The IACUC Handbook

Since its establishment by USDA regulation in the mid-1980s, the Institutional Animal Care and Use Committee (IACUC) has evolved as the premier instrument of animal welfare oversight within research institutions in the United States. As biomedical and wildlife research continues to grow, the role and impact of the IACUC has increased in scope and complexity. *The IACUC Handbook* has become an invaluable resource for individuals when the time comes for them to serve on their institution's IACUC. It provides a foundation for understanding and implementing the many and varied responsibilities of this committee.

This fourth edition comprehensively addresses the significant changes in the pertinent regulatory environment and interpretation of applicable federal laws, regulations, and policies. It includes modifications to regulations, as well as best practices, and new chapters on basic IACUC operations; animals used in agricultural research and teaching; invertebrates; pandemic and disaster planning; the Pacific Rim ethical review framework; operational communications and workflow; reduction of regulatory burden; and research reproducibility and the role of the IACUC. It also adds new authors and editors widely respected in the field.

An essential resource for IACUC members, university administrators and compliance personnel, laboratory animal veterinarians, and scientific investigators whose work involves animal research, this new edition ensures the book's continued acceptance as the standard reference in the field.

The IACUC Handbook

Fourth Edition

Edited by
Mark A. Suckow, Sreekant Murthy, Julie M. Sharp and Troy Hallman

CRC Press
Taylor & Francis Group
Boca Raton London New York

CRC Press is an imprint of the
Taylor & Francis Group, an **informa** business

Fourth edition published 2026
by CRC Press
2385 NW Executive Center Drive, Suite 320, Boca Raton FL 33431

and by CRC Press
4 Park Square, Milton Park, Abingdon, Oxon, OX14 4RN

CRC Press is an imprint of Taylor & Francis Group, LLC

ISBN: 9781032224978 (hbk)
ISBN: 9781032225203 (pbk)
ISBN: 9781003272861 (ebk)

DOI: 10.1201/9781003272861

Typeset in Times
by Deanta Global Publishing Services, Chennai, India

Dedication: Tribute to Dr. Jerald Silverman, MS, DVM, MPS

It is indeed an honor and a privilege to dedicate this fourth edition of the *IACUC Handbook* to Dr. Jerald Silverman, fondly known as "Jerry" in the IACUC world. I (Sreekant Murthy) had known Jerry for over a decade as a colleague and a friend. When I first met him, he was the Attending Veterinarian at Hahnemann University and Drexel University, Philadelphia. He was a fixture in his office daily from 9 a.m. to 9 p.m. Any time after 5 p.m. was dedicated to his professional writing or to his family. I always wondered how he balanced his professional life with his family life, but I know he was very adept at both. Jerry's wife, Dr. Lisa Barber, attributes gathering for dinner at the kitchen table every night as the glue that kept them connected. His daily hour-long drives to the skating rink served as the catalyst for all manner of conversation with his daughter. He reviewed protocols or drafted manuscripts while sitting in the bleachers during a practice session.

Mark Suckow, Jerry, and I had mutual respect and admiration for each other. We stayed connected with Jerry, even after he joined the University of Massachusetts (UMass) in 2002. The reasons for his move to Massachusetts were twofold. One, to change his professional venue, and the other to follow the dreams of his beloved wife who joined the faculty of Tufts University's veterinary school. At the time of his death, he was an emeritus professor at UMass and adjunct professor at Tufts. After retirement, he continued to write his IACUC-focused columns in *Lab Animal* until his death. Jerry passed away in 2020 at the age of seventy-eight succumbing to complications of COVID-19.

Jerry had reverence for all animals, a passion that was cultivated by an uncle who was a veterinarian, and which led Jerry to study ornithology at Cornell University and later to veterinary school, also at Cornell. Like many people engaged in biomedical research, he felt the emotional and ethical tug regarding the necessity for animal research to advance human and animal health.

Jerry actively advocated for implementation of three Rs in animal research, and he saw the IACUC as a powerful force in that regard. While he could be outspoken in his views, Jerry recognized that bringing experts together to share their wisdom is the key to the positive evolution of laboratory animal care and research. He was not shy from directly addressing controversies. He believed dialogue to be the foundation on which to build consensus for dealing with the challenges facing many laboratory animal veterinarians, researchers, and IACUCs.

His monthly column, "Protocol Review," in *Lab Animal* (NY) made Jerry a well-known figure in the research animal community. He used hypothetical issues for each of the columns, using a fictitious university name "Great Eastern University," where those issues arose. In the column he would seek opinions from the regulatory agencies and experts in the field as a way to provide insights into the dynamics of IACUC operations and protocol review. Jerry had the innate ability to ask questions and seek answers to unlock the complexities of IACUC management. Thus, one could assume why *The IACUC Handbook* was shaped as a book full of questions, answers, and opinions.

In 2004, at the request of CRC, Jerry, Mark Suckow, and I met to lay the foundation for publishing a handbook with an overall objective of producing a compendium of IACUC management. It was meant to bring consistencies in best practices across all institutions and IACUC reviews. We decided then that the handbook should be in the form of open-ended practical questions and answers that an IACUC would generally ask to understand regulations, obtain guidance, and seek expert opinions. Jerry took charge of developing surveys that would encompass every aspect of the animal welfare program. We are satisfied to know that *The IACUC Handbook* has become a widely used resource.

His idea for the book was not about asking random questions; rather, it was to craft guided questions that are appropriate for specific situations, whether it is to explain IACUC functions or the humane care and use of animals or occupational health. It is simply unimaginable how tirelessly and masterfully Jerry developed pointed questions in each of the chapters as a lever for freely seeking opinions from subject matter experts and guidance on regulations from the regulatory agencies. He also developed over 1,500 pointed survey questions in *The Handbook* that one could construe as conversations with representatives of many institutions as a way to gauge how varied institutions manage their programs and tackle many of the common, difficult issues. Further, this approach allows readers insight into how institutions are fulfilling their obligations and responsibilities in the care and use of animals in research, education, and testing.

Jerry was an exceptionally good managing editor. He took the primary responsibility of interacting with regulatory officials and the publisher. Instead of micromanaging us (Mark Suckow and I), he clearly defined our co-editor's roles to match our knowledge and capabilities to recruit authors, provide guidance to authors, stay focused on the topic, encourage authors to freely express their opinions, and explain flexibilities to meet the unique needs of the institution.

There is no doubt that Jerry left an immense legacy to the IACUC world by authoring and editing the first three editions of *The IACUC Handbook*. It is unfortunate that Jerry passed away at the time when we were planning the fourth edition of the book. Without question, the foundation he laid in the first three editions is the impetus for this, the fourth edition of *The IACUC Handbook*. This fourth edition has retained most of Jerry's survey questions, added two new editors, and followed the solid foundation Jerry established to add additional chapters and authors to help guide readers on contemporary issues common to IACUCs.

Mark, I, and the new editors have profound admiration for the legacy Jerry left behind. His broad vision to understand the complexities of IACUC regulations, his own views, and his commitment to bringing experts and regulatory agencies together to educate the IACUC community and to bring a balance between the humane use and care of animals in research and compliance were extraordinary. He truly was an icon in the IACUC world and of rare vintage in an era that values flexibility and self-governance in allowing animal welfare programs to meet the needs of institutions.

Finally, I thought the following few appreciative words from his wife would be a fitting tribute to Jerry.

"On behalf of my husband who I dearly loved and admired, I extend Jerry's appreciation and gratitude to the original and new editors for their unwavering dedication to the spirit and vision for this authoritative text and long hours in creating the new edition, particularly considering the challenges during the pandemic. The new chapters reflect their insights into refining the traditional roles of the IACUC as well as addressing ongoing challenges for researchers and exploring new directions for the IACUC. There is no greater tribute than to build on the foundation that he, Dr. Suckow and Dr. Murthy forged with an eye to leading the IACUC and the laboratory animal community into the future."

Thank you, Jerry, our gratitude to you has no bounds!
Sincerely,

Lisa Barber, DVM, DACVIM-Oncology
Adjunct Professor
Cummings School of Veterinary Medicine
Tufts University

Sreekant Murthy, Ph.D.
Emeritus Professor of Medicine
Drexel University College of Medicine
Adjunct Professor of Medicine
Rowan-Virtua School of Osteopathic Medicine

Contents

Preface

This fourth edition of *The IACUC Handbook* builds upon the foundation laid by the first (2000), second (2006), and third (2014) editions. Over time, both regulations and best practices have evolved, and we have attempted to capture current information and perspective. For example, NOT-OD-24-076, released by the U.S. National Institutes of Health, Office of Animal Welfare in March 2024 describes changes that allow flexibility regarding IACUC semiannual evaluations with respect to animals that are regulated by the U.S. Department of Agriculture. It can reasonably be anticipated that regulatory expectations, guidelines, and best practices will continue to undergo change, thereby supporting future revisions of this book.

Beginning with the third edition of *The IACUC Handbook*, consideration was given to research animal oversight expectations and practices on an international level, specifically Europe. The fourth edition of the book expands this perspective to include the Asia-Pacific region. This addition reflects the rising global acknowledgement that a strong program of animal care and use oversight not only ensures humane treatment of animals, but also supports excellence in science.

This fourth edition of *The IACUC Handbook* has expanded content that addresses features of IACUC operation and process. These include chapters on basic IACUC operations, disaster and pandemic planning, operational communications and workflow, and reduction of regulatory burden. Further, the expanded content addresses topics of growing focus and includes chapters on animals used in agricultural research and teaching, invertebrates, and research reproducibility and the role of the IACUC.

As in previous editions, the fourth edition is presented as items in a question-and-answer format that provides a regulatory basis relevant to the issue being discussed, followed by the opinion of the chapter author(s). Many of the items also include survey questions and summarized response data, as a way for readers to understand approaches that are commonly followed by IACUCs around the nation. In this way, readers benefit from the combined experience of numerous IACUCs and some sense of "best practices" can be ascertained. In some chapters we have permitted the repetition of items that are similar to those found in other chapters. These have been cross-referenced to provide a broader perspective of expert opinions. We hope all of this will be of value to you.

Outside review of manuscripts was conducted beyond the editorial review we provided. In this regard, we wish to express our sincere gratitude to the representatives of USDA/APHIS and OLAW who diligently reviewed and assessed the chapters to ensure agreement of content with prevailing regulations and policies.

The IACUC serves as a key operational hub in ensuring that research, teaching, and testing involving animals is conducted in a manner that is both humane and consistent with expectations of regulatory and accrediting agencies, as well as with institutional policies. A well-functioning IACUC will serve as a resource for investigators and the institution to navigate through an environment with ever-evolving expectations and best practices. Success depends on the collaboration of the attending veterinarian, the Institutional Official, the IACUC and animal researchers, together forming a foundation to ensure excellence in animal care and use and the science that it supports.

Mark A. Suckow, D.V.M., D.A.C.L.A.M.
Sreekant Murthy, Ph.D
Julie M. Sharp, D.V.M., C.P.I.A., D.A.C.L.A.M.
Troy Hallman, M.S., V.M.D., D.A.C.L.A.M

Disclaimer

Any content, opinions, findings, conclusions, or recommendations expressed in this publication are those of the authors and do not necessarily reflect the views of the U.S. Government, the National Institutes of Health (NIH) nor the Office of Laboratory Animal Welfare (OLAW). Neither the U.S. Government, the NIH nor OLAW assumes any legal liability for the accuracy, completeness, or usefulness of any information, products, or processes disclosed herein. For questions pertaining to the Public Health Service Policy on Humane Care and Use of Laboratory Animals (PHS Policy), particularly those regarding compliance, please contact OLAW directly.

Acknowledgment

Acknowledgment to Authors of Earlier Editions

The editors and the authors of *The IACUC Handbook*, fourth edition, acknowledge the contributions of the following authors to the earlier editions of this handbook. Their efforts were extremely important and served as a valuable basis for this new edition. Where appropriate, the authors in this edition have adopted the interpretations and opinions of earlier editions of the book or have made revisions based on current knowledge and future directions pertinent to the topic.

Chapter Authors

David R. Archer, Ph.D
Robert M. Bigsby, Ph.D.
Marylin Chimes, D.V.M., Dipl. A.C.L.A.M., J.D.
Stephen K. Curtis, D.V.M., Dipl. A.C.L.A.M.
Peggy J. Danneman, V.M.D., M.S., Dipl. A.C.L.A.M.
Elizabeth J. Dawe, D.V.M. Dipl. A.C.L.A.M.
Ralph B. Dell, M.D.
Kathryn A. Donohue, J.D.
Laura Bachich Ergin, J.D.
Paul Flecknell, M.A., Vet. M.B. Ph.D., D.L.A.S.., Dipl. E.C.V.A., M.B.C.V.S.
Diane J. Gaertner, D.V.M., Dipl. A.C.L.A.M.
Cynthia S. Gillett, D.V.M., Dipl. A.C.L.A.M.
Larry Iten, D.V.M.
Karen M. James, B.S.
Alicia Z. Karas, M.S., D.V.M., Dipl. A.C.V.A.A.
Kathy Laber, D.V.M., Dipl. A.C.L.A.M.
Angela Mexas, D.V.M., Ph.D., Dipl. A.C.V.I.M.
Kathleen D. Moody, V.M.D., M.S., Dipl. A.C.L.A.M.
Christian Newcomer, V.M.D., M.S., Dipl. A.C.L.A.M.
Lester R. Rolf, Jr., Ph.D., D.V.M.
Harry Rozmiarek, D.V.M., Ph.D. (deceased)
Howard Rush, D.V.M., M.S., Dipl. A.C.L.A.M.
Sallie Thiem Sanford, J.D.
Priya Shankar, J.D.
Mary Jo Shepherd, D.V.M., C.P.I.A.
Jerald Silverman, D.V.M., M.P.S. (deceased)
Jane Simpson
Alison C. Smith, D.V.M., Dipl. A.C.L.A.M.
Joseph Spinelli, D.V.M.
Harold F. Stills, Jr., D.V.M., Dipl. A.C.L.A.M.
Farol Tomson, D.V.M., M.B.A.
Richard C. Van Sluyters, O.D., Ph.D.
Lisa A. Vincler, J.D.
Stefan Wagener, Ph.D., C.B.S.P.
Patricia A. Ward, B.S., M.P.A., L.A.T.G.

Acknowledgment to Reviewers

The Editors of *The IACUC Handbook*, fourth edition, would like to express our sincere gratitude to colleagues at the USDA/APHIS and NIH/OLAW for their tremendous and diligent efforts in reviewing manuscripts to ensure concordance of content with relevant regulations and guidelines. These individuals are:

Lance Bassage, V.M.D., USDA APHIS
Patricia Brown, V.M.D., NIH OLAW (retired)
Louis DiVincenti, D.V.M., USDA APHIS
Neera Gopee, D.V.M., NIH OLAW
Nicole Lukovsky-Akhsanov, D.V.M., M.P.H., NIH OLAW
Eileen Morgan, NIH OLAW (retired)
Brent Morse, D.V.M., NIH OLAW
Jane Na, D.V.M., C.P.I.A., NIH OLAW
Nicolette Petervary, V.M.D., NIH OLAW
Carolyn Reneski, M.S., C.P.I.A., NIH OLAW
Angelina Williams, D.V.M., NIH OLAW
Axel Wolff, D.V.M., NIH OLAW

The editors wish to specifically express our deep gratitude to Catherine Lavinski for her exceptional volunteer efforts in contributing to the preparation of this book. Her work in analyzing survey data, compiling the survey results, and integrating the survey findings into the various chapters has been invaluable and played a crucial role in bringing this book to fruition. Cat's meticulous attention to detail and her dedication to ensuring the accuracy and clarity of the data significantly enhanced the quality of this publication. Her contributions have been instrumental, and we are deeply appreciative of her hard work and commitment.

Editors

Mark A. Suckow, D.V.M., D.A.C.L.A.M., is a private consultant to animal care and research programs and serves as Professor (Emeritus) of Biomedical Engineering at the University of Kentucky, where he previously served as Associate Vice President for Research and Attending Veterinarian. Earlier, he served as Attending Veterinarian and Assistant Vice President for Research at the University of Notre Dame. He received his D.V.M. from the University of Wisconsin in 1987 and completed a post-doctoral residency program in laboratory animal medicine at the University of Michigan in 1990. He is a Diplomate of the American College of Laboratory Animal Medicine.

Dr. Suckow has published more than 140 scientific papers, patents, books, and book chapters. He has created a company to commercialize technology based on intellectual property he developed. Dr. Suckow is active in professional organizations and has served as President of both the American Association for Laboratory Animal Science (AALAS) and the American Society of Laboratory Animal Practitioners (ASLAP). He currently serves as an emeritus member of the AAALAC Council on Accreditation and as Editor in Chief for the *Journal of the American Association for Laboratory Animal Science.*

Sreekant Murthy, Ph.D., is an Emeritus Professor of Medicine at Drexel University College of Medicine, Drexel University, Philadelphia, and Adjunct Professor of Medicine, Rowan-Virtua School of Osteopathic Medicine. He served as the Senior Associate Vice Provost for Research Compliance (1997–2013) and Chief Research Compliance Officer at Drexel University and Rowan University (2013–2020), respectively. He retired from Drexel University in 2013 and from Rowan University in 2020. Since his retirement, he started a consulting agency, and he is providing consultations in the regulated research arena, export control, and research security to universities.

Dr. Murthy is a well-established and recognized researcher in the area of peptic ulcer disease. He studied the role of regulatory peptides in regulating gastrointestinal functions and he eventually engaged in inflammatory bowel diseases research until his retirement. He developed and popularized an animal model of experimental colitis that researchers and the industry across the world are extensively using even today. He has published over 70 papers in peer-reviewed journals, authored chapters in nine books, edited two books, published nearly 150 abstracts at conferences and symposia, and made nearly 200 presentations nationally and abroad. His published work has been cited 5,500 times by other scientists in his field and read by over 12,000 scientists across the world. He is in the top 3% of all published biological scientists and is in the top 5% of researchers published in gastrointestinal inflammation. He served as a peer reviewer for several journals and grant applications.

Julie M. Sharp, D.V.M., C.P.I.A., D.A.C.L.A.M., is a private consultant and the Executive Director of the Institutional Animal Care and Use Program at the University of California, San Francisco. Previously she served in animal program leadership roles at Duke University, the University of Pennsylvania, and the Research Foundation for the State University of New York. She has expertise in regulatory compliance and IACUC administration in state, private, and government-based animal research facilities. She earned her veterinary medical degree from North Carolina State University, Raleigh, North Carolina. Dr. Sharp is a Diplomate of the American College of Laboratory Animal Medicine, an ad hoc Specialist for AAALAC International, and actively participates in the Federal Demonstration Partnership's Compliance Unit Standard Procedures initiative as a member of the Steering Committee and Co-Chair of the Help Desk Team.

Troy Hallman, M.S., V.M.D., D.A.C.L.A.M., serves as the Director of the Office of Animal Research Support at Yale University. Earlier, he served as the Director of the Office of Animal Welfare and a clinical laboratory animal veterinarian at the University of Pennsylvania. Dr. Hallman completed a master's degree in zoology from the University of British Columbia in 1997. He earned his veterinary medical degree from the University of Pennsylvania in 2000 and completed a post-doctoral residency program in laboratory animal medicine at the University of Pennsylvania in 2002. He has been a Diplomate of the American College of Laboratory Animal Medicine since 2008. Dr. Hallman currently serves as a member of the AAALAC Council on Accreditation. His expertise includes regulatory oversight of veterinary, biomedical, agricultural programs, and wildlife research programs, as well as laboratory animal welfare regulations and guidance.

Contributors

Kenneth P. Allen, D.V.M., Dipl. A.C.L.A.M.
Assistant Provost for Research and Director
Biomedical Resource Center
and
Professor
Microbiology & Immunology
Medical College of Wisconsin
Milwaukee, Wisconsin

David R. Archer, Ph.D.
Associate Professor
Division of Hematology/Oncology
Department of Pediatrics
School of Medicine
Emory University
Atlanta, Georgia

Cecile Baccanale, D.V.M.
University Veterinarian
Associate Vice President, Animal Care
and
Professor
Division of Clinical and Translational Research
Faculty of Medicine and Health Sciences
McGill University
Montreal, Quebec, Canada

Ron E. Banks, D.V.M., D.A.C.A.W., Dipl. A.C.L.A.M., Dipl. A.C.V.P.M., Fellow N.A.P., C.P.I.A.
Attending Veterinarian
Office of Research and Sponsored Programs
University of Denver (Colorado Seminary)
Denver, Colorado

Szczepan W. Baran, M.S., D.V.M.
Founder & CEO
Veterinary Innovation Center (VIC)
and
Principle
Baran Café
and
Chief Barks and Purrs Officer
Tech4Pets
and
Research Associate
Division of Animal and Human Health Engineering
M3-BIORES (Measure, Model & Manage Bioresponses)
KU Leuven
Leuven, Belgium

Kathryn Bayne, M.S., Ph.D., D.V.M., Dipl. A.C.L.A.M., Dipl. A.C.A.W., C.A.A.B. (Emeritus)
Consultant

Christine A. Boehm, M.Ed., D.V.M. M.S., Dipl., A.C.L.A.M., Dipl. E.C.L.A.M.
CEO
The IACUC Vet, LLC.

John F. Bradfield, D.V.M., Ph.D., Dipl. A.C.L.A.M.
Attending Veterinarian (Emeritus)
and
Professor (Emeritus)
Pathology
University of Nebraska Medical Center
Omaha, Nebraska

Marcy Brown, M.A., C.M.A.R., L.S.S.B.B.
IACUC Program Consultant
Program Consultant
Boston, Massachusetts

Madeline L. Budda, D.V.M., M.S., Dipl. A.C.L.A.M.
Director
Animal Welfare Assurance
and
Associate Professor of Cell Biology
University of Oklahoma Health Sciences Center
Oklahoma City, Oklahoma

Jennifer Camacho, A.S. (Veterinary Technology), M.S.
Environmental Enrichment Program Manager
Massachusetts General Hospital
Boston, Massachusetts

Larry Carbone, D.V.M., Ph.D., Dipl. A.C.L.A.M., Dipl. A.C.A.W.
Director (former)
Institutional Animal Care and Use Program
University of California, San Francisco
San Francisco, California

Bart Carter, D.V.M., M.S., Dipl. A.C.L.A.M.
Assistant Dean for Animal Resources
University of Texas Southwestern Medical Center
Dallas, Texas

Kristina Cataline, D.V.M., M.P.V.M., Dipl. A.C.V.P.M.
Associate Director (former)
University of California San Francisco Institutional Animal Care and Use Program
SanMateo, California

Sonja T. Chou, V.M.D., M.S., Dipl. A.C.L.A.M.
Director
Laboratory Animal Resources
and
Clinical Assistant Professor
New York University Shanghai
Shanghai, China

Susan Stein Cook, D.V.M., M.S., Dipl. A.C.L.A.M.
Consultant in Laboratory Animal Medicine
Williamston, Michigan

Robin Crisler, D.V.M., M.S., Dipl. A.C.L.A.M.
Laboratory Animal Veterinarian Specialist and Consultant
Indianapolis, Indiana

Erin Czarniak, C.P.I.A., L.A.T.G.
Associate Director – Quality Assurance
and
Animal Care & Use Office
University of Michigan
Ann Arbor, Michigan

Lauren Dandridge, C.P.I.A., L.S.S.S., S.S.G.B., C.P.B.C.A.
Associate Director
IACUC/IBC/PAM for the Research Integrity and Assurance
Princeton University
Princeton, New Jersey

Bernard J. Doerning, D.V.M., Dipl. A.C.L.A.M., M.B.A.
Director
Division of Laboratory Animal Resources
University of Kentucky
Lexington, Kentucky

Carolyn M. Doerning, D.V.M., Dipl. A.C.L.A.M., M.B.A.
Assistant Director and Assistant Professor
Cincinnati Children's Hospital Medical Center
Cincinnati, Ohio

Melissa C. Dyson, D.V.M., M.S., Dipl. A.C.L.A.M., C.P.I.A.
Senior Director
and
Clinical Professor
Unit of Laboratory Animal Medicine
University of Michigan
Ann Arbor, Michigan

Deborah Frolicher, B.S., C.P.I.A.
IACUC Consultant
San Diego, California

Alan Gelperin, Ph.D.
Professor of Neuroscience
Princeton, New Jersey

Rose Gillesby, D.V.M.
Vice President
Global Animal Science and Welfare
Zoetis
Kalamazoo, Michigan

Neera Gopee, D.V.M., Ph.D., Dipl. A.C.L.A.M., Dipl. A.B.T.
Associate Director
Animal Welfare Policy
Office of Laboratory Animal Welfare (OLAW)
National Institutes of Health (NIH)
Bethesda, Maryland

Edward J. Gracely, Ph.D.
Associate Professor
Family, Community, & Preventive Medicine
College of Medicine, and Epidemiology and Biostatistics
Dornsife School of Public Health
Drexel University
Philadelphia, Pennsylvania

William G. Greer, B.S., C.P.I.A., L.A.T., C.M.
Assistant Vice President for Research
Animal Program Compliance Oversight
University of Michigan
Ann Arbor, Michigan

Javier Guillén, D.V.M.
Senior Director for Europe and Latin American
AAALAC International
Pamplona, Spain

Anna Hampton, D.V.M. Dipl. A.C.L.A.M., Dipl. A.C.A.W.
Director
Office of Animal Welfare Assurance
Duke University
Durham, North Carolina

F. Claire Hankenson, D.V.M., M.S., M.F.A., Dipl. A.C.L.A.M.
Professor of Pathobiology (CE)
and
Associate Vice Provost for Research
and
Executive Director
University Laboratory Animal Resources
University of Pennsylvania
Philadelphia, Pennsylvania

Tracy M. Heenan, D.V.M., C.P.I.A.
Director
Office of Animal Care and Use
and
Professor
Department of Pathology and Laboratory Medicine
University of North Carolina at Chapel Hill
Chapel Hill, North Carolina

JoAnn Henry, A.A.S.
Assistant Director
Division of IACUC and IBC Administration
New York University, Langone Health
New York University Grossman School of Medicine,
New York, New York

Debra L. Hickman, D.V.M., M.S., Dipl. A.C.L.A.M., Dipl. A.C.A.W.
Global Director
AAALAC International
Frederick, Maryland

Michael J. Huerkamp, D.V.M., Dipl. A.C.L.A.M.
Emeritus Professor of Pathology and Laboratory Medicine
Emory University
Atlanta, Georgia

Tanise Jackson, D.V.M., Dipl, A.C.L.A.M., C.P.I.A.
Director
Animal Welfare Research and Integrity
Florida A&M University
Tallahassee, Florida

Vickie Jarrell, Ph.D.
Private Consultant
Gates, Oregon

Elaine K. Kim, B.S., C.P.I.A.
Senior Technical Professional – Biology
KBR Wyle Services LLC
Houston, Texas

Jennifer A. Klahn, M.A., C.P.I.A.
Director
Research Safety & Animal Welfare
University of California
Los Angeles, California

Thomas A. Leach, Esq.
Executive Director
New Jersey Association for Biomedical Research
and
Pennsylvania Society for Biomedical Research
Atco, New Jersey

Stuart E. Leland, D.V.M., Dipl. A.C.L.A.M.
Director of Research Integrity and Assurance
Princeton University
Princeton, New Jersey

Neil S. Lipman, V.M.D., Dipl. A.C.L.A.M.
Executive Director
and
Professor
Center of Comparative Medicine and Pathology
Memorial Sloan Kettering Cancer Center
Weill Cornell Medicine Hospital for Special Surgery
New York, New York

Carolyn M. Malinowski, M.S., D.V.M., Dipl. A.C.L.A.M., C.M.A.R., C.P.I.A.
Assistant Professor
MD Anderson Cancer Center
The University of Texas
Houston, Texas

Michael D. Mann, Ph.D.
Professor (Emeritus)
Cellular and Integrative Physiology
University of Nebraska Medical Center
Omaha, Nebraska

Tara Martin, D.V.M., Dipl. A.C.L.A.M., M.S.
Clinical Associate
and
Professor of Laboratory Animal Medicine
Unit for Laboratory Animal Medicine
University of Michigan
Ann Arbor, Michigan

Natalie L. Mays, B.A., L.A.T.G., C.P.I.A.
Director
Division of IACUC and IBC Administration
NYU Grossman School of Medicine
NYU Langone Health
New York, New York

Emily L. Miedel, V.M.D., Dipl. A.C.L.A.M.
Associate Professor of Laboratory Animal Medicine
Division of Comparative Medicine and Research
University of Pennsylvania
Philadelphia, Pennsylvania

Deborah Mook, D.V.M., Dipl. A.C.L.A.M.
Executive Director of the Division of Animal Resources
Emory University
Atlanta, Georgia

Sreekant Murthy, Ph.D.
Professor of Medicine (Retired)
and
Chief Research Compliance Officer
Rowan University
and
Adjunct Professor of Medicine
Rowan University School of Medicine,
Stratford, New Jersey

Gwenn S. F. Oki, M.P.H.
Compliance Director (Retired)
Van Andel Institute
Grand Rapids, Michigan

Esther J. Pearl, Ph.D.
Program Manager
Experimental Design
National Centre for the Replacement, Refinement and Reduction of Animals in Research (NC3Rs)
London, United Kingdom

Nathalie Percie du Sert, Ph.D.
Director of Research Practice
National Centre for the Replacement
Refinement Reduction of Animals in Research (NC3Rs)
London, United Kingdom

Scott E. Perkins, V.M.D., M.P.H., Dipl. A.C.L.A.M.,
Senior Director
Tufts Comparative Medicine Services
Tufts University
Boston, Massachusetts

Marcel Perret-Gentil, D.V.M., M.S.
University Veterinarian
and
Director
Laboratory Animal Resources Center
University of Texas at San Antonio
San Antonio, Texas

Kiirsa Pokryfke, M.S., C.M.A.R.
Managing Director
Training Core and Technical Services
Unit for Laboratory Animal Medicine
University of Michigan
Ann Arbor, Michigan

Ernest D. Prentice, Ph.D.
Professor (Emeritus)
Department of Genetics, Cell Biology and Anatomy
University of Nebraska Medical Center
Omaha, Nebraska

Mark J. Prescott, Ph.D.
Director of Policy and Outreach
National Centre for the Replacement, Refinement and Reduction of Animals in Research (NC3Rs)
London, United Kingdom

Kevin Prestia, D.V.M., M.B.A., M.P.H., Dipl. A.C.L.A.M.
Director
Division of Comparative Medicine
New York University, Langone Health
New York University Grossman School of Medicine
New York, New York

Stacy L. Pritt, D.V.M., M.S., M.B.A., C.P.I.A., C.H.R.C., E.X.C.S., E.Co.P., Dipl. A.C.A.W.
Associate Vice Chancellor
and
Chief Research Compliance Officer
Texas A&M University System
College Station, Texas
and
Adjunct Professor
College of Veterinary Medicine & Biomedical Sciences
Texas A&M University
College Station, Texas

Jennifer K. Pullium, M.V.B., Dipl. A.C.L.A.M.
Senior Director
Division of Comparative Medicine
New York University Langone Health
New York University Grossman School of Medicine
New York, New York

Penny S. Reynolds, Ph.D.
Assistant Professor
Department of Anesthesiology
University of Florida College of Medicine
Gainesville, Florida

Karen Rogers, D.V.M., Dipl. A.C.L.A.M., C.P.I.A.
Laboratory Animal Professional
Bloomington, Indiana

Dewi K. Rowlands, MPhil., M.B.A., Ph.D.
Director
Centre for Comparative Medicine Research
Li Ka Shing Faculty of Medicine
The University of Hong Kong
Hong Kong, China

Aubrey Schoenleben, Ph.D., C.P.I.A.
Director
Regulatory Affairs and External Partnership
University of Washington
Seattle, Washington

Sally Thompson-Iritani, D.V.M., Ph.D., C.P.I.A.
Assistant Vice Provost
Animal Care, Outreach, & 3Rs
and
Associate Professor
School of Public Health
University of Washington
Seattle, Washington

Joseph D. Thulin, D.V.M., M.S., Dipl. A.C.L.A.M.
Professor (Emeritus)
Physiology
Medical College of Wisconsin
Milwaukee, Wisconsin

Noé Tirado-Muñiz, D.M.V., D.V.M., M.S., C.P.I.A.
Associate Attending Veterinarian
and Associate Director
Office of the Attending Veterinarian (OAV)
University of Kentucky
Lexington, Kentucky

Robin Lyn Trundy, M.S., R.B.P., C.B.S.P.
Institutional Biosafety Office & Biosafety Team Lead
Environmental Health and Safety
Vanderbilt University
Nashville, Tennessee

Joanne L. Zahorsky-Reeves, D.V.M., Ph.D., Dipl. A.C.L.A.M.
Associate Director
Division of Laboratory Animal Medicine
University of California
Los Angeles, California

Abbreviations

Several abbreviations and acronyms are frequently used throughout this book. Because they are not defined at every use, this list provides a summary of meanings for these items.

21CCA: 21st Century Cures Act
AAALAC: AAALAC, International; previously, the Association for Assessment and Accreditation of Laboratory Animal Care
AALAS: American Association for Laboratory Animal Science
ACLAM: American College of Laboratory Animal Medicine
ACUP: Animal care and use protocol
Ag Guide: *Guide for the Care and Use of Agricultural Animals in Research and Teaching*; 4th edition (2020)
APHIS: USDA Animal and Plant Health Inspection Service
APHIS/AC: Animal and Plant Health Inspection Service/ Animal Care
AV: Attending Veterinarian
AVMA: American Veterinary Medical Association
AWA: Animal Welfare Act
AWAR: Animal Welfare Act Regulations (note that the plural use is intended)
BAHSCR: Board on Animal Health Sciences, Conservation, and Research within the Division on Earth and Life Studies of the U. S. National Academies; replaces the former ILAR (Institute for Laboratory Animal Research)
BARDA: Center for the Biomedical Advanced Research and Development Authority
BMBL: *Biosafety in Microbiological and Biomedical Laboratories*, 6th edition (2020)
CCAC: Canadian Council on Animal Care
CDC: United States Centers for Disease Control and Prevention
CFR: Code of Federal Regulations
COGR: Council on Governmental Relations
DoD: United States Department of Defense
DMR: Designated member review
FCR: Full committee review
FDA: United States Food and Drug Administration
FEMA: Federal Emergency Management Agency
Guide: *Guide for the Care and Use of Laboratory Animals*; typically, the 8th edition (2011)
HREA: Health Research Extension Act of 1985
IACUC: Institutional Animal Care and Use Committee
IBC: Institutional Biosafety Committee
IO: Institutional Official
IRAC: Interagency Research Animal Committee
MOU: Memorandum of Understanding
NASA: National Aeronautics and Space Administration
NIH: United States National Institutes of Health
NIOSH: National Institute for Occupational Safety and Health
OHSP: Occupational health and safety program
OLAW: Office of Laboratory Animal Welfare
Opin.: Opinion of the chapter's author(s)
OSHA: Occupational Health and Safety Administration
PAM: Post-approval monitoring
PHS: United States Public Health Service
PHS Policy: Public Health Service Policy on Humane Care and Use of Laboratory Animals
PPE: Personal protective equipment
Reg.: Regulatory information offered as guidance to the reader.
SOP: Standard operating procedure
Surv.: Survey performed by the chapter's author(s).
USDA: United States Department of Agriculture
VVC: Veterinary Verification and Consultation

Survey Methods and Statistics

The survey in the current edition is structured as in the previous editions, designed to collect information from a broad representation of IACUCs regarding the approaches they take to manage their programs and to remain in compliance with regulations. Most of the questions in the previous editions were generated by Dr. Jerald Silverman with input from the other editors. For the current edition of the book, the authors of each of the chapters were asked to review all the questions from the previous editions and to add additional questions where necessary to draw additional insights into their respective chapters. Questions for the new chapters were developed entirely by the respective authors.

Survey questions were sent to approximately 1,500 NIH/OLAW-accredited institution's Program leaders across the world. We thank Dr. Patricia Brown, former Director for OLAW for providing the list of assured institutions.

An anonymous survey was conducted using the Qualtrics survey platform. Nearly 350 institutional representatives responded to approximately 580 comprehensive survey questions over the period of approximately three months. We thank the IT Department at the University of California, San Francisco, for their guidance on hosting the Qualtrics survey.

The responses were collated and analyzed using the same method of analysis consistent with previous editions. As in the previous editions, the survey questions were multiple choice, though some questions also included text boxes that allowed respondents to provide more specific input. The responses were calculated using the number of respondents (institutions) as the denominator and the number of individuals who chose the answer as the numerator. As shown below in the demographic questions, the percentage of responses was calculated as a quotient (Numerator/Denominator) multiplied by one hundred. Some items allowed multiple answers as a way to tease out more specific information.

Ms. Catherine Lavinski, MSc (Wildlife Conservation and Management), ALAT, formerly an Animal Technician at Yale University, performed the statistical analysis and painstakingly analyzed the data and collated the responses for each of the questions in the respective chapters.

The following demographic questions were asked for each of the institutions.

Primarily academic	203/349 (58%)
Primarily pharmaceutical (with drug discovery)	19/349 (5%)
Primarily private non-pharmaceutical research or testing including CROs	48/349 (14%)
Primarily zoological or other exhibitor with associated research	1/349 (0%)
Primarily laboratory animal breeding	4/349 (1%)
Primarily government research and testing	45/349 (13%)
Other*	29/349 (8%)

*Other ranged from hospital or health-care systems, private medical/nonprofit research institutions, military facility, and natural history museum.

2. What is the approximate size of your organization? If your organization has multiple sites, consider only the site(s) for which your IACUC has regulatory oversight responsibility?

- About 1 – 100 full-time-equivalent employees — 81/349 (23%)
- About 101 – 500 full-time-equivalent employees — 64/349 (18%)
- About 501 – 1,000 full-time-equivalent employees — 36/349 (10%)
- About 1,001 – 5,000 full-time-equivalent employees — 87/349 (25%)
- Over 5,000 full-time-equivalent employees — 81/349 (23%)

3. Does your organization currently have any level of AAALAC accreditation?

- No — 116/349 (33%)
- Yes — 233/349 (67%)

4. Is your organization legally required to comply with the PHS Policy for all or some of its animal-related work?

- No — 29/349 (8%)
- Yes — 320/349 (92%)

5. Provide the approximate number of approved IACUC protocols at your institution.

- < 50 — 133/349 (38%)
- 50 – 100 — 66/349 (19%)
- 100 – 300 — 82/349 (23%)
- 300 – 500 — 32/349 (9%)

- 500 – 1,000 19/349 (5%)
- 1,000 – 2,000 11/349 (3%)
- 2,000 – 5,000 1/349 (0%)
- 5,000 – 10,000 0/349 (0%)
- > 10,000 0/349 (0%)
- I don't know 5/349 (5%)

6. Is your organization legally required to comply with the Animal Welfare Act for all or some of its animal related work?

- No 43/349 (12%)
- Yes 306/349 (88%)

*Other responses reflected use of various electronic and paper versions of record maintenance.

1

Origins of the IACUC

Neera V. Gopee

Introduction

The need to provide regulatory oversight to ensure animal welfare in the research laboratory was recognized in Victorian England over 120 years ago, which led to the first anti-vivisection law in 1876, known as the Cruelty to Animals Act. Modifications of this law exist today as the Animals (Scientific Procedures) Act. Research with animals was subject to specific licensing requirements overseen by the United Kingdom Government Home Office, which incorporated ethics committees called Animal Welfare and Ethical Review Bodies.[1] In the United States (U.S.), the earliest national law addressing animal welfare was the Twenty-Eight Hour Law, which was enacted in 1873 and enforced by the United States Department of Agriculture (USDA).[2] This law set humane standards for the transportation of livestock and did not specifically address animals in the research laboratory.

The New York Anticruelty Bill of 1866 and the formation of Humane Societies in New York (1866), Pennsylvania (1867), Massachusetts (1868), and Washington, D.C. (1870) addressed the use of animals in research, as did the formation of the American Anti-Vivisection Society in Jenkintown, Pennsylvania, in 1883. Nevertheless, none of these actions had national authority or scope. The first official law addressing the care and use of laboratory animals in the U.S. was the Laboratory Animal Welfare Act of 1966. An amendment in 1970 changed its name to the Animal Welfare Act, by which it is known today.

Background

Prior to the 20th century, the responsibility for animals used in research in the U.S. was placed directly in the hands of the research investigator. The quality of animal care and animal welfare varied tremendously among research laboratories; even laboratories within the same school or institution had different animal care policies and standards of care. Animal care was frequently the responsibility of a diener, who provided food, basic care, and much of the animal manipulation. Basic nutrition and sanitation were often inadequate with no available environmental or housing standards. In many instances, the animal handling staff meant well, but they were neither adequately trained nor qualified. In other instances, animal welfare and even adequate care were of low priority. Rodents were commonly fed cereal grains and dogs and cats received table scraps. Backyard and part-time breeders provided rodents of undetermined genetic background, and the animals harbored a variety of parasites, pathogenic bacterial agents, and viruses. Long-term studies with rodents were impossible, as animals suffered from malnutrition and clinical disease and accurate research data were difficult to obtain. Dogs and cats were acquired individually, often provided in crates or gunny-sacks, and obtained from undocumented sources. Breeders of good-quality, standardized animals for research did not exist.

Central Management and Organization of Laboratory Animal Care

One of the first major research institutions in the U.S. that provided central management and care for research animals was the University of Chicago. Dr. Nathan Brewer, a highly trained and qualified individual, was hired in 1945 to direct this institution's centrally managed program of laboratory animal care. Dr. Brewer provided animal facility management at the University of Chicago as early as 1930. He subsequently received formal training as a veterinarian and obtained a Ph.D. in Physiology. Several other institutions in the country were following similar paths as the need for improved and more standardized animal care and welfare was being recognized. However, it was not until the 1940s that several professionals with a major interest in laboratory animal care began meeting in the Chicago area, which was undoubtedly stimulated and encouraged by Dr. Brewer. In 1949, veterinarians responsible for laboratory animal medicine at five Chicago area institutions met at least monthly to discuss animal care problems.[3] They were Drs. Nathen Brewer, University of Chicago; Elihu Bond, University of Illinois; Bennett Cohen, Northwestern University; Robert Flynn, Argonne National Laboratory; and Robert Schroeder, Hektoin Institute. Special meetings were also scheduled when colleagues from other institutions visited Chicago. Dr. W.T.S. Thorpe, Director of the newly organized Laboratory Aids Branch at the National Institutes of Health (NIH), which eventually developed into the National Center for Research Resources, was among one of the most esteemed visitors. Recognizing the need for other institutions and individuals around the country to exchange information on what was referred to as "laboratory animal care," a forum or panel was held on November 28, 1950, which led to the formal establishment of the Animal Care Panel (ACP). In 1967, the ACP

DOI: 10.1201/9781003272861-1

became the American Association for Laboratory Animal Science (AALAS). Dr. Brewer served as the first national president of the ACP, and, with other colleagues, he inspired many additional activities and organizations instrumental in shaping animal care and use policies and laws in the U.S.[4]

Officially established, and conducting annual conferences on animal care, the ACP appointed the Professional Standards Committee (PSC) in 1961 to determine and establish a professional standard for laboratory animal care and facilities under a contract awarded by the National Institutes of Health. The result of this effort was the 1963 publication of the first edition of the *Guide for Laboratory Animal Facilities and Care* (now known as the *Guide for the Care and Use of Laboratory Animals*), which was, and still is, widely accepted and serves as a primary reference for laboratory animal care and use. Future editions of this publication were supported by the NIH and guided by the Institute for Laboratory Animal Research (ILAR, now known as the Board on Animal Health Sciences, Conservation, and Research). In 2011, the National Academy Press, under the auspices of the National Research Council (NRC), published the current *Guide for the Care and Use of Laboratory Animals: Eighth Edition (Guide).*[5] This single document serves as the "bible" for laboratory animal care and use policies and guidelines in the U.S. and several countries worldwide by assisting institutions in caring for and using animals in ways judged to be scientifically, technically, and humanely appropriate. It is excerpted from, and referenced in, all major guidelines and regulations and has been translated into and published in many other languages. The Public Health Service (PHS) Policy on Humane Care and Use of Laboratory Animals (Policy) incorporates the *Guide*, requiring institutions to use this resource as a basis for developing and implementing an institutional program for activities involving animals.[6] The *Guide* expects every institution with an animal care and use program to have an Institutional Animal Care and Use Committee (IACUC) or institutional equivalent. This was based on federal law, which created a statutory basis for the Institutional Official (IO), Attending Veterinarian (AV), and IACUC to have shared responsibility and authority for the animal care and use program,[7] with the IACUC being responsible for the assessment and oversight of the program.

AAALAC International

Amid the extensive growth in biomedical research with animals and concurrent anti-vivisection activity in the mid- to late 1940s, five veterinarians from major Chicago institutions pioneered the movement to address issues related to the emerging field of laboratory animal science. They spearheaded the national organization called the ACP to serve as a forum for the exchange of information on the care and management of animals, as well as the design of animal facilities. However, the ACP struggled to implement a program for standards, certification, and accreditation, and in 1957 an *ad hoc* Committee for the Consideration of Animal Regulatory Activities of "Accreditation" Committee was appointed.[8] The ACP became the Professional Standards Committee (PSC) in 1959 to reflect its new focus on developing standards of care in response to allegations by the Humane Society of the United States. The PSC, later renamed the Animal Facilities Certification Committee (AFCC), outlined a certification program focused on determining professional standards for laboratory animal care and developing an accreditation program, for approval by the Animal Facilities Certification Board (AFCB).[9]

In 1963, the Animal Facilities Accreditation Board (AFAB) (previously the AFCB) saw a need to assist institutions in evaluating their animal facilities through a voluntary accreditation program. It was believed that this could be accomplished through the conduct of institutional site visits by peers to promote quality animal research and achieve the highest standards of care, in accordance with the first publication of the *Guide*. This was accomplished in 1964, when two-person teams from the AFAB visited 26 major institutions in the U.S. that conducted research with animals and evaluated their animal care policies and programs using the standards in the new *Guide*. The National Advisory Committee (NAC) to the AFAB was appointed in 1964 to represent the general scientific community interests in this program and the classes of institutions likely to be affected, as well as to advise the AFAB on conducting and implementing the program nationally.

On September 15, 1964, the ACP Board of Directors reviewed the detailed report "Accreditation of Laboratory Animal Facilities and Care," and acknowledged that the feasibility of a privately operated autonomous and national voluntary accreditation program for animal care would require the endorsement, cooperation, participation, and financial support of the scientific community. The AFAB, directed by representatives of member societies within the biomedical community interested in excellence in laboratory animal care and use, served as the body charged with the administration of the nonprofit accreditation corporation. This board soon saw the need to function independently, and, on April 30, 1965, the articles of incorporation for the American Association for Accreditation of Laboratory Animal Care (AAALAC) were filed and incorporated in the state of Illinois. The first two institutions accredited by AAALAC were the University of Louisville in Kentucky and Howard University in Washington, D.C. AAALAC was renamed the Association for Assessment and Accreditation of Laboratory Animal Care International (AAALAC International) in 1996, reflecting the organization's growth in other countries and its commitment to enhancing life sciences and quality animal care globally.

The Association for Assessment and Accreditation of Laboratory Animal Care International changed its name in 2016 to AAALAC International. Around the globe, AAALAC International continues to improve the welfare of animals in science and education through the accreditation of organizations meeting high standards of humane and responsible animal care and use. In addition to compliance with all applicable local and national regulations, AAALAC International assesses programs using principles outlined in the Three Primary Standards: (1) the *Guide*, (2) the Guide for the Care and Use of Agricultural Animals in Research and Teaching, and (3) the European Convention – ETS 123. In 2022, more than 1,000 institutions and companies in over 50 countries have earned AAALAC accreditation. AAALAC International envisions a world where excellence in animal welfare and science converge.

Animal Welfare Act

Prior to 1966, the U.S. did not have a national law addressing the acquisition and care of animals in biomedical research. Local humane societies were actively lobbying for the protection of pets and occasionally for farm animals. At the same time, the scientific community was improving the quality of animal care and well-being in the research laboratory, as evidenced by the creation of the ACP in 1950 and AAALAC in 1965. The increasing need for research dogs and cats was partially fulfilled by animal dealers who obtained these animals in various ways, including alleged theft for sale to research laboratories. A series of articles and news reports featuring photos of animal neglect and abuse at the hands of animal dealers and the theft and sale of pets to research laboratories culminated in two major articles in *Life* (February 4, 1966) and *Sports Illustrated* (November 29, 1965) magazines. These articles served as the impetus for legislative intervention, suggesting a need for regulation and a system of enforcement. In response to a wave of public sentiment, catalyzed in part by these articles, the Laboratory Animal Welfare Act (LAWA) was passed by Congress and signed into law by President Lyndon Johnson on August 24, 1966 (Public Law 89–544).[10] For the first time in the U.S., there were legal standards for laboratory animal care and use to ensure that certain animals intended for use in research facilities were provided humane care and treatment. The U.S. Department of Agriculture (USDA) was delegated the authority for implementing and enforcing this new law, and it promptly began providing applicable regulations.[11] Although the 1966 Act set minimum standards for the handling, sale, and transport of dogs, cats, nonhuman primates, guinea pigs, hamsters, and rabbits held by animal dealers or pre-research in laboratories, it was enacted primarily to protect dogs and cats and to counteract the business of pets being stolen by dealers and subsequently sold for research purposes. Research laboratories and dealers were required to register their facilities, be licensed, and undergo unannounced inspections by USDA's Animal Health Division, known since 1988 as Animal and Plant Health Inspection Service (APHIS), Animal Care (AC). Reports were subsequently prepared, and citations issued for noncompliance identified during the inspections. However, the law limited the USDA's authority within the research facility to only animals being held "pre-research," and these early inspections did not extend into the research laboratory, where animal care and use remained under the direction of the research investigator.

On December 24, 1970, President Nixon signed the 1970 amendments and enacted Public Law 91-579, commonly referred to as the Animal Welfare Act (AWA).[12] The definition of "animal" was changed to expand the scope of the animals covered, to include all warm-blooded animals used or intended for use in research, testing, experimentation, or exhibition, with the exception of horses not used in research and farm animals used or intended for use as food or fiber. The amendments also addressed an increased understanding of the importance of animal welfare. Research facilities were required to demonstrate, at least annually, professionally acceptable standards governing the care, treatment, and use of animals, including adequate veterinary care, as it related to the appropriate use of anesthetics, analgesics, and tranquilizing drugs during experimental procedures. The USDA was also required to compile a report to Congress, and make publicly available, data reported by research facilities on the number of animals used by species, inclusive of whether pain-relieving drugs were administered or justified for withholding.[13]

Public organizations, worried about the acquisition and welfare of animals destined for research, continued their activities. Such groups ranged from local and national humane societies concerned about animal welfare and well-being to radical animal rights activists opposed to the use of animals for any reason. Participation in the more radical elements of these groups gained momentum in the U.S., becoming more vocal in the early 1980s. Their activities peaked in a series of illegal break-ins and vandalism and moved to the forefront of public opinion soon after two incidents involving allegations of "animal cruelty" in two well-known research institutions.[14,15] This volatile climate raised public concern and visibility of animals in research, serving as a catalyst for amendments and clarifications to guidelines and regulations to provide additional protection for animal welfare.

A series of legislative bills and meetings in the early to mid-1980s, spearheaded by Senators Robert Dole, George Brown, John Melcher, and others, stimulated interest in including a mandate for the establishment of an Institutional Animal Committee. Such a committee would oversee animal care and use at each registered institution, and requirements for the committee's roles, composition, and responsibilities would be included as part of the USDA regulations. Public Law 99-198, the Food Security Act of 1985, Subtitle F –Animal Welfare, also called the Improved Standards for Laboratory Animals Act, was enacted on December 23, 1985.[16] This Act expanded the scope of USDA's authority over the welfare of animals in the laboratory, during testing, when utilized in higher education, and on exhibit, including captive marine mammals. The law required that each research facility have an Institutional Animal Committee composed of not fewer than three members. It included a Doctor of Veterinary Medicine and a member not affiliated in any way with the facility, other than as a member of the committee, and to provide representation for general community interests in the proper care and treatment of animals. A quorum of the committee was required for all formal actions, including review of the practices involving pain and distress, and review of the health and welfare of the animals documented during inspections conducted as part of the semiannual review of all animal study areas and animal facilities.

This amendment also addressed the care and management of laboratory animals by clarifying the meaning of *humane care* to include descriptions of sanitation, housing, and ventilation. It directed the Secretary of Agriculture to establish regulations to provide exercise for dogs and an adequate physical environment to promote the psychological well-being of nonhuman primates. This bill specified the IACUC's responsibility for ensuring that alternatives to animal use were explored when experimental procedures involved pain or suffering and that experiments were not unnecessarily duplicative. In addition, the Act defined practices considered to be painful, and it stated that no animal could be used in more than one major operative procedure from which it recovered (with listed exceptions). An additional mandate included the formation of an information service at the National Agricultural Library (currently known as the Animal Welfare Information Center) to provide resources to the general public and the research community on the care and welfare of animals and alternatives, as defined by reduction, refinement, and replacement. The final section of the amended Act described the penalties for release of trade secrets by regulators and the regulated community.

In the August 31, 1989, *Federal Register* publication, the USDA published final rules for Title 9, Code of Federal Regulations (CFR) Parts 1, 2, and 3 of the Animal Welfare Act Regulations (AWAR).[17] For the first time, each registered research facility was required to appoint an IACUC "for the purpose of evaluating the care, treatment, housing, and use of animals, and for certifying compliance with the Act by the research facility." Although there were no significant changes from the 1985 mandate to the composition and responsibilities of the committee, each research facility was required to provide training for all personnel involved in animal care.

Subsequent amendments to the AWA have led to regulations that included an expansion of the term "animal" to include more species and provisions for additional protections for dogs. The Farm Security and Rural Investment Act of 2002, Subtitle D, further redefined "animal" under the AWA, and although it continued to exclude common laboratory rats (*Rattus*), mice (*Mus*), and farm animals used in production agriculture, it narrowed the scope of the exclusion for birds to only those bred for use in research.[18] It is this amendment that firmly established that birds (not bred for use in research) were included with the warm-blooded animals covered under the AWA. Later, the 2008 Farm Bill (Public Law 110–246) included a number of amendments to the AWA prohibiting the importation of dogs into the U.S., unless they were at least six months of age, in good health, and had all necessary vaccinations.[19] All live dogs imported into the U.S. for resale, research, or veterinary treatment were also required to be accompanied by an import permit issued by APHIS.

Some of the most notable recent additions to the AWARs as they relate to registered research facilities include the final rule published on November 24, 2021, eliminating the requirement in § 2.30(a)(1) to update the research facility registration every three years after the facility's initial registration.[20] In addition, § 2.31(d)(5) was amended to remove the annual continuing review requirement and add the requirement for a complete review of activities at appropriate intervals as determined by the IACUC, but not less than every three years. The 2021 Congressional Appropriations Act was of particular interest since it required the USDA to lift the July 31, 2013, stay of a final rule mandating contingency planning. Consequently, the USDA published a final rule in the *Federal Register* on December 3, 2021, requiring all entities responsible for AWA-regulated species to meet certain requirements for contingency planning and training of personnel to safeguard the health and welfare of the animals in the event of possible emergencies or disasters (9 CFR § 2.134).[21] Perhaps of greatest significance in terms of the scope of protection and enforcement under the AWA and its regulations was the publication of a final rule on February 21, 2023, that implemented regulations and standards governing the humane handling, care, treatment, and transportation of birds, excluding birds bred for use in research.[22] USDA APHIS AC continues to ensure the humane treatment of animals covered by the AWA through quality inspections and guidance and by providing excellent customer service, learning opportunities, and assistance through sound financial investments and by virtue of the veterinary medical officers and specialists who have expertise with many species, including marine mammals, exotic cats, nonhuman primates, and birds.[23]

Public Health Service Policy

The first Public Health Service (PHS) Policy on Humane Care and Use of Laboratory Animals (Policy) went into effect in 1973, with revisions in 1979, 1986, 1996, 2002, and 2015.[24] All PHS policies on this subject evolved from an NIH policy published in 1971 under the auspices of the NIH's Institutional Relations Section of the Division of Research Grants.[25,26] Each successive revision was intended to ensure that awardee institutions provided appropriate care and ensured the humane treatment of animals involved in research by placing more specific oversight responsibilities on the animal care committee.

The 1971 NIH Policy "Care, and Treatment of Laboratory Animals" (NIH 4206) stated that institutions or organizations using warm-blooded animals in projects or demonstrations supported with funds by NIH grants, awards, or contracts would "assure the NIH that they will evaluate their animal facilities in regard to the maintenance of acceptable standards for the care, use, and treatment of such animals."[27] The NIH-funded institution could demonstrate that it was either accredited by a recognized professional laboratory animal accrediting body or had established its own committee to carry out its assurance function. The minimal number of committee members was not stated, but at least one member had to be a Doctor of Veterinary Medicine. Guidelines to be followed included the 1968 *NIH Guide for Laboratory Animal Facilities and Care* (PHS Publication No. 1024), all applicable portions of the AWA, and an appended set of guidelines known as the NIH Guidelines for Use of Experimental Animals. The committee was required to inspect the animal facilities of the institution at least once a year and report its findings and recommendations to responsible officials of the institution. Additionally, records of activities and recommendations must be made available for inspection by NIH representatives. Under this policy, AAALAC accredited institutions were not required by NIH to have an animal care committee.

The first PHS Policy replaced the NIH Policy on July 1, 1973, and continued to accept AAALAC accreditation in lieu of an institutional committee.[26,28] The minimum number of committee members was set at three, and unaccredited or partially accredited institutions were required to have a committee. A veterinarian was required on the committee only when institutions used a "significant number of animals." If the number of animals used in activities supported by all components of the Department of Health, Education, and Welfare (DHEW) was not "significant," a veterinarian was not required, but one of the committee members must then be "a scientist with demonstrated expertise in the care and use of laboratory animals." If such a person was not available, a consultant veterinarian was an acceptable alternative. A "significant number of animals" was not quantitatively defined, and the final determination was made by the DHEW from animal inventory information provided as part of the institution's Assurance statement.

The policy for institutional review of applications and proposals stated, "Grantee and contractor institutions are encouraged to review their applications and proposals in the light of the pertinent provisions of the Animal Welfare Act, the standards set by the Institute of Laboratory Animal Resources (the NIH *Guide*), and the DHEW Principles for the Use of Laboratory Animals, and to familiarize their staff with these provisions, standards, and principles." However, there was no requirement that institutional committees perform review of individual proposals or regularly provide to the DHEW summaries or certifications of such committee actions. There was also a requirement to keep records of committee activities for at least three years after the budget period. Assurance statements had to list the facilities and components of the program and required an "evaluation committee" to conduct periodic facility inspections and report to responsible officials at least once a year. For the first time, restrictions could be imposed for non-compliance with the terms of the Policy and the institution's eligibility to receive future DHEW grants or contracts involving the use of animals could be withdrawn.

Unlike previous versions of the PHS Policy, the third Policy on record was applicable to all grants, contracts, or other awards from all PHS components of the Department of Health and Human Services (formerly DHEW) involving the use of live vertebrate animals used or intended for use in research, experimentation, testing, training, or related purposes.[29,30] The 1979 revised PHS Policy was adopted as NIH Manual Chapter 4206 and the PHS Grants Manual Chapter 1-43, "Animal Welfare."[26] It required institutions to submit an Assurance signed by the Institutional Official (IO) to the NIH Office for Protection from Research Risks (OPRR), currently known as Office of Laboratory Animal Welfare (OLAW), and must be deemed acceptable before receiving an award for studies involving animals or animal facilities. OPRR was responsible for the compliance, general administration, and coordination of the Policy and not for scientific merit review or the purposes for which PHS-supported projects were awarded. The Assurance committed the institution to follow the edition of the *Guide* then in use, the AWA, DHEW Principles for the Use of Animals, and other applicable laws and regulations.

All grantee or contractor institutions with PHS awards involving animals or animal facilities were required "to appoint and maintain an institutional committee to maintain oversight of its animal facilities and procedures." A minimum of five members, including at least one veterinarian, knowledgeable in the care and use of animals used in research, was required to be on the committee. AAALAC accreditation was again recommended as the best means of demonstrating conformance with good animal care and use provisions of the *Guide*. An alternative to AAALAC accreditation was annual review of the animal facilities and procedures by the institution's committee. Institutions, as in the past, were obligated to keep records of committee activities or accrediting body determinations, including recommendations and determinations, with those records being available for inspection by authorized PHS officials. Although review of individual proposals or projects by the institution's committee was encouraged, it was still not required. However, proposals were required to indicate the rationale for using animals, confirmation that the species and numbers of animals were appropriate, that unnecessary discomfort and injury to animals would be avoided, and that analgesic, anesthetic, and tranquilizing drugs would be used where indicated to minimize stress to the animals. Even though the review remained merely a recommendation, the suggestion that the committee act as a reviewing body further strengthened its role. The institution was required to provide the names, position titles, and credentials of the committee members, as well as any changes in the membership on an annual basis to OPRR. A sample statement on the Assurance was preceded by the following:

> The samples given below for committee members are not intended to dictate numbers of or qualifications for committee members. However, except in unusual circumstances, the committee should be of at least five members and include at least one veterinarian. Any such unusual circumstances should be explained in a statement accompanying the assurance.

The following was contained in the sample statement:

> We have appointed and will maintain a committee of at least five members to maintain oversight of our animal care program. The members have appropriate education and experience to perform their duties with respect to the types of animals and species used and the kinds of projects to be undertaken. If the conduct of a specific project is to be reviewed, the quorum will not include any member having an active role in the project. Changes in membership will be reported annually to the Office for Protection from Research Risks (OPRR), National Institutes of Health (NIH), and PHS.

Three options were provided for reporting of an institution's degree of compliance: AAALAC accreditation, certification by its own institutional committee, or committee recommendations of improvements needed for compliance. In the event noncompliance with the *Guide* was identified by the committee or OPRR, the institution was required to submit an annual report to OPRR describing progress toward full conformance. Each institution was required to submit a completely new Assurance to OPRR every five years.

After allowing five years to study the effectiveness of the Assurance system, in a March 1984 report to the Director of the NIH, it was concluded that the committees frequently seemed "less than fully assertive" in carrying out their responsibility.[24,31] The report was based on random site visits to ten Assured institutions that were compliant with the *Guide*, but not AAALAC accredited. These ten institutions were selected from the more than 800 institutional Assurances on file with OPRR and from each of the ten geographic regions of DHHS.

It was evident that all ten institutions shared a common concern that the care and use of laboratory animals must be in accordance with good science and consideration of animal welfare. The two conclusions drawn from the report were that:

1. Reliance upon voluntary compliance with PHS policy and recommendations in the *Guide* is a realistic approach to fostering proper care and use of laboratory animals in biomedical research. There is no reason to believe that regular NIH inspections are needed or would be more effective than the traditional assurance process.
2. The present assurance system should be strengthened by modifying the current PHS policy on animal welfare to promote more conscientious involvement by both the NIH and its awardee institutions.

A recommendation after this study suggested that "the PHS Policy should be further modified to define more precisely the responsibilities of the awardee institutions, particularly the role of the animal care committee. It is imperative that the experience and expertise of the members of such committees be used to conduct full and effective reviews of proposals involving research with animals. The appointment of a nonscientist and an individual unaffiliated with the institution should be given serious consideration."[27,31] This study resulted in the revision of the PHS Policy, which further defined and outlined the requirements of an animal care committee. The new "Policy on the Humane Care and Use of Laboratory Animals by Awardee Institutions" was developed under the authority and chairmanship of OPRR and published in a special edition of the NIH Guide for Grants and Contracts, "Laboratory Animal Welfare," on June 25, 1985.[26]

That same year, the Health Research Extension Act (HREA), was enacted on November 20, 1985, as Public Law 99-158, "Animals in Research," and it became the statutory authority for the PHS Policy.[32] As a result of the HREA, at least four provisions had to be added to the June 1985 revised PHS Policy, which went into effect September 16, 1986.[26] OPRR was responsible for the general administration and coordination of the PHS Policy. The Policy endorsed the "US Government Principles for the Utilization and Care of Vertebrate Animals Used in Testing, Research and Training" which were promulgated in 1985 by the Interagency Research Animal Committee (IRAC). It also required that methods of euthanasia be consistent with the American Veterinary Medical Association (AVMA) Panel on Euthanasia unless a deviation is justified for scientific reasons in writing by the investigator. The PHS Policy also continued to require institutions to use the existing *Guide* as a basis for developing and implementing an institutional program for activities involving animals. In addition, all institutions were required to comply, as applicable, with the AWA, and other federal statutes and regulations relating to animals. For the first time, the PHS Policy applied to NIH intramural research and the IACUC was appointed by the Chief Executive Officer (CEO) of the institution.[24] The Policy required that each institution receiving PHS funds for research involving live vertebrate animals provide a written assurance acceptable to the PHS. It also identified the IO as having ultimate responsibility for the institution's program for the care and use of animals, with the authority to sign the institution's Assurance and commit on behalf of the institution that the requirements of the PHS Policy would be met. Footnote 1 of the PHS Policy stated that "The name Institutional Animal Care and Use Committee (IACUC) as used in this policy is intended as a generic term for a committee whose function is to ensure that the care and use of animals in PHS-supported activities is appropriate and humane in accordance with this policy. However, each institution may identify the committee by whatever name it chooses." The committee consisted of not less than five members, including an individual unaffiliated with the institution; a veterinarian with program responsibilities and who has training or experience in laboratory animal science and medicine; a practicing scientist experienced in research involving animals; and a member whose concerns are in a non-scientific area. Among the seven functions described, IACUCs were responsible for conducting program evaluations at least annually and reviewing and approving those sections of applications for PHS funds that relate to the care and use of animals before PHS funds may be awarded. Institutions not accredited by AAALAC were required to conduct a self-assessment of the institution's program based on the *Guide*.

The March 1996 reprint of the PHS Policy contained minor corrections and did not change significantly from the previous 1986 version. In 2000, the Animal Welfare Division of OPRR was renamed the Office of Laboratory Animal Welfare (OLAW) and this Office assumed the responsibility for the general administration and coordination of the 2002 PHS Policy by providing specific guidance, instruction, and materials to institutions that must comply with the Policy.[33] The PHS Policy incorporated the HREA and IRAC's U.S. Government Principles for the Utilization and Care of Vertebrate Animals Used in Testing, Research, and Training. It continued to use the most current *Guide* as a basis for developing and implementing an institutional program for activities involving animals. It also required that the program description include an explanation of the training or instruction available to scientists, animal technicians, and other personnel involved in animal care, treatment, or use. This training or instruction must include information on the humane practice of animal care and use and the concept, availability, and use of research or testing methods that minimize the number of animals required to obtain valid results and minimize animal distress. The IACUC was required to evaluate and prepare twice-yearly reports on all of the institution's programs and facilities (including satellite facilities) for activities involving animals instead of once each year. The IACUC, through the IO, was responsible for reporting requirements, and minority views filed by members of the IACUC must be included in reports to OLAW. The revised policy permitted institutions with PHS Animal Welfare Assurances to submit verification of IACUC approval for competing applications or proposals after peer review but prior to an award. Also, new footnotes 6 and 12 of the PHS Policy provided institutions with the option of coding the names of the IACUC members in materials routinely submitted to OLAW.

OLAW continues to monitor compliance of Assured institutions with the most recent 2015 reprint of the PHS Policy by the authority of the HREA of 1985.[34] The PHS Policy applies to research involving live, vertebrate animals used or intended for use in research, research training, experimentation, or biological testing or for related purposes conducted or supported by the PHS. The PHS agencies include the NIH, Biomedical Advanced Research and Development Authority (BARDA), Centers for Disease Control and Prevention (CDC), and Food and Drug Administration (FDA). Through memoranda of understanding with OLAW, the PHS Policy also applies to animal activities supported or conducted by the National Aeronautics and Space Administration (NASA), National Science Foundation (NSF), and Department of Veterans Affairs (VA). The current PHS Policy reflects minor updates to the 1985 reprint, including adoption of the Eighth Edition of the *Guide*, which was released by the NAS ILAR in 2011. Footnotes 2, 7, 9, 11, and 13 were modified to require PHS Assured institutions to comply with USDA regulations that are applicable to their programs. All institutions are also required to comply with applicable provisions of the AWA and other federal statutes and regulations relating to animals.

21st Century Cures Act

In 2016, Title II, Section 2034(d) of the 21st Century Cures Act directed the NIH, in collaboration with the USDA and Food and Drug Administration (FDA), to complete a review of applicable regulations and policies for the care and use of laboratory animals and to make revisions, as appropriate, to reduce administrative burden on investigators while maintaining the integrity and credibility of research findings and protection of research animals.[35] The Act instructed NIH to: (1) seek the input of experts, if appropriate; (2) identify ways to ensure applicable regulations and policies are not inconsistent, overlapping, or unnecessarily duplicative; (3) take steps to eliminate or reduce identified inconsistencies, overlap, or duplication among such regulations and policies; and (4) take other actions, as appropriate, to improve the coordination of regulations and policies with respect to research with laboratory animals. A Working Group of federal subject matter experts convened and reviewed published reports, communications, and surveys, highlighting the regulations and policies that contribute to researchers' administrative burden. Listening sessions were conducted and requests for information seeking input were issued before releasing the final report, Reducing Administrative Burden for Researchers: Animal Care and Use in Research, on August 28, 2019. The Report describes the efforts of the Working Group and the collaborative decisions of NIH, USDA, and FDA on the recommendations to protect animal welfare while reducing unnecessary administrative burden on researchers.

Both the USDA and the OLAW have made substantial progress on the steps and actions since the release of the final report in this report and will continue to identify additional areas to protect animal welfare while reducing unnecessary administrative burden on researchers. USDA APHIS published a final rule in the *Federal Register* on November 24, 2021, that amended § 2.31(d)(5) by removing the continuing review requirement and adding the requirement for a complete review of activities at appropriate intervals as determined by the IACUC, but not less than every three years.[20] Additional amendments to § 2.30(a)(1) and § 2.36(a) eliminated the requirement for research facility registration updates at three-year intervals and for CEO and IO signatures on Annual Reports, respectively. Some of the guidance OLAW has released since 2020 include changing the reporting period for the Annual Report to OLAW to harmonize with the reporting period of the USDA.[36] Institutions also have the option to use sections of the AAALAC Program Description when preparing a Domestic Animal Welfare Assurance.[37] Both federal agencies clarified existing flexibilities for conducting semiannual facility inspections.[38] Two equally important guidance published by OLAW on October 18, 2021, clarified the requirements, responsibility, timing, and conduct of grant and contract to protocol congruency review for NIH grant applications and contract proposals, respectively.[39,40] Harmonization of regulations from other federal agencies, e.g., Department of Defense, National Science Foundation, National Aeronautics and Space Administration, Department of Veterans Affairs, continues to be an important and ongoing initiative in reducing burden as part of the 21st Century Cures Act.

Department of Defense (DoD)

While the scientific community and the public sector were gradually evolving guidelines and policies to assure animal welfare and good animal care,[41–50] the Department of Defense (DoD) had similar concerns for applicable research at the agency. As early as 1961, the DoD issued the *Policy on Experimental Animals*,[51–53] which directed that "all aspects of investigative programs involving the use of laboratory animals and sponsored by Department of Defense agencies will be conducted according to the Principles of Laboratory Animal Care as promulgated by the National Society for Medical Research." While this early Policy provided few specifics, several revisions followed and included all animals used in DoD laboratories, both in the U.S. and abroad.

The 1982 DoD Directive 3216.1, "Use of Laboratory Animals in DoD Programs," established policy on the humane treatment and appropriate care of animals used in research. In general, it followed the AWA and the *Guide*, with an attempt to incorporate alternatives to animal use in the form of replacement, reduction, and refinement. A Joint Service regulation 70-18 was issued by the Army, Navy, Air Force, Defense Nuclear Agency, and Uniformed Services University on June 1, 1984, *Use of Animals in DoD Programs*, to implement the directive's policies uniformly for all DoD components. This directive required that "all DoD organizations having animals (other than military working, recreational, and ceremonial) will seek accreditation by AAALAC."[54] It further required that "the local commanders of each DoD organization conducting or sponsoring activities involving animals in Research, Development, Teaching and Education (RDTE), clinical investigations, diagnostic procedures, or instructional programs will form a committee to oversee the care and use of animals." Such committees were to be appointed by the local commander and consisted of at least three members, including a Doctor of Veterinary Medicine, with the responsibility for the review of protocols and assuring compliance with policies, standards, and regulations. The concept and practice of such committees to review and assure appropriate animal care and use, while not known by the acronym *IACUC*, were in place at many military installations prior to their being regularly formed at most academic institutions. Army Regulation (AR) 40-33, "The Care and Use of Laboratory Animals in DOD Programs" superseded AR 70-18 on February 16, 2005.[55]

Both the 1995[56] and 2005[57] DoD Directives required review and approval of proposals or protocols by a duly constituted IACUC consisting of a minimum of five members. These Directives required both a primary and an alternate nonaffiliated member intended to facilitate community representation at each IACUC meeting, facility inspection, and semiannual program review. DoD organizations or facilities maintaining animals for use in DoD programs were still required to attain and maintain AAALAC (International) accreditation. All animal research conformed to the requirements of the AWA, *Guide*, U.S. Government Principles for the Utilization and Care of Vertebrate Animals Used in Testing, Research, and Training, and the requirements of the applicable regulations of the USDA. The 1995 Directive defined "animal" as any dog, cat, nonhuman primate, guinea pig, hamster, rabbit, or any other live vertebrate animal, including birds, rats of the genus *Rattus* and mice of the genus *Mus* bred for use in research, training, testing, or experimentation purposes. In contrast, the 2005 Directive included any live vertebrate animal used or intended for use in research, training, or testing, or for experimentation purposes (other than military working, recreational, and ceremonial and intended for use as livestock and poultry as food or fiber).

The September 13, 2010, DoD Directive[58] once again redefined "animals" as any living or dead vertebrate animal, including birds, cold-blooded animals, rats of the genus *Rattus*, and mice of the genus *Mus*. The offspring were considered animals only after hatching for avian and other egg-laying vertebrate species and the larval offspring were considered animals for fish and amphibians. The IACUC membership was still set at a minimum of five members; however, at least one veterinarian and one nonaffiliated member (or his or her alternate) were required to be present for a quorum.

In the most recent March 10, 2019, DoD Directive, IACUCs are still required to have a minimum of five members, including a Chair, veterinarian, non-scientific member, and nonaffiliated member.[59] DoD IACUCs should also consider the designation of an alternate member(s) for the nonaffiliated member to facilitate community representation in the IACUC's activities. Although the definition of "animal" did not change from the 2005 Directive, the requirement for a quorum was redefined "when a majority of the voting members of the IACUC are present." The Animal Care and Use Review Office (ACURO) is a component of the United States Army Medical Research and Development Command (USAMRDC), Office of Research Protections (ORP), responsible for the oversight of the implementation of USAMRDC, U.S. Army, and DoD policies regarding the use of animals in research, development, testing, evaluation, and training.[60] The DoD is firmly committed to ensuring that all funded or supported research involving animals is conducted in compliance with federal, DoD, USAMRDC, and international regulatory requirements.

National Science Foundation (NSF)

The National Science Foundation (NSF) is an independent federal agency created by the National Science Foundation Act of 1950, as amended (42 USC 1861-75), with a mission to promote the progress of science; advance the national health, prosperity, and welfare; and secure the national defense.[61] Dating as far back as October 1977, Section 713 of the NSF Grant Policy Manual (GPM) (NSF 77-47) included provisions requiring any grantee performing research on warm-blooded animals to comply with the Laboratory Animal Welfare Act of 1966, with the expectation that the guidelines described in the *Guide* be followed.[62]

About a decade later, the 1989 GPM NSF 88-47[63] described the responsibility of grantees performing activities that involved vertebrate animals to comply with the AWA and AWAR as they relate to the care, handling, and treatment of vertebrate animals held or used for research, teaching, or other activities. In addition, grantees were required to ensure that the recommendations in the *Guide* were followed, and that they complied with the PHS Policy and the U.S. Government Principles for the Utilization and Care of Vertebrate Animals Used in Testing, Research, and Training. Each proposal submitted by grantees was also required to be reviewed by an IACUC and accompanied by a statement from the grantee that assures the grantee's compliance with the PHS Policy. Sufficient information had to be provided to allow evaluation of the appropriateness of the choice of species, the number of animals to be used, and any unnecessary exposure of animals to discomfort, pain, or injury.[64]

Although the 1995 NSF GPM 95-26[65] animal welfare requirements did not vary significantly from the 1989 Policy Manual, the July 2002 NSF GPM[66] included significant changes. Animal coverage included not only vertebrate animals covered by the AWA, but it was extended to rats, birds, and mice. Apart from requiring IACUC review and approval prior to award, additional IACUC approval was needed for protocols that were changed substantively from those originally proposed and approved, or if the previous approval was more than three years old. Awards to U.S. grantees for projects involving the care or use of vertebrate animals at a foreign institution or a foreign field site also required approval of research protocols by the U.S. grantee's IACUC. However, for projects awarded to a foreign institution, NSF required a statement of compliance indicating that animal activities would be conducted in accordance with all applicable laws in the foreign country and that the International Guiding Principles for Biomedical Research Involving Animals would be followed.

In April 2007, the NSF GPM, which previously was a standalone compendium of basic NSF policies and procedures for use by the grantee community and NSF staff document, was renamed the NSF Award & Administration Guide (AAG) and was incorporated as Part II of the new NSF Proposal & Award Policies & Procedures Guide (PAPPG).[67] The animal welfare requirements, however, mirrored those described in the 2002 GPM. About five years later, the October 2012 PAPPG[68] included provisions to follow applicable taxon-specific guidelines, as appropriate. The memorandum of understanding (MOU) between NSF and NIH OLAW on August 10, 2015, represented a major milestone in ensuring consistent and effective oversight of the welfare of animals used in activities funded by the NSF.[69] Institutions that received NSF support were required to include NSF-supported activities with live vertebrate animals as covered activities in their Assurance and promptly report situations involving NSF-supported animal activities to OLAW as required by Section IV.F.3. of the PHS Policy. Consequently, both the 2017 (NSF 17-1)[70] and October 4, 2021 (NSF-22-1)[71] PAPPGs clarified the use of vertebrate animals to include rats, mice, and birds for research or education on NSF-supported projects by stating that "any project proposing the use of vertebrate animals for research or education must comply with the provision in PHS Assurance for Institutional Commitment (Section II) that requires the submitting organization to establish and maintain a program for activities involving animals in accordance with the *Guide for the Care and Use of Laboratory Animals* (*Guide*)." The 2021 PAPPG requires grantees include adequate information in the Project Description of the application to enable reviewers to evaluate the: (1) rationale for involving animals; (2) choice of species and number of animals to be used; (3) description of the proposed use of the animals; (4) exposure of animals to discomfort, pain, or injury; and (v) description of any euthanasia methods to be used. NSF remains committed to ensuring appropriate policies and procedures are implemented aimed to ensure the "humane care and treatment of any [live] vertebrate animal used or intended for use in such activities as field or laboratory research, development, training, experiments, biological testing, or for related purposes supported by NSF grants."

National Space and Aeronautics Administration (NASA)

Although the National Aeronautics and Space Administration (NASA) began operating as the nation's civilian space agency in 1958 after passage of the National Aeronautics and Space Act,[72] documentation of the first animal in space dates as far back as June 11, 1948, when a V-2 Blossom launched from White Sands, New Mexico, carried a rhesus monkey named Albert I.[73] NASA's policies governing the care and use of its research animals initially required compliance predominantly with federal law, most notably the 1966 AWA, but they were later expanded in the 1980s with the development of additional policies and regulations. In the early to mid-1980s, NASA's animal research policy existed as a proposed NASA Management Instruction (NMI) requiring AAALAC accreditation for all NASA installations.[64] The establishment of an Animal Care and Use Committee (ACUC) in each animal facility consisting of a research veterinarian, a biomedical scientist, a nonscientist, and a person not affiliated with NASA was also a required element of the NMI. Responsibility of the ACUC included overseeing the animal care facility, establishing specific guidelines, reviewing proposals, and making recommendations for approval or disapproval of funding. The "Animal User's Guide" for studies at the Ames Research Center in California exemplified the implementation with NASA's policy by maintaining an animal care facility with appropriate animal housing and monitoring of all animal care and experimentation progress by the ACUC.

The August 30, 1989, 14 Code of Federal Regulations (CFR) Part 1232 titled "Care and Use of Animals in the Conduct of NASA Activities," established the policy; implementation procedures; and management authority and responsibility for the care and use of vertebrate animals in the conduct of NASA activities.[74] This rule required NASA's laboratories and the institutions performing NASA-supported activities using any live vertebrate animals to comply with the AWA, PHS Policy, *Guide*, and the U.S. Government Principles for the Utilization and Care of Vertebrate Animals Used in Testing, Research, and Training. NASA also codified a formal rule known as the "NASA Principles for Ethical Care and Use of Animals" in the Sundowner Report of 1996.[75,76] Akin to the Belmont Report, the Sundowner Report expounds on the premise that the use of animals in research involves responsibility, not only for the stewardship of the animals, but also to the scientific community and society and adherence with the tenet of three basic principles: (1) respect for life, (2) societal benefit and (3) nonmaleficence. As a result, the March 28, 1998, NASA Policy Directive (NPD) 8910.1 titled "Care and Use of Animals" required that NASA conduct all activities involving vertebrate animals at AAALAC International accredited centers and in compliance with the PHS Policy, *Guide* and AWA, and NASA Principles for the Ethical Care and Use of Animals.[77]

In 2003, the International Space Life Sciences Working Group (ISLSWG) recommended that the Committee on Space Research (COSPAR) develop an international policy on animal care and use for space-borne research.[78] The resulting COSPAR "Policy and Guidelines for the Use and Care of Animals in Space-borne Research" applied to the conduct of all space-based and associated ground-based procedures involving vertebrate animals that directly support flight activities. It outlined ten principles focused on replacement, reduction, and refinement, consistent with the Council for International Organizations of Medical Sciences (CIOMS) "International Guiding Principles for Biomedical Research Involving Animals." Consequently, for ground-based activities performed by an international partner, the December 2, 2011, NASA Procedural Requirements (NPR) 8910.1c[79] required compliance with applicable local and national laws and regulations of the country where the activities occur, with a minimum standard being the CIOMS "International Guiding Principles for Biomedical Research Involving Animals" and the COSPAR "Policy and Guidelines for the Use and Care of Animals." This NPR defined "animal" as any live or dead vertebrate animal used or intended for use in research, teaching, testing, or experimentation or hardware development, including wildlife and agricultural animals when used exclusively in nonagricultural research activities. Each institution using animals in research, testing, or training activities must also establish an IACUC in accordance with the requirements of the AWA and PHS Policy.

About a decade later in 2019, a MOU between NASA and NIH OLAW was established to obtain services in the administration and coordination of the PHS policy by providing assistance with negotiating Animal Welfare Assurances; evaluating allegations of noncompliance; advising the implementation and interpretation of the PHS Policy; authorizing waivers to the PHS Policy; and conducting site visits.[80] The most recent January 20, 2022, NPD established policy for the care and use of vertebrate animals and higher order cephalopods.[81] This Directive required adherence to the NASA Principles for Ethical Care and Use of Animals and compliance with all applicable federal laws, regulations, policies, and guidelines regarding the care and use of vertebrate animals and cephalopods, including the AWA, PHS Policy, *Guide*, and COSPAR Guidelines for the Use and Care of Animals in Space-borne Research. All members of the animal care and use program are instrumental to ensuring NASA's ongoing commitment to humane and responsible use of live vertebrates and higher order cephalopods in agency-supported research, testing, teaching, and hardware development activities. This is accomplished part through the compliance with applicable federal, state, and local animal welfare laws and regulations, policies, and guidelines, and ensure adherence to the NASA Principles for the Ethical Care and Use of Animals, as described in the *Sundowner Report*.

Principles for Animal Care in Research

The *International Guiding Principles for Biomedical Research Involving Animals* (International Guiding Principles) were first developed and first published in 1985 by the Council for International Organizations of Medical Sciences (CIOMS). These principles were developed as a result of extensive international and interdisciplinary consultations spanning over a three-year period between 1982 and 1984.[82] Not surprisingly, a revision of the 1985 International Guiding Principles was warranted, taking into account, among other things, the exponential growth in research with animals, concomitant with technological advancements and societal expectations for exhibiting a culture of care for the animals. In 2012, CIOMS, in partnership with the International Council for Laboratory Animal Science (ICLAS), updated the International Guiding Principles.[83] The revised 2012 International Guiding Principles serves as an international benchmark, providing an ethical framework for the development of laws and policies to ensure the appropriate use of animals. The *U.S. Government Principles for the Utilization and Care of Vertebrate Animals Used in Testing, Research, and Training* (U.S. Government Principles) were promulgated in 1984 by the U.S. Interagency Research Animal Committee and were, to a large extent, based on the CIOMS International Guiding Principles.[84] The U.S. Government Principles are currently published in both the PHS Policy and the *Guide*. The need for authorization and ethical review is described in Principle X of the International Guiding Principles.

> X. While implementation of these Principles may vary from country to country according to cultural, economic, religious, and social factors, a system of animal use oversight that verifies commitment to the Principles should be implemented in each country. This system should include a mechanism for authorization (such as licensing or registering of institutions, scientists, and/or projects) and oversight which may be assessed at the institutional, regional, and/or national level. The oversight framework should encompass both ethical review of animal use as well as considerations related to animal welfare and care. It should promote a harm-benefit analysis for animal use, balancing the benefits derived from the research or educational activity with the potential for pain and/or distress experienced by the animal. Accurate records should be maintained to document a system of sound program management, research oversight, and adequate veterinary medical care.

This need for an IACUC is specifically addressed in the U.S. Government Principles by the following.

> IX. Where exceptions are required in relation to the provisions of these Principles, the decisions should not rest with the investigators directly concerned but should be made, with due regard to Principle II, by an appropriate review group such as an institutional animal care and use committee. Such exceptions should not be made solely for the purposes of teaching or demonstration.

Evolution of the IACUC

The 1986 PHS Policy first described the IACUC in its present form and membership composition, and it provided only general guidance on how such a committee should be formed and operated. With an amendment to the AWA effective October 30, 1989, USDA regulations required a similar IACUC. Although membership requirements differ, most institutions have a single oversight committee that satisfies both PHS and USDA requirements. The Scientists Center for Animal Welfare (SCAW) was instrumental in providing early guidance to institutions on IACUC functions and organization through regional conferences and workshops, culminating in a special AALAS publication, *Effective Animal Care and Use Committees*.[85] Since 1983, training, workshops, and regular conferences of this type have been available to the research and animal care and use community. Several of the current initiatives and organizations are supported, in part, by OLAW's National Animal Welfare Education Program, which includes the Interagency Collaborative Animal Research Education (ICARE) Project, SCAW, IACUC Administrators Association (IAA), Public Responsibility in Medicine and Research (PRIM&R), and IACUC 101 Series.

As described in the AWAR (§ 2.31,b,2), the IACUC "shall be composed of a chairman and at least two additional members" and must "prepare reports of its evaluations" … "and submit the reports to the Institutional Official of the research facility."[18] While originally borrowed from the human Institutional Review Board structure, the establishment of local animal care committees (i.e., IACUCs) to review and assure animal welfare and well-being is a common practice, if not a requirement, in the animal research community. The goal of each committee is compliance with guidelines and regulations while allowing for the flexibility of individual institutions to tailor their programs to effectively meet the unique needs of the institution. Active participation by research scientists allows for the unique needs of research investigators to be voiced, participation by members unaffiliated with the institution protects the public conscience, and veterinarians assure appropriate medical care and provisions. This committee has become the primary oversight mechanism for animal care and use, responsible through an IO and reporting directly to regulatory and granting agencies on animal care and use matters. OLAW, USDA, and AAALAC International continue to provide additional guidance and clarification of the expectations and responsibilities of the IACUC through respective online guidance and "frequently asked questions." There is a continuing need for education to ensure that this concept works as well as possible.

Acknowledgments

The author would be remiss for not acknowledging, first and foremost, the legacy and contributions of the author of the previous edition of this chapter, Dr. Harry Rozmiarek, upon which the foundation of the current chapter rests. The author would also like to acknowledge the valuable contributions and insightful comments offered by the following reviewers for their technical expertise related to their associations with the various agencies: Dr. Patricia Brown, OLAW; Dr. Axel Wolff, OLAW; Drs. Lance Bassage and Louis DiVincenti, USDA APHIS; Dr. Helen E. Diggs, AAALAC International; Major Michael Bencivenga, U.S. Army; Dr. Edda Thiels, NSF; and Dr. Richard Simmonds, Nevada System of Higher Education and University of Nevada, School of Medicine.

REFERENCES

1. McLeod, C. and Hartley, S. 2018. Responsibility and laboratory animal research governance. *Science, Technology, & Human Values*. 43: 723–741.
2. National Agricultural Library, U.S. Department of Agriculture. *Twenty-Eight Hour Law*. https://www.nal.usda.gov/animal-health-and-welfare/twenty-eight-hour-law. Accessed March 1, 2024.
3. Middleton, C.C. and La Regina, M. 2007. The history of the American College of Laboratory Animal Medicine: The first fifty years 1957–2007. Available at: https://www.aclam.org/about/college-history. Accessed March 1, 2024.
4. Rozmiarek, H. 1986. Current and future policies regarding animal welfare. *Investigative Radiology* 22: 175.
5. National Research Council. 2011. *Guide for the Care and Use of Laboratory Animals: Eighth Edition*.
6. U.S. Department of Health and Human Services, National Institutes of Health, Office of Laboratory Animal Welfare. 2015. Public health service policy on humane care and use of laboratory animals.
7. Van Sluyters, R.C. 2008. A guide to risk assessment to animal care and use programs: The metaphor of the 3-legged stool. *ILAR Journal* 49: 372–378.
8. AAALAC International. History (long version). https://www.aaalac.org/about/history/long-version/. Accessed March 1, 2024.
9. Miller, J.G. and Clark, J.D. 1999. The history of the Association for Assessment and Accreditation of Laboratory Animal Care International, in *50 Years of Laboratory Animal Science*, pp. 39–43.
10. 89th Congress, H. R. 13881. 1966. *Public Law 89-544*.
11. U.S. Department of Agriculture, Animal and Plant Health Inspection Service, Animal Care. https://www.aphis.usda.gov/aphis/home. Accessed March 1, 2024.
12. 91st Congress 1970. H. R. 19846, Public Law 91-579.
13. Schwindaman, D.F. 1999. The history of the Animal Welfare Act, in *50 Years of Laboratory Animal Science*, pp. 147–151.
14. Fraser, C. 1993. The raid at Silver Spring. *The New Yorker*, p. 66, 19.
15. McCabe, K. 1986. Who will live, who will die. *The Washingtonian*, pp. 113–157.
16. 99th Congress. 1985. H.R. 2100, *Public Law 99-198*.
17. U.S. Department of Agriculture, Animal and Plant Health Inspection Service. 1989. *Final Rule: Animal Welfare; 9 CFR Parts 1, 2, and 3*, vol. 54.
18. 107th Congress. 2002. H.R. 2646, *Public Law 107-171*.
19. 110th Congress. 2008. H.R. 6124, *Public Law 110-246*.
20. U.S. Department of Agriculture, Animal and Plant Inspection Service. 2021. *AWA Research Facility Registration Updates, Reviews, and Reports*, vol. 86.
21. U.S. Department of Agriculture, Animal and Plant Inspection Service. 2021. *Handling of Animals; Contingency Plans*, vol. 86.
22. U.S. Department of Agriculture, Animal and Plant Health Inspection Service. 2023. *Final Rule: Standards for Birds Not Bred for Use in Research under the Animal Welfare Act, 9 CFR Parts 1, 2, and 3*, vol. 88.
23. U.S. Department of Agriculture, Animal and Plant Inspection Service. Animal Welfare Act. https://www.aphis.usda.gov/aphis/ourfocus/animalwelfare/awa/ct_awa_program_information. Accessed March 1, 2024.
24. National Institutes of Health, Office of Laboratory Animal Welfare. 2002. *Institutional Animal Care and Use Committee Guidebook*, 2nd edition.
25. Whitney, R.A., Jr. 1987. Animal care and use committees: History and current national policies in the United States. *Laboratory Animal Science*, no. Special Issue, pp. 18–21.
26. Gordon, H.S. 1999. The history of the public health service policy on the humane care and use of laboratory animals, in *50 Years of Laboratory Animal Science*, pp. 152–155.
27. U.S. Department of Health, Education, and Welfare, National Institutes of Health. 1971. Care and treatment of laboratory animals (NIH 4206), *NIH Guide for Grants and Contracts*, pp. 3–4.
28. Department of Health, Education, and Welfare. 1973. *Grants Administration Manual*, Chapters 1–43, Animal Welfare.

29. U.S. Congress, Office of Technology Assessment. 1986. Institutional and self-regulation of animal use, in *Alternatives to Animal Use in Research, Testing, and Education*, pp. 333–355.
30. U.S. Department of Health and Human Services, National Institutes of Health, Office for Protection from Research Risks. 1979. *Public Health Service Policy on Humane Care and Use of Animals.*
31. U.S. Department of Health and Human Services, National Institutes of Health. 1984. *Laboratory Animal Welfare*, 5th ed., vol. 13.
32. 99th Congress. 1985. H.R. 2409, *Sec. 495, Animals in Research.*
33. U.S. Department of Health and Human Services, National Institutes of Health. 2002. *Public Health Service Policy on Humane Care and Use of Laboratory Animals.*
34. U.S. Department of Health and Human Services, National Institutes of Health. 2015. *Public Health Service Policy on Humane Care and Use of Laboratory Animals.*
35. National Institutes of Health, HHS, Animal and Plant Health Inspection Service, USDA, and Food and Drug Administration, HHS. 2019. Reducing administrative burden for researchers: Animal care and use in research.
36. National Institutes of Health, Office of Laboratory Animal Welfare. 2020. *Notice of Update to the Reporting Period of the Annual Report to OLAW,* 2020.
37. National Institutes of Health, Office of Laboratory Animal Welfare. 2021. *Notice to Encourage Using AAALAC International Program Description Sections to Complete Parts of the OLAW Domestic Animal Welfare Assurance.*
38. National Institutes of Health, Office of Laboratory Animal Welfare. 2024. *Guidance on Flexibilities for Conducting Semiannual Inspections of Animal Facilities.*
39. National Institutes of Health, Office of Laboratory Animal Welfare. 2021. *Notice of Clarification of Institutional Responsibilities Regarding NIH Grant to Protocol Congruence Review.*
40. National Institutes of Health, Office of Laboratory Animal Welfare. 2021. *Notice of Clarification of Offeror Responsibilities Regarding Contract to Protocol Congruence Review.*
41. American Societies of Animal Science Societies. 2020. *Guide for the Care and Use of Agriculture Animals in Agricultural Research and Testing*, 4th Edition.
42. American Psychological Society, Committee on Animal Research and Experimentation. 2022. *Guidelines for Ethical Conduct in the Care and Use of Nonhuman Animals in Research.*
43. U.S. Department of Health and Human Services. 1988. *National Institutes of Health, Institutional Administrators' Manual for Laboratory Animal Care and Use.*
44. ILAR Committee on the Use of Laboratory Animals in Neuroscience and Behavioral Research. 2003. *Guidelines for the Use of Mammals in Neuroscience and Behavioral Research.*
45. ILAR Committee on the Use of Animals in Precollege Education. 2006. *Principles and Guidelines for the Use of Animals in Precollege Education.*
46. Joint AAMC-AAU Committee. 1985. *Recommendations for Governance and Management of Institutional Animal Resource.*
47. U.S. Department of Health and Human Services, NIH Publication No. 91-3207. 1991. *Preparation and Maintenance of Higher Mammals during Neuroscience Experiments.*
48. The Universities Federation for Animal Welfare. 2010. *The UFAW Handbook on the Care and Management of Laboratory Animals*, 8th edition.
49. International Air Transportation Association. Live animal regulations. https://www.iata.org/en/publications/store/live-animals-regulations/. Accessed March 1, 2024.
50. National Research Council. 2004. *Science, Medicine, and Animals.* Washington, DC: The National Academies Press.
51. U.S. Department of Defense. 1959. *DoD Directive 5129.1.*
52. U.S. Department of Defense. 1961. *DoD Directive 51225.5.*
53. U.S. Department of Defense. 1961. *Policy on Experimental Animals in Department of Defense Research.*
54. U.S. Department of Defense, United States Army. 1984. *The Use of Animals in DOD Programs.*
55. Departments of the Army, Navy, Air Force, Defense Advanced Research Projects Agency, and the Uniformed Services University of the Health Sciences. 2005. *The Care and Use of Laboratory Animals in DOD Programs.*
56. Department of Defense Directive. 1995. *Use of Animals in DoD Programs.*
57. Department of Defense Directive. 2005. *Use of Animals in DoD Programs.*
58. Department of Defense Directive. 2010. *Use of Animals in DoD Programs.*
59. Department of Defense Directive. 2019. *Use of Animals in DoD Programs.*
60. United States Army Medical Research and Development Command. Animal Care and Use Review Office (ACURO). https://mrdc.health.mil/index.cfm/collaborate/research_protections/acuro. Accessed March 1, 2024.
61. National Science Foundation. About the National Science Foundation. https://www.nsf.gov/about/. Accessed March 1, 2024.
62. National Science Foundation. 1977. *NSF Grant Policy Manual, NSF 77-47.*
63. National Science Foundation. 1989. *NSF Grant Policy Manual, NSF 88-47.*
64. Office of Technology Assessment. 1986. Alternatives to Animal Use in Research, Testing , and Education, in *Appendix B: Regulation of Animal Use within Federal Departments and Agencies*, pp. 386–394.
65. National Science Foundation. 1995. *Grant Policy Manual NSF 95-26.*
66. National Science Foundation. 2002. *Grants Policy Manual NSF 02-151.*
67. National Science Foundation. 2007. *NSF Proposal & Award Policies & Procedures Guide (PAPPG) NSF 07-140.*
68. National Science Foundation. 2012. *Proposal and Award Policies and Procedures Guide (PAPPG) NSF 13-1.*
69. National Institutes of Health, Office of Laboratory Animal Welfare. 2015. *Notice of Memorandum of Understanding between NIH and NSF Concerning Laboratory Animal Welfare.*
70. National Science Foundation. 2017. *Proposal and Award Policies and Procedures Guide (PAPPG) NSF-17-1.*
71. National Science Foundation. 2021. *Proposal and Award Policies and Procedures Guide (PAPPG) NSF 22-1.*

72. National Aeronautics and Space Administration. 1958. *National Aeronautics and Space Act of 1958.*
73. Gray, T. 1998. A brief history of animals in space. National Aeronautics and Space Administration. https://www.nasa.gov/history-of-animals-in-space/.
74. National Aeronautics and Space Administration. 1990. *Title 14 CFR Ch. V, Part 1232 – Care and Use of Animals in the Conduct of NASA Activities*, vol. 5.
75. National Aeronautics and Space Administration, *The Sundowner Report, NASA Principles for the Ethical Care and Use of Animals*, 1996.
76. Van Hoosier, G.L. 2000. Principles and paradigms used in human medical ethics can be used as models for the assessment of animal research. *Comparative Medicine*, vol. 50, no. 2, pp. 103–106.
77. U.S. Department of Agriculture, Animal Welfare Information Center. 1999. *Information Resources for Institutional Animal Care and Use Committees 1985–1999.*
78. Stabekis, P. 2007. Policy and guidelines for the use and care of animals in space-borne research. *Space Research*, vol. 169, pp. 3–6.
79. National Aeronautics and Space Administration. 2011. *NASA Procedures and Guidelines, Care and Use of Animals.*
80. National Institutes of Health, Office of Laboratory Animal Welfare. 2020. *Notice of Memorandum of Understanding between NIH and NASA Concerning Laboratory Animal Welfare.*
81. National Aeronautics and Space Administration. 2022. *NASA Policy Directive, Care and Use of Animals.*
82. Council for International Organizations of Medical Sciences. 1985. *International Guiding Principles for Biomedical Research Involving Animals.*
83. Council for International Organization of Medical Sciences and the International Council for Laboratory Animal Science. 2012. *International Guiding Principles for Biomedical Research Involving Animals.*
84. Office of Science and Technology Policy. 1985. *U.S. Government Principles for the Utilization and Care of Vertebrate Animals Used in Testing, Research, and Training*, vol. 50.
85. Orlans, F.B., Simmonds, R.C., and Dodds, W.J. 1987. Effective animal care and use committees (published in collaboration with the Scientists Center for Animal Welfare. *Laboratory Animal Science (Special Issue).*

2

Circumstances Requiring an IACUC

Marcy Brown and Deborah Frolicher

Introduction

APHIS/AC and NIH/OLAW have overlapping regulatory functions codified in the AWA, the AWAR, the Health Research Extension Act of 1985 (HREA), and the PHS Policy. These documents specify the various situations in which an IACUC is required. Although all components of an animal care and use program—the IO, the AV, the animal care staff, the research community and the IACUC—have important roles to play, unquestionably, federal law, regulation, and policy place the IACUC in a pivotal position for ensuring animal welfare. In addition, the *Guide* and the *Institutional Animal Care and Use Committee Guidebook*[1] describe the key role IACUCs have in the assessment and oversight of the animal care and use program. This chapter explains the various circumstances in which institutions are required to have an IACUC and summarizes the regulatory requirements of the AWA, AWAR, HREA, and PHS Policy for an IACUC. Confusion may arise regarding the requirement to have an IACUC since the mandates concerning the creation of an IACUC differ between the AWA and the HREA. In an effort to alleviate this confusion, the reader is directed to the specific sections of the AWA, AWAR, and PHS Policy that describe the regulatory requirements for the IACUC. The remainder of the chapter provides the answers to a series of questions designed to identify the circumstances under which IACUCs are required at research institutions, and whether IACUCs are required at elementary and secondary schools, research and teaching hospitals, institutional exhibits, science fairs, zoos, aquaria, animal shelters, or local humane organizations.

2:1 What federal agencies require the appointment of an IACUC?

Reg. There are two: the USDA and the PHS (see 2:2). APHIS/AC, through the AWAR, requires an IACUC at any institution that uses animals in research, testing, or teaching, provided the institution meets the definition of "research facility" and the species is among those listed under the definition of "animal" in §1.1 of the AWAR. The PHS, through NIH/OLAW, requires an IACUC at any institution where activities involving any live vertebrate animal, supported by the PHS or an agency that requires compliance with the PHS, are conducted (PHS Policy IV,A,1).

2:2 What specific documents describe the regulatory requirements for the IACUC?

Reg. The specific documents that describe the regulatory requirements for the IACUC are the AWA, Title 7 of the U.S. Code (7 U.S.C. Chapter 54) §2131–2159 (Public Law 89-544), and the HREA (Public Law 99-185). Detailed regulations and standards, which the IACUC must consider when evaluating an institution's Animal Care Program, are set forth by the USDA in the AWA Title 9 of the Code of Federal Regulations (9 CFR), Chapter 1, Subchapter A, Parts 1, 2, and 3 and by the NIH in the HREA as set forth in PHS Policy IV,A,3; IV,B; IV,C; and IV,E–F.

See below for specific sections within the AWAR and PHS Policy that include references to the IACUC.

PART 2 — REGULATIONS

Subpart C — Research Facilities

• §2.31	Institutional Animal Care and Use Committee
• §2.33,a,3	Attending Veterinarian and Adequate Veterinary Care
• §2.35,a–§2.35,f	Recordkeeping Requirements
• §2.36,b,3	Annual Report
• §2.37	Federal Research Facilities
• §2.38,f,2,11; §2.38,k,1	Miscellaneous

DOI: 10.1201/9781003272861-2

PART 3 — STANDARDS

Subpart A — Specifications for the Humane Handling, Care, Treatment, and Transportation of Dogs and Cats

- §3.6,d Primary Enclosures
- §3.8,b,1; §3.8,c; §3.8,d,2 Exercise for Dogs
- §3.13 Veterinary Care for Dogs

Subpart B — Specifications for the Humane Handling, Care, Treatment, and Transportation of Guinea Pigs and Hamsters

- §3.28,c,3 Primary Enclosures

Subpart C — Specifications for the Humane Handling, Care, Treatment, and Transportation of Rabbits

- §3.53,c,3 Primary Enclosures

Subpart D — Specifications for the Humane Handling, Care, Treatment, and Transportation of Nonhuman Primates

- §3.80,b,2,iii; §3.80,c Primary Enclosures
- §3.81,c,3; §3.81,d; §3.81,e,ii Environment Enhancement to Promote Psychological Well-Being
- §3.83 Watering

Subpart G – Specifications for the Humane Handling, Care, Treatment, and Transportation of Birds Not Bred for Use in Research

- §3.154 Environment Enhancement to Promote Psychological Well-Being

The PHS Policy also sets forth detailed requirements governing IACUCs.

References to IACUCs are found primarily in the following sections:

- IV,A,3 Institutional Animal Care and Use Committee
- IV,B Functions of the IACUC
- IV,C Review of PHS-Conducted or PHS-Supported Research Projects
- IV,E Recordkeeping Requirements
- IV,F Reporting Requirements

The PHS Policy (IV,A,1) requires institutions to use the *Guide* as a basis for developing and implementing an institutional program for activities involving animals.

2:3 What is the difference between the Animal Welfare Act and the Animal Welfare Act Regulations in regard to requirements for an IACUC?

Reg. The AWA (Public Law 89-544, U.S. Code 7 §2131–2159) is a statute enacted by Congress in 1966 and governs the care and use of animals in research at both federal and nongovernment facilities. The AWAR are found in the Code of Federal Regulations (CFR) Title 9, Chapter 1, Subchapter A – Animal Welfare, Parts 1–4. Every regulation listed in the CFR must have an enabling statute (in this case, the AWA). The AWAR describe the standards and guidance adopted by the USDA to implement and enforce the Animal Welfare Act.[2]

Opin. Both the AWA and the AWAR require research facilities using regulated species to establish an IACUC (see 2:1; 2:2). However, the AWAR set forth a more extensive list of functions of the IACUC (§2.31,c).

2:4 What is the difference between the HREA and the PHS Policy regarding the requirements for an IACUC?

Opin. The HREA provides the legislative mandate for the PHS Policy. Both the HREA and the PHS Policy require institutions using animals in research, teaching, or testing activities conducted or supported by the PHS or an agency that requires compliance with the PHS Policy to establish an IACUC (see 2:1; 2:2). The PHS Policy sets forth an extensive list of functions of the IACUC (IV,B).[3] One notable difference between the HREA and PHS Policy is that the HREA states that the animal care committee "shall be composed of not fewer than three members" but the PHS Policy states that the Institutional Animal Care and Use Committee "shall consist of not less than five members."

2:5 What are the regulatory charges of the IACUC?

Reg. The regulatory charges of the IACUC are described in the AWA §2143,b and §2143,c; the AWAR §2.31,c; HREA §495,b,3; and PHS Policy IV,B,C,F.

Opin. In summary, depending on the species and source of funding involved, the above require the IACUC to*:

	REQUIREMENT	PHS	USDA
1.	As applicable: PHS Animal Welfare Assurance; USDA Registration	IV.A	§2.30.a
2.	Establish a Committee Membership, orientation, resources, and training; Provide resources for an Animal Care and Use Program	IV.A3.a–c; B.1–8	§2.31.a
3.	Semiannual review of programs and facility inspections Report to the IO	IV.B.1–3	§2.30.c.1-3
4.	Review activities (protocols) involving animals Review of new and continuing activities Review/approve modifications to activities Conduct post-approval monitoring	IV.B.4, 6–7; C.1–7; D.1–3	§2.31.c-e
5.	Review concerns involving animals	IV.B.4	§2.31.c.4
6.	Make recommendations to the IO	IV.B.5	§2.31.c.5
7.	Suspend activities	IV.C.7	§2.31.d.6-7
8.	Training and instruction for personnel	IV.A.1.g	§2.32.a-b
9.	Reporting and Recordkeeping Requirements External annual reports, if applicable: Minutes of IACUC meeting: Records regarding protocol review; and Records of semiannual review and inspections	IV.F.1–4	§2.336.a-b
10.	Occupational health and safety program	IV.A.1.	
11.	Security, disaster, emergency, crisis plans	See OLAW website (https://olaw.nih.gov/resources/disaster-planning.htm)	Contingency Planning and Training of Personnel Rule (APHIS-2020-0101)

*excerpted in part from the Institutional Animal Care and Use Guidebook[1]

2:6 How might records for IACUCs be effectively maintained?

Reg. Institutions are required to maintain various records related to the animal care and use program. The AWAR (§2.35) and PHS Policy (IV,E) require each institution to maintain:

- Minutes of IACUC meetings, including records of attendance, activities of the committee, and committee deliberations (see 7:23 and 7:26).
- Records of applications, proposals, and proposed significant changes in the care and use of animals, and whether IACUC approval was given or withheld
- Records of semiannual IACUC reports and recommendations, including minority views, that have been forwarded to the Institutional Official (IO)

In addition, PHS Policy (IV,E) requires each institution to maintain:

- A copy of its Animal Welfare Assurance
- Records of accrediting body determinations

The AWAR (§2.35,f) and PHS Policy (IV,E) stipulate that all records must be maintained for at least three years and that records relating to applications, proposals, and proposed significant changes must be maintained for three years after completion of the activity. All records must be accessible for inspection and copying by authorized representatives of APHIS/AC, NIH/OLAW, or federal funding agency representatives at reasonable times (AWA §2140; AWAR §2.31,c,3; AWAR §2.35,f); PHS Policy (IV,E,2) adds "and in a reasonable manner."

Opin. The IACUC or IACUC staff should either maintain the records listed above or ensure that they are maintained in a secure and accessible location at the institution. Additional records that should be kept include:

- IACUC correspondence, including e-mail records
- Institutional policies, guidelines, and SOPs for animal care and use
- Reports of IACUC investigations of concerns involving animals

Methods of maintaining records will vary, depending upon the organization of the institution and the size and complexity of its animal use program. However, any method that is used to maintain records must be efficient, reliable, and secure.

While some institutions choose to use traditional paper records, most institutions, as evidenced by the survey results below, use electronic protocol management systems to maintain records of review and approval of animal use protocols. Such systems conduct the entire protocol workflow electronically, including authorship, submittal, review, comments, modifications, and approval, as well as the processing of amendments, annual reviews (if required by the institution), and triennial reviews. Systems may also include IACUC meeting agenda and minutes automation, e-mail notifications, information alerts, and historical reporting, thereby facilitating regulatory compliance. Electronic systems have many advantages, including the ability to sort and retrieve information quickly by various parameters such as species, type of procedure, protocol expiration date, and location of the procedure or animal housing. Many systems also manage the purchase of animals to ensure congruence between number of animals ordered and number of animals approved on a protocol. Information stored in an electronic protocol system may also facilitate preparation of any required annual reports (e.g., to APHIS/AC, NIH/OLAW, and/or AAALAC) as well as preparation of the NIH/OLAW Animal Welfare Assurance and the AAALAC Program Description.

Surv. What method does your institution use to maintain required records? For example,

IACUC meeting minutes; records of attendance; activities of the committee and committee deliberations; records of applications, proposals, and proposed significant changes in the care and use of animals; whether IACUC approval was given or withheld; records of semiannual IACUC reports and recommendations, including minority views that have been forwarded to the IO; and, if applicable, a copy of the Animal Welfare Assurance and records of accrediting body determinations? Select all that apply.

- Electronic protocol/data management systems — 264/309 (85%)
- Paper copies — 138/309 (45%)
- E-mail correspondence — 198/309 (64%)
- Other* — 45/309 (15%)

*Other responses reflected use of various electronic and paper versions of record maintenance

Surv. Aside from records that are required to be maintained, what additional documents are maintained as part of your institution's Animal Care Program? Select all that apply.

- IACUC meeting agenda — 278/309 (90%)
- IACUC correspondence, including e-mail records — 247/309 (80%)
- Institutional policies, guidelines, and SOPs for animal care and use — 305/309 (99%)
- Reports of IACUC investigations of concerns involving animals — 292/309 (94%)
- Reports of post-approval monitoring activities — 252/309 (82%)
- Records of the number of animals approved on a protocol and the number of animals ordered — 268/309 (87%)
- Other* — 22/309 (7%)

*Other responses included items that are likely maintained by many, if not most institutions, though not reflected as such in the number of respondents.

- Animal health records
- Semi-annual inspection and review information
- Annual reports (OLAW/AAALAC)

2:7 Is an IACUC *always* required where animals are used for research, teaching, testing, or product safety evaluation?

Reg. No. An IACUC is required only if one of the following applies:

- The species used for research, teaching, or testing is covered by the definition of animal given in §1.1 of the AWAR. The AWAR definition of animal specifically excludes certain animals bred for use in research, such as birds, rats of the genus *Rattus*, and mice of the genus *Mus*; horses not used for research purposes; and other farm animals used in agricultural research. Birds, rats of the genus *Rattus*, and mice of the genus *Mus* not bred for use in research would require IACUC oversight.
- The facility meets the definition of research facility given in §1.1 of the AWAR.
- The research, teaching, or testing is supported by the PHS, or the activity will be performed at an institution with an Animal Welfare Assurance, committing that institution to comply with the PHS Policy.
- The institution is receiving support for animal research, teaching, or testing from an agency that requires compliance with PHS Policy [e.g., National Institutes of Health (NIH), the National Science Foundation (NSF), Centers for Disease Control and Prevention (CDC), Food and Drug Administration (FDA), National Aeronautics and Space Administration (NASA), Biomedical Advanced Research and Development Authority (BARDA), and the Veterans Administration (VA)].[4]
- The institution is subject to state or local ordinances that require institutions using animals in research, teaching, or testing to establish an IACUC.

Opin. If an institution is, or desires to be, accredited by AAALAC, it must have an IACUC.

2:8 A pet dog is used in a research or teaching hospital as nothing more than a model in an art therapy class. Is it necessary for the hospital to have (or establish) an IACUC to approve the use of this animal?

Reg. See 2:7.

Opin. No. An IACUC is required only when AWAR-regulated species are being used in research, teaching, or testing. A hospital that allows dog owners to bring their pet animals into the hospital for use in an art therapy class is not conducting research, teaching, or testing with the animals. Therefore, the AWAR do not apply and an IACUC review is not required. An IACUC review is not required under the PHS Policy, unless the activity is part of a research project receiving support for animal research, teaching, or testing from an agency that requires compliance with PHS Policy (see 2:7).

2:9 A pet dog is used in a research or teaching hospital to evaluate the impact of pet-assisted therapy on the recovery of patients. The resulting data will be summarized into a manuscript to be submitted for publication. Is it necessary for the hospital to have (or establish) an IACUC to approve the use of this animal?

Reg. See 2:7.

Opin. No. An IACUC is required only when AWAR-regulated species are being used in research, teaching, or testing. In this case, the research or teaching hospital is not conducting research or teaching **with** the pet animals, but rather with the patients. An IACUC review is not required under PHS Policy, **unless** the activity is part of a research project receiving support for animal research, teaching, or testing from an agency that requires compliance with PHS Policy (see 2:7).

2:10 Do elementary or secondary schools need to have an IACUC to keep pet animals?

Reg. No, the AWA does not regulate the keeping of pets, whether in a private home or by a school. In addition, both the AWA (§2132,e) and the AWAR (§1.1, Research facility) specifically exclude elementary and secondary schools from being designated as research facilities. Therefore, the regulations do not apply and an IACUC is not required.

Opin. An elementary or secondary school that keeps pet animals is also not considered an exhibitor under the definitions in §1.1 of the AWAR because it does not purchase animals and exhibit them to the public for compensation. An IACUC is not required under PHS Policy, unless the activity is part of a research project receiving support for animal research, teaching, or testing from an agency that requires compliance with the PHS Policy (see 2:8).

2:11 Is it necessary to have IACUC oversight of elementary or secondary school science fair projects or other educational activities involving the use of vertebrate animals?

Reg. No. Elementary and secondary schools are specifically excluded from the definition of research facility in §2132,e of the AWA and §1.1 of the AWAR. Accordingly, the AWAR do not apply and an IACUC is not required. An IACUC also would not be required under PHS Policy, since it applies only to institutions that receive funds for vertebrate animal activities from PHS or from an agency that requires compliance with the PHS Policy (see 2:7).

2:12 What guidelines exist for the use of animals in elementary or secondary school educational activities?

Opin. Although elementary or secondary schools are generally not covered by the AWA, AWAR, HREA, or the PHS Policy, they must follow the PHS Policy if the activity is part of a research project supported by the PHS or from an agency that requires compliance with the PHS Policy. Otherwise, educational institutions may choose to follow the guidance contained in the *Guide* and AALAS Position Paper *Use of Animals in Precollege Education*.[5] The latter document presents guidelines and resources for the humane care and responsible use of animals in precollege education and offers recommendations on classroom dissection and on the use of animals in science fair projects. As part of its broader educational mission to ensure that all animal use is performed responsibly and humanely, AALAS has also developed *Caring for Animals: A Guide to Animals in the Classroom*, which provides information about species commonly found in classrooms settings such as mice, rats, hamsters, guinea pigs, gerbils, rabbits, reptiles, and amphibians.[6] The National Science Teachers Association has also developed a Position Statement on Responsible Use of Live Animals and Dissection in the Science Classroom[7] and supports including live animals as part of instruction in the K-12 science classroom. The Society for Science provides requirements, information, and resources for studies involving vertebrate animals in precollege studies conducted as part of an educational activity.[8]

Other guidelines that may be useful in elementary or secondary school educational activities are:

- The *Use of Animals in Biology Education* developed by the National Association of Biology Teachers[9]
- *Guidelines for Ethical Conduct in the Care and Use of Animals* from the American Psychological Association[10]
- *Principles and Guidelines for the Use of Animals in Precollege Education* from the Institute of Laboratory Animal Resources[11]

In addition to these guidelines, several states have laws or regulations governing the use of animals in the classroom and individual state or local governments or local school districts may have standards that are either guidelines or requirements. In many states there are nonprofit biomedical research societies that schools can consult for information on guidelines for the use of animals in educational activities.[12] In general, these societies promote public education and understanding in support of biomedical research.

2:13 Do colleges or universities with live mascots need to have the care and use of these animals approved by the IACUC?

Reg. Under the AWA the care and treatment of a regulated species mascot would come under APHIS/AC jurisdiction *if* the animal is used in research (§2143,a,1) or exhibited (§2132,h) to the public during football games, etc. In the event the animal is used in research, an IACUC would be required per AWAR §2.31. The animal would not fall under APHIS/AC jurisdiction if it is merely a pet.

Opin. Institutions receiving PHS funding that use live animals as mascots are technically not required to have the care and use of these animals approved by the IACUC since the PHS Policy applies only to animal activities supported by PHS funds or funding from an agency that requires compliance with PHS Policy. However, to the extent the mascot is housed at university facilities, it should be cared for and housed in a manner consistent with the *Guide* and AWAR, and the housing and care of an AWAR-regulated species is subject to APHIS/AC inspection if it is exhibited to the public.

Therefore, the question of whether the care and use of a mascot must be approved by the IACUC depends upon an institution's policies since it is not required under the AWAR or PHS Policy. For example, it is the policy at some institutions that all uses of live vertebrate animals must be approved by an IACUC. In these cases, it is important for the institution to have clear written policies or SOPs regarding animals used as mascots, exhibits, or for purposes other than research, teaching, or testing. Such policies, or SOPs, should include a provision for IACUC review and approval if the institution requires it, as well as stipulations for animal care (e.g., stating who provides housing and husbandry) and veterinary services. For institutions participating in AAALAC's accreditation program, if institutional policies require some degree of IACUC oversight of animals used for a purpose other than research, testing, or teaching, then AAALAC would expect adherence to those policies.[13]

2:14 Animals are included as part of institutional exhibits for the purpose of educating the public about the use of animals in research or providing entertainment. Would use of these animals need to be reviewed and approved by the IACUC?

Reg. See 2:7.

Opin. No. An IACUC is required only when AWAR-regulated species are being used in research, teaching, or testing or the institution is receiving support for animal research, teaching, or testing from PHS or an agency that requires compliance with the PHS Policy. Individuals or businesses with AWAR-regulated species that are only on display, perform for the public, or are used in educational presentations must be licensed as exhibitors with APHIS but there is no requirement for IACUC review and approval unless, as in 2:13 (above), an institution has a policy or SOP requiring that all use of live vertebrate animals must be approved by an IACUC.

2:15 Do zoological gardens and aquariums fall under the jurisdiction of the AWAR or PHS Policy?

Reg. Yes. Zoos and aquariums that maintain AWAR-regulated species (e.g., marine mammals, exotic rodents, etc.) are defined as exhibitors under §1.1 of the AWAR and must comply with all regulations governing exhibitors, as described in §2.1–2.5 of the AWAR, as well as the applicable standards for species housed in zoos and aquariums as described in §3.1-3.168 of the AWAR.

Opin. Strictly speaking, only zoos and aquariums receiving support for animal research, teaching, or testing from PHS or an agency that requires compliance with the PHS Policy are required by NIH/OLAW to comply with PHS Policy (see 2:7).

2:16 Under the AWAR, are zoological gardens and aquariums expected to adhere to the same research criteria as biomedical research institutions? (See 2:15)

Reg. See 2:7.

Opin. If a licensed exhibitor such as a zoo or aquarium is conducting biomedical research, teaching, or testing activities with AWAR-regulated species, the licensee is *not* exempt from the need for registration with APHIS/AC and compliance with the AWAR (for further information on this topic, see reference[14]). Similarly, zoos and aquariums receiving support for animal research, teaching, or testing from PHS or an agency that requires compliance with the PHS Policy are required by NIH/OLAW to comply with the PHS Policy.

2:17 Is research conducted at zoological gardens and aquariums required to come under the purview of an IACUC?

Reg. See 2:1, 2:3, and 2:7. If a zoological garden or aquarium is required to be registered as a research facility under the AWA, research conducted is required to be reviewed by an IACUC, just as any other AWA-registered research facility.

Opin. If the activity involves an AWAR-regulated species and is considered regulated research by APHIS/AC (see 2:13), IACUC review and approval is required. Similarly, if the activity is supported by the PHS or an agency that requires compliance with the PHS Policy, it must be reviewed and overseen by an IACUC (see 2:7 and 9:4–9:6).

2:18 Do private and municipal animal shelters and humane societies fall under the jurisdiction of the AWA and its regulations or the PHS Policy and therefore require an IACUC?

Reg. The legal requirements for businesses regulated by the AWA are described in the AWA (§2133; §2136) and the AWAR (§1.1, Dealer; §2.1,a,3,i–viii).

Opin. Generally, animal shelters and humane societies do not fall under the jurisdiction of the AWAR unless they sell dogs or cats to a research facility, licensed dealer, exhibitor, or sell animals wholesale as pets (AWAR §1.1, Dealer). If a private shelter provides animals to a research facility, it must be licensed with APHIS/AC as a dealer and must comply with all AWA regulations governing dealers, which do not include a requirement for an IACUC. A municipal animal shelter or humane society that provides animals to a research facility is not defined as "a person" by APHIS/AC and, consequently, does not meet the definition of dealer under §1.1 of the AWAR. Accordingly, municipal shelters and humane societies are exempt from the regulations governing dealers.

Animal shelters and local humane organizations fall under the jurisdiction of the PHS Policy only if they receive support from the PHS or other agency that requires compliance with the PHS Policy (see 2:7), which would include the requirement for an IACUC.

2:19 The AWAR (§2.31,d,1) exempts studies defined as field studies from the requirement for IACUC review. However, the AWAR definition of a field study *excludes* any study that involves an *invasive procedure*, *harms*, or *materially alters the behavior of the animals under study*. In determining whether a field study is exempted from the requirement for IACUC review, what is meant by an "invasive procedure" and "materially alters the behavior of the animals"?

Reg. The AWAR (§1.1) state: "Field study means a study conducted on free-living wild animals in their natural habitat. However, this term excludes any study that involves an invasive procedure, harms, or materially alters the behavior of an animal under study." The AWAR do not further define the phrases "invasive procedure," "harms" or "materially alters the behavior" (see 2:18).

Opin. Although the phrases are not further defined, the context in which they appear in the AWAR gives guidance as to their intended meaning. Clearly, the intent of this regulation is to require IACUC review of any projects that have the potential to disrupt animals or to alter their environments. These projects include field studies that require animals to be trapped or confined in any way, handled (see AWAR §1.1 for a definition of handling), anesthetized, tagged or marked in any other way, or subjected to blood or tissue sampling. In other words, only strictly observational studies—ones in which neither the animals nor their environments are disturbed or permanently altered—are exempt from the requirement for IACUC review.

2:20 Is an institution that conducts only field studies required to have an IACUC?

Reg. See 2:19.

Opin. It depends; if the field studies involve vertebrate animals and are supported by PHS or an agency that requires compliance with PHS Policy, an IACUC is required. According to NIH/OLAW FAQ A.6, the IACUC is responsible for oversight of vertebrate animal activities supported by PHS or an agency that requires compliance with the PHS Policy. To approve such activities in the field, IACUCs must know where field studies will be located, what procedures will be involved, and be sufficiently familiar with the nature of the habitat to assess the potential impact on the animal subjects. If the activity alters or influences the activities of the animal(s) that are being studied, the activity must be reviewed and approved by the IACUC (e.g., capture and release, banding). If the activity does not alter or influence the activity of the animal(s), IACUC review and approval is not required (e.g., observational, photographs, collection of feces)."[15] A study of free-living, wild, AWA-regulated species that involves invasive procedures, harms, or otherwise materially alters the biology, behavior, or ecology of an animal under study is covered by USDA AWAR and requires IACUC review and approval.

2:21 The AWAR (§2.31,d,1) exempt studies defined as field studies from the requirement for IACUC review. However, the AWA definition of a field study (§1.1) includes the phrase "free-living wild animals in their natural habitat." In determining whether a field study is exempted from the requirement for IACUC review, can an animal be kept for any length of time outside of its natural habitat before it is no longer considered free living?

Reg. See 2:19. The AWAR definition of a field study makes it clear that any study that involves an invasive procedure or has the potential to harm or materially alter an animal's behavior is not exempt from IACUC review.

Opin. Removing an animal from its natural environment for any length of time would appear to always materially alter its behavior. Such removal would typically involve capture, trapping or herding, handling, restraint, and/or physical confinement. These activities, no matter how carefully they are performed, would carry a likelihood of altering an animal's behavior in a significant fashion, at least in the short term.

REFERENCES

1. Applied Research Ethics National Association/Office of Laboratory Animal Welfare. 2002. *Institutional Animal Care and Use Committee Guidebook*. Bethesda, MD: National Institutes of Health. https://grants.nih.gov/grants/olaw/GuideBook.pdf. Accessed March 3, 2024.
2. USDA National Agricultural Library, Animal Welfare Act, Current Law and Regulations. https://www.nal.usda.gov/animal-health-and-welfare/animal-welfare-act. Accessed March 3, 2024.
3. National Institutes of Health, Office of Laboratory Animal Welfare. 2015. *Public Health Service Policy on Humane Care and Use of Laboratory Animals*. Section IV. Implementation by Institutions. National Institutes of Health. https://grants.nih.gov/grants/olaw/references/phspol.htm. Accessed March 3, 2024.
4. National Institutes of Health, Office of Laboratory Animal Welfare. *Obtaining an Assurance*. https://olaw.nih.gov/guidance/obtaining-an-assurance.htm. Accessed March 3, 2024.
5. American Association for Laboratory Animal Science. *Use of Animals in Precollege Education*. https://www.aalas.org/about-aalas/position-papers/use-of-animals-in-precollege-education. Accessed March 3, 2024.
6. American Association for Laboratory Animal Science Caring for Animals in the Classroom. https://www.aalas.org/media/c03a4cff-16d7-4656-8959-b947c93fc202/t2Jerw/Public%20Outreach/animals-in-classroom.pdf. Accessed March 3, 2024.
7. National Science Teachers Association. 2008. NSTA position statement: Responsible use of live animals and dissection in the science classroom. https://www.nsta.org/about/positions/animals.aspx. Accessed March 3, 2024.
8. Society for Science-Vertebrate Animals. https://www.societyforscience.org/isef/international-rules/vertebrate-animals/. Accessed March 3, 2024.
9. National Association of Biology Teachers. *The Use of Animals in Biology Education*. https://nabt.org/Position-Statements-The-Use-of-Animals-in-Biology-Education. Accessed March 3, 2024.
10. American Psychological Association. 2022. *Guidelines for Ethical Conduct in the Care and Use of Nonhuman Animals in Research*. https://www.apa.org/science/leadership/care/guidelines.aspx. Accessed March 3, 2024.
11. National Academy of Sciences. 2004. Institute of Laboratory Animal Resources. *Principles and Guidelines for the Use of Animals in Precollege Education*. https://nabt.org/files/galleries/Principles_and_Guidelines_for_the_Use_of_Animals_in_Precollege_Education.pdf. Accessed March 3, 2024 .
12. States United for Biomedical Research. https://statesforbiomed.org/. Accessed March 3, 2024.
13. Bayne, K. Global Director, AAALAC International, personal communication, 2013.
14. Kohn, B. and S.L. Monfort. 1997. Research at zoos and aquariums: Regulations and reality. *J. Zoo Wildlife Med*. 28: 241–250.
15. National Institutes of Health, Office of Laboratory Animal Welfare. Frequently Asked Questions A.6, Does the PHS Policy apply to animal research that is conducted in the field? https://olaw.nih.gov/faqs#/guidance/faqs?anchor=question50290. Accessed March 3, 2024.

3

Creation of an IACUC

Marcy Brown and Deborah Frolicher

Introduction

This chapter provides guidance on how institutions may form an IACUC. It describes whose interests the IACUC should serve and whose responsibility it is to appoint the members. Since the nonaffiliated member of the committee is the general public's link to the federally mandated oversight process for animal welfare, information is provided on the critical role of the nonaffiliated member and how institutions might go about finding an effective one. For an IACUC to fulfill its regulatory requirements properly, its members must be informed of their responsibilities under federal laws, regulations, and policies. In addition to these federally mandated obligations, IACUCs may be charged with additional responsibilities by their state, city, or institution. Given the complexity of this environment, institutions have a responsibility to provide their IACUC members with specialized training; therefore, this chapter also describes methods for IACUC member training. The latter part of the chapter suggests means by which larger institutions may utilize multiple IACUCs to reduce the workload associated with reviewing hundreds of animal use protocols and inspecting dozens of animal facilities. Finally, there is a discussion regarding whether a small institution may choose to use the services of the IACUC at a larger institution with a more extensive animal care and use program.

3:1 In general, what is expected of the IACUC, and whose interests does it represent?

Reg. The AWA (§2143,b,1) charges the members of the IACUC with representing "society's concerns regarding the welfare of animal subjects." In addition, the nonaffiliated member is specifically charged with representing general community interests in the proper care and treatment of animals (AWA §2143,b,1,B,iii; AWAR §2.31,b,3,ii; *Guide* p. 24, *Ag Guide* pp. 1). The PHS Policy (IV,B) references the IACUC as "an agent of the institution" that will "oversee the institution's animal program, facilities, and procedures" (PHS Policy IV,A,3,a).

Opin. The IACUC is expected to oversee and evaluate the institution's animal care and use program to ensure consistency with the requirements of the AWAR, and PHS Policy, as well as the recommendations of the *Guide* and *Ag Guide*, as applicable to the institution. The authority, composition, and functions of the IACUC are also described in the *Institutional Animal Care and Use Committee* Guidebook[1] (pp. 11–18). The Committee represents multiple interests, including those of the institution, the research community, the animals, and the public. It also serves as the self-monitoring agent of federal agencies and accrediting bodies, such as APHIS/AC, NIH/OLAW, and AAALAC, International, respectively.

3:2 Who is responsible for appointing the IACUC?

Reg. The Chief Executive Officer (CEO) is charged with appointing an IACUC (AWA §2143,b,1; AWAR §2.31,a; PHS Policy IV,A,3,a). The CEO is defined in PHS Policy (IV,A,3,a, footnote 5) as the highest operating official of the organization, such as the President, Provost, or Chancellor of a university. PHS Policy allows the CEO to delegate the authority to appoint the IACUC but requires that such delegation be specific and in writing. The *Guide* (p. 12) requires institutions to comply with the AWAR and PHS Policy (see 5:13).

Opin. At larger academic institutions, it is not unusual for the CEO to delegate authority for appointment of the IACUC to a senior administrator who is more directly responsible for the institution's research program, such as a Vice Chancellor, Dean, or Vice President for Research. This individual frequently serves as the organization's Institutional Official (IO) (PHS Policy III,G) and signs and submits documents, such as the PHS Animal Welfare Assurance. It is important to note that the CEO and the IO do not have to be the same individual.

3:3 For practical purposes, should the IACUC Chair be given the option of accepting or rejecting a proposed appointment to the committee?

Reg. Both the AWAR (§2.31,a) and the PHS Policy (IV,A,3,a) mandate that the IACUC be appointed by the CEO (see 4:5).

Opin. No; however, input from the Chair, as well as other IACUC members, the IACUC staff, the AV, and even members of upper management or the research community may be a useful method of informing the CEO of specific needs of the IACUC in order to ensure it possesses the expertise necessary to fulfill its responsibilities.

DOI: 10.1201/9781003272861-3

3:4 Are there specific membership requirements for an IACUC?

Reg. Yes, the AWAR (§2.31,b,2) state that the committee shall be composed of a Chair and at least two additional members. The AWAR (§2.31,b,3) further require that at least one member shall be a Doctor of Veterinary Medicine, with training or experience in laboratory animal science and medicine, and at least one member shall not be affiliated in any way with the facility other than as a member of the Committee.

PHS Policy (IV,A,3,b), on the other hand, states that the committee shall include at least one veterinarian, one practicing scientist experienced in research involving animals, one member whose primary concerns are in a nonscientific area, and one individual who is not affiliated with the institution in any way and is not a member of the immediate family of a person who is affiliated with the institution. An individual may fulfill more than one requirement; however, no IACUC constituted under the PHS Policy may consist of fewer than five members (PHS Policy IV,A,3,b).

The *Guide* (pp. 24–25) echoes these requirements and adds that nonaffiliated members "should not be laboratory animal users." The *Ag Guide* (p. 1) further requires that IACUC membership include "a scientist who has experience in agricultural research or teaching involving agricultural animals; an animal, dairy or poultry scientist who has training and experience in the management of agricultural animals, and a veterinarian who has training and experience in agricultural animal medicine and who is licensed or eligible to be licensed to practice veterinary medicine." Specific for DoD (DoD Instruction 3216.01, section 3.1.c), membership must have both a primary and an alternate nonaffiliated member.

Opin. The final composition and number of members on an IACUC will depend on a number of factors, including the nature of the research and funding, the size of the institution, and the species involved.

Surv. How many regular voting members are on your IACUC? (see 5:1)

• 3 to 5	27/296 (9%)
• 6 to 10	136/296 (46%)
• 11 to 15	86/296 (29%)
• 16 to 20	29/296 (10%)
• >20	1/296 (<1%)
• I don't know	1/296 (<1%)

3:5 What is the definition of the nonaffiliated member of the IACUC?

Reg. The nonaffiliated member is defined as an individual who represents general community interests in the proper care and use of animals, is not a laboratory animal user, is not affiliated with the institution, and is not an immediate family member of a person who is affiliated with the institution (AWA §2143,b,1,B; AWAR §2.31,b,3,ii; PHS Policy IV,A,3,b,4; *Guide* pp. 24–25; *Ag Guide* p. 1) (see 5:32).

Opin. The nonaffiliated member, also sometimes referred to as the public or community member, is intended to provide representation for general community interests in the proper care and treatment of animals.

3:6 How might IACUCs find individuals to serve as nonaffiliated members?

Opin. No specific regulations prescribe how nonaffiliated members must be selected. Nonaffiliated members may be identified in several ways. Based on survey responses the most common source is through contacts of the present members, including the current nonaffiliated member. Other useful sources for recruiting nonaffiliated members may include the CEO; the AV; personnel in departments at the institution, such as the Public Relations office; professional societies (e.g., lawyers, ethicists, clergy, teachers, librarians, health-care professionals, police, firefighters); local humane organizations; state biomedical advocacy organizations (see 5:34–5:36); local veterinarians; and nonprofit service organizations.

Surv. Identify methods used at your institution to find and recruit effective nonaffiliated IACUC members (NAM). Select all that apply.

• Recommendation from a current NAM or IACUC member	273/298 (92%)
• Recommendation from local professional societies (lawyers, veterinarians, teachers, librarians, etc.)	90/298 (30%)
• Recommendations from state biomedical advocacy groups	21/298 (7%)
• Recommendations from local humane societies or animal rescue groups	20/298 (7%)
• Other*	62/298 (21%)

*Other responses reflected variations of the choices above; however, the following were also offered.

- Online search for clergy members
- Posting of advertisement online
- Recommendation of PIs, staff, IO, other institutional employees
- Recommendation from volunteer services at the institution
- Recommendation from veterinary associations, veterinary technology teaching programs
- Recommendation from local teachers union
- Recommendation from other institutional committees

3:7 What is the suggested duration of IACUC membership for the Chair and committee members?

Opin. No specific regulations require minimum or maximum terms for IACUC members, including the Chair. Serving as an effective member of an IACUC is a difficult task. To perform their duties competently, the members of an IACUC must become familiar with a large body of regulations, policies, and guidelines. They also must learn about the history, special problems, research strengths, goals, objectives, and idiosyncrasies of their institution and its investigators. Experience has shown that a new IACUC member requires six months to a year to gain the experience required to function as a fully effective member of an IACUC. In light of this, most institutions ask members to serve for at least two years, and many for three years. Some institutions also ask capable and willing members to serve multiple two- or three-year terms. While NIH/OLAW does not address length of service for IACUC members, they do point out that if a required member (e.g., scientist, nonscientist, veterinarian, or unaffiliated) leaves the committee so that position is no longer filled, the IACUC is not properly constituted and may not conduct official business until a member who fulfills the required position is appointed.[2,3] To avoid vacancies in these critical membership positions, many institutions have more than one scientist, nonscientist, veterinarian, or nonaffiliated member serving on their IACUC, or they appoint alternates to these membership positions.

Serving as an effective IACUC Chair requires all of the knowledge and expertise needed to become an effective committee member, plus leadership skills to manage the committee effectively and to interact with the institution's administration. For these reasons, the person selected to Chair the IACUC has usually served at least one term as a committee member. Many institutions ask the IACUC Chair to serve multiple terms in order to retain someone with a high level of regulatory and institutional knowledge in this position. (See Chapter 5 for additional information.)

Surv. What is the average length of service of current IACUC members (non-chairperson) at your institution?

- We have not yet had any turnover 12/296 (4%)
- 1–2 years 10/296 (3%)
- 3–4 years 80/296 (27%)
- 5–10 years 151/296 (51%)
- 10–15 years 29/296 (10%)
- >15 years 5/296 (2%)
- I don't know 9/296 (3%)

3:8 What is the typical length of service for an IACUC chairperson at your institution?

Opin. No regulations require a specific term of service for the IACUC Chair (see 3:7). However, in addition to serving as an IACUC member, the Chair has the added responsibility of keeping abreast of regulatory trends and interpretations, ensuring that the IACUC carries out its required mandates, educating and supporting the IACUC members, serving as the IACUC spokesperson and leader, and communicating regularly with the IO, AV, IACUC Administrator, and researchers. For these reasons, many institutions ask the IACUC Chair to serve for several years in order to retain someone with a high level of regulatory and institutional knowledge in this position. This is evident in the survey results where the greatest percentage of responses (40%) indicated that the typical length of service for the IACUC Chair is three to six years and 47% noted that years of service prior to being appointed Chair was greater than four years. Transitioning a Vice Chair to Chair status may be an effective way of ensuring continuity of the IACUC. Approximately 59% of survey respondents indicated that their IACUC has an appointed Vice Chair.

Surv. What is the typical length of service for an IACUC chairperson at your institution?

- 1–3 years 34/296 (11%)
- 3–6 years 117/296 (40%)
- 7–10 years 62/296 (21%)
- >10 years 59/296 (20%)
- I don't know 24/296 (8%)

Surv. How many years of IACUC service did your Chair have prior to being appointed as the Chair?

- None 36/298 (12%)
- 1–2 years 42/298 (14%)
- 3–4 years 59/298 (20%)
- > 4 years 139/298 (47%)
- I don't know 22/298 (7%)

Surv. Does your IACUC have an appointed Vice Chair?

- No 84/296 (28%)
- Yes 174/296 (59%)
- An IACUC member is delegated as Chair only when needed. 38/296 (13%)

3:9 Does your institution provide compensation of any kind (other than reimbursed parking fees, lunch during a meeting, or similar minor compensation) to either the IACUC chairperson or IACUC members, including the nonaffiliated member? (see 5:9)

Opin. There are no specific regulations regarding compensation for IACUC members. However, both OLAW and APHIS/AC suggest that if financial compensation is provided it should not be so substantial as to be considered an important source of income or to influence voting on the IACUC. NIH/OLAW guidance in its FAQ B.11, also states: "It is acceptable for the institution to pay for IACUC training of nonaffiliated members (e.g., attendance at IACUC 101)."[4] Lastly, the *Guide* (pp. 24–25) echoes this sentiment: "The public member may receive compensation for participation and ancillary expenses (e.g., meals, parking, travel), but the amount should be sufficiently modest that it does not become a substantial source of income and thus risk compromising the member's association with the community and public at large."

The majority of survey respondents indicated that their IACUC members are *not* compensated. A total of 45% of respondents indicated that the Chair receives direct or indirect compensation. When provided, compensation was described as a stipend or salary supplement, funds for professional activities or laboratory research, effort credit distribution, travel to continuing education, lunch as part of the meeting, free parking, donation to a designated charity, or distribution of annual gift cards. None of these methods appear to be substantial enough to influence voting or be an important source of income.

Surv. Does your institution provide compensation of any kind (other than reimbursed parking fees, lunch during a meeting, or similar minor compensation) to either the IACUC chairperson or IACUC members (other than the nonaffiliated member? Select all that apply and briefly describe the method of compensation in the text box.

- I don't know about compensation to the Chairperson. 12/296 (4%)
- I don't know about compensation to IACUC members. 7/296 (2%)
- No, the chairperson is not compensated. 113/296 (38%)
- No, IACUC members are not compensated. 179/296 (60%)
- Yes, the Chairperson receives direct compensation. 93/296 (31%)
- Yes, the Chairperson receives indirect compensation. via a credit to his/her home department. 41/296 (14%)
- Yes, IACUC members receive direct compensation. 16/296 (5)
- Yes, IACUC members receive indirect compensation via a credit to their home department. 11/296 (4%)
- Please briefly describe method of compensation. 73/296 (25%)

Variable methods of compensation include the following:

- Stipend or salary supplement provided to Chair and/or Vice Chair
- Partial salary recovery provided to administrative or academic unit of Chair and/or Vice Chair
- Funds provided to Chair and/or Vice Chair for professional activities, research
- Distribution of effort credit, relief from other duties given to Chair and/or Vice Chair
- Stipend or salary supplement provided to members other than Chair or Vice Chair
- Travel to continuing education provided
- Lunch provided at meetings or gift cards

Surv. Does your institution provide financial compensation to the nonaffiliated member(s)?

- I don't know about compensation to the nonaffiliated member(s). 16/296 (5%)
- No, the nonaffiliated member(s) isn't/aren't financially compensated. 170/296 (57%)
- Yes, the nonaffiliated member(s) is/are financially compensated. Method and amount of compensation is:* 110/296 (37%)

*Respondents indicated a variety of compensation strategies for the nonaffiliated member, with some receiving monetary compensation on a per meeting, monthly, hourly, or annual basis. Some respondents indicated specific amounts; others did not state specifics. Some institutions provided non-monetary compensation to nonaffiliated members, and some provided a combination of monetary and non-monetary compensation.

- $25–$50 per meeting
- $75 per meeting
- $100 per meeting
- $150–$200 per meeting
- $50–$75 per hour
- $200 per year
- $300 per year
- $500 per year
- $600 per year
- $800 per year
- $1,000–$1,200 per year
- $2,400 per year
- Annual gift card
- Travel, mileage expenses
- Parking
- Travel and costs to attend external CE events
- Donation to designated charity

3:10 Are institutions required to train their IACUC members?

Reg. Both the AWAR and the PHS Policy are silent on a specific method or type of training that should be provided to IACUC members to ensure they are qualified. However, the AWAR (§2.31,a) and the PHS Policy (IV,A,3,a) state that the IACUC must be qualified through the experience and expertise of its members to assess the research facility's animal program, facilities, and procedures, while the AWA (§2143,b) states that members shall possess sufficient ability to assess animal care, treatment, and practices in experimental research as determined by the needs of the research facility. In addition, the AWAR (§2.32,a) state: "It shall be the responsibility of the research facility to ensure that all scientists, research technicians, animal technicians, and other personnel involved in animal care, treatment, and use are qualified to perform their duties." IACUC members are "involved in animal care, treatment, and use" because they oversee it, so they are included in the training requirements.

Recognizing the complexity of the responsibilities that IACUC members are asked to fulfill, the *Guide* (p. 17) states: "It is the institution's responsibility to ensure that IACUC members are provided with training opportunities to understand their work and role. Such training should include formal orientation to introduce new members to the institution's Program; relevant legislation, regulations, guidelines, and policies; animal facilities and laboratories where animal use occurs; and the processes of animal protocol and program review. Ongoing opportunities to enhance their understanding of animal care and use in science should also be provided." The *Guide* (p. 15) also states: "All Program personnel training should be documented." Accordingly, PHS-Assured and AAALAC-accredited institutions are expected to provide adequate training for IACUC members and to document that such training has been provided.

Opin. There are many ways that institutions can train IACUC members. Although no formal regulations require this training, it is common best practice to provide training that ensures IACUC members are qualified to perform their duties. A common method is to provide copies of, or Internet links to, relevant regulations, policies, and guidelines, as well as copies of the institution's forms, guidelines, and SOPs. If an electronic protocol system is used, training on the system features should be provided. For those institutions using paper protocol forms, a copy of the institution's forms should be provided. New members typically have one or more orientation sessions under the guidance of the IACUC staff, the Chair, and/or the AV. New committee members may also be assigned a mentor who is a more experienced member of the IACUC, and they may be invited to attend one or two IACUC meetings as observers before their membership term starts. In addition, members may attend an ever-growing number of meetings, workshops, and webinars that address animal care and use issues of importance to IACUCs. Other useful information for training new IACUC members includes a tutorial on the PHS Policy (https://olaw.nih.gov/resources/tutorial).

AALAS sponsors a variety of online animal care and use courses in the AALAS Learning Library,[5] which is an excellent source of training materials for IACUC members. The IACUC 101™ series, supported in part by OLAW, presents programs designed to provide IACUC members, administrators, veterinarians, animal care staff, researchers, regulatory personnel, compliance officers, and others with information on the role and responsibilities of IACUCs and an understanding of federal policies and regulations governing the care and use of animals in research and education.[6] Public Responsibility in Medicine and Research (PRIM&R) offers webinars throughout the year and other educational opportunities in conjunction with its annual conference. The Scientists Center for Animal Welfare (SCAW)[7] sponsors an annual educational Winter conference as well as a variety of IACUC Training workshops useful for IACUC members and administrators, Principal Investigators, Attending Veterinarians, regulatory personnel, and animal care staff. The Massachusetts Society for Medical Research (MSMR)[8] offers "Introduction to Institutional Animal Care and Use Committees," a six-week online training course for prospective and new IACUC members, staff, and community members. The Collaborative Institutional Training Initiative (CITI) offers member institutions an online "Essentials for IACUC Members" course.[9] The IACUC Administrators Association (IAA) holds "Best Practices" meetings for IACUC Administrators where common programmatic issues and potential solutions, as well as successful processes and ideas, are informally discussed.[10]

Several Institute for Laboratory Animal Research (ILAR) publications also contain information of value to IACUC members.[11–16]

Surv. Identify methods used at your institution to train IACUC members. Select all that apply.

- Provide hard copies of regulatory documents — 146/298 (49%)
- Provide electronic links to regulatory documents — 285/298 (96%)
- One-on-one mentoring with a current IACUC member — 145/298 (49%)
- One on one meeting with the IACUC Chair — 132/298 (44%)

- One-on-one meeting with a member of the IACUC support staff 168/298 (56%)
- One-on-one meeting with the AV 72/298 (24%)
- One-on-one meeting with the IO 20/298 (7%)
- Periodic educational topics at scheduled IACUC meetings 243/298 (82%)
- Other* 97/298 (33%)

*Other responses reflected variations of the choices above, though the following were also offered:

- Attendance at external meetings/training events
- Instruction from animal facility manager
- Provision of an IACUC member handbook or training materials
- E-mail updates of relevant information
- Self-guided PowerPoint training
- Webinars
- Formal orientation program

3:11 Should the performance of individual IACUC members be evaluated?

Opin. There are no specific regulations with regard to evaluating individual IACUC member performance. However, an effective IACUC, with engaged members, can help foster an institutional culture that values research quality and the humane care of research animals. Further, a goal of the IACUC is to find the optimal balance between achieving experimental aims while ensuring compliance with regulatory requirements and with accreditation and institutional policies. A self-evaluation of the IACUC and its members, at the time of program evaluation, can highlight support needs and/or process changes necessary to effectively meet the responsibilities of the IACUC within the institution's Animal Care Program. An IACUC mission statement may be useful and can contain language that details required performance standards for members. According to survey responses below, 44% of respondents indicated that they currently evaluate IACUC member performance. In addition, it should be noted that both OLAW and USDA will recommend a change in IACUC membership if documentation indicates low or nonparticipation in IACUC official business by a particular member.

Surv. Does your institution evaluate the performance (e.g., review timeliness, quality of reviews and inspections, etc.) of the IACUC members?

- No 157/298 (53%)
- Yes 131/298 (44%)
- I don't know 10/298 (3%)

3:12 What criteria might be used to evaluate the performance of individual IACUC members?

Opin. As indicated by the survey responses below, performance criteria of individual IACUC members may include tracking the number of meetings and/or facility inspections attended, the number of documents reviewed, and/or individual member participation in IACUC subcommittees or educational opportunities. This information can be maintained by IACUC office staff and also works as a method to benchmark IACUC activities. Specific to some institutions, providing this type of information to a member's department chairperson/director may be appreciated when a member is applying for tenure and/or a promotion.

Use of committee evaluation tools can provide a mechanism for individual member evaluation as well as determining committee functions that may need improvement. For example, an agree-disagree style survey could be sent to each member to address issues such as whether the number of members is adequate, the number of meetings held is sufficient, the day/time of meetings is appropriate, or the meetings themselves are well organized. Self-evaluation focuses on best practices and therefore compares the IACUC's performance with its mandated responsibilities.

Surv. If your institution evaluates the performance of your IACUC members, which of the following criteria are used for this evaluation? Select all that apply.

- Number of protocols/amendments reviewed each year 72/298 (24%)
- Timeliness of protocol/amendment reviews 91/298 (31%)
- Number of IACUC meetings attended 137/298 (46%)
- Quality of reviews and inspections 60/298 (20%)
- Number of semiannual facility inspections attended 93/298 (31%)
- Number of subcommittee assignments (e.g., policy development/revision, noncompliance investigations, PAM visits, etc.) in which the member participated 32/298 (11%)
- Number of external relevant educational program attended 27/298 (9%)
- Other* 142/298 (48%)

*Other responses reflected variations of the choices above, though the following were also offered:

- Performance of IACUC members is not regularly evaluated
- Overall engagement of the member, participation in meetings, informal assessment
- Ensure that members are up to date with respect to training

3:13 May an institution have more than one IACUC?

Reg. The AWA (§2143,b,1) states: "The Secretary shall require that each research facility establish at least one committee." The AWAR (§2.31,a), which implement the AWA read, "The Chief Executive Officer of the research facility shall appoint an Institutional Animal Care and Use Committee (IACUC)," language that is echoed in PHS Policy (IV,A,3,a).

Opin. No regulatory language precludes establishment of multiple IACUCs at one institution. The statements above suggest open interpretation for appointing more than one committee at the discretion of the institution. When institutions have separate physical areas of jurisdiction in their animal care program, they may opt to appoint more than one IACUC. Institutions with multiple IACUCs should contact their regional APHIS/AC office for guidance and those with activities supported by an agency that requires compliance with the PHS Policy should contact NIH/OLAW for advice on how to describe their IACUC structure and function in their Assurance. Multiple IACUCs are found most commonly at large institutions that have multiple campuses or locations, diverse scientific missions, many animal facilities, and/or large numbers of investigators and protocols.

3:14 Under what circumstances might multiple IACUCs be advantageous?

Opin. Multiple IACUCs may be advantageous at large institutions where the sheer volume of research activity exceeds the capacity of a single IACUC to review the associated protocols and programs in an effective and timely manner. Dividing the work among multiple IACUCs, each appointed to oversee a discrete portion of the Animal Care Program, may be an optional way to manage the workload, thereby reducing the administrative burden. An added advantage of dividing the workload is that it allows committees to gain expertise in reviewing protocols in particular research areas, since they usually are appointed within academic units or for coverage of specific research disciplines. The existence of more localized IACUCs also makes it easier for IACUC members to maintain a personal relationship with investigators and a sense of ownership and pride in the portion of the program they oversee.

Another situation in which the existence of more than one IACUC may be advantageous is in institutions that have disparate research programs, such as those with Colleges of Medicine, Veterinary Medicine, and Agriculture. In others, it occurs when the research program is physically divided between two locations. In both situations, the organization may find it effective to appoint individual IACUCs to review protocols from and oversee the separate components of its overall Program. In these cases, it is important that the responsibility for the ongoing review (e.g., amendment, annual if applicable, and/or triennial review) of an individual protocol remains with the original IACUC that approved it.

3:15 What might be the disadvantages of multiple IACUCs?

Opin. A major challenge for a multiple IACUC system is to develop mechanisms for maintaining consistency and quality in program review, facility inspection, and protocol review across the institution. If there is not an overarching coordinating group, the institution may become fragmented and inconsistent, or conflicting practices may arise. However, one IACUC may not have authority over another IACUC (see 3:17). Another disadvantage is the need for increased IACUC staffing and funding to sustain the necessarily redundant components of multiple committees. The cost of maintaining an IACUC may be considerable and, in the absence of centralized support, can place a large burden on smaller administrative units, such as colleges or departments.

Another potential problem with the multiple IACUC system is that each committee must meet all the membership requirements for an individual IACUC (e.g., veterinarian, nonaffiliated member, nonscientist), including the requirement that no more than three members from the same administrative unit may serve on the IACUC (AWAR §2.31,b,4; *Guide*, p. 25). It can be difficult to appoint a properly constituted committee from within a small administrative unit. Moreover, difficulties can arise with locally constituted IACUCs because no member may participate in the IACUC review or approval of a research project in which the member has a conflicting interest, nor can any member who has a conflicting interest contribute to the constitution of a quorum during the review of the activity in which the person has a conflict (PHS Policy IV,C,2; AWAR §2.31,d,2). Even when there is no obvious conflict of interest, it is important to recognize that there can appear to be a conflict when IACUC members drawn from a small administrative unit review each other's protocols. Finally, when IACUC members are drawn from a limited pool, individuals may find that they have an excessive time commitment to the IACUC.

Some of the difficulties cited, with issues related to consistency of actions, can be mitigated by having one or more of the same individuals serve on each IACUC and/or by having an experienced IACUC staff that supports all the IACUCs. However, according to the language in PHS Policy, an individual would be disqualified for serving as a nonaffiliated member on one IACUC if they served in any other capacity for the institution (e.g., service on a different IACUC).

3:16 What are some methods for multiple IACUCs to effectively interact?

Opin. In institutions where multiple IACUCs report to a single Institutional Official (IO), the IO may appoint an IACUC advisory committee to address general issues that affect the entire animal care and use program. The committee could consist of the IO or the IO's designee, the Chairs of the individual IACUCs, the AV, and the IACUC Administrator. Some simple strategies for multiple IACUCs to effectively interact can include:

- Having some of the same members, other than the nonaffiliated member(s), serve on more than one committee remembering that no more than three voting members associated with a single administrative unit may be on the same committee
- Using a standardized animal use protocol form
- Sharing the development/modification of common guidelines, SOPs, procedures, and policies
- Sharing an IACUC administrator and/or IACUC staff
- Developing a common IACUC training program
- Coordination of an AAALAC site visit, including shared responsibility for the Program Description
- Developing and maintaining centralized information resources for all the IACUCs

3:17 Should multiple IACUCs all report to a central IACUC, advisory committee, or other authority?

Opin. According to language in the OLAW "Sample Assurance" document and the AAALAC "Instructions for Completing the Program Description," the lines of authority and responsibility for an institution's animal care program must demonstrate a direct reporting structure and lines of communication from the IACUC to the IO.[17,18] The IO must be the same individual for all IACUCs under one NIH/OLAW Assurance. Parts of the organization that have different Assurances could have different IOs. As stated above in 3:16, an overarching advisory group may be utilized to support multiple IACUCs; however, this group must carefully confine its activities to policy guidance and may not assume the functions of the IACUC as described in the PHS Policy (IV,B,1–8) and AWAR (§2.31,c).

3:18 What strategies can be used to ensure that protocol review is consistent throughout an institution with multiple IACUCs?

Opin. The most important step in ensuring consistent protocol review is to develop a single animal use protocol form that can be used throughout the institution. The form should be designed to allow sufficient flexibility in describing proposed procedures, so as to encompass the full range of activities at the institution (e.g., antibody production, survival surgery, product safety testing, field research, breeding) while requiring all investigators to provide mandated information for proposed uses of animals (e.g., rationale for animal use, justification for animal species and numbers, alternatives to painful procedures, humane endpoints). NIH/OLAW has posted a comprehensive animal study proposal sample document on their website at https://grants.nih.gov/grants/olaw/sampledoc/animal_study_prop.htm. AALAS's "IACUC Central" has posted sample protocol forms from various institutions on its website at https://www.aalas.org/iacuc/iacuc_resources/protocol-forms. The sample forms from both resources may be downloaded and customized by institutions at their discretion for use in their program.

Another important strategy to ensure consistent protocol review is for each IACUC to share the same guidelines, SOPs, and policies, especially those that pertain to the protocol review process and special considerations for IACUC review, such as those described in the *Guide* (pp. 27–31) on experimental and humane endpoints, physical restraint, multiple survival surgical procedures, food and fluid regulation, and use of non-pharmaceutical-grade chemicals.

It is also important for the IO to ensure that the membership of each IACUC not only meets the minimal regulatory requirements but also includes individuals who have sufficient expertise to competently evaluate the protocols they review. Each committee needs to have scientists familiar with the kinds of animal use they will be asked to review and veterinarians knowledgeable in the appropriate care and husbandry needs of the proposed animal species. If a committee is asked to review a protocol in an area outside of the expertise of its members, it may use non-voting *ad hoc* consultants to assist in its evaluations. Depending on the nature of the activities to be reviewed, sufficient technical expertise must be available for each IACUC to evaluate potential risks associated with protocols (e.g., biosafety, chemical and radiation safety, occupational health) and to ensure that appropriate safeguards are proposed to mitigate these risks. This can be accomplished by including a health and safety specialist on every committee or by providing a centralized health and safety resource for use by all IACUCs.

Lastly, a formal, coordinated IACUC member training program for the multiple IACUCs, using methods described in 3:10, will enhance consistency of protocol review.

3:19 What mechanism might be used to ensure consistent IACUC semiannual inspection and review of animal care facilities and programs at institutions that have multiple IACUCs?

Opin. One way for institutions that have multiple IACUCs to ensure that semiannual inspections and program reviews are consistent is to utilize an institution-wide committee composed of at least two members from each individual IACUC to perform these functions. Remember, a majority of the members of each individual IACUC would have to review and sign the report for it to be a valid semiannual report under the AWAR. Also, any individual IACUC member wanting to participate in the inspections or program review may not be excluded. The institution could also have an institution-wide committee to review the reports of the individual IACUCs that perform semiannual reviews and to make recommendations to improve consistency. However, the content of the semiannual report is the responsibility of the IACUC with jurisdiction over that area. An institution-wide committee may not conduct oversight editing or influence the content of the reports. The AWAR (§2.31,c,3) require the IACUC to perform this function. The development of institution-wide forms to be used in conducting inspections and preparing reviews can also help ensure consistency. One tool to assist IACUCs in conducting semiannual facility and program reviews is the NIH/OLAW *Semiannual Program Review and Facility Inspection Checklist.*[19] Another strategy is for personnel from a central unit (e.g., AV, institutional compliance officer) to participate as *ad hoc* consultants in the facility inspections and program reviews conducted by each IACUC and participate in all IACUCs to ensure consistency. The PHS Policy (IV,B,3, footnote 8) states that the IACUC "may, at its discretion … invite *ad hoc* consultants to assist in conducting the evaluation. However, the IACUC remains responsible for the evaluation and report." The AWAR (§2.31,c,3) have similar wording. Again, a formal, coordinated IACUC member training program will ensure all members understand the semiannual inspection and program review process.

Lastly, institutions that require all of their components to be accredited by AAALAC should have an independent method for ensuring uniform compliance with applicable regulations, policies, and guidelines.

3:20 What is the least common denominator (e.g., school, college, department, individual investigator) around which an IACUC can form?

Opin. Institutions with multiple IACUCs might be formed based on the following criteria, keeping in mind whether these institutions have separate physical areas of jurisdiction in their animal care and use program. Separate jurisdiction prevents overlap of oversight among the committees, thereby preventing conflict.

- Discrete units within the institution's administrative structure (e.g., medical school, veterinary school, graduate school, college of agriculture)
- The nature of the proposed research (e.g., agricultural, biomedical, field research, product testing)
- The animal species being used (e.g., farm animals, laboratory animals, nonhuman primates, exotic species)
- Geographically separate facilities (e.g., if an institution has more than one IACUC, each IACUC might oversee different physical areas. See 3:14)

Surv. If your institution has more than one IACUC, provide the rationale for the multiple IACUCs.

• NA—we have only one IACUC.	278/298 (93%)
• Our IACUCs differ by department, school, or college.	2/298 (1%)
• Our IACUCs differ based on the type of work conducted, i.e., biomedical, agricultural, field studies, etc.	6/298 (2%)
• Our IACUCs differ based on geographical location.	5/298 (2%)
• Other*	7/298 (2%)

*Other responses mostly reflected variations of the choices above, though some offered additional information as shown below.

- Respondents stated that the species involved determined the need for an additional IACUC.

3:21 What needs to be considered when selecting the size of a unit upon which to base an IACUC?

Opin. When selecting the size of a unit upon which to base an IACUC, consideration needs to be given to:

- The potential workload of the IACUC (e.g., number of protocols per year; number, size, and location of animal facilities).
- The requirement that each IACUC must have a direct reporting structure to the IO (see 3:17).
- Whether the unit has the resources to support an IACUC.
- Whether the unit has sufficient breadth of animal use to prevent frequent potential conflicts of interest within the IACUC.

- Whether the pool of animal users from which IACUC members can be drawn is sufficiently large.
- The AWAR requirement (§2.31,b,4) that no more than three members of the IACUC can be from the same administrative unit. The *Guide* (p. 25) echoes this language.

3:22 Can a small institution without an IACUC use the services of a nearby institution with an IACUC to review and approve research supported by a PHS grant?

Reg. PHS Policy (IV,A,3) stipulates that the IACUC must be appointed by the CEO (or the CEO's designee) and mandates the number and qualifications of the IACUC members (see Chapter 5). The AWAR (§2.31,a; §2.31,b) essentially echoes PHS Policy, although the minimal number of IACUC members required differs between these two documents. Neither the PHS Policy nor the AWAR stipulate that any of the *non*-veterinary IACUC members must also be employees of, or in any other way affiliated with, the institution (see 5:19). However, PHS Policy (IV,A,3,b) and the AWAR (§2.31,b,3,i) state that the IACUC veterinarian must have "direct or delegated program authority and responsibility for activities involving animals at the institution (PHS Policy) [or] research facility (AWAR)." The AWAR (§2.33,a,1) further stipulate, "Each research facility shall employ an attending veterinarian under formal arrangements. In the case of a part-time attending veterinarian or consultant arrangements, the formal arrangements shall include a written program of veterinary care and regularly scheduled visits to the research facility." The AWAR note (§2.33,a,3), "The attending veterinarian shall be a voting member of the IACUC; Provided however, that a research facility with more than one Doctor of Veterinary Medicine (DVM) may appoint to the IACUC another DVM with delegated program responsibility for activities involving animals at the research facility." Per PHS Policy (IV,A,3,B) the veterinarian with direct or delegated program authority, named in the Assurance as such, is a member of the IACUC.

Opin. Although there are some potential pitfalls to this approach (e.g., see 5:19), it might be worth considering in certain situations, such as when a small institution with only a few animal users has as its neighbor a larger institution with a more extensive and experienced animal care and use program. The CEO (or the IO, if the authority to appoint the IACUC has been delegated) of the smaller institution could agree to appoint the larger institution's AV to serve as the smaller institution's AV and the larger institution's IACUC to serve as the smaller institution's IACUC, in each case with the proper program responsibilities. Other variations of this approach are also possible, but, in each situation, care must be taken by both institutions to ensure that the arrangements they make do not compromise their overall compliance with the requirements of the PHS Policy and the AWAR regarding the responsibilities of the IACUC, AV, CEO, and IO (see also Chapter 5). Institutions considering any such relationship are well advised to consult representatives of NIH/OLAW and APHIS/AC about the appropriateness of their proposed arrangements.

3:23 How might an IACUC be effectively constituted at a very small institution where all or most personnel are likely to be directly or indirectly involved with the research to be reviewed?

Opin. As pointed out in 3:22, the use of persons not employed by the institution as members on the IACUC is compatible with AWAR and PHS Policy. In this situation, supplementing the affiliated members with several nonaffiliated members whose expertise is well suited to the needs of the institution would prevent the problems inherent in trying to populate an IACUC with the institution's entire staff of animal users.

REFERENCES

1. Office of Laboratory Animal Welfare, Applied Research Ethics National Association. 2002. *Institutional Animal Care and Use Committee Guidebook*. Bethesda, MD: National Institutes of Health. https://olaw.nih.gov/guidance/ARENA-OLAW-IACUC-guidebook.htm. Accessed March 9, 2024.
2. OLAW Guidance No. 97-03. Maintenance of properly constituted IACUCs. https://olaw.nih.gov/guidance/articles/dc97-3.htm. Accessed March 9, 2024.
3. National Institutes of Health, Office of Laboratory Animal Welfare. Frequently asked questions B.1, What are the IACUC membership criteria? https://olaw.nih.gov/faqs#/guidance/faqs?anchor=50294.Accessed March 9, 2024.
4. National Institutes of Health, Office of Laboratory Animal Welfare. Frequently asked questions B.11, May the institution pay or reimburse expenses incurred by nonaffiliated members? https://olaw.nih.gov/faqs#/guidance/faqs?anchor=50293. Accessed March 9, 2024.
5. American Association for Laboratory Animal Science. AALAS Learning Library. https://www.aalaslearninglibrary.org. Accessed March 9, 2024.
6. IACUC 101™ Series. https://iacuc101.org/. Accessed March 9, 2024.
7. Scientists Center for Animal Welfare. https://www.scaw.com/iacuc-training-workshops.html. Accessed March 9, 2024.
8. Massachusetts Society for Medical Research. *Introduction to Institutional Animal Care and Use* Committees On-Line Training. https://www.msmrcourses.org/courses/introduction-to-animal-care-and-use-committees/. Accessed March 9, 2024.
9. Collaborative Institutional Training Initiative. https://about.citiprogram.org. Accessed March 9, 2024.

10. The IACUC Administrators Association. https://iacucaa.org/. Accessed March 9, 2024.
11. van Zutphen, B. 2007. Education and training for the care and use of laboratory animals: An overview of current practices. *ILAR J.* 48(2):72–74.
12. Dobrovolny, J., J. Stevens, and L.V. Medina. 2007. Training in the laboratory animal science community: Strategies to support adult learning. *ILAR J.* 48(2):75–89.
13. Anderson, L.C. 2007. Institutional and IACUC responsibilities for animal care and use education and training programs. *ILAR J.* 48(2):90–95.
14. Medina, L.V., K. Hrapkiewicz, M. Tear et al. 2007. Fundamental training for individuals involved in the care and use of laboratory animals: A review and update of the 1991 NRC Core Training Module. *ILAR J.* 48(2):96–108.
15. Greene, M.E., M.E. Pitts, and M.L. James. 2007. Training strategies for Institutional Animal Care and Use Committee (IACUC) members and the Institutional Official (IO). *ILAR J.* 48(2):131–142.
16. Foshay, W.R., and P.T. Tinkey. 2007. Evaluating the effectiveness of training strategies: Performance goals and testing. *ILAR J.* 48(2):156–162.
17. OLAW Sample Domestic Assurance. https://olaw.nih.gov/resources/documents/assur.htm. Accessed March 9, 2024.
18. AAALAC Program Description Instructions. https://www.aaalac.org/program-description. Accessed March 9, 2024.
19. National Institutes of Health, Office of Laboratory Animal Welfare. Semiannual Program Review and Facility Inspection Checklist. https://olaw.nih.gov/resources/documents/cheklist.htm. Accessed March 9, 2024.

4

Reporting Lines of the IACUC

Marcy Brown and Deborah Frolicher

Introduction

Federal law, regulation, and policy stipulate that the IACUC must report to the individual who has been designated to serve as the Institutional Official (IO). This chapter clarifies what it means to be the IO by reviewing this individual's responsibilities under the PHS Policy, the AWAR, and the *Guide*. Examples of common job titles for IOs in various types of institutions are used to provide the reader with a sense of the level of authority that this person should have within an institution. The chapter concludes with brief discussions of useful methods for communicating with and educating the IO about the functions of the IACUC. The chapter also provides suggestions on which individuals may contact NIH/OLAW, APHIS/AC, or AAALAC for guidance on IACUC related issues.

4:1 To what person is the IACUC required to report?

Reg. The IACUC must report to the institution's IO. According to PHS Policy (IV,B,3; IV,B,5) and the AWAR (§2.31,c,3–5), the IACUC is required to submit reports of its semiannual program evaluations and facility inspections to the IO and to make recommendations to the IO regarding any aspect of the institution's program, facilities, or personnel training. PHS Policy (IV,F,1; IV,F,2) requires that the IACUC, through the IO, submit an annual report to NIH/OLAW. In addition, PHS Policy (IV,C,7; IV,F,3) requires the IACUC, through the IO, to promptly provide NIH/OLAW with a full explanation of the circumstances and actions taken with respect to any serious or continuing noncompliance with PHS Policy, any serious deviation from the *Guide*, or any IACUC suspension of an activity. In its *Guidance on Prompt Reporting to OLAW under the PHS Policy on Humane Care and Use of Laboratory Animals*,[1] NIH/OLAW provides examples of situations that must be reported by the IACUC, through the IO, including the "failure to correct deficiencies identified during the semiannual evaluation in a timely manner."

The AWAR (§2.31,c,3) require the IACUC to report through the IO to APHIS/AC and any federal funding agency, in writing, within 15 business days whenever the institution fails to adhere to a reasonable plan and schedule for correcting a significant deficiency in the institution's program or facilities (see 29:12; 30:16). In addition, if the IACUC suspends an activity involving animals, PHS Policy (IV,C,7) and the AWAR (§2.31,d,7) require that the IO, in consultation with the IACUC, review the reasons for the suspension, take appropriate corrective action, and report that action with a full explanation to NIH/OLAW and any federal agency funding the project and/or APHIS/AC if the animal activity is regulated under the AWAR.

4:2 What methods for communication with and keeping the IO involved should the IACUC employ?

Opin. Given that the IO is the individual at a research facility who is authorized legally to commit on behalf of the research facility that it will meet the requirements of the AWAR and PHS Policy (see 4:4), an effective method for communicating with the IO is crucial for keeping the IO involved and informed on program issues or concerns. To ensure awareness of the various Animal Program components, periodic in-person meetings with the IO are valuable. Scheduling these meetings in combination with delivery of the required Semiannual Evaluation Reports is a good first step toward strengthening institutional commitment and oversight of the Animal Care Program.

Surv. What methods are used at your institution to communicate with your IO and keep him/her involved in the Animal Care Program? Select all that apply.

• No direct communication with the IO to ensure involvement in the Animal Care Program	3/285 (1%)
• Our IO attends IACUC meetings	62/285 (22%)
• Provision of periodic written reports, such as semiannual program documents with results of facility inspections	252/285 (88%)
• Provision of IACUC meeting minutes	139/285 (49%)
• Regular, scheduled face-to-face/virtual meetings with IACUC Chair	93/285 (33%)
• Regular, scheduled face-to-face/virtual meetings with the IACUC support staff	83/285 (29%)
• Regular, scheduled face-to-face/virtual meetings with the IACUC membership	16/285 (6%)

DOI: 10.1201/9781003272861-4

• Regular, scheduled face-to-face/virtual meetings with faculty/investigator oversight committee	14/285 (5%)
• *Ad hoc* meetings between the IO and relevant individuals (e.g., IACUC Chair, IACUC support staff, IACUC members) only as needed	138/285 (48%)
• Other	37/285 (13%)

4:3 How is the Institutional Official defined?

Reg. The AWAR (§1.1, Institutional Official) define the IO as the individual at a research facility who is authorized to legally commit on behalf of the research facility that it will meet the requirements of the AWAR. Similarly, PHS Policy (III,G) defines the IO as the individual who signs and has the authority to sign the institution's Assurance, which commits the institution to meet the requirements of PHS Policy. If not done by the Chief Executive Officer (CEO), the IO (or designated facility representative[2]) signs and submits the annual report to APHIS/AC (§2.36,a). The CEO may or may not be the IO as defined by PHS Policy (IV,A,3,a, footnote 5). The *Guide* (pp. 13–14) notes that the IO is "The individual who, as a representative of senior administration, bears ultimate responsibility for the Program and is responsible for resource planning and ensuring alignment of Program goals with the institution's mission."

Opin. NIH/OLAW FAQ G.5[3] also describes the differences between the CEO and IO and states that "In some institutions, the IO and the CEO may be one and the same." If the CEO does not designate an IO, it is the authors' opinion that the CEO becomes the IO by default and must be identified as such on the NIH/OLAW Assurance and any other pertinent documents. NIH/OLAW will not approve an Assurance without the signature of an individual identified as the IO.

4:4 What are the responsibilities of the IO?

Reg. The IO is the individual who is responsible for ensuring that an institution complies with all applicable animal welfare laws, regulations, and policies. The IO signs forms, reports, and letters on behalf of the institution and interacts with the IACUC in overseeing the institution's animal care and use program.

According to PHS Policy, the IO makes commitments on behalf of the institution that the requirements of the Policy will be met:

- Signs and has the authority to sign the Assurance (III,G)
- Receives semiannual evaluation reports and recommendations from the IACUC (IV,B,3; IV,B,5), including minority views (IV,E,1,d)
- In consultation with the IACUC, determines whether deficiencies are significant or minor (IV,B,3)
- On behalf of the institution, receives notification of the IACUC's decision to approve or withhold its approval of animal activities (IV,C,4)
- Receives and signs annual reports to NIH/OLAW (IV,F)
- Consults with the IACUC regarding suspensions and corrective actions; reports to regulatory and funding agencies (IV,C,7)
- May subject protocols that have been approved by the IACUC to further review and approval but may not approve an activity that has not been approved by the IACUC (IV,C,8)
- Ensures that the institution maintains required records for the specified period (IV,E)

According to the AWAR (§1.1, Institutional Official), the IO legally commits the institution to meet the requirements of the AWAR as follows:

- Signs and submits the registration form (§2.30,a)
- If not done by the CEO or a designated facility representative,[2] signs and submits the annual report to APHIS/AC (§2.36,a)
- Receives semiannual evaluation reports along with recommendations and minority views, from the IACUC (§2.31,c,3; §2.35,a,3)
- In consultation with the IACUC, determines whether deficiencies are significant or minor (§2.31,c,3)
- Reports to APHIS/AC and any federal agency funding the research, in writing, within 15 business days, any significant deficiency the IACUC found uncorrected within the prescribed time that was indicated on the semiannual inspection report (§2.31,c,3)
- Consults with the IACUC regarding suspensions and corrective actions; reports to regulatory and funding agencies (§2.31,d,7)
- May subject protocols that have been approved by the IACUC to further review and approval but may not approve an activity that has not been approved by the IACUC (§2.31,d,8)
- Notifies APHIS/AC, in writing via certified mail, of any change in the name, address, or ownership, or other change in operations affecting its status as a research facility, within 10 days after making such changes (§2.30,c)

4:5 Who can serve as the IO if that role is delegated by the Chief Executive Officer?

Reg. The Health Research Extension Act of 1985 requires the IACUC to be appointed by the Chief Executive Officer (CEO) of the entity for which the committee is established. The PHS Policy requires the CEO to appoint the IACUC in accord with specified qualifications and membership criteria, although the CEO may delegate this authority in writing. In some institutions, the IO and the CEO may be one and the same, whereas in other institutions, particularly large ones, the CEO may be further removed from the day-to-day program oversight.[3] If the CEO delegates authority to appoint the IACUC then that delegation must be specific and in writing.

Opin. The AWAR and PHS Policy describe no restrictions nor do they stipulate any minimum qualifications for individuals who serve as IOs. However, to be effective, an IO must have the level of authority described in 4:7 (see 3:2; 4:4).

4:6 Who typically fills the role of IO in academia and in industry?

Opin. Some common titles of individuals who serve as IOs in academia and industry include Chancellor; President; Provost; Vice Provost; Vice Chancellor for Health Affairs; Vice Chancellor, Administration; Vice Chancellor for Academic Affairs; Vice Chancellor for Research; Executive Director, Research Resources; Director, Research Administration; Director of Sponsored Research; Dean, School of Medicine; Associate Dean, Research and Sponsored Programs; Chief Executive Officer; Executive Vice President for Health Affairs; Senior Vice President and Vice Provost for Health Affairs; Vice President for Health Affairs; Vice President for Research; Vice President for Research and Graduate Studies; Vice President for Clinical Affairs; and Vice President for Worldwide Toxicology.

Surv. Identify the job title for the individual at your institution who fulfills the role of IO.

• Chancellor or Vice Chancellor	21/285 (7%)
• Provost or Vice Provost	27/285 (9%)
• Vice President for Research	90/285 (32%)
• Executive Director	31/285 (11%)
• Chief Operating Officer	14/285 (5%)
• Dean or Associate Dean	16/285 (6%)
• Other (please specify)	86/285 (30%)

*Responses reflected the diversity of institutions represented in the survey, with a variety of titles, many of which are similar in scope to those listed above. In addition, the following were offered.

- Acting Director
- Associate Executive Director
- Associate Laboratory Director
- Assistant/Associate Vice President for Research
- Associate Vice President for Academic Facilities and Planning
- Chief Executive Officer
- Center Director
- Chief Science Officer
- Research Integrity Officer
- Chief Veterinarian
- Chief Animal Welfare Officer
- Chief of Staff
- Deputy Director
- General Counsel
- Director of Sponsored Research
- Director of Research Compliance
- Medical Center Director
- President

4:7 What institutional power or authority should the IO have?

Reg. See 4:3.

Opin. The IO must be authorized to legally commit on behalf of the institution that the requirements of the AWAR and/or PHS Policy will be met. To be effective, the IO must have sufficient administrative authority to promulgate, implement, and enforce policies across departmental lines. In addition, the IO must have sufficient fiscal authority to approve and fund a level of staffing and resources that are adequate to meet "the goals of quality animal care and use" (*Guide*, p. 14) and any needed program improvements, facility repairs, and renovations (see 4:3).

4:8 How involved does the IO need to be in the general activities of the IACUC?

Opin. The IO's level of involvement in the activities of the IACUC varies widely from institution to institution. Some factors that can influence the IO's involvement include the size and complexity of the institution's research program, whether the IO has a background in biomedical research, and the management styles of the organization and the IO. In some institutions, the IO or a representative from the IO's office serves as a member of the IACUC and is directly involved in the general activities of the committee. However, NIH/OLAW recommends that the IO not serve as a voting member on the IACUC, since the IACUC reports to the IO.[4] At a minimum, the IO must understand the functions of the IACUC as they are defined by the AWAR and PHS Policy. Furthermore, IOs that are organizationally distant from the IACUC must ensure that there is a mechanism by which they are promptly informed of any potential threats to animal welfare, or any violations of the AWAR, PHS Policy, the *Guide*, or the institution's PHS Assurance.

Surv. Practically speaking, how involved is the IO at your institution? Select all that apply.

- The IO attends all IACUC meetings. 38/285 (13%)
- The IO meets regularly with IACUC Support staff. 109/285 (38%)
- The IO meets regularly with the IACUC Chair. 129/285 (45%)
- The IO meets regularly with the Attending Vet. 125/285 (44%)
- I am not sure who the IO is. 2/285 (1%)
- Other: (Please specify) 100/285 (35%)

*Responses reflected a variety of means of engagement with the IO, including some that are similar to the above responses. Some respondents provided multiple answers. Other responses included:

- Communication with the IO via email, including reports sent.
- Meeting with a designate of the IO.
- The IO attends some, but not all, IACUC meetings
- The IO meets with the IACUC as needed.
- The IO reviews semi-annual reports.
- The IO reviews IACUC meeting minutes.
- The IO meets with an advisory group.
- The IO is not very engaged with the IACUC.
- The IO attends semi-annual inspections and/or reviews.
- The IO meets with the Associate Vice President for Research Administration.

4:9 What means might be useful for educating the IO with respect to the functions of the IACUC and related policies of the institution?

Opin. The IO should be aware of the laws and regulations governing the use of animals in research, teaching, or testing; understand and support the key components of the animal care and use program; and know where to find information when needed. Training of the IO may be provided by the IACUC Chair, the IACUC Administrator, and/or the AV and may be formal, informal, or a combination of both. Training should include information on the specific IO responsibilities as well as IACUC and AV responsibilities, a broad overview of the program and key reference documents. Useful information regarding training for both the IACUC and the IO may be found in the article "Training Strategies for Institutional Animal Care and Use Committee (IACUC) Members and the Institutional Official (IO)."[5] The NIH/OLAW online tutorial on the PHS Policy is also a useful training resource.[6] Finally, NIH/OLAW conducts a free online webinar series to help both IACUCs and IOs "explore their responsibilities in the oversight of PHS-funded research that involves the use of live vertebrate animals."[7] Webinar recordings are available to the public through the NIH/OLAW website (see also 3:10). Numerous workshops and meetings that provide IACUC training to IOs, members, and administrators are also listed on the NIH/OLAW website.

4:10 Is it appropriate for a research facility to establish guidelines on who may contact either NIH/OLAW or APHIS/AC for guidance on IACUC-related issues, or must this always be the IO?

Opin. It is prudent for institutions to have designated spokespersons for communicating with external entities such as federal regulating agencies and the media. Such designations help to prevent inadvertent errors and miscommunications. In situations in which an institution is making an official report of an animal welfare incident to APHIS/AC, NIH/OLAW, or AAALAC, the AWAR (§2.31,d,7), PHS Policy (IV,F), and AAALAC's Rules of Accreditation (Sect. 2)[8] stipulate the mechanisms for submitting a formal report. Adherence to these requirements is, of course, not optional. Efforts to designate official spokespersons must not, however, infringe upon the rights granted to individuals to independently report alleged animal welfare incidents under institutional, state, and federal whistleblower and whistleblower protection policies.

Surv. Who at your institution may contact NIH/OLAW or APHIS/AC for guidance on IACUC-related issues and is this described within an institutional guidance/policy document? Select all that apply.

- The Director of our IACUC support staff or their primary designee and IS described in a guidance/policy document. 71/285 (25%)
- The Director IACUC support staff or their primary designee and is NOT described in a guidance/policy document. 105/285 (37%)
- The IACUC Chair/Vice Chair and IS described in a guidance/policy document. 71/285 (25%)
- The IACUC Chair/Vice Chair and is NOT described in a guidance/policy document. 96/285 (34%)
- The IO and IS described in a guidance/policy document 94/285 (33%)
- The AV or designee and IS described in a guidance/policy document. 60/285 (21%)

- AV or designee and is NOT described in a guidance/policy statement. 84/285 (29%)
- The IO and is NOT described in a guidance/policy document. 71/285 (25%)
- Other. 27/285 (9%)

REFERENCES

1. National Institutes of Health, Office of Laboratory Animal Welfare. 2005. Notice NOT-OD-05-034, Guidance on Prompt Reporting to OLAW under the PHS Policy on Humane Care and Use of Laboratory Animals. https://grants.nih.gov/grants/guide/notice-files/not-od-05-034.html. Accessed March 2, 2024.
2. USDA APHIS website: Submit Research Facility Annual Reports. https://www.aphis.usda.gov/aphis/ourfocus/animalwelfare/SA_Obtain_Research_Facility_Annual_Report. Accessed March 2, 2024.
3. National Institutes of Health, Office of Laboratory Animal Welfare. 2012. Frequently Asked Questions, G.5. What Is the difference between the Institutional Official and the Chief Executive Officer? https://olaw.nih.gov/faqs#/guidance/faqs?anchor=question50380. Accessed March 2, 2024.
4. National Institutes of Health, Office of Laboratory Animal Welfare/Applied Research Ethics National Association. 2002. *Institutional Animal Care and Use Committee Guidebook*. Bethesda, MD: National Institutes of Health. https://olaw.nih.gov/guidance/ARENA-OLAW-IACUC-guidebook.htm. Accessed March 2, 2024.
5. Greene, M.E., M.E. Pitts and M.L. James. 2007. Training strategies for Institutional Animal Care and Use Committee (IACUC) members and the Institutional Official (IO). *ILAR J.* 48(2):131–142.
6. National Institutes of Health, Office of Laboratory Animal Welfare. 2011. Tutorial on PHS Policy. https://olaw.nih.gov/resources/tutorial. Accessed March 2, 2024.
7. National Institutes of Health, Office of Laboratory Animal Welfare. OLAW Online Webinars and Podcasts. https://olaw.nih.gov/education/webinars-and-podcasts. Accessed March 2, 2024.
8. AAALAC, International. Rules of Accreditation. https://www.aaalac.org/accreditation-program/rules-of-accreditation. Accessed March 2, 2024.

5

General Composition of the IACUC and Specific Roles of the IACUC Members

William G. Greer and Erin Czarniak

Introduction

The general composition requirements for IACUCs, established by federal regulations since 1985, give institutions considerable latitude in fashioning their IACUCs to reflect the scientific expertise and meet the needs of their animal care and use programs. Although the specific IACUC composition requirements vary according to the regulatory oversight agency involved, both the AWAR and the PHS Policy require the IACUC to have a diversified membership, including a veterinarian and a member who is unaffiliated with the institution. The PHS Policy further extends the diversification by requiring the institution to include a nonscientist on the IACUC. Membership diversification on the IACUC is intended to broaden the perspective and add depth to the important IACUC review processes. For example, in institutions conducting agricultural animal research programs, IACUCs should endeavor to incorporate members with appropriate agricultural expertise as recommended by the *Guide for the Care and Use of Agricultural Animals in Research and Teaching*[1]; other specialty areas of research animal use should be similarly addressed.

Most institutions recognize that developing and sustaining an effective IACUC requires planning and an ongoing commitment. The selection of individuals to meet specific membership requirements, new IACUC member orientation or education activities, and successful integration of new IACUC members as contributing members demand considerable effort. Even in a fully functional and effective IACUC, self-assessment and periodic adjustments are necessary for continued high-quality service to the institution and for maintenance of an appreciation for and understanding of the emerging trends in animal care and use programs. The *Guide*[2] (pp. 12–13) emphasizes this point stating, "The body of literature related to animal science and use of animals is constantly evolving, requiring programs to remain current with the information and best practices."

To assist in IACUC development and self-review efforts, this chapter is intended to examine the overall composition of the IACUC and provide information on the roles, responsibilities, and issues involving the IACUC and its various member categories. Some of the trend information provided was acquired through the authors' personal experiences as IACUC members and/or administrators at more than a dozen institutions, and from over 300 IACUC Administrators that attended past IACUC Administrators Best Practice Meetings.

5:1 How many members must an IACUC have?

Reg. The number of IACUC members required depends upon whether the institution is a recipient of funding from, or supported by, the PHS or other cooperating federal agencies, such as the Department of Veterans Affairs, the National Science Foundation, or the National Aeronautics and Space Administration. In these instances, the institution must have at least five IACUC members (PHS Policy IV,A,3,b). Institutions that use animal species regulated under the AWAR and funded through internal or private sources are required to have at least three IACUC members (AWAR §2.31,b,2). The *Guide* (p. 5) states: "The size of the institution and the nature and extent of the Program will determine the number of members of the committee and their terms of appointment." However, the minimum number of members allowable must also have the appropriate roles on the IACUC as described in 5:2.

Opin. In the authors' experience, organizations usually exceed the minimal membership requirements set by the federal regulations and this impression is affirmed by the survey data provided by 61 institutions queried in 2023 about the characteristics of their IACUC membership in this survey. When responders were asked "How many regular voting members are on your IACUC?" the majority of institutions (47%) reported that the committee size was between ten and twenty members. Several institutions (23%) reported larger committees of over twenty members. They comprised the second most prevalent size category followed by smaller committees of less than ten members in 31% of institutions. Thus, it remains uncommon (as reported in the previous edition of this book)[3] for institutions to operate with a committee size hovering at the regulatory minimums. Although a minimum of five IACUC members may be adequate for a small organization with a focused research program and a limited spectrum of animal use, generally a

DOI: 10.1201/9781003272861-5

minimum of ten to twenty members may be necessary in larger, more complex programs and to meet the growing demands of responsible oversight (see 3:4.)

5:2 What specific members are required for the IACUC?

Reg. The AWAR (§2.31,b) state that the committee shall be composed of a Chair and at least two additional members. The AWAR do not describe the qualifications of the Chair, but it states that the other two members of the committee will include at least one veterinarian and at least one member shall not be affiliated in any way with the facility and shall not be a member of the immediate family of a person who is affiliated with the facility. The AWAR (§2.33,a,3) require the veterinarian appointed to the IACUC to be either the Attending Veterinarian or, at facilities with more than one Doctor of Veterinary Medicine, another veterinarian with delegated program responsibility for activities involving animals at the research facility.

The AWAR (§2.31,b,4) state that if the committee consists of more than three members, not more than three members shall be from the same administrative unit of the facility. The administrative unit is defined as the organizational or management unit at the departmental level of a research facility (AWAR §1.1, Administrative Unit), which, for example, may include the Office of Research Administration and the university (institutional) laboratory animal resources.

The *Guide* (p. 24) states that the IACUC should have a veterinarian "either certified (e.g., by ACLAM, ECLAM, JCLAM or KCLAM) or with training and experience in laboratory animal science and medicine or in the use of the species at the institution." In addition, there should be "at least one practicing scientist experienced in research involving animals." Finally, there should be "at least one member from a nonscientific background, drawn from inside or outside the institution" and "at least one public member to represent general community interests in the proper care and use of animals. Public members should not be laboratory animal users, affiliated in any way with the institution, or members of the immediate family of a person who is affiliated with the institution."

PHS Policy (IV,A,3,b), on the other hand, states that membership must consist of not fewer than five members. The PHS Policy (IV,A,3,b,1–IV,A,3,b,4; IV,A,3,c) states that the committee shall include at least one veterinarian with training or experience in laboratory animal science and medicine who has authority and responsibility for activities involving animals at the facility, one practicing scientist experienced in research involving animals, one member whose primary concerns are in a nonscientific area (for example, ethicist, lawyer, member of the clergy), and one individual who is not affiliated with the institution in any way other than as a member of the IACUC and is not a member of the immediate family of a person who is affiliated with the institution. One individual may fulfill more than one requirement. Also note that PHS Policy (II; IV.A.1) requires institutions to comply with the AWAR, as applicable, and to follow the *Guide*. Readers should refer to the NIH/OLAW FAQ B.1 for a succinct discussion of IACUC membership criteria.[4]

Opin. Organizations involved in agricultural research and teaching should be aware that there are guidelines for the composition of animal care and use committees overseeing agricultural research activities.[1] The Agricultural Guide[1] recommendations are used as a standard by AAALAC but have no regulatory empowerment. They closely parallel the PHS Policy, differing only by specifying two types of scientific individuals on the committee. One should be a scientist from the institution who has experience in agricultural research or teaching involving agricultural animals, and the other should be an animal, dairy, or poultry scientist who has training and experience in the management of agricultural animals. This document further recommends that the composition of the IACUC be modified according to the recommendations just noted to provide for the centralized and uniform oversight of the institution's animal care and use program. In an institution that conducts both agricultural research and PHS-funded research in nonagricultural species, the recommendation for two committee members to have the agricultural expertise specified would likely necessitate exceeding the five-person minimal membership requirement of the IACUC under PHS Policy. However, in the authors' experience, the IACUCs serving these dual arenas have taken this approach.

5:3 Must an IACUC report to NIH/OLAW and APHIS/AC the names and titles of all members of the IACUC?

Reg. The general requirements to NIH/OLAW in an institution's Animal Welfare Assurance include the provision for the reporting of IACUC members' names, position titles, and credentials; however, a footnote in the report template clarifies that institutions may, at their discretion, represent the names of members other than the IACUC Chair and veterinarian with numbers or symbols (PHS Policy IV,A,3,b, footnote 6)[6,7] with the understanding that the coding system must be made available to NIH/OLAW upon request. The ability to withhold the names of IACUC members may be beneficial for the individuals and institutions in the event that The Freedom of Information Act 5 U.S.C. 552 requests are later pursued through NIH/OLAW. The AWAR do not contain a comparable provision and therefore, there is no reporting requirement to APHIS/AC.

5:4 Can one person fill more than one position on the IACUC?

Reg. The PHS Policy (IV,A,3,c) provides a written assent to dual representation as long as the IACUC totals at least five members.[3,8] The AWAR offer no direct discussion of this matter and do not preclude this practice.

Opin. It is permissible for an individual to fill more than one of the membership categories on an IACUC under both the AWAR and the PHS Policy. While it is permissible, the authors' identify the significant IACUC responsibilities, workload, benefits of diversification, and challenge in replacing an exiting individual occupying a dual role as reasons to not consolidate roles in an individual. Very few institutions have exploited dual category representation as a long-term strategy for their IACUC, and, when it occurs, it frequently involves the unaffiliated member serving in the dual capacity as a nonscientist. In the foregoing circumstance, the objectivity and independence of the IACUC's decisions could be compromised, particularly if the committee is small. In the authors' experience, dual-capacity appointments as the non-scientist and nonaffiliated member are most commonly used only transiently to bridge a gap in the mandated membership after the departure or resignation of a member.

5:5 What problems can occur if the number of IACUC members is too large or too small?

Opin. There can be significant problems related to the size of the IACUC. IACUCs that have a small membership often have a narrower base of expertise and, depending upon the size of program they service, may encounter an onerous administrative workload resulting in issues such as meeting regulatory requirements and deadlines. Also, small IACUCs may often have a more difficult time making a quorum, and, even in the presence of a legitimate quorum, member absences can have a marked effect on decision making. The challenges for IACUCs that have large memberships are very different. In large IACUCs, the principal problem is that members may become disengaged from the activities of the IACUC. This might be due to impediments to open communication in the IACUC or to the daunting scope and size of the program and its challenges. Thus, members have no sense that their efforts are having any impact on program quality or progress. In the depths of ennui, these members contribute to the quorum in number only. The challenges for the IACUC Chair are to ensure that all members have the opportunity for real contribution and an opportunity to be heard on issues, to help members feel a sense of accomplishment, and, when necessary, to inform the IO that the IACUC may be too large for the tasks at hand. Training opportunities are important for the maturation and effective operation of all IACUCs and can help keep committee members engaged.

5:6 What institutional constituencies should be represented on the IACUC?

Opin. The majority of organizations go beyond the federally mandated composition of the IACUC (see 5:2) to diversify the IACUC membership. The priority in this effort generally is to enlarge the complement of scientific members on the IACUC to match the predominant areas of scientific expertise in the program. In addition to the potential political benefits this has for the IACUC within the scientific community it serves, it confers tangible benefits for the IACUC's deliberations on protocol matters. It also may improve the quality of correspondence exchanged between the IACUC and PIs.

Other institutional constituencies that the authors find to be favored are the following:

- IACUC administrators
- Health and safety personnel (including occupational health and safety, and medical professionals)
- Senior animal management or supervisory personnel
- Statisticians
- Research laboratory technicians (with extensive animal involvement)
- Communications and marketing professionals
- Librarians
- Legal counsel
- Public relations personnel
- Grants and financial personnel
- Student body representatives

The merits of having these people on the IACUC should be evaluated in the context of the program for which they are intended. In the authors' opinion, IACUC administrators, health and safety personnel, statisticians, librarians, legal counsel, communication professionals, and public relations personnel should be given priority. IACUC administrators can provide consistency, serve as a source for regulatory knowledge, enhance institutional memory and action on important issues/decisions, improve IACUC processes and efficiency, offer principal investigator training and assistance and solidify the IACUC's reputation in "customer service." Health and safety personnel are comfortable working at the interface of science and society and can make a significant contribution to the IACUC by keeping it apprised of the status of health and safety regulations, and recommendations relating to the animal care and use environment.[9] Statisticians and librarians can help the IACUC and investigators address the appropriateness of animal numbers and availability of alternatives. Communication specialists can perform a critical role by extending necessary knowledge and outreach to critical animal care and use program members. Legal counsel and public relations personnel also can be of great assistance in the consideration of issues relating to science, policy, perception, and the public.

5:7 Should the selection of IACUC members depend upon the institution's research goals and expertise?

Opin. Two of the most important functions of the IACUC, protocol review and programmatic review, should be major driving forces for the IACUC to select members with a share in the institution's research portfolio. Protocol review should be insightful, thorough, and objective. Members who have scientific expertise in the areas under review can help focus the discussion of relevant issues, improving the quality of the review. IACUC programmatic review is a mandated responsibility of the IACUC (PHS Policy IV,B,1–IV,B,5; AWAR §2.31,c,1–§2.31,c,5) but it really is best performed as a bidirectional process involving the community served. This implies that the IACUC should foster conduits for the flow of information from all participants in the animal care and use program. The appointment of IACUC members who represent specialty interests or areas of expertise can prove advantageous to this process. Most institutions make a concerted effort to staff their IACUCs with members who reflect the research expertise of the program, at the same time making sure to prevent possible conflicts of interest.

The research portfolios amongst institutions are diverse. Arguably, an IACUC lacking the necessary expertise may be unable to appropriately evaluate and approve certain animal activities. For example, a medical school's research portfolio may focus on research and scholarship specific to human health, and the expertise of its IACUC may be biomedical. The problem occurs when the Biomedical IACUC is asked to review field research originating from, for example, its Department of Environmental Sciences. For the IACUC to thoroughly assess the proposed activities, it must have members or ad hoc consultants with expertise in animal activities that involve wildlife. The list of issues that may require discussion can be exhausting, and unique to the situation. Consequently, a committee member with wildlife expertise should be able to fulfill two primary responsibilities:

1. Identify, understand, and reference the wildlife taxonomy guidelines on, for example, the use of wild mammals, birds, fish, reptiles, and amphibians.
2. Depending on the circumstances, identify specific activity nuances that require an IACUC to seek outside guidance from an ad hoc consultant with specialized expertise to participate in the deliberation and approval process.

PHS Policy (IV.C.3) recognizes that the diversity of projects to be reviewed may pose a challenge and therefore states that "the IACUC may invite consultants to assist in the review of complex issues. Consultants may not approve or withhold approval of an activity or vote with the IACUC unless they are also members of the IACUC."

5:8 What is the typical IACUC composition in a large versus a small institution, and are the members' roles different in these settings?

Opin. There are different trends in the composition of the IACUC and in the ways the IACUC uses its members in large versus small institutions. In small commercial institutions or independent research organizations based around a particular area of scientific inquiry, the IACUCs tend to be smaller (fewer than ten members) because the number and diversity of issues requiring consideration by the IACUC are limited. Small academic institutions tend to have slightly larger IACUCs (seven to ten members). Although the research portfolio may be limited compared to that of a large academic institution, the relative diversity of the research activities can be quite high even though each department may only have a few active faculty researchers. The intimacy of the small institutional environment sometimes imposes a higher level of expectation that the IACUC members will be able to speak authoritatively about the research and programmatic needs of their colleagues. IACUC members in small institutions also are more likely to be active participants in developing and implementing programmatic components (e.g., biosafety review and IACUC educational efforts) that may fall under the IACUC's purview.

Large institutions sometimes have 20 or more IACUC members simply to provide ample representation of the animal user groups, scientific areas under study and number of investigators (e.g., 150 to 200) in the program. Data from the 2023 survey of institutions indicated that almost 23% operated with IACUCs of 20 or more members compared to 9% in the previous edition. In the authors' experience, these IACUCs tend to be more formal, impersonal, and insular. Frequently, institutionally provided administrative support obviates the need for the personal involvement of the IACUC members in ushering new initiatives forward. As noted in 5:5, a significant challenge for IACUC members in this position is to combat these tendencies and convince the faculty that the IACUC is a receptive, helpful, and responsive body that works for the institution in support of the scientific mission by encouraging improvement of the animal care and use program. Another issue germane to extremely large IACUCs is the number of veterinarians appointed to that committee. Although federal law has stipulated that an IACUC need only have one veterinary member, most institutions that have large IACUCs (greater than 15 persons) have increased the number of veterinarians appointed. This action helps ensure that the veterinary perspective will be represented consistently and affords the veterinarians an opportunity to divide the workload and concentrate on specialty areas of interest and needs.

5:9 Should the IACUC Chair, individual IACUC members, or their departments receive compensation for their efforts?

Reg. Although PHS Policy has not commented on the general topic of IACUC member compensation, the *Guide* (pp. 24–25) does suggest that compensation of the public member should, "be sufficiently modest that it does not become a substantial source of income and thus risk compromising the member's association with the community and the public at large." NIH/OLAW FAQ B.11 is also instructive on this matter. The AWAR are silent on this issue.

Opin. In contrast to earlier comments on this topic in previous editions of this book which were based upon anecdotal information, compensation for the IACUC Chair has not been established as a common practice and occurs relatively infrequently. For those who have received this acknowledgment of their efforts, salary support, funding to support professional development, research program development, research technical support, and a variety of other approaches have emerged from faculty–administrator negotiations. With less success than the Chair to date, other IACUC members with faculty appointments continue to pursue and gain some forms of compensation for their efforts. Although the primary compensation received is reduction in instructional commitments, some committee members continue to receive, for example, travel support to national meetings or release from ancillary academic obligations such as service on other committees. Members may find that documentation of their time and effort on IACUC activities may help advance the argument for reciprocity of benefits.

Surv. According to the 2012 survey of 296 respondents, approximately 16% acknowledged providing direct (12%) or indirect (4.5%) support to the IACUC Chairperson. In contrast, 7% of institutions reported compensating IACUC members through direct (6%) or indirect (1%) mechanisms. The 2012 survey clarified that compensation such as parking fees, local travel, lunch during the meeting, and other minor compensations should be excluded in responding (see 3:9.)

5:10 How can the performance of the IACUC be evaluated?

Opin. The overall performance of the IACUC is heavily influenced by the degree of each individual member's level of participation. As a result, some institutions evaluate the effectiveness of each IACUC member. The most common method reported to the authors involves a type of scoring system that is associated with a member's agreement form. The agreement form is thoroughly discussed at the time of appointment and clearly outlines and stresses the importance of the individual member's duties (e.g., regular meeting attendance, timely submission reviews, and participation in semiannual inspections and reviews). Throughout the year, adherence to the members agreement is documented by, for example, the IACUC Administrator and subsequently translated into an annual performance review. The IACUC Chair and Administrator then perform an annual performance review with each committee member, either providing a "pat on the back" or offering suggestions for increased engagement.

Several other mechanisms are available and should be utilized to evaluate the performance of the IACUC on an ongoing basis. The IO should meet periodically with the Chair of the IACUC to discuss the content, style, and timeliness of IACUC reports and correspondence with members of the institution's research community. This is also a process commonly used, although few institutions couch such meetings in terms of an evaluation. Also, without the inducement of a particular incident, the IO should consider soliciting the comments of the faculty or staff who use laboratory animals as well as of the members of the IACUC. The assessment of the IACUC as a working group seems particularly important because of the profound potential impact of group dynamics on IACUC morale, attention to detail, and, ultimately, decision making.

Programmatic review contains the essential element of self-assessment that should be practiced by every IACUC. The *Guide* (p. 13) sets the expectation that responsible resource planning by the IO will provide the IACUC with the authority and resources to fulfill its responsibilities including, "specific training to assist IACUC members in understanding their roles and responsibilities and evaluating issues brought before the committee." Such training ordinarily entails member attendance at national conferences or webinars, followed by the communication and discussion of pertinent topics by the entire committee. Simple measures such as the quality of meeting minutes and the attendance records for meetings (i.e., just making or failing to make a quorum repeatedly portends disaster) can be useful. Also, the level of member participation in post-approval monitoring activities and the overall record and severity of indiscretions found during post-approval monitoring indirectly reflects on the performance of the committee. Lack of programmatic oversight, inadequate training programs for personnel, lack of rigor in protocol review and/or failure to provide guidance to PIs through post-approval monitoring activities can increase the number of reportable events, a metric that reflects negatively on the IACUC. Alternatively, lack of findings can also be an indicator of a disengaged IACUC and/or a need for greater IACUC training. Post-approval monitoring findings collected by, for example, an IACUC Administrator, that indicates a contrary programmatic state than indicated by the findings of semi-annual IACUC inspections may indicate inadequate IACUC inspections.

Participation in the voluntary peer-review process of AAALAC entails a substantial review of IACUC function through interviews and the review of written materials. Generally, AAALAC site visits are conducted at triennial intervals and should be supplemented by the institution's internal evaluation mechanisms. In some instances, expert consultants in the area of IACUC function also have been used by institutions.

5:11 Who can remove the IACUC chairperson or other IACUC members?

Reg. The PHS Policy (IV,A,3,a) and the AWAR (§2.31,a) require the Chief Executive Officer (CEO) to appoint the IACUC members. Although the IACUC possesses autonomous review and reporting functions once established, service on the IACUC remains at the discretion of the CEO. Under the PHS Policy (IV,A,3,a, footnote 5), the CEO may delegate the authority to appoint the IACUC if the delegation is specific and in writing.

Opin. It is very fortunate that institutions are rarely, if ever, confronted with this quandary because very few institutions have clearly written bylaws for their IACUCs that stipulate who has jurisdiction or what the process is in this matter, and no clarification is offered in the PHS Policy or the AWAR. For the most part, informal discussions originating from the Chair, in the case of other IACUC members, or from the CEO, in the case of the Chair, apparently have proved adequate for the silent and amicable departure of IACUC members in the vast majority of cases. In the absence of bylaws clearly delineating who has authority in this matter, the CEO retains that authority. In a related matter, without bylaws the extent to which, and the situations in which parliamentary procedure should be applied in IACUC meetings and deliberations are also questionable. This includes, for example, the ability of the IACUC to self-regulate through the use of censure or expulsion procedures[11] (see 6:12).

It is reasonable and advisable for institutions to establish performance standards, i.e., participation expectations, for the Chair and other IACUC members to ensure that discussions with individuals about termination of service are not regarded as arbitrary and capricious. Attendance at meetings, participation in other IACUC activities and ability to perform work assignments adequately in a timely manner are fair performance measures. In recognition of workload and the demanding schedules of IACUC members, the appointment of alternate members on the IACUC, as well as the formal designation of a Vice Chair may have value as an aid to IACUC function and in succession planning.

5:12 What criteria can be applied for removing the IACUC Chairperson or other IACUC members?

Opin. Most organizations agree that inappropriate personal conduct, poor attendance or lack of participation in IACUC activities, or repeated inadequate preparation for assigned IACUC duties might constitute sufficient reason to seek the removal of a member. Ideally, these broad areas would be included in the bylaws developed for the IACUC, and the IACUC members would be informed during their initial IACUC training of the general performance criteria for committee members. The IACUC Chair or IO should inquire whether there are any mitigating circumstances for those members who cannot regularly attend a high percentage (e.g., 75%) of the IACUC meetings/activities held annually and consider the replacement of these individuals if better attendance is not forthcoming. Committee membership agreements can also be used effectively for this purpose. These agreements define expectations of service in each of the areas of IACUC responsibility and serve as a tangible measure of member engagement and contribution and can be used by some programs for the periodic review of IACUC member contribution (see 3:11–3:12).

5:13 What is the procedure for selecting and appointing the IACUC Chair?

Opin. Many IACUCs attempt to prepare for a transition in the position of Chair through the appointment of a Vice Chair or an *ad hoc* Chair-in-Training. In this regard, the IACUC usually has the opportunity to serve in an advisory capacity to the CEO, who ultimately has the responsibility of appointing the Chair (AWAR §2.31,a; §2.31,b). (See 3:2). Other groups that can prove useful in identifying and supporting candidates for the position of IACUC Chair in academic institutions include the faculty senate and faculty research advisory committees. The CEO can also solicit the input of an IACUC administrator who may have valuable insights on the members who have been especially conscientious and made valuable contributions to the IACUC.

5:14 What traits or qualifications are desirable for the IACUC Chair?

Opin. In most institutions, a concerted effort is made to recruit a Chair who is a well-respected individual who has served in a leadership role. Regardless of whether or not the Chair has a scientific background, this individual should be regarded as a good colleague in the context of the institution's scientific mission and have exhibited strong leadership qualities. The Chair should be patient, tolerant, diplomatic, tactful, and efficient in the handling of the IACUC's business and sensitive issues. This person should be able to dedicate the time necessary with sufficient flexibility to plan, oversee, and/or participate in all critical IACUC functions and activities. The notion of adequate participation in critical IACUC functions presupposes that the Chair will be knowledgeable and conversant in current regulatory interpretations, contemporary practices, and trends pertaining to animal use and welfare and committed to leading a successful institutional animal care and use program. In addition to providing a continuing and stabilizing presence, the Chair is likely to encounter occasional urgent matters that require immediate attention.

Some institutions have been slow to recognize that they have appointed a Chair who has not been performing adequately or may be constitutionally unfit for this important position. A partial list of problems indicative of such an appointment includes disorganization of IACUC meetings and activities; inattention to issues raised by the faculty, IACUC members, or veterinary staff in IACUC meetings; a lack of understanding of the regulations and standards and reluctance to consult with other knowledgeable IACUC members or staff or members who do; resistance to the adoption of needed programmatic changes as new standards emerge; proclivity for unbalanced communications (written or verbal) with investigators without the collaborative input or endorsement of the IACUC; or inappropriate and unilateral supplication, or alternatively, rigidity in dealing with investigators.

5:15 What is the Chair's role in the management of IACUC functions?

Opin. The Chair should serve as a primary point of contact and play an active role in the management of all IACUC activities regardless of the individual's role as an actual participant. The Chair serves six important constituent groups:

- The senior administration (embodied in the CEO and IO)
- The scientific community
- Other members of the IACUC
- The Office of Ethics and/or Program Compliance
- The federal government
- The public

The immense commitment within the institution of time, resources, and interest to the IACUC's activities should compel the Chair's enthusiastic involvement in IACUC oversight (see 3:1).

5:16 Can the AV serve as the IACUC Chair?

Opin. While there is no prohibition in the PHS Policy or AWAR of the AV serving as the IACUC Chair, this practice is strongly discouraged. Only a few institutions among the hundreds known to the authors have taken this approach. NIH/OLAW had issued an explanation that strongly suggested that having either the AV or IO serve as the IACUC Chair may be inappropriate because of real or perceived conflicts of interest, the consolidation of important leadership roles and the disruption of the necessary checks and balances intended in the institutional reporting structure and IACUC review processes.[12] Consequently, it is important for the AV and the IACUC maintain independent reporting lines to the IO to preserve the intent of checks and balances. The programmatic review activities of the IACUC encompass laboratory animal management and veterinary care, both of which are often under the direct purview of the AV. Thus, the appointment of the AV as the Chair affords the AV an opportunity to influence the IACUC openly or subtly. To reinforce the conflict of interest (COI) perspective, it is important to remember the IACUC oversees and ensures ongoing compliance within those areas directly overseen by the AV, including the vivarium. Efforts to stimulate institutions to reflect upon the advisability of continuing to use this approach have remained common during AAALAC site visit discussions. AAALAC has often observed that the responsibilities of the AV were already sufficiently demanding and considerably compounded by adding the complex job of the IACUC Chair. Moreover, AAALAC has recognized some instances in which the AV did not carry the title of IACUC Chair but was functioning *de facto* as the Chair and carrying the workload of the IACUC.[13] Institutions should be wary of this devolutionary development.

5:17 Should the manager of the institution's laboratory animal resource unit serve as the IACUC Chair?

Opin. The appointment of the animal facility manager as the IACUC Chair is much more common than the appointment of the AV in this capacity but still quite rare. Facility managers have a broad understanding of the various aspects of the animal care and use program but generally have a narrower scope of responsibilities than the AV. This lessens the COI issue discussed in 5:16. The issue is not entirely eliminated, however, since the review of facilities and the diverse provisions for laboratory animal management are critical elements of the IACUC's programmatic review that generally fall within the domain of the facility manager. The involvement of the facility manager in the financial management of the resource (e.g., husbandry supplies and staffing) may be another COI related deterrent to the appointment of this individual as the Chair in some institutions. Additionally, the animal facility manager may lack the knowledge necessary to be conversant in the overarching scientific enterprise at the institution.

5:18 In an academic setting, what should be the faculty rank of the IACUC Chair? Is a person who has academic tenure preferred?

Opin. The faculty appointment status of IACUC members in an academic setting has been observed by the authors to be both tenured and non-tenured faculty. Institutions have favored appointing tenured faculty members due to nuances associated with a faculty member seeking tenure. Faculty members placed on the IACUC that are pursuing tenure have been observed to have competing interests. For example, the responsibilities associated with IACUC membership are significant and time consuming whereas achieving tenure may require considerable time to be focused on research activities, grant writing and publishing. In addition, non-tenured IACUC members may have an unconscious bias should they be involved in deliberations that involve, for example, senior faculty members on their tenure committee.

Surv. We queried the faculty appointment status of the Chair in 297 institutions. Resulting data are summarized in Table 5.1.

Academic institutions demonstrated an overwhelming preference for the appointment of a faculty member to Chair the IACUC and a majority of the faculty appointed had achieved a rank of Full Professor; tenure status was not solicited.

TABLE 5.1

Faculty Status of the Chair in Reporting Institutions

Academic Title	Total (n = 297)
Non-faculty title or unreported[a]	108
Instructor/Assistant Professor	19
Associate Professor	61
Full Professor	109

[a] Non-academic institutions and Chairs without faculty title are combined.

In an earlier edition of this book,[3] we established that the percentage of tenured Full Professors serving as the Chair was higher than non-tenured Professors and that regardless of faculty rank, 62% of Chairs had attained tenure. Also, more Associate Professors occupy the position of Chair than Assistant Professors (20.5% versus 6.4%), according to the 2012 survey.

Opin. The data support the interpretation that most academic institutions believe that high-ranking faculty will best command the respect and wield the authority necessary to lead the IACUC through its federally mandated review activities impacting institutional resources and faculty satisfaction. In many large academic institutions, the appointment of a Full Professor (and one who has an active research program or prior history of laboratory animal use) is deemed desirable. Academic rank, *per se*, is a less critical factor than an individual's scientific stature, collegiality, and reputation for fairness within the institution's scientific community. Institutions should recognize that the faculty often directs its antagonism toward adverse decisions made by the IACUC to the Chair. Hence, the appointment of a non-tenured faculty member (in a tenure track position) as the IACUC Chair conceivably might enhance that individual's vulnerability during tenure review decisions. Most institutions (or faculty members) appear to recognize this and avoid placing lower-ranking, untenured academic faculty in this precarious position.

5:19 Can an IACUC be composed entirely of persons not employed by the university (or other institutions)?

Reg. Both the AWAR and the PHS Policy (see 5:2) require that at least one member of the IACUC not be affiliated with the institution in any way other than as a member of the IACUC. Other members are not required to be affiliated except that the veterinarian must have direct or delegated program authority (PHS Policy IV,A,3,b,1) and responsibility (AWAR §2.31,b,3,i) for activities involving animals at the institution. Also, pursuant to PHS Policy (IV,B), the IACUC functions as an "agent of the institution."

Opin. The AWAR (§2.33,a,1 and §2.33,a,3) state that the research facility must employ an AV under formal arrangements and that the AV must be a voting member of the IACUC. Under the PHS Policy, the AV must have programmatic authority, so it is unlikely that the research facility will have an animal care and use program without some financial arrangements with the AV. Even if the AV is not employed by the organization, he/she would be affiliated by acceptance of critical responsibilities within the program. However, neither the PHS Policy nor the AWAR stipulate that some (or any) of the remaining members appointed must be institutional employees. Nevertheless, the authors have never encountered an IACUC that did not include at least one member who was an employee. However, the authors are aware of companies that have contracted with contract research organizations or other animal program service providers to conduct their animal research studies without representation on the IACUC. A more common occurrence is that an organization participating in a cooperating research partnership with another organization may elect to co-appoint a knowledgeable IACUC that serves the research missions of both organizations, irrespective of the employment associations of the membership. The pivotal issue in the matter of IACUC appointment is not whether the CEO holds any financial sway over the committee members through employment, but rather, whether the CEO is capable of disbanding the IACUC if it proves to be incompetent, inattentive to animal welfare issues or fails to discharge its federally mandated responsibilities with due diligence in a competitive scientific environment. The growth of inter-institutional collaborations has prompted a *Guide* recommendation (p. 15) that organizations develop a formal written understanding of the responsibilities of each party in the partnership including IACUC oversight.

5:20 Can persons who do not have prior experience or expertise with research animals be appointed as members of an IACUC?

Reg. The regulatory requirements for personnel having experience with laboratory animals are minimal (PHS Policy IV,A,3,a; IV,A,3,b; AWAR §2.31,a; §2.31,b). Both sets of regulations require that the veterinarian have experience with laboratory animals and the PHS Policy stipulates that, in addition, one practicing scientist has experience in research involving animals. Members who do not have laboratory animal experience can be called into IACUC service.

Opin. Most organizations easily exceed the regulated number of those with laboratory animal experience, but IACUCs often have many members who have laboratory animal experience related to one or a few species involving applications in a particular area of research. Much is to be said for the vicarious experience these individuals garner from talking with colleagues, participating in active research programs, reading the literature, and gaining experience on the IACUC. However, introspection by the IACUC is also desirable to allow the self-identification of areas of IACUC weakness and foster the recruitment of internal or external consultants in areas of need. For example, an IACUC overseeing wildlife work would be better served by wildlife specialists than laboratory animal ones. Training should also be provided to individuals with no prior experience with research animals, such as lay members.

5:21 Can or should an outside consultant serve as the IACUC Chair?

Reg. The regulatory references for an IACUC Chair appear in PHS Policy (IV,A,3,b) and in the AWAR (§2.31,c,3; §2.31,b,2). The regulatory references for consultant(s) also appear in the PHS Policy (IV,C,3) and the AWAR (§2.31,c,3; §2.31,d,3). No guidance in PHS Policy or the AWAR is offered on whether a consultant can or should serve as the IACUC Chair.

Opin. This practice is not prohibited by the PHS Policy or the AWAR and the authors are aware of organizations that have used outside expert consultants in this leadership position. Although an outside consultant may initially lack an appreciation for the scientific landscape and key players of the institution, this is not an insurmountable deficit and might be easily offset by the consultant's superior knowledge of the regulations and best practices. This may also afford some institutions a fiscal advantage in the final analysis and help preserve valuable time in the laboratory for a scientist who would otherwise be assigned this important duty.

5:22 What are the specific duties of the IACUC Chair?

Reg. The PHS Policy (IV,A,3,b) indicates that the IACUC Chair designates at least one member of the IACUC, qualified to conduct the review of proposed animal activities, if full committee review is not requested.

Opin. It is important to note that the role of the IACUC Administrator has significantly evolved. It is the administrator who frequently serves as the compliance specialist at many institutions, and he or she needs to develop a close working relationship with the IACUC Chair.

The IACUC Chair and Administrator work in concert to ensure the institution's animal care and use program maintain operations that are consistent with the applicable standards. The Chair, through consultation with, and implementation through, the administrator has the responsibility for overseeing the coordination and implementation of effective, efficient systems for protocol review by the IACUC in compliance with the PHS Policy and the AWAR. These review activities can only be performed at a properly convened meeting of the IACUC unless the review to approve or require modifications (to secure approval) or request Full Committee Review (FCR) is done using Designated Member Review (DMR). Thus, the Chair through collaboration with the administrator, should:

- Ensure that a quorum of the IACUC is present
- Declare the loss of a quorum, resulting in the end of official business if a sufficient number of members depart

- Prepare or oversee the preparation of meeting minutes and reports and submit these documents to the IO in accordance with PHS Policy (IV, E; IV,F) and the AWAR (§2.31,c,3; §2.31,d,2; §2.31,d,4; §2.35,a)
- Report to the IO any activities that have been suspended by the IACUC for noncompliance with PHS Policy or the AWAR (see Chapter 33)
- Establish a sound system of written communication for the IACUC with investigators concerning the approval status of protocols and the steps necessary to secure approval
- Designate, as specified in the PHS Policy (IV,C,2), at least one qualified member of the IACUC to conduct designated member review if full committee review is not requested

Beyond these relegated duties that the Chair performs on behalf of the committee, most institutions expect the Chair and the IACUC office compliance professionals to keep abreast of new regulatory trends and interpretations and evaluate and champion policy and practice initiatives (e.g., new training and educational programs) to improve the animal care and use program. This process involves the Chair's regular interaction with other areas of expertise within the organization, ranging from the research compliance office to other institutional committees to occupational health and safety, human resources, and the physical plant. The Chair may also provide guidance to, and field concerns from, investigators. Moreover, because these onerous diverse duties intensively consume time and energy and require consistent relationship-building within and outside the organization, organizations are increasingly relying on the recruitment of professional IACUC administrators to partner with the Chair to ensure the detailed, competent, and timely completion of many tasks related to IACUC function.

5:23 What is meant by an "alternate member" of the IACUC?

Opin. The regulatory standards specifically define the composition of a qualified IACUC to ensure institutions develop impartial committees that can effectively satisfy public concerns associated with research animals. Since neither the PHS Policy, nor the AWARs codify the use of alternate members, their use was not originally intended. Given that the use of alternate voting members provided institutions the human resources to better manage their animal care and use programs, the regulatory agencies provided standards and expectations regarding alternate IACUC members.

Reg. The NIH, USDA, and FDA confirm agreement through a memorandum of understanding[5] and discuss the appointment and use of alternate IACUC members for official business in NIH notice NOT-OD-11-053, Guidance to Reduce Regulatory Burden for IACUC Administration Regarding Alternate Members and Approval Dates (March 18, 2011).

Alternate members are fundamentally the same as a voting member with the caveat that each can serve in an official capacity only in the absence of a voting member with their same designated role (e.g., nonaffiliate, scientific member, non-scientific member, or veterinarian). Consequently, an alternate scientific member may serve in the absence of any voting scientific members. Institutions can appoint as many alternates as they wish to the committee, but a single alternate member can serve in the capacity only of a single voting member. In other words, if two scientific members are unavailable to serve during an IACUC meeting a scientific alternate can only serve in place for one of the missing members (i.e., one vote and one member toward the quorum).

Alternate members are appointed to the IACUC consistent with the practice used to appoint voting members and must receive the same training as voting members and maintain their qualifications to effectively serve as a voting member (provided in NOT-OD-11-053).

- The Chief Executive Officer, or designee, must appoint alternates to the IACUC in writing. Appointment of the IACUC is described in the PHS Policy at IV.A.3.a and the AWAR at 2.31(b). This procedure applies to appointment of both regular members and alternates.
- Alternates may serve as an alternate only in the membership category(s) for which they are qualified. Membership categories are described in the PHS Policy at IV.A.3.b and the AWAR at 2.31(b). The specific name(s) of the regular member(s) for whom the alternate serves are not required in the appointment record.
- Alternate members are permitted and encouraged to attend IACUC meetings and other IACUC activities even when the regular member is present.
- Alternates should receive IACUC training and orientation identical to that which is provided to regular members.
- Alternates must be listed as alternates on the IACUC roster submitted to OLAW as part of the Animal Welfare Assurance and annual reports. Institutions may, at their discretion, represent the names of alternates by using

numbers or other symbols in submissions to OLAW. Sufficient information for OLAW to determine that alternates are appropriately qualified must be provided, including the alternate's degree(s)/credentials, position title, and membership category as described in the PHS Policy at IV.A.3.b. The identity of alternates must be readily ascertainable by the institution and available to authorized OLAW or other PHS representatives upon request.

5:24 When a quorum is lost because of the mandated non-participation of a member that results from a conflict of interest, may that person's alternate reconstitute the quorum and vote?

Opin. Situations such as this should provide a strong impetus for institutions to articulate the role, responsibility, and authority of alternate members clearly in a charter document. The NIH has issued guidance on this issue (NOT-OD-11-053), indicating that a qualified alternate can be appointed to represent one or more IACUC members and/or more than one alternate can be appointed to represent a particular IACUC member.[15] However, alternate members cannot serve simultaneously with the member they represent in any committee activity.

Notice NOT-OD-11-053 accommodates the instance where an IACUC loses a quorum due to a member's unavailability (i.e., disqualification) by virtue of conflict of interest; a qualified alternate member may be used to reconstitute the quorum to proceed with a vote. The authors agree that such a substitution to make a quorum would be appropriate because the alternate member represents a redundancy in position and not of perspective. The underlying assumption should be that an alternate member will be free from the conflict of interest; knowledgeable about the facts under review and evaluate them independently and autonomously; understand all relevant IACUC issues; and, vote his or her conscience. The AWAR do not address alternate members, but APHIS/AC does not preclude this policy.

5:25 Can alternate members of the IACUC perform semiannual inspections that involve AWA-regulated species?

Guidance. APHIS/AC and NIH/OLAW have agreed in a guidance document (NIH/OLAW FAQ B.2) that alternate members can participate by sharing their expertise and perspectives in IACUC meetings and review activities even when their primary IACUC member is present. However, they may only contribute to a quorum in place of, and not as a supplement to, the primary member they are intended to represent.

Opin. Appropriately appointed alternate members would be eligible to participate in semiannual inspections involving APHIS/AC-regulated species provided that the member and his or her alternate are not participating simultaneously. Notice NOT-OD-11-053[15] (which provides guidance from both NIH/OLAW and APHIS/AC) encourages alternate member attendance at IACUC meetings and other activities even when the regular member is in attendance, however, this does not expand the opportunity for simultaneous participation during semiannual inspections, even if the member and his or her alternate are inspecting different facilities. Similarly the member and his or her alternate cannot constitute the two-person team required under the AWAR (§2.31,c,3) during the facility inspection process, because they represent only a single vote on the IACUC. As suggested in 5:23, creation of a charter document governing these activities would be advisable to clarify the institution's expectations of its IACUC members (see 23:17).

5:26 An IO appoints a scientist as an alternate member of the IACUC, to fill in when any scientific member is "unavailable." Under what conditions can this individual participate in IACUC functions? Should institutions place any conditions on what constitutes the "unavailability" of a member?

Opin. According to NIH Notice NOT-OD-11-053[15] unavailability encompasses instances in which the member is unable to attend the meeting, needs to leave the meeting early or arrive late, or is recused from participating due to a real or potential conflict of interest. The Notice represents the growing recognition that obligations in regulatory oversight compete with scientific productivity and scholarship in many instances. Institutions may find it useful to communicate a performance standard indicating that members who are regularly "unavailable" at the last moment or otherwise appear to be avoiding participation are not meeting their institutional and IACUC commitments. Alternate IACUC members should resist the capricious or compromised decision making that might be associated with short notice circumstances.

5:27 Must the institutional AV also serve as the veterinary member of the IACUC?

Reg. The PHS Policy (IV,A,3,b,1) and the AWAR (§2.31,b,3,i) state that the veterinarian on the IACUC shall have training or experience in laboratory animal science and medicine and have direct or delegated program authority and responsibility for activities involving animals at the institution. The AWAR (§2.33,a,3) state: "The attending veterinarian shall be a voting member of the IACUC; Provided, however, that a research facility with more than one Doctor of Veterinary Medicine (DVM) may appoint to the IACUC another DVM with delegated program responsibility for activities involving animals at the research facility."

Opin. Written guidance jointly provided by NIH/OLAW and APHIS/AC[16] recognizes that institutions with large and complex programs may wish to appoint more than one AV to address the needs of the research program and the specialty care of animal species involved. Not all of the AVs need to serve on the IACUC and service on the IACUC implies nothing about the authority hierarchy in veterinary care in the institution as a whole. Therefore, programs may appoint several AVs but only one has direct or delegated program authority and responsibility and *must* be a voting member of the IACUC (see below).

The wording of AWAR §2.31,b,3,i, as indicated in the Regulatory paragraph, describes the IACUC veterinarian with the wording found in AWAR §1.1, Attending Veterinarian. Although this suggests that the AV must be an IACUC member, AWAR §2.33,a,3 states that if the research facility has more than one veterinarian, then a veterinarian other than the AV can serve on the IACUC in place of the AV as long as that other veterinarian has delegated program responsibility for animal activities. PHS Policy IV,A,3,b,1 appears to offer concurrence with the AWAR by stating that the IACUC veterinarian is one who has "direct or delegated program authority and responsibility." Nevertheless, NIH/OLAW, in reviewing this chapter, provided guidance by stating that "For PHS Assured institutions, the veterinarian with program authority must serve on the IACUC. Other veterinarians with delegated clinical responsibilities may also serve, but the veterinarian with program authority and responsibility for activities involving animals at the institution must serve on the IACUC."

NIH/OLAW and APHIS/AC have jointly issued a very helpful clarification on the appointment and expected role of the AV on the IACUC and the opportunities for how to approach the matter of appointment in complex institutions.[16] Clear lines of authority and responsibility should be sought in the appointment of the AV to best effect programmatic oversight, essential programmatic changes if needed and quality protocol procedures. Institutions with complex programs may wish to discuss their options with NIH/OLAW and APHIS/AC.

5:28 What training or experience is useful for the veterinary member of the IACUC?

Reg. The PHS Policy (IV,A,3,b,1) and AWAR (§2.31,b,3,i) stipulate that the veterinary member of the IACUC should have training or experience in laboratory animal medicine and science (see 27:2). The AWAR (§1.1) further clarify, under the definition of the attending veterinarian, that this individual should have either graduated from a veterinary school accredited by the American Veterinary Medical Association Council on Education, acquired a certificate issued by the American Veterinary Medical Association's Education Commission for Foreign Veterinary Graduates, or received equivalent formal education as deemed appropriate by the APHIS administrator.

Opin. The intent of the PHS Policy and AWAR is to help ensure that the IACUC can rely upon the veterinary member not only for competent clinical insights, but also for information on ancillary areas such as unusual animal models, zoonoses, other occupational health and safety concerns, hazard containment, genetics, and unique nutritional and husbandry requirements of laboratory animal species. Proficiency in these diverse subject areas can be demonstrated formally by board certification in the American College of Laboratory Animal Medicine (ACLAM), and it is advisable for one or more of the veterinarians responsible for a large and complex program of research animal use to be board certified by ACLAM or comparable national laboratory animal medicine specialty (e.g., ECLAM, JCLAM or KCLAM). However, the needs of research animal care and use programs that are smaller or of a more limited scope may be met by veterinarians who are not ACLAM board certified. Regardless of their ACLAM certification status, all veterinary personnel involved in the oversight of research animal use should be aware and stay abreast of the central research, clinical, and regulatory concerns pertaining to the species under their care. This can be accomplished by reviewing the literature and attending the national meetings sponsored by various organizations, including American, Association for Laboratory Animal Science, ACLAM, American Society of Laboratory Animal Practitioners, Institute for Laboratory Animal Research, Association of Primate Veterinarians, American Veterinary Medical Association, AAALAC and other international laboratory animal science/medicine organizations.

5:29 What is the role of the veterinarian on the IACUC?

Reg. The role of the veterinarian within the context of the IACUC is defined under the AWAR (§2.31,d,iv,B; §2.31,d,vi; §2.31,d,vii; §2.31d,ix; §2.31,d,x,B) and by the *Guide* (p. 14) which serves as an extension and amplification of the PHS Policy. The *Guide* (p. 14) and PHS Policy (IV,A,3,b,1) indicate that the AV is responsible for the health and well-being of all the laboratory animals used in the institution.[15] In summary, the AWAR require consultation with the AV or their designee for procedures that cause more than momentary pain or distress and includes provisions for

- Direction of animal care and use, specifically housing, feeding and nonmedical care of the animals (a specific role of the AV)

- Medical care, including appropriate methods to prevent, control, diagnose and treat disease and injuries
- Aseptic surgery, use of anesthetics, analgesics, paralytics, pre- and postoperative care, and euthanasia
- Oversight of multiple major survival surgery resulting from a veterinary condition in an animal that also had experimental surgery

Specific provisions within the *Guide* (pp. 27–28, 30) include

- Advising the IACUC on new procedures or procedures with the potential to cause pain and distress that cannot be reliably controlled
- Ensuring that veterinary care is available to mitigate the illnesses, lesions, or behavioral abnormalities associated with animal restraint

Opin. Veterinarians play a unique role on IACUCs, as a result of their expertise in laboratory animal medicine and science and broader understanding of the physiological, behavioral, nutritional, and husbandry needs of laboratory animal species. Although by no means a lone voice on the IACUC on matters of animal health and welfare, the veterinarian should make the concerted effort to address these areas with precision and authority. Special areas of emphasis for the IACUC veterinarian should include the discussion of the use of proper anesthesia, analgesia and euthanasia in laboratory animals in the relief of pain and distress; assurance of veterinary involvement in pre-surgical planning and surgical skills/program assessment; the analysis of possible iatrogenic complications to the procedures used or the disease model proposed; the selection of humane endpoints and response to unexpected outcomes; and a review of the plans for appropriate and timely medical intervention.[16]

5:30 Under what conditions can the veterinarian delegate some of his or her responsibilities to other IACUC members?

Reg. According to the AWAR (§2.31,d,1,iv,B) the AV may delegate his or her responsibility for playing a role in the planning and consultation on procedures that may cause more than momentary or slight pain or distress.

Opin. Depending on the background, training, and experience of the other IACUC members, and the size, complexity, and maturity of the animal care and use program, many of the veterinarian's responsibilities can be delegated to another veterinarian or other members for defined periods (e.g., a particular IACUC meeting or semiannual review). For the most part the appointment of an alternate for the veterinarian is used to address this issue in most institutions. Very few of the institutions known to the authors would proceed with an IACUC meeting in the absence of a veterinarian, particularly if this involved neglecting significant issues because of insufficient veterinary input. The strategy most often used when the IACUC veterinarian is absent is to invite another veterinarian with relevant skills and knowledge, to participate as a guest in an advisory capacity or require evidence of veterinary approval before proceeding. Secondarily, IACUCs can table issues requiring the veterinarian's special analysis and comment.

5:31 Does the attending veterinarian have any mandated training responsibilities under the AWAR or PHS Policy?

Reg. Training is not assigned as a specific requirement to the AV in either the AWAR or the PHS Policy. The AWAR (§2.32,a; §2.32,b) indicate that training "shall be the responsibility of the research facility." The PHS Policy (IV,C,1,f) places the responsibility of assuring that personnel are appropriately qualified and trained to conduct the procedures proposed on research animals on the IACUC, through its protocol review process. AVs, or their designee, however, must provide "guidance to principal investigators or other personnel involved in the care and use of animals regarding handling, immobilization, anesthesia, analgesia, tranquilization and euthanasia; and ... adequate pre-procedural care and post-procedural care in accordance with current established veterinary medical and nursing procedures," as stipulated in the AWAR (§2.33,b,4; §2.33,b,5). IACUCs should work collaboratively in support of the program of veterinary care to ensure that ample resources are available to the veterinarians to meet this critical objective. The *Guide* (p. 106) has a requirement for the AV to provide guidance "to investigators and all personnel involved in the care and use of animals to ensure appropriate husbandry, handling, medical treatment, immobilization, sedation, analgesia, anesthesia, and euthanasia. In addition, the AV should provide guidance... to surgery programs and perioperative care involving animals." With regard to surgical training, the *Guide* (p. 116) states the "IACUC, together with the AV, is responsible for determining that personnel performing surgical procedures are appropriately qualified and trained in the procedures."

Opin: In most institutions, species specific training initiatives are run by the veterinarian(s) or under the direct management of the veterinary team operating either through the animal resource or the IACUC administrative office. Collaborative ventures with the veterinary staff augment, or in some cases supplant, this effort in some institutions. In these cases, most use a multidisciplinary approach involving institutional scientists, capable technicians with advanced skills, dedicated training personnel within the IACUC, or the veterinary resource group as well as the veterinarian(s). In the authors' view, verification of training and documentation of proficiency continue to be areas of vulnerability in many animal research programs.

5:32 What is meant by the nonaffiliated (outside) member of the committee, and what is the intent of including such an individual on the committee?

Reg. The intent of including a nonaffiliated member on the IACUC is to ensure that someone who is not affiliated with (or beholden to) the institution in any manner is involved in the review of the institution's animal care and use activities. This individual also may not be a member of the immediate family of a person affiliated with the institution (AWAR §2.31,b,3,ii; PHS Policy IV,A,3,b,4; *Guide* p. 24). According to the AWAR (§2.31,b,3,ii), this individual is intended to represent the "general community interests in the proper care and treatment of animals." (See 3:5; 3:6.) The language in the PHS Policy is the most restrictive, providing for an individual "who is not affiliated with the institution in any way other than as a member of the IACUC, and is not a member of the immediate family of a person who is affiliated with the institution."[7]

Opin. The IACUC membership must always include a nonaffiliated member for it to be legally constituted. In addition, each nonaffiliate must be readily available to serve in an official capacity. As a result, many institutions choose to identify more than one individual to serve as a nonaffiliated member of the IACUC.

According to the language in PHS Policy, an individual would be disqualified as an unaffiliated member if they served voluntarily in any other capacity for the institution (e.g., service on an IBC). NIH/OLAW does not allow individuals who have used laboratory animals in the past to be nonaffiliated members. Ostensibly, because the nonaffiliated member is not expected to benefit personally from any of the activities proposed and is not dependent on the institution for his or her livelihood, she or he enhances the public's confidence in the unfettered objectivity of the IACUC review processes. Institutions should consider other situations that might be perceived as conflicts of interest in the appointment process, however tenuous, such as a retiree receiving a pension from the institution and should be prepared to explain their decision process in these cases. The unaffiliated member should be regarded as a full IACUC member and should be afforded opportunities to participate equivalent to those of all other IACUC members. However, in the authors' view, the unaffiliated member should not be asked to take the lead in any areas in which he or she may not be comfortable (e.g., as the primary reviewer in the Designated Member Review process for a protocol).

5:33 As an extension and example of 5:32, if two colleges, A and B, reciprocate in course admissions for students and exchange faculty for course instruction, would it be appropriate for the nonaffiliated IACUC member at College A to be a spouse of a faculty member at College B?

Opin. The institution must realize that the value of a nonaffiliated member is to document an opinion that is completely autonomous of the institution. Consequently, it behooves institutions to be conservative when defining what is considered an affiliation. The best practice is to be extremely inclusive when identifying points that disqualify an individual from serving as a nonaffiliate. In addition to the circumstances discussed in Section 5:32, an institution may choose to disqualify an individual from serving as a nonaffiliated member if any of the following circumstances, for example, exist.

- Any member of the candidate's family is an employee of the institution.
- The spouse or child of the candidate is a student at the institution.
- The candidate serves as a nonaffiliate on another of the institution's compliance committees (e.g., the Institutional Biosafety Committee).
- The candidate or their family member is receiving a pension that is completely supported by the institution.
- The candidate is participating in collaborative research with members of the institution.

In the authors' view, institutions should exercise common sense in the application of appointments when the "affiliation" is extremely tenuous as it is in this case. If the faculty member at College B is not personally involved in some aspect of the animal care and use program at College A or is not dependent upon the success of College A for salary and livelihood, there is no compelling reason to believe that the spouse would not be able to perform this important public service admirably.

5:34 What are the typical backgrounds of individuals who serve as nonaffiliated members of IACUCs?

Opin. There is no typical background of these individuals. Institutions have chosen a wide variety of individuals to serve in this capacity. Without implying that the following categories are mutually exclusive, the range of individuals includes broadly educated, erudite humanists; businesspersons; public servants; educators; persons involved in other types of animal care and use; physicians, veterinarians, scientists, and technicians who do not use laboratory animals; and concerned citizens who seek to make a public contribution[7] (see 3:6).

5:35 Can the nonaffiliated member be a scientist, veterinarian, ethicist, or biostatistician who is not affiliated with the institution?

Opin. Any of these individuals would qualify as long as he or she is not involved in the care and use of laboratory animals at the institution where they serve as an IACUC member, either directly or through consulting and similar tangential service.

5:36 Can a retiree from an institution serve as the unaffiliated member of the same institution's IACUC?

Opin. In the authors' experience, very few institutions exercise the opportunity to appoint either retirees or alumni as unaffiliated members, perhaps because they are disinclined to create the perception of a conflict of interest (see 5:32). Even in the case of pensioners, the retirement income was earned previously, and the unaffiliated member's livelihood is not in immediate jeopardy if they offer opinions at odds with institutional interests. Appointments of this type are not prohibited by federal regulations nor discouraged in guidance documents. In the few cases of this type known to one of the authors, fidelity to the institution appeared to engender robust participation, yielding very honest and critical questions, review, and commentary.

Neither the PHS Policy nor the AWAR prohibit a retiree from an institution from serving as the unaffiliated member, although cautionary guidance in NIH Notice NOT-DO-15-109[15] does state that "in evaluating the qualifications of an individual to serve as a nonaffiliated member, the CEO should confirm the appointee has no discernible ties or ongoing affiliation with the institution. Regarding service of former employees or students as nonaffiliated members, the appointing official must be assured that the person is not in any way obligated to the institution. Real or perceived conflicts of interest must be avoided to ensure the IACUC's and the institution's integrity. Appointment of an individual who is unambiguously unaffiliated is the most effective way to fulfill the intent of the Policy."

5:37 Can or should the nonaffiliated member be compensated for service on the IACUC?

Opin. Neither the PHS Policy nor the AWAR prohibit the compensation of the outside member. NIH/OLAW has stated that nominal compensation is permissible without jeopardizing a member's nonaffiliated status if compensation is only in conjunction with service on the IACUC and is not so substantial as to be considered an important source of income or to influence voting on the IACUC[17] (see 5:9). The *Guide* (pp. 24–25) also endorses this approach.

In accordance with the findings of an earlier survey on this subject, respondents offered narrative comments indicating that the nonaffiliated members were compensated minimally. Support for the travel and parking for the nonaffiliated member was frequently mentioned and several organizations indicated that the nonaffiliated members were compensated for their time or were given honoraria using the descriptor "small." Only one institution offered information on the level of compensation, noting that the member received a payment per meeting and per protocol. Other types of compensation mentioned included support for attendance at meetings on IACUC function and provision of gift cards periodically. For example, an institution may make a charitable contribution on behalf of the nonaffiliated members. These types of innovative approaches illustrate the wide latitude institutions can exercise to acknowledge the valuable contributions and important public service provided by the unaffiliated member. Institutions should be cautioned that compensation of their nonaffiliated members at or exceeding the rate of the member's primary income source is likely to be perceived by the lay public and by regulatory agencies as a conflict of interest and cannot be recommended.

5:38 Can the nonaffiliated member serve on more than one IACUC, in the same or in different institutions?

Opin. There is no proscription against nonaffiliated members' serving on more than one IACUC if they have an interest in that level of involvement, and the authors are aware of individuals who have done this with laudable interest and commitment. PHS-Assured institutions with two or more IACUCs should contact NIH/OLAW for advice if they intend to utilize one nonaffiliated member for service on more than one of these. However, NIH/OLAW, in reviewing this chapter, has indicated that the same nonaffiliated member may not serve on multiple IACUCs at the same institution. Skeptics may speculate that the excessive involvement of an individual as a nonaffiliated member on several IACUCs is an indication of a shift in the individual's neutrality and objectivity, which might be a reason for institutions to avoid the "overzealous" contributor. Also, scientific staff may be concerned that the individual could be a potential source for the leakage of scientific ideas to other institutions. This matter could be readily addressed by assuring that the nonaffiliated member has received training in issues concerning confidentiality and nondisclosure of proprietary information. Institutions may also want to consider asking their nonaffiliated members to sign formal agreements to this effect.

5:39 What is the role of the nonaffiliated member of the IACUC?

Opin. The role of the nonaffiliated member is not really defined other than as described in 5:32. Most institutions are hopeful that this individual will play an active role in all IACUC activities and will be comfortable with and become adept at making persistent, straightforward, and disarming inquiries about matters that are undetected by the institutional members of the IACUC. The nonaffiliated member also projects the institution's commitment to transparency by affording the member a full understanding of the organization's scientific programs involving animals and ensuring that control measures for research posing a risk to the public are adequate.

5:40 What are the background and qualifications of individuals who typically fill the role of the nonscientist?

Reg. The PHS Policy (IV,A,3,b,3) indicates that the "primary concerns" of individuals serving in this capacity should be "in a nonscientific area," and the specific examples cited (e.g., ethicist, lawyer, member of the clergy) have no obvious connections to any area of science.[7] NIH Notice NOT-OD-15-109[15] reiterates the necessity to appoint individuals possessing "a naïve attitude with regard to science and scientific activities," and provides further examples for persons without scientific training (e.g., English instructors and those working in finance). The AWAR do not specify the need to appoint an IACUC member in this particular category.

Opin. Selecting a nonscientist can be difficult and one could argue that in many institutions, an individual who has been chosen as the nonscientist has some scientific training but clearly does not qualify as a practicing scientist. For example, NIH Notice NOT-OD-15-109 identifies those working in finance as an option for filling the nonscientist role even though finance can be viewed as a mix of art and science with a strong emphasis on mathematics.[15] The authors suggest an extensive analysis of potential nonscientists to ensure the chosen individual lacks scientific training to meet the intent of the policy.

The types of individuals seen in this role among the institutions surveyed for the previous edition[3] included lawyers, clergy, health and safety personnel, business and human resources personnel, public relations personnel, quality assurance/control personnel, and technicians not involved with animal care or use.

5:41 Is a biostatistician considered a nonscientist?

Opin. A FAQ issued by NIH/OLAW describes a nonscientist as an individual who had a "naive attitude with regards to science and scientific activities."[8] Clearly, statisticians are scientists because statistics is a branch of mathematics, a pure science. On the bases of promoting a variation in perspectives, a biostatistician would not fill the role of nonscientist.

5:42 What interests does the nonscientist represent and what role does he/she play on the IACUC?

Opin. This individual serves to diversify the IACUC membership further by bringing a perspective lacking scientific training and adding balance for the scientific members who may be regarded as having a vested interest in the promotion of animal studies. As with the nonaffiliated member, the nonscientist should participate in and contribute to all of the IACUC's mandated activities. With regard to protocol review, this member can be especially valuable by working to ensure that the IACUC meets its obligation under the *Guide* (p. 25) to provide a "clear and concise sequential description of the procedures involving the use of animals that is easily understood by all members of the committee."

5:43 Should the scientist on the IACUC have animal research experience?

Reg. PHS Policy (IV,A,3,b,2) and the *Guide* (p. 9) but not the AWAR, indicate that an IACUC should include in its membership at least one practicing scientist who has laboratory animal experience.

Opin. Appointing a scientist with animal research experience is a universal practice among the institutions known to the authors. In large institutions, typically several scientists are appointed to the IACUC to reflect the interests of different user groups in the organization. The appointment of scientists who have laboratory animal experience aids the IACUC's discussion of relevant issues during protocol and program review. It helps the IACUC better understand the selection, use, and limitations of animal models and certain aspects of experimental design. Institutions conducting research and teaching in agricultural animals should also be aware of and follow recommendations of the *Ag Guide* concerning the experience of the scientific members. The *Ag Guide* recommends two scientists should be included in a basic IACUC: one is a scientist with experience in conducting research or teaching with agricultural animals and the other is an animal, dairy or poultry scientist with experience in the management of agricultural animals.

5:44 Should the scientific member of the IACUC be a senior-level scientist, or is a junior-level scientist acceptable?

Opin. Scientists at any level are appropriate, but many institutions feel that senior scientists confer more authority and credibility on the IACUC. However, most academic organizations have found the recruitment of senior scientists to the IACUC to be difficult because these individuals usually have already made significant committee contributions to the institution in other areas and have other significant obligations within the institution. In academic institutions, junior faculty appointed to the IACUC may find the extensive time commitment an impediment to their efforts to develop a viable portfolio in research and scholarship for promotion and tenure consideration. Also, regarding promotion and tenure, institutions should be aware of the conflicts of interest that may arise when a junior faculty member reviews the protocols of senior faculty members involved in salary, promotion, and tenure decisions.

5:45 What is the role of the scientist on the IACUC?

Opin. The principal roles of the scientist on the IACUC are to ensure that the interests of scientific colleagues are being fairly represented in the review process and to aid in the IACUC's assessment of the relevance, validity, and technical aspects of the studies proposed. Of course, the scientist recognizes the confluence of animal health, welfare, and scientific interests in research animal studies and plays an important role as a proponent for the prudent, ethical, and humane use of animals. In broader issues involving program development and implementation, the scientist can give the IACUC perspectives on how best to launch new initiatives to engender the support of the scientific community and others involved in the care and use of laboratory animals.

5:46 Should the IACUC include an ethicist as a member?

Opin. Some institutions have been successful in recruiting one or more individuals knowledgeable in ethics with an education and work focus in the arts and humanities or in business, but few have been able to identify a bona fide ethicist. Clearly, issues in ethics may be important in the IACUC's deliberations and there may be occasions when an ethicist can be helpful in leading this process.

5:47 Should the IACUC include a biostatistician as a member?

Opin. Most IACUCs have not pursued the option of including a biostatistician to ensure that investigators are using "the minimum number (of animals) to obtain valid results" (PHS Policy IV,D,1,a–IV,D,1,b; U.S. Government Principle III; AWAR §2.31,e,1; §2.31,e,2) because they expect the scientist(s) and veterinarian on the IACUC to evaluate this area. The *Guide* (p. 25) emphasizes that animal care and use protocols should include a discussion of statistical justification for the numbers of animals and the experimental group sizes. Scientific, veterinary, or other IACUC members should be capable of performing this function if they can commit the effort and are given sufficient information by the investigator about the assumptions on the data anticipated in the experimental study. However, confirming that the investigator is proposing to use the minimal number of animals necessary can involve an extensive and potentially redundant effort of the IACUC. If an IACUC routinely encounters weak or dubious explanations for the numbers of animals requested in protocols, the addition of a biostatistician to the IACUC may be helpful (see 5:40).

5:48 Should animal care technicians be considered for IACUC membership?

Opin. In the authors' experience, several institutions have appointed animal care technicians to their IACUCs in an effort to develop a better sense of how the basic animal care program functions. An added benefit is that it may promote a sense of empowerment and inclusion in the IACUC review processes for the entire animal care staff.

5:49 If an institution is performing Good Laboratory Practices (GLP) studies, can a quality assurance (QA) person also serve on the IACUC?

Opin. In the authors' opinion, it would be permissible to have a QA person serve on the IACUC. QA personnel have responsibility for all processes influencing the integrity, quality, validity, and documentation of data. IACUC functions are not included in the QA bailiwick, but some IACUCs might benefit from a heightened scrutiny of their processes as well as the general contributions an individual who has a QA background would be expected to add as his or her experience accumulates.

5:50 Should individuals who have specialty expertise be included as voting, non-voting, or *ad hoc* IACUC members?

Reg. PHS Policy (IV,C,2; IV,C,3) and the AWAR (§2.31,d,2; §2.31,d,3) clearly indicate that voting privileges are reserved for full IACUC members who are appointed by the CEO. Nevertheless, the PHS Policy and AWAR do not preclude the use of non-voting consultants.

Opin. Having individuals available who have specific expertise to bring the best decision making to the IACUC in different areas is always in the institution's interest. Individuals who have specialty expertise might include properly appointed *ad hoc* members who are willing to participate in the full range of IACUC activities. Individuals who have specialty expertise and are not interested in assuming all of the responsibilities of full membership but who would be helpful periodically to the IACUC can be invited to participate in an advisory capacity as consultants without voting privileges. The types of *ad hoc* appointees vary by function and are diverse. Institutions might consider including librarians for alternative searches; environmental or occupational health and safety personnel for hazardous studies; and ichthyologists, herpetologists, or ornithologists for studies involving species in their respective disciplines. Some institutions maintain a list of consultants willing to participate at any time as experts in different areas of importance to rigorous and thoughtful protocol review. These contingencies should be addressed in the charter developed by the institution for the IACUC.

5:51 Can non-voting members count toward the quorum needed for IACUC meetings?

Reg. OLAW FAQ B-5, to which the USDA agrees, indicates that a quorum of the IACUC is a majority (i.e., half of the voting members plus one) of the total number of voting members. Consequently, only voting members can count toward the required quorum, which must be satisfied before a meeting can be called to order for the purpose of conducting official business.

5:52 Should the IACUC seek the advice of consultants for issues that may require special expertise, such as pain and distress concerns?

Reg. PHS Policy (IV,C,3) and the AWAR (§2.31,d,3) describe the use of consultants and how they may not "approve or withhold approval of an activity or vote with the IACUC." The AV or their designee must also be consulted on pain and distress concerns as this is a specific role given to the AV by the AWAR (§2.31,d,1,iv,B and (§2.33,b,4 and 5).

Opin. IACUCs have the freedom to exercise their ability to use consultants whenever necessary. However, in most institutions, their use is reserved for truly extraordinary issues as opposed to any areas in which the IACUC feels it might improve its decision making. In the authors' opinion, many IACUCs would potentially benefit from the more liberal use of consultants.

5:53 Should consultants be from inside or outside the institution, and should they be anonymous?

Opin. Consultants can be selected either from inside or outside the institution, depending upon the availability of relevant expertise and the resolution of concerns about objectivity. Most organizations have not found it practical, necessary, or desirable to maintain the anonymity of the consultant. However, if the need for a consultant is likely to involve a highly contentious and disputed high-stakes issue, anonymity may be elected by the IACUC or imposed by the consultant as a condition of participation. Regardless of the circumstances, the IACUC has the obligation to foster an environment conducive for the investigator to respond to the consultant's critique.

5:54 Should consultants to the IACUC be compensated?

Opin. External consultants frequently are compensated for their efforts, and it seems reasonable that these individuals should be compensated according to the institutional policies established for other types of external academic review.

5:55 Should there be a confidentiality agreement between the consultant and the institution?

Opin. It is advisable for institutions to have a confidentiality agreement with consultants to prevent the disclosure of important, and potentially patentable, scientific or technical information. In addition, confidentiality agreements are important because they establish or reaffirm a precedent for the manner in which all materials relevant to the IACUC's deliberations are handled. This issue becomes particularly important from the legal standpoint if the institution is faced with the prospect of retrieving documents that may have been obtained illegally by adversarial parties. In a broader sense, this issue is also important for all members of the IACUC. It is highly advisable that all institutions assure that IACUC members are aware of and adhere to the requirements for confidentiality pertaining to protocol review and other committee business. Training of the IACUC should encompass this important area.

5:56 What support staff is useful for the IACUC?

Opin. Many institutions have developed an animal care and use compliance office to provide the administrative, monitoring, and training activities of the IACUC. The composition of this office depends upon the size and scope of the animal care and use program and the extent to which the IACUC members are willing and able to become personally involved in the fine details of committee function. In general, the growth of IACUC administrative offices nationally affirms that scientific (and other) members of IACUCs have increasingly found significant support from their IOs for divesting the responsibility for program implementation in many areas of IACUC function. In the opinion of the authors, an effective IACUC office can greatly benefit an institution ensuring timely efficient IACUC administrative processes, creating a partnership with the scientific community/programs and building a reputation for customer service by the IACUC. A survey from within the IACUC administrator community indicated a 24% growth in IACUC offices from 2005 (44%) to June 2022 (68%).[18] The types of positions that have proved to be very useful on the IACUC office staff in different settings include a director/administrator or senior assistant to the IACUC Chair; training, monitoring (e.g., post-approval monitoring) and compliance personnel; computer/database specialists; biostatisticians; and clerical staff. In small institutions with limited programs, a small staff of 0.5 to 1 full-time equivalent (FTE) or less may be sufficient, but in large, complex programs it is not uncommon for the IACUC office staff to include ten to fifteen FTEs.

5:57 Who typically provides the administrative "support" staff to the IACUC?

Opin. The support staff is generally recognized and provided as an administrative cost that is provided under the authority of the CEO or IO of the organization.

5:58 What are some of the typical responsibilities of the administrative office?

Opin. The role of IACUC Office staff has evolved into that of compliance specialists. They often are involved in conducting facility inspections and preparing the correspondence related to semiannual reviews, protocol matters, post-approval monitoring activities, policies and procedures development of the IACUC, and maintaining the records thereof. The staff also may be responsible for making the arrangements for all IACUC activities. If staff members have appropriate training and technical expertise, they may be involved in conducting institutional training seminars and visiting laboratories and animal use areas to provide instruction and monitor ongoing activities. This staff may maintain databases of IACUC protocol approval and of the appropriate ancillary approvals in other areas such as use of biohazards, and the enrollment in occupational health and safety activities. In many cases they may be a key liaison with other important institutional functions that support the animal care and use program such as occupational health and safety and physical plant personnel. In a few institutions, IACUC office staff has been available to conduct literature searches for alternatives to animal use and to provide a statistical analysis to confirm that the investigator has properly justified the number of animals requested.

Although many institutions draw their IACUC support staff from technical backgrounds involving experience with laboratory animals, others have had success recruiting from diverse disciplines without this experience. Institutions should promote the development activities of this staff by supporting their attendance at workshops and conferences dedicated to IACUC-related topics, animal care and use program improvements, and/or the training of technical staff. These types of programs are offered by AALAS, AAALAC, Public Responsibility in Medicine and Research (PRIM&R), IACUC Administrators Association, the Laboratory Animal Welfare Training Exchange (LAWTE), IACUC 101, Scientists Center for Animal Welfare (SCAW) and NIH/OLAW in partnership with regional academic sponsors, and commercial enterprises. This investment in staff training will make the staff more effective in general. Training by these programs can efficiently give staff who do not have prior experience in the laboratory animal arena a quick and balanced understanding of key issues in this new environment as well as enhance the effectiveness of IACUC professionals that already have experience.

Surv. For academic institutions, what is the faculty rank of your IACUC chairperson? Include persons with annotated titles such as "clinical assistant professor."

• Not applicable as we are not an academic institution	76/284 (27%)
• Instructor	2/284 (1%)
• Research scientist (non-tenure track)	1/284 (4%)1
• Assistant professor	10/284 (4%)
• Associate professor	57/284 (20%)
• Full professor	112/248 (39%)
• I don't know	1/284 (0%)
• We are academic but our chair does not have a faculty title.	15/284 (5%)

REFERENCES

1. Guide for the Care and Use of Agricultural Animals in Agricultural Research and Teaching. 2020. *Guide for the Care and Use of Agricultural Animals in Agricultural Research and Teaching*, 4th ed. Champaign: Federation of Animal Science Societies. https://www.asas.org/docs/default-source/default-document-library/agguide_4th.pdf?sfvrsn=56b44ed1_2.
2. Committee for the Update of the Guide for the Care and Use of Laboratory Animals. 2011. *Guide for the Care and Use of Laboratory Animals*, 8th ed. Washington, D.C.: National Academies Press.
3. Newcomer, C.E. 2012. General composition of the IACUC and specific roles of the IACUC members. In *The IACUC Handbook*, 3rd ed., eds. J. Silverman, M. Suckow and S. Murthy, 37–60. Boca Raton: CRC Press.
4. National Institutes of Health, Office of Laboratory Animal Welfare. 2012. *Frequently Asked Questions B.1, What Are the IACUC Membership Criteria?* https://olaw.nih.gov/faqs#/guidance/faqs?anchor=50294.
5. Memorandum of Understanding among the Office for Laboratory Animal Welfare, APHIS Animal Care and the FDA (https://olaw.nih.gov/guidance/finalmou.htm)
6. National Institutes of Health, Office of Laboratory Animal Welfare. 2015. *Public Health Service Policy on Humane Care and Use of Laboratory Animals*. Bethesda: National Institutes of Health. https://olaw.nih.gov/resources/documents/annual-report.htm#download .
7. National Institutes of Health, Office of Laboratory Animal Welfare. 2015. *Public Health Service Policy on Humane Care and Use of Laboratory Animals*. Bethesda: National Institutes of Health. https://olaw.nih.gov/policies-laws/phs-policy.htm#FOOTNOTES.
8. National Institutes of Health, Office of Laboratory Animal Welfare. 2012. *Frequently Asked Questions B12, How Does OLAW Define Nonscientific IACUC Member?* https://olaw.nih.gov/faqs#/guidance/faqs?anchor=50304.
9. U.S. Department of Agriculture, Animal and Plant Health Inspection Service. 2011. *Policy 15, Institutional Official*

and IACUC Membership. https://www.aphis.usda.gov/animal_welfare/downloads/AC_BlueBook_AWA_508_comp_version.pdf .
10. Committee on Occupational Safety and Health in Research Animal Facilities. 1997. *Occupational Health and Safety in the Care and Use of Research Animals.* Washington, D.C.: National Academy Press.
11. Robert, H.M. and S.C. Robert. 1990. *The Scott, Foresman Robert's Rules of Order Newly Revised*, eds. H.M. Robert III and W.J. Evans. Glenview: Scott, Foresman.
12. National Institutes of Health, Office for Protection from Research Risks. 1993. Frequently Asked Questions about the Public Health Service Policy on Humane Care and Use of Laboratory Animals. *ILAR News* 35(3–4):47.
13. Newcomer, C.E. 1997. Behold! The Animal Care and Use Magician! *Connection* [Newsletter of AAALAC]: 10–11.
14. National Institutes of Health, Office of Laboratory Animal Welfare. 2009. *Notice NOT-OD-09-035, Guidance to IACUCs Regarding Use of Designated Member Review (DMR) for Animal Study Proposal Review Subsequent to Full Committee Review (FCR).* http://grants.nih.gov/grants/guide/notice-files/NOT-OD-09-035.html.
15. National Institutes of Health. 2011. *Notice, NOT-OD-11-053, Guidance to Reduce Regulatory Burden for IACUC Administration Regarding Alternative Members and Approval Dates.* http://grants.nih.gov/grants/guide/notice-files/NOT-OD-11-053.html.
16. Brown, P. and C. Gibson. 2009. Multiple campuses, one IACUC; How many AVs? A word from OLAW and USDA. *Lab Animal* (New York) 38:114.
17. National Institutes of Health, Office of Laboratory Animal Welfare. 2012. *Frequently Asked Questions B.11, May the Institution Pay or Reimburse Expenses Incurred by Nonaffiliated Members?* https://olaw.nih.gov/faqs#/guidance/faqs?anchor=50293.
18. Greer, W.G. 2022. Personal Communication. *Summary of Data Obtained from a Survey Conducted by the IACUC Administrator Association.*

6

Basic IACUC Operations

Jennifer A. Klahn and Joanne L. Zahorsky-Reeves

Introduction

The regulatory expectations are very clear with respect to the functions of the IACUC in an animal care and use program (ACUP), but they are silent on the specifics of how those functions are supported; this is intentional. ACUPs vary dramatically as far as the size and focus of their programs. The IACUC for a small, contract research organization, for example, will have a very different support structure compared to the IACUC for a large university with a veterinary school and nonhuman primate center. While a smaller program may involve the IACUC members more directly in administrative operations, larger organizations are likely to employ an administrative office of certified staff. These staff handle everything from answering general questions to scheduling facility inspections to drafting correspondence on behalf of the IACUC to filing formal reports—all this while serving as institutional memory for the IACUC and subject matter experts on regulatory matters for the investigators.

This chapter will cover administrative operations of the IACUC, predicated on the idea that there is an independent and organized office providing support for those operations. A theme that will repeat throughout this chapter is that the resources required to support the program are driven by the needs and operational scope of the ACUP but dictated by the institutional investment in the program. An institution that is unwilling or unable to invest in IACUC support will struggle with its animal care and use program.

6.1 What is the appropriate staffing level for an IACUC support office?

Opin. There is no "one size fits all" approach for the IACUC support office. The appropriate staffing level is widely variable and based on institutional needs, program size, complexity, and support. When evaluating the number of staff needed to assist the IACUC, consideration should be given to the following:

- Number of protocols approved by the IACUC
- Number of Principal Investigators (PIs) and researchers working under them
- Number of centralized and satellite facilities in which animals are housed and used
- The different species used at the institution
- Complexity/diversity of the research portfolio
- Institutional risk tolerance and history of incidents and/or citations

It stands to reason that a smaller program with fewer protocols, PIs, and facilities may require a smaller administrative staff to support the IACUC, while larger, more elaborate programs may necessitate more staff; however, this is not always the case. For example, a modest program at a private institution with sufficient funding may see value in expanding the administrative support to their IACUC, perhaps to assist the research community with quick protocol turnaround or to ensure institutional compliance with regulatory oversight. Similarly, a program that perceives greater threat from past compliance failures may employ additional individuals to aid the IACUC with inspections and post-approval monitoring (PAM).

The IACUC support office may be a stand-alone entity, in which staff are dedicated solely to IACUC and ACUP operational activities; this model may be employed more frequently by large programs. Alternatively, the staff supporting the IACUC may be cross-trained on other committee support, for example Institutional Biosafety Committee (IBC), Institutional Review Board (IRB), and/or Conflict of Interest (COI). This latter model may be used more often by smaller institutions out of budgetary necessity. Regardless of the size of the institution, cross-training allows for greater coverage in the case of staff outages and can benefit a program with different perspectives.

In summary, offices may exist with as few as one staff member or be a vast organization of many people. Nevertheless, an entity of any size can function to support its IACUC efficiently and well.

6.2 What distinct areas of responsibility are there within an IACUC office?

Opin. IACUC support staff cover a variety of responsibilities and may be dedicated to specific distinct areas, depending on program size and needs.

DOI: 10.1201/9781003272861-6

Responsibilities vary and may include:

- Answering general inquiries from researchers and providing information on the role of the IACUC and need for IACUC review
- Managing general animal user training and recording completion of that training in a centralized database
- IACUC meeting management
  - Scheduling meetings, preparing agendas, ensuring meeting quorum, assigning reviewers to protocols,* taking notes, tracking discussion and action items, providing historical and regulatory perspectives, drafting meeting minutes
- Involvement in protocol review:
  - Checking for application completeness, conducting administrative pre-review, performing more in-depth review, processing reviews through Designated Member Review (DMR) and Full Committee Review (FCR), handling communications from the IACUC, evaluating for completeness based on regulatory and institutional requirements, finalizing review and issuing approvals, managing follow-on submissions and use of different review processes (specifically: administrative review, veterinary verification and consultation [VVC], DMR, and FCR).
- Post-approval monitoring†
  - May be basic PAM (e.g., continuing review, semiannual program reviews and facility inspections, daily caretaker/technician/veterinarian observations) or more developed/formalized program (e.g., protocol audits, meetings with research labs, procedure observations, review of records)
- Support for semiannual program reviews and facility inspections
  - Scheduling reviews and inspections, assisting members with performing the program review and conducting inspections, taking notes, identifying/tracking findings, corresponding with responsible parties to address findings, reporting program review and facility inspection results to the IACUC, reporting to the IO
- Compliance investigations
  - Receiving allegations of non-compliance or departures/deficiencies, supporting IACUC investigations by gathering information and/or communicating on behalf of the IACUC, ensuring corrective actions are identified and implemented, providing guidance regarding when reports to external agencies are required
- Preparing reports to the IO, OLAW, USDA, other entities
  - Working with the IACUC Chair, AV, and/or IO to prepare content for and submit required regulatory reports (e.g., Animal Welfare Assurance, USDA registration, AAALAC program description, annual reports to OLAW/USDA/AAALAC, reports of noncompliance and/or uncorrected significant deficiencies, etc.)

6.3 How are staff best divided across the distinct areas of responsibility?

Opin. How staff are divided across the areas of responsibility depends on the size and scope of the program. A smaller program may be well served by a single staff member who performs all elements of IACUC operational support, from general inquiries to regulatory reporting on behalf of the IACUC and IO. That same program, however, may feel it advantageous to add individuals to the support team, to ensure coverage not just over vacations but in the case of prolonged staff absence or unexpected departure (see 6:1). When multiple staff members support an IACUC, having a clear understanding of responsibilities, and an individual who supervises or manages the distribution of work, will be beneficial.

Offices with more than one individual supporting the IACUC may choose to train those individuals on all responsibilities, so that all staff are equally prepared to complete a given task or advise the IACUC. As the number of staff in the administrative office increases, having clear expectations of responsibilities will ensure that these responsibilities are adequately covered and executed. A supervisor or manager would do well to continue to check-in with their staff to ensure ongoing understanding of their defined roles as a way to avoid straying into an area better suited to a subject matter expert (e.g., presenting comments in an administrative review of an animal-use protocol that are of a veterinary nature).

An office may choose to dedicate some individual(s) to protocol review only. Such staff may be assigned to specific PIs, departments, types of research, species, etc. These staff would serve as the source of information for their assigned area

* Readers are reminded that, if full committee review is not requested, the IACUC Chair is responsible for designating at least one IACUC member who is appropriately qualified to perform the review. This responsibility is codified in both the PHS Policy C.2 and the Animal Welfare Regulations 2.31(d)(2).

† See OLAW FAQ G6 (Is post approval monitoring required?) for more information. https://olaw.nih.gov/faqs#/guidance/faqs?anchor=50381

and manage all associated inquiries and reviews. When multiple individuals perform such duties, coordination is essential to ensure that reviews are conducted consistently.

Similarly, an office may choose to assign IACUC staff solely to the semiannual inspection program, ensuring that the individual(s) with this responsibility handle the entire inspection workflow, from identification of needed inspections through completion and filing of reports.

Programs that elect to have a dedicated PAM program may find it advantageous to assign staff solely to PAM activities, to best manage the various elements of PAM. The individual(s) involved in PAM must have excellent communication skills to foster a respectful and supportive relationship with the research community, and they should work closely with the veterinary team. For such personnel, a background in research and some direct knowledge of science and veterinary procedures can be tremendously helpful to instilling confidence.

6.4 What attributes, skills, education, and/or certification(s) are needed to be a successful IACUC administrator?

Opin. The successful IACUC administrator will have a combination of attributes and skills, not the least important of which are patience and good customer service, but also critical thinking, organization, and excellent communication. For administrators who work away from the animal facilities (e.g., telework, off-site office), strong and frequent communication, technology skills, and self-motivation are necessary to ensure that distance does not impact quality of service.

It is not uncommon for the IACUC to be perceived as a hurdle standing between the researcher and their ability to do their science. Understanding that the IACUC might be seen less than favorably, the administrator can do a lot of good for their program by offering superior customer (researcher) support, demonstrating empathy for the multiple burdens (including regulatory—see also 6.15) on investigators, facilitating reviews and information when urgent deadlines are identified, conveying complicated messages with clarity and regulatory information (when available), and educating the community on the role of the IACUC as a partner in research.

A background in the sciences and prior experience working with animals can be invaluable to the IACUC administrator's ability to be successful in their role. However, with training and time, those individuals with no such background or experience can still do well. Academic institutions may require their IACUC staff to minimally have a bachelor's degree due to the community of researchers they will serve; however, experience working in the field can replace a degree in some cases.

Whether other credentials such as the American Association for Laboratory Animal Science (AALAS) Technician Certification Program or state licensure as a Registered Veterinary Technician (RVT) are beneficial to the IACUC administrator may depend on the IACUC's expectations for how this individual will contribute to the program. If an employee with an RVT will participate in PAM and may, for example, assess laboratory member's abilities to perform surgery, then that credential would be highly desirable. It should of course go without saying that more training and education is never a bad thing and an individual who holds the Certified Professional IACUC Administrator (CPIA) certification,* for example, will be well positioned to understand the needs of the IACUC.

6.5 How best to determine a candidate's fitness for the job of IACUC administrator during the interview process?

Opin. The authors encourage development of structured interview questions that explore different categories to evaluate fitness to work in IACUC administration. For example, questions may be developed related to the following topics:

- Regulatory knowledge
- Research experience
- Veterinary or animal husbandry skills/knowledge
- Interpersonal skills
- Writing skills
- Public speaking
- Animal welfare versus animal rights

In each category, the hiring party should develop general and specific questions to elicit responses that will help judge how well prepared an individual may be for a role in IACUC administration and at what level. For example, a general question seeking knowledge about IACUC regulations and policies might be followed by a scenario question aimed at understanding how a candidate would apply that knowledge to work through a problem.

An individual with no regulatory knowledge may be an excellent candidate for an entry-level job if they have strong interpersonal and writing skills, while individuals with strong knowledge of IACUC regulations would be critical for higher level jobs that require interpretation and enforcement of those regulations.

* The CPIA credential is offered through PRIM&R and "constitutes formal recognition of an IACUC professional's broad knowledge of IACUC functions and expertise about animal care and use programs and signals that its holders are highly qualified to discharge their duties pursuant to United States rules and regulations, prevailing ethical codes, and administrative best practices." See https://primr.org/cpia for more information.

Individuals who have experience working in a research environment, or performing husbandry or technician duties, are likely to bring an important, broad perspective to the IACUC administrative team. When these individuals have regulatory knowledge and good administrative skills, they are likely to be considered strong applicants to the IACUC office.

Different insights into a candidate's potential can be gleaned by using an interview panel. In addition to the IACUC administrator or director, participants on such a panel could include the IACUC Chair, a member of the veterinary team, other members of the IACUC staff, and/or a representative from the institutional Environment Health & Safety (EH&S) group.

6.6 How much animal or laboratory/research experience is necessary to work in IACUC support?

Opin. While direct animal/research experience is valuable to help understand the different challenges faced in the IACUC oversight process, it is not a requirement to be successful to work for the IACUC. Junior-level positions supporting the IACUC, for example, are excellent entry points into the field and provide sufficient exposure to help an individual determine whether they want to learn more about IACUC administration. Personnel who do not have research or research animal experience can still make valuable contributions, so long as they have administrative skills, inquisitive minds, and a desire to learn and help.

6.7 How much assistance should an IACUC office offer researchers in the creation of animal use protocols?

Opin. The IACUC office has no regulatory obligation to assist researchers with the creation of animal use protocols and, in fact, doing so may blur the understanding of the role the staff have in the institution's animal care and use program. That's not to say that some IACUC offices do not provide some level of assistance with protocol preparation. However, there is a valid concern that if the IACUC staff assist too much with writing protocols, they run the risk of investigators not fully understanding the contents and requirements of what they are submitting. If PIs themselves do not realize what they have been approved to do, they cannot appropriately instruct their lab staff and this may create potential compliance issues down the road, and even stray into the realm of legal implications for scientific misconduct, misrepresentation, etc. It must be remembered that an initial offer of assistance may also lead to subsequent expectations of support and over-reliance on the staff, who in the future may not have the bandwidth to routinely offer such assistance.

That being said, the IACUC staff generally have the best familiarity with the IACUC's protocol forms/online system, review process, and general expectations, including the institution's minimum training requirements (such as enrollment in and compliance with the institutional occupational health and safety program). It is recommended that the willingness to assist investigators with completing their protocol be limited to gently guiding provision of information and educating investigators on why the information is needed, and how and where to provide it within an application. For reasons already stated, programs are cautioned against IACUC staff themselves completing protocol applications on behalf of investigators, although they may "prime the pump" by getting an investigator started, for example, creating an application in an online protocol format in which the new PI is not proficient, so that they may continue writing the protocol once the bare outline has been started.

6.8 Is an online system required to effectively manage protocol submission, review, and approval?

Opin. An online system is not required to manage protocol submission, review, and approval. However, an online system will greatly improve the IACUC's ability to track and report on protocol processes, particularly for larger programs with numerous protocols. Online systems facilitate the involvement of required and *ad hoc* reviewers and allow for concurrent reviews when multiple individuals are tasked with contributing at the same time. In the absence of an online system to coordinate the workflow, an IACUC office may find themselves using several programs to conduct a review: e-mail to receive and distribute applications as well as communicate about the status and send/receive committee correspondence; a program (such as MS Word or Adobe) in which those applications are actually written and saved (including multiple versions as a protocol is revised and later amended); another program to track the different assigned reviewers, review processes, and status; and other programs that might be needed to track user training, facility inspection dates and findings, and so on. A robust online protocol management system will allow for each element of the protocol review process, from submission to approval, including subsequent amendments and triennial reviews, as well as communication with investigators, to be contained within it—a true benefit when it comes to audits, accreditation site visits, and regulatory inspections.

6.9 What are some desirable features and capabilities of an effective electronic protocol management system?

Opin. An electronic protocol management system should facilitate transparency into the review process across different stakeholder groups. Investigators should be able to communicate with the IACUC/

office, access IACUC documents related to their submissions, and clearly understand what activities are approved and for what timeframe. The veterinary staff should be able to easily identify approved procedures, anticipated outcomes and potential complications, planned humane endpoints, and criteria for earlier than scheduled euthanasia. For the IACUC office, integration with the institution's learning management system or user training database will make it easier to evaluate whether users have completed minimum training requirements to participate in animal activities.

A good electronic system will allow the IACUC office to generate reports on topics like protocol review turnaround times, which protocols are approved for certain activities, which investigators are approved to work with particular substances; and information obtained from such reports will aid communication to the research community when changes to IACUC-approved procedures are made and/or if certain substances become unavailable for procurement and must be replaced with alternatives. The electronic system should further aid the IACUC office and research community by sending notifications to investigators about upcoming and required protocol actions, such as the need for a triennial review.

6.10 What are the relative advantages of a vendor product versus a home-grown electronic protocol management system?

Opin. Working with a vendor can be advantageous, as the institution will not be starting from scratch to build a robust electronic program and will benefit from the collective knowledge the vendor has gained from working with prior clients. Purchasing a vendor product often allows access to vendor support for system bugs and enhancements, as well as a community of other users who can advocate for updates and provide insights into implementation of the product. Further, vendor products are regularly upgraded to improve customer experience and address changing regulatory requirements.

The flip side, of course, is that the institution may struggle to find an off-the-shelf product that works with its IACUC business processes, thus necessitating changes to those processes. This can cause confusion and frustration for everyone who is comfortable with the previous way of doing business. Further, vendor support—both in scope and in longevity—is not guaranteed; the institution would be wise to interview other clients to learn about their experiences with the product and customer service. How well does the vendor listen to and address client concerns? How often is the product updated and are those updates done thoughtfully, with the client base in mind? Does the vendor allow users to configure and/or customize the product locally, and how much does the vendor charge the client for these changes? Do such customizations impact future support? Does the company producing the software have a long track record, or is it otherwise likely to still be in business and able to provide ongoing support a few years down the road?

A home-grown system has the advantage of being more precisely tailored to the business needs of the institution. So long as the institution's IT support can be ensured after the initial development and launch, going the home-grown route ensures that the institution is not competing with other clients for vendor support. However, if IT support cannot be ensured from the institution, a home-grown system may quickly become outdated and/or difficult to maintain. Institutions may find it challenging to develop as robust a system as a vendor can offer, limiting integrations and requiring continued reliance on external programs to support the program.

Surv. Do you use software designed for scheduling, conducting, and communicating about facility inspections and/or PAM?

• No	236/284 (83%)
• Yes. The software we use is*	48/284 (17%)
• Not sure	0/284 (0%)

*Software we use is:

- System developed by institution/home-grown
- Doodle
- Filemaker
- Huron product/system
- IRBNet
- IRBManager
- InfoEd
- Key Solutions
- Microsoft product
- Outlook
- REDCap
- Topaz Technologies
- Atune Tick@lab
- Google
- Workday Enterprise Management

Surv. Does your IACUC track performance metrics as a way to assess efficiency?

• No	170/284 (60%)
• Yes. The software we use is*	92/284 (32%)
• Not sure	22/284 (8%)

*The responses to software used are reflected in responses to previous survey question.

Surv. If yes to above question, which metrics are tracked? Select all that apply.

- Overall turnaround time from protocol submission to approval 85/91 (93%)
- Overall turnaround time from amendment submission to approval 75/91 (82%)
- Time for individual IACUC members to submit reviews 45/91 (49%)
- Other* 21/91 (23%)

*Other:

- Number of protocols processed by various review methods (e.g., FCR, DMR, VVC))
- Non-compliance events
- Post-approval monitoring activity
- IACUC administrative actions
- Number of protocols and amendments approved each month
- Adverse events
- Time for the AV to review and approve submissions via VVC
- Time for other steps in the review process (e.g., response from PI)
- Number of review rounds until approval
- Number of submissions by individual departments or units

6.11 Should new researchers be oriented to the institutional animal care program, policies, and requirements? How is this done?

Reg. Regulatory and oversight bodies expect that institutions will ensure that personnel conducting procedures on animals will be "appropriately qualified and trained in those procedures" (PHS Policy IV,C,f). The Animal Welfare Regulations reinforce this expectation and indicate that the institution shall fulfill this requirement "in part through the provision of training and instruction to those personnel" (§2.32,a). The *Guide* further expands on this topic by highlighting the expectation that the institution "ensure that [researchers] have the necessary knowledge and expertise for the specific animal procedures proposed and the species used" (p. 16–17). And the Health Research Extension Act[4] describes the requirement for grant applications and contract proposals to include assurances that "scientists, animal technicians, and other personnel involved with animal care, treatment, and use by the applicant have available to them instruction or training in the humane practice of animal maintenance and experimentation, and the concept, availability, and use of research or testing methods that limit the use of animals or limit animal distress."

Opin. The institution can best fulfill expectations by providing training specific to the program of animal care and use at the institution. While training offered through professional organizations and national training programs will provide a solid foundation for IACUC expectations, researchers will be better equipped to fulfill their local duties if they receive institution-specific orientation. This training should cover such topics as IACUC policies and guidelines, required training for animal users, veterinary and EH&S requirements, and more. The institution may offer this training formally, via required in-person or online training, or informally, via optional meetings and existing online resources, such as the AALAS Learning Library (https://www.aalaslearninglibrary.org). The formal approach to this type of institution-specific training will require additional resources for development and implementation, as well as enforcement and compliance; however, this approach ensures that researchers have been oriented to a minimum standard. The informal approach requires minimal resources beyond maintaining a list of accessible online resources; however, this approach is likely to be less successful at disseminating information as compared to an institution that maintains a robust mandatory training program.

6.12 What content is useful for the IACUC to include on its website?

Opin. A healthy IACUC website will cover the following topics:

- Role/authority of the IACUC at the institution
- Local IACUC policies and guidance
- Meeting dates and submission deadlines
- Application and review processes, including links to protocol forms and/or online programs
- Animal user training program and links to required classes
- Occupational health program information for animal users
- Description of the semiannual inspection program
- Procedures for reporting allegations of mistreatment and noncompliance
- Regulatory resources, including institutional OLAW Assurance, USDA Registration, and/or AAALAC accreditation information
- IACUC office contact information

Institutions may consider including additional information on their IACUC website to aid in education, outreach, and transparency efforts. Some topics may include:

- Overview of the type of animal research at the institution
- IACUC news and announcements
- IACUC metrics
- Oversight body inspection reports (summaries or the reports themselves)

6.13 Should the IACUC website be password protected?

Opin. The type of information made available on the IACUC website and the level of concern about negative attention will inform whether an institution decides to make information on their site password protected. Requiring credentials to access the IACUC website may prompt the IACUC to provide greater detail about sensitive information, including IACUC staff names, specific procedural information, and facility locations. However, some individuals who legitimately need access to IACUC information may be deterred by the requirement to login to the site. A combination of approaches, by which more general information is publicly available, while other information requires a password, can also be used; this may be the solution to allow some initial website visitors (such as PIs new to the institution) to obtain the information they need, without jeopardizing more sensitive data.

Surv. Does your institution have an IACUC website?

- No 86/284 (30%)
- Yes 197/284 (69%)
- Not sure 1/284 (0%)

Surv. If your institution has an IACUC website, is this site password protected or publicly available?

- Password protected 94/197 (48%)
- Publicly available 101/197 (51%)
- Not sure 2/197 (1%)

6.14 By what methods might the IACUC communicate important policy and program updates to the institutional research community?

Opin. The IACUC has an obligation to communicate policy and program news and updates clearly, effectively, and promptly to the research community. Such communications may take the form of listserv announcements, postings to an electronic message board or website (these may be public-facing or require credentials to access), townhall meetings (in person or via video conferencing), small group meetings with departments and/or individual research groups, regularly scheduled training sessions, and more. The method used for communication should be appropriate to the type of message being distributed. For example, policy updates must be broadly distributed and communicated, and they can be done using all the above suggested means. Program updates of a sensitive nature (e.g., describing specific techniques for euthanasia procedures) may be more appropriately communicated on a password-protected website or in targeted small group meetings and training sessions.

Surv. Does your IACUC have a method for communicating important policy and program updates to the research community?

- No 28/284 (10%)
- Yes 251/284 (88%)
- Not sure 5/284 (2%)

Surv. If yes to the above question, how is this done? Select all that apply.

- Via a listserv 98/247 (40%)
- Via a newsletter 67/247 (27%)
- Via an institutional IACUC website 133/247 (54%)
- Via group presentation (in person or recorded) 89/247 (36%)
- Via contact with individuals who might be specifically affected by changes to specific policies or programs. 182/247 (74%)
- Other* 54/247 (22%)

*Other responses largely reflected communication via e-mail listserv, though the following responses were also offered:

- Communication directly to lab personnel during PAM visits or semiannual inspections
- Communication via protocol submission and tracking system
- Communication via document control software

6.15 What is meant by "regulatory burden," and what are some examples with respect to the IACUC?

Opin. An entire book could be dedicated to the topic of regulatory burden and the role of the IACUC. Perhaps the most striking recent effort to alleviate regulatory burden is the 21st Century Cures Act (21CCA), which was signed into law in 2016. As it pertains to animal research and IACUCs, Title II, Section 2034(d) of the 21CCA directed the NIH to collaborate with USDA and FDA "to complete a review of applicable regulations and policies for the care and use of laboratory animals and to make revisions, as appropriate, to reduce administrative burden on investigators while maintaining the integrity and credibility of research findings and protection of research animals" (https://olaw.nih.gov/sites/default/files/21CCA_final_report.pdf).

The establishment of a workgroup, led by OLAW, to evaluate and address regulatory burden sparked important conversations in the regulated community about regulatory burden. A 2008 article offered perspective on regulatory burden, identifying three sources of regulatory burden: "specific requirements in law and regulation, interpretive requirements or 'guidance' by regulatory agencies, and self-imposed regulatory burden resulting from institutional interpretations."[5] In 2016, the Council on Governmental Relations (COGR) released a tool to aid institutions in evaluating actions to reduce administrative burden: https://www.cogr.edu/sites/default/files/Administrative%20Burden%20Checklist%20Revised%20June%2015%202016.pdf

The COGR checklist is an excellent starting place for institutions that wish to evaluate local policies and practices that exceed regulatory requirements.

Some examples of IACUC practices that increase regulatory burden include:

- Applying USDA requirements to non-regulated species such as purpose-bred mice and rats (e.g., use of USDA pain categories in these protocols requiring a search for alternatives to proposed painful/distressful procedures, requiring two IACUC members to participate in all semiannual facility inspections, veterinary consult for all protocols)
- Requiring full committee review for all protocol submissions, including amendment applications
- Requiring very specific information, such as exact drug doses as opposed to acceptable ranges, exact needle gauges for injections, etc.
- Requiring annual review of all animal care and use activities

The institution may have given thought to their practices and made the decision to go above and beyond minimum requirements; in such cases, there may have been a risk/benefit calculation that tipped toward reducing risk to the institution or a consideration of how differing processes might be perceived by the public. In cases where consideration has not been given to such practices, the authors encourage the IACUC administrator/director to discuss with the IACUC Chair and Attending Veterinarian the Animal Research/IACUC items included in the COGR checklist and to determine the value of making such changes versus not doing so.

6.16 How might the IACUC reduce regulatory burden?

Opin. As noted above, the IACUC can identify programmatic elements that exceed regulatory requirements by using the COGR checklist. Once the self-evaluation has been completed, it is up to the IACUC to determine whether to pursue programmatic changes. Just because an opportunity has been identified does not mean an IACUC is obligated to make changes. Local reasons may exist for exceeding regulatory requirements, for example, a prior history of compliance issues related to a particular topic. It is perfectly acceptable to conclude that, institutionally, the IACUC remains risk averse on certain topics and, therefore, sees value in additional requirements. Simply completing a self-evaluation for regulatory burden does not require change be made.

Surv. Has your institution made efforts to reduce regulatory burden associated with animal research?

- No 76/280 (27%)
- Yes 174/280 (62%)
- Not sure 30/280 (11%)

6.17 What documents are required to be maintained as part of IACUC business operations?

Reg. The PHS Policy (IV.E) states that the following records shall be maintained:

1. A copy of the Assurance which has been approved by the PHS
2. Minutes of IACUC meetings, including records of attendance, activities of the committee, and committee deliberations
3. Records of applications, proposals, and proposed significant changes in the care and use of animals and whether IACUC approval was given or withheld
4. Records of semiannual IACUC reports and recommendations (including minority views) as forwarded to the Institutional Official
5. Records of accrediting body determinations

All records shall be maintained for at least three years; records that relate directly to applications, proposals, and proposed significant changes in ongoing activities reviewed and approved by the IACUC shall be maintained for the duration of the activity and for an additional three years after completion of the activity.

The USDA (9 CFR §2.35) requires the following IACUC records to be maintained and made available to USDA inspectors:

1. Minutes of IACUC meetings, including records of attendance, activities of the Committee, and Committee deliberations
2. Records of proposed activities involving animals and proposed significant changes in activities involving animals,

and whether IACUC approval was given or withheld

3. Records of semiannual IACUC reports and recommendations, including minority views
4. Records that relate directly to proposed activities and proposed significant changes in ongoing activities reviewed and approved by the IACUC shall be maintained for the duration of the activity and for an additional three years after completion of the activity.

Opin. The PHS Policy and AWA/AWAR clearly detail IACUC and ACUP recordkeeping requirements, with overlap between the two seen in maintaining records of the following:

- IACUC meeting minutes, attendance records, IACUC activities and deliberations
- Proposed activities involving the care and use of animals, including proposed significant changes to those activities, and the IACUC determination for each
- Semiannual IACUC reports and recommendations (including minority views)

Additionally, the PHS Policy requires maintenance of the approved Assurance document for the institution and records of accrediting body determinations, while the AWAR include specific requirements for maintenance of records related to the acquisition and disposition of dogs and cats (§2.35,b,c).

In both cases, maintenance of such records is minimally required for at least three years; for ongoing activities, records are to be maintained for the duration of the activity, then another three years after termination.

In addition, the animal program is expected to maintain records to document the following:

- Training for researchers, IACUC members, husbandry staff, etc.
- Annual review of the institutional disaster/emergency response plan
- IACUC-approved departures from the *Guide* and the PHS Policy
- Animal health checks, procedures, administration of medications, and monitoring

6.18 Where should the aforementioned documents be maintained?

Opin. Records of IACUC operations and business are best maintained by the IACUC administrative team, for example, on a shared computer or cloud drive that is accessible to all relevant staff. Animal health, procedure, acquisition, and disposition records are best maintained by the veterinary team, but maintenance of procedure records may be shared with relevant research staff so long as they are easily accessible for inspection.

Maintenance of training records in a single location may be challenging, depending on how many sources of training the institution has. A centralized learning management system may be used to record locally provided training, but if professional agency training is required (e.g., AALAS, CITI), the institutional learning management system may not be equipped to maintain external records. If the IACUC has established minimum training requirements for researchers, integration of the protocol application with training records, or tracking such records within the system, will make it infinitely easier for IACUC staff to verify completion of these requirements for IACUC approval.

For any records that are cloud-based, issues of accessibility must be balanced with those of security. IT professionals should be consulted regarding security firewalls and procedures to ensure, as much as possible, that outside entities will not be able to access sensitive information regarding any institution's ACUP without authorization.

6.19 What aspects of institutional reporting (e.g., annual reports to OLAW/USDA/AAALAC, noncompliance reports, etc.) should the IACUC office coordinate?

Opin. The IACUC office, as the most likely entity at the institution to maintain relevant records, will be best positioned to coordinate institutional reporting of the annual report and any noncompliance to OLAW. Annual reports to USDA and, if relevant, AAALAC, may be better coordinated through the veterinary team and/or animal procurement office, but with input from the IACUC office when needed. Ultimately, providing reports to oversight bodies requires coordination among those who maintain relevant records and primary responsibility and processes should be established.

Surv. Does your institution have a policy for record retention?

• No	28/280 (10%)
• Yes	242/280 (86%)
• Not sure	10/280 (4%)

6.20 How should meetings be scheduled?

Opin. To ensure maximum participation among busy members, meetings should be scheduled so that they occur on an expected schedule. For example, the IACUC may choose to hold their monthly meeting on the first Wednesday of each month. Another IACUC may schedule meetings the second and fourth Monday of every month. The same start time should be used for regularly occurring meetings, again, to ensure that members have enough advance notice to plan their own calendars accordingly. It is beneficial to set the meeting schedule for a given calendar year a couple months before the end of the current year, so that adequate time is available for rescheduling if a regular meeting date occurs on a holiday and enough time is available to book meeting rooms, if the IACUC meets in person.

If the IACUC meets so infrequently that mandated activities are unreasonably delayed, the institution should reevaluate the meeting schedule and consider increasing the frequency. Members may initially balk at the idea of more meetings; however, more frequent meetings means that business can be handled in a timely fashion. Fewer meetings result in longer individual meetings due to the buildup of items for discussion.

When scheduling meetings, consideration should be given to whether the IACUC works better when members are in-person, or if a remote format using a web-conferencing program is an effective means for handling business. Remote meetings may improve member attendance and more easily allow for guests and observers to participate. Those IACUCs that utilize a remote format for meetings are cautioned to consider and routinely monitor whether this presents any barriers to member engagement. If members do not comment, remain off-camera, or appear distracted, the IACUC would be well advised to weigh the convenience of calling in from anywhere with their ability to have robust discussion on matters of concern.

6.21 How is the meeting agenda set and how far in advance of a convened meeting?

Opin. The IACUC members can expect certain categories of items to appear on the agenda for each meeting, such as review of meeting minutes and discussion of protocols that have been selected for full committee review. Other agenda items may occur less frequently but also with some regularity, for example, facility inspection reports, allegations of noncompliance, policy review, and continuing education.

Members should be provided adequate advance notice of items to be discussed at an upcoming meeting to allow for sufficient time to review materials and prepare for discussions. A point person in the IACUC office (e.g., the Administrator or Director) should work with the IACUC Chair to establish guidelines for regular agenda items (e.g., protocol review, inspection reports) and *ad hoc* items (e.g., allegations of non-compliance, policy discussions). Deadlines for agenda item submissions should be publicly available and kept as regularly as possible. Late additions to the meeting agenda should be infrequent and based on urgent need, to respect members' time and contributions. Late assignments should always be cleared with the member to which an agenda item is assigned as well as the IACUC Chair.

6.22 What topics should be discussed by the IACUC at a convened meeting?

Opin. The IACUC agenda may include the following topics:

- Discussion of new protocols and continuing reviews, based on local policy/requirements for use of DMR versus FCR
- Discussion of significant changes to approved activities, based on local policy/requirements for use of VVC, DMR, and/or FCR
- Reports for semiannual program reviews and facility inspections, to include findings and proposed corrective actions
- Reports of concerns/non-compliance, including proposed corrective actions when appropriate
- Regulatory updates
- IACUC policy updates

Additional topics may include:

- Proposed suspension of activities, remembering that such action must be reviewed and approved at a convened meeting of the IACUC
- Concerns about the animal care and use program, to report to the IO
- Updates from the Attending Veterinarian
- Updates from EH&S, occupational health personnel
- Updates from the post-approval monitoring program

6.23 How should items be prioritized on the IACUC meeting agenda?

Opin. The IACUC meeting agenda should not be so stacked with discussion items that the members cannot commit the time necessary to work through that agenda. For heavy agendas, the IACUC may be inclined to prioritize protocol reviews and postpone less seemingly consequential discussions, for example, continuing education or policy review. However, this approach places greater emphasis on one element of the IACUC's charge at the expense of managing the animal care and use program as a whole.

If a meeting must be ended early, for example, due to loss of quorum, the IACUC may choose to prioritize those items of business that have a due date, such as the semiannual program evaluation and facility inspections. However, the next scheduled meeting agenda should be organized such that any postponed or deferred items from the prior meeting are given priority.

6.24 How does the IACUC office staff support or engage with the committee during convened meetings?

Opin. The authors believe that the IACUC administrative staff should participate in meetings to the extent necessary to understand and implement committee decisions. The staff provide critical institutional memory and regulatory knowledge and, when welcome, can help guide the IACUC in fulfilling their mandated tasks. An IACUC that does not seek engagement from the administrative staff may experience a disconnect between the actions they vote on and the execution of those actions.

At the very least, administrative staff should assist the IACUC with tracking member attendance, ensuring that quorum is maintained (particularly if/when a conflict is identified), recording votes, and taking notes. Staff who were present during facility inspections, for example, can often help answer member questions or provide clarifications on findings. Staff who have been working with PIs can sometimes provide context on protocol amendments or insight into intended changes. Information that is beneficial to a discussion in a convened meeting should be shared.

6.25 How are committee decisions and subsequent action items executed after the meeting? Who tracks such items and how is this done to ensure resolution or return to FCR?

Opin. It is critical that there be a clear understanding of who or what entity will take responsibility for implementation of/action on IACUC decisions after a convened meeting. It may seem to go without saying that the IACUC administrative staff would follow up on protocol discussions and communicate concerns back to the PI. However, assuming that is the institution's expectation, are any others involved in that process? Does the Chair or primary reviewer, for example, participate in crafting correspondence to the PI? At what point would the Chair, primary reviewer, veterinarian, and/or other ancillary reviewers be tasked with participation in protocol review after discussion at a meeting, and in what capacity? These are decisions that should be made in advance, or at least shortly after the meeting is adjourned and before any actions are undertaken.

Different staff roles may take primary responsibility for IACUC meeting follow-up. The staff member who supports the semiannual inspection program, for example, would be the likely individual to finalize inspection reports. If an institution has a dedicated post-approval monitoring coordinator, that individual may be the best person for any "for-cause" meetings or audits.

An online program or centralized database, with shared files, will provide visibility into IACUC tasks as they are conducted. Reviewing meeting minutes will help determine if committee decisions have been implemented.

IACUC staff would be well served to develop their own checklists to track review assignments and other IACUC tasks to ensure that they are completed as voted on by the IACUC, and that any matters that require re-review at a convened meeting are, thus, returned to the full committee.

6.26 Should the IACUC staff perform any type of administrative protocol pre-review?

Opin. The IACUC staff are well positioned to identify trends in the type of questions raised by the IACUC and the areas of the protocol form that are typically less well understood or completed. Consequently, a preliminary review by a member of the administrative office can help obtain clarification in advance of the committee review, thus improving the ability of the IACUC to focus on animal welfare, scientific explanations or justifications, and procedural concerns. The IACUC office staff generally have greater exposure to regulatory and oversight body communications and updates, for example, through listservs and webinars, and they can aid the IACUC with a first pass of protocol applications to ensure that the minimal required content is included. It is the authors' position that there is real value in IACUC staff performing protocol preview of administrative and regulatory portions of the protocol application that are likely to distract or which may be inadvertently overlooked during the committee review. That said, we caution against expanding the scope of the staff review beyond administrative, operational, and regulatory items, and into the realm of veterinary and clinical concerns, as this creates the potential for stepping on toes and providing information that is not the most current or correct based on institutional veterinary practices. While those trends in questions we mention above can certainly include those of a veterinary or clinical nature, related pre-review comments are either best left to a veterinarian reviewer or heavily caveated to refer the researcher to the veterinarian for confirmation.

Based on the use of "should" in the above question, the authors believe that the ultimate response for this item is no. Rewording the question to ask whether there is value in IACUC staff performing pre-review, we offer a resounding YES.

6.27 How should the IACUC office manage the requirement for veterinary review of protocols with possibility of animal pain/distress?

Reg. §2.31,d,1,iv,B of the AWAR describes the requirement for involving in their planning, consultation with the attending veterinarian (or his or her designee) when procedures may cause more than momentary or slight pain or distress to animals. Additionally, §2.33,b,4 requires a program of adequate veterinary care that includes guidance to principal investigators and other personnel involved in the care and use of animals regarding handling, immobilization, anesthesia, analgesia, tranquilization, and euthanasia.

Opin. As gatekeepers for the IACUC protocol review process, the IACUC administrative staff should have a clear understanding of the type(s) of submissions that require a veterinary review based on regulation and institutional policy.

For example, protocols with USDA-regulated species undergoing procedures defined as Pain Category D* or E† would be required, by regulation, to undergo veterinary consultation prior to IACUC approval. However, an institution may choose to extend the requirement for veterinary consultation to include all species, not just those regulated by the USDA, for which more than momentary or slight pain or distress is expected to occur. The IACUC may further expand the veterinary review process to include activities limited to no more than momentary or slight pain or distress if there is the likelihood for potential adverse effects or complications related to the procedure, such as with administration of drug via injection.

Consequently, it is critical that the IACUC has a clear process for conducting the veterinary review and that the IACUC staff understand the process, particularly at which point in the review the veterinary part is expected to occur. An internal guidance document, including a workflow and checklist, may be a useful tool to train IACUC staff and ensure that the process of veterinary review is managed correctly.

6.28 What types of outreach, education, and training programs are commonly offered by the IACUC office to the research community?

Opin. The nature and extent of IACUC staff involvement in outreach, education, and training programs will depend on the experience of the staff and their bandwidth. Smaller programs may have less capacity to offer regular training or educational workshops, particularly if staff are tasked with supporting or liaising with additional regulatory committees beyond the IACUC (e.g., IBC, IRB, see Chapter 37, "Operational Communications and Workflow," for more on this topic). Capacity issues aside, those topics that may be covered by the IACUC office in training sessions include:

- Regulatory requirements for activities involving the use of animals
- How to use the institutional IACUC online system
- Tips and tricks for protocol submission
- Institutional IACUC policies and processes
- Common mistakes and compliance issues
- Training requirements

6.29 Should IACUC staff be assigned to support specific researchers or types of protocols?

Opin. There can be great value in establishing relationships between individual IACUC staff and specific research teams or departments. Creation of a dedicated IACUC coordinator or researcher/department liaison program should be undertaken with the intent of facilitating understanding and information, such that the IACUC can better guide researchers through the protocol development and review process. Having a dedicated person assigned to specific types of research or species, particularly if that person previously worked with animals or on the technician side of the program, can further facilitate protocol development and review in advance of IACUC evaluation.

6.30 Should the IACUC staff be authorized to review any non-significant changes to protocols? If yes, what types?

Reg. NOT-OD-14-126, *Guidance on Significant Changes to Animal Activities*,[8] permits administrative processing of the following changes:

a. Correction of typographical errors
b. Correction of grammar
c. Contact information updates
d. Change in personnel, other than the PI

Opin. The OLAW Guidance on Significant Changes to Animal Activities allows for administrative handling of a few types of changes, including to the list of protocol personnel other than the PI. Through omission, the Guidance tacitly permits for administrative processing of changes in funding associated with a protocol (absent any procedural changes) and use of additional locations that are already part of the animal program but have not previously been identified on the protocol. These types of changes—personnel, funding, and locations—are likely to occur frequently and will benefit from a rapid turnaround from submission to approval, in order to not delay research operations. With established processes, appropriately trained

* Column D (also referred to as Pain Category D) is defined on USDA Form 7023 as activities "involving accompanying pain or distress to the animals and for which appropriate anesthetic, analgesic, or tranquilizing drugs were used."

† Column E (also referred to as Pain Category E) is wdefined on USDA Form 7023 as activities involving "pain or distress to the animals and for which the use of appropriate anesthetic, analgesic, or tranquilizing drugs would have adversely affected the procedures, results, or interpretation of the teaching, research, experiments, surgery, or tests."

IACUC administrative staff can greatly assist the committee by being authorized to perform these reviews, thus reducing the IACUC's workload.

6.31 To what extent should the IACUC staff offer to edit or author protocols for investigators?

Opin. If the IACUC uses an online system that permits doing so, it can be helpful for IACUC staff to assist investigators by directly editing text within the protocol when minimal changes are necessary, for example, correcting a typo or adjusting a dosage based on PI request. Changes or edits made by staff should be confirmed with the PI. When more extensive updates are required, however, the authors believe that investigators have primary responsibility for authoring and editing their own protocols, primarily to ensure that the content of the protocol accurately reflects the work as it will be done and includes all necessary information and justifications. When the IACUC staff have too large of a role in the protocol writing process, there is a potential for later issues, including noncompliance, due to decreased familiarity with the description of activities. See also 6.7.

6.32 Should the IACUC administrative staff have a role in the semiannual inspection process?

Opin. The authors feel strongly that the IACUC administrative staff should have a role in the semiannual inspection process, for example:

- Scheduling inspections to ensure that they are performed on schedule as required
- Planning the inspection route for order of entry (e.g., if certain locations need to be visited prior to others)
- Verifying that responsible parties will be available during the inspection if needed
- Ensuring that the minimum number of IACUC members required for an inspection will be present
- Providing checklists to IACUC members for inspections
- Taking notes of observations and findings
- Communicating findings to responsible parties
- Preliminary evaluation of responses to all findings
- Preparing reports of inspections and findings to the IACUC
- Conducting required follow-up on inspection findings as directed by the IACUC
- Finalizing IACUC reports and obtaining member signatures when required
- Filing final reports with the Institutional Official
- Maintaining records of locations, species and purpose of use, responsible parties, and any specialized equipment
- Tracking repeat findings and corrective actions
- Ensuring proper documentation is on file and current for PI-maintained locations

Surv. Does your IACUC staff support the semiannual inspection process?

• No	18/280 (6%)
• Yes	261/20 (93%)
• Not sure	1/280 (< 1%)

Surv. If your IACUC staff supports the semiannual inspection process, please indicate the type of support provided. Select all that apply.

• Scheduling	242/272 (89%)
• Ensuring that rooms are visited in proper order (if order of entry exists)	174/272 (64%)
• Ensuring that participants attendees are current on occupational health requirements (e.g., visiting NHP areas)	155/272 (57%)
• Recordkeeping (e.g., locations visited, when visited, who attended, what was observed, presence and type of any deficiencies)	251/272 (92%)
• Writing report to the IACUC	224/272 (82%)
• Other*	50/272 (18%)

*Other responses included aspects of the activities described above, though the following were also offered:

- Communicating deficiencies to laboratory and facility personnel, following up on resolution of deficiencies
- Provide transportation to sites to be inspected
- Educational outreach to laboratory personnel during inspection
- Serve as an inspector in at least some circumstances
- Send reports to the IO

6.33 Should the IACUC administrative staff have a role in post-approval monitoring?

Opin. Post-approval monitoring (PAM) refers to every follow-up review or action that takes place after the IACUC approves the use of animals and facilities in research, teaching, or testing activities. When semiannual inspections are conducted, PAM is being performed. When triennial review of a previously approved activity is conducted, PAM is being performed.

The aforementioned activities are required by the PHS Policy[9,10] and AWAR[11,12] and are the minimum requirements for PAM. It goes without saying that the IACUC administrative staff will have a role in these PAM activities, though the extent of that role may vary by program. For example, a program may have dedicated staff who perform facility inspections and, by virtue of this role, are appointed to the IACUC. Absent a formal role on the IACUC, administrative staff are routinely involved in scheduling inspections, communicating with researchers about upcoming inspections, and handling preliminary communications about any inspection findings.

A program that has expanded their PAM activities beyond those that are required by policy or regulation will benefit from involving the IACUC administrative team in their efforts. Staff who have laboratory animal or veterinary technician training will be well equipped to verify researcher proficiency at handling animals and performing procedures and may have a role in observing procedures as requested by the IACUC. Absent technician training, IACUC administrative staff can easily review training and surgery records, so long as IACUC expectations for these documents are pre-established. Simply being an accessible presence in the animal facility is an excellent way for IACUC staff to establish relationships with their researchers and may end up being more beneficial than performing a formal review of approved activities, for example, via a protocol audit.

Surv. Does your institution have personnel dedicated to performing post-approval monitoring activities?

- No 109/280 (39%)9
- Yes 160/280 (57%)
- Not sure 11/280 (4%)

REFERENCES

1. National Institutes of Health, Office of Laboratory Animal Welfare. 2015. *Public Health Service Policy on Humane Care and Use of Laboratory Animals.* Section IV.C.2, and 9 CFR Animal Welfare Regulations, Part 2, Subpart C, Section 2.31(d)(2).
2. NIH Office of Laboratory Animal Welfare, Frequently Asked Questions. G6: Is Post Approval Monitoring Required? https://olaw.nih.gov/faqs#/guidance/faqs?anchor=50381
3. Certified Professional in IACUC Administration. *About the CPIA Credential.* https://primr.org/cpia
4. Health Research Extension Act of 1985. *Public Law 99-158,* November 20, 1985, "Animals in Research." Sec. 495.
5. Haywood, J. R. and M. Greene. 2008. Avoid an overzealous approach: A perspective on regulatory burden. *ILAR Journal* 49(4):426–434.
6. U.S. Department of Agriculture, Animal and Plant Health Inspection Service. *Annual Report of Research Facility* (Form 7023). https://www.aphis.usda.gov/library/forms/pdf/APHIS_7023.pdf
7. U.S. Department of Agriculture, Animal and Plant Health Inspection Service. *Annual Report of Research Facility* (Form 7023). https://www.aphis.usda.gov/library/forms/pdf/APHIS_7023.pdf
8. National Institutes of Health, Office of Laboratory Animal Welfare. 2014. *Notice NOT-OD-14-126, Guidance on Significant Changes to Animal Activities.* https://grants.nih.gov/grants/guide/notice-files/NOT-OD-14-126.html
9. National Institutes of Health, Office of Laboratory Animal Welfare. 2015. *Public Health Service Policy on Humane Care and Use of Laboratory Animals.* Section IV.B.2
10. National Institutes of Health, Office of Laboratory Animal Welfare. 2015. *Public Health Service Policy on Humane Care and Use of Laboratory Animals.* Section IV.C.5
11. 9 CFR Animal Welfare Regulations, Part 2, Subpart C, Section 2.31(c)(2).
12. 9 CFR Animal Welfare Regulations, Part 2, Subpart C, Section 2.31(d)(5).

7

Frequency and Conduct of Regular IACUC Meetings

Sreekant Murthy

Introduction

The goal of the IACUC is to ensure the humane care and use of animals in research, testing, and teaching. Additional goals are for the IACUC to remain in compliance with regulations while maintaining flexibility to meet the unique needs of the institution's academic and research mission. There are four fundamental components for the effective and efficient functioning of the IACUC and these components are:

- Active participation of the IACUC members to meet the investigator's needs in conducting research
- Participation of community members to bring transparency and public conscience
- Participation of AVs to provide appropriate medical care and monitor the well-being of animals, and
- Firm commitment from the institution to comply with regulations and their Assurance (if applicable) by establishing a committee to oversee the institution's animal program, facilities, and research and educational projects involving the use of animals

The IACUC derives its authority from the 1985 HREA and the AWA. These laws mandate the proper constitution and functions of the IACUC. The institution and researchers play a significant role in the composition and functioning of the IACUC through self-governance and monitoring. These laws require an adequate veterinary care program and the provision of an appropriate environment and housing for animals. A critical part of this program is a requirement by the *Guide* for establishing an occupational health and safety program[1] as an essential part of the overall program of animal care and use.

A comparison of the general issues for IACUC composition and functions shows that the PHS endorses the *U.S. Government Principles for the Utilization and Care of Vertebrate Animals Used in Testing, Research, and Training.* Principle I states: "The transportation, care, and use of animals should be following the Animal Welfare Act (7 U.S.C. 2131 *et. seq.*) and other applicable Federal laws, guidelines, and policies." The *Guide* (Eighth Edition) also takes into account the U.S. Government Principles for the Utilization and Care of Vertebrate Animals Used in Testing, Research, and Training and endorses several principles (the *Guide*, p. 12).

The AWA requires that proposed activities using animals be conducted following the AWAR. Based on these acts, regulations, policies, and principles, IACUCs are constituted to protect the health and well-being of the animals used in research, research training, education, and biological testing. The AWAR (§2.31,a) state that "nothing in this part shall be deemed to permit the Committee or IACUC to prescribe methods or set standards for the design, performance, or conduct of actual research or experimentation by a research facility." This statement, and Principle II of the U.S. Government Principles noted above, ensures animal care and use without compromising research for the good of society. Therefore, a carefully composed IACUC that oversees the humane use and well-being of the animals and facilitates animal research is fundamental to a research facility's teaching and research objectives. This chapter addresses critical issues that are pertinent to the functioning of the IACUC.

7:1 How is the frequency of meetings of the IACUC decided?

Reg. IACUCs are required by the AWAR (§2.31,c,1; §2.31,c,2) and PHS Policy (IV,B,1-IV,B,3) to conduct a review of its animal care and use program every six months. APHIS/AC requires that a majority of the IACUC review and sign the ensuing semi-annual inspection report (AWAR §2.31,c,3); likewise, NIH/OLAW requires that the semiannual report has the signatures of a majority of the IACUC members. However, there is no requirement to do this at a convened meeting of a quorum of the IACUC.[2]

The *Guide* (p. 25) requires that "the IACUC must meet as often as necessary to fulfill institutional responsibilities and records of committee meetings and results of deliberations should be maintained." The AWAR provide flexibility to determine the frequency of meetings, leaving it to the needs of the institution. Thus, some large institutions may meet every month and other smaller institutions may meet when necessary. However, there is a requirement that if full committee review of an agenda item is required, then approval of such items must be made at a convened meeting of a quorum of the IACUC with the approval vote of a majority of the quorum present.[3]

The IACUC is empowered by PHS Policy IV, B,8, and AWAR §2.31,c,8 to suspend activity if it finds noncompliance with the PHS Policy, *Guide*, and Assurance, and if the activity is not being

DOI: 10.1201/9781003272861-7

performed following the description of activity provided by the investigator and approved by the IACUC. Suspension may occur only after review of the matter at a convened meeting of a quorum of the IACUC, and with the suspension vote of a majority of the quorum present. The AWAR (§2.31,d,5) require a complete review of all IACUC-approved protocols no less than every three (3) years. The PHS Policy (IV,C,5) requires that the IACUC conduct a complete review at least once every three (3) years. Thus, the number of convened meetings may vary from none to as many as needed to meet regulatory and institutional requirements.

Opin. There is a fundamental responsibility for the IACUC to meet as often as necessary to oversee the coordination and implementation of effective, efficient systems for protocol review and program review to remain in compliance with PHS Policy and the AWAR. Each institution must evaluate its responsibilities to the investigator and funding organizations. Based on the number of IACUC applications, the institution should conduct as many meetings as necessary to review protocols. Many institutions decide on the frequency based on NIH or other sponsor deadlines that may include industry-related and internal applications. Full committee reviews and suspension of an activity must be done in a convened meeting. It is this author's opinion that animal welfare is best served by having the semiannual review of the program of animal care and use (including facility inspections) at a convened meeting of the IACUC with a quorum present.

Surv. What are the circumstances that changed the frequency of meetings?

- Frequency never changed 189/277 (68%)
- Changed because FCR volume decreased 25/277 (9%)
- Changed because it was difficult to attain quorum if the meetings were held too often 5/277 (2%)
- Changed because of pandemics or other similar situations 18/277 (6%)
- Other* 40/277 (14%)

*Other responses reflected variations of the choices above, though the following were also offered:

- *Ad hoc* meetings convened for circumstances such as significant non-compliance and adverse events, as well as for the review of protocols with an urgent need for approval
- Increased volume of submissions
- Additional meetings convened for continuing education of IACUC members and development of policies
- Meetings may be canceled if there is no new business or protocols to review
- Meeting frequency increased to improve oversight of the animal care and use program

7:2 Who determines the frequency of IACUC meetings?

Opin. There is no regulation or policy requiring specific individuals to determine the frequency of meetings. The author recommends that the scheduling of a meeting be done by the IACUC after consultation with the Chair. It is good practice to write a policy on the frequency of meetings and publish the schedule of meetings and the deadlines for the submission of protocols. This will allow the investigators to be aware of schedules and prepare for submissions on time. If a meeting is canceled for any reason, the IACUC Chair or administrator should inform the members of the IACUC and the investigators as early as practicable that the regularly scheduled meeting will not be held and provide a date for the next meeting.

7:3 What is the frequency and conduct of Full Committee Review (FCR) IACUC meetings?

Opin. Regulations, policies, or guidance by OLAW or AWAR do not prescribe how often the IACUC should meet as a full committee. In circumstances where the IACUC is conducting FCR or the IACUC is contemplating suspending an activity, it requires the use of a full committee. FCR is required when any member of the IACUC requests FCR during the process of preparing to conduct DMR. PHS Policy IV.C.2. states that "each IACUC member shall be provided with a list of proposed research projects to be reviewed," with written descriptions available. Any member may obtain, upon request, a full committee review of those research projects. It is important to reiterate here that IACUCs must conduct as many convened meetings as necessary to comply with federal regulations in the situations stated above and to ensure that the institution's animal welfare program is fulfilling its responsibility to provide humane care and that the procedures used are supported by practical, ethical, and scientific principles.

Surv. How frequently does your IACUC have a full committee meeting to review protocols?

- Never 2/277 (1%)
- Only when requested by the chairperson or an IACUC member 31/277 (11%)
- Approximately monthly or more often 156/277 (56%)
- Approximately 6 times a year 15/277 (5%)
- Approximately 4 times a year 15/277 (5%)
- Approximately 2 times a year 26/277 (9%)
- Approximately once a year 4/277 (1%)
- Other* 28/277 (10%)

*Other responses mostly reflected slight variations of the choices above or did not directly address the frequency of meetings. The following were also offered:

- Weekly
- Twice per month

7:4 What parliamentary rules are used in conducting IACUC meetings?

Reg. The PHS Policy (e.g., IV,C,2; IV,C,6) and AWAR (e.g., §2.31,d,2; §2.31,d,6) address some aspects of parliamentary procedure, such as conducting FCR of protocols at a convened meeting of a quorum of the IACUC and voting in case of a suspension. NIH/OLAW has noted in a publication that the validity of IACUC activities is always predicated on the existence of a properly constituted IACUC, although there is no requirement that all of the members be present at all meetings.[3,4]

Opin. None of the regulatory agencies or the *Guide* prescribe regulation for conducting the IACUC parliamentary procedures It is imperative that institutions not allow *Robert's Rules of Order* or other parliamentary or internal rules to supplant those procedures in PHS Policy or the AWAR.

7.5 What information should be provided to IACUC members prior to a meeting?

Reg. The AWAR (§2.31,d,2) and PHS Policy (IV,C,2) state: prior to IACUC review, each member of the committee shall be provided with a list of proposed activities to be reviewed. Written descriptions of all proposed activities that involve the care and use of animals shall be available to all IACUC members, and any member of the IACUC may obtain, upon request, full committee review of those activities.

Opin. Some IACUCs provide their members with all correspondence concerning the review of a protocol. This includes amendments, correspondence concerning allegations of protocol violations or animal mistreatment, and results of subcommittee investigations (particularly those that may lead to suspensions and therefore require review and approval by the full committee). Additional material that the members can receive are reports of any task forces or subcommittees that the IACUC has requested to review or modify a policy or procedure and updates on institutional policies and procedures that affect the conduct of an IACUC meeting. Other reports include membership changes, updates on APHIS/AC regulations and NIH/OLAW policies and guidance on the PHS Policy and relevant changes to the NIH Grants Policy Statement, copies of all pertinent publications, and meeting notices that may enhance the review process.

Many institutions are transitioning their IACUC submission and review process from paper to using databases driven by software. This review process allows all members to have full access to new protocol submissions, active approved protocols, amendments, minutes of the meetings, minority views, correspondence concerning allegations of protocol violations or animal mistreatment, and results of subcommittee investigations (particularly those that may lead to suspensions; therefore, requiring review and approval by the full committee). There is no regulation to prevent or allow software-driven reviews so long as reviewers have the information about all the proposed activities. The entire electronic submission and review process depends upon the robustness of the system to permit the reviewer to view all pertinent documents specific to an existing continuing review, initial review submission, or a third-year complete review of a protocol renewal. The system must also have a built-in process to prevent members with conflicts of interest from having access to protocol deliberations. The system should permit viewing specific information about the investigator and study personnel to ensure that there is no reason for concern whether they have an outstanding noncompliance issue. The system should also be sufficiently robust to document the process of approval especially when conducting designated member review (DMR), amendments, and training updates.

7.6 What additional verbal or written information might be given to IACUC members at the time of a regular meeting, but before research protocols are discussed?

Opin. Verbal information presented at a convened meeting may include reports from the Chief Executive Officer, Institutional Official (IO), Director of the Office of Research, Chair of the IACUC, AV, supervisor of the animal care facilities, reports from other committees, and presentations from guests. Other written materials may include opinions by external consultants and any other reports that the Chair or the institution deems necessary for the proper conduct of IACUC functions. Written information may include minutes of a previous meeting for approval, the agenda for the convened meeting summaries of protocol reviews, external consultant's reports, post-approval monitoring reports, subcommittee reports, updates on policies and procedures, educational materials, regulatory updates, and any other handouts that are pertinent for the conduct of IACUC functions. Very few IACUCs conduct separate quarterly, semiannual, or annual business meetings to discuss various issues that have been of general concern to the IACUC. When held, these meetings may discuss the IACUC administrative procedures such as:

- Evaluation of surveys conducted by the IACUC
- Recruitment of new members
- Review of attendance records
- Replacing a member
- Training new members

- Creating and updating standard operating procedures (SOPs)
- Creating/reviewing the institution's emergency/disaster (contingency) plans
- Policy on whistleblowers
- Updating IACUC forms
- Educational matters
- Methods to train and assess qualifications of animal users
- Creation and dissemination of newsletters for animal users
- Security issues
- Legal opinions
- Ethical issues for the appropriate use of animals in research and education
- Post-approval monitoring reports
- Review of a disaster plan
- Review of concerns about the animal care and use program

An additional purpose of business meetings is to give the IACUCs undivided attention for the review of the institution's animal care and use program, protocols for animal use, and flexibility to meet the unique needs of the institution.

7:7 What specific business should be conducted by the IACUC? Is a formal vote required for any business conducted?

Reg. According to the AWAR (§2.31,c,1–8; §2.31,d,1; §2.31,d,4–6), the IACUC functions include the following. Some of the IACUC functions listed below are abbreviated for brevity; therefore, readers are advised to read the full text under §2.31,(c)(1) – (C)(8).

1. Review at least once every six months the institution's program for humane care and use of animals
2. Inspect at least once every six months all of the institution's animal facilities
3. Prepare reports of evaluations conducted as required by paragraphs in AWAR (§2.31,c,1–2) and submit the reports to the Institutional Official of the facility; *provided however*, that the IACUC may determine the best means of conducting evaluations of the research facility's programs program and facilities; and *Provided further;* That no committee member wishing to participate in any evaluations conducted under this subpart may be excluded. Readers are directed to follow additional requirements as enumerated in this Section §2.31,(c)(3).
4. Review and, if warranted, investigate concerns involving the care and use of animals at the research facility resulting from public complaints received and from reports of non-compliance received from all sources
5. Make recommendations to the IO on any aspects of institutional animal programs, facilities, and training
6. Review and approve or require modifications in (to secure approval) or withhold approval of those components of proposed activities related to the care and use of animals, as specified in paragraph (d) of AWAR §2.31,(d)(1), (d)(4) – (d)(6).
7. Review and approve, or require modifications in (to secure approval) or withhold approval of, proposed significant changes regarding the care and use of animals in ongoing activities, and
8. Be authorized to suspend an activity involving animals in accordance with specifications set forth in paragraph in the AWAR (§2.31,d,6)
9. In addition, the IACUC must review activities enumerated in AWAR §2.31,(d) (1)–d)(8).

Although the IACUC is responsible for many functions listed above, only formal voting in a convened meeting is required for activities reviewed as FCR and suspensions. Therefore, the number of convened meetings may vary depending on the number of FCRs periodically submitted and the suspensions the IACUC is required to review.

PHS Policy (IV,B, 1–8) reiterates some of the requirements enumerated in AWAR (§2.31,c and d); and emphasizes the *Guide* as a basis for IACUC functions. PHS Policy IV.B. states that as an agent of the institution, the IACUC shall with respect to PHS-conducted or supported activities:

1. Review at least once every six months the institution's program for humane care and use of animals, using the *Guide* as a basis for evaluation
2. Inspect at least once every six months all of the institution's animal facilities (including satellite facilities) using the *Guide* as a basis for evaluation
3. Prepare reports of the IACUC evaluations conducted as required by IV.B.1. and 2. of this Policy, and submit the reports to the Institutional Official

   (The reports must meet the following criteria:

   a. The reports shall be updated at least once every six months upon completion of the required semiannual evaluations.

b. The reports shall be maintained by the institution and made available to OLAW upon request.

c. The reports must contain a description of the nature and extent of the institution's adherence to the *Guide* and this Policy and must identify specifically any departures from the provisions of the *Guide* and this Policy and must state the reasons for each departure.

d. The reports must distinguish significant deficiencies from minor deficiencies. A significant deficiency is one that, consistent with this Policy, and, in the judgment of the IACUC and the Institutional Official, is or may be a threat to the health or safety of the animals. If program or facility deficiencies are noted, the reports must contain a reasonable and specific plan and schedule for correcting each deficiency.

e. If some or all of the institution's facilities are accredited by AAALAC or another accrediting body recognized by the PHS, the report should identify those facilities as such.)

4. Review concerns involving the care and use of animals at the institution
5. Make recommendations to the Institutional Official regarding any aspect of the institution's animal program, facilities, or personnel training
6. Review and approve, require modifications in (to secure approval), or withhold approval of those components of, PHS-conducted or supported activities related to the care and use of animals as specified in IV.C. of this Policy
7. Review and approve, require modifications in (to secure approval), or withhold approval of proposed significant changes regarding the use of animals in ongoing activities, and
8. Be authorized to suspend an activity involving animals in accordance with the specifications set forth in IV.C.6. of this Policy which states that the IACUC may suspend an activity which it previously approved if it determines that the activity is not being conducted in accordance with applicable provisions of the Animal Welfare Act, the *Guide*, the institution's Assurance, or IV.C.1.a.–g. of this Policy. The IACUC may suspend an activity only after review of the matter at a convened meeting of a quorum of the IACUC and with the suspension vote of a majority of the quorum present. A formal voting in a convened meeting is required for activities reviewed as FCR and Suspensions. Thus, the number of convened meetings may vary depending on the number of FCRs and Suspensions IACUC is required to review.

The *Guide* (p. 25) requires the IACUC to be responsible for oversight and evaluation of the entire program and its components, protocol review and approval, proposed significant changes to animal use, inspection of facilities and animal use areas, and regular review of the program. The *Guide* recommends the IACUC to meet as often as it is necessary to fulfill its responsibilities.

Opin. Many of the above items are pertinent for the semiannual review of an institution's animal care and use program. Any institutional program reviewed at a convened meeting, whether the program does or does not have deficiencies (minor or significant) and with a plan and schedule for corrections recommended by the committee, should be formally approved by a majority of the quorum present and all minority opinions must be recorded. Regular meetings are usually reserved for reviewing research protocols. However, there is no regulatory reason for discussing policies and procedures, review of training, discussion of animal welfare concerns, safety/occupational health updates, etc. that typically occur during regular meetings in addition to protocol review.

Many IACUCs have agendas that include discussions of the institutional program of animal care and use, review of protocols, and other general business of the IACUC. All business matters that pertain to institutional program evaluations, protocol reviews, sanctions (if the institution has authorized the IACUC to do so), and any other matters that are part of the AWAR and PHS Policy (or institutional policy) can be approved in a convened meeting. However, to suspend an animal activity or to conduct a full committee review of a protocol with a subsequent vote, the IACUC must meet in a convened meeting of a quorum of the IACUC with a majority of the quorum present.[3] A majority of the quorum present must vote to approve or suspend a project. All other business matters should be discussed. For example, if the committee is developing a policy, such policies should be approved by a majority of members. However, the IACUC must reserve the committee's time to conduct protocol reviews to aid the investigator and meet at a separate meeting to discuss business matters, unless the business matter is related to an activity being discussed.

7:8 At a full committee meeting, does the IACUC require a formal vote for items other than the approval of a protocol or the suspension of an animal activity?

Opin. PHS Policy requires a majority vote of the quorum present at a convened meeting to conduct a full committee review and approval of a proposed project or a significant change to a project and suspend an activity. However, the IACUC may review other activities that are not always protocol specific. Examples of these include allegations of mistreatment or non-compliance, suspension of researchers' other animal activities unrelated to the protocol in question, substantiation, or rejection of a whistleblower's allegations, and changes in policies and standard operating procedures. Such discussions may have to be approved by the IACUC as they may impact animal welfare. Many institutions take a formal vote on business items discussed at a convened meeting. Some institutions, in their internal SOPs, require approval of the minutes at a subsequent meeting. To use DMR subsequent to FCR, the quorum of members must approve the use of DMR.[5]

Surv. At a full committee meeting, does your IACUC require a formal vote for items other than the approval of a protocol or the suspension of an animal activity? Example: A formal vote to approve meeting minutes.

- No 19/277 (7%)
- Yes, for just about all administrative items 122/277 (44%)
- Yes, for some but not all administrative items 130/277 (47%)
- I don't know. 1/277 (0%)
- Not applicable 2/277 (1%)
- Other 3/277 (1%)

7:9 For a full committee meeting to occur, what should your IACUC's policy be about the attendance of specific members?

Reg. The AWAR (§2.31,b,1-§2.31,b,4) require that the IACUC consist of at least three members: a Chair, a veterinarian, and a person not affiliated with the institution. PHS Policy (IV,A,3,b; IV,A,3,c) and the *Guide* (p. 24) require a minimum of five members, including a veterinarian, a scientist, a nonscientist, a nonaffiliated member, and one other. One of those five will be designated as the Chair. For the PHS it is not one of the membership roles but rather a designation for one of the five members. The *Institutional Animal Care and Use Guidebook*[6] also states that one member can fulfill more than one of the roles in the IACUC membership. However, for a quorum, the minimum attendance is two under the AWAR and three under the PHS Policy. (See 7:14 for a definition of quorum.)

Opin. Ideally, nonaffiliated and nonscientist (under PHS Policy) members (or their respective alternates) should be present at meetings to address ethical issues and appropriate use of animals from non-institutional and non-scientific perspectives. Likewise, it is ideal for the attending veterinarian (AV) to be present at meetings since the AV is responsible for the health and well-being of all laboratory animals and other aspects of animal care, such as husbandry, and housing to ensure that the program complies with the *Guide* (p. 14). It is ideal to have the IACUC membership at more than the minimum number with additional alternate members so that the quorum and member's expertise can be maintained if one or two members are not available for the meeting. This will also help provide the expertise necessary for a thorough and efficient review of protocols. The presence of a statistician will provide additional strength in evaluating one of the 3Rs (Reduction). However, it is not legally required for any specific member to be present at all meetings if the committee membership is properly constituted and a quorum is present.[7]

Surv. For a full committee meeting to occur, what is your IACUC's policy about the attendance of specific members? Select all that apply.

- Not applicable as we do not have full committee meetings. 2/277 (1%)
- All voting members (or their alternate) must be present. 21/277 (8%)
- The Attending Veterinarian (or designee) must be there. 65/277 (23%)
- At least some of the members (or alternates) required by regulation or policy (scientist, non-affiliated, etc.) must be there. 69/277 (25%)
- As long as there is a quorum, we will have the meeting. 228/277 (82%)
- We may meet without a quorum, but then we will not take votes. 49/277 (18%)
- I don't know. 1/277 (0%)
- Other 7/277 (3%)

7.10 What should be your IACUC's expectation about the frequency of attendance of an IACUC member at meetings?

Opin. Attendance of a committee member at the scheduled meetings is an important consideration in selecting a member. The absence of specific members may result in unnecessary delays in the approval of a study or an inadequate/poor review due to a lack of reviewer expertise. Determination of the validity of the absence should be at the discretion of the institution's officials and particularly the committee Chair. The IO or the Chair may set guidelines for members missing consecutive or too many meetings. Likewise, the institution in its bylaws and standard operating procedures may set up policies when and how far in advance of a committee meeting a member should notify the committee of his/her absence, a policy for unacceptable absences, and reporting absences to the institution's officials. A point to consider at the time of appointing the committee is to select appropriate alternate members for each of the members so that there will be availability of expertise to review protocols in case the primary member is absent. This will also help to prepare alternates to become full members when an opportunity arises to change or replace a member. The IACUC must always strive for 100% attendance so that the quality of discussions is enhanced to the extent that the committee can speak with one voice or differing opinions are heard to take appropriate actions. While 100% attendance may not be achievable or possible, absences from meetings should be kept at a minimum and, when they occur, they are clearly unavoidable.

7:11 What actions are taken if an IACUC member does not attend an adequate number of meetings?

Opin. No regulation requires actions by the IACUC when members do not attend an adequate number of meetings. NIH/OLAW FAQ B.4 states: "Attention should be paid to attendance at IACUC meetings to ensure that an appropriate mix of members attends meetings. Chronic nonattendance by IACUC members, especially those explicitly required by PHS Policy or USDA regulations, implies a lack of participation in the oversight responsibility of the IACUC."[8] Thus, the institution and the IACUC should pay considerable attention to the attendance of PHS- and USDA-specified members to ensure robust and effective oversight of the research program by the IACUC. The institution or the IACUC must establish a policy on member attendance and what actions may be taken toward a member who fails to attend two consecutive meetings or groups of meetings without an explanation acceptable to the committee Chair. Institutions should establish policies on what constitutes grounds for removal of a member following a request by the Chair or a recommendation by the Chair or the IACUC to the institution's appointing official to remove such members. It is not unusual to recognize certain members typically attending only those meetings in which their own or their colleague's (departmental) protocol is being reviewed. In such cases, careful attention should be paid to ensure that there is no direct conflict in that member attending and voting on those protocols. In those cases, it may be better for the member to voluntarily abstain or for the Chair to request the member to abstain from voting. An abstention does not affect the outcome of a vote for approval made by a properly constituted quorum unless the abstention is due to a conflict of interest. A conflict of interest abstention (more commonly called a recusal) decreases the size of the quorum present.[9] PHS Policy (IV,C,2) states that "no member of the IACUC may participate in the IACUC review or approval of a research project in which the member has a conflicting interest (e.g., is personally involved in the project) except to provide information requested by the IACUC; nor may a member who has a conflicting interest contribute to the constitution of a quorum." In the author's opinion, the Chair or IO needs to send an informal notification or a memo requesting that the member attend more meetings before any formal action is taken.

7.12 If a member is asked to leave the committee, who would formally take this action?

Opin. Removing a committee member is also uncharted and controversial territory unless the evidence is clear that the member has violated certain IACUC policies of attendance requirements or breached confidential committee information. Often the Chair or the IACUC may make a recommendation to the appointing official (e.g., the Chief Executive Officer, CEO, IO) to remove or replace a member. From a regulatory perspective, an IACUC member appointed by the IO can only be "unappointed" by the IO. In general, the institutional officer appointing members honors such requests if the requests are well substantiated. Having the IO take formal action may be acceptable as long as the institutional policy allows to remove a member from the committee formally gives that responsibility to an administrative official such as the IO.

Surv. If an IACUC member (or alternate) were to be asked to resign from the committee, who would formally take this action?

- We have no policy. 59/277 (21%)
- The Institutional Official 128/277 (46%)
- The IACUC chairperson 50/277 (18%)
- I don't know. 14/277 (5%)
- Other* 26/277 (9%)

*Other responses mostly reflected the choices above, though the following were also offered:

- Director of the IACUC/Research Compliance
- Vice President for Research Administration
- Dean
- Decided by vote of the IACUC.

7:13 When does your IACUC veterinarian first evaluate pain and distress on a protocol and provide any input back to the investigator?

Reg. The AWAR (§2.33,b,4; §2.33,b,5) states that the AV must provide "Guidance to principal investigators and other personnel involved in the care and use of animals regarding handling, immobilization, anesthesia, analgesia, tranquilization, and euthanasia" and "Adequate pre-procedural and post-procedural care in accordance with current established veterinary medical and nursing procedures."

The AWAR (§2.31,d,1,iv,B) require that for procedures involving more than momentary or slight pain or distress, either the veterinarian or his or her designee be consulted in the planning of the research project. The *Guide* (p. 106) states that the AV "should guide investigators and all personnel involved in the care and use of animals to ensure appropriate husbandry, handling, medical treatment, immobilization, sedation, analgesia, and euthanasia." It also states that the AV should provide guidance and oversight of surgery programs and perioperative care.

Opin. When research projects involve pain or distress, veterinary consultation on analgesia or anesthesia is best provided prior to the IACUC review. It is also a good practice for the institution to establish a policy when this consultation should occur. It is preferable to have this consultation in the planning stages of the protocol. If for extraordinary reasons a veterinarian is not available, another approach to this requirement is to have the AV identify one or more designees who can be involved in planning stages of the protocol. Another veterinarian with knowledge in this area is an appropriate choice for such a designation. Non-veterinarians also may be involved in the consultation at the discretion of the AV.

The author and the *Guide* endorse the recommendations of the American College of Laboratory Animal Medicine (ACLAM) regarding adequate veterinary care.[9] "The institution bears responsibility and must assure, through authority explicitly delegated to the veterinarian or to the IACUC, that only facilities appropriate for the intended surgical procedures are utilized and that personnel are adequately trained and competent to perform the procedures. The veterinarian's inherent responsibility includes monitoring and providing recommendations concerning pre-, intra-, and post-operative procedures, surgical techniques, the qualifications of institutional staff to perform surgery, and the provision of postoperative care."[9]

ACLAM also recommends, "The veterinarian must be involved in the review and approval of all animal care and use in the institutional program. This includes advising on the design and performance of experiments using animals as related to model selection, collection and analysis of samples and data from animals, and methods and techniques proposed or in use. This responsibility is usually shared with investigators, the IACUC, and external peer reviewers."[9] It is important to note here that in some institutions both the AV and the IACUC members could evaluate the provision of anesthesia and analgesia at the same time during the protocol review process. In the author's opinion, the PI should involve the AV or the AV's designee in planning the type of care and appropriate use of drugs to relieve pain or distress to avoid possible delays in securing approval.

The NIH Grants Policy Statement[10] and the Worksheet for Review of the Vertebrate Animal Section (VAS) provided to NIH applicants[11] require that the investigator describe the procedures to ensure that the discomfort, distress, pain, and injury will be limited to that which is unavoidable in the conduct of scientifically sound research. The policy requires investigators to describe the use of analgesic, anesthetic, and tranquilizing drugs and/or comfortable restraining devices, where appropriate, to minimize discomfort, distress, pain, and injury. The author recommends that investigators involve the AV before submitting an NIH grant. By doing this, pain or distress and the appropriate use of drugs are properly discussed before an IACUC review. This allows the PI to secure timely approval, especially for "just-in-time" reviews.

7:14 What constitutes the quorum needed to conduct an IACUC meeting?

Reg. A quorum is an assembly of a majority (more than 50%) of the voting members of the IACUC. For the AWAR (§2.31,b), the minimum number is two, and for PHS Policy (IV,A,3,b), the minimum number is three (provided the IACUC has only three and five members, respectively). Otherwise, it requires more than 50% of the members to be present in a convened meeting to constitute a quorum. Both the AWAR and the PHS Policy state that a member who has a conflict of interest may not contribute to the constitution of a quorum. No member may participate in the IACUC review of an activity in which that member has a conflicting interest (e.g., is personally involved in the activity), except to provide information requested by the IACUC; nor can a member who has conflicting interest contribute to the constitution of a quorum at which a proposed activity relevant to the conflict is being considered for approval (AWAR §2.31,d,2; PHS Policy IV,C,2). The IACUC may invite consultants to assist in the review, but they may not approve or withhold approval of an IACUC activity; therefore, they cannot vote and be counted toward meeting a quorum.[6]

NIH/OLAW and APHIS/AC[12,13] have accepted the practice of using designated alternates for IACUC members, provided they are included in the NIH/OLAW Animal Welfare Assurance. There is no clear direction from the USDA on whether an IACUC membership list needs to be provided to USDA at any time for the registered facility. In these instances, the individual(s) serving as alternate(s) do not add to the total number of committee members unless the primary member is not at the meeting.

Opin. Often, institutions face the problem of attaining a quorum of the IACUC. Some institutions apply a variety of methods to attain or maintain quorum at a convened meeting. These methods include calling members of the committee ahead of the meeting date to alert them about an upcoming meeting or arranging a meeting on a day that is most suitable for all or a majority of the members. If electronic methods are used,[14,15] they present a convenient method to remind the members about the upcoming meeting, including setting up calendar reminders for the meeting date, time, and location. Other ideas include designating alternative members or providing food at the meetings so that the members can devote their lunch, breakfast, or dinnertime for the meeting. All of these methods are workable, and there are no AWAR or PHS Policy limitations for applying these methods.

Generally, it is advisable to have the IACUC membership at more than the minimum number mandated by either the AWAR or the PHS Policy or appoint alternate members with similar expertise to a primary member. For example, some members may leave a meeting early due to prior commitments or unavoidable circumstances. This might result in the loss of a quorum if only the minimum number of people are on the IACUC and present at the meeting. Persons with conflicting interests may affect the quorum composition if IACUC membership is small as they cannot be used to constitute a quorum (AWAR §2.31,d,2; PHS Policy IV,C,2). The problem of a member leaving a convened meeting can often be avoided if protocols reviewed by that individual are placed on the top of the agenda. The simple polling of IACUC members for their votes does not satisfy the definition of a meeting of a convened quorum.[14] Likewise, polling is not a voting method for the "full committee review" of protocols. Furthermore, polling is not permitted for suspension of an activity of a previously approved protocol/activity. However, e-mail, polling, and conference calls with stipulations are allowed by OLAW anytime that an in-person or teleconference/videoconference meeting is required. conference calls, etc., are not normally acceptable methods for full committee review (see 7:19, 7:20 for exceptions). A vote on a suspension of an animal activity requires an "in-person or teleconference/video conference meeting of a quorum of the IACUC (see 7:11). It is acceptable to poll members to see if any want a full committee review before relegating a protocol to Designated Member Review.

IACUCs often encounter problems when members abstain from voting due to a conflict of interest or other valid reasons. In such cases, the quorum will be lost if the abstention is due to a conflict of interest.[9] Regulations do not require all members mandated by PHS Policy to be present at each meeting. The AWAR requires only three members to constitute an IACUC. However, the majority of institutions that submit federal grants must adhere to the PHS Policy, which requires the institution to constitute an IACUC consisting of at least five members.

7:15 If a quorum is in place at the beginning of an IACUC meeting, but the quorum is lost during the meeting, can the meeting continue?

Opin. There are no regulations that stipulate whether a meeting be closed or suspended if the quorum is lost in the middle of a full committee meeting. However, the AWAR and PHS Policy (see 7:14) state that a quorum of members must be present to approve or withhold approval during full committee reviews or perform other specific IACUC actions such as suspending an animal activity. Therefore, all IACUC activities that require voting by a majority of the quorum are suspended when the quorum is lost. However, the protocols that were not reviewed at the meeting can be reviewed by DMR if members are all notified, and no one calls for FCR. If any of those protocols on the agenda were referenced to FCR, those could not be sent out for DMR; rather, they need to be rescheduled for another full committee meeting.[16] In other words, DMR subsequent to FCR is for use after a protocol outcome has been decided by a majority of a quorum of the IACUC present at a convened meeting. It cannot be used if there is no quorum.

In such exceptional circumstances, according to PHS Policy[16] the IACUC with a quorum of members present may unanimously vote to require modifications to secure approval and have the revised protocol reviewed and approved by DMR or returned for FCR at a convened meeting. If all members of the IACUC are not present at a meeting, the committee may use DMR subsequent to FCR by utilizing the following stipulations.[16]

a. All IACUC members agree *in advance in writing* that the quorum of members present at a convened meeting may decide by unanimous vote to use DMR subsequent to FCR when modification is needed to secure approval. However, any member of the IACUC may, at any time, request to see the revised protocol and/or request the FCR of the protocol.
b. In order to conduct reviews by DMR subsequent to FCR, the institution should specify its intention to conduct reviews in this manner in its Assurance with

OLAW. (IACUCs that newly elect to utilize a standard operating procedure for DMR subsequent to FCR should provide information about this program change to OLAW in the next Annual Report.)

If all members are not present and the IACUC lacks written standard procedures as described above, the committee has the option to vote to return the protocol for FCR at a convened meeting or to employ DMR. If electing to use DMR, all members, including the members not present at the meeting, must have the revised research protocol available to them and must have the opportunity to call for FCR. A DMR may be conducted *only* if all members of the committee have had the opportunity to request FCR and none have done so (PHS Policy IV,C,2).[16]

Alternatively, to avoid issues related to the maintenance of a quorum, the institution may amend its NIH/OLAW Assurance to conduct a review by tele or audiovisual communication means; however, even such reviews require stringent application of policies that are described in 7:17 and 7:18. In the event of pandemic circumstances, for example, or in similar situations where meetings are held by electronic methods most institutions had the foresight to include electronic means of conducting the meeting in their OLAW Assurance. They also had an easy way of conducting IACUC meetings to approve, secure approval for a modification, or halt the study. In March 2020, OLAW issued a notification (NOT-OD-20-088)[17] regarding the conduct of IACUC meetings. According to this notice, there are three specific points offered as social distancing measures to prevent the spread of COVID-19. However, this notice has been rescinded and replaced with NOT-OD-22-207.[18] This notice now refers to OLAW's Disaster Planning and Response Resources webpage and NOT-OD-21-164[19] of the Semiannual Facility Inspections webpage. The notification also affirmed that OLAW is the only entity authorized to make determinations regarding waivers to the provisions of the PHS Policy. Therefore, any modifications that infringe on the PHS Policy must be approved by OLAW and the IACUC before the change is implemented. Thus, the author recommends that institutions are well advised to include tele or audiovisual communication means of conducting meetings in their OLAW Assurance. This will avoid situations imposed by pandemics or any other disasters, as well as situations when an impromptu meeting is essential to conduct a review of a protocol or amendment; for matters related to suspension of a protocol; or for discussions and action that might need to be taken in response to non-compliance issues. APHIS/AC does not provide any direction for audiovisual conferencing as an alternative to face-to-face meetings. However, APHIS/AC concurs with NIH/OLAW that "the functional equivalent of a convened quorum in the full committee mode might be emulated through real-time, interactive, videoconferencing added to the end of the described electronic review process."[17]

7:16 Should any or selected investigators be invited to the IACUC meeting?

Opin. It is customary for some IACUCs to invite investigators to respond to questions in a face-to-face meeting or when meetings are held using tele/audiovisual methods. Since the committee has the dual responsibility of thoroughly reviewing protocols and executing decisions, the PI can answer specific questions raised by the IACUC. This often saves time for the IACUC and the PI. The PI should leave the meeting before a vote is taken. If the protocol involves a committee member who has a conflicting interest, the member is usually asked to leave the meeting while the committee is discussing that protocol (although the person may give information to the committee if requested to do so). The member with a conflicting interest should leave the meeting when the committee votes. In the author's opinion, it is the best practice for the chair to ask members at the beginning of the meeting whether any of the members have any conflict implicit, or explicit, so that only members who do not have a conflict are present at the time when a protocol is reviewed and voted upon.

7:17 What methods do IACUCs use to formally review a protocol?

Reg. To approve an animal activity the criteria to be met are covered under PHS Policy (IV,c,1,a-IV,c,1,g; IV,d,1,a-IV,d,1,e) and the AWAR (§2.31,d; §2.31,e). The method to accomplish this is left up to each individual institution. According to these regulations, there are only two lawful methods of IACUC review allowed by PHS Policy and AWAR: (1) full committee review (FCR) by a convened quorum of the IACUC and (2) designated member review (DMR) by one or more members chosen by the chairperson, employed only after all voting members have been provided an opportunity to call for full committee review. The NIH Guide for Grants and Contracts NOT-OD-06-052[15] provides guidance regarding the use of tele- and videoconferencing when a convened meeting is required. The guidance is also provided in an article published by NIH/OLAW concerning IACUC functions and the use of technological advances in communication technology.[17]

Opin. Many institutions use FCR to review protocols. FCR protocols require a convened meeting of a quorum of members, and protocols can be approved only by an affirmative vote of a majority of the quorum present in a meeting. Some institutions also use an informal pre-review process. During this pre-review, a primary member or a team of members may generate questions, and the PI can respond ahead of a meeting; thereafter, the modified protocol is reviewed by the IACUC. Pre-review in any institution is a voluntary process that aids the investigator and the IACUC in resolving some of the key issues related to the protocol, reduces the committee time in reviewing the protocol, and improves the turnaround time. However, even though a protocol has gone through a pre-review process, the IACUC must still provide written approval prior to the research team initiating the activities described in the protocol.

According to the NIH Guide for Grants and Contracts NOT-OD-06-052,[15] methods of telecommunication (e.g., telephone or video conferencing) that are acceptable for the conduct of official IACUC business requiring a quorum are as follows:

- All members are given notice of the meeting. Some IACUC software allow for an automatic e-mail notice to be sent to members.
- Documents normally provided to members during a physically convened meeting are provided to all members in advance of the meeting. If an electronic IACUC management system is used, the members are often automatically informed about the documents in advance of the meeting.
- All members have access to the documents and the technology necessary to fully participate.
- A quorum of voting members is convened when required by PHS Policy.
- The forum allows for real-time verbal interaction equivalent to that occurring in a physically convened meeting (i.e., members can actively and equally participate and there is simultaneous communication).
- If a vote is called, the vote occurs during the meeting and is taken in a manner that ensures an accurate count of the vote. A mail ballot or individual telephone polling cannot substitute for a convened meeting.
- Opinions of absent members that are transmitted by mail, telephone, fax, or e-mail may be considered by the convened IACUC members but may not be counted as votes or considered as part of the quorum.
- Written minutes of the meeting are maintained in accordance with the PHS Policy (IV,E,1,b). It should be noted that APHIS/AC concurs with Notice NOT-OD-06-052.[15]

Surv 1. What formal method is used to review a protocol?

Entirely full committee review	43/277 (16%)
Mostly full committee review but some designated member review	79/277 (29%)
Entirely designated member review (unless subsequently called to full committee)	41/277 (15%)
Mostly designated member review but some full committee review	88/277 (32%)
an even mix of full committee review and designated member review	26/277 (9%)

Surv 2. How does your IACUC perform Full Committee Reviews (FCR)?

• We do not use FCR.	8/276 (3%)
• We have 1 or 2 members (non-Chairperson) present the protocol and then the remainder of the committee can join the discussion before we vote.	121/276 (44%)
• The Chair presents the protocol and then remainder of the committee can join the discussion before we vote.	33/276 (12%)
• Most members of the committee routinely fully review all FCRs and most members are meaningful participants in the ensuing discussion.	86/276 (31%)
• Other*	28/276 (10%)

*Other responses include a mix of the choices above with a variety of specific approaches, though the following were also offered:

- The veterinarian who reviewed the protocol leads the FCR discussion.
- The PI presents the protocol at the meeting and then leaves prior to deliberation.
- A subcommittee of 3 to 4 members reviews each protocol and then presents it for FCR.

Surv 3. Concerning using Designated Member Review (DMR) after a Full Committee Review (FCR) what will the IACUC do? Select all that apply.

• We do not use DMR after FCR.	29/276 (11%)
• We occasionally or often use DMR after FCR.	112/276 (41%)
• The quorum present votes to send the quorum present votes to send the protocol to DMR and subsequently the full committee is given the opportunity to agree to use DMR.	164/276 (59%)
• I don't know.	5/279 (2%)

7:18 What are the requirements for conducting other than face-to-face full committee review IACUC meetings ("exceptional circumstances" meetings)?

Reg. NOT-OD-06-052[15] states that although "the traditional meeting is still seen as the optimum environment for fulfilling the intent of the PHS Policy, OLAW recognizes the new communications tools available and the need for flexibility in the ways that institutions may comply with the PHS Policy in the many diverse settings encountered."[15] APHIS/AC and NIH/OLAW concur that a traditional full committee meeting with a convened quorum of the IACUC leads to a review in real time. However, the functional equivalent of a convened quorum may be emulated through real-time, interactive video conferencing.[16] For these reasons, several criteria have been provided for establishing alternate methods that may be considered functionally equivalent to meetings of a convened quorum either face to face or tele/audiovisual methods. Two of those criteria are that the alternate approach must include a high degree of interactivity and allow for careful deliberation of sensitive issues. Additionally, a quorum of IACUC members must be in direct communication with each other and be given full opportunity to participate for the duration of the meeting. Regulations remind institutions that details of their IACUC procedures, especially those that may vary from those outlined in the PHS Policy, be thoroughly described in the institutional Assurance submitted for OLAW review. Thus, the requirements are the same as face-to-face meetings.

Opin. Telephone and audiovisual conferencing may be acceptable alternatives to face-to-face meetings for full committee review of protocols and other IACUC matters. The requirements for audiovisual meetings are[17]:

- All members must be given ample prior notice by paper or electronically to participate.
- At least a quorum of the voting members (>50%) must be convened on the same conferencing line whether it is a telephone or an audiovisual line.
- The quorum of IACUC members must be in direct communication with each other and be given full opportunity to participate for the duration of the meeting.
- The minutes of the meetings must be compiled and maintained on file as required by federal regulatory agencies.

7:19 How can a designated member review (DMR) of protocols be accomplished using the electronic media meeting format?

Reg. The PHS Policy (IV,C,2) states: "If full committee review is not requested, at least one member of the IACUC, designated by the chairperson and qualified to conduct the review, shall review those research projects and have the authority to approve, require modifications in (to secure approval) or request full committee review of those research projects." The AWAR (§2.31,d,2) have identical language. In both cases, prior to IACUC review, each member of the committee must be provided with a list of proposed activities to be reviewed. Written descriptions of all proposed activities that involve the care and use of animals shall be available to all members. Any member of the IACUC may obtain, upon request, a full committee review of those activities.[5] The regulations do not specify how the proposed activities list needs to be disseminated so long as the list of activities is provided or made accessible to the IACUC members.

Opin. A DMR means that the IACUC Chair appoints a reviewer(s) to review an IACUC protocol. DMR can be conducted using IACUC software to facilitate the review process if the requirements in the PHS Policy (IV,C,2) are met. (See chapter 9 for additional information.) Many institutions have a submission webpage interface that allows researchers to submit new protocols, continuing reviews, amendments, protocol closures, and three-year protocol renewals electronically to the IACUC. Investigators can expect faster service with electronic submission. Some institutions use electronic submission methods to automatically streamline protocols destined for FCR whenever protocols involve unalleviated pain and distress, surgery, food or water restriction, paralytics, or prolonged restraint. All other protocols are sent to DMR. Some recent IACUC software is equipped for primary and secondary reviewers to electronically enter their review comments at the time of the review. Such review comments are fully visible to all members of the committee prior to IACUC meetings. The process allows designated reviewers to either make unanimous decisions or call for a full committee review. Certainly, using IACUC-software for DMR serves the intended purpose of speeding up the protocol review process. It may also decrease the overall workload relevant to full committee review.

NIH/OLAW and APHIS/AC have endorsed the use of electronic communications for certain IACUC functions based on 2006 guidance.[15] According to this publication, "The conveyance, by fax or email, of information such as the institutional Assurance; animal study proposals; agendas and minutes of meetings; institutional policies and standard operating procedures; reports, announcements or correspondence from oversight or regulatory agencies; and other matters related to the institutional animal care and use program for consideration and review by IACUC members would be regarded as appropriate." Institutions should exercise caution when electronic methods are used to prevent security breaches and should ensure that privileged and confidential information is communicated through a secure, encrypted site that has controlled access to IACUC members and staff.

7:20 What procedures are appropriate for using designated member review (DMR) after a full committee review (FCR)?

Reg. NIH/OLAW has provided guidance to PHS awardee institutions and IACUCs concerning the acceptable actions of an IACUC to meet the requirements of PHS Policy.[5] This guidance relates to IACUC required modifications to secure approval, and actions the IACUC may take for the use of DMR subsequent to FCR. APHIS/AC concurs with NIH/OLAW on this guidance.[5] The guidance indicates that the IACUC may take one of the following three actions:

1. If all members of the IACUC *are present* at a meeting, the committee may vote to require modifications to secure approval and have the revised research protocol reviewed and approved by DMR or returned for FCR at a future convened meeting. The DMR is designated by the IACUC Chair.
2. If all members of the IACUC *are not present* at a meeting, the committee may use DMR subsequent to FCR according to the following stipulations:
   a. All IACUC members agree in advance, in writing, that the quorum of members present at a convened meeting may decide by unanimous vote to use DMR subsequent to FCR when a modification is needed to secure approval. However, any member of the IACUC may, at any time, request to see the revised protocol and/or request FCR of the protocol.
   b. To conduct reviews by DMR subsequent to FCR, the institution should specify its intention to conduct reviews in this manner in its Assurance with OLAW. (IACUCs that newly elect to utilize a standard operating procedure for DMR subsequent to FCR should provide information about this program change to OLAW in the next Annual Report.)
   c. The DMR is designated by the IACUC Chair.
3. If all members *are not* present and the IACUC lacks written standard procedures as described above, the committee has the option to vote to return the protocol for FCR at a convened meeting or to employ DMR. If electing to use DMR, all members, including the members who are not present at the meeting, must have the revised research protocol available to them and must have the opportunity to call for FCR.[5] DMR may be conducted only if all members of the committee have had the opportunity to request FCR (PHS Policy IV,C,2). The DMR is designated by the IACUC Chair.

Opin. Federal regulations allow only two methods of protocol review, FCR and DMR. It is entirely up to the institution to decide whether DMR is to be used subsequent to FCR. The institution, in its NIH/OLAW Assurance, must include that they will be using DMR subsequent to FCR. Regardless of what is in the Assurance, the IACUC, under specific circumstances where animal welfare issues are of significant concern, should exercise the option of using FCR. Even though the Assurance allows DMR subsequent to FCR, under no circumstances does the guidance or the Assurance prevent the use of FCR. While DMR after FCR speeds up the review time, in both forms of review the threshold for animal welfare concerns is the same.

7.21 Can IACUC reports be endorsed by electronic methods and can signatures of members be obtained by using electronic methods?

Opin. The semiannual program review and inspection reports require IACUC approval (see 7:1). The use of IACUC software or any other methods of capturing electronic signatures for semiannual program review and inspection reports outside of the convened meeting are acceptable methods to approve program review and semiannual reports by IACUC members. FCR meetings may not be conducted via e-mail or polling, as this does not provide the opportunity for real-time communication. IACUC activities require documenting votes or verifying committee approval. They also may require signatures to be legally binding. Electronic methods of voting are conducted quite rapidly, but signatures are difficult to obtain by this method. Therefore, in a meeting using telecommunication, the Chair calls for a vote, the IACUC coordinator or a designated member records the vote, and then prepares and distributes the final document by mail to solicit original signatures for the permanent record. Some electronic methods can validate and authenticate the credentials of investigators and committee members. If such methods are available, the IACUC can develop an SOP to solicit signatures. In this author's opinion, if electronic signatures are used for IACUC administrative activities, the institution's NIH/OLAW Assurance should specify such use.

7:22 Must the minutes of IACUC meetings be recorded? If so, what precautions should be taken to maintain confidentiality, and what is the impact on Freedom of Information Act (FOIA) or pertinent state laws?

Reg. PHS Policy (IV, E,1,b) and the AWAR (§2.35,a,1) require records of IACUC meetings to be kept, whether the meeting is face to face or held alternatively using electronic communications (see 6:18; 6:19). Records must be compiled and maintained on file for three years after the study ends (AWAR §2.35, f; PHS Policy IV,E,2). Minutes must include records of attendance, activities of the committee, and committee deliberations (PHS Policy IV,E,1,b). The federal FOIA, at the present time, applies only to documents in the possession of a federal agency.

Opin. Meeting minutes should be written carefully with appropriate usage of words so that the lay public and any outside auditors would not be alarmed by reports obtained through the FOIA or pertinent state laws (see Chapter 26). The *Institutional Animal Care and Use Committee Guidebook*[6] cautions that the person assigned to prepare reports should be aware of the FOIA and their state's open record laws. The FOIA does not require a private organization or business to release any information directly to the public, whether it has been submitted to the government or not. However, information submitted to the government by such organizations is made available through a FOIA request. There are some exemptions to FOIA requests such as trade secrets and confidential business information.[20]

While recording meeting minutes satisfies regulatory needs, particular care must be taken to ensure proper usage of words since meeting minutes can, under certain circumstances, be obtained through a state's FOIA or any state open record laws (the federal FOIA is at present only applicable to documents in the possession of a federal agency) (see Chapter 26). The best practice is not to include extraneous information in the minutes so that the information is not taken out of context by any readers. The minutes should contain sufficient detail for an outside person to determine the discussions and conclusions on routine and controversial issues discussed by the committee. It is not required to specifically identify a member involved in the deliberations. Follow your state and federal laws and regulations on recording minutes to remain in compliance. Institutions should seek advice from their legal counsel regarding what information needs to be included in the minutes and reports.

It is important that each IACUC account for potential security problems of any privileged and trade-secret confidential information and release of confidential information by members as required by AWA §2157. Since minutes are often generated using computers, particular care should be taken to prevent unauthorized access to computer records or the IACUC database. The mandatory enforcement of passwords, controlled access, and encryption should be practiced to prevent unauthorized access. IACUC members should be advised to return copies of the minutes to the IACUC secretary once the committee ratifies the minutes. Unwanted and used copies of the minutes should be destroyed using appropriate methods.

Many IACUCs record meetings on audiotape. This raises the question as to whether these recordings should also be saved as part of recordkeeping. This is an internal policy unique to each IACUC whether to save or erase the tape. In general, these tapes also contain irrelevant discussions; therefore, it is customary for the IACUC to transcribe portions of the discussion that are relevant and then erase the tape. There are no regulations that explicitly deal with audio taping and archiving of such tapes. (See 7:22)

Surv. Does the federal Freedom of Information Act (FOIA) or a similar state FOIA or "sunshine" law impact on what is placed in your IACUC minutes? Select all that apply.

• We do not record minutes.	3/276 (1%)
• FOIA has no impact on our minutes.	101/276 (37%)
• FOIA may have an impact, but it depends on the topic being discussed.	99/276 (36%)
• FOIA seems to have a major impact on our IACUC minutes.	47/276 (17%)
• Sunshine law has no impact on our minutes.	55/276 (20%)
• Sunshine law may have a major impact on our minutes.	40/276 (14%)
• I don't know.	23/276 8*%)

7:23 What voting information should be recorded by your IACUC?

Reg. The AWAR (§2.35,a,1) and PHS Policy (IV,E,1,b) state that the minutes of IACUC meetings include records of attendance and activities of the committee and committee deliberations. The AWAR (§2.31,d,2) and PHS Policy (IV,C,2) do not allow an IACUC member who has a conflicting interest to review, approve, or contribute to the constitution of quorum. The AWAR and the PHS Policy do not state whether the names of individuals who voted, approved, opposed, or abstained need to be recorded.

Opin. It is a customary practice in many institutions to record the names of individuals attending a meeting. It is customary to record the actual number of individuals approving, withholding approval, or abstaining in a vote, but not the names of members.

7:24 Should each IACUC protocol be voted on individually?

Opin. In general, each protocol is unique, and it is a best practice to approve each protocol individually on a project-specific basis, taking into consideration a number of factors such as the aims of the study, consideration of alternatives, and minimizing pain and distress. AWAR (§2.31,d,2) state: "If full Committee review is requested for a proposed activity, approval of that activity may only be granted after review, at a convened meeting of a quorum of the IACCU, and with the approval vote of a majority of the quorum present." To ensure that IACUC members clearly approve of the specific activity in question, an individual vote is necessary.

7:25 Who records IACUC meeting minutes?

Opin. The institution should assume responsibility for selecting a person to record meeting minutes, based on input from the CEO, IO, the Office of Research, and the Chair of the IACUC. In some institutions, the IACUC administrator or his/her designee records the minutes. Other examples are the Chair, a recording secretary, or a member of the committee recording the minutes. It is important that people assigned to this task be knowledgeable of federal and other requirements and the institution's animal care and use program. If not, someone who is knowledgeable of IACUC regulations, policies, and procedures should read the minutes before minutes are presented to the committee or administrative officials.

Surv. In your institution who records meeting minutes?

• We have a designated administrator/coordinator who records the meeting minutes.	199/276 (72%)
• IACUC Chair records the meeting minutes.	25/276 (9%)
• A designated and rotating IACUC member will record the meeting minutes.	5/276 (2%)
• Meeting minutes are taped and transcribed by someone in the office of research administration or the IACUC Office.	31/276 (11%)
• Other*	16/276 (6%)

*Other answers reflected variations on the above choices but also included:

- The Attending Veterinarian
- Head of the animal facility

7:26 What should be included in the meeting minutes?

Reg. AWAR (§2.35,a,1) and PHS Policy (IV,E,b) state that the minutes require the records of attendance, activities of the committee, and committee deliberations. The minutes should include sufficient detail of the discussions and language that is generally understood by the non-scientific and lay members of the committee and people from outside agencies.

Opin. In the author's opinion, it is reasonable that the following items may be included in IACUC meeting minutes:

- Program reviews, and inspection reports distinguishing significant and minor deficiencies with plans and schedules for correcting each deficiency.
- Investigations of complaints and issues of non-compliance received from the public or research facility personnel
- Approval, conditions such as modifications and clarifications, withheld approvals, and suspensions of protocols
- Approval or withheld approval of proposed significant changes to protocols
- Training and certification guidelines approved by the committee
- Consultants' reports on protocols
- New policies or guidance voted on by the committee

7:27 How does your IACUC use the minutes of its meetings?

Reg. The IACUC minutes must be maintained for the duration of the project plus at least three years and be accessible for inspection and copying by authorized NIH/OLAW or other PHS representatives at reasonable times and in a reasonable manner (PHS Policy IV,E,2). The AWAR (§2.35,a,1) lists the IACUC minutes as a record and the AWAR (§2.35,f) require all records and reports to be maintained for a minimum of three years.

Opin. There are no guidelines for the way the recorded minutes of the meetings are to be used. In general, minutes are maintained as the official and primary record for NIH/OLAW, APHIS/AC, and AAALAC inspections and site visits to document animal welfare issues, pain or distress issues, and programs and facilities issues discussed at the meeting. Minutes can be used for bringing consistency to program management, review procedures, confirming or modifying policies, and general procedures of the committee.

7:28 IACUCs often review many protocols at a meeting. The concern is that those protocols considered at the beginning of the meeting may be reviewed more thoroughly than those at the end of the meeting. How can IACUCs avoid this problem?

Opin. The AWAR (§2.31,c,4-§2.31,c,7) and PHS Policy (IV,C) provide appropriate guidelines for reviewing research projects, but the guidelines do not provide a time schedule for discussing reports and research projects in a meeting. The *Guide* (p. 25) suggests that each IACUC should meet as often as necessary to fulfill its responsibility and provides topics that should be considered in the review of animal care and use protocols (*Guide*, pp. 24–34). Various practices used to ensure a thorough review of protocols, irrespective of the number of protocols and the agenda sequence, include lengthening the meeting time to accommodate review of all protocols in a single convened meeting without losing a quorum and postponing the review of some protocols while making sure that the review is not delaying proposed research. This happens in large institutions with many protocols to review. However, for small institutions with few protocols and those institutions that effectively and efficiently use DMR, this is not an issue.

7:29 Should an IACUC have a written policy on what constitutes a conflict of interest which may lead to a member being excused from voting?

Reg. The AWAR (§2.31,d,2) and the PHS Policy (IV,C,2) state that no IACUC member may participate in the IACUC review or approval of an activity in which that member has a conflicting interest (e.g., is personally involved in the activity), except to provide information as requested by the IACUC.

Opin. Conflicts of interest may arise in many ways. A common potential or apparent conflict of interest is when a PI comes from the same department as the IACUC member. It is not acceptable to serve as a designated reviewer in those circumstances as the member may have sole say on approving the protocol. In full committee consideration a member must make up her/his own mind whether to vote or abstain. However, if there is any apparent, perceived, or potential conflict of interest, the member should declare so before joining any review, discussion, or vote. If assigned as one of a team of reviewers, the member can either participate in the review or ask to be relieved from that duty. A member from the same discipline can often provide the best review of a protocol but may then have to abstain from any vote.

The responsibility of determining if a member has a conflict with the protocol to be reviewed may fall on the Chair. Therefore, it is a best practice for the Chair to start the meeting by asking members whether they have a conflict with any of the protocols or animal issues that will be discussed. This will help determine how the agenda needs to be navigated while paying attention to the quorum.

The overall guiding principle is that members should act professionally and impartially in considering IACUC matters. Likewise, resolution of a conflict of interest may arise in many ways. An SOP to resolve the conflict is critical for the growth of a research enterprise. The SOP should describe the procedures for resolving conflicts arising from the investigator as well as a committee member. Investigator-initiated conflicts arise at the time of submitting a protocol. Here the investigator may request that a certain member be excluded from the review of the protocol, providing evidence that substantiates the claim that a conflict exists with that member. If that member's expertise is essential for the review, the IACUC can exercise the option of using a consultant for the review. If a member thinks that he/she has a conflict with a protocol, the member should notify the IACUC Chair and not participate in the IACUC review or approval except to provide information when requested by the committee. Potential conflict is implicit if the member has competing research, access to funding information which may give an unfair advantage to the member, and occasionally personal bias against an investigator's research.

7:30 Should the IACUC have a policy on maintaining confidentiality?

Reg. USDA/APHIS/AC and OLAW offer the following guidance regarding confidentiality as it relates to IACUC matters.

The AWA's provision on protecting confidential information is broad and applies to IACUC members regardless of whether a principal investigator elects to share information about his/her research. IACUC members must maintain confidentiality and not disclose trade secrets or proprietary information about activities and species regulated under AWA, or they risk exposure to penalties and damages for disclosing any such information. Penalties include removal from such Committee, and a fine of not more than $1,000 and imprisonment of not more than one year, or if such violation is willful, a fine of not more than $10,000 and imprisonment of not more than three years (§ 2157 (c)). The principal investigator and/or research facility may also pursue action to recover damages incurred as a result of the violation (§ 2157 (d); ref. 1).

An institution is required to ensure all persons involved in animal care and treatment are qualified and provide training where necessary (§ 2.32). Considering this, IACUC member training must ensure the attendees understand their responsibilities under the AWA, including the requirement to maintain confidentiality.

Public Law 99-158, "Animals in Research," requires that research institutions not disclose publicly trade secrets or commercial or financial information that is privileged or confidential (Sec. 495 (e)). Therefore, the Public Health Service (PHS) requires that all material reviewed by the IACUC be treated as confidential. The institution should ensure that new IACUC members understand and meet their responsibility to maintain confidentiality.

Opin: It is important to protect the confidential material pertinent to any protocol reviewed by the IACUC.

The IACUC should take the general approach that any material presented to the IACUC be considered confidential information and that the members must, therefore, ensure confidentiality of the material provided to them. Best practices to maintain confidentiality are:

1. Do not disclose any confidential material or data that are specific to an investigator to non-IACUC members.
2. Do not discuss with third parties any confidential material or information such as protocol reviews, matters of noncompliance, suspensions, and opinions to third parties without the explicit written permission from the IACUC Chair or institutional authorities.

Any compromise or breach of confidentiality may directly affect the integrity or the intellectual property of the researcher and the institution. Therefore, some institutions choose to have IACUC members sign confidentiality agreements when beginning IACUC service and, in some instances, annually. This allows documentation, in writing, an acknowledgement by individuals that they will abide by the institution's expectation for confidentiality. The acknowledgement document requires members to not disclose or utilize directly or indirectly any confidential or proprietary information belonging to the researcher, as well as any confidential information discussed at an IACUC meeting. In this author's opinion, it is important that institutions develop a policy on the definition of confidentiality, clearly describing how confidentiality should be maintained with respect to the research protocol or the data without compromising the integrity, the intellectual property, or other proprietary information of the researcher or the institution. The policy might detail proper methods to dispose of confidential material after meetings; specific circumstances under which confidential information can be discussed with other researchers or their collaborators; circumstances requiring the principal investigator's explicit permission to discuss confidential information with outsiders; and the IACUC's position on sharing information for purposes of providing examples to other researchers to help them understand IACUC policies and review processes. Some institutions may combine conflict of interest and confidentiality information into a single policy or SOP.

REFERENCES

1. Committee for the Update of the Guide for the Care and Use of Laboratory Animals. 2011. *Guide for the Care and Use of Laboratory Animals*, 8th ed., 17–23. Washington, D.C.: National Academies Press.
2. Office of Laboratory Animal Welfare. 2011. *Semiannual Report to the Institutional Official*. Accessed January 7, 2024. https://olaw.nih.gov/resources/documents/ioreport.htm.
3. Wolff, A. 2002. Correct conduct of full-committee and designated-member review. *Lab Animal (NY)* 31:28–31.
4. Office of Laboratory Animal Welfare. 2012. *Frequently Asked Questions B.3. Must Certain Members Be Present in Order to Conduct Official Business?* Accessed January 7, 2024. https://olaw.nih.gov/faqs#/guidance/faqs?anchor=50296.
5. Office of Laboratory Animal Welfare. 2009. *Guidance to IACUCs Regarding Use of Designated Member Review (DMR) for Animal Study Proposal Review Subsequent to Full Committee Review (FCR)*. Notice Number: NOT-OD-09-035. Accessed January 7, 2024. https://grants.nih.gov/grants/guide/notice-files/NOT-OD-09-035.html.
6. Office of Laboratory Animal Welfare. 2002. *Institutional Animal Care and Use Committee Guidebook*, 2nd ed., 32–33. Bethesda: National Institutes of Health. Accessed April 28, 2023. https://olaw.nih.gov/guidance/ARENA-OLAW-IACUC-guidebook.htm.
7. Ellis, G., and N.L. Garnett. 1997. *Maintenance of Properly Constituted IACUCs*. OPRR Reports, No. 97–03. Accessed January 7, 2024. https://olaw.nih.gov/guidance/articles/dc97-3.htm.
8. Office of Laboratory Animal Welfare. 2012. *Frequently Asked Questions B.4. Is a Certain Level of Meeting Attendance Required of IACUC Members?* Accessed December 1, 2022. https://olaw.nih.gov/faqs#/guidance/faqs?anchor=50297. Office of Laboratory Animal Welfare. 2012. *Frequently Asked Questions B.5. What Is a Quorum and When Is a Quorum Required?* https://grants.nih.gov/grants/olaw/faqs.htm#IACUC_5.
9. American College of Laboratory Animal Medicine, Public Statements. *Adequate Veterinary Care*. Accessed April 28, 2023. http://www.aclam.org/Content/files/files/Public/Active/https://www.aclam.org/media/576c987a-4380-4e7c-b439-4892ac7ab6ce/GhFYxA/ACLAM/About%20Us/Position%20Statements/ACLAM%20Position%20Statement%20-%20Adequate%20Veterinary%20Care.pdfposition_adeqvetcare.pdf.
10. NIH Grants Policy Statement-Dec 2022. Accessed January 7, 2024. https://grants.nih.gov/grants/policy/nihgps/HTML5/introduction.htm.
11. Office of Laboratory Animal Welfare. *Worksheet for Applications Involving Animals (VAS)*. Accessed April 28, 2022. https://olaw.nih.gov/sites/default/files/VASchecklist.pdf. Office of Laboratory Animal Welfare. 2002. *Institutional Animal Care and Use Committee Guidebook*, 2nd ed., 85. Bethesda: National Institutes of Health. Accessed January 7, 2024.
12. Animal and Plant Health Inspection Service. 2011. *Animal Care Resource Guide*. Policy 3. Pre- and Post-Procedural Care. https://www.aphis.usda.gov/animal_welfare/downloads/policy/Policy%203%20Final.pdf.
13. Office of Extramural Research. 2001. *Guidance Regarding Administrative IACUC Issues and Efforts to Reduce Regulatory Burden*. Notice NOT-OD-01-017. Accessed January 7, 2024. https://grants.nih.gov/grants/guide/notice-files/NOT-OD-01-017.html.

14. Garnett, N., and S. Potkay. 1995. Use of electronic communications for IACUC functions. *ILAR Journal* 37: 190–192. Accessed January 7, 2024. https://olaw.nih.gov/guidance/articles/ilar95.htm.
15. Office of Laboratory Animal Welfare. 2006. *Guidance on Use of Telecommunications for IACUC Meetings under the PHS Policy on Humane Care and Use of Laboratory Animals.* Accessed January 7, 2024. https://grants.nih.gov/grants/guide/notice-files/not-od-06-052.html.
16. Office of Laboratory Animal Welfare. 2012. *Frequently Asked Questions D.19. May an IACUC Use Designated Member Review (DMR) to Review an Animal Study Protocol Subsequent to Full Committee Review (FCR) When Modifications Are Needed to Secure Approval?* Accessed January 7, 2024. https://olaw.nih.gov/faqs#/guidance/faqs?anchor=50340.
17. Flexibilities for Assured Institutions for Activities of Institutional Animal Care and Use Committees (IACUC) Due to Covid-19 Notice NOT-OD-20-088.
18. OLAW Waivers to Conduct Semiannual Facility Inspections Due to COVID-19 Will Expire October 1, 2022. Accessed January 7, 2024. https://grants.nih.gov/grants/guide/notice-files/NOT-OD-22-207.html.
19. Guidance on Flexibilities for Conducting Semiannual Inspections of Animal Facilities. Accessed January 7, 2024. https://grants.nih.gov/grants/guide/notice-files/NOT-OD-21-164.html.
20. Freedom of Information Act. *5 U.S.C. § 552, As Amended by Public Law No. 104-231, 110 Stat. 3048.* Accessed January 7, 2024. https://www.justice.gov/oip/foia_updates/Vol_XVII_4/page2.htm.

8

General Format of IACUC Protocol Forms

William G. Greer and Lauren Danridge

Introduction

IACUC review and approval of proposed animal activities was mandated by federal law for most institutions using animals in research, teaching, and testing by the passage of an amendment to the AWA known as the Improved Standards for Laboratory Animals Act (Public Law 99-198) and by a component of the legislative reauthorization for the NIH, known as the Health Research Extension Act of 1985 (Public Law 99-158). In advance of this requirement, and as early as 1980, several academic institutions already had begun to implement the basic concepts of the IACUC review and approval of animal activities. The mechanism(s) by which IACUCs obtained information on proposed animal activities has developed over the years into what is commonly known as the IACUC protocol. In the context of the contemporary standards for protocol review in most institutions, the initial forms used were rudimentary, and the information retrieved in these pioneering efforts was extremely scant and generally would now be insufficient for the issuance of an IACUC approval. Nevertheless, these initial efforts were important because they introduced the setting, players, and ethos of an evolving process that eventually would be adapted and embellished at diverse institutions across the country as the federal mandates for protocol review took effect.

Many of the factors that influenced the early approaches to protocol review continue to be important today. These have shaped the character and content of the forms used in the protocol review process according to institutional preference and stimulated organizations to meet regulatory and ethical considerations with clarity, consistency, and completeness of all documents. For example, most institutions and investigators want the protocol review process to be objective, consistent, and efficient. This orientation fosters the support of research investigators and IACUC members who must commit precious time to fulfill this institutional requirement. However, the actual protocol review forms used to facilitate this process are markedly different from institution to institution. Institutional philosophy, the types of research conducted, pragmatic considerations related to administrative and regulatory detail, and many other ancillary factors can influence the format of the IACUC protocol review form.

The purposes of this chapter are to examine the various strategies used by institutions in the information collection phase of the protocol review process and to highlight the strategies that appear to be the most successful and widely adaptable. The chapter also discusses alternative approaches that have proved to be well suited to particular institutional environments. Information for the preparation of this chapter was acquired through insights into the practices of numerous institutions from conducting of AAALAC site visits or reviewing other materials related to AAALAC activities, from over 300 institutions during the conduct of IACUC Administrators Best Practice Meetings (see Chapter 5) and from the authors' personal experiences as IACUC members and/or administrators at more than a dozen institutions.

A well-designed protocol form can assist investigators in their efforts to provide the clear, concise, and comprehensive information necessary for the review process, and can also serve as a resource for programmatic expectations of various aspects of animal use (e.g., institutionally specific procedures for aseptic surgery). It is critical to the smooth functioning of the IACUC. Most IACUCs have revised their protocol forms several times since the inception of the protocol review process in response to new or augmented regulatory requirements, the need to improve the quality and detail of the information provided, or the necessity to enhance the retrieval and sharing of information contained in these documents. While the protocol template should be considered a living document and an IACUC should not be reluctant to change the protocol review format for due cause, format changes are likely to be disruptive and burdensome during a transition period. For this reason, format changes should be carefully planned and coupled with investigator outreach and education efforts and limited to the minimum amount of information required to successfully complete the review.

8:1 Should there be a standard protocol form for submission to the IACUC?

Opin. The large majority of institutions require investigators to submit their animal care and use protocols to the IACUC for review on an institutionally developed, standardized form. This approach is usually regarded as an essential first step in the fulfillment of the IACUC's mission to establish a review process that is perceived as unbiased, consistent, efficient, and, under optimal conditions, user-friendly. Most institutions have PIs who file multiple studies with the IACUC, and the use of a standardized form allows these individuals to develop a familiarity with the type of information

DOI: 10.1201/9781003272861-8

required by the IACUC to ensure successful completion of the review process. Most institutions have found that the quality of information provided by PIs using standardized protocol forms improves over time. Nevertheless, when changes are incorporated into existing forms to meet newly defined needs, a period of coaching and re-acclimatization may be necessary. Additionally, standardization is generally regarded as helpful and critical to the IACUC members in meeting their review obligations. The compartmentalization of information into predictable areas enables IACUC members to focus their review and discussion on sensitive areas with greater alacrity and facilitates the retrieval of information pertinent to any subsequent monitoring and compliance activities.

It was quite unusual for institutions to conduct IACUC protocol review without the use of a standardized form, but this practice was used by some institutions at the time of the first edition of this book. The authors are not aware of any institutions that have continued to use this method. It was used mostly by academic institutions with smaller animal care and use programs and a limited research profile, usually serving ten or fewer PIs. In those instances, the IACUC reviewed the PI's grant application and a supplemental narrative concerning specific areas deemed necessary for IACUC review. The narratives were responsive to a list of queries provided by the IACUC and veterinarian to ensure that all the federally mandated aspects of protocol review were covered. While it is conceivable that this approach could occasionally produce a reasonably integrated, relevant, and coherent document for the IACUC review process, in practice this rarely happened. In addition to requiring IACUC members to demonstrate a unique flexibility and commitment, this approach to protocol review challenged IACUC members to wade through much extraneous information and maintain a running inventory of the quality and location of the investigator's responses (or non-responses) to important issues. In order for the information gleaned from the grant application and protocol at the time of the review to remain conveniently accessible to the IACUC for later use, a synopsis of the review becomes necessary. For these reasons, a free-form protocol IACUC review process will likely remain limited, if indeed it continues, to animal care and use programs in small institutions imbued with a sense of collegial courtesy, reciprocity, and the good faith principle.

8:2 Would it be possible for oversight agencies, such as the USDA/APHIS, and PHS, to endorse a universal protocol form (UPF) that could be used across institutions?

Reg. The Guide for the Care and Use of Laboratory Animals (*Guide*) recognizes the use of the "IACUC Protocol"; however, there are no federal mandates that specifically identify the method(s) by which the IACUC must collect the needed information—only that certain information is needed for the committee to successfully conduct a review.

Opin. The short answer is yes; OLAW does, offer a sample "animal study proposal" and notes that it "is **not** required and is provided for the convenience of IACUCs at Assured institutions. Sections may be added, deleted, or modified to meet the needs of individual programs."* Such a universal protocol template (UPT) could provide institutions with education on the minimum information required for IACUCs to meet the expectations of the federal mandates for IACUC review. The value of a UPF is the inherent guidance that can help institutions ensure the only information gathered through the protocol template is that needed for the IACUC to successfully review and approve proposed animal activities. The most optimal format for an IACUC application form is one that satisfies the federal mandates and gathers the information that can only be acquired from the Principal Investigator (PI), such as animal activities details, locations animals will be utilized, and funding information.

In practice, and as OLAW has pointed out in their sample animal study proposal, the protocol form is expected to be adjusted to meet the needs of individual programs. The type of information requested in the IACUC protocol depends on each institution's research portfolio, the structure of their compliance units, and other institutionally specific requirements. For example, information required by an Institutional Biosafety Committee (IBC) may be included in the IACUC protocol and used by the IBC for review, while other institutions maintain separate IBC application forms. Similarly, information required by the Safety officer for risk assessment of the use of hazards in animal activities may be included in the IACUC protocol application or could be obtained outside of the IACUC protocol. Some institutions may utilize IACUC- approved, standardized procedure descriptions that are referenced within the protocol application, and other institutions may include these descriptions or SOPs as part of the IACUC approved documents (e.g., as appendices to the approved protocol). Some institutions may require very specific details of the research activities, such as which specific strains of mice will be used, while others may require information on only genetically modified animals and expected negative phenotypes. Still, there are organizations, such as the U.S. Department of Veterans Affairs (VA), that require a single, standardized protocol to be used, without modification, across all campuses.

Another variable affecting the content of the IACUC protocol form is the institutionally specific

* https://olaw.nih.gov/resources/documents/animal-study-prop.htm

philosophy toward the oversight and, dare we say, the amount of trust that the Animal Care and Use Program puts in the PIs and their animal users. This philosophical approach to oversight will dictate the level of detail required in the protocol (e.g., needle size, that the surgical site will be shaved) and the amount of flexibility in procedural descriptions (e.g., exact dates of blood collection vs. up to 1% of the animal's body weight in blood in a 14-day period). Another institutionally specific philosophical variable relates to the reduction of burden, for example many institutions require the application of USDA Pain and Distress category assignments for species that are not USDA-regulated. Protocol forms will also vary across institutions due to the degree of required justifications for topics such as deviations from institutionally specific policies or guidelines (e.g., the use of an analgesic not included in the veterinarian's list of the most used analgesics), for deviations from the *Guide* (e.g., single housing), and overall animal numbers (provision of an estimate vs. lengthy statistical narratives). Similarly, while some institutions prefer to have policies, guidelines, and other helpful resources as stand-alone documents, other institutions may prefer to include some of the educational information and ACUP expectations within the IACUC protocol form, thereby reducing the number of these standalone documents.

In addition to these programmatic influences, the inherent flexibility in the federal mandates provides additional opportunities for institutionally specific questions or formats. For example, the species utilized and type of animal activities conducted can influence the protocol format, which is most pronounced in the differences between laboratory and field research.

8:3 What general information is necessary for the IACUC to review a protocol adequately?

Opin. Several authoritative documents reiterate the basic components of an animal care and use activity that should be assessed by the IACUC during the protocol review process. These documents include the AWAR, the PHS Policy, U.S. Government Principles (VI to VIII), and the *Guide* (pp. 25–35). The AWAR impose specific review requirements for regulated animal species, and many institutions have chosen, as a matter of policy, to extend these requirements to all vertebrate species. Although the practice appears to be becoming rare, other institutions have maintained a dichotomy in the review criteria based upon exclusions provided in the definition of the term *animal* in the AWAR (§1.1) (see 13:1). The authors endorse the following list of elements delineated in the *Guide* as a complete, comprehensive, and convenient resource for achieving a sound protocol review process serving the researcher and the IACUC:

- Rationale and purpose of the proposed use of animals (i.e., "Procedures involving animals should be designed and performed with due consideration of their relevance to human or animal health, the advancement of knowledge, or the good of society"*)
- A clear and concise sequential description of the procedures involving the use of animals that is easily understood by all members of the committee
- Availability or appropriateness of the use of less-invasive procedures, other species, isolated organ preparation, cell or tissue culture, or computer simulation
- Justification of the species and number of animals proposed
- Whenever possible, the number of animals and experimental group sizes should be statistically justified (e.g., provision of a power analysis)
- Unnecessary duplication of experiments
- Non-standard housing and husbandry requirements
- Impact of the procedures performed on the animals' well-being
- Appropriate sedation, analgesia, and anesthesia (indices of pain or invasiveness might aid in the preparation and review of protocols)
- Conduct of surgical procedures, including multiple operative procedures
- Post-procedural care and observation (for example, inclusion of post-treatment or post-surgical animal assessment forms)
- Description and rationale for anticipated or selected endpoints
- Criteria and process for timely intervention, removal of animals from a study, or euthanasia if painful or stressful outcomes are anticipated
- Method of euthanasia or disposition of animal, including planning for care of long-lived species following study completion
- Adequacy of training and experience of personnel in the procedures used, and roles and responsibilities of the personnel involved
- Use of hazardous materials and provision of a safe working environment
- Weighing study objectives against potential animal welfare concerns when, for example, the animal activities include multiple survival surgical procedures, physical restraint,

* https://olaw.nih.gov/resources/documents/animal-study-prop.htm

food and fluid regulation, and use of non-pharmaceutical-grade chemicals and other substances

Surv. What information does your IACUC require regarding personnel safety to be included in the protocol? Select all that apply.

- Description of PPE to be worn 175/266 (66%)
- Method for scavenging waste anesthetic gas 120/266 (45%)
- Animal biosafety level at which work is to be conducted 211/266 (79%)
- Description of animal handling method 200/266 (75%)
- Other* 67/266 (25%)

*Other:

- General methods for risk mitigation
- Training and experience related to safe handling of hazardous agents
- Standard operating procedures (SOPs)
- Description of hazardous waste disposal
- Any deviations from standard practices or PPE for surgical procedures
- Evidence of IBC/safety committee review and/or approval
- Specification of the hazardous agent(s) to be used
- Evidence of enrollment of personnel in the occupational health program
- Safety data sheet (SDS) for any hazardous materials
- Contact information for personnel working with hazardous agents
- Description of signage/hazard identification to be used
- Location of the use of hazardous agents

Surv. Does your IACUC require descriptions of all procedures or are references to descriptions provided in other documents (e.g., standard operating procedures) acceptable in lieu of detailed descriptions?

- We require detailed descriptions. 119/266 (45%)
- We require references to descriptions in other documents. 5/266 (2%)
- We accept either detailed descriptions or references to descriptions in other documents. 136/266 (51%)
- We do not expect protocols to include or refer to documents containing descriptions for procedures. 6/266 (2%)

8:4 Should the standard protocol form be used to gather other information needed for the institution to ensure program compliance with federal guidance documents on other topics?

Opin. Documents governing the development and operation of animal care and use programs typically identify two responsible parties: the IACUC and the institution. By law and policy, the IACUC is typically responsible for those activities directly related to the proposed care and use of vertebrate animals for research, teaching, or testing. Although the IACUC must review and assume responsibility for the entire animal program, the institution carries the ultimate responsibilities for establishing a functional occupational health and safety program, ensuring federal funding to support animal research is administered according to the terms and conditions of the grant or contract, establishing an effective training program, and ensuring complete and accurate records are maintained.

The protocol review form and review process are convenient and adaptable for capturing the information needed by the institution in fulfillment of these other responsibilities. For example, a submission to the IACUC may not be considered complete unless the PI's submission includes a copy of his/her NIH grant application to allow the administrative office to complete the protocol and grant congruency process.[2] Institutions also can use the submission to ensure everyone listed on the protocol has completed federally required training in humane animal care and use. The submission form may include a question that asks the PI to verify the current training status of laboratory personnel and approval by other institutional compliance units, such as the IBC and radiation safety. Final protocol approval may not be given until successful completion of the training and approval by, for example, the IBC and radiation safety committee has been verified. In addition, organizations often facilitate the distribution and completion of personal risk assessment forms (e.g., medical surveillance) in conjunction with the protocol review process to aid the program's occupational medicine service.

8:5 Is there an optimal outline for the content of the IACUC protocol template form?

Opin. A central question in the protocol review process is how to optimize information retrieval from the areas identified in 8:1 and 8:4, in order for the IACUC protocol review process to be sufficiently rigorous to ensure high-quality animal care, safety, and other considerations and still facilitate approval on a first attempt. There is no apparent consensus on this matter as evidenced by the many different approaches taken by IACUCs, which, in turn, are mirrored by the variety of forms used by different IACUCs. The forms used by institutions generally fall into the following categories:

- Those that state a requirement or use a question (regarding the areas defined in 7:3) to elicit a free-form narrative response and consolidate all relevant information onto a single form
- Those that use a free-form narrative approach but are divided into modules allowing the form to be tailored to the particulars of the submission (e.g., those not involving surgery would delete the sections aligned with this activity)
- Those that emphasize a checklist and fill-in-the-blank approach whenever possible, either in the modular or in the consolidated format: forms generally intended to expedite completion and review of straightforward protocols. The use of specific protocol templates developed to accommodate specific research activities common in the institution are also aligned with this approach.
- Those that include references or assurance of adherence to policies, guidelines, SOPs, and/or IACUC approved procedures, without repeating the details of the activities in the protocol

Institutions using a free-form narrative for protocol review simply ask questions related to each of the items in 8:3, eliciting the investigator's response. The protocol form includes questions covering all areas, and answers are typically provided only for the segments of the protocol that apply. Thus, in studies involving minimal animal procedures, the majority of the form may not be used. However, in protocols involving complicated studies with extensive animal use and entailing specialized expertise (e.g., surgery, post-procedural care, or the use of biohazardous agents), all information deemed necessary for the IACUC review is provided. Most institutions realize that investigators may not recognize the kind of qualitative and quantitative information necessary for protocol approval without assistance, even though this subject may have been extensively covered in institutional training and education activities. Therefore, many institutions using this format generally make an effort to enhance the PI's appreciation for the "hot button" issues or nuances particular to each area of review. This is accomplished through the distribution of guidelines for the completion of a protocol review form, for example (mock) protocols, IACUC position statements, regulatory announcements on issues, or articles emphasizing the IACUC's interest or perspective on a topic. In a variant of this approach, some IACUCs incorporate this type of information into sidebar or parenthetical comments in the protocol form to channel the investigator in the right direction. As noted in section 8:2, the most optimal format for an IACUC application form is one that requests all the information required by the federal mandates and gathers the information that can be acquired only through the PI such as animal activities details, locations animals will be utilized, and funding information.

Institutions using the modular version of the free-form narrative protocol work toward paperwork reduction by capitalizing on the fact that many studies only involve a limited, and somewhat predictable, spectrum of activities around which the "core" protocol form is designed. This core form is linked to appendices that are designed for specific topics such as surgery and postoperative care, prolonged restraint, multiple major survival surgery, other procedures potentially involving significant pain and distress for laboratory animals, and the use of biohazards. The psychological benefits attendant to the reduction in paperwork for investigators and IACUC members are additional reasons given by institutions that use this approach.

Some IACUCs use checklists and "fill-in-the-blanks" formats as the predominant modes of information collection to expedite completion of the forms through a series of discrete, informative inquiries into aspects of the protocol. Some areas of inquiry and some research settings are better suited to this approach. For example, organizations having a narrow research mission in which many of the studies follow a predictable format and involve techniques that are well described and repetitive often use this format. In some institutions specific protocol templates are developed for specific activities. For example, some institutions create templates that focus on the use of animals in educational exercises if they have large teaching programs, or on breeding colony husbandry and management if they maintain substantial populations of genetically modified animals. Pharmaceutical companies and commercial laboratories involved in testing compounds in conformance with federal regulations often have extensive standard operating procedures for animal care and use. Many aspects of their IACUC protocol forms are addressed simply by citing these procedures or other documents used in project development if these primary sources are relevant and have been reviewed and approved by the IACUC. Some larger animal care and use programs, with diversified research endeavors, have incorporated this approach into their protocol form to service research activities that can benefit. A significant pitfall in this approach, when it is applied to complicated studies, is that it tends to disperse the information about the different types of procedures being performed on the animals and can limit the reviewer's appreciation for the temporal sequencing and impact of procedures on the animals. Flowcharts illustrating the various uses of animals in an application often can remedy this problem. A second problem is that these forms tend to be lengthy because they are structured to capture numerous small bits of information. If the forms are poorly worded, the information gathered may be voluminous but disconnected and confusing. Also, when SOPs are changed in the course of a study, the PI must assure that research personnel are following the SOP as currently written to achieve compliance with the IACUC-approved protocol.

Many institutions have blended the narrative, checklist, and fill-in-the-blank strategies in their protocol review efforts to optimize these efforts. This is the approach to protocol review that the authors deem optimal for most animal care and use programs. The experience of the IACUC often leads to identifying portions of the protocol form that investigators repeatedly do not answer correctly or adequately. These areas can be transformed into a checklist format or narrative blocks with supplemental instruction on the appropriate content of the blocks. For example, in the segment on the search for alternatives to painful procedures, the form might include a list of commonly used databases, along with prompts for the keywords used and the date of the search. As another example, in the discussion on the provisions for postoperative care, the questions might stimulate responses for each important aspect of care (e.g., physiological monitoring, analgesic therapy, fluid and antibiotic therapy, recordkeeping). Similarly, in polyclonal antibody production protocols, the questions could channel the investigator into methods that conform to IACUC guidelines.

An approved IACUC protocol form should be regarded as a component of the investigator's obligation to the ethical and humane care and use of animals. It is hypothetically a contractual agreement with the institution to conduct animal care and use activities in conformance with applicable laws, regulations, and sound clinical and scientific practices. In summary, any approach, or mix of approaches, has merit if it satisfies the investigators and IACUC members and produces a well-defined contract through a thorough IACUC review and approval process.

8:6 How specific should the information on the protocol form be to secure an approval? Is it acceptable for an IACUC to approve a generic protocol?

Reg. For PHS Policy purposes, IACUC approvals must ensure that the animal activities will be conducted in accordance with federal mandates (e.g., to address the required review criteria at PHS Policy IV,C,1). The use of a more generic protocol, with project-specific amendments to cover significant changes, is one method that meets this requirement.

Opin. There are many investigators who mistakenly view the IACUC protocol review and approval function only as an approval of the animal procedures. Thus, if they use the same animal procedures over an extended period, they believe that working under a generic approval is appropriate. However, most IACUCs have not acquiesced to this preference because protocol approval is not simply an approval of procedures. In general, and as identified in 8:1–8:4, IACUCs require PIs to submit very specific information on their protocol forms to allow the IACUC to link the scientific objectives with the justification for the proposed animal activities. Accordingly, in the few IACUCs that do use this generic protocol approach, the PIs are required to submit supplementary information for each study to define its objectives, assure the validity of the animal model and the lack of alternatives, and justify the number of animals requested. Information pertaining to any new animal welfare or personnel safety concerns that may have been introduced by new variables in the studies proposed also must be provided and reviewed.

There are times when the approval of a generic protocol, abiding by the contingencies stated, may be appropriate and efficient. An example is a protocol with a standard set of procedures that is used repetitively to test compounds and materials in a particular animal model with recognized validity of the study of a class of compounds. The only variable in these protocols is the material or compound under investigation. Also, scientists may reduce some of the specificity in their protocol details if this is acceptable to the IACUC based upon their assessment that satisfactory animal welfare outcomes will not be compromised. Since PIs are expected to conduct the procedures approved by the IACUC as written, they may afford themselves some latitude when preparing their submissions for committee review if their IACUC is receptive to this approach. For example, if an IACUC approved blood collection from a dog at 2 milliliters every three hours for a 24-hour period, a fixed and discrete expectation has been established in the protocol. In this example, if the PI later determined that 3 milliliters of blood was better suited to thoroughly complete their laboratory assay and provide for sample back-up storage, a protocol modification would need to be submitted and approved before additional blood could be collected. Had the PI prepared his submission and been approved by the IACUC to allow for some minor adjustment of study parameters (i.e., bleeding intervals and blood volumes) to optimize the data and assay by requesting permission to collect up to 4 milliliters of blood at 12 or fewer times in a 24-hour period and by providing the proper scientific rationale for the request, he/she would have had the flexibility to proceed without a protocol modification. In addition, some institutions may permit the approval of a class of drugs as opposed to an exhaustive list of all, for example, specific anti-depressants that will be examined in a mouse model of anxiety.

8:7 Should the protocol form be used as a vehicle for the continued instruction of investigators in aspects of the PHS Policy, AWAR, and ethical considerations in animal research?

Opin. As discussed in sections 8:1, 8:3–8:5, most IACUCs use standardized protocol forms for all types of animal activities and that refer to the PHS Policy, AWAR, and ethical issues in one or more areas. Some include a statement signed by the investigators assuring that all the activities described will be conducted in conformance with the aforementioned regulations. These references to PHS Policy and the AWAR serve to remind PIs of the regulatory underpinnings of the protocol review process. They also emphasize the contractual nature of an approved protocol. Similarly, many IACUC protocol forms attempt to evoke the discussion of ethical considerations in response to a variety of experimental practices such as prolonged restraint, food and water restriction, use of aversive stimuli, surgery, multiple major survival surgery, disease induction, post-procedural care, and the selection of study end points. While using the protocol form in this fashion to give these issues continued visibility to the investigators is well intended, in the authors' opinion that the use of the protocol form as a significant tool for instruction on ethical considerations of the care and use of animals is a superficial approach with little productivity. Most IACUCs would likely agree that investigators tend to focus their attention on filling out protocol forms for the purpose of securing of approval and not to be educated about potentially complicated matters.

8:8 Could inclusion of a questionnaire on the form on animal welfare considerations be useful to the IACUC?

Opin. Fewer than half of the IACUCs, with which the authors are familiar, have attempted to evaluate animal welfare considerations in a separate section or questionnaire. When this is done, it usually is limited to querying the investigator about the allocation of animals to the various APHIS/AC pain categories (if used) and to aid the PI's understanding of the AWAR requirement for the search for alternatives and scientific justification for conducting APHIS/AC Category E animal studies. (See 17:4.) Even at this basic level, most institutions encounter occasions in which the IACUC does not agree with the PI's assessment or the IACUC is not able to reach a full consensus on an animal welfare matter. On most IACUC protocol forms, this type of information is integrated into the sections involving the discussion of procedures performed and the overall management of the animal model. If the information given in these areas is insufficient for the IACUC to identify and assure the PI's attention to animal welfare issues, changes in the protocol form or the training of investigators is warranted. It seems very likely with increased attention given to animal welfare issues in the *Guide* (see 8:3), the statement in the *Guide* (p. 27) that the "IACUC is obliged to weigh the objectives of the study against potential animal welfare concerns," and AAALAC's discussion of the "harm-benefit" analysis[3] that institutions will be prompted to use the protocol form to capture animal welfare information. It is important to note that the assignment of USDA-regulated species into the Categories of B, C, D, or E is, ultimately, a retrospective assessment of what the animal experienced and is not to be limited to what is expected to occur at the time of IACUC protocol preparation and review. Therefore, regardless of the method of data capture at the front end of project design, animal welfare considerations are not stagnant, and institutions should have a mechanism for updating protocols to address issues that may arise as the research is conducted (e.g., amending the protocol to adjust humane endpoint criteria and responses to unexpected outcomes).

8:9 Should IACUC protocol forms require the PI to include a statement written in language understandable to the lay public about the purpose and relevance of the proposed studies?

Reg. This inquiry has become standard on most IACUC protocol forms. The basis for making this request is U.S. Government Principle II, which states: "Procedures involving animals should be designed and performed with due consideration for relevance to human or animal health, the advancement of knowledge, or the good of society." This is also strengthened through the 8th edition of the *Guide* that specifies protocols include "clear and concise descriptions" that non-scientists and unaffiliated members of the IACUC understand.

To be properly constituted, an IACUC must have at least one member "whose primary concerns are in a nonscientific area"* and who provides "representation for general community interests in the proper care and treatment of animals."† It follows, then, that for the non-scientific/community member to participate in the evaluation of the

* Public Health Service. Policy on Humane Care and Use of Laboratory Animals – Frequently Asked Questions. Applicability of the PHS Policy, Question No. A.6. Does the PHS Policy apply to animal research that is conducted in the field? (US Department of Health and Human Services, Bethesda, MD, USA). online at: https://olaw.nih.gov/faqs#/guidance/ faqs?anchor=question50290.

† Animal Welfare Act Regulations 9 CFR Part 1.1 https://www.ecfr.gov/current/title-9/chapter-I/subchapter-A

proposed animal activities, the individual must be able to understand the statement addressing U.S. Government Principle II. Further, OLAW's Sample Animal Study Proposal (refer to section 8:2) includes guidance on the provision of the study objectives, stating "Briefly explain the aim of the study and why the study is important to human or animal health, the advancement of knowledge, or the good of society in language that a layperson can understand."*

Opin. In the age of highly specialized research, most IACUCs regard this type of statement as essential for the lay member of the IACUC, institutional public relations or development personnel, and scientists on the IACUC who work in areas unrelated to the protocol under consideration. In the authors' experience, weak responses to this inquiry on the IACUC protocol form often reflect the general inexperience of scientists in addressing a scientifically unsophisticated public. Some institutions do not provide this type of inquiry on the form because all animal protocols are subsumed under an organizational or corporate research mission that assures this matter.

8:10 What information is appropriate to be included in the nontechnical description of the project(s) and the statement on the project(s) relevance to human or animal health, the advancement of knowledge, or the good of society?

Opin. The *Guide* (p. 25) states that the PI should provide "a clear and concise sequential description of the procedures involving the use of animals that is easily understood by all members of the committee" to help ensure, among other things, the participation of all IACUC members in the review process. Writing a cogent lay summary for the proposed research can be a challenging task, and the effort required is well invested in the authors' view. The function of the lay summary is to provide the nonscientific members of the IACUC, and particularly the nonaffiliated member, with an understanding of the scientific questions being asked and how the animals will be used to meet the project goals. The lay summary should indicate how the animal model contributes to the hypothesis being tested, the exploration of a new and potentially important area of science, or the resolution of an important scientific dispute. Obscure scientific jargon and the lack of effort to connect the studies to conditions or phenomena that are widely known or easily appreciated by the lay audience are common weaknesses that have been observed by the authors. To alleviate problems related to the use of scientific terms that are unfamiliar to lay members of the committee, some institutions have elected to provide a glossary of lay terms as a reference (e.g., https://az.research.umich.edu/animalcare/glossary).[6]

8:11 Should IACUC protocols submitted for educational activities have the same basic structure as a research protocol?

Opin. To provide for uniformity in oversight, most institutions use the same protocol form for all types of activities involving animal care and use (sections 8:1, 8:3–8:5). Alternatively, in institutions using animals in teaching activities that do not necessarily involve painful or distressful procedures, a unique and streamlined protocol template is often used. For example, life sciences and agricultural animal husbandry courses are frequently taught and may entail only approved methods of euthanasia or common farm animal husbandry practices. In situations such as these, institutions often create a protocol template to capture what routine, "non-invasive" techniques will be conducted, the number of animals needed per student, and the number of students participating in the activity. With this information, and the understanding that the course content encompasses this educational practicum, IACUCs can still approve and oversee the activity as they would any other project. This practice often optimizes the IACUC's administrative efforts and efficiency in the protocol review process and aids PIs who may be laden with the preparation of protocols from many other active research endeavors. Institutions also have found the development of special attachments to the protocol form to address areas regarded as unique to the use of animals in educational exercises necessary to fulfill adequate IACUC review requirements. The areas of special emphasis include a review of

- Course syllabus to understand how and why animal studies are integrated into the course
- All relevant handouts provided to students pertaining to animal use
- Student profile information (e.g., preparatory coursework and degree major)
- Summary of animal care and use educational discussion or materials provided to the students
- Backup provisions for students not fulfilling their obligations to laboratory animals
- Occupational health and safety provisions
- Summary of students' written critiques of the educational activity from the previous year
- Opportunities or plans for the course director to develop alternatives to the use of animals for subsequent sessions of the course

* National Institutes of Health, Office of Laboratory Animal Welfare. Sample Animal Study Proposal. http://grants2.nih.gov/grants/olaw/sampledoc/oacu3040-2.htm#admindata. Accessed June 30, 2022.

- The type of training, especially safety training, students receive as part of the overall activities

A similar approach allowing some modification of the protocol review template might also be applicable to studies for medical professionals to provide instruction or develop proficiency on particular procedures or the management of clinical emergencies such as laparoscopic surgery, endoscopy, bronchoscopy, endotracheal intubation, and Advance Trauma Life Support Training. The use of animals in these procedures is invasive and requires highly competent procedural and post-procedural clinical management of the animal model, and the assessment of the team providing the clinical care is clearly crucial. Also, the incorporation of alternative methods and the use of simulacra to prepare trainees for *in vivo* components would be important to ascertain. Further, it is essential that measures are taken to ensure that these trainees, who typically have only transient animal contact, are afforded knowledge of, for example, the ethical considerations of animal care and use, the hazards to which they may be exposed, and the practices used to achieve risk mitigation.

8:12 Is a PI required to provide information on the protocol review form that may be proprietary in nature?

Reg. Under the AWA (§2143,a,6,B), "No rule, regulation, order, or part of this chapter shall be construed to require a research facility to disclose publicly, or to the Institutional Animal Committee during its inspection, trade secrets or commercial or financial information which is privileged or confidential." §495,e of the HREA imparts a similar position to the PHS Policy.

Opin. PHS Policy and the AWA do not require IACUCs to review proprietary information. Most IACUCs have designed their protocol forms and implemented procedures to prevent the unwitting and unnecessary disclosure of proprietary information by investigators. On the other hand, IACUCs must have sufficient information to make an informed decision about the assurances for animal welfare during all phases of animal care and use. Where the PHS Policy is concerned, the IACUC must be assured that the occupational health and safety program has identified compounds and other materials, equipment, practices, and procedures that may have an impact on occupational health and safety considerations for personnel involved in the project (*Guide*, pp. 18–23). Hence, most IACUCs do not require the disclosure of the name of a proprietary compound but do require information about the chemical class of the compound, toxicity data in the target animal species or other animals, special handling requirements, and the basis for dosage selection. The development of new techniques and the use of new instrumentation in animals for application in humans pose more of a quandary. For example, if a new surgical technique is an integral component of a proprietary package involving a new device, an IACUC would be remiss to waive its responsibility to assess the impact of the procedure on experimental animals. Also, most IACUCs require sufficient information about the purported advantages of the new equipment to support the claim of potential benefit (see 26:3–26:6). The training of IACUC members should include the expectation and importance of nondisclosure of proprietary information available to them through the review process.

Protections for nondisclosure of confidential information are also built into the AWAR (§2.35, f), which state: "APHIS inspectors will maintain the confidentiality of the information and will not remove the materials from the research facilities' premises unless there has been an alleged violation, they are needed to investigate a possible violation, or for other enforcement purposes. Release of any such materials, including reports, summaries, and photographs that contain trade secrets or commercial or financial information that is privileged or confidential will be governed by applicable sections of the Freedom of Information Act." Institutions often have policies about confidentiality and nondisclosure as a condition of employment, and IACUCs often create additional nondisclosure agreements or contracts with each voting member and other, non-voting staff that may attend IACUC meetings. See Chapter 26 for additional information on propriety and confidentiality issues.

8:13 Should information such as the PI's name, title of the application, and research sponsor be included on the IACUC protocol form, or could that information bias the IACUC's decisions?

Opin. While there is some variation in the content of the protocol signalment among the institutions known to the authors, all identified the PI by name and included a title for the proposed activities. PHS Policy (IV,C,1,f) and the AWAR (§2.31,d,1,viii) require the IACUC to verify that personnel conducting procedures are appropriately qualified and trained. This would be difficult, if not impossible, without the identification of the PI. Most protocol forms also asked the PI to provide information about the funding source, particularly when the IACUC was responsible for sending the letters of protocol approval to funding agencies. Nevertheless, in some institutions, a response to this inquiry is at the discretion of the PI. In institutions that fund their research programs internally, this inquiry is not relevant. Virtually none of the institutions believe that this type of information is likely to compromise their objectivity, notwithstanding the admission by institutions that internally funded projects have to provide evidence of peer scientific review via an internal mechanism or consultants, whereas externally funded projects do not. Other information about the PI and associated laboratory personnel often collected in the protocol form includes position or academic title; office, laboratory, and home (emergency) phone numbers; and office and laboratory location. In the authors' view, emergency contact information is of paramount importance for critical correspondence on animal welfare matters and should always be provided. In many institutions, the ability to serve as a PI is reserved for individuals with particular positions or academic titles (see 26:6; 26:7).

8:14 What are some methods used by IACUCs to identify protocols for the purpose of correspondence, timely periodic reviews, and aiding of the oversight of ongoing animal-based research activities?

Opin. Most institutions use a numerical or alphanumerical identification system for the protocols that are submitted to and approved by the IACUC. This type of system offers several advantages. It provides a brief, content-neutral reference that links both the animals and, when necessary, the grant application or award to a set of approved animal care and use procedures. Many institutions indicate the year of protocol submission in their log or accession system. This method provides rapid identification of those protocols that are due for annual (only for Veterans Administration Protocols) or triennial renewal. Other numbers are added to reflect the sequence of protocol submission within the year, and sometimes letter or decimal extensions are used to indicate the renewal year within the triennial period. As an example, 13-38.16 might indicate the 38th protocol submitted to the IACUC in 2013, with a full review of the protocol scheduled in the year 2016.

It is not uncommon for the approval of protocols submitted at or near the end of the calendar year to be delayed until the following year. Computer databases are used routinely to track the actual dates of submission and approval to ensure that the requirements for renewal are met in a timely fashion. Also, organizations vary in their requirements for the number of animal species that are permitted in a particular protocol application (e.g., some allow multiple species, whereas others allow only one) or the association of a single IACUC protocol to a single grant (some permit multiple grants on a single IACUC protocol and others restrict to a 1:1 match).

Organizations also vary in their decision to implement the species-dedicated protocol approach to allow ease of activity tracking and post-approval monitoring efforts. Institutions that allow only one species per protocol application sometimes retain the same number but follow it with a letter designator for species (e.g., *R* = rat, *Rb* = rabbit, *M* = mouse) to associate the procedures described with a particular grant application or award. The same tracking number may be retained, along with a modifying extension, to indicate that animal care and use procedures are identical to those described in the parent protocol, but an alternate project title or funding source is associated with the protocol (see 8:13). Regardless of the protocol identification system developed, it is critical that all members of the ACUP are familiar with and understand the system.

8:15 When protocols are due to expire, should the IACUC retain the original number assigned or assign a new number to the protocol?

Reg. The AWAR (§2.31, d,5) and the PHS Policy (IV, C,5) require a complete review of animal activities at least every three years.

Opin. Most institutions prefer to retire protocol numbers when the protocols have reached the limit of their three-year approval period.[5] (This reference may change depending on PHS Policy and AWAR). This approach is convenient in that it ensures individuals involved in animal care and use (and research oversight) that the review process is current and provides an easy indicator of recent IACUC approval (e.g., the protocol number for a protocol approved in 1990 will differ significantly from one approved in 2022). It is becoming common in the authors' experience for institutions to retain the original protocol number through the life of the project for efficiency as well as to indicate the continuity and longevity of a research program. This approach is also satisfactory as long as the IACUC has performed a rigorous review at the appropriate review cycle. PIs and animal facility managers have found that changing their protocol numbers in the course of a project necessitates, for example, updating hundreds of cage cards on mouse cages or developing methods to ensure procedure and training details are consistently and adequately documented under the new protocol number. Retaining the protocol number might reduce the need to reaffirm and update details with unnecessary repetition. On the other hand, maintaining protocol numbers in perpetuity can affect other processes, such as records retention.

8:16 Are there advantages to using an "electronic" protocol e-submission format?

Opin. The use of an electronic protocol format requires institutions to either develop a homegrown system or purchase a commercially available product. In both cases, the process can be found to be cost prohibitive. Institutions using e-protocol management systems usually have large animal care and use programs and budgets that can support the associated expenses. However, regardless of size, the need for electronic systems may become a necessity due to government requirements trending toward paperless submissions and retention as mandated by the Paperwork Reduction Act (PRA).*

Institutions using e-protocol templates are afforded many benefits. Typically, the database used to manage the system is accessible to IACUC members, thereby giving committee members unrestricted access to all protocols, including those under review. In addition, the e-submission template is typically programed to gather only the relevant information the IACUC needs to conduct the project review. For example, a PI may be asked to check a series of boxes identifying the procedures they will conduct. For conducting surgeries, the survival surgery box would be checked and PIs performing survival surgery would need to answer all questions relating to the topic. As e-submission programs become more affordable they likely will help maximize the efficiency of the PIs' protocol completion process, the IACUC's review process, and document management. Institutions should conduct regular audits of e-protocol systems to avoid software issues, such as exceeding the three-year expiration date and approved animal numbers.

Surv. Does your IACUC allow an investigator to use veterinary drug menus on protocol forms or reference a standard operating procedure or policy? (A drug menu is a list of drugs and associated dosages that are nearly equally efficacious for a stated purpose.)

- No 121/266 (45%)
- Yes 125/266 (47%)
- Other* 20/266 (8%)

*Other:

- Not applicable to the work at our institution
- Depends on the species and complexity of the study

Surv. Does your IACUC have a written policy on what constitutes a conflict of interest for reviewers of protocols (either by Full Committee Review or Designated Member Review).

- No 100/266 (38%)
- Yes 156/266 (59%)
- Other 10/266 (4%)

Surv. Does your IACUC use a separate protocol form for each species proposed for use in a study?

- No, we use the same protocol form for all species to be used. 192/266 (72%)
- Yes, there is a separate form for each species. 54/266 (20%)
- Yes, but only for species covered under the Animal Welfare Act. 4/266 (2%)
- Other 16/266 (6%)

REFERENCES

1. National Institutes of Health, Office of Laboratory Animal Welfare. *Sample Animal Study Proposal.* http://grants2.nih.gov/grants/olaw/sampledoc/oacu3040-2.htm#admindata. Accessed April 3, 2013.
2. National Institutes of Health, Office of Laboratory Animal Welfare. 2012. *Frequently Asked Questions D.10, Is the IACUC Required to Review the Grant Application?* http://grants.nih.gov/grants/olaw/faqs.htm#proto_10.
3. Association for Assessment and Accreditation of Laboratory Animal Care International. 2013. *Frequently Asked Questions C.3, Harm-Benefit Analysis.* http://www.aaalac.org/accreditation/faq_landing.cfm#B3.
4. National Institutes of Health, Office of Laboratory Animal Welfare. 2012. *Frequently Asked Questions D.1, How Frequently Should the IACUC Review Research Proposals?* https://grants.nih.gov/grants/olaw/faqs.htm#proto_1.

* https://www.govinfo.gov/content/pkg/BILLS-104s244enr/pdf/BILLS-104s244enr.pdf

9

Submission and Maintenance of IACUC Protocols

Anna Hampton

Introduction

Use of animals in research, testing, teaching, and/or experimentation is a privilege that should not be taken lightly. To help mitigate the potential for misuse of laboratory animals, specific standards have been established that may be required for certain animal activities. These include the Animal Welfare Act (AWA) and Animal Welfare Regulations (AWAR), the Health Research Extension Act of 1985, the U.S. Government Principles for the Utilization and Care of Vertebrate Animals Used in Testing, Research, and Training (the *Principles*), the Public Health Service Policy on Humane Care and Use of Laboratory Animals (the PHS Policy) and select Department of Veterans Affairs (VA) and Department of Defense (DoD) **requirements** (*Reg.*).[1–5] In addition, the *Guide for the Care and Use of Laboratory Animals* (the *Guide*) is captured as a requirement because compliance with it is obligatory by the PHS Policy.[6] Furthermore, if an institution is AAALAC accredited, then AAALAC International standards would be considered a requirement.[7] In contrast, there sometimes are common practices among IACUCs, laboratory animal professionals, or voluntary oversight organizations and these are considered **opinions** (*Opin.*). All requirements must be followed over any opinion. If the reader seeks additional clarification on a topic, he or she is encouraged to reach out to the specific federal, grant, or oversight group that may be involved for guidance.

Because IACUC requirements are generally specific to certain characteristics of the animal activity, readers are encouraged to approach each situation they may encounter with a broad perspective to decipher the actual oversight needs and requirements. This can help streamline the IACUC's approach and reduce unnecessary burden on those who care for and/or use laboratory animals at the institution. Appropriate considerations include the species; the funding source used to conduct or support the animal activity; the location of the animal activity; the specific description of processes as stated in the institution's OLAW Animal Welfare Assurance; whether the animal care and use program is AAALAC accredited; any state and local requirements regarding research animal care and use; and institutional policies, SOPs, and precedents.

9:1 When does the IACUC need to approve an animal activity?

Reg. Per the AWAR (§2.31,c,6–7) and PHS Policy (B,6–7), the IACUC is required to "Review and approve, require modifications in (to secure approval), or withhold approval of those components of proposed activities related to the care and use of animals" as well as any significant changes to ongoing activities.

The AWAR (§1.1) define activity as "those elements of research, testing, or teaching procedures that involve the care and use of animals" and defines animal, in brief, as "any live or dead warm-blooded animal, which is being used, or is intended for use for research, teaching, testing, experimentation, or exhibition purposes. ... This term excludes birds, rats of the genus Rattus, and mice of the genus Mus, bred for use in research." The AWAR (§1.1) define "bred for use in research" as "an animal that is bred in captivity and used for research, teaching, testing, or experimentation purposes." Further, the AWA and AWAR definitions specifically exclude elementary or secondary schools; and, interestingly, they define "research facility" as a school, institution, or organization that uses live animals in "research, tests, or experiments" while making no specific mention of dead animals. Furthermore, the AWA (§2132) limits the scope of the USDA's authority to the United States, the District of Columbia, the Commonwealth of Puerto Rico, the Virgin Islands, Guam, American Samoa, or any other territory or possession of the United States, and the AWAR (§2.31,d,1) exempt field studies from IACUC review requirements. Taken together, the AWA and AWAR require that the IACUC approve animal activities that involve the use of certain live, warm-blooded animals in research, testing, or teaching at a level higher than elementary/secondary school levels.[8] However, they cover these activities if performed only in the United States or its territories and exclude specific animal uses (see Section 9:2).

The PHS Policy (III,A) applies to "any live, vertebrate animal used or intended for use in research, research training, experimentation, or biological testing or for related purposes." The PHS Policy covers animal activities supported or conducted only with PHS funds (PHS Policy III,H), which includes: the Agency for Healthcare Research and

DOI: 10.1201/9781003272861-9

Quality; the Agency for Healthcare Research and Quality, the Biomedical Advanced Research and Development Authority (BARDA), the Centers for Disease Control and Prevention (CDC); the Food and Drug Administration (FDA); the Health, Resources and Services Administration; the Indian Health Service; the National Institutes of Health (NIH); and the Substance Abuse and Mental Health Services Administration. The PHS Policy also applies to other federal funding sources that have an established Memorandum of Understanding (MOU) or agreement with the Office of Laboratory Animal Welfare (OLAW) (i.e., National Aeronautics and Space Administration (NASA), National Science Foundation (NSF), and U.S. Department of Veterans Affairs (VA)).[9–12] Per PHS Policy (II), the requirements of the PHS Policy extends to anywhere in the world that the activity is conducted.

The *Guide* (pg. 2) defines animal use as "The proper care, use, and humane treatment of laboratory animals produced for or used in research, testing, or teaching" and notes "[The IACUC's] oversight functions include review and approval of proposed animal use (protocol review) and of proposed significant changes to animal use" (p. 25).

It should also be noted that some animal activities conducted or supported with U.S. military funding also requires IACUC approval.[13–17]

Opin. AAALAC International's Council on Accreditation uses three primary standards to evaluate animal programs. Of these three, the two that are relevant to institutions in the United States are the *Guide for the Care and Use of Laboratory Animals* (the *Guide*) and the *Guide for the Care and Use of Agricultural Animals in Research and Teaching* (*Ag Guide*).[5, 18, 19] Both the *Guide* (p. 25) and *Ag Guide* (p. 2) have an expectation for the IACUC to review and approve proposed activities and proposed significant changes for animals, including agricultural animals, used in research and teaching, with the *Guide* adding testing as an additional method of animal use.

AAALAC International covers all institutionally owned animals, including traditional laboratory animals, farm animals, wildlife, and aquatic animals. Like the *Guide*, AAALAC does not exclude invertebrates and specifically notes that nontraditional animals, including invertebrate species, will be included in AAALAC's evaluation where relevant to the institution's mission.[7, 20–21] The AAALAC International definition of animal can broaden even further to include the evaluation of animals not owned by the institution, if that animal(s) impacts the health or welfare of the institutionally owned animals or the occupational health and safety of personnel. Relevant examples might include maintenance of feral cat colonies near vivarium entry points or personally owned animals housed in terrarium/aquariums at the institution (e.g., lab pet fish). In both examples, the animals may pose a biosecurity risk to the animals being used as research subjects, as well as an occupational health risk to personnel (see section 9:35). AAALAC International suggest consulting with AAALAC about specific types of animals and consult legal counsel regarding animal "ownership" at the institution or at collaborative sites.[7]

IACUC approval may also be required by other funding sources, specific municipalities, or be an overarching institutional requirement (e.g., described in the institution's OLAW Animal Welfare Assurance). In addition, many journals require confirmation of IACUC review and approval of live animal activities. Furthermore, many well-known reference documents indicate the need for the IACUC to approve specific aspects of animal activities.[18, 22–24]

9:2 What animal activities can be performed without an IACUC-approved protocol?

Reg. The requirement for IACUC approval prior to initiation is dependent on the type and species of animal, funding used to conduct or support the animal activity, location of use, and/or type of animal activity.

Type and species of animal: Section 9.1 covers the definitions of animals that need IACUC approval. The definition of animal in the PHS Policy (III,H), AWAR (§1.1), and by the DoD includes only vertebrate species.[14–16] The *Guide* (p. 2) specifically states it is not species specific but focuses on vertebrates. Thus, in general, the use of invertebrates does not require IACUC approval. Furthermore, AWAR (§1.1) exclude the following types of animals in its definition of animal, thus IACUC approval would not be required under this regulation for: (1) cold-blooded species or ectotherms, and (2) birds, rats of the genus *Rattus*, and mice of the genus *Mus*, bred for use in research. Where "bred for use in research" is defined as "an animal that is bred in captivity and used for research, teaching, testing, or experimentation purposes." Of course, the most stringent applicable requirement for an individual institution would need to be followed.

Funding used to conduct or support and location of use: The AWA oversight of "Research Facilities" applies regardless of funding used to conduct or support the activity (§ 2132); however, this requirement applies only to animal use in the United States and other territories in possession of the United States. The PHS Policy (Section II) applies only to national and international locations that conduct or support live animal activities with funds from the PHS, including BARDA, CDC, FDA, or funds associated with NIH OLAW MOUs (i.e., NASA, NSF, VA).[9–12] Military funds used to conduct or support animal activities are similar to the PHS policy requirements.[13–16]

Type of activity: The requirement for IACUC approval is specific to animals being used in research, teaching, testing, or experimentation (AWA and AWAR); or research, research training, experimentation, or biological testing or for related purposes (PHS Policy). The AWAR (§1.1) specifically exclude the following animal activities from their oversight requirements: (1) horses not used for research purpose; and (2) other farm animals used or intended for use for improving animal nutrition, breeding, management, or production efficiency, or for improving the quality of food or fiber. The AWAR(§2.31,d,1) state, and OLAW has clarified in their Frequently Asked Question (FAQ) A.6, that the IACUC is responsible for oversight of research conducted in the field and must evaluate the animal use to determine if protocol review and approval by the IACUC is required. If the activities alter or influence the biology, behavior, or ecology of the study animals or other species or "when capture, handling, confinement, transportation, anesthesia, euthanasia, or invasive procedures are involved, IACUC review and approval is required, and the IACUC must ensure that proposed studies are in accord with the PHS Policy and *Guide*" (OLAW FAQ A.6.).[6, 25] In addition, OLAW FAQ A.1. states "Institutions must implement the PHS Policy for all PHS supported activities involving animals, and must ensure that any standards that might not be consistent with PHS Policy do not affect or pose risks to PHS supported activities"; thus, animal activities that may not require IACUC approval may need to be included in IACUC oversight if it could pose a risk to PHS supported activities.[26] IACUC approval is not required for DoD funded activities that involve: (1) ceremonial, recreational, and working animals; (2) livestock or poultry used or intended for use as food or fiber; (3) livestock or poultry used or intended for use for improving animal nutrition, breeding, management, or production efficiency, or for improving the quality of food or fiber; (4) animals used in disease surveillance or in field studies; and (5) activities performed as part of environment, safety, and occupational health programs or as part of pest management programs.[14, 15]

Opin. As mentioned in 9:1, AAALAC International may expect the IACUC to review the use of invertebrates in some circumstances.[7, 20, 21] In addition, there may be additional funding, state, or local requirements that are stricter than those mentioned in this chapter and, if the animal activity is occurring at a non-U.S. performance-site, it is important to remember that oversight standards and requirements can vary considerably in other countries.[27]

The use of the term "teaching" as it relates to the AWA and AWAR can sometimes be hard to interpret. For example, the use of agriculture animals to teach methods that improve farm or ranch management (e.g., hoof trimming or shearing) would not require IACUC approval under the AWA and AWAR; however, use of farm animals in biomedical teaching that involves the training of human or veterinary medical personnel (e.g., veterinary technicians) in medical practices and methods (e.g., surgery, diagnostic techniques, anesthesia, and analgesia) would require IACUC review and approval.[28]

9:3 What information is required on an IACUC protocol?

Reg. The PHS Policy does not use the term "protocol," while the AWA and AWAR use the term "protocol" in a non-specific manner that could mean a variety of written procedures or agreements. Although there is a requirement for IACUC review and approval of certain animal activities (PHS Policy B,6.; AWAR §2.31,c,7), there is no specific requirement for the use of an IACUC protocol.

The term "protocol," as commonly used, is captured in the *Guide* (p. 25) and is defined as a detailed description of the proposed use of laboratory animals. The terms protocol, animal protocol, animal use protocol, or IACUC protocol are also captured in other documents and are ubiquitously used to describe the document that captures the animal activity components that require IACUC review and approval.[13–16, 29]

The AWAR(§2.31,d,1) specify that the IACUC conduct a review of the proposed activity to ensure that it (1) meets the requirements of AWAR, (2) provides an acceptable justification for any departures from AWAR, in writing, and (3) that the animal activity meets the following requirements:

- §2.31,d,1,i Procedures involving animals will avoid or minimize discomfort, distress, and pain to the animals.
- §2.31,d,1,ii The principal investigator has considered alternatives to procedures that may cause more than momentary or slight pain or distress to the animals and has provided a written narrative description of the methods and sources.
- §2.31,d,1,iii The principal investigator has provided written assurance that the activities do not unnecessarily duplicate previous experiments.
- §2.31,d,1,iv,A-C Procedures that may cause more than momentary or slight pain or distress to the animals will:
  - Be performed with appropriate sedatives, analgesics, or anesthetics, unless withholding such agents is justified for scientific reasons, in writing, by the principal

investigator and will continue for only the necessary period of time

- Involve, in their planning, consultation with the attending veterinarian or his or her designee
- Not include the use of paralytics without anesthesia

- §2.31,d,1,v Animals that would otherwise experience severe or chronic pain or distress that cannot be relieved will be painlessly euthanized at the end of the procedure or, if appropriate, during the procedure.
- §2.31,d,1,vi The animals' living conditions will be appropriate for their species in accordance with Part 3 (Standards) of the AWAR and contribute to their health and comfort. The housing, feeding, and nonmedical care of the animals will be directed by the attending veterinarian or other scientist trained and experienced in the proper care, handling, and use of the species being maintained or studied.
- §2.31,d,1,vii Medical care for animals will be available and provided as necessary by a qualified veterinarian.
- §2.31,d,1,viii Personnel conducting procedures on the species being maintained or studied will be appropriately qualified and trained in those procedures.
- §2.31,d,1,ix Activities that involve surgery include appropriate provision for pre-operative and post-operative care of the animals in accordance with established veterinary medical and nursing practices. All survival surgery will be performed using aseptic procedures, including surgical gloves, masks, sterile instruments, and aseptic techniques. Major operative procedures on non-rodents will be conducted only in facilities intended for that purpose which shall be operated and maintained under aseptic conditions. Non-major operative procedures and all surgery on rodents do not require a dedicated facility but must be performed using aseptic procedures. Operative procedures conducted at field sites need not be performed in dedicated facilities but must be performed using aseptic procedures.
- §2.31,d,1,x,A-C No animal will be used in more than one major operative procedure from which it is allowed to recover, unless:
  - Justified for scientific reasons by the principal investigator, in writing
  - Required as routine veterinary procedure or to protect the health or well-being of the animal as determined by the attending veterinarian, or
  - In other special circumstances as determined by the Administrator on an individual basis
- §2.31,d,1,xi Methods of euthanasia used must be in accordance with the definition of the term set forth in 9 CFR part 1, § 1.1 of this subchapter, unless a deviation is justified for scientific reasons, in writing, by the investigator.

The PHS Policy (C,1.). indicates that the IACUC shall review "those components related to the care and use of animals and determine that the proposed research projects are in accordance with (the PHS) Policy. In making this determination, the IACUC shall confirm that the research project will be conducted in accordance with the AWA insofar as it applies to the research project, and that the research project is consistent with The Guide unless acceptable justification for a departure is presented." It is the responsibility of the IACUC to ensure that the animal activity follows the institution's OLAW Animal Welfare Assurance and meets the following requirements outlined in PHS Policy (C,1,a–g):

a. Procedures with animals will avoid or minimize discomfort, distress, and pain to the animals, consistent with sound research design.
b. Procedures that may cause more than momentary or slight pain or distress to the animals will be performed with appropriate sedation, analgesia, or anesthesia, unless the procedure is justified for scientific reasons in writing by the investigator.
c. Animals that would otherwise experience severe or chronic pain or distress that cannot be relieved will be painlessly killed at the end of the procedure or, if appropriate, during the procedure.
d. The living conditions of animals will be appropriate for their species and contribute to their health and comfort. The housing, feeding, and nonmedical care of the animals will be directed by a veterinarian or other scientist trained and experienced in the proper care, handling, and use of the species being maintained or studied.
e. Medical care for animals will be available and provided as necessary by a qualified veterinarian.
f. Personnel conducting procedures on the species being maintained or studied will be appropriately qualified and trained in those procedures.

g. Methods of euthanasia used will be consistent with the recommendations of the American Veterinary Medical Association (AVMA) Panel on Euthanasia, unless a deviation is justified for scientific reasons in writing by the investigator.

The *Guide* (p. 25) provides a comprehensive list of topics that should be considered by the researcher and reviewed by the IACUC. These include:

- Rationale and purpose of the proposed use of animals
- A clear and concise sequential description of the procedures involving the use of animals that is easily understood by all members of the committee
- Availability or appropriateness of the use of less invasive procedures, other species, isolated organ preparation, cell or tissue culture, or computer simulation
- Justification of the species and number of animals proposed; whenever possible, the number of animals and experimental group sizes should be statistically justified
- Unnecessary duplication of experiments
- Nonstandard housing and husbandry requirements
- Impact of the proposed procedures on the animals' well-being
- Appropriate sedation, analgesia, and anesthesia
- Conduct of surgical procedures, including multiple operative procedures
- Post-procedural care and observation
- Description and rationale for anticipated or selected endpoints
- Criteria and process for timely intervention, removal of animals from a study, or euthanasia if painful or stressful outcomes are anticipated
- Method of euthanasia or disposition of animals, including planning for care of long-lived species after study completion
- Adequacy of training and experience of personnel in the procedures used, and roles and responsibilities of the personnel involved
- Use of hazardous materials and provision of a safe working environment.

In addition, the *Guide* (p. 27) mentions Special Considerations for IACUC Review. These considerations are listed below, but readers are encouraged to reference the *Guide* for more specific information on each topic.

- Experimental and Humane Endpoints, including precise definition of the humane endpoint, frequency of observation, training of personnel responsible, and response upon identification
- Unexpected Outcomes, highlighting that genetically modified animals may be at increased risk for these due to uncharacterized phenotypes
- Physical Restraint, outlining specific guidance to justify and refine restraint procedures
- Multiple Survival Surgical Procedures, noting when they are acceptable regardless of the surgery category (major vs. minor)
- Food and Fluid Regulation, focusing on IACUC review of necessity for such procedures, potential adverse complications, and methods of monitoring the health of the animal
- Use of Non-Pharmaceutical-Grade Chemicals and Other Substances, stating that they should only be used when necessary to meet the scientific goals and when a pharmaceutical-grade product is not available
- Field Investigations, encouraging IACUCs to consult with a qualified wildlife biologist when appropriate
- Agricultural Animals, stressing that animal well-being and justification for the scientific need to use animals still applies

The *Principles* specifically state that the Institutional Official is responsible to ensure that the *Principles* are adhered to; however, Principal 9 indicates that any exceptions to the *Principles* should be reviewed by an IACUC. In addition, with the exception of animal transportation considerations, most of the *Principles* are captured in the components that require IACUC review and approval as described in the AWA, AWAR, PHS Policy, and/or the *Guide*.[4]

It is important to keep in mind that other sources of funding used to conduct or support animal activities may have their own specific requirements with respect to IACUC review and approval of activities using animals. This may include additional reviews by other oversight bodies (e.g., U.S. Army Animal Care and Use Review Office, Navy Bureau of Medicine and Surgery, etc.) before initiation of the activity.[14, 16, 29, 30]

The Principal Investigator is responsible for the protocol and animal use activity; thus, it is reasonable to regard listing that person on the protocol as a requirement (see section 9:19 and 9:26). In addition, the IACUC should approve required departures, exceptions, or exemptions from the AWAR

(§2.38,k,1; §3.8,d,2; and §3.81,e,2) and/or the (PHS Policy (IV,C,1).

Opin. The use of an IACUC to ensure appropriate animal use at an institution is required by the AWA and PHS Policy. Other professional references and/or guidance documents mention the importance of IACUC review and approval of specific aspects of animal use. These are generally specific to the topic of the document (e.g., international animal use, use of agriculture or nonhuman primate species, biosafety, neuroscience and behavioral research, occupational health and safety, etc.) and can serve as valuable resources for more details on those specific aspects of animal care and use.[18, 22–24, 31, 32]

Although one can debate if a single, comprehensive IACUC protocol with the description of the animal activity is required, it has become accepted as a common approach to ensure that the IACUC is reviewing the necessary animal activity components.[33, 34] However, other approaches may be more appropriate for the activity and may include a written standard of care or standard operating procedure (see Section 9:27-30). Regardless, the IACUC has an obligation to review the most critical components related to the care and use of laboratory animals to ensure that considerations for the animal's health and well-being are met, the activity is consistent with application regulations, and the safety of personnel involved is ensured. Regardless of the method used to capture the required information, a standard approach should be utilized by the institution to avoid noncompliance. When developing IACUC protocol forms and workflows, institutions should consider the user experience to minimize frustrations and ensure the process is intuitive, easy to complete, promotes allowed flexibility, and limits unnecessary confusion or burden.[33]

Additionally, optional questions or items may be placed in a protocol form to help streamline the review and capture program critical aspects of animal use for other processes; however, these extra items should be judiciously evaluated by the institution to ensure they serve a legitimate purpose and do not just create additional work. In addition, institutions should remember to follow their OLAW Animal Welfare Assurance, which, in some cases, may be more stringent than the PHS Policy and the *Guide*.[35] For example, IACUC protocols might capture information such as:

- Administrative items (dates of approval and expiration, institution-specific identification numbers or tags, type of study, and title)[33]
- Lay summary[33]
- Listing all personnel who may handle animals as described in the protocol (see 9:26)[33]
- Description of protocol personnel's skills and experience (see 9:26)[33]
- Listing all locations of animal use (see 9:27)
- Capturing detailed information on hazardous agents, biologic materials, and/or controlled substances[22, 33, 36]
- Placing the animals in an anticipated USDA pain/distress category[37]
- Aims of the study and study design[34]
- Literature search for alternatives (AWA §2143,a,3,B; §2143,d,7; §2143,B,i; AWAR §2.31,d,1,ii)[38, 39]
- Scientific merit review (see 9:27)
- Harm-benefit analysis (see 9:27)
- Methods for animal care, housing, feeding, etc.[35]
- Movement of animals (see 9:27)[33]
- Species-specific enrichment[33]
- Use of non-pharmaceutical grade substances[33, 40]
- External collaborations or use of contract research organizations (See 9:18)

9:4 Do government agencies conducting, supporting, or funding animal research require IACUC-approved protocols?

Reg. The AWA and AWAR requirements are species-, activity-, and location-dependent but independent of funding source. If a USDA-regulated animal is being used in research, teaching, testing, or experimentation in the United States or its territories, then IACUC review and approval of the animal activity is required prior to initiation. Some exceptions may apply; see sections 9:1–9:3 and 9:31–32.

As mentioned, the PHS policy does not use the term protocol; however, the PHS Policy (IV,A,1) requires institutions to use the *Guide* as the basis in their animal care and use program, and the *Guide* (p. 25) defines and describes the need for a protocol. Furthermore, OLAW FAQ's A.3., A.6., A.7., and the FAQ section D: Protocol Review, use the term protocol regarding aspects of IACUC review and/or approval.[25, 41–43] With this, one can extrapolate that when PHS funds are used to conduct or support animal activities, an IACUC-approved protocol is required. In addition, the PHS Policy (V,B) defines the absolute requirement that "PHS awarding units may not make an award for an activity involving animals unless the prospective awardee institution and all participating institutions have approve Assurances on file with OLAW, and the awardee institution has provided verification of approval by the IACUC of those components of the application or proposal related to the care and use of animals."[5] OLAW has similar agreements with other government agencies (i.e., NASA, NSF, and VA) and the same expectation would apply to research funded by these agencies.[9–12]

The DoD and the VA require IACUC protocols that have been reviewed and approved by the IACUC. If the animal activity is being conducted at the government agency, the government agency will have their own required protocol form.[14, 16, 29, 30] However, if the government agency is funding an animal activity that is being conducted or supported at another performance site, local IACUC approval and oversight is required. In these situations, a protocol appendix that captures component specific requirements may be required.[14, 16, 29, 30]

The *Principles* apply to all federal agencies conducting or supporting testing, research, or training procedures involving vertebrate animals. Although the *Principles* do not specifically require an IACUC protocol, it is a common way of ensuring that the institution is adhering to the *Principles*.

Opin. The specific requirements for each funding source should be evaluated by the institution and followed.

9:5 Do private agencies funding biomedical research require IACUC-approved protocols?

Reg. As mentioned in 9:4, the AWA and AWAR requirements are species, activity, and location dependent, and compliance is required independent of funding source. If a USDA-regulated animal is being used in research, teaching, testing, or experimentation in the United States or its territories, then IACUC review and approval of the animal activity is required prior to initiation. Some exceptions may apply; see Sections 9:1–9:3 and 9:31–32.

Opin. It is beyond the scope of this chapter to capture the requirements for all potential funding sources that could be used to conduct or support animal activities. Institutions should be proactive and ensure that all funding requirements are met prior to the initiation of animal activities. For example, the American Heart Association Award Agreement Terms and Conditions states that "research involving animals has been approved by an appropriate animal care and use committee or equivalent and adheres to the U.S. Animal Welfare Act, and any applicable regulations and guidelines by the National Health Council, U.S. Department of Agriculture, or National Institutes of Health."[44]

Other considerations when conducting animal activities that are supported with private funds, with regard to IACUC approval and oversight:

- Institutional responsibility toward animals: If the oversight of the animal activities at the institution varies due to species, funding, activity type, or location of use, it may be advantageous to develop one standard institutional requirement for IACUC review and approval of an animal use protocol. This may prevent unintentional non-compliance with requirements.
- The institution's OLAW Animal Welfare Assurance: If the institution has an approved OLAW Animal Welfare Assurance it must be followed. If the Assurance indicates that all animal activities will receive IACUC-approval via a protocol, then all animal activities should have an associated, approved IACUC protocol, regardless of funding.
- AAALAC International accreditation status: AAALAC, International uses the *Guide*, the *Ag Guide*, and the European Convention for the Protection of Vertebrate Animals used for Experimental and Other Scientific Purposes (ETS 123) as primary standards for evaluation; and both the *Guide* (pg. 25) and *Ag Guide* (pg. 2) state that an essential IACUC function is to approve animal use (i.e., protocols).[6, 18, 19, 45]

9:6 How can institutions require IACUC approval of activities funded from agencies that do not require it, or that involve animals that are not covered by either the AWA/R or the PHS Policy?

Opin. This question is not specific to an external requirement but reflects on a specific institution's approach to requiring IACUC approval for animal activities regardless of species or funding source. That said, if an institution's OLAW Assurance indicates that all animal activities must receive IACUC approval via review of a protocol, then all animal activities should have an associated, approved IACUC protocol, regardless of funding.

In situations where IACUC review and approval of animal activities are not required, the institution is left to determine the local expectations and requirements for animal use. This can be further complicated, in rare situations, when the institution is not required to have an IACUC (e.g., no USDA-regulated species used, no PHS funding for animal research). OLAW FAQ B.6. states that "Institutions have discretion to subject animal activities to IACUC oversight regardless of the source of funding. This practice ensures uniform standards, appropriate oversight and accountability, and therefore is often in the best interest of the institution."[46]

When determining the need for IACUC approval outside the requirement, institutions should consider their goals and missions toward animal use, the benefit to the institution and animals, and the utility of being able to state that an animal activity is IACUC approved (e.g., journals, collaborations, etc.). In addition, a consistently applied standard for IACUC approval of all animal activities at an institution may enhance internal and external credibility by ensuring ethical use of animals and consideration for animal welfare. These benefits should then be weighed against the increase in burden to researchers and additional administrative burden. Maximizing flexibilities and creating a process that captures and oversees the animal activity information needed can create a manageable burden/oversight balance.[47]

If an institution decides to require IACUC approval for all animal activities, the IACUC should have clear, established policies and/or SOPs on the requirement. This includes creating an institutional mechanism to capture all animal use that requires IACUC approval before implementation. Possible considerations include:

- Documenting institutional support (CEO, IO, chairs, deans, etc.)
- Consulting with public relations as it relates to assuring to the public that the institution's use of animals is done responsibly and ethically
- Defining what requires IACUC approval: type and species of animal, type of animal activity, location of the animal activity, etc.
- AAALAC International accreditation status and requirements
- Animal acquisition/capture points to verify IACUC approval (e.g., the time of animal ordering, purchasing, housing, etc.)
- Publication requirements for IACUC approval of the research that is being reported. Some journals require a statement of IACUC approved animal activities, or similar.

9:7 Must the IACUC review and approve protocols using animal carcasses and parts of carcasses?

Reg. It depends. However, if the carcass or parts of the carcass will be used in research, teaching, training, or testing, and the animal was acquired alive or there were any antemortem procedures, alterations, or manipulations to the animal prior to death, the use of the animal, while it was alive, requires IACUC review and approval.[8, 41]

The PHS Policy Section III specifically mentions live vertebrates and, although the AWAR (§1.1) definition of animal includes "live or dead," the AWA definition of Research Facility (§2132,e) is specific to use of "live animals."

The DoD includes "living or dead" in their definition of animals but then defines dead as "animals killed for the direct purpose of conducting (research, development, test and evaluation) or training. However, it does not include dead animals or parts of dead animals purchased at grocery stores or slaughterhouses."[15]

The *Guide* (p. 2) defines animal as "Laboratory animals or animals: Any vertebrate animal (e.g., traditional laboratory animals, agricultural animals, wildlife, and aquatic species) produced for or used in research, testing, or teaching." Although the *Guide* does not specifically distinguish between live or dead animals, the PHS Policy does exclude dead animals. AAALAC International does not distinguish between live or dead animals and covers institutionally owned animals.[20]

Opin. Generally, if a live animal is euthanized specifically for the purpose of using the carcass for research or teaching, or for the purpose of harvesting tissues or other samples to be used in research or teaching, IACUC review and approval is expected. Regardless of the need for IACUC review and approval of carcass use, the IACUC may be contacted or consulted regarding this type of animal use and should be prepared to advise. The IACUC can then provide guidance on carcass use that could include:

- Additional information on why IACUC approval is or is not required
- Importation, transportation, and handling requirements (e.g., federal, state, local, institutional)
- Human safety considerations (e.g., zoonotic diseases, PPE)[33]
- Animal safety considerations (e.g., biocontainment)[33]
- Appropriate carcass disposal (the *Guide*, pp. 20, 73, 136, and 142)
- Possible publication requirements that would require IACUC or ethics body approval
- Other legal or local requirements (e.g., CITES, Endangered Species Act)

9:8 Must the IACUC review and approve a protocol for a study in which the investigator will use only tissues from animals found dead alongside the road (i.e., roadkill)?

Reg. No. Because the live animal was (1) not used in research, teaching, testing, training, or experimentation, (2) was not killed for the purposes of obtaining or using tissues, and (3) not manipulated antemortem for project-specific purposes, IACUC review and approval would not be required.[41]

Opin. IACUC review and approval of research and teaching activities using animal carcasses and parts of carcasses is not required, unless the live, vertebrate animal was obtained for a PHS conducted or supported activity and was euthanized to use the biological samples or carcass (see section 9.7). Given the scenario presented, animals found dead along the road would not require IACUC review or oversight. However, the IACUC may be contacted or consulted regarding this type of animal use and should be prepared to advise.

9:9 Must the IACUC review and approve protocols using animal tissues, cells, or biological fluids derived from living animals?

Req. Understanding the intended use of the material is important. If the animal-derived biological material will be used for a purpose unrelated to research or teaching, then it is likely that the activity does not need IACUC review and approval, if live animals are not manipulated in any way to obtain these materials.

With regard to the PHS Policy requirements, OLAW FAQ D.15. specifically states the "collection of biological samples from any live vertebrate animal for the purpose of a PHS-supported activity is covered by the PHS Policy." Thus, approval of a protocol that includes descriptions of animal use for the purpose of obtaining biological materials is a typical practice by IACUCs.[48]

Opin. Institutions should be proactive in ensuring that collection of the animal tissues, cells, or biological fluids from the live animal is done so in a manner that fits the institution's mission and expectations. To accomplish this goal, institutions may choose to ensure that the use of all live animals is reviewed and approved by the IACUC (see section 9:6). In addition, the Vertebrate Animals Section of an NIH grant proposal needs to describe the use of live animals in the generation of antibodies, tissues, or other materials, and use of PHS/NIH funds for work involving vertebrate animals requires that descriptions of the animal use in the grant proposal are congruent with the IACUC protocol.[49] Institutions should also consider occupational health and biosecurity issues around the use of biological materials.[22, 33]

9:10 Must the home institution's IACUC review and approve protocols using tissues, cells, or biological fluids acquired from sources other than the IACUC's institution?

Reg. See Section 9.9. The collection of biological materials from live animals, which are intended for research purposes, with few exceptions, requires IACUC review and approval (See Section 9:1–9:2).

Opin. If these substances are acquired from a location outside of an IACUC's home institution, both institutions should ensure all responsibilities related to the live animal use is defined, including which IACUC provides the review and approval (see Section 9:17).

9:11 Does obtaining antibodies from commercial sources require IACUC approval?

Reg. Obtaining antibodies, a biological material, from live animals would be analogous to, and have the same requirements as, the use of live animals to derive animal tissues, cells, or biological fluids (see Section 9:9). The VA has similar requirements.[16]

Opin. There are a few additional items to consider with the acquisition of commercial sources of antibodies. First, does the PHS Policy apply to the production of these antibodies? OLAW FAQ A.2. specifically addresses this issue and states that "the generation of custom antibodies is an activity involving vertebrate animals and is covered by the PHS Policy."[50] This requirement is specific to custom antibodies which should be distinguished from off-the-shelf antibodies. This distinction can be determined by evaluating the method of creation. If the investigator is providing an antigen or the antibody is being created at the specific request of the investigator, this would be considered as custom. If the antibody is for general sale and the creation process is not altered due to the request or requirements of an investigator this would be considered off the shelf. Second, if IACUC approval is required, whose IACUC should approve the animal activity? The use of live animals to produce antibodies to be used in PHS supported activities obligates the awardee institution (the purchaser) to ensure the producer (commercial source) is covered by an OLAW Animal Welfare Assurance (either their own or a component of the awardee's Assurance) and that the use of live animals in the creation of the antibodies has project-specific IACUC approval.[49, 51, 52] As with the use of live animals to create biologic materials, the Vertebrate Animals Section of an NIH grant proposal needs to describe the use of live animals in the generation of antibodies, and when PHS/NIH funds support the use of live vertebrate animals, the institution is responsible for ensuring the proposal is congruent with the IACUC protocol.[49, 53–55] The IACUC(s) that provide(s) approval can vary (see Section 9:17). If a species being used to generate the antibody to be used in research is regulated by the AWA and AWAR, the producer (commercial source) must be registered as a "Research Facility" with the U.S. Department of Agriculture.[51]

Because AAALAC International assesses care and use of animals owned by the accredited institution, if the live animals used to create the antibodies are considered owned by the home institution, then AAALAC would consider these animal activities as part of the home institution's accreditation.[20]

9:12 If an investigator wishes to obtain a surgically altered animal for research use at their home institution, does the pre-acquisition surgery require IACUC approval?

Reg. The requirement would be similar to that for the acquisition of antibodies, biological materials, or tissues derived from live animals (see Section 9:10–9:11). Once at the acquiring institution, any additional procedures performed on the animal would also need IACUC approval, if required (see Section 9:1).

Opin. In line with the opinion of 9:11, there are a few additional items to consider with the procurement of animals that have undergone pre-acquisition surgery. First, does the PHS Policy apply to the surgical procedure? OLAW FAQ A.2. states that "surgery conducted in response to a specific, custom request is covered by the PHS Policy."[50] This requirement applies to surgeries that are performed at the specific, custom request of the investigator and are not off the shelf (e.g., animals previously modified and available before the request was made). Second, if IACUC approval is required, whose IACUC should approve the animal activity? As stated with the acquisition of antibodies, all the PHS Policy requirements would apply and would be the responsibility of the awardee institution. This includes ensuring the institution performing the pre-acquisition surgery has an OLAW Animal Welfare Assurance and verifying that this project-specific surgery has been approved by an IACUC (see Section 9:17).[52] Like with the use of animal derived biological materials, the pre-acquisition, custom surgery would need to be described in the Vertebrate Animals Section of the corresponding NIH grant proposal, since it is part of the scope of work, and it should be congruent with the IACUC protocol (PHS Policy V,B).[54] If the species being used in research is regulated by the AWAR (§2.30), the source providing the animals must be registered as a "Research Facility" with the U.S. Department of Agriculture.[51]

9:13 Does the use of fertilized eggs (birds, reptiles, and amphibians) or fetuses (mammalian) require an approved protocol?

Reg. Although the definition of animal in the AWA (§2132,g) and AWAR (§1.1) includes birds not bred for research, the definition of bird in AWAR (§1.1) specifically excludes eggs prior to the commencement of the hatching process. These regulations do not apply to ectotherms (e.g., fish, reptiles, amphibians, etc.). The PHS Policy applies only to egg-laying vertebrates after hatching, thus IACUC approval would not be required for fertilized eggs. This is also true for fish—the PHS Policy would only apply to larval fish after hatching. That said, the egg laying female would be subject to the requirements of the PHS Policy, if applicable.[56, 57]

The DoD considers egg laying, vertebrate species to be an animal only after hatching. The DoD does not specify the use of fetuses; however, use of the live dam carrying the fetuses would require an IACUC approved protocol and if the fetuses were born alive they would require one as well. The DoD does cover dead animals if the animal was killed for the direct purpose of conducting research, development, test, evaluation, or training.[15]

Opin. Even though fertilized eggs are not currently covered by the AWA, AWAR, or PHS Policy, institutions that use fertilized eggs in research, teaching, testing, training, and experimentation should have plans in place to address the potential for inadvertent hatching of fertilized eggs. This should include the expectations for care, IACUC approval, and likely euthanasia of these animals.[47] In addition, determining when an animal is born is important for animal counting and tracking. This is especially important for USDA regulated animals to ensure appropriate reporting of animal use numbers on the USDA Annual Report of Research Facilities.[58]

9:14 Is an IACUC protocol required for breeding colonies, regardless of planned use in animal activities?

Reg. Both the PHS Policy (III,A) and AWA/AWAR (§2132/§1.1) definition of animals includes "animals … intended for use"; thus, breeding colonies and their offspring that are intended for use in research would require IACUC review and approval. If livestock or poultry are being bred to improve breeding efficiency, then they would be excluded from AWA/AWAR requirements.[14–16]

Opin. Even though these animals may not be associated with a specific animal research activity, according to OLAW, if the breeding and subsequent offspring are part of, and directly support, the institution's research program, the use of these animals requires IACUC review and approval.[59]

Surv. Does your IACUC require an approved protocol for maintenance of a breeding colony?

• No.	11/263 (4%)
• Yes.	217/263 (83%)
• I am unsure.	4/263 (2%)
• We do not maintain breeding colonies at my institution.	31/263 (12%)

9:15 What information should be included for a protocol describing the maintenance of a breeding colony?

Reg. See question 9.3. Some items might not be applicable.

Opin. Consistent with application of one of the 3 R's, reduction, the *Guide* (p. 107) states that the establishment of breeding colonies should be based on need and should be managed in a way that minimizes animal numbers. It would be advantageous for IACUCs to confirm the need to maintain a breeding colony, how breeding will be managed to limit the number of unneeded animals, how the offspring will be used, the final disposition of offspring that are not needed, and how animal use will be tracked.[60] In addition, information on colony management should be included, such as estimated number of breeding adults and offspring, cage/tank density, breeding procedures and system, weaning plans, animal identification/genotyping, etc.[36, 60]

9:16 Do sentinel animals or blood donors need IACUC approval? Since these animals are not used in specific research projects, is it necessary to develop protocols for the review and approval of these activities by the IACUC?

Reg. Neither the AWA nor AWAR cover animals owned by the institution that are used to support clinical care under a veterinary-client-patient relationship (e.g., hospital blood or sample donors).[61] OLAW FAQ D.16. concisely answers this question by stating: "animals used as sentinels, breeding stock, blood and blood product donors, or for other similar purposes may not be part of specific research protocols, their use is part of the institutional research program and directly supports research activities. Consequently, the IACUC should review protocols and SOPs that involve animals for such purposes."[59]

Opin. If research animals are also being used as blood donors or sentinels, this type of use can be captured in the research protocol or activity specific protocol. In addition, both the OLAW and the USDA allow institutions to utilize Standard Operating Procedures (SOPs) that are reviewed and approved by the IACUC to capture required animal use activities in lieu of an IACUC protocol (see Section 9:3 and 9:28).[62]

9:17 If an investigator wishes to conduct or support research using animals at another institution, must the IACUC at the investigator's home institution approve the protocol? Must the other institution's IACUC approve it?

Reg. When USDA-registered research facilities collaborate, it is the responsibility of both registered parties to ensure they meet the requirements of the AWA and AWAR. This can be accomplished through clear delegation of responsibilities for the functions of the IACUC, animal care and handling, and reporting of the animals on the Annual Report that is captured in a contract or similar formal agreement. If these responsibilities are not delegated, all parties involved bear the responsibility of performing the required IACUC functions. Only one of the collaborating institutions should report the animals on the Annual Report to USDA.[38, 63]

The PHS Policy applies to animal activities conducted or supported by the PHS and through MOU or other agreements with OLAW (NASA, NSF, and VA).[9–12] This requirement is based on the source of funding and not location of use. The awardee institution holds the responsibility of meeting the PHS Policy requirements (IV,A). Thus, the home institution would be responsible for ensuring IACUC review and approval of the animal activity occurring at another institution. More than one IACUC can review and approve an animal activity; however, it is not a federal requirement. If one IACUC will be used, discretion can be used to determine which IACUC provides the review and approval.[63–65]

The DoD and VA requirements also apply to animal activities conducted or supported by the associated funding source.[15, 16]

Opin. As eloquently stated by the *Ag Guide* (p. 3), "the fact that research is conducted off site does not lessen the responsibility of the Institution to assure appropriate and humane animal care and use."[18] It is imperative that collaborating institutions define respective responsibilities, including which IACUC(s) will review and approve the activity, to ensure compliance with animal use requirements. This can be accomplished through written agreements between the institutions supporting and/or conducting the animal activity and the performance site where the animal work is actually occurring. This could be captured in a standard document template like a formal, written MOU, contract, or agreement that unambiguously defines respective responsibilities for animal ownership, and animal care and use;, IACUC review, approval, and oversight; and/or other federal, state, or local requirements (*Guide*, p. 15).[38, 64, 65] The Federal Demonstration Partnership developed a sample MOU and associated guidance that is available for use by collaborating institutions.[66]

Conduct of PHS-supported activities obligates the awardee institution to ensure the performance site is covered by an OLAW Animal Welfare Assurance (either their own or a component of the awardee's Assurance).[64] The NIH requires that awardees have a formal written agreement with each consortium participant that includes compliance with applicable public policy requirements, in this case, the PHS Policy, and thus all the provisions of the PHS Policy (IV.B.2.) for required functions of the IACUC.[54] Consideration can also be given for the collaborating institution's status as it relates to OLAW Assurance, USDA license, USDA registration, AAALAC International Accreditation, or other accreditations and possible permits or other requirements (e.g., those associated with the Endangered Species Act, Convention on International Trade in Endangered Species of Wild Fauna and Flora, etc.).[33]

It is advisable for the IACUC to have established processes in place that enable identification of collaborations which involve the conduct of work using live vertebrate animals. An effective process will be one that informs and allows the initiating investigator to understand and obtain the required documents of the home institution in an efficient manner and in advance of beginning any activities using animals.[27]

9:18 If an investigator wishes to conduct or support research using animals at a foreign institution, must the

IACUC at the investigator's home institution approve the protocol? Must the other institution's IACUC approve it? What about animal samples that are obtained by citizens of the foreign country and sent to a domestic institution?

Reg. Application of the AWA is limited to the "state"; and the AWA (§2132,d) defines "state" as the United States, the District of Columbia, the Commonwealth of Puerto Rico, the Virgin Islands, Guam, American Samoa, or any other territory or possession of the United States. Thus, animal activities conducted at institutions in these defined locations would require IACUC approval prior to initiation. (See Sections 9:2 and 9.17.)

The PHS Policy requirements stated in 9:10 and 9:18 would apply to both foreign and national collaborations. If the domestic institution is the prime awardee, the domestic institution's IACUC is responsible for verification of IACUC approval. The Awardee may accept the animal use protocol from the foreign institution, but the domestic awardee IACUC is responsible for the review and approval. It is important to note that if the animal activity will be conducted or supported with PHS, NSF, or NASA funds and the performance site is a foreign institution, the foreign institution must have an Animal Welfare Assurance for Foreign Institutions with OLAW, regardless of the prime awardee's assurance status.[67]

Collection of animal samples is analogous to the scenario addressed in 9:7–11. Specific to samples collected outside the United States, the home institution should consider if the acquisition of the animal sample would require IACUC review and approval (see Section 9:2). If IACUC review and approval is required and animals will be used for a PHS-supported activity, then the awardee institution would be responsible for ensuring the activity was consistent with the PHS Policy Requirements (see above and Section 9:17).[48]

Opin. Institutions that collaborate with international performance sites should evaluate the arrangement to ensure it meets all funding requirements, their standards of animal use, and that the benefit of the collaboration outweighs the potential risk (e.g., reputation damage if there is animal harm, less direct oversight by the home institution, ethical considerations, etc.). Understanding the prevailing standards of the country in which the collaborating institution is located, as well as any associated internal (Ethics Review Committee, Institutional Animal Care and Use Committee, etc.) and/or external (government agencies, OIE member, AAALAC International, other accrediting bodies) mechanisms of oversight can help the home institution determine if the collaboration meets their expectations.[27] IACUCs can also consider following OLAW's requirement for work performed at foreign institutions by requiring international collaborators to follow the International Guiding Principles for Biomedical Research Involving Animals.[31, 68] As indicated above, having a written agreement between the institutions that are supporting and those conducting the animal activity, which captures the details concerning the specific animal care, use, and handling requirements, is valuable (see Section 9:17). Flexibility can be considered as long as such accommodations meet the requirements and institutional mission.[69]

Although it is beyond the scope of this chapter, institutions should be aware of, and follow, other applicable federal laws, importation requirements, permits, and occupational health concerns related to importation of animal-derived samples.[22, 48, 70]

9:19 What minimal qualifications does a person need in order to be the Principal Investigator (PI) of an IACUC protocol?

Reg. The AWAR (§1.1) define Principal Investigator (PI) as "an employee of a research facility, or other person associated with a research facility, responsible for a proposal to conduct research and for the design and implementation of research involving animals."

PHS Policy does not specifically define the term, "principal investigator." The U.S. Government Principle VIII uses the term "Investigator" and the requirement indicated is that this person should be appropriately qualified and experienced for conducting procedures on living animals. The NIH Grants Policy Statement defines the PI as "The individual(s) designated by the applicant organization/recipient to have the appropriate level of authority and responsibility to direct the project or program to be supported by the award."[54]

The DoD and VA have specific requirements of the PI but do not mention specific minimal qualifications.[13, 14, 16, 29]

Opin. The minimum qualifications for a PI are left to the institution to define.[71] The PI has responsibilities with regard to the activities the IACUC must review and approve as outlined in section 9:3. As stated in the *Guide* (p. 27), the PI has precise knowledge of the animal activity. In addition, the PI holds responsibility within the IACUC protocol over the use of animals on the protocol, to the personnel in the laboratory, and within the institution to ensure compliance with local and federal requirements.[33] Given the PI's obligations, it is advisable that IACUCs confirm that anyone in this role has the authority, oversight, and knowledge of the IACUC-approved animal activity to ensure that it can be managed and implemented as approved. Furthermore, the PI needs to be able to meet the training and qualification requirements and, hopefully, be available to respond to concerns in a timely fashion. An IACUC could, for example, set minimum qualifications for one to qualify as a PI.[34, 49, 72] If an IACUC protocol has more than one principal investigator, it should be understood that the responsibility and accountability of an individual PI is not diminished.[54, 61] For IACUC protocol purposes identifying one responsible individual for the oversight of the activities in the IACUC protocol would be important to prevent ambiguities in responsibilities.

9:20 What minimal qualifications does a person need to submit a protocol to the IACUC? For example, can students, postdoctoral fellows, medical residents, or visiting scientists submit a protocol?

Reg. There are no specific regulatory expectations or guidelines that define minimal qualifications for one to submit a protocol to the IACUC.

Opin. It is recommended that the institution have established requirements (i.e., policy) outlining who can submit a protocol to the IACUC. Principal investigators have significant responsibilities with regard to the IACUC protocol and should be able to submit the protocol for IACUC review and consideration of approval (see Section 9:19). If the submitter is not the PI, IACUCs may want to consider who can submit on the PI's behalf, how the PI is made aware of the submission, if the submitter holds any other authority over the protocol (e.g., can they respond to modifications to secure approval or IACUC questions/clarifications), etc. Considering the potential impact if the protocol is submitted and managed by an individual who is not the PI, IACUCs may consider consulting with general counsel on their submission requirements to ensure there are no ambiguities with respect to institutional responsibilities.

9:21 Who typically maintains IACUC protocols and for how long should they be kept?

Reg. Both the AWAR (§2.35) and the PHS Policy (IV,E) require the research facility or awardee institution, respectively, to maintain records of protocols and the IACUC review outcome (e.g., IACUC approval given or withheld) for three years after completion of the activity. Longer retention may be required to comply with other applicable laws or if requested by an APHIS Administrator.[38]

The VA and DoD are similar and require record retention for at least three years after the end of the animal activity.[15, 16]

Opin. The institution is responsible for ensuring IACUC records are maintained as required. This responsibility is generally delegated to the IACUC administration office. It would also be prudent for the PI to maintain copies of the approved protocol for the required retention period. Final disposition of the protocol after the required retention period is not required, but advised. Any records in the institution's possession, regardless of the retention period, in either digital or physical format, may be subject to subsequent release to a FOIA request and/or for legal reasons (e.g., subpoena or discovery demands).[36, 73, 74]

9:22 Who should have access to IACUC protocols?

Reg. The AWAR (§2.31,d,2) require that written descriptions of all proposed activities that involve the care and use of animals be available to all IACUC members. Additionally, the AWAR (§2.35,f) require that authorized APHIS or federal funding agency representatives have access to protocols for inspection. The PHS Policy (IV,E,2) is similar and requires authorized OLAW or other PHS representatives to have access to records for inspection.[38]

Opin. To the author's knowledge, there is no requirement for specific institutional representatives, other than IACUC members, to have access to protocols; however, it is important that key personnel have access to help ensure protocol compliance. This may include providing IACUC protocol access to:

- IACUC administrative personnel: Allows them to provide assistance with protocol management, track deadlines, and answer questions.[74]
- The Attending Veterinarian and veterinary support personnel: Allows them to view approved animal activities; assist in post-approval monitoring; review IACUC approved exemptions, exceptions, and departures; and reference established experimental and humane endpoints.[74]
- Animal husbandry personnel: Allows them to verify protocols for animal procurement; view approved animal activities; assist in post-approval monitoring; review IACUC approved exemptions, exceptions, and departures; reference approved experimental and humane endpoints; and review study specific husbandry needs.[36, 74]
- Occupational health and safety personnel: Allows them to view approved species and hazardous substance use and, possibly, the location of use if that is a component of the protocol.
- The Institutional Official (IO): Enables one possible mechanism by which the IACUC can notify the IO regarding the outcome of IACUC protocol review (approve, withhold approval, or modifications required to secure approval) as required by the PHS Policy (IV,C,4) and AWAR (§2.31,d,4).
- IACUC members: Provides them written descriptions of all proposed animal activities and aides IACUC members in completing semiannual facility inspections and program reviews as required by the AWAR (§2.31) and PHS Policy (IV,B)

- Principal Investigator: Provides an unambiguous understanding of the final approved protocol and what activities are approved (see Section 9:19).[36]
- Research personnel: Allows those performing the animal activity to have the final IACUC-approved version of the animal procedures, thus, aiding in maintaining compliance (see Section 9:26).[36]
- Other: Other individuals may need access to help fulfill the institution's research mission and facilitate responsibilities such as post-approval monitoring, protocol compliance, and/or grant/funding requirements (e.g., grants and contracts staff).[74]

Although it is likely advantageous to provide key personnel access to IACUC protocols, IACUCs should consider document security and confidentiality. Having limited and controlled access to protocols can provide a mechanism to protect against release of sensitive or confidential information. IACUCs are encouraged to develop policies and processes related to IACUC protocol access (see Section 9:23). As mentioned in Section 9:22, records in the institution's possession may be subject to release due to FOIA requests and/or for legal purposes (e.g., subpoena or discovery demands), even after the required retention period has passed.

Surv. Who has ready access to an IACUC protocol form other than the Institutional Official, the IACUC office, the IACUC chairperson, the Principal Investigator, and the Attending Veterinarian? Select all that apply.

• Nobody else	15/263 (6%)
• Certain high level institutional officials	50/263 (19%)
• Certain animal facility administrators	147/263 (56%)
• Certain animal care or animal health technicians	157/263 (60%)
• The Principal Investigator and the research team	191/263 (73%)
• USDA/Animal Care inspector (for Animal Welfare Act covered species)	74/263 (28%)
• Other IACUC members or alternates	169/263 (64%)
• There is open access to anybody	22/263 (8%)
• I don't know.	0/263 (0%)

9:23 What are some recommendations for maintaining confidentiality if electronic systems are used for protocol submission and review?

Reg. The AWA (§2157) states that it is unlawful for an IACUC member to release any confidential information. In addition, the PHS requires IACUC members to treat all items they review as confidential.[75]

Opin. Storing confidential material online provides consistent and convenient retrieval to those that need it, but it can make the information more vulnerable to remote, unnoticed, and/or un-approved access. It is imperative that institutions take the necessary steps to ensure information within an IACUC protocol is protected, secure, and treated as confidential. Appropriate due diligence in this regard will not only ensure that the requirements are met, but will also safeguard the institution from potential electronic breaches and possible targeting by animal rights activism. Some items that might be considered with respect to maintaining protocol confidentiality include[36, 76]:

- Coordination between the IACUC and the information technology team to ensure any online electronic systems that are used for protocol submission, review, and storage meet all applicable institutional requirements for security of proprietary, confidential, trade secrets, intellectual property, business information, etc.
- Coordination with institutional public relations representatives to ensure there is appropriate security for information that could be used by animal rights activists or other individuals/groups that would use the information to harm the institution, personnel, animals, property, or reputations. Protocols may include descriptions of animal procedures, personal information (names, phone numbers, addresses), animal use locations, substance storage locations (e.g., select agent, controlled substances), etc. that require protection.
- Providing secure access to online protocol information and having a gatekeeping process that verifies and limits those with access to the protocol system (see Section 9:22).[76]
- Consideration for requiring confidentiality agreements, which have been reviewed by institutional general counsel, for those with access to protected/confidential information.
- Further protective measures to consider for online systems include:
  - Consistent and routine personnel training
  - Secure login, passwords, encryption[77]
  - Multifactor authentication
  - Access control[77]
  - Role control (e.g., who can edit, delete, submit, upload, etc.)
  - Regular and secure back-ups
  - Ensuring that software is up to date, and all security patches are in place

- Audit trail and recordkeeping
- Use of institutional intranets[36]

Surv. Does your IACUC currently use a reasonably sophisticated electronic protocol submission and/or review system? This refers to commercial or in-house developed systems other than basic word processing programs.

• Yes, but for submissions only.	9/263 (3%)
• Yes, but for review only.	1/263 (0%)
• Yes, for submissions and review.	154/263 (59%)
• No, we do not use a sophisticated submission or review system.	99/263 (38%)

9:24 What is meant by "just-in-time" for NIH grants/awards and how is this related to the IACUC protocol submission and review process?

Reg. NIH Grants Policy Statement 1.2 describes just-in-time (JIT) as:

> "NIH policy allows the submission of certain elements of a competing application to be deferred until later in the application process, after review when the application is under consideration for funding. Within the Status module of the eRA Commons, users will find a feature to submit Just-In-Time information when requested by NIH. Through this module, institutions can electronically submit the information that is requested after the review, but before award."

Use of the just-in-time (JIT) process is optional, and part of the NIH's JIT procedures is to verify that IACUC approval of the use of live vertebrate animals has been granted.[54] If protocols are used as the IACUC's mechanism to review and approve animal activities, the JIT process is linked to the institutional IACUC protocol submission and review process for any NIH grant with a Vertebrate Animals Section.

Opin. The JIT process requires verification of IACUC approval of the project's proposed use of live vertebrate animals. To complete this verification, the grant applicant must submit to NIH the date the IACUC approved the vertebrate animal activities described in the grant proposal. The NIH uses a virtual meeting space for this submission called Electronic Research Administration Commons, or eRA Commons. The approval date must be within three years of the date the verification is provided through eRA commons. The NIH will not accept approval dates that are beyond the three-year complete review cycle, and NIH funds cannot be used to conduct or support live vertebrate animal use until there is documentation of IACUC approval for the animal activity.[54] In addition to capturing the IACUC approval date, institutions may have additional verifications associated with the JIT process to ensure that all requirements are satisfied, such as verification of congruence between descriptions of animal use in the grant application with descriptions offered in the IACUC protocol. It is important for Institutions to understand that when an IACUC approval date is provided to the NIH, the NIH considers this a verification by the institution that there is congruence between the grant/contract and the IACUC-approved protocol.

Because the JIT process is optional, if an institution chooses to use this deferred verification process, the institution is required to support the decisions of the IACUC and cannot pressure or overrule an IACUC decision. In addition, the PI is required to be forthcoming and provide timely communication of any changes to animal usage that are a result of the NIH review and award process.[78]

Because of the time and effort involved in the IACUC protocol review and approval process, the IACUC protocol review and approval is usually one of the elements deferred to the JIT stage. Not all grant applications that are submitted are funded; thus, this deferral process saves considerable administrative burden on the PI and the IACUC by allowing animal activities that will not be funded or performed from having to undergo IACUC review and approval. PIs may already have an approved IACUC protocol in place that describes the activities in the JIT-related grant and providing a date of IACUC approval is relatively easy. In addition, they may be able to capture the animal activities described in the grant by modifying an existing protocol (e.g., protocol amendment; see Chapter 11). However, if there is no existing protocol for the animal activity then PIs can be caught off-guard by the short turnaround time for submission of the required elements related to the JIT. The NIH Grants Management Officer or Grants Management Specialist provides a submission deadline when notifying investigators of a JIT. This deadline may be just a few days after the initial JIT notification. In general, this is not enough time for the PI to draft and submit a protocol and have the protocol reviewed and approved by the IACUC.[79] In the author's experience, providing an IACUC memo to the NIH stating that a protocol has been submitted and is undergoing IACUC review may appease the NIH and provide additional time to complete the required items of the JIT.

9:25 Investigator Jones has a PHS grant and permanently moves from Institution A to Institution B. The PHS allows the grant to be transferred to Institution B. Dr. Jones has

an IACUC-approved protocol at Institution A but will complete the animal-related research at Institution B. Must Institution B's IACUC review and approve the proposed research before it can begin?

Reg. OLAW FAQ F.3. states:

> "The transfer of PHS-supported research to a different Institution requires the prior approval of the funding component. The proposed new grantee Institution must have or obtain an Animal Welfare Assurance and possess all the resources necessary to fulfill the conditions of the grant, and its IACUC must review and approve the animal activities. The original IACUC approval is void when the original grantee formally relinquishes the award. The receiving Institution must provide verification of IACUC approval prior to receiving funding. Note that the conditions of approval by the IACUC at the receiving Institution may differ from those required by the original grantee's IACUC."[80]

Opin. The NIH Grants Policy 8.1.2.7 uses the term "Change of recipient organization" and defines it as a transfer of the legal and administrative responsibilities of the grant.[54] It is advisable for institutions to have a mechanism to capture these situations and engage the IACUC to help ensure adherence to the requirement.

9:26 Which people should be listed on an IACUC protocol?

Opin. Many of the items that should be in an IACUC protocol are reliant on the PI providing documentation of, and consideration for, aspects of the animal activity (see Section 9.3). Also, in some cases there is a requirement that the PI be named.[16] In any case, it is common practice that the PI(s) be listed on each IACUC protocol (see Section 9.19).

It is common for institutions to use the IACUC protocol to capture other relevant information beyond that which is required (see Section 9:3). For example:

- The name of each person who will be conducting live animal procedures on the protocol. This practice can be advantageous because it can help capture, document, and verify the qualifications, training, and experience of these individuals with respect to the animal activities that specific personnel will perform[7, 18, 33, 36, 81–83] This approach can also be used to verify that the protocol participants are enrolled in a program of risk assessment as described in the *Guide* (p. 22).[22, 23, 17, 33, 84]
- The name of each person who will be using hazardous substances, entering restricted areas, or involved in other activities that may pose a health or safety risk (e.g., handling nonhuman primates, radioisotopes, lasers, etc.). This can help ensure occupational health and safety requirements related to these activities are met.[18, 22, 23, 33, 84]
- The names of personnel who can be contacted in case of an animal emergency and/or other urgent IACUC need relevant to the protocol.

9:27 What is a reasonable sequence of time and events that begins when a PI submits a protocol and ends when the protocol is ready for a definitive disposition (approve or withhold approval)?

Reg. The AWAR (§2.31,d,2–4) and PHS Policy (IV,C,2–4) describe a very specific sequence that must occur for all IACUC protocols:

- Prior to the review, each IACUC member shall be provided with a list of proposed research projects to be reviewed.
- Written descriptions of research projects that involve the care and use of animals shall be available to all IACUC members, and any member of the IACUC may obtain, upon request, full committee review of those research projects.
- If full committee review is not requested, at least one member of the IACUC, designated by the Chairperson and qualified to conduct the review, shall review those research projects and have the authority to approve, require modifications (to secure approval), or request full committee review of those research projects.
- If full committee review is requested, approval of those research projects by the full committee may be granted only after review at a convened meeting of a quorum of the IACUC and with the approval vote of a majority of the quorum present. Alternatively, designated member review may occur subsequent to full committee review when specific criteria are met, see Chapter 11.[85]
- The IACUC may invite consultants to assist with the review of complex issues. Consultants may not approve or withhold approval of an activity or vote with the IACUC unless they are also members of the IACUC.
- The IACUC shall notify investigators and the institution in writing of its decision to approve or withhold approval of those activities related to the care and use of animals, or

of modifications required to secure IACUC approval. If the IACUC decides to withhold approval of an activity, it shall include in its written notification a statement of the reasons for its decision and give the investigator an opportunity to respond in person or in writing.

Opin. The workflow of events that occur once a protocol is submitted can vary widely depending on the needs of the program. As long as the sequence indicated in the requirements are met, the IACUC can use its discretion related to other steps in the review process. However, OLAW Assured institutions must be cautious to ensure that they are following the path agreed to in their Animal Welfare Assurance.[12] If workflow is modified, it must meet the minimum requirements, and PHS Assured Institutions should report these changes on their OLAW Annual Report.[86]

Items that may be additionally included in a protocol review workflow to ensure all IACUC and/or other Animal Program expectations are being met (see Section 9:3) include:

- Review the consideration of alternatives, if needed[33]
- Review of the statistical justification of animal number and/or group sizes, if needed[33]
- Administrative verification that protocol personnel are enrolled in a risk assessment program or similar occupational health and safety program requirement[22, 23, 33, 84]
- Administrative verification that personnel have completed all required training and are qualified to conduct the work described in the protocol (see Section 9:26)[7, 18, 33, 81, 83, 87]
- Review of appropriateness and availability of the proposed animal care and use facilities for the proposed procedures[22, 23, 33, 36]
- Verification that scientific merit has been established[16, 18, 33, 74, 88]
- Verification that an appropriate harm-benefit evaluation has been performed[47, 89]
- Verification that any collaboration or contracted services external to the home institution meets institutional and external requirements (See Sections 9:10–12 and 9:17–18)
- Safety review of hazardous agent or equipment use (radiation, biological, chemical, IBC, etc.)[18, 22, 23, 33, 84]
- Review of the animal transportation plan within the facility, between the facilities, and off-campus[33, 90, 91]
- Consultation with the Attending Veterinarian or designee related to painful or distressful procedures, established humane endpoints, surgery, euthanasia, etc.[33, 36, 74]
- Administrative pre-review for completeness[33, 36, 74]
- Evaluation of efforts undertaken to enhance reproducibility of the work[33, 47, 92]
- Review from other consultants to assist with the review of complex issues[36]

Surv. Does your IACUC have a general administrative review to ensure that the protocol form is filled in correctly before the protocol is either pre-reviewed by a veterinarian or sent for formal IACUC review?

- No 44/263 (17%)
- Yes 213/263 (81%)
- Other 6/263 (2%)

Surv. If your IACUC uses a veterinary pre-review process, how long does that process typically take from the time the veterinarian receives the protocol until s/he returns it to either the IACUC or the investigator?

- We do not use a veterinary pre-review. 92/263 (35%)
- Less than 1 week. 93/263 (35%)
- 1–2 weeks. 70/263 (27%)
- 3–4 weeks. 8/263 (3%)
- More than 4 weeks. 0/263 (0%)
- I don't know. 0/263 (0%)

Surv. If your IACUC ever uses a Designated Member Review (DMR) for protocol review, how long does it take from the time the designated reviewer receives the protocol until the protocol is first returned to either the investigator or the IACUC office?

- We do not use DMR. 19/263 (7%)
- Less than 1 week. 100/263 (38%)
- 1–2 weeks. 130/263 (49%)
- 3–4 weeks. 11/263 (4%)
- More than 4 weeks. 1/263 (0%)
- I don't know. 2/263 (1%)

Surv. If your organization ever uses Full Committee Review (FCR), how long does it take from the time committee members first receive a protocol until it is discussed at a full committee meeting? This includes the time it may require to have back-and-forth discussions with the investigator. Please respond to this question only if your IACUC meets approximately once a month as a full committee. Otherwise, respond as "No/not applicable."

- No/not applicable. 45/261 (17%)
- We do not use FCR. 8/261 (3%)
- Less than 1 week. 25/261 (10%)

- 1–2 weeks. 100/261 (38%)
- 3–4 weeks. 71/261 (27%)
- 5–6 weeks. 8/261 (3%)
- More than 6 weeks. 3/261 (1%)
- I don't know. 1/261 (0%)

9:28 If the exact same procedure is repeatedly used (such as a surgical procedure), can the IACUC or PI have an approved "procedure bank" of standard animal procedures so a PI can reference the approved procedure without having to write it out in detail every time? (see 9:30)

Reg. Animal activities should be reviewed and approved by the IACUC through one of the two established and required mechanisms, Designated Member Review or Full Committee Review (See 9:27), and receive a complete review by the IACUC at least once every three years as outlined in the AWAR (§2.31,d) and PHS Policy (IV,C).

OLAW FAQ D.14. states: "The IACUC must review and approve activities on a project-specific basis, taking into account a number of factors, e.g., the aims of the study, consideration of alternatives, minimization of pain and distress. For routine aspects of research (e.g., species specific techniques for immunization and titer determinations during antibody production), IACUCs may approve SOPs that can be cited by investigators in their protocols in order to avoid needless repetition. SOPs should be reviewed by the IACUC at appropriate intervals (at least once every three years) to ensure they are up-to-date and accurate."[93]

Opin. It is assumed that the use of the word "approved" in this scenario is referring to IACUC approval. How IACUCs choose to use "Procedure Banks" that contain approved standard animal procedures to reduce administrative burden can vary widely. The IACUC needs to ensure that standard procedures are available to both the PIs and the individuals who will perform the procedure. In addition, there should be a mechanism in place to ensure that the IACUC provides the initial review and approval and the required triennial review and approval as described in the OLAW FAQ D.14. Institutions with OLAW Animal Welfare Assurances will need to ensure they are following the approval process agreed to in their Assurance.[12, 93]

One approach to utilizing approved standard animal procedures is to allow PI's to simply reference the standard animal procedure in their IACUC protocol without requiring a detailed description within the protocol. With this approach, it is imperative that the PI and individuals who are carrying out the procedures have access to the final, approved, detailed, written description of the standard animal procedure.[18, 52, 93–95] The approved standard animal procedure, like any animal activity approved by the IACUC, may undergo modifications (e.g., significant changes) throughout its three-year lifespan. In addition, modifications may be required at the time of the triennial protocol review and approval. If a PI simply references the standard animal procedures in the IACUC protocol, how will these changes be managed? Will the PI need to commit to follow a specific version in their IACUC protocol; if not, does the IACUC have a mechanism to notify them of changes to ensure they remain compliant? It is important to ensure that the PI and personnel performing the procedures have working knowledge of the exact version of the approved standard animal procedure so that any potential changes with new versions of standard animal procedures do not create ambiguities and confusion, which can lead to non-compliance.[74]

One could also combine multiple projects or investigators that perform similar, routine procedures under a single IACUC protocol to avoid needless repetition.[52]

9:29 If a procedure is repeatedly performed, but a single variable is altered (e.g., a new drug to be tested), can the IACUC have an approved "procedure bank" of standard animal procedures so a PI can reference the approved procedure without having to write it out in detail every time? If yes, how should the IACUC address the issue of the single variable?

Reg. See Section 9:28.

Opin. Flexibility in procedure descriptions is allowed and encouraged by OLAW and the USDA. For example, PIs can describe dose ranges, drug formularies, and procedure frequency ranges in the IACUC protocol. In addition, IACUCs can develop policies to permit the use of veterinary verification and consultation for some significant changes.[38, 96–99] How to manage the change of a single variable in an approved standard animal procedure depends on the type of variable that may change. If the different variables (e.g., drugs) that may be tested are known at the time of protocol creation, they can all be mentioned as potentially being administered in the approved protocol. Alternatively, if the variables share a common characteristic (e.g., specific type of drug) the protocol may be able to capture all administration options under that drug class (e.g., alkylating chemotherapeutics, specific agonist, etc.) or structural analogs of the substance, which will allow flexibility to evaluate compounds. It is important to ensure that variations within the drug class or analog would not change any component of the approved protocol (e.g., pharmaceutical grade, preparation of the drug, route or frequency of administration, volume of administration, biosafety level, animal handling, etc.) or animal impact (potential complications, adverse events, or pain/distress).

If there is not enough known about the variable to include flexibility in the approved standard animal procedure, the PI could reference the standard procedure and list all the variables they anticipate using in the IACUC protocol. A change in the variable would require a protocol modification that would need to be reviewed and approved by the IACUC. Significant changes to the approved IACUC protocol must be approved by one of the valid IACUC approval methods allowed by the PHS Policy and/or AWAR (see Chapter 10).[38, 97]

9:30 Do inter-institutional "procedure banks" of standard animal procedures exist that can be used for the purposes of descriptions included in protocols?

Reg. There is no requirement related to inter-institutional "procedure banks" of standard animal procedures; however, should an IACUC adopt the use of "procedure banks" they should ensure that the animal activity is reviewed and approved by the IACUC through one of the two established and required mechanisms, Designated Member Review or Full Committee Review (see Section 9:27).

Opin. IACUCs may develop "procedure banks" consisting of standard animal procedures for their own internal use. Such a bank of standard animal procedures might be placed, for example, on the IACUC's website or be shared by other means. To the author's knowledge, there are currently no "procedure banks" which are marketed as being inter-institutional and freely available for other institutions to use. That said, the Federal Demonstration Partnership currently has a Compliance Unit Standard Procedure (CUSP) Working Group which is a multi-institutional initiative to develop and share standard animal procedures that can be used in IACUC protocols.[100, 101]

If an IACUC chooses to use a "procedure bank" available at another institution, it is advisable that the local IACUC perform the complete review and approval of the standard animal procedures versus accepting the other institution's IACUC approval. This recommendation stems from the requirement that the home institution's IACUC review and approve the animal activities to be conducted at the IACUC's institution (see Section 9:25).

Surv. Does your IACUC have a collection of IACUC approved research techniques (i.e., standard operating procedures) or policies that an investigator can reference in his or her application without having to rewrite the entire technique/policy? Examples: Tail snips for genotyping, tattooing, common surgical procedures, animal restraint policies.

• No	88/261 (34%)
• Yes	173/261 (66%)

9:31 A PI wishes to use a very small amount of blood (for research) that will be taken for clinical purposes by a veterinarian from a privately owned pet dog. Is IACUC approval needed for the veterinarian to give the blood to the researcher?

Reg. If the dogs are not being handled in response to the research requirements, and the blood is obtained as part of care that was pursuant to a valid veterinarian-client-patient relationship, then this animal activity qualifies as veterinary clinical care rather than research. Veterinarian-Client-Patient Relationship is generally a fee-for-service arrangement between the pet owner and the veterinarian and is regulated by state veterinary licensing boards.[102, 103] Veterinary clinical care of a privately owned animal is not considered a research activity by the USDA or OLAW and does not require IACUC approval. However, if any part of the veterinarian's participation in the blood draw from the dog is for the purpose of the research activity or influenced in any way due to the study parameters (e.g., more frequent blood draws or collection of a larger blood volume) then both the USDA and the OLAW would require IACUC review and approval of the animal activity, where applicable.[42, 102, 104–107] Furthermore, OLAW states in FAQ A.8: "When a privately owned animal is recruited, with the owner's consent, for participation in a research study or veterinary clinical trial and the activity includes collection or generation of data for research purposes, such activities are considered research and are subject to IACUC oversight."[102]

If it is determined that IACUC review and approval is required for this animal activity, the PI is responsible for obtaining an informed client consent form from the pet owner for the research activity. If the animal activity is overseen by OLAW through PHS Policy or OLAW MOU, then the performance site must secure an OLAW Animal Welfare Assurance before initiation of the animal activity. In addition, if performed at another institution or veterinary clinic, the home institution should evaluate the need for a written agreement of animal care and oversight (see Section 9:17).[42, 102]

Opin. It may be advantageous for IACUCs to establish formal processes and/or procedures to determine when IACUC approval is required for animal use situations that are not very clear (see Sections 9:6–14, 9:16, 9:18)[102, 108] Such an approach can help establish consistent, established mechanisms within the program to provide a definitive response regarding the need for IACUC approval on a case-by-case basis. Avoiding miscommunication or misunderstandings related to animal use activities is imperative. Furthermore, having documentation of such decisions may be advantageous to the investigator if they hope to publish results from the work and IACUC approval is expected by the journal. The IACUC can also use this formal process to communicate additional information regarding the animal activity that may lie outside of the IACUC's purview. This might include, for example, import-export requirements, occupational health and safety contacts, and requirements of other compliance committees (e.g., IBC or IRB) (see Section 9:7).

If it is determined that IACUC review and approval is not required, it is critical that both the researcher and the veterinarian understand that the sample collection by the veterinarian cannot be modified in any way as a result of the study design. Use of already collected blood that was to be discarded would be a good way to ensure the veterinarian would have no study bias.

9:32 A veterinarian draws blood from a cow that is to be slaughtered the next day. The blood is specifically destined for use in an in-vitro biomedical research project. Is IACUC approval needed to draw and use the blood?

Reg. Yes. As mentioned in Section 9:31, if any part of the veterinarian's participation in the sample collection (e.g., blood) from a client-owned animal is for the purpose of the research activity or influenced in any way due to the study parameters (e.g., more frequent blood draws or collection of a larger blood volume), then IACUC approval would be required prior to any activities involving an animal(s).

Opin. See Section 9:31.

9.33 An investigator receives Institutional Review Board approval to compare humans who receive animal assisted therapy to those who do not. Is IACUC approval needed for the animals involved in the therapy?

Reg. If the animals are not the subjects of the mentioned study and will not be manipulated in any way as a result of, or for, the study activity, the use of animals in this scenario would not be regulated by the AWA or AWAR. The author recommends that the IACUC review the institution's OLAW Animal Welfare Assurance and/or contact OLAW to determine if OLAW would require the animal activity to have IACUC review and approval.[59, 109]

Opin. See the opinion in Section 9:31.

9:34 Live mice are fed to snakes as a source of food, and the snakes are the research subject supported by PHS funds. The mice arrive in the morning and are fed by the end of the day. Does the use of the mice as food require IACUC approval?

Reg. OLAW states "in situations where the PHS Policy is applicable, the feeding of live vertebrate animals to other animals is also subject to IACUC oversight either as a part of the protocol review process or as a covered component of the institutional program of animal care and use."[110]

Opin. Although there is no clear guidance on this topic, if the live mice that are acquired by the institution are used in the course of research, teaching, or testing, then one could assume that IACUC review and approval would be required.

9:35 Is IACUC approval required for institutional animals not used in research, teaching, testing, or experimentation (e.g., mascots, feral cats, office pets, therapy animals).

Reg. If the animal is not being used for research, teaching, training, testing, or experimentation as defined by the definition of animal in the AWAR (§1.1) or the PHS Policy (III,A), then these activities would not require IACUC review and approval (see Section 9:1 and 9:3).

Opin. IACUC knowledge and/or oversight of institutionally owned animals that do not require IACUC review and approval is advisable. What this oversight consists of might be circumstance and species dependent but could include obtaining a consent agreement with the owner/responsible individual; ensuring all human occupational health and safety concerns are addressed; ensuring the animal has appropriate veterinary care; and establishing animal care standard operating procedures while at the institution. IACUC consultation with the institution's general counsel about potential risk and liabilities related to these animals may be appropriate. The institution will need to follow their OLAW Animal Welfare Assurance, and if the Assurance is written to require IACUC approval of all animal activities regardless of type or funding source, then IACUC review and approval would be required.[111]

The IACUC should ensure that the presence of such animals does not impact the health status of animals being maintained at the institution and which are being used for research, teaching, training, testing, or experimentation. In addition, it is imperative that the institution comply with all federal, state, and local laws associated with service animals. A process that includes the IACUC and Attending Veterinarian in the review of requests to allow service animal entry into restricted animal use areas is imperative to ensure health and safety of existing colony animals.[112]

REFERENCES

1. U.S. Department of Agriculture. 1966 (Last amended 2013). *Animal Welfare Act (Public Law 89-544).* https://www.aphis.usda.gov/animal_welfare/downloads/AC_BlueBook_AWA_508_comp_version.pdf. Accessed January 28, 2023.
2. U.S. Department of Agriculture. *Animal Welfare Regulations 9 CFR* (U.S. Department of Agriculture, 2021). https://www.aphis.usda.gov/animal_welfare/downloads/AC_BlueBook_AWA_508_comp_version.pdf. Accessed January 28, 2023.
3. Health Research Extension Act of 1985. *Public Law 99-158. Sec. 495.* https://grants.nih.gov/grants/olaw/references/phspolicylabanimals.pdf. Accessed January 28, 2023.
4. Interagency Research Animal Committee. *US Government Principles for the Utilization and Care of Vertebrate Animals Used in Testing, Research, and Training* (US

Department of Health and Human Services, Washington, DC, 1985). https://grants.nih.gov/grants/olaw/references/phspolicylabanimals.pdf. Accessed January 28, 2023.

5. Office of Laboratory Animal Welfare, National Institutes of Health. 2015. *Public Health Service Policy on Humane Care and Use of Laboratory Animals* (US Department of Health and Human Services, Bethesda, MD). https://grants.nih.gov/grants/olaw/references/phspolicylabanimals.pdf. Accessed January 28, 2023.
6. Institute for Laboratory Animal Research. 2011. *Guide for the Care and Use of Laboratory Animals,* 8th ed. (National Academies Press, Washington, DC). https://grants.nih.gov/grants/olaw/guide-for-the-care-and-use-of-laboratory-animals.pdf. Accessed January 28, 2023.
7. AAALAC International. *Accreditation Program – Rules of Accreditation.* May 2022. https://www.aaalac.org/accreditation-program/rules-of-accreditation/. Accessed January 28, 2023.
8. Garnett, N.L. and W.R. DeHaven. 1997. OPRR and USDA Animal Care response on applicability of the AWA/R and the PHS Policy to dead animals and shared tissues. *Lab Animal* 26(3): 21. https://olaw.nih.gov/guidance/articles/laba97v26n3p21.htm. Accessed on March 19, 2024.
9. National Institutes of Health, Office of Laboratory Animal Welfare, 2022. MOU between NIH and NASA. https://olaw.nih.gov/nasa.html. Accessed August 31, 2022.
10. National Institutes of Health, Office of Laboratory Animal Welfare. 2022. MOU between NIH and NSF. https://olaw.nih.gov/NSF.html. Accessed August 31, 2022.
11. National Institutes of Health, Office of Laboratory Animal Welfare. 2022. MOU between NIH and VA. https://olaw.nih.gov/DVA.html. Accessed August 31, 2022.
12. National Institutes of Health, Office of Laboratory Animal Welfare. 2022. Obtaining an Assurance. https://olaw.nih.gov/guidance/obtaining-an-assurance.htm. Accessed August 31, 2022.
13. 59th Medical Wing Instruction 40-402, Medical Command. April 4, 2021. *Animal Care and Use in Clinical Investigations, Training, and Research & Development.* https://static.e-publishing.af.mil/production/1/59mdw/publication/59mdwi40-402/59mdwi40-402.pdf. Accessed August 31, 2022.
14. Army Regulation 40-33, SECNAVINST 3900.38.C, AFMAN 40-401 (I), DARPAINST 18, USUHSINST 3203. Medical Service. February 16, 2005. *The Care and Use of Laboratory Animals in DOD Programs.* Departments of the Army, Navy, Air Force, Defense Advanced Research Projects Agency, and the Uniformed Services University of the Health Sciences Washington, DC. https://armypubs.army.mil/epubs/DR_pubs/DR_a/pdf/web/r40_33.pdf. Accessed August 31, 2022.
15. Department of Defense. March 20, 2019. DoD Instruction 3216.01. *Use of Animals in DoD Conducted and Supported Research and Training.* https://www.esd.whs.mil/Portals/54/Documents/DD/issuances/dodi/321601p.pdf. Accessed August 31, 2022.
16. Department of Veterans Affairs. November 23, 2011. Veterans Health Administration, Washington, DC, 20420. *VHA Handbook 1200.07.* https://www.va.gov/vhapublications/ViewPublication.asp?pub_ID=2464. Accessed August 31, 2022.
17. Office of Naval Research, Science and Technology. 2022. *Animal Use Research Requirements.* https://www.nre.navy.mil/work-with-us/how-to-apply/compliance-and-protections/research-protections/animal-use. Accessed August 31, 2022.
18. American Dairy Science Association, American Society of Animal Science, and Poultry Science Association. 2020. *Guide for the Care and Use of Agricultural Animals in Research and Teaching,* 4th ed. Champaign, IL. https://www.asas.org/docs/default-source/default-document-library/agguide_4th.pdf?sfvrsn=56b44ed1_2. Accessed January 28, 2023.
19. AAALAC International. *Resources – The Three Primary Standards.* 2022. https://www.aaalac.org/resources/the-three-primary-standards/. Accessed August 31, 2022.
20. AAALAC International. June 2022. *Accreditation Program – Frequently Asked Question B.1. Animal Ownership.* https://www.aaalac.org/accreditation-program/faqs/. Accessed March 21, 2024.
21. AAALAC International. June 2022. *Accreditation Program - Frequently Asked Question C.10. Client-Owned Animals in Veterinary Colleges.* https://www.aaalac.org/accreditation-program/faqs/. Accessed August 31, 2022.
22. Center for Disease Control and Prevention. June 2020. *Biosafety in Microbiological and Biomedical Laboratories,* 6th ed. National Institutes of Health. https://www.cdc.gov/labs/pdf/SF__19_308133-A_BMBL6_00-BOOK-WEB-final-3.pdf. Accessed August 31, 2022.
23. National Research Council of the National Academies. Institute for Laboratory Animal Research Division of Earth and Life Studies. Committee on Occupational Health and Safety in the Care and Use of Nonhuman Primates. 2003. *Occupational Health and Safety in the Care and Use of Nonhuman Primates.* Washington, DC. National Academy of Sciences. 2003. https://www.ncbi.nlm.nih.gov/books/NBK43451/pdf/Bookshelf_NBK43451.pdf. Accessed August 31, 2022.
24. National Research Council of the National Academies. Institute for Laboratory Animal Research Division of Earth and Life Studies. Committee on the Use of Animals in Neuroscience and Behavioral Research. 2003. *Guidelines for the Care and Use of Mammals in Neurosciences and Behavioral Research.* Washington, DC: National Academy of Sciences. 2003. https://www.ncbi.nlm.nih.gov/books/NBK43327/pdf/Bookshelf_NBK43327.pdf. Accessed August 31, 2022.
25. National Institutes of Health, Office of Laboratory Animal Welfare. 2023. *Frequently Asked Questions A.6., Does the PHS Policy Apply to Animal Research That Is Conducted in the Field?* https://olaw.nih.gov/faqs#/guidance/faqs. Accessed January 28, 2023.
26. National Institutes of Health, Office of Laboratory Animal Welfare. 2024. *Frequently Asked Questions A.1., Should Institutions Apply the PHS Policy to All Animal Activities Regardless of the Source of Funding?* https://olaw.nih.gov/faqs#/guidance/faqs. Accessed March 19, 2024.
27. Bayne, K. and P. Turner. 2019. Animal Welfare Standards and International Collaborations. *ILAR Journal* 60: 86–94.
28. Gipson, C. and C. Wigglesworth. 2006. A word from OLAW and the USDA. *Lab Animal (NY)* 35:4, 17. https://olaw.nih.gov/sites/default/files/35_4_0704.pdf. Accessed August 31, 2022.

29. Office of Naval Research. Navy Bureau of Medicine and Surgery. 2016. *Animal Use Appendix for Research Involving Animals*. https://www.nre.navy.mil/media/document/bumed-animal-use-appendix-research-involving-animals-full-version. Accessed August 31, 2022.
30. U.S. Army Research and Development Command. Animal Care and Use Review Office. *Appendix for Animal Use in Research, Development, Testing, Evaluation, and Training*. 2020. https://mrdc.amedd.army.mil/assets/docs/orp/acuro/ACURO_Appendix.pdf. Accessed August 31, 2022.
31. Council for International Organization of Medical Sciences and The International Council for Laboratory Animal Science. 2012. *International Guiding Principles for Biomedical Research Involving Animals*. https://olaw.nih.gov/sites/default/files/Guiding_Principles_2012.pdf. Accessed August 31, 2022.
32. National Research Council. Commission on Life Sciences. Institute of Laboratory Animal Resources. Committee on Occupational Safety and Health in Research Animal Facilities. 1997. *Occupational Health and Safety in the Care and Use of Research Animals*. https://www.aaalac.org/pub/?id=E9017004-C6F8-5B42-C5D3-451EAA7B6E2D. Accessed August 31, 2022.
33. Mohan, S. and P. Foley. 2019. Everything you need to know about satisfying IACUC protocol requirements. *ILAR Journal* 60(1): 50–57.
34. Preisig, P., J. Silverman, P. Brown, and N. Petervary. 2017. Correspondence: Tough decisions about protocol review. *Lab Animal (NY)* 46(7): 290–294. https://archives.nih.gov/asites/grants/09-18-2017/grants/olaw/references/laban46_07_0717p290.pdf. Accessed August 31, 2022.
35. Wolff, A. and B. Morse. Office of Laboratory Animal Welfare. Webinars and Podcast. December 9, 2010. *IACUC Review of Proposed Animal Activities*. https://olaw.nih.gov/education/educational-resources/webinar-2010-12-09.htm. Accessed August 31, 2022.
36. Applied Research Ethics National Association/Office of Laboratory Animal Welfare. 2002. *Institutional Animal Care and Use Committee Guidebook*, 2nd ed. Bethesda: National Institutes of Health. https://olaw.nih.gov/sites/default/files/GuideBook.pdf. Accessed August 31, 2022.
37. U.S. Department of Agriculture, Animal and Plant Health Inspection Service. 2020. *APHIS Form 7023: Annual Report of Research Facility*. https://www.aphis.usda.gov/library/forms/pdf/APHIS_7023.pdf. Accessed August 31, 2022.
38. U.S. Department of Agriculture. 2021. Animal Welfare Inspection Guide. *Animal Care Inspection Guide*. https://www.aphis.usda.gov/animal_welfare/downloads/Animal-Care-Inspection-Guide.pdf. Accessed August 31, 2022.
39. National Institutes of Health, Office of Laboratory Animal Welfare. 2022. *Frequently Asked Questions D.7., Should the IACUC Consider the Three "Rs" of Alternatives When Reviewing Protocols?* (Refinements to research, Reduction of animal numbers, and Replacement with non-animal models). https://olaw.nih.gov/faqs#/guidance/faqs. Accessed January 28, 2023.
40. National Institutes of Health, Office of Laboratory Animal Welfare. 2022. *Frequently Asked Questions F.4., May Investigators Use Non-Pharmaceutical-Grade Substances in Animals?* https://olaw.nih.gov/faqs#/guidance/faqs. Accessed January 28, 2023.
41. National Institutes of Health, Office of Laboratory Animal Welfare. 2023. *Frequently Asked Questions A.3., Does the PHS Policy Apply to Use of Animal Tissue or Materials Obtained from Dead Animals?* https://olaw.nih.gov/faqs#/guidance/faqs. Accessed January 28, 2023.
42. National Institutes of Health, Office of Laboratory Animal Welfare. 2023. *Frequently Asked Questions A.7., Does the IACUC Need to Approve Research Studies That Use Privately Owned Animals, Such as Pets?* https://olaw.nih.gov/faqs#/guidance/faqs. Accessed January 28, 2023.
43. National Institutes of Health, Office of Laboratory Animal Welfare. 2023. *Frequently Asked Questions Section D., Protocol Review*. https://olaw.nih.gov/faqs#/guidance/faqs. Accessed March 21, 2024.
44. American Heart Association. 2022. *AHA Award Agreement Terms and Conditions*. https://professional.heart.org/-/media/PHD-Files/Research/Research-Policies-and-Awardee-Hub/aha_award_agreement.pdf. Accessed August 31, 2022.
45. Council of Europe. 1986. *European Convention for the Protection of Vertebrate Animals Used for Experimental and Other Scientific Purposes*. Strasbourg, 18.III.1986 (ETS123). https://www.coe.int/en/web/conventions/full-list/-/conventions/treaty/123. Accessed January 28, 2023.
46. National Institutes of Health, Office of Laboratory Animal Welfare. 2023. *Frequently Asked Questions B.6., Does the IACUC Have Authority over Activities Not Supported by PHS?* https://olaw.nih.gov/faqs#/guidance/faqs. Accessed January 28, 2023.
47. Leland, S., P. Straeter, and B. Gnadt. 2019. The role of the IACUC in the absence of regulatory guidance. *ILAR Journal* 60(1): 95–104.
48. National Institutes of Health, Office of Laboratory Animal Welfare. 2022. *Frequently Asked Questions D.15., Is IACUC Approval Required for the Collection of Samples in Foreign Countries from Captive Wild Animals or Research Colonies?* https://olaw.nih.gov/faqs#/guidance/faqs. Accessed January 28, 2023.
49. Silk, S., L. Hampton, and P. Brown. 2014. What investigators need to know about the use of animals. *ILAR Journal* 54(3): 324–328.
50. National Institutes of Health, Office of Laboratory Animal Welfare. 2023. *Frequently Asked Questions A.2., Does the PHS Policy Apply to the Production of Custom Antibodies or the Purchase of Surgically Modified Animals?* https://olaw.nih.gov/faqs#/guidance/faqs. Accessed January 28, 2023.
51. National Institutes of Health. *Office of Laboratory Animal Welfare. OPRR Reports Number 95-03*. March 8, 1995. Sources of Custom Antibody Production. https://olaw.nih.gov/guidance/articles/dc95-3.htm. Accessed August 31, 2022.
52. Wolff, A., N. Garnett, S. Potkay, D. Wigglesworth, and V. Thornton. 2003. Frequently Asked Questions About the Public Health Service - Policy on Humane Care and Use of Laboratory Animals. *Lab Animal (NY)* 32(9): 33–36. https://olaw.nih.gov/guidance/articles/lab_animal2003v32n9_wolff.htm. Accessed August 31, 2022.
53. National Institutes of Health. Office of Laboratory Animal Welfare. 2022. *Vertebrate Animal Section*. https://olaw.nih.gov/guidance/vertebrate-animal-section.htm. Accessed August 31, 2022.

54. U.S. Department of Health and Human Services, National Institutes of Health. December 2021. *NIH Grants Policy Statement.* https://grants.nih.gov/policy/nihgps/index.htm. Accessed March 19, 2024.
55. National Institutes of Health, Office of Laboratory Animal Welfare. 2023. *Frequently Asked Questions D.10., Is the IACUC Required to Review the Grant Application?* https://olaw.nih.gov/faqs#/guidance/faqs. Accessed January 28, 2023.
56. National Institutes of Health, Office of Laboratory Animal Welfare. 2022. *Frequently Asked Questions A.4., Does the PHS Policy Apply to Live Embryonated Eggs?* https://olaw.nih.gov/faqs#/guidance/faqs. Accessed January 28, 2023.
57. National Institutes of Health, Office of Laboratory Animal Welfare. 2022. *Frequently Asked Questions A.5., Does the PHS Policy Apply to Larval Forms of Amphibians and Fish?* https://olaw.nih.gov/faqs#/guidance/faqs. Accessed January 28, 2023.
58. U.S. Department of Agriculture, Animal and Plant Health Inspection Service, Animal Care. *Memo RE: Annual Report.* 2021. https://www.aphis.usda.gov/animal_welfare/downloads/reports/fy21-annual-report-packet.pdf. Accessed August 31, 2022.
59. National Institutes of Health, Office of Laboratory Animal Welfare. 2022. *Frequently Asked Questions D.16., Is IACUC Approval Required for the Use of Animals in Breeding Programs, as Blood Donors, as Sentinels in Disease Surveillance Programs, or for Other Non-Research Purposes?* https://olaw.nih.gov/faqs#/guidance/faqs. Accessed January 28, 2023.
60. National Institutes of Health, Office of Laboratory Animal Welfare. 2022. *Frequently Asked Questions F.2., Is the IACUC Responsible for Tracking Animal Usage?* https://olaw.nih.gov/faqs#/guidance/faqs. Accessed March 21, 2024.
61. National Association for Biomedical Research. *Q&A with the USDA – The Tenth Edition.* Originally aired on July 19, 2022. https://www.youtube.com/watch?v=xy_hpm6JonU. Accessed August 31, 2022.
62. Brown, P. and C. Gipson. 2013. A word from OLAW and USDA. *Lab Animal* 42(9): 318.
63. Juarez, B. and P. Brown. 2019. A word from APHIS and OLAW. *Lab Animal* 48(01): 8
64. National Institutes of Health, Office of Laboratory Animal Welfare. 2022. *Frequently Asked Questions D.8., When Institutions Collaborate, or When the Performance Site Is Not the Awardee Institution, Which IACUC Is Responsible for Review of the Research Activity?* https://olaw.nih.gov/faqs#/guidance/faqs. Accessed January 28, 2023.
65. Office of Extramural Research. *NOT-OD-01-017: Guidance Regarding Administrative IACUC Issues and Efforts to Reduce Regulatory Burden.* February 12, 2001. https://grants.nih.gov/grants/guide/notice-files/not-od-01-017.html. Accessed August 31, 2022.
66. The Federal Demonstration Partnership. 2024. *Animal Subjects Subcommittee: FDP MOU Sample and FDP IACUC MOU User Guide.* https://thefdp.org/committees-subcommittees/research-compliance-committee/animal-subjects-subcommittee/. Assessed March 19, 2024.
67. National Institutes of Health, Office of Laboratory Animal Welfare. 2022. *Frequently Asked Questions D.13., If an Animal Activity Will Be Performed Outside of the US (Either by a Foreign Awardee or by a Foreign Institution as a Subproject for a Domestic Awardee), Is the Awardee's IACUC Required to Review and Approve That Activity?* https://olaw.nih.gov/faqs#/guidance/faqs. Accessed January 28, 2023.
68. National Institutes of Health. *Office of Laboratory Animal Welfare. Foreign Assurance.* 2022. https://olaw.nih.gov/resources/documents/foreign.htm. Accessed August 31, 2022.
69. Holthaus, K., D. Goldberg, C. Connelly, B. Corning, C. Nascimento, E.Witte, and B. Bierer. 2020. Facilitating collaborative animal research: The development and implementation of a master reciprocal institutional agreement for animal care and use. *Journal of Clinical and Translational Science* 4(2): 96–101
70. Potkay, S., N. Garnett, J. Miller, C. Pond, and D. Doyle. 1997. Frequently asked questions about the public health service policy on humane care and use of laboratory animals. *Contemporary Topics* 36(2): 47–50. https://grants.nih.gov/grants/olaw/references/faq_labanimals1997.htm. Accessed August 31, 2022.
71. Juarez, B. and P. Brown. 2018. A word from APHIS and OLAW. *Lab Animal (NY)* 47(7): 172.
72. Brown, P. and C. Gipson. 2007. A word from OLAW and the USDA. *Lab Animal* 36(1): 17
73. American Bar Association. November 28, 2021. *How Courts Work, Steps in a Trial.* https://www.americanbar.org/groups/public_education/resources/law_related_education_network/how_courts_work/discovery/. Accessed August 31, 2022.
74. Greer, W. and R. Banks. 2020. *The IACUC Administrator's Guide to Animal Program Management.* CRC Press. Taylor and Francis Group, Boca Raton, London, New York.
75. Brown, P. and B. Juarez. 2016. A word from USDA and OLAW. *Lab Animal* 45(9): 328.
76. Coventry L. and Branley, D. 2018. Cybersecurity in healthcare: A narrative review of trends, threats and ways forward. *Maturitas* 113: 48–52.
77. Garnett, N. and S. Potkay. 1995. Use of electronic communications for IACUC functions. *ILAR Journal* 37(4) 190–192.
78. National Institutes of Health. *NOT-OD-10-128: Clarification on the Roles of NIH Scientific Review Groups (SRG) and Institutional Animal Care and Use Committees (IACUC) in Review of Vertebrate Animal Research.* September 1, 2010. https://grants.nih.gov/grants/guide/notice-files/NOT-OD-10-128.html. Accessed August 31, 2022.
79. Mann, M. and E. Prentice. 2007. Verification of IACUC approval and the just-in-time PHS grant process. *ILAR Journal* 48(1): 12–28.
80. National Institutes of Health, Office of Laboratory Animal Welfare. 2023. *Frequently Asked Questions F.3., May an Investigator Transfer Animals and Research to an Institution Different than the Grantee Institution?* https://olaw.nih.gov/faqs#/guidance/faqs. Accessed January 28, 2023.
81. American Veterinary Medical Association. 2013. *AVMA Guidelines for the Euthanasia* of *Animals*: 2020 edition. https://www.avma.org/sites/default/files/2020-02/Guidelines-on-Euthanasia-2020.pdf. Accessed August 31, 2022.

82. Maloney, S. Office of Laboratory Animal Welfare. *Webinars and Podcast.* June 20, 2013. Training Issues (Transcript). https://olaw.nih.gov/education/educational-resources/webinar-2010-12-09.htm. Accessed August 31, 2022.
83. National Institutes of Health, Office of Laboratory Animal Welfare. 2022. *Frequently Asked Questions G.1., What Kind of Training Is Necessary to Comply with PHS Policy, and How Frequently Should It Be Provided?* https://olaw.nih.gov/faqs#/guidance/faqs. Accessed January 28, 2023.
84. National Institutes of Health, Office of Laboratory Animal Welfare. 2022. *Frequently Asked Questions G.2., What Is Required for an Occupational Health and Safety Program?* https://olaw.nih.gov/faqs#/guidance/faqs. Accessed January 28, 2023.
85. National Institutes of Health, Office of Laboratory Animal Welfare. 2023. *Frequently Asked Questions D.19., May an IACUC Use Designated Member Review (DMR) to Review an Animal Study Protocol Subsequent to Full Committee Review (FCR) When Modifications Are Needed to Secure Approval?* https://olaw.nih.gov/faqs#/guidance/faqs. Accessed March 20, 2024.
86. National Institutes of Health. Office of Laboratory Animal Welfare. 2022. *Annual Report to OLAW.* https://olaw.nih.gov/resources/documents/annual-report.htm. Accessed August 31, 2022.
87. Brown, P. and B. Goldentyer. 2021. A word from the USDA and OLAW. *Lab Animal* 50(11): 298
88. National Institutes of Health, Office of Laboratory Animal Welfare. 2022. *Frequently Asked Questions D.12., Is the IACUC Responsible for Judging the Scientific Merit of Proposals?* https://olaw.nih.gov/faqs#/guidance/faqs. Accessed March 21, 2024.
89. AAALAC International. *Accreditation Program – Frequently Asked Question C.3. Harm-Benefit Analysis.* June 2022. https://www.aaalac.org/accreditation-program/faqs/. Accessed March 21, 2024.
90. AAALAC International. *Accreditation Program – Frequently Asked Question C.11. Transportation of Animals Used in Research.* June 2022. https://www.aaalac.org/accreditation-program/faqs/. Accessed August 31, 2022.
91. National Institutes of Health, Office of Laboratory Animal Welfare. 2022. *Frequently Asked Questions F.12., What Are the Institution's Responsibilities in Ensuring That Animals Are Shipped Safely and in Reporting Adverse Events That Occur in Shipment of Animals to or from the Institution?* https://olaw.nih.gov/faqs#/guidance/faqs. Accessed January 28, 2023.
92. National Centre for the Replacement, Refinement, & Reduction of Animals in Research. *The ARRIVE Guidelines 2.0.* July 2020. https://arriveguidelines.org/sites/arrive/files/documents/ARRIVE%20guidelines%202.0%20-%20English.pdf. Accessed August 31, 2022.
93. National Institutes of Health, Office of Laboratory Animal Welfare. 2022. *Frequently Asked Questions D.14., May Standard Operating Procedures (SOPs) or Blanket Protocols That Cover a Number of Procedures Be Utilized in Lieu of Repeating Descriptions of Identical Procedures in Multiple Protocols?* https://olaw.nih.gov/faqs#/guidance/faqs. Accessed January 28, 2023.
94. Brown, P. 2016. A word from OLAW. *Lab Animal* 45(8): 288.
95. Brown, P. 2018. A word from OLAW. *Lab Animal* 47(5): 115.
96. Brown, P. 2015. A word from OLAW. *Lab Animal* 44(3): 86.
97. National Institutes of Health. NOT-OD-14-126: *Guidance on Significant Changes to Animal Activities.* August 26, 2014. https://grants.nih.gov/grants/guide/notice-files/NOT-OD-14-126.html. Accessed August 31, 2022.
98. National Institutes of Health. *NOT-OD-20-088: Flexibilities for Assured Institutions for Activities of Institutional Animal Care and Use Committees (IACUCs) Due to COVID-19.* March 16, 2020. https://grants.nih.gov/grants/guide/notice-files/NOT-OD-20-088.html. Accessed August 31, 2022.
99. Pritchard, C., N. Petervary, and N. Gopaee. July 2020. *Laboratory Animal Science Professional.* 48–50.
100. The Federal Demonstration Partnership. 2022. *Compliance Unit Standard Procedure (CUSP) Working Group.* https://thefdp.org/default/committees/research-compliance/iacuc/compliance-unit-standard-procedure-cusp/. Accessed August 31, 2022.
101. The Federal Demonstration Partnership. *Virtual Meeting Research Compliance Committee May 11, 2022. Research Compliance Committee (RCC) Updates.* https://thefdp.org/default/assets/File/FDP_21CCA_and_RCC_Updates_May%202022.pdf. Accessed August 31, 2022.
102. National Institutes of Health, Office of Laboratory Animal Welfare. 2022. *Frequently Asked Questions A.8., How Can the IACUC Determine If Activities Involving Privately Owned Animals Constitute Veterinary Clinical Care or Research Activities?* https://olaw.nih.gov/faqs#/guidance/faqs. Accessed January 28, 2023.
103. American Veterinary Medical Association. 2022. *Veterinary-Client-Patient Relationship (VCPR) FAQ.* https://www.avma.org/resources-tools/pet-owners/petcare/veterinarian-client-patient-relationship-vcpr-faq. Accessed August 31, 2022.
104. Gipson, C. and C. Wigglesworth. 2006. A word from USDA and OLAW. *Lab Animal* 35(6): 17.
105. Brown, P. and C. Gipson. 2010. A word from OLAW and USDA. *Lab Animal* 39(3): 68.
106. Brown, P. and B. Juarez. 2016. A word from USDA and OLAW. *Lab Animal* 45(7): 252
107. Brown, P. and B. Goldentyer. 2022. A word from the USDA and OLAW. *Lab Animal* 51(1): 2.
108. Brown, P. 2009. A word from OLAW. *Lab Animal* 38(6): 186.
109. Brown, P. and C. Gipson. 2014. A word from USDA and OLAW. *Lab Animal* 43(4): 122.
110. Garnett, N. 2003. A word from OLAW. *Lab Animal* 32(10): 19.
111. Brown, P. 2019. A word from OLAW. *Lab Animal*, 48(3): 69.
112. Code of Federal Regulations (CRF): 28 CFR parts 35 (title II) and 36 (title III). *The Americans with Disabilities Act of 1990 and Revised ADA Regulations Implementing Title II and Title III.*

10

General Concepts of Protocol Review

John F. Bradfield, Ernest D. Prentice, Gwenn S. F. Oki, and Michael D. Mann

Introduction

Reviewing protocols involving the use of animals is one of the most important responsibilities of the IACUC. Although the format employed for this review varies across institutions, the criteria used in reviewing and approving protocols should be as consistent as possible. A thorough and up-to-date review is a particularly important consideration with regard to animal welfare issues, such as the use of non-animal model alternatives and the application of refinement techniques to reduce animal pain, discomfort, and distress.

The purpose of this chapter is to present general concepts of protocol review that will help IACUC administrators, IACUC members, Attending Veterinarians, and other interested individuals gain an understanding for how IACUCs conduct protocol review. Particular attention is given to issues such as scientific merit review, pre-review, use of consultants, designated member review (DMR) versus full committee review (FCR), and IACUC actions. Because the answers to the questions posed in this chapter are designed to be as succinct as possible, the reader is encouraged to consult the PHS Policy, the AWAR, and the *Guide* for necessary further guidance.

10:1 What review and approval criteria do the PHS Policy and AWAR require with respect to IACUC review of protocols?

Reg. PHS Policy (IV,C,1,a–g; IV,D,1,a–e) and the AWAR (§2.31,d; §2.31,e) specify criteria that must be met before an IACUC can approve a proposed activity involving the use of animals (e.g., research) as described in a protocol.

The PHS Policy (IV,A,1) requires that Assured institutions also base their IACUC review of protocols on the *Guide* and that they comply with the AWAR which apply to APHIS/AC regulated species. The *Guide* (pp. 25–26) lists 15 topics that IACUCs should review.

The AWAR (§2.31,d; §2.31,e) list a total of 16 requirements that must be met before the IACUC can approve a protocol. Some of the requirements mirror the PHS Policy, whereas others are not found in the PHS Policy. For example, the AWAR require an investigator to consider alternatives to procedures that may cause more than momentary pain or distress and that the investigator has provided a written narrative description of the methods and sources used to determine that alternatives are not available in such circumstances. The AWAR also require the investigator to provide written assurance that the animal-related activities do not unnecessarily duplicate previous experiments. The PHS Policy requires that the Investigator consider alternatives but does not require a description of the methods and sources used in that consideration. However, it should be noted that if the activity is both PHS funded and APHIS/AC regulated, compliance with the AWAR is an absolute requirement (PHS Policy II).

Whereas the AWAR are silent regarding the U.S. Government Principles, the PHS Policy includes the nine *U.S. Government Principles for the Utilization and Care of Vertebrate Animals Used in Testing, Research, and Training* that provide additional guidance concerning the use of alternatives that embrace the concepts of Replacement, Reduction, and Refinement (3 Rs). The *Guide* (p. 4) also endorses these Principles and further emphasizes the 3 Rs. Of particular note is Principle II that states: "procedures involving animals should be designed and performed with due consideration of their relevance to human or animal health, the advancement of knowledge, or the good of society." Principle II represents the scientific and societal justification for using animals in research, which is also addressed as "Scientific Review" in the *Guide* (p. 26) (see 10:13).

Opin. It should be noted that neither the PHS Policy nor the AWAR provide specific guidance in all areas of protocol review. Consequently, it is left to the institution to develop specific guidelines for implementation of federal requirements. Therefore, many institutions have developed investigator handbooks that include information concerning local policies for anesthetics, analgesics, blood sampling, tail sampling, antibody production, and preoperative and postoperative care for non-rodent animals, and other aspects of animal research.

Finally, it should be mentioned that NIH periodically publishes guidance on selected topics, as notices in the NIH *Guide for Grants and Contracts*, that provide information concerning NIH/OLAW's

DOI: 10.1201/9781003272861-10

interpretation of the requirements of the PHS Policy.[1] It is particularly useful to access NIH/OLAW's FAQs, which provide guidance on many topics, including protocol review.[2] In the interest of harmonization, APHIS/AC concurs with NIH/OLAW guidance provided in these FAQs, where applicable. In addition, in Protocol Review articles in *Lab Animal*, NIH/OLAW and APHIS/AC often respond to questions posed in various scenarios.[3]

10:2 What are some useful ways to disseminate protocols for initial and continuing review once they have been submitted to the IACUC?

Reg. PHS Policy (IV,C,2) requires that each IACUC member be provided with a list of proposed projects to be reviewed, and that written descriptions of projects be available to all IACUC members. The AWAR (§2.31,d,2) have the same requirement. Neither the AWAR nor the PHS Policy require that IACUC members be given a complete protocol, or how protocols should be distributed to the IACUC. In addition, the AWAR and PHS Policy do not specify when FCR and DMR should be used.

Opin. It is a preferred practice, adopted by many institutions, to provide complete protocols for initial and continuing review by the IACUC. This is also true for significant changes that have a major impact on the protocol (e.g., addition of a major survival surgery). Suffice it to say, the IACUC members must have all the information necessary to perform the review.

There are many methods used to disseminate protocols for review, including hard copies that are mailed, distributed by fax, standard electronic mail, and other electronic methods; or via IACUC member access to secure IACUC websites/servers (see 6:8). When a protocol is assigned to full committee review (FCR), it is common practice for IACUCs to assign each protocol to one reviewer or to a primary and secondary reviewer (chosen based upon their expertise in the subject matter of the protocol) who are given principal responsibility for in-depth review. With this method, the other members of the IACUC receive either a complete copy of the protocol or a protocol summary for review. Some IACUCs, particularly at institutions that have a large amount of research using animals, utilize a pre-review system where protocols are clarified and modified as necessary prior to full committee review (see 10:4–10:11).

Many IACUCs also utilize a designated member review (DMR) system, sometimes incorrectly referred to as "expedited review." The use of the latter term is strongly discouraged by APHIS/AC, NIH/OLAW, and AAALAC because it has been incorrectly interpreted as "light review." DMR helps to decrease the workload of the IACUC during full committee meetings. In general, the method used to disseminate protocols for review is dictated by the size, diversity of research, electronic capability, and specific needs of the institution (see 10:20–10:32).

10:3 What are the IACUC considerations for reviewing a protocol not associated with a PHS grant?

Reg. If an institution has elected, in its NIH/OLAW Assurance, to apply the requirements of the PHS Policy to all activities involving animals (regardless of the source of funding), then the IACUC obligations for reviewing internally funded projects are the same as those obligations under the PHS Policy. Activities involving animals regulated under the AWA require adherence to the AWA and AWAR regardless of funding source.

Opin. Internally funded projects, or other projects not subject to external peer review, may pose special problems for the review process, particularly with respect to scientific merit review. Even IACUC members who maintain steadfastly that the committee has no responsibility or right to review scientific merit agree that the IACUC has some obligation to do so when the proposal will not be subjected to external peer review. A minimal level of merit that justifies animal use must be assured before protocols are approved. Needless to say, animal welfare considerations should be the same, and the same rigorous review process with regard to regulatory compliance should be uniformly applied to both internally and externally funded projects.

10:4 Is pre-review of a protocol required by AWAR or the PHS Policy?

Reg. Pre-review of protocols is not a PHS Policy requirement; however, the AWAR (§2.31,d,1,iv,B) require PIs to consult with the attending veterinarian or his or her designee when procedures are planned that may cause more than momentary or slight pain or distress.

Opin. Pre-review refers to the informal review of a protocol by one or more individuals prior to formal review by the IACUC, which can be conducted only by FCR or DMR. The use and scope of pre-review varies by institution. In some institutions, pre-review is conducted by an IACUC administrator or staff who checks for typographical errors and to verify that the form is complete. In other institutions, veterinarians or subject matter experts conduct a rigorous review of animal welfare concerns, raising questions that they anticipate will be asked by the IACUC. A pre-review by the veterinarian can also be used to meet the AWAR consultation requirement. Pre-review is not a substitute for FCR or DMR. Pre-review is optional; a review, not a pre-review, must be conducted before an action can be taken (approval, seek modifications to secure approval, or withholding approval). A good pre-review can certainly help reduce the workload of the IACUC. It should be noted that scientific merit review by an external body prior to IACUC review can be considered a specialized form of pre-review (see 10:12).

Surv. Does your committee ever do veterinary review concomitant with protocol review?

• Yes. Specify under what circumstance*	111/257 (43%)
• No	146/257 (57%)

*Yes:

- In circumstances in which the protocol describes procedures that involve pain/distress or use of anesthesia during invasive procedures
- In circumstances in which a veterinary pre-review was not performed
- When there is urgency for review that would not allow for a pre-review prior to FCR
- Sometimes, if the procedures described involve no or minimal pain and veterinary pre-review has not been performed

10:5 What are the reasons for a pre-review of protocols?

Opin. Protocols are sometimes incomplete when they are submitted for IACUC consideration. Information that the IACUC needs to effectively review the protocol might be absent or unclear. For example, the investigator may be proposing to use a method of anesthesia involving injectable agents without specifying the dose. Clearly, this information must be provided before the IACUC can approve the protocol. Pre-review saves the IACUC time because IACUC members do not have to read protocols that are obviously not ready to be reviewed. Pre-review is also helpful to the investigator in terms of speeding up the review process because the protocol is less likely to be referred back for review by the full committee pending further information.

Surv. Regarding veterinary pre-review of protocols: Does your IACUC require review of all protocols by the veterinarian prior to dissemination to the full committee? Select all that apply.

• Yes.	120/257 (47%)
• No, we do not conduct veterinary pre-reviews.	67/257 (26%)
• Not all protocols, pre-review depends on species.	20/257 (8%)
• Not all protocols, pre-review depends on procedure.	37/257 (14%)
• Not all protocols, pre-review depends on pain/distress category.	35/257 (14%)
• Other*	27/257 (11%)

*Other:

- Conducted for protocols involving new or unusual procedures
- Conducted if requested by the PI

10:6 Who can conduct a general pre-review of a protocol?

Opin. Because pre-review is not a mandated function of an IACUC, the pre-review process can be structured in a way that is most helpful to the IACUC and investigators. Pre-review can be carried out by designated IACUC members, qualified IACUC office staff, campus veterinarians, or any other experienced individuals who are able to adequately review the protocol and do not have a conflict of interest.

10:7 What is a workable mechanism for the pre-review of protocols?

Opin. The pre-reviewer reviews the proposal, notes any points of inadequacy in the document, and communicates with the PI. The pre-reviewer may suggest that the PI make corrections or clarifications to the protocol (e.g., changing the route, dose, or type of anesthetic agent) or suggest adding an additional detail to the protocol (e.g., indicating the volume of blood to be withdrawn). In the case of more substantial corrections, the pre-reviewer interactively discusses the problems with the PI and may recommend that the investigator rewrite the protocol and submit a corrected version. There is the potential problem that the PI may interpret pre-review as actual IACUC review, and changes made may be interpreted as the only action that will be necessary to obtain IACUC approval. The PI should clearly understand that pre-review is not full committee or designated member review and that additional changes may be required by the IACUC.

10:8 What happens to a protocol after it is pre-reviewed?

Opin. After a protocol is pre-reviewed, the investigator submits a modified or unmodified protocol that must be processed for formal review by the IACUC. The formal review may be accomplished by DMR (see 10:20–10:27) or by FCR, which is reviewed at a convened meeting of a quorum of the IACUC (see 10:28–10:32).

10:9 Are the results of the pre-review presented to the full IACUC? If so, by whom?

Opin. In institutions that still use paper submissions, the IACUC may be presented with a completely revised application form, including all the changes made by the PI as a result of the pre-review process. Alternatively, the changes may be presented as a separate document along with the original application form. The former method is preferred because it yields a "clean," complete protocol application, which is a benefit to both the IACUC and the investigator and his or her staff. In institutions that use electronic submissions, the PI can simply alter the online form, especially if the submission software is able to track all the changes. From the point of view of minimizing the workload of the IACUC, it is preferable to have a single updated document for review. A single accurate copy of the approved protocol facilitates protocol compliance. Conducting work that has not been approved by the IACUC is a frequently occurring form of non-compliance reported to NIH/OLAW. NIH/OLAW has a webinar on reporting non-compliance which provides helpful guidance.[4]

10:10 Are pre-review comments binding on investigators?

Opin. No. Pre-review is simply a method for giving investigators advice on their protocols. A possible exception may arise when the AV determines during pre-review that a particular anesthetic, analgesic, or surgical procedure is not appropriate (AWAR §2.31,d,iv,B; §2.33,b,4). Investigators may occasionally disagree with the pre-reviewer's suggestions and choose to provide an explanation and send an unmodified or a partially modified protocol to the IACUC for formal review. However, an investigator who insists on presenting an inadequate document to the IACUC is likely to find that the IACUC will not approve the protocol, and the protocol will be returned for modifications or clarification for the same reasons identified by the pre-reviewer. This will result in unnecessary delays in obtaining IACUC approval of the project.

10.11 In an effort to reduce administrative burden, is it beneficial for the IACUC to provide investigators with assistance in writing protocols?

Opin. The preparation, review, revision, and re-review of protocols takes a significant amount of time for the investigator, the IACUC, and the IACUC staff. Even in modest-sized programs, the workload burden is not trivial. There may be obvious benefits to the IACUC aiding with writing protocols. Such assistance would likely improve the clarity and detail of the submitted protocols, thereby streamlining the subsequent review process. This would also likely decrease the number of protocols that require modification and re-review; all of which would save significant time and effort. However, such assistance requires that there is sufficient highly trained and experienced IACUC staff available. This would require a potentially substantial allocation of institutional resources that may not be practical for all programs.

Surv. Does your IACUC office provide assistance to investigators in drafting protocols? Select all that apply.

• No.	34/256 (13%)
• Yes, we offer dedicated staff support.	86/256 (34%)
• Yes, we offer one-on-one sessions with investigators.	145/256 (57%)
• Yes, we provide extensive written instruction sets.	75/256 (29%)
• Yes, we offer periodic training sessions.	61/256 (24%)
• Other*	49/256 (19%)

*Other:

- Veterinary staff/attending veterinarian provides assistance.
- Assistance provided upon request or as needed by the IACUC Chair, IACUC office staff, or IACUC members.
- Assistance provided by senior investigators to junior investigators.
- Support offered on a fee-for-service basis.

10:12 Should the IACUC review protocols for scientific merit? Do the AWAR or the PHS Policy specifically require (or prohibit) review of scientific merit?

Opin. In our experience, many IACUCs review protocols for some level of "scientific merit" (e.g., value, importance, scientific relevance). It is not clear, however, that these terms have the same or similar meaning to all IACUCs. Whether IACUCs should perform such review is open to debate. Prentice et al.[5] and Mann and Prentice[6] commented on the PHS Policy, in which the term "scientific merit" is not used. Rather, it refers to "relevance" in U.S. Government Principle II in a manner that strongly suggests the terms are synonymous. The PHS Policy also uses the terms "sound research design" (IV,C,1,a) and "scientifically valuable research" (IV,D,1,d). When all of these terms are considered together, they support the position of NIH/OLAW for PHS-funded research activity that an IACUC should consider the scientific relevance of a proposal. Indeed, NIH/OLAW has written in an FAQ, the following guidance which makes it clear that it is seldom possible to separate considerations of scientific value from animal welfare:

> Peer review of the scientific and technical merit of an application is considered the purview of the NIH Scientific Review Groups (SRGs), which are composed of scientific experts from the extramural research community in a particular area of expertise. However, SRGs also have authority to raise specific animal welfare concerns that can require resolution prior to a grant award.
>
> Although not intended to conduct peer review of research proposals, the IACUC is expected to include consideration of

> the U.S. Government Principles in its review of protocols. Principle II calls for an evaluation of the relevance of a procedure to human or animal health, the advancement of knowledge, or the good of society. Other PHS Policy review criteria refer to sound research design, rationale for involving animals, and scientifically valuable research. Presumably a study that could not meet these basic criteria is inherently unnecessary and wasteful and, therefore, not justifiable.[7]

The AWAR appear inconsistent in their reference to "scientific merit review." In the Public Comment Section of the regulations, APHIS/AC stated, "we added the term 'animal care and use procedure'... to avoid any misunderstanding or implication that APHIS intends to become involved in the evaluation of the design, outlines, guidelines, and scientific merit of proposed research."[8] Also, the AWAR (§2.31,a) and the AWA itself (§2143,a,6,A,i–ii) state that "except as specifically authorized by law or these regulations, nothing in this part shall be deemed to permit the Committee or IACUC to prescribe methods or set standards for the design, performance, or conduct of actual research or experimentation by a research facility." One way to interpret this apparent inconsistency in the regulatory language is to accept that there may be many scientific aspects of a study that do not relate to the animal activities, and are, therefore, outside the boundaries of IACUC review. On the other hand, the AWAR (§2.31,e,4), like the PHS Policy, make reference to "scientifically valuable research." It is certainly possible that a flawed research design could negatively impact animal welfare and may, in turn, violate requirements in the AWAR. In this case, a required change in the experimental design would seem to be permitted by AWAR.

The *Guide* (p. 26) indicates that the IACUC, "should evaluate scientific elements of the protocol as they relate to the welfare and use of animals. For example hypothesis testing, sample size, group numbers and adequacy of controls." Whereas IACUC's may have varying interpretations of what constitutes "scientific elements," clearly, according to the *Guide*, the IACUC is tasked with some level of responsibility for evaluating fundamental aspects of scientific methodology as they pertain to animal use and welfare.

Whereas it appears that neither NIH/OLAW nor the *Guide* explicitly require IACUC review of scientific merit, nevertheless, according to NIH/OLAW,[9] an institution cannot defer scientific relevance review to the funding agency. Therefore, approval by an IACUC of a proposed activity that is conditional upon successful peer review by the funding agency is not consistent with the PHS Policy requirements. Such conditional action does not constitute IACUC approval required by the PHS Policy either prior to review by the NIH Scientific Review Group (SRG) or under just-in-time[10] requirements. The NIH review of merit should, therefore, be viewed as additional assurance, rather than the only assurance that the research has value.[5,9] Based on this, Prentice et al.[5] concluded that the IACUC does have a responsibility to review the scientific relevance of animal projects that are subject to the requirements of the PHS Policy. There is not, however, general agreement with this conclusion. Black[11] argues that merit and relevance are not the same thing; that IACUC review does not constitute peer review, the consequences of IACUC review are different from those of external review, and IACUC review may constitute a violation of researchers' academic freedom, which may be considered a questionable idea. Prentice et al.[12] and Mann and Prentice[6] offer counterarguments. In recent years, however, the issue of the IACUC's role in review of the science has become far less contentious.

The advent of just-in-time certification of IACUC approval of research proposals (see 9:24) seems to add another level to the arguments above. In reference to the just-in-time process, the NIH wrote, "The fundamental PHS Policy requirement that no award may be made without an approved Assurance and without verification of IACUC approval remains in effect. This change only affects the timing of the submission of the verification of that review."[13] Given this, and the information in the *Guide* (p. 26), it seems that the PHS holds the IACUC responsible for review of scientific relevance regardless of whether the just-in-time process was used or not. However, prior review of the grant application by the NIH SRG is clearly beneficial to the IACUC in its consideration of the science.

Surv. Does your IACUC consider the scientific merit of a protocol application that has had external scientific review by a peer group?

- Not applicable. 60/260 (23%)
- No, we believe it is the external peer reviewers who have that responsibility. 71/260 (27%)
- Yes, we ask our members to provide at least a cursory review of scientific merit. 104/260 (40%)
- Yes, but only for species covered by the Animal Welfare Act. 5/260 (2%)
- I don't know. 11/260 (4%)
- Other. 9/260 (3%)

Surv. Does your IACUC provide a scientific review for internally funded research proposals (i.e., research proposals that have not undergone external peer review)?

• Yes.	121/260 (47%)
• No, but they are reviewed by another group within our organization such as a departmental or other committee.	64/260 (25%)
• No, and they are not reviewed by any other group within our institution.	38/260 (15%)
• I don't know.	12/260 (5%)
• Other.	11/260 (4%)
• Not applicable. We do not have any internally funded proposals; all proposals are supported by external funds that undergo external peer review.	14/260 (5%)

10:13 Should the IACUC approve protocols without consideration of scientific merit?

Opin. Prentice et al.[5] note that there are two levels of review for scientific merit. They refer to a "fundamental level" of review in which scientists form "basic judgments about the adequacy and appropriateness of experimental design in terms of the ability to test the hypothesis, use controls, sample size, statistical analysis, and the training and experience of investigators." This first-level merit review is necessary for the IACUC to approve the judicious use of animals. The other level of review, "knowledge-based level," requires "an assessment be made of the scientific importance of the study." An assessment of merit at the fundamental level could be made by any appropriately constituted IACUC, even in the absence of special expertise in the topic of the protocol. It would seem that a judgment of the merit at the knowledge-based level would be necessary to determine the appropriateness of the proposed animal use. Of course, the latter judgment requires that an IACUC, involved in reviewing proposals over a range of subjects, have suitable expertise in that range of subject matters or employ *ad hoc* consultants (see 10:12).

10:14 The AWAR (§2.31,a) state that the IACUC shall not prescribe methods or set standards for the design, performance, or conduct of actual research or experimentation by a research facility. What does this actually mean given that the IACUC has the authority to approve or withhold approval of animal research activities?

Reg. AWAR (§2.31,a) state: "Except as specifically authorized by law or these regulations, nothing in this part shall be deemed to permit the Committee or IACUC to prescribe methods or set standards for the design, performance, or conduct of actual research or experimentation by a research facility" (see 10.12).

Opin. This apparent contradiction in the AWAR was discussed in detail by Prentice et al.[5] and Mann and Prentice.[6] Consider an animal use protocol that did not contain suitable controls. One could argue that if such a protocol contained no other animal welfare issues, it should be approved by the IACUC because the quoted text above seems to say that scientific design should not be considered. On the other hand, the AWAR (§2.31,e,2) say that a protocol should contain "A rationale for involving animals and for the appropriateness of species and numbers to be used." In this case, the numbers would be inappropriately low because of the missing controls. Lack of controls is an animal welfare issue. This is a poorly designed experiment that will not produce usable data and therefore leads to the unnecessary use of the experimental animals with no scientific benefit. Therefore, the IACUC should not approve the protocol.

10:15 When is it appropriate for the IACUC to use consultants to aid in protocol review?

Reg. PHS Policy (IV,C,3) and the AWAR (§2.31,d,3) allow the use of consultants for the review of a protocol. Consultants may not vote with the IACUC, and the IACUC remains responsible for its actions and decisions when consultants are used.

Opin. The use of consultants is highly variable among IACUCs. In some institutions, it is common practice for IACUC reviewers to use colleagues within the institution as consultants for clarification of issues raised during prereview of a protocol (see 10:4; 10:5). This pre-review clarification can greatly speed the final review process. Outside consultants may be engaged, for example, when the animal proposal involves highly novel and controversial, cutting-edge research or for research using wildlife when the institution has no other expert in the field. Outside consultants may be used when there is disagreement among the IACUC members as to the scientific relevance of a given proposal or if unresolved concerns exist regarding animal welfare. Consultants also are useful if an investigator asks the committee to reconsider its decision (see 10:56; 33:45).

Surv. How frequently does your IACUC enlist the participation of non-member consultants to provide specific expertise in the review of a protocol?

• Frequently, at least several times per year	16/256 (6%)
• Occasionally, such as once or twice per year	23/256 (9%)
• Rarely, such as less than once per year	135/256 (53%)
• Never	82/256 (32%)

10:16 Is it appropriate for the IACUC to use consultants to review scientific merit?

Reg. See 10:12.

Opin. It is appropriate for the IACUC to use consultants, particularly if the IACUC does not have the requisite expertise to assess scientific merit. Also, outside reviewers would likely be most useful when there is disagreement among the IACUC members as to the scientific merit of a protocol or if an investigator appeals the decision of the IACUC (see 10:13; 10:56; 33:45). Indeed, the IACUC could seek advice from consultants regarding any aspect of the project that is problematic.

10:17 What types of consultants might be useful for review of scientific merit? Who picks them?

Opin. Consultants can be:

1. Experts on the topic of the protocol under review
2. Experts in the use of animals, with respect to a particular procedure proposed in the protocol
3. Both

An example of the first use specified above is an expert who is asked to review a protocol because the IACUC is uncertain whether the use of the ascites method for making monoclonal antibodies is required in a particular project, as the PI claims. Alternatively, the consultant can be asked to judge whether any useful information will be forthcoming once the antibody is made. An example of the second use is an expert in primate behavior and care who is asked to review a protocol because the IACUC is uncertain whether a rhesus monkey may reasonably be restrained continuously for a specified period of time. The expert could be asked whether any useful information could be derived from the experiments that require the monkey to be restrained for that period.

Consultants should be selected by the IACUC, but it may also be reasonable to ask the PI to suggest possible experts. This allows the committee to maintain objectivity in the review process while considering the PI's point of view. Also, should the Institutional Official (IO) have the need to review (and potentially disallow) a protocol that had been approved by the IACUC, he/she could appoint a consultant to assist in the review. However, it must be remembered that if an IACUC has withheld approval of a protocol, no one, including the IO, can overturn that decision (AWAR §2.31,d,8; PHS Policy IV,C,8). As mentioned above, the IO may disallow IACUC-approved research, but this is usually done for reasons unrelated to the criteria used by the IACUC when it approved the protocol. Finally, some institutions require the consultant to sign a Confidentiality Agreement to protect the proprietary nature of the proposed study. This agreement will be useful in terms of protecting the anonymity of the consultant and the confidential nature of the study.

10:18 Should the identity of consultants be kept anonymous to the investigator?

Opin. The same arguments apply to the anonymity of consultants and to the anonymity of grant proposal reviewers. It has been argued that reviewers refrain from making negative comments about a protocol if their identity is known to the investigator and may fear reprisals. It also is possible to argue that reviewers may make unfair statements about a protocol because their identity is withheld. Whether this will happen likely depends on the individual reviewers and consultants and on the culture of animal care and use at the institution. In any case, if the reviewers of a protocol remain anonymous, then certainly consultants should be as well. On the other hand, anonymity may be difficult if the investigator is asked to suggest possible experts. Regardless, the IACUC must determine whether the consultant is conflicted before appointment.

10:19 An IACUC retains the services of a consultant suggested by the PI to help evaluate a complex protocol. Based in part on the consultant's recommendation, the protocol is approved. The IACUC subsequently learns that the consultant was a former student of the PI who submitted the protocol. Should the IACUC revisit the protocol's approval?

Opin. The PI should have informed the IACUC of this consultant's conflict of interest. But, even if consultants are used in the review process, it is still the IACUC that must make the decision for approval. In this case, the IACUC determined that the protocol should be approved. The occurrence of the conflict of interest would not necessarily change the fact that the IACUC had found the protocol acceptable. However, if the IACUC's decision to approve the protocol was largely based on the consultant's review, then the IACUC might well want to review the protocol again and perhaps retain the services of another consultant.

10:20 What is "designated member review," and what is its intent?

Reg. Both the PHS Policy (IV,C,2) and the AWAR (§2.31,d,2) describe a method of IACUC review termed "designated member review" (DMR). Some institutions inaccurately and inappropriately refer to this review process as expedited review, but designated member review is the correct term. DMR must conform to the following process: written descriptions of research projects that involve the care and use of animals must be made available to all IACUC members, and any member of the IACUC must have the opportunity to request full committee review of those research projects. If full committee review (FCR) is not requested, at least one member of the IACUC, designated by the Chair and qualified to conduct the review, shall review those activities, and shall have the authority to approve, require modifications in (to secure approval), or request full committee review of any of those activities. (PHS Policy IV,C,2; AWAR §2.31,d,2).

Opin. Although the exact procedures frequently vary between institutions, the DMR process may enable IACUCs to review and approve protocols faster than those presented for FCR. It is important to mention, however, that DMR in no way implies that the quality of review is less stringent than a protocol reviewed by the full committee. A successful DMR process allows institutions, particularly larger institutions that process many protocols, the opportunity to reduce the workload of the IACUC at convened meetings, thereby, allowing members to focus on protocols that may warrant more time and attention. It should also be noted that the DMR process does not reduce the IACUC office staff workload that may include follow-up reminders to the designated reviewer, correspondence with the PI, etc.

There is no federal requirement that limits the number of protocols that may be reviewed by the DMR process or specifies the number of full committee meetings. The *Guide* (p. 25), however, states "the committee must meet as often as necessary to fulfill its responsibilities." Additionally, in accordance with best practices the expectation is that an IACUC will meet in a convened meeting at least twice per year to approve the semiannual program review and facility inspections, but this is not an NIH/OLAW or APHIS/AC requirement. Clearly, progressive IACUCs meet on a regular basis to conduct whatever business is appropriate to their programs. This would, of course, be described in the PHS Assurance.

It is possible that all protocols could be reviewed, at least initially, by DMR. This is made more likely by use of an effective pre-review process. FCR would then be applied only to protocols for which FCR was requested by an IACUC member. However, such a review system would not be considered best practice. On the other hand, some IACUCs review all protocols by the full committee at convened meetings. This also does not reflect best practice unless the number of protocols is very small.

In the DMR process, the members of the IACUC do not vote. If a FCR is not called for, the designated reviewer has the authority to approve a protocol, require modifications (to secure approval), or refer it to the full committee. If there is more than one designated reviewer, the reviewers must agree on a course of action or refer the protocol to the full committee.

Surv. For the initial formal review of IACUC protocols (designated member review or full committee review), which review method does your IACUC use?

• We use designated member review 100% of the time,	26/260 (10%)
• We use designated member review about 75%–99% of the time,	75/260 (29%)
• We use designated member review about 50%–74% of the time,	40/260 (15%)
• We use designated member reviews about 25%–49% of the time.	17/260 (7%)
• We use designated member reviews about 1%–24% of the time.	39/260 (15%)
• We only use full committee review.	63/260 (24%)

10:21 Under what circumstances might a protocol be assigned to a designated member review process?

Reg. A protocol may be assigned to one or more designated reviewers only after all IACUC members have been provided the opportunity to call for full committee review (PHS Policy IV,C,2; AWAR §2.31,d,2) (see 10:20; 10:22).

Opin. Because the AWAR and the PHS Policy do not define the criteria for assigning protocols to the DMR process, each institution should develop an internal policy, tailored to the individual institution's needs and idiosyncrasies. Even if an institution develops criteria to determine the kinds of activities or protocols that it will permit to conduct review by DMR, all IACUC members must still be provided with the opportunity to call for full committee review of each individual project.

An institution should first decide whether it is advantageous to establish a DMR process and whether it is advisable for the IACUC to establish a list of well-defined criteria to determine the types of protocols qualifying for DMR versus FCR. For example, some institutions restrict DMR to protocols involving noninvasive or acute procedures that do not cause more than momentary pain or distress to the animals (e.g., the procedures involve euthanizing the animals to collect tissues) or to protocol amendments that involve minor changes to an approved study (see 10:22).

If an institution elects to adopt a DMR process, the IACUC should document the assignment criteria and the procedures for processing the protocols. The IACUC should develop a standardized process or set of SOPs which describe the DMR process. The description of the DMR process and its use should be included in the institution's Animal Welfare Assurance, approved by NIH/OLAW and the SOP should be included in the institution's IACUC policy and procedures.

10:22 Under what circumstances is the designated member review not appropriate?

Opin. Although each institution must determine the criteria appropriate for its own program, DMR may be unacceptable under many circumstances, including the following:

- IACUC members are not provided with sufficient information concerning the proposed research activities.
- IACUC members are not given sufficient opportunity to call for FCR prior to the approval of the proposed protocol via the DMR process (PHS Policy IV,C,2; AWAR §2.31,d,2).
- Any IACUC member requests FCR (PHS Policy IV,C,2; AWAR §2.31,d,2).
- One or more of the designated reviewers has a conflict of interest (PHS Policy IV,C,2; AWAR §2.31,d,2).
- The proposed protocol does not meet the requirements for DMR delineated in the institution's NIH/OLAW-approved Animal Welfare Assurance or the institution's IACUC policy and procedures. For example, the invasive nature of the research may not permit the protocol to be reviewed by the DMR method according to the institution's policies and its Animal Welfare Assurance. Other examples include procedures that involve unalleviated pain or distress in APHIS/AC-regulated species or major changes in ongoing research with APHIS/AC-regulated species (see 10:30).
- Any request for FCR made prior to the time of protocol approval requires FCR, even if the request is made after the time period to call for FCR has elapsed.

10:23 Should the entire IACUC be involved in the designated member review process, or can a subcommittee conduct designated member review?

Reg. See 10:20.

Opin. The entire committee must have the opportunity to review the protocol and to request FCR, if warranted. If there is no call for FCR, one or more members of the IACUC who are qualified to review the protocol can be assigned by the Chair to review the protocol in its entirety. Questions and concerns raised by any member of the IACUC should always be addressed prior to approval by DMR. Even though an individual IACUC member is not assigned as a designated reviewer, if he or she raises an issue without calling for FCR, that issue should be brought to the attention of the designated reviewer(s) or cause the protocol to be brought to FCR, if that is the desire of the member who raised the issue. If a subcommittee is being utilized for this review process, all designated reviewers *must* agree on the decision. It is not acceptable to use a majority vote.[14] If one of the reviewers disagrees, the protocol must be referred to the full committee for review.

10:24 What is required for approval of a protocol via the designated member review process?

Reg. The designated reviewer(s) must use the same criteria that are applicable to protocols undergoing FCR. Neither the PHS Policy nor the AWAR make any distinction regarding approval criteria for DMR vs. FCR.[15] The designated reviewer(s) must ensure that the review method and criteria for approval are in full compliance with all of the applicable requirements of the PHS Policy (IV,C,1,a–g; IV,D,1,a–e) and the AWAR (§2.31,d,1,i–xi; §2.31,e,1–5).

Opin. One of the intents of the DMR process is to conduct rapid review of routine protocols, but by no means does this process require less information than the FCR process. Another intent could be to redistribute IACUC workload or to enable IACUCs to focus on thorny issues without the distraction of routine protocols.

10:25 What is an effective means for administering a designated member review?

Opin. One significant difference between DMR and FCR is that the DMR process does not require a convened meeting of a quorum of the IACUC. Despite this fact, after a protocol is assigned to DMR, at least one or more IACUC members (designated reviewers) should review the entire protocol with the same degree of thoroughness that is given to a protocol reviewed by the full committee.

The following is an example of a DMR process. This particular example goes beyond the requirements of the AWAR and PHS Policy in that it requires more protocol specific information to be sent to all members and documentation of each designated reviewer's decision with regard to DMR:

- The IACUC administrator reviews the protocol submitted to the committee and decides whether it qualifies for DMR based upon the IACUC's predetermined criteria. In some instances, additional information may be included (e.g., supporting grant materials).
- Each IACUC member reviews the protocol to determine whether clarification or changes are needed and has the opportunity to call for FCR.
- If no member calls for FCR in a reasonable, specified time period, the Chair appoints one or more designated reviewers. For example, a subcommittee, composed of the IACUC Chair and the AV, is authorized to review and approve the protocol. In a case where the Chair or the AV has a conflict of interest (e.g., is personally involved in the project),

he or she must be replaced by another member qualified to conduct the review (PHS Policy VI,C,2; AWAR §2.31,d,2) (see 7:11; 7:14; 7:29). In our experience, 3–5 days is a sufficient time for the members to decide whether the protocol should be referred to FCR. The protocol can be reviewed and approved, or modifications may be required to secure approval, or the protocol can be referred to the full committee. However, if a request for FCR occurs after the designated time period and before protocol approval, the protocol must be referred to FCR.

- Documentation, including review sheets, is maintained as evidence of each designated reviewer's decision regarding these protocols. The designated reviewer sheets are kept in the protocol file in the event the DMR is ever questioned. *No matter what process an institution adopts, the DMR procedure must be documented from assignment to approval each time it is used.*

10:26 An IACUC Chair appoints designated member reviewers. Can an IACUC member who was not chosen as a designated reviewer demand to be made a designated reviewer?

Reg. Both AWAR (§2.31,d,2) and PHS Policy (IV,C,2) specify that at least one member of the IACUC be designated *by the Chairperson* and be qualified to conduct the review.

Opin. There is no requirement for the Chair to honor such a request. Although this may be an unusual request, an IACUC Chair would be wise to find out why an IACUC member has made this demand and perhaps refer the protocol to full committee review. In fact, the IACUC member who made the demand can call for full committee review (see 10:23).

10.27 What is the IACUC's obligation to ensure consistency of protocol review, particularly for protocols that are reviewed by DMR?

Opin. The AWAR and PHS Policy require that the designated member reviewer(s) must be qualified to conduct the review (AWAR 2.31,d,2 PHS Policy IV,C,2). Qualifications might include a reviewer's expertise in a particular scientific discipline, or experience with the specific species or the procedures involved. Also, a qualified designated member reviewer is one who is properly trained and experienced in IACUC review procedures and the required review criteria so that an appropriate review of the protocol is achieved. All members of the committee should be well versed in the protocol review process and the criteria that are reviewed to ensure a consistent level of review for all DMR-reviewed protocols. The training of new IACUC members should be sufficient to ensure their proficiency and consistency of review. Many IACUCs will assign DMR-protocols to experienced members who work with new members as a method to teach new members the intricacies of protocol review, thereby improving overall consistency in review.

Surv. How does your IACUC ensure consistency in reviews of protocols by committee members? Select all answers that apply.

• We train committee members on how to review protocols.	171/256 (67%)
• We provide guidelines regarding IACUC protocol review policies.	177/256 (69%)
• We intentionally keep the pool of reviewers small to decrease variability (we assign the same expert reviewers, when possible, to the same type of protocol.	72/256 (28%)
• Other approaches*	42/256 (16%)

*Other:

- During IACUC meetings, review expectation and importance of review consistency.
- Encourage members to review the comments provided by other reviewers.
- Use of a review checklist.
- IACUC staff monitor reviews and discuss inconsistencies at meetings or on an individual basis with reviewers.
- Periodic audits of protocol review.

10:28 What is "full committee" review?

Reg. Full committee review (FCR) of an IACUC protocol is one that is conducted by a quorum of the IACUC at a regularly scheduled or specially convened meeting (PHS Policy IV,C,2; AWAR §2.31,d,2).

Opin. Due to the volume of protocols reviewed at many institutions, a good portion of the actual work involved in protocol review may be done before the meeting via a mechanism such as pre-review (see 10:4–10:14). Thus, at the meeting, the committee members are able to concentrate on the proposed protocol, its "scientific merit," experimental design, and the humane care and use of laboratory animals without the necessity of obtaining clarifications related to incomplete information and/or confusing points.

10:29 What is the purpose of a full committee review?

Opin. The purpose of FCR is to have all IACUC members involved in reviewing and making decisions regarding the disposition of protocols during an interactive meeting. This allows the IACUC to utilize the expertise of its members in a discussion-based format, thus facilitating the resolution of protocol-related issues.

10:30 When is full committee review of a protocol warranted?

Reg. Full committee review is one mechanism of review described by the PHS Policy and AWAR.[14]

Opin. There are no federal requirements specifying when a protocol should receive FCR or the type of protocol that should receive FCR. However, many IACUCs have developed criteria to determine the types of protocols or changes that require FCR. Items that may be considered in developing such criteria include:

- Invasiveness of procedures
- Level of pain or distress
- Species and number of animals requested
- Experimental design
- The nature of the animal use, e.g., research project versus educational use
- "Controversial" procedures, e.g., death as an endpoint
- Procedures that request exceptions to regulations, e.g., multiple survival surgeries that require appropriate justification and which the IACUC may want to review only at the time of a fully convened meeting
- Whether the protocol will receive peer review by the funding agency or other group prior to funding
- Major protocol amendments involving procedures that would require FCR at initial review

10:31 What is an effective full committee review process?

Reg: PHS Policy (IV,C,1,a–g; IV,D,1,a–e) and AWAR (§2.31,d; §2.31,e). Whereas the PHS Policy and AWAR require only a quorum (simple majority) to officially convene a full committee meeting, it is obviously advantageous to have a greater number of IACUC members present, particularly key individuals such as the AV and, of course, the nonaffiliated member. It should also be noted that consultation with the AV is required during design of the study, prior to IACUC review, for APHIS/AC-regulated research that involves procedures that may cause more than momentary or slight pain or distress to the animals (AWAR §2.31,d,1,iv,B). NIH/OLAW provides excellent guidance on the correct conduct of full committee review as well as designated member review.[14]

Opin. Each IACUC should establish SOPs regarding the conduct of FCR of protocols. One effective method of full committee review includes the following process:

- Pre-review (see 10:4–10:12).
- Review by IACUC staff or designated member of the committee for completeness of the application and compliance with information requirements, as indicated in PHS Policy (IV,C,1,a–g; IV,D,1,a–e; AWAR §2.31,d,1,i–xi; §2.31,e,1–5).
- Review by the AV in addition to the consultation required during design of the study (AWAR §2.31,d,1,iv,B).
- Review by outside consultants if the IACUC members do not possess relevant expertise.
- Review by committee members assigned as primary and secondary reviewers.
- Primary and secondary reviewers communicate with the investigator prior to the meeting, during which the reviewers may suggest modifications in the protocol which may facilitate subsequent review (PHS Policy IV,C,2; AWAR §2.31,d,2).
- Every member of the committee should be provided with a copy of, or access to, the protocol.
- Presentation of the protocol involving all the IACUC members in attendance by the primary and secondary reviewers followed by a general discussion of the protocol.
- Following discussion at a convened quorum of the committee, a vote is taken to determine final disposition of the protocol.

10:32 What are the PHS Policy and AWAR requirements for approval of a protocol by full committee review?

Reg.

- Both the PHS Policy (IV,C,1,a–g; IV,D,1,a–e) and the AWAR (§2.31,d; §2.31,e) have specific requirements about information that must be included in the protocol as well as the items the IACUC must consider (see 10.1).
- No member of the IACUC may participate in the review or approval of a research project in which the member has a conflict of interest (e.g., is personally involved in the project or has a financial interest) except to provide information requested by the IACUC (PHS Policy IV,C,2; AWAR §2.31,d,2). IACUCs should require this member to leave the room during final discussion and voting on the protocol in question. If an IACUC meeting occurs using virtual technology, a means of excluding members with a conflict of interest should be similarly implemented (see 10:67).
- Approval of a protocol considered by the full committee "may be granted only after

review at a convened meeting of a quorum of the IACUC and with the approval vote of a majority of the quorum present" (PHS Policy IV,C,2; AWAR (§2.31,d,2).

- A member who has a conflict of interest in a protocol under review may not contribute to the constitution of a quorum (PHS Policy IV,C,2; AWAR §2.31,d,2). This person may not vote on the protocol in question.
- Institutions must maintain written documentation of committee deliberations (PHS Policy IV,E,1,2; AWAR §2.35,a,1–2).
- Both the investigator and the institution must be notified in writing of the committee's decision (PHS Policy IV,C,4; AWAR §2.31,d,4).

Opin. Whereas most IACUCs include documentation of committee deliberations in the minutes, such documentation can also be maintained on file with the protocol. Filing records of particularly sensitive deliberations along with the protocol is of importance to institutions subject to state open records laws. (See Chapter 30.) It is up to the IO, IACUC Chair, or IACUC staff to explain to APHIS/AC, NIH/OLAW, or AAALAC, as appropriate, the rationale for the institution's recordkeeping methods.

10:33 What is veterinary verification and consultation (VVC)?

Reg. In 2014, OLAW issued updated guidance on significant changes to animal activities (NOT-OD-14-126) with concurrence by APHIS/AC that listed specific significant changes to a protocol that can be handled through consultation with an IACUC-authorized veterinarian (veterinary verification and consultation or VVC).

Opin. Prior to 2014 there were only two mechanisms by which the IACUC could review and approve significant changes to a protocol, either by DMR or by FCR. The guidance issued in 2014 by OLAW created another mechanism, VVC, for handling some significant changes. Changes that can be handled by this process include changes in:

- anesthesia, analgesia, sedation, or experimental substances
- euthanasia to any method approved in the AVMA Guidelines for the Euthanasia of Animals[16]
- duration, frequency, type, or number of procedures performed on an animal

The process for VVC includes IACUC-approved policies that allow for consultation with an IACUC-authorized veterinarian. The investigator can then directly contact the veterinarian who determines whether a particular proposed change, in his or her judgment, qualifies for VVC. If so, the veterinarian can then authorize implementation of the change if it is justified and congruent with OLAW requirements and the IACUC policy that specifies the types of changes eligible for review by VVC. Alternatively, the veterinarian can refer the requested change to the IACUC for review. If the proposed change does not meet the above criteria, it must be referred to the IACUC for review by either DMR or FCR. When VVC is used, the veterinarian is not conducting DMR, but rather is verifying that a proposed change is appropriate according to existing IACUC policy using IACUC-approved guidance, references, and/or formularies in conjunction with his or her veterinary judgment. This allows the change to be implemented. All significant changes handled by VVC must be documented and the protocol updated.

OLAW describes categories of significant changes that must be approved only by DMR or FCR (i.e., changes that do not qualify for VVC). These include the following:

a. Those that describe procedures changing from non-survival to survival surgery
b. Procedures resulting in greater pain, distress, or degree of invasiveness
c. Changes in housing and or use of animals in a location that is not part of the animal program overseen by the IACUC
d. Change in species
e. Change in study objectives
f. Change in Principal Investigator (PI)
g. Any modifications that negatively impact personnel risk and safety

Surv. Does your IACUC use the Veterinary Verification and Consultation (VVC) process for review of protocol modifications?

• No	82/257 (32%)
• Yes	175/257 (68%)

Surv. If your IACUC uses VVC, which of the following changes usually qualify for review by VVC? Select all that apply.

• Change in anesthetic	170/216 (79%)
• Changes in species	31/216 (14%)
• Change in analgesic	169/216 (78%)
• Change in method of euthanasia	156/216 (72%)
• Change of the PI	16/216 (7%)
• Change in method of blood collection	135/216 (63%)
• Other*	84/216 (39%)

*Other: generally contingent on substitution of one approach or procedure for another that is already approved by the

IACUC as part of a policy or SOP. Some of these additional responses are below.

- Change to a similar experimental compound
- Change to a different cell line for implantation
- Change in route/dosage/volume/frequency of already approved compounds
- Addition of compounds to offer physiological support
- Change in volume/schedule/or method of blood sampling
- Collection of additional tissues following euthanasia
- Increase in animal number of less than 10%
- Change in aseptic surgical technique
- Change in animal breed or strain
- Change in location
- Change in source of animal
- Change in experimental timeline
- Change in the number of times a procedure is performed
- Change in the weight range of animals used
- Change in anesthesia/analgesia from one approved method to another approved method
- Addition of procedures that are already approved in a policy or SOP
- Change in the method of animal identification
- Change in species for field studies only
- Change in approved euthanasia method
- Change in sampling for genotyping
- Institution does not utilize VVC

10:34 An IACUC protocol was properly approved. Can an investigator begin research with animals based on an oral approval from the IACUC office or must there be written documentation of the approval?

Reg. PHS Policy IV, C, 4 states: "The IACUC shall notify investigators and the institution in writing of its decision to approve or withhold approval of those activities related to the care and use of animals, or of modifications required to secure IACUC approval."

Opin. Once approval of the protocol has been documented in the IACUC records, which can mean written in the IACUC minutes by the recording secretary or recorded on a summary check sheet or some other process approved by the IACUC, the research may begin but not until the date of written notification of approval that has been issued to the investigator. Once all requirements are satisfied and documented accordingly, IACUC written notification of approval should be issued as soon as possible. Oral conveyance of IACUC approval is not sufficient.

10:35 How should the IACUC review a change to a protocol that involves only a change in title and/or funding source?

Opin. If a protocol has been previously reviewed and approved, then the title or funding source may be changed without re-review by the IACUC if it is the policy of the institution to handle changes in title and or funding source administratively. If there is no institutional policy to delineate changes that are eligible for administrative review, then a proposed change in title warrants review by the IACUC.

10:36 What procedures can the IACUC use to determine if a protocol is new or a resubmission with a change in title?

Opin. If a protocol is determined to be a resubmission with a change in the title only, the IACUC may decide to verify this by comparing the resubmission with the previously approved protocol on file. Investigators sometimes modify protocols during the course of resubmission without recognizing that the IACUC must approve all proposed significant changes (PHS Policy IV,B,7; AWAR §2.31,c,7). Indeed, most IACUCs require approval of any change prior to implementation. It is perhaps a better procedure to make changing the title of a protocol a separate process with its own form or treat it as just one of several types of amendments to a protocol. That way the changes in the protocol would be obvious.

10:37 Should IACUC decisions be influenced by the source or size of a research grant?

Reg. The PHS Policy and the NIH Grants Policy Statement (Part II, Terms and Conditions) require the institution to verify, before award, that the IACUC has reviewed and approved those components of grant applications and contract proposals related to the care and use of animals.

Opin. On reflex, most IACUC members would probably answer "no" to this question. In general, all protocols should be reviewed with the same rigor regardless of how much money is involved or the source of funding. The size of the grant should not impact IACUC decisions.

Whereas political pressures for IACUC approval of large or prestigious grants are real, studies should be approved based on appropriateness of the proposed research. Certainly, animal welfare should never be compromised in the interest of increasing grant funds. IACUCs must be constantly vigilant for inconsistencies in their deliberations and decisions, especially those affected by such pressures or perceptions.

Institutions may specify in their Animal Welfare Assurance that research not funded by the PHS will be conducted according to the PHS Policy and NIH/OLAW guidance. Institutions choose to do this for several reasons:

- Having a single set of standards for the animal program simplifies operations oversight.
- They wish to provide state-of-the-art care to all their research animals—a choice that appeals to students, donors, the public, and one that strengthens recruitment efforts.
- An outside authority regulates the behavior of difficult animal users.

10:38 Should the decisions of the IACUC be influenced by the potential scientific importance of a project proposed in a protocol?

Reg. The IACUC has the authority to approve a proposed project. NIH/OLAW and APHIS/AC have stated that under no circumstances is an IACUC required to approve a project against its will[17] (see 17:7).

Opin. IACUC decisions should not be influenced by the investigator, species, funding source, or "hotness" of the research topic. Review should be based on animal welfare issues in consideration of the regulatory requirements. Nevertheless, both the dollar value of the grant and a species-specific view of an animal's societal worth potentially could, but should not, affect IACUC deliberations[18] (see 10:68).

10:39 If a protocol is completely novel and contains untested surgical or experimental procedures, how could such a protocol be reviewed and approved when the procedures cannot be referenced?

Opin. Few protocols actually contain untested surgical procedures. In some cases, the IACUC could recommend that a pilot study be conducted involving only the part of the protocol using the untested procedures. With these pilot data derived from successful use of the previously untested procedures, the investigator could then obtain IACUC approval for the complete protocol. Pilot studies, however, must also be reviewed and approved by the IACUC. (See NIH/OLAW FAQ D.11.)

Alternatively, the IACUC could approve the protocol as submitted but with a reduced number of animals, the remainder being approved when the investigator has tested the new procedure. Both of these actions allow the researcher to proceed with the study while protecting the animal subjects. The protocol could also be reviewed based on knowledge of the technique most closely related to the proposed procedure—the ones from which this was extended. Also, the IACUC may factor in the effect this procedure might have on human pain and distress and extrapolate to animals (see 14:9; 14:11; 14:12).

10:40 What are the possible decisions an IACUC can make after protocol review?

Reg. Both the PHS Policy (IV,B,6) and the AWAR (§2.31,c,6; §2.31,d,4) allow the IACUC to approve, require modifications (in order to secure approval), or withhold approval of proposed activities related to the care and use of animals. This includes significant changes to ongoing animal activities (AWAR §2.31,c,6; §2.31,d,4; PHS Policy IV,B,7). The IACUC can also vote to suspend a protocol (AWAR §2.31,c,8; §2.31,d,4; PHS Policy IV,C,6).

Opin. In determining the disposition of a protocol submitted for review by the full committee review method, an IACUC has multiple options including:

- Approval
- Require modifications to secure approval (see 10:42)
- Refer back to the full committee for re-review
- Withhold approval
- Defer decision—pending additional information.

When the DMR method is used, the only options available are

- Approval
- Require modifications to secure approval
- Refer to the full committee for review

It should be noted that, during re-review of a PI's protocol and/or response to the IACUC's review, the committee or designated reviewer may identify additional concerns or even change the requirements in consideration of new information.

10:41 What constitutes "approval" of a protocol? Does this action necessarily mean that no further changes or information are needed?

Reg. Per PHS Policy and the AWAR, IACUCs either approve, require modifications in (to secure approval), or withhold approval of protocols (AWAR §2.31,c,6; PHS Policy IV,B,7). The same holds true for designated reviewers with an additional option of referring the protocol for FCR but without the option of withholding approval. Anything short of final written approval is not adequate for initiation of animal activities or submission of an IACUC approval date to the NIH as part of a grant application.

Opin. Generally, an approved protocol is one that contains all the required information, has been judged by the IACUC to be acceptable, has satisfied all approval criteria, and for which approval has been recorded in the IACUC minutes or other records.

10:42 What constitutes "requires modifications to secure approval"?

Reg. See 10:41.

Opin. The IACUC, during a full committee meeting, may require modifications of a protocol in order for the investigator to secure approval. This action may be inappropriately referred to as "approval with conditions," which will be addressed shortly. In any event, often the Chair, the AV, or another member is assigned by the IACUC Chair as a designated reviewer to review the investigator's response or the revised protocol. He or she is empowered by the IACUC to approve the revised protocol without further review by the full committee. Alternatively, a subcommittee may be formed and empowered to review the response and approve the revised protocol via the DMR process; however, depending upon the conditions imposed, the entire committee may wish to see the response.

In 2009, NIH/OLAW issued guidance on DMR subsequent to FCR.[19, 20] This guidance states that if *all* members of the IACUC are present at a meeting they may vote to require modifications to secure approval, and the protocol will then be reviewed by DMR. However, if not all members are present at a meeting, the guidance allows a quorum of the members to decide by unanimous vote to use DMR subsequent to FCR providing *all* IACUC members agree in advance, in writing, to a policy which allows this review procedure (see 10:47).

Before the study can be initiated, the PI must first comply fully with all conditions arising from the IACUC's review, and then final approval can be granted. The reader is cautioned that terms such as "conditional approval," "provisional approval," or "approved pending clarification" frequently cause confusion. NIH/OLAW and APHIS/AC strongly advise IACUCs to avoid these terms, as they may lead to non-compliance (see 10:47).

IACUCs might determine that a protocol is approvable, contingent on receipt of a specific modification or confirmation (e.g., receipt of assurance that the PI will conduct the procedure in a fume hood). The IACUC can handle this modification (or clarification) as an administrative detail that is documented in the protocol record. On the other hand, protocols that are missing substantive information necessary for the IACUC to make a judgment (e.g., justification for withholding analgesics in a painful procedure) are incomplete. If the protocol is incomplete, it is not possible to satisfy the protocol review criteria. IACUCs should devise effective ways of differentiating between substantive omissions and administrative issues.

10:43 What constitutes "tabling a protocol" for reconsideration by the full committee?

Opin. "Tabling" of a protocol is a term often used by IACUCs, but it is not an action described in the PHS Policy, AWAR, or in any related guidance. The IACUC may "table" or "postpone" further review of a protocol pending receipt of additional substantive information or a significant revision of the protocol. The action of tabling a protocol is generally used during the process of FCR when the IACUC decides it is necessary for the whole committee to re-review the protocol before further action can be taken. Tabling a protocol is usually reserved for proposals that do not contain sufficient information or those in which the IACUC has identified a serious animal welfare concern. The term "tabling," however, may be perceived as negative. A more diplomatic term is "refer back to the full committee" for re-review by the IACUC.

10:44 What might cause an IACUC to withhold approval of a protocol?

Reg. Approval may be withheld if any of the PHS Policy (IV,C,1) or AWAR (§2.31,d,1; §2.31,e) criteria are not met.

Opin. Withholding approval of protocols is infrequent. Generally, approval is withheld when the PI and the IACUC are not able to come to an agreement on issues in which the IACUC requires modifications to secure approval.

Both the PI and the IO must be notified in writing of the committee's decision. This written notification should include a statement of the reasons for the decision and provide the PI with an opportunity to respond in person or in writing. Certainly, the IACUC may reconsider its decision with documentation in the committee minutes, in light of new information provided by the PI (PHS Policy IV,C,4; AWAR §2.31,d,4). In some cases, the IACUC may want to consult an outside expert in order to obtain a supplemental review.

10:45 How should an investigator respond to questions or conditions from the IACUC?

Opin. Responses from the PI to questions or concerns raised by the IACUC should be in such a format as to result in a protocol that contains a complete, easy-to-discern description of the proposed activities. The response from the PI should clearly answer each issue raised. This can be achieved via a point-by-point letter that serves as a means for the PI to respond to the committee's questions and concerns accompanied by a revised protocol that incorporates all required changes and clarifications. Requiring the PI to provide a revised protocol has the advantage of ensuring that a complete and up-to-date protocol is embodied in one document. Copies of the final approved IACUC protocol should be maintained in the IACUC administrative office, in the animal facility, and in the investigator's laboratory. This final, approved protocol also has the advantage of facilitating the management of the protocol by the animal care staff and for compliance review purposes. In the case of electronic protocol submissions, such responses by the investigator are probably best made directly to the online protocol document. Finally, if PIs submit clear protocols and IACUCs perform equally clear reviews, then the overall efficiency of the review process improves accordingly.

Surv. Approximately what percentage of your IACUC protocols are approved the first time they undergo either full committee or designated member review after they have received a pre-review?

• None	15/260 (6%)
• About 1%–10%	65/260 (25%)
• About 11%–25%	28/260 (11%)
• About 26%–50%	38/260 (15%)
• Greater than 50%	109/260 (42%)
• I don't know	5/260 (2%)

10:46 Should there be a time limit to receive responses to queries raised by the IACUC?

Opin. Establishment of time limits and other constraints should be considered when the IACUC develops an SOP. Many IACUCs have found it useful to set specific expectations for time constraints and deadlines which can be of benefit to both the investigator and the committee. Ironclad deadlines and inflexible staff, however, can present a major point of contention for faculty already juggling multiple projects and attempting to comply with much paperwork and many deadlines.

Surv. What does your IACUC do if an investigator does not respond to IACUC queries in a reasonable length of time?

• We have never encountered this problem.	50/260 (19%)
• The protocol remains open for a response, forever.	26/260 (10%)
• Our IACUC has a time limit, after which the protocol is considered invalid.	86/260 (33%)
• We have no firm policy.	87/260 (33%)
• I don't know.	0/260 (0%)
• Other.	11/260 (4%)

10:47 After full committee review of a protocol in which modifications are required to secure approval, can the IACUC Chair or a designated reviewer approve the protocol once the modifications are verified?

Opin. This should be determined at the meeting when the decision is made to grant further review and approval authority via the DMR process (see 10:42). That is, after subsequent review, approval may be granted if the revised protocol has sufficiently addressed the concerns identified on initial FCR review. For DMR subsequent to FCR, a formal process must be followed as described in 10:42. Any member of the IACUC, including the Chair, can serve as the DMR reviewer of the revised protocol upon designation by the Chair as long as he or she is qualified to do so.

10:48 Is a majority vote required for any IACUC actions related to protocol approval?

Reg. Approval of a protocol being reviewed by FCR "may be granted only after review at a convened meeting of a quorum of the IACUC and with the approval vote of a majority of the quorum present" (PHS Policy IV,C,2; AWAR §2.31,d,2). This is also true for votes to suspend an activity. (See 10:20 for designated-member review requirements.)

Opin. The PHS Policy and AWAR represent the minimum requirements, and an institution may implement more stringent requirements. These should be clearly stated in the institution's policy and procedures and its PHS Assurance (if applicable). It is important to consider the decision to exceed the requirements carefully, as the procedures stated in the institution's PHS Assurance become the legal requirements for PHS funding upon NIH/OLAW approval of the Assurance. APHIS/AC also expects adherence to institutional policies. It is also important to remember that all approval criteria, including institutional specific requirements, must be satisfied for the protocol to be approved and authorized for implementation.

10:49 At a full committee meeting of a 20-member IACUC, 15 members are present. Six vote to approve a protocol, six abstain, and three vote against approval. Is the protocol considered approved according to the AWAR and PHS Policy?

Opin. No. In this example, the 15 members present constitute the quorum. To approve a protocol, a simple majority of those members present, i.e., eight, would have to vote for approval of the protocol. Because only six members voted to approve, the vote count fell short by two votes, and the protocol cannot be approved according to the AWAR and PHS Policy[21] (see 10:48).

10:50 How should the IACUC handle minority opinions to IACUC actions on the review of protocols?

Reg. PHS Policy (IV,E,1,d; IV,F,4) and AWAR (§2.31,c,3) refer to semiannual facility inspections and program evaluations and annual reports and for those purposes minority views must be included. For minority views expressed during the review of an animal use protocol, NIH/OLAW has provided guidance on a dissenting vote versus a minority view:

> "Both protocol approval and suspension of animal study protocols by the IACUC require a majority vote of a quorum of the IACUC. Although an IACUC member's dissenting vote on these issues must be recorded in the minutes, this does not constitute a minority view for reporting purposes. Any IACUC member may submit a minority view to OLAW addressing any

aspect of the institution's animal program, facilities, or personnel training. Whether OLAW receives a minority view as part of an annual report, renewal Assurance document materials (i.e., the most recent Semiannual Report to the IO for Category 2 institutions that lack AAALAC accreditation), or directly from the dissenting IACUC member, it carefully reviews the information provided in accordance with requirements of the PHS Policy and provisions of the Guide."[22]

Opin. Because neither the PHS Policy nor the AWAR address this issue relative to specific IACUC protocols, it should be considered when the IACUC develops an SOP. A common practice is to record dissenting opinions in the minutes. However, another option is to allow the dissenter to write a letter expressing his/her opinion. This is then filed along with the IACUC protocol. When deciding on the method of dissent documentation, the IACUC should review any applicable open records laws for its state. (See Chapter 26.)

10:51 A protocol undergoes a congruency check with the corresponding NIH grant application, and the IACUC requires protocol modifications which result in a change of scope in the grant. Does the granting agency need to be notified?

Reg. The PHS Policy (IV,D,2) requires "verification of approval (including the date of the most recent approval) by the IACUC of those components related to the care and use of animals. If verification of IACUC approval is submitted subsequent to the submission of the application or proposal, the verification shall state the modifications, if any, required by the IACUC."

For changes of scope that may include modifications that impact the scientific aims of the grant, NIH Grants Policy 8.1.2.5 requires the recipient to obtain prior approval from the NIH awarding institute or center.

Since 2002, NIH Grants Policy has allowed the verification of IACUC approval to be submitted later in the application process to reduce the time to award and ensure the accuracy and timeliness of the information.[10] Such a request for information later in the review cycle is referred to as "just-in-time" (see 9:24; 10:12.) NIH/OLAW has described the oversight role of the IACUC by stating:

"An institution that elects to proceed according to "just-in-time" procedures for IACUC approval bears the responsibility for supporting the decisions of the IACUC. Under no circumstances may an IACUC be pressured to approve a protocol or be overruled on its decision to withhold approval. The PHS Policy requires that modifications required by the IACUC be submitted to the NIH with the verification of IACUC approval, and it is the responsibility of institutions to communicate any IACUC-imposed changes to NIH staff. It is incumbent upon investigators to be totally forthcoming and timely in conveying to the IACUC any modifications related to project scope and animal usage that may result from the NIH review and award processes. Should an institution find that one of its investigators disregards his/her responsibilities, the institution may, for example, determine that all animal protocols from that investigator be subject to IACUC approval prior to allowing that investigator to submit an application."[23]

Surv. For institutions that receive grants from the U.S. Public Health Service (PHS), does your IACUC inform the granting institute or center when a procedure or species that is not on a funded grant is substituted for one that is on the funded grant?

• Not applicable, we do not receive PHS funding.	42/260 (16%)
• Yes we do so routinely.	28/260 (11%)
• Yes, we do so on occasion for any such substitution.	37/260 (14%)
• Yes, we do so but only when species covered by the Animal Welfare Act are involved.	6/260 (2%)
• No, we do not.	56/260 (22%)
• I don't know.	51/260 (20%)
• Other.	40/260 (15%)

10:52 To check for congruence, should the IACUC review just the IACUC protocol or should the IACUC also review the animal care and use sections of an associated NIH grant proposal?

Reg. "It is the *institution's* responsibility, not the IACUCs, to ensure that the information the IACUC reviews and approves is consistent with that contained in the application to be funded. While the IACUC may perform the grant to protocol congruence review, it is not a required IACUC function and may be performed by other qualified institutional personnel with responsibility designated to a specific office or position (e.g., sponsored programs or compliance office)."[24] Verification of IACUC approval submitted to the NIH means that the institution certifies to the NIH that the IACUC has reviewed and approved those components of the grant application related to the care and use of animals. "Applications or proposals (competing and non-competing) covered by this Policy from institutions which have an approved Assurance on file with OLAW shall include verification of approval ... by the IACUC of those components related to the care and use of animals" (PHS Policy IV,D,2). The signature of the institutional representative on the grant application (PHS 398 form) is a legally binding statement that the IACUC has approved all animal activities covered in the grant application. The submission of just-in-time verification also constitutes a binding statement. It should be noted that the signature on the grant application is an attestation of assurance provided by the institution and not the IACUC.

NIH/OLAW in its webinar on "Grants Policy and Congruence" states: "The institution must document the IACUC's approval of the animal activities proposed in the grant. The institution, when asked by NIH, must be able to associate each grant or grants with a relevant protocol or protocols. If the institution uses a protocol numbering system, it must be able to link protocol numbers to grant numbers. But a 1:1 ratio is not required."[25]

Opin. If the IACUC protocol accurately reflects the information contained in the grant application, then the verification is valid. The IACUC, however, cannot be assured of such validity unless it also reviews the vertebrate animal sections of the associated grant application, or the committee relies on another mechanism such as a valid comparison of the protocol(s) with the grant performed by a qualified administrator such as a grant specialist, compliance officer, or personnel within the IACUC office. With the advent of just-in-time (see 9:24; 10:12; 10:51), review of the grant proposal need not take place until it appears that the proposal will be funded.

Some institutions do not review the grant proposals at all. They simply rely upon a statement by the PI or department chair that the protocol accurately reflects what is in the grant proposal. However, if the PI's or chair's certification is inaccurate, the resulting institutional verification to the PHS will be invalid. If such a discrepancy were to be discovered, there are potential civil and criminal penalties for submitting false statements.[25]

Finally, it should be noted that all animal use on a specific grant must be approved by the IACUC but does not have to be contained in only one IACUC protocol. It is an institutional decision as to whether the procedures will be covered in one or several (many) IACUC protocols.

Surv. For institutions that receive grants from the U.S. Public Health Service (PHS), how does your institution assure that what is written on a grant application is covered by one or more IACUC protocols?

- Not applicable, we do not receive PHS funding. 39/257 (15%)
- We make no such determinations. 5/257 (2%)
- The Principal Investigator attests to congruency between the grant and protocol(s). 47/257 (18%)
- The IACUC office compares the grant to one or more IACUC protocols. 111/257 (43%)
- Our institution, but not the IACUC office, makes the determination. 20/257 (8%)
- Members of our IACUC are asked to make the determination. 19/257 (7%)
- I don't know. 8/257 (3%)
- Other. 8/257 (3%)

10:53 Should the IACUC accept an approval statement from an IACUC at another institution?

Reg. An IACUC may accept the approval of an IACUC at another institution if that institution has an approved NIH/OLAW Animal Welfare Assurance. This practice is normally limited to collaborations or subawards involving performance sites outside the awardee institution. In most instances, the IACUC of the performance site assumes responsibility for animal activities in its facilities. Both institutions should have a clear understanding and a written agreement of what their respective responsibilities are in this situation, particularly if one IACUC agrees to abide by the determinations of the other[26] (see 9:17).

Opin. This is an area for which the IACUC should establish a SOP. The IACUC should have some "comfort level" about the other institution, its committee, and the quality of its research. To avoid problems, a collaborative protocol approved by an IACUC at another institution may warrant some level of review by the local IACUC (e.g., DMR). In some instances, FCR may be considered depending on the potential for animal pain or distress and protocol complexity (see 10:54). Issues to consider in accepting an approval statement from an IACUC at another institution include:

- Does the institution have an approved NIH/OLAW Animal Welfare Assurance?
- Is the institution a USDA-registered research facility?
- Is the institution accredited by AAALAC? AAALAC accreditation may provide an additional level of confidence in the IACUC's review.

When animal studies involve experiments at two different institutions a memorandum of understanding (MOU) between the two institutions should be implemented. The MOU should detail areas of responsibility by both IACUCs, AVs, and Programs.

Acceptance of an external IACUC review and approval also includes the responsibility for continuing review. The external institution should agree to supply all continuing review documents relative to the protocol as they are reviewed and approved. Failure to do this can result in withholding a notice of approval to the granting agency and possible loss of grant funds. In some cases, the home institution's sponsored programs office may require an assurance of continuing IACUC approval for the grant funds to be released. This can be another source of problems for the PI's research.

10:54 Can an investigator rely on a protocol reviewed and approved by an IACUC at another institution in order to initiate his or her research? What should be the conditions and limitations for this reciprocity of IACUC approval?

Reg. See 9:17; 10:53.

Opin. An IACUC may choose to accept approval for a protocol by an IACUC at a different institution. However, any conditions regarding acceptance of an approval statement from an IACUC at another institution should be described in the IACUC's written policies and in the written agreement between the two institutions. For example, in order to facilitate research, an IACUC may choose to allow animals to be purchased or transferred to its institution, but procedures that involve animals cannot be performed until the local IACUC has reviewed and approved the protocol (see 10:53). Because local institutional requirements vary, automatic approval is not advisable.

10:55 An investigator subcontracts part of a PHS-funded research project to another institution where animals will be used. What oversight and paperwork responsibilities does the awardee institution have relative to the PHS Policy and the AWAR?

Reg. NIH/OLAW requires that any subcontracted work involving animals that is supported by PHS funds be conducted only at other NIH/OLAW-Assured institutions (PHS Policy V,B). If the performance site (collaborating institution or subcontractor) is not NIH/OLAW-Assured, then NIH/OLAW requires that an Assurance for the performance site be negotiated in order for the work to go forward (PHS Policy V,B). Alternatively, an institution can choose to accept responsibility for the work under its own Assurance and include the performance site as a covered component with a written agreement between the institutions. It should be recognized, however, that the latter arrangement means that the NIH/OLAW-Assured institution assumes full responsibility for ensuring that all the animal work conducted at the performance site is in full compliance with the PHS Policy and the AWAR. This arrangement necessitates that the NIH/OLAW-Assured institution's IACUC conduct semiannual program reviews and facility inspections of the performance site (see 9:17).

In addition, the NIH Grants Policy Statement on written agreements requires that awardees have a formal written agreement with consortium participants that includes agreement for meeting the PHS Policy requirement for review and approval of proposed animal activities and significant changes to animal activities and semiannual facilities review by an IACUC.[26]

The AWAR do not address subcontracting. The institution that conducts the animal activity will be held accountable for animal care; hence, a protocol approved by the IACUC of the institution conducting the animal activity must be in place. The approval process is outlined in AWAR (§2.31,c,6; §2.31,c,7).

Opin. When an institution subcontracts part of a research project to another institution, there should be a clearly defined written agreement (*Guide*, p. 15) concerning the responsibilities of each institution to comply with the PHS Policy and AWAR regardless of whether both institutions have approved Assurances on file with NIH/OLAW. If a serious compliance problem arises at a subcontract site, it will likely impact the primary institution. However, if the subcontract site is registered with APHIS/AC and is the facility whose IACUC approved the protocol, that site would be held primarily responsible. Written agreements can help minimize and resolve potential problems.

Surv. If your IACUC subcontracts or otherwise collaborates for animal care or use with another institution, do you require a Memorandum of Understanding to delineate areas of responsibility for animal care and use?

• Not applicable	58/257 (23%)
• No	6/257 (2%)
• Rarely	8/257 (3%)
• Yes, but only if we are the primary grant or contract holder	40/257 (16%)
• Yes, but only if we are dealing with species covered by the Animal Welfare Act	2/257 (1%)
• Yes, under most circumstances	127/257 (49%)
• I don't know	8/257 (3%)
• Other	8/257 (3%)

10:56 Can an investigator appeal the decision of the IACUC relative to a protocol review decision?

Reg. The PHS Policy (IV,C,4) and the AWAR (§2.31,d,4) require that written notification of withholding approval include a statement of the reasons for the decision, "to give the investigator an opportunity to respond in person or in writing." Nevertheless, IACUC decisions to withhold approval may not be overturned by a higher institutional authority (PHS Policy IV,C,8; AWAR §2.31,d,8).

Opin. The above regulatory information suggests that a reconsideration by the IACUC should be an option. Furthermore, the "Supplementary Information" that accompanies the August 31, 1989, Federal Register containing 9 CFR Parts 1, 2, and 3 Final Rules (p. 36132), states that "on the basis of the response, the committee may reconsider its decision." This is another area for which the IACUC is advised to establish a written SOP for exactly how an investigator appeal will be handled with attention to detail and due process. Certainly, investigators should be invited to meet with the committee during the appeal process (see 10:57).

10:57 What is an appropriate and effective mechanism for appeal of IACUC decisions (see 10:56; 33:45).

Opin. Some suggestions to consider in developing an SOP for appeals include:

- Assignment of an IACUC member to work with the investigator to facilitate the process
- Involvement of an expert consultant to advise the IACUC

It should be remembered that according to both the AWAR and PHS Policy an IO may not overturn a decision of the IACUC to withhold approval (see 10:56). Therefore, the IACUC must alter its own decision, which it can do by a vote of a majority of a quorum at a convened meeting or via the DMR review process, although the latter method may not do justice to an appeal that really should be considered by the full committee.

10:58 Can a protocol for which approval has been withheld be resubmitted with modifications?

Reg. Yes, neither PHS Policy nor the AWAR preclude resubmission.

Opin. Resubmission of a modified protocol should be encouraged. A resubmission provides the investigator with an opportunity to respond to the IACUC's review as allowed by PHS Policy (IV,C,4) and the AWAR (§2.31,d,4). The IACUC also may consider assigning a member to work with the investigator to develop a protocol that can be approved by the committee and allow the investigator's research program to progress (see 10:57).

10:59 Is there any institutional authority that can reverse the decision of the IACUC to withhold approval?

Reg. No. The AWAR (§2.31,d,8) state that "officials of the research facility … may not approve an activity involving the care and use of animals if it has not been approved by the IACUC." The PHS Policy (IV,C,8) is effectively identical (see 10:56; 33:38).

10:60 At times, PIs who do drug testing may receive a test drug, but the manufacturer does not want to reveal what the drug is. Can the IACUC approve the use of the drug in animals?

Opin. It is not uncommon that a pharmaceutical company contracts the testing of a new drug to an investigator in a research institution. Because such drug development is highly competitive, the company may not want to divulge the exact identity of the drug. This poses a problem for the IACUC charged with evaluating the effect of the drug on the welfare of animals. In order to approve the use of the drug, the IACUC must know at least the general class of the drug as well as its dose and route of administration, and any known risks to the animals. With this much information it is usually possible for the IACUC to approve use of the drug. It is often advantageous to have all IACUC members sign a binding confidentiality agreement (see 10:61; 31:20; 31:23).

10:61 Sometimes PIs do not know which of a class of drugs they will use, either because there are many or because they don't know which will be available. Can the IACUC approve the use of a class of drugs in animals?

Opin. Sometimes an investigator wants to use a particular class of drug in a study, but he or she is unsure if or when a given drug will be used. The IACUC can approve the use of a general class of drugs if a list of drugs that might be used along with dose range, frequency, and route of administration for each one is provided. It would also be helpful for the committee to know the conditions under which each would be used, the mechanism of action, potential side effects, and toxicity. The IACUC may choose to require that the PI notify the committee of which drugs are actually used either at the time of annual or continuing review or in a special interim or periodic report, as deemed appropriate by the IACUC (see 10:60).

10:62 The IACUC has a responsibility for initial review and continuing oversight of animal research projects. Sometimes projects involve adverse events or unexpected problems. What actions should the IACUC take, if any, when notified of these adverse events?

Opin. If the unexpected problem results in morbidity or the death of animals, the IACUC may require the PI to submit interim reports, i.e., to report more often than triennially as required by AWAR and PHS Policy. The nature and duration of such reporting would be determined by the severity of the problem and its frequency. The IACUC may also ask the PI to alter the procedures of the project and to submit a revised protocol for review.

10:63 An IACUC member requests that the IACUC reopen the discussion of a previously approved protocol. He was away when the protocol was approved, and he has concerns that were not considered during the review process. Can

an IACUC reexamine the approval of a protocol under these circumstances?

Opin. The IACUC is charged with continuous oversight of projects and to review concerns involving the care and use of animals at the institution. Therefore, the IACUC *can* rereexamine a protocol at any time it thinks it is necessary. If an absent IACUC member raises valid concerns that were not addressed at the time of initial or continuing review, these should be brought before the full IACUC. It should be noted, however, that the project may already have been started and therefore the IACUC should include this fact in their reconsideration of a project. Requiring major changes in the project may invalidate research that has already been conducted.

Finally, any IACUC member at any time should express any concerns he or she has regarding animal welfare or compliance. NIH/OLAW and APHIS/AC have advised: "Review of concerns is not a matter of choice; concerns must be brought to the IACUC's attention.... Concerns about animal activities at an institution must be reviewed at a convened meeting of the IACUC."[27] All concerns warrant referral to the full committee and consideration for re-review of the protocol as necessary.

10:64 Can an IACUC develop and/or accept SOPs in lieu of the investigator describing procedures in detail within a protocol? What special concerns might be generated by doing so?

Opin. The AWAR and the PHS Policy do not specifically preclude the use of SOPs. Nevertheless, there are some issues that should be considered. First, it is incumbent on the IACUC to be familiar with the SOP in order to conduct an appropriate review of the protocol. This is an important facet of the review process and clarifies what will happen to an animal from the beginning to the end of the study. If not, then all aspects of the protocol have not been reviewed by the IACUC. Second, when the SOP is modified, this would constitute a change in every protocol that references its use, and all associated protocols should be amended each time the SOP is modified. In addition, every investigator who uses the SOP must be aware of the change and adopt it into their research protocol. Changing an SOP without adoption by the investigator might well have the effect of putting him/her into non-compliance.

It might be possible for the institution to maintain different versions of the same SOP, but it appears that keeping track of which version goes with which protocol would pose a difficult process. The investigator of each study must take responsibility for use of any SOP in their research. NIH/OLAW and APHIS/AC issued an NIH/OLAW FAQ D14[28] on SOPs or blanket protocols that provides guidance on this issue. Additional guidance has been published elsewhere.[29]

10:65 For commonly used, well-established procedures, such as collecting a blood sample from a dog's cephalic vein, should the IACUC request details of how the procedure itself will be performed?

Reg. The PHS Policy (IV,C, 1,f) and AWAR (§2.32,a) require the personnel conducting procedures on animals to be appropriately qualified and trained. It is the responsibility of the institution to provide necessary resources for training and it is the responsibility of the IACUC to ensure that personnel are, in fact, qualified. Even if a procedure is well established, the protocol should contain sufficient detail about the procedure(s), which, in turn, helps the IACUC in its evaluation of personnel qualifications.

Opin. The IACUC should ask for information about parameters of preparation of the skin for the sample, a range of the volume of blood to be drawn, the interval between draws, and the range of frequency of draws. It would not be necessary to ask for the angle of needle insertion and such items unless the IACUC was not convinced that the person doing the blood draw knew how to do it. Training should be required if personnel lack sufficient expertise prior to initiation of the procedure(s).

10:66 Rather than providing details of an experimental procedure on an IACUC form, a PI provides literature references that clearly describe the details of what he will do. Should the IACUC consider approving this study with just the literature reference or should the actual procedure be part of the IACUC protocol?

Opin. The actual procedure should be part of the IACUC protocol so that it is clear what will be done as part of the experimental procedures. As necessary, the IACUC reviewer may choose to read the literature references (see 10:65).

10:67 Can the IACUC use teleconferences or other electronic communication in its review of animal use protocols? (see 7:17–7:19).

Reg. According to PHS Policy, "If full committee review is scheduled, or requested by a DMR reviewer, approval of those research projects may be granted only after review at a convened meeting of a quorum of the IACUC and with the approval vote of a majority of the quorum present." The AWAR use essentially the same language, and neither document addresses how the convened meeting may occur.

Opin. Traditionally, a quorum of the IACUC members meets together in a room to discuss and vote on a protocol. According to Garnett and Potkay, "The traditional meeting provides the kind of environment which is most conducive to thoughtful deliberation and interaction and is still regarded as the optimum forum for many of the IACUC functions."[30] But, this is not the only possibility. With modern technology, virtually identical face-to-face meetings are possible without the participants being in the same room.

What should be the criteria for such electronic meetings? NIH/OLAW has listed, and USDA/APHIS has concurred, with what are appropriate criteria.[31]

Methods of telecommunications (e.g., telephone or video conferencing) are acceptable for the conduct of official IACUC business requiring a quorum, provided the following criteria are met:

1. All members are given notice of the meeting.
2. Documents normally provided to members during a physically convened meeting are provided to all members in advance of the meeting.
3. All members have access to the documents and the technology necessary to fully participate.
4. A quorum of voting members is convened as required by PHS Policy and AWAR.
5. The forum allows for real-time verbal interaction equivalent to that occurring in a physically convened meeting (i.e., members can actively and equally participate and there is simultaneous communication).
6. If a vote is called for, the vote occurs during the meeting and is taken in a manner that ensures an accurate count of the vote. A mail ballot or individual telephone polling cannot substitute for a convened meeting.
7. Opinions of absent members that are transmitted by mail, telephone, fax, or e-mail may be considered by the convened IACUC members but may not be counted as votes or considered as part of the quorum.
8. Written minutes of the meeting are maintained in accord with the PHS Policy (IV, E,1,b) and the AWAR (§2.35,a,1).

Clearly, it is possible to accomplish FCR in electronic meetings. There is no real need to use such techniques for designated reviewer methods. The conveyance of information required for the latter method can clearly be handled electronically. There is no voting involved in the process.

10:68 Should the decisions of the IACUC be influenced by the species proposed for use in the project?

Opin. "Speciesism," the treatment of animals differently solely because they are members of a different species, can drift into the IACUC deliberation process. Some species-specific care and use criteria must be part of the protocol, but it is assumed by some individuals that some species have a greater moral standing than others. This can come about because they are more like human beings than other species or because they often serve as pets. Some people justify distinctions between species on the basis of supposed sentience. However, there is little solid evidence for a difference in sentience between a number of species, such as a rat and a mouse or a cat and a dog. According to one survey, only a minority of IACUCs let the species to be used influence their decisions.[32]

No one has reported tracking the time spent in deliberation of protocols for different species, but many will assert that protocols involving nonhuman primates often take longer for the IACUC than those of other species for the same level of invasiveness. There is nothing in the AWAR, the PHS Policy, or the *Guide* to suggest that regulatory agencies expect this. Species selection and IACUC deliberations should be made on the basis of scientific appropriateness for the proposed project (see 10:37).

10:69 Is it possible to set up a commercial IACUC that hires out to review and approve protocols for an institution that doesn't want to set up its own IACUC?

Reg. The AWAR (§2.33) require the AV to be employed under a formal contract with the institution and for the AV to be a voting member of the IACUC. The PHS Policy (IV,A,1) requires compliance with the AWA and, like the AWAR, states that the IACUC veterinarian must have programmatic authority and is responsible for animals at the institution. As such, it seems nearly impossible to have an IACUC-for-hire.

NIH/OLAW states: "Domestic Assurances are negotiated with institutions that: control their own animal facilities, conduct animal research onsite, have an animal care and use program with an IO, an IACUC and a veterinarian with program authority and access to all animals."[33]

Opin. There are commercial/independent Institutional Review Boards (IRBs, independent of research institutions) that perform the jobs of institutional IRBs on a fee-for-service basis. There is no mention in the PHS Policy or AWAR of a commercial IACUC. An institution that receives PHS funding must submit an Assurance outlining its program of animal care and use. It must also name the IO, IACUC Chair, and AV. Both the PHS Policy and the AWAR specify a line of authority that would be very difficult to establish with such an external IACUC. In addition, the AV must be a member of the IACUC. That would mean that the AV of every institution that used such a commercial IACUC would have to be a member of that IACUC. This appears to be logistically problematic even using electronic technologies. In addition, according to the PHS Policy and the AWAR, the IACUC must conduct semiannual evaluations of the institution's animal care and use program along with inspections of animal facilities. This would also be difficult to achieve with a commercial IACUC. Finally, IACUCs have responsibilities that can be very institution-specific, and these include more than protocol review. IACUCs are expected to address immediate animal concerns, non-compliance, and formulate case-specific corrective actions, and having an outside entity to do this would not be advisable or feasible.

10:70 Can polling be used to obtain approval for a protocol in lieu of having a full committee meeting?

Reg. The word "polling" or the word "electronic" (associated with protocol review) do not exist in the PHS Policy or AWAR.

Opin. Whereas it is not referred to directly in the PHS Policy, polling has been addressed by the following NIH/OLAW guidance[29]:

> "Polling is defined as sequential, one-on-one communication, either in person or via telephone, e-mail, fax, U.S. mail, or by other similar means. Polling is an appropriate mechanism for providing all committee members with the opportunity to call for full review of a protocol prior to initiating the "designated reviewer" method of protocol review described below. It may also be appropriate as a mechanism for distributing and reviewing drafts of meeting minutes or reports.
>
> The simple polling of IACUC members does not, however, satisfy the definition of a meeting of a convened quorum and should not be used for conducting IACUC business that requires the vote of a convened quorum of the committee. For example, polling should not be considered a valid method of voting under the "full committee" review method of protocol review and is not an acceptable substitute for having a vote of a convened quorum on the suspension of a previously approved activity involving animals."

The bottom line is that polling can be used for setting up meetings and for collecting responses from members as to whether a protocol should be referred for full committee review as a part of the designated reviewer process. It is not acceptable to employ polling to establish a quorum to conduct business or to review or vote on protocols as a part of a full committee review process. Polling is not conducive to the dynamic exchange of ideas that is important in the full committee deliberation process. As a result, polling precludes due consideration of all the aspects of the full committee review process.

Acknowledgment

The authors wish to thank Rebecca Bogatz for her unfailing assistance in typing the chapter, checking references, and providing valuable editorial comments.

REFERENCES

1. National Institutes of Health, Office of Laboratory Animal Welfare. 2013. *Notices*. https://grants.nih.gov/grants/olaw/references/notices.htm. Accessed January 3, 2023.
2. National Institutes of Health, Office of Laboratory Animal Welfare. 2013. *Frequently Asked Questions (FAQs)*. https://olaw.nih.gov/faqs#/guidance/faqs. Accessed January 3, 2023.
3. National Institutes of Health, Office of Laboratory Animal Welfare. 2013. *Commentary*. https://www.grants.nih.gov/grants/olaw/references/commentary.htm. Accessed January 3, 2023.
4. National Institutes of Health, Office of Laboratory Animal Welfare. 2009. *Reporting Noncompliant Events to OLAW*. https://grants.nih.gov/grants/olaw/educational_resources.htm. Accessed January 3, 2023.
5. Prentice, E.D., D.A. Crouse, and M.D. Mann. 1992. Scientific merit review: The role of the IACUC. *ILAR News* 34:15–19.
6. Mann, M.D. and E.D. Prentice. 2004. Should IACUCs review scientific merit of animal research projects? *Lab Animal (NY)* 33:26.
7. National Institutes of Health, Office of Laboratory Animal Welfare. 2013. *FAQ D.12, Is the IACUC Responsible for Judging the Scientific Merit of Proposals?* https://olaw.nih.gov/faqs#/guidance/faqs?anchor=50327. Accessed January 3, 2023.
8. Final Rules: Animal Welfare. *9 CFR Parts 1 and 2, Federal Register*, Vol. 54. No. 168, August 31, 1989, P. 36114. https://www.nal.usda.gov/legacy/awic/final-rules-animal-welfare-9-cfr-parts-1-2-and-3. Accessed August 16, 2022.
9. National Institutes of Health, Office for Protection from Research Risks. 1991. The Public Health Service responds to commonly asked questions. *ILAR News* 33:68. https://grants.nih.gov/grants/olaw/references/ilar91.htm. Accessed January 3, 2023.
10. National Institutes of Health. April 20, 2012. *Notice NOT-OD-12-101, Notice of Requirement for Electronic Submission of Just-In-Time Information and Related Business Process Changes Beginning 2012* https://grants.nih.gov/grants/guide/notice-files/NOT-OD-12-101.html. Accessed August 16, 2022.
11. Black, J. 1993. Letter to the editor. *ILAR News* 35:1. Accessed August 16, 2022
12. Prentice, E.D., D.A. Crouse, and M.D. Mann. 1993. Letter to the editor. *ILAR News*, 35:2.
13. Federal Register. August 7, 2002. *67 FR 51289*.
14. Wolff, A. 2002. Correct conduct of full-committee and designated-member protocol review. *Lab Animal (NY)* 31(9):28–31. https://olaw.nih.gov/guidance/articles/laba02v31n9.htm. Accessed June 16, 2023.

15. National Institutes of Health, Office of Laboratory Animal Welfare. 2013. *FAQ D.6, What Criteria Should the IACUC Consider When Reviewing Protocols. FAQ D.7, Should the IACUC Consider the Three "Rs" of Alternatives When Reviewing Protocols?* https://olaw.nih.gov/faqs#/guidance/faqs?anchor=50319. Accessed January 3, 2023.
16. American Veterinary Medical Association. 2020. *AVMA Guidelines for the Euthanasia of Animals.* 2020 edition. https://www.avma.org/KB/Policies/Documents/euthanasia.pdf. Accessed January 3, 2023.
17. Garnett, N.L. and W.R. DeHaven. 1998. A word from the government. *Lab Animal (NY)* 27(3):19. https://olaw.nih.gov/guidance/articles/laba98v27n3.htm. Accessed June 16, 2023.
18. Silverman, J. 1997. Do pressure and prejudice influence the IACUC? *Lab Animal (NY)* 26(5):23–25.
19. National Institutes of Health, Office of Laboratory Animal Welfare. 2009. *Notice NOT-OD-09-035, Guidance to IACUCs Regarding Use of Designated Member Review (DMR) for Animal Study Proposal Review Subsequent to Full Committee Review (FCR).* http://grants.nih.gov/grants/guide/notice-files/NOT-OD-09-035.html. Accessed January 3, 2023.
20. National Institutes of Health, Office of Laboratory Animal Welfare. *FAQ D.19, May an IACUC Use Designated Member Review (DMR) to Review an Animal Study Protocol Subsequent to Full Committee Review (FCR) When Modifications Are Needed to Secure Approval?* https://olaw.nih.gov/faqs#/guidance/faqs?anchor=50340. Accessed January 3, 2023.
21. Silverman, J. 1996. Majority rules? *Lab Animal (NY)* 25(4):22. https://grants.nih.gov/grants/olaw/references/laba96v25n5.htm. Accessed January 3, 2023.
22. National Institutes of Health, Office of Laboratory Animal Welfare. *FAQ C.6, What Are PHS Requirements for Recording and Reporting Minority Views?* https://olaw.nih.gov/faqs#/guidance/faqs?anchor=50311. Accessed January 3, 2023.
23. National Institutes of Health. 2010. *Notice NOT-OD-10-128, Clarification on the Roles of NIH Scientific Review Groups (SRG) and Institutional Animal Care and Use Committees (IACUC) in Review of Vertebrate Animal Research.* https://grants.nih.gov/grants/guide/notice-files/NOT-OD-10-128.html. Accessed January 3, 2023.
24. National Institutes of Health. 2021. *Notice NOT-OD-22-005, Notice of Clarification of Institutional Responsibilities Regarding NIH Grant to Protocol Congruence Review.* https://grants.nih.gov/grants/guide/notice-files/NOT-OD-22-005.html. Accessed March 18 2024.
25. National Institutes of Health, Office of Laboratory Animal Welfare. June 7, 2012. *Transcript of OLAW Webinar. Grants Policy and Congruence.* https://grants.nih.gov/grants/olaw/120607_seminar_transcript.pdf. Accessed August 8, 2022.
26. National Institutes of Health, Office of Laboratory Animal Welfare. 2013. *FAQ D.8, When Institutes Collaborate, or When the Performance Site Is Not the Awardee Institution, Which IACUC Is Responsible for Review of the Research Activity?* https://olaw.nih.gov/faqs#/guidance/faqs?anchor=50320. Accessed January 3, 2023.
27. Brown, P. and C.A. Gipson. 2010. A word from OLAW and USDA. *Lab Animal (NY)* 39:167. https://grants.nih.gov/grants/olaw/references/39_6_0610.pdf. Accessed January 3, 2023.
28. National Institutes of Health, Office of Laboratory Animal Welfare. *FAQ D.14, May Standard Operating Procedures (SOPs) or Blanket Protocols That Cover a Number of Procedures Be Utilized in Lieu of Repeating Descriptions of Identical Procedures in Multiple Protocols?* https://olaw.nih.gov/faqs#/guidance/faqs?anchor=50329. Accessed January 3, 2023.
29. Gipson, C. and P. Brown. 2012. A word from OLAW and USDA. *Lab Animal (NY)* 41:43. https://grants2.nih.gov/grants/olaw/references/41_02_0212.pdf. Accessed January 3, 2023.
30. Garnett, N. and S. Potkay. 1995. Use of electronic communications for IACUC functions. *ILAR Journal* 37:190–192. https://olaw.nih.gov/guidance/articles/ilar95.htm. Accessed June 16, 2023.
31. National Institutes of Health, Office of Laboratory Animal Welfare. 2006. *Notice NOT-OD-06-052, Guidance on Use of Telecommunications for IACUC Meetings under the PHS Policy on Humane Care and Use of Laboratory Animals.* https://grants.nih.gov/grants/guide/notice-files/NOT-OD-06-052.html. Accessed January 3, 2023.
32. Silverman, J., S.A. Baker, and C.W. Lidz. 2012. A self-assessment survey of the Institutional Animal Care and Use Committee, Part 1: Animal welfare and protocol compliance. *Lab Animal (NY)* 41:230–235.
33. National Institutes of Health, Office of Laboratory Animal Welfare. 2012. *Obtaining an Assurance.* https://grants.nih.gov/grants/olaw/obtain_assurance.htm. Accessed January 3, 2023.

11

Amending IACUC Protocols

Cecile Baccanale

Introduction

The IACUC approves protocols for a period of three years and ensures that the requested animal use is compliant with the PHS Policy, AWAR, and institutional policies. Throughout the three-year period, investigators may need to modify protocols as research evolves, and changes may be required to improve the study.

Reasons for an investigator to request modifications to a previously approved protocol may include, but are not limited to, the following:

- Based on initial experimental data
- Based on unanticipated outcome(s)
- Based on new, emerging, published research

In addition, practices in veterinary care may be refined as new drugs are introduced, dosages modified, or procedures are refined. Some examples may include:

- A change in analgesia to a sustained release compound
- A change from a laparotomy to a laparoscopy surgical approach
- A change from an injectable anesthesia to an inhalant one

As stated in the NIH/OLAW Guidance NOT-OD-14-126,[1] changes to approved research projects must be conducted in accordance with the institution's PHS Assurance, PHS Policy (IV,C,1,a–g), the USDA AWAR, and must be consistent with the *Guide,* unless an acceptable justification for a departure is presented. Significant changes to previously approved animal activities are granted by the IACUC after Full Committee Review (FCR) or Designated Member Review (DMR). An institution may establish IACUC-approved policies for the conduct of animal activities and define what it considers a significant change. Some specific significant changes may be handled administratively based on IACUC-approved policies such as an increase in previously approved animal numbers (specific policy)[2] or an authorized veterinarian who verifies compliance with approved policies, Veterinary Verification and Consultation (VVC). Other changes can be handled administratively without approved policies such as typographical and grammatical errors, contact information updates, and changes in personnel other than the Principal Investigator (PI). The use of fewer animals does not require approval. Procedures for handling significant changes in accordance with the AWAR are described in the *Animal Welfare Inspection Guide* (Section 7.1.7 Significant Changes to Animal Activities).[3]

Therefore, changes to protocols are submitted to the IACUC through an amendment process for the committee's review and approval. These changes must be approved before their implementation. The chapter will raise and respond to some questions related to amendments and present the survey results.

11:1 What is the purpose of a protocol amendment?

Reg. The PHS Policy (IV, B,7) and the AWAR (§2.31, c,7) require Principal Investigators (PIs) to seek IACUC approval of proposed significant changes regarding the care and use of animals in ongoing activities. The IACUC may approve an amendment, require modifications to secure approval, or withhold approval after review of the proposed changes. According to the *Guide* (p. 25), "The committee is responsible for oversight and evaluation of the entire Program. Its oversight functions include review and approval of proposed animal use (protocol review) and of proposed significant changes to animal use."[4]

Opin. The purpose of an amendment is to allow investigators to make changes to their approved protocol(s). Changes to an approved protocol must be submitted using an amendment request to the IACUC for review and approval before their implementation. An investigator cannot initiate the changes prior to receiving IACUC approval (NIH/OLAW FAQ D.5).[5] Investigators may wish to modify their protocol based on experimental data or emerging research. Changes may also be recommended by the veterinary staff because of refinements in animal and veterinary care. Additionally, IACUC members may suggest changes based on their continued oversight of animal activities. The IACUC can approve, withhold approval, or require modifications to secure approval. There are no regulatory limitations on the number of amendments that can be submitted.

DOI: 10.1201/9781003272861-11

11:2 What are the regulatory requirements for protocol amendment?

Reg. PHS Policy (IV, B,7) and the AWAR (§2.31, c,7) require that the IACUC review and approve, require modifications to secure approval or withhold approval of proposed significant changes regarding the care and use of animals in ongoing activities. PHS Policy (IV, C,1, a–g) states that the IACUC should review animal-related components of any proposed protocol and determine if the proposed research is in accordance with PHS Policy, the AWA, the *Guide*, and the institution's PHS Assurance. The AWAR (§2.31, d, 1) describe how to conduct the review and states that the IACUC shall determine that the proposed significant changes meet requirements as detailed in AWAR (§2.31, d, 1, i–xi). These requirements include minimizing pain and distress, considering alternatives to painful procedures, and assuring activities are not unnecessarily duplicative, animal housing and veterinary care are appropriate, personnel are qualified, surgical standards are met, and appropriate euthanasia techniques are used.[4]

Opin. Initiating significant changes to IACUC-approved protocols without prior IACUC approval is a noncompliance with PHS Policy and should be promptly reported to OLAW (NIH/OLAW Guidance NOT-OD-05-034[6]). For APHIS/AC-regulated species, it is also considered a noncompliance although it is not reportable to APHIS/AC unless protocol suspension occurs. However, it can be cited as a finding during an inspection. Establishing an effective and time-efficient review by the IACUC of amendments supports compliance.

11:3 What changes constitute a significant modification to an approved animal use protocol?

Reg. In 2014, NIH published a notice on significant changes to animal activities,[2] which is consistent with the USDA Animal Care Inspection Guide 7.1.7.[3] Significant changes include changes that have, or have the potential to have, a negative impact on animal welfare. In addition, some activities that may not have a direct impact on animal activities are also considered significant (please see below). Furthermore, in support of the use of performance standards and professional judgment and to reduce regulatory burden, the notice allowed IACUCs some discretion to use approved policies to define significant changes.

The IACUC remains responsible for confirming that all IACUC review criteria are maintained and documented (PHS Policy IV, C,)1; AWAR (§2.31, d; §2.31, e; §2.35, a, 2).

Significant changes described below must be approved by one of the valid IACUC approval methods described in PHS Policy IV.C.2, that is FCR or DMR. Changes:

- from non-survival to survival surgery
- resulting in greater pain, distress, or a degree of invasiveness
- in the housing and or use of animals in a location that is not part of the animal program overseen by the IACUC
- in the species
- in the study objectives
- in the Principal Investigator
- that impacts personnel safety

The following specific significant changes can be handled administratively according to IACUC policies in consultation with a veterinarian authorized by the IACUC (VVC). The changes include:

- Anesthesia, analgesia, sedation, or experimental substances
- Euthanasia to any method approved in the AVMA Guidelines for the Euthanasia of Animals[7]
- Duration, frequency, type, or number of procedures performed on an animal

The IACUC policy on VVC must include four components:

- a list of the specific significant changes eligible for VVC
- the IACUC-approved policies or references that can be used by the veterinarian to verify the proposed changes
- which veterinarian(s) or group of veterinarians are authorized to conduct VVC (e.g., all Division of Laboratory Animal Medicine veterinarians, institutional veterinarian only)
- a method to document the veterinary consultation and verification, reflected in the protocol

Finally, a significant change to increase previously approved animal numbers may be handled administratively according to an existing IACUC policy[2] (see 11:11).

Opin. Any proposed changes to an animal-related protocol should be submitted for review and approval by the IACUC before their implementation. The IACUC or the IACUC administrator (under the direction of the IACUC) must distinguish between protocol changes to determine the course of action required for the review and implementation of such changes in a timely manner.[3] There are five categories of changes to be considered:

1. Changes that significantly alter the scope of the project would best be achieved by a complete protocol submission rather than a protocol amendment.
2. Changes that have, or have the potential to have, a negative impact on animal welfare require both submission of a protocol amendment and FCR or DMR prior to their implementation. These also include new procedures that are added to a protocol.
3. Changes that are specific and detailed in an IACUC-approved policy require submission of a protocol amendment and verification by a subject matter expert (e.g., a veterinarian) authorized by the IACUC to verify compliance with the approved policy (VVC).
4. An increase in previously approved animal numbers requires submission of a protocol amendment and may be handled administratively according to an existing IACUC policy.
5. Changes in contact information, personnel other than the Principal Investigator (PI) (review ensures that all IACUC required criteria such as training, enrollment in the Occupational Health and Safety Program (OHSP), etc. are met), and typographical or grammatical errors require submission of a protocol amendment and may be approved administratively without IACUC-approved policies, consultations, or notifications.

Each IACUC should establish a list of guidelines and examples to clarify within their institution as to what types of significant changes should be handled via protocol amendment for review by the IACUC or VVC, and which changes require that a new protocol be submitted. As an example, in Table 11.1 below, we categorize significant protocol changes into those we think would require a new protocol, those that would require an amendment with review by the committee, those that would require a VVC (see 11:3), and those that could be handled through administrative review. While this table is intended to help provide some examples, it represents only the author's opinion. The IACUC is responsible for carefully delineating how the changes will be processed, and IACUC administrative staff should be trained accordingly. Therefore, the IACUC approves all methods of review.

Additional significant changes that meet NIH/OLAW criteria for requiring IACUC review, but are not in the NIH/OLAW guidance,[1] include the following:

- Requests to house animals outside of the vivarium (e.g., in a lab) for more than 12 consecutive hours for species covered by the

TABLE 11.1

Methods of IACUC Approval of Protocol Changes

Changes Requiring a New Protocol FCR/DMR	**Changes Requiring an Amendment**
Changes in the purpose, aim, or objective	Change in PI
Change from peer-reviewed to non-peer-reviewed funding sources	Departures from the *Guide*
	Addition of a new procedure
	Addition of a new animal use site (lab)
	Addition of a new housing site
	Addition of a new species without changing the study objectives
	Addition of a procedure that increases the level of invasiveness
Changes Requiring an Amendment to Be Handled through VVC Based on IACUC Policies (see 11:3)	**Changes Requiring an Amendment to Be Handled Administratively Based on IACUC Policy**
Change in euthanasia method (consistent with AVMA Guidelines on Euthanasia)	Increase in animal numbers
Change in analgesia (e.g., to a SR compound)	Change in housing to another room already approved as a housing site
Changes in anesthesia method (e.g., from an injectable to an inhalant)	Addition or removal of personnel
Changes in the number of samples to be collected	Addition of a new strain of mice with no debilitating phenotype
Changes in the volume of a sample	
Changes in the number or volume of an experimental compound if level of invasiveness is not changed	
Addition of procedures already approved in the protocol (e.g., additional number of imaging or gavage procedures)	

AWA or 24 consecutive hours for other species (NIH/OLAW FAQ. E.1.)[8]

- Addition of any procedure that may result in unexpected death or other complications not described in the original protocol
- Changes that would render immune competent animals immunocompromised[4]

While there is no limit to the number of amendments that can be submitted for a specific protocol, complex protocol changes or submission of multiple subsequent changes to one protocol are not always suited for the amendment mechanism. The use of an electronic IACUC protocol management system, which can produce an integrated version of the protocol after each amendment is approved, may allow more complex protocol changes to be documented and may avoid some occasions where complex amendments require the writing of a new protocol to achieve clarity.[4] Alternatively, IACUCs without an electronic system may choose to accept track changes within the body of the approved protocol.

Surv. Members of the laboratory animal community were asked to categorize individual types of changes to a protocol into those that would require no amendment, an amendment, or a new protocol at their respective institutions. Most of the responses were consistent with the regulatory requirements except for a change in the study objectives. The results were almost evenly divided at 46% requiring a new protocol and 49% an amendment. The shift appeared significant from the results of *The IACUC Handbook* (3rd edition),[4] where the majority indicated that a new protocol would be required.

Surv. Does your IACUC require an amendment or submission of a new protocol if changes to the objectives of a study are going to be made?

- No amendment needed 5/247 (2%)
- New amendment needed 122/247 (49%)
- New protocol needed 113/247 (46%)
- Not applicable 7/247 (3%)

Surv. Does your IACUC require an amendment or submission of a new protocol if addition of minor surgery is added to an approved study?

- No amendment needed 0/247 (0%)
- New amendment needed 227/247 (92%)
- New protocol needed 14/247 (6%)
- Not applicable 6/247 (2%)

Surv. Does your IACUC require an amendment or submission of a new protocol if a major surgery is added to an approved study? Note: Major operative procedure means any surgical intervention that penetrates and exposes a body cavity or any procedure which produces permanent impairment of physical or physiological functions (AWAR (§1.1).

- No amendment needed 0/245 (0%)
- New amendment needed 206/245 (84%)
- New protocol needed 34/245 (14%)
- Not applicable 5/245 (2%)

Surv. Does your IACUC require an amendment or submission of a new protocol if changing from nonsurvival to survival surgery?

- No amendment needed 0/245 (0%)
- New amendment needed 198/245 (81%)
- New protocol needed 41/245 (17%)
- Not applicable 6/245 (2%)

Surv. Does your IACUC require an amendment or submission of a new protocol if changing the degree of invasiveness of a procedure?

- No amendment needed 0/245 (0%)
- New amendment needed 226/245 (92%)
- New protocol needed 15/245 (6%)
- Not applicable 4/245 (2%)

Surv. Does your IACUC require an amendment or submission of a new protocol if changing the expected pain or distress of an experimental animal?

- No amendment needed 0/245 (0%)
- New amendment needed 221/245 (90%)
- New protocol needed 20/245 (8%)
- Not applicable 4/245 (2%)

Surv. Does your IACUC require an amendment or submission of a new protocol if performing a new procedure or changing a procedure being used on mice or rats?

- No amendment needed 0/245 (0%)
- New amendment needed 216/2545 (88%)
- New protocol needed 22/245 (9%)
- Not applicable 7/245 (3%)

Surv. Does your IACUC require an amendment or submission of a new protocol if performing a new procedure or changing a procedure being used on an USDA species?

- No amendment needed 0/245 (0%)
- New amendment needed 180/245 (73%)
- New protocol needed 26/245 (11%)
- Not applicable 39/245 (16%)

Surv. Does your IACUC require an amendment or submission of a new protocol if changing personnel involved in animal procedures?

- No amendment needed 27/244 (11%)
- New amendment needed 211/244 (86%)

- New protocol needed 2/244 (1%)
- Not applicable 4/244 (2%)

Surv. Does your IACUC require an amendment or submission of a new protocol if changing anesthetic agent or agents?

- No amendment needed 3/244 (1%)
- New amendment needed 234/244 (96%)
- New protocol needed 5/244 (2%)
- Not applicable 2/244 (1%)

Surv. Does your IACUC require an amendment or submission of a new protocol if changing to the use of or withholding of analgesics?

- No amendment needed 0/244 (0%)
- New amendment needed 229/244 (94%)
- New protocol needed 11/244 (5%)
- Not applicable 4/244 (2%)

Surv. Does your IACUC require an amendment or submission of a new protocol if changing the analgesic agent?

- No amendment needed 6/244 (2%)
- New amendment needed 234/244 (96%)
- New protocol needed 2/244 (1%)
- Not applicable 2/244 (1%)

Surv. Does your IACUC require an amendment or submission of a new protocol if changing the methods of euthanasia?

- No amendment needed 3/244 (1%)
- New amendment needed 231/244 (95%)
- New protocol needed 6/244 (2%)
- Not applicable 4/244 (2%)

Surv. Does your IACUC require an amendment or submission of a new protocol if changing the duration, frequency, or number of procedures performed on an animal?

- No amendment needed 4/244 (2%)
- New amendment needed 223/244 (91%)
- New protocol needed 15/244 (6%)
- Not applicable 2/244 (1%)

Surv. Does your IACUC require an amendment or submission of a new protocol if addition of a new strain or stock of the same species?

- No amendment needed 39/244 (16%)
- New amendment needed 193/244 (79%)
- New protocol needed 5/244 (2%)
- Not applicable 7/244 (3%)

Surv. Does your IACUC require an amendment or submission of a new protocol if a change that increases the numbers of mice or rats?

- No amendment needed 6/244 (2%)
- New amendment needed 231/244 (95%)
- New protocol needed 1/244 (0%)
- Not applicable 6/244 (2%)

Surv. Does your IACUC require an amendment or submission of a new protocol if a change that increases the numbers of a USDA-regulated animal?

- No amendment needed 2/244 (1%)
- New amendment needed 192/244 (79%)
- New protocol needed 9/244 (4%)
- Not applicable 41/244 (17%)

Surv. Does your IACUC require an amendment or submission of a new protocol if a change that increases the number of animals by 5% or less?

- No amendment needed 28/244 (11%)
- New amendment needed 212/244 (87%)
- New protocol needed 1/244 (0%)
- Not applicable 3/244 (1%)

Surv. Does your IACUC require an amendment or submission of a new protocol if change that increases the number of animals by 5%–10%?

- No amendment needed 19/244 (8%)
- New amendment needed 221/244 (91%)
- New protocol needed 3/244 (1%)
- Not applicable 1/244 (0%)

Surv. Does your IACUC require an amendment or submission of a new protocol if a change that increases the number of animals by 10% or more?

- No amendment needed 5/244 (2%)
- New amendment needed 230/244 (94%)
- New protocol needed 8/244 (3%)
- Not applicable 1/244 (0%)

Surv. Does your IACUC require an amendment or submission of a new protocol if a change of Principal Investigator?

- No amendment needed 4/244 (2%)
- New amendment needed 187/244 (77%)
- New protocol needed 48/244 (20%)
- Not applicable 3/244 (1%)

Surv. Does your IACUC require an amendment or submission of a new protocol if change or addition of a new species?

- No amendment needed 0/244 (0%)
- New amendment needed 164/244 (68%)
- New protocol needed 73/244 (30%)
- Not applicable 5/244 (2%)

Surv. Does your IACUC require an amendment or submission of a new protocol if a change from a peer-reviewed funding source to a non-peer reviewed funding source?

- No amendment needed 85/242 (35%)
- New amendment needed 112/242 (46%)
- New protocol needed 9/242 (4%)
- Not applicable 36/242 (15%)

Surv. Does your IACUC require an amendment or submission of a new protocol if changed to a vendor, such as a producer of polyclonal antibody?

- No amendment needed 100/244 (41%)
- New amendment needed 113/244 (47%)
- New protocol needed 6/244 (2%)
- Not applicable 23/244 (10%)

Surv. Does your IACUC require an amendment or submission of a new protocol if changed to a location where all or part of the study will be done?

- No amendment needed 11/242 (5%)
- New amendment needed 206/242 (85%)
- New protocol needed 10/242 (4%)
- Not applicable 15/242 (6%)

Surv. Does your IACUC require an amendment or submission of a new protocol if a request for housing conditions that are not typical of vivarial housing or husbandry?

- No amendment needed 3/242 (1%)
- New amendment needed 212/242 (88%)
- New protocol needed 17/242 (7%)
- Not applicable 10/242 (4%)

Surv. Does your IACUC require an amendment or submission of a new protocol if a change to reflect more clinical signs or animal deaths than previously anticipated?

- No amendment needed 15/240 (6%)
- New amendment needed 202/240 (84%)
- New protocol needed 16/240 (7%)
- Not applicable 7/240 (3%)

Surv. Does your IACUC require an amendment or submission of a new protocol if addition of a prolonged restraint procedure?

- No amendment needed 0/240 (0%)
- New amendment needed 217/240 (90%)
- New protocol needed 15/240 (6%)
- Not applicable 8/240 (3%)

Surv. Does your IACUC require an amendment or submission of a new protocol if the addition of or a change to the use of hazardous agents?

- No amendment needed 4/240 (2%)
- New amendment needed 223/240 (91%)
- New protocol needed 15/240 (6%)
- Not applicable 2/240 (1%)

When should a change to the stated study objectives be categorized as a significant change that requires an amendment? Most IACUCs categorized an increase in animal numbers greater than 10% as a significant change, while < 10% was considered as a minor amendment. The specific responses regarding when a change to the protocol objectives are needed are noted below (See survey responses 1–28).

Surv. Changes to the protocol objectives

- Minor amendment 21/240 (9%)
- Significant/major amendment 203/240 (85%)
- Not applicable 11/240 (5%)
- I don't know 5/240 (2%)

Surv. Addition of minor surgery

- Minor amendment 50/240 (21%)
- Significant/major amendment 184/240 (77%)
- Not applicable 5/240 (2%)
- I don't know 1/240 (1%)

Surv. *Addition of major surgery*

- Minor amendment 0/240 (0%)
- Significant/major amendment 233/240 (97%)
- Not applicable 6/240 (3%)
- I don't know 1/240 (0%)

Surv. Changing from non-survival surgery to survival surgery.

- Minor amendment 6/239 (3%)
- Significant/major amendment 225/239 (94%)
- Not applicable 8/239 (3%)
- I don't know 0/239 (0%)

Surv. Changing the degree of invasiveness of a procedure.

- Minor amendment 8/240 (3%)
- Significant/major amendment 224/240 (94%)
- Not applicable 4/240 (2%)
- I don't know 3/240 (1%)

Surv. Changing the expected pain or distress of an experimental animal.

- Minor amendment 8/239 (3%)
- Significant/major amendment 225/239 (94%)
- Not applicable 3/239 (1%)
- I don't know 3/239 (1%)

Surv. Addition of a new procedure or changing a procedure being used on mice or rats.

- Minor amendment 30/239 (13%)
- Significant/major amendment 200/239 (84%)
- Not applicable 7/239 (3%)
- I don't know 2/239 (1%)

Surv. Addition of a new procedure or changing a procedure being used on the USDA species.

- Minor amendment 18/239 (8%)
- Significant/major amendment 180/239 (75%)
- Not applicable 35/239 (15%)
- I don't know 6/239 (3%)

Surv. Changing personnel involved in animal procedures.

- Minor amendment 211/239 (88%)
- Significant/major amendment 17/239 (7%)
- Not applicable 11/239 (5%)
- I don't know 0/239 (0%)

Surv. Change of anesthetic agent.

- Minor amendment 101/239 (42%)
- Significant/major amendment 134/239 (56%)
- Not applicable 2/239 (1%)
- I don't know 2/239 (1%)

Surv. Changing to the use of, or withholding of, analgesics.

- Minor amendment 17/239 (7%)
- Significant/major amendment 215/239 (90%)
- Not applicable 4/239 (2%)
- I don't know 3/239 (1%)

Surv. Change of the analgesic agent.

- Minor amendment 104/239 (44%)
- Significant/major amendment 130/239 (54%)
- Not applicable 3/239 (1%)
- I don't know 2/239 (1%)

Surv. Change to the method of euthanasia.

- Minor amendment 82/239 (34%)
- Significant/major amendment 153/239 (64%)
- Not applicable 2/239 (1%)
- I don't know 2/239 (1%)

Surv. Change to the duration, frequency or number of procedures performed on an animal.

- Minor amendment 36/239 (15%)
- Significant/major amendment 199/239 (83%)
- Not applicable 2/239 (1%)
- I don't know 2/239 (1%)

Surv. Addition of new strain or stock of the same species.

- Minor amendment 141/239 (59%)
- Significant/major amendment 67/239 (28%)
- Not applicable 28/239 (12%)
- I don't know 3/239 (1%)

Surv. A change that increases the number of mice or rats (non-USDA regulated).

- Minor amendment 108/239 (45%)
- Significant/major amendment 119/239 (50%)
- Not applicable 8/239 (3%)
- I don't know 4/239 (2%)

Surv. A change that increases the number of a USDA-regulated species.

- Minor amendment 56/239 (23%)
- Significant/major amendment 144/239 (60%)
- Not applicable 38/239 (16%)
- I don't know 1/239 (0%)

Surv. A change that increases the number of animals by 5% or less.

- Minor amendment 170/239 (71%)
- Significant/major amendment 62/239 (26%)
- Not applicable 7/239 (3%)
- I don't know 0/239 (0%)

Surv. A change that increases the number of animals by 5%–10%.

- Minor amendment 148/239 (62%)
- Significant/major amendment 83/239 (35%)
- Not applicable 8/239 (3%)
- I don't know 0/239 (0%)

Surv. A change that increases the number of animals >10%.

- Minor amendment 48/239 (20%)
- Significant/major amendment 187/239 (78%)
- Not applicable 4/239 (2%)
- I don't know 0/239 (0%)

Surv. A change of the Principal Investigator.

- Minor amendment 37/239 (15%)
- Significant/major amendment 188/239 (79%)
- Not applicable 12/239 (5%)
- I don't know 2/239 (1%)

Surv. A change or an addition of new species.

- Minor amendment 21/239 (9%)
- Significant/major amendment 198/239 (83%)
- Not applicable 19/239 (8%)
- I don't know 0/239 (0%)

Surv. A change from a peer-reviewed funding source to a non-peer reviewed funding source.

- Minor amendment 112/239 (47%)
- Significant/major amendment 30/239 (13%)
- Not applicable 91/239 (38%)
- I don't know 5/239 (2%)

Surv. Change of a vendor, such as a producer of polyclonal antibody.

- Minor amendment 110/239 (46%)
- Significant/major amendment 38/239 (16%)
- Not applicable 82/239 (34%)
- I don't know 8/239 (3%)

Surv. Change to a location where all or part of the study will be done.

- Minor amendment 125/239 (53%)
- Significant/major amendment 85/239 (36%)
- Not applicable 26/239 (11%)
- I don't know 2/239 (1%)

Surv. A request for housing conditions that are not typical of vivarium housing or husbandry.

- Minor amendment 31/239 (13%)
- Significant/major amendment 196/239 (82%)
- Not applicable 11/239 (5%)
- I don't know 0/239 (0%)

.Surv. A change to reflect additional or more severe clinical signs or animal death deaths than previously expected.

- Minor amendment 9/239 (4%)
- Significant/major amendment 212/239 (89%)
- Not applicable 13/239 (58%)
- I don't know 4/239 (2%)

Surv. The addition of, or change to, the use of hazardous agent.

- Minor amendment 35/239 (15%)
- Significant/major amendment 193/239 (82%)
- Not applicable 4/239 (2%)
- I don't know 2/239 (1%)

11:4 What format should be used to submit an amendment?

Reg. The format for protocol amendments is determined by the IACUC. The required content for a significant amendment is outlined in the AWAR (§2.31, e).

Opin. Amendments may be submitted through an amendment form or tracked changes to the approved protocol/electronic form or by email. All members of the IACUC must be provided with the opportunity to call for review at a convened meeting. If the amendment is a significant change, it may be reviewed by either the FCR or the DMR or for specific changes verified through VVC (see 11.3). An increase in animal numbers may be handled administratively based on IACUC-approved policy. The use of an electronic form facilitates the amendment review process, as it highlights the changes in each section. Alternatively, for programs without an electronic form, a Word document of the approved protocol can be used with amended changes tracked. Some institutions still prefer to use an amendment form that is smaller than the original protocol form. The form should be developed by the IACUC and ensure that all essential questions required for review and approval are met. For clarity, it may be necessary to require that only a single amendment be under current review at any given time. Questions to be addressed should include the following:

- What is the purpose or rationale for requesting the amendment?
- Is there a proposed change in personnel associated with the protocol?
- If adding personnel to the protocol or changing the roles of personnel on the approved protocol, the amendment should list their qualifications and training.
- Is there a proposed change to the species, number, sex, or strain of animal used in the protocol? If so, justification should be provided.

- Will specialized housing be required?
  - Is housing available?
  - Will the housing meet the experimental requirements and comply with the *Guide*'s and AWAR standards?
- Will a new procedure be added?
  - The description of the new procedure must be detailed and justified.
  - If the new procedure causes more than momentary pain or distress the following should be provided. Documentation demonstrating that alternatives have been considered; the activities do not unnecessarily duplicate previous experiments; and the attending veterinarian or his or her designee was involved in the planning (AWAR §2.31,d,1,ii-iv).
  - How often and for what duration will the procedure be performed?
  - Will the animal experience pain or distress before, during, or after the procedure? How will this be minimized and treated?
- Are you proposing a change in the procedures already described in the original protocol (any change in drugs, methods, anesthesia, analgesia, or euthanasia)?
- Will prolonged restraint of conscious animals be required?
- Will new hazardous materials be used?

Concerning the submission of amendments, most respondents either used a short amendment form or integrated changes directly into the existing protocol (>60%). The majority also did not accept changes submitted through an e-mail or a memo (87%), required a new literature search for a significant change (53%), and allowed only one amendment at a time per protocol (54%).

Surv. What would your institution require when amending the protocol? Select all that apply.

• Not applicable (we do not amend protocols).	1/251 (0%)
• We require a special amendment form to be used.	156/251 (62%)
• An amendment form is not mandatory, but we prefer it to be used.	5/251 (2%)
• Modifications are integrated directly into the content of the protocol.	151/251 (60%)

Surv. *Can protocols* be amended by e-mail without the use of an amendment form?

• Yes	21/251 (8%)
• No	216/251 (86%)
• Not applicable	8/251 (3%)
• Other	6/251 (2%)

Surv. Can protocols be amended by a memo without the use of an amendment form?

• Yes	15/251 (6%)
• No	218/251 (87%)
• Not applicable	14/251 (6%)
• Other	4/251 (2%)

Surv. Is this special form simpler/shorter than your "regular" protocol form?

• Yes	139/251 (55%)
• No	32/251 (13%)
• Not applicable	75/251 (30%)
• Other	5/251 (2%)

Surv. Do all amendments (major or minor) go through either DMR or FCR?

• Yes	137/251 (55%)
• No	98/251 (39%)
• Not applicable	2/251 (1%)
• Other	14/251 (6%)

Surv. Is a new or additional literature (or related) search required for minor amendments?

• No additional experimental work is allowed under any circumstance	7/251 (3%)
• Yes	29/251 (12%)
• No	183/251 (73%)
• Not applicable	6/251 (2%)
• Other	26/251 (10%)

Surv. Is a new or additional literature (or related) search required for significant (major) amendments?

• Yes	131/247 (53%)
• No	58/247 (23%)
• Not applicable	8/247 (3%)
• Other	50/247 (20%)

Surv. At your institution, can there be more than one amendment be reviewed at the same time to amend a single protocol?

• Not applicable	6/247 (2%)
• Yes	100/247 (40%)
• No	134/247 (54%)
• I don't know	7/247 (3%)

11:5 Do the designated member review and full committee review processes also apply to amendments?

Reg. PHS policy (IV,C,2) and the AWAR (§2.31, d, 2) require the same review procedure for proposed significant changes in ongoing activities (protocol amendments) as they do for new activities (new protocols), with the exceptions detailed in NIH/

OLAW Guidance NOT-OD-14-126[1] and *Animal Welfare Inspection Guide* 7.1.7,[3] regarding consultation by an IACUC authorized veterinarian who acts as subject matter expert to verify compliance with IACUC-reviewed and IACUC-approved policy.

Opin. Except for significant changes that qualify for VVC, all IACUC members must have the opportunity to evaluate proposed significant changes and request full committee evaluation at a convened meeting of the IACUC, as in full committee review. If no member requests full committee review, the IACUC Chair designates one of the more qualified individuals for designated member review. Whether a full committee review or designated member review[9,10] mechanism is used, all reviewers must have access to both the amendment document and the original protocol. Although members should be informed of significant changes by VVC, there is no requirement that they are informed of those changes at the time of the VVC review request. IACUC approval of protocol amendments must be documented in the protocol file with a dated signature.[4]

The following questions dealt with the mechanism used for amendment review. The majority indicated that significant amendments underwent FCR at a convened meeting (60%), while minor ones were reviewed through DMR (72%).

Surv. Please select the mechanism(s) below that best summarizes your mechanism for review of SIGNIFICANT amendments.

- Not applicable 0/247 (0%)
- Review and approval by IACUC staff member(s) 20/247 (8%)
- Review and approval by a single member of the IACUC (designated member review) 81/247 (33%)
- Review and approval by two members of the IACUC, with or without review by IACUC staff 79/247 (32%)
- Review and approval by the Full Committee at a convened IACUC meeting 147/247 (60%)
- Other 21/247 (9%)

Surv. Please select the mechanism(s) below that best summarizes your mechanism for review of MINOR amendments. Select all that apply.

- Administrative review and approval 162/247 (66%)
- DMR by one or more IACUC members 179/247 (72%)
- VVC or veterinary review and approval 129/247 (52%)
- FCR at a convened IACUC meeting 47/247 (19%)
- Other* 19/247 (8%)

*Other: some noted that the precise mechanism used for review depended upon aspects of the change, such as species and type of modification being proposed in the amendment.

11.6 What is the purpose of Veterinary Verification and Consultation (VVC)?

Reg. As defined in the NIH/OLAW Guidance NOT-OD-14-126,[1] and the *Animal Welfare Inspection Guide* 7.1.7,[3] in support of the use of performance standards and professional judgment and to reduce regulatory burden, IACUC-reviewed and IACUC -approved policies such as guidance documents, standard operational procedures, drug formularies, etc., may be used for the administrative handling of protocol change in consultation with a veterinarian authorized by the IACUC. The veterinarian acts as a subject matter expert to verify, in consultation with the research team, compliance with the IACUC-approved policy. VVC is documented and changes are reflected in the protocol.

Opin. The purpose of VVC (see 11.3) is to reduce the regulatory burden for PIs and IACUC members. NIH/OLAW is careful to state that this process is not considered a DMR but a verification by a subject matter expert. The IACUC is required to have approved policies or SOPs to be used as references for each specific significant change included in the VVC Policy. Examples of policies or SOPs may include the volume, number, and interval of blood collections on a species; duration, frequency, type, or number of procedures performed on an animal; formularies to cover anesthesia, analgesia, sedation, euthanasia (AVMA Guidelines for the Euthanasia of Animals[7]), and experimental substances. If these policies or SOPs or formularies are used by the veterinarian for the VVC program, they must be reviewed and approved by the IACUC. VVC cannot be used to add new procedures to an approved protocol.[11] While this removes regulatory burden on the PIs, it increases the level of responsibility assigned to the veterinarian. The veterinarian may refer any request to the IACUC for review for any reason and must refer any request that does not meet the parameters of the IACUC-reviewed and IACUC -approved policies.

Surv. What is the purpose of a Veterinary Verification and Consultation (VVC)? Select all that apply.

- Approve amendments faster 124/234 (53%)
- Reduce burden on the IACUC 153/234 (65%)
- Approve significant changes on a case-by-case basis in accordance with PHS Policy 106/234 (45%)
- I don't know 15/234 (6%)
- Other 31/234 (13%)

Most of the respondents indicated that the purpose of VVC was to reduce regulatory burden and accelerate approval of amendments based on policy, consistent with the regulations.

11.7 How does VVC differ from other methods for review of significant changes?

Reg. As defined in the NIH/OLAW Guidance NOT-OD-14-126,[1] VVC can be used by an IACUC to review specific significant amendments based on IACUC-reviewed and IACUC -approved policies for a select group of changes. The veterinarian authorized by the committee acts as a subject matter expert to verify compliance with the policy. If the request does not meet the parameters of the policies, the veterinarian refers the request to the IACUC for review (see 11:3). If a new procedure is added to the protocol, VVC cannot be used.

Opin. While VVC can be used in general to verify significant changes against formularies or methods of euthanasia consistent with the AVMA Guidelines for the Euthanasia of Animals,[7] it can also be used for changes in procedures. However, the veterinarian must be cautious that the significant change does not result in greater pain or distress to the animal. Changes that result in increased pain, distress, or degree of invasiveness must be reviewed and approved by the IACUC, by either full committee review or DMR. One could argue that increasing the number of procedures could potentially result in more distress to an animal. The veterinarian may choose to use the USDA categories of invasiveness for guidance to assist in making such determinations.

As shown in the survey results below, most respondents indicated that changes requested through VVC would be consistent with approved IACUC SOP or Policy. However, a minority allowed VVC without IACUC-approved SOP or Policy under certain circumstances when animal welfare would not be unduly impacted, such as with a change in behavioral testing methods (21%), at the discretion of the AV for a minor procedure (30%), enhancement of enrichment (41%), or a change in stock/strain when special housing was not required (38%).

Surv. In your institution, when does the IACUC allow significant changes based on VVC? Select all that apply.

• The change must be part of an SOP or policy already approved by the IACUC. The veterinarian is verifying that the requested change is consistent with a procedure already approved by the IACUC.	162/234 (69%)
• A change in dosage, route, frequency, or duration within acceptable veterinary practices and/or consistent with an IACUC-approved SOP or policy.	167/234 (71%)
• When there is a switch of analgesics, anesthetics, or sedative agents to another that is part of an IACUC-approved SOP or policy.	160/234 (68%)
• Changes in the dosage, timing, or route of an experimental agent if the change will not increase the potential for pain and distress	135/234 (58%)
• Changes to experimental agent, including a change in the test agent, dose, or route of administration when changes do not result in the study objective, pain, and distress.	91/234 (39%)
• Changes to euthanasia to any method approved by the AVMA Guidelines on Euthanasia. The Guidelines has been approved by the IACUC.	151/234 (65%)
• When the AV under his/her discretion and judgment authorizes a minor procedure that will not unduly impact on animal welfare.	70/234 (30%)
• Changes related to blood collection with respect to volume, interval, frequency, and access to the route of collection and number of samples that is consistent with an IACUC-approved SOP or policy.	136/234 (58%)
• Changing biopsy from one method to another that is consistent with an IACUC-approved SOP or policy.	78/234 (33%)
• Changing the method of identifying animals that is consistent with an IACUC-approved SOP or policy.	110/234 (47%)
• Changing behavioral testing methods.	48/234 (21%)
• Changing methods to enhance enrichment.	96/234 (41%)
• Changes to stock, strain, or genetic modification unless the change requires special housing or care.	90/234 (38%)

11.8 Must the VVC be done by the Attending Veterinarian (AV), or can the process be delegated to another veterinarian who is not a voting member of the IACUC?

Reg. As defined in the NIH/OLAW Guidance NOT-OD-14-126,[1] the person conducting the verification is a veterinarian authorized by the IACUC.

Opin. As written in the NIH/OLAW Guidance the veterinarian is authorized by the IACUC. The Guidance does not specify that it must be the AV or even a voting member. IACUCs may authorize more than one veterinarian, who need not be an employee, to conduct VVC. However, the veterinarian should be familiar with the animal care and use program—in large programs with multiple veterinarians this could be a clinical veterinarian or a compliance veterinarian.

Surv. In your institution, what is your process for VVC? Select all that apply.

- All significant changes to the approved protocol must be submitted as an amendment to the IACUC. 185/234 (79%)
- The IACUC office and the AV conduct a pre-review of all amendments to determine whether the amendment qualifies for VVC. 110/234 (47%)
- If a protocol does not qualify for administrative review or VVC, the amendment will be sent to FCR or DMR. 159/234 (68%)
- All administratively reviewed amendments and VVC are documented, and the PI will receive written confirmation when approved. 145/234 (62%)
- A list of administratively and AV reviewed amendment is provided to the IACUC at its next meeting 119/234 (51%)

11:9 Can the IACUC Chair or an IACUC subcommittee review and approve significant changes or must the full committee review and approve such submissions?

Reg. PHS Policy (IV,C,2) and the AWAR (§2.31,d,2) identify full committee review and designated member review as the only approved methods to evaluate animal protocols and proposed significant changes to ongoing protocols. As previously discussed, NIH/OLAW Guidance NOT-OD-14-126[1] allows a verification (not approval) of limited defined significant changes under certain circumstances by an IACUC-authorized veterinarian who acts as a subject matter expert. In addition, an increase in animal numbers may be handled administratively, based on an IACUC-approved policy. IACUC subcommittees are not authorized to approve proposed changes to an ongoing protocol, unless they operate as designated members for review and are appointed as such by the Chair. In instances when a full committee review is conducted, it should be done at a convened meeting of a quorum, either electronically or in person, with a majority vote of the quorum present.[12] APHIS/AC concurs with PHS on this matter, which is addressed in the *Animal Welfare Inspection Guide* under Chapter 7.1.8.[3] Real-time interactive electronic methods of protocol review are acceptable only as described by NIH/OLAW in FAQ B.8.[13] In general, the methods typically used for protocol review should also be used to review amendments.

Opin. Institutional policy will determine which type of protocol amendments are eligible for designated member review, full committee review, VVC, or administrative review. Any member of the IACUC, including the Chair, may serve as a designated reviewer. IACUC members determine which amendments will need to be reviewed by the full committee since they have the option of requesting FCR for any amendment. An institution may establish a policy to have all protocol amendments reviewed by a full committee. On the other hand, the IACUC may allow personnel changes (other than the PI) or other administrative changes, such as typographical errors, to be approved by the Chair or an IACUC administrator without review by FCR or DMR.

11:10 How should the IACUC review a protocol amendment?

Opin. Each institution must develop its own policies regarding significant protocol modifications. To determine the significance of a proposed change, the IACUC should consult NIH/OLAW Guidance NOT-OD-14-126.[1] The investigator and the IACUC should consider the potential impact on animal welfare and the overall ethical cost-benefit ratio of the research proposed, as they would when conducting a complete, three-year protocol review. Careful consideration of the basic areas of concern in primary protocol review should be revisited.

- The potential for pain or distress
- The level of invasiveness of procedures described in the amendment
- Consideration of the cost-benefit
- Consideration of alternatives
- The potential for unexpected outcomes
- Humane experimental endpoints
- Use of non-pharmaceutical grade chemicals in an animal protocol
- Physical restraint
- Methods or indications for euthanasia or removal from the protocol

For institutions that separated amendments into major (significant) and minor ones, the review process for minor amendments was shorter (< 1 week) versus (1 to 2 weeks) for major ones.

Surv. If your institution separates minor from significant (major) amendments, do they undergo different courses of protocol review by IACUC members or IACUC staff?

- Not applicable. We do not separate minor from significant amendments. 58/234 (25%)
- Minor and significant amendments have the same course of protocol review. 57/234 (24%)

- Minor and significant have a different course of protocol review. 115/234 (49%)
- I don't know. 1/234 (0%)
- Other 3/234 (1%)

Surv. If your review separates minor from significant (major) amendments, of the total number of amendments reviewed per your institution, how many are minor amendments per year?

- 0–100 amendments 99/234 (42%)
- 100–500 amendments 39/234 (17%)
- More than 500 amendments 16/234 (7%)
- Don't know 14/234 (6%)
- Do not distinguish 39/234 (17%)
- Not applicable 27/234 (12%)

Surv. If you review separate minor from significant (major) amendments, of the total number of amendments reviewed per year how many are significant amendments/year?

- 0–100 amendments 105/233 (45%)
- 100–500 amendments 34/233 (15%)
- More than 500 amendments 10/233 (4%)
- Don't know 18/233 (8%)
- Not applicable 66/233 (28%)

Surv. If your review separates minor from significant (major) amendments, of the total number of amendments reviewed, how long does it take to review a major amendment/year?

- Less than one week 20/233 (9%)
- One or two weeks 93/233 (40%)
- Up to one month 77/233 (33%)
- Other 43/233 (18%)

Surv. If your review separates minor from significant (major) amendments, of the total number of amendments reviewed, how long does it take to review a minor amendment?

- Less than one week 121/233 (52%)
- One to two weeks 62/233 (27%)
- Up to one month 9/233 (4%)
- Other 41/233 (18%)

11:11 How many animals may be added to an ongoing study using an amendment mechanism?

Reg. PHS Policy (IV,D,1,a; IV,D,1,b) and the AWAR (§2.31,e,1; §2.31,e,2) require that proposals to the IACUC specify and include a rationale for the approximate number of animals proposed for use. Individual institutions, therefore, must establish mechanisms to monitor and document the actual number of animals acquired and used in approved activities. PHS Policy (via U.S. Government Principle III) also requires that the minimum number of animals needed to obtain valid scientific results be approved for use. NIH/OLAW Guidance NOT-OD-14-126[1] states that an amendment requesting an increase in animal number may be handled according to existing IACUC policy. Such a policy should include the identification of the individual(s), by title or name, who are authorized to approve the increase in animal numbers relative to the IACUC- approved policy; address whether the requested change is in keeping with the original rationale for the number of animals approved; take into consideration if there is a change in study objectives; and take into consideration if the change may result in a negative impact on animal welfare.

Opin. It is up to the individual IACUC to establish such policies, for example, some institutions may allow for a small percentage (5%–10%) more than the approved number of animals to be approved administratively based on approved policy.

11:12 Can the IACUC approve an amendment administratively?

Reg. According to NIH/OLAW Guidance NOT-OD-14-126[1] and APHIS/AC *Animal Welfare Inspection Guide*,[2] changes that can be handled administratively without IACUC-approved policies, consultations, or notifications include:

- Correction of typographical errors
- Correction of grammar
- Contact information updates; and
- Changes in personnel other than the PI (the review must ensure that personnel are appropriately identified, adequately trained and qualified and meet other criteria as required by the IACUC)
- An increase in previously approved animal numbers (only APHIS/AC *Animal Welfare Inspection Guide*[2])
- Use of fewer animals than approved; does not require IACUC approval, notification, consultation, or administrative handling

Opin. Administrative approval of changes as described above contributes significantly to the reduction in regulatory burden for the PIs and IACUC members.

Consistent with regulations, most respondents were administratively approving addition of personnel (other than the PI) (90%) and changes in contact information (85%). Responses for a survey question for administratively approving the changes are given below.

Surv. Can your IACUC approve an amendment administratively for any of the reasons below? Select all that apply.

• Addition of personnel other than the PI, after verification of training and enrollment in the OHSP or deletion of personnel	209/233 (90%)
• Procedures that do not cause more than momentary pain or distress (USDA category C or lower)	19/233 (8%)
• Decrease in number of sample collection time points or sample volume	59/233 (25%)
• Addition of a new mouse strain	55/233 (24%)
• Addition of a new enrichment device	43/233 (18%)
• Change in location	94/233 (40%)
• Contact information	198/233 (85%)
• Change in sex of the animals to be used	37/233 (16%)
• Addition of a specified number or percentage of non-USDA-regulated animals beyond that originally approved	71/233 (30%)
• Addition of any number or percentage of non-USDA-regulated animals	14/233 (6%)
• Addition of a specified number or percentage of USDA-regulated animals beyond that originally approved	37/233 (16%)
• Addition of any number or percentage of USDA-regulated animals.	6/233 (6%)

11:13 After two years of conducting survival surgeries, the PI on an IACUC-approved protocol decides to change the procedure to non-survival. Must the PI submit an amendment? If so, how can this amendment be reviewed?

Reg. NIH/OLAW and USDA state that a change from a non-survival surgical procedure to a survival surgical procedure or a change resulting in greater pain, distress, or degree of invasiveness both constitute a significant change and thus require that an amendment be submitted for full committee review or DMR. They are silent on a procedure resulting in less distress. However, a change in the type of procedures performed on an animal may be handled administratively according to IACUC-reviewed and IACUC -approved VVC and referenced policies in consultation with a veterinarian authorized by the IACUC (see 11:3).

Opin. The amendment may be reviewed by full committee review (or DMR) or, depending on the nature of the procedure, VVC may be considered. As a subject matter expert, the veterinarian will determine if an anticipated change in the level of pain or distress is anticipated, although with a non-survival procedure, this is highly unlikely.[15] The verification and consultation must be documented. After evaluating the request, the veterinarian may choose to refer it to the full committee.[16]

11:14 A PI is requesting to change the method of rodent anesthesia from an injectable (ketamine/xylazine) to an inhalant (isoflurane) in a study. How can the IACUC handle the request?

Reg. According to NIH/OLAW and USDA, the change can be handled through VVC if the institution has developed, reviewed, and approved an IACUC VVC policy. The veterinarian must determine that the change will not negatively affect animal welfare, and the consultation must be documented (see 11:3).

Opin. The IACUC can handle the request through VVC provided that the committee has approved policies, SOPs, and formularies for the species, and the veterinarian determines the change will not negatively affect the animal's welfare. If the change to gas anesthesia was requested by a PI that uses inhaled anesthesia and is trained in its use and the surgery will be performed in a location that has a scavenger table that has been safety tested, then the veterinarian could verify and approve the change.[14]

11:15 A protocol is approved for non-survival surgery on swine. The PI submits an amendment to add a tracheostomy. The proposed procedure would be done on animals already anesthetized for the non-survival surgery. Must the amendment include a literature search for the new procedure?

Reg. The AWAR (S2.31, d,1,ii) state that the PI must consider alternatives to procedures that may cause more than momentary or slight pain or distress to the animals and provide a written narrative description of the methods and sources used to determine that alternatives were not available.

Opin. A literature search should be done as it is a surgical procedure and therefore more than momentary pain or distress is expected. Since the protocol already consists of a surgical procedure, the addition of another procedure would not change the USDA category of invasiveness. However. the literature search should be revised to incorporate the new procedure.

11:16 A protocol is approved for behavioral studies; the testing chambers are located within the animal facility. The facility is scheduled to undergo renovations, and the PI worries that the resulting noise and vibration could have a negative impact on his study. A colleague graciously offers the use of her chambers located in her laboratory; this space is already approved by the IACUC. Does the PI need to submit an amendment to use the chambers in the colleague's laboratory?

Reg. Both NIH/OLAW Guidance NOT-OD-14-126[1] and USDA *Animal Welfare Inspection Guide*[3] classify a change in the use of animals in a location that is not part of the animal program overseen by the IACUC to constitute a significant change that requires review by a full committee. The *Guide* (pp.149–150) details several important considerations related to physical plant and behavioral studies. Furthermore, the *Guide* (p. 25) states that the IACUC is responsible for oversight and evaluation of the entire Program and its components; this includes regular inspection of animal facilities and animal use areas. Both the PHS Policy (IV,C,1,d) and the AWAR (§2.31,d,1,vi) state that the living conditions of animals will be appropriate to their species and contribute to their health and comfort.

Opin. If the colleague's chambers in her laboratory are already part of the animal care program and inspected regularly by the IACUC, then adding the new location would not necessarily constitute a significant change in the Program. However, it may constitute a significant change to the investigator's protocol. The PI should submit an amendment for the new site. The IACUC should then inspect the site to confirm that the work will be conducted in accordance with the approved use as described in the protocol.

11:17 A dog is anesthetized and prepared for surgery. The PI then recognizes that the dosage of an experimental drug that she believed to be safe might cause a problem. The potential problem is easily avoided by lowering the dosage. Should she make the correction or obtain IACUC approval before proceeding?

Reg. Both the PHS Policy (IV,B,7) and AWAR (§2.31,c,7) require the IACUC to review proposed significant changes in ongoing activities using animals before implementation of such changes. AWAR (§2.31, d,1, iv, B) requires that the AV be involved in planning and consultation regarding procedures that cause more than momentary or slight pain and distress. NIH/OLAW Guidance and USDA both support changes in experimental substance as a significant change that may be handled administratively if the institution has developed, reviewed, and approved an IACUC VVC policy (see 11:3).

Opin. There should be an immediate consultation with a veterinarian to determine if the change in dosage is warranted or permissible. If a VVC Policy was in place and this specific significant change is included in the IACUC-approved VVC Policy as eligible for verification by the veterinarian of an IACUC reviewed and approved policy regarding dosage changes of experimental substances, VVC could be used (see 11:3). If not, the AV has authority to intervene and change dosages for individual animals for medical reasons and "an integral component of veterinary medical care is the prevention or alleviation of pain associated with procedural and surgical protocols," (*Guide*, p. 120). This could be considered such a case since the animal is already anesthetized and intervention by the AV to adjust the dose would avoid the need for an additional anesthetic event. In such a case, documentation of both the change and the consultation, along with the veterinarian's advice and/or consent, should be followed by submission of a protocol amendment.[3]

11:18 During a semiannual inspection, the IACUC members meet an investigator in his laboratory. The PI describes the need to change a surgical approach to facilitate liver biopsies. Should the members inform the PI that an amendment must be submitted to cover this change?

Reg. The *Guide* (p. 25) states that the following should be considered in a protocol for review by the IACUC.

- a clear and concise sequential description of the procedures
- use of the less invasive procedure,
- impact on the animals' well-being
- appropriate sedation, analgesia, and anesthesia
- conduct of surgical procedure
- Post-procedural care and observation
- humane endpoints
- adequacy of training and experience of personnel in the procedure used

Based on NIH/OLAW Guidance NOT-OD-12-126,[1] if the procedure causes greater pain, distress, or degree of invasiveness, it should be reviewed by the IACUC. However, if the new surgical approach does not increase the level of pain, distress, or degree of invasiveness, VVC may be considered if the following is in place. The IACUC has approved a VVC Policy, and this specific significant change is included as eligible for verification of an IACUC-reviewed and IACUC-approved policy regarding dosage changes of experimental substances (see 11:3).

Opin. A change in a surgical approach may be a refinement for the animal, such as a change from a laparotomy to a laparoscopy to collect a liver biopsy. However, even if this is the case, there are several factors for the IACUC to consider, such as:

- What is the reason for the change?
- What is the expertise of the surgeons on the new procedure?
- Is a new literature search required?
- Will additional animals be needed?
- Are the humane endpoints the same?
- Does the survival rate change?
- Does the recovery change?
- Do the anesthetic and analgesic regimes remain the same?

The investigator should be encouraged to contact a veterinarian to discuss the change and then submit an amendment accordingly.

REFERENCES

1. National Institutes of Health, Office of Laboratory Animal Welfare. 2014. *Notice NOT-OD-14-126, Guidance on Significant Changes to Animal Activities.* Accessed December 14, 2022, https://grants.nih.gov/grants/guide/notice-files/NOT-OD-14-126.html.
2. National Institutes of Health, Office of Laboratory Animal Welfare. 2014. *Guidance on Significant Changes to Animal Activities.* https://olaw.nih.gov/guidance/significant-changes.htm
3. United States Department of Agriculture, Animal and Plant Inspection Service. 2021. *U.S. Department of Agriculture.* Accessed December 15, 2022. www.aphis.usda.gov.
4. Mexas, A.M. and D.J. Gaertner. 2014. Amending IACUC Protocols. In the *IACUC Handbook*, ed. J. Silverman, M. Suckow, and S. Murthy, 3rd ed., 177–198. Boca Raton, FL: CRC Press.
5. National Institute of Health, Office of Laboratory Animal Welfare. 2022. *Frequently Asked Questions D.5, May the Investigator Begin Animal Work before Receiving IACUC Approval?* Accessed December 15, 2022, https://olaw.nih.gov/faqs#/guidance/faqs?anchor=question50316.
6. National Institutes of Health, Office of Laboratory Animal Welfare. 2005. *Notice NOT-OD-05-034, Guidance on Prompt Reporting to OLAW under the PHS Policy on Humane Care and Use of Laboratory Animals.* Accessed December 15, 2022, https://grants.nih.gov/grants/guide/notice-files/not-od-05-034.html.
7. AVMA Guidelines for the Euthanasia of Animals: 2020 Edition. *American Veterinary Medical Association.* Accessed December 15,2022. https://www.avma.org/sites/default/files/2020-02/Guidelines-on-Euthanasia-2020.pdf.
8. National Institute of Health, Office of Laboratory Animal Welfare. 2022. *Frequently Asked Questions E.1, Should the IACUC Inspect Laboratories or Other Sites Where Investigators Use Animals?* Accessed December 15, 2022. https://olaw.nih.gov/faqs#/guidance/faqs?anchor=question50347.
9. National Institute of Health, Office of Laboratory Animal Welfare. 2022. *Frequently Asked Questions D.3, What Are the Possible Methods of IACUC Approval?* Accessed December 15, 2022. https://olaw.nih.gov/faqs#/guidance/faqs?anchor=question50314.
10. Garnett, N., and S. Potkay. 1995. Use of electronic communications for IACUC functions. *ILAR Journal*, 37:190. Accessed December 15, 2022, http://grants.nih.gov/grants/olaw/references/ilar95.htm.
11. Silverman J. 2017. Making changes: When is VVC appropriate? *Lab Animal (NY).* 46(3):59. doi: 10.1038/laban.1197. PMID: 2821186 Accessed December 15, 2022. https://olaw.nih.gov/sites/default/files/laban46_03_0317.pdf .
12. National Institutes of Health, Office of Laboratory Animal Welfare. 2006. *Notice NOT-OD-06-052, Guidance on Use of Telecommunications for IACUC Meetings under the PHS Policy on Humane Care and Use of Laboratory Animals.* Accessed December 15, 2022. https://grants.nih.gov/grants/guide/notice-files/not-od-06-052.html.
13. National Institute of Health, Office of Laboratory Animal Welfare. 2022. *Frequently Asked Questions B.8.* Accessed December 15, 2022. https://olaw.nih.gov/faqs#/guidance/faqs?anchor=question50301.
14. Silk, S., G. Babcock, J. Collins, B. Goldentyer, M.L. James, and C. Gillett. Transcript and Submitted Questions. page 13 for *OLAW's Webinar "Guidance on Significant Changes to Animal Activities."* https://olaw.nih.gov/sites/default/files/140821_seminar_transcript.pdf at: https://olaw.nih.gov/education/educational-resources/webinar-2014-08-21.htm.
15. Kendall, L.V., E.K. Kim, J.R. Owiny, and J.D. Ayers. 2017. Veterinary Verification and Consultation: Implementation of OLAW guidance on significant changes. *Lab Animal* (NY) 46(3):63–65. doi: 10.1038/laban.1201. PMID: 28211868. Accessed December 15, 2022. https://pubmed.ncbi.nlm.nih.gov/28211868/.
16. Kendall, L.K. and E.K. Kim. 2016. Implementing Guidance on Significant Changes: One Institution's Experience. *OLAW Online Seminar.* Accessed December 15, 2022. https://grants.nih.gov/grants/olaw/educational_resources/webinar_09082016.htm.

12

Continuing Review of Protocols

John F. Bradfield, Ernest D. Prentice, and Gwenn S. F. Oki

Introduction

An IACUC's initial review and approval of a project represents the committee's informed judgment that when the protocol is performed as approved, it will comply with all applicable federal requirements contained in the Animal Welfare Act and Regulations (AWAR) and the Public Health Service (PHS) Policy, as well as institutional requirements. However, the IACUC's review responsibilities do not end with this initial approval. The IACUC is obligated by both the AWAR and the PHS Policy to conduct ongoing protocol reviews.

The purpose of this chapter is to present general concepts of continuing review that will be useful to IACUCs. It should be noted that continuing review is just as important as the initial review because it serves to confirm the IACUC's decisions relative to the status of the protocol and helps the investigator maintain compliance.

12:1 What is meant by continuing review?

Opin. Until recently, the Animal Plant Health Inspection Service/Animal Care (APHIS/AC) required that a continuing review be performed no less often than annually. However, as of December 27, 2021, the USDA made a rule change by amending the annual requirement to once every three years.[1] This was done in response to requirements under the 21st Century Cures Act to review regulations and policies involving the care and use of laboratory animals, in collaboration with the FDA and NIH. This change in USDA requirements, now coincides with the longstanding PHS Policy requirement to conduct Complete Triennial *(de novo)* Review. Complete Triennial Review is a monitoring process to determine that the study remains in compliance, that the activities have been "conducted in accordance with the approved protocol," that modification to the protocol have received prior IACUC approval, and to ensure that the investigator is informed of new institutional, state, or federal requirements.[2] It should be noted that the *Guide* (p. 34) endorses APHIS/AC and NIH/OLAW requirements and further states that "some institutions use the review as an opportunity for the investigator to submit proposed amendments for future procedures, to provide a description of any adverse or unanticipated events, and to provide updates on work progress."[3] Both the USDA and NIH/OLAW require Complete Triennial Review to be a *de novo* process (every three years),[1,4] meeting all the review criteria as set forth in the AWAR's (§2.31,d) and PHS Policy (IV,C,1, a–g). However, it may be prudent for IACUC's to consider a more frequent review (e.g., Periodic Continuing Review) of some protocols depending on the nature of the activity. Protocols that involve areas of investigation that are novel to the investigator or the IACUC, or that involve potential for significant pain or distress, may warrant more frequent IACUC oversight and review. In such cases, the IACUC would require a more frequent review of activities based on need, rather than by regulatory requirement. This more frequent review, which is not required by NIH/OLAW or the USDA, would be termed "Periodic Continuing Review" to distinguish it from the required Complete Triennial Review. A decision by the IACUC to conduct Periodic Continuing Review is usually made at the time a protocol is initially approved or upon approval of a significant change.

12:2 What is the intent of Complete Triennial Review?

Opin. APHIS/AC in alignment with NIH/OLAW, interprets triennial (*de novo*) review as a comprehensive process which allows the IACUC to evaluate the status of the project and whether it qualifies for reapproval.

12:3 Is a protocol initially considered approved on the date when the research funding begins or on the date when the IACUC has taken final action on the protocol?

Opin. The PHS Policy and AWAR are silent regarding the date of approval of an animal activity. However, NIH/OLAW and APHIS/AC offer the following guidance on this topic. In its guidance document (NOT-OD-11-053)[5] NIH/OLAW (with APHIS/AC concurrence) stated: "Date of [IACUC] approval is the date that appears in the written notification of approval to the principal investigator. IACUCs have flexibility to develop a system that functions well for their institution, provided that the approval date occurs within a reasonable period

DOI: 10.1201/9781003272861-12

of time after review and approval. For example, the IACUC may designate an approval date as the last working day of the month in which the proposal was reviewed, allowing time to make minor corrections or confirm other activities which may affect the approval (e.g., safety committee review, research staff training)."

If the IACUC approves a protocol and submits a letter to the funding agency certifying that the protocol was approved, the actual date of the IACUC's approval must be indicated as required by the PHS Policy (IV,D,2). This is not a future date with final approval contingent on actual funding. In addition, the just-in-time process for verification of IACUC approval for the animal use portions of the grant application and any IACUC imposed changes must be noted on the IACUC's verification of approval prior to the grant award. The institution should have a mechanism and procedures established to verify congruence of the grant application with the approved IACUC protocol.[6] Despite the availability of the just-in-time process, it should be noted that institutions have the prerogative to require investigators to obtain IACUC approval of the grant application prior to NIH peer review, thereby not utilizing the just-in-time process.

12:4 What is the maximal life of an approved protocol?

Opin. The AWAR and PHS Policy do not limit the life of a protocol; however, both require that the IACUC conduct continuing reviews of approved studies at specified intervals determined by the IACUC, but no less often than once every three years (see 12:11). The triennial review must be completed within three years from the last complete review (either initial or triennial review) and cannot be extended.[5] Therefore, many institutions require a new submission at least triennially since it is very likely that the science has evolved, that the current protocol may include multiple amendments that have been approved over the three-year period, and that the IACUC may have new policies or a revised protocol application form that asks new questions that need to be specifically addressed. Since the triennial review constitutes a *de novo* review, it may be more efficient to require submission of a new protocol.

12:5 What information is necessary to conduct appropriate Complete Triennial Review of a protocol?

Opin. Most IACUCs require investigators to utilize an institution-specific continuing review form which is used in conjunction with the complete, up-to-date protocol. This standardizes the information required for all ongoing studies and streamlines the review process. The form should be designed in consideration of the intent of continuing review described previously. One model developed in 1996 in collaboration with NIH/OLAW and APHIS/AC has been described, which is still relevant today.[7]

A relatively simple mechanism can be implemented using a standard review form which contains basic protocol information. This could include, PI, protocol title and number, species, progress report, problems/adverse events, and "other" category the PI can use to provide additional information to be considered/required by the IACUC. During the Complete Triennial Review process the IACUC should prompt the investigator to address any animal use/welfare concerns that may have occurred since the last review.

12:6 Is full committee review (FCR) by the IACUC necessary in order to conduct Complete Triennial Review?

Opin. FCR is not required by AWAR or the PHS Policy. However, IACUCs may decide that full committee review is necessary for studies that involve sensitive species or high-profile controversial activities or serious animal welfare concerns. Protocols involving no changes or minor proposed changes that present no animal welfare issue during continuing review can be most efficiently handled by designated member review (DMR) process.

12.7 Are IACUC approval criteria any different when Complete Triennial Review is conducted by designated member review (DMR) vs. full committee review (FCR)?

Opin. The IACUC approval criteria specified in the AWAR and PHS Policy apply equally regardless of whether the Complete Triennial Review is conducted by DMR or FCR.

12:8 What action should the IACUC take if an investigator does not respond to a request for information concerning the Complete Triennial Review of their protocol?

Reg. If an investigator allows IACUC approval to expire because of failure to respond to the continuing review notification, and PHS-funded and/or USDA-regulated animal work is ongoing, then the research is no longer in compliance with PHS Policy or AWAR. IACUCs are required to report this non-compliance to NIH/OLAW (PHS Policy IV,F,3,a). There is no requirement to report this non-compliance to APHIS/AC, although it would be cited as a non-compliant item (NCI) if identified during a USDA inspection. However, the expiration of a protocol where all animal activities have ceased does not require a report to NIH/OLAW.[8]

Opin. The AWAR (§2.31,d,5) and PHS Policy (IV,C,5) require that reviews of animal protocols be conducted at least triennially. If the IACUC's request for information is necessary for the continuation report to be approved, the IACUC has no choice but to halt further use of any animals until the issue is resolved and the protocol is either reinstated or terminated. If the continuation reapproval date has passed with no response from the investigator, appropriate individuals in the institution should be notified (e.g., the animal facility management, the institution's AV, the PI's department head) with a copy of the notification to the PI. Operationally, IACUCs should have established policies and procedures for handling these types of situations. It should be noted that both the PHS Policy and the AWAR are silent on the disposition of animals already in the process of experimentation when the protocol lapses and a continuation report has not been submitted (see 12.9).

12:9 If a protocol expires, can the investigator continue to use animals that have already undergone some of the research procedures and are in the queue for completing study-related interventions?

Reg. See 12.8.

Opin. The AWAR and PHS Policy prohibit continuation of a study if the investigator has not secured approval to continue the study. Many IACUCs have established procedures for the transfer of the animals onto a "holding protocol" managed by the animal care facility or AV. The IACUC-approved holding protocol does not permit any experiments to be conducted but will allow the animals to be maintained until the investigator secures continuation approval. This holding protocol is an example of a procedure that will temporarily maintain the animals so that once the investigator secures approval, the research interventions can be completed and wasting of animals is minimized. If a protocol involves maintenance of animals that require non-standard husbandry procedures (e.g., insulin administration for diabetic animals), this may be permitted with IACUC approval. It should be noted that financial charges cannot be made to a PHS grant during lapses in IACUC approval.[8]

Survey Questions:

Surv 1. If a protocol expires, which of the following statements are applicable to your IACUC? Select all that apply.

• No additional experimental work is allowed under any circumstances.	203/231 (88%)
• We allow an investigator to conclude any ongoing work.	13/231 (6%)
• We allow the animal facility personnel to conclude any ongoing work on the protocol.	9/231 (4%)
• We only allow the investigator to continue to breed animals.	12/231 (5%)
• We only allow animal facility personnel to continue to breed animals for the investigator.	40/231 (17%)
• We only allow the investigator to euthanize animals for the investigator.	13/231 (6%)
• We only allow animal facility personnel to euthanize animals.	77/231 (33%)
• Other.	34/231 (15%)

Surv 2. With regard to continuing review of protocols, what level of detail should the IACUC consider? Select all that apply.

• Only the previously approved protocol with no additional details needed	39/231 (17%)
• The number of animals used	174/231 (75%)
• Which objectives/ aims have been achieved	146/231 (63%)
• A description of any unanticipated animal welfare problems that may have occurred	189/231 (82%)
• Scientific progress (peer-reviewed publications, presentations, intellectual property developed, patents, etc.	119/231 (52%)
• Other*	41/231 (18%)

*Other responses reflected the choices above, as well as noting the distinction between the triennial continuing protocol review by OLAW and annual reviews, a process that is no longer required by USDA. In addition, the following were offered:

- Changes/amendments to the protocol
- Modifications made to address unanticipated animal welfare concerns
- Change in study objectives
- Animal health issues
- Personnel and training updates
- Updated literature review

12:10 If a protocol is near completion and the continuation report due date is imminent, can the IACUC vote to extend the protocol for a specified short period without submitting a formal report for continuing review?

Reg. See 12:8.

Opin. No. Both APHIS/AC and NIH/OLAW are clear that the IACUC may not extend the protocol approval dates without the appropriate triennial reviews within the specified period.[9]

12:11 When should an entirely new protocol be submitted? Why?

Opin. Many institutions require investigators to submit a new protocol every three years. Although this procedure conforms with AWAR and the PHS Policy (IV,C,5) requirements for Complete Triennial Review (*de novo*), it should be emphasized that there is no limit to the life of a protocol. Therefore, it is possible to allow a protocol to continue indefinitely without a new protocol submission as long as all amendments are included in the protocol, and both the AWAR and the PHS Policy requirements for complete triennial (*de novo*) review are met. The triennial review must include a review of all current information relevant to the protocol, although submission of an entirely new protocol is not required (see 12:4).

12:12 How much information concerning results of a study at the time of Complete Triennial Review should be provided to the IACUC?

Opin. Reporting the results of an ongoing study is not an AWAR or PHS Policy requirement. However, at the time of triennial review, the IACUC is required to perform a complete (*de novo*) review of the continuing study (see 12:4). This necessitates that the protocol meet all review criteria and study conduct requirements as stipulated in PHS Policy (IV,C,5) and AWAR (§2.31,d,i–xi; §2.31,e,1–5). It, therefore, makes sense for the IACUC to require that the investigator provide an informative summary of the results of the study to date that would facilitate IACUC review and, if necessary, IACUC-required information. Indeed, the only way the IACUC can justify use of the animals already included (to date) in the research project is to obtain information on the progress of the study. Admittedly, investigators encounter "blind alleys" and other problems that may affect study results. In some cases, animal experiments must be repeated. The IACUC should be advised of such issues that arise during the course of the study and at the time of continuing review in order to fulfill its responsibilities.

12:13 Is a literature search required at the time of Complete Triennial Review?

Reg. The AWAR (§2.31,d,1,ii) require documentation that alternatives were considered. This applies to both initial review and triennial continuing review. Although the need for a similar literature search is not specifically indicated, PHS Policy (IV,A) requires that applicable AWAR requirements also be met as part of the institution's Assurance.

Opin. Since the PHS Policy has no literature search requirement for research not subject to AWAR, it is up to the institution's IACUC to decide whether to require a literature search for non-USDA covered species.

REFERENCES

1. Office of the Federal Register, 9 CFR Parts 1, 2, and 3, *Federal Register* 54(168), 36133. https://www.federalregister.gov/documents/2021/11/24/2021-25614/awa-research-facility-registration-updates-reviews-and-reports.
2. Applied Research Ethics National Association/Office of Laboratory Animal Welfare. 2002. *Institutional Animal Care and Use Committee Guidebook*, 2nd ed., 96. Bethesda, MD: National Institutes of Health.
3. Committee for the Update of the Guide for the Care and Use of Laboratory Animals. 2011. *Guide for the Care and Use of Laboratory Animals*, 8th ed., 34. Washington, DC: National Academies Press.
4. Office of Laboratory Animal Welfare. 2013. *Frequently Asked Questions D.1, How Frequently Should the IACUC Review Research Protocols?* Accessed December 15, 2022. https://grants.nih.gov/grants/olaw/faqs.htm#IACUC_2.
5. Office of Extramural Research. 2011. *Notice NOT-OD-11-053, Guidance to Reduce Regulatory Burden for IACUC Administration Regarding Alternate Members and Approval Dates.* Accessed December 15, 2022. https://grants.nih.gov/grants/guide/notice-files/NOT-OD-11-053.html.
6. National Institutes of Health, Office of Laboratory Animal Welfare. 2013. *Frequently Asked Questions D.6, What Criteria Should the IACUC Consider When Reviewing Protocols?* Accessed December 15, 2022. https://grants.nih.gov/grants/olaw/faqs.htm#.
7. Oki, G.S.F., E.D. Prentice, N.L. Garnett, et al. 1996. Model for performing Institutional Animal Care and Use Committee continuing review for animal research. *Contemporary Topics in Laboratory Animal Science* 35(4):53. Accessed December 15, 2022. https://grants.nih.gov/grants/olaw/references/contop96.htm.
8. National Institutes of Health, Office of Extramural Research. 2010. *Notice NOT-OD-10-081, Guidance on Confirming Appropriate Charges to NIH Awards during Period of Noncompliance for Activities Involving Animals.* https://grants.nih.gov/grants/guide/notice-files/NOT-OD-10-081.html.
9. National Institutes of Health, Office of Laboratory Animal Welfare. 2013. *Frequently Asked Questions D.2, May the IACUC Administratively Extend Approval of a Project That Has Expired?* Accessed December 15, 2022. https://grants.nih.gov/grants/olaw/faqs.htm#.

13

Justification for the Use of Animals

Kristina Cataline and Larry Carbone

Introduction

Legally and ethically, scientists must justify their need for animals if they are to use them in their studies. PHS Policy and the AWA and its regulations call on the IACUC to review the investigators' proposal to use laboratory animals, including the investigator's justification of the type and number of animals. The IACUC must also review exceptions to the *Guide*, PHS Policy, and the Animal Welfare Act, including multiple major survival surgery, use or withholding of painkilling drugs, and choice of any non-standard euthanasia method.

Legal and accreditation guidance and regulations use several terms but do not provide definitions for some key terms, including *justification*, *harm*, *ethical*, *pain*, *distress*, *benefit*, *merit*, *animal welfare concerns*, and *suffering*. This can lead to conflict and confusion in the IACUC if members are unclear about what kinds of research, and quality of research methods, they should review. Must they restrict themselves to harm reduction via application of "The Three Rs" (replacement, reduction, and refinement) or does review of justification require a broader review of the proposed science?

Regulatory guidance presumes that scientists use animals only for important, well-designed studies, and that harming animals requires evidence that experiments are particularly well designed, necessary, and important. Whether the IACUC can and should make the evaluation is controversial.

Regulatory guidance sets minimum standards for animal housing and animal use to limit pain and distress to animals. Institutions and their IACUCs are free to go beyond the minimum, such as setting a higher bar for confining animals in ways that limit their behavioral opportunities, promoting programs to minimize killing—even painless killing—of animals, and reviewing every experiment's scientific and ethical justification.

In this chapter, we present information on what IACUCs must do to meet regulatory standards, but we also emphasize (in Sections labeled *Opin.*) what IACUCs may consider in reviewing animal use for ethical justification. In the years since the last updates of the relevant laws, the *Guide*, and the AWR, we look to emerging literature in animal welfare science and ethics to update this chapter for this edition.

13:1 What is the regulatory definition of an *animal*?

Reg. The HREA does not define "animal." PHS Policy (III,A) defines an *animal* as "any live, vertebrate animal used or intended for use in research, research training, experimentation, or biological testing or for related purposes." NIH/OLAW explicitly states that unhatched embryos of egg-laying species may have vertebrae, but they are not covered by PHS Policy, and that presumably goes for fetal mammals *in utero* as well. Larval forms of amphibians and fish are considered vertebrates.[1] Zebrafish hatch from their chorion at approximately 72 hours post-fertilization, at which point they become legal animals in PHS policy.[2] PHS Policy only applies to animals and institutions with certain classes of federal funding unless an institution includes all its animals in its OLAW-approved Assurance.

> The AWA (§1.1, Animal) and AWAR (§2132,g) define an *animal* as "any live or dead dog, cat, nonhuman primate, guinea pig, hamster, rabbit, or any such other warm-blooded animal, as the secretary may determine is being used or is intended for use for research, teaching, testing, experimentation, or exhibition purposes, or as a pet. This term excludes birds, rats of the genus *Rattus*, and mice of the genus *Mus*, bred for use in research, and horses not used for research purposes and other farm animals, such as, but not limited to, livestock or poultry used or intended for use as food or fiber, or livestock or poultry used or intended for use for improving animal nutrition, breeding, management, or production efficiency, or for improving the quality of food or fiber. With respect to a dog, the term means all dogs, including those used for hunting, security, or breeding purposes."

Thus, under the AWA, the overwhelming majority of animals used in laboratories (common laboratory rats, mice, and, increasingly, zebrafish) are not regulated under the AWA. Fetuses are not regulated under the AWA. Birds, rats, and mice not bred for research (bred in the pet industry, for example, or wild caught) are covered by AWAR. The definition of "animal" is in the Animal Welfare Act itself,

DOI: 10.1201/9781003272861-13

which only the U.S. Congress, not the USDA, is empowered to change.

AAALAC International considers all members of the kingdom Animalia to be animals, including invertebrates such as cephalopods, nematodes, and zooplankton, though it does not expect institutions to apply the same "intensity of oversight" to a nematode that it would to a mouse.[3]

The regulatory landscape for octopuses and other more cognitively developed (or sentient) invertebrates is in flux. AAALAC now considers them animals that require consideration in an animal care program. The AWAR exclude them and will continue to do so without a congressional amendment. The NIH put out a Request for Information in 2023 and so the NIH could choose to modify the PHS Policy to include animals beyond the presently covered vertebrates. At this point, OLAW guidance regarding cephalopods restates their exclusion from PHS Policy, from any required reporting of their use, and from a requirement that they be covered in a Vertebrate Animals Section of grant applications. OLAW writes: "Though cephalopods are not covered under the PHS Policy, institutions are expected to adhere to the U.S. Government Principles … for humane care and use of all animals in their care. In addition, because cephalopods are not vertebrate animals, they are not currently subject to the requirements for reporting and there is no requirement to submit a Vertebrate Animals Section for cephalopod use in funding applications." In 2023, OLAW proposed some changes to this, including that IACUCs include cephalopods in their reviews, post-approval monitoring, and semiannual evaluations.[4]

Opin. Regulatory definitions of "animal" map very roughly onto categories of animals thought to have significant sentience, which we loosely define here as the ability to experience pain and pleasure.[5] The AWA's inclusion of hamsters while it excludes laboratory-bred mice, and its total exclusion of non-mammals, is inconsistent with the principle of sentience-based species coverage. Likewise, exclusion of highly sentient invertebrates such as octopuses is not rooted in neurobiology or evidence of sentience, and, thus, the NIH has this issue under review. While amending the AWA to include all sentient animals, or even all mammals, would require congressional action, it is within NIH's scope of authority to define or refine "animal" to include all sentient animals, including some invertebrates.

Institutions are free to expand upon the regulatory definition of animal and to charge their IACUC to cover more animals in more settings than the minimum required by regulation. They may review protocols for non-AWAR-regulated vertebrates, whether the animals are on publicly funded research or not. They may include embryos or invertebrates such as cephalopods or decapods at their discretion. Going beyond the regulatory definitions of "animal" brings them closer to the ethical principle that harming sentient animals requires justification.

Surv. Does your IACUC review protocols for species not covered by the Animal Welfare Act or the PHS Policy (e.g., cephalopods, agricultural species used in non-biomedical research)?

- Not applicable 61/226 (27%)
- Yes, routinely 53/226 (23%)
- Yes, but only upon request 22/226 (10%
- No 86/226 (38%)
- Other 4/226 (2%)

13:2 What is the regulatory definition of "justification"?

Reg. The PHS Policy, the AWA, the AWAR, and the *Guide* call for justification in the use of animals, but they do not explicitly define "justification."

13:3 In what ways is it required that the use of animals be justified on an IACUC protocol?

Reg Institutions using animals regulated by the AWAR must ensure the following:

- That all proposals to the IACUC contain a rationale for involving animals, and for the appropriateness of the species and numbers of animals to be used (AWAR §2.31,e,2).
- A description of procedures designed to assure that discomfort and pain to animals will be limited to that which is unavoidable for the conduct of scientifically valuable research, including provision for the use of analgesic, anesthetic, and tranquilizing drugs where indicated and appropriate to minimize discomfort and pain to animals (AWAR §2.31,e,4).
- That the PI has considered alternatives to procedures that may cause more than momentary or slight pain or distress to the animals (AWAR §2.31,d,1,ii) and has provided a written narrative description of the methods and sources (e.g., the Animal Welfare Information Center) used to determine that alternatives to animal use are not available (AWAR §2.31,d,1,ii).
- That the PI has provided written assurance that the research activities do not unnecessarily duplicate previous experiments (AWAR §2.31,d,1,iii).

The HREA (Sect. 495,c,2) and the Vertebrate Animals Section (VAS) of the NIH grant application require applicants for grants, contracts, or cooperative agreements involving research on animals to submit "a statement of the reasons for the use of animals in the research to be conducted with funds provided under such grant or contract." Applications and proposals (competing and non-competing) for awards submitted to PHS that involve the use of animals are required to specify the species and approximate number of animals to be used, the rationale for involving animals, and the appropriateness of the species and numbers to be used.

The VAS states: "Investigators must justify the use of animals in the proposed research. U.S. Government Principles require grantees to consider mathematical models, computer simulation, and in vitro biological systems. The justification should:

- Indicate why the research goals cannot be accomplished using an alternative model (e.g., computational, human, invertebrate, in vitro)
- Provide rationale for the choice of species (e.g., advantages of the species chosen and why alternative species are not appropriate)
- Discuss why less highly evolved or simpler animal models are not appropriate. For example, the use of nonhuman primates, dogs, or cats should be thoroughly justified.
- The justification for the number of animals, sex, ages, and other biological variables should be described in the Research Strategy (PHS Policy IV,D,1,a; IV,D,1,b).

The *Guide* (p. 25) states: "The following topics should be considered in the preparation and review of animal care and use protocols:

- Rationale and purpose of the proposed use of animals.
- Justification of the species and number of animals requested. Whenever possible, the number of animals requested should be justified statistically.
- Availability or appropriateness of the use of less-invasive procedures, other species, isolated organ preparation, cell or tissue culture, or computer simulation."

13.4 What types of justification does an IACUC review?

Opin. The word "justification" may imply moral or ethical analysis to some, or at least some sort of cost/benefit, risk/benefit, or harm/benefit analysis. Regulations and guidelines do not define justification, so it is not always clear if the IACUC is charged with making a technical or an ethical evaluation. In the relevant regulations and guidelines, one finds reference to several potential types of justification, including scientific necessity, economics, convenience, "conservation of scarce animal resources," and ethics.

"Scientific justification" is synonymous with the scientific rationale for an action. In the AWAR, for example, using animals in more than one major survival operative procedure (AWAR §2.31d,1,x,A) must be "justified for scientific reasons," i.e., the experiment requires multiple surgeries per animal or would otherwise fail. The *Guide* (pp. 30, 64, 124) also requires scientific justification of certain practices, such as singly housing social species, performing multiple surgeries or painful procedures on the same animal, or using physical methods of euthanizing animals. For example, a scientist explains to the IACUC why they need to perform sequential surgeries on the same animal subject, why, for example, only rabbits can yield the data the scientist needs, or why anti-inflammatory analgesics will render the experiment useless. The IACUC can review the data that supports those claims.

"Ethical justification" is not synonymous with scientific justification, and procedures that have sound scientific justification may yet be unethical. If withholding analgesics is the only route to data the scientist seeks, the scientist, IACUC, or others may yet decide that the project is not valuable enough ethically to justify the pain to be experienced by the animals. Regulatory language could be read to imply that justification is all or nothing; matters are rarely so simple. IACUCs must deliberate how strong a justification is required to allow the use of animals, especially for animals at risk of significant unalleviated pain or distress. For projects of questionable ethical justification, the IACUC may approve closely monitored pilot work that establishes the degree of animal suffering and the robustness of data and then readdress the justification.

Cost savings and efficiency could justify some practices but are explicitly ruled out in some specific cases. For example, cost savings *alone* [emphasis added] is cited in the *Guide* (p. 30) as an inadequate justification for performing multiple major survival surgical procedures on animals. Conversely, cost savings is not mentioned as adequate or inadequate for other issues. As a general rule, OLAW writes: "Animal welfare and the integrity of research findings, rather than cost alone, should be the primary factors in decisions related to assuring compliance with the recommendations in the *Guide* in PHS-funded research."[6] That FAQ goes on to state: "In many instances, institutions and IACUCs elect to exceed the standards described in the *Guide*. This is not required and can add expense to the program. OLAW does not discourage or encourage institutions from exceeding the standards." We believe that IACUCs should encourage going beyond the minimum regulatory standards and seek refinements in animal care to improve animal welfare even when those require some material purchases or staff time. Examples might include running post-use adoption programs or purchasing caging systems that exceed the space recommendations in the *Guide* when emergent literature shows them to be superior.

The *Guide* (p. 30) states that "conservation of scarce animal resources may justify the conduct of multiple major surgeries on a single animal," but does not elaborate on a standard of what might constitute a "scarce resource." This could refer to animals in short supply in labs or in captivity, as well as to founder animals of new genetic lines might qualify as "scarce."

Presently, the numbers of available nonhuman primates do not meet the demand, with finite NIH or other resources to support domestic breeding in the National Regional Primate Centers, increased demand (with COVID-19 responsible for some of this), restrictions of exports from other countries, and legal and conservation concerns regarding importation of wild-caught (or potentially wild-caught) nonhuman primates. The Association of Primate Veterinarians cautions that this shortage could lead institutions to "become more likely to re-use animals for multiple projects, resulting in increased long-term costs/harms to the animals."[7–9]

Note that the AWAR do not contain a similar allowance for conservation of scarce resources, so the *Guide*'s flexibility on this matter could lead to AWA non-compliance, at least in the case of multiple major survival surgeries on primates or other AWA-covered species.

13:5 Should the IACUC perform an ethical review of animal-use protocols?

Reg. The AWA, AWAR, and PHS Policy do not explicitly define or call for an *ethical* or *moral* review. The *Guide* (pp. 25–33) lists several functions of an IACUC; ethical review is not one of those. The *Guide* encourages scientists and institutions "to give careful and deliberate thought to the decision to use animals, taking into consideration the contribution that such use will make to new knowledge, ethical concerns, and the availability of alternatives to animal use."

The U.S. Government Principle II states: "Procedures involving animals should be designed and performed with due consideration of their relevance to human or animal health, the advancement of knowledge, or the good of society."[10] The *Guide* endorses this principle, but neither the Principles nor the *Guide* assign this task to anyone in particular, whether the IACUC, funding reviewers, the investigator, or institutional administrators.

Opin. IACUCs operate within a societal consensus ethic that accepts the use of animals for human purposes, including food, entertainment, work, and research. IACUCs are free to recommend members for appointment by the chief executive officer (CEO) who do not share this ethic. The societal ethic is not *laissez-faire*: some animal uses are permitted, but they must be justified. IACUCs thus may begin from the ethical premise that harming animals in research is permissible, but that scientists are constrained, ethically, in what they can do to animals. The governmental/societal belief that animal use can be ethically justified leads not just to permitting, but to active funding, animal research. The belief that such use *must be* ethically justified leads to the establishment of IACUCs and other safeguards to animal welfare.[11]

Most IACUCs devote themselves to minimizing animal pain and distress, using the Three Rs framework (replace, reduce, and refine; see below) to guide their efforts. A full ethical review requires going further. Once pain and distress have been reduced to their minimum achievable for a given project, does even that amount of animal pain and distress warrant performing the experiment? A full ethical review requires that someone—the IACUC, the research investigator, a funding agency, a research administrator—determines that a project's worth is sufficiently high to justify animal use.[11–13]

A full ethical review would go beyond assessing how potential harm to the animals balances against potential benefits to humans. It should consider other costs of conducting animal research, such as money, environmental impacts, and potential harm and risks to the health and safety of workers and other human beings. It should consider possible benefits to animals and environmental health. It should consider the harms of not performing projects that could benefit human, animal, or environmental health.[13] Whether the IACUC can and should make these determinations is controversial. Currently, neither funding agencies—for projects receiving funding—nor IACUCs typically do an in-depth ethical review, leaving only the investigator as the person making an ethical evaluation of their own research.[14]

Surv. Does your IACUC consider the balance between animal welfare and scientific (or educational) importance while reviewing a proposed animal activity?

• Yes, this is routinely incorporated into the protocol review.	213/226 (94%)
• Yes, but only for species regulated under the USDA Animal Welfare Act.	2/226 (1%)
• No.	5/226 (2%)
• I don't know.	2/226 (1%)
• Other.	4/226 (2%)

13:6 Must the IACUC perform a harm-benefit, risk-benefit, or cost-benefit analysis?

Reg. None of these terms exist in PHS Policy, in the AWA, or the AWR.

For certain procedures, the *Guide* (p. 27) states that "the IACUC is obliged to weigh the objectives of the study against potential animal welfare concerns," perhaps its most explicit call for IACUC ethical review. AAALAC International refers to this as a "'harm-benefit' [quotes in the original] analysis" and states that "site visitors will assess how the Committee conducts this analysis," without further guidance or restriction on how the IACUC's harm-benefit analyses should be performed or how they will evaluate it.[15]

Opin. Some authors use these three terms, "harm-benefit," "risk-benefit," and "cost-benefit," interchangeably, while others draw distinctions.[16–18] Here we use them interchangeably, considering "risk" to mean risk of animal suffering. The *Guide*'s call to weigh both objectives of the study and animal welfare concerns cannot be met solely by reviewing alternatives to pain and distress, with the 3Rs framework as guidance but somehow evaluating the value of the scientific proposal.

13:7 What is *harm to animals* in the context of animal research?

Reg. The *Guide* and PHS Policy do not use the word "harm," choosing instead "unrelieved pain or distress" or "potential animal welfare concerns." In the AWAR, harm denotes physical injuries in trapping or handling animals. AAALAC uses the *Guide*'s "animal welfare concerns" to imply the need for a harm-benefit analysis, though without defining "harm."

Opin. The word "harm" can vary in its usage, with differing implications for animal care practices. At its most restrictive, harm signifies pain and distress, which are clearly in the IACUC's legal purview, at least in their more severe manifestations. We favor a more expansive definition, in which the word "harm" encompasses pain, distress, and suffering, as well as depriving animals of opportunities for happiness and pleasure.[19]

Performed correctly, killing animals in accord with the AVMA Guidelines for the Euthanasia of Animals should not cause significant pain or distress.[20] The AVMA Guidelines note that "debate exists about whether *euthanasia* appropriately describes the killing of some healthy animals at the end of biological experiments." Is it a harm to kill healthy young animals, depriving them of whatever positive experiences their lives could hold? The prevalence of post-research adoption programs, along with IACUCs' concern for painless euthanasia of very large numbers of unwanted or unusable genotypes of mice, suggest that many people consider painless death at least somewhat of a harm to be avoided.

Animals may be harmed by being deprived of the good things in life, even if they are not actively experiencing deprivation as distress.[19, 21] Laboratory-bred dogs can learn to enjoy running, chasing tennis balls, and lying in the sun, all of which current regulation-size cages deprive them of. We believe that IACUCs should consider confinement and limited opportunities—even in enclosures that meet the size guidelines of the *Guide* and the AWRs—as animal harms that they should try to minimize. Requirements to enrich animals' lives serve not just to prevent distress and improve study outcomes but also to promote positive welfare and well-being.[22, 23]

IACUCs, veterinarians, and investigators must also remember outliers, animals that fare worse in some experiments than their cohorts. Carefully developed and monitored intervention endpoints are essential to minimize harm to these animals.[21]

13.8 Are suffering and harm the same?

Reg. PHS Policy and the *Guide* do not use these words. The U.S. Government Principles call for euthanasia of animals that would otherwise *suffer*, and veterinarians in the U.S. swear an oath to prevent and relieve animal *suffering*.[24] The AWA and AWAR require prevention of suffering—more or less synonymous with injury or disease—but as part of animal holding, not in the context of research procedures under the IACUC's oversight.

Opin. We propose that *suffering* best captures the spectrum of animal harms that the IACUC's must strive to prevent or minimize suffering. Animals in significant unalleviated pain and distress are suffering and may have a life not worth living.[21] These animals are placed in the highest severity band ("severe") in most European systems, and in Category E of the AWR.

IACUCs must remain aware of projects, genotypes, and conditions that cause chronic and/or intense pain that is beyond the reach of available analgesic therapies to manage, no matter how willing an investigator would be to try treatments. Animal housing, management, lack of enrichment, and other non-experimental factors may also cause animals to suffer distress. The IACUC should evaluate these situations with the same caution they give Category E experimental procedures in which pain medicines are actively withheld out of concern for how drugs could harm the experiments. They must evaluate with the investigator whether refinements will adversely or beneficially affect data.

13.9 Should the IACUC set an upper limit to animal harms?

Reg. Regulations do not require setting upper limits on animal pain and distress, beyond which an IACUC may not approve a protocol. Instead, the greater the potential for pain and distress, the greater the obligation for undertaking literature searches, setting intervention and humane endpoints, and scrutinizing any plans to withhold anesthetics, analgesics, or other refinements that could mitigate animal pain and distress.

Opin. Some scholars cite the challenge of a true harm-benefit or cost-benefit analysis as reason to avoid them. Can one truly predict where a given body of well-conducted research will read in the decades that follow it? As an alternative, they recommend setting upper limits on the duration or intensity of suffering an animal might endure.[13, 25] Most such proposals include a provision, comparable to the European Union's special provisions for procedures of "Severe Severity," to allow procedures that breach this upper limit with heightened scrutiny, justification, and monitoring. As long as there remains flexibility to breach a proposed upper limit to suffering when there is perceived scientific necessity, we question whether this is truly an upper limit.

13:10 What benefits should an IACUC consider as justification?

Reg. Advancing human health is the most obvious potential benefit of most animal research. Scientific studies aimed to improve animal health, as well as environmental and conservation research, may require animal use. Classroom use ranges from high school labs to advanced surgical training for medical and veterinary practitioners. This use is less likely to generate new scientific knowledge, but it may impart important knowledge and skills, after adequate training with non-animal models.

Research benefits are on a spectrum from "applied" to "basic" research, and the justification for using animals for imminent important benefits is most obvious. Testing a product for release to the public is an example of applied research, while studies of some cell functions with no currently known or obvious connection to a particular disease might be labeled "basic." Basic biological knowledge can be crucial in future studies, and projects should not be hastily dismissed for lack of an obvious near-immediate application.[18]

13:11 Should the IACUC review scientific merit as a component of the justification of use of animals?

Reg. The AWAR do not specifically call for scientific merit review by the Institutional Official (IO), the IACUC, or other bodies. The regulations require that pain be limited to that "which is unavoidable for the conduct of scientifically valuable research" (AWAR §2.31,e,4). Additionally, the AWAR require that no animal be used in more than one major operative procedure from which it is allowed to recover unless justified for scientific reasons by the Principal Investigator (PI). However, the regulations do not explicitly charge the IACUC with making these determinations.

NIH/OLAW's stance is more complicated. PHS Policy requires compliance with the *Guide*'s call for the IACUC to weigh study objectives (one element of merit) against potential animal welfare concerns, as well as U.S. Government Principle II's call for an evaluation (though not necessarily by an IACUC) of the relevance of a procedure to human or animal health, the advancement of knowledge, or the good of society.

NIH/OLAW has provided guidance on this topic in FAQ D.12: "Peer review of the scientific and technical merit of an application is considered the purview of the NIH Scientific Review Groups (SRGs), which are composed of scientific experts from the extramural research community in a particular area of expertise. However, SRGs also have authority to raise specific animal welfare concerns that can require resolution prior to a grant award.

"Although not intended to conduct peer review of research proposals, the IACUC is expected to include consideration of the U.S. Government Principles in its review of protocols. Principle II calls for an evaluation of the relevance of a procedure to human or animal health, the advancement of knowledge, or the good of society. Other PHS Policy review criteria refer to sound research design, rationale for involving animals, and scientifically valuable research. Presumably a study that could not meet these basic criteria is inherently unnecessary and wasteful and, therefore, not justifiable."

> "The primary focus of the SRG is scientific merit and the primary focus of the IACUC is animal welfare. The two bodies have differing constitutions, mandates, and functions. However, since it is not entirely possibly to separate scientific value from animal welfare some overlap is inevitable. SRGs may raise concerns about animal welfare and IACUCs may question the scientific rationale or necessity for a procedure."[26]

Opin. Scientific merit includes both the importance of the scientific question being asked and the likelihood that the proposed project will answer it.[11, 27] Scientific merit certainly is an inherent component of the justification of the use of animals, as non-meritorious research can hardly justify harming animals.

Institutions and their IACUCs must decide how to evaluate proposals that will not have undergone review by an SRG or do not have funding sources that require OLAW oversight. They may use internal funds or grants from small foundations. Most use by private industry of animals does not receive funding over which OLAW has oversight or SRG review. Moreover, even those grant applications that receive SRG review are often less detailed than the procedures the IACUC must review. This would suggest a need to compare the peer-reviewed application in detail with the IACUC protocol and to consider experiments not detailed in the grant application as not having had a merit review. Animals in training labs will not have received NIH peer review, nor will much research or testing in pharmaceutical companies.

An additional challenge at a multidisciplinary institution is that the IACUC may not have the expertise to review either the importance of the research question being addressed or the soundness of the hypotheses being tested, or the quality of the methods proposed for the work.

In such circumstances, some in-house departmental review committee or *ad hoc* expert consultation may be sought. The IACUC may decide to require and accept such reviews rather than try to assess the scientific merit on its own. While the IACUC may not be the body that assesses the scientific merit of a project, it should know that peer review has been performed. The IACUC needs to know that this review was favorable and done by a body the IACUC finds credible. The IACUC needs to know if it is reviewing a project that has not had scientific merit review. It needs to know if a project was judged meritorious but not meritorious enough to beat out the competition for funding, or if it was externally reviewed and found to be non-meritorious. Should the IACUC approve animal use for studies that were deemed insufficiently robust to garner funding?

Best practice would be for IACUCs to review the merit review comments that investigators have received and, where there are none, conduct a review themselves if they have the necessary expertise, or require external review, with written review comments, not simply a department chair's signature.

13:12 How can the IACUC evaluate the justification for withholding pain medicines, nonpharmaceutical pain management techniques, or other welfare refinements for procedures that cause animals significant unalleviated pain or distress?

Reg. The AWAR (§2.31, d), based on the AWA (§2143,a,3,A), require the IACUC to ensure that "Procedures that may cause more than momentary or slight pain or distress to the animals will be performed with appropriate sedatives, analgesics or anesthetics, unless withholding such agents is justified for scientific reasons, in writing, by the principal investigator and will continue for only the necessary period of time." This information must be provided to the IACUC (AWAR §2.31,e,4), and the investigator must have considered alternatives to these procedures, provide a written description of the methods and sources used to determine alternatives were not available, and assure that the activities do not unnecessarily duplicate previous experiments (AWAR §2.31,d,1,i–iii). Furthermore, the IACUC must ensure that these activities involved, in their planning, consultation with the attending veterinarian or his or her designee (AWAR §2.31,d,1,iv,B).

The USDA Animal Welfare Inspection Guide requires that "endpoints are described that will be used to terminate the study and/ or used as the basis for when treatment or euthanasia will be performed."[28]

IACUCs must answer a "threshold question" of whether a procedure's pain or distress is significant enough to warrant alleviation, in which case it is reported in Column D (USDA's "With Pain With Drugs") or in Column E (USDA's "With Pain Without Drugs"). For pain, the threshold is "pain in excess of that caused by injections or other minor procedures" (AWAR §1.1).

The AWAR (§2.36, b,7) require research institutions to file an annual report, including in Column E: "the common names and the numbers of animals upon which teaching, experiments, research, surgery, or tests were conducted involving accompanying pain or distress to the animals and for which the use of appropriate anesthetic, analgesic, or tranquilizing *drugs would have adversely affected the procedures, results, or interpretation of the teaching, research, experiments, surgery, or tests* [emphasis added]. An explanation of the procedures producing pain or distress in these animals and the reasons such drugs were not used shall be attached to the annual report."

PHS Policy (IV,C,1,b) similarly calls on the IACUC to determine that "Procedures that may cause more than momentary or slight pain or distress to the animals will be performed with appropriate sedation, analgesia, or anesthesia, unless the procedure is justified for scientific reasons in writing by the investigator." It does not set thresholds for "momentary or slight" pain or distress. It does not require categorizing animals into "pain categories" or reporting types of animal use or pain categories to the NIH as the USDA requires for AWA-covered species.

The *Guide* (p. 5) states that studies that will cause severe or chronic pain or distress should include descriptions of appropriate humane endpoints or provide science-based justification for not using a particular, commonly accepted humane endpoint. The *Guide* no longer specifically addresses situations in which the IACUC might review an investigator's plans to withhold perioperative analgesics or other such sources of transient (but not momentary) pain.[29]

Opin. Approximately 8% (rough estimate) of AWA-covered animals were reported in Column E in 2022.[30] There is no reason to assume rats and mice are used less in painful studies than are the AWAR-regulated species, and whether IACUCs use the USDA's pain categories or not, they should pay special attention to any protocols in which pain management and other welfare refinements that could reduce suffering will be withheld.

The USDA system requires that justification rest on the adverse effects that medications (and other welfare refinements) could have on study validity, not on cost, convenience, reluctance to obtain controlled substances, or other challenges.

The IACUC and investigator face both a technical and an ethical determination. IACUCs should ask what it means that analgesic or other drugs might "adversely affect" data. Experiments performed with analgesics administered or withheld may yield different results, but "in studies where the use of certain analgesics appears to be contraindicated, investigators should be mindful that unwanted variables from pain-induced perturbation of homeostatic mechanisms can affect the animal model."[31] There may be times when treating inflammation and pain may result in a truer model with more valid data that translate better to human health. Only occasionally will IACUCs go into protocol review with clear model-specific data on how pain medications and refinements will affect outcomes. A literature search or expert consultation examining the potential adverse effects of medical and nonpharmaceutical pain treatments on the data should likewise explore the potential effects of untreated pain or distress on study outcomes before deciding that withholding pain management is justified. The *Guide* advises that "nonpharmacologic control of pain may be effective and should not be overlooked as an element of the post-procedural or perioperative care provided to research animals." As with pain medicines, the IACUC, veterinarian, and investigator should examine nonpharmacologic pain control options for their efficacy in alleviating animal pain and for good and bad effects on research outcomes.

Thus, the questions to consider are:

- Will analgesics, nonpharmaceutical pain management techniques, and welfare refinements result in altered data?
- Are these data less reliable than data in studies with unalleviated pain/distress?
- How large is the effect?
- If welfare refinements will have a significant negative impact on the data, do the study objectives and the scientific merit outweigh the potential animal suffering?

Specific data will not be available for every model. The IACUC must decide its standard of evidence in evaluating such protocols: whether to approve such studies only when interference by analgesics has been demonstrated or when such interference seems plausible but uncertain. IACUCs make an ethical decision regarding what level of evidence, or scientific justification, they need to conclude that the use of analgesics or other refinements will have a significant deleterious effect on an experiment.[29] Those requiring a lower standard of evidence will allow more animals with untreated pain and distress than those with more stringent requirements.

Finally, once IACUCs have determined the likelihood of significant unalleviated pain and distress for a project or procedure, they may evaluate whether the objectives of the study and its merits justify harming the animals as proposed.

Surv. When an investigator proposes a study in which pain-relieving medications will be withheld during potentially painful procedures, what does your IACUC typically require as scientific justification that analgesic or anesthetic drugs would adversely affect the scientific project?

- Not applicable. 14/226 (6%)
- No references are required but the investigator must make a statement that analgesic or anesthetic drugs may interfere with the scientific project. 16/226 (7%)
- No references are required but the investigator must make a statement describing how all specific classes of analgesic or anesthetic drugs that could alleviate pain or distress will interfere with the scientific project. 24/226 (11%)
- The investigator must make a statement describing how all specific classes of analgesic or anesthetic drugs that could alleviate pain or distress will interfere with the scientific project, and provide references 107/226 (47%)
- Any of the above justifications, plus the investigator must include parallel information on his/her review of how untreated pain or distress may affect the scientific project. 58/226 (26%)
- None of the above. 1/226 (0%)
- Other. 6/226 (3%)

13:13 What are alternatives in the context of animal research?

Russell and Burch provided a useful framework for considering ways to minimize the suffering of live sentient laboratory animals. The framework covers animal housing and care, as well as experimental use. It covers animal distress (they use the word "inhumanity") that is the focus of the research ("direct" distress, for example, as may occur in some pain studies) and distress that is essentially a side effect of animal care and use practices ("indirect" distress, such as keeping an animal in an enclosure that restricts species-typical behaviors.).[32] Russell and Burch elaborated the sometimes overlapping, sometimes contradictory "3Rs" framework of replacement, reduction, and refinement alternatives.

Reg. The AWA (§2143,a,3,B) requires "that the principal investigator considers alternatives to any procedure likely to produce pain to or distress in an experimental animal" and provide "a written narrative description of the methods and sources, e.g., the Animal Welfare Information Center, used to determine that alternatives were not available" (AWAR §2.31,d,1,ii). The *Guide* similarly "endorses" the principle of "Consideration of alternatives (in vitro systems, computer simulations and/or mathematical models) to reduce or replace the use of animals" (p. 12) and calls the Three Rs "a practical strategy for decision-making" (p. 3). NIH-OLAW FAQ D.7 highlights the similarities of U.S. Government Principles III and IV to the 3Rs. U.S. Government Principle III requires the use of the minimum number of animals necessary to obtain valid results, synonymous with a requirement to reduce animal numbers.[33] It also requires that mathematical models, computer simulation, and *in vitro* biological systems should be considered, synonymous with a requirement to replace non-animal models wherever possible. U.S. Government Principle IV requires proper use of animals to avoid or minimize discomfort, distress, and pain consistent with sound scientific practices, synonymous with a requirement to implement refinements.

Opin. The 3Rs alternatives framework is a useful matrix for systematically minimizing pain and distress of any given set of experiments, though not all alternatives neatly fit one of the Rs or another. Though originally written to reduce distress in animal laboratories, institutions, IACUCs, and scientists can find the framework helpful in identifying ways to maximize animal well-being.[34] The 3Rs alternatives framework is not an ethical system; it contains no answer for what a scientist or IACUC should do when rigorous pursuit of alternatives still yields an experimental plan with considerable pain or distress.[13]

13:14 What is meant by replacement alternatives?

Opin. Replacement alternatives are conceptually the most straightforward: finding ways to generate research data without harming sentient animals. Candidates for consideration include studying cells in tissue culture (*in vitro* techniques), microorganisms, computer simulations, human epidemiological data, or using inanimate models. Replacement can include replacing sentient animals, such as mammals, with animals presently thought to be essentially nonsentient, such as insects or nematodes. Replacement might include using slaughterhouse material or cadavers of animals euthanized for purposes other than research or teaching.

Replacement can overlap with reduction and refinement, and there is no need to over-analyze how to characterize an alternative under consideration. Many people working today would consider it a refinement to conduct invasive procedures on live animals under general anesthesia, whether the animal awakens later from the procedure or not. Russell and Burch saw this as a replacement, as the (temporarily) insentient, unconscious animal has replaced the sentient, conscious creature.[34] Likewise, people can argue, unnecessarily, whether using live animals as a source of tissues and cells for *in vitro* studies is better seen as replacement or reduction.

For want of a better "R," we include studies of human volunteers, companion animals, and observation of unconfined animals as replacements to studying confined, non-consenting sentient laboratory animals. Cognitive biology uses spontaneous behaviors, as well as neuroimaging of cooperating human volunteers or trained dogs for some studies, while clinical studies of naturally occurring conditions in companion animals (cancer models, pain, others) are valuable in medical research.

Replacements are constantly evolving, especially, but not exclusively, in the field of regulatory testing. Paradigmatic examples are the production of monoclonal antibodies in cell culture, as opposed to growing them in mouse ascites fluid, and replacing live-rodent colony sentinels with testing the animal cages and bedding directly for pathogens. An investigator needs a strong justification to choose these outdated methods in the face of available replacements.

More problematic is the attempt to replace more sentient animals with less sentient animals (the authors resist the terms "higher animal" and "lower animal" or reference to a "phylogenetic scale," as none of these terms have precise biological definitions). Despite this, "lower on the phylogenetic scale" is a term used, without definition, in the *Guide* (p. 5). Are frogs, fish, or octopuses reliably known to be less sentient than mammals? Among the mammals, are there any data that rodents are significantly less sentient than dogs, swine, or even primates? For the purposes of this chapter, *sentience* is defined as the capacity to experience pain, happiness, and suffering as affective, or emotional, states. Sentient animals are those that can experience a range of good or bad welfare.[5] Sentience is more than simply the ability to sense stimuli (even noxious stimuli) and less than the complex self-awareness and self-consciousness possessed by humans. There is no biological reason to assume that sentience is an all-or-nothing trait, possessed equally, in full, by all animals that possess it. Rather, it exists on a continuum across species, with monkeys, apes, and dolphins, as far as we know, more sentient than fish or squid, and adult animals likely more sentient than larval or embryonic forms.[5] Among mammals, while some species are clearly more self-conscious and more intelligent than others, the authors are less confident stating that rodents differ in sentience from dogs or swine or even primates and cetaceans.

There are valid ethical and welfare reasons to want to replace monkeys in an experiment with pigs, to replace dogs with mice, even though all mammals are likely to be about equal in sentience. For example, a National Academies review of use of dogs at the U.S. Department of Veterans Affairs states: "Some animal research protocols may be less harmful when conducted in dogs than in other species such as pigs. Others might be the reverse. For example, anxiety associated with blood draws, ureteral catheterizations, or other procedures might be less burdensome for trained dogs than for pigs."[35]

Surv. It is generally considered preferable in research to choose "less sentient" species whenever possible. How does your IACUC evaluate the sentience of various species of animals?

- Rodents are less sentient than dogs, cats, or swine, which are less sentient than primates. 94/226 (42%)
- Mammals are considered to be equally sentient and more sentient than non-mammals. 15/226 (7%)
- All vertebrates (fish, amphibians, reptiles, birds, and mammals) are considered equally sentient. 54/226 (24%)
- We do not have a consistent way to assess sentience. 63/226 (28%)

13:15 What is meant by reduction alternatives?

Opin. Reduction includes all the efforts to lower the numbers of animals used to the minimum necessary for sound statistical analysis. This often means rethinking statistical tests to use just the number necessary for a statistically valid result. Reduction attempts may rely on refining the study, as when the use of healthier, more genetically homogeneous animals lowers in-group variability, allowing a smaller number of animals in each group. Reduction efforts may rely on careful planning of statistical analyses, use of power analyses or other methods for projecting the minimum number of animals needed, and use of historical controls, when appropriate. When concurrent controls are required (as is most typical), it means carefully thinking how many control groups are needed. Sharing tissues across several projects when euthanizing animals is another form of reduction.

Many breeding regimens generate more animals than will be used in experiments, especially if breeding for complex genotypes in heterozygous animals. This problem is compounded if multiple investigators at an institution are maintaining their own small colonies of the same strains in parallel. OLAW FAQ F2 requires tracking animal use, including the generation and culling of unwanted or unusable genotypes of rodents; thus, IACUCs should oversee breeding practices in the interests of reduction.

Reduction does not mean conducting underpowered studies with too few animals per treatment to yield robust, statistically valid results. Not only would this be a waste of the animals' lives, but also it can produce misleading information when such substandard studies are published.[36]

13:16 What is meant by refinement alternatives?

Opin. Refinement alternatives are the most varied, as they comprise all the myriad ways to rethink animal care and use to reduce the potential for pain or distress. Some examples are listed below. The authors recommend that investigators seeking refinements in their studies and committees reviewing protocols conduct a birth-to-death welfare inventory. To do this exercise, visualize every step in the animal's life from the initial point of contact (birth at or arrival into the facility) until the animal's euthanasia or departure from the institution. List every reasonable potential source of pain or distress, whether related to housing, routine husbandry, or the experimental procedures. Consider going beyond this, listing ways to enhance the animal's positive experiences during and after research. Each of those potential sources of pain or distress and opportunities to maximize well-being should then be addressed, whether in a targeted literature search; in consultation with peers, veterinarians, and animal care specialists; and/or in the conduct of pilot studies, and with as much creative thinking as the investigator and committee members can muster.

A small sampling of refinement possibilities includes:

- Choice of experimental endpoints that precede onset of disease or mortality
- Setting intervention endpoints for animals that develop problems during the study
- Improved use of anesthetics and pain treatments, including preventive pain treatment and multimodal analgesic regimens
- Using telemetry devices to replace tethers or rigid restraint devices

- Replacing open surgery with endoscopic techniques
- Providing supportive veterinary and nursing care
- Maintaining infection-free animal colonies
- Choosing enclosures that go beyond regulatory minimum standards, allowing animals to dig, run, climb, socialize, or hide as they are motivated to do
- Using positive reinforcement training so that animals cooperate with research procedures
- Certifying staff competence before allowing them to conduct experiments
- Seeking alternatives to killing animals when studies are completed

13:17 How do the "3Rs" of alternatives complement or conflict with one another?

Often, the 3Rs are synergistic, and it is not important to define changes as one or the other "R." For example, replacing live animals with cells in culture may still require some live animals as sources for cells, but at a reduced number. Controlling for some extraneous variability, such as clinical disease with adventitious viruses, in study populations is a refinement that improves animal health and reduces animal numbers.

The *Guide* (p. 5) cautions: "Refinement and reduction goals should be balanced on a case-by-case basis. In other words, reduction should not serve as a rationale for reusing an animal or animals that have already undergone experimental procedures, especially if the well-being of the animals would be compromised."

13:18 What is meant by the search for alternatives?

Reg. The *Guide* (p. 25) calls on investigators and IACUCs to consider "less invasive procedures, other species, isolated organ preparation, cell or tissue culture, or computer simulation," but it does not provide guidance or set standards for how to make or evaluate that consideration. The AWA (§2143,a,3,B; §2143,d,3; §2143,e,3) and the AWAR (§2.31,d,1,ii) go one step further by mandating that the investigator not just consider, but actively search for such alternatives, that the IACUC determine that the investigator has considered such alternatives, and, if identified, that alternatives will not be used, review how the investigator determined alternatives could not be used. The AWAR requirement for alternatives is limited to "procedures that may cause more than momentary or slight pain or distress to the animals." It is not simply a requirement to search for replacement alternatives.

Opin. An investigator searches for alternatives *to* animal research by looking for non-animal models that would replace the use of sentient animals and for alternatives *in* animal research by searching for ways to reduce animal numbers and to reduce animal pain and distress.

13:19 How can the IACUC evaluate the investigator's search for and consideration of alternatives?

Reg. Under the AWAR (§2.31,d,1,ii) and the AWA (§2143,a,3,B), the IACUC is required to ensure that "the principal investigator has considered alternatives to procedures that may cause more than momentary or slight pain or distress" and "has provided a written narrative description of the methods and sources, e.g., the Animal Welfare Information Center, used to determine that alternatives were not available."

The *Guide* (p. 25) is less detailed on this topic. It specifies that justification of animal numbers should be statistically justified, whenever possible. The availability of refinements and replacements are two of the topics that "should be considered in the preparation of the protocol by the researcher and its review by the IACUC."

Opin. Institutions must comply with AWA requirements for AWA-regulated species but should also recognize the limitations. IACUCs should not assume that the relevant information is to be found only in published, indexed articles. Very useful information can be found through consultation with others performing similar work and with veterinarians and animal care specialists who focus on laboratory animals. That said, the authors believe it would be a rare project in which consultation with experts was the sole or even primary route to alternatives and would advise most investigators to include the electronic resources described later.

Surv. What does your IACUC accept as an appropriate means of searching for alternatives to painful or distressful procedures? Select all that apply.

• Not applicable	10/226 (4%)
• A single database search	64/226 (28%)
• At least two database searches	155/226 (69%)
• Consultation with experts in the field	114/226 (50%)
• Attendance at scientific meetings or study sections	62/226 (27%)
• Other*	12/226 (5%)

- A written narrative from the PI describing prior work
- Listing of search terms and strategy used
- Consultation with the veterinarian
- I don't know

13:20 What sources, resources, and methods are useful when searching for alternatives to animal use?

Reg. The AWAR (§2.31,d,1,ii) mention the Animal Welfare Information Center (AWIC) as a useful resource but does not mandate its use.

Opin: AWIC provides detailed step-by-step guidance for constructing a comprehensive literature search for alternatives. Although designed for AWA-regulated species, the search strategies apply to any animals used in research. Some highlights:

- Ask targeted questions to guide selection of databases and keywords
- Search at least three databases
- Use multiple search strings and keywords (both scientific and 3Rs related terms)
- Include specific terms and synonyms as keywords (for example, "environmental enrichment" or "enriched housing" or "nesting material" or "bedding" or "cardboard" or "tube" or "wheel" or "toy").[37]

Appointing a librarian as an IACUC member or regular consultant can help establish an appropriate framework for literature searches and assist investigators who need assistance with a complex set of scientific questions. This additional committee expertise can provide a IACUC member continuing education that establishes a baseline level of knowledge about how to search for alternatives. IACUCs could also consider building AWIC's guidance into the protocol form or a pre-review checklist to ensure that targeted questions are addressed prior to IACUC review.

The IACUC cannot thoroughly review literature searches if the only information it receives are the names of the databases, the key words, and the dates of the search. The AWAR specify a *narrative*, which we believe should include discussion of what information the alternatives searches yielded, and why the protocol is in its current form. The AWAR require (§2.31,d,1,ii) that the narrative explain how the investigator determined "that alternatives were not available." Replacements are not always available, certainly, but reduction and refinement are always possibilities that must be considered in an alternatives search.

A word of caution: smaller institutions may not subscribe to all the major journals (print or electronic) that would allow an investigator to see more than just the abstracts of relevant publications. They can take the delays and costs of requesting access or skew their search solely to open-access journals that may not have the articles they most need for their alternatives search. A complete alternatives narrative should include a description of how the investigator managed this limitation.

Surv. If a database search is typically used by researchers/educators at your institution to help find alternatives to painful/distressful procedures, what does your IACUC require for a "written narrative description of the methods and sources used to determine that alternatives were not available?"

- Not applicable 22/225 (10%)
- Date of search(es), database(s) searched, and search terms used 65/225 (29%)
- Date of search(es), database(s) searched, search terms used, plus a description of what the search(es) produced 52/225 (23%)
- Date of search(es), database(s) searched, and search terms used, a description of what the search(es) produced, and rationale for rejecting some/all of the alternatives that the search produced 82/225 (36%)
- Other 4/225 (2%)

13:21 Must the IACUC ensure that the animal use activity does not unnecessarily duplicate previously conducted work? If so, how is this done?

Reg. The AWAR (§2.31,d,1,iii) require that "the principal investigator has provided written assurance [to the IACUC] that the activities do not unnecessarily duplicate previous experiments." The *Guide* (p. 26) lists unnecessary duplication of experiments as a topic the investigator should consider in preparation of the protocol for IACUC review. Neither the AWAR nor the *Guide* require that the IACUC make this assessment directly.

Opin. The IACUC must not solely accept the investigator's claim that the proposal does not entail unnecessary duplication but apparently should review how the investigator knows this to be true. Though the search strategy for non-duplication differs from the search for alternatives, a simple approach to documenting compliance is to include a statement on the investigator's signed protocol application that the work does not entail unnecessary duplication; in addition, the documentation for an alternatives search should include information on how that search would uncover unnecessary duplication.

The key word here is *unnecessary*. Duplication and replication are essential components of hypothesis-driven research and are permitted if they serve a valid scientific purpose, such as confirming another's work or validating laboratory assay systems. To determine whether the duplication is or is not necessary, the IACUC needs to evaluate the justification provided by the investigator who proposes the duplication. In addition to evaluating if a project unnecessarily duplicates others' works, IACUCs must evaluate whether an investigator's plans (or a journal's requirements) for in-lab replication is acceptable.

13:22 Is the use of animals in teaching protocols considered necessary duplication?

Opin. There are no clear federal laws, regulations, or policies pertinent to this question. A genuine consideration of alternatives, not a simple statement that the protocol is for teaching, is warranted. The important question is not whether duplication (i.e., teaching a new cohort of students using procedures employed in previous semesters) is necessary, but whether animal use, especially invasive animal use, is necessary. Careful alternatives search on this topic would include consideration of published studies on the educational effectiveness of live animal versus other teaching methods and consultation with faculty at other institutions who are employing alternative methods. Though animal laboratories may be retained in the curriculum, refinements and reductions may be appropriate. These could include videotaping some animal laboratories to show future generations of students, requiring preliminary skills development with nonliving models prior to handling live animals, or delegating the most challenging part (e.g., pithing or anesthetizing an animal) to an experienced professor rather than an inexperienced roomful of students.

13:23 Is a requirement by a funding or regulatory agency to use animals, or a manuscript reviewer's requirement for additional experiments for a publication, considered adequate justification?

Reg. The USDA's animal care inspection guide states: "Protocols must meet the following criteria. … Provide a rationale for the number of animals to be used [2.31(e)(2)].

Examples of rationale that may be used include, but are not limited to, as required by the Food and Drug Administration (FDA) or other federal agency (federal code, regulation, or standard, etc., should be cited) [or] Required by international testing requirements (code, regulation, standards, etc., should be cited)."[28]

NIH/OLAW has posted guidance clarifying the relative roles of Scientific Review Groups (SRGs) serving NIH funding agencies and IACUCs.[38] Briefly, if the SRG raises a specific "concern" about the vertebrate animal section of a grant application or contract proposal, the NIH informs the applicant of that concern. The applicant must resolve the concern in conjunction with the NIH Program Officer and OLAW. This guidance may be relevant for manuscript reviews of PHS-funded projects, in which the journal requires experiments that were not initially in the grant application or the IACUC approval. Once that happens, the PI should convey to the IACUC any modifications related to animal usage that may have emanated from the NIH or the journal review. Only in select instances are product safety tests mandated by federal agencies as part of PHS-supported grants or contracts, so the posted NIH guidance is less relevant to such situations.

Opin. Adjudication as to what is proper use of animals is not dictated *solely* by granting or regulatory agencies or by journal reviewers. The institution, through its IACUC, must determine whether the justification for using animals is appropriate. If a granting or regulatory agency requires the use of animals, or a particular species or strain, their reasons for doing so should be submitted to the IACUC for its consideration. However, the IACUC must base its decision on the parameters previously noted, not just on the policy of the granting agency. Other details of a protocol may also be mandated either by a granting agency or by a regulatory agency that will review the data. For example, approval of veterinary vaccines may require submission of vaccine trial data that use death as an end point. Though efficacy may be well demonstrated long before mortality, or even before significant morbidity, animal lives would be wasted on such a study if the efficacy data would not be accepted by the Department of Agriculture, which licenses such vaccines. Approving such a study therefore means approving potentially painful procedures when alternatives have been rejected not for scientific but for regulatory reasons.

There may be occasions when the IACUC finds itself in a bind, asked to approve agency-mandated procedures that the IACUC considers unjustified. An IACUC might strike a balance in which it approves some animal use but also initiates its own correspondence to persuade the agency in question to change its mandate or, at least, to better explain and justify it. In such a situation, it seems the IACUC may have little choice but to defer to the agency requirements, even if it disagrees with the agency on the justification.

13:24 Are there any federal requirements dictating that animals or certain species of animals must be used in certain forms of research or product safety testing?

Reg. The AWA and the HREA do not include mandates for animal testing of any sort.

Several federal agencies, including the FDA and the EPA, are responsible for oversight of safety testing for human and animal drugs, medical devices, biologics, pesticides, and other compounds. For the most part, these agencies provide general guidance on the types of testing data required for licensing a product. Guidance documents may reference traditional animal models, or those used to support previous approvals, but do not contain any specific requirements to use animals. Though their regulatory requirements for validation of alternative methods may slow recognition of some non-animal assays, these federal agencies nonetheless have institutional commitments to replace animal tests with non-animal tests when possible. The IACUC can promote development and validation of alternative methods through comprehensive literature searches, consultations with other experts in the field, and review of agency correspondence with investigators to better understand options for regulatory flexibility.

The 2022 FDA Modernization Act 2.0 gives congressional authority to the Food and Drug Administration to consider non-animal safety and efficacy data. The FDA can accept

1. Cell-based assays
2. Organ chips and micro-physiological systems (MPS)
3. Computer modeling, i.e., in silico research
4. Other nonhuman or human biology-based test methods e.g., human tissue testing
5. Animal models[39]

Surv. Is a federal requirement to use animals for a safety test considered to be adequate justification for your IACUC to approve animal use?

- Not applicable 54/225 (24%)
- Yes 47/225 (21%)
- No 35/225 (16%)
- Depends on the "requirement" because some requirements are actually suggestions 57/225 (25%)
- I don't know 28/225 (12%)
- Other 4/225 (2%)

13:25 Since the AWA definition of an animal includes dead animals (see 13:1), should the IACUC review a justification for the use of dead animals?

Reg. The PHS Policy does not require IACUC review and approval of the use of dead animals, unless they are being killed for use in PHS-supported activities. In such a case, the IACUC is really reviewing the use of live animals being euthanized for research, teaching, or testing and the focus is on animal care and handling before and during euthanasia, as well as the justification that there is merit in the planned use of the cadaver or tissues, and that the numbers are appropriate to the task at hand.

While the statutory definition of *animal* includes dead animals used or intended for use in research, the definition of a *research facility* is one using or intending to use live animals (AWAR §1.1, Research Facility). Therefore, reference to committee review of use of animals applies solely to use of live animals, though the actual regulations specify "animals," not "live animals" (§2.31). Facilities are nonetheless required to maintain acquisition and disposition records of any dead dogs or cats—but only dogs or cats—it receives.[39]

Opin. From a legal point of view, justification does not appear to be required. It is an excellent application of alternatives to make maximal use of euthanized animals by using them as teaching cadavers or distributing needed tissues to several investigators, rather than requiring each to purchase animals separately. IACUC review could slow the process and increase the burden of in-house tissue sharing with no improvement in animal welfare and should not be required. A middle position between total oversight and no oversight would be to allow in-house tissue sharing without case-by-case IACUC oversight but to require oversight if dead animals are being received through purchase or donation from outside the institution. Professors who use animal cadavers in classrooms might want the prior approval and support of the IACUC.

Health and safety oversight is necessary if the cadaveric material comes from animals likely to be harboring zoonotic infections (e.g., macaques, sheep, and wild animals) or from animals on infectious disease or other hazardous studies. State and federal permits to possess cadavers or parts may be necessary; oversight of permits may be assigned to the IACUC by the institution but is not a federally mandated IACUC role.

13:26 How much detail should the IACUC require relative to strain and genotype in the IACUC protocol? Is it sufficient to simply list the species?

Reg. The *Guide* (p. 25) lists choice of species, not strain, stock, genotype, or phenotype, as an object of IACUC review. Other sections of the *Guide* (pp. 27, 31, 121) note times when the strain or stock of animals should be considered and note (p. 75) that this strain, stock, and genotype information should be on cage cards. The AWAR (§ 2.31,e,1 and 2) require "identification of the species" and "rationale … for the appropriateness of the species" in animal protocols, but no further description of strain or breed.

Opin. IACUCs should ask for more detailed information on the strains, breeds, and genetic lines the investigator plans to use, though a detailed list of genotypes on its own may be of limited use. An investigator should identify strains and lines prone to unhealthy genotypes and needing special care versus those that should remain healthy through life. Knowing the expected phenotypes can help focus the search for refinements.

If the investigator plans to compare several genotypes in an experiment, the IACUC should understand the rationale, which can guide efforts at reduction.

13:27 Should the IACUC request a justification for euthanasia of animals?

Reg. The AWAR (§1.1 Euthanasia; §2.31,d,1,i; §2.31,d,1,v), PHS Policy (IV,C,1,c and g) and the *Guide* (pp. 123–124) require the IACUC to review the proposed method of euthanasia and stress the need to minimize pain and distress. The IACUC must ensure that methods of euthanasia produce rapid unconsciousness and subsequent death without evidence of pain or distress, unless scientifically justified, and the Attending Veterinarian should provide guidance regarding euthanasia as a component of adequate veterinary care (AWAR §2.33,b,4). Euthanasia methods likely to cause animal pain or distress must be explicitly approved by the IACUC. Beyond this, there is no special justification required for killing animals, though the numbers of animals killed must be justified. While euthanasia need not be justified to the IACUC, refusal to euthanize animals that experience untreatable pain or distress does require very special justification. The AWA (§2143,A,3,c,v) requires "that the withholding of tranquilizers, anesthesia, analgesia, or euthanasia when scientifically necessary shall continue for only the necessary period of time," and, as previously discussed, these animals must be reported under Category E on the facility's annual report. U.S. Government Principle VI states: "Animals that would otherwise suffer severe or chronic pain or distress that cannot be relieved should be painlessly killed at the end of the procedure or, if appropriate, during the procedure."

In states with dog and cat adoption laws, healthy animals must be offered for adoption when their research use has ended. OLAW FAQ F.11 covers many of the considerations relevant to adopting out animals purchased for OLAW-covered projects:

> "The PHS Policy and the *Guide* are silent on the issue of private adoption of research animals for pets after a study has been completed and the animals are no longer required. For purposes of property standards under the uniform administrative requirements for HHS grant awards, animals purchased for research purposes are considered to be supplies. Generally, the title to supplies vests with the grant recipient upon acquisition (2 CFR § 200.314). OLAW supports the safety and protection of animals and reminds institutions that their policies must clarify the disposition of animals acquired for research once the research has ended, which may include adoption. The PHS will not assume legal or financial responsibility for any adoption program, or any matters arising from or related to the adoption of research animals. The institution should ensure that its policy meets pertinent state and local laws and regulations for the transfer of animal ownership and is encouraged to coordinate with local animal shelters."[40]

Opin. While not a source of pain or distress, many consider death to be a harm to animals insofar as it deprives them of potential positive experiences of living their lives. We recommend that IACUCs support well-run, responsible, post-use retirement and adoption programs. Coordinating an adoption program is not typically an IACUC role, but the IACUC can contribute by using a protocol form that lists adoption and retirement as options alongside euthanasia. Even rodent and rabbit protocols might include IACUC approval of allowing adoption of animals and reviewing criteria of which animals might be suitable adoption candidates.

13:28 Should the IACUC request justification for the use of animals in field studies?

Reg. The AWAR (§1.1, Field Study) state that a field study is "a study conducted on free-living wild animals in their natural habitat. However, this term excludes any study that involves an invasive procedure, harms, or materially alters the behavior of an animal under study." Under the AWAR (§2.31,d,1), animals involved in activities meeting that definition of field studies are exempt from IACUC review, but they require the same standards as laboratory-based research if invasive techniques will be performed or the animals' behavior is altered. USDA has a Tech Note summarizing the requirements for IACUCs working with field biologists.[41]

PHS Policy allows the investigator to submit basic information about a proposed field study. FAQ A6 states: "if the IACUC determines that the proposed activity is purely observational *and* will not alter or influence the biology, behavior or ecology of the study animals or other species, IACUC review and approval is *not* required." Otherwise, the IACUC must provide oversight of field studies, including purely observational work, if it has the potential to impact human health and safety or to impact the observed animals, the local environment, or other species.[42]

The *Guide* (p. 32) lists field investigations as one of the eight topics that are "the most common requiring special IACUC consideration" and calls attention to special welfare concerns (including return to wild or other disposition), to permitting requirements, and to human health and safety concerns.

Opin. Field investigations vary from hands-off observation to capturing, holding, marking, surgically altering, or tissue collection. If invasive procedures occur in field studies, the AWAR exemption does not apply, and these activities must be reviewed and approved by the IACUC. "Materially altering" animal behavior is left undefined, including duration of any alteration. Studies that require capture and brief restraint for marking or tagging animals may not be invasive, but they may affect animal behavior (especially if anesthesia is used), and the marking or tagging method may itself affect the animals in various ways. Depending on the species and season, even walking through a habitat or playing a recorded call to observe behavior could adversely affect animals. Vertebrate animals in PHS-funded field studies are not excluded from IACUC consideration in institutions covered by the PHS Policy, regardless of the invasiveness or behavioral effects of the study. For consistency, it is a good policy to grant the same review of justifications for field studies as for non-field studies. Institutional officials should also remember that many tagging systems identify the institution placing the tag and will want a record that the study was duly reviewed if animals with problems are later found and identified as institutional research subjects. The authors would recommend that the IACUC, not just the investigator, determine whether a particular field study might pose significant risk to animals to warrant IACUC review. If the committee is insufficiently familiar with the species and the experimental methods, it should consult appropriate, knowledgeable experts.

13:29 Must the IACUC review and approve the use of animals destined to be used as a food source for other animals being used in IACUC-approved studies?

Reg. The AWAR and PHS Policy do not explicitly call for IACUC approval for animals maintained as dietary items for other animals.

Opin. Studies of animal predation that involve feeding live or euthanized animals to others are, of course, research projects that must receive IACUC review and approval. In addition to the welfare of the prey species, it is essential to rule out their potential as infectious vectors as well.

Institutions may decide to include animals used for food in an SOP, on their own IACUC protocol, or on the protocol for the species that consumes these animals. The *Guide* (p. 91) advises, "Live food sources need to be maintained and managed to ensure a steady supply and the health and suitability of the organism as a food." Though the *Guide*'s examples are all invertebrates, such as crickets or brine shrimp, the same principle should apply to vertebrates fed as live food.

13:30 Must the IACUC review other non-research uses of live vertebrates, such as sentinel programs, breeding programs, or as blood donors?

Reg. Yes. NIH OLAW's posted guidance says, "Although animals used as sentinels, breeding stock, blood and blood product donors, or for other similar purposes may not be part of specific research protocols, their use is part of the institutional research program and directly supports research activities. Consequently, the IACUC should review protocols and SOPs that involve animals for such purposes, initially and at appropriate intervals (at least once every three years)."[43]

REFERENCES

1. NIH Office of Laboratory Animal Welfare. 2011. *FAQ A.5: Does the PHS Policy Apply to Larval Forms of Amphibians and Fish?* Accessed September 15, 2021. Available from: https://olaw.nih.gov/faqs#/guidance/faqs?anchor=50289.
2. Moorman, S.J. 2001/ Development of sensory systems in zebrafish (Danio rerio). *ILAR J,* 42(4): 292–298.
3. AAALAC International. 2021. *FAQ: Animals Included in the AAALAC International Accredited "Unit" 2. Invertebrates.* Accessed February 20, 2021. Available from: https://www.aaalac.org/accreditation-program/faqs/#A2.
4. National Institues of Health. 2023. *Cephalopods in Research.* Accessed May 10, 2024. Available from: https://olaw.nih.gov/guidance/cephalopods-in-research#:~:text=Though%20cephalopods%20are%20not%20covered,use%20of%20all%20animals%20in.
5. Mellor, D.J. 2019. Welfare-aligned sentience: Enhanced capacities to experience, interact, anticipate, choose and survive. *Animals (Basel)* 9(7).
6. NIH Office of Laboratory Animal Welfare. 2022. *Frequently Asked Questions: May Cost Be Used as Justification for Not Implementing Animal Welfare Standards?* Accessed October 31, 2022. Available from: https://olaw.nih.gov/faqs#/guidance/faqs?anchor=questionhttps://olaw.nih.gov/guidance/faqs.
7. Association of Primate Veterinarians. 2022. Association of Primate Veterinarians Lifetime Use Guidelines for nonhuman primates in biomedical research. *J Am Assoc Lab Anim Sci* 61(4): 320–321.
8. National Academies of Sciences, E., and Medicine. 2023. *Nonhuman Primate Models in Biomedical Research: State of the Science and Future Needs.* Washington, D.C.: National Academies Press.
9. Eissinger, R. 2022. *Panel 1: NIH Perspectives.* Accessed April 1, 2024. Available from: https://www.nationalacademies.org/event/08-25-2022/docs/DAEBA66EF288778A40DC37CDCEECEA999A3086EE1C65?noSaveAs=1.
10. United States Interagency Research Animal Committee. 1985. Principles for the Utilization and Care of Vertebrate Animals Used in Testing, Research, and Training. *Federal Register* 50(97): 20864–20865.
11. Prentice, E.D. 2021. *OLAW Online IACUC Staff Seminars: Ethics and IACUC Responsibility.* Accessed October 15, 2022. Available from: https://olaw.nih.gov/sites/default/files/100610_seminar_transcript.pdf.
12. National Air and Space Administration. 1996. *NASA Principles for the Ethical Care and Use of Animals*, p. 25.
13. Beauchamp, T. and D. DeGrazia. 2019. *Principles of Animal Research Ethics.* New York: Oxford University Press.
14. Carbone, L. 2019. Ethical and IACUC considerations regarding analgesia and pain management in laboratory rodents. *Comp Med* 69(6): 443–450.
15. AAALAC International. 2022. *Harm-Benefit Analysis.* Accessed December 13, 2021.2cited 2018 December 13,]; Available from: https://www.aaalac.org/accreditation/faq_landing.cfm.
16. Laber, K., et al. 2016. Recommendations for addressing harm-benefit analysis and implementation in ethical evaluation—Report from the AALAS-FELASA Working Group on Harm-Benefit Analysis – Part 2. *Lab Anim* 50(1 Suppl): 21–42.

17. Grimm, H., I.A.S. Olsson, and P. Sandoe. 2019. Harm-benefit analysis—What is the added value? A review of alternative strategies for weighing harms and benefits as part of the assessment of animal research. *Lab Anim* 53(1): 17–27.
18. Niemi, S.M. 2020. Harm-benefit analyses can be harmful. *ILAR J* 60(3): 341–346.
19. Makowska, I.J. and D.M. Weary. 2020. A good life for laboratory rodents? *ILAR J* 60(3): 373–388.
20. AVMA Panel on Euthanasia. 2020. *AVMA Guidelines for the Euthanasia of Animals: 2020 Edition*. Available from: www.avma.org/KB/Policies/Documents/euthanasia.pdf.
21. Nunamaker, E.A., et al. 2021. Developing recommendations for cumulative endpoints and lifetime use for research animals. *Animals (Basel)* 11(7).
22. Gaskill, B.N. and K. Gouveia. 2020. "The Mouse," in *Animal-centric Care and Management*, B. Gaskill and S. Cloutier, eds. CRC Press, pp. 103–119.
23. Cait, J., et al. 2022. Conventional laboratory housing increases morbidity and mortality in research rodents: Results of a meta-analysis. *BMC Biol* 20(1): 15.
24. American Veterinary Medical Association. 2003. *Veterinarian's Oath (Updated November 1999)*, in *AVMA Membership Directory and Resource Manual*. American Veterinary Medical Association, p. 398 (insert).
25. Olsson, I.A.S., et al. 2020. From unpleasant to unbearable: Why and how to implement an upper limit to pain and other forms of suffering in research with animals. *ILAR J*.
26. NIH Office of Laboratory Animal Welfare. 2022. *Frequently Asked Questions: Is the IACUC Responsible for Judging the Scientific Merit of Proposals?* Accessed October 31, 2022. Available from: https://olaw.nih.gov/faqs#/guidance/faqs?anchor=questionhttps://olaw.nih.gov/guidance/faqs.
27. Bateson, P. 1986. When to Experiment on animals, in *New Scientist*. 1986. p. 30–32.
28. United States Department of Agriculture. 2021. *Animal Welfare Inspection Guide*. [Includes the annual report form, and instructions, and pain categories.] Accessed September 12, 2022. Available from: https://nationalawa.org/files/Animal_Care_Inspection_Guide_Jan_2021.pdf.
29. Carbone, L. and A. Shriver. 2023. Ethical considerations in anesthesia and analgesia in laboratory animals. In *Anesthesia and Analgesia in Laboratory Animals*, 3rd ed. P.J. Melissa Dyson, Jennie Lofgren, Elizabeth Nunamaker, Daniel Pang, eds. Academic Press, pp. 3–17.
30. U.S. Department of Agriculture, A.a.P.H.I.S. 2024. *Research Facility Annual Usage Summary Report*. Available from: https://www.aphis.usda.gov/awa/research-facility-report/annual-summary.
31. Committee on Recognition and Alleviation of Pain in Laboratory Animals. 2009. *Recognition and Alleviation of Pain in Laboratory Animals*. Washington, DC: National Academies Press, p. 270.
32. Russell, W.M.S. and R.L. Burch, 1959. *The Principles of Humane Experimental Technique*. 1959. Methuen & Co., p. 238.
33. National Institues of Health, O.o.L.A.W. Undated. *FAQ D.7 Should the IACUC Consider the Three "Rs" of Alternatives When Reviewing Protocols? (Refinements to Research, Reduction of Animal Numbers, and Replacement with Non-Animal Models)*. Accessed May 15, 2024. Available from: https://olaw.nih.gov/faqs#/guidance/faqs?anchor=50319.
34. Tannenbaum, J. and B.T. Bennett. 2015. Russell and Burch's 3Rs then and now: The need for clarity in definition and purpose. *J Am Assoc Lab Anim Sci* 54(2): 120–132.
35. National Academies of Sciences, E., and Medicine. 2020. *Necessity, Use, and Care of Laboratory Dogs at the U.S. Department of Veterans Affairs*.
36. Advisory Committee to the Director, N. *ACD Working Group on Enhancing Rigor, Transparency, and Translatability in Animal Research*. 2021. Accessed November 4, 2022. Available from: https://acd.od.nih.gov/documents/presentations/06112021_RR-AR%20Report.pdf.
37. National Agricultural Library, A.W.I.C. 2022. *Literature Searching: How to Find Animal Use Alternatives*. Accessed November 6, 2022. Available from: https://www.nal.usda.gov/services/literature-searching-animal-use-alternatives#toc-step-1-identify-areas-to-potentially-implement-alternatives.
38. National Institutes of Health, O.o.L.A.W. 2021. *Vertebrate Animals Section*. Accessed April 1, 2024. Available from: https://olaw.nih.gov/guidance/vertebrate-animal-section.htm.
39. Adashi, E.Y., D.P. O'Mahony, and I.G. Cohen. 2023. The FDA Modernization Act 2.0: Drug testing in animals is rendered optional. *Am J Med* 136(9): 853–854.
40. National Institutes of Health, O.o.L.A.W. Undated. *FAQ F.11 Can IACUCs Authorize the Adoption of Research Animals as Pets after the Animals Are No Longer Needed for Study?* Accessed May 15, 2024. Available from: https://olaw.nih.gov/faqs#/guidance/faqs?anchor=50374.
41. USDA APHIS Animal Care. 2024. *Research with Free-Living Wild Animals in Their Natural Habitat and the Animal Welfare Act*. Accessed April 5, 2024. Available from: https://www.aphis.usda.gov/sites/default/files/ac-tech-note-research-free-living-wild-animals.508.pdf.
42. National Institues of Health, O.o.L.A.W. Undated. *FAQ A.6 Does the PHS Policy Apply to Animal Research That Is Conducted in the Field?* Accessed May 15, 2024. Available from: https://olaw.nih.gov/faqs#/guidance/faqs?anchor=50290.
43. NIH Office of Laboratory Animal Welfare. 2022. *Frequently Asked Questions: Is IACUC Approval Required for the Use of Animals in Breeding Programs, as Blood Donors, as Sentinels in Disease Surveillance Programs, or for Other Non-research Purposes?* Accessed October 31, 2022. Available from: https://olaw.nih.gov/faqs#/guidance/faqs?anchor=question50331.

14

Justification of the Number of Animals to Be Used

Ed J. Gracely

Introduction

Using an appropriate number of subjects is an important part of any research project. A proper sample size is essential to obtaining valid results and to minimizing the number of individuals exposed to the potential risks and harms of research. For this reason, the IACUC is interested in ensuring that a sufficient but not excessive number of animals are used, consistent with the 3 R's. A study using too few animals is ethically problematic because the animals are being subjected to potentially painful procedures or loss of life with relatively little likely benefit to the advancement of knowledge. On the other hand, a study using more animals than are truly needed is also problematic, because it unnecessarily exposes some of them to the same harms.

This chapter discusses key issues related to IACUC review of animal numbers. The existing regulations and policies are based on the use of performance standards empowering IACUCs to use professional judgment in weighing questions about animal numbers. Although IACUCs differ on many of these issues, experience, common sense, and the results of an informal survey all help provide direction. The survey results emanate from a questionnaire sent primarily to academic institutions.

14:1 What do the AWAR and PHS Policy state with respect to justifying the number of animals requested in the protocol?

Reg. The AWAR (§2.13,e,1; §2.31,e,2) state that the proposal to use animals must include "identification of the species and the approximate number of animals to be used" and "a rationale for involving animals and for the appropriateness of the species and numbers of animals to be used." U.S. Government Principle III states: "The animals selected for a procedure should be of an appropriate species and quality and the minimum number required to obtain valid results." The PHS Policy (IV,D,1,a; IV,D,1,b) also states that all applications and proposals that are submitted to the PHS and that involve animals must include "identification of the species and approximate number of animals to be used" and "rationale for … appropriateness of the species and numbers used."

The *Guide* (p. 25) states that "whenever possible, the number of animals and experimental group sizes should be statistically justified (e.g., provision of a power analysis…)." This is the most specific statement of how animal numbers should be justified in any of these documents. The *Guide* also notes that, "hypothesis testing, sample size, group numbers, and adequacy of controls can relate directly to the prevention of unnecessary animal use or duplication of experiments." The Worksheet for Review of the Vertebrate Animal Section (VAS) 2017 at the NIH/OLAW has largely removed prior details about animal number justifications and now states: "The justification for the number of animals, sex, ages, and other biological variables should be described in the Research Strategy."[1]

NIH/OLAW FAQ F.2 adds a bit more detail.

> "Although the PHS Policy does not explicitly require a mechanism to track animal usage by investigators, it does require that proposals specify a rationale for the approximate number of animals to be used and be limited to the appropriate number necessary to obtain valid results.[2] This implicitly requires that institutions establish mechanisms to document and monitor numbers of animals acquired and used, including any animals that are euthanized because they are not needed. Monitoring should not exclude the disposition of animals inadvertently or necessarily produced in excess of the number needed or which do not meet criteria (e.g., genetic) established for the specific study proposal. Institutions have adopted a variety of administrative, electronic, and manual mechanisms to meet institutional needs and PHS Policy requirements."

Opin. The phrase "approximate number" appears in both the AWAR and the PHS Policy, without further explanation. All animal number estimations are approximate in some sense. Animals die, procedures fail, unexpected events occur, and new research directions emerge. Furthermore, it is always possible to amend a protocol to deal with

DOI: 10.1201/9781003272861-14

such eventualities or with data more variable than had been anticipated, thus requiring additional animals per group. If the wording in the law and the policy is interpreted as primarily intended to encompass these sorts of uncertainties, then that wording is consistent with requiring an initial estimate that is specific but understood implicitly to be approximate.

It is possible that the regulations were intended to allow a broader type of flexibility (such as "from 200 to 250 animals will be required"). As this type of request is not prohibited, if the IACUC finds it to be reasonable and scientifically justified, then it may approve an approximate number of animals.

NIH/OLAW FAQ F.2 makes clear that researchers cannot decide to monitor and record only animals that are actually used in studies. Animal welfare regulations (and ethical considerations) apply to any animal produced in the process of doing the research, even if it has the wrong genotype or sex, or is merely surplus.[2] The IACUC is responsible for ensuring that these animals are accounted for and humanely disposed of if they are not used.

14:2 What constitutes sufficient justification for a requested number of animals?

Reg. See 14:1.

Opin. First, note that almost all animal protocols require some justification for animal numbers. This obviously includes not only those involving actual group comparisons and statistics, but also those involving students and teaching, breeding protocols, *in vitro* work, and wildlife studies. Even pilot studies—why did you want six pilot animals, rather than four or eight? Of course, the form of the justification will vary by design.

Institutions may want to develop their own Standard Operating Procedures guidelines for number justification.

As for official guidelines, the AWAR and PHS Policy do not address this issue. Thus, it is left to the institution to develop appropriate policies. The IACUC must be able to determine how all the requested animals will be utilized, which research questions the number to be used will potentially answer, and why those questions could not be answered with fewer animals or no animals at all. Describing animal use is generally straightforward, except in studies that are open ended and have evolving goals. In the latter instance, the investigator should attempt to provide a sequence of events and a desired endpoint. A justification of numbers based solely on an ongoing need for a certain number of animals per year without an explanation of what is likely to be accomplished in each specific period is not acceptable.

An interesting set of three comments by researchers and veterinarians on the justification of numbers for an open-ended study has been published.[3–5] Four invited commentators (in three articles) responded to a scenario in which an established scientist planned a three-year series of studies requiring a total of 5,500 mice. The project had already been funded (pending IACUC approval) by the NIH. The discussion centered on the fact that the justification of numbers is brief for such a large number of animals, being basically a statement of animal needs per week multiplied by 52 weeks x 3 years = the total number. To provide some context, note that an APHIS/AC reviewer who read the present chapter (Chapter 14) before publication agreed with the chapter's author: "This justification would not be adequate for AWAR purposes. A scientific justification is needed; i.e., an explanation of why x number of animals per week is needed." Interestingly, only one of the responses to the scenario (the last one) took a similar position. One commentator was fairly satisfied with the justification, especially given that the NIH grant review panel was satisfied. This respondent wanted only a few examples of the kinds of studies to be done. The second comment (written by two respondents) listed a variety of questions, mainly about ways the researcher might reduce the required animals or better justify the weekly details (for example, could more cells be obtained from each animal by using improved techniques?). It was not clear whether this pair of commentators would have accepted the unspecified research to be done if these more practical questions had been satisfactorily answered. The third respondent took objection to the whole process, arguing that if the researcher could not explicitly lay out the studies to be done over three years, then approval should be requested only for that series that could be planned, with amendments to request more animals as needed. Clearly, this topic generates strong and diverse opinions!

Justifying the specific animal numbers for individual research questions is often difficult as well, and there is no perfect way to do it. Previous research, while sometimes helpful, is only a crude guide to the number of animals needed in the present study. Even statistical power analysis requires assumptions (such as the magnitude of effects worth looking for) that are often subjective or that must themselves be estimated from other data.

This author believes that sample size justification should be seen to ensure that researchers and IACUCs are aware of the need for an adequate but not excessive sample size and are attempting to prevent serious miscalculations. A moderate amount of detail is sufficient to accomplish this. Consider three specific approaches, each of which may be adequate when it is applicable:

- Report of a power analysis, with enough information provided to show that the researcher knows how to analyze the data and how to use power analysis techniques. Question 14:5 lists key considerations of a thorough power analysis. A briefer approach may be acceptable, but this author's experience suggests that attempts at an abbreviated power analysis often produce descriptions that are garbled or lack key information.
- Citation of previous research, with sufficient information provided to indicate that the previous research is similar enough in concept and methodology to the present proposal as to make it a reasonable model for sample sizes in the latter.
- Derivation of animal numbers from material needs, when the study has no statistics (e.g., histological characterization of tissues), with a clear indication why the specific amount of material is needed, and why the number of animals requested is appropriate to provide that amount of material.

One of the survey questions asked if IACUCs view justification of animal numbers as part of their responsibility:

Surv.t Does your IACUC regard reviewing sample size (animal numbers) justification as part of its responsibility?

- Yes, we review animal number justifications and may critique their content. 176/222 (79%)
- Yes, but we primarily ensure that a justification is provided. We let scientific reviewers verify the justification if there is a grant or contract. 36/222 (16%)
- No. It is a scientific question outside of the purview of the IACUC. 4/222 (2%)
- Our decision to review or not depends on the funding source or species of animal being used. 2/222 (1%)
- Other. 4/222 (2%)

Clearly, the large majority (79%) believe it is their responsibility to verify and validate the sample size justification. A smaller number (16%) merely checks for the presence of a justification, without checking its validity. Only a few do not do either.

Opin. Just verifying the presence of a justification, as 16% do, may be problematic if it is meant literally. PI's make numerous, fixable, errors in their protocols, like stating that beta was set to 80% (they mean power = 80%), or in which no group sample size is offered (just a total sample size). Attrition calculations can be done incorrectly. And sometimes a protocol states, say, that the behavioral studies need 10 per group, but one of the behavioral studies has 12 or 8. Basic error checking is essential and does not constitute questioning the science.

How much further to go is a matter of judgment, based partly on the type of review the protocol will receive and the expertise available on the IACUC.

Another survey question asked about studies that do not require any statistical analysis.

Surv.t Some protocols may not require any statistical analysis (e.g., histological analysis). When presented with such a protocol, what is the minimum your IACUC will accept for sample size justification?

- Few details, such as a statement that "n" animals are needed to provide enough tissue for all histological analyses 74/222 (33%)
- Simple calculation of the total amount of tissue needed and the amount available from one animal 66/222 (30%)
- A more detailed justification of the number of histological observations required and the amount of tissue available from one animal 71/222 (32%)
- Other 11/222 (5%)

Respondents clearly divided themselves along a range from basically taking the PI's word that, say "200 animals suffice" to a detailed calculation.

Opin. IACUC's should be prepared to answer the question, "How were you able to ensure that the animal numbers requested were appropriate and justified?" Would they be comfortable if the only justification they had was a non-specific statement that "n" were needed? Requesting some details beyond that seems prudent.

Is a power analysis required when appropriate? There is another relevant survey question for that.

Surv.t When the experimental design allows, does your IACUC request statistical justification (i.e., power analysis) for the number of animals being requested?

- Yes, almost always 136/222 (61%)
- Only sometimes 28/222 (13%)
- Not usually 11/222 (5%)
- We always leave the choice of how to justify animal numbers to the investigator 43/222 (19%)
- Other 4/222 (2%)

A simple majority generally require a power analysis if it is relevant, but close to 40% are less consistent if they require one at all.

Opin. Power analyses are the most sophisticated method of sample size justification, and one could argue that we should move in the direction of mandating them, but they are tricky to do well, require assumptions to be met, and a good understanding of the statistical model to be used. Power analyses are clearly the gold standard and should be employed whenever the researcher's understanding of the effect sizes in a complex design plus technical expertise allow them. On the other hand, a good, well-described, and delineated justification of sample size based on prior research may be more reliable and believable than a poorly done power analysis, especially for a complicated design.

Several other survey questions touch on aspects of this issue and clarify how IACUCs approach the problem:

Surv.t If an investigator uses inferential statistics to justify the number of animals in a particular protocol, what is the minimum statistical information your IACUC would usually accept?

- Details (e.g., test to be used, standard deviation, effect size, alpha, power, expected differences), plus the results of the power analysis. 61/222 (27%)
- A basically correct, brief statement of a power analysis with enough information to indicate competence in performing the analysis 112/222 (50%)
- We will accept most anything that looks like a statistical response 21/222 (9%)
- A simple statement that a power analysis has been done 17/222 (8%)
- Other 11/222 (5%)

The commonest answer does not require heavy details but does require a complete statement with a good sense that the PI knows what they are doing. But 17% accept almost anything! As above, that may be too little for proper evaluation.

A common alternative to power analysis is to cite relevant previous research, by the PI or others. Several survey questions address this topic.

Surv.t Would your IACUC routinely accept sample size justification in a statistical-type study that is based on a statement of previous experience of the investigator, and nothing else?

- Yes, if details of study comparability are provided 120/222 (54%)
- Yes, if that is what the investigator chooses to provide 28/222 (13%)
- No 60/222 (27%)
- Other 14/222 (6%)

Surv.t Does your IACUC accept sample size justification in a statistical-type study that is based on references to similar work in published literature?

- Yes, if adequate details are provided to demonstrate comparability 163/222 (73%)
- Yes, if that is what the investigator chooses to provide 36/222 (16%)
- No 18/222 (8%)
- Other 5/222 (2%)

Surv.t If your IACUC accepts sample size justifications based on references to similar work in the published literature, does any person check to see if the literature citation(s) actually supports the contention of the animal user?

- Yes, usually 94/217 (43%)
- Yes, occasionally 35/217 (16%)
- Yes, but rarely 35/217 (16%)
- No 24/21 (11%)
- I don't know 20/217 (9%)
- Other 9/217 (4%)

As with other questions, responses vary widely. Most IACUCs accept some version of previous work or published studies as a sample size justification, with published references preferred to previous work by the PI. The majority clearly want sufficient details provided.

Opin. In this author's experience, the amount of detail provided is often a weak spot. We get a list of references and a simple statement that the studies were similar, without details. Since a change of dependent variable, manipulation, or analysis could drastically affect the power, it would seem prudent to require fairly strict details and comparability. Thus, while previous experience and literature publications may, as noted above, sometimes be a good way to justify numbers, especially with designs for which a good power analysis would be challenging, the citation must be done *right.*

Note that fewer than 50% report that someone routinely checks the references to see if they actually support the sample size in the present protocol.

14.3 What issues arise if animals are added after a study is completed or if a study is terminated early?

How should the IACUC approach a scenario in which a study is almost statistically significant, albeit not quite, and the PI asks for a few more animals to try to achieve statistical significance?

First let's look at a relevant survey question.

Surv.t How does your IACUC approach an amendment request to add additional animals to a study for the purpose of reaching statistical significance if the original number approved fell just short of reaching significance? Select all that apply.

• We request statistical justification for the additional number requested.	121/217 (56%)
• We usually approve a request if it is for 5% or fewer animals.	68/217 (31%)
• We are stricter about approving animals covered by the Animal Welfare Act.	28/217 (13%)
We consider any such request to be a significant amendment.	63/217 (29%)
We typically consider such a request to be a minor amendment.	45/217 (21%)
We consider such a request to be scientifically inappropriate and do not allow it.	4/217 (2%)
Other.	21/217 (10%)

Opin: This is, of course, a problematic situation since adding a few extra subjects to a non-significant result can distort the p-value and lead to an increased risk of type 1 (false positive) error. Sample size should normally be decided in advance and stuck to. If you have a non-significant result and think it would be significant with a bigger sample size, a new study is called for, not a bit-by-bit increase in N until the desired results are obtained. Unfortunately, few IACUC's seem to understand this basic principle. Note that only 2% of respondents said that they would consider the request scientifically invalid, but this is statistically the correct answer! Well-planned sequential study designs in which stopping rules are established *a priori* are an exception to this rule.

Stopping early can also be problematic, for the same reason—it increases the probability of finding a significant result by chance.

In human subjects' research, it is common to design long studies with large sample sizes using sequential stopping rules. The researchers plan to perform interim analyses at pre-defined points and to stop the study if the results merit that. In some cases, the data is analyzed after every new subject is completed. Sequential testing is a specific procedure, with its own appropriate modifications to data analysis. For example, if the study is not stopped early, the analysis of the final dataset may require a smaller p-value to attain significance than would otherwise have been needed, in order to compensate for the multiple "looks" along the way.

A recent publication by Douglas Fitts[6] provides an easily applied mechanism for using such methods in small studies. For example, suppose a researcher does a preliminary power analysis which suggests 10 animals per group and decides that 6 per group is the smallest number worth trying, while 18 per group is the largest they would even consider. The study proceeds by doing the analysis after 6 animals have been completed in each group, and then after 1 more animal has been added to each group in turn. The trick is that the p-value for significance at each step is required to be less than or equal to 0.0165 (the new alpha level). Interestingly, if the p-value is 0.24 or greater at any step, the study also stops. The details of the significance and non-significance cutoffs vary with other decisions—see the reference for details. Such methods can be powerful and may, on average, use fewer animals than a single fixed sample-size model, but the PI must be prepared to continue to the end if neither stopping criterion is reached.

14:4 Relative to the number of animals used, are there different regulatory requirements for different species?

Reg. While the AWAR exclude mice, rats, and birds bred for use in research and cold-blood animals, the PHS Policy includes all vertebrate animals.

Opin. The definition of an animal by different regulatory authorities varies based on the species or where it was bred, but the definitions make no distinctions in terms of animal number justification. For example, there is no wording that distinguishes different phylogenetic levels of equally sentient animals in terms of animal number justification.

Researchers sometimes speak of wanting to use the "lowest phylogenetic level" that will answer a question as a method of replacement or refinement. However, other than a species (such as an insect or worm) that is presumably less subject to pain than "higher" animals, there is nothing in the regulations to mandate this, nor to suggest that more justification for a particular number of subjects is required for "higher" species than for "lower" ones among sentient animals.

One distinction of interest involves the study of endangered species as covered under the Endangered Species Act of 1973[7] and other relevant legal documents such as the Convention on International Trade in Endangered Species of Wild Flora and Fauna (CITES) treaty.[8] It seems likely that the total number of such animals to be studied, as well as the number to be euthanized, would be subject to stricter standards for endangered species than would otherwise prevail for other animals. Any IACUC dealing with such a situation will have to review all the relevant regulations.

Other than endangered species, there are few species-specific distinctions relative to justifying the number of animals proposed for use. From ethical, humanitarian, and regulatory standpoints, it is as important to justify the use of 100 mice properly as it is to justify the use of 100 dogs. Invertebrates are not included in the definition of an animal noted above, but it seems reasonable that certain apparently sentient and intelligent invertebrate groups (such as cephalopods) be given the same consideration as vertebrates. Naturally, the appropriate species for the research must be used.

Is there a level of animal intelligence that renders killing animals less justifiable? Is it important to provide more justification for the utilization or euthanasia of ten apes compared to ten mice? There are no such distinctions in the regulations (at least in terms of justifying the numbers of animals used), but implicit in all animal number justifications is the idea that the knowledge gained must be adequate to offset the impact on the individual animals involved. Each IACUC will need to determine whether highly intelligent species require special consideration in maintaining this balance. The statistical analysis is not affected, but the decision as to whether the research should occur at all may be.

14:5 What are the key considerations in using a statistical power analysis to justify animal numbers?

Opin. There are several key considerations when a researcher wants to perform and present a fully usable and appropriate power analysis. The researcher should have a thorough understanding of the appropriate statistical analysis for the design or must be willing to seek out and collaborate with someone who does. Much basic science research that uses statistics involves several groups or conditions and often several time points. Such designs require advanced analytic techniques, such as the analysis of variance (ANOVA) and specific follow-up tests (such as the Tukey test). In some instances, the major question involves the interaction of two independent variables, as tested in a two-way ANOVA.

These analytic decisions have important implications for power analysis. For example, a researcher with a complex design who estimates power merely for a two-group comparison may reach a substantially incorrect conclusion. Often, a thoughtful analyst can reduce a complex question to one for which sample sizes can be estimated with simpler methods, but this process requires a good understanding of the statistical issues.

The study design must be spelled out in sufficient detail for IACUC (or other) reviewers to understand it and to see how the power analysis is configured to match the analysis that will be employed. The researcher must present and justify the specific parameters and assumptions required by the power analysis. Typically, this will involve estimates of effect sizes (such as the mean difference between two groups) and estimates of variability, such as the standard deviation.

Analytic methods have different requirements that must be determined before a power analysis can be run. Prior experience, the theory underlying the research, and pilot data may be useful.

14:6 How should sample size for field studies be determined?

Opin. Field research is a different kind of research from lab research. In many cases (albeit not always), there is no control condition, no randomization of subjects, and a high degree of unpredictability as to what will be seen.

In the present author's experience, wildlife researchers often argue that such considerations make standard sample size justifications irrelevant. They commonly ask for IACUC approval to observe as many animals as possible given the available time and resources.

Field research is not necessarily unobtrusive observation, for which the number observed would be of little concern. Often wild animals are captured in ways that may cause disruption to their daily lives or physical harm to them. Electroshock seems to be a standard way of sampling fish, for example, and some studies of birds may need to sample species that rarely leave the upper canopy, thus necessitating that they be shot in order to, say, evaluate their parasite load or other medical conditions. In addition, some field research involves euthanizing captured specimens for further analysis and/or adding to a permanent collection. Thus, the IACUC has a real need to understand why the total number to be studied is optimal.

If a field researcher is going to argue for sampling as many as possible, there needs to be a clear and adequate explanation for why that is optimal. By comparison, we would not allow a lab researcher to justify a large sample size simply because it produces a more precise estimate. Rather, the lab scientist is required to estimate how much precision is needed, or what a reasonably important effect size would be (for a power analysis) then determine sample size based on those considerations. This ensures that the number of animals requested is adequate without wasting animals.

Wildlife studies of multiple species are different, however. There may be no upper limit to the number of rare and even previously unknown species to be found in a sample that would be considered a wasteful use of animals to obtain. Thus, it may be acceptable to make that point and then justify the expected numbers based on time and resources, in those limited situations.

Nevertheless, not all animal research fits this model. In many studies, the goal is to see how the numbers of a single species are changing over time or between locations. Sometimes the goal is to estimate some quantitative characteristic, such as mean parasite load, body weight, eggs per nest, and so on.

In these cases, where the outcome is quantitative, field researchers may overstate the difficulty of using standard techniques of precision or power to estimate an appropriate number. They probably do not need the mean number of eggs per nest to a margin of error of 0.1, for example. A reasonable margin can be stated, from which an appropriate sample size can be determined. Alternatively, if the goal is to see whether that mean number has changed from 2020 to 2023, a power analysis can be used.

In some cases, wildlife studies merge into lab studies, when wild animals are captured then subjected to various physiological procedures. In many such cases, standard power analysis can be employed.

Note that all wild animal research involves permits that should be verified by the IACUC.

It is also important that field researchers consider and report on, not only the animals to be studied, but also any non-study animals likely to be incidentally caught or killed. Estimates of their numbers should be included in the protocol.

The survey question on this topic was:

Surv.t When studies are done on animals living in the field (whether they are captured and released or euthanized), how does your IACUC approach sample size justification? Select all that apply.

N/A since we do not conduct animal studies in the field.	91/217 (42%)
We require power analysis.	35/217 (16%)
We require sample size justification based on previous experience showing that the requested numbers are adequate for the purpose.	93/217 (43%)
We accept sample size justification based on the researcher having used the sample size without further justification.	21/217 (10%)
We allow the researcher to make a judgment if conventional methods of sample size justification do not apply.	34/217 (16%)
We generally approve whatever number is approved on the relevant collection permits, if applicable.	28/217 (13%)
Other*	8/217 (4%)

*Other responses generally reflected the choices above, though some respondents noted that the precise expectation could depend on circumstances, such as species being used and invasiveness of procedures.

While interpreting this survey, note that respondents were permitted to choose more than one option and that 42% of respondents indicated that they never evaluate field studies, leaving only 126 that do. Thus 25% (35/126) of IACUCs that evaluate field studies require a power analysis and over 50% require a justification based on previous experience.

14:7 How can protocols with breeding, Mendelian ratios less than 1, significant attrition, and other complicating factors be clearly presented?

Commonly, researchers have complex designs to describe. This refers not so much to the multiple factors of the research itself, but to secondary considerations that affect the sample size, notably breeding that will produce substantial numbers of animals with unusable genotypes or breeding schemes that must produce more animals than needed to account for variability in breeding success. Attrition of animals can sometimes be severe and affect the numbers as well.

Opin. When the structure of the study is complex, it is best to start with the actual number of animals needed for the research, then work backwards to the total to be obtained. At the first stage of explanation, avoid talking about Mendelian ratios, breeding strategies, attrition, or otherwise unused animals. Just derive and explain the number that must be successfully operated on, injected, or used for tissue—in other words, the number of animals required to actually contribute to the research. In most cases, the key sample size consideration will be the number needed per study group. Then derive the total number from that starting point.

In the simplest cases, text wording like the following is suitable,

"Based on the power analysis described above, we need 6 successful rats in each of our three groups, 18 in total. Since we expect 75% of animals to be successfully trained, we need 24 to yield 18. Approximately ¼ of the offspring have the correct genotype, thus we need to breed 96 offspring to obtain 24 usable animals. A total of 8 parent rats will be required, for a total of 104 rats."

However, this example represents a comparatively simple case. Nowadays many protocols have specific strains used only for a step in the breeding process, none of which are actually suitable research subjects. In such cases, a table is very helpful with columns for each strain showing how many are produced, the Mendelian ratio for suitable genotypes (if any) from a given cross, the total number of each genotype to be produced, and the number suitable for research.

The use of a letter abbreviation for each strain and variant is also very helpful. Few reviewers will appreciate trying to tell the difference between two strains with complex names that differ by one superscript symbol! Define them early, in a table. Then say, "Strains A and B will be bred, from which 25% will be strain C, to be used in the next breeding cycle." And so on.

Another complex case is one in which the analysis will be on cells or tissues, rather than animals, with the number of cells or tissues from each animal often expanded via laboratory methods.

The first critical consideration is to avoid violating the assumption of independence. The assumption is violated if observations from the same animal are treated as independent, thus increasing the sample size without actually adding any animals. While widely practiced, few statisticians would accept this approach. The experimental unit here is the animal, not the observation within animals.

An example of an acceptable method is obtaining enough cells from each animal to expand into five plates and then assigning each plate to a different treatment to be analyzed in a repeated measures fashion across animals.

An example of a problematic approach would be taking the five plates from each of three animals, subjecting all fifteen to the same treatment and counting that as fifteen subjects. This is not appropriate. The experimental unit here is again the animal, for an n here of three, not fifteen.

A simple analytic approach is to average the five plate results within each animal. More sophisticated models like "Generalized Estimating Equations" can be used to include all fifteen observations. Nevertheless, simply counting them as fifteen is not appropriate. Very likely, the plates within an animal behave more like each other than they would to a random plate from another animal. They are not "independent."

Here again, the best description drills down to the most basic level and works up to the total number needed. For example:

"We need at least 5 x 10^6 cells per plate to do a biochemical analysis under one condition. Each animal provides enough cells for three plates, each of which will be randomized to a different condition to be statistically analyzed as repeated measures. Power analysis indicates that we need six observations in each condition, thus six mice will suffice. Specifically, 3 conditions x 6 observations per condition / 3 plates per rat = 6."

Be sure the reader can see exactly how the total number follows from the steps. The actual calculation at the end ties it all together.

14:8 What role can a biostatistician play in evaluating the justification of the number of animals requested?

Opin. Biostatisticians are trained to perform the kind of calculations used to determine sample size. They can advise the researcher how to analyze the data properly and can help to write the analysis and animal number sections of the protocol. To be most helpful, biostatisticians also should be familiar with the IACUC and its task. They should understand the audience that their comments will reach and understand enough of science to be able to recognize what is central and what is not. Having a statistician either on the committee or available to help researchers could be very useful. Since the development of statistical methods and the performance of power analyses are both lengthy tasks, IACUCs should not expect a statistician (whether a member of the committee or not) to perform these tasks routinely without making appropriate collaborative, and possibly financial, arrangements.

14:9 What exactly is a pilot study? How do pilot studies differ from other studies in animal number justification?

Opin. A pilot study is one involving a small number of animals (rarely more than ten) used to demonstrate that a technique can work or to estimate the variability in the data before performing a statistical analysis. A pilot study does not require the extent of sample size justification required by "full" studies. Nevertheless, a pilot study is part of an IACUC protocol and, therefore, does require a justification that is sufficient to indicate why that number of animals is needed to achieve the results of the pilot. If pilot studies are to be exempted from power analysis or other relatively stringent methods for sample size justification, then common sense suggests that they be narrowly defined and limited to a few special cases. Every institution should develop and disseminate specific guidelines, or an SOP, concerning what will be accepted as a pilot study.

It is important to note that pilot studies are much more useful for estimating variability in the data than for determining the likely magnitude of effects. The latter requires a much larger study than most pilots. Furthermore, if a pilot study is used to estimate magnitudes of differences or to determine whether or not a particular manipulation is worth pursuing further, then the pilot animals should not be used in the succeeding full study. If pilot study animals serve only to estimate variability or to verify that a technique works in your laboratory (e.g., that you can actually get reliable results on a particular piece of equipment), it may be acceptable to include the pilot study animals in the full study.

14:10 Can a study using four groups of ten animals each be considered a pilot study?

Opin. In the author's opinion, it is difficult to see how so many animals could be considered a pilot study or why a pilot study would require four groups. Variability can normally be adequately estimated with one group (or, at most, two groups) of animals. The viability of a technique also can typically be tested on one or two groups.

Not every possible situation can be anticipated. For example, consider a researcher who has four technically difficult manipulations, each of which must be shown to work before a larger study can be undertaken utilizing all four. Conceivably, a pilot study with four groups could be justified here.

14:11 If animals are requested both for a pilot study and for the full study to follow, should the IACUC approve both at the same time?

Opin. Assuming the number of animals used in the pilot study is justified and until the results of the pilot study are known, the IACUC should not approve the entire number of animals requested.

The pilot may show that the basic technique is flawed and the whole study should not progress.

In other instances, a pilot has as its purpose the estimation of variability in the data upon which a power analysis will be based. Until these estimates are in hand, any number of animals requested for the full study must be considered unjustified.

It is important to note that the regulations also do not allow a "conditional" approval of a protocol. Even the word "conditional" pending the pilot is not acceptable and sends the wrong message. Only the pilot animals are approved. The rest will be separately requested and approved when the pilot is completed (if it is consistent with continuing). Avoid talking about a preliminary report (implicitly to be followed by approval). PIs could misunderstand this as an effective "approval but for a technicality," which may not be the intent. It is, in this author's opinion, best to grant approval only for the number of animals fully and initially justified.

The survey responses suggest that IACUCs do not routinely follow these rules.

Surv.t If a pilot study is approved by your IACUC, how does your IACUC proceed with approving the total number of animals needed for the larger study?

Not applicable.	10/217 (5%)
We approve all the animals at once.	14/217 (6%)
We approve all the animals at once but only allow the investigator to order the animals for the pilot study. Once the pilot study is reviewed by the IACUC we may release the remaining animals to the investigator.	24/217 (11%)
We approve animals only for the pilot study and will only approve (via amendment) the remaining animals once the IACUC has evaluated results of the pilot study.	94/217 (43%)
Our decisions vary, based on specific protocols.	73/217 (34%)
Other.	2/217 (1%)

Opin. If the purpose of a pilot is to see whether something works, what is the point to approving all the animals at once? And approving all but only releasing some later sounds dangerously like a conditional approval. The correct answer is to approve for the pilot then approve the rest as part of an amendment.

Surv.t If the number of animals required for the latter part of a study is dependent upon findings from an earlier part of the same study (on the same protocol), how does your IACUC approach this situation? Select all that apply.

Not applicable.	18/217 (8%)
We require an estimate of the number of animals needed based on the most likely outcome of the first part of the study.	103/217 (47%)
We approve the initial number needed then the investigator submits an amendment with a justification for the remaining number needed.	110/217 (51%)
We allow the investigator to make a best guess of the number needed without additional justification.	15/217 (7%)
Other.	14/217 (6%)

14:12 Can the IACUC demand that a pilot study be done?

Opin. The IACUC has the right and responsibility to require adequate justification of the number of animals requested (see 14:1). If a researcher is unable to provide that, the IACUC may withhold its approval. If a pilot study can break the impasse, it may be an appropriate choice for both the IACUC and the investigator.

The IACUC, by exercising its responsibility to assure that the number of animals to be used is appropriate, may withhold approval of a project until a pilot study is completed. If a pilot study is agreed to, it requires IACUC approval. There is a brief reference in NIH/OLAW FAQ D.11 suggesting that IACUCs may be able to demand a pilot, without details, "Whether proposed by investigators or required by the IACUC, pilot studies require review and approval by the IACUC in accordance with the PHS Policy."[2] Since the IACUC will grant approval of the larger study only after the pilot study is completed (if it follows the preceding advice), it will necessarily need to receive a report on the pilot, with a revised request for the remaining animals. These must be seen as two separate steps since, as noted above, "conditional approval" of a protocol is not permitted by regulations. The IACUC approves the pilot. That's all. The PI then returns to the IACUC with a new request for the full number (if justified by the pilot).

A survey question on this topic was not asked for in this edition of *The IACUC Handbook* but in the second edition survey 69% of 162 respondents agreed that IACUCs could require a pilot study, 24% had no policy, and only 6% said no.[9]

14:13 Should the IACUC accept the argument that three replications are needed to ensure reliability?

A survey question addresses a closely related question:

Surv. Does your IACUC allow for the repeat of an experiment (as part of the same protocol) in order to demonstrate experimental repeatability prior to publication? Select all that apply.

Often yes, for *in vitro* studies using animal tissues	65/217 (30%)
Often yes, for *in vivo* studies	91/217 (42%)
Not often for *in vitro* studies	23/217 (11%)
Not often for *in vivo* studies	55/217 (25%)
Other	54/217 (25%)

Opin. This assertion is common in IACUC protocols in this author's experience. Fitts [10] says that looking through the Instructions for Authors in several journals he was unable to find any that require a certain number of replications, although reviewers might!

He argues, and the author agrees, that this rule is problematic. The author believes that the issue needs to be considered on a case-by-case basis. Fitts[10] accepts it for things like demonstrating that an antibody binds to a specific ligand.

The author thinks that one could also defend the three times rule if there was a real possibility that a given run through of an experiment could be biased in some way. For example, if an occasional batch of the substance of interest is contaminated or if a given set of rats may have some trait that affects the results, it may be necessary to perform a completely different replication to show that biasing factors do not underlie the result.

However, if the issue is mainly just chance and the risk of type I error (falsely finding an effect that does not exist), the author, with Fitts, would argue that adjusting the significance level is a better approach and will generally require far fewer than three times the number of animals in one study. Indeed, some statisticians have stated that using 0.05 as a cutoff is actually too weak, even with just a single analysis. 0.01 provides a more convincing significance level.

Consider a simple two-group comparison to be done via unpaired t-test with six animals per group yielding 80% power at alpha = 0.05. Keeping the other parameters the same, increasing the number of subjects to 10 per group gives 88% power at alpha = 0.01. Twelve subjects per group provides over 90% power for alpha = 0.005. Either of these high-powered analyses capable of producing very convincing p-values uses fewer subjects than three experiments with six per group each time (18 per group in all).

If there are several analyses, any one of which is important, the researcher could start with alpha = 0.01 and divide by the number of comparisons. With five comparisons, the new alpha becomes 0.002. Any result in five that is significant at $p < 0.002$ is quite convincing. If several are significant at that level, chance is virtually ruled out as an explanation.

14:14 Can an IACUC withhold approval of a protocol that requests fewer animals than the IACUC believes are needed for a valid study?

Reg. See 14:1.

Opin. Yes, if the researcher does not adequately justify the use of that number. The primary question is whether the sample size justification itself is adequate (see 14:1). Section 14.2 above discusses methods to adequately justify sample size. If the justification is invalid or inappropriate, the researcher should be asked to improve it. If the researcher fails to do this acceptably, then approval of the protocol is withheld until the justification is satisfactory to the IACUC. The use of too few or too many animals can be equally deleterious. A researcher who requests fewer animals than one might have expected is not excused from showing how that number is sufficient for research purposes.

A survey question on this topic was not asked for this edition of *The IACUC Handbook*; however, in the second edition survey 69% of 161 respondents agreed that their IACUC "would withhold approval" of a protocol that appears to be requesting too few animals, 24% said they would not, and 7% replied "other."[9]

14:15 When justifying the number of animals requested should fetuses and neonates be counted as vertebrate animals?

Reg. The AWAR and PHS Policy do not address this issue in detail. However, NIH/OLAW has provided guidance on this issue in FAQ A.4 and A.5 that state: "Although avian and other egg-laying vertebrate species develop backbones prior to hatching, OLAW interprets the PHS Policy as applicable to their offspring only after hatching."[2] Similarly, the AWAR define "bird" as "any member of the class Aves, excluding eggs, but including birds once the hatching process commences" (AWAR 1.1). Moreover, that "larval forms of fish and amphibians have vertebrae and are covered by the PHS Policy." These statements make clear that hatched chicks or other vertebrates should be counted. It would be reasonable to infer that the same logic would apply to neonate mammals. Thus, it seems that born and hatched vertebrates should all be counted. The handling of fetuses is less clear (see *Opinion*).

Opin. Authorities are divided on whether fetal mammals and unhatched birds can feel pain, and to some extent even neonates.

For example, the AVMA Guidelines for the Euthanasia of Mammals, 2020 states: "Mammalian fetuses are unconscious in utero due to a combination of factors, such as low oxygen tension and hormonal influences in the uterus suppressing consciousness. Rat and mouse pups are born neurologically immature when compared with humans, and their afferent pain pathways are not well developed until after postnatal day 5 to 7, with cortical development occurring later."[11]

But this same document says, a few lines later, "As cold surfaces can cause tissue damage and presumably pain, the animals should not come in direct contact with ice or precooled surfaces." For neonates exposed to inhaled anesthesia, they state: "Adequate exposure time should be provided, or an adjunctive method (e.g., cervical dislocation, or decapitation) should be performed after a neonate is nonresponsive to painful stimuli." Both latter quotes refer to pain perception in fetal animals, neonates, or both.

The Canadian Council on Animal Care in Science states: "during the final third of the gestation, fetuses should be given the same ethical considerations as apply to the fully mature animal."[12] As for eggs, chickens are considered a precocial species, since the hatched young need to be able to walk about almost immediately after hatching. According to the Opinion of the Scientific Panel on Animal Health and Welfare, "Precocial oviparous species, some of which are breathing at the time of hatching present much evidence of being aware before hatching and during the last days before hatching, perhaps for as much as the last third of their development."[13] They make a similar comment about mammals, "For most vertebrate animals, the stage of development at which there is a risk of poor welfare when a procedure is carried out on them is the beginning of the last third of development within the egg or mother."

In this author's opinion, late-stage fetuses should be considered as if they are able to feel pain and thus should be protected from it. The purpose of the IACUC is to help assure high standards of animal welfare in the institution. If a fetus can feel pain, then its welfare matters, and it should become an IACUC concern. Nevertheless, it is difficult to develop comprehensive and systematic guidelines for preventing fetal pain, because so little is known about many species and their precise neural development.

In order for the IACUC to keep track of how many animals are experiencing particular kinds of manipulation, some accounting for all of them is necessary. If late-stage fetal or neonatal animals are directly the subjects of research studies, accounting for them follows naturally from the sample size justification itself. The number of fetuses needed will critically determine the justification of animal numbers in such instances.

The *Institutional Animal Care and Use Committee Guidebook* section on breeding does not discuss fetuses but suggests that suckling animals be subdivided into those to be used in any way in the research versus those merely to be euthanized before weaning.[14] Both groups are then counted and reported, as recommended here.

14:16 To reduce the number of animals used, how might an animal be used in more than one biomedical research, teaching, or testing protocol?

Reg. The AWAR (§2.31,d,1,x) and the *Guide* (p. 30) limit the situations under which animals can be subjected to a second major survival surgery or related operative procedure. This happens primarily in instances of scientific necessity and is clarified in NIH/OLAW FAQ F9.[2]

Opin. Any sacrificed animal should be utilized in as many studies as can benefit from the use of its tissues. It is appropriate to do so at any time several protocols can use an animal without subjecting it to any additional pain or distress. A second use of a live animal in a non-painful or non-distressful study is also appropriate and can reduce the total number of animals involved in research. For example, after a survival surgery study in which animals need not be euthanized, a simple non-stressful behavioral study might be appropriate.

A real problem concerns the reuse of an animal in a painful or distressing way after it has already been used once in this way. Federal guidelines focus on surgery, but common sense suggests that the painful or distressful reuse of an animal that previously was subjected to a distressful or painful procedure requires careful justification by the investigator and careful consideration by the IACUC Both NIH/OLAW FAQ F9 (on multiple major surgical procedures in a single animal) and NIH/OLAW FAQ F13 (on laparoscopic procedures) are relevant here.[2, 9] The latter recognizes that the IACUC has some discretion on how to classify laparoscopic procedures (as major or minor), but not unlimited discretion. Notably, "Any laparoscopic surgery that produces substantial impairment of physical or physiological function must be considered a major operative procedure."

14:17 How else can animal numbers be reduced?

Opin. It is worthwhile considering ways that animal numbers can be reduced carefully. A table of possible methods for reducing the number of animals used is provided in the *Institutional Animal Care and Use Committee Guidebook*.[14] Some of what is recommended is already customary practice for most researchers, such as standardized procedures; use of "healthy, genetically similar animals"; and good postoperative care to minimize losses. Other suggestions include sharing of tissue (as noted in 14:16), several terminal procedures per animal, appropriate choice of control groups, and maximization of the use of information with good statistical software.

Morton describes some additional methods.[15] For example, he notes that if a certain drug would be considered as too toxic for use if even one of six subjects had severe adverse reactions, it would make sense to test the six in sequence and then stop the study if one of them, in fact, had such a reaction.

Morton further notes that pilot studies can help to prevent problems and the data may be able to be used in the main study.[15] Nevertheless, as noted in 14:9, one needs to be cautious with this concept. If pilot studies are used merely to test procedures and estimate variance, it is true that one can use the data in the full study. If, however, pilot studies are used to determine, for example, which drugs from a large set of drugs have an effect, one needs independent replication of the results with entirely new data. Anyone proposing to mix pilot data with full study data should consult a statistician.

Gaskill and Garner note that effective use of analysis and design can also help reduce animal numbers. They discuss two scenarios.[16]

First, consider a two-group comparison of an intervention versus control in which it is desired to test the effect within male and female groups. If a power analysis suggests eight animals per group, and if the analysis is done by unpaired t-tests comparing the treatment groups for male and female animals separately, a total of 2 x 16 = 32 animals is needed.

A more powerful analysis is a 2 x 2 analysis of variance in which the treatment effect is estimated as an average across the two sexes. This design also produces a test for the sex by group interaction, assessing directly whether the intervention effect differs by sex.

Another example Gaskill and Garner provide is an interesting way to structure a study.[16] Again, consider a simple two-group comparison, and let us ignore sex as a factor. From each cage of four animals, randomly select two to get the treatment and two to get the control condition. Now do the analysis as an ANOVA in which cage is a blocking factor. Assuming animals in a single cage are more similar to each other than to animals in different cages, the blocked design helps reduce unwanted random variation and may allow fewer animals to be used.

14:18 What role can veterinarians and animal facility personnel play in reducing the number of animals used?

Opin. The IACUC has a general responsibility to oversee the appropriate use of animals in biomedical research. It closely interacts with the AV and the animal care staff. Veterinarians and other animal facility personnel have responsibilities that generally include assuring that animals are obtained from high-quality vendors, reviewing the health status of incoming animals based on a vendor's reports, assuring quarantine and testing of animals when appropriate, and facilitating the re-derivation of animal strains when appropriate. They are also responsible for developing the entire program of preventive medicine and often oversee all animal care operations. The use of healthy, well-treated, and well-maintained animals helps decrease the numbers of animals needed by decreasing the need to repeat studies and reducing variability.

Veterinarians and animal facility personnel also can assist with preoperative, operative, and postoperative planning to help assure animal survival, further decreasing the numbers needed.

REFERENCES

1. National Institutes of Health. 2017. *Worksheet for Review of the Vertebrate Animal Section (VAS) 2017.* Accessed on December 12, 2022. https://olaw.nih.gov/sites/default/files/VASchecklist.pdf .
2. PHS Policy on Humane Care and Use of Laboratory Animals. 2022. Accessed December 12, 2022. https://olaw.nih.gov/policies-laws/phs-policy.htm.
3. Silverman, J. 32004. Animal use for in vitro work: How much justification is enough? *Lab Animal (N.Y.)* 33(5): 15.
4. Doyle, R. 2004. Give some examples. *Lab Animal (N.Y.)* 33(5): 15.
5. Matthews, M. 2004. A little short. *Lab Animal (N.Y.)* 33(5): 16.
6. Fitts, D. 2010. Improved stopping rules for the design of efficient small-sample experiments in biomedical and biobehavioral research. *Behavior Research Methods* 42: 3–22.
7. U.S. Fish and Wildlife Service. 2013. *Endangered Species Act (P.L. 93-205, 87 Statute 884).* Accessed December 26, 62022. https://www.fws.gov/endangered/laws-policies/index.html.
8. Convention on International Trade in Endangered Species. www.cites.org.
9. Gracely, E.J. 2007. Justification of the number of animals to be used. In *The IACUC Handbook,* 2nd ed., eds. J. Silverman, M. Suckow and S. Murthy, Chapter 13. Boca Raton, FL: CRC Press.
10. Fitts, D. 2011. Ethics and animal numbers: Informal analyses, uncertain sample sizes, inefficient replications, and type I errors. *Journal of the American Association for Laboratory Animal Science* 50(4): 445–453.
11. AVMA Guidelines for the Euthanasia of Animals: 2020 Edition. https://www.avma.org/sites/default/files/2020-02/Guidelines-on-Euthanasia-2020.pdf.
12. CCAC Guidelines on Euthanasia of Animals Used in Science. https://ccac.ca/Documents/Standards/Guidelines/Euthanasia.pdf.
13. EFSA Panel on Animal Health and Welfare. 2005. Opinion of the Scientific Panel on Animal Health and Welfare on a request from the Commission related to "Aspects of the biology and welfare of animals used for experimental and other scientific purposes." *The EFSA Journal* (2005) 292: 1–46
14. Applied Research Ethics National Association/Office of Laboratory Animal Welfare. 2002. *Institutional Animal Care and Use Committee Guidebook,* 2nd ed., *98,* 130–9133. Bethesda, MD: National Institutes of Health. Accessed December 12, 2022. https://olaw.nih.gov/sites/default/files/GuideBook.doc.
15. Morton, D. 1998. The importance of non-statistical design in refining animal experiments. ANCCART News Fact sheet. Accessed on December 12, 2022. (researchgate.net).
16. Gaskill, B.N. and J.P. Garner. 2020. Power to the people: Power, negative results and sample size. *Journal of the American Association for Laboratory Animal Science* 59 (1): 9–16.

15

Animal Acquisition and Disposition

Deborah Mook, Michael J. Huerkamp, and David R. Archer

Introduction

Millions of animals are acquired and used for scientific research annually in the U.S. Because animal acquisitions for research and the eventual disposition of these animals are under the stringent regulation of federal law and the oversight of funding agencies, there is a lot at stake. If animals are not acquired and used in accordance with appropriate laws and standards, institutions risk punitive measures and/or public relations dilemmas. Complicating these issues are myriad details related to the potential sources of animals or research specimens (e.g., farms, stockyards, commercial breeders, USDA Class B dealers, foreign import, pet shops, wild caught, donations or loans from citizens, institutional breeding colonies, reanimation from sperm or embryo cryopreserved repositories, transfer from other research projects, interstate movement, and abattoirs), the type of research for which animals are acquired, the animal species to be used, stage of development/maturation, and eventual disposition (e.g., euthanasia, slaughter for food, donation to rehabilitation program or sanctuary, and adoption). After animals are acquired, the tracking and accounting of their use are arguably among the major administrative challenges at a research institution. Consequently, to act in a legal, fair, consistent, and rational manner, it is important for IACUCs to be cognizant of the many issues related to acquisition, records, tracking, and disposition of research animals.

15:1 What are the usual and ordinary sources of animals for research?

Opin. The most conventional way to obtain animals for research is to purchase them from outside the institution from a commercial vendor, licensed dealer, farm, or researcher from another institution. Within an institution, sources include breeding or stock colonies and those transferred from or exchanged with another research project. Other sources, usually requiring a certain level of justification, include pet shops, private pets (such as those used in veterinary clinical studies with owner consent) (see 15:42), and those transferred from other research institutions. Animals may also be studied in, or collected from, the wild.

In some instances, scientists may require tissues, rather than live animals, for research. As an alternative to purchasing animals to be immediately euthanized, scientists may obtain biological materials from slaughtered livestock, captured wildlife, harvested marine animals, or, perhaps, transfer from a colleague from within or outside the institution who has euthanized animals for another purpose.

It is important to understand that the breadth of species defined as *animal* varies with the regulating body and does not necessarily adhere to the classic biological definition. PHS Policy (III,A) and the *Guide* (p. 2) include all vertebrates in their definition of animal, while the AWAR (§1.1, Animal) specifically exclude mice, rats, and birds bred for research, horses not used specifically for research purposes, and livestock or poultry used for food, fiber, or agricultural research.

15:2 What are the legal requirements governing procurement of animals and tissues for research?

Reg. The AWAR require that animals used for biomedical research.

- Be acquired by lawful means and in compliance with recordkeeping requirements (§2.35,b). Note that there are specific requirements for the acquisition and recordkeeping of dogs and cats (§2.35,b–c).
- Have a rationale for involving animals, and for the appropriateness of the species to be used (§2.31,e,2).
- Have appropriate identification of the species and the approximate number of animals to be used (§2.31,e,1).

Opin. For animals used for food, fiber, and in agricultural research, the AWAR do not apply, but recommendations are given in the *Guide for the Care and Use of Agricultural Animals in Agricultural Research and Teaching.*[1] For research funded by entities of the PHS, the salient features of the AWAR are reinforced by the PHS Policy (IV,D,1,a; IV,D,1,b), U.S. Government Principle III, and the *Guide* (pp. 11–35, 106–107).

The provisions in the *Guide*, including those that relate to justifying and tracking animal use,

DOI: 10.1201/9781003272861-15

are also used by AAALAC in accrediting research institutions.[2] While the AWAR exclude poikilothermic animals and typical laboratory rats and mice from these considerations, all vertebrates are covered under PHS Policy and, by extension, AAALAC. Consequently, in most research institutions in the U.S., federal regulations, AAALAC standards, and granting agencies require that vertebrate animals be acquired lawfully, used judiciously, and disposed of appropriately. Non-PHS-funded research may be included in PHS requirements if an institution, via its Animal Welfare Assurance with NIH/OLAW, indicates that it will include activities funded by sources other than the PHS. Institutions may exclude non-PHS-funded or agricultural animal activities from its Assurance yet still empower the IACUC to oversee the use of animals in those areas. Areas that are not directly supported by the PHS, but are functionally, programmatically, or physically in an area that could affect PHS supported activities, would not be considered excluded from the Assurance. In the case of field studies, local, state-provincial, federal-national, or international laws, regulations, and permit requirements likely will pertain to scientific collecting, transport, possession, sale, purchase, barter, exportation, and importation of specimens or parts thereof, or other activities involving native or non-native species of animals, whether endangered or not,[3] and fall under IACUC purview minimally for mammals regulated under the AWAR (§1.1, Animal).

The determination of the use of tissue versus the use of animals would depend on how the tissue is obtained. Specifically, for PHS-funded research, if animals are purposely obtained to collect tissue, which would be considered use of vertebrate animals and the requirements of the PHS Policy, including an Assurance and IACUC approval, would apply. If tissues are provided as extra material from another IACUC-approved study that would not be considered animal use under the PHS Policy. The AWAR function similarly here and consider the use of tissues from animals in unrelated studies as a way to promote the 3Rs.

Also impacting research animal acquisition and disposition are regulations and standards covering importation of animals and safety and efficacy of biologicals,[4] endangered and threatened species,[5,6] the safety of the American food supply,[7] the use of infectious agents (biohazards, bloodborne pathogens),[8–10] radioisotopes (usually state regulated), and chemicals and certain toxins.[11] Finally, beyond these bounds, an institution must also act and set policy in ways that serve to protect its good reputation.

15:3 Is a land grant college that uses animals only in agricultural research and training required to follow the AWAR, PHS Policy, or both?

Reg. The AWAR (§1.1, Animals, §2.1,a,3,vi) and the *Guide* (pp. 32–33) specifically exclude livestock bought, sold, or used in agriculture production, agricultural research, or agricultural training. PHS Policy also does not normally cover animals acquired and studied for agricultural purposes; however, it does apply if the agricultural research and teaching component are part of a larger institution with an institutional Animal Welfare Assurance that states that all institutional entities will adhere to PHS Policy. This becomes a germane issue when PHS-funded research is done on institutional farms or other agriculture-related facilities.

Opin. Regardless of the category of research, institutions are expected to provide oversight of all research animals and ensure that pain and distress are minimized (*Guide*, p. 33). Whether the research is interpreted as falling within the realm of biomedical or agricultural research, and the concomitant standards for care and use, is a decision best made by the IACUC in consideration of the goals of the investigator and the well-being of the study subjects and not necessarily as defined by the setting (i.e., farm versus laboratory) for the holding of animals or conduct of experiments (*Guide*, p. 33). Where animals are used in research strictly for agricultural purposes, the *Guide* does not apply. For animals maintained in a farm setting, the *Guide for the Care and Use of Agricultural Animals in Research and Teaching* becomes the applicable guidance document.[1] In order to preserve a good reputation and to obtain and maintain AAALAC accreditation, any institution doing agricultural research with food and fiber species should endorse the principles in the agricultural guide.[1] The standards of the "*Ag Guide*," similar to the *Guide*, advocate roles for the IO, AV, and a fully authorized IACUC and promote thorough protocol review and semiannual inspections of housing and study areas. Adherence to standards in that document also promotes good science, enhances animal well-being, maintains credibility with the public, provides standardization of research conditions across institutions (and hence reduces variability), and ultimately protects the institutional privilege of conducting research with animals.

15:4 How should the use of amphibians, reptiles, fish, and other ectotherms in research be monitored by the IACUC?

Reg. PHS Policy (III,A) covers the use of all vertebrates in research, including reptiles, amphibians, and fish, and the *Guide* (pp. 77–88) addresses husbandry of aquatic species in research to a significant degree covering areas of environmental conditions and enrichment, water quality, life support systems, habitat enclosure characteristics, behavior and social management, husbandry, sanitation, identification, and recordkeeping.

Opin. Although the PHS Policy is intentionally broad in scope and does not prescribe specifics about the care and use of any species, it assigns that task to the IACUC and allows for professional judgment. Many of the principles it advocates generally can be adapted to animal care and use programs for animals of any phylogenetic class. Recommendations for specific species are often available from organizations that have an interest in the appropriate care and use of these species in laboratory and field studies.[12–20] It is clear, however, that individual requirements for these three classes of vertebrates, which contain more than 28,000 species, cannot be addressed in a single set of guidelines. Consequently, NIH/OLAW recommends that the advice of experts be obtained to design and develop studies and suitable housing and care procedures for species not commonly used in research.[21,22]

15:5 For the purpose of IACUC review, are birds used in scientific research covered by the AWAR or PHS Policy?

Reg. Historically, all birds were excluded from the AWA. In 2002, Congress passed the Farm Security and Rural Investment Act and amended the definition of an animal in the AWA to exclude only those birds bred for use in research from the AWA. USDA/APHIS finalized standards and regulations concerning birds not bred for use in research in 2023 (AWAR Subpart G).[23] The PHS Policy covers the use of all vertebrates in research, and this category obviously includes birds (PHS Policy III,A).

Opin. As is the case for other vertebrate species, PHS Policy assigns oversight for the care and use of birds to the IACUC and then allows for professional judgment. There are numerous scenarios in which birds may be used in research. Depending upon specific need, birds may be acquired from the wild or from domestic commercial or private sources. Commercial purveyors of birds may provide them for regulated (research) use or for non-research purposes. At our institution, purchased society finches and chicks are used in neuroscience experiments. Canaries and zebra finches may be acquired from the pet trade for similar use. Broad spectra of wild indigenous species are captured for sample collection in studies of enzootic ectoparasites' and West Nile virus prevalence. Domestic chickens, under some circumstances, have proved useful for antibody production, and quail may be used in cardiac research. Where avian species are not commonly used in research, NIH/OLAW recommends that expert advice be obtained to design and develop studies and suitable housing and care procedures.[21,22]

15:6 Are there any special legal requirements regarding accounting and tracking of specific species used in research?

Reg. The federal government specifically requires that detailed records be kept on individual dogs and cats acquired and used in research (AWAR §2.35,b,1–§2.35,b,8) and that annual use of regulated species be reported by pain/distress category (AWAR §2.36,b,5–§2.36,b,8). The species required to be reported are listed under Animal in §1.1, Definitions of the AWAR (see 15:1). Federal and state laws generally require special permits for the use of endangered species in captivity or in field research. Specific recordkeeping can aid the researcher in proving that animals were born in captivity and in compliance with applicable laws.

Opin. Annual tracking of animal use[24] is a potentially valuable management resource. Such data can be used to show trends in program size and changes in focus that may be useful in program justification and projections. Such information may also be requested by institutional administrators or public officials and may be helpful in clarifying facts about animal research for the public. Some institutions, ours for example, may require accounting of animal use in an internal annual report.

15:7 Should investigators be allowed to acquire, but not use, animals prior to IACUC approval of the protocol?

Opin. It is best practice not to allow advance purchase, capture, or acquisition by other means for several reasons. First, there may be unintentional use or temptation to use or to prepare to use the animals without appropriate authorization. The conduct of any animal activities that have not been reviewed and approved by the IACUC is considered by APHIS/AC and NIH/OLAW as non-compliance, and for NIH/OLAW, it must be reported to that office.[25] (As OLAW requires prompt reporting, and as submitted documents are subject to the Freedom of Information Act, we recommend a preliminary consultation with OLAW be conducted prior to the submission of a final report.) Second, if animals are acquired prior to consideration of a research proposal and the proposed use of the animals is subsequently rejected by the IACUC or the number of animals approved by the IACUC is reduced from the original request, this situation may present problems in managing the population and would constitute violation of legal requirements to use the minimal number of animals necessary. Potential exceptions to this might include fish maintained and studied in large numbers, those kept at marine or freshwater experiment stations for prospective use, and those trapped in the wild (see 15:8; 15:25). Another possible exception could include a scenario where an investigator moves from one institution to another and must transfer animals prior to IACUC approval. In this instance, the institution would need a pre-established mechanism for management and oversight of these animals, perhaps via memorandum of understanding (MOU). Several institutions utilize holding protocols where animals can reside, but no research can be done, until IACUC approval.[26]

15:8 Is it necessary to have IACUC approval for animals that are just being maintained (no research being performed)?

Opin. Yes. In addition to animals directly used in research, NIH/OLAW and APHIS/AC implicitly require that institutions establish mechanisms to monitor and document all animals acquired or produced.[24] This includes any animals that are intended for use in research, e.g., kept for breeding purposes, nonhuman primates awaiting assignment in core pools at primate centers, dogs and cats in centralized institutional conditioning programs, gift rodents secured in quarantine programs, and even animals held in abeyance in the case of a lapsed or suspended protocol. Without IACUC approval or at the expiration of a protocol, animal use is not valid. Continuation of animal activities beyond the expiration or use of animals without approval is a serious violation of the PHS Policy and AWAR and is reportable to NIH/OLAW.[25] In the case of an expired protocol, especially where the likelihood exists that it will be legally renewed or the animals will be transferred to another protocol, rather than the wasteful and internecine approach of animal depopulation, the IACUC should establish a mechanism for temporary oversight and management of these animals. The holding protocol may be of use in this situation. Even in the case of fish caught and held for prospective use, a husbandry and holding protocol is appropriate. While other mechanisms may be contemplated to manage these circumstances generating a protocol would be consistent with standard practices (see 15:7).

15:9 What institutional entity should procure animals for research?

Opin. One of the critical responsibilities of an IACUC is to assure that the institution lawfully and ethically procures appropriate animals for research. This is best accomplished through the institution's animal resource program or other appropriately designated office.[27] Veterinarians involved in the animal resource program must have specialized training in laboratory animal medicine (AWAR §2.31,b,3,i; PHS Policy IV,A,3,b,1; *Guide* p. 15) and have the expertise to evaluate animal sources and identify those that are appropriate for research given the institution, its resources, and its needs.

It is well within the jurisdiction of the IACUC, especially in consideration of its responsibility to ensure an adequate program of veterinary care and a safe work environment, to provide guidance to the animal resources program concerning the species, legal sources, genotypes, and microbiological background of animals acquired for research use.

In smaller academic or industrial programs with part-time or consultant veterinary staff, the IACUC may well want or need to have considerable input into this process. Coordination of veterinary care, vendor approval, and a centralized purchasing system will aid the tracking of animal use as discussed in 15:6; 15:10; 15:11; 15:13; and 15:14.

15:10 Should the IACUC become involved in approving an "acceptable" animal vendor?

Opin. The loss of animals or research to preventable infectious diseases has ramifications for the institutional requirement to ensure that research is not unnecessarily duplicative (AWAR §2.31,d,1,iii; *Guide*, p. 26) and is not in keeping with the principle of the 3Rs. Consequently, it is within the purview of the IACUC to work with the professional veterinary staff to identify acceptable animal health standards. In most instances, IACUCs defer to the professional judgment of the veterinary staff to identify vendors that meet appropriate health and legal standards. Often, this is the most effective means of addressing this obligation. Nevertheless, when rodent use is high, when certain animals are scarce, or when funds are inadequate, veterinarians may be under pressure to permit the entry of animals with suboptimal health status into the institution. In this situation, it may be worthwhile for the IACUC, as the *de facto* representative of the community of scientists, to work with the veterinary staff to develop and use standards that are in the best interests of the community of scientists within the institution.

15:11 What constitutes an acceptable animal vendor? What procedures can the IACUC or animal resources program use to assess vendor quality?

Opin. In general, animal vendors should have a practice of producing consistently healthy animals of a specific genotype and health status in compliance with applicable laws, statutes and regulations, and institutional needs. Under the AWAR (§2.40,a), dealers are required to have a program of veterinary care. Specifically, the AWAR require licensing of any of the following that supply warm-blooded animals, other than common laboratory rats and mice, to research institutions:

- Pet shops (§2.1,a,3,i)
- Breeders selling more than 25 dogs or cats annually that were born and raised on their premises (§2.1,a,3,iv)
- Persons or commercial enterprises deriving more than $500 annually from such sales (§1.1, Dealer; §2.1,a,3,ii)

- Any person or entity (Class B Dealer) selling any dog or cat not born and raised on its premises (§2.1,a,3,iv)

Additionally, institutions should make a dedicated effort to ensure that all transactions and related procedures involving animal procurement are done in a lawful manner (*Guide*, pp. 106–107). Other considerations are whether the vendor has an adequate program of veterinary care and a history of consistently meeting consumer needs and expectations. Ideally, vendors should be AAALAC accredited or provide assurance that they meet standards for their industry (i.e., health quality, animal care, customer service) and applicable federal laws. For rodent vendors, the health status of the animals should be documented through a regular program of health surveillance and verified by a rational means that includes sampling technique, sampling strategy, appropriate diagnostic tests, sufficiently frequent testing, a reliable laboratory, and prompt reporting of changes in health status. In most instances, the evaluation and consideration of animal vendors are handled within the program of veterinary medical care, with IACUC oversight.

15:12 What precautions should the IACUC and animal resources program require if animals of unknown health status must be acquired? And what precautions should the IACUC and animal resources program require if animals of known poor health status must be acquired?

Opin. Acquisition of animals of unknown health status should be limited to those instances when the animals are available only from one source. Nevertheless, the use of this class of animals has several risks. The primary threat is the transmission of infectious diseases to pathogen-free animals. In addition to causing otherwise preventable pain and distress, research may be compromised or invalidated and therefore become unnecessarily duplicative (see 15:10). Consequently, the IACUC should ensure that the institution has the components of a program of veterinary medical care that provide for the stabilization, isolation, and health characterization of animals of an undefined or poor health status (*Guide*, pp. 109–112).

Animals of unknown health status can often be tested for specific agents prior to shipment, commonly using relatively rapid PCR techniques. Once they have arrived at the receiving institution, one relatively safe approach is to isolate and use the animals for rapid experiments whereby they do not enter the general population. If they are to remain long term, they must undergo quarantine. Quarantine programs should be suitably robust to detect infection through the expression of clinical signs or developing antibody titers or permit the incubation or subclinical stage of a disease to become manifested via culture, the presence of parasites and/or their eggs, pathogen nucleic acids, or pathognomonic lesions and thereby minimize the risk of introduction of pathogens into established colonies. Often bacteriology or polymerase chain reaction (PCR) technology can be employed and results can often be obtained quickly; however, responsive antibody production may not be detectable for 1 to 3 weeks or more, depending on the amount and timing of the pathogen inoculum, husbandry factors, and characteristics of the host. Although investigators may feel a stringent quarantine and testing program is a hindrance to their research, data is readily compromised when using potentially infected animals.

Health background information from trustworthy sources should be sufficient to enable the veterinarian to establish appropriate quarantine parameters (e.g., duration, diagnostic procedures, therapeutic or prophylactic interventions, etc.) and to enable the physiological, nutritional, and behavioral adaptation of new arrivals to any social or environmental novelty.

An effective quarantine program should also provide for the re-derivation of diseased or suspect animals by surgical or other appropriate means and other relevant services such as pathogen testing for cells and cell lines that may potentially affect the health of animals within the program.

15:13 Should the IACUC attempt to track the number of all animals acquired or only those used under approved protocols?

Reg. The AWAR (§2.36,b,5–§2.36,b,8) require that the common names and numbers of warm-blooded animals (other than common laboratory rats and mice and birds bred for use in research) used for teaching, research, experiments, or tests or held for all such purposes except applied production agriculture research be included in the annual report to APHIS/AC. For this reason, facilities must track the number of animals regulated under the AWAR acquired and used during the reporting period. Additionally, PHS Policy requires tracking and accounting for all animals used or intended for use, thereby ensuring that the animal numbers that were approved by the IACUC are not exceeded.

Opin. Although neither the PHS Policy nor the AWAR explicitly require an institutional mechanism to track animal usage by investigators under IACUC-approved activities, both require that applications to the IACUC specify and include a rationale for the approximate number of animals proposed to be used (AWAR §2.31,e,1; §2.31,e,2; PHS Policy IV,D,1,b; *Guide*, p. 25). In addition, AAALAC expects institutions to provide the number of animals used at the time of annual report. These provisions implicitly require that institutions establish mechanisms to monitor and document the number of animals acquired or produced and used in approved activities.[24] This includes any number of animals that are kept for breeding purposes or culled prior to research use and not subjected to any experimental manipulations.[27] Tracking is particularly important for institutions that maintain large breeding colonies of nonhuman primates where animals produced from this population may be held for lengthy periods before assignment to a research protocol. Institutions that use dogs and cats must maintain acquisition and disposition records in compliance with federal law (AWAR §2.35,b) and should invoke steps to verify that pets are not being received (*Guide*, p. 106). These resources may also operate, depending on institutional need, by maintaining pools of unassigned animals and allotting them to research protocols only after a suitable stabilization period. In some cases, however, tracking animal numbers may be difficult or counterproductive, as in the case of aquatic animals in large numbers (see 15:15).

15:14 How should numbers of animals acquired and used be tracked by the IACUC?

Reg. As above, the AWAR (§2.36,b,5–§2.36,b,8) require that the common names and numbers of warm-blooded animals (other than common laboratory rats and mice and birds bred for use in research) used for teaching, research, experiments, or tests or held for all such purposes except applied production agriculture research be included in the annual report to APHIS/AC. NIH/OLAW requires only reporting of average daily inventories of animals, by species, in new and renewal Animal Welfare Assurances.[28] PHS Policy implicitly requires that institutions establish mechanisms to document and monitor numbers of animals born, acquired, and used.

Opin. In order to complete the required reports, institutions must establish mechanisms to monitor and document the number of animals acquired and used in approved activities (see 15:13).[24] The most effective approach combines a methodology for tabulating animal use with a policy regarding the accuracy or precision of the process. Methods of tabulation can be either manual or automated. The use of commercial animal ordering software and census tabulation by bar code scanning or radio-frequency identification (RFID) allows for most integrated animal resources programs to track, with relative ease, the number of animals purchased from commercial sources and imported into the institution through a quarantine program. This process provides accurate figures with relatively little investment in time and personnel. Admittedly, accounting to produce rodents bred and weaned onsite may be more challenging. When acquisitions are linked to IACUC approval numbers, investigators, depending upon the capabilities of the technology, can be automatically informed when their use has reached a preset percentage (e.g., 80%–90%) of the animals approved on that protocol and with a request to provide a specific justification if it is anticipated that the number of animals ultimately required will exceed the number approved.[21] Small institutions that use limited numbers of animals may find it most effective to maintain a log of each IACUC- approved activity, merely subtracting the number of animals acquired for each order or breeding and add to the census from the number approved, with verbal or written notification to the investigator as the number of animals approved is approached.[21]

With respect to the precision of the tabulation process, IACUCs in aggregate historically have approached this consideration with some lack of unity.[29] While most committees attempt to track all animal use with precise accuracy, others attempt to track use with approximate accuracy and apparently roughly one in twelve do not do anything at all (see the first survey item, below). More flexibility is often found in practices concerning rodents and ectotherms than those concerning nonrodent mammals. The major reason for this diversity in accounting essentially relates to discrepancies in tracking and reporting requirements between the AWAR and PHS Policy. Species not regulated by the AWAR (e.g., rats, mice, birds bred for use in research, fish) are often accounted for by using approximate means (PHS Policy IV,D,1,a), appropriate amounts (*Guide*, p. 12), or the minimum number to obtain valid results (U.S. Government Principle III) rather than prescribed precise figures. However, even when a species may not be regulated under the AWAR, where the experimental use is highly invasive or has considerable potential for significant pain or distress, it may, at the discretion of the IACUC, be subjected to higher scrutiny and require precise justification and tracking.

Institutions that track and report the approximate number of certain species used in research may be in full compliance with PHS Policy; however, the AWAR require accurate and precise reporting of animals used. Additionally, many internal and external factors (including financial management, public opinion, and the spirit of the law) support an accurate accounting of the animals used in research. Whatever approaches and processes an institution chooses, it must satisfy the PHS Policy requirement that the number of animals used be limited to the appropriate number necessary to obtain valid results.[24,30] The AWAR (§2.31,e,2) also have a requirement to provide a rationale for the appropriateness of the number of animals to be used.

Surv. Does your IACUC (or an administrative office) track the number of animals acquired for research with precise or approximate accuracy?

• We track all animals acquired with precise accuracy.	125/211 (59%)
• We track all animals acquired with approximate accuracy.	57/211 (27%)
• We do not track the number of animals acquired.	17/211 (8%)
• Other.	12/211 (6%)

Surv. If your IACUC tracks the number of animals acquired for research with only approximate accuracy, what species are being tracked? Select all that apply.

• All species.	80/211 (38%)
• USDA regulated species.	13/211 (6%)
• Rodents.	35/211 (17%)
• Birds.	9/211 (4%)
• Aquatic species and other cold-blooded species.	18/211 (9%)
• Wildlife vertebrate species.	12/211 (6%)
• Not applicable, we track the number of animals with precise accuracy.	74/211 (35%)
• Other*	26/211 (12%)

*Other:

- We track animals that are housed in the central housing units. Studies occurring at farms are not tracked.
- We do not track the number of animals acquired.

15:15 What factors prevent the precise prediction and accounting of the number of animals acquired and used for research?

Opin. Although, with the exception of the annual report required by the AWAR (see 15:14), precise accounting of animals is not always required, investigators must still satisfy the requirement that the number of animals used be limited to the appropriate number necessary to obtain valid results (AWAR §2.31,e,1–§2.31,e,2; PHS Policy IV,A,1,g; PHS Policy IV,D,1,a; U.S. Government Principles III). This requirement often produces conflicting pressures, on the one hand to justify animal numbers (including the use of statistical analysis and power calculations (see Chapter 14), on the other, to account for the inherent unpredictability of science. Foremost, it is fundamentally difficult to forecast animal requirements accurately for the three- to five-year funded lifetime of all but the smallest and simplest endeavors, especially given that results and their impact on subsequent research direction cannot typically be predicted. This difficulty has a virtually exponentially greater effect on sizable, active laboratories with multiple projects and numerous technicians, graduate students, and postdoctoral fellows carrying out various aspects of research. These conditions can often lead to discrepancies between the number of animals approved and the number necessary for the completion of the work. In addition to changes in research direction or increases in experimental complexity, failed experiments, unforeseen technical challenges, unexpected disease, and other complications may cause more animals to be used than originally expected (or requested). Given these circumstances, it is not reasonable to expect scientists to be able to project the exact number of animals needed to do the work. It is reasonable, however, to require a valid estimation and, where additional animals must be added to enable ongoing work to continue via the protocol modification process, to expect a logical explanation of why previous approximations were incorrect or why more animals are needed.

The real challenge in animal tracking is found in large breeding colonies of prolific animals such as ectotherms (e.g., fish), chickens, or rodents. Dealing with the sheer volume of animals produced can be a full-time job for animal caretakers and users and attempts to count the animals accurately are often overwhelming. Activities involving the production of genetically manipulated animals are of low yield but require the production of large numbers of animals, many of which are unsuitable for research[27] (see 15:18.). Avian and rodent breeding colony accounting may be facilitated if the populations are managed by a consolidated animal resources program and/or centralized software, but there are trade-offs in this practice, such as insulation of researchers from their animals and risks of inequity between production and demand.

Additional difficulties are presented when animals are used or collected in the field or bred in fisheries or when animals are transferred from one investigator to another. In large-scale pond and tank production systems, where fish are propagated or held for subsequent research projects or used in production aquaculture research, there may be tens to hundreds of thousands of fish involved.[31] In such studies, it is not ordinarily possible to count the precise number of animals used and an estimation of the number of fish acquired and maintained should be sufficient. Additionally, handling of fish, simply for purposes of counting, may be distressful and counterproductive to the maintenance of good health.[32] As such, IACUCs should consider whether the precise tabulation of fish is critical to the experiment itself or important to realistic minimization of fish pain or distress, or whether it constitutes an otherwise groundless exercise. In the context of protocol review involving fish, it is often reasonable for the IACUC to review the general number of animals involved, the rationale for that number, and justification for any vagueness.

While it is not outside the realm of possibility to count every single animal in an enormous population, the cost-benefit ratio is not conducive to the practice, and doing so arguably would have little impact on the humane care or use of animals except in identifying situations where a disease, environmental variability, or genetic contamination may be influencing the numbers of pups born and weaned. Although of less significance, receiving procedures may not account for extra animals included in a shipment to be counted against ceilings approved by the IACUC (see 15.21).

Other accounting challenges include instances in which animals may be used in more than one study or where animals are kept for several years for the same study. Animal use in more than one study usually occurs in one of two ways. First are studies in which the animal is killed for a specific purpose, but tissues are taken for another study (discussed in 15.33). Next are studies in which the longevity or rarity of the animal, the duration (e.g., aging) or innocuous nature of the primary study (e.g., behavioral), or the need to study disease pathogenesis or response to treatment over a lengthy period would allow for the animals to be used in long-term or justified sequential or serial experiments. The latter scenarios would usually involve nonhuman primates or higher vertebrates; thus, it is especially important that each use of an animal is justified in an application and that multiple experiments do not jeopardize the quality of the data or unduly impact the welfare of the animal. For reporting purposes, animals should be included only once except when specific procedures are required to be reported. For purposes of annual reporting of species regulated by the AWAR, which require reporting of accurate (not approximate) numbers of animals, animals are to be counted only once, regardless of the number of studies in which they are used, and those used in more than one protocol should be counted in the most painful or distressful category.[4,33] When animals such as dogs, cats, or monkeys are studied for several years or alternate between periods of study and disuse, they should be counted annually.[34]

In the context of the overall research program, scientists often fail to see precise accounting as valuable and thus do not commit precious human resources to the activity. There may also be a disincentive to accurate reporting if it brings the risk of extra time and effort required to amend protocols and add more animals. Peer review groups of most funding agencies do not micromanage projects down to details such as animal numbers. Instead, most rely on the powerful influence of funding, an influence that is often underestimated by the public and largely ignored by the vocal opponents of research. IACUCs, in turn, may stringently enforce what has been conveyed as a loose regulatory standard for animal number approximation (AWAR §2.31,e,1–§2.31,e,2; PHS Policy IV,A,1,g; PHS Policy IV,D,1,a; U.S. Government Principles III). Sometimes, an "us and them" relationship can develop between investigators and IACUCs and can lead to less-than-optimal accounting of animal usage. Generally, IACUCs need to communicate their role to investigators in an effective way and be responsive to the sometimes rapidly changing needs of the investigators. An effective mechanism of communication is the institutional training program (particularly for new investigators), supported by periodic workshops or communiqués to the established community of scientists explaining the relevant laws and the obligation of the investigators.

Technological tools save labor and enhance process consistency, such as electronic census acquisition by bar code scanning or RFID. The use of these technological tools ordinarily forces institutions to address the issue of accounting for carryover animals as old protocols and associated cage cards expire, renewed protocols are initiated, and replacement cage cards must be produced. The identification or estimation of the number of carryover animals will be integral to the accuracy of this process and must be considered in the renewal along with any new animals to be added. Otherwise, investigators will run the risk of exhausting the number of animals approved by the IACUC before the anticipated conclusion of the project or the IACUC risks approving substantially more animals than necessary.

The deployment of census management technology is, however, not a cure-all, as there remain limitations related to most products and their applicability. Despite all these considerations, the expectation of most organizations with research oversight responsibility is that there will be a good faith effort made in specifying and tracking the number of animals used.[24,28]

15:16 Should the number of animals generated by breeding colonies be tracked?

Reg. The AWAR require reporting of the species and number of regulated animals that are bred, conditioned, or held for use (§2.36,b,8). NIH/OLAW implicitly requires that institutions have mechanisms and processes in place to document and monitor the numbers of animals acquired and used, including those produced from in-house breeding colonies.[24]

Opin. The breeding of animals for research purposes, stated simply, constitutes research. Consequently, animals produced by breeding colonies should be counted. The expectation of legal, regulatory, and accrediting bodies is that research facilities make a good faith effort at counting not only the number of progeny from these colonies that are used directly in research, but also the total number of animals produced.[27] A previous survey from 2014 showed that IACUCs consistently, and in overwhelming numbers, require the tabulation of rodents and other mammalian species produced from in-house colonies.[29] Consistently, the survey conducted more recently (see below) reflects the same practices. When a breeding colony is a centralized core or supports multiple projects (or even individual investigators), animal production may be tracked by a dedicated breeding protocol; breeding management software is also virtually essential. In the case of dogs and cats, federal law requires precise accounting at the time of weaning (AWAR §2.35,b; §2.38,g,3; §2.50,d).

The survey below shows that institutions vary in their timing of accounting for pre-weanling animals, with 26% of institutions tracking at or around birth and 48% at weaning. NIH/OLAW expects that animals born or hatched be accounted for at the first opportunity without compromising the health and welfare of the animals, e.g., at the time of first cage manipulation. The authors caution institutions to weigh the pros and cons of tasking someone with the requirement of tabulating animals as they are born and debiting them promptly from an IACUC-imposed limitation. Such intrusive activity seemingly creates an animal welfare threat (e.g., the possibility of litter cannibalism or abandonment), may disrupt normal behavior (e.g., nursing care, sleeping), may add variability to the environment, and may not provide much of a benefit beyond the rational and comprehensive consideration of animal numbers done in the context of IACUC review. Beyond these considerations, there is also an annoyance risk in this process for the investigator. For example, assume that the IACUC approves 100 male and female rats and some of the females are pregnant. Their progeny is not only born, but also weaned and counted. In this situation the scientist will be in a deficit (over the approved limit) situation that must be administratively reconciled. These undesirable predicaments can be obviated when the policies and procedures are clear and understood by all parties.

With this said, an institution may choose to debit animals at birth. This practice keeps the institution in compliance with the expectations of NIH/OLAW and may be beneficial possibly in situations where a high level of neonatal monitoring is required and a medical record is needed, for example. In the case of either tracking at birth or at first cage change, the authors recommend a system of approximating animal numbers to ameliorate the animal welfare concerns described above; rather than disturb a nest of mouse pups, assign a standard number of animals to a litter. Fundamentally, the expectation is that all progenies are tracked.[35] Regrettably, the unavoidable production of unusable offspring is an integral part of the research enterprise[27] (see 15.18).

Surv. At what age does your IACUC include neonatal mice in the animal census?

• Not applicable	28/211 (13%)
• At or very soon after birth (i.e., within a day of birth)	54/211 (26%)
• Within 48–72 hours of birth	7/211 (3%)
• Within 4–5 days after birth so as not to disturb the litter	10/211 (5%)
• At weaning	101/211 (48%)
• Other	11/211 (5%)

15:17 What information is useful and how should the IACUC address the acquisition of new research animals from breeding colonies?

Opin. The IACUC's role for oversight regarding breeding colonies includes ensuring that the need for a breeding colony has been established on the basis of scientific or animal welfare concerns; that the procedures for care, genetic authenticity, genotyping, and use employed in the breeding colony are consistent with the *Guide* and evaluated and approved by the IACUC on a regular basis (i.e., as part of both protocol review and the semiannual program review); and that there is a mechanism for tracking animals.[27,29,33,36] The IACUC will need information about colony management. These practices should be briefly described in the investigator's animal protocol and justification provided for departure from any standard institutional practices.

To reduce the potential for unnecessary animal wastage and assure the validity of data derived from experiments using animals from the colony, the IACUC should assess the specific mouse breeding colony management knowledge and skills of persons managing the colonies. Special emphasis should be placed on this individual's background in rodent genetics, breeding practices, genetic monitoring, and understanding of nomenclature.

Examples of other information that might prove useful to the committee include the typical number of breeders and number of young per cage, the breeding system (including number of females per male or continuous versus interrupted mating), the intended weaning age, methods for identification of individual animals, and even the breeding scheme (e.g., cross–intercross, backcrossing). This sort of information is especially important with polygamous or continuous mating systems when there is the prospect of large numbers of pups from several litters to be present in any cage, causing overcrowding. The estimated number of animals that are kept for breeding purposes and not subject to any experimental manipulations should be part of the animal protocol.[27] Intensively bred rodent colonies should be managed in ways that demonstrate successful production outcomes and preserve animal well-being.

Determining which animals to include in the estimated number of animals on an animal protocol can be especially daunting to the investigator and the IACUC in the absence of IACUC-developed guidelines. Studies involving genetic analysis are particularly animal intensive. It is possible for the investigator to estimate the number of animals required, but difficult for the IACUC to evaluate this estimate in the absence of experience.[27] A reasonable estimate is for the requirement of about 1,200 mice to map a single gene with recessive inheritance and full penetrance derived from a nucleus of 10 to 12 monogamously mated mice.[27] Similar numbers of mice may be required for quantitative trait loci analysis using the F_2 progeny derived from an initial cross of four to six pairs. Using "speed" congenic technology, up to 750 mice may be necessary to derive a congenic strain, providing the homozygous mutant is fertile.[27] Even in the straightforward case of established colonies in which all offspring consistently meet criteria to be suitable for research, the exact number of animals can only be approximated because it is impossible to predict the exact number and sex of offspring. After founder transgenic or "knock-out" mice have been identified, between 80 and 100 mice may be needed to maintain and characterize a line and assure heritability under the assumptions of five breeder pairs per line, regular breeder replacement prior to senescence, no unusual infertility, and adequate numbers of weanlings for genotyping and phenotyping.[27]

In the preparation of a protocol, the estimated number of animals should clearly distinguish among breeders, young that cannot be used in experiments because they are of the wrong genotype or gender, and animals that will ultimately be subject to experimental manipulations. If a study requires fertilized one-cell eggs, embryos, or fetuses, the protocol should indicate the number of eggs, embryos, or fetuses that are required for proposed studies.[27] The estimated number of experimental animals may be limited to the number of female animals that are mated and euthanized or surgically manipulated to collect the required eggs, embryos, or fetuses. In this situation, males might be listed as breeders if they are not subject to any experimental manipulation. Likewise, when breeding colony animals will be used as sources of tissues, cells, or other biological materials, the total number to be used for experiments should be estimated on the basis of the typical amount of biological material needed for an experiment, the approximate number of animals necessary to yield the material, and the number of experiments projected.

The accounting of animal production can be done by animal care personnel recording births and weaning. On a large scale, this is an intrusive and labor-intensive process often best done electronically. As a new cage of mice is populated with weanlings removed from a production cage, either the exact number of mice or an average (typically 3.25–3.5 mice/cage) can be deducted from the maximal number of mice approved from the protocol. Regardless of approach, the process must support the mandate to the IACUC to limit the number of animals produced necessary to support the work and not exclude those animals produced in excess of need, either inadvertently or necessarily, or genetic culls.[28] Litter sizes also should be assessed with sufficient frequency and in a manner ensuring that space recommendations of the *Guide* are routinely heeded.[28] As an alternative, investigators report to the designated office, at regular intervals, the number of animals born, weaned, or used in studies.[27] This report can be tallied against the numbers in the approved protocol.

15:18 Is there a percentage of unused breeding colony animals that are euthanized (without having been used as breeding stock or experimental subjects) that should be tolerated by the IACUC?

Opin. Substantial numbers of animals may be required to maintain a breeding colony. For example, while trying to establish a breeding colony for a new mutant rodent or ectotherm model, the investigator will also be working to determine phenotype, to identify affected physiological system(s), and to define inheritance pattern in addition to breeding founder and subsequent generational animals. Under these circumstances, the production of unused animals that are subsequently euthanized is almost inescapable. This is a significant issue for commercial and core rodent production colonies that must operate at a production excess (approximately 30% or more) in order to meet unexpected increases in market demand or that produce animals at an excess when demand unexpectedly takes a downturn. A number of user factors such as newly funded projects, changing experimental initiatives, failure to plan, cessation of standing orders, and periods of relative research dormancy alone or collectively may have a dynamic impact on production. For animals from a production colony, matching supply with need is a moving target, and, therefore, production cannot reasonably be expected to match demand.

The case of creating genetically unique animals by using homologous or nonhomologous recombination presents many examples of animals that are produced and euthanized without further use. In the creation of a transgenic (nonhomologous insertion) mouse using the pronuclear microinjection technique, only 10%–30% of the mice born will carry the transgene; thus, 70%–90% cannot be used. [38] Although most transgenic founders will subsequently transmit the foreign gene in 50% of their offspring, approximately 20%–30% are mosaic and transmit the gene at a lower frequency (i.e., 5%–10%).[39,40] Consequently, in the breeding of transgenic founder animals anywhere from 50% to 95% of the offspring will not carry the gene of interest, will have no research value in many instances, and must be euthanized.

In the case of genetically unique mice or breeding of small colonies used by a single investigator during periods when progeny may not be needed for experiments and where sperm or embryo cryopreservation techniques may not be available, mating must continue in order to perpetuate the colony. When this happens, the offspring that are not selected as breeders must be euthanized. In managing inbred animals by a single-line system, genetic divergence is prevented by euthanizing breeders, their progeny, and their grand-progeny that cannot be traced within a minimal number of generations to a common ancestor. As one might imagine, in a large production operation involving multiple genotypes of animals, the euthanasia of large numbers of animals results.

For these reasons, it is recommended that IACUCs permit animals to be produced from breeding colonies in reasonable excess of projected or historical need.

15:19 Should unseamed animals (particularly rodents), or animals used or harvested *in utero*, be accounted for in the protocol review process?

Reg. There is no requirement to account for unborn fetuses. OLAW/NIH directs that if a fetus is removed early for experimental purposes and survives, it must be accounted for.

Opin. If painful procedures are to be performed on animals, including those that are immature but at a stage of development able to perceive pain (i.e., late *in utero* or neonatal), it is clearly within the area and responsibility of the IACUC to address this issue (*Guide*, pp. 25–26). If suckling rodents will be subject to any manipulation, such as thymectomy, toe clip, ear notch for identification, tail tip excision for genotyping, or behavioral tests, the estimated number of manipulated sucklings must be included in the number of animals used.[27] Essentially all IACUCs adhere to this doctrine and, although no specific guidance exists, most also oversee the use of fetuses in research (see survey below). NIH/OLAW clarifies that institutions must appropriately monitor and document numbers of all animals acquired (through breeding or other means) and used in approved activities and to include genetic culls or those produced in excess of need.[28]

Surv. Does your IACUC ever consider a mammalian fetus to be a vertebrate animal subject to the oversight of your IACUC?

- Not applicable 55/211 (26%)
- Yes, at any fetal age 36/211 (17%)
- Yes, after a certain fetal age (species dependent) 52/211 (25%)
- No 60/211 (28%)
- Other 8/211 (4%)

15:20 Should preweanling rodents be counted and debited from approved protocols especially if procedures are done, and at what age?

Reg. NIH/OLAW expects that all preweaning rodents will be accounted for at the first opportunity without compromising the health and welfare of the animals. These animals should be counted and tracked at the "first manipulation," such as cage change or study-related handling.

Opin. While this expectation seems reasonable, it may be difficult to manage logistically; especially, it may add confusion to the process of monitoring and tracking animal use with accuracy and may be possibly inversely associated with the size of the mouse population and census and influenced by other animal accounting processes. Be that as it may, NIH/OLAW stresses that "institutions need to appropriately monitor and document numbers of animals acquired (through breeding or other means) and used in approved activities" and that such practices "not exclude the disposition of animals inadvertently or necessarily produced in excess of the number needed or which do not meet criteria (e.g., genetic) established for the specific study proposal."[28] In particular, it is difficult to tabulate precise numbers of rodents born without disturbing them at critical times at or around birth and during rearing. The animal resources program's engaging in this process adds disruption and expense to the process without any apparent animal welfare benefit. While it is possible for the research staff to report these numbers to the animal resources program or IACUC office as they are generated by the manipulative event, a high degree of dedication to the process of all parties is required. After protocol review and depending upon the accuracy of information conveyed from the IACUC, there is a risk for the person responsible for tracking to mix, confuse, and interchange the total number of animals born, the subset of any used for a distressful manipulation, the accounting of weanlings, and any parental animals that may be purchased (see 15:19 and 15:21). In institutions where it is applicable, approximately 55% of IACUCs attempt to tabulate the precise number of rat or pups born (see survey below). Clearly, if painful procedures are to be done on animals, including presumably young ones capable of pain perception, or experiencing distress, it is within the aegis of the IACUC to address this issue (*Guide*, pp. 25–26).

Surv. With reference to indicating the number of pups born in a rat or mouse litter what does your institution do?

- Not applicable. 39/211 (18%)
- We try to count individual animals. 117/211 (55%)
- We estimate the number of pups in the litter. 42/211 (20%)
- Other. 13/211 (6%)

15:21 A commercial animal vendor sends two more rats than were ordered. How should the two extra animals be accounted for in the IACUC's and PI's records?

Opin. On one hand, it may not be necessary to tabulate the extra rodents given that the animals were not ordered, were received unexpectedly, and could not reasonably be considered by using statistical analysis or other means in the formulation of a protocol. Conversely, the animals do represent a windfall that could be used in manners or for purposes outside IACUC oversight. If the approved use involves highly invasive procedures, considerable pain or distress by design, or species of interest for any variety of reasons (e.g., regulated by the AWAR), the IACUC may be particularly interested in regulating any access to individual animals in excess of the IACUC-reviewed and IACUC-approved figures. In the end, the housing density, census tabulation methodology, and protocol debiting procedures at the operative level in the animal research facility are likely to determine whether the extra animals are counted. For example, at our institution the rat census is done by counting each individual animal. Each rat has a corresponding individual cage card, and electronic census methods are used. Therefore, the extra rats would be debited from the protocol when new cage cards were made at the time of initial housing immediately after receipt. If the census is done by the cage, as for mice at our institution, extra mice in a shipment may or may not be counted. Capture of these animals for accounting purposes would hinge upon the research instructions for housing density and whether additional cage cards must be generated to accommodate all of the cages. In the context of the overall numbers of animals received by the typical investigator or acquired annually by the institution, these extra animals, included in shipments to buffer against any incidental mortality, are generally insignificant. The authors recommend that the IACUC work with the institutional animal resources program to develop a written policy and procedure regarding the disposition of animals received unexpectedly.

15:22 An IACUC approves the use of one litter of rodent pups (approximately ten pups) that will be born of one dam. What can the IACUC do to prevent a person from purchasing a total of 11 animals, rather than just the one dam?

Opin. Human and technological resources must be used rationally to apply constraints that in the end allow only the acquisition of the number of animals approved by the IACUC to conduct valid experiments. For example, when timed-pregnant female rats are purchased as the source of pups to be used in a study, to prevent excessive ordering it is necessary to segregate the number of sucklings projected for experimental use from the number of dams estimated to produce them, and to convey the appropriate information to the purchasing entity. If one thousand nursing rat pups of either sex have been justified for use in acute experiments and one hundred outbred, pregnant female rats have been estimated to produce that number, the total number of rats approved to be *purchased* for the study would be 100 (the timed-pregnant females). If the IACUC was to approve 1,100 rats (1000 + 100) and convey this to the purchasing entity (and even injudiciously to the investigator), this might create an error whereby up to 1,100 timed-pregnant females could be ordered. The requirement to monitor, document, and scientifically justify the acquisition of all 1,100 rats required by the study can be met in the context of the IACUC protocol review. After all, the pivotal regulatory requirements to approve a certain number of animals for a study are somewhat conflicting and ambiguous. The various charges are to consider the approximate (AWAR §2.31,e,1; PHS Policy IV,D,1,a), appropriate (AWAR §2.31,e,2; PHS Policy IV,D,1,b), or minimal (PHS Policy IV,A,1,g) number of animals needed. Arguably, this is not necessarily the precise number. As a facet of protocol review, and in an attempt to meet these requirements, it may be helpful to specify the number of adult animals to be acquired versus those to be subsequently born at the institution.

15:23 An investigator uses avian embryos in his research. What is the responsibility of the IACUC, if any, in overseeing this activity?

Opin. Although PHS Policy does not discuss considerations attendant to embryonated bird eggs (or mammalian fetuses, for that matter), nonvertebrate animals (chick embryos within eggs) have the potential to develop into hatched vertebrate animals. While embryonal states of avian species develop vertebrae at a stage in their development prior to hatching, NIH/OLAW has interpreted *live vertebrate animal* to apply to avians only after hatching,[41] and the AWAR define bird for the purposes of regulation as "any member of the class Aves, excluding eggs, but including birds once the hatching process commences." However, the risk of eggs hatching and producing chicks (requiring food, water, proper housing, and veterinary care and placing them under the purview of PHS Policy) dictates that IACUCs consider developing policies for different aged avian embryos, newly hatched birds, the point at which bird embryos are considered vertebrate animals, and interventions that address the care or euthanasia of animals that hatch unexpectedly.[41–43] For chickens, the last three days of incubation (incubation days 18–21) represent the last stage of embryo development and coincide with the chick's drawing the yolk sac into the body and having sufficient pulmonary maturation to handle oxygen and carbon dioxide exchange.[44] During this period, some chicks may hatch normally and some prematurely hatched chicks may survive outside the egg with little additional care.

15:24 Zebrafish are born with their yolk sac attached. At what stage of development are they considered vertebrate animals under the PHS Policy?

Reg. The PHS Policy has been interpreted by NIH/OLAW to apply to fish and amphibians at the time of hatching.[45]

Opin. As larval forms of fish and amphibians have vertebrae, PHS Policy applies to the offspring of egg-laying vertebrates upon hatching. For zebrafish larvae, for example, this typically occurs three days post-fertilization. This definition should cover all reproductive strategies found in fish, including oviparous and viviparous species.[45] Fish are not regulated by the AWAR.

15:25 How, and with what precision, should animals be accounted for when used in field studies?

Reg. The AWAR (§1.1, Field Study) grant an exemption to field studies in which free-ranging wild animals are observed in their natural environment and without the involvement of invasive procedures or intrusions that harm or materially alter behavior. Where federal funds are used to support field studies, PHS Policy and the U.S. Government Principles, as specified in the *Guide* (pp. 199–200), dictate that the approximate number of animals to be captured be addressed.

Opin. Although the AWAR grant an exemption to observational field studies, few such studies can be done without interventions or some degree of invasiveness.[46] Given the variety of procedures that could be done in the field, the associated stress and risk of untoward outcomes associated with capture, and variation in researcher experience and skills, virtually any procedure, even simple handling, has the potential to cause harm. Because the tabulation of animal numbers revolves around the issue of invasiveness, it is important for the IACUC and involved researchers to address the specific procedures, potential refinements, and qualifications and training of personnel. After appropriate deliberation, it should be possible to define the relative invasiveness of the field study. If the study is determined to involve more than detached observation, the IACUC and investigator must address not only the acquisition of animals by capture, but also any allowable rate of attrition related to capture and the experimental interventions. IACUCs also should be aware that the federal and state wildlife permitting processes are usually rigorous and involve an assessment of hazard of the procedures and overall risk to the population to be studied. However, this does not necessarily include the implications for individual animals. Investigators should define the potential number and species of targeted and untargeted species that might be captured. Additionally, the protocol should address, and the IACUC should consider the merit of allowing investigators to use excess animal numbers. The numbers of animals required in studies of wildlife will vary greatly depending on study design, species' life-history characteristics, and questions posed. The needs for behavioral studies range from the capture of only a few animals where the focus is on a specific behavior to an entire population to mark all individuals.[3] Genetic, taxonomic, ecological, and other studies require a minimum sample size for statistical analyses. Too few animals might not allow the investigator to address research questions with sufficient scientific robustness and, subsequently, will result in a waste of animals if the results cannot be applied to test a hypothesis.[3] Statistical tools, such as power analysis, however, still have utility for field studies in estimating the number of animals required to obtain statistical significance for a given level of variance and a minimum difference between samples.

It is reasonable for the IACUC to review and approve a field study protocol with only a general description of animal numbers provided there is a rationale for that number and explanation for any vagueness.

While in small-scale, well-defined experiments in the laboratory using only a few fish or birds or wild-caught rodents, it is possible to limit animal acquisition to a specific number, this is not ordinarily the situation in the field. For example, in the capture of fish, netting of birds, or live trapping of small mammals, IACUCs should appreciate that while investigators can define the number and identity of targeted species in the field required to meet their research objectives, it is not reasonable to expect them to know how many will actually be captured or caught. This is especially true in survey work and in other types of exploratory investigations.[3] If captured, targeted, and accidental species are of such rarity as to be unlikely to be encountered, or if the captured animals are highly stressed and have a strong likelihood of dying upon release, some argue that it is irrational not to use them to advance the knowledge base to benefit society.[47] In the situation in which capture has an unexpected high yield, this serendipitous increase in animal numbers may provide a more robust database for evaluating specific hypotheses. Finally, the IACUC and field investigators should establish a mechanism to gain IACUC approval for protocol modification expeditiously from the field. A DMR process that allows for rapid review and approval or allowing a broad yet justifiable range for the number of animals to be collected would be best practices. Anticipating and describing the rare occurrences in the protocol allows for their review and approval and is preferable to addressing a noncompliance situation after the fact.

The investigator should also provide assurance to the IACUC that collection permits necessary for the use of wild mammals have been issued for the proposed project and copies of permits might be requested by a prudent IACUC.[3] Many states require an annual report as a condition of permitting. These reports can be submitted to the IACUC to satisfy the APHIS/AC annual reporting requirements of the number of animals used. (Animals that are part of field studies, as defined in the AWAR (§1.1, Field Study), are not required to be listed in the APHIS/AC annual report.) Investigators should additionally obtain permission of the owner, operator, or manager of any private property to be used or of the management authority for public lands involved as a site before commencing fieldwork thereon.[3]

15:26 Should the IACUC allow additional animals to be acquired for approved studies to compensate for animal losses due to random deaths, experimental error, husbandry-caused mortality, surgical complications, and the like?

Opin. It is reasonable to assume that not all animals that begin on a study will be alive at the end, as a result of causes other than those related to the experiment. In a breeding colony of healthy C57Bl/6J mice, for example, the mortality rate of adult mice in otherwise good health and not subjected to experimental manipulation has been documented at 0.13% per week.[48] At our institution, the tracking of mouse mortality by our animal resources program from 2012 to 2023 and for an approximate average daily census by head count over that period of 63,115, the mortality rate on a weekly basis was at least 0.15% +/– 0.01% (mean +/– 1 standard deviation). This was for an uncharacterized and mixed population of inbred, outbred, mutant, and various crossbred mice used for breeding, aging, invasive and noninvasive experimental purposes, and other applications. The number most likely represents an underestimate to some degree as it comprised only mortality reported by animal care personnel and, with rare exception, not those animals found by researchers. Separate from research factors, the husbandry program can be a source of unanticipated morbidity or mortality such as from cage flooding or dehydration if, for example, an automated watering system fails. The risk of not allowing compensation for some degree of unplanned attrition is that data will be collected from a population with insufficient animals, statistical analysis will lose robustness, some opportunities to arrive at significant conclusions will be forfeited, and animals may be used wastefully. The jeopardy in allowing overly excessive animal numbers may be preventable animal pain and distress and undue loss of animals that may be undetected by the IACUC for a considerable period.

The important consideration for the IACUC is to determine a "normal" or "acceptable" rate of spontaneous mortality for the genotype and that associated with a given procedure or aspect of research. Excluding iatrogenic causes, attrition ordinarily from natural causes should not be very high and in the range of 5%–7% annually for many populations of healthy, non-aged adult inbred mice.[40,48] and substantially less for non-inbred species higher on the phylogenetic scale. However, in situations in which the animal phenotype predisposes to a general lack of vigor or even death, and inexperienced persons are being trained in a new procedure, mortality above nominal rates may be observed. Especially with surgical models, IACUCs are often asked to allow for additional numbers of animals to account for technical mistakes, experimental errors, and other complications.

Mortality related to surgery may be influenced by the genetic and health background of the animal, the anesthetic regimen, competency of the surgeon, programs of intraoperative and postoperative care, and the invasiveness of the surgical procedure. High-risk procedures such as thoracotomy, bowel transplant, and three-quarter and five-sixth nephrectomy models will have ostensibly higher associated death rates than uncomplicated procedures such as embryo transfer, orchiectomy, or cranial implant. Mortality should not be very high in surgery, however, when well-trained and experienced persons use healthy animals and established procedures that meet contemporary veterinary standards, including appropriate asepsis and postoperative care. In the absence of readily available published or internal data applicable to the model, a reasonable expectation for mortality in association with surgery in such cases might be 5% or less. When highly invasive, painful, or distressful experiments are planned, especially for species regulated by the AWAR, the IACUC may choose to exercise appropriate oversight by approving only the minimal quantity of animals rationalized as being necessary. Additional animals might be allowed to be added only by amendment, after IACUC review. Most IACUCs (see survey below) proceed with no guidelines regarding unforeseen losses. Instead, the rationale for extra animals is considered on a case-by-case basis in the context of protocol review. When IACUCs allow for a mortality "cushion" to be built into experiments, the general practice is to allow for a 10% or lower mortality rate without additional review.

Given that it is reasonable for new technical employees (not to mention students engaged in research in academic institutions) to be learning procedures, the IACUC might prudently focus on training by initially allowing a number of animals to be used for non-survival training preparations rather than artificially increasing the sizes of experimental groups.

Surv. What percentage of animals does your IACUC typically approve of for unexpected deaths or related research problems?

- Not applicable 21/211 (10%)
- Each request handled on a case-by-case basis 147/211 (70%)
- 10% or less 25/211 (12%)
- More than 10% 2/211 (1%)
- Typically, whatever the investigator requests 10/211 (5%)
- Other 6/211 (3%)

15:27 How can the IACUC prevent the excessive acquisition or breeding of animals beyond approved limits?

Opin. Many institutions have automated systems that will alert an appropriate individual when an investigator has reached a preset percentage (e.g., 80%–90%) of the number of animals approved for a specific project and can prevent ordering of animals in excess of the number approved.[27] The circumstances are more problematic when animals are produced to excess from an in-house breeding colony. In many instances, it may be not only impractical, but also draconian, to cease all breeding immediately, euthanize all animals that are in excess of the IACUC allowance, separate all breeders, and the like. It is incumbent upon the IACUC to articulate a clear policy and work closely and collegially with the investigators and animal resources program to nurture a culture of compliance. In the rare and unfortunate case of recalcitrance, the IACUC can resort to the options of protocol suspension or denial of access to facilities and notification of the proper regulatory authorities and funding entities.

15:28 Is IACUC approval needed for the use of animal tissues acquired from a slaughterhouse?

Reg. Ordinarily, neither the AWAR nor the PHS Policy protocol review requirements apply to dead animals in the context of research, including those obtained from abattoirs[49] (see 15:51).

Opin. The use of shared tissues and slaughterhouse material is an effective application of the "three Rs" and is encouraged where scientifically appropriate.[49] IACUC approval is not required if the collection takes place postmortem and as a byproduct of the commercial enterprise.[49] (If AWA regulated species are euthanized specifically for the purposes of a research study, this is regulated under the AWA.) The use of dead animals or parts of animals in research, however, is an area for which a clear institutional policy can help prevent serious misunderstandings and possible compliance problems.[49] It is in the institution's best interest to ensure that issues such as occupational health and safety, potential disease transmission to colony animals, institutional and personal liability, and related matters are addressed. A few IACUCs delegate oversight for this to some combination of the AV, safety officer, and IACUC Chair or other official designee of the committee. Whatever the mechanism, it is recommended that the IACUC and investigators maintain appropriate documentation, and institutions tailor their policy to meet their needs.[49] If the samples are obtained antemortem and are not incidental to slaughter (e.g., blood), or if they dictate the slaughter procedures in any way, then a protocol may be necessary, depending on the nature of the study. Examples of scenarios of when a protocol is likely to be needed include where biological fluids (e.g., blood, urine, milk, saliva, semen) or tissue specimens or biopsies are collected from animals prior to death or where live fetuses are obtained.

PHS grant applicants using shared animal tissues or slaughterhouse materials are advised to specify the origins of the tissues when describing their proposed use as appropriate in an application.[49] Otherwise, any reference to the use of animal tissues in the application is likely to trigger questions related to IACUC review. Therefore, an explanation in the application that the tissues will be harvested from dead animals at a slaughterhouse will help prevent complications in the peer review process.

Surv. Does your IACUC review the use of animal tissues acquired from a slaughterhouse?

- Not applicable. 98/211 (46%)
- We have no policy regarding that type of acquisition. 23/211 (11%)
- IACUC review is not required. 55/211 (26%)
- A formal IACUC review is required. 9/211 (4%)

- We require notification via a notice of intent, memorandum of understanding, informal review by an IACUC member, or other such means of notification. 18/211 (9%)
- Other. 8/211 (4%)

15:29 What should be required by the IACUC if a PI wishes to use blood or other body parts from an animal that was killed at a slaughterhouse, animal control institution, veterinary clinic, or for reasons otherwise not related to research?

Opin. It is recommended that IACUCs have a pro forma policy and procedure established for this sort of eventuality. Many IACUCs find it effective to have short, one-page forms for the purpose of specifically documenting animal tissue acquisition and use. A useful mechanism is to have a form for tissues from any source including abattoirs, intramural projects in which animals are euthanized, and tissues are made available to others, or extramural transfers. Completed forms can be given designated member review by the committee chairperson or designee, acting under the auspices of the IACUC, although this is not required.

15:30 Is IACUC approval needed for the use of animal tissues acquired from animals euthanized as part of an unrelated experiment? For example, without having an IACUC-approved protocol, can a scientist remove body parts immediately after euthanasia from an animal on a colleague's IACUC protocol?

Opin. NIH/OLAW considers the use of shared tissues to be an effective application of reduction, refinement, and replacement and encourages this practice.[49] Assuming the animals and the method of euthanasia were covered under an approved IACUC protocol, IACUC approval is not needed.[50] Likewise, the harvest of tissues from dead animals itself is not considered research under the AWA and therefore is not a regulated activity.[49] However, if animals are killed for purposes of obtaining or using their tissues or other materials or if there must be distinct antemortem manipulation necessary to obtain the tissues, IACUC review and approval are required.[49] Many institutions choose to require protocol review for all activities involving living or dead animals. This places the decision-making control at the IACUC level, and where the practice may best serve the interests of the institutions for a variety of regulatory and nonregulatory reasons (e.g., public relations, liability, occupational health and safety).[49] Committees that elect not to oversee this activity generally do so as long as the tissues collected do not alter the approved procedures in any way (including addition of extra animals), that the animals were used originally for the research of others (and not for the individual receiving the tissue), that hazard assessment showed no risk of injury or infection for personnel, and that the tissues were collected after the animal was dead. In these instances, it is important for investigators and IACUCs to avoid the temptation to misuse the dead animal exemption to circumvent policy and regulation. For example, the idea that the sole determining issue is whether or not whole animals are involved (making oversight of tissue acquisition unnecessary since tissues do not constitute whole animals) is incorrect.[50] IACUCs should consider this issue and have a policy and practice. Some committees may abdicate this responsibility because the animal is dead or only parts are taken, but this has the potential for creating problems if safeguards or understandings are not in place to ensure personnel safety and oversight of the ultimate use(s) of the animal. In institutions where IACUC approval is required, the process often involves the use of a brief form and a designated member review by the IACUC Chair, AV, or other designee. In all cases, a "paper trail" is recommended to show that an institution has applied the appropriate standards to the acquisition, use, and disposition of animals.[50] Likewise, to prevent peer review complications, PHS grant application procedures should be assiduously observed.

Surv. Does your IACUC require approval for the use of animal tissues acquired from dead animals that were euthanized as part of an unrelated experiment?

- Not applicable 33/211 (26%)
- Yes 41/211 (19%)
- No 115/211 (55%)
- I don't know 5/211 (2%
- Other 17/211 (8%)

15:31 Are there reasons why IACUCs should review the acquisition of animal tissues acquired from animals euthanized as part of an unrelated experiment?

Opin. Assuming the animals and the method of euthanasia were covered under an approved protocol, IACUC approval is generally not needed.[49] However, approval is obviously necessary if the investigator receiving the tissue, in the course of meeting research needs, causes more animals to be used or different procedures to be performed on live animals (see 15:30). Review or policy establishment by the IACUC should also be considered when there is a risk of pathogen transmission from one facility to another within a research institution. For example, several rodent viruses may be transferred into tissues. Direct or indirect contact with such tissues may infect previously disease-free rodents in another animal research facility, invalidate the research in which the animals were used, and cause additional animals to be used unnecessarily as a replacement for those infected. Occupational health and safety issues also dictate IACUC consideration of tissue transfer and use. Some rodent pathogens, such as lymphocytic choriomeningitis virus, are transmissible to humans. Tissues from nonhuman primates, wild-caught animals, livestock, and unconditioned dogs or cats may also serve as a source of pathogens for humans. The authors suggest institutions implement a policy for management of biosecurity in the situation where tissues are used in animals. Additionally, there may be experiments done that could lead to public relations challenges for the institution, such as research using fetal tissue that may not be consistent with the institutional mission or philosophy, or research on tissues acquired from animals euthanized at an animal shelter.

Surv. Is IACUC approval needed for the use of animal tissues acquired from animals euthanized as part of an unrelated experiment? For example, without having an IACUC-approved protocol, can a scientist remove body parts immediately after euthanasia from an animal on a colleague's IACUC protocol?

- Yes 87/211 (41%)
- No 76/211 (36%)
- We do not have a policy 48/211 (23%)

Surv. Does your IACUC have a policy or SOP for the use of postmortem tissues?

- No 139/211 (66%)
- Yes, and require an end user to operate under IACUC oversight. 26/211 (12%)
- Yes, but we do NOT require an end user to operate under IACUC oversight 46/211 (22%)

Surv. If there is a policy or SOP, what does the policy say?

- A protocol is required. 29/211 (14%)
- A memorandum is required. 13/211 (6%)
- No review or documentation is required. 71/211 (34%)
- Other* 98/211 (46%)

*Other:

- Notification by means such as a tissue use form.
- Documentation of the source of the tissue.
- A protocol is required if the animals are euthanized solely for the purpose of tissue harvest.
- Statement that the tissue has been tested for pathogens if the tissue is from an external source.
- A protocol is required only when tissue harvest involves a species regulated by the USDA.
- A protocol that is not reviewed by the IACUC is required as a way to document the activity.
- Safety review.
- We do not have a relevant policy.

Surv. Does your IACUC assess potential biosecurity risks associated with animal carcasses?

- No 41/211 (29%)
- Yes 170/211 (81%

15:32 Without having an IACUC-approved protocol, can a scientist remove body parts immediately before euthanasia from an anesthetized animal on a different scientist's IACUC protocol?

Reg. NIH/OLAW and APHIS/AC stipulate that any procedure involving antemortem manipulation of live animals must be subject to protocol review.[49] PHS policy does not apply to dead animals (see 15:31).

Opin. In the context of the tissue-receiving investigator, it should be noted that NIH/OLAW has issued guidance that identifies a change, including addition of a species, as an example of a "significant change" requiring prior IACUC approval.[51] Likewise, IACUCs must have administrative mechanisms to ensure that all participating personnel are adequately trained and qualified, participate appropriately in the occupational health and safety program, and meet all other applicable criteria.[52] Acquisition of body parts from a live animal by a person not approved to be engaged in the procedure represents potentially serious non-compliance with the AWAR, PHS Policy, and federal granting procedures. NIH/OLAW has indicated its determination that conducting unauthorized animal research activities is a serious and reportable violation.[25] This position is also consistent with the NIH Grants Policy Statement (Rev. 04/21) that identifies the substitution of one animal model for another (or presumably addition of an animal model where none existed previously) as an example of "change of scope" likely to require prior NIH (grants management officer) approval.[53,54] The signature block on the face page of the PHS form 398 grant application requires the institution to declare that it meets and will maintain compliance with all applicable terms and conditions of award. Consequently, IACUCs should have clearly articulated and well-known policies in this regard. When the situation of tissue sharing has been encountered by IACUCs, survey of IACUC Chairs (below) shows that the majority require an IACUC review. Given that prudent sharing of animals is encouraged as a means of reducing the number of animals used in research,[50,55] it may not be necessary for the second scientist to have an independent protocol. Nevertheless, the "piggybacked" scientist should be added to the personnel list and meet credentialing qualifications, and the IACUC should specifically review any new or modified procedures for the existing protocol.

Surv. Does your IACUC require its approval for the use of tissues acquired from an anesthetized animal that is about to be euthanized (i.e., before the animal is dead) to be used in an unrelated experiment?

• Not applicable.	13/208 (6%)
• Our IACUC would require its prior approval before tissues could be removed.	176/208 (85%)
• Our IACUC would not require its prior approval before tissues could be removed.	8/208 (4%)
• I don't know.	5/208 (2%)
• Other.	6/208 (3%)

15:33 Is IACUC approval needed for acquisition of animal tissues from abroad?

Opin. Investigators who collaborate with scientists in other countries may receive animal tissues from these foreign sources and may sometimes go abroad to collect samples from captive wild animals maintained in zoological collections and in research colonies. Just as animal-related activities supported by the PHS require review by the IACUC of the domestic awardee institution, foreign institutions that serve as performance sites must have a Foreign Assurance on file with NIH/OLAW.[21,56,57] NIH/OLAW considers institutions whose scientists are engaged in such collaborative work accountable for the animal-related activities from which they receive animals or animal parts. When a foreign institution holds a PHS Assurance, it also is expected to conduct the study in accordance with the applicable host nation's policies and regulations. In the specific case of sample collection, the review should consider the species involved, the nature of the specimen, and the degree of invasiveness of the procedure and should give appropriate consideration to the use of anesthetics and analgesics. When samples are to be obtained directly by citizens of a foreign country for subsequent shipment, the recipient PHS supported investigator should determine the proposed methods of collection and present that information to his or her IACUC for review. Prior to sample collection, and regardless of whether specimens are obtained by an awardee institution's investigator directly or by persons in a foreign country, NIH/OLAW strongly recommends that each awardee institution consult other agencies of the U.S. government concerning importation requirements.[21,58] Depending on the species involved and the nature of the specimen, the following may be of assistance: the U.S. Fish and Wildlife Service, Department of the Interior (for compliance with the International Convention on Trade in Endangered Species of Fauna and Flora [CITES]),[6] USDA/APHIS (regarding potential animal pathogens),[4] and Centers for Disease Control and Prevention (concerning importation of nonhuman primates and potential pathogens of human beings).[8]

15:34 Using PHS grant monies, a research institution purchases surgically modified mice from a commercial vendor. Must the vendor have NIH/OLAW Assurance?

Reg. The PHS Policy is applicable to all PHS-supported activities involving animals, whether the activities are performed at a PHS agency, an awardee institution, or another institution.[59]

Opin. The determining issue is whether the surgery is conducted in response to a specific custom surgery request or whether the animals were previously surgically modified and available for sale "off the shelf" (e.g., from a catalog) before the PI's need was anticipated.[59] If an investigator requests that a specific custom surgical procedure or procedures be performed on an animal for use in activities funded by the PHS, then the organization or person who conducts the surgical procedure(s) is considered to operate from a performance site and must either have on file with NIH/OLAW an approved Animal Welfare Assurance or be included as a component of the applicant organization's Assurance.[59] If the animals are commercially available as surgically modified models, such as in the case of hypophysectomized rats or mice with implanted devices, then the supplier is not considered a performance site and is not required to file an Assurance with NIH/OLAW.

15:35 Under what conditions can the IACUC permit the use of animals purchased from a retail pet store or the wholesale pet trade?

Reg. The AWAR require licensing of pet shops that supply warm-blooded animals, other than captive-bred birds and domesticated laboratory rats and mice, for research purposes (§1.1, Retail Pet Store; §2.1,a,3,i).

Opin. As a general rule, this is a practice that should be avoided. It may be acceptable for ectotherms, such as fish; for rare types of rodents or rabbits; or for certain birds or reptiles when such animals cannot be acquired from a more traditional or conventional source. Where regulated species (see 14:1) are to be obtained from a pet shop, the institution has an obligation to obtain the informed consent of pet shop management and to procure the animals in a lawful manner (*Guide*, pp. 106–107). The purchase of animals from pet shops should also fall under the auspices of the IACUC because of the potential for abuse and the risk to the institutional reputation should this practice become known to the public. Animals that are obtained from pet stores may not be acclimated to a laboratory environment and are of unknown genetic and health backgrounds. With respect to the latter, the veterinary staff should be called upon to evaluate the health status of the animals and quarantine, isolate, and otherwise manage these animals to protect other research colonies.

Surv. Does your IACUC permit the acquisition of research animals from a pet store or commercial pet supplier? Select all that apply.

- Not applicable. 40/208 (29%)
- Yes, and we require a justification for the source of the animals. 57/208 (27%)
- Yes, and we do not require justification for the source of the animals. 2/208 (1%)
- Yes, but only for species not covered by the Animal Welfare Act regulations. 18/208 (9%)
- Yes, for any species. 7/208 (3%)
- No, pet stores are prohibited as a source of animals. 73/208 (35%)
- This question has not been encountered by our IACUC. 45/208 (22%)

Surv. Should the IACUC permit donations of pets or sporting animals from their legal owners for research purposes?

- No 168/206 (82%)
- Yes 38/206 (18%)

Surv. If yes to above question on donation of sporting animals, how can this best be promoted to legally serve the donor and recipient?

- Please specify 30/30 (100%)
    - Donation form/MOU with signatures
    - Formal agreement describing how the animals will be used and for what purpose
    - Formal agreement that specifies who the legal owner is at all times
    - Formal legal document

15:36 Are there legal requirements that must be met for purchase of animals from a commercial farm?

Reg. Farm animals used in non-agricultural research are regulated under the AWAR, as are their dealers and those that transport them. Farm animals that are exempt from the AWAR include those intended for use as food or fiber or used or intended for use improving animal nutrition, breeding, management, or production efficiency, or for improving the quality of food or fiber (see 15.2). Any dealer selling farm animals must be licensed to do so, unless they have been formally exempted by the Secretary as being "de minimis" (AWA § 2133).

Opin. AWAR regulate farm animals produced and used for biomedical research purposes and their sale. The research facility becomes responsible at the point where they assume custody of the animal. Therefore, if an employee of the research facility goes to a livestock market or farm and purchases livestock, then that person, as an agent of the research facility, must satisfy the AWAR transportation requirements (AWAR §3.136–§3.142) and the institution must satisfy health and husbandry standards (AWAR §3.125–§3.133).

15:37 Can animals be transferred between studies that have IACUC-approved protocols?

Opin. Provided certain requirements are met, it is permissible to transfer animals between studies. Survey data from previous editions[29,33,60] and this volume show this to be a universal and standard practice. Unrestricted trading of animals between research protocols, however, could lead to unauthorized use, defeat attempts at tracking animal utilization, put legal records pertaining to dogs and cats in jeopardy (AWAR §2.35,c,1–§2.35,c,2), and be potentially confusing to veterinarians who try to determine the ownership of animals requiring medical care. As such, transfer procedures should include informed IACUC consent for the transaction; approval by the IACUC of the use of the species in the research; a means to assure that multiple, unrelated major survival surgical procedures are prevented; and a mechanism to credit animals to one protocol and transfer them to another via a debit transaction. IACUCs may choose to discriminate in terms of the stringency of process control. For example, the process could be managed through an IACUC administrative function or by the animal resources program on behalf of the IACUC or deferred to the scientists involved. Important considerations in this regard might be whether or not the species is AWA regulated or the degree of invasiveness of the intended use. For those species that are not regulated under the AWA, such as rats and mice bred for use in research and fish, IACUCs might allow less restricted transfer. The technological innovation of census management by automated processes allows transfer procedures by crediting the animals to the donor's account and debiting the exchanged animals from the total approved for the recipient collaborator.

Surv. Does your IACUC permit animals to be transferred between studies having IACUC-approved protocols?

• No applicable.	7/208 (3%)
• Yes, we permit transfers.	181/208 (87%)
• No, we do not permit transfers.	14/208 (7%)
• Other.	6/208 (3%)

15:38 Under what circumstances or conditions may animals be transferred between studies and how can the IACUC track this activity?

Opin. At institutions where there are breeding colonies of animals supporting more than one research project, there must be a mechanism to transfer animals from the breeding colony to IACUC-approved research protocols. An example of the latter is an institution that has a core facility to produce genetically engineered animals. Those of a desired genotype are produced under one IACUC-approved activity (the core) and are then transferred to the IACUC-approved activity of a specific scientist for research use. Other cases involve simple transfers in which one investigator receives a few breeders of one-of-a-kind genetically manipulated mice from the breeding colony of a colleague. Nonhuman primates used in operant conditioning or behavioral paradigms present another example. These animals are usually purchased from a commercial vendor and are quarantined in the animal research facility. During the quarantine phase (or shortly thereafter), the investigator may become aware of behavioral considerations that render the animal less appropriate for the intended research and more appropriate for use by a colleague. In this instance, the investigators may wish to exchange animals. The same may be true for dogs or cats obtained for use in research involving surgical procedures. In this circumstance, demeanor and size may lead scientists to trade animals between acute and chronic studies, or where the animals have anatomical or other attributes that are better suited for certain studies than others. Two investigators, one with expertise in a specific experimental manipulation and a second with animals of a unique genotype, may collaborate by transferring the animals from the original owner to the party doing the manipulation.

The most important consideration is that the IACUC employs a mechanism to ensure that the total number of animals used and the species of animal are appropriate in terms of the approved protocol and that there are not multiple major operative procedures in unrelated research projects (AWAR §2.31,d,1,x,A; *Guide*, p. 30). The IACUC should develop some guidelines for review of historical research use, the transfer of animals between approved protocols, and the subsequent monitoring of well-being and outcomes if this is a circumstance that might occur repeatedly. An important consideration in setting guidelines is obtaining the surgical history of the animals from either the investigator or the veterinary staff. This is necessary to prevent animals from being unintentionally subjected to multiple major survival operative procedures. When applicable, the Institutional Official (IO) must submit a request to APHIS/AC and receive approval in order to allow a regulated animal to undergo multiple survival surgical procedures in unrelated experiments (*Guide*, p. 30; AWAR §2.31,d,1,x,C). The movement of animals from one protocol to another ideally should involve debiting or crediting the number of animals against those approved for the study. IACUCs often refer these activities to the animal resources program, which often has the staffing and computer resources to handle the task. Some IACUCs, however, find it advantageous to require the Principal Investigator (PI) to inform the committee of the practice by memorandum or other mechanism.

Surv. Assume animals are bred on Protocol A and then used on Protocols B and C. Does your IACUC have an effective system of tracking the number of animals born and then transferred to the other protocols?

• Not applicable.	24/208 (12%)
• We do not have a tracking system.	13/208 (6%)
• We have a tracking system, but it is not very accurate.	16/208 (8%)
• We have a reasonably accurate tracking system.	131/208 (63%)
• We have a reasonably accurate tracking system, but it is difficult to use or is not used effectively.	13/208 (6%)
• I don't know.	2/208 (1%)
• Other.	9/208 (4%)

15:39 If a protocol is being renewed, how can the IACUC ensure that it is intentionally renewing the original number of animals requested?

Opin. This is an important issue from the perspective of preventing unnecessarily duplicative experiments and providing a justification for the total number of animals used in research. An example of this situation is useful for illustrative purposes.

An investigator is funded for a five-year study involving thirty macaques. He purchases them over the first three years of the study and then submits a renewal after three years, requesting 30 macaques. How can the IACUC be sure that the 30 animals requested on the renewal are new monkeys, and not just a relisting of those already on the study and still in the investigator's colony?

As an extension of this problem, what if the investigator originally was approved for 30 animals but acquired only 20 by the time of the three-year renewal? If 10 animals are now requested in the renewal, how can the IACUC determine whether these are the remaining 10 from the original approval, or 10 more above the original request of 30 (for a total of 40)?

These problems can be addressed with careful formulation of the renewal and modification requests. IACUC review may be facilitated when requests are supported by progress reports of animal usage on the previous iteration. For an annual renewal (end of first or second year for PHS Policy regulated species) the application should not include animal numbers, as the request has already been approved for the first three years. The annual accounting of animals for APHIS/AC should be monitored at the level of acquisition and then once annually for individual animals kept over multiple years.

Requests for additional animals before the end of the original protocol should be handled in a different manner than new or three-year renewals and specifically indicate that additional animals are required. One mechanism to help eliminate the issues raised in the examples is to close out the original protocol regardless of the number of animals used or not used and to consider the continuation as an essentially new proposal, with a new set of experiments and requirement for justification of animal numbers. Obviously, there are circumstances, as in the example, where further questions, such as the following, should be asked:

- Is this a new protocol or a three-year renewal?
- If the protocol is a renewal, what is the previous protocol number?
- If the protocol is a renewal, are there *existing* animals that are already assigned to the study?
- If so, how many? (Do not include this number in your answer to the next question.)
- Justify the number and use of animals specific to this application.

Many IACUCs do not consider this issue with meticulous detail. Instead, they rely on faith that good science is not going to be repetitious or wasteful, that the vast majority of scientists can be trusted, that economic factors (as much as anything else) have a limiting effect on animal use and dictate judiciousness, and that the IACUC application will show that the research requires a certain number of animals and is novel and important.

When an animal resources program performs census management using electronic technology, it dictates downstream processes for the IACUC, the precise details of which vary depending on the software. Generally speaking, for protocols that will be deactivated due to expiration, three groups of animals must be taken into consideration: 1) the animals that have been used and are no longer on census on this protocol, 2) animals that are actively being used and are on census, and 3) those that have not yet been ordered and are available (theoretically) for purchase. Once the expiring protocol is deactivated, the animals that previously were available to purchase no longer will be and a new cohort of animals must be approved for use on the new protocol. The animals that are on census will be moved to the new protocol and be debited from it. In some systems, they then are immediately issued a new cage card (with new protocol and billing account numbers) and considered as a new acquisition A second option is to transfer the animals from the expiring protocol to the new one, treating them as an internal receipt. To reduce confusion during consideration of the three-year renewal, some IACUCs find it useful for the remaining animals to be added to the new protocol to be listed or described on a supplementary document separate from the renewal application. At our institution, the electronic protocol and census management systems have allowed for immortal protocol numbers. This strategy alleviates the need to expend significant effort creating new cage cards at the time of three-year renewal. The above mechanisms of transferring animals from old to new protocols apply and will be illustrated with an example: An investigator is approved for the use of 50 pigs on a protocol and, over the course of the three-year period, uses 45. At time of renewal, she has 10 pigs (of the 45 used) on census. There are theoretically 5 pigs still available for her to purchase, but these 5e will not transfer to the new protocol. Our investigator needs 50 additional pigs to continue her work for the next three-year period. In her submission to the IACUC, she should describe the 10 pigs existing on census and explain that she needs 50 additional, so she should request 60 pigs. Upon approval, the current census is transferred to the new protocol and debited. Thus, the 10 pigs already on census are tracked and 50 are available for her work to continue.

In summary, special attention should be paid to the type of application (new or renewal), the numbers of animals used, the number requested, and the number to be carried over from the previous protocol. Although it is beyond the jurisdiction of individual IACUCs, in the opinion of the authors many of these problems could be alleviated if periods covered by IACUC protocols and extramural grants could be matched in duration.

Surv. Assume a study is being renewed for an additional period of time. What actions does your IACUC take to assure that animals that were not used during the original approval period are being included in a request for additional animals to be used during the renewal period? Select all that apply.

- Not applicable. 19/208 (9%)
- Our IACUC does not have a policy on this matter. 46/208 (22%)
- If there are any "leftover" animals from the initial approval they must be transferred to the renewed protocol and counted as part of the number of animals approved for the renewed protocol. 110/208 (53%)
- Animals carried over to the renewed protocol are added to the total number of animals to be used via a separate line item. They are then debited from the grand total when the renewal is activated (such as when updated cage cards are made). 15/208 (7%)
- The expired protocol is closed out and any remaining animals are deleted from the census. They can be added onto the renewed protocol via a special form, but they are debited from the total. 14/208 (7%)
- Carryover animals are specifically identified in the renewal. No additional consideration is given except that only the total number of new animals needed over the next three years is divulged to those who purchase the animals. 16/208 (8%)
- Investigators are encouraged to identify carryover animal numbers but doing so is not enforced. 9/208 (4%)
- Other. 10/208 (5%)

15:40 What guidelines can the IACUC establish for the use of endangered species in research?

Opin. The use of endangered species in research must conform with all federal and state laws and international conventions to which the U.S. subscribes, including the Lacey Act, the Endangered Species Act, the Marine Mammal Protection Act, and the CITES treaty. It is desirable for IACUC guidelines to investigators to repeat this requirement.

15:41 Should institutional programs related to veterinary care (such as a rodent sentinel program, quarantine programs involving diagnostic sample collection from animals, necropsy service, rodent colonies for sustaining predatory species (e.g., snakes) or raptor rehabilitation programs, and therapeutic surgical interventions) be approved via the protocol review system?

Reg. The AWAR (§2.31,a) require that all animal activities for regulated species be reviewed by the IACUC. PHS Policy (IV,B,1) and the *Guide* (p. 9) state that the entire "program" must be reviewed by the IACUC.

Opin. PHS Policy applicability is not limited to research; it also includes all activities involving animals, such as testing and teaching. NIH/OLAW has determined that although animals used as sentinels, breeding stock, chronic donors of blood and blood products and for other like needs may not be part of specific research protocols, their use for these purposes contributes significantly to the institutional research program and constitutes activities involving animals.[61] Consequently, the IACUC must receive and approve protocols, have appropriate systems to monitor the use of animals prior to the commencement of such activities, and then perform reviews at the appropriate intervals.[61] The authors also interpret this to include rodent colonies used as a source of food for raptors in rehabilitation programs animals used for training purposes, core rodent breeding services, and those held on expired protocols. All clinical activities (e.g., diagnostic, and therapeutic interventions) and the necropsy service involving existing research or instructional animals ordinarily are under the supervision of the AV, should be covered under the semiannual institutional program assessment, and do not require approval under the protocol review system.

Surv. At your institution, which of the following programs requires an IACUC approved protocol? Select all that apply.

- Our rodent sentinel program. 140/208 (67%)
- We do not have a rodent sentinel program. 55/208 (26%)
- Animal (mammal) quarantine program. 85/208 (41%)
- We do not have an animal quarantine program. 53/208 (25%)
- Rodent surgical re-derivation program. 66/208 (32%)
- We do not have a rodent surgical rederivation program. 119/208 (57%)
- Animal adoption program. 42/208 (20%)
- We do not have an animal adoption program. 100/208 (48%)

15:42 Does the IACUC need to approve studies that use privately owned animals, such as those in a clinical trial in a school of veterinary medicine? (See 14:50: 29:64; 15:7; 15:8.)

Reg. The AWAR (§1.1) include all regulated species used, or intended for use, in research, teaching, and testing. The PHS Policy (III,A) includes any live vertebrate animals used or intended for use in research, research training, experimentation, or for related purposes (if the work is PHS funded, or the institution voluntarily includes all animals under its Animal Welfare Assurance). Neither the AWA, AWAR, nor PHS Policy distinguishes between privately owned animals and those owned by the institution.

Opin. The determining factor as to whether PHS Policy and the AWAR are applicable is whether the veterinary engagement with the animal is in response to the requirements for research or in the usual care of a patient.[62–64] Given that the animals are enrolled in a clinical trial in this scenario, the IACUC must provide oversight. The law and regulations give the IACUC room to deliberate justifiable exemptions, particularly where the use is not covered in the AWAR or PHS Policy (see bulleted examples that follow). Ultimately this involves doing what is appropriate by all parties and documenting the exemption and its rationale. If the treatment plans and monitoring elected by the client-owners are not influenced by the study, then the research use would be incidental and may not fall under the AWAR or PHS Policy.[64] In this case, some entity would need to determine whether or not IACUC oversight applied and the IACUC likely would be in the most qualified position to decide. Beyond that, it is strongly recommended for the protection of the institution that the IACUC has a written policy with respect to the research use of client-owned animals and, at the least, an informed consent is obtained from the client as the representative of the animal (the subject).

As noted in the Regulatory comments to this question, when federal oversight is mandated, neither the AWAR nor PHS distinguishes privately owned animals from those owned by the institution. Several factors are to be considered, including but not limited to the following:

- The specific type of work
- Species to be used
- The need to board or retain the animals for a period
- The nature and resources of the facility
- Involvement of federal funding
- Requirement of IACUC review by a nonfederal granting agency
- The content of the institutional Animal Welfare Assurance

Pets used in clinical trials are privately owned animals recruited from volunteers and, as such, usually remain the property of the owners, who maintain responsibility for their upkeep, housing, and transportation. However, once these animals appear on institutional property, as noted previously, rules should be established as to how the animals are managed and temporarily housed. Such rules should take into consideration facility operations, informed consent, institutional liability, public safety, and animal welfare. When client-owned animals are donated to the institution, the full range of compliance issues takes effect. At those veterinary schools where the IACUC does not review the research use of client-owned animals (e.g., does not have an NIH/OLAW Assurance), teaching hospitals typically have policies governing the conduct of research. Where the veterinary school has an Animal Welfare Assurance or the species to be used is covered by the AWAR or the Assurance of the parent university, the IACUC clearly has full oversight responsibility. In situations not covered by Assurance or the AWAR, the proposed research may be reviewed by the department and then a hospital board or committee akin to an institutional review board for human subject experimentation. In practice, the overwhelming preponderance of IACUCs elect to oversee the experimental use of client-owned animals.

Surv. With reference to veterinary clinical trials at a veterinary or agricultural school: Does your IACUC approve these clinical trials? Select all that apply.

- Not applicable. 175/206 (85%)
- No, a separate clinical trials committee approves these trials. 3/206 (1%)
- Yes, the IACUC approves these trials because there is no other committee to do so. 6/206 (3%)
- Yes, the IACUC approves these trials because they are considered to be research. 17/206 (8%)
- A clinical trials committee approves clinical trials but IACUC approval may also be required in certain cases. 4/208 (2%)
- I don't know. 2/208 (1%)
- Other. 5/208 (2%)

15:43 An institution, a technical college, does not own or house animals but in the veterinary technology program, students encounter animals in classroom demonstrations of noninvasive procedures and during externships. Is the IACUC required to follow the AWAR, PHS Policy, or both, and is it required to monitor the acquisition and disposition of the animals?

Reg. The AWAR apply to the use of animals in teaching, including those that might be encountered in a veterinary technology program such as dogs, cats, guinea pigs, hamsters, or rabbits (§1.1, Animals).

Opin. Although not applicable where such animal use is supported by tuition or other institutional funds, for purposes of consistency and simplicity all instructional use of animals should be reviewed by the IACUC.[27]

The American Veterinary Medical Association requires that the instructional use of animals in accredited veterinary technology programs, regardless of funding source, species, or nature of use, be under the authority of an IACUC.[65] This logically extends even to instances when student-owned animals are used for the classroom demonstration of noninvasive procedures such as physical restraint for examination, vital signs recording, and positioning of conscious patients for radiography and that remain on the premises for less than 12 hours. Under this scenario at the teaching institution, it is within the purview of the IACUC to monitor and track the numbers of animals procured or participating, the frequency of use, and the eventual disposition of the animals. When students work with client-owned animals during an externship at a community veterinary private practice establishment for purposes of acquiring practical skills in specimen collection, medication administration, surgery assistance, or postoperative care, the institution has a stake in assuring that these experiences are meaningful and that the students will be appropriately supervised, mentored, and trained.

Techniques learned in this environment will be carried into subsequent careers. While the level at which oversight occurs is a matter of institutional decision, it is logical to assign the responsibility to the IACUC. In this case, the IACUC should at least consider whether or not these experiences should be covered by protocols.

15:44 An institution's veterinarians or physicians occasionally are asked to provide veterinary services to a pet at no or reduced cost. How should the IACUC address these situations?

Opin. Although research institutions have great resources and clinicians with unique skills, engaging in these activities is a legal "slippery slope" fraught with risk. There is great potential for both institutional and veterinary liability should something go awry.

All states have veterinary practice acts that regulate the practice of veterinary medicine in the state and protect veterinary private practice as a business. When unlicensed specialists, including physicians, undercut fees or provide veterinary services for individuals who otherwise would seek care from private practitioners, the institution and medical professional assume a legal risk.

Some institutions and individuals believe that they can protect themselves by rendering free service or requiring that clients sign consent forms. Nevertheless, the judicial system most likely would not enforce provisions of consent forms in which clients agree not to hold an institution or individual liable for the negligent treatment of an animal.[66] Nor will any consent by a client to have a veterinary procedure done by a non-veterinarian protect the non-veterinarian from prosecution for practicing veterinary medicine without a license.[66]

In some instances, veterinarians and institutions must be pragmatic. Acutely ill or dying animals on or near institutional property, such as stray cats in end-stage disease, dogs that have been hit by cars, or injured wildlife, may require humane intervention and cooperation with local animal control and/or natural resources management authorities. Whatever the reason, veterinarians are bound by oath to relieve animal suffering, but, as noted previously, they must also act in ways that protect research animals from those that have undefined health status. Consequently, institutions may want to develop policies that address situations in which emergencies involving animals that are not university property require prompt and humane intervention. At the author's institution, the preferred course of action is to first request assistance from local nonprofit groups (e.g., county animal enforcement or wildlife groups). If state, local, or private resources are unavailable to address an emergency situation, any person acting in good faith may offer care to animals at the scene of an accident or emergency without incurring liability provided that his/her actions are not grossly negligent. Should care be provided, the expectation is to stabilize the condition of animals with a favorable prognosis that require critical care with the understanding that a county animal control officer will transport them immediately to a local veterinary hospital for subsequent care. In an extreme situation, a severely injured animal may require euthanasia.

15:45 A laboratory animal facility is associated with a human hospital. What policy, if any, should the IACUC have relative to housing service or assistance animals (e.g., Seeing Eye dogs) or other animals that might accompany patients admitted into the hospital?

Opin. At the authors' institution, with a large medical center and several affiliated hospitals, Seeing Eye dogs and animals involved in pet-facilitated therapy programs are permitted in the wards and public areas of the hospital. These issues fall under the jurisdiction of hospital policy and practice and not under the oversight of the IACUC as the animals are privately owned and not used for research, testing, teaching, or training. Under the Americans with Disabilities Act, hospitals, as public accommodations, are expected to provide reasonable access to services and facilities (health care) for those who use service animals,[43] but they are not required to provide care or supervision while they are onsite.[67] Likewise, at our medical center, pet-facilitated therapy is made available to patients through a relationship with a bonded, outside provider. The provider ensures that the privately owned animals participating in the program are of the appropriate age, size, and disposition; are acclimated to patients and the hospital environment; and are in apparently good health (including parasite control and full immunizations) as certified annually by a contract veterinarian.

15:46 What are the legal requirements regarding the disposition of research animals?

Reg. The AWAR require euthanasia methods to produce rapid unconsciousness and subsequent death without evidence of pain or distress, unless otherwise scientifically justified (§2.31,d,1,xi). Further, animals that would otherwise experience severe or chronic pain or distress that cannot be relieved will be painlessly euthanized (§2.31,d,1,v). The AWAR cite specific identification requirements for identification of dogs and cats disposed of by a research facility (§2.38,g,1).

Opin. Animals used in research may be disposed of by euthanasia, adoption, retirement (e.g., endangered species), safe and responsible return to production agriculture (food and fiber species), sale for slaughter, release to the wild, or transfer to another institution or research project. In most instances, animals used in research are euthanized. Considerations when disposing of these animals include the experimental history (e.g., method of euthanasia and exposure to chemicals, toxins, radiation, infectious agents, recombinant DNA, human tissues, and drugs) and may involve burial in a landfill, incineration, alkaline hydrolysis with discharge into the sewage system, or rendering into animal feeds. In some instances, particularly in the case where radioisotopes have been given to an animal, the carcass must be held until the radioactivity decays to safe levels.

15:47 Can animals used in research, teaching, or testing be adopted as pets once they have been used in an IACUC-approved project? What is an appropriate mechanism for facilitating this process?

Opin. Provided the animals are emotionally and physically healthy with appropriate socialization and have not been disfigured or disabled by the research, adoption of animals that make suitable pets is appropriate at the discretion of the institution.[68] This practice, as have many others, may have legal ramifications. Guidelines for adoptions should be formally developed in consultation with legal counsel. Important considerations relevant to adoption include adoption to a reliably suitable and permanent home, consent of the researcher, approval of the request by the IACUC, and confirmation of the good health of the animal by a veterinarian. There may be other restrictions, such as limiting adoptions to employees and students of the institution and not to the general public. Some public institutions do not permit adoption *per se* as adoption entails a change in ownership that involves divestiture of public property. Pragmatic approaches to circumvent this challenge may include permitting animals to leave under the status of a long-term loan or allowing retired animals to remain in possession of the laboratory with care provided by staff. This may lead, however, to questions of whether such "institutional pets" are covered in a protocol and how the program is funded. At the author's institution, healthy animals are released for adoption after an approval process (e.g., the committee has a chance to vote) and the completion of relevant records (i.e., USDA Veterinary Services Form 18-6: Record of Disposition of Dogs and Cats). Copies of all correspondence related to adoptions and the permanent medical records of adopted animals are maintained in the animal resources program office. This allowance comes with the risk that the institution forsakes control of the narrative around the life of the animal as well as the suitability of subsequent possessors. A circumstance may arise, for example, in which an adopter subsequently wishes to surrender the animal. IACUCs should be aware of this potential eventuality and include the possibility in their institutional guidelines.

Surv. Does your IACUC allow animals used in research, teaching, or testing to be adopted as pets once they have been used in an IACUC-approved project?

- Yes 109/206 (53%)
- No 97/206 (47%)

Surv. What are the responsibilities of the IACUC in assessing a potential adopter of research animals? Does the IACUC vet potential adopters of animals?

- Yes 81/206 (39%)
- No 125/206 (61%)

Additional Comment:

- No adoption program, no animals adopted out
- Veterinary staff reviews potential adopters
- Informal process
- External group facilitates adoptions
- Legal department reviews potential adopters

Surv. Does the IACUC vote to allow or disallow a particular adoption?

- Yes 12/206 (6%)
- No 123/206 (60%)
- Other* 71/206 (34%)

*Other:

- No adoption program, no animals adopted out
- No, the veterinary staff approves adoptions
- Yes, if the animal has undergone a surgical procedure other than sterilization
- IACUC approved adoption in the protocol as a possible method of disposition, but does not provide any input on specific adoptions

Surv. Should the IACUC have any role in monitoring the animal post-adoption?

- Yes 22/206 (11%)
- No 184/206 (89%)

Surv. Does your institution allow adoption of research animals only to certain individuals?

- My institution does not allow adoption of research. animals. 81/206 (39%)
- Yes, only to employees. 14/206 (7%)
- Yes, but only to employees and close associates. 26/206 (13%)
- My institution allows adoption to anyone who undergoes an approval process to do so. 53/206 (26%)
- Other* 32/206 (16%)

*Other:

- Adoption to specific groups of people depends on the species being adopted
- Adoption to pre-approved organizations

Surv. Are there restrictions on species?

- Yes 66/206 (32%)
- No 71/06 (34%)
- Only non-ACT animals 2/206 (1%)
- Other restrictions* 67/206 (33%)

*Other:

- No adoption program, no animals adopted out
- Nonhuman primates are not adopted out (other than for release to sanctuary)
- Immunocompromised or genetically modified animals are not adopted out
- Animals exposed to hazardous agents are not adopted out
- Animals that have undergone surgical procedures are not adopted out
- Animals that are a non-native (potentially invasive) species are not adopted out
- Venomous animals are not adopted out

Surv. Does your IACUC play any role in following animals' post-adoption? For example, if the adopter chooses to surrender the animal.

- Yes 10/206 (5%)
- No 138/206 (67%)
- Other* 58/206 (28%)

*Other responses largely reflected respondent's institution as not having an adoption program. Other respondents noted that the IACUC must be notified if the adopter chooses to surrender the animal. Others noted that the veterinary staff conducts follow-up with adopters.

15:48 Should the IACUC permit farm animal species that have been used in teaching, testing, or research to be sold for slaughter? What guidelines would be useful to the IACUC in this regard?

Opin. Domestic livestock, usually raised for food or fiber, are also commonly used in research, teaching, and testing. In some instances, it may be appropriate to return live animals that have been used for research or training purposes to production agriculture or to render carcasses into livestock feed. However, if this is done, it should be done with caution and with full awareness of the potential for institutional and individual liability.

When animals used in research are subsequently slaughtered for human or animal food, the major concern is adulteration of the meat by residues (e.g., anesthetics, antibiotics, hormones, toxins, radioisotopes, carcinogens) or infectious agents that would render the meat or its food products unwholesome. The agencies that regulate slaughter are the USDA's Food Safety and Inspection Service (for federally inspected facilities) or state agencies for state inspected slaughter facilities. Endangerment of the nation's food supply by failure to adhere to food safety standards constitutes a felony in the U.S. that could be referred to the appropriate legal jurisdiction. Given these potential legal ramifications, an institutional policy on the sale, slaughter, consumption, or rendering of research animals is strongly encouraged. The policy should be developed with appropriate legal counsel and should be consistent with the institution's Animal Welfare Assurance. Veterinary care of animals sent to market or slaughter and drug withdrawal periods must be consistent with the Animal Drug Use Clarification Act of 1994 (AMDUCA).[7] In the event that the animals are given a new animal drug as defined by the Food and Drug Administration (FDA),[69] no meat, eggs, or milk from those animals may be processed for human food without FDA approval. If animals are being slaughtered at the research facility for subsequent consumption, the method of euthanasia must be stated in the approved protocol and should be in compliance with the AWAR (§1.1, Euthanasia) and the American Veterinary Medical Association's guidelines on euthanasia.[43] For animals that have had invasive surgical procedures, the general practice should be for the level of surgical asepsis and postoperative care to be consistent with current veterinary standards for the species. Where IACUCs allow this practice, FDA restrictions are adhered to, and the animals must have no lingering evidence of manipulation such as the presence of a cannula or other permanent implant.

Surv. Does your IACUC permit farm animals that have been used for teaching, testing, or research to be sold for slaughter or released back to production agriculture?

- Not applicable. 121/206 (59%)
- This has not been discussed by our IACUC. 13/206 (6%)
- Yes, we always allow slaughter or release back to production agriculture. 8/206 (4%)
- Yes, but under specified conditions. 23/206 (11%)
- No, we do not allow these animals to be slaughtered or released back to production agriculture. 37/206 (18%)
- Other. 4/206 (2%)

15:49 Is IACUC approval required (either formally or informally) to donate surplus or culled rodents to zoos or similar institutions to be used as a food source for reptiles or birds?

Reg. The *Guide* (p. 73) stipulates that conventional, biologic, and hazardous waste, including carcasses, should be removed, and disposed of regularly and safely, including rendering hazardous wastes innocuous by an appropriate means and in accordance with all applicable regulations.

Opin. The AWAR and PHS Policy are silent with respect to the disposal of culled or euthanized research animals with the exception of requirements related to humane euthanasia practices (AWAR §2.31,d,v; §2.31,e,5; PHS Policy IV,C,1,c). The AWAR apply only to rodent species other than rats of the genus *Rattus* and mice of the genus *Mus* (e.g., hamsters, voles, squirrels; §1.1 Animal). Be that as it may, the diversion of carcasses from research sources to raptor rehabilitation programs, zoological parks, other exhibitors, or as a food source for the reptile pets of employees may represent a cost-effective and beneficial arrangement particularly for those recipient operations with expenses functioning on slim financial margins and, in the case of employee pets, as a component of the facility biosecurity program (i.e., preventing employees from maintaining rodent colonies at home of a lesser health status). Recipients may become highly dependent upon the arrangement and, superficially, the public relations benefits might seem to be positive for all of those involved. The authors recommend, however, that these undertakings be entered into with mutual caution. Cells immortalized with viral agents, cells carrying viral genomic material, tumorigenic human cells, and cell lines of human or nonhuman primate origin inoculated into animals present potential hazards beyond the research environment particularly if disposed via digestion by other animals. Disposal of potentially hazardous carcasses by alimentation would not be consistent with the conditions of the *Guide* nor applicable bloodborne pathogen and biosafety standards.[9,10] With respect to genetically engineered mice, regardless of how harmless or inconsequential the insert or targeted mutation, one must be cognizant of the public relations challenge created by the terms "transgene," "genetically engineered/modified," and "mutant" and the difficulties in prevailing in a 15-second sound bite debate on the news with a partially informed concerned party. Institutions clearly have an obligation to create internal processes ensuring that pathogenic materials, chemicals, radioactivity, and other known or potential hazards properly stream for disposal as hazardous waste or are rendered into a harmless state. The simplest solution, where surplus or culled rodents are to be donated, is to establish programs allowing the release for this purpose only of animals of wild-type genetic background (currently, a small fraction of any census) known to be free of hazards. This topic is highly complex and potentially dynamic from the aspect of safety regulation with exceptions and exemplars for safe diversion of carcasses for food use beyond the limited purvey of the treatment of this subject here. As such and at a minimum, responsible institutional review is recommended with the IACUC ordinarily the best body to coordinate this.

Surv. Is your IACUC's approval required (either formally or informally) to donate surplus or culled rodents to zoos or similar institutions to be used as a food source for reptiles or birds of prey?

- Not applicable. 72/206 (35%)
- We have no policy on this topic. 41/206 (20%
- We do not allow such donations. 44/206 (21%)
- We allow such donations, but only of dead animals. 7/206 (3%)
- We allow such donations of dead or live animals. 0/206 (0%)
- We allow such donations, but only of dead animals known to be free of potential hazards (e.g., infectious organisms, transgenes, radioactivity, etc.). 28/206 (14%)
- We allow such donations of live or dead animals known to be free of potential hazards (e.g., infectious organisms, transgenes, radioactivity, etc.). 9/206 (4%)
- Other* 5/206 (2%)

*Other responses reflected the choices above, but some respondents noted that approval for such donations was made not by the IACUC but by the veterinary staff or by animal facility management.

15:50 Is the IACUC required to approve the use of privately owned animals that are used for teaching general physical examinations at a veterinary school?

Reg. The AWAR (§1.1) apply in the case of regulated species used for teaching purposes. The PHS Policy (III,A) includes any live vertebrate animals used or intended for use in research, research training, experimentation, biological testing, or for related purposes. Neither the AWA, AWAR, nor PHS Policy distinguishes between privately owned animals and those owned by the institution.

Opin. The discriminating factors to consider as to whether PHS Policy and the AWAR apply are the species, source of funding for the activity, and whether the supervised teaching of veterinary skills and engagement with the animal is truly for instructional purposes only and does not involve research. The AWAR would apply for instructional activities with regulated species and IACUC oversight would be necessary given the use of the patients for training purposes. The PHS Policy would not apply given that experimentation is not a component of the activity and financial support most likely would derive from tuition and client fees and ordinarily would not be derived from PHS. For additional discussion, see 15:42.

15:51 Do IACUCs have to approve the use of cadavers on animals obtained from a pound?

Reg. The AWAR include in the definition of "Animal" both live subjects and dead specimens of regulated species (§1.1 Animal), including those obtained from pounds or shelters (§1.1 Pound or Shelter, Random Source). The *Guide* (p. 106) requires that all animals be acquired lawfully.

Opin. The inclusion of both live and dead specimens of regulated species under the definition of "Animal" in the AWAR includes those obtained from pounds or shelters (§1.1 Pound or Shelter, Random Source); however, the definition of "research facility" in the AWAR is limited to the use of live animals, and research facility recordkeeping requirements under the AWAR pertain only to live animals (§2.35,b, Recordkeeping requirements). While the *Guide* requires lawful acquisition of animals, its definition of laboratory animal (p. 2) carried with it the practical interpretation and associated historical application that attention would be directed to cases of live animals. While it is safe to conclude that the IACUC would not be required to approve the acquisition of cadaver animals donated from a pound, the involvement of a financial transaction may have different implications. IACUCs may want to carefully review and consider situations of cadaver donation from pounds given the animal welfare, public relations, and other implications related to the care, handling, and humane euthanasia practices at the source (see 15:28).

Acknowledgment: We acknowledge the contributions of Dr. Larry Iten to this chapter.

REFERENCES

1. Societies FoAS. 2020. *Guide for the Care and Use of Agricultural Animals in Agricultural Research and Teaching.* 4th ed., 1–9.
2. AAALAC. *Three Primary Standards.* 2022
3. Sikes, R.S. and W.L. Gannon. 2016. Guidelines of the American Society of Mammologists for the use of wild mammals in research. *J Mammol* 97:663–688.
4. Animal and Plant Health Inspection Service – Veterinary Services. In: USDA, editor. Title 9; Part 104. Washington, DC: U.S. Government Printing Office.
5. Permits for Scientific Purposes, Enhancement of Propagation or Survival, or for Incidental Taking of Endangered Wildlife. In: DOI, editor. Title 50; Part 17. Code of Federal Regulations. Washington, DC: U.S. Government Printing Office; 1975 (Revised 1997).
6. Endangered Species Convention. In DOI, editor. Title 50; Part 23. Code of Federal Regulations. Washington, DC: U.S. Government Printing Office; 1977 (Revised 1997).
7. Extralabel Drug Use in Animals. In: DHHS, editor. Title 21; Part 530. Code of Federal Regulations. Washington, DC: U.S. Government Printing Office; 1997 (Revised 1998).
8. OFR. Foreign Quarantine; Importations; Dogs and Cats; Turtles, Tortoises, and Terrapins; Nonhuman Primates; Etiologic Agents, Hosts, and Vectors. In: DHHS, editor. Title 42; Part 71; Subpart F. Washington, DC: U.S. Government Printing Office; 1985 (Revised 1997).
9. Biosafety in Microbiological and Biomedical Laboratories Appendix H: Working with Human, NHP, and other Mammalian Cells and Tissues (U.S. Government Printing Office) (2020).
10. Occupational and Health Standards, Bloodborne Pathogens. In: OHSA, editor. Title 29; Part 19101030. Washington, DC: U.S. Government Printing Office.
11. OFR. Occupational Safety and Health Standards; Toxic and Hazardous Substances; Chemical Hygiene Plan. In: OHSA, editor. Title 29; Part 1910 Subpart Z. Washington, DC: U.S. Government Printing Office; 1990 (Revised 1998).
12. Pough, F.H. 1991. Recommendations for the Care of Amphibians and Reptiles in Academic Institutions. *ILAR J.* 33(4).
13. Schaeffer, D.O., K.M. Kleinow, L. Krulisch, Welfare SCfA, Medicine LSUSoV. 1992. *The Care and Use of Amphibians, Reptiles, and Fish in Research.* Scientists Center for Animal Welfare.
14. UFR (Use of Fishes in Research) Committee. 2013. Guidelines for the use of fishes in research. *American Fisheries Society.* Bethesda, Maryland. p 13. Available at: https://www.asih.org/s/asf-guidelines-use-of-fishes-in-research-2013.pdf
15. Committee HACau. 2004. *Guidelines for Use of Live Amphibians and Reptiles in Field and Laboratory Research.* 2nd ed. American Society of Ichthyologists and Herpetologists.
16. Cartner, S.C., J.S. Eisen, S.C. Farmer, K. Guillemin, M.L. Kent, and G.E. Sanders. 2020. *The Zebrafish in Biomedical Research: Biology, Husbandry, Diseases, and Research Applications.* American College of Laboratory Animal Medicine series. Academic Press.

17. Alworth, L.C., S.B. Harvey. 2007. IACUC issues associated with amphibian research. *ILAR J.* 48(3):278–89. doi:10.1093/ilar.48.3.278.
18. Green, S.L., 2010. *The Laboratory Xenopus sp.* The laboratory animal pocket reference series. Boca Raton, FL: CRC Press.
19. Greene, H. 1995. Nonavian reptiles as laboratory animals. *ILAR.* 37(4):182–186.
20. McNamara, S., M. Wlizla, and M.E. Horb. 2018. Husbandry, general care, and transportation of Xenopus Laevis and Xenopus Tropicalis. *Methods Mol Biol.* 1865:1–17.
21. Potkay, S., N.L. Garnett, J.G. Miller, C.L. Pond, and D.J. Doyle. March 1997. Frequently asked questions about the public health service policy on humane care and use of laboratory animals. *Contemporary Topi Lab Anim Sci.* 36(2):47–50.
22. OLAW. 2012. *Frequently Asked Questions: PHS Policy on Humane Care and Use of Laboratory Animals, D17: What Guidelines should IACUCs Follow for Fishes, Amphibians, Reptiles, Birds, and Other Nontraditional Species Used in Research?* In: OLAW, editor. Bethesda, MD.
23. APHIS. Standards for Birds Not Bred for Use in Research under the Animal Welfare Act; 2022.
24. OLAW. 2012. *Frequently Asked Questions: PHS Policy on Humane Care and Use of Laboratory Animals, F2: Is the IACUC Responsible for Tracking Animal Usage?* In: OLAW, editor. Bethesda, MD.
25. OLAW. 2012. Guidance on Prompt Reporting to OLAW under the PHS Policy on Humane Care and Use of Laboratory Animals. In: DHHS, editor. Notice NOT-OD-05-034. Bethesda, MD.
26. Silverman, J. July 20, 2012. Management of a holding protocol. *Lab Anim (NY)* 41(8):219.
27. ARENA/OLAW. 2002. Institutional Animal Care and Use Committee Guidebook. 2nd ed. U.S. Government Printing Office.
28. OLAW. Animal Welfare Assurances: Update of Animal Welfare Assurance. Bethesda, MD; 2008.
29. Huerkamp, M.J., L. Iten, D. Archer. 2012. Animal Acquisition and Disposition. In: Silverman J, Suckow MA, Murthy S, eds. *The IACUC Handbook.* 3rd ed. Boca Raton, FL: CRC Press.
30. OLAW. 2012. U.S. Government Principles for the Utilization and Care of Vertebrate Animals Used in Testing, Research, and Training. In: OLAW, editor. Bethesda, MD.
31. Yu, X., Y. Wang, D. An, and Y. Wei. 2022. Counting method for cultured fishes based on multi-modules and attention mechanism. Article. *Aqua Eng.* 96:102215.
32. Ashley, P.J. 2007. Fish welfare: Current issues in aquaculture. *Appl Anim Behav Sci.* 104(3–4):199–235.
33. Huerkamp M.J. and D. Archer. 2007. Animal Acquisition and Disposition In: Silverman J, Suckow MA, Murthy S, eds. *The IACUC Handbook.* 2nd ed. Boca Raton, FL: CRC Press; 188–225.
34. APHIS. 2020. Guidelines for Preparing USDA Annual Reports and Assigning USDA Pain and Distress Categories. In: USDA, editor. Washington, DC.
35. Mohan, S. and P.L. Foley. December 31, 2019. Everything you need to know about satisfying IACUC Protocol Requirements. *ILAR J.* 60(1):50–57.
36. Behringer, R., M. Gertsenstein, K.V. Nagy, and A. Nagy. 2014. *Manipulating the Mouse Embryo: A Laboratory Manual.* 4th ed. Cold Spring Harbor Laboratory Press, xxii, 814.
37. Nagy, A., M. Gertsenstein, K. Vinterstein, and R. Behringer. 2003. *Manipulating the Mouse Embryo: A Laboratory Manual.* 2nd ed. Cold Spring Harbor Laboratory Press; 152.
38. Wilkie, T.M., R.L. Brinster, and R.D. Palmiter. November 1986. Germline and somatic mosaicism in transgenic mice. Research Support, U.S. Gov't, P.H.S. *Devel Biol.* 118(1):9–18.
39. OLAW. 2012. Frequently Asked Questions: PHS Policy on Humane Care and Use of Laboratory Animals, A4: Does the PHS Policy Apply to Live Embryonated Eggs? In: OLAW, editor. Bethesda, MD.
40. Saif, Y.M. and W.L. Bacon. 1996. Protocol review: Simply stated. *Lab Anim (NY)* 25(5).
41. Association AVM. AVMA Guidelines for the Euthanasia of Animals: 2020 Edition. 2020. Accessed August 12, 2022. https://www.avma.org/sites/default/files/2020-02/Guidelines-on-Euthanasia-2020.pdf.
42. Swayne, D.E. 1996. Protocol Review: Preventative measures. *Lab Anim (NY)* 25(5).
43. OLAW. 2012. Frequently Asked Questions: PHS Policy on Humane Care and Use of Laboratory Animals, A5: Does the PHS Policy Apply to Larval Forms of Amphibians and Fish? In: OLAW, editor. Bethesda, MD; 2012.
44. Mulcahy, D.M. December 15, 2017. The Animal Welfare Act and the conduct and publishing of wildlife research in the United States. *ILAR J.* 58(3):371–378.
45. Borski, R.J. and R.G. Hodson. 2003. Fish Research and the Institutional Animal Care and Use Committee. Research Support, U.S. Gov't, Non-P.H.S. *ILAR J.* 44(4):286–294.
46. Reeb-Whitaker, C.K., B. Paigen, W.G. Beamer, R.T. Bronson, G.A. Churchill, I.B. Schweitzer, and D.D. Myers. January 2001. The Impact of Reduced Frequency of Cage Changes on the Health of Mice Housed in Ventilated Cages. Evaluation Studies Research Support, U.S. Gov't, P.H.S. *Lab Anim (NY)* 35(1):58–73.
47. OLAW. 2012. Frequently Asked Questions: PHS Policy on Humane Care and Use of Laboratory Animals, A3: Does the PHS Policy Apply to Use of Animal Tissue Obtained from Dead Animals? In: OLAW, editor. Bethesda, MD.
48. Garnett, N. and W.R. DeHaven.1997. OPRR and USDA/Animal Care Response on Applicability of the Animal Welfare Regulations and the PHS Policy to Dead Animals and Shared Tissues. *Lab Anim (NY)* 26(3).
49. OLAW. 2013. Frequently Asked Questions: PHS Policy on Humane Care and Use of Laboratory Animals, D9: What is Considered a Significant Change to a Project that Would Require IACUC Review? In: OLAW, editor. Bethesda, MD.
50. OER. 2003. Revised Guidance Regarding IACUC Approval of Changes in Personnel Involved in Animal Activities. In: DHHS, editor. *Notice NOT-OD-03-046.* Bethesda, MD.
51. NIH. 2012. NIH Grants Policy Statement, Part II: Terms and Conditions, Subpart A, Change in Scope. In: NIH, editor. Bethesda, MD.

52. OLAW. 2012. Frequently Asked Questions: PHS Policy on Humane Care and Use of Laboratory Animals, B13: Does OLAW Expect the IACUC to Notify NIH When there is a Change in an Animal Activity Supported by PHS Funds? In: OLAW, editor. Bethesda, MD.
53. Brown, P. 2007. Response to Protocol Review Scenario: A word from OLAW. *Lab Anim (NY)* 36(3).
54. OER. 2012. Change in Criteria for Renewal of Animal Welfare Assurances for Foreign Institutions. In: DHHS, editor. Notice NOT-OD-12-081. Bethesda, MD.
55. OLAW. 2012. *Frequently Asked Questions: PHS Policy on Humane Care and Use of Laboratory Animals, D13: If an Animal Activity Will Be Performed Outside of the US (Either by a Foreign Awardee or by a Foreign Institution as a Subproject for a Domestic Awardee), Is the Awardee's IACUC Required to Review and Approve that Activity?* In: OLAW, editor. Bethesda, MD.
56. OLAW. 2012. Frequently Asked Questions: PHS Policy on Humane Care and Use of Laboratory Animals, D15: Is IACUC Approval Required for the Collection of Samples in Foreign Countries from Captive Wild Animals or Research Colonies? In: OLAW, editor. Bethesda, MD.
57. OLAW. 2012. Frequently Asked Questions: PHS Policy on Humane Care and Use of Laboratory Animals, A2: Does the PHS Policy Apply to the Production of Custom Antibodies or to the Purchase of Surgically Modified Animals? In: OLAW, editor. Bethesda, MD.
58. Huerkamp M.J. and D. Archer. 2000. Animal Acquisition and Disposition. In: Silverman J., Suckow M.A., Murthy S., eds. *The IACUC Handbook.* 1st ed. Boca Raton, FL: CRC Press.
59. OLAW. 2012. Frequently Asked Questions: PHS Policy on Humane Care and Use of Laboratory Animals, D16: Is IACUC Approval Required for the Use of Animals in Breeding Programs, as Blood Donors, as Sentinels in Disease Surveillance Programs, or for Other Non-Research Purposes? In: OLAW, editor. Bethesda, MD.
60. OLAW. 2012. Frequently Asked Questions: PHS Policy on Humane Care and Use of Laboratory Animals, A8: How Can the IACUC Determine if Activities Involving Privately Owned Animals Constitute Veterinary Clinical Care or Research Activities? In: OLAW, editor. Bethesda, MD.
61. OLAW. 2012. Frequently Asked Questions: PHS Policy on Humane Care and Use of Laboratory Animals, A7: Does the IACUC Need to Approve Research Studies that Use Privately Owned Animals, Such as Pets? In: OLAW, editor. Bethesda, MD.
62. Brown, P. and C. Gipson. 201o. A word from OLAW and USDA. *Lab Anim* 39(3):68.
63. Education AVMACoVT, Activities. 2005. Accreditation Policies and Procedures of the AVMA Committee on Veterinary Technician Education and Activities (CVTEA). American Veterinary Medical Association.
64. Tannenbaum, J. (Personal Communication). 1998.
65. OFR. Nondiscrimination since Disability by Public Accommodations and in Commercial Facilities; Service Animals. In: DOJ, editor. Title 28; Part 36. *Code of Federal Regulations.* Washington, D.C.: U.S. Government Printing Office; 1991 (Revised 1998).
66. OLAW. 2012. Frequently Asked Questions: PHS Policy on Humane Care and Use of Laboratory Animals, F11: Can IACUCs Authorize the Adoption of Research Animals as Pets after the Animals Are No Longer Needed for Study? In: OLAW, editor. Bethesda, MD.
67. OFR. New Animal Drugs for Investigational Use. In: DHHS, editor. Title 21; Part 511. *Code of Federal Regulations.* Washington, DC: U.S. Government Printing Office; 1975 (Revised 1998).

16

Animal Housing, Use Sites, and Transportation[*]

Robin Crisler and Karen Rogers

Introduction

This chapter focuses on examples of acceptable and unacceptable housing or holding locations for animals used in research, teaching, and testing. The Animal Welfare Act (AWA) and Animal Welfare Act Regulations (AWAR),[1] the *Guide for the Care and Use of Laboratory Animals (Guide)*,[2] the Public Health Service (PHS) Policy,[3] Office of Laboratory Animal Welfare (OLAW), and AAALAC websites are the primary references utilized. There are few absolutes when considering animal housing and use sites. Professional judgment, institutional policy, and protocol goals and needs must all be considered when determining what is and is not an acceptable location for the care and use of animals.

IACUC Chairs from throughout the United States and Canada responded to a questionnaire about animal housing and use sites. Their responses are tabulated and summarized where appropriate to provide further insight into the breadth of policies currently in practice.

16:1 What is the definition of an animal facility?

Reg. The AWAR (§1.1) do not provide a definition for "animal facility"; however, animal housing is described within separate definitions for each of the following terms: *housing facility, indoor housing facility, outdoor housing facility, sheltered housing facility*, and *study area*. For example, a housing facility is any land, premise, shed, barn, building, trailer, or other structure intended to house animals.

PHS Policy (III,B) defines an *animal facility* as "any and all buildings, rooms, areas, enclosures, or vehicles, including satellite facilities, used for animal confinement, transport, maintenance, breeding, or experiments inclusive of surgical manipulation."

The *Guide* (pp. 135–136) states that an animal facility consists of functional areas for animal housing, care, and sanitation; receipt, quarantine, and separation of animals, separation of species, or isolation of individual projects; and storage. Some facilities may encompass space for surgery, intensive care, necropsy, irradiation, diet preparation, procedural space, imaging, clinical treatment, containment or barrier facilities, supply, receiving, cage washing, carcass storage, facility administration, training, locker rooms, break rooms, and maintenance areas.

Opin. The authors' working definition of an animal facility is an integrated concept that is inclusive of the many programmatic aspects that relate to animal housing, care, and use in an institutional setting. However, for the purposes of this chapter, the term *animal facility* shall refer to the primary housing location and support space for animals and animal studies overseen by the IACUC.

16:2 What are the circumstances and length of time that animals may be kept at a location other than their primary housing site?

Reg. Any space outside of the vivarium to which animals are moved for study, testing or observation must be listed in the IACUC-approved and subsequently modified protocol, must be maintained in accordance with all applicable policies, regulations, and guidelines, and is subject to periodic inspection by the IACUC. There are two basic scenarios in which this can occur: for research purposes (lab, surgical suite, testing area) or for medical reasons (veterinary or surgical treatment area). In both cases, if housed for longer than 12 hours, these locations are considered *study areas* as per the AWAR §1.1 definition. According to AWAR §1.1, a *study area* means any building room, area, enclosure, or other containment outside of a core facility or centrally designated or managed area in which animals are housed for more than 12 hours, and therefore must be inspected by the IACUC as per AWAR §2.31,c. The AWAR (§2.31,c,2) require that any location where animals are housed for more than 12 hours be inspected by the IACUC at least once every six months.

PHS Policy (III, B) states that "a satellite facility is any containment outside of a core facility or centrally designated or managed area in which animals are housed for more than 24 hours." Nevertheless, PHS Policy (II) requires compliance

[*] This revised Chapter 16 builds on the contributions of Dr. Cynthia Gillett, author of this chapter in the 1st and 2nd editions, as well as Dr. Robert M. Bigsby, in the 3rd edition of *The IACUC Handbook*.

DOI: 10.1201/9781003272861-16

with the AWA where applicable. Therefore, the PHS Policy definition of a *satellite facility* and the duration of stay requirements may be applied to species that are not regulated by the AWAR, such as typical laboratory rats and mice (see NIH/OLAW FAQ E.1).[4,5] However, the more restrictive AWA regulation (§2.31,c,2) requiring IACUC semiannual inspection of locations where animals are housed for more than 12 hours and areas where any form of surgical manipulations (minor, major, survival, non-survival) are performed must be applied to all AWA-regulated species.

The *Guide* (p. 25) states that the IACUC is responsible for oversight of animal use areas; however, it does not define that term. It also states (p. 134): "Animals should be housed in facilities dedicated to or assigned for that purpose, not in laboratories merely for convenience. If animals must be maintained in a laboratory to satisfy the scientific aims of a protocol, that space should be appropriate to house and care for the animals and its use limited to the period during which it is required." Likewise, if animals are held under prolonged restraint outside of animal housing areas, if lesions, illness, or severe behavioral change occur, the animal should temporarily or permanently be removed from the animal restraint (*Guide*, p. 30). Animals that must be restrained for procedures should be restrained for the minimum required time period to accomplish research objectives.

Opin. The authors' interpretation of the term *animal use area*, as stated in the *Guide* (pp. 25, 34), is that it implies research procedural areas such as a laboratory or testing site. Neither the *Guide*, the AWAR, nor the PHS Policy prohibits removal of animals from the animal holding facility for procedures or for housing elsewhere, such as in an investigator's laboratory or a satellite facility. Additionally, NIH/OLAW indicates: "Institutions are responsible for oversight of all animal-related activities regardless of how long or where the activity occurs" (NIH/OLAW FAQ E.1.).[4] However, doing so may trigger additional regulatory oversight responsibility of the IACUC or bring into play individual institutional requirements. Although one could apply different time standards on the basis of whether or not an AWAR-regulated species is involved, applying the most restrictive criterion (12 hours) will result in compliance in all situations.

Surv. For how long does your IACUC allow animals to be removed from their primary housing site without an IACUC-approved exception to federal regulations? Select all that apply.

• Not applicable	65/204 (32%)
• Up to 12 hours for species covered by the Animal Welfare Act	54/204 (26%)
• Up to 12 hours for all species	73/204 (36%)
• Up to 24 hours for species covered only by the PHS Policy	49/204 (24%)
• Up to 24 hours for all species	7/204 (3%)
• Greater than 24 hours for all species	1/204 (0%)

16:3 What are examples of acceptable circumstances for keeping animals outside the primary animal facility?

Reg. The *Guide* (pp. 134–135) refers to a centralized animal facility as one in which support, care, and use areas are adjacent to the animal housing space, while in a decentralized system animal housing and use may occur in space that is not solely dedicated to animal care and support, e.g., laboratories, treatment, behavioral testing or imaging facilities, or surgical suites physically separated from animal housing/support areas. The *Guide* (p. 135), states: "The opportunity for exposure to disease agents is much greater in these (decentralized) situations and special consideration should be given to biosecurity, including transportation to and from the site, quarantine before or after use of the specialized research area, and environmental and equipment decontamination." The *Guide* also indicates that animals should not be housed in laboratories outside of a dedicated animal facility merely for convenience.

Opin. The AWAR and PHS Policy, while providing definitions and oversight recommendations for animal facilities (including satellite facilities; see 16:1), do not provide examples of acceptable circumstances or rationale for keeping animals outside the primary animal facility. However, the *Guide* (pp. 146–150) does provide examples of circumstances where decentralization of animal housing and use may be preferred, such as facilities for imaging or irradiation or laboratories for behavioral studies.

As the *Guide* indicates, there are other considerations when housing animals outside a centrally managed facility. For reasons of disease control, some facilities have a policy stating that if animals leave the facility they may not be returned. This is particularly applicable to microbiologically defined (specific pathogen free) rodents. The method and route of transportation to the study area also must be monitored to minimize potential for disease transmission and to ensure conformance with transportation guidelines. Facility construction, environmental monitoring, security, consistency of animal care, and veterinary availability should also be considered when housing outside of the primary facility.

Surv. What are some circumstances that are typically acceptable to your IACUC for keeping animals outside of the primary animal facility? Select all that apply.

• Not applicable, we do not permit animals outside of the primary facility.	49/204 (24%)

- Biocontainment requirements. 55/204 (27%)
- Needed research equipment cannot be placed within the animal facility. 132/204 (65%)
- Behavioral or similar testing would be affected by daily movement of animals to and from their usual holding room. 104/204 (51%)
- Postoperative monitoring. 52/204 (25%)
- Need for frequent observations or sampling. 71/204 (35%)
- Convenience for the laboratory. 7/204 (3%)
- Other* 16/204 (8%)

*Other:

- Circumstances in which housing of a species cannot be accommodated by the primary animal facility
- When animals are to be used in terminal procedures
- When a shared procedure room is full or unavailable
- When zebrafish embryos are used in a teaching exercise
- For purposes related to acclimation of the animals to the testing environment

16:4 Are there regulatory considerations regarding the actual act of transportation that the IACUC must consider when animals are moved from their primary housing site to a secondary site?

Reg. The AWAR (Part 3, Standards) have extensive transit regulations for each regulated species. They include primary enclosures; primary conveyances; animal identification; space, food, and water requirements; care in transit; handling; sanitation; ventilation; shelter; temperature; and the terminal facilities. The AWAR (§2.38,f) also cover the transit of animals as part of an adequate "handling" consideration, which implies animal handling during transportation is included as an item the IACUC must consider, though this is not explicitly stated. PHS Policy II requires compliance with the AWAR, where applicable, and PHS Policy III defines an animal facility to include vehicles used for transport.

The *Guide* (pp. 107–109) states: "Careful planning for all types of transportation should occur to ensure animal safety and well-being. The process of transportation should provide an appropriate level of animal biosecurity while minimizing zoonotic risks, protecting against environmental extremes, avoiding overcrowding, providing for the animals' physical, physiologic, or behavioral needs and comfort, and protecting the animals and personnel from physical trauma" (*Guide*, pp. 107). In addition, "special considerations may be necessary for transporting animals during certain phases of their life or in certain conditions, such as pregnant, perinatal, and geriatric animals; animals with preexisting medical conditions (e.g., diabetes mellitus); and animals surgically prepared by the supplier." Importantly, for the sake of a controlled experimental environment, "biosecurity during transportation is always important, it is of particular importance for immunocompromised, genetically modified and specific pathogen-free rodents." Measures such as decontaminating the surfaces before placement of containers in clean animal facilities should be taken. "Transportation of animals in private vehicles is discouraged because of potential animal biosecurity, safety, health, and liability risks for the animals, personnel, and institution" (*Guide*, p. 108).

Transport of animals that have been infected with an infectious agent or ones that are carrying grafted tissue that is suspected to be infected with such an agent is subject to national and international rules regulating the transport of hazardous materials. The website for the Centers for Disease Control and Prevention (CDC)[6] contains links to relevant regulatory agencies.

Opin. The AWAR on the transport of animals are designed to ensure humane animal care. These regulations should be consulted when planning intra-facility and inter-facility institutional animal transfers. While animal transport is not required to be directly reviewed and approved by the IACUC, it is a best practice if the IACUC is aware of such transport and what standard operating procedures may be used. In the authors' opinion, transport activities should be reviewed during the semiannual program evaluation to ensure appropriate procedures are being followed. Since they can be considered animal use areas, transport vehicles used routinely by the institution should be included in the semiannual facility inspection, assessing sanitation and climate control systems just as with any other animal housing unit. Animal transport issues also should be raised during protocol review when justification for off-site housing is being requested. International transport is outside the scope of this chapter; however, all applicable national and international standards and laws should be followed. The National Research Council's publication *Guidelines for the Humane Transportation of Research Animals* and other resources provide a comprehensive review of issues related to transportation of animals.[7–10]

Biosafety concerns are paramount when considering the transport of animals carrying infectious agents or tissues (human or animal xenografts) that cannot be reasonably considered to be non-infectious. According to the Dangerous Goods website for the International Air Transport Association (IATA)[10]: "A live animal that has been intentionally infected and is known or suspected to contain an infectious substance must not be transported by air unless the infectious substance contained cannot be consigned by any other means. Infected animals may only be transported under terms and conditions approved by the appropriate national authority." Although there are guidelines for packaging tissue specimens containing potentially hazardous material, there are no readily available guidelines for securing a live animal for such transport and the reader is advised to contact the CDC and IATA for more information.

Surv.1 Does your IACUC allow small animals (e.g., rodents) to be transported from one animal housing site to another by private vehicles?

• No	91/204 (45%)
• Yes, with IACUC approval	84/204 (41%)
• Yes, and IACUC approval is not necessary	8/204 (4%)
• Other*	21/204 (10%)

*Other:

- Yes, for circumstances related to veterinary treatment.
- Yes, in the event of a disaster, part of the facility's contingency plan.

Surv. *2* If your IACUC allows the use of privately owned passenger vehicles for animal transport, what environmental stipulations have been established for that purpose? Select all that apply.

• Not applicable.	98/204 (48%)
• Cages must be secured within the vehicle (e.g., via seat belts).	82/204 (40%)
• Vehicles must maintain a temperature that is appropriate for the species.	92/204 (45%)
• Primary enclosures must be contained in a secondary enclosure (e.g., a large plastic tote).	64/204 (31%)
• Enclosures must be transported on a sanitizable surface (e.g., cargo liner and not carpet or seats).	48/204 (24%)
• Even if approved by the IACUC we discourage this practice as there may be animal allergens or other factors left behind from the transported animals.	61/204 (30%)
• Additional review and approval is also needed by Environmental Health and Safety.	16/204 (8%)
• Other*	12/204 (6%)

16:5 Are there other practical considerations besides regulatory requirements the IACUC must consider regarding animal transport?

Reg. The AWAR (§2.38,f,1) state that handling of all animals shall be done as expeditiously and carefully as possible in a manner that does not cause trauma, overheating, excessive cooling, behavioral stress, physical harm, or unnecessary discomfort. The *Guide* (p. 107) indicates that transportation should provide for animal biosecurity, minimize zoonotic risks, protect from environmental extremes, avoid overcrowding, provide for animals' needs and comfort, and protect people and animals from trauma.

Opin. The practical, non-regulatory day-to-day considerations that can arise during intra- and inter-institutional animal transport tend to be situational and are usually best handled by discussions with the laboratory animal facility professionals that oversee the transportation processes. Local on-site examples include situations such as whether or not a box of rodents can be taken outside to transport them to a laboratory in another building; which elevator can be used to transport animals to a laboratory; or which travel path may be best to bring animals into areas where imaging equipment is shared for human clinical use. Other considerations may include which commercial vendors are considered acceptable to use for ground or air shipping to off-site locations, how to evaluate shipping vendors and what standards should be used to do so, and ensuring the feasibility of the shipment with the receiving institution. The answers to these and similar issues surrounding the appropriateness of transportation procedures can be complex and depend on factors such as characteristics of the animals (species, age, size, health status, etc.) climate, time of year, facility design, and institutional policies. Experience, knowledge, and good judgment around situations commonly encountered during animal transportation are necessary to achieve successful outcomes. Excellent organization, communication, and recordkeeping are also key factors. Institutional plans should be in place to ensure that animal transportation occurs in a safe and controlled manner and utilizes advance planning to avoid potential pitfalls. It is recommended that institutions provide and require training for all individuals who participate in transportation activities. The institution should ensure that the transporters are trained and qualified to perform the work and understand the expectations and regulatory needs. If researchers perform export functions in addition to the animal facility staff, researchers should be included in this training requirement as well. Personnel transporting animals should be trained to perform risk assessments and to identify animal welfare issues at any point in the transportation process, including those that may occur mid-shipment. Personnel should be trained to contact the Attending Veterinarian in case a concern arises. If an animal health or welfare concern happens related to transport, the IACUC must also be notified. The IACUC, during the course of its semiannual inspections, program review, or through reports of concern about transportation, may find evidence of unacceptable transportation practices that will need to be addressed.

It is advisable that an institution has a transportation policy or standard operating procedure to describe processes, to provide continuity, and to help establish expectations between investigators and animal facility staff. This policy should be reviewed by the IACUC as part of the program evaluation even though aspects of the policy may contain non-regulatory elements. Guidelines and training for the transport of animals that have been administered biologic, chemical, or radiologic hazards should be established with the help of safety experts and the appropriate institutional committees (Institutional Biosafety Committee, Radiologic Safety Committee, Environmental Health and Safety). If the infectious disease or hazard safety status of the animals may differ between shipping locations, caution must be used to ensure that appropriate quarantine procedures are followed, and that animals are not shipped until confirmation of acceptance is expressed by the recipient.

Exposure of persons to allergens through the transportation process must also be considered. Individuals who do not have direct animal contact may not be covered by the institution's Occupational Health and Safety Program yet still need information about the risks of laboratory animal allergy and instructions on what to do should they experience symptoms. Transportation practices should be designed to reduce and limit exposure of all persons who may be exposed to airborne allergens along the transport route, such as the general public and employees in laboratories adjacent to animal use areas.

16:6 If animals are being housed in a research laboratory setting, are there species differences to consider?

Reg. The housing requirements for various APHIS/AC-regulated species are addressed in Part 3 of the AWAR. These standards apply if the animals are approved by the IACUC to be "housed" in a laboratory. The IACUC may make exceptions to these standards if there is scientific justification to do so (AWAR §2.36,b,3), and these exceptions must be included in the facility's annual report. The Animal Welfare Inspection Guide pp. 7–6 – 7.7 clarifies which exceptions require reporting and which do not.[25] Likewise, PHS Policy (IV,C,1,d) requires the same standards of care for animals housed in or away from the primary care facility, ensuring that "living conditions of animals will be appropriate for their species and contribute to their health and comfort." The *Guide* (p. 74) states: "Animals should be cared for by qualified personnel every day, including weekends and holidays, both to safeguard their well-being and to satisfy research requirements. Emergency veterinary care must be available after work hours, on weekends, and on holidays." Note that this requirement is imperative when having animals housed in labs or other satellite areas. The *Guide* (pp. 41–103) provides additional examples of species-specific environmental considerations, such as temperature, humidity, ventilation, air quality, illumination, noise, vibration, social needs, environmental enrichment, space, husbandry, and sanitation. The use of laboratory space for housing should be limited to the period during which it is required.

Opin. The need for laboratory housing should be protocol driven and not motivated by convenience or economics (*Guide*, p. 134). Not all laboratory facilities will be structurally or environmentally appropriate for housing all types of animal species. The suitability of laboratory housing and the expertise, training, and availability of scientists to provide nonmedical care seven days a week must be carefully evaluated on a case-by-case basis by the IACUC. A comprehensive husbandry plan for the animals should be submitted to the IACUC for review. The IACUC should keep records of approved laboratory housing sites and perform, or delegate, periodic inspections of these areas when animals are present, in addition to the scheduled semiannual facility inspections. Situations where animals are infected with, or dosed, hazardous materials should generally not be considered for laboratory housing due to the many special requirements for containment of the hazard and protection of both humans and the animals.

Surv. How does your IACUC inspect animal housing and use sites if they are at geographically distant locations? That is, locations that would require unique travel plans were the IACUC to do an on-site inspection. Select all that apply.

- Not applicable, we do not have geographically distant locations. 111/204 (54%)
- At a geographically distant institution with a collaborating institution, we depend on that institution's IACUC. 46/204 (23%)
- At a geographically distant field site we ask nearby colleagues to inspect the site. 9/204 (4%)
- At a geographically distant field site we depend on videos or similar means to evaluate the site. 45/204 (22%)
- We inspect all sites by ourselves, whether near or distant. 40/204 (20%)
- We do not inspect our geographically distant sites. 5/204 (2%)
- Other. 12/204 (6%)

16:7 May animals be housed at sites not belonging to an institution? Under what circumstances?

Reg. The AWAR (§2.38,i) state: "If any research facility obtains prior approval of the APHIS AC Regional Director, it may arrange to have another person hold animals, *provided* that.

1. The other person agrees, in writing, to comply with the regulations in this part and the standards in part 3 of this subchapter, and to allow inspection of the premises by an APHIS official during business hours.
2. The animals remain under total control and responsibility of the research facility; and
3. The Institutional Official agrees, in writing, that the other person or premises is a recognized animal site under its research facility registration."

APHIS/AC suggests that APHIS Form 7009 should be used for approved off-premises housing sites. The APHIS/AC Regional Office can be contacted for help and guidance on this topic should questions arise.

PHS Policy (IV,A) requires that awardee institutions and other participating institutions that receive PHS support have an approved Assurance on file with NIH/OLAW. NIH/OLAW has stated that vendors that supply custom antibodies to investigators, or perform a surgery conducted in response to a specific request, must either have an Animal Welfare Assurance on file with NIH/OLAW or be included in the receiving research institution's Assurance statement (see NIH/OLAW FAQ A.2.)[4,12] Additionally, if AWAR-regulated species are used to provide blood or serum at the vendor's site or are used in laboratory tests, quality-control studies, or the manufacture or testing of biologics, then the vendor must be registered with APHIS/AC as a research facility.[13] The USDA publications "Licensing and Registration Under the Animal Welfare Act," along with the "Animal Welfare Inspection Guide" provide information on the responsibilities of institutions and vendors for contracts or projects that involve multiple research housing sites.[13]

The *Guide* (p. 15) along with OLAW (FAQ D8) indicate that collaborating institutions should have a formal written understanding that addresses responsibilities of animal care and use, ownership, and IACUC review and oversight.[2,4]

Opin. The legal necessity for the non-institutional site to be APHIS/AC registered must be evaluated along with a separate verification that an Assurance is on file with NIH/OLAW if one is legally required, as outlined above. AAALAC accreditation status for the non-institutional site may also be highly desirable. These considerations are dependent on the primary institution's status and on the individual housing situation. Affiliate Veterans Administration Medical Center facilities are an example where housing animals at a non-institutionally owned site might be routinely permitted. In the latter example, this should be listed as a "site" of the research institution via a memorandum of understanding (MOU) put into place for assurance that compliance with applicable portions of the VA Research with Animals (VHA Directive 1200.07) will be met.[14] Other examples of off-site housing may include when institutions outsource rodent quarantine or embryo rederivation to commercial vendors specializing in those services. Also, during animal transport long distances, a brief temporary or overnight housing stay in an off-site animal holding area may be required mid-shipment as arranged through a commercial transport vendor (for example between airline flights).

AAALAC follows animal ownership in terms of defining who is responsible for animals at an off-site program for purposes of accreditation. During the accreditation process, AAALAC considers the outside contractor's accreditation standing as well as that of the institution seeking accreditation.[15]

Surv. Under what circumstances would your IACUC approve the housing of institution-owned animals at sites not belonging to your institution? Select all that apply.

• Not applicable	80/204 (39%)
• If we would run out of general housing space	14/204 (7%)
• Need for specialized housing (e.g., for nonhuman primates)	29/204 (14%)
• Collaborative research arrangements	109/204 (53%)
• For long-term housing of animals on study	22/204 (11%)
• Other*	20/204 (10%)

*Other:

- In the event of an emergency or disaster, as part of the facility's contingency plan
- In circumstances in which specialized equipment or procedures related to the research are needed

16:8 Are there any circumstances under which animals may be housed in a person's home or on a private farm?

Reg. Both NIH/OLAW and APHIS/AC will consider the use of the animal, the housing, and the animal's ownership to determine the appropriate level of regulatory oversight that is required. As far as the use of the animal is concerned, APHIS/AC specifically does not regulate pets or animals owned by private citizens in a veterinarian–client-patient-relationship (VCPR) because the owner's consent is present, and clinical work is outside of the original intent of the AWA. Animals in a veterinarian–client-patient-relationship are not regulated even when they are being treated at a research institution. In addition, as per the definition of animal in the AWA (§2132,g), APHIS/AC has no regulatory authority over animals used in agricultural research that are housed on private farms.

The PHS Policy (PHS Policy I) covers live vertebrate animals used or intended for use in research, research training, and biological testing activities conducted or supported by the PHS. OLAW states in its FAQ A.7, 8 that pets used in PHS-funded research must be covered under an IACUC-approved protocol. The institution must have an OLAW-approved Animal Welfare Assurance covering all performance sites. If the study is being conducted in collaboration with a private clinical veterinary practice, the operational components of the practice associated with the research activity should be a covered component of an Assured institution. When doing research on a pet, the institution is responsible for obtaining informed consent from the pet owner for the research activity. If the research activity is being conducted in collaboration with a private veterinary practice, the institution should consider the use of a memorandum of understanding (MOU) agreement. The institutional legal counsel may be involved in the development of the document. The veterinary clinical care of a privately owned animal is not a research activity and does not require IACUC approval. Veterinary clinical care is typically offered as a fee-for-service activity and is regulated by state veterinary licensing boards.

In a guidance statement provided for this question, APHIS/AC (with NIH/OLAW concurrence) indicated that pet and client-owned animals participating in an animal study proposal for educational and teaching purposes (e.g., as part of a veterinary technical program) are regulated during the activity. Although the pet owner's home would not be inspected, the animal study protocol utilizing these animals must be reviewed and approved by the IACUC.

In the event an institution elects to house its animals used in biomedical research in a private home or private farm for greater than 12 hours, the home or farm would be considered a study area (AWAR §1.1, Study Area) and therefore subject to IACUC inspection. Under the PHS Policy (III,B), the same is true for animals housed in a private home or farm for more than 24 hours. Regarding PHS policy, pets and livestock used in research must be covered under an IACUC-approved protocol and the institution must have an NIH/OLAW-approved Assurance covering all performance sites.

Opin. The origins and intent of the AWAR are rooted in the need for protection of pets from theft and resale, especially for the unauthorized sale of stolen animals for use in research institutions (AWA §2131), which was a national concern first documented in the November 29, 1965, issue of *Sports Illustrated* magazine regarding Pepper, a Dalmatian dog, and helped pave the way for Congress to enact the AWA in 1966.

Prior approval by the IACUC for private housing is considered by surveyed respondents to be essential. The authors feel that IACUC-approval is needed for these animals, even when PHS funding does not support the research. When animals remain privately owned, as opposed to institutionally owned, they most often are housed in that person's home or farm (e.g., blood donor animals or veterinary clinical studies on companion and farm animals).

Surv. Under what circumstances would your IACUC allow animals to be housed in a person's home or on a private farm? Select all that apply.

• Not applicable, we do not allow animal housing at privately owned locations.	155/203 (76%)
• Study subjects are privately owned animals (e.g., clinical trials at veterinary schools).	35/203 (17%)
• On weekends and holidays if intensive (around-the-clock) monitoring is needed.	1/203 (0%)
• IACUC approval is required.	34/203 (17%)
• Other*	6/203 (3%)
• For long-term housing of agricultural animals on study (e.g., private farm).	13/203 (6%)

*Other:

- Housing would be allowed under emergency/disaster circumstances.

16:9 Do all wildlife or field studies need to be reviewed by the IACUC?

Reg. The AWAR definition of *animal* includes wild birds and most mammals; therefore, wild mammals and birds are subject to IACUC protocol review. However, the regulations specifically exempt the IACUC from the requirement to review field studies that are conducted on free-living wild animals in their natural habitat, and which do not involve an invasive procedure, any harm to the animal, or any material alteration of the behavior of an animal under study (AWAR §1.1; §2.31,d,1).

PHS Policy II requires compliance with the AWA, as applicable, and (III,A) defines an *animal* as any live, vertebrate animal. Therefore, all vertebrate animals used in field studies fall under the PHS Policy.

In FAQ A.6 by NIH/OLAW, the IACUC is responsible for oversight of vertebrate animal activities supported by the PHS and those supported by NSF and VA in accord with PHS Policy.[4] To conduct such activities in the field, the investigator must provide the IACUC with the following information: where the activity will be conducted; what procedures will be involved; a brief description of how those procedures are likely to affect the biology and ecology of the study animals by describing the relationship of that species to the habitat and to other species found in the study area (including the nature and duration of potential effects); and assurance that permit requirements of pertinent local, state, national, and international wildlife regulations will be obtained before work begins. If the IACUC determines that the proposed activity is likely to alter or influence the biology, behavior, or ecology of the study animals or other species found in the study area, then protocol review and approval is required. However, if the IACUC determines that the proposed activity will not alter or influence the biology, behavior, or ecology of the study animals, IACUC review and approval is not needed.

The *Guide* (p. 32) states: "Investigations may involve the observation or use of non-domesticated vertebrate species under field conditions," and that the IACUC should assure the field study "does not compromise the health and safety of either animals or persons in the field." The *Guide* (p. 32) indicates that field studies need assessment for issues, including zoonotic disease and safety, as well as any laws and regulations that may apply.

Opin. If an institution's NIH/OLAW Assurance covers all vertebrate animals, not just those funded by PHS, the IACUC must review wildlife field studies of all vertebrates that include activities other than simple unobtrusive observation. It is the authors' opinion that applying uniform policies to all animals involved in the institution's animal care and use program regardless of the type of research or teaching activity involved promotes consistent oversight.

APHIS/AC has indicated that it is implied through the AWA (§2143,a,7,A) regarding "professionally acceptable standards governing the care, treatment, and use of animals" that the *Guide for the Care and Use of Laboratory Animals*, the *Guide for the Care and Use of Agricultural Animals in Research and Teaching*, as well as peer-reviewed taxon-specific journals and publications are considered acceptable standards of care to APHIS/AC. These resources may be helpful to IACUCs when evaluating wildlife or field studies.

16:10 If animals are wild-trapped, are there conditions the IACUC should place on that activity in terms of type of trap, frequency of checking of the trap, and euthanasia of injured or to-be-collected animals?

Reg. A *field study*, as defined in the AWAR (§1.1), is a study conducted on free-living wild animals in their natural habitat. In accordance with the AWAR (§2.31,d,1), a field study is exempt from IACUC review because it does *not* harm or materially alter the behavior of an animal under study. If an IACUC determines that a study does not meet this definition, the research cannot be classified as a field study under the AWAR, and the IACUC must review the study. The trapping of wild animals covered by the AWAR would not be consistent with the definition of field study and would require IACUC review and approval to determine the details of how the work could be performed to address each of these points. There must be procedures in place for checking traps frequently to prevent unnecessary distress and animal injury, as well as provisions for the treatment or euthanasia of injured animals.

The *Guide* (p. 32) points out that field investigations may require international, state, and/or local permits and that the permitting agencies may require a review of the scientific merit of the proposed study. The *Guide* states: "When species are removed from the wild, the protocol should include plans for either a return to their habitat or their final disposition, as appropriate." In addition, potential occupational health and safety issues should be reviewed by the appropriate institutional committee(s), "with assurances to the IACUC that the field study does not compromise the health and safety of either animals or persons in the field."

Opin. To help ensure appropriate animal care and use, the IACUC is required to review field research that involves animal trapping. Thus, IACUC protocols must provide sufficient detail on methods to be used (e.g., trapping, tagging, collaring, blood collections, euthanasia), frequency of observations, and a contingency plan for adverse events that address animal and human safety. Traps must be frequently checked to prevent unnecessary distress and animal injury, provisions for treatment or euthanasia of injured animals must be in place. IACUCs also have a responsibility to ensure compliance with the appropriate entity regarding permits, handling, and transportation.

Appendix E of the *Institutional Animal Care and Use Committee Guidebook*[16] describes local, national, and international regulations requiring permits for conduct of field studies. *The Guidebook* and publications of several professional societies provide guidelines to the capture and care of wild animals.[17–23] Capture of many wildlife species are recently relying on available video technology to mount cameras at the trapping site to monitor traps remotely.[24]

Surv. How does your IACUC evaluate appropriate means of performing euthanasia for field studies? Select all that apply.

- Not applicable, we do not have field studies. 90/203 (44%)
- An IACUC representative or AV designee travels to the field location to observe the procedure being performed. 7/203 (3%)
- The IACUC reviews a written description of the methods to be used in the field. 102/203 (50%)
- The IACUC reviews a video of the procedures. 10/203 (5%)
- The IACUC consults with an external subject matter expert to determine the appropriateness of the method. 39/203 (19%)
- Other* 16/203 (8%)

*Other:

- The IACUC ensures that training is provided to personnel conducting the procedure.
- The IACUC reviews a written description of the training and experience of personnel with the specific procedure to be used.

16:11 Should field sites be included in the semiannual IACUC inspections?

Reg. The AWAR (§2.31,c,2) state that "animal areas containing free-living wild animals in their natural habitat need not be included in such inspections." This statement agrees with the APHIS/AC *Animal Welfare Inspection Guide* to facility inspections.[25] Neither the *Guide* nor the PHS Policy addresses the issue of inspecting field sites; however, PHS Policy II requires compliance with the AWA as applicable. Although the PHS Policy does not specifically discuss wild animals or field studies, the definition of *animal* in PHS Policy III,A applies to all live vertebrate animals. Therefore, under strict interpretation, field sites could be considered animal study areas by NIH/OLAW and require semiannual inspection by the IACUC. NIH/OLAW clarifies in their FAQ E.4[4] that "While semiannual IACUC inspections of field study sites are not required and in many circumstances are impractical, IACUCs should be apprised of the circumstances under which studies are conducted so that they can consider risks to personnel and impact on study subjects. This may be partially accomplished by written descriptions, photographs, or videos that document specified aspects of the study site (and can include assessments by study personnel located there). The IACUC should also ensure that appropriate permits are in place."

Opin. It is the authors' opinion that the question of whether field study sites are to be included in the IACUC semiannual inspection must be decided on a case-by-case basis. Certainly, field study sites that include housing enclosures should be inspected, as the animals involved are no longer truly free-living and are dependent on adequate monitoring and oversight. The IACUC should inspect such sites when they are in use. However, practical issues must also be considered. If the field site is geographically distant, sending an inspection team may not be feasible. It is recommended that IACUCs be sensitive to public concerns regarding the trapping, restraint, and capture of animals under wildlife and field conditions. IACUC monitoring of these studies may help avoid instances of negative public perception.

Surv. Does your IACUC include field sites in its semiannual inspections? Select all that apply.

- Not applicable, we do not have field studies 90/203 (44%)
- No 72/203 (35%)
- Yes, if not geographically distant 15/203 (<1%)
- Yes, using "surrogate inspections" if not close by (e.g., consultants, nearby institutions) 8/203 (4%)
- Yes, but only if the protocol includes an invasive procedure 11/203 (5%)
- Other* 16/203 (8%)

*Other:

- In some cases, we do virtual inspections or review reports of activities at the sites provided by personnel working at those sites.
- On-site inspections are conducted only if animals are held in captivity.

16:12 Are animals that are used in field studies included on the APHIS/AC annual report?

Reg. The AWA (§2143 a,7,A) requires research facilities to annually report that the provisions of the Act along with professionally acceptable standards governing the care and use of animals in research have been followed. The AWAR (§2.36) define this as the Annual Report and provides greater detail on the requirements.

A regulated species, in accordance with the definition of the word *animal* (AWAR §1.1), not only includes free-living and wild animals but also includes free-living birds. This definition does not include rats (of the genus *Rattus*) and mice (of the genus *Mus*), because they were bred for research. As a result, all animals used in field investigations that *do not* meet the definition of a *field study* in AWAR §1.1 must be listed on the APHIS/AC Annual Report. Research on APHIS/AC-regulated animals in the field is a regulated activity when the work fails to meet the AWAR definition of *field study*. In contrast, animal use that falls under the AWAR definition of field study (§1.1) need not be included on the APHIS/AC annual report because this is not a regulated activity.

Opin. Including the animals in the APHIS/AC annual report was occasionally not done by some survey respondents. However, the institution and the IACUC must provide a mechanism for reporting (on the APHIS/AC annual report) the AWAR-regulated species used in field investigations. For example, the IACUC can send each investigator who has an approved protocol involving research in the field an annual request for summary information on the number and species used during the reporting period (October 1 to September 30). These numbers can then be compiled for the institutional annual report, which is due before December 1 of each year.

Special note: The authors wish to highlight that the use of the terminology "field study" and "field investigation" are not interchangeable and have entirely different meanings in the regulatory context of the AWAR definitions in section §1.1. This wording is often confused, and, if incorrectly used, it would make a distinct difference in an institution's regulatory compliance. Similarly, if the term "wildlife research" is used, this needs to be further described as either a "field study" under the AWAR definition or a regulated activity conducted in the field.

16:13 With reference to housing requirements, should the IACUC make an attempt to distinguish proactively between biomedical research and agricultural research when reviewing proposals that use large farm animals?

Reg. The definition of an *animal* in the AWAR (§1.1) excludes "farm animals, such as, but not limited to, livestock or poultry used or intended for use as food or fiber, or livestock or poultry used or intended for use for improving animal nutrition, breeding, management, or production efficiency, or for improving the quality of food or fiber."

The *Guide* (pp. 32–33) states: "The use of agricultural animals in research is subject to the same ethical considerations as for other animals in research. … The protocol, rather than the category of research, should determine the setting (farm or laboratory). Housing systems for farm animals used in biomedical research might or might not differ from those in agricultural research; animals used in either biomedical or agricultural research can be housed in cages, stalls, paddocks, or pastures. The *Guide* applies to agricultural animals used in biomedical research, including those maintained in typical farm settings."

PHS Policy II does not refer to a particular type of research. It states that it is applicable to all PHS-conducted or PHS-supported activities involving animals. The Health Research Extension Act (§495a), on which the PHS Policy is based, does refer to "animals to be used in biomedical and behavioral research," a reference that, by inference, might be construed as not applicable to agricultural research. NIH/OLAW FAQ A.1 indicates: "There are many valid reasons for institutions to perform program oversight institution-wide using uniform and consistent standards for animal care and use. Institutions must implement the PHS Policy for all PHS supported activities involving animals and must ensure that any standards that might not be consistent with PHS Policy do not affect or pose risks to PHS supported activities." OLAW advises that uniform and consistent standards are essential in a quality animal care and use program, and that public perception of a potential double standard should also be considered.[9] This admonition is elaborated on in NIH/OLAW FAQ A.1.

Opin. The AWAR language has become the working definition of agricultural research. The unstated converse is that farm animals used in studies where the goal is the advancement of biomedical science are regulated by the AWAR. Therefore, the practical reason to distinguish between biomedical and agricultural research is that the latter need not be included in AWAR requirements, such as annual reporting and search for alternatives, and need not conform to recommendations in the *Guide* with particular reference to the space and physical plant recommendations that are focused on biomedical research animals (see 16:14). It is the authors' opinion that applying uniform policies to all animals involved in the institution's animal care and use program regardless of the type of research or teaching activity involved promotes consistent oversight; however, decisions on housing farm animals used in biomedical research under conditions typical of an agricultural setting should be made by the IACUC on a case-by-case basis. There is leeway for institutions to delineate animal areas that are programmatically and functionally distinct and do not support PHS activities via the institution's Assurance per NIH/OLAW FAQ A.1; however, the institution must exercise the utmost in care and diligence in doing so.

APHIS/AC has indicated that it is implied through the AWA (§2143,a,7,A) regarding "professionally acceptable standards governing the care, treatment, and use of animals" that the *Guide for the Care and Use of Laboratory Animals*, the *Guide for the Care and Use of Agricultural Animals in Research and Teaching*, as well as peer-reviewed taxon-specific journals and publications are considered acceptable standards of care to the USDA.

16:14 What criteria are used to distinguish biomedical from agricultural research? How are the standards different for biomedical research housing sites versus agricultural research housing sites?

Reg. See 16:13. The Guide states: "The use of agricultural animals in research is subject to the same ethical considerations as for other animals in research, although it is often categorized as either biomedical or agricultural because of government regulations and policies, institutional policies, administrative structure, funding sources, and/or goals" (*Guide*, pp. 32). Because of this, different criteria for evaluating housing and care of animals of the same species exist based on whether they are used in biomedical or agricultural research. "For example, animal models of human diseases, organ transplantation, and major surgery are considered biomedical uses; and studies on food and fiber production, such as feeding trials, are considered agricultural uses." "Decisions on categorizing research uses of agricultural animals and defining standards for their care and use should be made by the IACUC based on both the researcher's goals and concern for animal well-being" (*Guide*, pp. 33). Therefore, the *Guide for the Care and Use of Laboratory Animals* applies to agricultural animals used in biomedical research, whereas for animals used in agricultural research, the *Guide for the Care and Use of Agricultural Animals in Research and Teaching* is a useful reference.

Opin. It is clear that farm animals such as sheep, cows, goats, swine, and horses used in biomedical research are covered by both the AWAR and the *Guide*. It is also clear that the same animals used for research on livestock production are not ordinarily regulated by the AWAR. PHS-funded agricultural projects (which would be covered by the *Guide*) are very unlikely, given that the focus of PHS funding is biomedical research.

What is not clear are the circumstances under which farm animals used in biomedical research must be housed according to the *Guide*'s recommendations as compared to when they may be housed in more farm-like settings (such as paddocks, pastures, and barns). The *Guide* (p. 33) states that the housing system chosen should be protocol driven. The authors interpret that to mean that farm animals used in studies in which the minimization of variables is vital to the study's outcome (e.g., transplant research) should be housed in the more environmentally controlled setting of a traditional laboratory animal facility. Conversely, farm animals intended for use in studies for which each animal's environment need not be finely controlled (e.g., antibody production) may be housed in farm-type settings. As stated in the *Guide* (p. 33), decisions on the type of enclosure and density of housing for farm species should be made with due consideration to both the researcher's goals and the animal's well-being. For animals housed at an accredited facility in the agricultural setting, AAALAC's perspective on this topic is that "the housing and care for farm animals should meet the standards that prevail on a high-quality, well-managed farm. The collective professional judgment of the responsible oversight body (i.e., IACUC, Ethics Committee), principal investigator and veterinarian should determine which standard(s) applies best regarding the care and welfare of agricultural animals, based on a performance approach in the context of the requirements of the study and the species used. The rationale for making this determination should be documented."[15]

Surv. What criteria does your IACUC use to distinguish biomedical from agricultural studies? Select all that apply.

- Not applicable, we only have research in one of these categories. 164/203 (81%)
- If the end goal is to benefit human or animal health, we consider it biomedical research. 25/203 (12%)
- If the end goal is to benefit food/fiber/etc. production practices, we consider it agricultural research. 28/203 (14%)
- We look at the funding, e.g., NIH funding means biomedical research while the Pork Producers Council means agricultural research. 6/203 (3%)
- We have no specific criteria. 6/203 (3%)
- Other. 2/203 (1%)

Surv. Which standards does your IACUC use for agricultural research housing sites? Select all that apply.

- Not applicable, we do not have agricultural research 172/203 (85%)
- Those of a well-managed farm 12/203 (6%)
- Those in the *Guide for the Care and Use of Agricultural Animals in Research and Teaching* 28/203 (14%)
- Housing standards are protocol driven 5/203 (2%)
- I don't know 0/203 (0%)
- Other 0/203 (0%)

16:15 Is it acceptable to house different animal species in the same holding room?

Reg. The AWAR (§2.33,b,1–§2.33,b,2) state that adequate veterinary care standards require appropriate facilities for housing animals and appropriate methods for prevention of disease. Regarding keeping in the same animal holding/housing room, the AWAR (§2.33,b,5) indicate adequate veterinary care allows for care in accordance with currently accepted veterinary medical and nursing procedures.

The AWAR (§1.1) define a *primary enclosure* as "any structure or device used to restrict an animal or animals to a limited amount of space, such as a room, pen, run, cage, compartment, pool, or hutch." The AWAR do not address housing of different species in the same holding room but do address housing of different species in the same primary enclosure. The AWAR (§3.33,b) prohibit housing hamsters and guinea pigs in the same primary enclosure and housing any other species in the same primary enclosure with guinea pigs and hamsters. Housing rabbits in a primary enclosure with other species is prohibited, "unless required for scientific reasons" (AWAR §3.58,a). Furthermore, the AWAR (§3.33,c; §3.58,b) require the separation of hamsters, guinea pigs, and rabbits that are under quarantine or treatment for a communicable disease from same-species animals and other susceptible species of animals so as to minimize dissemination of such disease. The AWAR (§3.7,d) prohibit cohabitation of dogs or cats with other species in the same primary enclosure unless they are compatible. Nonhuman primates may not be housed with other species of primates or animals unless they are compatible, do not prevent access to food, water, or shelter by individual animals, and are not known to be hazardous to the health and well-being of each other (§3.81,a,3). For all other regulated species, the AWAR (§3.133) state: "Animals housed in the same primary enclosure must be compatible. Animals shall not be housed near animals that interfere with their health or cause them discomfort."

PHS Policy is silent on the subject of mixing different species in the same primary enclosure or holding room, although PHS Policy II requires compliance with the AWAR, where applicable.

The *Guide* (p. 52) states: "Housing should provide for the animals' health and well-being while being consistent with the intended objectives of animal use." The *Guide* (p. 111) also states that "Physical separation of animals by species is recommended to prevent interspecies disease transmission and to eliminate the potential for anxiety and physiologic and behavioral changes due to interspecies conflict. … it may also be acceptable to house different species in the same room, for example, two species that have a similar pathogen status and are behaviorally compatible."

Opin. Professional judgment usually dictates that mixing certain species in the same room or same primary enclosure is inappropriate to their life histories (e.g., cats and dogs), behavioral characteristics, or predator-prey relationships. The potential for disease and parasite transmission is an additional consideration regarding housing different species in the same room or enclosure. As examples of recommended separate housing by species, the *Guide* (pp. 111–112) notes that rats and mice should be housed separately where risk of interspecies transmission with disease-causing bacteria or viruses may exist, and nonhuman primates should be separated by geographical origin due to differing viral disease susceptibility. Our opinion (supported by AAALAC)[15] is that the practice of co-housing differing species in the same room or primary enclosure should be specifically reviewed and approved (or have approval withheld) by the IACUC on a case-by-case basis for all the factors mentioned above.

Surv. Which species do you house together (in separate cages) in the same holding area? Select all that apply.

• Not applicable, we work with only one species.	11/203/ (5%)
• Not applicable, we never mix species in the same holding room.	95/203 (47%)
• Species the veterinary or animal care specialists allow.	29/203 (14%)
• Rodent species that are kept in microisolator-type cages.	67/203 (33%)
• Amphibians and reptiles (same or different species)	28/303 (24%)
• Farm species.	12/203 (6%)
• Nonhuman primate species (although separating Old and New World).	8/203 (4%)
• Other*	16/203 (8%)

*Other:

- Fish species together
- Fish with amphibians
- Agricultural animals when compatible conspecifics are not available.

16.16 What considerations for animal transportation should be taken into account when national or international travel has become limited, such as during a pandemic or with airline restrictions?

Reg. Several federal agencies in the United States regulate the transportation of research animals or of species that might be used for research. Only two laws identify standards for humane care during transportation: the AWA, which details specific standards for animal care during transportation, and the Lacey Act, which provides that wildlife be transported in accordance with the International Air Transportation Association (IATA) Live

Animals Regulations (LARs), which are updated annually.[10] The AWAR provide standards for the humane handling, care, treatment, and transportation of all warm-blooded animals intended for use in research, teaching, or testing and is implemented by the USDA. There are separate rules for transporting dogs and cats (§3.14–3.20), guinea pigs and hamsters (§3.35–3.41), rabbits (§3.60–66), nonhuman primates (3.86-3.92), marine mammals (§3.112–118), all other covered warm-blooded animals (§3.136–142), and birds (§3.161–168). The act contains standards for consignment (delivery of animals to an entity for transport), primary transportation enclosure, primary conveyance, food and watering requirements during transportation, terminal facilities, care in transit, and handling. For international travel, institutions should contact appropriate authorities to ensure compliance with any relevant statutes and other animal transportation requirements that must be met for animals to cross international boundaries, including those of the final destination.

The *Guide* provides performance standards for research animal transportation. It states: "The process of transportation should provide an appropriate level of animal biosecurity while minimizing zoonotic risks, protecting against environmental extremes, avoiding overcrowding, providing for the animals' physical, physiologic or behavioral needs and comfort, and protecting the animals and personnel from physical trauma" (*Guide*, pp. 107). Animals being transported for research should also be accompanied by appropriate documentation, including health certificates, sending and receiving institutions' addresses and contacts, emergency procedures and veterinary contact information, and agency permits as needed. If adverse events occur in the shipping of animals to or from an institution, reporting to OLAW must occur as per NIH/OLAW FAQ F.12.[4] Receiving institutions should notify the shipping institution if problems are found upon arrival.

Opin. When travel is restricted, such as during a pandemic or other type of emergency, planning becomes the most significant step in arranging research animal travel. Normal flights can become delayed or cancelled, potentially leaving shipments of animals delayed at an intermediate stop in the journey. Therefore, it is extremely important to include additional feed and water sources in the shipping container to account for these delays. Ground transportation may be more reliable but can be limited if couriers are understaffed or otherwise impacted. The background, training, and experience of the courier becomes an important factor to verify before deciding to use the courier. If travel options are extremely limited, animal shipments may have to be delayed until transportation becomes more readily available and reliable. Technology that makes use of the global positioning system, and data loggers which monitor environmental parameters, are recommended for use with all animal transport. Many commercial shippers offer these monitoring services.

The reduced number of commercial airlines available and willing to transport animals intended for research, especially nonhuman primates, is a growing concern. The ability to safely transport all animals, including those associated with research, is critical and necessary for animal movement and scientific advancements across the globe. Sharing knowledge and resources worldwide is important to achieve reduction, refinement, and the highest quality of scientific outcomes. Continued advocacy and education of the public, the airlines, and government legislators on the importance of biomedical research and the need for safe transport for research animals is key. Without the capacity to safely transport research animals, the essential work being done to treat, cure, and prevent disease that impacts humans and animals is at risk. Support for the National Association for Biomedical Research and other global policy organizations by the general public and the worldwide biomedical research community is paramount to safeguard responsible, humane, and ethical study of animals in research.

Surv. How does/did your IACUC handle performing semiannual inspections of facilities and program review during a pandemic? Select all that apply.

- We performed them as normally scheduled and with the number of inspectors typical at our institution (i.e., there were no changes). 71/203 (35%)
- We performed them with fewer than normal participants to increase social distancing. 117/203 (58%)
- We performed some portion of them using live streaming/by video. 74/203 (36%)
- We were required to delay some inspections from their usual inspection schedule. 41/203 (20%)
- Other/not listed* 14/203 (7%)

*Other:

- More extensive record review in lieu of in-person inspection
- Relied on investigators for self-evaluation of satellite spaces
- No animals were on-site during the pandemic

Surv. How does/did your IACUC handle performing semiannual inspections of facilities and program review during a pandemic? Select all that apply.

- We performed them as normally scheduled and with the number of inspectors typical at our institution (i.e., there were no changes). 71/203 (35%)

REFERENCES

1. Title 9 - Animals and Animal Products, Subchapter A – Animal Welfare, Code of Federal Regulations. https://www.nal.usda.gov/animal-health-and-welfare/animal-welfare-act Accessed August 11, 2025.
2. National Research Council (NRC). 2011. *Guide for the Care and Use of Laboratory Animals*, 8th edition. Washington, DC: National Academies Press. https://grants.nih.gov/grants/olaw/guide-for-the-care-and-use-of-laboratory-animals.pdf. Accessed August 11, 2025.
3. National Institutes of Health (NIH), Office of Laboratory Animal Welfare (OLAW). *PHS Policy on Humane Care and Use of Laboratory Animals*. https://olaw.nih.gov/policies-laws/phs-policy.htm Accessed August 11, 2025.
4. National Institutes of Health (NIH), Office of Laboratory Animal Welfare (OLAW). Frequently Asked Questions (FAQs). *PHS Policy on Humane Care and Use of Laboratory Animals*. https://olaw.nih.gov/faqs#/guidance/faqs Accessed August 11, 2025.
5. National Institutes of Health (NIH), Office for Protection from Research Risks (OPRR). 1991. The Public Health Service Responds to Commonly Asked Questions. *ILAR News* 33(4):69. Available at: https://olaw.nih.gov/guidance/articles/ilar91.htm which is working. Accessed August 11, 2025.
6. Centers for Disease Control and Prevention (CDC). Biosafety in Microbiological and Biomedical Laboratories—6th Edition, Available at: https://www.cdc.gov/labs/pdf/SF__19_308133-A_BMBL6_00-BOOK-WEB-final-3.pdf Accessed August 11, 2025.
7. National Research Council (NRC). 2006. *Guidelines for the Humane Transportation of Research Animals*. Washington DC: National Academies Press. https://nap.nationalacademies.org/catalog/11557/guidelines-for-the-humane-transportation-of-research-animals Accessed August 11, 2025.
8. USDA, APHIS, Animal Health, National Veterinary Accreditation Program, NVAP Reference Guide, Animal Movement. https://www.aphis.usda.gov/nvap/reference-guide/animal-movement/interstate Accessed August 11, 2025.
9. Grandin T. Behavior of Cattle, Pigs, Buffalo and Antelope during Handling and Transport. http://www.grandin.com/behaviour/transport.html. Accessed August 11, 2025.
10. International Air Transport Association. 2025. https://www.iata.org/en/publications/store/live-animals-regulations/ Accessed August 11, 2025.
11. Silverman, J., M.A. Suckow and S. Murthy. 2014. *The IACUC Handbook*, 3rd edition. Boca Raton: CRC Press.
12. Potkay, S., N.L. Garnett, J.G., Miller et al. 1995. Frequently asked questions about the Public Health Service Policy on Humane Care and Use of Laboratory Animals. *Lab Anim. (NY)* 24(9): 24–26.
13. U.S. Department of Agriculture (USDA), Animal and Plant Health Inspection Service (APHIS). 2018. Licensing and Registration Under the Animal Welfare Act Guidelines for Dealers, Exhibitors, Transporters, and Researchers. https://www.aphis.usda.gov/sites/default/files/graybook.pdf Accessed August 11, 2025.
14. Department of Veterans Affairs. VHA Directive 1200.07 https://www.research.va.gov/programs/animal_research/documents.cfm Accessed August 11, 2025.
15. Association for Assessment and Accreditation of Laboratory Animal Care International (AAALAC). 2022. Frequently Asked Questions. https://www.aaalac.org/accreditation-program/faqs/ Accessed August 11, 2025.
16. Applied Research Ethics National Association/Office of Laboratory Animal Welfare. 2002. *Institutional Animal Care and Use Committee Guidebook*, 2nd edition. Bethesda: National Institutes of Health. http://grants.nih.gov/grants/olaw/GuideBook.pdf Accessed August 11, 2025.
17. Use of Fishes in Research Committee (joint committee of the American Fisheries Society, the American Institute of Fishery Research Biologists, and the American Society of Ichthyologists and Herpetologists). 2014. Guidelines for the use of fishes in research. American Fisheries Society, Bethesda, Maryland. Available at: https://fisheries.org/docs/wp/Guidelines-for-Use-of-Fishes.pdf. Accessed August 11, 2025.
18. Herpetological Animal Care and Use Committee. 2004. *Guidelines for the Use of Live Amphibians and Reptiles in Field and Laboratory Research*, 2nd edition. American Society of Ichthyologists and Herpetologists. Available at: https://ssarherps.org/wp-content/uploads/2014/07/guidelinesherpsresearch2004.pdf. Accessed August 11, 2025.
19. Sikes, R.S., and the Animal Care and Use Committee of the American Society of Mammalogists. 2016. 2016 Guidelines of the American Society of Mammalogists for the use of wild mammals in research and education. *J. Mammal.* 97(3):663–688.
20. Braun, C.E. (ed). 2005. *Techniques for Wildlife Investigations and Management*, 6th edition. Bethesda: The Wildlife Society.
21. Captive Care Committee. 2007. *IPS International Guidelines for the Acquisition, Care, and Breeding of Nonhuman Primates*, 2nd edition. International Primatological Society. http://internationalprimatologicalsociety.org/wp-content/uploads/2021/10/IPS-International-Guidelines-for-the-Acquisition-Care-and-Breeding-of-Nonhuman-Primates-Second-Edition.pdf Accessed August 11, 2025.
22. The Ornithological Council. 2023. *Guidelines to the Use of Wild Birds in Research*, 4th edition. https://birdnet.org/wp-content/uploads/2023/09/Guidelines-September-2023.pdf Accessed August 11, 2025.

23. Powell, R.A., Prouix, G. 2003. Trapping and Marking Terrestrial Mammals for Research: Integrating Ethics, Performance Criteria, Techniques and Common Sense. *ILAR J.* 44(4): 259–276.
24. Meek, P.D., Fleming, P.J.S., Ballard., G., Claridge, A.W., Banks, P., Sanderson, J., Swann, D. (eds). 2014. *Camera Trapping: Wildlife Management and Research.* Collingwood, Australia: CSIRO Publishing.
25. U.S. Department of Agriculture (USDA) , Animal and Plant Health Inspection Service (APHIS). June, 2025. Animal Welfare Inspection Guide. Available at: https://www.aphis.usda.gov/sites/default/files/animal-welfare-inspection-guide.pdf Accessed August 11, 2025.

17

Considerations for Mitigation of Pain and Distress

Emily L. Miedel and F. Claire Hankenson

Introduction

Pain and manifestations of stress and distress are important considerations in the well-being of laboratory animals. Overall, the tenet that pain and distress should be minimized, if not eliminated, is a federal and ethical expectation for enrollment of animals in research and testing procedures.[1–3] The mitigation of potential pain in animals is highlighted in the U.S. Government Principles (1985), specifically in the fourth statement, "IV. Proper use of animals, including the avoidance or minimization of discomfort, distress, and pain when consistent with sound scientific practices, is imperative. Unless the contrary is established, investigators should consider that procedures that cause pain or distress in human beings may cause pain or distress in other animals."

The intention to minimize pain and distress for animals is a direct correlate to one of the 3Rs terms, that of "refinement," as described in Russell and Burch's classic text, *The Principles of Humane Experimental Technique*.[4] The interpretation of potential refinements has expanded from their original intent and is now reflected in the contemporary interpretation of refinement as "Advancing research animal welfare by exploiting the latest *in vivo* technologies and by improving understanding of the impact of welfare on scientific outcomes."[5] More specifically, refinements in experimental design include species-specific training of animal care personnel, acclimation of animals after arrival to new sites, provision of social contact, veterinary supportive care, environmental enhancement, increased monitoring for pain and distress in order to intervene as necessary, and objective humane endpoints. Our community continues to investigate how best to assess animal behaviors as they relate to detection of pain and distress; therefore, refinement has become an increasingly important aspect of animal research aimed to enhance animal welfare and improve the quality of scientific data obtained.

Approval of research projects with significant potential for pain or distress is one of the most serious ethical judgments an IACUC makes. The critical IACUC requirement that research investigators are to provide relevance, cost-benefit consideration, and objective validation prior to the conduct of any animal studies cannot be overstated. Veterinary consultations are encouraged, and pilot studies can be undertaken that require an investigator to validate how experimental interventions might impact the animal model. The decision to potentially withhold analgesia should be made based upon relevant published literature, in consultation with veterinary staff and with the agreement that other methods of supportive care are to be provided.[6,7] Supportive measures, by their very nature, are intended to reduce the amount of distress endured by the animal patient. Supportive or palliative measures will require some frequency of monitoring and care beyond the daily cageside check. While the incorporation of critical care equivalent to human medical emergency care may be challenging, whether due to animal size or to complexity of species, the authors contend that synergistic refinements in veterinary care of research animals are achievable and imperative.

Historically, the study of pain to better understand neurologic mechanisms underlying sensory transmission through the peripheral and central nervous systems has relied extensively on animal modeling. Certain research models may necessitate that some degree of stress or distress experienced by animals is fundamental to the model's success. When working with models that induce or evoke painful responses, there is an art and science to performing therapeutic interventions. The use of anesthesia and analgesia are vital steps to be employed over the time course of a research project. The authors emphasize that all therapeutic agents delivered to alleviate pain and distress should be used according to the approved research protocol with documentation detail to the extent sufficient to meet the legal, ethical, and research reproducibility obligations.[6,7]

Ultimately, consideration of methods for minimizing or eliminating pain and distress should be of central concern to the IACUC when evaluating the benefit of the studies. World Organization for Animal Health defines animal welfare as "the physical and mental state of an animal in relation to the conditions in which it lives and dies." The effort to maintain positive animal welfare with measures deployed to reduce or remove any negative impact on physical, mental, and overall state remains paramount. Rather than simply avoiding negative states, a more contemporary approach to animal welfare is promoting positive experiences, through social housing, consistency of husbandry and enrichment, and an animal's agency over its environment.[8] Buchanan-Smith and colleagues[9] have proposed an enhanced definition of refinement, one that incorporates measures that amplify elements of well-being through proactive approaches. The public will continue to seek understanding about the relevance and involvement of animals for scientific discovery; therefore, all stakeholders in the research regulatory oversight community should stay abreast of relevant veterinary and scientific literature, medical advancements,

DOI: 10.1201/9781003272861-17

alternatives to animal use, and related approaches that will balance the intended *benefit to society* with *harm/cost* to the animal subjects.

17:1 What is the difference between stress and distress?

Reg. Distress is defined in the *Guide* (p. 121) as an aversive state in which an animal fails to cope or adjust to various stressors with which it is presented. Although the Animal Welfare Act (AWA) requires that animal pain and distress are minimized, neither the AWA nor the associated regulations (AWAR) provide a definition of distress.

Opin. It is important to recognize that there is a difference between stress and distress. The most recent report from the Committee on Recognition and Alleviation of Distress in Laboratory Animals[10] stated that the adaptive *stress response* and a *state of distress* can be differentiated based on "good scientific evidence" but that "distress remains a complex and still poorly understood phenomenon." Attempts have been made to define and differentiate stress and distress, and these vary in their approach.[10,11] The Committee on Recognition and Alleviation of Distress in Laboratory Animals further clarifies that based on current scholarly thinking "stress denotes a real or perceived perturbation to an organism's physiological homeostasis or psychological well-being." The role of the stress response is to make the organism able to adapt to a given situation, theoretically to regain a state of homeostasis, or balance, which will allow it to function well. The ability to adapt physiologically or psychologically to a disturbance is of benefit to the individual animal; however, the end result is not always to restore the animal to its exact state prior to the stress-inducing event.[12] In addition, the events that are *stressors* and the *condition* of mounting a stress response are not always unpleasant. As an example, strenuous play may trigger a stress response but could be highly rewarding and enjoyable to the animal—this is defined as "eustress," where a stressor initiated an activity (playing) that ultimately enhanced well-being. Recent work clarifies that the condition of *stress* is present only when the individual cannot predict a stimulus or control its environment (i.e., adaptation is not possible); otherwise, a living being's physiologic response to perturbations resulting in the constellation of events called "the stress response" is simply that—a response.[11]

When the effort required to adapt to some stressors or the aversive nature of the stressor increases and reaches a point where the animal fails to be able to adapt adequately, this is considered to represent "distress." In other words, it is in the "failing to cope with" severe, sustained, or multiple combined stressors that propel the animal's homeostasis into distress. Stressors often cause a low level unpleasant affective or emotional state (e.g., thirst) and the degree of distress caused by an experimental procedure (e.g., fluid deprivation) will increase as the animal's affective and physiologic states become overwhelmed. Unless manifested by overt external physical signs or behaviors, states of distress may be difficult to recognize in laboratory animals. In other words, it might be challenging to determine when a mouse becomes distressed by fluid deprivation, yet renal failure and hypovolemic shock are adverse clinical outcomes that can be measured objectively. McMillan[13] gives a simple explanation for why certain experiences (such as hypoxia, hunger, nausea, urinary bladder distention, or hypothermia) cause an increasingly unpleasant state; put more simply, these individually or collectively represent a threat to survival. It is important to recognize that a poor physiologic state does not always mean that distress is present: an animal may have undiagnosed cancer or other subclinical disease that may indicate a diminished state of health, although the experience of the animal is not one of distress. Judgments and conclusions made during an ethical review process should be based on objective criteria, such as percentage of weight loss or diagnostic evidence of a tumor. Following the example above, merely seeing an animal reach a measurable threshold (objective criterion) does not confirm that distress is present, and distress may be experienced by an animal in situations where we cannot (or do not) assess their mental state. Lastly, individuals may experience distress at different stages in a disease process, highlighting the importance of individual monitoring and serial assessments.

Pain is only one of several aversive states that can cause distress, and although the regulations governing laboratory animal use tend to favor scrutiny of protocols for occurrence of pain and its alleviation, all procedures that have the *potential* to cause pain and distress should be thoroughly reviewed. Pain itself is complicated as a cause of distress; the degree of pain a stimulus causes might differ between individual animals, between species, and it may depend on the situation. For example, housing on wire-bottom cages might be more distressful for rodents that have injured footpads than for clinically healthy rodents; as well, standing on a heated surface is likely to be more uncomfortable and painful to the footpads of a mouse than to the hoofs of a horse. It might be said that an aversive state has dimensions of time and magnitude and becomes "distress" when the product of the two exceeds a certain threshold. In any case, the IACUC must take responsibility for estimating the degree of averseness of an experience or research condition, which in some cases may have been studied scientifically (e.g., by preference and motivation studies) or otherwise may require a subjective interpretation (e.g., this is what a human would experience in an equivalent state) in order to guide attempts at refinement.[14]

17:2 What is the difference between pain and nociception?

Reg. *Pain* is not directly defined in the AWAR, but a *painful procedure* (§1:1, Painful Procedure) is defined as one "that would reasonably be expected to cause more than slight or momentary pain or distress in a human to which that procedure is applied." The example provided is pain in excess of that caused by injections or other minor procedures. The *Guide for the Care and Use of Laboratory Animals* (the *Guide*) states that pain is a complex experience that typically results from stimuli that damage or have the potential to damage tissue (p. 120). *Nociception* is defined in neither the AWAR nor the Public Health Service (PHS) Policy. The definition of a *painful procedure* in U.S. Government Principle IV (which is incorporated into the PHS Policy) and the *Guide* is similar to the AWAR definition.

Opin. Nociception represents only the detection of certain stimuli by sensory nerves and the transmission of signals to the central nervous system. In other words, nociception represents the way that noxious (potentially tissue damaging) stimuli are transmitted through the spinal cord and brain. Noxious input of a sufficient magnitude will cause depolarization of sensory nerve endings; examples of such stimulus types are applications of heat or cold, focal pressure, and chemical or biological irritants. The stimulatory signal travels through the sensory nerve tracts to the spinal cord and potentially to the brain. Noxious stimuli may concomitantly trigger both involuntary (e.g., reflex limb withdrawal, autonomic nervous responses) and voluntary (purposeful) aversive movements. Pain is the subjective experience of nociceptive signal input. The nociceptive signals arrive at different levels of the brain (depending on the phylogenetic rank of the species) and may be processed or interpreted as a pain sensation. When the overall interpretation of the stimulus is that it is painful, an affective-motivational response occurs, involving prior experience and contextual information, and thus the *pain experience* of the individual depends upon many factors.

Although reflex motor and autonomic responses may occur in the unconscious or anesthetized animal in response to noxious stimuli, they do not result in pain or distress because the animal is unaware of the stimulus. The administration of anesthetics to the point where both voluntary and involuntary movements are suppressed, and autonomic responses are blunted in response to a noxious input, is presumed to be sufficient to eliminate the experience of pain for the duration of the anesthesia. Upon recovery to a conscious state, the animal's capacity to experience or perceive pain will return.

Pain in humans is described as having both a sensory and an emotional component. For example, the International Association for the Study of Pain (IASP) definition is as follows: "Pain is an unpleasant sensory and emotional experience associated with actual or potential tissue damage."[15] Defining animal pain is challenging because of uncertainties relating to the emotional (affective) component of pain in nonhuman animals. This has been avoided in some definitions by interpreting pain in relation to its effects on animal behavior. For example: "Pain in animals is an aversive sensory experience that elicits protective motor actions, results in learned avoidance, and may modify species-specific traits of behavior, including social behavior."[16]

Nociception and pain serve the organism by protecting it from further injury. One neuroscientist's view of pain is that it evolved as part of the function of homeostasis (ensuring that the body regulates its own well-being).[17] Water and salt balance, for example, are regulated and involve internal physiologic compensations as well as behavioral drives. "The behavioral drive that we call pain usually matches the intensity of the sensory input, but it can vary under different conditions, and can become intolerable or, alternatively, disappear, just as hunger or thirst.[17] Regardless of whether the reader agrees that animals experience emotions, it is generally accepted by veterinary staff, scientists, and the public that common and domesticated animal species experience pain. Animals exhibit the behavior-motivational component in response to pain, which can vary depending on the circumstances of what else is happening at the time, and, if the experience continues with sufficient magnitude, it can lead to adverse health consequences.

A useful elucidation of nociception, pain, pain assessment, and pain recognition and treatment is given in the Institute of Laboratory Animal Resources report, titled *Recognition and Alleviation of Pain in Laboratory Animals*.[10] Definitions, importance, recognition, and treatment goals for pain are also reviewed in the American College of Veterinary Anesthesia and Analgesia position paper on pain in animals[18] and the American College of Laboratory Animal Medicine (ACLAM) Position Statement on Pain and Distress in Laboratory Animals.[19]

17:3 What should be the minimal expectations of the IACUC related to the relief of pain and distress?

Reg. The IACUC, under the AWA (§2143,a,3,A; §2143,a,3,C) and AWAR (§2.31,d,1,iv,A–C), must ensure that procedures causing more than momentary or slight pain or distress will be performed with appropriate sedatives, analgesics, or anesthetics unless certain specific criteria are met (e.g., IACUC-approved scientific justification to withhold analgesia for only the necessary period of time). These procedures must involve the Attending Veterinarian (AV) or the AV's designee in their planning, and investigators are not permitted to use paralytics without anesthesia. Animals experiencing severe or chronic pain or distress that cannot be relieved must be humanely euthanized during or at the end of the study. The AWAR (§2.31,d,1,ii) require the consideration of alternatives to procedures that may cause more than momentary or slight pain or distress to animals and a written narrative description of the methods and resources used to determine that alternatives are not available. To aid researchers with their database searches, the AWA (§2143,e) specifies the establishment of the National Agricultural Library (NAL) to serve as a resource on the 3Rs to help identify means to limit pain and distress.

The PHS Policy (IV,C,1,a; IV,C,1,c) is worded similarly to the AWAR with respect to avoiding or minimizing pain and distress, the need to use appropriate means to alleviate pain and distress, and the need to euthanize animals during or at the end of a procedure if severe pain or distress cannot be alleviated. PHS Policy (IV,C,1,b) states: "Procedures that may cause more than momentary or slight pain or distress to the animals will be performed with appropriate sedation, analgesia, or anesthesia, unless the procedure is justified for scientific reason in writing by the investigator." The onus to clearly demonstrate the exclusion of analgesics is on the investigator and must be based on actual scientific reasons grounded in physiology, not just on a reference to a previous publication where analgesia was not given. PHS Policy (IV,D,1,d) notes that applications and proposals to the PHS must contain a description of the procedures that are designed to assure that discomfort and injury to animals will be limited to that which is unpreventable in the conduct of scientifically valuable research, and that proper drugs will be used where indicated to minimize animal pain and discomfort. The U.S. Government Principles (specifically IV–VI) are the basis for the language found in the AWAR and PHS Policy

The *Guide* (p. 120), U.S. Government Principle IV, and the AWAR (§1.1, Painful Procedure) assume that procedures that cause pain in humans will likely cause pain in nonhuman animals. The *Guide* (p. 121) states that the selection of analgesia and anesthesia for animals should reflect appropriate veterinary judgment. It also notes (p. 120) that the recognition of pain in different species is a key to its prevention or alleviation. The *Guide* (pp. 122–123) warns that certain drugs, such as sedatives, anxiolytics, and neuromuscular-blocking agents, are not analgesics or anesthetics.

Opin. Pain and distress are undesirable from a scientific and ethical perspective and form the basis for much public debate about the use of animals in research. The role of the IACUC is to thoroughly review all protocols for elements that might lead to pain or distress and to assess the appropriateness of methods designed to minimize these occurrences. In doing so, there should be scrutiny of the requisite search for alternatives. Alternatives in this context range from replacement of animals altogether to the refinement of methods of animal treatment so that less total pain or distress are experienced (see "3Rs" as described in the Introduction to this chapter). Perhaps it is in this arena that the IACUC will have its most important oversight, reviewing not only the proposed alleviation of pain or distress, but also the relevant and timely monitoring strategies, as well as verification of the training and competency of those involved in conducting the procedures. For example, an inexperienced or novice animal research employee may handle an animal inappropriately and cause unanticipated pain that could otherwise have been prevented by proper training. It also is likely that a person skilled in surgical procedures will carry out a project using fewer animals (because of fewer technical failures), leading to less postoperative pain (because of less tissue trauma), and effect a more rapid recovery (because of shorter anesthesia time) than an inexperienced or less competent surgeon. The total amount of pain or distress from all aspects of animal use, including not only experimental procedures themselves, but also such things as marking/identification and general husbandry procedures, should also be considered.

For an IACUC to decide whether the Principal Investigator (PI) has sufficiently considered pain and distress alternatives, all potential sources of these experiences should be considered and alleviation or preventative measures sought. Although the regulations use broad terms such as *appropriate* (drugs) or *unavoidable* (discomfort) (AWAR §2.31,d,iv,A; §2.31,e,4; PHS Policy IV,C,1,b), some methods to reduce pain or distress might be possible but unfeasible (e.g., requiring years of surgical training prior to permitting a new surgeon to operate). The minimization of stressors that might lead to pain and distress can also be addressed through environmental supportive care, such as by providing additional bedding, more easily reachable food or water, fluid therapy, or other improvements in comfort and homeostasis (*Guide*, p. 122). The IACUC, especially in consult with veterinary staff members, plays an important role in reviewing and encouraging all avenues of pain relief in the protocols under review.

Finally, the IACUC should ensure that animals are observed for documentation of species-specific behaviors and health condition at regular intervals, using generally agreed upon criteria (e.g., body condition scoring, active behaviors such as food consumption, activity levels, and attitude) in both the normal state and during experimental conditions when pain or distress is expected. Monitoring of animal well-being can ensure that measures to reduce pain or distress are deployed rapidly when needed and are effective in action. The frequency of observations should usually be greater when a significant abnormal state is anticipated. All these factors should be holistically considered by the IACUC when assessing the efforts an investigator intends to make in order to reduce pain and distress experienced by animals.

17:4 What alternatives must be considered for protocols that will involve pain or distress? How can this be done?

Reg. The AWAR (§2.31,d,1,ii) require the investigator to consider alternatives to procedures that may cause more than momentary or slight pain or distress to the animal. The investigator must provide a written description of the methods and sources used to determine whether alternatives are available. The Animal Welfare Information Center (AWIC) of the National Agricultural Library offers training in performing acceptable literature searches for alternatives and will perform searches upon request.[20] The *Guide* (p. 5) states that consideration should be given to the use of inanimate systems or nonvertebrate animals. The U.S. Government Principle III states that mathematical models, computer simulation, and *in vitro* biological systems should be considered.

Opin. From 17:3, all protocols that have the potential to cause pain and distress should be reviewed with reference to reduction of animal numbers to only those necessary to produce statistically relevant results, replacement with alternatives to animals, and refinements in best practices to improve animal health. This requires significant effort from both the IACUC and the investigator. The investigator should explain, in detail, how an animal is to be manipulated, handled, and studied and justify their selection of animals over inanimate modeling and the use of more invasive methods over less invasive ones. The investigator must describe recent literature sources/searches that have been reviewed to demonstrate that there are no available alternatives that will minimize pain or distress. Within the IACUC protocol, procedural and experimental details should be selected as terms to search for alternatives. As the number of databases on alternatives grows, reference to searches on these can be included within the IACUC application (see 17:20).

Investigators also should demonstrate to the IACUC an appropriate experimental design that maximizes the information gained and minimizes the number of animals necessary to achieve the expected outcome. It is helpful to indicate whether expert advice from a statistician or power analyses have been included, particularly when the experimental design is complex (*Guide*, pp. 26–27). Consultation on algorithms for estimating appropriate animal numbers is encouraged.[21] Finally, a range of options are available for reducing pain and distress, and those appropriate to the investigation should be considered. For example, it might be possible to conduct a study entirely while the animal is under general anesthesia, reduce pain by using appropriate analgesics, or, as technology expands, to use less invasive methods of collecting data so that the overall numbers of animals are reduced. It also may be possible to limit the distress experienced by animals, or the duration of that distress, by requiring defined endpoints for a study (e.g., placing upper limits on the total tumor burden in studies of carcinogenesis or defining a set of criteria that will be used to determine when an animal should be removed from study and humanely euthanized). The *Guide* (p. 26) recommends the use of pilot studies to evaluate the potential for pain or distress from procedures that have not been previously performed (see 17:55).

17:5 Since the actual assessment of pain cannot be made until after a procedure is performed, is it appropriate for the IACUC to attempt to assess and categorize the potential for pain and distress when first reviewing a protocol?

Reg. The AWA (§2143,a,7,A) and the AWAR (§2.36) require an annual report for regulated species, encompassing animal use activity in testing, research, and teaching from October 1 through September 30 (the federal fiscal year). The annual report (AWAR §2.36,b,5–§2.36,b,7) categorizes total animal numbers used on the basis of the level of pain or distress they experienced. Humane Use Category C (on APHIS/AC Annual Report form 7023) is used to enumerate animals which experience no more than momentary or only slight pain or distress and did not require pain-alleviating drugs; Humane Use Category D is used to enumerate animals which experienced pain or distress alleviated by anesthesia, drugs, and supportive care; and Humane Use Category E is used to enumerate animals in which pain or distress could not be alleviated due to research needs. The PHS Policy has no requirement for the categorization of pain and distress. (See 17:2 for the definition of a *painful procedure*.)

Opin. Many IACUCs use the APHIS/AC Humane Use Categories (pain and distress categories) for all animals, not only those regulated by the AWAR. Nevertheless, the AWA requirement applies to USDA-regulated species only. The wording of AWAR §2.36,b,5–§2.35,b,7 is in the past tense, and the information provided to APHIS/AC in the annual report should be based on the pain or distress actually experienced by the animal, not solely on an educated estimate made by the IACUC or the investigator prior to the initiation of the study. While prospective estimates of the number of animals that will fall into each of the pain or distress categories can simplify the work of the IACUC, the reality is that research needs and experiences often change during the course of a study. If the IACUC uses a prospective system of assigning pain or distress categories, the IACUC should have an established process that allows an investigator (or the IACUC) to reassess and change the categorization retrospectively, on the basis of the animals' actual experience of and alleviation of painful sensations. A good clarification of prospective vs. retrospective categorization of pain and distress can be found in the *Reducing Administrative Burden for Researchers: Animal Care and Use in Research* report in response to the 21st Century Cures Act (p. 20).[22] Although the estimation of pain may not be fully calculable for a particular animal until after a procedure has been performed, it is frequently possible to make an informed estimate of the likely degree of pain that may be experienced. If the procedure, or a similar procedure, has been carried out previously at the institution, then this can be used to help predict the likely pain characterization, regardless of animal species. If the technique or species/strain is new to the institution, then colleagues at other institutions where the procedure has been performed should be consulted. If possible, individuals who actually performed or cared for animals undergoing the novel procedure should provide input. Extrapolations of pain experienced by humans undergoing similar procedures can also act as a guideline. There should be few occasions when some indication cannot be obtained as to the likely consequences of pain and distress. This initial assessment should certainly be sufficient to determine, for example, what level of analgesic use would be appropriate, and what type of periprocedural care and support the animal might require. In all instances, measures for the control of pain require increased monitoring of animals to ensure the interventions are effective; subsequently, if interventions are not effective, these should be adjusted accordingly with veterinary oversight.

17:6 Can the IACUC approve protocols in which animals will experience pain or distress that is not relieved by the use of anesthetics or analgesics? Under what circumstances?

Reg. If an investigator can justify, in writing, that withholding anesthesia or analgesia is required for scientific reasons and will continue only for the necessary period, then the IACUC can review and approve such an activity (AWAR §2.31,d,1,iv,A). The PHS Policy (IV,C,1,b) has similar wording. The *Guide* (p. 27) indicates that for studies with the potential for unalleviated pain or distress, the IACUC is obligated to weigh the objectives of the study (the benefits) against animal welfare concerns (the cost to the animal).

Opin. IACUCs may be asked to approve protocols in which animals experience pain or distress that, due to experimental needs, cannot be alleviated by the use of anesthetics or analgesics. For example, in studies of pain pathophysiology and treatment, animals are subjected to noxious stimuli or tissue injury and may experience pain for which analgesic administration may interfere with the protocol to such an extent that its use would invalidate the data obtained. Pain and distress may also occur in animal models such as those involving neoplasia, chronic organ failure, infectious diseases, inflammatory responses, toxicity testing, and a wide range of other circumstances. In many of these studies, alleviation of pain and distress by pharmacological means may not be possible because of interactions between the drugs used and the research objectives. However, methods to reduce pain or disease-related disability, such as providing environmental support (e.g., nondrug therapies that deliver additional comfort, provision of soft surfaces and substrates for resting, deep bedding, treats, and social housing with familiar conspecifics) may act to reduce overall distress (*Guide*, p. 122). Humane endpoints and minimizing (or eliminating) the amount of time when pain is experienced are crucial components of these protocols. Defining interventions or the criteria for conducting euthanasia when preestablished endpoint criteria are reached needs to be included in these IACUC applications. In protocols requiring surgeries to induce models that will subsequently experience pain associated with disease progression, alleviation from surgical pain should be considered separately and in addition to pain resulting from the disease model. IACUCs can request a pilot study or validation study to demonstrate that short term therapeutic perioperative analgesic use significantly alters the research model studied. Decision trees, such as those found in Peterson et al.,[7] are available that may benefit IACUC members in deciding whether a Principal Investigator is justified in withholding analgesics. When pain or distress is unpreventable, IACUCs should ask the researcher whether the potentially serious impact of such experiences on animal physiological processes will detrimentally influence study results. Pain and stress cause metabolic and immunological alterations in animals and humans. If a control group does not experience the equivalent amount of pain or stress as the experimental group, then results of certain types of studies might potentially be biased to a greater degree than if both groups were treated with analgesics. Another consideration is whether the benefit of the information or results outweighs the total degree of animal distress imposed by the study (*Guide*,

p. 27). Whenever there is unalleviated pain or distress, efforts must be made to reduce the number of animals used in such a study, replace animals in the study with alternative techniques that do not cause pain, and refine the study design so that pain and distress are minimized. Although these basic tenets should be applied to all protocols, they clearly are of particular importance in circumstances in which significant pain and distress can be anticipated and would otherwise be relieved in humans with the same disease/health condition as those under study in animals.

17:7 If the AV and an investigator differ on the pain or distress potential for a particular procedure, is it the responsibility of the AV, IACUC, investigator, or all to provide supporting evidence for their point of view?

Reg. The AWAR require research facilities to assure that the AV has appropriate authority to ensure the provision of adequate veterinary care and to oversee the adequacy of other aspects of animal care and use (§2.33,a,2). Under the AWAR, the AV provides guidance to the PI relative to the use of anesthesia and analgesia (§2.33,b,4; §2.31,d,1,iv,B), while the IACUC maintains its responsibility to ensure studies are performed with proper sedation, anesthesia, or analgesia (AWAR §2.31,d,1,iv,A). The PHS Policy itself is silent on the authority of the AV to change post-procedural care; however, the *Guide* (p. 114) notes that the AV and the investigator should make all efforts to work together to determine the appropriate treatment or course of action. In the event of a pressing health problem where consensus between the AV and the research team cannot be resolved, the AV must have the authority to institute appropriate measures to relieve severe pain or distress (*Guide*, p. 114).

Opin. Veterinarians and veterinary care staff are trained professionals knowledgeable in animal physiology and medical care with respect to species-specific differences in behavior and husbandry needs. In some instances, the veterinarians, veterinary technicians, and animal care staff are in a position to observe animals more often and with greater insight than research personnel. Many researchers also have expertise in animal husbandry or behavior or may have acquired substantial experience over the course of time. If the research team can offer evidence to the IACUC about the impact of a study on animal well-being, then this should be considered. However, as the science of animal welfare advances and other improvements in research refinement are published, veterinarians and animal welfare scientists may be able to contribute new knowledge to the assessment of the potential for pain or distress and suggest remedies for their alleviation. Any decision made by an IACUC should incorporate input from all stakeholders involved in animal care and experimentation.

17:8 If an investigator provides, in writing, a scientific rationale for withholding the provision of analgesia during and after a painful procedure, is the IACUC obliged to accept this rationale?

Reg. The AWAR (§2.31,d,1,iv,A) require the use of analgesia during a procedure that causes more than momentary pain or distress unless the investigator provides the IACUC written scientific justification for withholding such agents. The PHS Policy (IV,C,1,b) includes the same wording. The AWA (§2143,a,6,A) and the AWAR (§2.31,a) state that except as authorized by law or by the AWAR, the Secretary (of USDA) or the IACUC, respectively, cannot prescribe methods or set standards for the design, performance, or conduct of actual research or experimentation by a research facility. The HREA (§495,a,2,B) states that the guidelines established in the PHS Policy "shall not be construed to prescribe methods of research" but also indicates that this statement does not override compliance with the PHS Policy (HREA §495,b,1). Both the AWAR (§2.31,c,6) and the PHS Policy (IV,B,6) provide the IACUC with the authority to approve or withhold approval of the use of animals in research or other animal activities.

Opin. It has been argued that the scientific justification presented to the IACUC must have some substance; otherwise, it is no better than no justification at all.[23] The same reference stated that if a logical justification could be provided, even if "reasonable minds may differ as to whether the justification suffices, the IACUC must defer to the investigator." In contrast, both APHIS/AC and NIH/OLAW then responded that it is their interpretation "that the AWA and PHS Policy do not require an IACUC to approve a proposed project against its will."[24] This is reaffirmed in NIH/OLAW FAQ D.6,[25] which states that IACUCs must confirm that the protocol is consistent with the *Guide* unless a scientific justification for a departure is presented and is acceptable to the IACUC.

Any justification to withhold analgesia should be clearly reasoned and, whenever possible, documented by scientific data or literature references. Generic statements, such as "Analgesia will interfere with the collection of useful data" may be correct, but the IACUC should not accept such statements at face value (see 17:7). The IACUC needs to consider what analgesics have been considered and needs to be provided requisite literature or data to determine if the justification to withhold analgesics is comprehensive and is not contradictory or inconclusive. IACUCs should not base judgments on extrapolations or assumptions from different models or different analgesics (even when within the same class of drugs). PIs should explore options, in consultation with veterinary staff, for all classes of analgesics in their protocol to determine if another option for pain relief could be considered (e.g., could the study incorporate a long acting local anesthetic if an NSAID or opioid was contraindicated in study outcomes. Furthermore, the IACUC needs to be presented with justification that the proposed lack of analgesic use would negatively impact the model more so than unalleviated pain and its characterized downstream effects. Additionally, every attempt should be made to assist with supportive care to provide comfort to the animal during the study phases, whether through nutritional, bedding, thermal, or environmental enrichments.

17:9 What are clinical and behavioral criteria an IACUC, veterinary staff, or an investigator can use to determine whether an animal is in pain or distress?

Reg. The *Guide* (pp. 120–121) provides examples of some species-specific manifestations of pain (such as nasal porphyrin discharge in rodents) and quotes U.S. Government Principle IV, which states that unless the contrary is known, it should be considered that conditions causing pain in humans also may cause pain in other animals. The AWAR (§1.1, Painful procedure) use similar wording, in addition to stating "that is, pain in excess of that caused by injections or other minor procedures."

Opin. Attempts to recognize pain or distress in animals rely on understanding the potential causes of pain or distress, observing animals using a combination of behavioral and physiological variables, and looking for deviations over time from clinical health and normality of behaviors. In the case of pain, for example, an animal may change its spontaneous behavior so that it becomes less active, decreases grooming or nest-building activity, and/or reduces its food and water consumption. It also may change its responses to handling, for example, by exhibiting increased aggression. Conversely, some animals may become apathetic and unresponsive to handling. Grimace scales of pain have been developed for several species which are helpful objective tools, but it is important to note that the absence of a grimace does not necessarily mean an animal is not in pain, especially if the observer is in the room.[26–29] For rodents, evaluating nest building, burrowing activity, or lack of grooming can help identify animals that may be in pain or distress.[30] Body weight changes may occur as a result of acute or chronic pain. The criteria used to assess pain and distress will depend on the species, the type of pain, and the degree of socialization, with the overarching constant of behavioral assessment as compared to expected behaviors. There is no purely objective way to assess "clinical" pain or distress in animals, and so the most common way to evaluate this is to observe behavior or representations of behavior (such as nest-building quality or utilization of enrichments) within the home environment/cage/pen. Species-specific signs of pain have been reported, and the literature continually expands as refinements are researched and implemented to improve animal well-being.[10,30–47] Importantly, clinical signs of pain and distress may be difficult to detect and their recognition almost always requires a detailed knowledge of the normal behavior of the animal species within the confines of the provided enclosure. Recognition also requires that sufficient time be allocated for observation of the animal, potentially at numerous intervals; in some circumstances, it may be necessary to observe the animal's behavior in such a way that it is unaware of the presence of the observer (e.g., by using a video camera) so as to eliminate the animal's instinct to mask displays of pain.

17:10 An IACUC protocol indicates that a procedure is moderately painful to its canine subjects for approximately 1 hour, then never repeated. During that time, pain-relieving drugs cannot be used, but an alternative pain relief mechanism will be used. That is, calming music, a darkened room, and human petting of the animal will occur. The technique has not been previously attempted. Should the IACUC consider this to be alleviated or unalleviated pain?

Reg. See 17:2. The definition of a *painful procedure* is available (see 17:2) along with related regulatory information (see 17:3). For painful procedures in which appropriate anesthetic, analgesic, or tranquilizing drugs would have adverse effects on procedures, results, or the interpretation of teaching, research, experimentation, or tests, the AWAR (§2.36,b,7) specifically require an explanation of the procedures and reasons such drugs were not used. The *Guide* (p. 122) notes that in addition to anesthetics, analgesics, and tranquilizers, nonpharmacological control of pain and proper nursing support is sometimes effective (see 17:6).

Opin. When the likely outcome of nonpharmalogical intervention is not known from a review of the literature or from other sources, a pilot study may prove helpful (*Guide* p. 26). (See 17:55.) However, from a regulatory standpoint, withholding analgesic drugs for a procedure in which providing them is considered the current standard of veterinary care, regardless of alternative methods that might be used to minimize it, is considered unalleviated pain.

Evidence for benefit of a variety of nonpharmacologic methods of pain relief can be found in the literature; substantiation depends on the modality, situation, and species involved. Social contact can offer alleviation of anxiety levels, as pain may be intensified under anxiety-inducing situations.[10] Documented experience in clinical veterinary practice has shown that many socialized animals, especially dogs and cats, respond positively to human presence and touch. The mechanisms behind the beneficial effects of social contact and touch-based therapies have been investigated and may involve the production of oxytocin or other neuropeptides that modulate pain.[48,49]

17:11 Can mammals that have a cerebral cortex feel pain if made decerebrate?

Opin. When addressing this question, it is important to differentiate between pain and nociception (see 17:2). Nociception is the response to damaging or potentially damaging stimuli, whereas pain has a subjective component and is interpreted by most humans (and theorized that a functioning forebrain is required for pain perception or the experience of pain) such that decerebration should remove that capacity.[50] Since thalamic structures are believed to play a role in pain perception, it is usually considered that removal of the forebrain, including the thalamic nuclei, ensures that pain cannot be perceived. The position in regard to decorticate animals is less clear. Given the higher levels of organized behavior shown by decorticate or decerebrate animals of some species (e.g., rats) and that subcortical structures for pain perception are present, some caution is required in making extrapolations that decorticate animals have completely diminished pain perception, as rats that are decerebrate still respond to noxious stimuli in a manner that is nearly identical to non-decerebrate rats.[51–53]

17:12 Can nonmammalian vertebrates lacking or having only a primitive cerebral cortex (e.g., frogs, fish) feel pain?

Opin. IACUCs typically adopt an approach that presumes an animal with a certain level of central nervous system development can experience pain, and research involving these species should be conducted in a way that reduces the likelihood of causing pain or distress.

It seems unlikely that frogs experience pain in the same manner as humans, but since we can demonstrate nociception in amphibians[54] and since their degree of cerebral development cannot be said to preclude the possibility of pain perception, we should assume that these species can experience pain. Despite the assumption that these taxa experience pain, pharmacologic alleviation of pain in nonmammalian species is not well studied (although studies of nociception in amphibians and less cerebrally complex animals are available), and clinical signs of pain are not well characterized. When considering the aquatic environment that fishes inhabit, it is unlikely that their nociception and pain system is identical to terrestrial animals. However, several studies have identified nociceptors similar to that of mammals, and fish express molecular and physiological changes during potentially painful stimulation in higher brain areas. Fish also exhibit prolonged behavioral responses to noxious stimuli.[55] Zebrafish larvae as early as five days (days post-fertilization) exhibit behavioral changes when exposed to noxious heat, CO2-infused water and acetic acid.[56] When experiencing pain, fish do not demonstrate expected fear or anti-predator responses, which may be detrimental in their natural environment if predation risk is present. Changes in behavior are prevented by drugs that deliver effective analgesia, providing evidence that these changes are driven by nociception and pain mechanisms.[57–59] Taken together, this combined evidence suggests a pain experience that warrants attention in nonmammalian vertebrates, and thus it is vital that the IACUC and research teams seek to minimize and alleviate pain in these species when possible.

17:13 Which drugs can be considered adequate anesthesia for major surgery, such as abdominal surgery in laboratory rodents, rabbits, or cats?

Reg. The AWAR require the AV or AV designee to be consulted, particularly during the planning stages, of procedures that may cause more than momentary or slight pain or distress to these animal species and that the AV provide guidance regarding immobilization, anesthesia, analgesia, tranquilization, and euthanasia (§2.31,d,1,iv,B; §2.33,b,4).

Opin. Institutional drug formularies and published veterinary resources (textbooks and scientific literature) will provide the most current and effective drug types and dose ranges and combinations for consideration of premedication, anesthesia, analgesia, and post-surgical support for these animal species.

17:14 Should the use of urethane or alpha chloralose be allowed as anesthesia for recovery surgery?

Reg. The AWAR require the AV or AV designee to be consulted, during the planning, of procedures that may cause more than momentary or slight pain or distress to the animals and that the AV provide guidance regarding immobilization, anesthesia, analgesia, tranquilization, and euthanasia (§2.31,d,1,iv,B; §2.33,b,4).

Opin. Urethane had been historically used for the long-lasting immobility, anesthesia, and reported cardiovascular stability that it provides. A review of its use and impacts is available.[60,61] Both the availability of modern and refined injectable and inhalant anesthetic techniques and many undesirable features of urethane suggest that it should be used rarely, if at all, in contemporary research practice and only under careful consideration and consultation with veterinary experts. Personnel health and safety is a major consideration and exposure to urethane may cause neoplasia in some species. From a regulatory standpoint, pharmaceutical grade urethane is not available, which indicates that any nonpharmaceutical grade substance would need to be scientifically justified in the IACUC protocol application.

Alpha Chloralose has also been used historically as an anesthetic agent. It is a hypnotic and is still occasionally used in certain types of studies (e.g., cardiovascular, respiratory, neurovascular physiology, and neurotransmission) because of its reputation for creating less interference with the processes being studied than other anesthetic agents and because it can produce long-lasting immobility. Chloralose is compared to pentobarbital; however, modern anesthetic techniques may provide acceptable hypnosis or anesthesia without the potential for pain or distress associated with chloralose. Certain methods of administering chloralose may cause injury and subsequent pain or other morbidity during the postoperative period. Problems attributed to chloralose include adynamic ileus and intestinal injury when administered intraperitoneally, phlebitis associated with accidental extravascular injection, and a stressful recovery phase.[62] In many types of studies, alternative modern anesthetic techniques have been shown to provide equal or superior methods of immobilization.[63] Anesthetics have variable effects on experimental results, and an investigator should carefully scrutinize the literature for refined, contemporary, less noxious anesthetics and techniques, rather than relying on historical drugs based on long-standing practices. If, for example, a researcher uses modern anesthetics in one species and chloralose in another to measure the same parameters, it could be assumed the modern anesthetic would be appropriate for both species. Many experts do not recommend chloralose be used for survival surgery, and, in addition, pharmaceutical grade chloralose is not commercially available which indicates any nonpharmaceutical grade substance would need to be scientifically justified in the IACUC protocol application.[60–62]

17:15 Should the IACUC allow for the use of a nonpharmaceutical grade drug for anesthesia or analgesia if pharmaceutical grade alternatives are available, but may be in limited (back-ordered) supply or significantly financially burdensome to procure?

Reg. The AWAR do not comment directly on the use of nonpharmaceutical grade drugs; however, the AWA (§2143,a,3,A) states that adequate veterinary care with the appropriate use of analgesic or tranquilizing drugs is required. The AWAR (§2.33,a) also require adequate veterinary care. The NIH/OLAW FAQ site states in F.4:

> "The IACUC may use a variety of administrative methods to review and approve the use of non-pharmaceutical-grade substances. For example, the IACUC may establish acceptable scientific criteria for the use of these substances within the institution, rather than on a case-by-case basis. Investigators and IACUCs should consider relevant animal welfare and scientific issues including safety, efficacy, availability of pharmaceutical-grade substances, and the inadvertent introduction of new variables. Cost savings alone are not an adequate justification for the use of non-pharmaceutical-grade substances in animals. However, unavailability or shortages of pharmaceutical-grade substances may lead to cost increases and the IACUC may determine that this justifies the use of the non-pharmaceutical-grade substitution. Procedures that may cause more than momentary or slight pain or distress to animals must be relieved by sedation, analgesia, or anesthesia using veterinary or human pharmaceutical-grade substances, unless the use of a non-pharmaceutical-grade substance is scientifically necessary, appropriately justified, and approved by the IACUC. The use of a non-pharmaceutical-grade euthanasia agent must meet the same criteria."[64]

Opin. The preparation of nonpharmaceutical grade drug solutions can lead to inconsistent product quality and the potential for adverse effects, including tissue damage, pain, or lack of sterility or of a sufficient clinical effect. For this reason, their use may constitute inadequate veterinary care. NIH/OLAW has clarified that a nonpharmaceutical grade anesthetic agent can be used when scientifically necessary, appropriately justified, and approved by the IACUC.[65] The IACUC should ensure that investigators have fully considered the utility of alternative pharmaceutical grade drugs, and, if the use of nonpharmaceutical grade drugs is approved, these solutions are to be prepared, stored, and used with appropriate consideration of sterility, purity, stability, and related safety factors. A more detailed discussion about the use of nonpharmaceutical quality drugs or chemicals is found in a Position Statement by OLAW.[66]

Surv. Should the IACUC allow for the use of a nonpharmaceutical grade drug for anesthesia or analgesia if pharmaceutical grade alternatives are available, but may be in limited (back-ordered) supply or significantly financially burdensome to procure?

- Not applicable 17/201 (8%)
- Yes, almost always 11/201 (5%)
- Yes, but only with a science-based explanation 106/201 (53%)
- No (unless the nonpharmaceutical grade drug itself is being evaluated) 57/201 (28%)
- Other* 10/201 (5%)

*Other:

- Scientific need or supply chain issues would be acceptable rationale for use of such compounds.
- No support was expressed for allowing such use due to financial constraints.
- Noted that use of such compounds should be approved by the IACUC.
- Acceptable purity standards should be expected by the IACUC.

17:16 The AWA, the AWAR, and the *Guide* have restrictions on performing multiple major survival surgeries (MMSS) on the same animal. Are there any such restrictions on performing multiple painful procedures on the same animal?

Reg. AWAR (§2.31,d,x) prohibit more than one major operative procedure from which it is allowed to recover unless A) justified for scientific reasons by the investigator in writing; B) required as routine veterinary procedure or to protect the health or well-being of the animal as determined by the AV; or C) in other special circumstances as determined by the APHIS Administrator on an individual basis. PHS Policy (IV,D,1,d) notes that applications and proposals to the PHS must contain a description of the interventions that will be used to ensure that discomfort and injury to animals will be limited and that appropriate drugs will be used to minimize animal pain and discomfort. U.S. Government Principle IV emphasizes the "proper use of animals, including the avoidance or minimization of discomfort, distress, and pain when consistent with sound scientific practices, is imperative." NIH/OLAW, in reviewing 17:19, has added that "Allowing transfer of an animal to another study in which additional major painful procedures are performed does not comply with the AWAR or the PHS Policy. OLAW does not condone the use of such a transfer to circumvent guidance and would consider this activity to be noncompliant with PHS Policy and U.S. Government Principles." NIH/OLAW FAQ F.9[64] directly addresses MMSS, indicating that the practice is acceptable only if the following criteria are met:

1. The major survival surgery is included in and essential to a single research project or proposal.
2. The major survival surgery is scientifically justified as a benefit to the health of the animal or society.
3. The major survival surgery is necessary for clinical reasons.

The AWAR (§2.31,d,1,i) state that procedures involving animals are to avoid or minimize discomfort, distress, and pain to the animals. The AWAR (§2.31,d,1,ii) also state that the investigator must consider alternatives to procedures that may cause more than momentary or slight pain or distress. Further, the AWAR (§2.31,e,4) note that the procedures should be designed to limit pain and discomfort to that which is unavoidable for the conduct of scientifically valuable research.

Opin. The intent of the federal regulations restricting multiple major survival surgeries is to reduce the total amount of pain and distress experienced by any individual animal and to prevent animals from being transferred at the end of one study to another study where the animal may be subjected again to major survival surgery or invasive procedures. Therefore, it seems prudent that studies in which animals are potentially subjected to multiple painful procedures should be reviewed by IACUC using a paradigm similar to that used for multiple major survival surgeries (AWAR §2.31,d,1,x; *Guide*, p. 30) (see 17:19). A review for this purpose is implied in the *Guide* (p. 30), which states that "some procedures characterized as minor may induce substantial post-procedural pain or impairment and should similarly be scientifically justified if performed more than once in a single animal." If a procedure causes significant tissue damage (e.g., skin burns, osteoarthritis, or nerve damage) but is not a major surgery, there is still a potential that the animal is left with a chronic state in which pain or distress thresholds may be lowered and subsequent uses might result in more total pain than if another animal were used. On the other hand, a distinction should be made between *potentially* painful procedures and those that involve *unalleviated* pain. Techniques to reduce pain and distress include local and systemic analgesics, anesthetics, and operant conditioning/acclimation. Other supportive care modalities can be used alone or in combination and should be aggressively pursued if they will not specifically interfere with the study.

If a study requires that an animal be subjected to repeated painful or distressful manipulations, then the investigator should outline for the IACUC the impact, magnitude, and duration of the pain; the pain's potential effects on experimental outcome; methods to monitor the animal's ability to adapt to or recover from the procedures; and reasons why the pain cannot be relieved, if this is proposed (see 17:6). Without such information, any attempt to set a limit on the frequency with which such an event can be repeated is arbitrary.

Surv. If an animal is transferred from one study to another, does your IACUC have an effective means of determining if prior major survival surgery was performed on that animal?

- Not applicable 33/201 (16%)
- Yes 134/201 (67%)
- No 8/201 (4%)

- No, unless the issue is specifically brought to the committee's attention 20/201 (10%)
- Other 6/201 (3%)

Surv. If an animal is transferred from one study to another, does your IACUC have an effective means of determining if an animal experienced significant pain or distress in the original study?

- Not applicable 31/201 (15%)
- No 10/201 (5%)
- No, unless the issue is specifically brought to the committee's attention 42/201 (21%)
- Yes 114/201 (57%)
- Other 4/201 (2%)

17:17 Is it a good idea to have an institution-wide policy that some form of analgesia must be provided for any major survival surgical procedure?

Reg. The AWAR (§2.31,d,1,iv,A) state that procedures involving more than momentary pain or distress will be performed with appropriate sedatives, analgesics, or anesthetics unless their withholding is justified, in writing, by the PI. U.S. Government Principle IV speaks of avoiding or minimizing discomfort, distress, and pain when consistent with sound scientific practices.

Opin. Analgesics, and often more than one class of analgesic drug, and nondrug approaches aimed at providing pain relief should be used for any potentially painful survival surgical procedure, whether major or minor; this is expected for the provision of appropriate veterinary medical care. The type and amount of analgesia will depend on the relative invasiveness of the procedure and the degree of pain that is caused. Published estimates of the degree of pain expected from various types of surgery[30,35,37–39] (as well as for various models of disease, injury, or health conditions) can be consulted to determine the amount, type, and combinations of analgesics to be used. In many cases, the amount of pain an animal will be expected to perceive is comparable to that felt by humans in the similar situation (AWAR §1.1, Painful Procedure; U.S. Government Principle IV; *Guide*, p. 120). In order to get an idea of the incidence and magnitude of pain that a surgery or condition causes in humans, the authors find it instructive to consult human clinical literature databases. Studies of analgesic interventions for a particular procedure (e.g., hip replacement surgery) or a disease state (e.g., ulcerative colitis) commonly include a description of the severity of experienced pain. To best comply with both the regulatory and the ethical aspects of humane animal care, the institution should have a policy stating that analgesia for any survival surgery (or other pain-inducing procedure or disease state) should be provided to the animal, and that the type of analgesia should be commensurate with the degree of expected pain, unless adequate scientific justification is provided for not doing so.

17:18 Should the IACUC require the use of appropriate analgesia after embryo transfer surgery in mice?

Reg. See 17:3; 17:17. PHS Policy (IV,C,1,b) states that procedures that may cause more than momentary or slight pain or distress to the animals will be performed with appropriate sedation, analgesia, or anesthesia, unless the procedure is justified for scientific reasons in writing by the investigator. U.S. Government Principle IV speaks of avoiding or minimizing discomfort, distress, and pain when consistent with sound scientific practices.

Opin. Embryo transfer surgery, which is a major operative procedure, is expected to cause post-surgical pain. The degree of pain from a flank incision and subsequent visceral manipulations depends upon surgical finesse, the type of anesthetic agent used, and other factors. There are few studies investigating the amount of pain that this commonly performed procedure causes in mice; at least one indicated that it had an impact on behavior,[67] and the equivalent surgery in humans would require analgesia. In recent years, studies exploring a variety of analgesics alone or in combination for this procedure have confirmed there is no reduction in viability of transferred embryos in mice and rats.[68–71] These studies assessed representative strain/stocks of mice; the results and findings should be applicable to all laboratory strains that might undergo major surgery for embryo transfer and subsequent pregnancy and delivery of offspring. Future refinements that are alternatives to surgery for embryo transfer, such as using the noninvasive NSET (non-surgical embryo transplantation) device should also be considered as this model continues to undergo refinement.[72]

Surv. Does your IACUC require the use of appropriate analgesia after embryo transfer surgery in mice?

- Not applicable. 118/201 (59%)
- We do not require an analgesic. 2/201 (1%)
- We do not require an analgesic but we encourage its use. 10/201 (5%)
- We usually require the use of an analgesic. 66/201 (33%)
- Other. 5/201 (2%)

17:19 AWA-regulated animals arrive at an animal facility, already having undergone major surgery (under anesthesia) at the originating site. The animals have healed before arrival, with no clinical demonstration of pain or distress upon arrival. There will be no painful or distressful procedures when used at the receiving institution. On the

APHIS/AC annual report form (Form 7023), are these animals placed in Column C (no pain or distress) or Column D (pain or distress alleviated by drugs)?

Reg. The AWAR (§2.31,d,1,x) and the *Guide* (p. 30) restrict multiple survival surgeries to certain circumstances. (See 17:5 for general considerations of pain categories.)

Opin. On the APHIS/AC annual report form (Form 7023) of the receiving facility these animals should be listed in Column C because no surgery was performed by the facility. The originating site, however, would need to place the animals in Column D on its annual report. If these animals were to be transferred to another protocol at the receiving facility where a major operative procedure is to be done, the facility must get approval from APHIS/AC (as a special circumstance study) as well as IACUC approval because of the multiple major procedures that the animal will undergo in its research role per the AWAR (§2.31,d,1,x).

Vendors typically have their own APHIS/AC registration; in those circumstances, the vendor's annual report states the appropriate pain or distress category for the animal while at the vendor's facility. If the vendor does not have an independent APHIS/AC registration, the receiving institution must arrange to have the vendor's animal use operations covered under the receiving institution's registration. How the receiving institution categorizes animals in the study should refer to the specific way that the animals are used at the institution. In the scenario presented above, there was no apparent pain detected upon arrival. If there is no need from that point onward to administer analgesics or anesthetics to minimize pain or distress, then the study animals are placed in column C (Humane Use Category C).

17:20 When an IACUC reviews a literature search strategy for alternatives to painful or distressful procedures, must each potentially painful or distressful procedure be listed as a keyword in the search?

Reg. See 17:4.

Opin. See 17:4. For an investigator to determine whether the proposed use of laboratory animals will be conducted with contemporary refinements and updated contemporary techniques (i.e., refined approaches), a search of each potentially painful or distressful procedure should be done. This type of literature search can be challenging yet can be aided by use of resources such as the USDA's Animal Welfare Information Center[20] or the Animal Welfare Institute database for refinement.[73] In addition to searching for alternative models (which may be found and deemed unsuitable), the investigator is strongly encouraged to search for methods by which animal well-being can be maintained or improved. For example, researchers in unrelated disciplines may have developed behavioral assessment tools or otherwise refined animal use methods that can be applied to the new study protocol as it is developed. In order to address the mitigation of any pain or distress, each type of painful or distressful exposure should be listed as search terms. Importantly, use of the word "alternatives" or "non-animal" as search terms may unduly limit the number of useful hits in a literature search, whereas using terms such as *severity, assessment*, combined with terms such as *model* and *animal*, and with *pain, nociception*, or *illness*, may reveal contemporary approaches for assessment of pain, disability, or humane endpoints. It should also be noted that such papers may not have as their primary focus the reduction of pain or distress, and that the methods section of the papers may have to be examined for the applicable and specific experimental details.

Surv. Have animal users at your institution performed literature searches that resulted in the implementation of alternatives to painful or distressful procedures that were in the original IACUC application?

- Not applicable 13/201 (6%)
- No 45/201 (22%)
- Yes, but infrequently for basic research needs 90/201 (45%)
- Yes, and somewhat often for basic research needs 14/201 (7%)
- Yes, but infrequently for educational needs 4/201 (2%)
- Yes, and somewhat often for educational needs 4/201 (2%)
- I don't know 31/201 (15%)
- Other (Specify examples of alternatives that have migrated pain and distressful procedures in animals) 0/201 (0%)

17:21 Is a measured decrease in food or water consumption a reasonable indicator of pain in laboratory rodents?

Reg. See 17:3.

Opin. Numerous factors can influence food and water consumption in rodents; the experience of pain being only one potential reason. If a rodent reduces its food and water intake, pain should be included as a possible cause along with oral or dental disorders, infectious disease, changes in environment, alteration in the husbandry regimen, experimental phenotype, and other stressors. Nevertheless, if animals have undergone an operative procedure, immediate changes in food and water consumption in the postoperative period have been cited as indicators of procedural impact, discomfort, and pain.[30,37] Since rodents are expected to be socially housed, measuring individual food and water consumption may be difficult. Therefore, assessments of body condition scores along with body weight measurements are a standard of care for rodent medicine. It is important to note that many studies are carried out on growing animals, so establishing average weight gain and growth curves for a few days before an operative procedure is beneficial. After surgery, rats and mice almost invariably demonstrate a reduction in body weight (between 5% and 15%, depending upon the nature of the surgical procedure) as a consequence of reduced appetite leading to reduced food and water consumption. In some cases, perioperative analgesia may either exacerbate or attenuate this change in bodyweight compared with untreated control animals. It is reported that the use of buprenorphine in both operated and control rodents may result in weight loss.[74]

Although measuring body weight is said to be useful as an indicator of pain or distress, it is essentially a retrospective index of the amount of pain or discomfort that was produced over the preceding 12 to 24 hours. Thus, weight loss probably indicates that unalleviated pain or distress has already occurred, making it a more useful barometer to guide changes in future animal surgeries. Behavioral assessments should be carried out routinely in the immediate days following surgery to allow for adjustments of pain medications and other interventions, to align with the animals' responses (see 17:22).

17:22 What percentage of weight loss over what period might indicate that an animal is experiencing pain or distress?

Opin. See 17:21. Loss of body weight does not necessarily indicate that an animal is experiencing pain and distress but typically is related to a health condition or environmental stressor that might evolve into distress. Chronic weight loss, particularly if unexpected, should be investigated carefully. In many institutions, limits are placed on weight loss in advance of reaching a humane endpoint where the animal should be removed from study. Brief periods of inappetence will cause weight loss from baseline weight of 5%–10% in rodents; this can happen within a 24-hour period following an operative procedure in mice. Many institutions choose to limit weight loss for rodents to a maximum of 15%, which is clinically apparent, would be reflected in a loss of body condition, and would reasonably cause distress to the animal. However, weight loss limits should be appropriate for the model used, as some papers have recently reported modest weight loss of as low as 5%–10% correlating with pain/distress in certain models.[30,75,76] Most guidelines are based on professional judgment regarding animal health and well-being, and it is advisable to examine the issue on a case-by-case basis, taking the species and overall body condition scoring into consideration. Veterinary staff should be consulted to aid in determination of the root cause of weight loss, for example, to determine if it is the outcome of lowered feed intake, increased metabolism, or decreased absorption of nutrients. Further consultations with the research team should review if the weight loss is an adverse side effect of the research protocol, and if supplemental feeding, administration of analgesics, sedatives, or environmental changes will prevent or ameliorate the loss. If cachexic states associated with disease (e.g., cancer) are not the research focus, then supportive measures may enhance animal survival to reach the intended experimental endpoints. It also is important to compare the animal's weight with that of an untreated, normally growing control animal. Note: within the research animal environment, adults of several animal species when fed *ad libitum* may become obese (body condition score assessed as 5 out of 5). This emphasizes the importance of adopting a reasoned, logical approach to interpreting the significance of weight loss in an animal.

Surv. When weight loss is accepted by your IACUC as one indicator that an animal may be in pain or distress, what percentage of weight loss is typically considered to be significant (as compared to control animals or a previous maximum weight)? Assume the average recovery time for any procedure that might have been performed has passed.

- Not applicable 16/201 (8%)
- Usually 5%–10% 21/201 (10%)
- Usually 10%–15% 38/201 (38%)
- Usually 15%–20% 59/201 (29%)
- Greater than 20% 58/201 (29%)
- I don't know 9/201 (4%)

17:23 What are useful guidelines for the IACUC to consider when the use of neuromuscular blocking (NMB) agents is requested? (See 17:24.)

Reg. The AWAR (§2.31,d,1,iv,C) state that NMB agents (paralytics) will not be used without anesthesia. The PHS Policy (U.S. Government Principle V) states that surgical or other painful procedures should not be performed on unanesthetized animals paralyzed by chemical agents. The *Guide* (p. 123) states that when these agents are used, it is recommended that the appropriate amount of anesthetic first be defined based on results of a similar procedure that used the anesthetic without a blocking agent.

Opin. NMBs prevent voluntary muscle activity so that an animal cannot respond to painful or other stimuli with reflexes or movement, including eye position changes and respiratory efforts. It is essential to preclude the very real possibility that the level of anesthesia is inadequate in a paralyzed animal and the animal is aware of the surgical procedure being performed, which may be both painful and a source of distress in certain species. Lack of movement in response to a surgical stimulus or pinch of extremities are the most commonly used methods to assess the adequacy of anesthetic depth; therefore, when using NMBs as part of an anesthetic protocol, alternate methods of assessment and monitoring are required. In the management of anesthesia for human patients, NMBs are often used to facilitate endotracheal intubation and major surgery and estimates of "awareness" are based on postsurgical reports. Feedback from human patients indicates that approximately 4 of every 1,000 patients can recall intraoperative events that occurred while they were under anesthesia.[77,78] Put into perspective, these human cases were anesthetized by professionally certified medical anesthetists, while most animals are anesthetized by non-medically trained staff and non-veterinary investigators within a research setting. Thus, the use of NMBs in research animals should only be approved in consultation with veterinary medical staff and with awareness of the risks of NMB causing unnecessary pain or distress if inadequate anesthetic monitoring occurs.

Before examining what constraints might reasonably be placed on the use of NMBs, the IACUC should first establish why it is necessary to include them. In some instances, investigators may simply have taken a human anesthetic protocol and applied it to their animal model. This is often inappropriate, since the rationale for NMB use in humans is not readily applicable to other species.[79] Veterinary anesthesiology textbooks discuss mechanical ventilation during anesthesia, and these chapters do not mention use of neuromuscular blockers. In a discussion of respiratory management of anesthetized animals, three options for preventing spontaneous breathing are listed:

1. Slight hyperventilation
2. Achieving adequate anesthetic levels
3. Use of NMB

Option 1 is deemed to be the easiest to execute.[80] Thus, NMBs are not necessary for mechanical ventilation and the IACUC should not accept this as the sole reason for requiring NMBs. Neither are NMBs needed to produce sufficient muscle relaxation for the majority of surgical procedures. It is true that electrosurgery and cautery cause skin and muscle to twitch in the absence of NMBs, which can make the limb or head move slightly for brief periods. Nevertheless, this is only critical if the structures being accessed (e.g., eye, brain, spinal cord) are sensitive to tiny movements. Under those conditions, neuromuscular blockade is only necessary during specific portions of surgery (not, for example, for incision of skin and cranium). For the most part, patient movement can be prevented by ear bars and other security measures. Certain intraocular surgeries may require a central eye position and an NMB may be needed to facilitate this positioning. Typically, human intraocular surgery is performed on awake patients using a local anesthetic block of the orbit. Presumably, in the anesthetized animal, a local block of orbital structures will ensure the central eye position as well as provide excellent preemptive analgesia, without the need for NMBs. If, however, NMBs are required, then the following points should be considered:

- The investigators should have experience with anesthesia and surgery in the species under study and should be incorporating an anesthetic regimen that is effective for inducing insensibility to pain and lack of awareness *on its own for that surgery type* in the absence of an NMB.
- The NMB should not be administered until after the anesthetic has reached a stable level and surgery has commenced. If practical, the NMB should be allowed to wear off periodically, so that somatic reflex responses can be assessed before additional doses of NMB are administered.
- The heart rate and blood pressure should be monitored continuously and elevations (an increase in either of 15%–20%) in response to painful stimuli indicate the need for additional anesthesia. It is important to note that this monitoring technique is not fully reliable and awareness in humans can occur without major changes in these variables.[81] With some anesthetic regimens (e.g., a volatile anesthetic such as isoflurane), monitoring the electroencephalogram can be of value, but this technique also can prove unreliable.[50]
- If inhalant anesthesia is used, an end tidal agent monitor (which monitors the exhaled anesthetic gas concentration as an indication of the central nervous system concentration) is extremely useful to evaluate and document that the animal's anesthetic level was in the range in which insensibility is typically present. End tidal agent measurements are currently only feasible in larger animals (the approximate size of a rabbit or larger).
- The IACUC should incorporate post-approval monitoring of anesthetic records of cases wherein NMBs are used to confirm documentation of all relevant assessments and physiologic parameters under anesthesia and paralytics as an indication that appropriate and humane use of these agents occurred.

17:24 If an NMB is used along with an anesthetic, what might the IACUC request to help assure that the animal is anesthetized and not simply immobilized?

Reg. See 17:23.

Opin. Some common situations in which NMBs are used include neurophysiological and imaging studies in animal models. In many of these studies, a very light plane of anesthesia must be maintained to minimize interference between the anesthetic regimen and the study protocol. It is precisely in these circumstances that delivery of inadequate anesthesia is most likely. Some investigators have suggested that after completion of surgery, the surgical wounds can be infiltrated with local anesthetic, while general anesthesia is discontinued, and the animal remains immobilized due to ongoing effects of the NMBs (although the animal must still receive ventilatory support when under the influence of NMBs). This situation raises two concerns: first, it is difficult to ensure the adequacy of local anesthetic blockade—either initially or after the NMB has taken effect; and second, perhaps more importantly, even if the animal is pain free, it may become distressed if exposed to other stimuli (e.g., odors, sounds) and is unable to react due to the paralysis caused by the NMB. General anesthesia is used in animals not only to provide insensibility to pain, but also to induce unconsciousness, which helps prevent the distress caused by physical restraint (immobility) and experimental manipulations. A paralyzed, conscious animal, even if pain free, is likely to experience considerable distress. If NMBs are to be used during surgery or painful procedures, then the monitoring techniques that can be applied are those described in 17:23.

17:25 Is the use of a local anesthetic (e.g., an agent applied to a specific region of the body) to perform a minor procedure considered to be alleviation of pain by anesthesia?

Reg. See 17:3.

Opin. Yes. Local anesthetics offer a valuable alternative to general anesthesia for a range of procedures. In some species, the use of regional anesthesia produced by local anesthetics can be considered the method of choice (e.g., a paravertebral block to carry out abdominal surgery in cattle).[82] Provided the anesthetic is administered and dosed appropriately, pain can be prevented by drugs like bupivacaine either by local infiltration, by infiltration around nerve trunks, or by epidural or intrathecal administration. Experience with humans and nonhuman animals suggests that these techniques are useful, provided any distress caused by physical restraint or other (nonpainful) procedures can be controlled. The duration of local anesthetic action varies with the agent, and additional doses of local anesthetic or other analgesic administration may be necessary to maintain alleviation of pain. The development and availability of long-acting (sustained-release) local anesthetic formulations in recent years has provided tremendous advancement in analgesia control and should be used when available. Local anesthetics can cause an uncomfortable sensation (e.g., stinging or burning) on injection. Techniques to reduce this pain involve the use of sedatives or tranquilizers (which by reducing anxiety or the level of consciousness can act to raise the pain detection threshold), buffering of acidic solutions, using the smallest needle size possible, and cold or warm compresses applied to the injection site.

17:26 What records related to the use of anesthetics and analgesics should be maintained and reviewed by the IACUC?

Reg. The AWAR (§2.35,a; §2.35,f) require that all records and reports directly relating to proposed activities involving animals (i.e., those elements of research, testing, or teaching procedures that involve the care and use of animals [AWAR §1.1, Activity]) and proposed significant changes in ongoing activities reviewed and approved by the IACUC are to be maintained for at least three years after the completion of the activity. PHS Policy (IV,E,2) has the same requirement. Although anesthesia and analgesia records for most covered species are not specifically required by the AWAR (with marine mammals as the exception [AWAR §3.110]), the concept of adequate veterinary care (AWAR §2.33; *Guide*, pp. 12, 14, 105) strongly implies the need for maintenance of detailed medical records that would include anesthesia and analgesia records. The *Guide* (p. 115) states that medical records are a key element of the veterinary care program and ACLAM has published guidance on medical records.[83]

Opin. The extent of medical record documentation, that includes anesthetic and analgesic use, will depend upon the type and complexity of the species and procedures. At a minimum, the anesthetic or analgesic drugs, dosage, time delivered, route of administration, and effect should be recorded, together with details about the experimental animal (age, weight, sex, strain, etc.). Increasingly, provision of this information will also be required as part of the materials and methods by editorial boards and scientific journals.[84] If surgical procedures are to be carried out, the investigator should note the adequacy of anesthetic depth and the way this was assessed. If additional doses of drugs are given, the time delivered, route of administration, and the effect should be recorded for each dose. It can be beneficial to the welfare of the animal to have records of analgesic administration available at the cage side for review by veterinary staff and the IACUC. Also, be sure to adhere to all DEA documentation requirements for any controlled drugs.

In addition, research and veterinary teams should record the duration of anesthesia, the recovery time to sternal recumbency, any morbidity, any unexpected adverse effects (e.g., vomiting), and when full recovery was achieved (e.g., return to normal activity, eating, drinking, etc.) and indicate how this assessment was made. As discussed in 17:23 and 17:24, it may be desirable to keep a continuous recording of heart rate and blood pressure, if feasible, which is particularly important if neuromuscular blocking agents are used. These data permit a critical review of anesthetic practices and should be considered an essential part of any scientific protocol and medical record for laboratory animals as part of veterinary care standards.

In many instances the research projects extend for many years with certain animals, with additional time necessary for manuscript completion and publication of work with animal models. For USDA-covered species, review of medical records is expected at the time of unannounced annual inspections and should provide a clear timeline of routine health assessments and interventions (physical examination, dental prophylaxis) and experimental procedures (surgery, imaging events). Therefore, it is prudent to maintain all medical records in accessible formats while animals are at the institution and for at least three years after the euthanasia or other disposition (transfer to another project or protocol or adoption) of the animal.

Surv. Does your IACUC require investigators to maintain anesthesia and analgesia usage records (other than notes on cage cards) for rats and mice? This refers to animal/cage identification, dates, dosages, drug used, effects, and any other items considered important by your IACUC.

• Not applicable	11/200 (6%)
• No	13/200 (7%)
• Yes, always	135/200 (68%)
• Sometimes, depending on the details of the procedure	36/200 (18%)
• Other	5/200 (3%)

17:27 Should the use of anesthetics and analgesics be based on a strict dosing schedule or on varying dosages as determined by sound clinical judgment?

Reg. The AWAR (§2.33,b,4) require the AV and AV designees to guide the research team on the appropriate use of anesthesia, analgesia, and immobilization. The AWAR (§2.33,b,5) empower the AV with additional authority, stating that adequate pre- and post-procedural care shall be in accordance with current established veterinary medical and nursing procedures. The PHS Policy (IV,A,3,b,1) states that a veterinarian will have direct or delegated authority and responsibility for activities involving animals. The *Guide* (p. 120) notes that an integral component of veterinary medical care includes effective programs for the alleviation of pain.

Opin. Anesthetics and analgesics should be given, at least initially, according to dose schedules. These are determined from scientific literature and veterinary formularies but are often modified because of variations in response by different species of animals. Administering a standard dose from a recommended range of anesthetic may produce the desired effect, but it might instead cause animals to be too deeply anesthetized or inadequately anesthetized. After assessing the response to an anesthetic for the particular species, strain, age, and sex of animal to be used in a study, a more appropriate dose and administration schedule can be developed. One must consider that in rodent strains, anesthesia variation can be substantial. For example, when using the duration of unconsciousness as an indicator of anesthetic efficacy, doubling of sleep time has been observed in diverse strains of mice.[85]

It is relatively easy to adjust anesthetic drug doses to suit particular groups of animals, but this is more difficult with analgesics because of the limited ability to assess postoperative pain. The authors do not recommend that an investigator propose to administer analgesics "as needed" to treat pain. "As needed" dosing strategies are effective in humans, only when they can articulate their pain levels and self-administer medications by means of patient-controlled analgesia devices. "As needed" analgesic dosing strategies in animals have two major limitations: first, whether it will be possible to detect and quantify pain in an individual animal, and second, whether timing of the monitoring assessments will be frequent enough that lengthy periods of substantial pain are prevented (see 17:51). In summary, pain assessments should be performed frequently and with knowledge of species-specificity to ensure that appropriate vigilance will ameliorate pain in that animal.

Literature and experience with preventative and multimodal analgesic techniques are rapidly expanding. The administration of analgesics prior to incision (preemptive analgesia) is an expected aspect of a balanced anesthetic regimen, with the continuation of analgesic coverage throughout and beyond the postoperative period. Fixed dose administration for up to 72 hours postoperatively, combined with (a) monitoring for evidence of pain between doses, (b) extending the duration of treatment for procedures that are felt to cause moderate or severe pain, and (c) using modalities that overlap in terms of analgesic coverage are highly encouraged in current veterinary textbooks on pain management.[82] As a caveat, any fixed dose schedule may lead to insufficient pain relief. In fact, to be evaluated properly, an animal's pain should ideally be assessed in the interval between doses and just prior to redosing in order to detect whether an adequate duration and magnitude of effect has been achieved. One benefit of a fixed dosing schedule, with monitoring and ability to intervene between doses within the first 24–48 hours post-surgery, is that this method rarely leads to complications of excessive analgesic side effects. It also has the benefit of assuring that if personnel are less confident in their pain assessment skills, the animal will be given at least a baseline of coverage for pain.

A general inadequacy of postoperative pain relief in humans, among other factors, was often attributed to the use of rigid dosing schedules with no inherent flexibility. From the veterinary standpoint, individual animals (like humans) will vary in their response to surgery and postoperative outcomes. Therefore, dose rates obtained from the scientific and veterinary literature are a helpful starting point, and every attempt should be made to assess the adequacy of analgesia in each individual animal in case a particular patient requires more analgesia than others. Even if retrospective measurements such as body condition scoring and body weight are used, this approach encourages flexibility in dose ranges for adequate analgesia in subsequent studies.[86] In some species (e.g., dog, goat) complex behaviors, human–animal interactions, and animal–animal interactions are relatively well understood, and these behaviors can be used as a foundation for postoperative observation to design objective scoring algorithms for the quantification of pain levels. Evaluation of pain in rodent and other "prey" species solely through observation of animal behavior is more challenging, as these species may be relatively inactive because of photoperiod, observer presence, or masking of pain behaviors.

One major barrier to the provision of analgesia in the immediate postoperative period is that the shorter dosing interval for opioid analgesics (4–6 hours) is shorter than the period where animals are typically left without monitoring or treatment (e.g., overnight, up to 16 hours). This should be recognized, with the recommendation of longer-duration (sustained-release) formulations that can be used (e.g., transdermal fentanyl in combination with long-acting local anesthetics) with the overlap of other agents (e.g., nonsteroidal anti-inflammatory drugs) to achieve sufficient analgesic coverage throughout overnight periods. Longer acting formulations of opioids and NSAIDS are becoming increasingly more available, which will help with gaps in dosing (although these should not be a substitute for decreasing intervals to check/assess postoperative animals), Heightened frequency of animal monitoring and other scheduling modifications will be necessary in order to meet regulatory and ethical requirements for mitigation of pain and distress in animal models.

Surv. Does your IACUC require overnight monitoring of animals when a procedure has been performed and dosing or observations are necessary at 4–6 hour intervals for more than 24 hours post-procedure?

• Not applicable	30/200 (15%)
• No, overnight monitoring not required	37/200 (19%)
• Yes, for all species	68/200 (34%)
• Yes, but only for species covered by the Animal Welfare Act	8/200 (4%)
• Yes, but only if requested by our veterinarian and approved by the IACUC	45/200 (23%)
• Other*	12/200 (6%)

*Other:

- IACUC expectation for documentation of monitoring and/or dosing
- None of the respondents indicated a specific method for documentation nor for verification of monitoring and/or dosing

17:28 If the use of anesthesia might impair the survival ability of an animal to be released in the wild as part of a field study, can the IACUC appropriately approve procedures that use either no anesthesia or approve anesthetic regimens that do not produce complete pain alleviation?

Reg. Neither the AWAR nor the PHS Policy have pain relief exemptions for animals used in field studies. A field study is defined in the AWAR (§1.1) as any study conducted on free-living wild animals in their natural habitat. It excludes a study that involves an invasive procedure or harms or materially alters the behavior of an animal under study; therefore, any study that includes procedures that require the use of anesthetics or analgesics does not meet the definition of a field study under the AWAR. The PHS Policy does not have a specific definition of a field study but notes that veterinary and IACUC input may be needed for field studies involving anesthesia (*Guide*, p. 32). Guidance from the NIH/OLAW (FAQ A6)[87] states: "[i]f the activities are PHS supported and involve vertebrate animals, then the IACUC is responsible for oversight in accordance with PHS policy. IACUCs must know where field studies will be located, what procedures will be involved, and be sufficiently familiar with the nature of the habitat to assess the potential impact on the animal subjects. Studies with the potential to impact the health or safety of personnel or the animal's environment may need IACUC oversight, even if described as purely observational or behavioral. When capture, handling, confinement, transportation, anesthesia, euthanasia, or invasive procedures are involved, the IACUC must ensure that proposed studies are in accord with the *Guide*."[87] When anesthesia is required for a field study, the IACUC must assure that the proposed study is in accord with the *Guide*. IACUCs are encouraged to consult the Guidelines of the American Society of Mammalogists for the use of wild mammals in research as a reference for guidance on field studies.[88]

Opin. The use of animals in field studies poses particular difficulties, since it is usually intended that the animal will survive the study and resume its normal activities; application of anesthesia may impair this process. However, the introduction of reversible anesthetic regimens has greatly improved management of wildlife anesthesia, although IACUCs must be cognizant of anesthetic withdrawal times and potential for consumption by humans or predators. Similarly, in smaller species, use of modern inhalational agents (e.g., isoflurane) can result in very rapid recovery with few significant aftereffects. While the use of potent inhalational agents, such as isoflurane delivery via induction chambers, is not generally recommended, publications have shown safe inhalant use in a variety of field conditions.[89] Finally, the use of local anesthetics can be evaluated as a means of minimizing the experience of pain for an animal in the field. If all these options have been considered and rejected as impracticable or ineffective, then an ethical judgment must be made as to whether the aims of the project outweigh the pain or distress potentially induced by the required manipulations. Alternatively, if the project seeks to obtain only tissue samples from an animal, a judgment must be made as to whether the procedure should be allowed without adequate anesthesia and the animal allowed to recover, or the procedure carried out with anesthesia and the animal euthanized rather than released back to its natural environment.

17:29 What is the maximal practical length of time that the IACUC should allow a rat to be regulated or restricted from food or fluid for behavioral motivation studies? How do these suggestions change for a mouse, dog, nonhuman primate, or other common species?

Reg. The AWAR (§2.38,f,2,ii) state that short-term restriction of access to food or fluid from animals is allowed when specified in an IACUC-approved activity that includes a description of monitoring procedures. PHS Policy (IV,A,1) requires institutions to follow the *Guide* (pp. 30–31), which addresses food or fluid restrictions. The *Guide* (pp. 30–31) specifically notes that restriction for research purposes should be the least restriction necessary to achieve the scientific objective while maintaining animal well-being. The *Guide* (p. 31) also indicates that records should be established to monitor physiologic or behavioral indices, including objective criteria for temporary or permanent removal of an animal from the experimental protocol. NIH/OLAW FAQ F.19 also discusses food and fluid restrictions[90] (see 17:30). In brief, the term "deprivation" indicates a complete withholding of food or water; this differs from regulation/restriction, which is the act of controlling the amount of food or fluid an animal receives and scheduling when the animal receives it.

Opin. Food and fluid restriction may be carried out for many reasons and attempting to establish maximal periods of restriction without reference to the aims of a particular study is undesirable. For example, a project might be judged so important in terms of its potential benefits that prolonged periods (up to 48 hours) of food deprivation (defined as complete withholding of food) could be permissible if this were considered a necessity.

Often, only relatively minor periods of food and fluid restriction are required, for example, to ensure an empty stomach or gastrointestinal tract prior to anesthesia, to motivate behavior, or to induce a catabolic state in the animal. In each study, it is important to determine the minimal period of restriction needed to achieve the desired objective. Often 16- or 24-hour periods of feed restriction are chosen, as these are a convenient interval for removal of the food at the end of the typical workday, followed by experimental use of the animal the following morning. In the case of small rodents, this length of time for fasting/restriction may be excessive and may induce unnecessarily severe effects because of the high metabolic rate of these smaller species. It has been shown that when presented with a limited quantity of food (e.g., half the amount normally consumed overnight), rats will eat normally until the food is exhausted. Thus, by reducing the amount of food placed in their food hopper, effective food removal (e.g., from 3:00 A.M.) can be achieved, which is sufficient to produce an empty stomach by the following morning.[91] Investigators should note, of course, that coprophagy occurs in rodents and other species, as well as the potential of pica (eating of bedding and nesting substrates) so complete gastric emptying may be challenging and require instead a combination of fasting the animal within an empty cage and the use of an anal cup to prevent ingestion of feces.[92]

A second aspect of the animal's normal biological processes also should be considered when determining food and fluid restriction. Some species (e.g., mice, rats) typically eat the majority of their calories in the dark phase of their photoperiod, which then triggers fluid consumption during this phase. If food is withdrawn overnight and a procedure is carried out the next day, and if that procedure causes adverse effects, the rat may not eat the following night or until >36 hours have passed. As a result, up to 48 hours of fasting may have inadvertently occurred for the study animal, which may have consequences for the particular study, as well as for the welfare of the animal.

When dealing with larger species, longer periods of food withdrawal might be needed and, generally, are better tolerated than in smaller animals. If food is being withdrawn solely to reduce the risk of emesis during induction or recovery after anesthesia, a period of 8 to 16 hours is appropriate for dogs, cats, ferrets, or nonhuman primates, but unnecessary in rabbits and rodents as these latter species do not vomit. Withholding of fluid is not recommended for more than a few hours in monogastric species. Withholding food in pigs can be helpful in reducing the volume of gut contents for abdominal surgery and in ruminants a minimum of 12-hour withholding of food and water may reduce some of adverse consequences of anesthesia, such as regurgitation, and rumenal tympany.[60]

17:30 Are food or fluid restrictions considered pain or distress?

Reg. The AWAR (§1.1, Painful procedure) define a painful procedure as one that would reasonably be expected to cause more than slight or momentary pain or distress in a human being to which that procedure was applied, that is, pain in excess of that caused by injections or other minor procedures. The *Guide* (pp. 30–31) states that when the availability of food or fluid is regulated/restricted animals should be closely monitored to ensure that food and fluid intake meet their nutritional needs and that written records should be maintained for each animal to document daily food and fluid consumption, hydration status, and any behavioral and clinical changes (see 17:29).

Opin. NIH/OLAW, in a position statement, has written: "The IACUC must evaluate the level of restriction and potential adverse consequences in regulating food or fluid. The IACUC must also evaluate the methods for assessing the health and wellbeing of animal activities that involve regulation of food or fluid. The IACUC has the authority to approve scientific justifications for departures from the recommendations in the *Guide*. For instance, using scheduled access to food or fluid sources may be justified by describing procedures based on performance standards that ensure adequate maintenance of hydration, body weight, and behavioral and clinical health. (See also 17:29). It may be necessary to monitor both food and fluid intake if regulation of one influences consumption of the other."[66] NIH/OLAW FAQ F.19 has similar wording.[90]

Many studies report the withholding of water for physiologic studies. There is no question that dehydration causes distress that increases with time. A brief search of methods sections of papers using water deprivation in rats indicates that water deprivation in rats may be imposed for periods of 24–96 hours.[93] One refinement study reports that rats exhibit signs of dehydration (lethargy, hunched posture, myosis (constriction of pupil) and decreased skin elasticity) after 18–22 hours of water restriction.[94] Food deprivation is also a cause of distress, and the effects will depend on the species and environmental conditions. IACUCs are encouraged to explore alternatives to food and fluid regulation that utilize positive reinforcement to motivate behaviors to achieve scientific goals.

17:31 Is it appropriate to use electric shock to stimulate animals to run or walk on a treadmill?

Reg. PHS Policy (IV,C,1,a) requires that procedures with animals avoid or minimize discomfort, distress, and pain, consistently with sound research design. The AWAR (§2.31,d,1; §2.31,d,1,i) state that unless acceptable justification for a departure from the standards of care in the AWAR is presented in writing, procedures involving animals will avoid or minimize discomfort, distress, and pain to the animals. The AV must be consulted on procedures which cause greater than slight or momentary pain or distress (AWAR§2.31,d,1,iv,B; §2.33,b,4).

Opin. When using any conditioning stimulus, it is desirable to use a reward system rather than a mild or moderate noxious stimulus such as electric shock (*Guide*, p. 31). When examining a proposal for the latter, an IACUC should require evidence to show that an aversive stimulus is the only technique that can be used to produce the required behavior in the animal. If it is concluded that aversive stimuli are the only practical conditioning stimuli, then the least aversive and most tissue-sparing stimulus used to motivate animals should be chosen (e.g., a puff of air, sound, or vibration) over one that is overtly designed to result in a negative or pain-producing stimulus.

17:32 Can a mammalian fetus feel pain? If so, at what age can a rodent fetus be presumed to feel pain? Should the rodent fetus be included as a vertebrate animal when the IACUC considers the number of animals requested for the study?

Opin. The concern relating to fetal and neonatal pain has emerged largely from studies in human infants. Prior to the mid-1980s, procedures were undertaken on human infants without effective anesthesia or analgesia; later studies indicated that human neonates could experience pain.[95] Investigations in neonatal rodents have shown not only that pain (or at least responses to noxious stimuli) can be demonstrated, but that these experiences produce long-term changes in the nervous system. The stage at which these abilities become functional during fetal and neonatal development is still under debate, and case-by-case interpretation may be needed as the extent of mammalian neurophysiological development differs depending on the species. In rodents and other species, anatomical studies suggest that *nociceptive* responsiveness is present in the second half of gestation (see 17:2).

Fetal direct anesthesia is now widespread accepted practice among human fetuses undergoing surgery, and increasing evidence suggests human fetuses feel pain after the age of 12 weeks gestation.[96–100] Neurologic maturation, indicated by EEG "differentiation" may dictate the point at which pain can be perceived; however, this remains hypothetical.[101–104] The *AVMA Guidelines for the Euthanasia of Animals: 2020 Edition* state: "scientific data indicate that mammalian embryos and fetuses are in a state of unconsciousness throughout pregnancy and birth." Reviews of potential for pain sensation in the fetus and neonate may be of value to IACUC members.[100]

17:33 Can unhatched avian embryos be presumed to feel pain? Is IACUC approval needed for the use of avian embryos?

Reg. The AWAR definition of animal excludes eggs until the pipping process commences (AWAR§1.1 Definitions – Animal and Bird). NIH/OLAW has interpreted the term *live vertebrate animal* to apply to avians (e.g., chick embryos) only after hatching.[105] See also NIH/OLAW FAQ A.4.[106]

Opin. Although it seems illogical to conclude that the capacity to experience pain emerges at the instant of hatching, we cannot as yet determine the stage of development at which this capacity is sufficiently well developed to warrant concern.[107] The *AVMA Guidelines for the Euthanasia of Animals: 2020 Edition* state that "bird embryos that have attained >80% incubation demonstrate EEG activity that is sustained, with increases in amplitude suggesting the potential for pain perception in conscious embryos; therefore they should be euthanized by similar methods used in avian neonates such as anesthetic overdose, decapitation, or prolonged (>20 minutes) exposure to CO_2. Eggs at < 80% incubation may be destroyed by prolonged exposure (>20 minutes) to CO_2, or to cooling (<4°C for 4 hours) or freezing."[108] IACUCs should acknowledge that pain may occur, acknowledge the collective uncertainty as to the stage of development at which this occurs, and set some initial guidelines that can be reviewed as the understanding of avian neurobiology improves. Some institutions may choose to permit early embryonic avian stages to proceed without IACUC oversight, but request IACUC protocols once embryos reach a particular developmental stage (see 17:32).

17:34 Is hypothermia an acceptable form of anesthesia for fetal or neonatal homeothermic animals?

Opin. Hypothermia produces immobility and deep hypothermia (cryoanesthesia) is often used as general anesthesia during surgery in neonatal rodents. The rationale for this approach in rodents is due to the significant physical development that occurs post-birth in these species and the leveraging of slowing nerve conduction using low temperatures to depress body systems to the point of unconsciousness. Study results demonstrate that neonatal cryoanesthesia (< 12 minutes) has no adverse effects on the body weight of mice during development or on their cognition into adulthood.[109] Many U.S. institutions permit hypothermia as part of their adopted guidelines and policies for brief anesthesia in rodent pups. However, it has been suggested that since rewarming from hypothermia is associated with pain in humans,[110] the technique may be undesirable even if it produces a state of insensibility. Perhaps a more constructive approach is to examine the alternative anesthetic techniques available for use in neonates. Because volatile and injectable anesthetics can be used to produce safe and effective anesthesia in older rodents, it seems reasonable for an IACUC to ask for justification of this technique over other options to euthanize very young neonates. Investigators who have used all of these techniques often find the use of conventional anesthesia more convenient, especially for prolonged surgical procedures, since the neonates do not need to be maintained on an ice pack for the duration of surgery. Importantly, application of focal cooling and freezing sprays (e.g., ethyl chloride) are not recommended for biopsy or other tissue sampling procedures in rodents due to more widespread tissue damage and inflammation that are sequalae to these spray exposures.[111,112]

Surv. Does your IACUC typically accept hypothermia as an acceptable form of anesthesia for neonatal mice or rats? Select all that apply.

- Not applicable 55/200 (28%)
- Yes, up to 5 days of age 46/200 (23%)
- Yes, up to 10 days of age 14/200 (7%)
- Yes, up to 14 days of age 1/200 (1%)
- Yes, and requires an explanation as to why other anesthetics are not acceptable 34/200 (17%)
- No 54/200 (27%)
- Other 13/200 (7%)

17:35 Is hypothermia an acceptable form of anesthesia for poikilothermic animals?

Reg. PHS Policy (IV,C,1,b) and the AWAR (§2.31,d,iv) state that procedures causing more than slight or momentary pain or distress will be performed with appropriate sedation, analgesia, or anesthesia unless otherwise justified in writing by the investigator.

Opin. Similar concerns to those described in 17:34 relate to the use of hypothermia as a means of "anesthesia" in poikilotherms, which are organisms with a variable body temperature that tends to fluctuate with, and is similar to or slightly higher than, the temperature of its environment. As with mammalian neonates, well-established alternative anesthetic regimens are available[113]; therefore, hypothermia may not offer any refinement against other options for anesthesia of poikilotherms. Gradual chilling of water will slow movements and reduce the general metabolic activity of warm-water species of fish (like zebrafish). Hence, oxygen consumption is reduced while, at the same time, the capacity for holding dissolved oxygen in water is increased. Depending upon the species of poikilotherm, nerve conduction is reduced but not completely blocked, compared to effective block of conduction with the use of MS-222. In general, hypothermia for cooling anesthesia is not recommended for invasive procedures, and although it is a convenient method to immobilize fish, it is assumed to be an ineffective anesthetic. In comparison, for euthanasia of these species, rapid chilling induces a lethal shock presumably due to rapid disruption of osmoregulatory functions.[114,115]

17:36 An investigator proposes to administer a small amount of nonirritating fluid daily for 1 month by intraperitoneal injection into hamsters. The injection is performed by a skilled person and the sites will be rotated. Is this APHIS/AC pain/distress Category C or Category E?

Reg. See 17:5 for a discussion of the APHIS/AC pain/distress categories. The *Guide* (p. 26) states that "If little is known about a specific procedure, limited pilot studies … conducted under IACUC oversight, are appropriate" (see 17:55).

Opin. The essential determination is whether the procedure causes momentary pain or distress or causes more cumulative and substantial pain that is unrelieved. The investigator and IACUC are unlikely to be able to answer this question at the outset. Secondary determinations may include how to gauge if the fluid is truly nonirritating. In other words, any data, like gross histology findings and histopathologic exam of the peritoneum, should be reviewed, as well as observations of animal behavior after the injection procedure to see if behavioral responses are evident, transient, or persistent. Unintended adverse events should also be evaluated, for example, should there be inadvertent penetration of an abdominal organ from the injection needle, this could lead to complications that would elevate the level of pain and distress in the animal. Animals should be evaluated in a prospective manner, by observing animals after a single injection and then after multiple days of injections, by behavioral responses, by body condition scoring, by weight assessments, and by comparisons to age-matched (and ideally sham-handled) animals. The degree to which a relatively minor noxious stimulus can lead to pain depends on whether inflammation occurs and on the overall health and well-being of the animal. It has been reported that daily intraperitoneal injections (n=30) into CD-1 mice resulted in no differences in complete blood counts and behaviors, and there were minimal focal histopathological lesions from the needle punctures that were not deemed to be clinically significant.[116] Serial assessment of animals by daily monitoring of behavior and weight can aid in detecting the magnitude of the cumulative procedure's impact. Once this information is known, future categorization can be completed with more objective evidence, and any initial or pilot studies can be retrospectively characterized as Humane Use Category C or E. As a reminder, Humane Use Category D categorization would apply only if animals received an alleviation of the potential for painful and distressful procedures.

17:37 An investigator proposes to administer a small amount of nonirritating fluid daily for 1 month by oral gavage to unanesthetized and untranquilized hamsters. A skilled person performs the gavage. Is this APHIS/AC pain/distress Category C or Category E?

Reg. See 17:5; 17:36.

Opin. Oral gavage has the potential to cause irritation and inflammation of the esophagus, which might be expected to result in lower weight gain and food consumption. Even with extensive training, there is a potential for perforation of the esophagus during oral gavage procedures in small mammals, which may be difficult to evaluate at the time of the procedure and may be discovered upon postmortem examination. Body weight curves, posture, and swallowing behavior during voluntary oral intake are likely to be indicative of the presence of this pain. If there is a well trained and experienced person performing the gavage, Category C would be appropriate, particularly if techniques are used to provide sweet enticements for the hamsters to voluntarily ingest the fluid without having to use gavage; this would be an excellent refinement in animal procedures.[117,118] Post-gavage responses and behaviors should be closely evaluated (see 17:36).

17:38 An inhalation anesthetic is used to restrain an animal briefly for a painless procedure (such as an intranasal inoculation). Does this fall under APHIS/AC annual report Category C or D?

Reg. See 17:5. See 17:1 and 17:2 for definitions of distress.

Opin. The primary advantage of using sedatives or anesthetics to immobilize animals for the performance of nonpainful procedures is that these drugs are used in lieu of more forceful or stress-inducing physical restraint methods, which might lead to unintended pain and distress for the animal. The IACUC should evaluate the rationale presented by the investigator for using any anesthetic. For example, if the IACUC agrees with the investigator that the extent of physical restraint needed to deliver accurate dosing through intranasal inoculation is likely to distress the animal more than momentarily, then the use of an anesthetic (or perhaps a sedative) to reduce the need and consequences of that manual or physical restraint is appropriate. Alternately, the researcher might argue that anesthesia is required to have the animal deeply inhale the intranasal inoculation and that other means of assuring lung delivery would be more distressful or painful. Regardless of the explanation, the IACUC will have to support with an approval of the stated need. Although using an anesthetic or sedative might be considered a refinement of the experimental technique, it does not change the fact that regulated animals that receive drugs for the purpose of alleviating pain or distress are required under the AWAR to be categorized as Humane Use Category D.

To help assure there is no confusion as to when animals can receive general anesthesia yet still be included in Category C, APHIS/AC has provided the following clarification:

> The IACUC has the latitude to consider anesthetizing or tranquilizing an animal for radiographs, MRI, or similar procedures as Category C if it determines that only minimal pain or distress is involved with the procedure. If, however, the IACUC determines that any needed restraint process is itself painful or very distressful requiring medications (sedatives, tranquilizers, anesthetics) for alleviation, then Category D is appropriate. (IACUC Handbook, 3rd Edition)

17:39 How long is the time length defining "prolonged restraint" in animal studies, regardless of species? At what time point should an animal be provided with relief from prolonged restraint? Should prolonged restraint be considered as a distressful procedure for which anesthesia or supportive care should be administered?

Reg. The PHS Policy (IV,C,1,a; IV,C,1,b; U.S. Government Principle IV) and the AWAR (§2.31,d,1,i; §2.31,d,1,ii) note that the IACUC should determine that a proposed activity prevents or minimizes discomfort and distress, and that alternatives to painful or distressful procedures have been considered. The *Guide* (p. 29) is more specific. It states that prolonged restraint (including chairing of nonhuman primates) should be avoided unless it is scientifically essential and is approved by the IACUC. It suggests the use of less restrictive systems, such as a backpack-fitted infusion pump for nonhuman primates and free-stall housing for farm animals. The *Guide* (pp. 29–30) provides advice for restraint, including: restraint time should be the minimum required to accomplish the research objectives; animals should be acclimated to the restraint device; animals that fail to acclimate should be removed from the study; restraint devices should not be used as a convenience when handling animals. The AWAR describe long-term restraint for nonhuman primates as lasting more than 12 hours and contains specific requirements for the use of restraint devices in this taxon (§3.81,d).

Opin. Chronic (prolonged) restraint can be required for a variety of purposes but is usually relied upon when administering compounds or sampling body fluids via implanted catheters or obtaining continuous recording of physiological variables, although it is acknowledged some induced-stress models will utilize a prolonged restraint procedure. The need for prolonged physical restraint has been offset by the development of harness-and-swivel devices that allow an animal some degree of movement, while implantable telemetry devices or ambulatory infusion systems allow complete freedom of movement. These mobile-restraint devices are refinements that may reduce the need for maximal restraint and mitigate the potential for accompanying pain and distress; however, there are certain caveats. The weight of, and the surgical procedure required to insert, telemetry implant devices can inflict a metabolic burden on the animal and impart postoperative pain. Tethers and metabolism cages allow freedom of movement but require animals to be isolated from conspecifics and maintained in barren (non-enriched) environments. Boredom and social isolation can contribute to stress and distress. The effects of environmental stressors on many experimental variables, including toxicity of substances, prenatal developmental biological processes, and cardiovascular, neurologic, and immunologic parameters have been well documented.[119–122]

When considering protocols that require restraint (or any of the refinements mentioned), an IACUC should confirm whether physical restraint is needed or whether an alternate approach can be adopted. This must be balanced by an appreciation that these alternate systems may still cause some pain, distress, or discomfort to an animal, potentially during the conditioning period and after implantation. Ideally and hypothetically, if a study can be completed with acclimated physical restraint lasting <2 hours, this might be preferable to the use of a longer-term tethering system.

As the period of restraint increases, the balance between different restraint options may shift. It is important to determine whether the animal can be readily trained to acclimate to physical restraint. Measurement of stress-sensitive indices, such as blood glucose concentration, heart rate, or behavioral stereotypy, suggests that animals can be acclimated or conditioned to accept restraint in slings or other restraint devices.[123] Problems arise, however, when physical restraint is carried out for prolonged periods in such a manner that the animal is unable to eat, drink, or carry out species-specific clinically normal behaviors, such as self-grooming.

Many IACUCs choose to define prolonged conscious restraint (e.g., 15 min of restraint) quantitatively within institution-specific policies so that those protocols using restraint can be identified and appropriately reviewed. The amount of time an animal should be provided with relief from conscious prolonged restraint is dependent on the model; however, the AWAR require at least one continuous hour of unrestrained activity daily for nonhuman primates maintained in restraint devices for more than 12 hours, unless continuous restraint is required by the research proposal approved by the IACUC (§3.81,d). For all species, IACUCs should rigorously evaluate the proposed restraint period to ensure that it is the minimum time necessary.

17:40 A research project with hamsters requires 0.5 ml of blood once a week for 6 months. Postorbital blood collection, under anesthesia, is approved by the IACUC. Even with a skilled technician, is it reasonable for the IACUC to assume pain or distress will occur over time from so many collections?

Reg. Neither the AWAR nor the PHS Policy directly address postorbital blood collection (see 17:3–17:5; 17:41). However, investigators are required to consult with the AV or AV designee during the planning of procedures that have the potential for more than slight or momentary pain or distress (§2.31,d,1,iv,B).

Opin. This historical blood sampling technique is optimized if personnel undergo coordinated hands-on training to provide consistent and sufficient instruction to best perform retroorbital blood collection with competence. It has been shown in anesthetized laboratory mice that 79 of 80 animals (98.8%) showed normal body condition, posture, and behavior throughout a study where eyes were alternated every other week for blood sampling. There may be occurrence of transient ocular clinical signs in animals after sampling.[124] Abnormal behavior indicative of pain was studied in rats after a single postorbital puncture under anesthesia. Using intensive monitoring, the duration of the impact of the procedure appeared to be about five hours.[125,126] Repeated tissue injury may cause altered threshold to pain sensitivity, and repeated sampling might cause increased pain over time. The severity of impact on an individual animal might be measured by alterations in body condition score, body weight, and, of course, gross appearance of the eye. Alternating between the left and right eyes reduces the number of times a single eye is sampled, allowing time for tissue healing and recovery and lowering the overall potential for induction of a painful state. IACUCs may choose to require a dose of analgesics (perhaps topical anesthesia to the eye, as well as systemic analgesics) if repeat retroorbital blood collection is permitted.

17:41 Does blood collection from rodents via retro-orbital route normally require anesthesia, analgesia, or tranquilization?

Reg. Neither the AWAR nor the PHS Policy directly address postorbital blood collection.

Opin. See 17:40. Postorbital (retroorbital) blood sampling in rodents can provide moderate quantities of blood quickly and easily and when carried out expertly. In trained hands, this collection technique seems to cause minimal long-term complication. As mentioned in the prior answer, appropriate training to conduct the procedure, with gentle handling and collection approaches, is critical for success and to minimize adverse outcomes.[124] When retro-orbital sampling is carried out less competently, injuries to the globe can occur and may be severe.[127] The procedure itself may be no more painful than peripheral venipuncture (e.g., collection from the tail vein). Many research units carry out this sampling technique with the animal under general anesthesia, both to reduce any pain and to minimize inadvertent ocular injury should the animal struggle during the procedure. The IACUC should encourage veterinary consultations and consider alternatives and refinements, like sampling from the submental, tail, or saphenous vein instead.[128–131] (See 17:4; 17:5; 17:40.)

17:42 Relative to pain or distress, how many episodes of blood collection should the IACUC reasonably allow on a single animal?

Opin. Blood collection as a method of routine health analysis in domestic species is a tenet of veterinary practice. Often collections are deemed to be momentary in terms of pain or distress, similar to the human experience of having blood sampled for diagnostic testing. It is extremely common for IACUCs to have blood collection guidelines that outline how much blood can be collected in a particular timeframe. Rigorous review of these practices and a balance of animal health against data collection is necessary if there is a concern about excessive sampling that might cause undue harm to the animal (see 17:5).

17:43 For performing procedures such as genetic monitoring, can the tail of a rat or mouse be clipped without causing more than momentary pain to the animal? If yes, what length of tail on mice?

Opin. Removal of a small portion of the tail has become an established procedure in many institutions that work with transgenic animals. There is little doubt that removing the tip (1-2mm) of a sensitive structure causes pain. A number of studies have been done to examine the short- or long-term effects on rodents and whether anesthesia or analgesia is effective.[130–140] The conclusions vary, but there is evidence for an effect that is more than momentary as animals age; however, handling of animals and anesthetizing of animals with isoflurane also causes short term effects on behavior and physiology. Anesthesia without lasting analgesic effects may have more detrimental impact on mice than performing the procedure without anesthesia.[130,133,134,139] A consensus may shortly begin to emerge as the results of studies of mice of varying age and of technique (i.e., amount of tail removed) accumulate. These studies have used an assortment of assessment methods, such as physiological, behavioral, and persistent pain hypersensitivity as well as histological evaluation. Currently, different institutions vary in their opinion as to whether the procedure requires anesthesia and whether anesthesia should be local or general. An alternative technique described by Meldgaard and associates,[139] which uses swabs of buccal cells, could be attractive in that it replaces a surgical procedure with potential for pain; collection of buccal cells, or potentially rectal mucosal swabs, should be performed only in adult rodents as the swab is too invasive in neonates. A subsequent study found that the physiologic and behavioral changes that result from both invasive (i.e., cutting tissue) and noninvasive DNA sampling methods (i.e., hair or mucosal swabs) did not differ; however, the quantity of DNA from hair and buccal samples is much less than what can be obtained by ear punch, tail biopsy, and rectal mucosal sampling.[140,141]

Surv. Does your IACUC usually permit tail clipping of mice as a means of obtaining tissue for genotype?

• Not applicable	36/200 (18%)
• No	11/200 (6%)
• Yes, but only 2 mm or less	82/200 (41%)
• Yes, but only 5 mm or less	57/200 (29%)
• Other*	14/200 (7%)

*Other:

- Animals must be 21 days of age or younger unless otherwise justified.
- No tail length maximum is specified.
- PIs are strongly encouraged to use ear punching to obtain tissue for genotyping.

Surv. If your IACUC permits tail clipping for genotyping, at what age is it allowed to be performed without local or general anesthesia? Assume no more than a 3 mm sample will be removed.

• Not applicable	49/200 (25%)
• Less than 7 days of age	22/200 (11%)
• Less than 10 days of age	13/200 (7%)
• Less than 17 days of age	15/200 (8%)
• Less than 21 days of age	78/200 (39%)
• Less than 28 days of age	9/200 (5%)
• At any age	10/200 (5%)
• Other	4/200 (2%)

Surv. We encourage the use of alternatives to toe clipping, including the following. Select all that apply.

• Buccal mucosal swabbing	24/113 (21%)
• Hair pluck	24/113 (21%)
• Ear punch/snip	100/113 (88%)
• Tail snip	98/113 (87%)
• Rectal mucosal swabbing	11/113 (10%)
• Other tissue source*	3/113 (3%)

*Other: blood sampling.

17:44 What are the important issues and what are some general guidelines for the IACUC to consider relative to methods of animal identification?

Reg. The *Guide* (p. 75) states that toe clipping, as a method of identification of small rodents, should be used only when no other individual identification method is feasible although it may be the preferred method for neonatal mice up to seven days of age. Although the *Guide* lists other methods of identification (e.g., tattooing, ear notching) that can potentially be painful or distressful to an animal, only toe clipping is singled out for additional discussion. Fin clipping, identification tags, etc. can be used to identify aquatic species (*Guide,* p. 87). The AWAR (1:1, Painful Procedure) define a *painful procedure* as one that causes more than momentary pain or distress (e.g., as can occur from a needle prick). PHS Policy (IV,C,1,a; IV,C,1,b) has similar wording and the same intent. Identification requirements for dogs and cats are noted in the AWAR (§2.38,g) (see 17:45).

Opin. Accurate and reliable identification of laboratory animals is essential for most research projects. Methods of identification should:

- Be reliable and simple to perform
- Cause minimal pain or distress to an animal
- Be standardized, so that all can readily interpret the concerned identification system
- Not impact the well-being of non-captive (e.g., field study) animals

The IACUC should balance the consequences of failure of an identification system (loss of animals from a study and the need to repeat the investigation using additional animals) against the impact on the animal of the marking technique. For example, the investigator should consider whether the technique would impact the ability to thrive in the animal's environment. These issues are often debated when an investigator wishes to use relatively low-cost (in economic terms) techniques such as ear tagging, ear punching, or toe clipping, compared to tattooing or the more refined methods such as individual microchip implantation. Aside from concerns that the former methods are less reliable (fighting can result in loss of tags or tearing of aural openings), methods of physical marking result in a physical change to the animal that may cause more than momentary pain. Checking the identification of an animal may cause stress due to circumstances of physical restraint. In contrast, the placement of microchips might be considered to cause only momentary pain, and identification can be carried out more easily, often using remote sensing devices without the need to restrain the animal. Recently, however, a few studies have supported that toe clipping in very young mice does not cause appreciable indications of persistent pain and that a microchip transponder was more aversive than toe clipping.[140,142–144] The authors of one study suggested that toe clipping might be an alternative that allows identification of animals but also provides a sample for DNA testing. The impact of a variety of identification procedures in rodents has been reviewed.[137]

Means of identification of aquatic species are also unique and should be included in IACUC guidance to researchers. Fin clipping in zebrafish is a common means of identification, although there is evidence that this may be a painful procedure and alternatives should be evaluated as they develop and refinements should be applied if feasible (e.g., skin swabbing).[145–147]

Surv. Does your IACUC allow toe clipping of young rats or mice as a means of obtaining tissue for genotyping?

• Not applicable	35/200 (18%)
• No, we do not	70/200 (35%)
• Yes, with adequate scientific justification	81/200 (41%)
• Yes, with or without adequate scientific justification	7/200 (4%)
• Other	5/200 (3%)

Surv. Does your IACUC encourage the use of alternatives to toe clipping for genotyping?

• Not applicable because our researchers do not perform such procedures.	77/200 (39%)
• Yes.	116/200 (58%)
• No. We permit toe clipping, but we do NOT encourage an alternative.	7/200 (4%)

17:45 What refinement efforts have been undertaken at institutions to promote rodent identification practices that limit pain and distress?

Reg. See 17:44.

Opin. A similar approach to that of 17:44 can be applied to this question. Investigators should state why they consider it necessary to remove a piece of tissue if less invasive methods are available. These issues may seem relatively trivial when weighed alongside questions of end points in carcinogenesis studies or subjecting animals to major survival surgery, but they should be addressed in a way that recognizes the overarching intention to minimize any pain or distress caused to animals. In particular, as a refinement, collection of a tissue sample (e.g., ear punch from the pinna) for identification can serve a dual purpose of providing a sample for genotyping, so that additional invasive tissue samples are not necessary.

Surv. What refinement efforts have been undertaken at your institution to promote rodent identification practices that limit pain and distress? Select all that apply.

• Tattooing.	99/198 (64%)
• Microchipping.	64/198 (32%)
• RFID tags.	30/198 (15%)
• Limited use of metal ear tags.	72/198 (36%)
• Limited to no use of tissue removal.	29/198 (15%)
• We do not permit toe clipping as a means of identification.	73/198 (37%)
• Identification by means of non-toxic marker.	134/198 (68%)

17:46 What justification should an IACUC expect if an investigator proposes that the death of animals or the death of 50% of study animals is the most relevant end point for the study?

Reg. Neither the AWAR nor the PHS Policy address the use of the lethal dose 50% (LD_{50}) test. The U.S. Food and Drug Administration Modernization Act of 2021 does not require LD_{50} test data and now allows an applicant for market approval for a new drug to use methods other than animal testing to establish the drug's safety and effectiveness. Under this bill, these alternative methods may include cell-based assays, organ chips and micro-physiological systems, sophisticated computer modeling, and other human biology-based test methods.[148] A search for alternatives to potentially painful or distressful procedures is required under the AWAR (§2.31,d,1,ii).

Opin. There are two primary points to consider for this question. First is the justification for the use of death as an endpoint, and second is the use of a 50% mortality criterion. The use of death as an endpoint has been regarded as essential in some investigations, but rapid developments and refinements are being made in this field, and it is an area in which both investigators and the IACUC should make particular efforts to keep abreast of the current literature. Regulatory testing requirements for acute toxicity do not require LD_{50} data and allow for animals that are moribund or exhibiting clinical signs to be euthanized rather than die spontaneously.[149] The usual reason offered for selecting death as an endpoint is the difficulty of reliably differentiating animals that will die from those that will recover, despite their showing severe clinical signs of illness or toxicity. Surrogate humane endpoints have been proposed to intervene when death can reliably be predicted from clinical signs or test results. In infectious disease, disease prevention, or therapeutic studies, death indicates a lethal infectious burden or failure of therapy. Often, these studies involve species for which detailed monitoring is not typically performed (e.g., rodents). If there is not a rapid onset of morbidity, animals gradually develop progressively more severe abnormalities, commencing with mild depression of normal activity and signs associated with lack of grooming activity (e.g., ruffled fur coat) and finally progressing to coma and death. When the clinical signs include subjectively distressing changes such as convulsions, weakness, cachexia, or severe dyspnea, there is pressure to euthanize an animal rather than allow further deterioration of its condition. As previously mentioned, the difficulty is the fear that some of these animals may recover, and early euthanasia might invalidate a test result; yet, data that animals can recover to clinical health from moribund states is lacking.

Several constructive suggestions have been proposed to reduce the need to use death as an endpoint in studies. There are several alternatives to using a strict 50% lethality criterion for acute toxicity studies that can reduce animal numbers used by 60%–70%.[150] Furthermore, it is important to keep in mind that the results of a given test or exposure may be markedly affected by the handling or environment of the animals. This means that careful attention to laboratory conditions, training of personnel, and animal monitoring will play a vital role in experimental validity. Without a complete necropsy, a conclusion that the agent, disease, or intervention caused death could be erroneous. Instead, the animals may have died because of weakness and subsequent inability to eat and maintain caloric needs. Recognition of what is happening physiologically is missed when death is the only criterion evaluated; more detailed tracking of clinical signs or parameters might lead to better information, thereby increasing the discriminatory power of the experiment. Guidelines on the management of animals being used for the assessment of novel antibacterial or antifungal agents have been proposed by representatives of organizations involved in this type of investigation.[151] In some circumstances, simple clinical indices such as the development of hypothermia from a controlled baseline of a healthy cohort of animals can be used to predict death reliably.[152] There remain many studies, however, in which no validated criteria have been devised. In these circumstances, an IACUC should carefully assess whether criteria can be developed during the progress of the study. In many instances, failure to develop these criteria may be due to infrequent or inadequate observations of the animals. At other times critical events may occur when personnel are not usually available (e.g., dark cycle/overnight periods), thereby precluding detailed observation. There seems no doubt that progress in refining endpoints has only occurred because of investigators carefully evaluating their own particular models. It may be that even after careful assessment, no progress is made, and animals must be allowed to die if the aims of the study are regulated for a specific testing outcome. It is essential, however, for an IACUC to be proactive, require genuine attempts to develop nonlethal endpoints, and not allow investigators to use death simply because it is an unambiguous criterion that can be applied. Finally, the adoption of nonlethal endpoints can have beneficial effects on a study because they allow blood and tissue samples to be taken that may enable crucial intermediary data points, prior to spontaneous death, to be obtained.

17:47 A mammal is born with a genetic defect that leads to an abnormal but nonlethal health condition very soon after birth (e.g., blindness). The animal adapts as well as can be expected to that condition and requires no significant additional levels of husbandry or veterinary care. Should the IACUC consider this condition as nonpainful and nonstressful (Category C on the APHIS/AC annual report) or as unalleviated pain or distress (Category E on the APHIS/AC annual report)?

Opin. Although animals may have no apparent abnormality on clinical examination and require no special husbandry, one might consider that the lack of ability to move about or sensory deprivation causes an increase in the barrenness of the environment to that animal, and therefore a form of deprivation distress. It is advantageous to provide blind animals with increased enrichment and/or socialization whenever possible to mitigate the effects of sensory deprivation. On the other hand, one could argue that since we are unable to detect any apparent adverse effect, except by neurological examination, we should classify this as nonpainful or nondistressful, and therefore Category C. It also may be necessary to reconsider the classification as the animal ages and all aspects of the life cycle are considered.

17:48 When a condition such as loss of sight or Parkinson's disease is acutely induced in a laboratory animal should this be considered distressful to that animal?

Reg. See 17:1, 17:2 and 17:55. US Government Principle IV states that unless the contrary is established, procedures that cause distress in humans should be considered as potentially being able to cause distress in other animals. Similar wording is used in the AWAR (§1.1, Painful procedure) (see 17:9).

Opin. Induced conditions in laboratory animals may have different clinical presentations depending on the species of animal, the method used to induce the condition, the rapidity of onset, associated research procedures, and other variables. For example, the induction of Parkinson's disease with MPTP (1-methly-4-phenyl-1,2,3,6-tetrahyd ropyridine) in nonhuman primates typically is presented as a clinically more severe condition than when this chemical is used to induce disease in the mouse. The determination as to whether an animal is distressed by the induced condition can be subjective but as with the definition of pain (and pain is a form of distress) it is helpful to consider if the same condition would be distressing if it occurred in a human. Thus, if a human became acutely blinded, even if there was no associated pain, one could extrapolate that a degree of distress will be experienced; similarly, if a clinically healthy human were to become acutely Parkinsonian, even with the best of supportive care, that person would most likely experience stress that could evolve to distress. Given the U.S. Government Principle IV, it is reasonable therefore to assume that an animal would also be distressed, perhaps more so since an animal cannot comprehend the rapidity of the disease process and rapid loss of sight or mobility.

Surv. If your IACUC approves a study in a mouse model that usually results in overt discomfort (i.e., Parkinson's model) where no pain medication is given but the animal is euthanized upon reaching humane endpoint (i.e., weight loss), what APHIS/AC pain/distress category does your institution place these experimental animals in?

- Not applicable. 35/198 (8%)
- Mice are not assigned pain/distress categories at my institution. 16/198 (8%)
- Mice are placed in Category D (pain/distress alleviated by euthanasia). 54/198 (27%)
- Mice are placed in Category E (unrelieved pain/distress). 93/198 (47%)

17:49 Ferrets and pigs are used to teach tracheal intubation techniques. They are anesthetized, recovered, and used again. What is a reasonable number of times, and at what intervals, might a ferret be used for this purpose? How does the IACUC determine this?

Opin. It is more appropriate to state that an animal can be used as often as required for such a procedure, provided it shows no behavioral changes that increase in severity over time and is treated appropriately for any potential pain and is allowed to recover from any pain or disability that might result from the procedure. If the animal continues to feed, drink, and grow normally and does not develop clinical abnormalities after the intubation procedure, this indicates that appropriate handling and techniques were used that prioritized health and safety. For teaching protocols, there is a degree of subjective fatigue that should be ascertained if the animal is excessively handled, in addition to consideration of the length of time the ferret remains under anesthesia for the practice procedures. Animals should always be thoroughly examined for evidence of laryngeal, oral, or other trauma at the end of repeated technical procedures.

Familiarizing an animal to the procedure and providing rewards (positive reinforcement training; PRT) for appropriate behavior and an enriched environment, including social housing of social species can lead to improved welfare of the animals involved. This welfare enhancement could not easily be achieved when animals are used on a few occasions, euthanized, and a naive animal used for the next series of manipulations. The availability of nonsentient alternatives (e.g., inanimate plastic or 3D printed models for training intubation) should continuously be explored. Training with the animal species involved is a prerequisite and the cumulative potential for pain or distress from handling by numerous trainees should be assessed to determine when to recuse animals from further training procedures.

17:50 A drug to be used in a research project is known to cause tremors or mild seizures that occur once and last for approximately 2 minutes. Should the IACUC classify this as unalleviated distress? In this example, if seizures occur frequently, should the IACUC classify this as unalleviated distress?

Reg. The AWAR (§2.36,b,7) specifically require an annual report indicating painful or distressful procedures for which the use of appropriate anesthetic, analgesic, or tranquilizing drugs would adversely affect the procedures, results, or interpretation of the teaching, research, experiments, surgery, or tests.

Opin. Many drugs produce mild neurological side effects, and when these include seizures, particular concern over the welfare of the animals involved should be a priority. It is generally perceived that seizure activity can cause postictal pain and distress, particularly when marked muscle spasm or other bodily injury occurs. In the case of seizures that cause a loss of consciousness, there is no evidence that the seizure itself is painful, although it should be considered that anxiety or fear may occur in the subject in anticipation of the seizure (pre-ictal phase) or in animals co-housed with others that might react to the seizure episode. It is the opinion of APHIS/AC[153] that seizures such as this constitute unalleviated distress.

As in other areas of research using animals, we have no clear evidence regarding the degree of pain or distress caused by seizures. As reported for humans, some seizure activity may occur while the person is consciously experiencing the event, although there are reports of rare cases where individuals experience severe pain during the seizure and a relatively high reported incidence of post-seizure headache.[154,155] It could be argued that if the seizure is of a type that results in loss of consciousness and is not associated with potentially distressing aftereffects, then it should not be considered to produce unalleviated distress. There is no way to assess whether an animal is aware during a seizure, thus the IAUC may want to classify it as unalleviated distress unless detailed observations of the animal during and after the seizure can objectively aid in determining the extent of impact to animal health.

17:51 An investigator performs a research procedure and states that he will euthanize an animal, as soon as any pain or distress is detected, using pentobarbital. Under those circumstances, would the IACUC place those animals in the APHIS/AC annual pain and distress report Category C or D?

Reg. See 17:5; 17:57. See 17:1–17:2 for definitions of *pain* and *distress*.

Opin. In this question, pentobarbital will be used for euthanasia, not anesthesia. The central issue is the timing of observations relative to the onset of pain or distress. If, for example, the animal is continuously observed until it shows the first signs indicative of *slight or momentary pain or distress* and then is immediately euthanized, the study might be placed in Category C. The fact that euthanasia was performed does not change the APHIS/AC category to Category D (pain or distress alleviated by drugs) because APHIS/AC requests the most severe category of pain or distress *actually experienced* by the animal. In this case, as it was only slight or momentary discernable pain, then Humane Use Category C.

However, most species of animals, and especially prey species (mice/rats) may be experiencing pain/distress for prolonged times before actually exhibiting overt signs that are measurable to an observer. In the research environment, animals are observed once daily such that signs may not be immediately detected if there is ongoing pain, nor are there consistent quantitative parameters available that can be instituted to detect the exact moment an animal feels pain. An animal that is observed only twice a day might experience several hours of *more than* slight pain before observation. Animals that experience breakthrough pain, despite receiving appropriate anesthetic and analgesics, or experience pain or distress before the need for anesthetics or analgesics is detected and addressed, should be placed in Category D. This Humane Use Category is further supported as long as the animals are appropriately monitored; the type, dose, and frequency of analgesics being administered are appropriate for the procedure and species; and the intent is to alleviate the pain and distress as needed.

Euthanasia is an appropriate form of pain relief, so animals that are euthanized when pain is detected are Category D.

17:52 Should an IACUC allow cardiac puncture without prior anesthesia?

Reg. Cardiac puncture as a means to effect euthanasia by exsanguination (unless there is prior anesthesia, stunning, or sedation) is considered unacceptable by the *AVMA Guidelines for the Euthanasia of Animals: 2020 Edition* because of anxiety associated with extreme hypovolemia.[119] PHS Policy (IV,C,1,g) requires compliance with the AVMA *Guidelines for the Euthanasia of Animals* unless the IACUC approves a deviation.

There are no specific regulatory statements concerning cardiac puncture for purposes other than euthanasia, assuming the volume of blood removed is sufficiently small as not to cause hypovolemia. Personnel performing cardiac puncture must be appropriately trained (AWAR §2.31,d,1,viii; U.S. Government Principle VIII; PHS Policy IV,C,1,f).

Opin. Cardiac puncture is discouraged in laboratory animal medicine, unless deployed solely as a terminal blood sampling access point or site for injection of euthanasia solution, in the absence of availability of other central routes. The entry point on the body surface is likely to be the factor that may result in the sensation of pain. In veterinary patients that require pericardial fluid removal (pericardiocentesis), this would be conducted using only a localized anesthetic block or under deep anesthesia to avoid painful stimuli of the needle placement. If cardiac puncture is used to deliver euthanasia solution, then there should be a justification for why it cannot be replaced by a less invasive euthanasia technique that will minimize the pain of intercostal injection or by peritoneal injection. If the investigator proposes to use this technique in an unanesthetized or unsedated animal to sample blood, the anatomical approach and justification should address these concerns. In general, cardiac sampling is primarily justified only as a terminal procedure under deep anesthesia to maximize blood sample volume and avoid any risk of injury and pain to the animal.

17:53 Is tattooing of animals a painful procedure?

Reg. The AWAR (§1.1, Painful procedure) define a painful procedure as one that would reasonably be expected to cause more than slight or momentary pain or distress in a human being to which that procedure was applied, that is, pain in excess of that caused by injections or other minor procedures. The AWAR (§2.36,b,5) indicate that tattooing should be categorized on its annual report as a procedure involving no pain, distress, or use of pain-relieving drugs.

Opin. Based on the experiences of humans, the size, location, and manner of application of a tattoo are all factors in determining whether the application of a particular tattoo is painful. All permanent methods of tattooing cause a certain amount of tissue damage, and it is therefore likely that a certain amount of pain will occur although it may be minimal and momentary. The use of new sterile needles and nontoxic pigments on disinfected skin, while under local or central anesthesia (if appropriate), all contribute to lessening any immediate or subsequent pain from tattooing. Experience suggests that the pain caused by tattooing of rodents, particularly neonatal animals, is usually minimal. The impact of a variety of identification procedures in rodents is reviewed, and a comparison of three types, including tattooing, in adult rats has been reported.[137]

17:54 Is death at the end of a normal lifespan considered a painful experience?

Reg. NIH/OLAW Notice NOT-OD-05-034[156] states that under certain conditions the mortality of animals may *not* have to be reported to NIH/OLAW, based on the IACUC's consideration of the surrounding circumstances.

Opin. Examples provided in the above NIH/OLAW notice include:

- Death of animals that have reached the end of their natural life spans
- Death of neonates when husbandry and veterinary medical oversight of dams and litters was appropriate
- Death from spontaneous disease when appropriate quarantine, preventive medical, surveillance, diagnostic and therapeutic procedures were in place and followed
- Death from manipulations that fall within parameters described in an IACUC-approved protocol

Death, even at the end of a normal lifespan, may involve pain or other distress. If an animal has a sudden event and dies acutely, there is possibly no experience of pain or distress. Most animals that die spontaneously do so as a result of organ dysfunction, chronic disease, or lack of ability to maintain homeostasis and there is a high likelihood that pain or distress will occur. Intervention with euthanasia per the *AVMA Guidelines for the Euthanasia of Animals: 2020 Edition* would ideally be used to prevent moderate to severe pain.

Surv. If an untreated control animal dies a so-called natural death due to old age, what APHIS/AC pain/distress category is assigned by your IACUC (assume a APHIS/AC category is to be used)?

- Not applicable 39/198 (20%)
- Category B – if the animal was just being held and not used for experimentation 49/198 (25%)
- Category C – if the animal was part of a study or for teaching and the study/teaching resulted in minimal or no pain and distress 78/198 (39%)
- Category E – unalleviated pain/distress due to experimental needs 4/198 (2%)
- I don't know 16/198 (8%)
- Other 12/198 (6%)

17:55 Can the IACUC suggest pilot studies to determine the extent of pain or distress when they are not known?

Reg. Although pain or distress is not formally defined in the AWAR (see 17:2 for the definition of a painful procedure and 17:1 for a discussion of distress), the *Guide* (p. 120) defines pain as a complex experience that typically results from stimuli that damage or have the potential to damage tissue. The *Guide* (p. 121) defines distress as an aversive state in which an animal fails to cope or adjust to various stressors with which it is presented. The *Guide* (p. 26) states that "if little is known about a specific procedure, limited pilot studies, designed to assess both the procedure's effects on the animals and the skills of the research team and conducted under IACUC oversight, are appropriate." NIH/OLAW in FAQ D.11 has stated: "Pilot studies may be appropriate to determine the technical feasibility of larger studies or to make initial assessments of the effect of procedures on animals."[157]

Opin. Pilot studies, or the conduct of the experiment in a limited number of animals (e.g., 2–6) will allow the research team, veterinary staff, and IACUC to determine whether monitoring and treatment strategies are adequate to maintain well-being and mitigate pain and distress. Pilot studies may involve focused involvement of the veterinary team and consultation about refinements and revision of procedures if unanticipated issues or clinical anomalies develop.

17:56 If during a study an animal covered by the AWAR accidentally fractures a leg bone and the animal receives appropriate veterinary treatment, which APHIS/AC annual report pain and distress category should the animal be assigned to?

Reg. The AWAR (§2.36,b,1) require an annual report to assure, in part, that professional standards of treatment have been used, including the use of anesthetic, analgesic, and tranquilizing drugs when required by the IACUC. Animals are placed into one of three Humane Use Categories (C, D, or E) as part of the mandatory annual report (AWAR §2.36,b,5–7). The AWAR (§2.33,b,2) require adequate veterinary care to treat diseases and injuries. The PHS Policy IV,C,1,e requires medical care for animals by a qualified veterinarian (see 16:5).

Opin. Although the AWAR (§2.36,b,5–§2.36,b,7) require an annual report categorizing regulated species used during experimentation, teaching, or testing into one of three categories, this categorization does not apply to accidental injuries to an animal. The APHIS/AC *Animal Welfare Inspection Guide*[158] states that the animal should be reported in the pain category appropriate to its experiences related to the research study as determined by the research facility. The accident and treatment do not affect the reporting category. Therefore, if the animal did not experience any pain or distress as part of the approved study, it would be reported in Category C (no pain or distress or use of pain-relieving drugs) even though the unanticipated bone fracture may have caused initial pain, and anesthesia and analgesics may have been used during its appropriate treatment under the guidance of veterinarian.

17:57 If an animal covered by the AWAR experiences unanticipated pain as part of a research procedure and the pain is treated in a timely manner, which of the APHIS/AC annual report categories for pain and distress is appropriate to use?

Reg. See 17:56.

Opin. The key phrase in the above query is that the pain was treated "in a timely manner." Although neither the AWAR nor the APHIS/AC Inspection Guide define that phrase, it is reasonable to assume it implies that the animal is successfully treated with analgesia as soon as the pain is recognized and that observations have been made with a frequency approved by the IACUC. With those caveats, it is suggested by the APHIS/AC Inspection Guide to assign Humane Use Category D to the animal (pain or distress alleviated using drugs).

17:58 If an animal experiences unanticipated pain/distress during a research procedure, and is euthanized in a timely manner, which of the APHIS/AC annual report pain categories is appropriate to use?

Reg. Animals that are unable to have chronic pain or distress alleviated will be euthanized at the end of the procedure or, if appropriate, during the procedure (AWAR §2.31,d,1,v; PHS Policy IV,C,1,c).

Opin. As noted in 17:57, the key phrase is "in a timely manner" and the phrase assumes that the pain is recognized during observations made within the defined criteria approved by the IACUC and the animal is euthanized soon thereafter. Technically, euthanasia should be accomplished by the use of a drug to comply with the wording of Category D; however, it is the opinion of the authors that nondrug means of euthanasia complying with the AWAR's definition of euthanasia (the humane destruction of an animal accomplished by a method that produces rapid unconsciousness and subsequent death without evidence of pain or distress, or a method that utilizes anesthesia produced by an agent that causes painless loss of consciousness and subsequent death) (§1.1) likely would be found acceptable to APHIS/AC.

17:59 An animal on a study experiences unanticipated pain from an infection that is entirely unrelated to the study. The pain cannot be treated due to interference with the goals of the study. Nevertheless, the animal can be treated with antibiotics. Which of the APHIS/AC annual report categories for pain and distress is appropriate to use?

Reg. See 17:56. The AWAR (§2.33,b,2) require adequate veterinary care to treat diseases and injuries. PHS Policy (IV,C,1,e) requires medical care for animals overseen by a qualified veterinarian.

Opin. Under the circumstances described in the above query, Humane Use Category E is most appropriate (unalleviated pain or distress because pain relieving drugs would have adversely affected the procedure, results, or interpretation of the study). In this situation, the needs of the study, not the infection *per se*, dictated the withholding of analgesia. Even if the antibiotics that were used eventually resolved the infection and mitigated the pain, it is unlikely that the pain was alleviated in a timely manner (see 17:57).

17:60 Can the creation of intentional ascites with peritonitis be a justification for antibody production in rodents and rabbits? If ascites is allowed in rodents, how is the pain and distress categorized by the IACUC?

Opin. Often called an ascites peritonitis model, the intent of ascites induction is to attract inflammatory cells to the site of sterile inflammation. The procedure typically involves a single intraperitoneal injection of sterile 4% thioglycolate medium, time for the immunologic response to the medium, and subsequent euthanasia of the animal to collect fluid and cells from the peritoneal cavity. Although non-sterile peritonitis is often a painful condition in humans, numerous empirical observations of mice used for this sterile peritonitis procedure suggest that there is no pain experienced during the relatively short (typically 48 hours) period of time prior to euthanasia. Frequent monitoring of the animals is necessary to ensure that ascites does not develop into an uncomfortable state; in addition, alternatives to a live animal model for this purpose have been developed and may replace the ascites model.

Surv. Can the creation of intentional ascites with peritonitis justify antibody production in rodents?

- No, there are alternative *in vitro* methods to ascites that should be encouraged. 86/198 (43%)
- Yes, the alternative *in vitro* methods may not be as robust as *in vivo* ascites. 43/198 (22%)
- Other* 69/198 (35%)

*Other:

- Use of the ascites method for antibody production does not apply to many institutions.
- Scientific justification is required for the use of the ascites method at their institutions.

Surv. How is the pain and distress with respect to antibody production by means of ascites categorized?

- Not applicable 105/198 (53%)
- Category C – minimal or no pain/distress 6/198 (3%)
- Category D – Pain/distress alleviated by drug use because our IACUC requires an analgesic to be used 36/198 (18%)
- Category E – Unalleviated pain/distress due to experimental requirements 42/198 (21%)
- Other 9/198 (5%)

17.61 Should rodent cancer models be considered "unrelieved pain and distress"?

Opin. The degree of pain resulting from cancer induction in animals is likely quite variable, based on type, location, and burden, and certainly challenging to quantify. Pain or discomfort in a rodent cancer model could arise from two sources: 1) surgery—typically associated with the implantation of cells or tissue samples instead of therapy, and 2) local tumor growth, metastasis, and comorbidity, with acute pain arising from implantation and chronic pain accompanying tumor development and metastasis. Tumor inoculation can often be accomplished by using fairly noninvasive methods that require only short-term physical or chemical restraint (for example, when subcutaneous injection is used). Conversely, some orthotopic models of brain, bone, pancreatic, and liver cancer, for example, may involve a surgical procedure for implantation. In these models, management of acute pain in the post-surgical period is often practiced, with analgesic agents administered for 24 to 48 hours after the procedure but not necessarily continuing for subsequent days and weeks throughout the course of disease development. Sufficient evidence is available to support the idea that rodent cancer models likely engender some degree of pain that, if left unrelieved, serves as a confounding study variable, primarily by inducing a stress response. In addition, sufficient evidence exists to support the conclusion that analgesics influence cancer biology and likewise may serve as confounding factor to experimental outcomes. In the face of supporting evidence that analgesic agent provision will invalidate or confound a study as cancer progresses, withholding could be justified. Conversely, human patients are routinely provided with analgesics that may make a similar animal model more readily translatable.[159] Depending on the tumor model, objective assessments will be necessary to determine which, if any, cancer studies would most accurately be placed in Humane Use Category E.

Surv. If your IACUC approves a study in a mouse model that may not result in overt discomfort (i.e. subcutaneous tumor model) where no pain medication is given but the animal is euthanized upon reaching a humane endpoint (i.e. weight loss or tumor ulceration), what APHIS/AS pain/distress category does your institution place these experimental animals in?

- Not applicable. 33/198 (17%)
- Mice are not assigned pain/distress categories at my institution. 15/198 (8%)
- Mice are placed in Category D. 123/198 (62%)
- Mice are placed in Category E. 27/198 (14%)

17.62 At what age should an IACUC assume pain can be experienced by zebrafish embryos and what is done to alleviate pain?

Reg. Larval forms of fish and amphibians have vertebrae and are covered by the PHS Policy. As noted in OLAW FAQ, the PHS Policy applies to the offspring of egg-laying vertebrates only after hatching. Zebrafish larvae, for example, typically hatch three days post-fertilization (dpf).

Opin. Larval stages of zebrafish possess a similar organization of molecular nociceptive circuits compared with mammals and display comparable behavioral responses to adult fish after a potentially painful procedure. Zebrafish larvae as early as five days perform behavioral changes when exposed to noxious heat, CO_2-infused water, and acetic acid, which are alleviated when fish are immersed in lidocaine, buprenorphine, or morphine at appropriate doses. While many IACUCs do not require protocol approval for embryos under three days, proper euthanasia consistent with the *AVMA Guidelines for the Euthanasia of Animals: 2020 Edition* should be followed after this developmental stage. There is increasing evidence that young zebrafish larvae do experience distress and should be maintained on an IACUC protocol either at time of fertilization (if an IACUC chooses to do so) or within 72 hours (as OLAW mandates) to ensure painful procedures are addressed accordingly and euthanasia is performed according to AVMA standards.[56,160]

17.63 How do you assess pain and distress in zebrafish?

Opin. An important aspect of methods used to identify potential pain or distress is the ability to detect changes in typical behavior for that species. Identifying normal behaviors and then recognizing abnormal behaviors that may be indicative of distress is still under investigation for fish utilized in laboratory settings. Preliminarily, the use of standard behavioral measures such as time spent in the bottom half of the tank, swimming speed, position in the tank, or percentage of tank explored can help identify possible criteria to be utilized for a pain/distress scoring system for zebrafish after common procedures are performed. Overt distress (buoyancy issues, rapid opercular movement, sores, lesions, etc.) can be identified on daily health checks and should be evaluated by veterinary staff.[55,146]

17.64 How do you assess pain and distress in cephalopods?

Opin. On the basis of currently available evidence, cephalopods possess (or are likely to possess) a number of attributes that are believed to be necessary for the ability to experience and suffer from pain. Evaluation of the global regulations under development for cephalopods, in addition to review of current literature regarding pain identification and assessment in cephalopods, is encouraged. The monitoring system implemented should recognize normal, desirable, behaviors such as locomotion, foraging or hunting, reproductive behavior, territory inspection, and social behaviors, including communication. Animal observation should similarly document elimination of normal behaviors and identify the presence of damaging or otherwise unexpected behaviors, like self-mutilation or injuring behaviors, stereotypies, withdrawal, anxiety-related behaviors, or aggression. These will often be species-specific, e.g., an apathetic or withdrawn octopus may have arms that resemble corkscrews or are curled into a ball, whereas cuttlefish and squid may have their arms hanging loose. Examples of abnormal body positions include octopus floating mantle-upwards or positive buoyancy (floating) in cuttlefish.[161] IACUCs may choose to place cephalopods within the animal care program using traditional protocol review and approval, despite their invertebrate status, to best evaluate wellbeing, care, and health status during research involving these animals.

Surv. Does your IACUC require non-Animal Welfare Act-regulated species to be placed into APHIS/AS pain categories as part of protocol review?

• Not applicable	23/198 (12%)
• No – only USDA species are placed into pain/distress categories	16/198 (8%)
• Yes – all species (including invertebrates)	47/198 (24%)
• Yes – but limited to vertebrates	108/198 (55%)
• Other	4/198 (2%)

Acknowledgments: former authors Alicia Z. Karas and Jerald Silverman; Norman D. Wiltshire, chapter review.

REFERENCES

1. Lund, T.B., M.R. Morkbak, J. Lassen, *et al.* 2012. Painful dilemmas: a study of the way the public's assessment of animal research balances costs to animals against human benefits. *Public Underst. Sci.* doi:10.1177/0963662512451402.
2. Nuffield Council on Bioethics Report. 2005. The ethics of research involving animals, pp. 35–36. https://www.nuffieldbioethics.org/publication/the-ethics-of-research-involving-animals/. Accessed August 11, 2025.
3. Swami, V., A. Furnham, A.N. Christopher. 2008. Free the animals? Investigating attitudes toward animal testing in Britain and the United States. *Scand. J. Psychol.* 4:269–276.
4. Russell, W.M.S., R.L. Burch. 1959. *The Principles of Humane Experimental Technique*. London: Methuen.
5. National Centre for the Replacement Refinement & Reduction of Animals in Research. https://www.nc3rs.org.uk/who-we-are/3rs. Accessed August 11, 2025.
6. Carbone, L. 2019. Ethical and IACUC considerations regarding analgesia and pain management in laboratory rodents. *Comp. Med.* 69:443–450.
7. Peterson, N.C., E.A. Nunamaker, P.V. Turner. 2017. To treat or not to treat: the effects of pain on experimental parameters. *Comp. Med.* 67:469–482.
8. Mellor, D.J., E. Patterson-Kane, K.J. Stafford. 2009. *The Sciences of Animal Welfare*. Hoboken: Wiley-Blackwell.
9. Buchanan-Smith, H.M., A.E. Rennie, A. Vitale, *et al.* 2005. Harmonising the definition of refinement. *Anim. Welfare.* 14:379–384.
10. Committee on Recognition and Alleviation of Distress in Laboratory Animals. 2008. *Recognition and Alleviation of Distress in Laboratory Animals*. Washington, DC: National Academies Press.
11. Koolhaas, J.M., A. Bartolomucci, B. Buwalda, *et al.* 2011. Stress revisited: a critical evaluation of the stress concept. *Neurosci. Biobehav. Rev.* 35:1291–1301.
12. Korte, S.M., B. Olivier, J.M. Koolhaas. 2007. A new animal welfare concept based on allostasis. *Physiol. Behav.* 92:422–428.

13. McMillan, F.D. 2003. A world of hurts—Is pain special? *J. Amer. Vet. Med. Assoc.* 223:183–186.
14. Orlans, F.B. 1987. Review of experimental protocols: classifying animal harm and applying "refinements." *Lab. Anim. Sci.* 37:50–56.
15. International Association for the Study of Pain. 2011. IASP taxonomy. http://www.iasp-pain.org/AM/Template.cfm?Section=Pain_Definitions. Accessed August 11, 2025.
16. Zimmerman, M. 1984. Neurological Concepts of Pain, its Assessment and Therapy. In *Neurophysiological Correlates of Pain*, ed. Bromm, B. Amsterdam: Elsevier.
17. Craig, A.D. 2003. A new view of pain as a homeostatic emotion. *Trends Neurosci.* 26:303–307.
18. American College of Veterinary Anesthesiologists. 1998. Position paper on the treatment of pain in animals. *J. Am. Vet. Med. Assoc.* 213:628–630.
19. ACLAM Position Statement on Pain and Distress in Research Animals. November 2016. https://www.aclam.org/about/position-statements. Accessed August 11, 2025.
20. National Agriculture Library, Animal Welfare Information Center. https://www.nal.usda.gov/programs/awic. Accessed August 11, 2025.
21. Khamis, H.J. 1997. Statistics and the issue of animal numbers in research. *Contemp. Topics Lab. Anim. Sci.* 36:54–59.
22. Reducing Administrative Burden for Researchers: Animal Care and Use in Research Report in response to the 21st Century Cures Act, p. 20. https://olaw.nih.gov/sites/default/files/21CCA_final_report.pdf. Accessed August 11, 2025.
23. Francione, G.L., A.E. Charlton. 1998. Bound by law. *Lab. Anim. (NY).* 27(3):20–21.
24. Garnett, N.L., W.R. DeHaven. 1998. A word from the government. *Lab Anim. (NY).* 27(3):21.
25. National Institutes of Health, Office of Laboratory Animal Welfare. 2012. Frequently asked questions D.6. *What Criteria should the IACUC Consider When Reviewing Protocols?* https://olaw.nih.gov/faqs#/guidance/faqs?anchor=50318. Accessed August 11, 2025.
26. Dalla, C.E, M. Minero, D. Lebelt, D. Stucke, E. Canlai, M.C. Leach. 2014. Development of the Horse Grimace Scale (HGS) as a pain assessment tool in horses undergoing routine castration. *PLoS ONE* 9:e92281.
27. Langford, D.J., *et al.* 2010. Coding of facial expressions of pain in the laboratory mouse. *Nat. Methods* 7:447–449.
28. Sotocinal, S.G, R.E. Sorge, A. Zaloum A.H. Huttle, L.J. Martin, J.S. Wieskopf, J.S. Mogil. 2011. The Rat Grimace Scale: a partially automated method for quantifying pain in the laboratory rat via facial expressions. *Mol. Pain* 7:55.
29. Viscardi, A.V., M. Hunniford, P. Lawlis, M.C. Leach, P.V. Turner. 2017. Development of a Piglet Grimace Scale to evaluate piglet pain using facial expressions following castration and tail docking: a pilot study. *Front. Vet. Sci.* 4:51.
30. Turner, P.V., D.S.J. Pang, J.L.S. Lofgren. 2019. A review of pain assessment methods in laboratory rodents. *Comp. Med.* 69:451–467.
31. Adrian, D., M.Papich, R. Baynes, J. Murrell, B.D.X. Lascelles. 2017. Chronic maladaptive pain in cats: a review of current and future drug treatment options. *Vet. J.* 230:52–61.
32. Balko, J.A., S.K. Chinnadurai. 2017. Advancements in evidence-based analgesia in exotic animals. *Vet. Clin.* 20:899–915.
33. Belshaw, Z., J. Yeates. 2018. Assessment of quality of life and chronic pain in dogs. *Vet. J.* 239:59–64.
34. Boesch, J.M. 2019. Advances in pain management: palliative care applications. *Vet. Clin.* 49:445–461.
35. Clutton, R.E. 2018. A review of factors affecting analgesic selection in large animals undergoing translational research. *Vet. J.* 236:12–22.
36. Flecknell, P. 2018. Analgesics in small mammals. *Vet. Clin.*, 21:83–103.
37. Flecknell, P. 2018. Rodent analgesia: assessment and therapeutics. *Vet. J.* 232:70–77.
38. Foley, P.L., L.V. Kendall, P.V. Turner. 2019. Clinical management of pain in rodents. *Comp. Med.* 69:468–489.
39. Gruen, M.E., B. Duncan, X. Lascelles, E. Colleran, A. Gottlieb, J. Johnson, P. Lotsikas, D. Marcellin-Little, B. Wright. 2022. 2022 AAHA pain management guidelines for dogs and cats. *J Am Anim Hosp Associat* 58:55–76.
40. Lascelles, B.D.X., D.C. Brown, M.G. Conzemius, M. Gill, M.L. Oshinsky, M. Sharkey. 2019. Measurement of chronic pain in companion animals: Discussions from the Pain in Animals Workshop (PAW) 2017. *Vet. J.* 250:71–78.
41. Perry, S.M., J.G. Nevarez. 2018. Pain and its control in reptiles. *Vet. Clin.* 21:1–16.
42. Pinho, R.H., M.C. Leach, B.W. Minto, F.D.L Rocha, S.P.L Luna. 2020. Postoperative pain behaviours in rabbits following orthopaedic surgery and effect of observer presence. *PloS One* 15:e0240605.
43. Plummer, P.J., M.N. Hempstead, J.K. Shearer, T.M. Lindquist. 2021. Evaluating the welfare of small ruminants: practical management advice. *Vet. Clin. North Am.* 37:33–54.
44. Reid, J., A.M. Nolan, E.M. Scott. 2018. Measuring pain in dogs and cats using structured behavioural observation. *Vet. J.* 236:72–79.
45. Ruel, H.L.M., P.V. Steagall. 2019. Adjuvant analgesics in acute pain management. *Vet. Clin.* 49:1127–1141.
46. Smith, J.S., J. Schleining, P. Plummer. 2021. Pain management in small ruminants and camelids: analgesic agents. *Vet. Clin.* 37:1–16.
47. Suckow, M.A., P.V. Turner. 2019. Pain as a clinical factor and experimental variable in research rodents. *Comp. Med.* 69:441–442.
48. Lund, I., L-C. Yu, K. Uvnas-Moberg, *et al.* 2002. Repeated massage-like stimulation induces long-term effects on nociception: contribution of oxytocinergic mechanisms. *Eur. J. Neurosci.* 16:330–338.
49. So, P.S., J.Y. Jiang, Y. Qin. 2008. "Touch therapies for pain relief in adults. *Cochrane Database Syst. Rev.* 4. Art.No: CD006535. https://iris.uniroma1.it/bitstream/11573/918178/1/TESI%20Dottorato%20SFinale.pdf. Accessed August 11, 2025.
50. Silverman, J., N.L. Garnett, S.F. Giszter, *et al.* 2005. Decerebrate mammalian preparations: unalleviated or fully alleviated pain? A review and opinion. *J. Am. Assoc. Lab. Anim. Sci.* 44:34–36.
51. Rose, J.D., F.W. Flynn. 1993. Lordosis response components can be elicited in decerebrate rats by combined flank and cervix stimulation. *Physiol Behav* 54:357–361.
52. Woods, J.W. 1964. Behavior of chronic decerebrate rats. *J Neurophysiol* 4:635–644.

53. Woolf, C.J. 1984. Long term alterations in the excitability of the flexion reflex produced by peripheral tissue injury in the chronic decerebrate rat. *Pain.* 18:325–343.
54. Stevens, C.W. 2004. Opioid research in amphibians: an alternative pain model yielding insights on the evolution of opioid receptors. *Brain Res. Rev.* 46:204–215.
55. Sneddon, L.U. 2015. Pain in aquatic animals. *J Exp. Bio.* 218:967–976.
56. Lopez-Luna, J., Q. Al-Jubouri, W. Al-Nuaimy, L.U. Sneddon. 2017. Impact of analgesic drugs on the behavioural responses of larval zebrafish to potentially noxious temperatures. *App. Anim. Behav. Sci.* 188:97–105.
57. Sneddon, L.U. 2003. The evidence for pain in fish: the use of morphine as an analgesic. *App. Anim. Behav. Sci.* 83:153–162.
58. Sneddon, L.U. 2019. Evolution of nociception and pain: evidence from fish models. *Philosoph. Trans. R. Soc. B*, 374:20190290. https://doi.org/10.1098/rstb.2019.0290. Accessed August 11, 2025.
59. Thomson, J.S., A.G. Deakin, A.R. Cossins, J.W. Spencer, I.S. Young, L.U. Sneddon. 2020. Acute and chronic stress prevents responses to pain in zebrafish: evidence for stress-induced analgesia. *J. Exp. Biol.* 223:224527.
60. Flecknell, P.A. 2015. *Laboratory Animal Anesthesia*, 4th edition. London: Academic Press.
61. Meyer, R.E., R.E. Fish. 2008. Pharmacology of injectable anesthetics, sedatives, and tranquilizers. In *Anesthesia and Analgesia in Laboratory Animals*, ed. R. Fish, P.J. Danneman, M. Brown, *et al.* San Diego, CA: Academic Press.
62. Silverman, J., W.W. Muir. 1993. A review of laboratory animal anesthesia with chloral hydrate and chloralose. *Lab. Anim. Sci.* 43:210–216.
63. Theodorsson, A., L. Holm, E. Therodorsson. 2005. Modern anesthesia and preoperative monitoring methods reduce pre- and postoperative mortality during transient occlusion of the middle cerebral artery in rats. *Brain Res. Protoc.* 14:181–190.
64. National Institutes of Health, Office of Laboratory Animal Welfare. 2012. Frequently Asked Questions. https://olaw.nih.gov/faqs#/guidance/faqs. Accessed August 11, 2025.
65. National Institutes of Health, Office of Laboratory Animal Welfare. 2012. Use of Non-Pharmaceutical-Grade Chemicals and Other Substances in Research with Animals. https://olaw.nih.gov/sites/default/files/120301_seminar_transcript.pdf. Accessed August 11, 2025.
66. National Institutes of Health. Office of Laboratory Animal Welfare. 2011. Position Statements. https://grants.nih.gov/grants/olaw/2011positionstatement_decarchive.pdf. Accessed August 11, 2025.
67. Goecke, J.C., H. Awad, J.C. Lawson, *et al.* 2005. Evaluating postoperative analgesics in mice using telemetry. *Comp. Med.* 55:37–44.
68. Goulding, D.R., P.H. Myers, E.H. Goulding, *et al.* 2010. The effects of perioperative analgesia on litter size in Crl:CD1(ICR) mice undergoing embryo transfer. *J. Am. Assoc. Lab. Anim. Sci.* 49:423–426.
69. Krueger, K.L., Y. Fujiwara. 2008. The use of buprenorphine as an analgesic after rodent embryo transfer. *Lab Anim. (N.Y.).* 37:87–90.
70. Parker, J.M., J. Austin, J. Wilkerson, *et al.* 2011. Effects of multimodal analgesia on the success of mouse embryo transfer surgery. *J. Am. Assoc. Lab. Anim. Sci.* 50:466–470.
71. Smith, J.C., T.J. Corbin, J.G. McCabe, *et al.* 2004. Isoflurane with morphine is a suitable anaesthetic regimen for embryo transfer in the production of transgenic rats. *Lab. Anim.* 38:38–43.
72. Steele, K.H., J.M. Hester, B.J, Stone, K.M. Carrico, B.T. Spear, A. Fath-Goodin. 2013. Nonsurgical embryo transfer device compared with surgery for embryo transfer in mice. *J. Am. Assoc. Lab. Anim. Sci.* 52:17–21.
73. Animal Welfare Institute. Additional Resources. http://awionline.org. Accessed August 11, 2025.
74. Navarro, K.L., K. Jampachaisri, M. Huss, C. Pacharinsak. 2021. Lipid bound extended release buprenorphine (high and low doses) and sustained release buprenorphine effectively attenuate post-operative hypersensitivity in an incisional pain model in mice (Mus musculus). *Anim. Mod. Exp. Med.* 4:129–137.
75. Chapman, K., F Sewell, L Allais, *et al.* 2013. A global pharmaceutical company initiative: an evidence-based approach to define the upper limit of body weight loss in short term toxicity studies. *Regul. Toxicol. Pharmacol.* 67:27–38.
76. Roughan, J.V., C.A. Coulter, P.A. Flecknell, H.D. Thomas, K.J. Sufka. 2014. The conditioned place preference test for assessing welfare consequences and potential refinements in a mouse bladder cancer model. *PLoS One* 9(8):e103362.
77. Samuelsson, P., L. Brudin, R. Sandin. 2007. Late psychological symptoms after awareness among consecutively included surgical patients. *Anesthesiol.* 106:26–32.
78. Sebel, P.S., T.A. Bowdle, M.M. Ghoneim. 2004. The incidence of awareness during anesthesia: a multicenter United States study. *Anesth. Analg.* 99:833–839.
79. Marsch, S.C.U., W. Studer. 1999. Guidelines to the use of laboratory animals: what about neuromuscular blockers? *Cardiovasc. Res.* 42:565–566.
80. Paddleford, R. Pulmonary dysfunction. 1996. In *Lumb and Jones Veterinary Anesthesia*, 3rd edition, ed. J.C. Thurmon, W.J. Tranquilli, J.G. Benson, 774. Baltimore: Williams and Wilkins.
81. Whelan, G., P.A. Flecknell. 1992. The assessment of depth of anesthesia in animals and man. *Lab. Anim.* 26:153–162.
82. Grimm, K.A., L.A. Lamont, W.J. Tranquilli, S.A. Greene, S.A. Robertson, S.A. 2015. *Lumb and Jones' Veterinary Anesthesia*, 5th edition. Baltimore: John Wiley and Sons.
83. Medical Records for Animals Used in Research, Teaching and Testing. ACLAM Position Statement. 2020. https://www.aclam.org/about/position-statements. Accessed August 11, 2025.
84. Percie du Sert N, Hurst V, Ahluwalia A, *et al.* 2020. The ARRIVE guidelines 2.0: Updated guidelines for reporting animal research. *PLoS Biol.* 18(7):e3000410. Accessed January 6, 2023.
85. Lovell, D.P. 1986. Variation in pentobarbitone sleeping time in mice. 1. Strain and sex differences. *Lab. Anim.* 20:85–90.
86. Blaha MD, Leon LR. 2008. Effects of indomethacin and buprenorphine analgesia on the postoperative recovery of mice. *J. Am. Assoc. Lab. Anim. Sci.* 47:8–19.

87. National Institutes of Health, Office of Laboratory Animal Welfare. 2012. Frequently Asked Questions A.6. *Does the PHS Policy Apply to Animal Research that Is Conducted in the Field?* https://olaw.nih.gov/faqs#/guidance/faqs?anchor=50290. Accessed August 11, 2025.
88. Sikes, R.S. Animal Care and Use Committee of the American Society of Mammalogists. 2016. 2016 Guidelines of the American Society of Mammalogists for the use of wild mammals in research and education. *J. mammal.* 97:663–688.
89. Desmarchelier, M., Cheveau, M., Imbeau, L., Lair, S. 2007. Field use of isoflurane as an inhalant anesthetic in the American marten (Martes americana). *J. Wildlife Dis.* 43:719–725.
90. National Institutes of Health, Office of Laboratory Animal Welfare. 2012. Frequently Asked Questions F.19. *Can Investigators Restrict Animals' Food and Fluid?* https://olaw.nih.gov/faqs#/guidance/faqs?anchor=52964. Accessed August 11, 2025.
91. Vermeulen, J.K., A.D. Vries, F. Schlingmann, *et al.* 1997. Food deprivation: common sense or nonsense? *Anim. Technol.* 48:45–54.
92. Waynforth, H.B., P.A. Flecknell. 1992. *Experimental and Surgical Techniques in the Rat.* London: Academic Press.
93. Bennett, T. Gardiner, S.M. 1987. Water deprivation: effects on fluid and electrolyte handling and plasma biochemistry in long-evans and brattleboro rats. *J. Physiol.* 385:35–48.
94. Kulick, L.J., D.J. Clemons, R.L. Hall, *et al.* 2005. Refinement of the urine concentration test in rats. *Contemp. Topics Lab. Anim. Sci.* 44:46–49.
95. Anand, K., W.G. Sippell, A. Aynsley-Green. 1987. Randomised trial of fentanyl anesthesia in preterm babies undergoing surgery: effects on the stress response. *Lancet.* 329(8524):243–248.
96. Anand, K.J. 2000. Effects of perinatal pain and stress. *Prog. Brain Res.* 22:117–129.
97. Anand, K.J.S., M. Maze. 2001. Fetuses, fentanyl and the stress response: signals from the beginnings of pain? *Anesthesiol.* 95:823–825.
98. Houfflin-Debarge, V., A. Delelis, S. Jaillard, *et al.* 2005. Effects of nociceptive stimuli on the pulmonary circulation in the ovine fetus. *Am. J. Physiol. Regul. Integr. Comp. Physiol.* 288:R547–R553.
99. Senat, M.V., C. Fischer, C., Y. Ville. 2002. Funipuncture for fetocide in late termination of pregnancy. *Prenatal Diagnosis.* 22:354–356.
100. Bellieni, C.V. 2021. Analgesia for fetal pain during prenatal surgery: 10 years of progress. *Pediat. Res.* 89:1612–1618.
101. Diesch, T.J., D.J. Mellor, C.B. Johnson, *et al.* 2007. Responsiveness to painful stimuli in anaesthetized newborn and young animals of varying neurological maturity (wallaby joeys, rat pups and lambs). *Proc. 6th World Congress on Alternatives & Animal Use in the Life Sciences,* August 21–25, Tokyo, Japan. AATEX 14(Special Issue):549–552.
102. Mellor, D.J., T.J. Diesch. 2006. Onset of sentience: the potential for suffering in fetal and newborn farm animals. *Appl. Anim. Behav. Sci.* 100:48–57.
103. Mellor, D.J. 2010. Galloping colts, fetal feelings, and reassuring regulations: putting animal-welfare science into practice. *J. Vet. Med. Educ.* 37:94–100.
104. Murrell, J.C., D.J. Mellor, C.B. Johnson. 2008. Anaesthesia and analgesia in the foetus and neonate. In *Anesthesia and Analgesia in Laboratory Animals,* ed. R. Fish, P.J. Danneman, M. Brown, *et al.* Waltham: Academic Press.
105. National Institutes of Health, Office for Protection from Research Risks. 1991. The Public Health Service responds to commonly asked questions. *ILAR News* 33(4):68.
106. National Institutes of Health, Office of Laboratory Animal Welfare. 2012. Frequently Asked Questions A.4. *Does the PHS Policy Apply to Live Embryonated Eggs?* https://olaw.nih.gov/faqs#/guidance/faqs?anchor=50288. Accessed August 11, 2025.
107. Mellor, D.J., Diesch, T.J. 2007. Birth and hatching: key events in the onset of awareness in the lamb and chick. *New Zeal. Vet. J.* 55:51–60.
108. American Veterinary Medical Association. 2020. AVMA Guidelines for the Euthanasia of Animals, 2020 edition. https://www.avma.org/sites/default/files/2020-02/Guidelines-on-Euthanasia-2020.pdf. Accessed August 11, 2025.
109. Janus, C., T. Golde. 2014. The effect of brief neonatal cryoanesthesia on physical development and adult cognitive function in mice. *Behav. Brain Res.* 259:253–260.
110. Davis, K.D., G.E. Pope. 2002. Noxious cold evokes multiple sensations with distinct time courses. *Pain* 98:179–185.
111. Braden, G.C., A.K. Brice, F.C. Hankenson. 2015. Adverse effects of vapocoolant and topical anesthesia for tail biopsy of preweanling mice. *J. Am. Assoc. Lab. Anim. Sci.* 54:291–298.
112. Paluch, L.R., C.C. Lieggi, M. Dumont, S. Monette, E.R. Riedel, N.S. Lipman. 2014. Developmental and behavioral effects of toe clipping on neonatal and preweanling mice with and without vapocoolant anesthesia. *J. Am. Assoc. Lab. Anim. Sci.* 53:132–140.
113. Mitchell, M.A. 2009. Anesthetic considerations for amphibians. *J. Exot. Pet Med.* 18:40–49.
114. Lillywhite, H.B., R. Shine, E. Jacobson, D.F. DeNardo, M.S. Gordon, C.A. Navas, *et al.* 2017. Anesthesia and euthanasia of amphibians and reptiles used in scientific research: should hypothermia and freezing be prohibited?. *BioScience.* 67:53–61.
115. Matthews, M., Z.M. Varga. 2012. Anesthesia and euthanasia in zebrafish. *ILAR J* 53:192–204.
116. Davis, J.N., C.L. Courtney, R.G. Hunter, *et al.* 2012. Behavioral, clinical, and pathologic changes seen in mice receiving repeated intraperitoneal injections. *Poster Presented at the Annual Meeting of the American Association for Laboratory Animal Science,* Minneapolis.
117. Turner, P.V., T. Brabb, C. Pekow, M.A. Vasbinder. 2011. Administration of substances to laboratory animals: routes of administration and factors to consider. *J. Am. Assoc. Lab. Anim. Sci.* 50:600–613.
118. Turner, P.V., C. Pekow, M.A. Vasbinder, T Brabb. 2011. Administration of substances to laboratory animals: equipment considerations, vehicle selection, and solute preparation. *J. Am. Assoc. Lab. Anim. Sci.* 50:614–627.
119. Cordero, M.I., J.J. Rodriguez, H.A. Davies, *et al.* 2005. Chronic restraint stress down-regulates amygdaloid expression of polysialylated neural cell adhesion molecule. *Neuroscience.* 133:903–910.

120. Costa, A., A. Smeraldi, C. Tassorelli, *et al.* 2005. Effects of acute and chronic restraint stress on nitroglycerin-induced hyperalgesia in rats. *Neurosci.* Lett. 383:7–11.
121. Colomina, M.T., M.L. Albina, D.J. Sanchez, *et al.* 2001. Interactions in developmental toxicology: combined action of restraint stress, caffeine, and aspirin in pregnant mice. *Teratology.* 63:144–151.
122. King-Herbert, A.P., T.W. Hesterburg, P.P. Thevenaz, *et al.* 1997. Effects of immobilization restraint on Syrian golden hamsters. *Lab. Anim. Sci.* 47:362–366.
123. Hassimoto, M., T. Harada, T. Harada. 2004. Changes in hematology, biochemical values, and restraint ECG of rhesus monkeys (Macaca mulatta) following 6-month laboratory acclimation. *J. Med. Primatol.* 33:175–186.
124. Fried J.H., D.B. Worth, A.K. Brice, F.C. Hankenson. 2015. Type, duration, and incidence of pathologic findings after retroorbital bleeding of mice by experienced and novice personnel. *J Am Assoc Lab Anim Sci.* 54:317–327.
125. Toft, M.F., M.H. Petersen, N. Dragsted, *et al.* 2006. The impact of different blood sampling methods on laboratory rats under different types of anaesthesia. *Lab. Anim.* 40:261–274.
126. Van Herck, H., V. Baumans, H.A.G. Boere, *et al.* 2000. Orbital sinus blood sampling in rats: effects upon selected behavioral variables. *Lab. Anim.* 34:10–19.
127. Van Herck, H., V. Baumans, C.J.W.M. Brandt, *et al.* 1998. Orbital sinus blood sampling in rats as performed by different animal technicians: the influence of technique and expertise. *Lab. Anim.* 32:377–386.
128. Heimann, M., H.P. Kasermann, R. Pfister, *et al.* 2009. Blood collection from the sublingual vein in mice and hamsters: a suitable alternative to retrobulbar technique that provides large volumes and minimizes tissue damage. *Lab. Anim.* 43:255–260.
129. Heimann, M., D.R. Roth, D. Ledieu, *et al.* 2010. Sublingual and submandibular blood collection in mice: a comparison of effects on body weight, food consumption and tissue damage. *Lab. Anim.* 44:352–358.
130. Holmberg, H., M.K. Kiersgaard, L.F. Mikkelsen, *et al.* 2011. Impact of blood sampling technique on blood quality and animal welfare in haemophilic mice. *Lab. Anim.* 45:114–120.
131. Regan, R.D., J.E. Fenyk-Melody, S.M. Tran, G. Chen, K.L. Stocking. 2016. Comparison of submental blood collection with the retroorbital and submandibular methods in mice (Mus musculus). *J. Am. Assoc. Lab. Anim. Sci.* 55:570–576.
132. Hankenson, F.C., L.M. Garzel, D.D. Fischer, *et al.* 2008. Evaluation of tail biopsy collection in laboratory mice (*Mus musculus*): vertebral ossification, DNA quantity, and acute behavioral responses. *J. Am. Assoc. Lab. Anim. Sci.* 47:10–18.
133. Morales, M.E., R.W. Gereau. 2009. The effects of tail biopsy for genotyping on behavioral responses to nociceptive stimuli. *PLoS One.* 4:e645.
134. Hankenson, F.C., G.C. Braden-Weiss, J.A. Blendy. 2011. Behavioral and activity assessment of laboratory mice (Mus musculus) after tail biopsy under isoflurane anesthesia. *J. Am. Assoc. Lab. Anim. Sci.* 50:686–694.
135. Arras, M., A. Rettich, B. Seifert, *et al.* 2007. Should laboratory mice be anaesthetized for tail biopsy? *Lab. Anim.* 41:30–45.
136. Sørensen, D.B., C. Stub, H.E. Jensen, *et al.* 2007. The impact of tail tip amputation and ink tattoo on C57BL/6JBomTac mice. *Lab. Anim.* 41:19–29.
137. Kasanen, H.E., H-M. Voipio, H. Leskinen, *et al.* 2011. Comparison of ear tattoo, ear notching and microtattoo in rats undergoing cardiovascular telemetry. *Lab. Anim.* 45:154–159.
138. Zhuo, M. 1998. NMDA receptor-dependent long term hyperalgesia after tail amputation in mice. *Eur. J. Pharmacol.* 349:211–220.
139. Meldgaard, M., P.J. Bollen, B. Finsen. 2004. Non-invasive method for sampling and extraction of mouse DNA for PCR. *Lab. Anim.* 38:413–417.
140. Cinelli, P., A. Rettich, B. Seifert, *et al.* 2007. Comparative analysis and physiological impact of different tissue biopsy methodologies used for the genotyping of laboratory mice. *Lab. Anim.* 41:174–184.
141. Dysko R.C., F.C. Hankenson, F.J. Benavides, L.A. Colby, P.C. Smith, D.K. Taylor, N. Nowak, D.E. Hoskins. 2020. Assessment of existing and novel tissue sample collection methods for standard and automated rodent genotyping. *J. Am. Assoc. Lab. Anim. Sci.* 59: 632.
142. Castelhano, M.J., N. Souas, F. Ohl, *et al.* 2010. Identification methods in newborn C57BL/6 mice: a developmental and behavioural evaluation. *Lab. Anim.* 44:88–103.
143. Schaefer, D.C., I.N. Asner, B. Seifert, *et al.* 2010. Analysis of physiological and behavioural parameters in mice after toe clipping as newborns. *Lab. Anim.* 44:7–13.
144. Frezel, N., G. Kratzer, P. Verzar, J. Bürki, F.A. Weber, H.U. Zeilhofer. 2019. Does toe clipping for genotyping interfere with later-in-life nociception in mice?. *Pain Rep.* 4:e740.
145. Breacker, C., I. Barber, W.H. Norton, J.R. McDearmid, C.A. Tilley. 2017. A low-cost method of skin swabbing for the collection of DNA samples from small laboratory fish. *Zebrafish.* 14:35–41.
146. Deakin, A.G., J.W. Spencer, A.R. Cossins, I.S. Young, L.U. Sneddon. 2019. Welfare challenges influence the complexity of movement: fractal analysis of behaviour in zebrafish. *Fishes.* 4:8.
147. Fu, C., Z.D. Cao, S.J. Fu. 2013. The effects of caudal fin amputation on metabolic interaction between digestion and locomotion in juveniles of three cyprinid fish species with different metabolic modes. *Comp. Biochem. Physiol. Part A Mol. Integr. Physiol.* 164:456–465.
148. FDA Modernization Act of 2021. H.R. 2565. 117th Congress (2021–2022). Congress.gov.
149. National Institutes of Health (US). Office of Laboratory Animal Welfare, & Applied Research Ethics National Association. 2002. Institutional animal care and use committee guidebook. Office of Laboratory Animal Welfare.
150. Rispin, A., D. Farrar, E. Margosches, *et al.* 2002. Alternative methods for the median lethal dose (LD50) test: the up-and-down procedure for acute oral toxicity. *ILAR J.* 43:233–243.
151. Acred, P., T.D. Hennessey, J.A. MacArthur-Clark, *et al.* 1994. Guidelines for the welfare of animals in rodent protection tests: a report from the Rodent Protection Test Working Party. *Lab. Anim.* 28:13–18.

152. Hankenson, F.C., N. Ruskoski, M., van Saun, G.S. Ying, J. Oh, N.W. Fraser. 2013. Weight loss and reduced body temperature determine humane endpoints in a mouse model of ocular herpesvirus infection. *J. Am. Assoc. Lab. Anim. Sci.* 52:277–285.
153. DeHaven, W.R., personal communication, 1999. Previous author.
154. Cavanna, A.E.M. Mula, S. Servo, *et al.* 2008. Measuring the level and content of consciousness during epileptic seizures: the Ictal Consciousness Inventory. *Epilepsy Behav.* 13:184–188.
155. Förderreuther, S., A. Henkel, S. Noachtar, *et al.* 2002. Headache associated with epileptic seizures: epidemiology and clinical characteristics. *Headache.* 2:649–655.
156. National Institutes of Health, Office of Laboratory Animal Welfare. 2005. Notice NOT-OD-05-034, Guidance on prompt reporting to OLAW under the PHS Policy on Humane Care and Use of Laboratory Animals. http://grants.nih.gov/grants/guide/notice-files/NOT-OD-05-034.html. Accessed August 11, 2025.
157. National Institutes of Health, Office of Laboratory Animal Welfare. 2012. Frequently Asked Questions D.11. *May the IACUC Approve Pilot Studies?* https://olaw.nih.gov/faqs#/guidance/faqs?anchor=50323. Accessed January 5, 2023.
158. U.S. Department of Agriculture. Animal and Plant Health Inspection Service. 2013. *Animal Welfare Inspection Guide.* https://www.aphis.usda.gov/sites/default/files/animal-welfare-inspection-guide.pdf Accessed August 11, 2025.
159. Taylor, D.K. 2019. Influence of pain and analgesia on cancer research studies. *Comp. Med.* 69:501–509.
160. Steenbergen, P.J., N. Bardine. 2014. Antinociceptive effects of buprenorphine in zebrafish larvae: an alternative for rodent models to study pain and nociception?. *Appl. Anim. Behav. Sci.* 152:92–99.
161. Andrews, P.L., A.S. Darmaillacq, N. Dennison, I.G. Gleadall, P. Hawkins, J.B. Messenger, *et al.* 2013. The identification and management of pain, suffering and distress in cephalopods, including anaesthesia, analgesia, and humane killing. *J. Exp. Marine Biol. Ecol.* 447:46–64.

18

Euthanasia

Debra L. Hickman

Introduction

This chapter addresses questions on the topic of euthanasia. The chapter is organized into eight sections: General Guidelines and IACUC Policy; Endpoints; Inhalant and Other Chemical Methods; Physical Methods; Adjunctive Methods—Stunning, Pithing, Exsanguination; Pre-natal and Neonatal Rodents; Field Studies; and Other Euthanasia-Related Questions. In addition to the *Guide*, PHS Policy, and AWAR, the major "regulatory" (see 18:2) reference for this chapter is the 2020 *AVMA Guidelines for the Euthanasia of Animals*. The reader is urged to consult the most recent guidance from NIH/OLAW, APHIS/AC, and 2020 version of AVMA Guidelines prior to making decisions regarding acceptable techniques and policies for euthanasia. The Canadian Council on Animal Care (CCAC) also has published several guideline documents that address various issues related to euthanasia. Although U.S. and other non-Canadian readers are not required to adhere to the CCAC guidelines, these documents include some excellent information that should be of interest to all readers. Where pertinent, this information is summarized in the Regulatory and/or Opinion parts of this chapter.

General Guidelines and IACUC Policy

18:1 What are some general guidelines for euthanasia?

Reg. Euthanasia of experimental animals may be performed for many reasons, including procurement of tissues or blood as part of the experimental design or as part of general breeding colony maintenance. Animals also may be euthanized to alleviate otherwise untreatable pain or distress. Both the AWAR (§2.31,d,1,v) and the PHS Policy (IV,C,1,c) require investigators to euthanize animals that would otherwise experience severe or chronic pain or distress that cannot be relieved by other means. In such instances, animals should be euthanized at the end of the procedure or, if appropriate, during the procedure. The *Guide* (p. 123) further indicates that animal use protocols should specify endpoint criteria that will "enable a prompt decision by the veterinarian and the investigator to ensure that the endpoint is humane and, whenever possible, the scientific objective of the protocol is achieved."

Opin. The term "euthanasia" is derived from Greek and means "easy death." According to definitions in the AWAR (§1.1, Euthanasia), the *Guide* (p. 123), and the *AVMA Guidelines for the Euthanasia of Animals*,[1] euthanasia involves killing in a humane manner that causes rapid unconsciousness and death with little or no pain or distress. These goals are met by selecting an agent or technique that is appropriate for the situation, considering the needs of the individual animal itself. It also is important that the person performing the euthanasia be compassionate, gentle, fully trained, and technically proficient in the technique to be used and experienced in the humane restraint of the species to be euthanized.

Aside from the all-important humane issues, four broad, overlapping criteria should be considered when selecting a method of euthanasia.

Animal: The method should consider the well-being of the individual animal being euthanized, based on an understanding of methods that can be utilized to minimize the potential pain or distress experienced by the animal. It is critical that the selected method minimizes the potential for pain or distress to be experienced by the animal, recognizing that it is not possible to fully prevent all pain or distress. An ideal method of euthanasia will result in a rapid and reliable irreversible death. Factors that should be considered include the species, its age and health status, temperament, availability of appropriate equipment for restraint, and the environment in which euthanasia will be performed. The environment can include considerations such as the macro-environment (e.g., light levels, presence or absence of conspecifics, comfort with handler) and micro-environment (e.g., home cage versus induction chamber, familiar or strange animals in the cage). Although well-being is studied at population levels, it should be recognized that it is experienced at the individual level and consideration should be given to the needs of an individual animal to ensure appropriate euthanasia.

Regulatory: The method should comply with all relevant regulations and guidelines. Deviations from these regulations and guidelines are

DOI: 10.1201/9781003272861-18

permitted under certain circumstances and should be approved based on sound written justification for scientific or medical reasons (*Guide*, p. 123; AWAR §2.31,d,1,xi; PHS Policy IV,C,1,g). It should be noted that some states and Canadian provinces also have specific laws related to euthanasia.

Human: The method should consider personnel and management issues, including the qualifications and training of the personnel administering euthanasia; the need to minimize emotional distress or compassion fatigue in human participants and observers by proper attention to training and to the aesthetic implications of the method; and attention to the health and safety of humans (and other animals).

Scientific: The method should consider the potential effect of the method on the scientific objectives of the research or planned postmortem evaluations. Any method of euthanasia should be viewed as having the potential to interfere with scientific objectives, and it is important to choose the method with care if any data will be collected from the animal after death.

Other criteria to consider are the additional time or procedures that may be required to assure that an animal will not recover after disposal, the expense and availability of drugs or equipment, the potential for human drug abuse, the ability to maintain euthanasia equipment in proper working order, safety for predators or scavengers that may eat the carcass, and potential environmental impacts of the method itself or animal's carcass following disposal.[1–3]

18:2 The *AVMA Guidelines for the Euthanasia of Animals* are not recognized as law. Must the IACUC use them when determining the appropriateness of the proposed method of euthanasia?

Reg. It is true that the AVMA euthanasia guidelines are presented as guidelines, not legal requirements. However, the AVMA Panel on Euthanasia acknowledges that the Guidelines will be used by some "as part of regulatory structures designed to protect the welfare of animals used for human purposes."[1] PHS Policy (IV,C,1,g) specifically states that the IACUC must determine that methods of euthanasia in a proposed research project are consistent with the AVMA euthanasia guidelines "unless a deviation is justified for scientific reasons in writing by the investigator." While this wording allows for deviation from the *AVMA Guidelines for the Euthanasia of Animals* under specific circumstances, it clearly indicates that IACUCs are required to give the Guidelines careful consideration when evaluating a proposed method of euthanasia. The AWAR (§2.31,d,1,xi) state that "Methods of euthanasia used must be in accordance with the definition of the term set forth in 9 CFR part 1, Sec. 1.1 of this subchapter, unless a deviation is justified for scientific reasons, in writing, by the investigator." The Animal Welfare Inspection Guide specifically notes that utilization of a method of euthanasia not found in the AVMA Guidelines is not necessarily a noncompliance with the Regulations provided that the method meets the AWAR definition of euthanasia (section 6.4.7.5).[4] A program of adequate veterinary care that includes guidance to principal investigators regarding euthanasia is also required (§2.33,b,4).

Opin. For the most part, adherence to the *AVMA Guidelines for the Euthanasia of Animals* should present little challenge to the IACUC, but situations may arise where judgments must be made within the framework or spirit of the Guidelines and regulations. The Guidelines are based on scientific evaluation of the pertinent literature but published findings and/or conclusions regarding a particular euthanasia method are not always consistent, nor is all pertinent information necessarily to be found in the published literature. In such situations, the IACUC may need to make its own determination regarding the best approach. In consultation with the AV, it must apply knowledge of the literature and other pertinent facts as they relate to the species in question along with sound professional judgment to achieve the goal of humane euthanasia. The IACUC also may encounter situations where an investigator wishes to use a method of euthanasia that is not addressed in the AVMA Guidelines or that has been classified as unacceptable. The potential for such a situation is acknowledged within the Guidelines, which go on to state that a veterinarian who is experienced with the species in question should apply both professional judgment and knowledge of clinically acceptable techniques to the situation. Finally, situations may arise where a deviation from the Guidelines is necessitated by scientific or clinical objectives. Both PHS Policy (IV,C,1,g) and the *Guide* (p. 123) state that the IACUC may approve a method of euthanasia that is inconsistent with the AVMA Guidelines provided that the investigator has justified the method for scientific (PHS Policy and *Guide*) or medical (*Guide*) reasons.

In the case of disaster situations, it is worth noting that the AVMA has recently released Guidelines for the Depopulation of Animals (https://www.avma.org/resources-tools/avma-policies/avma-guidelines-depopulation-animals) which may have additional guidance of use to an animal care and use program. Additionally, if the animal care and use program includes agricultural animals, the IACUC should be familiar with the *AVMA Guidelines for the Humane Slaughter of Animals*.

18:3 Is the performance of euthanasia considered pain relief when assigning APHIS/AC pain categories?

Opin. This question is not addressed directly in the AWAR. In the AWAR (§2.36,d,6–7), it is specifically the use of "appropriate anesthetic, analgesic, or tranquilizing drugs" that distinguishes animals placed in Category D from those in Category E. However, this question is partially addressed in APHIS/AC's *Animal Welfare Inspection Guide*.[4] In section 7.2.4.1, it is indicated that special attention should be paid to ensuring that the PI has considered alternatives to the painful or distressing procedure and that proper considerations for the humane timing of euthanasia are described.[4] However, the guidance suggests that in the presence of an advance plan with clear justification and endpoints that are followed, the assignment of Category D would be appropriate. If these plans are not followed and an animal experiences pain or distress that is not alleviated by euthanasia (or other methods), then the animal would be counted in Column E of the annual report.

18:4 Can the AV or a DVM with delegated authority demand or perform euthanasia of an animal that s/he believes is in significant pain or distress?

Reg. Both the AWAR (§1.1, Attending Veterinarian) and the PHS Policy (IV,A,3,b) state that the AV must have direct or delegated authority for activities involving animals at the facility, and the AWAR (§2.33,a,2) require that the AV have appropriate authority to ensure the provision of adequate veterinary care. However, although neither the AWAR nor the PHS Policy directly address the authority of the AV to euthanize an animal that s/he believes is experiencing significant pain or distress, failure to treat (or euthanize) an animal experiencing unanticipated pain or distress would constitute inadequate veterinary care. The *Guide* (p. 114) is more specific in this regard, stating that the AV has the authority to intervene, even perform euthanasia, to relieve pain or distress "if the responsible person (e.g., investigator) is not available or if consensus between the investigator and veterinary staff cannot be reached concerning treatment."

Opin. The regulations and guidelines clearly state that the AV, or their appropriate designee, is responsible for the health and welfare of animals at the institution. Less clearly acknowledged is the role of the AV in supporting animal research at the institution. These two roles—animal protector and research facilitator—must be balanced, a goal that sometimes requires walking a fine line. The AV must protect animal health and welfare and must remain vigilant to assure that their professional judgment is not compromised by institutional interests that are unrelated to the responsible conduct of research.[5] Nevertheless, every effort must be made to avoid unnecessarily compromising the use of the animal in research. Upon encountering an animal that is experiencing unanticipated significant pain or distress (i.e., not addressed in the approved animal use protocol), the first thought should be to identify an effective treatment that is consistent with the goals of the study. Only if this is not possible, or the treatment fails, should an intervention that is inconsistent with experimental objectives be considered. Among the more difficult situations is one in which the IACUC-approved protocol clearly states that animals will experience significant pain or distress prior to reaching the experimental endpoint. In those cases, it would not be appropriate for the AV to intervene and euthanize an animal that was exhibiting the expected signs of pain or distress but had not yet reached the IACUC-approved experimental endpoint.

18:5 If an animal is to be euthanized for clinical reasons that are unrelated to the research project (e.g., spontaneous disease, severe injury), must the method of euthanasia be one that was approved during review of the research protocol by the IACUC?

Opin. There are no specific guidelines that address this question. In the author's opinion, the primary concern should be the welfare of the animal to be euthanized, although the objectives of the research should be considered also. If the method of euthanasia is critical to research objectives and the data collected from the animal can still be included in the study, it would be desirable to euthanize the animal following the approved protocol. However, if the method of euthanasia is not critical to research objectives, if the animal is being removed from the study, or if use of the method of euthanasia specified in the protocol is contraindicated by the animal's clinical condition, the method used should be the one that would provide the most humane death for the animal. In any case, the method chosen should be consistent with recommendations of the *AVMA Guidelines for the Euthanasia of Animals*.[1]

Surv. If an animal is to be euthanized for clinical reasons that are unrelated to ongoing research (e.g., spontaneous disease, severe accident), which method of euthanasia listed below can be used? Select all that apply.

• Not applicable	2/196 (1%)
• Only the method specified in the IACUC protocol	108/196 (55%)
• Any method normally accepted by the IACUC (e.g., carbon dioxide, pentobarbital)	73/196 (37%)
• The method requested by the Principal Investigator	21/196 (11%)
• The method approved by our veterinarian	132/196 (67%)
• Other	4/196 (2%)

18:6 Is it acceptable to euthanize animals in the presence of other animals of the same species?

Reg. There are no regulatory prohibitions against euthanizing animals in the presence of conspecifics, although the *AVMA Guidelines for the Euthanasia of Animals* notes that, for sensitive species, it is desirable to avoid having other animals present when individual animals are being euthanized.[1]

Opin. It has long been believed that vocalizations, behaviors, or odors released by animals undergoing frightening or stressful experiences could cause distress in other animals, particularly animals of the same species. For this reason, it is often recommended that other animals not be present during euthanasia, especially euthanasia of a conspecific.[1,2,6] There is some evidence that rats are aroused, and perhaps stressed, when exposed to conspecifics that are undergoing stressful procedures.[7] In a series of studies, Sharp and colleagues[8–11] observed that heart rate and/or blood pressure were elevated in rats that witnessed other rats undergoing procedures such as decapitation, necropsy, routine cage change, restraint, vaginal lavage, or subcutaneous or tail vein injections. The animals also showed increases in heart rate and blood pressure when exposed to urine from stressed rats or dried rat blood. The magnitude of the cardiovascular changes observed in these studies was influenced by housing density and gender. In most cases, cardiovascular responses to witnessing decapitation were generally no greater than, and sometimes less than, the responses to witnessing procedures such as routine cage change. Boivin and colleagues demonstrated that mice observing the decapitation of other mice appeared to be responding to the sound of the guillotine instead of the death of the conspecifics.[12] While these studies suggest that rodents at least are stressed by witnessing a particularly violent euthanasia procedure, the evidence does not support the conclusion that witnessing euthanasia of conspecifics is any more stressful to other animals than is witnessing many other, more benign (at least from the human perspective) manipulations. As most of this concern is likely based on the extrapolation of the perspective of the human observer to the animal,[13,14] this is an area that would benefit from additional study.

Surv. Does your IACUC allow for the euthanasia of animals in the presence of animals of the same species (either in the visual or in the auditory range) that are not being euthanized?

- Not applicable, we do not euthanize animals. 1/196 (1%)
- Yes, we allow this for all species. 15/196 (8%)
- Yes, but this is only allowed for rodents and cold-blooded animals. 33/196 (17%)
- Yes, but this is only allowed for cold-blooded animals (e.g., Xenopus, zebrafish, etc.). 12/196 (6%)
- No, this is not allowed for any species. 119/196 (61%)
- Other* 16/196 (8%)

*Other:

- We do not have a relevant policy.
- This practice is discouraged but allowed if not avoidable.
- Acceptable if other animals are in IVCs and euthanasia is conducted in a ducted biosafety cabinet.

18:7 Can tissues that are removed post-euthanasia be used for research that has not been approved by the IACUC?

Reg. The AWAR pertain to both live and dead animals (AWAR §1.1, Definitions) and dead animals or parts that are sold for research may also be regulated under the AWAR. While the PHS Policy covers only live vertebrate animals, IACUC approval is required if: (1) the animals will be killed specifically for the purpose of obtaining or using their tissues; or (2) use of the tissues will require project-specific antemortem manipulation of the animals.[15]

Opin. As noted above, there are specific circumstances under which IACUC approval for the use of tissues from dead animals might be mandated under federal regulations or requirements. However, even under circumstances where IACUC approval might not be a regulatory requirement, it is generally appropriate for the IACUC to review all aspects of the proposed use of animals in research. The IACUC is expected to make a determination that studies using animals are scientifically valuable and consistent with the U.S. Government Principles. Principle II states that "procedures involving animals should be designed and performed with due consideration of their relevance to human or animal health, the advancement of knowledge, or the good of society." If the objectives of the study will be met in part using tissues collected postmortem, the IACUC must review these postmortem uses to fully understand the research and the contribution of animals to it. Consider a situation where it might be possible to make use of tissues from an animal that otherwise would be discarded following its use in another study. This would be a desirable means of reducing the total number of animals used, provided that previous use of the animal did not render the tissues unacceptable for the proposed postmortem study. For example, it would be important to verify that the animal had not been treated with any substances that might have resulted in contamination of the tissues with residues that would interfere with the postmortem use or pose a human safety risk. In the author's opinion, the IACUC should at least be notified prior to the "reuse" of tissues, even if they would simply have been discarded—this is essential if the IACUC is to be fully informed regarding all uses of animals at the institution.

18:8 Is perfusion under anesthesia considered euthanasia or a non-survival surgical procedure?

Reg. Perfusion fixation is not addressed in the federal regulations, but it is addressed as an adjunctive method of euthanasia in the *AVMA Guidelines for the Euthanasia of Animals.*[1] Provided that animals are fully anesthetized prior to perfusion, this technique is viewed as a fully acceptable adjunctive method of euthanasia.

Opin. Perfusion, or perfusion fixation, is a technique for replacing the blood in the animal's body with tissue fixative. Typically, this is done by initial infusion of saline to remove the blood, followed by infusion of the fixative. Death may occur relatively early or late in this process. Animals must be deeply anesthetized prior to starting the procedure.[1] Sedation would not be adequate for perfusion fixation, which involves pain or distress related to replacement of blood with fixative and, in most cases, some degree of surgical manipulation.

Whether to classify perfusion under anesthesia as a method of euthanasia or a non-survival surgical procedure has been a topic of debate. Because preparation of an animal for perfusion often requires opening the thorax or abdomen to provide access to the heart, many IACUCs have classified the procedure as non-survival surgery. However, preparation for perfusion may involve nothing more than cut down to a major vessel or simple venipuncture. Furthermore, if the animal is deeply anesthetized prior to starting the procedure, even if surgical preparation is involved, perfusion fixation meets the definition of euthanasia in the AWAR (§1.1, Definitions), AVMA Guidelines,[1] and *Guide* (p. 123), i.e., rapid loss of consciousness followed by death without pain or distress. On this basis, many IACUCs have classified the procedure as euthanasia, particularly if there is minimal or no surgical preparation involved. In the 2020 *AVMA Guidelines for the Euthanasia of Animals*, the AVMA Panel on Euthanasia has taken a definite stand in this debate by classifying perfusion as an adjunctive method of euthanasia. Thus, for most circumstances, it is most consistent with current guidelines for IACUCs to classify perfusion under anesthesia as euthanasia. The only exception might be a situation where the animal is subjected to a surgical procedure, at the termination of which it is perfused prior to regaining consciousness.

Surv. Does your IACUC consider perfusion to be a primary form of euthanasia or is it considered non-survival surgery?

- Not applicable, we don't perfuse animals 34/196 (17%)
- Euthanasia 39/196 (20%)
- Nonsurvival surgery 78/196 (40%)
- Nonsurvival surgery if a body cavity is opened; euthanasia if a peripheral vein is used 34/196 (17%)
- Other 11/196 (6%)

18:9 Should the IACUC review methods used to collect and remove escaped or verminous rodents from the animal facility?

Reg. Neither PHS Policy nor the AWAR address this question directly. The *Guide* (p. 74) states that any trapping methods used to control rodent pests should be humane, and also notes that live-catch traps require frequent observation and humane euthanasia after capture.[3] The *AVMA Guidelines for the Euthanasia of Animals* state in general terms that the same standards for humane euthanasia should be applied to killing wild or feral animals.[1] Wild-caught animals should be handled and euthanized in a manner that causes the least amount of stress. These Guidelines specifically address the acceptability of kill traps for euthanasia. These traps do not always cause death in a manner consistent with the criteria for euthanasia established by the AVMA Panel on Euthanasia and use of live traps followed by other methods of euthanasia is preferred. If used, kill traps should be checked daily and designed to kill captured animals swiftly; if a trapped animal is found alive, it should be killed quickly and humanely.

Opin. Nothing in the *Guide* or *AVMA Guidelines for the Euthanasia of Animals* specifically dictates IACUC involvement in determining methods for controlling escaped or verminous rodents. However, PHS Policy and the AWAR require the IACUC to review and approve the animal care and use program, of which pest control is an essential aspect. The IACUC, in close consultation with the AV, is also well qualified to ensure that the provisions of the *Guide* and AVMA Guidelines are followed with respect to humane euthanasia of wild/feral animals, in general, and trapping of wild/feral pests, in particular. There is an added element of ethical obligation for the welfare of escaped rodents. It is ironic that the status of research rodents can change so dramatically—from valued, even pampered, to scorned—when they escape. Nonetheless, the point could be made that the IACUC's responsibility to protect the welfare of these animals should continue despite this change in status.

18:10 If a trained person performs euthanasia improperly, resulting in injury to an animal or disposal of an animal that is not dead, what actions should the IACUC take?

Reg. IACUCs are required to ensure that personnel are appropriately qualified and trained (§2.31,d,1,viii), and research facilities are required to provide training in humane methods of animal maintenance and experimentation, including proper maintenance and utilization of euthanasia equipment (§2.32,c,1). Additionally, qualifications of personnel must be reviewed with sufficient frequency to fulfill this responsibility (§2.32,b). NIH/OLAW,[16] CCAC,[17] the *Guide* (p. 124),[3] and the *AVMA Guidelines for the Euthanasia of Animals*[1] emphasize that death must be verified after euthanasia and before disposal of the animal. This is viewed as particularly important when using carbon dioxide (CO_2) for euthanasia, as animals may appear dead while in the chamber, but recover when exposed to room air. In addition, PHS Policy (IV,B,4), the AWAR (§2.31,c,4), and the *Guide* (p. 23) stipulate that it is the responsibility of the IACUC and the Institutional Official (IO) to review and/or investigate animal welfare concerns at the institution.[3] Both the AWAR (§2.31,d,7) and the PHS Policy (IV,C,7) require that a report be submitted to APHIS/AC and/or NIH/OLAW if an activity is suspended by the IACUC as a result of such an investigation. NIH/OLAW views failure to ensure death of animals after euthanasia as a serious non-compliance with PHS Policy and requires PHS-Assured institutions to report the incident promptly—whether or not the incident results in suspension of an activity—with a full explanation of circumstances and actions taken.[15] Readers are encouraged to consult NIH/OLAW's Reporting Noncompliance webpage at Reporting Noncompliance | OLAW (nih.gov). for more information on this topic. OLAW also encourages the animal care program representatives to contact them as soon as possible for the preliminary report.

Opin. Improper performance of euthanasia, leading to injury of the animal or recovery of the animal following disposal, is a serious matter. The regulations and other guidelines are unambiguous in their requirements that: 1) death must be confirmed following euthanasia; and 2) animal welfare concerns must be reviewed by the IACUC. Disposal of a live animal after a botched attempt at euthanasia clearly would constitute a failure to confirm death, and either injury to an animal during a failed attempt at euthanasia or disposal of a live animal would fall within the definition of an animal welfare concern. Furthermore, failure to ensure death is identified by NIH/OLAW as an example of serious non-compliance with PHS Policy.[16] In a situation where an animal recovers following disposal, the following actions would be appropriate: investigate the situation to determine why it happened and what can be done to prevent it from happening again; and, for PHS-Assured institutions, report the incident to NIH/OLAW. A report needs to be made to APHIS/AC only if the corrective action includes suspension of an activity involving a covered species (see Chapter 29). Reporting to NIH/OLAW might not be necessary following a failed attempt at euthanasia provided that the failure is recognized immediately, and the animal is killed promptly before it regains consciousness. If for some reason the animal were not euthanized immediately, it would most certainly require a report to NIH/OLAW. Either way, this situation would qualify as an animal welfare concern, requiring some level of review to determine the appropriate corrective and preventive actions. Retraining might be in order to prevent future occurrences. Retraining should focus on proper performance of the technique in question; recognition of cessation of vital signs in the species in question; and/or effective application of an adjunctive method of euthanasia, e.g., thoracotomy, decapitation, cervical dislocation, or exsanguination.[1,3,6]

Surv. If a trained person performs euthanasia improperly, resulting in the injury or disposal of an animal that is unconscious but not dead, what actions would your IACUC typically take? Select all that apply.

• No action for a first-time offense.	2/196 (1%)
• We typically take no action, even for repeated offenses.	0/196 (0%)
• We would give a verbal reprimand.	58/196 (30%)
• We would give a written reprimand..	117/196 (60%)
• We may provide retraining	173/196 (88%)
• We would notify NIH/OLAW for covered species.	129/196 (66%)
• We would notify USDA/Animal Care for covered species.	87/196 (44%)
• I don't know.	8/196 (4%)
• Other.	24/196 (12%)

18:11 What is the difference between an experimental endpoint and a humane endpoint?

Reg. The *Guide* (p. 27) defines these terms as follows: "The experimental endpoint of a study occurs when the scientific aims and objectives have been reached. The humane endpoint is the point at which pain or distress is prevented, terminated or relieved in an experimental animal."[3] The *AVMA Guidelines for the Euthanasia of Animals* distinguish between euthanasia as an act undertaken to serve the animal's best interests vs. the use of humane techniques to induce the most rapid and painless and distress-free death possible, while further noting that these mandates are codependent.[1] The Guidelines also state that, in research situations, animal welfare considerations are balanced against the merits of the experimental design and the research. In these situations, the IACUC must apply the principles of refinement, replacement, and reduction, and ensure a respectful death for research animals.

Opin. The goal for all studies should be to identify an endpoint that is humane and supports the goals of the research. If animals are terminated before scientific objectives are achieved, the value of the study is compromised and animal lives—not to mention time and resources—are wasted or used to poor end. However, if animals are allowed to suffer beyond the point where scientific objectives have been achieved, their suffering is unnecessary, and the situation must be defined as inhumane. Ideally the endpoint should be selected before the study starts, and it should be clearly defined with objective criteria so that everyone can agree when it is reached. This process should involve the investigator, the AV, and the IACUC. In many studies, perhaps even the majority, identification of an appropriate endpoint is not difficult. However, for some studies—those involving cancer, infectious disease, or trauma are examples—identifying an endpoint that is both humane and scientifically sound is more challenging. If the investigator has significant experience with the model, they may be able to use existing data from previous studies to identify an appropriate endpoint. If not, a thorough review of the literature may reveal endpoints that proved useful for similar studies. In cases where there are no previous or published data upon which a decision can be based, a pilot study may be performed for the purpose of identifying a suitable endpoint. In any of these situations, further refinement of the endpoint may be possible over time.

Surv. Which description below is closest to how your IACUC typically approaches or defines a humane endpoint?

- Not applicable, we only have acute studies — 3/196 (2%)
- The point in a study in which an animal is killed. — 3/196 (2%)
- The point in a study in which an animal is to be killed if pain or distress cannot be alleviated. — 47/196 (24%)
- The point in a study in which pain or distress must be alleviated by some means, including euthanasia. — 74/196 (38%)
- We consider humane endpoints to be protocol specific and we do not set specific criteria or definitions. — 66/196 (34%)
- Other. — 3/196 (2%)

18:12 What is the moribund condition? Is it an acceptable endpoint in biomedical research studies?

Opin. *Moribund* has been variously defined as "being in the state of dying; approaching death," "in a dying state," and "dying; at the point of death." In studies where meeting scientific objectives requires knowing that a subject will die within a predictable timeframe and no other criterion that accurately predicts death in a more timely manner can be identified, euthanasia of the animal when it becomes moribund may be viewed as a more humane alternative to allowing the animal to die a natural death. Recognizing the moral and ethical responsibility to minimize potential pain or distress experienced by the animal, the question is whether an animal is likely to experience significant pain or distress in the interval between being recognized as moribund and natural death. Depending on the animal's level of conscious perception at the point where it is judged to be moribund, euthanasia at this point might or might not prevent additional suffering. It must be assumed that prompt euthanasia would prevent suffering in animals that are judged to be moribund but that still appear conscious and responsive. But what about animals that appear unconscious? Studies of human patients show that apparent unconsciousness may not be correlated with a complete lack of perception, suggesting that even apparently comatose animals might benefit from euthanasia in lieu of progression to natural death.[18] However, the same studies suggest that a significant percentage of moribund, unconscious animals would experience no further suffering if left to die naturally. Thus, generally, it is desirable to euthanize animals prior to the point where they enter a terminal comatose state. Recognizing that moribund animals might be past the point of suffering, the CCAC advises performing detailed observations of the animals to help set pre-moribund endpoints.[2] Similarly, ILAR's Committee on Pain and Distress in Laboratory Animals cautions that, unless euthanasia would interfere with experimental goals, animals should be killed before they become moribund. This report further recommends that the "criteria for determining the humane end to a study should be frequently reevaluated and revised as new information becomes available."[6]

Surv. With reference to the word "moribund," which description below is closest to how does your IACUC interprets its meaning?

- Not applicable, we only have non-invasive, non-painful, non-distressful protocols. — \6/196 (3%)
- Moribund means near death; the animal is not conscious. — 9/196 (5%)
- Moribund means near death; the animal may or may not be conscious. — 139/196 (71%)
- Moribund means very ill. — 23/196 (12%)
- Moribund means any animal experiencing pain or distress. — 7/196 (4%)
- I don't know. — 10/196 (5%)
- Other. — 2/196 (1%)

18:13 In situations in which natural death is not an acceptable endpoint, what clinical signs can be used to determine the appropriate time for euthanasia?

Reg. Federal regulations provide no specific guidelines as to which clinical signs might be most useful in determining an appropriate humane endpoint. The *Guide* (p. 27) states only that the PI should identify in the study protocol an endpoint that is both humane and scientifically sound. The CCAC provides the same general guidance along with some more specific suggestions regarding physiological and behavioral parameters that could be assessed as indicators of pain and/or distress associated with deterioration in the animal's condition.[2] The CCAC also provides some more detailed guidelines for selecting appropriate endpoints in cancer, toxicology, infectious disease, or pain research, as well as studies involving monoclonal antibody production.[2]

Opin. In general, allowing an animal to proceed to a natural death without intervention should be reserved for rare, highly specific research studies which have provided a scientific justification for the withholding of earlier interventions. When an IACUC is faced with one of these studies, there are additional considerations. Signs that can be used to determine the appropriate time for euthanasia fall into two general categories: (1) those indicating pain, distress, and or deterioration in clinical condition in an animal in which effective treatments are contraindicated or have failed; and (2) those that predict an irreversible progression to the outcome required to meet scientific objectives, e.g., death. General indicators of pain, distress, and overall clinical condition include body weight; body condition score; physiological indices such as heart rate, respiration, and body temperature; external appearance; unprovoked behavior; and behavioral responses to external stimulation. The value of these indicators is increased when each is evaluated according to a precisely defined scoring system, and combining the individual scores into a cumulative score provides a means of assessing the animal's overall deviation from normal.[2,19] Species-specific signs of chronic pain or distress also may be useful in identifying early humane endpoints.[20,21]

Signs that predict an irreversible progression to a scientifically valid outcome may be more variable depending on the research area or condition being studied. Potentially useful signs can be identified based on a literature search, previous experience with the model, and/or a pilot study. Before being adopted as definitive endpoints, potential endpoints should be evaluated in studies designed to demonstrate their reliability in predicting the desired experimental outcome (e.g., pilot studies). Humane endpoints that have proven useful as valid scientific endpoints include decrease in body condition score with or without changes in body weight and/or appearance or behavior for cancer research[19,20,22]; weight loss and/or changes in body temperature for infectious disease research[18,23]; and loss of muscular control or decrease in body temperature for vaccine potency testing.[24]

Surv. Which of the following methods has your IACUC approved in the past to help determine when an animal is to be euthanized? Select all that apply.

• Not applicable	2/196 (1%)
• A group of nonspecific clinical signs (e.g., ruffled fur, vocalizations, weight loss)	149/196 (76%)
• A group of clinical signs, each having a numerical value, and euthanasia is performed when a certain sum is reached (e.g., body condition scoring)	146/196 (74%)
• A disease severity index (e.g., when hind paralysis is first noticed in EAE models)	131/196 (67%)
• Study-specific signs developed by the investigator, with input from the veterinarian	177/196 (90%)
• Other	7/196 (4%)

Inhalant and Other Chemical Methods

18:14 What are the most appropriate conditions (e.g., concentration, flow rate) for euthanasia of adult rodents with carbon dioxide?

Reg. The *AVMA Guidelines for the Euthanasia of Animals*[1] view CO_2 as conditionally acceptable for euthanasia of those species in which aversion or distress can be minimized. A gradual-fill procedure, in which 30%–70% of the chamber volume is displaced per minute, is recommended. It is not acceptable to place conscious animals in a chamber that has been pre-filled with 100% CO_2. When gradual-fill methods cannot be used, the animal should be rendered unconscious prior to immersing it into 100% CO_2. CO_2 must be supplied in a "precisely regulated and purified form … typically from a commercially supplied cylinder or tank." Additional requirements include "an appropriate pressure-reducing regulator and flow meter or equivalent equipment with demonstrated capability for generating the recommended displacement rates for the size container being utilized." Mixing the CO_2 with oxygen is not recommended as this can allow the CO_2 to act more as an anesthetic and decrease its effectiveness as an agent for euthanasia (https://pubmed.ncbi.nlm.nih.gov/27648836/). These guidelines are similar to those of the CCAC, although these guidelines maintain that CO_2 is not an ideal method for humane killing in any species and that it should be used as the sole agent for euthanasia only under circumstances where other methods are impractical.[17]

Opin. CO_2 is an agent that creates a reversible anesthetic state by causing a respiratory acidosis.[1] The acceptability of CO_2 as an agent for euthanasia of adult rodents has been called into question in recent years due to concerns regarding the potential for pain or distress.[25] Even among those who support the use of CO_2 for this purpose, there is controversy regarding the conditions that are most likely to ensure a humane death.[5,26] The most controversial issues are the absence of alternatives that significantly reduce the potential for pain or distress as compared to CO_2 and the optimal delivery approach for the use of CO_2. While acknowledging that high concentrations of CO_2 could be distressful to some animals, the AVMA Panel on Euthanasia continues to conclude that CO_2 is acceptable with conditions for euthanasia of laboratory rodents.[1] The existing literature suggests that there is no apparent advantage to combining oxygen with CO_2, and the 2020 report recommends against this practice.[1]

There is evidence that a majority of animals may experience significant pain and/or distress when exposed to high concentrations of CO_2.[27–29] This is consistent with research indicating that carbon dioxide, which forms carbonic acid in a moist environment (e.g., the nasal mucosa), specifically excites small nerve fibers that transmit sensations of pain.[30,31] For this reason, it has been recommended that conscious animals should not be exposed to CO_2 concentrations greater than 70%.[27] On the other hand, unconsciousness and death occur more quickly when animals are exposed to high concentrations of CO_2.[32–34] Historically, some observers argued that prefilling the chamber with CO_2 before introducing animals may reduce anxiety and struggling[6] ; however, additional literature suggests that pain and distress can be reduced if the chamber is not prefilled.[5,26–28] The two most recent versions of the AVMA Guidelines for the Euthanasia of Animals (2013 and 2020) view the immersion of conscious animals into 100% CO_2 as unacceptable.[1] The American College of Laboratory Animal Medicine goes further, recommending that the chamber not be prefilled "because inspiration of high concentrations of CO_2 is both aversive and painful."[5]

Several sources suggest the use of low gas flow rates that lead to a gradual increase in the concentration of CO_2 in the chamber.[1,5,6,17,25,26] In 2013, the AVMA Panel on Euthanasia suggested that a volume per minute displacement rate that would displace 10%–30% of chamber volume should be utilized to allow the animal to lose consciousness prior to asphyxiation and with less pain than they might experience if exposed to high concentrations of CO_2 while still fully conscious.

After implementation of these changes, additional controversy was noted regarding the potential for the rodents to experience significant distress associated with the very low volume per minute displacement rates (such as <20%).[35] Studies of "air hunger" (an uncomfortable urge to breathe) supported this concern. This highly unpleasant, even frightening, condition can be induced experimentally in subjects rendered mildly hypercapnic by exposure to CO_2 concentrations of 7%–8%.[32,36] Although the phenomenon of air hunger has not been demonstrated conclusively in animals, there is evidence to suggest that the physiological basis for air hunger is well developed in small rodents.[32,37] Additional studies suggested that volume per minute displacement rates that are more moderate may provide the balance of minimizing potential pain or distress for laboratory rats and mice,[32–34,38] and the 2020 *AVMA Guidelines for the Euthanasia of Laboratory Animals* recommend volume per minute displacement rates of 30%–70%.[1] It should be noted that, while these recommendations are made for small rodents, there are significant differences in the literature regarding the welfare of mice versus rats with the volume per minute displacement rates. This suggests that caution should be taken in extrapolating the data to other rodent species.

Surv. Does your IACUC allow the use of CO_2 as a euthanasia agent for adult rodents? Select all that apply.

• Not applicable, we don't use rodents	4/196 (2%)
• Fully acceptable	169/196 (86%)
• Acceptable only with scientific justification	16/196 (8%)
• Acceptable only if the rodent is anesthetized (e.g. with isoflurane) or sedated prior to euthanasia	7/196 (4%)
• Not acceptable for rodents	5/196 (3%)

18:15 For euthanasia of rodents with carbon dioxide, are there any guidelines regarding proper use of the chamber?

Reg. The *AVMA Guidelines for the Euthanasia of Animals*[1] address general considerations for placing animals in the chamber before euthanasia with CO_2. Both emphasize the importance of avoiding overcrowding and also of minimizing distress that might be caused by mixing animals of different species or animals that are otherwise incompatible or simply unfamiliar with each other. If necessary, animals should be restrained or separated to avoid harm to themselves or others. The AVMA Guidelines further emphasize the need to keep chambers clean "to minimize odors that might cause distress in animals subsequently euthanized."

Opin. For practical purposes, it is generally recommended that the density in the chamber must not exceed the point where the CO_2 can circulate freely, ensuring that all animals will be equally exposed. Although there is no data to support this animal welfare perspective, it has been recommended that all animals have the ability to have all four feet on the floor at the same time. It is preferable to load the animals into the chamber immediately prior to administration of euthanasia. However, if it is necessary to hold animals in the chamber for any interval prior to euthanasia, density should be maintained below the point where the temperature, humidity, and levels of oxygen and CO_2 within the chamber would be adversely affected.

As noted by the AVMA Guidelines, density is only one of the factors that should be taken into consideration to protect animal welfare during CO_2 euthanasia.[1] Equally important are concerns related to social compatibility and other potential stressors to which animals might be subjected by humans seeking to complete an unpleasant task as quickly and efficiently as possible. Laboratory mice and rats exhibit evidence of fear, stress, and possibly distress in response to a variety of environmental conditions and procedures in the vivarium, including noises, pheromones and other odors, handling, and transportation.[1,7,9,11] Mice, in particular, also exhibit signs of stress when combined with unfamiliar conspecifics; depending on sex, strain, and other factors, they may or may not behave aggressively toward each other in such situations.[39,40] Stress due to these factors can be minimized by not mixing animals from different social groupings; transporting and euthanizing animals in the home cage; transporting them on carts that roll smoothly and quietly; and cleaning the euthanasia chamber between groups of animals.[1,5] However, the phenomenon of "stress-induced analgesia" may complicate these recommendations. Although the mechanisms of action have not been characterized, it is well understood that stress is associated with attenuated pain perception, most likely through the opioid pathways.[41] This raises the intriguing question that combination of unfamiliar animals may actually decrease the potential for animals to experience pain or distress, though this question has not been explored to date.

18:16 Is carbon dioxide an appropriate agent for euthanasia of rabbits?

Reg. The *AVMA Guidelines for the Euthanasia of Animals* list CO_2 as acceptable with conditions for the euthanasia of adult rabbits, noting that prior sedation will minimize distress associated with CO_2 euthanasia in this species. The Guidelines also observe that animals of the genus *Oryctolagus* may have prolonged survival times following CO_2 exposure.[1] In contrast, the CCAC does not view CO_2 as either fully or conditionally acceptable for euthanasia of rabbits.[17]

Opin. CO_2 is viewed by many as an effective, but often less preferred, method of euthanasia for rabbits.[1,6] The AVMA Guidelines,[1] for example, note that it is a useful agent but that it appears to cause apprehension or distress in rabbits, although more recent studies have suggested that it can be delivered at volume per minute chamber displacement rates of 28% or 58% without clear signs of distress.[42] In the author's opinion, CO_2 is an acceptable agent for euthanasia of sedated rabbits, but other methods are preferable. When rabbits are exposed to high concentrations of this gas, particularly at first, they tend to struggle and kick, and the potential for injuring themselves or an animal handler is significant. This is a common response of rabbits to any stimulus that causes pain or fear and it is probably not indicative that rabbits experience more discomfort than other animals exposed to CO_2. If scientific or other circumstances necessitate the use of CO_2 to euthanize a conscious rabbit, it is essential that the animal be sedated and properly restrained (e.g., in a commercial rabbit restrainer) to prevent it from kicking. The rabbit should never be placed loose in a euthanasia chamber, where the animal could injure itself if it were to panic during the procedure. If the rabbit is still conscious, the CO_2 should be introduced slowly into the chamber, and appropriate precautions should be taken to assure that the animal is dead prior to disposal.

Surv. Does your IACUC allow the use of CO_2 as a euthanasia agent for rabbits? Select all that apply.

• Not applicable, we don't use rabbits	94/196 (48%)
• Fully acceptable	11/196 (6%)
• Acceptable only with scientific justification	17/196 (9%)
• Acceptable only if the rabbit is anesthetized (e.g. with isoflurane) or sedated prior to euthanasia	19/196 (10%)
• Not acceptable for rabbits	64/196 (33%)

18:17 Is nitrogen or argon—alone, combined, or mixed with carbon dioxide—an appropriate agent for euthanasia of rodents or rabbits?

Reg. The AVMA *Guidelines for the Euthanasia of Animals* view nitrogen and argon as acceptable as sole agents for euthanizing chickens and turkeys, while the CCAC view them as acceptable for euthanizing poultry and swine.[1,17] However, the AVMA Guidelines indicate that high concentrations of nitrogen or argon (>98%) may be used to kill laboratory rodents after they have been rendered unconscious by other acceptable means, further noting that prolonged exposure may be required to ensure death.

Opin. Nitrogen and argon are inert gases that are available commercially in compressed gas cylinders. A mixture of argon (70%) and carbon dioxide (30%) is also commercially available for use in welding. These agents, alone or in combination, have been used to euthanize animals of several species, including dogs,[43,44] rats,[45] mink,[46] swine,[47] and poultry.[48,49] Disturbing reactions, including convulsions, were seen in some species after induction of unconsciousness. Animals of most species would willingly enter hypoxic chambers and overt signs of distress were seldom observed in conscious animals. However, only poultry and swine were willing to remain in hypoxic chambers to the point of unconsciousness. In contrast, several investigators[50–52] noted that rats and mice were reluctant to enter chambers filled with argon or that they left chambers with gradually rising argon concentrations before losing consciousness. Nonetheless, Leach et al.[50] concluded that rodents appeared to find 100% argon less aversive than either carbon dioxide or a mixture of carbon dioxide and argon.

Surv. Does your IACUC approve the use of other gases or gas mixtures (such as isoflurane, oxygen) alone or with carbon dioxide for small rodent euthanasia? Select all that apply.

- Not applicable. 10/196 (5%)
- We have no policy. 17/196 (9%)
- We allow carbon dioxide alone. 104/196 (53%)
- We allow isoflurane (or other halogenated anesthetics) alone. 97/196 (49%)
- We allow various mixtures of carbon dioxide, oxygen, and isoflurane (or other halogenated anesthetics). 43/196 (22%)
- We do not allow the use of other gases alone or in mixture. 17/196 (9%)

Physical Methods

18:18 What type of justification is needed if an investigator states that a small rodent is to be euthanized by cervical dislocation?

Opin. The *AVMA Guidelines for the Euthanasia of Animals* views cervical dislocation of small rodents and rabbits as acceptable with conditions, primarily due to concerns about the ability of personnel to consistently ensure a humane death.[1] This concern was confirmed by a study by Carbone et al, which demonstrated that 21% of the mice that underwent cervical dislocation continued to breathe, resulting in an unsuccessful euthanasia.[53] Indeed, this author also noted a lack of success (not quantified) that was disconcerting when performing a study comparing different methods of euthanasia, despite significant personal experience with cervical dislocation as a secondary method of euthanasia. Methods of euthanasia must be reviewed by the IACUC, which may require justification for use of this technique vs. one that is classified by the AVMA Guidelines as fully acceptable. The best justification would be provision of data—from the Principal Investigator's (PI) own laboratory or the published literature—showing how each acceptable method would interfere with specific scientific objectives using the model in question and also showing that cervical dislocation would not interfere. In lieu of such data on each acceptable method, a plausible rationale might be presented for extrapolating data demonstrating the undesirable effects of one particular method to other similar methods, e.g., from isoflurane to all inhalant anesthetics. In the absence of any existing data, the IACUC has the option of either: (1) accepting a plausible explanation of how fully acceptable methods would be likely to (or might) adversely affect scientific objectives; or (2) requiring the PI to perform a pilot study to generate data. In deciding whether to require the PI to perform a pilot study, the IACUC would need to assess whether the sacrifice of additional animal lives would outweigh the value of the information that would be gained. Regardless, the PI should describe clear experience with this technique and how to assess the success of the euthanasia process to ensure the potential for pain and distress is minimized. If the IACUC is unsure if the experience described is appropriate or sufficient, they can request that the individual demonstrate the technique to an experienced evaluator. In the author's opinion, the most humane option would be to explore alternative methods of euthanasia, but provide whatever training is necessary to ensure the proficiency of the personnel who will perform the procedure if scientifically justified.

Surv. What type of justification does your IACUC require if an investigator proposes to euthanize a rodent unanesthetized by cervical dislocation? Select all that apply.

- Not applicable, we do not use rodents. 4/196 (2%)
- We require no additional justification. 12/196 (6%)
- A general or a specific statement that other means of euthanasia or the use of anesthesia would interfere with the study but references are not required. 68/196 (35%)
- A specific statement that other means of euthanasia or the use of anesthesia would interfere with the study AND includes one or more published references or laboratory data. 103/196 (53%)
- Personnel must complete specialized training to ensure proficiency. 118/196 (60%)
- Other* 18/196 (9%)

*Other:

- Personnel must demonstrate proficiency as a condition for approval.
- We do not allow euthanasia of unanesthetized rodents by cervical dislocation.

18:19 Is cervical dislocation an acceptable method of euthanasia for large rats (> 200 gm) or rabbits (> 1 kg)?

Reg. The *AVMA Guidelines for the Euthanasia of Animals* classify manual cervical dislocation as a conditionally acceptable method for euthanasia of rabbits and for mice and rats weighing less than 200 g.[1] In contrast to the 2007 iteration of the *AVMA Guidelines on Euthanasia*, the 2020 Guidelines do not limit application of this technique to smaller rabbits, although the observation is still made that this procedure is physically more difficult with larger rabbits (and rats) due to their increased cervical muscle mass. The current Guidelines emphasize the importance of assuring technical competence in personnel who will perform this procedure on conscious mature rabbits, and they state that commercially available equipment designed to aid in this procedure should be evaluated for effectiveness. The CCAC classifies cervical dislocation as conditionally acceptable for rodents and rabbits in general, but further specifies that commercial dislocators/luxators must be used on heavier rats (>200 g) and rabbits (> 2 kg).[17]

Opin. It is generally acknowledged that it is physically challenging to perform manual cervical dislocation on larger, more heavily muscled animals. As a result, there is a high potential for severely injuring, without killing, such an animal. The likelihood of this outcome can be significantly reduced by use of a mechanical dislocator, which can help to ensure the technique is consistently applied, and it is only with use of this equipment that the CCAC views cervical dislocation as conditionally acceptable for larger rats or rabbits.[17] The AVMA Guidelines do not recognize any conditions under which cervical dislocation of conscious mature rats might be acceptable, but consider the procedure conditionally acceptable for conscious mature rabbits.[1] While not requiring the use of a mechanical dislocator for larger rabbits, the AVMA Guidelines advise institutions to evaluate the effectiveness of such equipment. An IACUC in the U.S. would have to grant an exception from AVMA Guidelines before an investigator could euthanize a mature rat (>200 g) by cervical dislocation. In the author's opinion, such an exception might be considered—with excellent scientific justification—if the skill of the person performing the technique were unquestioned, if a mechanical dislocator were to be employed, and/or if the animal were to be heavily sedated or anesthetized for the procedure. These same conditions must apply for euthanasia of mature rabbits by cervical dislocation to be performed in accordance with AVMA Guidelines.

Surv. Does your IACUC permit euthanasia of large rats (>200 grams) or rabbits (>1 kg) by cervical dislocation? Select all that apply.

• Not applicable, we don't use rats or rabbits	34/196 (17%)
• Yes, for rats, but not rabbits	38/196 (19%)
• Yes, for rabbits, but not rats	1/196 (1%)
• Yes, both rats and rabbits	5/196 (3%)
• Yes, both rats and rabbits, while awake	2/196 (1%)
• Anesthesia or sedation is required	136/196 (69%)

18:20 Must a moribund mouse be anesthetized prior to euthanasia by decapitation or cervical dislocation?

Reg. Anesthesia *per se* is not required prior to decapitation or cervical dislocation,[1,17] although the *AVMA Guidelines for the Euthanasia of Animals* specifies that animals must be unconscious or anesthetized prior to cervical dislocation in situations where the operator has not demonstrated technical proficiency.[11] On the other hand, the Guidelines specifically and repeatedly equate the moribund condition with unconsciousness or anesthesia, suggesting that the Guidelines would not require anesthesia of a moribund animal prior to cervical dislocation even by a less technically proficient operator.

Opin. The AVMA Guidelines require that small rodents be unconscious or anesthetized prior to cervical dislocation. If this procedure is to be performed on a conscious rodent for scientifically justified reasons, the IACUC must ensure that the individual is appropriately trained. As noted above (see 18:18), this training may be described or demonstrated. This is to protect the welfare of the animal, which might be left seriously injured but not dead following a bungled attempt at cervical dislocation.[53] This is less of a concern with decapitation, which is technically (though, for most people, not psychologically) easier to perform correctly, but concerns for both human safety and animal welfare dictate that animals be rendered unconscious prior to decapitation by untrained individuals. Could an untrained operator perform either cervical dislocation or decapitation as safely and humanely on a moribund mouse as on an anesthetized mouse? While a moribund animal is by definition near death, it is not necessarily unconscious (see 18:11) and might still be able to experience pain or distress. It might even be able to struggle, thereby posing a human safety risk during decapitation by an untrained individual. It is best to assume that the moribund mouse would be able to experience pain and distress, and to assign euthanasia by decapitation or cervical dislocation to a skilled operator. If for any reason the procedure must be performed by an inexperienced individual, the mouse should be deeply sedated or anesthetized beforehand.

18:21 Is it acceptable to use scissors, rather than a guillotine, for decapitation of adult mice?

Opin. The *AVMA Guidelines for the Euthanasia of Animals*[1] notes that "sharp blades"—presumably scissors—can be used for decapitation of neonatal rodents, but the use of scissors for decapitation of adult mice is not addressed by the Guidelines or in any of the U.S. regulations pertaining to euthanasia. The ILAR Committee on Pain and Distress in Laboratory Animals[2] states that decapitation of rodents may be accomplished by using a guillotine or heavy shears. The CCAC does not specifically address the use of scissors for decapitation, but may be alluding to—and tacitly approving?—their use in the following statement: "blades should be kept very sharp and guillotines should be well maintained and cleaned between uses."[18] In the author's opinion, humane euthanasia of adult mice using heavy, well-sharpened scissors is possible, provided that the operator is highly skilled in this technique. However, this procedure is exceedingly distasteful for most people, and this can lead all but the most resolute and experienced operator to perform the procedure less quickly and cleanly than is required for humane euthanasia. For this reason, use of scissors for decapitation is best limited to euthanasia of neonates or for adult mice that have been heavily sedated or anesthetized. People who wish to use this approach to euthanize fully conscious adult mice should be especially skilled and experienced.

18:22 When reviewing a protocol that involves euthanasia of mice by cervical dislocation or decapitation, how should the IACUC determine that personnel performing these techniques are properly trained?

Reg. PHS Policy (IV,C,1,f), the AWAR (§2.31,d,1,viii; §2.32), and the *Guide* (pp. 16–17, 124) emphasize that personnel performing procedures on animals must be properly trained. In all instances, euthanasia procedures are explicitly or implicitly included in these requirements (see Chapter 24. The *AVMA Guidelines for the Euthanasia of Animals* and CCAC specifically emphasize the need for personnel to be properly trained prior to performing euthanasia by decapitation or cervical dislocation.[1,17]

Opin. Although PHS Policy (IV,C,1,f), the AWAR (§2.31,d,1,viii), and the *Guide* (p. 15) specify that the IACUC is responsible for assuring that personnel are appropriately trained and that training programs are effective, none of these documents specifies the basis upon which the IACUC would determine that a particular individual has been properly trained. The *AVMA Guidelines for the Euthanasia of Animals* state that inexperienced persons should practice on dead or anesthetized animals until they are proficient in performing these procedures properly and humanely.[1] The CCAC also emphasizes the importance of assuring that personnel are trained in the humane performance of these procedures, as well as in the safe use of equipment for decapitation.[17] As for who should provide the training, the AWAR (§2.33,b,5) places guidance on euthanasia under the program of veterinary care and §2.32 makes "the research facility" responsible for providing training. The *Guide* (p. 16) states only that it is an institutional responsibility to assure that training is provided. The ILAR Committee on Pain and Distress in Laboratory Animals is bit more specific, maintaining that the IACUC and AV should provide for the training and supervision of personnel who will perform euthanasia.[6] Training by the AV or IACUC-approved institutional training program provides the IACUC with the highest degree of certainty that personnel are properly trained. An alternative approach is to have the PI or other skilled individual perform the training. In this case, direct observation of the trainee's skill by the AV, institutional trainer, or other experienced member of the IACUC is desirable prior to approving him/her to perform the procedure unsupervised.

Ideally, the skills of trained personnel would be reassessed or reapproved on a regular basis—every 2–5 years, for example. Trained personnel would be evaluated in a less formal manner as part of the post-approval monitoring process. Should any concerns arise regarding the proficiency of a particular operator, it would be incumbent upon the IACUC to require an assessment of their skill by the AV, institutional trainer, or other IACUC-approved individual.

Surv. When reviewing a protocol that involves the euthanasia of mice by cervical dislocation or decapitation, how does your IACUC determine that personnel performing the procedure are properly trained to do so? Select all that apply.

• Not applicable.	18/196 (9%)
• We rely on the investigator to provide assurance that personnel are properly trained.	123/196 (63%)
• We ask the veterinary staff to make appropriate observations.	93/196 (47%)
• We have a post-approval monitoring program that periodically checks techniques.	84/196 (43%)
• We require pre- or post-approval training.	71/196 (36%)
• Other.	17/196 (9%)

Adjunctive Methods—Stunning, Pithing, Exsanguination (and see 17:8)

18:23 An investigator states that she has many years of experience euthanizing rats by swinging them rapidly by the tail and hitting their heads against the edge of a table. Should the IACUC allow this procedure, if properly performed, as an acceptable form of euthanasia?

Reg. The *AVMA Guidelines for the Euthanasia of Animals* acknowledges that manually applied blunt force trauma to the head can be an effective and humane method of euthanasia when it is performed correctly on animals with thin craniums.[1] However, humane concerns can arise when it is performed repeatedly or on an inappropriate subject, and it can be aesthetically displeasing to human personnel. As a result, the AVMA Guidelines recommend that those who use this technique actively search for alternative methods. Concussion (crushing blow to the head) is recognized by the CCAC as conditionally acceptable under emergency conditions only.[17] The European Commission states that concussion by a trained individual is allowable for smaller animals, including rodents, to render the animal insensible prior to exsanguination or removal of the heart or brain (https://publications.europa.eu/resource/cellar/b0ecd88f-8c31-4a5f-9ebd-2c8ad4da4c14.0001.02/DOC_1#:~:text=The%20primary%20criteria%20for%20euthanasia,in%20the%20animal%2C%20be%20reliable%2C).

Opin. Cranial trauma created by swinging the animal and hitting the back of its head on the edge of a table is an old method that was used primarily to euthanize small laboratory animals with thin craniums; it is seldom mentioned in more recent literature. However, this technique has been described as a humane method of euthanasia or as a method of rendering animals unconscious prior to exsanguination or cervical dislocation in small animals.[17] The need for training of personnel is emphasized by the *AVMA Guidelines for the Euthanasia of Animals*, which also emphasize the need to monitor personnel for proficiency and assure that they are aware of the aesthetic implications of the technique.[1] Stunning by hitting the animal's head on the edge of a table requires a high level of concentration in addition to considerable technical proficiency. It takes only a small deviation in the force or angle of the blow to result in severe injury rather than instant death. Inconsistency in the performance of the technique is more likely as personnel become fatigued after performing the procedure repeatedly. Furthermore, swinging the animal by the tail is likely to produce distress, even if collision of the head were to produce rapid death. In the author's opinion, the IACUC should allow use of this method only when it is well justified and the committee is satisfied that other, preferable alternatives cannot be used.

Surv. Does your IACUC accept stunning of rodents, if properly performed, as an acceptable form of euthanasia?

- Not applicable, we do not use rodents. 4/196 (2%)
- We have no policy on stunning. 58/196 (30%)
- Acceptable with scientific justification. 9/196 (5%)
- Acceptable without scientific justification. 2/196 (1%)
- Not acceptable. 123/196 (63%)

18:24 Is pithing an acceptable method for euthanasia of conscious frogs?

Reg. The *AVMA Guidelines for the Euthanasia of Animals* strongly discourages the use of pithing for euthanasia of frogs and other amphibians unless the animal is anesthetized or otherwise unconscious; personnel must be properly trained in the technique.[1] Similarly, the CCAC indicates that pithing may be used to render frogs brain dead, but the animals must be at a surgical plane of anesthesia for the procedure.[17]

Opin. Simple pithing—destruction of the brain only—does not result in the death of the frog (the heart remains beating, respirations continue, and many reflexes remain intact) and is not viewed as an effective or acceptable method of euthanasia. Double pithing, which involves destruction of the brain followed immediately by destruction of the spinal cord, is required to assure death. Pithing requires considerable technical proficiency; if the person performing the technique is not skilled, the animal could experience considerable pain and suffering.

18:25 Is sedation (e.g., with diazepam or acepromazine) an acceptable alternative to anesthesia for pretreatment of animals prior to exsanguination?

Reg. The *AVMA Guidelines for the Euthanasia of Animals* state that exsanguination may be used as an adjunctive method to ensure death in animals that have been stunned or are otherwise unconscious.[1] The CCAC is less equivocal, stating that animals must be unconscious prior to exsanguination and that sedation is inadequate to ensure unconsciousness.[17]

Opin. While there is no reason to believe that animals would experience significant pain associated with competently performed exsanguination via a needle or cannula placed in a peripheral vein, it is likely that animals would experience considerable distress associated with hypovolemia-induced anxiety.[1,6] It is not clear whether sedation is adequate to eliminate this distress. In the author's opinion, anesthesia should be required unless compelling scientific justification is presented to explain why anesthetic drugs, but not a particular sedative drug, would interfere with scientific objectives. There is no question that an animal should be anesthetized prior to an exsanguination procedure that involves creation of a surgical incision. Exsanguination by cardiac puncture also requires anesthesia.[1]

18:26 Must a moribund rabbit be anesthetized prior to euthanasia by exsanguination?

Reg. See 18:20 and 18:25.

Opin. This question hinges in part on whether a moribund rabbit would be capable of perceiving pain and distress. As discussed in 18:12 and 18:20, it cannot be assumed that a moribund animal is incapable of perceiving pain and distress, and it is generally best to proceed under the assumption that a moribund animal could experience pain and distress. Were the rabbit healthy and fully alert, anesthesia would be required prior to exsanguination (see 18:25). However, there is an additional issue to consider in the case of a moribund animal. An animal near death might be unable to tolerate a general anesthetic and could die before or early in the exsanguination procedure, possibly limiting the amount of blood that could be collected. This could have a negative impact on the research if it were necessary to collect a large quantity of blood for use in the research project. In such a situation, the need to minimize distress in the rabbit would have to be balanced with the requirements of the study. Provided that the blood was to be collected via a needle or cannula placed in a peripheral vein, light anesthesia or sedation might be more appropriate than deep anesthesia. The justification would have to be more compelling to consider waiving the need for anesthesia prior to cardiac puncture, and even a moribund animal should be anesthetized prior to making any kind of surgical incision.

18:27 How should death be ensured prior to disposal of animals killed by methods such as carbon dioxide or inhalant anesthetics?

Opin. It is essential that death be verified before disposal of an animal that has been euthanized.[1,3,6,17] This is especially important when animals are euthanized by techniques that can cause loss of consciousness or cessation of visible respirations well before death occurs, as these animals are more prone to regain consciousness after disposal. The potential for regaining consciousness following disposal is a particular concern with CO_2 asphyxiation.[6,54] Maintaining the animal in the CO_2-rich environment for a period of time after the animal appears dead can help to minimize this potential, as can observation of the animal for a few minutes after re-exposure to room air.[1,5,6,54] Death can be verified by examining the animal for cessation of a combination of vital signs, including heartbeat, respiration, and corneal reflex,[1,5,6] but the only sign that definitively confirms death is rigor mortis.[1] Another way to assure that the animal is dead is to follow up with a physical method once the animal appears dead or unconscious. Reliable follow up methods include decapitation, cervical dislocation, thoracotomy, or exsanguination or removal vital organs.[1,5,6]

Surv.1 Does your IACUC require using a second "confirmatory" method of euthanasia to assure a euthanized animal is truly dead?

• Not applicable, we don't euthanize animals.	0/196 (0%)
• Yes, we have such a policy.	154/196 (79%)
• Yes, we have such a policy but only for certain animal species.	10/196 (5%)
• No, we do not have such a policy.	26/196 (13%)
• Other.	6/196 (3%)

Surv.2 If your IACUC requires a confirmatory physical method to assure that an animal is dead, which of the following methods are acceptable? Select all that apply.

• Not applicable, we do not require a second method	24/196 (12%)
• Bilateral thoracotomy	133/196 (68%)
• Decapitation	136/196 (69%)
• Cervical dislocation	154/196 (79%)
• Exsanguination	130/196 (66%)
• Removal of a vital organ	107/196 (55%)
• Other*	26/196 (13%)

*Other:

- Additional, prolonged exposure to carbon dioxide.
- Cardiac perfusion
- Confirmation of death determined by presence of rigor mortis or absence of breathing.
- Pithing for fish/amphibians/reptiles
- Freezing
- Thoracic compression

Peri-natal and Neonatal Rodents

18:28 What agents or techniques are most appropriate for euthanasia of neonatal altricial rodents (e.g., rats and mice)?

Reg. The *Guide* (p. 124) states that methods such as injection of a chemical agent, cervical dislocation, or decapitation should be considered as alternatives to CO_2 for euthanasia of neonatal rodents.[3] While CO_2 is not ruled out as a potentially acceptable method, neonatal rodents will require longer exposure times than adults to ensure death, especially if non-haired. The *AVMA Guidelines for the Euthanasia of Animals* echo these concerns, noting that non-haired neonatal mice may take up to 50 minutes to die from CO_2 exposure.[1] Intraperitoneal (IP) injection of pentobarbital is the recommended method for euthanasia pre-weanling small mammals. Conditionally acceptable methods include injectable dissociative anesthetics combined with α-adrenergic receptor agonists or benzodiazepines; nonflammable inhalant anesthetic gases; CO_2; hypothermia; immersion in liquid nitrogen; decapitation; and cervical dislocation.[1]

Opin. There are few published guidelines that specifically address euthanasia of neonatal animals. The few specific recommendations that do exist emphasize the fact that newborn animals are typically more resistant to hypoxia and more capable of coping with high environmental CO_2 than are older animals.[1,5,17] For this reason, it is often recommended either that inhalant agents (e.g., inhalant anesthetics, CO_2, carbon monoxide) not be used to euthanize non-haired neonates or that these agents be used only when the prolonged exposure times necessary to ensure death can be emplo yed.[1,5,17,55–57] Klaunberg and colleagues found that CO_2 was more effective than halothane for euthanasia of neonatal mice up to 7 days of age, and that CO_2 produced more rapid euthanasia of older (haired) preweanlings than either halothane or intraperitoneal pentobarbital.[58] Other methods that have been specifically identified as acceptable or conditionally acceptable for euthanasia of neonatal rats and mice include decapitation, cervical dislocation, anesthetic overdose (prolonged exposure times may be needed with inhalant anesthetics), or anesthesia followed by perfusion with chemical fixatives.[1] Hypothermia can be used instead of injectable or inhalant anesthetics to anesthetize pups <6 days of age prior to euthanasia by another method.[1]

When choosing a method of euthanasia for a neonate, it should be kept in mind that neonates, including neonatal altricial rodents, appear to be at least as capable of perceiving pain as adult animals.[6,17,59–61]

In the author's opinion, the most humane methods of euthanasia for neonatal mice or rats ≤10 days of age are IP injection of pentobarbital, cervical dislocation, or decapitation with heavy, sharp scissors. Sedation or anesthesia is not necessary prior to decapitation or cervical dislocation if the procedure is performed correctly by a skilled operator (see 17:20). Decapitation and cervical dislocation are aesthetically upsetting for many people, so CO_2 asphyxiation may be a more desirable alternative under circumstances where pentobarbital is not available or practical. Prolonged exposure of the pups to CO_2 (and maybe also to inhalant anesthetics) is essential, as is verification of death prior to disposal.

Surv. What methods of euthanasia are acceptable for the euthanasia of neonatal rats and mice? Select all that apply.

Carbon dioxide alone, with a prolonged well time	59/196 (30%)
Carbon dioxide followed with a physical method	137/196 (70%)
• Decapitation awake	86/196 (44%)
• Decapitation while under anesthesia	125/196 (64%)
• Cervical dislocation awake	57/196 (29%)
• Cervical dislocation while under anesthesia	83/196 (43%)
• Injectable anesthetic overdose	83/196 (42%)
• Hypothermia alone	16//196 (8%)
• Hypothermia followed by a physical method	77/196 (39%)
• Other*	24/196 (12%)

*Other:

- Not applicable to my institution.
- Anesthesia followed by exsanguination
- Prolonged exposure to inhalant anesthetic
- Rapid freezing in liquid nitrogen

18:29 Is carbon dioxide an appropriate agent for euthanasia of neonatal rodents? Are conditions for its use different for neonates than for adults?

Reg. Noting that neonatal rodents are resistant to the hypoxia-inducing effects of CO_2 and that they therefore require longer exposure times for euthanasia with this gas, the *Guide* (p. 124) states that alternative methods should be considered, e.g., injection with chemical agents, cervical dislocation, or decapitation. The *AVMA Guidelines for the Euthanasia of Animals* recommends IP pentobarbital as the preferred method for euthanasia of unweaned small mammals, but views CO_2 as conditionally acceptable for this purpose.[1]

Opin. As noted above, the *AVMA Guidelines for the Euthanasia of* Animals view CO_2 as not preferred but conditionally acceptable for the euthanasia of small mammals.[1] The need for a prolonged period of exposure is emphasized. The CCAC and ACLAM Task Force on Rodent Euthanasia also address the use of CO_2 in neonatal rats and mice; both define neonates as animals up to 10 days of age.[5,17] CCAC classifies CO_2 as only conditionally acceptable even for adult rodents and emphasizes that the time to death may be considerably longer in neonates.[17] The need for prolonged exposure times is equally emphasized by ACLAM, which also stresses the need to verify that the pups are dead prior to disposal.[1,5] In fact, time to death may exceed 35 minutes in neonatal rats and 50 minutes in neonatal mice exposed to 100% CO_2[55,56] and isoflurane.[57] Resistance to CO_2 decreases steadily during the first 10 days after birth, but even 10-day-old rats and mice require longer exposure times than adults of the same species.[55–58] During the first 7–8 days of life, inbred mice took longer than outbred mice to die following CO_2 exposure[55]; this difference was not demonstrated for rats.[56]

In the author's opinion, CO_2 is an acceptable but not preferred method of euthanasia for neonatal rats and mice ≤6 days of age (see 18:28). If CO_2 will be used to euthanize neonates of these species, prolonged exposure times are essential. To ensure death, recommended exposure times are 45 minutes for rat pups up to 6 days of age and 10–20 minutes for older pre-weanlings. For mice, recommended exposure times are 60 minutes for pups up to 6 days of age and 15–20 minutes for older pre-weanlings. As with adult animals euthanized with CO_2, it is essential to verify that the pups are dead prior to disposal.

Surv. If your IACUC allows the use of carbon dioxide for the euthanasia of mice, are there different requirements for neonates?

• Not applicable.	32/196 (16%)
• We have no different requirements.	13/196 (7%)
• We require a longer exposure time for neonates.	120/196 (61%)
• We recommend but do not require a longer exposure time for neonates.	6/196 (3%)
• I don't know.	8/196 (4%)
• Other.	17/196 (9%)

18:30 Is decapitation with sharp scissors an acceptable method for euthanasia of 1–2 day old rats or mice?

Reg. The *AVMA Guidelines for the Euthanasia of Animals* note that "sharp blades" can be used in place of a guillotine for decapitation of neonatal rodents.[1] The CCAC does not mention the use of scissors for decapitation of neonates, but does note that "sharp, well-maintained" scissors are acceptable for the decapitation of rodent fetuses.[17]

Opin. Decapitation is viewed as a conditionally acceptable method of euthanasia for rats and mice of any age.[1,17,18] While guillotines are generally recommended for decapitation of adults (see 18:21), scissors are an acceptable alternative for neonates and late-term fetuses.[1,5,17] In the author's opinion, decapitation with heavy, sharp scissors is a preferred method of euthanizing neonatal rats and mice (see 18:28). If scissors are used for this purpose, it is essential that they be kept clean and sharp. It is also essential that the operator be experienced in this procedure and able to decapitate the pup with a single, swift, decisive cut.

Surv. Does your IACUC approve decapitation with sharp scissors as an acceptable method of euthanasia of neonatal rats or mice?

• Not applicable, we do not use mice or rats.	9/194 (5%)
• Yes, it is acceptable, with age limits, without sedation.	104/194 (54%)
• Yes, it is acceptable, but only if the animal is sedated.	57/194 (29%)
• No, it is not acceptable at any age.	24/194 (12%)

18:31 If a pregnant mouse or rat dies or is euthanized, must additional steps be taken to ensure death of the fetuses?

Reg. The *Guide* (pp. 123–124) states that special consideration should be given to euthanizing fetuses depending on species and gestational age but provides no further guidance on what this special consideration might entail. The *AVMA Guidelines for the Euthanasia of Animals* and the CCAC suggest that no additional steps are needed to ensure death of fetuses during the first two trimesters of gestation; but, during the last trimester, the fetuses must be euthanized separately following euthanasia of the dam.[1,17]

Opin. Based on neural development, many have concluded that it is plausible, even likely, that humans and other mammals, including rats and mice, are able to perceive pain prior to parturition.[62–65] Although some question remains whether the capacity to experience pain truly exists before birth, there is accumulating support for the belief that human fetuses are able to experience pain during the third trimester of pregnancy, as early as 26 weeks of gestation.[64,65] The evidence is not as robust for rodents, but rat and mouse fetuses exhibit neural and behavioral responses to painful and other types of somatosensory stimulation during late gestation, as early as embryonic day 17 in the rat.[58,62,64,65] On the basis of this evidence, it is now widely recommended that late-term rat and mouse fetuses be handled in a manner that minimizes the potential for pain or distress. To quote the CCAC: "during the final third of gestation, fetuses should be given the same ethical considerations as apply to the fully mature animal."[17] While the AVMA Guidelines have taken a sharply different stand on this issue, many believe that additional steps should be taken following the death of the mother to protect the welfare of the fetuses. In the author's opinion, it is preferable to err on the side of caution until this issue can be decided more definitively. Appropriate methods of euthanasia include those recommended for neonatal rats and mice (see 18:28). There are no data or recommendations suggesting that special precautions are needed to minimize pain or distress in fetuses during the first two trimesters of gestation.

Field Studies

18:32 Is it acceptable to use inhalant gases to euthanize small rodents in the field?

Reg. The *Guide* (p. 32) is largely silent on this question except to comment that, in general, issues associated with euthanasia in the field are similar, if not identical, to those associated with euthanasia in the laboratory. The AWAR (Sect. 1.1) define a field study as a "study conducted on free-living wild animals in their natural habitat." Studies that involve invasive procedures or harm or alter the behavior of an animal are not field studies, according to the AWAR, so, §2.31,d,1,xi would apply, and the IACUC should consider the same issues as those associated with euthanasia in the laboratory.

The *AVMA Guidelines for the Euthanasia of Animals* describe the use of inhalant anesthetics, CO_2, carbon monoxide (CO), and other inert gases as conditionally acceptable for the euthanasia of free-ranging wild rodents, noting that the same conditions must be met for use of these agents as would apply for their use with domestic animals.[1] The CCAC also views CO_2 or CO as acceptable agents for use in the field, but emphasizes the need for a closed chamber and attention to personnel safety.[66] The CCAC also endorses the use of volatile anesthetic gases, except in the case of species that are able to hold their breath for long periods of time; however, ether and nitrous oxide are not recommended for euthanasia of wildlife.

Opin. Techniques of euthanasia that are preferred in the laboratory also may be applicable in the field. However, the *Guide* (p. 32), *AVMA Guidelines for the Euthanasia of Animals*, and CCAC all observe that there are situations in which it may be necessary to approach euthanasia in the field differently than in the laboratory.[1,66] Different methods may be mandated because of feasibility, human safety, animal welfare, or animal safety. Research objectives may limit the use of some methods or agents for wildlife species, but humane concerns must override mere convenience when selecting a method.[1] Provided that proper equipment is available and precautions are taken to protect personnel from harm, inhalant gases are acceptable for euthanasia of small rodents in the field. Note, however, that the CCAC specifically advises against the use of nitrous oxide or ether for this purpose as these gases are combustible and/or explosive and have potential for human toxicity and abuse.[66]

Surv. Does your IACUC allow an overdose of inhalant gases as a primary means of euthanasia for small rodents captured in field research?

• Not applicable, we do not conduct studies in the field.	99/194 (51%)
• Yes, we allow the use of isoflurane or other halogenated anesthetics.	61/194 (31%)
• Yes, we allow the use of diethyl ether.	2/194 (1%)
• No, we do not allow the use of inhalant gases in the field.	19/194 (10%)
• Other*	13/194 (7%)

*Other:

- Respondents also noted that secondary methods are required.
- Chloroform was used to euthanize small rodents in the field.

18:33 Is thoracic compression an acceptable method of euthanasia for wild rodents trapped in field studies?

Reg. The use of thoracic compression as a method of euthanasia is not addressed in the AWAR, PHS Policy, or the *Guide*. The *AVMA Guidelines for the Euthanasia of Animals* state that, under rare circumstances, thoracic compression might be used to euthanize animals that are deeply anesthetized or otherwise unconscious, or as a final confirmatory procedure to ensure death in cases where the animal's status is uncertain.[1]

Opin. In contrast to the *AVMA Guidelines for the Euthanasia of Animals*, the American Society of Mammalogists (ASM) views thoracic compression as an acceptable alternative for euthanasia of small mammals under circumstances where use of injectable pharmaceutical agents or inhalant gases may be stressful for the animals or pose risks to investigators.[67] ASM notes that thoracic compression has been used effectively for decades by practicing mammologists.[67] Sikes et al. note that thoracic compression appears to be painless and that it maximizes the use of the carcass.[67] Hickman and Johnson demonstrated that thoracic percussion (technique similar to thoracic compression) led to rapid exsanguination in anesthetized rats.[14] In this author's opinion, just as for the other physical methods of euthanasia, this method can be considered if scientifically justified and performed by individuals who have been trained in how to consistently perform the technique and recognize when the technique has failed so that a secondary method of euthanasia can be rapidly applied to minimize the potential for pain or distress.

Surv. Does your IACUC consider thoracic compression an ethical primary form of euthanasia for wild small rodents or birds trapped in field research?

Not applicable, we don't conduct field research 100/194 (52%)

Acceptable with scientific justification 43/194 (22%)

Acceptable without scientific justification 6/194 (3%)

Not acceptable 33/194 (17%)

Other* 12/194 (6%)

*Other:

Thoracic compression would be acceptable only for anesthetized animals.

18:34 Sometimes animals are unintentionally injured during field studies. Is it necessary for the investigator to euthanize these animals?

Reg. Except for situations in which live, wounded animals are found in kill traps, the issue of how to deal with animals that are injured during field studies is not specifically dealt with in the federal regulations or policies. However, the AWAR (§2.33,b,2) require "The use of appropriate methods to prevent, control, diagnose, and treat diseases and injuries." This requirement also applies to animals in those field studies that are not exempted from regulation (based on the AWAR (§1.1) definition of "field study"). Traps must be checked at least once daily and any trapped animals that are not dead must be killed quickly and humanely.[1] The CCAC addresses this issue more broadly, stating that the investigator must euthanize any animal in the field that is suffering unrelievable pain or distress as a result of capture, handling, or experimental manipulations.[66]

Opin. Both the AVMA Panel on Euthanasia and CCAC emphasize that challenges encountered during field research do not reduce the ethical obligation of the responsible individual to reduce pain and distress during the performance of euthanasia, in addition to consider the potential negative effects on other animals that may encounter the carcass.[1,66] This obligation is echoed by the American Society of Mammalogists, which further states that protocols involving fieldwork should explicitly describe circumstances and methods for euthanizing injured or distressed animals, even for studies where injury or death is not an anticipated outcome.[67] The *AVMA Guidelines for the Euthanasia of Animals* address in more general terms the performance of euthanasia to improve animal welfare.[1] For an animal that no longer enjoys good welfare—defined as feeling well, functioning well, and able to perform innate behaviors—the humane alternative is to give it a good death. While this discussion in the Guidelines centers primarily around companion animals, it could apply equally well to wild animals used in field studies. In the author's opinion, investigators have an ethical obligation to humanely kill any animals that would otherwise experience unrelieved pain or distress, or would be more vulnerable to predation, as a result of the investigative team's actions.

Surv. Does your IACUC require that injured animals (e.g., from investigator trapping) be treated or euthanized?

- Not applicable, we don't conduct field research. 94/194 (48%)
- Yes, in all cases. 36/194 (19%)
- Typically, yes, but the IACUC may consider not treating or euthanizing on a case-by-case basis (e.g., permit language). 62/194 (32%)
- No, the IACUC does not require injured wild animals to be treated or euthanized.

2/194 (1%)

Other Euthanasia-Related Questions

18:35 Which methods of euthanasia are acceptable for euthanizing zebrafish?

Reg. The *AVMA Guidelines for the Euthanasia of Animals* lists several methods as acceptable for the euthanasia of fish in general.[11] Of these, the following are acceptable for zebrafish in research facilities: immersion into buffered tricaine methane sulfonate (MS222, tricaine, TMS), buffered benzocaine, quinaldine sulfate, ethanol, or 2-phenoxyethanol; and rapid chilling in an ice-water slurry. In addition, rapid chilling followed by immersion in a dilute sodium hypochlorite or calcium hypochlorite solution is acceptable for zebrafish embryos and larvae. Methods that are conditionally acceptable for zebrafish in research settings include immersion in CO_2-saturated water, eugenol, isoeugenol, or clove oil; decapitation followed by pithing; cranial concussion followed by pithing; and maceration. The CCAC guidelines are similar in some respects but different in others (see below).

Opin. Both the *AVMA Guidelines for the Euthanasia of Animals* and CCAC guidelines on euthanasia recommend immersion in tricaine methane sulfonate (MS222, tricaine, TMS) or benzocaine as preferred methods for euthanizing fish.[1,18] Other methods that are viewed as acceptable or conditionally acceptable by both AVMA and CCAC are immersion in clove oil and maceration.[1,17] Cranial concussion also is listed as conditionally acceptable by both guidelines, but the AVMA Guidelines further specify that this must be followed by pithing.[1,17] Listed as acceptable or acceptable with conditions by AVMA but not addressed by CCAC are: immersion in quinaldine sulfate or 2-phenoxyethanol; rapid chilling in an ice-water slurry; rapid chilling followed by immersion in a dilute sodium hypochlorite or calcium hypochlorite solution is acceptable (for zebrafish embryos and larvae); immersion in CO_2-saturated water, eugenol, or isoeugenol; and decapitation followed by pithing. Conversely, immersion in etomidate is listed as acceptable by CCAC[17] but is not addressed by AVMA, and exsanguination under anesthesia is listed as conditionally acceptable by CCAC[68] but is not addressed by AVMA. The biggest difference in these two sets of guidelines is in relation to metomidate, immersion in which is listed as an acceptable method of euthanasia by CCAC,[17] but which is viewed as an unacceptable method by AVMA.[1]

Surv. 1 Which methods of euthanasia are generally acceptable to your IACUC for euthanizing zebrafish? Select all that apply.

• Not applicable, we do not use zebrafish	85/194 (44%)
• An overdose of MS-222	105/194 (54%)
• Rapid immersion in ice water	65/194 (34%)
• Decapitation	38/194 (20%)
• Sodium pentobarbital	14/194 (7%)
• Benzocaine HCl	16/194 (8%)
• Clove oil	10/194 (5%)
• Hypercarbonated water (CO2)	5/194 (3%)
• Ethanol	3/194 (2%)
• Other	9/194 (5%)

Surv. 2 Does your IACUC have a policy that describes expectations for the euthanasia of invertebrate animals?

- Not applicable, our institution does not use invertebrate animals (e.g., nematodes, fruit flies, etc.). 93/194 (48%)
- The IACUC does not have a policy regarding oversight of invertebrate species. 65//194 (34%)
- The IACUC does not have an official policy, but it would cover invertebrates if the institution ever procured any cephalopods. 16/194 (8%)
- We have such a policy, and it states the IACUC does not oversee any invertebrate species. 8/194 (4%)

We have such a policy, but it is limited to a subset of specific invertebrates (e.g. cephalopods). 12/194(6%)

18:36 Invertebrates are not regulated under the AWAR and PHS Policy. Is it necessary for the IACUC to review methods of euthanasia for invertebrates used in research?

Opin. Invertebrates are not regulated under the AWAR or PHS Policy. AAALAC states that invertebrates used in research are covered as part of an accredited program "where they are relevant to the unit's mission."[69] However, their use in research does not necessarily require formal IACUC review and approval. The level of IACUC oversight, and interest in review of the research by AAALAC site visitors, may vary depending on the species of invertebrate, with greater oversight expected for higher-level species such as octopi than for lower-level species such as sea slugs. Recognizing the importance of invertebrates in research, as display animals, and even as companion animals, the 2013 iteration of the *AVMA Guidelines for the Euthanasia of Animals* includes a section on euthanasia of invertebrates.[1] The Guidelines acknowledge the uncertainty regarding the ability of invertebrates to experience pain, distress, or compromised welfare, but assume that a conservative and humane approach to the care of any animal is warranted and expected by society. Guidelines also exist in the published literature for the provision of anesthesia and analgesia for invertebrates.[70–72]

In the author's opinion, IACUC review of activities involving invertebrates should, at a minimum, assure that the housing and care of the animals will be appropriate for their species; the maintenance and research use of these animals will not pose any threat to the health or safety of humans or other animals; and methods of euthanasia of these animals will be consistent with the AVMA Guidelines unless an exception is justified. In the case of a invertebrates, IACUC oversight might consist of nothing more than the review of a letter from the investigator describing the intended research use of the animals, the provisions for housing and care of the animals, any appropriate precautions to prevent escape of the animals or other unintended impact on humans or other animals, methods that will be taken to prevent unnecessary pain or distress in the animals, and the method of euthanasia that will be used at the end of the study.

18:37 On what basis might an IACUC approve use of a method of euthanasia that is not reviewed in the *AVMA Guidelines for the Euthanasia of Animals*?

Opin. It is acknowledged within the report itself that circumstances may arise that are not covered by the *AVMA Guidelines for the Euthanasia of Animals*.[1] In such situations, it is recommended that an appropriate method be chosen by a veterinarian who has experience with the species to be euthanized and who will apply professional judgment based on the animal's size, physiology, and behavior. Within a research setting, the method may actually be proposed by the investigator, but the IACUC should rely heavily on the advice of the AV (or other veterinary expert) in deciding whether to approve the method. Additional pertinent information on the method may be found in the published literature or in euthanasia guidelines developed by other professional bodies, e.g., the CCAC or International Council for Laboratory Animal Science (ICLAS). Insights might also be gained from colleagues at other institutions who are familiar with the method in question. As approval is being deliberated, all members of the IACUC should consider whether the method will satisfy the minimum requirements for a humane death: rapid loss of consciousness without pain, distress, or fear; and reliable progression to death following loss of consciousness. The availability of personnel with adequate expertise to apply the method skillfully should also be considered, as well as potential effects on the health, safety, and emotional responses of human personnel and other animals (see also 18:1).

REFERENCES

1. Association AVM. AVMA Guidelines for the Euthanasia of Animals: 2020 Edition. 2020. Available from: https://www.avma.org/sites/default/files/2020-02/Guidelines-on-Euthanasia-2020.pdf. Accessed September 28, 2022.
2. Care CCoA. Guidelines on: Choosing an Appropriate Endpoint in Experiments Using Animals for Research, Teaching, and Testing. 1998. Available from: https://ccac.ca/Documents/Standards/Guidelines/Appropriate_endpoint.pdf. Accessed December 31, 2022.
3. Guide for the Care and use of Laboratory Animals. 2011. Available from: https://grants.nih.gov/grants/olaw/guide-for-the-care-and-use-of-laboratory-animals.pdf. Accessed December 31, 2022.
4. Animal and Plant Health Inspection Service U. Animal Welfare Inspection Guide. 2021. Available from: https://www.aphis.usda.gov/animal_welfare/downloads/Animal-Care-Inspection-Guide.pdf. Accessed December 31, 2022.
5. Shomer NH, Allen-Worthington KH, Hickman DL, *et al.* Review of rodent euthanasia methods. *J Am Assoc Lab Anim Sci* 2020;59(3):242–253, doi:10.30802/aalas-jaalas-19-000084.
6. Recognition and Alleviation of Pain in Laboratory Animals. *National Research Council (US) Committee on Recognition and Alleviation of Pain in Laboratory Animals.* Washington (DC): National Academies Press (US); 2009. Recognition and Alleviation of Pain in Laboratory Animals - NCBI Bookshelf (nih.gov).
7. Castelhano-Carlos MJ, Baumans V. The impact of light, noise, cage cleaning and in-house transport on welfare and stress of laboratory rats. *Lab Anim* 2009;43(4):311–327, doi:10.1258/la.2009.0080098.
8. Sharp J, Zammit T, Azar T, *et al.* Does witnessing experimental procedures produce stress in male rats? *Contemp Top Lab Anim Sci* 2002;41(5):8–12.
9. Sharp JL, Zammit TG, Azar TA, *et al.* Stress-like responses to common procedures in male rats housed alone or with other rats. *Contemp Top Lab Anim Sci* 2002;41(4):8–14.
10. Sharp J, Zammit T, Azar T, *et al.* Are "by-stander" female Sprague-Dawley rats affected by experimental procedures? *Contemp Top Lab Anim Sci* 2003;42(1):19–27.
11. Sharp J, Zammit T, Azar T, *et al.* Stress-like responses to common procedures in individually and group-housed female rats. *Contemp Top Lab Anim Sci* 2003;42(1):9–18.
12. Boivin GP, Bottomley MA, Grobe N. Responses of male C57BL/6N mice to observing the euthanasia of other mice. *J Am Assoc Lab Anim Sci* 2016;55(4):406–411.
13. Baker BA, Hickman DL. Bias in rating of rodent distress during anesthesia induction for anesthesia compared with euthanasia. *J Am Assoc Lab Anim Sci* 2018;57(2):143–156.
14. Hickman DL, Johnson SW. Evaluation of the aesthetics of physical methods of euthanasia of anesthetized rats. *J Am Assoc Lab Anim Sci* 2011;50(5):695–701.
15. National Institutes of Health OoLAW. Frequently Asked Questions A.3, *Does the PHS Policy Apply to the Use of Animal Tissues Or Materials Obtained from Dead Animals?* Available from: https://olaw.nih.gov/faqs#/guidance/faqs?anchor=question50287. Accessed December 31, 2022.
16. National Institutes of Health OoLAW. NOT-OD-05-034, Guidance on Prompt Reporting to OLAW under the PHS Policy on Humane Care and Use of Laboratory Animals. 2005. Available from: https://grants.nih.gov/grants/guide/notice-files/NOT-OD-05-034.html. Accessed December 31, 2022.
17. Care CCoA. CCAC: Guidelines on: Euthanasia of animals used in science. 2010. Available from: https://www.ccac.ca/Documents/Standards/Guidelines/Euthanasia.pdf. Accessed December 31, 2022.
18. Toth LA. Defining the moribund condition as an experimental endpoint for animal research. *ILAR J* 2000;41(2):72–79, doi:10.1093/ilar.41.2.72.
19. Morton DB. A systematic approach for establishing humane endpoints. *ILAR J* 2000;41(2):80–86, doi:10.1093/ilar.41.2.80.
20. Wallace J. Humane endpoints and cancer research. *ILAR J* 2000;41(2):87–93, doi:10.1093/ilar.41.2.87.
21. Carstens E, Moberg GP. Recognizing pain and distress in laboratory animals. *ILAR J* 2000;41(2):62–71, doi:10.1093/ilar.41.2.62.
22. Paster EV, Villines KA, Hickman DL. Endpoints for mouse abdominal tumor models: refinement of current criteria. *Comp Med* 2009;59(3):234–241.
23. Olfert ED, Godson DL. Humane endpoints for infectious disease animal models. *ILAR J* 2000;41(2):99–104, doi:10.1093/ilar.41.2.99.
24. Hendriksen CF, Steen B. Refinement of vaccine potency testing with the use of humane endpoints. *ILAR J* 2000;41(2):105-13, doi:10.1093/ilar.41.2.105.
25. Hawkins P, Prescott MJ, Carbone L, *et al.* A good death? Report of the second Newcastle meeting on laboratory animal euthanasia. *Animals (Basel).* 2016.
26. Turner PV, Hickman DL, van Luijk J, *et al.* Welfare impact of carbon dioxide euthanasia on laboratory mice and rats: a systematic review. *Front Vet Sci* 2020;7:411, doi:10.3389/fvets.2020.00411.
27. Danneman PJ, Stein S, Walshaw SO. Humane and practical implications of using carbon dioxide mixed with oxygen for anesthesia or euthanasia of rats. *Lab Anim Sci* 1997;47(4):376–385.

28. Kirkden RD, Niel L, Weary DM. Aversion to carbon dioxide. *Lab Anim* 2005;39(4):453–455, doi:10.1258/002367705774286420.
29. Peppel P, Anton F. Responses of rat medullary dorsal horn neurons following intranasal noxious chemical stimulation: effects of stimulus intensity, duration, and interstimulus interval. *J Neurophysiol* 1993;70(6):2260–2275, doi:10.1152/jn.1993.70.6.2260.
30. Steen KH, Reeh PW, Anton F, *et al.* Protons selectively induce lasting excitation and sensitization to mechanical stimulation of nociceptors in rat skin, in vitro. *J Neurosci* 1992;12(1):86–95, doi:10.1523/jneurosci.12-01-00086.1992.
31. Thürauf N, Ditterich W, Kobal G. Different sensitivity of pain-related chemosensory potentials evoked by stimulation with CO2, tooth pulp event-related potentials, and acoustic event-related potentials to the tranquilizer diazepam. *Br J Clin Pharmacol* 1994;38(6):545–555, doi:10.1111/j.1365-2125.1994.tb04396.x.
32. Hickman DL, Fitz SD, Bernabe CS, *et al.* Evaluation of low versus high volume per minute displacement CO_2 methods of euthanasia in the induction and duration of panic-associated behavior and physiology. *Animals (Basel)* 2016;6(8), doi:10.3390/ani6080045.
33. Boivin GP, Bottomley MA, Dudley ES, *et al.* Physiological, behavioral, and histological responses of male C57BL/6N mice to different CO_2 chamber replacement rates. *J Am Assoc Lab Anim Sci* 2016;55(4):451–461.
34. Boivin GP, Bottomley MA, Schiml PA, *et al.* Physiologic, behavioral, and histologic responses to various euthanasia methods in C57BL/6NTac male mice. *J Am Assoc Lab Anim Sci* 2017;56(1):69–78.
35. Boivin GP, Hickman DL, Creamer-Hente MA, *et al.* Review of CO_2 as a euthanasia agent for laboratory rats and mice. *J Am Assoc Lab Anim Sci* 2017;56(5):491–499.
36. Banzett RB, Lansing RW, Binks AP. Air hunger: a primal sensation and a primary element of dyspnea. *Compr Physiol* 2021;11(2):1449–1483, doi:10.1002/cphy.c200001.
37. Schimitel FG, de Almeida GM, Pitol DN, *et al.* Evidence of a suffocation alarm system within the periaqueductal gray matter of the rat. *Neuroscience* 2012;200:59–73, doi:10.1016/j.neuroscience.2011.10.032.
38. Hickman DL. Wellbeing of alcohol-preferring rats euthanized with carbon dioxide at very low and low volume displacement rates. *J Am Assoc Lab Anim Sci* 2019;58(1):78–82, doi:10.30802/aalas-jaalas-18-000004.
39. Barabas AJ, Lucas JR, Erasmus MA, *et al.* Who's the Boss? Assessing convergent validity of aggression based dominance measures in male laboratory mice, mus musculus. *Front Vet Sci* 2021;8:695948, doi:10.3389/fvets.2021.695948.
40. Theil JH, Ahloy-Dallaire J, Weber EM, *et al.* The epidemiology of fighting in group-housed laboratory mice. *Sci Rep* 2020;10(1):16649, doi:10.1038/s41598-020-73620-0.
41. al'Absi M, Nakajima M, Bruehl S. Stress and pain: modality-specific opioid mediation of stress-induced analgesia. *J Neural Transm (Vienna)* 2021;128(9):1397–1407, doi:10.1007/s00702-021-02401-4.
42. Walsh JL, Van de Vegte J, Mercer B, *et al.* Euthanasia of meat rabbits with carbon dioxide: Behavioral and physiologic responses to gas chamber gradual- and fast-fill rates. *Can Vet J* 2019;60(7):770–778.
43. Herin RA, Hall P, Fitch JW. Nitrogen inhalation as a method of euthanasia in dogs. *Am J Vet Res* 1978;39(6):989–991.
44. Quine JP, Buckingham W, Strunin L. Euthanasia of small animals with nitrogen; comparison with intravenous pentobarbital. *Can Vet J* 1988;29(9):724–726.
45. Sharp J, Azar T, Lawson D. Comparison of carbon dioxide, argon, and nitrogen for inducing unconsciousness or euthanasia of rats. *J Am Assoc Lab Anim Sci* 2006;45(2):21–25.
46. Enggaard Hansen N, Creutzberg A, Simonsen HB. Euthanasia of mink (Mustela vison) by means of carbon dioxide (CO_2), carbon monoxide (CO) and nitrogen (N_2). *Br Vet J* 1991;147(2):140–146, doi:10.1016/0007-1935(91)90104-u.
47. Rault JL, McMunn KA, Marchant-Forde JN, *et al.* Gas alternatives to carbon dioxide for euthanasia: a piglet perspective. *J Anim Sci* 2013;91(4):1874–1883, doi:10.2527/jas.2012-5761.
48. Wang X, Zhao D, Milby AC, *et al.* Evaluation of euthanasia methods on behavioral and physiological responses of newly hatched male layer chicks. *Animals (Basel)* 2021;11(6), doi:10.3390/ani11061802.
49. McKeegan DE, Reimert HG, Hindle VA, *et al.* Physiological and behavioral responses of poultry exposed to gas-filled high expansion foam. *Poult Sci* 2013;92(5):1145–1154, doi:10.3382/ps.2012-02587.
50. Leach MC, Bowell VA, Allan TF, *et al.* Aversion to gaseous euthanasia agents in rats and mice. *Comp Med* 2002;52(3):249–257.
51. Niel L, Weary D. Rats avoid exposure to carbon dioxide and argon. *Appl Anim Behav Sci* 2008;107:100-109.
52. Makowska I, Niel L, Kirkden R. Rats show aversion to argon-induced hypoxia. *Appl Anim Behav Sci* 2008;114:572–581.
53. Carbone L, Carbone ET, Yi EM, *et al.* Assessing cervical dislocation as a humane euthanasia method in mice. *J Am Assoc Lab Anim Sci* 2012;51(3):352–356.
54. Hickman DL. Minimal exposure times for irreversible euthanasia with carbon dioxide in mice and rats. *J Am Assoc Lab Anim Sci* 2022;61(3):283–286, doi:10.30802/aalas-jaalas-21-000113.
55. Pritchett K, Corrow D, Stockwell J, *et al.* Euthanasia of neonatal mice with carbon dioxide. *Comp Med* 2005;55(3):275–281.
56. Pritchett-Corning KR. Euthanasia of neonatal rats with carbon dioxide. *J Am Assoc Lab Anim Sci* 2009;48(1):23–27.
57. Seymour TL, Nagamine CM. Evaluation of isoflurane overdose for euthanasia of neonatal mice. *J Am Assoc Lab Anim Sci* 2016;55(3):321–323.
58. Klaunberg BA, O'Malley J, Clark T, *et al.* Euthanasia of mouse fetuses and neonates. *Contemp Top Lab Anim Sci* 2004;43(5):29–34.
59. Zouikr I, Tadros MA, Clifton VL, *et al.* Low formalin concentrations induce fine-tuned responses that are sex and age-dependent: a developmental study. *PLoS One* 2013;8(1):e53384, doi:10.1371/journal.pone.0053384.
60. McLaughlin CR, Dewey WL. A comparison of the antinociceptive effects of opioid agonists in neonatal and adult rats in phasic and tonic nociceptive tests. *Pharmacol Biochem Behav* 1994;49(4):1017–103, doi:10.1016/0091-3057(94)90258-5.
61. Guy ER, Abbott FV. The behavioral response to formalin in preweanling rats. *Pain* 1992;51(1):81–90, doi:10.1016/0304-3959(92)90012-z.

62. Yi DK, Barr GA. Formalin-induced c-fos expression in the spinal cord of fetal rats. *Pain* 1997;73(3):347–354, doi:10.1016/s0304-3959(97)00119-x.
63. Barr GA. Formalin-induced c-fos expression in the brain of infant rats. *J Pain* 2011;12(2):263–271, doi:10.1016/j.jpain.2010.09.005.
64. Derbyshire SW, Bockmann JC. Reconsidering fetal pain. *J Med Ethics* 2020;46(1):3–6, doi:10.1136/medethics-2019-105701.
65. Bellieni CV. Analgesia for fetal pain during prenatal surgery: 10 years of progress. *Pediatr Res* 2021;89(7):1612–1618, doi:10.1038/s41390-020-01170-2.
66. Care CCoA. Guidelines on: The Care and Use of Wildlife. 2003. Available from: https://www.ccac.ca/Documents/Standards/Guidelines/Wildlife.pdf. Accessed December 31, 2022.
67. Sikes R, Mammalogists ACaUCotASo. Guidelines of the American society of mammologists for the use of wild mammals in research. *J Mammal* 2016;97:663–688.
68. Care CCoA. Guidelines on: The Care and Use of Fish in Research, Teaching and Testing. 2005. Available from: https://ccac.ca/Documents/Standards/Guidelines/Fish.pdf. Accessed December 31, 2022.
69. International A. Animals included in the AAALAC International Accredited "Unit". Available from: https://ar.aaalac.org/accreditation/faq_landing.cfm#A2. Accessed December 31, 2022.
70. De Sio F, Hanke FD, Warnke K, *et al.* E Pluribus octo - building consensus on standards of care and experimentation in cephalopod research; a historical outlook. *Front Physiol* 2020;11:645, doi:10.3389/fphys.2020.00645.
71. Cooper JE. Anesthesia, analgesia, and euthanasia of invertebrates. *ILAR J* 2011;52(2):196–204, doi:10.1093/ilar.52.2.196.
72. Lewbart G. *Invertebrate Medicine*. Hoboken, NJ: Wiley Blackwell; 2022.

19

Surgery

Szczepan W. Baran and Marcel Perret-Gentil

Introduction

The Institutional Animal Care and Use Committee (IACUC) and Attending Veterinarian (AV) have the responsibility to ensure the humane care and use of animals undergoing surgery and to ensure that individuals who perform surgery are qualified and trained. This chapter addresses several major considerations of surgery performed on animals, such as anesthesia, analgesia, aseptic technique, and the perioperative care associated with the conduct of survival and nonsurvival animal surgery in biomedical research and teaching environments.

19:1 What information regarding survival surgical procedures should be included by the investigator in the IACUC protocol? How detailed a description is necessary?

Reg. There are numerous regulatory requirements that relate to survival surgical procedures. In the broadest sense, they revolve around adequate veterinary care. Animal Welfare Act & Animal Welfare Regulations (AWAR) §2.33; Public Health Service (PHS) Policy IV,C,1,e). Specific areas of concern for the IACUC include the following:

- Use of appropriate anesthesia and analgesia (AWAR §2.31,d,1,iv,A; PHS Policy, IV,C,1,b)
- Use of aseptic technique (AWAR §2.31,d,1,ix)
- Conduct of multiple survival surgeries (AWAR §2.31,d,1,x; Animal Care Policy 14[1]) U.S. Department of Agriculture, Animal and Plant Health Inspection Service. 2019. Policy 14, Major Survival Surgery Dealers Selling Surgically Altered Animals to Research. Accessed December 31, 2022.
- Qualifications of surgical personnel (AWAR §2.31,d,1,viii)
- Perioperative care (AWAR §2.31,d,1,ix)
- Whether a major operative procedure or a non-major operative procedure will be performed (AWAR §2.31,d,1,ix)
- Surgical facilities (AWAR §2.31,d,1,ix)
- Description of the procedures to be performed (AWAR §2.31,d,2; §2.31,e,3; PHS Policy IV,C,2; IV,D,1) and compliance, as applicable, with the Animal Welfare Act (AWA) required (PHS Policy, II, IV,C,1)
- Availability or appropriateness of the use of less-invasive procedures (*Guide for the Care and Use of Laboratory Animals* (*Guide*) p. 25; AWAR §2.31,d,1,ii)

Opin. An investigator should provide sufficient surgical procedure detail in the IACUC protocol for the committee to confirm that acceptable surgical technique is proposed, and the AV can evaluate the perioperative care program pertaining to adequate veterinary care needed.[2]

In the authors' opinion, incision location, use of chronic instrumentation and implants (when applicable), and the method of wound closure (including size and type of suture), potential complications and how they will be addressed should be included with the detailed description of surgical procedures. In addition, the description of the method of anesthesia should include preanesthetic, anesthetic, and analgesic drugs and doses, routes, and frequency of administration. It is helpful for the protocol to include an appropriate range of agents, routes, dosages, frequencies, procedures, minor surgical repairs (e.g., re-suturing), etc. that will allow flexibility and enable the research team to use appropriate professional judgment and still provide enough detail for the IACUC to determine that the proposed care is humane and appropriate. If neuromuscular blocking agents are to be used, the agent, dosage, and administration route of the agonist and reversal agent should be listed along with methods to ensure that the animal remains in a surgical plane of anesthesia while the neuromuscular blockade is active. Procedures that may cause more than momentary pain or distress to the animals should not involve use of neuromuscular blocking agents unless anesthesia is also used (AWAR §2.31,d,1,iv,C). When neuromuscular blocking agents are used, they must be used with an anesthetic and it is appropriate to require the PI to document a lack of pain response, including parameters which are not affected by neuromuscular blockade (such as heart rate) from the animal

DOI: 10.1201/9781003272861-19

subjects during the procedure. This places the AV and IACUC in a much more informed position regarding these study paradigms.

Surv. Which of the following items relating to major survival surgical procedures are typically requested by your IACUC to be included in a protocol application? Check as many boxes as appropriate.

• Not applicable	12/193 (6%)
• Incision location	171/193 (89%)
• Incision length	133/193 (69%)
• Specific instruments to be used	74/193 (38%)
• Type of suture material to be used (e.g., absorbable vs. non-absorbable, barbed vs. nonbarbed)	152/193 (79%)
• Method of wound closure (e.g., in layers, one suture for all)	167/193 (87%)
• When (or if) skin sutures or clips will be removed	158/193 (82%)
• Justification for wound closure (e.g., continuous vs. simple interrupted)	54/193 (28%)

19:2 What information related to pharmaceuticals and other methods of surgical support used during surgery should be included in the IACUC protocol?

Reg. The Food and Drug Administration (FDA) provides guidance on drug compounding, but it does not address animal research.[3]

Opin. With respect to pharmaceuticals administered as part of a surgical procedure, it is the authors' opinion that an adequate description would include all pharmaceuticals used during the perioperative period at least until the animal is awakened and returned to its housing unit. Further, planned use of non-pharmaceutical grade compounds and their compounding should be indicated and appropriately justified (Animal & Plant Health Inspection Service/Animal Care (APHIS/AC) Policy 3; *Guide*, p. 31).

The anesthetic agent and its inhaled concentration (if a volatile agent is used) or dosage and route of administration (if a parenteral agent is used) should be indicated and evaluated by the AV and the IACUC. If delivered by drip and/or constant infusion, the concentration and infusion rate, or at least a range, should be included. A range of concentrations for volatile agents and the carrier gas should be detailed. Reversal agents, if applicable, their dosage and route of administration should also be included in the protocol. Institutional websites providing tables of possible agents from which PIs may "pick and choose" may add ease to writing of protocols, but it is the responsibility of the IACUC, especially with the input from the AV, for fine-tuning any medication protocols to maximize safety of the patient while under anesthesia and for a timely and uneventful recovery following anesthesia. It also is useful for the protocol to include a range of dosages, routes of administration, and appropriate agents such that the research team has flexibility to make changes based upon professional judgment during the course of the project.

If neuromuscular blocking agents are used as part of the proposed research, the drug, dosage, and route of administration must be included in the protocol. Further, the protocol should include a discussion regarding how the depth of anesthesia will be determined to ensure that the patient remains pain free while paralyzed with physiological monitoring recommended.

If injectable anesthesia is used, oxygen supplementation should be described and if not used, justification should be given.

If significant blood loss is likely to occur and/or extracorporeal perfusion will be used during the surgical procedure, the use of heparin, protamine, calcium, potassium electrolytes, and electrolyte fluids needs to be described in the protocol. Further, it should be clear whether the perfusionist or the anesthetist will be responsible for adjustment of fluids and medications during the course of the surgical event. Indecision in this regard or confusion during the surgical procedure can be detrimental for the animal patient.

Surv. What information is typically requested of an investigator performing a survival surgery procedure? Check as many boxes as appropriate.

• Not applicable	10/193 (5%)
• Preanesthetic drugs used	162/193 (84%)
• Anesthetic drugs used	182/193 (94%)
• Analgesic drugs used	180/193 (93%)
• The dose, frequency, and route of administration of anesthetic	180/193 (93%)
• Justification for wound closure (e.g., continuous vs. simple interrupted)	56/193 (29%)
• The dosage, frequency, and route of administration of all other drugs and chemicals used	171/193 (89%)
• The name (e.g., lactated Ringer's), dosage, and route of fluid therapy (if used)	138/193 (72%)
• Are fluids warmed before dosing?	62/193 (32%)
• Method of heat support to be provided (e.g., homothermic)	144/193 (75%)
• Will the temperature be monitored	126/193 (65%)
• Timing of heat support (e.g., pre-, post-surgery)	118/193 (61%)
• Justification for not providing heat support	89/193 (46%)

19:3 What information with respect to qualifications of the surgeon should be included in the IACUC protocol?

Opin. With respect to surgical qualifications, aseptic technique, and monitoring of the animal, the IACUC has the responsibility for implementation of SOPs, policies and training programs that will ensure that animal well-being and humane treatment are priorities during every surgical procedure, survival, or non-survival. Often this responsibility is met through employees specifically trained in certain aspects of general animal care and, more specifically, perioperative procedures. In many cases, such individuals are veterinarians, physician-surgeons, or technicians who have been tutored in specific surgical procedures by veterinarians and Principal Investigators (PIs). It is important that technical skills are kept current, and that the well-being of the animal patient is always of primary concern. Personnel, including veterinarians and surgeons, may require species-specific and procedure-specific training and assessment before being allowed to conduct survival or non-survival animal surgery. It is the responsibility of the IACUC to ascertain that only qualified individuals are allowed to perform surgery on animals. Having the AV or facility staff monitor new procedures and personnel; acknowledging publications in which personnel are identified as having performed specific procedures and outcomes; and mentoring personnel in laboratories where the proposed surgery is routinely performed are but a *few* of the means available to the IACUC to ensure competency in the surgical theater.

Surv. For major survival surgical procedures, which of the following information items are requested from an investigator on his/her IACUC protocol application? Check as many boxes as appropriate.

• Not applicable	15/193 (8%)
• Length of the surgeon's experience as a surgeon	81/193 (42%)
• Length of the surgeon's experience with the procedure in the species being used	94/193 (49%)
• Experience in surgical procedure(s) described in protocol	140/193 (73%)
• Published success rate or survival rate if available	29/193 (15%)
• Surgeon's success rate or survival rate	21/193 (11%)
• List of potential surgical complications	124/193 (64%)
• Type of surgical training received (e.g., firsthand surgical courses, in-house training)	121/193 (63%)
• Surgical certification received (e.g., Academy of Surgical Research Certification)	30/193 (16%)
• Qualifications of others directly involved in surgery (e.g., the anesthetist)	91/193 (47%)
• Methods of intraoperative monitoring of the animal (e.g., pulse oximeter, ECG, heart rate, respiratory rate, and temperature)	140/193 (73%)
• The methods of aseptic technique that will be followed	150/193 (78%)
• The frequency of postoperative observations until the animal is recovered from general anesthesia (the definition of recovery is to be acceptable to the IACUC)	159/193 (82%)

19:4 Is a cutaneous biopsy considered a survival surgical procedure?

Reg. The *Guide* (p. 117) identifies skin biopsy as one of a few examples of a *minor surgical* procedure, describing a minor surgical procedure as one in which there is minimal tissue trauma, no penetration of a body cavity, and little or no physical impairment from such a procedure. The *Guide* (p. 117) and National Institutes of Health (NIH) / Office of Laboratory Animal Welfare (OLAW) frequently asked question (FAQ) F.13[4] define major survival surgery as one which "penetrates and exposes a body cavity, produces substantial impairment of physical or physiologic functions, or involves extensive tissue dissection or transection." AWAR (§1.1) provide a definition of a major operative procedure as "any surgical intervention that penetrates and exposes a body cavity, or any procedure which produces permanent impairment of physical or physiologic function." In addition, AWAR §2.31,d,1,ix classify procedures as either major operative procedures or non-major operative procedures. In conventional veterinary clinical practice, a skin biopsy would usually be performed with at least a sedative and a local anesthetic; might require suturing or at least repair with tissue adhesive and would, therefore, be considered a surgical procedure.

Opin. There is a wide spectrum of procedures done under the umbrella of skin biopsy, particularly in human dermatology, where very superficial layers of the epidermis can be removed for histological evaluation, usually with the patient anesthetized using only local anesthesia. However, some of these procedures can reach the dermis. Despite the overall minimal nature of such procedures, all the attendant risks, though reduced, are those that would be associated with penetration into a body cavity. That is, there is the risk of infection; risk of non-healing and requirement for debridement and/or a secondary surgical repair; risk of hemorrhage and nerve damage and, depending upon the area and reasons for biopsy, the need for prophylactic antimicrobial use. Because living tissue has been incised, a cutaneous biopsy would be considered a minor survival surgical procedure.

19:5 What information regarding *non-survival* surgical procedures should be included by the investigator in the IACUC protocol?

Reg. The *Guide* (p. 118) indicates that while not all techniques that should be followed for survival surgery would need to be followed for non-survival procedures, that "at a minimum, the surgical site should be clipped, the surgeon should wear gloves, and the instruments and surrounding area should be clean."

Opin. As with survival surgical procedures (see 19:1), an investigator should provide sufficient detail in the IACUC protocol when describing non-survival surgical procedures so that the committee can confirm that acceptable surgical techniques are proposed and the AV can evaluate the preoperative and intraoperative management pertaining to adequate veterinary care, especially regarding anesthesia and analgesia. Aseptic techniques may need to be implemented when samples are being collected and when contamination may affect the quality of research outcomes. The method of anesthesia described should include preanesthetic and anesthetic drugs, ranges of dosages, routes and frequency of administration, and monitoring methods. If neuromuscular blocking agents are to be used, the agent, dosage range, and administration route should be listed. When neuromuscular blocking agents are used, it is appropriate to require the PI to document lack of pain response, including parameters such as heart rate, which are not affected by neuromuscular blockade, from the animal subjects during the procedure. This allows the AV and IACUC to carry out their assessment regarding these study paradigms (see 17:23). Some IACUCs choose to be more restrictive than current regulations (i.e., *Guide*, p. 118; AWAR §2.31,d,1,ix) with issues such as appropriate surgical attire, dedicated space requirements, and use of aseptic procedures. When this occurs, the IACUC should clearly define such requirements to the investigator.

Surv. For *non-survival* surgical procedures, what information is requested from an investigator? Check as many boxes as appropriate.

• Not applicable	10/193 (5%)
• Preanesthetic and anesthetic drugs, dosages, and route of administration	179/193 (93%)
• The anticipated length of the procedure	122/193 (63%)
• Whether or not aseptic technique will be used	122/193 (63%)
• If used, the dose and route of neuromuscular blocking agents	131/193 (68%)
• If used, the precautions to ensure an animal is not awake during the use of neuromuscular blocking agents	128/193 (66%)

19:6 How detailed a description is necessary in an IACUC protocol that contains *non-survival* surgical procedures?

Opin. The description should be adequate to assure the IACUC that, in fact, normal procedures are not abbreviated and that conduct of the surgery will be done with the same care and attention that is applied in survival procedures.

Some procedures require less detail than others, particularly those that do not involve extensive intraoperative management. For example, a tissue harvest conducted alone or, perhaps, after some physiological measurements might only require a "minimal" surgical description, such as "Midline laparotomy is performed in surgically prepped and draped animals. The renal artery and veins are ligated at the pelvis and the kidney extirpated. The animal is euthanized as described in item."

A more detailed description might be warranted when the intraoperative time and care are likely to be substantial, such that failure to provide detailed attention would likely result in pain or distress to or mishandling of the animal. Examples include a series of physiological measurements such as blood pressure, pressure volume curves of the ventricular chambers, and transmural pressures, in which maintenance of vascular cannulas is involved. In such cases, the physiological stability of the anesthetized animal is required for reliable data. Also, because the potential for hemorrhage and the need for volume replacement exist, the IACUC requires more detailed information including how complications will be addressed and what endpoints will be used.

19:7 If an animal is anesthetized and then undergoes open-chest terminal perfusion, should the procedure be counted as euthanasia or a non-survival surgical procedure?

Reg. The *Guide* (p. 118) simply describes a non-survival surgical procedure as one in which the animal patient is euthanized before recovery from anesthesia. Euthanasia is the act of humanely killing an animal by methods that induce rapid unconsciousness and death without pain or distress. Unless a deviation is justified for scientific or medical reasons, methods should be consistent with the *AVMA Guidelines for the Euthanasia of Animals*.[5]

Opin. In the authors' opinion, this particular procedure could be classified as either euthanasia or non-survival surgery. The final semantics depends upon the protocol description of the ways in which the animal is to be terminated and "why" this particular method is chosen. Importantly, the criteria for euthanasia have been met as part of the procedure, with the animal undergoing terminal perfusion under anesthesia.

If experimental physiological measurements were made as part of the procedure (e.g., blood pressure, ventricular contractile force, FEV_1; Purkinje fiber conduction velocity, etc.) and then the animal was perfused, it might be considered a non-survival surgical procedure. If, in fact, the only thing done was to perfuse the heart and lung so that tissue could be taken for histology or other analytical work, it might be called a tissue harvest and could be considered as a euthanasia procedure rather than a non-survival surgery.

19:8 Can the IACUC classify surgical procedures as minor versus major? By what criteria? How can this classification be useful to the IACUC when reviewing protocols?

Reg. The AWAR (§1.1) define a *major operative procedure* as "any surgical intervention that penetrates and exposes a body cavity or any procedure that produces permanent impairment of physical or physiological functions." The AWAR classify survival surgical procedures as either major or non-major operative procedures (§2.31,d,1,ix). The *Guide* (p. 117) states that surgical procedures are categorized as major or minor. A *major survival surgery* is defined as surgery that "penetrates and exposes a body cavity, produces substantial impairment of physical or physiologic functions, or involves extensive tissue dissection or transection." The *Guide* (p. 117) offers as a general guideline that procedures such as laparotomy, thoracotomy, joint replacement, and limb amputation are examples of major surgical procedures. The *Guide* (p. 117) defines *minor survival surgery* as surgery that "does not expose a body cavity and causes little or no physical impairment" and suggests that procedures such as wound suturing, peripheral vessel cannulation, percutaneous biopsy, routine agricultural animal procedures such as castration and most procedures routinely done on an "outpatient" basis in veterinary clinical practice are examples of minor survival surgery. Furthermore, the *Guide* (p. 117) states: "Laparoscopic surgeries and some procedures associated with neuroscience research (e.g., craniotomy, neurectomy) may be classified as major or minor surgery depending on their impact on the animal."[6–10]

PHS Policy (IV,A,1) requires institutions to use the *Guide* as a basis for their animal care and use program (see Chapter 1).

Opin. Classification of surgical procedures by an IACUC as major or minor is based on the preceding regulatory definitions, and the IACUC is empowered to make this decision on a case-by-case basis. For example, intraperitoneal injections with needles up to 18-gauge (1.27 mm in diameter) are routinely performed and are not considered minor or major procedures. Currently, available laparoscopes for rodents and small animals such as squirrel monkeys are available in a range as small as 0.9 mm in diameter. It is appropriate for the IACUC to consider the type of equipment and procedure being performed such as monitoring disease progression with laparoscopy versus laparoscopic cholecystectomy. Further, this classification is useful to an IACUC when reviewing a protocol because it is the basis for the IACUC's determination of the type of surgical facility and the degree of aseptic technique required. Major survival surgery on U.S. Department of Agriculture (USDA)-regulated species and non-rodents, must be performed only in facilities designed, operated, and maintained for that purpose. Minor survival surgery (non-major operative procedures as defined by APHIS/AC) and all surgery on rodents do not require separate dedicated facilities; however, aseptic technique must be used (AWAR §2.31,d,1,ix; *Guide*, p. 118). Further, for animals to be used in multiple major survival surgeries, such use must be scientifically justified in the protocol and approved by the IACUC (AWAR §2.31,d,1,x,A; *Guide*, p. 30; NIH/OLAW FAQ F.9[4]). There is no regulatory limitation to multiple *minor* survival surgical procedures. Nevertheless, an IACUC should use professional judgment to limit the number of minor surgical procedures performed on an animal.[11]

19:9 Should laparoscopy be considered a major operative procedure (major survival surgery)?

Reg. The AWAR (§1.1) define a *major operative procedure* as "any surgical intervention that penetrates and exposes a body cavity or any procedure that produces permanent impairment of physical or physiological functions." The *Guide* (p. 117) states that a major survival surgery is one that penetrates and exposes a body cavity or produces substantial impairment of physical or physiologic functions.[12] NIH/OLAW FAQ F13[4] provides general guidance to the IACUC about the classification of laparoscopic procedures as major or minor surgery. Furthermore, the *Guide* (p. 117) further states that laparoscopic surgeries and some procedures associated with neuroscience research (e.g., craniotomy, neurectomy) may be classified as major or minor surgery depending on their impact on the animal. For example, laparoscopic techniques with minimal associated trauma and sequelae (e.g., avian sexing and oocyte collection) could be considered minor, whereas others (e.g., hepatic lobectomy and cholecystectomy) should be considered major. Whether a laparoscopic procedure is deemed major or minor should be evaluated on a case-by-case basis by the veterinarian and IACUC.

Opin. The regulatory definition of major survival surgery is not presented with respect to the extent of entry or exposure of a specific body cavity. If the body cavity has been entered and exposed, a criterion has been met for defining the procedure as a major survival surgery. However, intraperitoneal injections with needles up to 18-gauge (1.27 mm in dimeter) are routinely performed but are not considered major or even minor surgical procedures. Currently, available laparoscopes for rodents and small animals such as squirrel monkeys are available in the range as small as 0.9 mm in diameter, a size of 20-gauge needle (0.908 mm). It is appropriate for the IACUC to consider the type of equipment and procedure being performed, such as monitoring disease progression with 20-gauge needle size laparoscope versus laparoscopic cholecystectomy.

Depending upon the laparoscopic procedure, there may be just one port or there may be two or more. The welfare of the animal patient should be the focus. While the laparoscopic technique minimizes trauma, it has the potential to place the animal at increased risk of hidden hemorrhage, which should be mitigated with an appropriate post-surgical monitoring plan.

Surv. Does your IACUC usually consider a laparoscopic procedure in the abdomen or thorax to be a major or minor operative procedure?

• Not applicable.	47/193 (24%)
• Generally, a major procedure.	77/193 (40%)
• Generally, a minor procedure.	6/193 (3%)
• We evaluate the overall invasiveness of the procedure before determining if it is major or minor.	63/193 (33%)

19:10 Does your IACUC usually consider a craniotomy (via one or more burr holes) to be a major or minor operative procedure?

Reg. The AWAR (§1.1) define a major operative procedure as "any surgical intervention that penetrates and exposes a body cavity or any procedures that produces permanent impairment of physical or physiological functions." The *Guide* (p. 117) states that surgical procedures are categorized as major or minor. A *major survival surgery* is defined as surgery that "penetrates and exposes a body cavity, produces substantial impairment of physical or physiologic functions, or involves extensive tissue dissection or transection" and offers that procedures such as laparotomy, thoracotomy, joint replacement, and limb amputation are examples of major surgical procedures. The *Guide* (p. 117) defines *minor survival surgery* as surgery that "does not expose a body cavity and causes little or no physical impairment" and suggests that procedures such as wound suturing, peripheral vessel cannulation, percutaneous biopsy, routine agricultural animal procedures such as castration, and most procedures routinely done on an "outpatient" basis in veterinary clinical practice are examples of minor survival surgery. PHS Policy (IV,A,1) requires institutions to use the *Guide* as a basis for their animal care and use program and NIH/OLAW recognizes the authority of the IACUC to determine whether specific manipulations used in research are major operative procedures (NIH/OLAW FAQ F13[4]).

Opin. For uniformity and clarity, the NIH/OLAW position seems congruent with the majority opinion of laboratory animal veterinarians and AVs. APHIS/AC has left this decision to the discretion of the IACUC. A request to classify a craniotomy, as well as any other manipulation involving penetration into a body cavity, as a minor surgical procedure would need to be evaluated by the IACUC with input from the PI and the AV. For instance, a common clinical procedure might puncture the thorax or abdomen with cannulated needles to withdraw fluid without exposing the corresponding body cavity. This is a relatively common clinical practice, and, in both instances, the surrounding tissues usually seal the opening. Thus, it might be argued that the body cavity has not been significantly exposed, and the procedure qualifies as a minor surgical procedure. In a similar vein, percutaneous puncture with a closed endoscope into the coelomic cavity of an avian species for the purpose of sexing, with the closure of the skin around the exiting endoscope, might be considered a minor procedure. In this regard, it is much more difficult to envision entering the cranium, even with a very fine trephine, without exposing the cavity, at least momentarily. The risks of significant hemorrhage are greater in the cranium than a similar penetration into either the chest or the abdomen. In the authors' opinion, the greater risk of hemorrhage, as well as the more severe consequences related to physical tissue damage and/or bacterial contamination of the brain warrant classifying craniotomy as a major surgery. On the other hand, if the craniotomy is made through a single burr hole to introduce a fine needle for injection of a noninvasive chemical into a specific portion of the brain, it may be considered by some as a minor procedure if it has already been established that such procedure produces minimal or no side effects. Whether major or minor, the IACUC should still consider the cumulative effect of multiple procedures on the animal and the scientific justification for the procedure(s).

Surv. Does your IACUC usually consider craniotomy (via one or more burr holes) to be a major or minor operative procedure?

• Not applicable.	34/193 (18%)
• We generally consider it to be a major procedure.	120/193 (62%)
• We generally consider it to be a minor procedure.	8/193 (4%)

- We evaluate the size of the burr hole(s) before deciding on major or minor. 1/193 (1%)
- We evaluate the overall invasiveness of the procedure before deciding on major or minor. 30/193 (16%)

19:11 Under what circumstances can an individual animal be used in multiple survival surgical procedures?

Reg. Although performing multiple major survival surgical procedures on an individual animal is discouraged, the *Guide* (p. 30) states that they may be permitted if scientifically justified by the investigator and approved by the IACUC; if necessary for clinical reasons; or if the surgeries are included in and are essential components of a research project or protocol. The AWAR (§2.31,d,1,x,A–C) and NIH/OLAW[11]) permit multiple major survival operative procedures for clinical reasons and when justified in the protocol (see NIH/OLAW FAQ F.9 for guidance on multiple survival surgeries on the same animal):

- When scientifically justified by the investigator in writing
- When needed as a routine veterinary procedure or to protect the health and well-being of an animal
- When other special circumstances are authorized by the Administrator, APHIS, USDA, on a case-by-case basis

Regulations that address multiple survival surgical procedures on an individual animal have exemption criteria where it is allowed. According to the AWA (§2143,a,3,D), exceptions for special circumstances may be made as determined by the Secretary of Agriculture. As specified by the AWAR (§2.31,d,1,x), requests for an exception are sent to the Administrator of APHIS/USDA.

The NIH/OLAW[13] and the *Guide* (p. 30) permits multiple major survival operative procedures on individual animals in the following circumstances:

- The procedures are included in, and are essential components of a single research project or proposal.
- The multiple surgical procedures are scientifically justified by the investigator.
- The multiple surgical procedures are necessary for clinical reasons.

Opin. Examples of special circumstances that may potentially justify performing multiple surgical procedures on an animal include procedures that are related components of a research project, involve conservation of scarce animal resources, or are needed for clinical reasons. The *Guide* (p. 30) and interpretation of PHS Policy[14] do not consider cost savings alone as an adequate justification for performing multiple major survival surgeries on a single animal.

19:12 How many survival surgical procedures to harvest oocytes from Xenopus frogs are acceptable for a single animal?

Reg. There are no explicit regulations or guidelines from NIH/OLAW, APHIS/AC, or the *Guide* regarding this specific procedure. Indeed, the AWAR (§1.1, Animal) exclude amphibians as a regulated species per the definition of "animal." However, guidance from NIH/OLAW and the *Guide* as described above (19:11) for multiple survival surgery might be applicable to this question. In this regard, the surgeries must be scientifically justified; part of one distinct study and protocol; be done under general anesthesia; be attentive to using the best aseptic techniques possible under the circumstances; and provide appropriate perioperative monitoring.

Opin. In the authors' opinion, multiple surgical oocyte collection should be allowed in these species, based primarily on existing experiences of those who have been doing this procedure for years with minimal apparent impact on the well-being of the frogs. There appears to be rapid postoperative healing at the surgical site. Proficiency through experience can limit the incision length and return the animal to normal eating and behavior quite quickly. The fact that subsequent harvests result in good quality eggs from a frog that maintains or increases weight during collection intervals suggests that physiologic recovery is quite rapid. These latter facts strongly suggest that perioperative pain and distress may be more subdued compared to that seen with mammalian species experiencing abdominal surgery. Reproductive tissues (ovaries) are not removed in this procedure and almost no manipulation or mechanical disturbance of other coelomic viscera are required, minimizing postoperative scarring, adhesions, and attendant pain. The IACUC may wish to consider that an adequate rest period before the next surgical collection is stated and that the two sides of the abdominal cavity might be alternated for surgery to access the oocytes. Acceptable ranges established by IACUCs vary from two (2) to five (5) collection surgeries. For example, the NIH Animal Research Advisory Committee (ARAC) approved oocyte collection states: "maximum of two recovery surgeries one on the left ovary and one on the right ovary per animal is recommended if multiple surgeries are approved. A third non-survival surgery collecting all oocytes may be performed after full recovery from the second surgery. Adequate recovery time should be allowed between laparotomies. A minimum of 1 month recovery between surgeries should be allowed when repetitive laparotomies are approved to ensure full recovery and healing of the incision site."[15]

Surv. When performing routine oocyte collections on *Xenopus* frogs, how many survival surgical procedures does your IACUC typically allow on any one animal?

- Not applicable 135/193 (70%)
- One surgery 10/193 (5%)
- Two surgeries 10/193 (5%)
- Three surgeries 9/193 (5%)
- Four surgeries 4/193 (2%)
- Five surgeries 11/193 (6%)
- Six surgeries 9/193 (5%)
- More than six surgeries 5/193 (3%)

19:13 A vendor surgically modifies an animal to meet a research need (e.g., ovariectomy of a rat). Must that vendor have its own PHS Assurance statement if the animals are sold to an institution and used in a project supported by PHS funds? Should the IACUC obtain a copy of that Assurance statement?

Opin. Purchase of animals that have been surgically modified by a vendor and are available for general sale is covered by PHS Policy if the surgery is conducted in response to a specific, custom request.[16] If an institution purchases an animal to be used in a PHS-supported study, and the animal has been surgically modified by the breeder in advance of sale, as a readily available "off-the-shelf" catalog item, then the vendor is not required to have his or her own Assurance on file with NIH/OLAW. On the other hand, if an investigator subgrants or subcontracts with a supplier to produce surgically modified animals according to specification for a PHS-supported study, the supplier must have on file an Animal Welfare Assurance with NIH/OLAW and must provide verification of IACUC approval for the animal activity. (NIH/OLAW FAQ A.2[4] discusses the applicability of the PHS Policy to the purchase of surgically modified animals.) If both institutions have Assurances, the IACUCs may choose which IACUC will review the protocol for the animal activities being conducted. It is recommended that if an IACUC defers protocol review to another IACUC, documentation of the review should be maintained by both committees.[17] The institution must obtain the supplier's Assurance number by contacting the supplier or from the NIH/OLAW website having a list of Assured institutions (http://grants.nih.gov/grants/olaw/assurance/300index.htm). Obtaining a copy of the Assurance statement or the establishment of a Memorandum of Understanding between the two institutions would be a local decision made at the institutional level.

19:14 Should the IACUC permit a farm animal to recover from major surgery so that it can be sold for slaughter once it has fully recovered?

Reg. The AWAR (§2.31,d,1,ix) and the *Guide* (pp. 115–120) do not prohibit recovery from a first major surgery provided that all related requirements are met. However, PHS Policy (IV,C,1,a) states that "procedures with animals will avoid or minimize discomfort, distress, and pain to the animals, consistent with sound research design." This is similarly stated in the AWAR (§2.31,d,1,i). The *Guide for the Care and Use of Agricultural Animals in Research and Teaching*[12] (p. 10) is similarly specific with allowance given (1) when the surgeries are justified and approved by the IACUC; and (2) if needed to ensure or maintain the health of the animal.

Opin. Assuming that the investigator's research or educational needs can be met with non-survival surgery, anesthesia without recovery supports the welfare interests of the animal by fully preventing potential pain and discomfort rather than *minimizing* potential postoperative pain with analgesics. If meat from the slaughtered animal was destined for human consumption, the IACUC must ensure that the animal received only FDA-approved medications and drug withdrawal periods were met. Ultimately, the IACUC's awareness of the public's expectations and perceptions regarding the use of animals in research and education should take precedence in making this decision. The use of animals in research is held to a higher ethical standard than many other uses. The IACUC needs to be aware of all federal, state, and local laws regarding sending research animals into the food chain (for either human or animal consumption). The potential negative public perception of returning farm animals to the food chain after use in research or education might be a basis for the IACUCs to discourage and decide not to permit this practice, particularly given the negative publicity that might result in the event of an adverse event following entry of such meat into the food chain.

19:15 Should the IACUC permit a hamster, ovariectomized as customer service by the vendor, to undergo another major survival surgical procedure? For regulatory compliance purposes, should this be considered as two major survival surgical procedures on the same animal?

Reg. The AWAR (§2.31,d,x,A–§2.31,d,x,C) stipulate that no animal should be used in more than one survival major operative procedure unless justified for scientific reasons, required as part of the veterinary care of the animal, or otherwise exempted by the APHIS/AC Administrator as a Special Circumstance. In the event another surgery is required for scientific purposes, APHIS/AC can be contacted with a request for Special Circumstance considerations (AWAR §2.31,d,1,x,C) since hamsters are regulated by the USDA. The *Guide* (p. 30) also clearly identifies the need for scientific justification if more than one major survival surgery is to be performed on an animal.

Opin. Ovariectomy, performed according to current veterinary practices, would have to be classified as a major surgery, since a body cavity has been entered, even when entered laparoscopically (see 19:9). Thus, it may be logical for the IACUC to regard this situation as one in which multiple major survival surgical procedures are being conducted. The IACUC may permit, as an exception based on scientific merit, multiple major survival surgical procedures as described above.

19:16 A major surgical procedure is performed by a veterinarian for a spontaneous medical condition that is unexpected and unrelated to any ongoing research. Have conditions been established for prohibiting a subsequent survival major operative procedure unless scientifically justified and IACUC approval is given for "multiple major operative procedures from which the animal recovers"?

Reg. This condition is specifically addressed by the AWAR (§2.31,d,1,x,B), which indicate that an animal that has a major operative procedure as part of a facility's routine veterinary care program (unrelated to research) or to protect the health or well-being of the animal as determined by the attending veterinarian may still be used in a research proposal that requires a major survival operative procedure.

Opin. It is the responsibility of the AV and the PI to advise the IACUC on the medical and scientific suitability of these animals for further study. The *Guide* (p. 30) suggests that additional consideration could be given to the species availability or other unique features of the animal that might make it a valuable resource for additional studies involving major survival surgery, though such use should be critically reviewed by the IACUC. While not specifically stated, the assumption is that the emergency procedure was such that the procedure did not interfere with the potential use of the animal in a dedicated study.

19:17 If an investigator dedicates a part of a laboratory bench top for performing major survival surgery on a hamster, is this considered as being compliant with the AWAR?

Reg. The AWAR (§1.1, Animal) specifically include hamsters as a regulated species. However, the AWAR (§2.31,d,1,ix) indicate that major operative procedures performed on rodents do not require dedicated facilities, though aseptic technique is required. The *Guide* (p. 144) indicates that space dedicated to surgery and related activities when used for this purpose and managed to minimize contamination from other activities within the room being conducted at the same time is acceptable for rodent surgery. NIH/OLAW has provided similar guidance on this topic in FAQ F.7, What are the requirements for conducting rodent survival surgery?

Opin. Conducting hamster survival surgical procedures on a laboratory bench is acceptable if there is no competing activity in the area at that time. Further, use of this area is compliant if conditions minimize contamination from other activities within the laboratory. The AWAR exception for rodents should not be interpreted to exempt the surgeon from the use of aseptic technique, nor from being appropriately garbed. Further, as the hamster is a USDA-regulated species, adequate intraoperative and postoperative records of observations, analgesic or other medication administration, pain assessments, and notes on surgical healing should be maintained.

19:18 What is aseptic technique?

Reg. The *Guide* (p. 118) states that aseptic technique includes:

- Preparation of the patient, with body hair removal and disinfection of the operative site
- Preparation of the surgeon, such as wearing appropriate surgical attire, face masks, and sterile surgical gloves
- The use of sterilized instruments, supplies, and implanted materials
- The use of operative techniques to reduce the likelihood of infection

The *Guide* (p. 118) also states that "aseptic technique is used to reduce microbial contamination to the lowest possible practical level. No procedure, piece of equipment, or germicide alone can achieve that objective: aseptic technique requires the input and cooperation of everyone who enters the surgery area. The contribution and importance of each practice varies with the procedure."

Opin. Surgical *aseptic technique* is the performance of a surgical procedure in a manner to minimize exposure of the patient to pathogenic organisms.[1,4]

19:19 Is aseptic technique necessary for performing survival surgery in animals?

Reg. The AWAR (§2.31,d,1,ix) require survival surgery to be performed on all regulated animals using aseptic procedures, which include surgical gloves, masks, sterile instruments, and aseptic techniques. The *Guide* (p. 118) states that "aseptic technique should be followed for all survival surgical procedures."

Opin. An IACUC has a responsibility to ensure the use of acceptable standards of aseptic technique. Semiannual inspections of areas where surgery is performed as well as observation of research staff performing surgeries as part of a post-approval monitoring program can help ensure this standard. Aseptic technique is used to reduce microbial contamination of a surgical wound and exposed tissues to the lowest level possible (*Guide* p. 118). To use less than optimal aseptic technique potentially increases bacterial contamination and subsequent risk of both clinical (apparent) and subclinical (unapparent) infection, which can compromise an animal's health postoperatively as well as research data. Importantly, lack of clinical evidence of postoperative infection does not rule out clinically unapparent (subclinical) infection. A study performed in rats showed that infection can be clinically unapparent and yet cause adverse physiologic, immunological, and behavioral responses that may affect research results and may not be recognized.[17]

19:20 Are the standards for aseptic technique different for rodents compared to other mammals?

Reg. Neither the AWAR (§2.31,d,1,ix) nor the *Guide* (p. 144) require a dedicated surgical facility for major survival rodent surgery as compared to non-rodent surgery. The AWAR (§2.31,d,1,ix) state that all survival surgery must be performed using aseptic procedures, including surgical gloves, masks, sterile instruments, and aseptic technique. The *Guide* (p. 118) states that "aseptic technique should be followed for all survival surgical procedures" and makes no distinction in this regard for rodents versus non-rodent mammals.

Opin. Because surgery involving rodents may involve surgery on multiple individuals during a single surgery session, modifications in standard aseptic techniques are often followed, and experience has shown that such practices are typically successful, including use of one sterile instrument pack for up to five rodents incorporating techniques to maintain sterility between animals and sterilization of instruments between surgeries using hot bead sterilizers,[18] utilization of non-commercial drapes such as Press'n Seal and similar products,[19] and optional wearing of a mask with cap and sterile gown.[20,21] Although modifications in standard aseptic techniques may be necessary or desirable for rodents, the performance standards to prevent or minimize exposure of the patient to pathogenic organisms to reduce the likelihood of infection must be met and the well-being of the animals should not be compromised.[22]

While the rationale for less stringent regulations governing rodent surgery appears obvious, there are reasons to be more restrictive, on an institutional basis, than required. Many laboratories work with both USDA-regulated and nonregulated species. The "lower key" atmosphere of rodent surgery can spill over into surgical activities involving surgical work with regulated species. If one is doing five sequential rabbit or guinea pig surgeries, five separate sterilized instrument packs will be required. It may be reasonable for an IACUC to have this same standard for rodents, even though it is not required by regulations. If a particular IACUC is more restrictive than the regulations, these additional requirements should be clearly defined so that the PI can perform studies in a compliant fashion.

19:21 Does aseptic technique need to be followed for non-survival surgery?

Reg. The AWAR (§2.31,d,1,ix) state: "Major operative procedures on non-rodents will be conducted only in facilities intended for that purpose which shall be operated and maintained under aseptic conditions." The area to be used should be clean, free of clutter, and prepared by acceptable veterinary sanitation practices that would be used in a standard examination/treatment room. The *Guide* (p. 118) does not require using aseptic technique for non-survival surgery; however, "the surgical site should be clipped, the surgeon should wear gloves, and the instruments and surrounding area should be clean."

Opin. In certain circumstances when performing non-survival surgery an investigator may adopt the same aseptic standards used for survival surgery when the lack of its use could compromise experimental results. This approach is frequently used when the non-survival surgical procedure is prolonged in time. Slattum and colleagues[22] reported results demonstrating a need for aseptic technique when performing non-survival surgery. They found that gram-negative bacteremia and septic shock developed in dogs during non-survival cardiopulmonary studies performed without using aseptic technique. Laboratory-prepared nonsterile intravenous solutions were found to be contaminated with gram-negative bacteria. Bacteremia and septic shock ceased to occur after initiating some components of aseptic technique such as using sterile commercial saline solution and other sterile intravenous injectables, disinfecting equipment and instruments, and using sterile gloves. If non-survival surgery is conducted in a dedicated "survival" surgery area, then aseptic technique should be used, or the surgical room be returned to an appropriate level of cleanliness before it is used for major survival surgery.

19:22 What level of aseptic technique should the IACUC require for field surgery involving wildlife?

Reg. The AWAR (§2.31,d,1,ix) state that survival surgery conducted at field sites does not require dedicated facilities but must be performed using aseptic procedures. The *Institutional Animal Care and Use Committee Guidebook*[23] suggests that aseptic practices are the best practice when performing survival surgery on wildlife in field studies. The *Guide* (p. 118) indicates that general principles of aseptic technique should be followed for all survival surgical procedures, with no distinction made between those performed in surgical facilities versus those performed under field conditions. The *Guide* (p. 115) also indicates that standard techniques may need to be modified under conditions such as field surgery, and, in such cases, a close assessment of surgical outcomes may require criteria other than morbidity and mortality.

Opin. An experimental animal surgical facility environment may be unnecessary when performing survival surgery on wildlife if it is not needed to improve animal well-being or surgical outcome as measured by lack of postoperative complications, improved surgical survival, or minimized pain and stress. Taking free-living wildlife to a dedicated facility to perform surgery could, in fact, be detrimental to the well-being and survival of the animals, though settings typical of clinical veterinary practice may also be suitable for field surgery involving wildlife.[11] Nevertheless, aseptic technique must be used to prevent or minimize exposure of the animal to pathogenic organisms. This includes preparation of the operative site (including hair removal and disinfection); surgical mask; sterile surgical gloves, instruments, supplies, and implanted materials; and use of operative techniques to reduce the likelihood of infection (*Guide*, p. 118). Steam sterilization or autoclaving is the preferred method of surgical instrument sterilization. However, chemical "sterilization" using liquid chemicals with appropriate contact time may be necessary in some field surgery settings. Ultimately, professional judgment must be used to optimize the circumstances for the environment where the surgery is performed and the aseptic techniques used.

19:23 What level of aseptic technique should be required for survival surgery performed on non-mammalian aquatic species?

Reg. As described in the *Guide* (p. 118) general principles of aseptic technique should be followed for all survival surgical procedures. There are no specific regulations for aseptic techniques regarding aquatic species.

Opin. The NIH Intramural Research Program provides internal guidelines[15] (National Institutes of Health, Office of Animal Care and Use, 2010, Guidelines for egg and oocyte harvesting in *Xenopus laevis*) for aseptic technique during *Xenopus* oocyte collection that may be useful best practices: "Surgeries should be done as aseptically as practical including the use of sterilized instruments and powderless gloves. Instruments should be sterilized by autoclaving or using a glass-bead sterilizer." To this point, the description is quite like that for aseptic techniques typically used for rodents or for non-survival surgical procedures. The NIH intramural guidelines also state: "The use of surgical drapes and preparation of the surgical site remains controversial for aquatic species. The use of a sterile drape and preparation of the surgical site with dilute povidone iodine solution has been recommended. The use of these chemical agents may disrupt the normal skin flora of the patient and the constant mucous production of *Xenopus* skin makes any sterilization effort transient. When chemical surgical preps are used, they should be limited to the immediate area around the incision site and should only be solutions, not scrubs containing soaps or detergents."

"Similarly, arguments have been presented regarding the use of surgical drapes. Drapes may be useful to keep mucus from getting on instruments and suture material and can be moistened to keep skin from drying during surgery. However, amphibian skin can be easily damaged and paper drapes that become wet pose no barrier to bacteria. NIH veterinarians report that the incidence of clinical complications following surgical oocyte harvesting is rare." It is also worth noting here that the skin of fish is also protected with antimicrobial rich mucus as well as scales that need to be protected for rapid healing and tissue repair.

In the authors' opinion, the surgeon should be wearing sterile, powderless, surgical gloves and surgical drapes need not be used. The authors are unaware of requirements by any research institution for survival surgery on non-mammalian aquatic species to be performed in a surgical suite. As with surgeries for rodents, it is advisable to perform such surgery in at least a dedicated space, since such an area keeps a focus on the surgery, needs of the patient and the proper environment for conducting the surgery with expediency given the less-than-ideal conditions.

A dilemma exists with respect to the use of organic iodine (povidone iodine) as a skin antiseptic as a means to not only improve aseptic conditions, but also to reduce the possible contamination of harvested oocytes from *Xenopus*. Any potential benefit must be balanced against the possibility of chemical harm to the skin. It is not clear at this point that any antiseptics prevent infections and/or enhance wound healing as part of surgical harvest of *Xenopus* oocytes, but, until shown otherwise, it might be assumed that use of such antiseptics offer benefit to the animal and to the harvested oocytes. Alternatively, one could express the egg mass through a dilute povidone solution the way neonates are treated for cesarean derived rodents, followed with a saline rinse.

Surv. Due to the limited options in the application of the aseptic technique in fish, prior to surgery, does your IACUC require water testing (e.g., pH, ammonia, microbial counts, etc.) to minimize wound contamination?

- Not applicable. 104/193 (54%)
- No, we do not require water testing (e.g., pH, ammonia, microbial counts, etc.) to minimize wound contamination. 52/193 (27%)
- Yes, we do require water testing (e.g., pH, ammonia, microbial counts, etc.) to minimize wound contamination. 37/193 (19%)

Compared to other species, the literature on fish analgesia is scant. Does your IACUC require the use of analgesics in fish undergoing surgical procedures?

- Not applicable. 103/193 (53%)
- No, we do not require the use of analgesics in fish undergoing surgical procedures. 36/193 (19%)
- Yes, we do require the use of analgesics in fish undergoing surgical procedures. 54/193 (28%)

Due to the limitations in physiological monitoring of fish during the anesthetic period, does your IACUC require minimum monitoring (e.g., opercular movement, water temperature, etc.)?

- Not applicable. 99/193 (51%)
- No, we do not require minimum monitoring (e.g., opercular movement, water temperature, etc.). 19/193 (10%)
- Yes, we do require minimum monitoring (e.g., opercular movement, water temperature, etc.). 75/193 (39%)

19:24 Would less than optimal aseptic technique be acceptable if a PI's records indicated excellent surgical success with a lack of postoperative infections?

Reg. The AWAR (§2.31,d,1,ix) require that all survival surgery on regulated species be performed using aseptic technique. The *Guide* (p. 118) similarly indicates that aseptic technique should be followed for all survival surgical procedures.

Opin. No. In the authors' opinion, the performance of survival surgery using less than optimal aseptic technique indicates that although the IACUC and AV reviewed and approved the procedures, the review was not thorough enough, or the investigator was inadequately trained in aseptic technique. In other words, the IACUC did not fulfill its responsibility to assure use of acceptable standards of aseptic technique.

Aseptic technique is used to reduce microbial contamination of a surgical wound and expose tissues to the lowest possible practical level[13] (*Guide*, p. 118). Using less than optimal aseptic technique potentially increases bacterial contamination and subsequent risk of clinical (apparent) and subclinical (unapparent) infection which could compromise the animal's health and research data. Lack of clinical evidence of postoperative infection does not rule out clinically unapparent (subclinical) infection. A study performed in rats showed that infection can be clinically unapparent and cause unrecognized, adverse physiologic, immunological, and behavioral responses that can affect research results.[18]

The IACUC should be cautioned not to focus on the product rather than on the process. The IACUC should regularly review approved aseptic techniques vis-à-vis currently acceptable standards of veterinary care.

19:25 Should the IACUC demand that aseptic technique be used when surgery is performed on farm animals in the field (e.g., standing rumenotomy performed in a barn)?

Reg. The AWAR (§2.31,d,1,ix), which apply to farm animals used in biomedical research, state that survival surgery conducted at field sites does not require dedicated facilities but must be performed using aseptic procedures. PHS Policy (IV,A,1) requires compliance with standards for survival surgery as outlined in the *Guide* when using farm animals in biomedical research. The *Guide* (p. 115) recognizes that modification of standard techniques might be necessary when performing field surgery; however, animal well-being should not be compromised. When modifications are implemented, a thorough assessment of surgical outcomes should be performed to ensure that appropriate procedures are followed. Surgical outcome assessment may require other criteria in addition to clinical morbidity and mortality.

When using farm animals in agricultural research, standards for survival surgery, as outlined in the *Guide for the Care and Use of Agricultural Animals in Agricultural Research and Teaching* (*Ag Guide*, 4th Edition, 2022),[24] should be applied. The *Ag Guide* (p. 12)[25] states: "Major survival surgeries should be performed in facilities designed and prepared to accommodate surgery whenever possible, and appropriate aseptic surgical procedures should be used. Good surgical practice includes the use of surgical caps, masks, gowns, and sterile gloves, as well as aseptic surgical site preparation and draping." Further, "Minor surgical procedures that do not penetrate a body cavity or produce substantial impairment (e.g., wound suturing, peripheral-vessel cannulation, certain standard agricultural practices) may be performed under less stringent conditions in accordance with standard agricultural practices. Therapeutic and emergency surgeries (e.g., caesarean section, treatment of bloat, repair of displaced abomasum) may sometimes need to be performed in agricultural settings that are not conducive to rigid asepsis. However, every effort should be made to conduct such surgeries in a sanitary or aseptic manner and to use anesthetics or analgesics commensurate with the risks to the animal's well-being."

Opin. When farm animals in the "field" require elective major survival surgery, it should be performed in facilities designed and maintained for surgery using appropriate aseptic procedures, including cap, mask, sterile gown and surgical gloves, and sterile instruments. Therefore, an elective rumenotomy should not be performed in a barn. Minor survival procedures may be performed under less stringent conditions than major procedures but require sterile instruments and aseptic technique. When therapeutic and emergency surgeries do not allow transport to dedicated surgical facilities, the facilities used should be clean and methods of aseptic technique used to prevent or minimize exposure of the animal to pathogenic organisms to reduce the likelihood of infection.

19:26 How can the IACUC ensure that personnel are qualified to perform surgical procedures?

Reg. The PHS Policy (IV,C,1,f) and AWAR (§2.31,d,1,viii) place responsibility with the IACUC to determine that personnel performing surgical procedures are qualified and trained in those procedures. NIH/OLAW further describes application of the PHS Policy by stating that institutions must "ensure that research staff members performing experimental manipulation, including anesthesia and surgery, are qualified through training or experience to accomplish such procedures humanely and in a scientifically acceptable fashion.[24,26]

The *Guide* (pp. 115–116) recognizes that personnel performing surgery on research animals have a wide range of educational backgrounds and might require various levels and kinds of training to ensure that good surgical technique is used. Research institutions should also perform continuing and thorough assessments of surgical outcomes to ensure that appropriate procedures are followed.

Opin. The IACUC's goal is to assess surgical competency before IACUC approval of a protocol. The IACUC should assess surgical skills and proficiency by some method prior to allowing individuals to participate in animal activities and this may need to include direct observation in addition to reviewing an individual's education, training, certification, and experience for assessment of general surgical competency and qualifications to perform the specific surgical procedure. To assist the AV and IACUC in developing appropriate training programs, the Academy of Surgical Research (ASR) developed and published training guidelines for research surgery commensurate with a person's formal education and training background.[19] The ASR training guidelines can be used as a tool when evaluating a person's educational background to determine what an individual may already know. The IACUC is empowered to suggest incorporation of objective assessment tools that can reduce subjectivity of the assessment process. For example, a physician trained in a surgical specialty may need less training to perform surgery on animals within his area of surgical expertise. However, he or she may require training in interspecies variations of anatomy, anesthesia, analgesia, and postoperative care methods. The need for formal training might be waived if the surgeon participates in a multidisciplinary team approach to perform the specific surgical protocol and includes additional experienced personnel qualified to work with animals.

To assure a surgeon's competence, the IACUC may require that a laboratory animal veterinarian observe or assist with at least the first surgery. Assistance could continue with subsequent procedures until competency is achieved and a specific surgical procedure is predictable to the satisfaction of the veterinarian. Although there is no single credential to ensure competency, the best credential is a documented record of previous successful performance of the proposed surgical procedure on the specified species, demonstrating minimal operative and postoperative complications. With such documentation, there may be no need for additional training. After IACUC approval of a surgical protocol, continuing objective assessment of surgical outcomes should be performed. Participation and input from a laboratory animal veterinarian, animal care staff, surgical technicians, and the investigator are needed to ensure ongoing use of appropriate procedures, to address complications and evaluate their rate of occurrence, and to initiate necessary corrective changes.

19:27 What is meant by perioperative care?

Reg. Activities that involve surgery must include appropriate provision for preoperative and postoperative care of the animals in accordance with established veterinary medical and nursing practices (§2.31,d,1,ix), and the program of adequate veterinary care must include adequate pre-procedural and post-procedural care in accordance with current established veterinary medical and nursing procedures (§2.33,b,5).

Opin. Perioperative care encompasses all events associated with a surgical procedure. A perioperative care program comprises three overlapping components.[25]

- Preoperative planning and management
- Intraoperative care
- Postoperative care

Detailed descriptions of the perioperative care program components have been published. Preoperative planning should include

- Identification of members of the multidisciplinary surgical team, all of whom should provide input into presurgical planning
- Roles and training needs of personnel
- Equipment and supplies required
- Facilities for conducting procedures

A balanced anesthetic protocol should be developed, including anesthetic agents, techniques, and methods of anesthetic monitoring to be used. Planning of the surgical procedure should include aseptic techniques to be used and assessment of indications for perioperative antibiotics. A postoperative care plan should be outlined. Preoperative management should include:

- A preoperative animal-health assessment with a physical examination and laboratory examination if indicated
- A period of stabilization to a new environment for animals before undergoing surgical procedures
- Preoperative fasting of a specified duration if indicated for the species to be used
- Administration of preoperative medications, analgesics, or antibiotics

Components of intraoperative care include:

- Monitoring of anesthetic level and vital organ function
- Provision of vital organ support such as parenteral fluid administration, supplemental oxygen, and maintenance of body temperature
- Proper surgical technique, which comprises:
- Gentle tissue handling
- Effective hemostasis
- Maintenance of sufficient blood supply to tissues
- Asepsis
- Tissue hydration maintenance
- Accurate tissue apposition
- Proper use of surgical instruments
- Appropriate use of monitoring equipment
- Expeditious performance of the surgical procedure

The postoperative period can be divided into three overlapping phases: recovery from anesthesia, acute postoperative care, and long-term postoperative care. Frequent assessment of thermoregulation and cardiovascular and respiratory function is required during anesthetic recovery and acute postoperative care. Additional care should also include:

- Monitoring of the surgical incision
- Thermal support to combat hypothermia
- Parenteral fluid administration to maintain hydration
- Administration of analgesics for postoperative pain
- Administration of prophylactic antibiotics and other drugs

Long-term postoperative care after anesthetic recovery and adequate physiologic stabilization requires at least once-daily monitoring until sutures are removed (usually at 10–14 days postoperatively) and any postoperative complications are resolved. Monitoring of vital signs, hydration, feed and fluid intake, feces and urine output, attitude and activity, surgical wound condition, body weight, and signs of postoperative pain and infection should be performed. Special diets, analgesics, bandaging, antibiotics, and other medications may be indicated. After suture removal, postoperative care required depends upon the species and surgical procedure. For example, chronic catheters or other partially exteriorized implants require ongoing monitoring and care. Monitoring of body weight should be scheduled throughout the postoperative period. More recently there has been an emphasis on non-pharmaceutical methods of pain control. As such, they should be considered during the postoperative period as ancillary methods of pain control.

19:28 What level of perioperative care is appropriate for rodents when compared to nonrodents?

Opin. The general components of the perioperative care programs are the same for rodents and nonrodents. Nevertheless, some elements of the programs differ. For rodent surgery, the surgical "team" may be reduced to one person, who serves as surgeon, anesthetist, surgical technician, and scrub nurse. When one person performs surgery on multiple animals at one sitting, as frequently occurs in rodent surgery, careful pre-surgical planning is required to ensure availability of all supplies and equipment required to perform surgery and support necessary modifications in standard aseptic technique. The preoperative animal-health assessment may include vendor-supplied colony health testing and a visual examination rather than a physical examination of each animal when received.[11] Serologic testing of sentinel animals in the facility or other health surveillance, such as environmental air dust results, also may be useful.

Although methods for intraoperative monitoring of anesthetized rodents similar to ones used with larger species are available when scientifically required, such methods might not be practical or possible in some research situations. Relying on observing chest wall movement to determine respiratory rate and palpating the apical pulse through the chest wall may be sufficient to assess cardiovascular and respiratory stability.[27] However, for complex procedures such as cardiac surgery, IACUCs should be empowered to request implementation of physiological monitoring with appropriate tools. Procedures used in larger animal species for intraoperative vital organ support such as intravenous fluid therapy can be difficult to use in rodents and may require other routes of administration of fluids or other strategies for supporting circulating blood volume.

The same intensity of monitoring and supportive care commonly provided for larger animals during recovery from anesthesia and acute postoperative care might require modification for rodents. In addition, performing surgery on multiple animals at one sitting, which frequently occurs, requires developing methods of supportive care for the anesthetic recovery of multiple animals simultaneously.

19:29 How much detail should be provided in the IACUC protocol relative to perioperative care?

Opin. In the authors' opinion, the perioperative care detail that should be provided in the IACUC protocol includes the following:

- Names, qualifications, and responsibilities of participating personnel, including the surgeon, the person who will be administering and monitoring anesthesia, and the person who will perform postoperative care
- Location where surgery will be done
- Duration of fasting, with rationale
- Perioperative medications, fluids, and antibiotics; dose, volume, route, and frequency of administration; duration of treatment
- Intraoperative monitoring (including temperature) and methods, including equipment to be used to assess objectively adequate anesthesia level
- Surgical anesthesia monitoring system if neuromuscular blocking agents are administered
- Aseptic techniques to be used
- Perioperative heat support to be provided
- Frequency of animal monitoring during anesthetic recovery and postoperative period
- Postoperative monitoring and care
- Indication of compliance with postoperative monitoring and care guidelines in the institution's IACUC policies
- Indication of expected health changes or possible postoperative complications and description of methods of monitoring and care
- Perioperative analgesics to be given; agent, dose, route, and frequency of administration; criteria for determining need for analgesics

19:30 Can expired pharmaceuticals be used for *survival surgery*?

Reg. The Animal Welfare Inspection Guide advises inspectors that expired medications or medical supplies may not be used for covered animals. NIH/OLAW states: "Euthanasia, anesthesia and analgesia agents should not be used beyond their expiration date, even if a procedure is terminal."[28]

Opin. In the authors' opinion, expired drugs and fluids of any kind should not be used in animals undergoing survival surgery. The use of expired anesthetics, analgesics, euthanasia, and emergency drugs should not be allowed regardless of the procedure intended. Loss of pharmacological integrity may lead to insufficient pain relief or other unexpected anesthetic problems. It is fundamental to adequate veterinary care that the user has confidence in the administered drugs; this is best assured by use of non-expired pharmaceuticals.

19:31 Is the use of expired pharmaceuticals acceptable for non-survival surgery?

Reg. NIH/OLAW states: "Euthanasia, anesthesia and analgesia agents should not be used beyond their expiration date, even if a procedure is terminal."[4]

Opin. Without written documentation of safety and efficacy from the manufacturer and IACUC approval of an investigator's written request for their use in non-survival surgery, use of expired drugs and fluids should not be allowed for animals undergoing non-survival surgery.

19:32 Should the IACUC request information on the type of suture materials that will be used?

Opin. The PI should describe the type of suture material and method of wound closure on the protocol form to assure that they are appropriate for the species and the surgical incision. To facilitate wound healing, the type of suture material used to close the skin should be chosen to minimize tissue reaction and potential for wound infection. For example, to produce a lesion on healthy skin, an inoculum of 10^7 or more *Staphylococcus aureus* organisms is required, but in the presence of a silk suture, the required inoculum is reduced to less than 10^3 bacteria.[29] The capillary action of braided suture material and the inherent difficulty in keeping a wound clean can combine to increase the probability of infection at the surgical site.[29] Appropriate wound closure using methods that bury sutures (e.g., subcuticular suture pattern) should be considered when animals may bite or pick at exposed skin sutures and/or reduce the amount of suture and knot requirement under the skin (e.g., barbed suture) when animals may bite or pick at the suture irritating the skin.

19:33 Should the IACUC request information about when sutures (or clips) will be removed, if that is necessary?

Opin. The appropriate time for suture, skin staple, or clip removal might be provided to investigators in an institutional IACUC policy, which should contain guidelines on postoperative care. Rather than request that the PI restate when sutures or clips will be removed, a written indication on the protocol form that the PI has read and will comply with the postoperative care guidelines provides adequate assurance that timely suture removal will be done. If written IACUC guidelines are not available to the PI, the IACUC should request that the PI state on the protocol form when sutures or clips will be removed. IACUCs should feel empowered to recommend methods to eliminate the need of suture/clip removal (e.g., utilization of subcuticular suture pattern).

19:34 Should the IACUC request information on how rodents or other small animals will be kept warm during and after surgery?

Reg. The AWAR (§2.33,b,4–5) require appropriate care in accordance with current accepted veterinary medical and nursing practices. The *Guide* (pp. 26, 120) notes the importance of postoperative care, including thermoregulation.

Opin. Because of the large surface area-to-body mass ratio of rodents and other small animals, heat loss resulting in hypothermia is likely to occur without adequate thermal support during surgery and recovery from anesthesia. Maintenance of normal body temperature significantly reduces anesthesia-related cardiovascular and respiratory disturbances (*Guide*, p. 119) and is often critical to the successful recovery of rodents from anesthesia. Maintenance of normothermia can greatly reduce morbidity and mortality. Methods to combat hypothermia can be suggested to investigators in an institutional IACUC policy, which should contain guidelines for rodent surgery and postoperative care. Perioperative observations and support are often considered to be important aspects of species-specific training, including that for rodents. If the PI is expected to follow institutional guidelines with respect to body temperature maintenance, the protocol form could have an item in which the PI can indicate his or her intent to comply with these guidelines. This endorsement provides adequate assurance that appropriate thermal support will be provided. If written IACUC guidelines describing methods to combat hypothermia are not available to the PI, the IACUC should request that the PI state, on the protocol form, how rodents or other small animals will be kept under normal body temperatures during and after surgery.

19:35 Should the IACUC request information about the frequency of postoperative observations?

Opin. The frequency of postoperative observation is determined by the nature of the surgical procedure and the stage of recovery.[11] Guidelines for minimal frequency of postoperative observation could be provided to investigators in an institutional IACUC policy for postoperative care. A written indication on the protocol form, by the investigator, that she or he has read and will comply with these guidelines provides adequate assurance that appropriate postoperative observation frequency will be carried out. However, if a surgical procedure necessitates more frequent observations than the guidelines recommend, the customized monitoring frequency and circumstances requiring monitoring should be described in the protocol. If there are no written IACUC guidelines available to the PI, the IACUC should request that the PI provide information about the frequency of postoperative observations on the protocol form.

19:36 When is an animal considered to be sufficiently recovered to return it to its home cage?

Opin. Many biomedical scientists and veterinarians believe that an animal should remain in a recovery area until it has recovered from anesthesia, physiological parameters are adequately stabilized, and it has regained ambulatory and protective reflexes before being moved to its home cage. The caveat to this is that application and practicality are at least species and size dependent. Physiological stability and ambulatory behavior might be very different for an animal that has undergone a thoracotomy versus one that has undergone arthroscopy. As continuous observation by caregivers is a current standard of care in veterinary medicine, this observation should be carried out until the animal can be extubated, the core temperature approaches normothermia, and the patient remains in sternal recumbency.

Preemptive analgesia may cause a postoperative animal to appear lethargic and postoperative pain may cause delayed recovery. Unlike a tranquilized animal, however, one treated with narcotic or non-narcotic analgesics will respond to non-pain stimulation (e.g., retching). The ability to stay sternal (the return of the righting reflex) is associated with the ability to raise the head as needed and allows the animal to cope with postoperative vomiting should it occur. The ability to raise the head is also an added safeguard when there is need to place water bowls that are shallow enough to prevent accidental immersion of the nostrils.

19:37 May animals that have undergone surgery be recovered in their home cage, or is there a need for a separate recovery area?

Reg. The *Guide* (p. 145) states that "a postoperative-recovery area should provide the physical environment to support the needs of the animal during the period of anesthetic and immediate postsurgical recovery and should be sited to allow adequate observation of the animal during this period." The species and type of surgical procedure will dictate the type of caging and support equipment required, which should be designed to support physiologic functions, such as thermoregulation and respiration.

Opin. During anesthetic recovery, an animal should be in an area that is warm, safe, quiet, comfortable, and appropriate for the needs of the individual animal and species. Dedicated recovery rooms are recommended for species such as rabbits, cats, dogs, and swine. A separate recovery area provides the necessary physical environment, ambient temperature control, and equipment for monitoring and supportive care to manage complications that may occur during recovery. Animals can be housed individually in appropriately sized cages designed to prevent injury to occupants. Individual housing also prevents potential trauma from cage mates. In addition, recovery areas can be located to facilitate the appropriate frequency of monitoring an animal by personnel.

Small ruminants that are too large to fit into recovery cages and large farm animals can recover in their own pen or stall if provisions are made for adequate observation, warmth, and animal safety. The added advantage of conspecific odors, sounds, and visual contact may enhance the animal's recovery. Foam- or air-filled padding placed on the floor in secure, warm areas are also useful for these species. The additional space afforded by a pen or stall also allows for assistance with changing position and moving larger species that may be unable to move themselves or may begin thrashing about during recovery. Regardless of the location, the criteria noted in 18:36 also apply: monitoring of the animal until sternally recumbent, a core body temperature approaching normothermia, and extubation of the animal.

Rodents often recover from anesthesia in the same laboratory where surgery is performed. Provisions to support body temperature are frequently necessary as thermal support is often critical to the successful recovery of rodents from anesthesia. Recovering rodents may need to be housed individually to prevent injury by cage mates. Although continuous monitoring may not be possible, the animals should be observed frequently until they recover from anesthesia and adequately stabilized before being returned to their home cages.

19:38 How long should investigators keep perioperative records and what type of records should be required?

Reg. Although the AWAR specifically require medical records only for dogs (§3.13,b) and marine mammals (§3.110,d), the AWAR (§2.33,b,5) indicate that a program of adequate veterinary care is required including adequate pre- and post-procedural care in accordance with current established veterinary medical and nursing procedures. Since medical records are a component of adequate veterinary care, the need for medical records may be inferred. The AWAR (§2.35,f) require that all records and reports must be maintained for at least three years past the end of the activity; hence, this might infer that medical and surgical records should also be kept for this period of time. The *Guide* (p. 115) states that medical records are a key element of the veterinary care program and are considered critical for documenting animal well-being as well as tracking animal care and use at a facility.

Opin. The specific information requested in medical records is detailed in The *Institutional Animal Care and Use Committee Guidebook*,[23] which suggests that, as best practice, "an intra-operative anesthetic monitoring record should be kept and included with the surgeon's report as part of the animal's records. This record should be available to the personnel providing postoperative care. Postoperative records at a minimum should reflect that the animal was observed until it was extubated and had recovered the ability to stand. These should be supplemented by records evaluating the animal's recovery, administration of analgesics and antibiotics, basic vital signs, monitoring for infection, wound care, and other medical observations."

Preoperative records should include a health profile and medical and vaccination history (when applicable) provided by the vendor and subsequently supplemented by the clinical care staff during quarantine or with annual updates, as appropriate. Examples of additional information include vaccinations for dogs or cats, quarterly or semiannual blood work, tuberculosis testing and serological profiles in nonhuman primates. A physical exam to assess an animal's health status should be performed. A preanesthetic evaluation may include records of body weight, body temperature, heart rate, and respiratory rate to establish baseline data, recognizing that these values might be influenced by preoperative or preanesthetic medications depending on the species being worked on. The need for diagnostic radiographic tests and laboratory evaluation of blood, urine, and feces will depend on the animal species, research protocol, and health history of the animal or colony. Preoperative administration of medications and antibiotics should be documented.

An anesthetic record should document anesthetic administration, monitoring of anesthetic depth (including reflexes), and frequent assessment of the physiologic status of an animal, including body temperature and cardiovascular and respiratory function.[30]

The extent of recordkeeping and monitoring depends upon the potential complications associated with the anesthetic regimen, the surgical procedure, and the animal species. Minimal documentation should include vital signs, core body temperature, heart rate, and respiratory rate, recorded at an average interval of every 15 minutes, and the anesthetic administered, dose, route, and time of administration.

The postoperative period can be divided into three overlapping phases: recovery from anesthesia, acute postoperative care, and long-term postoperative care.[11] The extent of recordkeeping and intensity of monitoring during the postoperative period depend upon the species and the surgical procedure. Generally, the most intensive and frequent monitoring is required during recovery from anesthesia, since it is then that the animal is most vulnerable to the combined adverse effects of anesthesia and hypothermia. Frequent assessment of thermoregulation and cardiovascular and respiratory function should continue during anesthesia recovery and acute postoperative care until the animal is adequately stabilized. Vital signs, including body temperature, pulse rate and rhythm, respiratory rate and rhythm, and oxygen saturation, should be recorded frequently during anesthesia recovery with less frequent monitoring possible as an animal stabilizes. Documentation of clinical observations, including mucous membrane color, capillary refill time, pain assessment, and surgical wound condition is recommended where applicable. Administration of analgesics, antibiotics, parenteral fluids, and other medications should be recorded and should include the drug name, dosage, route, and time given. A brief description of the surgery should be recorded with the postoperative care records or as a separate surgeon's report, with a notation about complications and/or unexpected need to deviate from the approved surgical plan.

During long-term postoperative care after anesthetic recovery and adequate physiologic stabilization, monitoring should continue at least once a day until suture removal (usually at 10–14 days postoperatively) and any postoperative complications are resolved. Vital signs, hydration, feed and fluid intake, feces and urine output, attitude and activity assessment, pain assessment, surgical wound condition and care, and monitoring for postoperative infection should be recorded. All postoperative care and medications administered as described with anesthesia recovery should be documented. After suture removal, frequency and parameters monitored depend on the species and surgical procedure. For example, chronic catheters or other partially exteriorized implants require documentation of ongoing monitoring and care. Monitoring of body weight should be scheduled throughout the postoperative period.

19:39 What is considered health record maintenance "consistent with current veterinary practice" and how are these records used?

Reg. The AWAR (§2.33,b,5; §2.35,f) indicate that adequate pre- and post-procedural care is expected and that all records and reports must be maintained for at least three years. Individual medical records are required for dogs and marine mammals. Medical records for dogs must include the identity of the animal; the date and description of any problem, examination findings, test results, plan for treatment and care, and treatment procedures performed; the names of all vaccines and treatments administered and the dates of administration; and, the dates and findings of all screening, routine, or other required or recommended test or examination (§3.13,b). Medical records for marine mammals must include animal identification, physical examination information, identification of all medical and physical problems with proposed plan of action, all diagnostic test results, and documentation of treatment (§3.110) (see 19:38). The *Guide* (p. 115) states: "Medical records are a key element of the veterinary care program and are considered critical for documenting animal well-being as well as tracking animal care and use at a facility."

Opin. The animal health record describes a problem and observations, action plan, and resolution or lack thereof, whether alone or superimposed on other problems. It is a communication tool that allows for a variety of trained persons to interact with the patient in a meaningful way and have others on the care team recognize exactly what was done, why it was done, and why an anticipated outcome may or may not have been realized. It is a solid mechanism for communication between the clinical care staff and similarly trained physicians/surgeons. In the research environment, especially with novel animal models, the reason for an unexpected death can frequently be resolved by consulting a well-documented animal health record. The well-organized and documented animal health record serves to put everyone on the same page, is critical for good research, and demonstrates documented attention to animal welfare.

19:40 Should the IACUC request information on how rodents or other small animals will be supplemented with oxygen during surgery?

Opin. Anesthesia depresses the cardio-respiratory centers, frequently leading to hypoxia. Although most inhalant anesthesia is vaporized with pure oxygen, the same cannot be said when administering injectable anesthesia (e.g., ketamine and xylazine). Instead, animals are allowed to breathe room air, which at sea level contains 21% oxygen. Therefore, an already compromised animal can become further hypoxic when breathing room air. A study showed that oxygen saturation levels in these animals may drop to a dangerous level, which is dose-dependent and results in higher post-op morbidity and mortality.[31] The study suggested that mice anesthetized with injectable (and inhalant) anesthesia without supplemental oxygen are routinely hypoxic. Animals that survive these hypoxic episodes may suffer damage to various organs, and particularly the brain, which could impact experimental results.

In the authors' opinion, IACUCs should consider requiring oxygen supplementation for animals undergoing anesthesia.

Surv. Does your IACUC require oxygen supplementation with injectable anesthesia?

- Not applicable. 45/193 (23%)
- No, we do not require oxygen supplementation with injectable anesthesia for any species. 88/193 (46%)
- Yes, we do require oxygen supplementation with injectable anesthesia for some (e.g., large animals) but not all (e.g., rodents). 60/193 (31%)

19:41 Can digital noninvasive real-time animal monitoring technologies (video, radio frequency identification (RFID), electromagnetic filed (EMF)) be used in place of in-person post-surgical assessments of animals?

Reg. The *Guide* (pp. 119–121) states: "An important component of postsurgical care is observation of the animal and intervention as necessary during recovery from anesthesia and surgery. The intensity of monitoring will vary with the species and the procedure and may be greater during the immediate anesthetic recovery period, and it continues. After recovery from anesthesia, monitoring is often less intense but should include attention to basic biologic functions of intake and elimination and to behavioral signs of postoperative pain, monitoring for postsurgical infections, monitoring of the surgical incision site for dehiscence, bandaging as appropriate, and timely removal of skin sutures, clips, or staples."

Opin. Digital noninvasive real-time objective animal monitoring technologies are in early development; therefore, they should not replace all in-person post-surgical assessments. They can, however, complement these assessments with behavioral and/or physiological information. Knowledge about an animal's activity, mobility, respiratory rate, and food and water intake provided by these systems can provide a more holistic post-surgical assessment. Real-time video assessment provides an opportunity to monitor animals more frequently without entering a facility. Digital technologies may enhance the sensitivity of assessing postoperative pain and analgesic intervention and may be used as a complementary monitoring tool to refine postoperative care.

Surv. If digital monitoring (e.g., video, Radio Frequency Identification (RFID), electromagnetic force (EMF)) is used to monitor animal health and provides alarms overnight, does your IACUC require personnel to assess animals in response to such alarms?

- Not applicable. 42/193 (22%)
- No, we do not refinement/alternatives consideration for surgical and anesthetic procedures. 27/193 (14%)
- Yes, we do require refinement/alternatives consideration for surgical and anesthetic procedures. 124/193 (64%)

19:42 Does your institution have a policy on the adoption of animals that have undergone surgery?

Reg. The FDA,[31] AVMA,[32] and VA[33] have policies addressing adoption of research animals. As the PHS Policy applies to animal research conducted by PHS agencies, such as NIH, OLAW's recommendations are also relevant to the disposition of NIH IRP animals.[34] The Code of Federal Regulations recordkeeping expectations offer institutions the option of developing and implementing a policy for transfer of ownership.[35] In 2014, the NIH DDIR authorized the transfer of ownership of NIH laboratory animals from the federal government to a private recipient with the following stipulations.[36]

- The approval authority for the transfer of ownership of a research animal resides with the Scientific Director or their designee.
- The IC (NIH Institutes and Centers) must complete a transfer of ownership agreement that identifies the terms of the transfer, health of the animal at the time of the transfer, and relinquishes NIH of responsibility for the animal once the transfer has occurred.

- The transfer of ownership agreement should be developed and reviewed by the NIH Office of General Counsel (OGC).

The USDA requires Form 7006 (Record of Disposition of Dogs and Cats) for adoption of dogs and cats and Form 7020 for all other animals (Record of Acquisition, Disposition or Transport of Animals (Other than Dogs and Cats).[37]

Opin. Does the regulatory or FDA guidance allow the adoption of animals that have undergone surgery? Adoption of animals is permitted including animals that have undergone a surgical procedure, and authors are aligned with the American College of Laboratory Animal Medicine's (ACLAM) support of "the concept of adoption of healthy, post-study, research animals into long-term, caring private homes or farms that can provide appropriate and humane living conditions for these animals as pets."[38]

It is highly recommended to develop an adoption policy that considers state and local laws and includes background checks of adoptees, Attending veterinarian's approval, and the species' suitability as pets. Adopted animals should be spayed or neutered and vaccinated, and their health status assessed. Genetically modified and/or animals that have received experimental compounds do not have to be automatically excluded from adoption; however, the impact of these factors on animal's long-term prognosis and adopter's safety should be considered and addressed within the adoption policy. Transparency about the animal and animal's history minimizes potential challenges.[39] The AALAS IACUC resource page provides links to adoption policies and forms from various institutions.[40]

Surv. Does your institution have an animal adoption program/policy for animals that have undergone surgery?

- Not applicable, we don't adopt animals and/or do not conduct surgery. 75/193 (39%)
- No, we do not have a policy specific to adoption of animals who have undergone surgery. 27/193 (14%)
- We have a policy specific to adoption of animals who have undergone surgery. 7/193 (4%)
- We have general policy on adoption of animals. 67/193 (35%)
- We do not have a general policy on adoption of animals. 16/193 (8%)
- Don't know. 1/193 (1%)

Acknowledgment

We would like to thank and acknowledge Lester L. Rolf, Jr., who authored the original version and the 2nd and 3rd editions of this book.

REFERENCES

1. Silverman, J. 1994. Protocol review: whose responsibility is it? *Lab Anim. (NY)* 23(2):22.
2. FDA Issues Final Guidance on Animal Drug Compunding from Bulk Drug Substances. https://www.fda.gov/news-events/press-announcements/fda-issues-final-guidance-animal-drug-compounding-bulk-drug-substancess. Accessed January 1, 2023.
3. National Institutes of Health. Office of Laboratory Animal Welfare. Sample Animal Study Proposal. http://olaw.nih.gov/sites/default/files/aacu3040-2.htm#surg. Accessed October 15, 2022.
4. Office Laboratory Animal Welfare FAQs.F.13 Are Laparoscopic Procedures Considered Major Surgery? https://olaw.nih.gov/faqs#/guidance/faqs. Accessed January 1, 2023.
5. American Veterinary Medical Association. 2020. AVMA Guidelines for the Euthanasia of Animals: 2020 Edition. https://www.avma.org/sites/default/files/2020-02/Guidelines-on-Euthanasia-2020.pdf. Accessed December 31, 2022.
6. Devitt, C.M., R.E. Cox, F.N. Hailey. 2005. Duration, complications, stress, and pain of open ovariohysterectomy versus a simple method of laparoscopic-assisted ovariohysterectomy in dogs. *J Am Vet Med Assoc* 15;227(6):921–7.
7. Hancock, R.B., O.I. Lanz, D.R. Wldron, *et al.* 2005. Comparison of postoperative pain after ovariohysterectomy by harmonic scalpel-assisted laparoscopy compared with median celiotomy and ligation in dogs. *Vet Surg.* 34(3); 273–282.
8. National Research Council. 2003. *Guidelines for the Care and Use of Mammals in Neuroscience and Behavioral Research.* Washington, DC: National Academy of Science Press.
9. Perret-Gentl *et al.* 1999. Videoendoscopy: an effective and efficient way to perform multiple visceral biopsies in small animals. *J Invest Surg.* 12:157–165.
10. Perrett-Gentil, M.I., M.N. Sinan, M.B. Dennis, *et al.* 2000. Videoendoscopic techniques for collection of multiple, serial intra-abdominal biopsy specimens in HIV-negative and HIV-positive pigtail macaques (Macaca nemestrina). *J Investigat Surg.* 13(4); 181–195.
11. National Institutes of Health. Office of Laboratory Animal Welfare. Sample Animal Study Proposal. http://olaw.nih.gov/sites/default/files/aacu3040-2.htm#surg. Accessed October 15, 2022.
12. Brown, M.J., P.T. Pearson, F.N. Tomson. 1993. Guidelines for Animal Surgery in Research and Teaching. *Am J Vet Res.* 54:1544–1549.
13. Policy on Humane Care and Use of Laboratory Animals - OLAW. https://olaw.nih.gov/guidance/articles/lab_animal2003v32n9_wolff.htm. Accessed on January 1, 2023.

14. National Institutes of Health, Office of Laboratory Animal Welfare. 2013. Frequently Asked Questions, A.2, Does the PHS Policy Apply to the Production of Custom Antibodies Or to the Purchase of Surgically Modified Animals? http://grants.nih.gov/grants/olaw/faqs.htm#App_2. Accessed January 1, 2023.
15. Bradfield, J.F., T.R. Schachtman, R.M. McLaughlin, *et al.* 1992. Behavioral and physiologic effects of inapparent wound infection in rats. *Lab Anim Sci.* 42:572–578.
16. Hodridge, J, M.Nichols, W. Dupont, *et al.* 2021. The effectiveness of hot bead sterilization in maintaining sterile surgical instrument tips across sequential mouse surgeries. *J Am Assoc Lab Anim Sci.* 2021;60(6):700–708.
17. Emmer K, N. Celeste, W. Bidot, *et al.* 2019. Evaluation of the sterility of Press'n seal cling film for use in rodent surgery. *J Amer Assoc Lab Ani Sci.* 58:235–239.
18. Hoogstraten-Miller, S.L., P.A. Brown. 2008. Techniques in aseptic rodent surgery. *Curr Protoc Immunol.* Chapter 1:Unit 1.12.1–1.12.14.
19. Sharp, P., J. Villano. 2012. *The Laboratory Rat*, 2nd edition. Boca Raton: CRC Press.
20. Slattum, M.M., L. Maggio-Price, R.F. DiGiacomo, *et al.* 1991. Infusion-related sepsis in dogs undergoing acute cardiopulmonary surgery. *Lab Anim Sci.* 41:146–150.
21. Applied Research Ethics National Association/Office of Laboratory Animal Welfare. 2002. *Institutional Animal Care and Use Committee Guidebook*, 2nd edition, Chapter C.3.g. Bethesda: National Institutes of Health.
22. National Institutes of Health, Office of Animal Care and Use. 2010. Guidelines for Egg and Oocyte Harvesting in Xenopus Laevis. http://oacu.od.nih.gov/ARAC/documents/Oocyte_Harvest.pdf. Accessed January 23, 2023.
23. Committees to revise the Guide for the Care and Use of Agricultural Animals in Research and Teaching. 2021. *Guide for the Care and Use of Agricultural Animals in Research and Teaching*, 4th Edition. Champaign, IL: Federation of Animal Science Societies.
24. National Institutes of Health, Office of Laboratory Animal Welfare. 2013. Frequently Asked Questions, G.1, What Kind of Training Is Necessary to Comply with PHS Policy, and How Frequently should it be Provided? http://grants.nih.gov/grants/olaw/faqs.htm#instresp_1. Accessed January 23, 2023.
25. Brown, M.J., J.C. Schofield. 1994. Perioperative care. In *Essentials for Animal Research: A Primer for Research Personnel*, eds. B.T. Bennett, M.J. Brown, J.`C. Schofield, 79. Washington, DC: National Agricultural Library.
26. Gaertner, D.J., T.M. Hallman, F.C. Hankenson, *et al.* 2008. Anesthesia and analgesia for laboratory rodents. In *Anesthesia and Analgesia in Laboratory Animals*, eds. R.E. Fish, M.J. Brown, P.J. Danneman, A.Z. Karas, 239. New York: Academic Press.
27. National Institutes of Health, Office of Laboratory Animal Welfare. 2012. Frequently Asked Questions, F.5, May Investigators Use Expired Pharmaceuticals, Biologics, and Supplies in Animals? http://grants.nih.gov/grants/olaw/faqs.htm#useandmgmt_5. Accessed January 23, 2023.
28. McCurnin, D.M., R.L. Jones. 1993. Principles of surgical asepsis. In *Textbook of Small Animal Surgery*, 2nd edition, ed. D.H. Slatter, 114. Philadelphia: W.B. Saunders.
29. Brown, M.J. 1994. Principles of anesthesia and analgesia. In *Essentials for Animal Research: A Primer for Research Personnel*, eds. B.T. Bennett, M.J. Brown, J.C. Schofield, 39. Washington, DC: National Agricultural Library.
30. Effects of Oxygen Supplementation on Injectable and Inhalant Anesthesia in C57BL/6 Mice. 2021. Caroline E Blevins,1,2 Natalie A Celeste,3, and James O Marx1,2, *JAALAS* 60(3):289–297.
31. U.S. Food and Drug Administration. Placement of Research Animals After Study Completion. Version 1. November 18, 2019. https://blog.whitecoatwaste.org/wp-content/uploads/2020/01/FDA-Placement-Policy_111819.pdf. Accessed January 23, 2023.
32. American Veterinary Medical Association. Research Dogs and Cats Adoption Policy. https://www.avma.org/resources-tools/avma-policies/research-dogs-and-cats-adoption. Accessed January 21, 2023.
33. Department of Veterans Affairs. Policy for Adoption of Research Animals. https://www.va.gov/files/2020-11/A-011_AdoptionPolicy2020F_508.pdf. Accessed January 21, 2023.
34. National Archives. Code of Federal Regulations Title 9 Chapter I Subchapter A Part 2 Subpart C Section 2.35. Recordkeeping Requirements. https://www.ecfr.gov/current/title-9/chapter-I/subchapter-A/part-2/subpart-C/section-2.35. Accessed January 21, 2023.
35. National Institutes of Health. Guidelines for Post-Study Research Animal Disposition. https://oacu.oir.nih.gov/system/files/media/file/2022-04/a4_research_animal_disposition.pdf#:~:text=The%20NIH%20Office%20of%20Research%20Services%2C%20Division%20of,study%20%28ORS-DVR%20Animal%20Adoption%2FRetirement%20Policy--OAM%20Policy%20Number%20P10. Accessed January 21, 2023.
36. U.S. Department of Agriculture. Animal and Plant Health Inspection Service Form 7006. https://www.aphis.usda.gov/aphis/ourfocus/animalwelfare/sa_publications/ct_publications_and_guidance_documents. Accessed January 21, 2023.
37. American College of Laboratory Animal Medicine. Adoption of Research Animals Position Statement. https://www.aclam.org/about/position-statements. Accessed January 21, 2023.
38. Carbone, L., L. Guanzini, C. McDonald, 2003. Adoption options for laboratory animals. *Lab Anim.* 32(9):37–41.
39. American Association for Laboratory Animal Science. Animal Adoption Forms and Policies. https://www.aalas.org/iacuc/iacuc_resources/animal-adoption-forms. Accessed January 21, 2023.
40. Bernal, J., *et al.* 2009. Guidelines for rodent survival surgery. *J Investigat Surg.* 22:445–451.

20

Custom Antibody Production and Blood Collection

Noé Tirado-Muñiz

Introduction

Modern biological research techniques often require the production of custom antibodies (Abs) as an essential component of the research protocol. Abs, which have a wide range of applications, including diagnostic tests, analytic/chemical uses and as therapeutic drugs, can be divided into polyclonal or monoclonal antibodies according to their cellular origin. Polyclonal Abs (pAbs) are produced by different populations of immune system cells (B cells) and can bind to a variety of sections (epitopes) of a foreign substance or antigen (substances which induce an immune reaction). Monoclonal Abs (mAbs), on the other hand, are produced by a single type of B cell and bind to a specific epitope of an antigen (Ag).

Production of both polyclonal and monoclonal antibodies may require the immunization of animals, which may cause pain and or distress. This presents the IACUC with the dilemma and duty of evaluating the immunization procedures and schedules with respect to their potential negative impact on animal welfare. Published guidelines from a number of sources are available to assist the IACUC in developing policies and procedures for animal use in antibody production. [1–6]

In vitro alternatives to the use of live animals in the production of antibodies are often appropriate (see 20:11) and investigators should explore such methods. The use of animals should not be considered the default approach for the production of Abs (especially when using the mouse ascites method) and justification for not using non-animal alternatives should be established. However, this chapter primarily focuses on those circumstances in which the IACUC reviews and approves the use of live animals for antibody production.

Production of a high-quality antibody requires an appropriate stimulation of the immune system. The initial immune activating event involves processing of the Ag by specialized immune system cells. Activation of these cells is often enhanced by the addition of various biologically active compounds (adjuvants). Adjuvants aid the immune stimulation process in several ways, including protecting the Ag from rapid degradation and permitting the slow release of the Ag to these specialized immune system cells. The result is an increase in the desired response to the Ag. However, this also brings about the production of a substantial inflammatory reaction with undesirable tissue damage and potential pain and distress.

Of the available adjuvants in use today, none has proven more effective or controversial than Freund's complete adjuvant (FCA).[7–14] First described in 1942,[7,15] it was originally composed of paraffin oil, mannide monooleate, and killed mycobacteria. The presently available formulations of FCA, which still contain killed bacteria and non-metabolizable oils among other components, produce significantly fewer severe inflammatory lesions.[8,9] Intradermal injection of FCA produces a chronic skin inflammatory reaction at the injection site, which usually ulcerates[10] along with granulomatous inflammation and focal necrosis.[10–14,16,17]

However, the pain and distress associated with these skin lesions is debatable as literature reports are conflicting (see 20:2). Severe pain was reported in five of nine humans accidentally injected with FCA.[18] and a researcher who accidentally injected FCA into a finger self-reported pain and discomfort of several weeks duration.[19] Taking into account the U.S. Government Principles[20] ("Unless the contrary is established, investigators should consider that procedures that cause pain or distress in human beings may cause pain or distress in other animals"), the use of such adjuvants may be considered to cause pain or distress in animals. Because the adjuvants' undesirable side effects may be influenced by the type used, dose, administration route, etc., additional details should be sought by the IACUC to ascertain that their use is properly justified and that alternative methods (which avoid or minimize animal pain and/or distress) have been considered.

20:1 Should investigators be required to use off-the-shelf Abs if available?

Opin. Following the spirit of the 3Rs as promulgated by Russell and Burch,[21] investigators should be encouraged to use off-the-shelf products when available. In doing so, the IACUC ensures the number of animals used and the number of painful procedures are reduced.

Another advantage for using off-the-shelf Abs (i.e. not produced specifically for PHS funded activities) is that the supplier is not required to hold an OLAW Assurance or be covered by another institutions' Assurance (even if the activity is PHS funded).

However, purchasing off-the-shelf products can come with its own set of potential issues[22–24] and may be a contributing factor in the so-called reproducibility crisis.[22–24] If the Ab is not commercially available, or if commercially available products are not suitable for the proposed activities,

DOI: 10.1201/9781003272861-20

strong justification for their production is needed (see 20:11).

20:2 Should IACUCs require that commercially available antibodies be validated before they are used in research specific studies?

Opin. Although validating the proposed antibodies to be used (i.e., specificity, selectivity, and reproducibility within the context of the specific experiments) may help reduce animal use, prevent pain and distress, and may avoid reproducibility issues, this requirement is not without burden to the investigator and may require additional IACUC oversight. Procedures for validation may involve additional time, money, and resources that researchers may not have at their disposal. Although a brief search of IACUC requirements for antibody production/use identified many institutional policies, the search failed to identify institutional policies that require antibody validation.

Surv. Does your IACUC require that commercially available antibodies be validated, such as for selectivity and specificity, before they are used in research specific studies?

- Yes 35/116 (30%)
- No 81/116 (70%)

20:3 Is the use of FCA a painful or distressful procedure?

Reg. The PHS Policy,[25] the U.S. Government Principles,[20] and the *Guide*[26] require the minimization of discomfort, distress, and pain in concert with good science. Additionally, the *Guide* requires that the investigator and the IACUC evaluate the "impact of the proposed procedures on the animal's well-being" in the preparation and approval of the animal use protocol.

Opin. There is much disagreement in the literature regarding the pain and distress elicited by the use of FCA. As described in the introduction, pain was reported by several individuals accidentally injected with FCA.[18,19] Taking U.S. Government Principle V[20] into account, the use of FCA may be considered to similarly cause pain or distress in animals. On the other hand, numerous reports have failed to document any pain or distress in animals injected with FCA[13,16,17,27,28] or in tuberculin-negative humans.[18,29] In mice and other rodents, however, the intraperitoneal injection of FCA and other adjuvants has been associated with the formation of granulomas, fibrous adhesions, abdominal fluid distension, and clinical signs indicative of pain and distress.[1,12,13,30–33] The use of intraperitoneal FCA in rodents as a model of acute inflammation[33–35] is an additional indication of the potential for pain and distress. Uncomplicated FCA–antigen emulsion intradermal and subcutaneous injections are not characterized by tenderness, irritation, or any other indication of greater than slight or momentary pain or distress in the animal.[36] Feed consumption and body weights are unaffected by uncomplicated FCA–antigen injections.

The choice of immunogen, regardless of the choice of adjuvant, may induce an autoimmune condition that is associated with pain and distress. The adjuvant-induced model of arthritis is a prime example and has even been recommended as a model of chronic clinical pain.[37]

Surv. Which USDA/AC pain classification does your IACUC usually assign to projects using Freund's Complete Adjuvant for antibody production? Select all that apply.

- Not applicable 95/191 (50%)
- Category C (no/minimal pain/distress) 9/191 (5%)
- Category D (pain/distress that will be alleviated) 42/191 (22%)
- Category E (no relief from pain/distress due to experimental needs) 41/191 (21%)
- Varies with the species 14/191 (7%)
- Varies with the site of inoculation 17/191 (9%)

20:4 Should the IACUC require the use of analgesics when FCA is used?

Reg. The AWA[38] (§ 2143) requires that practices should "ensure that animal pain and distress are minimized, including adequate veterinary care with the appropriate use of anesthetic, analgesic, tranquilizing drugs, or euthanasia." Similarly, U.S. Government Principles[20] (Principle V) requires that "procedures with animals that may cause more than momentary or slight pain or distress should be performed with appropriate sedation, analgesia, or anesthesia."

Opin. The use of analgesics with FCA immunization in mice has been recommended based upon attenuation of pain behaviors and lack of decrease on antibody responses.[39] Pain management modalities may include systemic non-steroidal and opioid anti-inflammatory drugs,[39] topical preparations,[40] and even electroacupuncture![41] Nonpharmacological modalities (e.g., additional nesting material, increased warmth, hi-value food items, etc.) may also be valuable. Some institutions strongly recommend that researchers consider the administration of analgesics to alleviate any associated pain/distress caused by FCA. However, substantiating such recommendation may be challenging in light of some literature, which establishes that the use of FCA produces no pain (or at least not more than other adjuvants). Although the potential pain and distress should be evaluated on a case-by-case basis, a possible approach to this apparent dilemma is to follow U.S. Principle IV, which states: "Unless the contrary is established, investigators should consider that procedures that cause pain or distress in human beings may cause pain or distress in other animals"[20] (see 20:3).

20:5 What type of justification should be required for the use of adjuvants in antibody production?

Reg. See 20:3.

Opin. The dramatic inflammatory reaction generated by the injection of the original formulation FCA has restricted its use to research antibody production and has also resulted in refinements in the original FCA composition to reduce the resulting inflammation. This has additionally stimulated the development and use of other adjuvants that purport to produce similar antibody responses without the lesions associated with FCA. Unfortunately, many of the alternative adjuvants selected to replace FCA, by nature of their mode of action, produce severe inflammatory reactions[11,13,42–44] or may produce lower antibody titers or no titers at all.[45] Additionally, because many antigens of interest are poorly immunogenic (i.e., elicit a weak or short-lived immune response), adjuvants remain essential. The choice of any adjuvant should be based upon a sound scientific rationale that considers the specific immunogen, the species being immunized, the route of administration, the impact on the animal's well-being, and the desired antibody.[4,36] Other considerations should include the experimental purpose for the adjuvant selected, the dose, the volume to be administered, and the distribution of such injections.

Surv. Does your IACUC require justification for the use of Freund's Complete Adjuvant in mice? Select all that apply.

• Not applicable – we do not use Freund's Complete Adjuvant in mice	84/191 (44%)
• Yes, when used by any route	95/191 (50%)
• Yes, if we were to approve subcutaneous use	8/191 (4%)
• Yes, if we were to approve intradermal use	7/191 (4%)
• Yes, if we were to approve intramuscular use	7/191 (4%)
• Yes, if we were to approve intraperitoneal use	8/191 (4%)
• Yes, if we were to approve intradermal use in the footpad	10/191 (5%)

20:6 Should the IACUC limit the number of injections of Freund's Complete Adjuvant (FCA) to any given animal?

Reg. The USDA AC Inspection Guide[46] lists Ab production as a high-risk procedure.

Opin. Although controversy still exists on whether the inflammation and ulceration subsequent to the administration of FCA causes pain and or distress,[10,36] the AWA requires that researchers consider "alternatives to any procedure likely to produce pain to or distress in an experimental animal."

Many institutions have developed policies regulating the administration of FCA and some go as far as discouraging its use. The number of injections on a single animal may be determined by institutional policies limiting the maximum allowable injectate volume/animal (e.g., maximum total volume on a mouse of 0.3 ml, with a maximum volume per injection site of 0.05 ml. for a total of six injection sites). The use of FCA is usually restricted to the initial inoculation and subsequent boosters are performed using Freund's Incomplete Adjuvant (FIA), which induces a lesser inflammatory reaction. Departures from this approach should be clearly stated in the protocol and closely evaluated by the IACUC.

20:7 Should the IACUC limit the anatomical distribution of Freund's Complete Adjuvant (FCA) injections or discourage specific routes of administration (e.g., intramuscular, intraperitoneal) to any given animal?

Reg. NIH OLAW FAQ F8[47] states that IACUCs are "expected to critically evaluate the proposed uses of the mouse ascites method. Prior to approval of such protocols, IACUCs must determine that (i) the proposed use is scientifically justified, (ii) methods that avoid or minimize discomfort, distress, and pain (including in vitro methods) have been considered, and (iii) the latter have been found unsuitable."

Opin. As previously alluded to, the negative results of the use of FCA may be related to route, dose, sterility and other factors. Many institutions have established specific requirements, including injection location, distribution (i.e., to avoid coalescing lesions), and volumes. Footpad injections are discouraged by some and intraperitoneal injections are limited to rodents only. The use of the intraperitoneal route requires clear monitoring and humane endpoint descriptions. Others prohibit intramuscular injections in mice and rats while allowing them in rabbits. Researchers should consult with their institution's Attending Veterinarian and review any institutional policies to ensure animal welfare and regulatory compliance.

20:8 What procedures should be required by the IACUC to minimize the potential for contamination of injection sites during the injection of adjuvant/antigens?

Opin. Contaminants or other impurities can increase the undesirable and painful side effects of FCA. For that reason, the antigen/adjuvant should be aseptically prepared and free of microbial or other types of contamination. Some institutions require that pharmaceutical-grade components be used in the production of adjuvants. Using aseptic procedures for antigen–adjuvant injections, including sterile preparation of the antigen–adjuvant mixture, clipping the injection site, and sterile scrubbing of the injection site with an antiseptic, have been recommended.[1,3,4,16,17] Additionally, the clipping of the animal's fur permits a more thorough examination of the injection sites post-immunization.

Many contract laboratories have specific requirements for the use of customer-supplied antigens in the production of custom Abs. These may include Ag purity, quantities, and concentration, among others. These requirements may decrease negative impact on animal welfare.

Surv. What procedures does your IACUC require to minimize contamination of injection sites used for adjuvant-antigen preparations? Select all that apply.

- Not applicable. 88/191 (46%)
- We have no policy. 37/191 (19%)
- We require a sterile preparation of the antigen-adjuvant mixture. 54/191 (28%)
- We require clipping the animal's fur prior to injection. 42/191 (22%)
- We require swabbing the area with a disinfectant prior to injection. 55/191 (29%)
- We require a surgical (or near surgical) prep of the area prior to injection. 21/191 (11%)

20:9 Should the amount of killed mycobacteria in the Freund's Complete Adjuvant (FCA) be limited?

Opin. Formulations of FCA containing no more than 0.1 mg/ml of dry mycobacterial cell mass have been recommended as a method to reduce the associated inflammatory response.[10] Nevertheless, commercially available formulations of FCA generally range from 0.4 to 1.0 mg/ml of dry mycobacterial cell mass.

Surv. Does your IACUC limit the amount of mycobacteria in Freund's Complete Adjuvant?

- Not applicable. 100/191 (52%)
- There is no limit specified. 71/191 (37%)
- Our limit is 0.5 mg/ml. 19/191 (10%)
- Our limit is up to 1.0 mg/ml. 1/191 (1%)

20:10 What type of justification or information should be required by the IACUC to approve a schedule of antigen–adjuvant injections?

Opin. The administration of booster injections prior to the optimal time not only increases the potential for pain and distress but also may be detrimental to the ultimate antibody level and affinity.[1,16,48–51] Animals mounting a robust immune response may develop an Arthus reaction (an acute inflammatory and painful reaction) at the booster injection site. The antigen, quantity of antigen, adjuvant used, injection route, and frequency are all important factors in determining the optimal schedule for immunization and, optimally, the individual animal's antibody titer should be followed to determine when the antibody level has plateaued and booster injections are indicated.[52]

Surv. Does your IACUC require investigators to monitor serum titers before booster injections are given?

- Not applicable. 100/191 (52%)
- Yes, we require monitoring. 17/191 (9%)
- No, we do not require monitoring. 49/191 (26%)
- It depends on the antigen used. 25/191 (13%)

20:11 What oversight actions should an IACUC consider if an investigator has contracted with a commercial laboratory for the production of custom antibodies in animals?

Reg. The AWAR (§2.31,d,1,ii; §2.31,d,1,iii) and the PHS Policy (IV,C,1,a) require that Principal Investigators (PIs) provide assurances that, in the case of painful animal procedures, alternatives have been searched for and found unacceptable and that the proposed activities are not duplicative. When USDA registered research facilities contract research out to be conducted at another facility (e.g., the production of custom antibodies in animals), it is the responsibility of the registrants to determine which party is responsible for the functions of the IACUC, animal care and handling, and reporting of the animals on the Annual Report. If no delegation of responsibilities occurs, then both registered parties are responsible, both IACUCs should perform all required functions, and both facilities may be cited for any non-compliance identified.[47] PHS Policy requires that any institution performing animal activities supported by PHS funds, even if the funds are awarded by subcontracting or subgranting, must either have an approved Animal Welfare Assurance[53] or be included as a component under the primary institution's Assurance. If both institutions have approved Assurances, it is necessary that only one of the institutions' IACUCs review and approve the specific activity. IACUCs may choose which IACUC will review the protocol for the animal activities being conducted. Verification of IACUC approval for the production of antibodies must be project specific. It is recommended that if an IACUC defers protocol review to another IACUC, both committees should maintain documentation of the review. The subcontracting or subgranting institution, however, retains accountability for "providing effective oversight mechanisms to ensure compliance with the PHS Policy."[54–56]

For PHS-funded projects involving the production of custom antibodies, the NIH Grants Policy Statement applies[57,58] (NIH GPS Part II, 4.1.1.3).

Under consortium (subaward) agreements in which the grantee collaborates with another organization, the grantee, as the primary recipient of NIH grant funds, is accountable for the performance of the project. Animal welfare requirements that apply to grantees also apply to subcontracts and subprojects. Each institution in the consortium must have an Assurance (Domestic, Interinstitutional, or Foreign).

Opin. The contracting of antibody production is a difficult problem for many institution's IACUCs. The availability of numerous commercial contracting companies along with the many alliances between peptide-generating and antibody-producing companies makes institutional oversight a nightmare. IACUCs must be diligent in their attempts to identify potential situations and be proactive in the education of PIs as to the necessity of notifying and working with the IACUC when contracting antibody production from a commercial firm.

The IACUC's duty in reviewing the contracted facilities protocols and procedures for antibody production protocols is less well defined. Some IACUCs require a complete review of contracted protocols in addition to the review by the IACUC of the contracted facility. Others limit their requirements to the documentation of IACUC approval from the contracted PHS Assured facility, while others do no review or have no requirements. The accessibility of a commercially available and acceptable antibody is a clear and definite example of an alternative to the use of additional animals and duplicative procedures; however, commercial production facilities may not necessarily perform such evaluations. Since the contracting institution retains partial accountability, it would seem prudent that IACUCs should review contracted antibody production protocols to assure that the procedures and policies involved adhere to those of the contracting institution and that the acceptability of available commercial antibodies has been evaluated.

Another issue frequently encountered is when the contracted company will not review/approve an animal use protocol until a contract is in place. This limits the ability of the contracting institutions' IACUC to approve activities if a complete review of the contracted company's protocol is required. This situation can lead to a difficult position in which the contracting institution will not approve the researcher's protocol pending review of the contracted company protocol and the contracted company not approving their internal protocol pending a contract from the contracting institution. In such cases a compromise could be reached where the contracting institutions' IACUC approves the researcher's protocol with the understanding that once the contracted company has an approved protocol this will be provided for additional review.

Surv. If an investigator uses a commercial laboratory to produce antibodies using Public Health Service funds, what information is required by your IACUC? Select all that apply.

• Not applicable.	89/191 (47%)
• We ask for a copy of the laboratory's PHS Assurance approval and its USDA registration (for USDA covered species).	51/191 (27%)
• We ask to see that laboratory's IACUC approval letter for raising the antibodies.	38/191 (20%)
• We review that laboratory's IACUC-approved protocol.	20/191 (10%)
• We have no policy.	35/191 (18%)
• Other*	12/191 (6%)

*Other:

- Confirmation of AAALAC accreditation
- Submission of an approved Memorandum of Understanding (MOU)
- Testing certificates
- We verify PHS Assurance of the laboratory but do not require a copy of the approval letter

20:12 Should additional justifications be requested by the IACUC for the production of monoclonal antibodies using the mouse ascites method?

Reg. NIH/OLAW issued a Frequently Asked Question (F8)[47] directing that IACUCs "critically evaluate the proposed use of the mouse ascites method.… IACUCs must determine that (i) the proposed use is scientifically justified; (ii) methods that avoid or minimize discomfort, distress, and pain (including in vitro methods) have been considered; and (iii) the latter have been found unsuitable."

Opin. The development of practical and reasonably priced *in vitro* methods for hybridoma growth, coupled with evidence that the mouse ascites method causes discomfort, has led many IACUCs to require that *in vitro* methods be utilized. Published comparisons of the *in vitro* methods and the ascites method have shown the results to be highly dependent upon the *in vitro* system used and the specific hybridoma.[59–65] In light of recent advances in monoclonal Ab production, there are few justifications for the use of the ascites method for monoclonal antibody production. Nearly all hybridomas can be propagated *in vitro* and the levels of monoclonal antibody production can be increased by the use of specifically designed growth chambers, hollow-fiber bioreactor systems, and other means.[66] In all instances, the burden of proof should be upon the PI to show that multiple *in vitro* methods have been attempted and were unsuccessful prior to the

IACUC's approving the ascites method of monoclonal antibody production.

A report[67] commissioned by the NIH summarizes the findings of the National Academy of Sciences as to the scientific necessity for producing monoclonal antibodies by the mouse ascites method, including ways to minimize any associated pain or distress. The report also discusses regulatory considerations for the mouse method and summarizes the current stage of development of tissue culture methods. In addition, the NIH Office of Intramural Research has established guidelines for the production of monoclonal antibodies in mice when using the ascites method.[68]

Surv. What justifications are typically acceptable to your IACUC for permitting the in vivo production of monoclonal antibodies by the mouse ascites method? Select all that apply.

- Not applicable. 120/190 (63%)
- We do not request any special justification. 4/190 (2%)
- *In vitro* production is too expensive. 2/190 (1%)
- The investigator tried *in vitro* methods and was not successful. 43/190 (23%)
- The literature states that *in vitro* methods were not successful and *in vivo* is required. 55/190 (29%)
- The structure or molecular weight of the protein suggests that *in vitro* production will fail. 23/190 (12%)
- Other. 11/190 (6%)

*Other:

- Statement that the ascites method will yield a higher concentration of antibody having the proper glycosylation
- Prior or ongoing successful experience using ascites-generated antibody
- Documented problems with yield and purity when *in vitro* methods are used

20:13 Should the IACUC require additional welfare safeguards for the production of monoclonal antibodies using the mouse ascites method?

Reg. See 20:12.

Opin. The use of the ascites method of monoclonal antibody production can cause pain or distress to the animals used.[4,68] Taking this into account, the IACUC is obligated to minimize any unnecessary suffering. Several steps can be taken to ensure the successful production of antibodies while maintaining animal welfare. A review of institutional websites reveals IACUC policies on the use of priming agents, required testing of hybridoma cells for rodent viruses, limits on volume and number of cells injected, aseptic preparation of the ascites fluid collection site, and many more. These can all be regarded as complying with the spirit of the 3Rs and every IACUC should have similar requirements commensurate with the expected impact on the health and welfare of the animals used.

20:14 Should the IACUC limit the number of allowable peritoneal taps for the collection of ascitic fluid?

Opin. Hybridomas are made by fusing antibody producing cells with myeloma (cancer) cells. The result is a hybrid cell with the Ab producing capabilities of the immune cell and the longevity and reproducibility characteristics of the myeloma cells. These are then injected into the mouse abdomen for production of monoclonal antibodies (the ascites method). The process of removing peritoneal fluid from mice used in monoclonal Ab production requires the use of a large-gauge needle and, often, some form of anesthesia. Many institutions limit the total number of peritoneal taps (with the last tap performed postmortem) while other institutions place limits on the animal's physical condition.[3] All hybridoma lines behave somewhat differently when growing in the animal's peritoneal cavity. Certain hybridomas are extremely invasive, resulting in bloody peritoneal fluid on the first tap. The tumor masses that hybridomas form in the peritoneal cavity may be either disseminated or solid and singular.[69] Likewise, the dynamics of monoclonal Ab production from specific hybridomas also varies.[70] The practice of placing absolute limitations on the number of taps is inappropriate as it fails to recognize the differences between hybridoma lines; however, there is little reason for the maximal number of taps to exceed three.

Surv. What limit does your IACUC place on the number of peritoneal taps when using the mouse ascites method for monoclonal antibody production?

- Not applicable. 128/190 (67%)
- We have no limits. 14/190 (7%)
- One tap only. 11/190 (6%)
- Two to three taps. 36/190 (19%)
- More than three taps. 1/190 (1%)

20:15 Should the IACUC require a monitoring plan with specified interventions for work that proposes polyclonal or monoclonal antibody production?

Reg. The Guide[26] includes the production of monoclonal Abs as one of the activities, which requires special consideration due to their potential for pain, distress, or other animal welfare concerns. The USDA AC Inspection Guide[46] lists Ab production as a high-risk procedure.

Opin. Some authors recommend monitoring animals daily after the initial immunization and to examine the animal at least three times a week for specific side effects. After intravenous or intraperitoneal booster injection hourly monitoring is important as there could be a potential for anaphylactic reactions.[4] A brief search of policies in U.S. institutions identified several approaches to monitoring animals used for Ab production. Some require very specific timelines, including how long the animals will be kept and prefer that a good Ab response is elicited early in the process followed by terminal blood collection (provided ongoing needs justify such approach) rather than keeping animals for undetermined lengths of times.

When using the ascites monoclonal Ab method, many institutions require that animals are observed at least three times per day for the first week, and then daily thereafter including weekends and holidays. Abdominal distension should also be monitored and no more than 20% of the pre-inoculation body weight should be gained without ascites fluid collection.

Signs to be monitored may vary between species and may be influenced by the product used, volume injected, number of injections, etc. and may include general appearance (coat, posture, grimace,[71] etc.), mentation, vital signs, weight, body condition, food and water intake, and others. It is important to recognize that under certain conditions symptoms can develop quickly and individuals should be proficient at identifying these and taking required steps to prevent or minimize pain and distress. Of similar importance is a close communication line between the research and veterinary staff to ensure provision of adequate veterinary care when needed.

Another important consideration is the definition of clear humane and experimental endpoints. Humane endpoints are those taken to alleviate pain or distress and may include euthanasia. Experimental endpoints are those reached when the study objectives are reached. Both humane and experimental endpoints can be intimately linked and striking a balance between the two ensures the best study outcomes while preserving animal quality of life and welfare. Experimental endpoints may include failure to mount a desirable Ab reaction, failure to produce ascites within a specified timeframe, while humane endpoints may include clinical signs of serious respiratory or circulatory stress, extreme weight loss (cachexia), production of cloudy/hemorrhagic ascites fluid and many others.

Surv. Does your IACUC require a monitoring plan as part of a protocol that describes production of polyclonal or monoclonal antibody production?

- No, a monitoring plan is not required in the protocol. 15/190 (8%)
- Yes, only for polyclonal antibody production. 3/190 (2%)
- Yes, only for monoclonal antibody production. 3/190 (2%)
- Yes, for both polyclonal and monoclonal antibody production. 58/190 (31%)
- There are no protocols involving monoclonal or polyclonal antibody production at our institution. 111/190 (58%)

20:16 What is the limit for an acceptable volume of blood withdrawal from a single collection?

Reg. The AWAR[39] require that procedures with the potential to cause more than momentary or slight pain or distress involve, in their planning, consultation with the Attending Veterinarian or his or her designee (§2.31,d,1,iv,B).

Opin. Numerous studies have associated severe hemodynamic changes and hemorrhagic shock with blood losses greater than 30% of the total blood volume.[72,73] Smaller volume losses in rats (15%–20% of the blood volume) reduce cardiac output by nearly 50%.[74–76] Some authors have recommended a maximal withdrawal of 15% of blood volume,[77,78] while others have reported acceptable results with blood collections of up to 25% of the total blood volume in mice.[79] Many institutions have developed policies regarding allowable blood collection volume and frequency. Of the institutional websites visited, a large number of institutions use 10% of the calculated total blood volume as a cutoff for single sample collection without fluid replacement. Fluid supplementation (e.g., 0.9% saline solution, lactated Ringer's solution, etc.) is required by many institutions when collecting larger amounts of blood (i.e., >10% total volume). This author has worked with several different species using 10% of the calculated total blood volume and has found no clinical untoward effect on animals' health.

Surv. What is the maximum blood volume withdrawal that your IACUC accepts as reasonable for a single survival withdrawal from a mouse?

- Not applicable 31/190 (16%)
- No policy in place 10/190 (5%)
- 0%–5% of the animal's total blood volume 52/190 (27%)
- 5.1%–10% of the animal's total blood volume 77/190 (41%)
- 10.1%–15% of the animal's total blood volume 16/190 (8%)
- More than 15% of the animal's total blood volume 4/190 (2%)

20:17 What limits should be set if repeated blood sampling is a component of the protocol?

Reg. The AWAR[38] (§2.31,d,1,i), the PHS Policy[25] (IV,C,1,a), and the *Guide*[26] (p. 12) all require that procedures used on animals be chosen to avoid or minimize discomfort, distress, and pain to the animals consistent with sound research design.

Opin. While standard hematology tests (hematocrit and hemoglobin concentration) may be used in larger animals to monitor the effects of multiple blood withdrawals upon an individual animal's well-being, they are rarely used in mice and rats. IACUCs are therefore left to prospectively evaluate the impact of proposed multiple blood collections upon animal well-being and to set limitations. Several studies with repeated blood collections have reported that collection of 250–500 µl daily from adult mice (approximately 10%–20% of total blood volume) results in a significant drop in the hematocrit and a profound reticulocytosis (elevation in the number of young/immature red blood cells).[80,81] The findings of several authors[77,82] suggest that weekly blood collection volumes exceeding 7.5% of the estimated blood volume results in significant hematologic changes supports a recommendation to limit weekly collections to under 10% of the total blood volume whenever maintaining hematologic parameters is necessary. If scientifically necessary, weekly blood collections of up to 15% of total blood volume in males and up to 25% in females has been reported to be sustainable in mice[79] (see 17:42).

An alternative to traditional blood sampling is automated micro-sampling technologies.[83] These techniques allow collection of smaller volumes on awake free-moving animals and are a refinement, which supports the 3Rs. However, these computer-controlled systems can be costly, and their use may not be feasible for smaller labs with limited budgets.

Surv. What is the maximum amount of blood your IACUC typically permits for once weekly blood collection in surviving adult mice?

- Not applicable 25/190 (13%)
- No policy in place 19/190 (10%)
- Up to 5% of the estimated blood volume 75/190 (39%)
- 5.1%–10% of the estimated blood volume 65/190 (34%)
- 10.1%–15% of the estimated blood volume 3/190 (2%)
- 15.1%–20% of the estimated blood volume 3/190 (2%)
- More than 20% of the estimated blood volume 0/190 (0%)

Surv. What is the minimum length of time until the next blood draw that is typically accepted by your IACUC when the maximum allowable amount of blood is withdrawn from an adult mouse?

- Not applicable 27/190 (14%)
- No policy in place 27/190 (14%)
- One week 41/190 (22%)
- Two weeks 68/190 (36%)
- Three weeks 15/190 (8%)
- More than three weeks 12/190 (6%)

20:18 What should the IACUC require regarding orbital sinus (retro-orbital, postorbital) and other blood collection procedures in rodents?

Reg. The AWAR[38] (§2.31,d,1,i), the PHS Policy[25] (IV,C,1,a), and the *Guide*[26] (p. 12) all require that procedures used on animals be chosen to avoid or minimize discomfort, distress, and pain to the animals consistent with sound research design. The AWAR (§2.31,d,1,iv,A) and the PHS Policy (IV,C,1,b) also require that procedures that may cause more than momentary or slight pain or distress to the animals will be performed with appropriate sedation, analgesia, or anesthesia unless scientific justification is provided by the investigator.

Opin. Orbital sinus or retro-orbital (postorbital) blood collection, a standard technique in mice and rats, is not without potential problems. Several studies have described the tissue trauma resulting from orbital sinus bleeding[84–86] and other studies have described behavioral changes (possibly indicating post-procedural pain) resulting from orbital sinus bleeding.[87,88] It is clear from the literature[89,90] and personal experience that the level of trauma associated with the orbital sinus bleeding technique is highly correlated with the skill of the technician performing the procedure. The variability in results and the potential for serious tissue trauma with resulting pain and distress have led many IACUCs and professional organizations to require the use of anesthesia or sedation for orbital sinus blood collections.[91] Unless the technician performing the procedure has demonstrated the requisite skill and experience to repeatedly perform orbital sinus blood collections without trauma to the animal, it is the author's opinion that anesthesia and analgesia should be required.

Surv. Does your IACUC usually require anesthesia or sedation for retro-orbital blood collection in mice?

- Not applicable. 49/190 (26%)
- No anesthesia or sedation is required. 26/190 (14%)
- Sedation along is sufficient. 19/190 (10%)
- We require general anesthesia. 85/190 (45%)
- Topical ocular anesthesia only is sufficient. 11/190 (6%)

Opin. Many IACUCs recommend alternative sites to the orbital sinus blood collection procedure. These alternative sites may include the facial vein (submandibular vein, superficial temporal vein, linguofacial vein), the sublingual vein, the femoral vein, the saphenous vein, the tail vein, and others. Each of these sites has advantages and disadvantages with the site chosen often determined by the familiarity of the staff with the procedure. The facial vein has become a popular alternative to orbital sinus blood collection permitting rapid blood collection without anesthesia in mice. This procedure, however, is not without potential complications with poor blood sample quality, tissue trauma, and reduced weight gains in comparison to the sublingual collection method.[92,93]

REFERENCES

1. Canadian Council on Animal Care. 2002. Guidelines on Antibody Production https://ccac.ca/Documents/Standards/Guidelines/Antibody_production.pdf.
2. Guidelines for the Use of Adjuvants in Research Special Emphasis on Freund's Adjuvant https://oacu.oir.nih.gov/system/files/media/file/2022-04/b8_adjuvants.pdf.
3. Jackson, L.R. and J.G. Fox, J.G. 1995. Institutional policies and guidelines on adjuvants and antibody production. *ILAR J.* 37:141–152.
4. Leenaars, M. and C.F.M. Hendriksen. 2005. Critical steps in the production of polyclonal and monoclonal antibodies: Evaluation and recommendations. *ILAR J.* 46:269–279.
5. Leenaars, P.P., C.F.M. Hendriksen, W.A. de Leeuw *et al.* 1999. The production of polyclonal antibodies in laboratory animals. *ATLA-Altern. Lab. Anim.* 27:79–102.
6. Workman, P., P. Twentyman, F. Balkwill *et al.* 1998. United Kingdom Co-Ordinating Committee on Cancer Research (UKCCCR) guidelines for the welfare of animals in experimental neoplasia (Second Edition). *Br. J. Cancer* 77:1–10.
7. Freund, J. and D. McDermott, 1942. Sensitization to horse serum by means of adjuvants. *P. Soc. Exp. Biol. Med.* 49:548–553.
8. Stewart-Tull, D. 1998. The future potential for the use of adjuvants in human vaccines. In *Biomedical Science and Technology: Recent Developments in the Pharmaceutical and Medical Sciences*, ed. A.A. Hincal and H.S. Kas, 129–136. New York: Plenum Press.
9. Stewart-Tull, D.E.S. 1995. Freund-type mineral oil adjuvant emulsions. In *The Theory and Practical Application of Adjuvants*, ed. D.E.S. Stewart-Tull, 1–19. New York: John Wiley and Sons.
10. Broderson, J.R. 1989. A retrospective review of lesions associated with the use of Freund's adjuvant. *Lab. Anim. Sci.* 39:400–405.
11. Leenaars, P.P., C.F. Hendriksen, A.F. Angulo *et al.* 1994. Evaluation of several adjuvants as alternatives to the use of Freund's adjuvant in rabbits. *Vet. Immunol. Immunopathol.* 40:225–241.
12. Leenaars, P. P., C.F. Hendriksen, M.A. Koedam *et al.* 1995. Comparison of adjuvants for immune potentiating properties and side effects in mice. *Vet. Immunol. Immunopathol.* 48:123–138.
13. Leenaars, P.P., M.A. Koedam, P.W. Wester *et al.* 1998. Assessment of side effects induced by injection of different adjuvant/antigen combinations in rabbits and mice. *Lab. Anim.* 32:387–406.
14. Stills, H.F., Jr. 1994. Polyclonal antibody production. In *The Biology of the Laboratory Rabbit*, 2nd edition, ed., P.J. Manning, D.H. Ringler, and C.E. Newcomer, 435–448. San Diego: Academic Press.
15. Freund, J. 1956. The mode of action of immunologic adjuvants. *Bibl. Tuberc.* 10:130–148.
16. Stills, H.F., Jr. 2012. Polyclonal antibody production. In *The Laboratory Rabbit, Guinea Pig, Hamster, and Other Rodents*, ed. M.A. Suckow, K.A. Stevens and R.P. Wilson, 259–274. Amsterdam: Elsevier.
17. Stills, H.F., Jr. and M.Q. Bailey. 1991. The use of Freund's complete adjuvant. *Lab Anim. (NY)* 20(4):25–30.
18. Chapel, H.M. and P.J. August. 1976. Report of nine cases of accidental injury to man with Freund's complete adjuvant. *Clin. Exper. Immunol.* 24:538–541.
19. Gould, Harry J. 2000. III Complete Freund's adjuvant-induced hyperalgesia: a human perception. 851):301–303.
20. U.S. Government Principles for the Utilization and Care of Vertebrate Animals Used in Testing, Research, and Training. https://olaw.nih.gov/policies-laws/gov-principles.htm.
21. Russell, W.M.S. and R.L. Burch. 1959. The principles of humane experimental technique. https://caat.jhsph.edu/principles/the-principles-of-humane-experimental-technique.
22. John, R. 2009. Couchman commercial antibodies: the good, bad, and really ugly. *J Histochem. Cytochem.* 57(1):7–8.
23. Saper, C.B. 2005. An open letter to our readers on the use of antibodies. *J. Comp. Neurol.* 493:477–478.
24. Jan, L.A. Voskuil, Anita Bandrowski, C. Glenn Begley *et al.* 2020 The antibody society's antibody validation webinar series. *MABS* 12(1).
25. PHS Policy (revised). https://olaw.nih.gov/policies-laws/phs-policy.htm.
26. Institute for Laboratory Animal Research. 2011. *Guide for the Care and Use of Laboratory Animals*, 8th edition. Washington, DC: National Academies Press.
27. Smith, D.E., M.E. O'Brien, V.J. Palmer *et al.* 1992. The selection of an adjuvant emulsion for polyclonal antibody production using a low-molecular-weight antigen in rabbits. *Lab. Anim. Sci.* 42:599–601.
28. Halliday, L.C. *et al.* 2000. Physiologic and behavioral assessment of rabbits immunized with freund's complete adjuvant. *Contem Top* 39(5):8–13.
29. Hughes, L.E., R. Kearney, and M. Tully. 1970. A study in clinical cancer immunotherapy. *Cancer* 26:269–278.
30. Giffen, P.S., J. Turton, C.M. Andrews *et al.* 2003. Markers of experimental acute inflammation in the Wistar Han rat with particular reference to haptoglobin and C-reactive protein. *Arch. Toxicol.* 77:392–402.
31. Lipman, N.S., L.J. Trudel, J.C. Murphy *et al.* 1992. Comparison of immune response potentiation and in-vivo inflammatory effects of Freund's and RIBI adjuvants in mice. *Lab. Anim. Sci.* 42:193–197.
32. Toth, L.A., A.W. Dunlap, G.A. Olson, G.A. *et al.* 1989. An evaluation of distress following intra-peritoneal immunization with Freund's adjuvant in mice. *Lab. Anim. Sci.* 39:122–126.

33. Wanstrup, J. and H.E. Christensen. 1965. Granulomatous lesions in mice produced by Freund's adjuvant; morphogenesis and phasic development. *Acta Pathol. Mic. Sci.* 63:340–354.
34. Geisterfer, M. and J. Gauldie. 1996. Regulation of signal transducer, GP13O and the LIF receptor in acute inflammation in vivo. *Cytokine* 8:283–287.
35. Olivier, E., E. Soury, J.L. Risler, F. Smith *et al.* 1999. A novel set of hepatic mRNAs preferentially expressed during an acute inflammation in rat represents mostly intracellular proteins. *Genomics* 57:352–364.
36. Stills, H.F., Jr. 2005. Adjuvants and antibody production: Dispelling the myths associated with Freund's complete and other adjuvants. *ILAR J.* 46:280–293.
37. Besson, J.-M.R. and G. Guilbaud. 1988. The arthritic rat as a model of clinical pain? *Proceedings of the International Symposium on the Arthritic Rat as a Model of Clinical Pain.* Saint-Paul de Vence, France, 6–8 June 1988. Amsterdam: Elsevier.
38. USDA AWAR (blue book) https://www.nal.usda.gov/legacy/awic/animal-welfare-act.
39. Kolstad, A.M., R.M. Rodriguiz, C.J. Kim *et al.* 2012. Effect of pain management on immunization efficacy in mice. *J. Amer. Assoc. Lab. Anim. Sci.* 51:448–457.
40. Iwaszkiewicz, K.S. and Susan Hua. 2014. Development of an effective topical liposomal formulation for localized analgesia and antiinflammatory actions in the complete Freund's adjuvant rodent model of acute inflammatory. *Pain Phys.* 17:E719–E735.
41. Xiang, X., Sisi Wang, Fangbing Shao, Junfan Fang et al. 2019. Electroacupuncture stimulation alleviates CFA-induced inflammatory pain via suppressing P2X3 expression. *Int. J. Mol. Sci.* 20:3248.
42. Hendriksen, C.F.M. and J. Hau, J. 2003. Production of polyclonal and monoclonal antibodies. In *Handbook of Laboratory Animal Science*, 2nd edition, ed. J. Hau and J. Van Hoosier, 391. Boca Raton: CRC Press.
43. Johnson, D.K. 1994. Adjuvant comparison in rabbits. *Sci. Anim. Care* 5(2):2–4.
44. Johnson, D.K. 1994. Adjuvant comparison in rabbits. In *Rodents and Rabbits: Current Research Issues*, eds. S.M. Niemi, J.S. Venable and H.N. Guttman, 77–80. Washington, D.C.: Scientists Center for Animal Welfare.
45. The Care and Feeding of an IACUC: The Organization and Management of an Institutional Animal Care and Use Committee, 1st edition.
46. USDA AC Inspection Guide. https://www.aphis.usda.gov/animal_welfare/downloads/Animal-Care-Inspection-Guide.pdf.
47. National Institutes of Health, Office of Laboratory Animal Welfare. 2021. Frequently Asked Questions F8, DIs the mouse ascites method an acceptable method of monoclonal antibody production? https://olaw.nih.gov/faqs#/guidance/faqs?anchor=question50367.
48. Herbert, W.J. 1968. The mode of action of mineral-oil emulsion adjuvants on antibody production in mice. *Immunology* 14:301–318.
49. Hu, J.G., A. Ide, T. Yokoyama *et al.* 1989. Studies on the optimal immunization schedule of the mouse as an experimental animal. The effect of antigen dose and adjuvant type. *Chem. Pharm. Bull.* 37:3042–3046.
50. Hu, J.G. and T. Kitagawa. 1990. Studies on the optimal immunization schedule of experimental animals. VI. Antigen dose-response of aluminum hydroxide-aided immunization and booster effect under low antigen dose. *Chem. Pharm. Bull.* 38:2775–2779.
51. Hu, J.G., T. Yokoyama, and T. Kitagawa. 1990. Studies on the optimal immunization schedule of experimental animals. IV. The optimal age and sex of mice, and the influence of booster injections. *Chem. Pharm. Bull.* 38:448–451.
52. Hanly, W.C., J.E. Artwohl, and B.T. Bennett. 1995. Review of polyclonal antibody production procedures in mammals and poultry. *ILAR J.* 37:93–118.
53. NIH OLAW Institutions with a PHS Approved Animal Welfare Assurance https://olaw.nih.gov/assured/app/index.html#DOMESTIC.
54. Potkay, S., N.L. Garnett, J.G. Miller *et al.* 1995. Frequently asked questions about the public health service policy on humane care and use of laboratory animals. *Lab Anim. (NY)* 24:24-26.
55. Wolff, A., N. Garnett, S. Potkay *et al.* 2003. Frequently asked questions about the public health service policy on humane care and use of laboratory animals. *Lab Anim. (NY)* 32(9):33–36.
56. National Institutes of Health, Office of Laboratory Animal Welfare. 2021. Frequently Asked Questions D8, When institutions collaborate, or when the performance site is not the awardee institution, which IACUC is responsible for review of the research activity? https://olaw.nih.gov/faqs/#/guidance/faqs?anchor=question50320.
57. NIH Grants Policy Statement https://grants.nih.gov/grants/policy/nihgps/nihgps.pdf.
58. National Institutes of Health, Office of Laboratory Animal Welfare. 2021. Frequently Asked Questions A2, Does the PHS Policy apply to the production of custom antibodies or to the purchase of surgically modified animals? https://olaw.nih.gov/faqs/#/guidance/faqs?anchor=question50286.
59. Bruce, M.P., V. Boyd, C. Duch *et al.* 2002. Dialysis-based bioreactor systems for the production of monoclonal antibodies—alternatives to ascites production in mice. *J. Immunol. Meth.* 264(1–2):59–68.
60. Jackson, L R., L.J., Trudel, J.G., Fox, and N.S., Lipman. 1996. Evaluation of hollow fiber bioreactors as an alternative to murine ascites production for small scale monoclonal antibody production. *J. Immunol. Meth.* 189(2):217–231.
61. Lipman, N.S. and L.R. Jackson. 1998. Hollow fibre bioreactors: An alternative to murine ascites for small scale (< 1 gram) monoclonal antibody production. *Res. Immunol.* 149:571–576.
62. Marx, U. 1998. Membrane-based cell culture technologies: A scientifically and economically satisfactory alternative to malignant ascites production for monoclonal antibodies. *Res. Immunol.* 149:557–559.
63. Peterson, N.C. and J.E. Peavey. 1998. Comparison of in vitro monoclonal antibody production methods with an in vivo ascites production technique. *Contemp. Topics Lab. Anim. Sci.* 37(5):61–66.
64. Trebak, M., J.M. Chong, D. Herlyn *et al.* 1999. Efficient laboratory-scale production of monoclonal antibodies using membrane-based high-density cell culture technology. *J. Immunol Meth.* 230:59–70.

65. Valdes Veliz, R. 2002. Alternative techniques to obtain monoclonal antibodies at a small scale: Current state and future goals. *Biotecnologia Aplicada* 19:119–131.
66. I. Jyothilekshmi and N. S. Jayaprakash 2021. Trends in monoclonal antibody production using various bioreactor systems. *J. Microbiol. Biotechnol.* 31(3): 349–357
67. National Research Council (U.S.). Committee on Methods of Producing Monoclonal Antibodies. 1999. *Monoclonal Antibody Production.* Washington, DC: National Academies Press. http://www.ncbi.nlm.nih.gov/books/NBK100199/.
68. Guidelines for Monoclonal Antibody Production in Mice https://oacu.oir.nih.gov/system/files/media/file/2021-02/b10_ascites_guide_archive.pdf.
69. Jackson, L.R., L.J. Trudel, J.G. Fox *et al.* 1999. Monoclonal antibody production in murine ascites - I. Clinical and pathologic features. Lab. Anim. Sci. 49(1):70–80.
70. Jackson, L.R., L.J. Trudel, J.G. Fox *et al.* 1999. Monoclonal antibody production in murine ascites - II. Production characteristics. *Lab. Anim. Sci.* 49(1):81–86.
71. Coding of facial expressions of pain in the laboratory mouse https://www.nature.com/articles/nmeth.1455.
72. Kaushansky, K. and W.J. Williams. 2010. *Williams Hematology*, 8th edition. New York: McGraw-Hill Medical.
73. Noble, R.P., M.I. Gregersen, P.M. Porter *et al.* 1946. Blood volume in clinical shock. 2. The extent and cause of blood volume reduction in traumatic, hemorrhagic, and burn shock. *J. Clin. Invest.* 25(2):172–183.
74. Ploucha, J.M. and G.D. Fink. 1985. Systemic hemodynamics during hemorrhage in the conscious rat and chick. *Fed. Proc.* 44:1349–1349.
75. Ploucha, J.M. and G.D. Fink. 1986. Hemodynamics of hemorrhage in the conscious rat and chicken. *Am. J. Physio.* 251:R846–R850.
76. Sapirstein, L.A., E.H. Sapirstein, and A. Bredemeyer. 1960. Effect of hemorrhage on the cardiac output and distribution in the rat. *Circ. Res.* 8:135–148.
77. Mitruka, B.M. and H.M. Rawnsley. 1977. *Clinical Biochemical and Hematological Reference Values in Normal Experimental Animals.* New York: Masson Publishing.
78. Raabe, B.M., J.E. Artwohl, J.E. Purcell *et al.* 2011. Effects of weekly blood collection in C57BL/6 mice. *J. Amer. Assoc. Lab. Anim. Sci.* 50:680–685.
79. Holm, T.M., A. Braun, B.L. Trigatti *et al.* 2002. Failure of red blood cell maturation in mice with defects in the high-density lipoprotein receptor SR-BI. *Blood* 99:1817–1824.
80. Macleod, J.N. and B.H. Shapiro. 1988. Repetitive blood-sampling in unrestrained and unstressed mice using a chronic indwelling right atrial catheterization apparatus. *Lab. Anim. Sci.* 38:603–608.
81. Diehl, K.H., R. Hull, D. Morton *et al.* 2001. A good practice guide to the administration of substances and removal of blood, including routes and volumes. *J. Appl. Toxicol.* 21:15–23.
82. McGuill, M.W. and A.N. Rowan. 1989. Biological effects of blood loss: Implications for sampling volumes and techniques. *ILAR News* 31(4):5–20.
83. L. David Hopper. 2016. automated microsampling technologies and enhancements in the 3Rs. *ILAR J.* 57(2): 166–177.
84. McGee, M.A. and R.R. Maronpot. 1979. Harderian gland dacryoadenitis in rats resulting from orbital bleeding. *Lab. Anim. Sci.* 29:639–641.
85. Vanherck, H., V. Baumans, N.R. Vandercraats *et al.* 1992. Histological changes in the orbital region of rats after orbital puncture. *Lab. Anim.* 26(1):53–58.
86. Le Net, J.E.L., D.P. Abbott, P.R. Mompon *et al.* 1994. Repeated orbital sinus puncture in rats induces damages to optic nerve and retina. *Vet. Pathol.* 31:621.
87. van Herck, H., V. Baumans, H.A.G. Boere *et al.* 2000. Orbital sinus blood sampling in rats: Effects upon selected behavioural variables. *Lab. Anim.* 34:10–19.
88. van Herck, H., V. Baumans, C. Brandt *et al.* 2001. Blood sampling from the retro-orbital plexus, the saphenous vein and the tail vein in rats: Comparative effects on selected behavioural and blood variables. *Lab. Anim.* 35:131–139.
89. Everds, N.E. 2007. Hematology of the laboratory mouse. In *The Mouse in Biomedical Research: Normative Biology, Husbandry, and Models*, 2nd edition, eds. J. G. Fox, S. W. Barthold, M. T. Davisson *et al.*, 133–170. Amsterdam: Elsevier.
90. van Herck, H., V. Baumans, C. Brandt *et al.* 1998. Orbital sinus blood sampling in rats as performed by different animal technicians: The influence of technique and expertise. *Lab. Anim.* 32:377–386.
91. Taylor, R., K.E. Hayes, and L.A. Toth. 2000. Evaluation of an anesthetic regimen for retroorbital blood collection from mice. *Contemp. Topics Lab. Anim. Sci.* 39(2):14–17.
92. Heimann, M., H.P. Kaesermann, R. Pfister *et al.* 2009. Blood collection from the sublingual vein in mice and hamsters: A suitable alternative to retrobulbar technique that provides large volumes and minimizes tissue damage. *Lab. Anim.* 43:255–260.
93. Heimann, M., D.R. Roth, D. Ledieu *et al.* 2010. Sublingual and submandibular blood collection in mice: A comparison of effects on body weight, food consumption and tissue damage. *Lab. Anim.* 44:352–358.

21

Animals Used in Agricultural Research and Teaching

Bart Carter, Vickie Jarrell, and Rose Gillesby

Introduction

Federal laws, regulations, and policies apply to the use of agricultural animals (e.g., sheep, swine, goats, cattle, horses) in biomedical research and teaching; however, agriculture animals used for improving animal nutrition, breeding, management, production efficiency, or for improving the quality of food or fiber, are exempt from these requirements. Institutions may voluntarily choose to include oversight of agricultural animals in their Animal Welfare Assurance, which would then apply PHS Policy requirements to these animals regardless of their use.

Agricultural animals used in biomedical research are regulated by the AWA, requiring oversight by an IACUC, and they are subject to the regulations and standards in Title 9 of the CFR. When agricultural animals are used in biomedical research funded by the Public Health Service (PHS), National Science Foundation (NSF), National Aeronautics and Space Administration (NASA), or Department of Veterans Affairs (VA), standards for care must be based upon the *Guide for the Care and Use of Laboratory Animals* (*Guide*). The *Guide* clarifies that "Decisions on categorizing the research uses of agricultural animals and defining standards for their care and use should be made by the IACUC based on both the researcher's goals and concern for animal well-being."

Institutions that use agricultural animals in food and fiber research and teaching may decide to assign responsibility for oversight of these activities to an IACUC; however, this is not required by law. Similarly, the decision to use the *Guide* and/or *Guide for the Care and Use of Agricultural Animals in Research and Teaching* (*Ag Guide*)[1] as reference resources in food and fiber research is at the discretion of the institution.

The apparent existence of two sets of standards for care can be complicated, presenting challenges to an oversight committee with respect to regulatory expectations and best practices for ensuring the welfare of farm animals used in research and teaching. Even the terminology for naming these animals is confusing with the *Guide* and the *Ag Guide* referring to them as agricultural animals and the AWA referring to them as farm animals. For consistency, in this chapter they will be referred to as agricultural animals except when specifically discussing AWA regulations. Both the *Guide* and the *Ag Guide* define the use of the words "must" and "should" indicating when a standard "must" be implemented to meet the minimal requirements and when an IACUC may accept justification for alternative approaches related to "should" statements. Adherence to "must" or "should" statements is mandatory for AWA regulated or federally funded research from the PHS, NSF, NASA, or VA, but voluntary when agricultural animals are used in food or fiber research. For the sake of consistency, many institutions opt to include agricultural animals, regardless of the goals of the research, under the oversight of the IACUC. It should be recognized that when U.S. institutions seek to have their agricultural animal programs accredited by AAALAC, International, the IACUC is expected to review and approve standards of care for agricultural animal activities.

Systems to produce food and fiber can differ substantially between various regions of the United States. The strategies for managing the animals, the breeds, and the environments in which they live depend largely on the local climate, topography, and markets. Production systems that work well in the northern tier of the Great Plains may not support the welfare of animals produced in the southwestern states. The strategies for housing, feeding, watering, handling, and delivery of clinical care may differ by necessity. Therefore, it is important that scientists with expertise on local production systems be involved as members of the IACUC. The *Ag Guide* provides recommendations for the care of agricultural animals; however, the IACUC is ultimately responsible for defining acceptable standards and for approving justifiable alternatives. OLAW has developed FAQ G.7,[2] that provides additional information that institutions with an approved Animal Welfare Assurance can utilize when determining when or if it is appropriate to utilize the *Ag Guide*.

The public has an increasing interest in the humane treatment of animals used for production of food and fiber. Proactive industry standards for humane care of animals in production are constantly evolving, and well-managed farms are frequently members of quality assurance programs. Public institutions that use farm animals, whether for research and teaching or strictly for production and sale, are held to a high expectation for humane and appropriate treatment. Oversight by an appropriately experienced IACUC can ensure standards are met and contribute to a positive reputation for the institution by implementing and teaching best practices for humane care and use.

Challenging ethical and regulatory considerations evolve when an institution decides to include food and fiber research activities under the oversight of an IACUC. Additional challenges may be presented when agricultural animals used for research also enter the food chain. This chapter explores

DOI: 10.1201/9781003272861-21

institutional decisions, regulatory requirements, and the roles and ethical considerations for the IACUC's oversight of the use of farm animals in biomedical and agricultural research and teaching.

21:1 What is the USDA's definition of a farm animal?

Reg. The Animal Welfare Regulations define farm animal as "any domestic species of cattle, sheep, swine, goats, llamas, or horses, which are normally and have historically, been kept and raised on farms in the United States and used or intended for use as food or fiber, or for improving animal nutrition, breeding, management, or production efficiency, or for improving the quality of food or fiber. This term also includes animals such as rabbits, mink, and chinchillas, when they are used solely for purposes of meat or fur, and animals such as horses and llamas when used solely as work and pack animals" (9 CFR § 1.1).

21:2 When do farm animals meet the requirements for regulation under the AWA?

Reg. The Animal Welfare Act (AWA) § 2132(g) specifically excludes from regulatory oversight "horses not used for research purposes, and …[…]other farm animals, such as, but not limited to livestock or poultry, used or intended for use as food or fiber, or livestock or poultry used or intended for use for improving animal nutrition, breeding, management, or production efficiency, or for improving the quality of food or fiber."

Opin. Whether an activity using farm animals is regulated under the AWA can be complex and sometimes difficult to determine. It is important to clearly define the goals of the research or teaching to interpret the requirements. Institutions should seek advice from the USDA/APHIS/AC when applying regulatory exemptions to the use of agricultural animals. The most impactful difference between exempt and regulated activities is the requirement for IACUC oversight and regular inspection by the USDA. When institutions assign oversight responsibilities for exempt farm animal use to an IACUC, the appropriate standards to support the research goals must be determined by the committee.

21:3 When are teaching activities involving farm animals regulated by the AWA?

Reg. The AWA does not directly address the regulation of farm animals for teaching activities.

Opin. Consistent with the definition of "animals" within the AWA, the goals of the teaching activity establish whether or not the AWA is applied. For example, while the use of farm animals in teaching basic science (anatomy, physiology, toxicology, etc.) and medical procedures to veterinary students is regulated, the use of farm animals to teach breeding, husbandry, and production efficiency is not regulated. Teaching of animal sciences and veterinary students enrolled in an animal sciences course to learn standard agricultural production practices. such as surgical castration or dehorning, is not a regulated activity, while teaching of animal sciences or veterinary students in appropriate aseptic techniques used in surgery is regulated. It is important for the goals of the activity to be clearly described in order for the IACUC to determine if a certain activity is covered under the AWA.

Surv. If your IACUC reviews activities involving agricultural animals, what categories of activities are reviewed? Select all that apply.

• Agricultural research	22/27 (81%)
• Biomedical research	23/27 (85%)
• Teaching of veterinary students	16/27 (59%)
• Teaching of animal sciences or biology students	18/27 (67%)
• Production of food and fiber for sale	13/27 (48%)
• Field trials where the animals are part of the applied treatment but are not themselves the focus of the research (i.e., when beef cattle graze forages but the experimental observations are on forage growth)	11/27 (41%)

21:4 What are some of the requirements for an IACUC providing oversight to an agricultural animal program?

Reg. An IACUC is required for review of activities with agricultural animals when used in basic science or biomedical research and teaching as these are subject to the AWA and AWAR. IACUC oversight is expected of agricultural animal care programs accredited by AAALAC, International as the *Ag Guide* (Chapter 1: Institutional Policies) has been adopted by the Council on Accreditation as a Primary Reference Resource. Additionally, the *Ag Guide* endorses the *US Government Principles for the Utilization and Care of Vertebrate Animals Used in Testing, Research, and Training* (*US Government Principles*)[3] as a basis for professional judgment about appropriate treatment and use of agricultural animals. *The US Government Principles (IX)* require that exceptions to the provisions of the Principles be reviewed by an appropriate committee such as an IACUC. When research activities are funded by the NIH or any affiliated agencies under the oversight of the Office of Laboratory Animal Welfare (OLAW), oversight by an IACUC is required, as described in the approved Animal Welfare Assurance.

Opin. The basic tenets of the ethical use of animals in research and teaching rest upon objective, professional judgment in providing for their best welfare during required use to further science and education. To exempt an animal from the protection of an objective review of the procedures they will experience is in direct conflict with responsible ethical consideration for their humane treatment. While laws and regulations by necessity are limited, the institution bears responsibility for the humane and ethical treatment of all animals used for research and teaching. A consistent standard of review of all animal activities by an IACUC not only simplifies the oversight process for the institution, but also, more importantly, provides objectivity to ensure the animals are treated ethically and humanely.

Surv. Does your institution use agricultural animals for non-biomedical research, teaching, or testing?

- Yes 27/190 (14%)
- No 163/190 (86%)

Surv. Does your institution's IACUC review activities that involve agricultural animals (used for any purpose)?

- No 0/27 (0%)
- Yes. They are reviewed by the same IACUC that reviews laboratory animal activities. 25/27 (93%)
- Yes. They are reviewed by a separate IACUC for review of activities involving agricultural animals. 2/27 (7%)

Surv. What records of on-farm activities with agricultural animals does your IACUC review? Select all that apply.

- Records of daily observations of the animals, whether in groups or as individual animals. 23/27 (85%)
- Records of husbandry activities such as cleaning, feeding, and checking watering systems. 22/27 (81%)
- Records of animal movements and transports. 9/27 (33%)
- Records of environmental parameters such as relative humidity and temperature for indoor housing facilities. 11/27 (41%)
- Records of individual animal treatments for minor injuries and illnesses. 20/27 (74%)
- Records of major clinical conditions and treatments requiring the attention of a veterinarian. 21/27 (78%)
- Records of euthanasia, including identity of the person conducting procedures and disposition of the carcass. 11/27 (41%)
- Records of the number of animals used on each IACUC protocol. 15/27 (56%)
- None of these type records are reviewed by the IACUC. 2/27 (7%)

21:5 How should an IACUC be constituted if part of their responsibility is to provide oversight of an agricultural animal program?

Reg. The AWAR (§2.31,a-b) define the required composition of an IACUC for activities regulated under the AWA. The members must be appointed by the Chief Executive Officer of the research facility, and the committee must be composed of a minimum of three members: a Chairman and at least two additional members, including a Doctor of Veterinary Medicine with training or experience in laboratory animal science and medicine and who has direct or delegated program responsibility for activities involving animals, and a member who is not affiliated in any way with the facility other than as a member of the committee. The nonaffiliated member represents the general community interests in the proper care and treatment of animals.

The Public Health Service Policy IV.A.3 establishes the membership of the IACUC for research or teaching supported by funds from the NIH or other agencies under the oversight of OLAW. The PHS requires five members on a legally constituted IACUC, adding one practicing scientist experienced in research involving animals and one member whose primary concerns are in a nonscientific area such as an ethicist, lawyer, or member of the clergy.

The *Ag Guide* (p. 1) also requires a minimum of five members, specifically a scientific member who should have experience in agricultural research or teaching involving agricultural animals. One member should also be an animal, dairy, or poultry scientist who has training and experience in the management of agricultural animals. The required veterinarian should have training and experience in agricultural animal medicine. Similar to the AWAR and the PHS Policy, the *Ag Guide* requires both a nonaffiliated member and a person whose primary concerns are in an area outside of science.

Opin. The specific requirements for expertise in agricultural animal management, research and teaching, and veterinary medicine are important if the IACUC is to make well-informed decisions regarding the appropriate use and standards of care for agricultural animals. While the basic principles of humane treatment apply to both laboratory and agricultural species, hypothesis-driven research in food and fiber production should remain relevant to the management systems. The investigators, in consultation with knowledgeable members of the IACUC, can best ensure animal welfare is maintained at a high level while addressing the challenges associated with agricultural animal housing, handling, and production.

Surv. If your institution has one IACUC that reviews the use of both laboratory and agricultural animals, what proportion of the membership of your IACUC has experience in the care and use of agricultural animals?

- More than 75% have agricultural animal experience. 4/27 (15%)
- 50% to 75% have agricultural animal experience. 5/27 (19%)
- 25% to 50% have agricultural animal experience. 9/27 (33%)
- Less than 25% have agricultural animal experience. 8/27 (30%)
- Not applicable, we have a separate IACUC for review of activities involving agricultural animals. 1/27 (4%)

21:6 Are the requirements for IACUC oversight of farm facilities different from those responsibilities for biomedical research programs?

Reg. The same oversight responsibilities that apply to the use of traditional laboratory animals used in research and teaching apply to farm animals when they are regulated by the USDA or overseen by the OLAW. These responsibilities include the required considerations during animal care and use protocol review. For those activities that are not directly overseen by government agencies, the *Ag Guide* (pp. 2–4) describes in detail the considerations for IACUC oversight. Protocol review by the committee includes standard agricultural animal husbandry practices that may cause temporary discomfort or pain. The *Ag Guide* indicates the IACUC must review and approve all operating procedures that have the potential to cause pain or distress but suggests that standard operating procedures may be used in lieu of, including detailed descriptions, in every agricultural animal care and use protocol.

Opin. The IACUC should review the use of agricultural animals regardless of the specific goals of the research or teaching activity. When standard agricultural procedures that may cause pain or distress are involved, the review should involve the expertise of those familiar with improvements in production industry practices. If an IACUC lacks expertise, it should seek the appropriate consultation. It is imperative to ensure the animals are humanely treated but equally important to ensure they are managed so results of the research remain relevant to industry challenges.

Surv. Does your IACUC inspect institutionally owned farm facilities when used solely for production of agricultural animals not intended for research and teaching?

- No. 3/27 (11%)
- Yes, at the same frequency as the rest of the animal care and use program. 16/27 (59%)
- Yes, but a lesser frequency than the rest of the animal care and use program. 1/27 (4%)
- Not applicable. We do not have institutionally owned farm facilities used solely for production of agricultural animals not intended for research and teaching. 7/27 (36%)

21:7 What ethical considerations should the institution review when determining if it wants to require IACUC oversight of all agricultural animals?

Reg. The *Guide* (p. 32) states: "The use of agricultural animals in research is subject to the same ethical considerations as for other animals in research, although it is often categorized as either biomedical or agricultural because of government regulations and policies, institutional policies, administrative structure, funding sources, and/or user goals. … Decisions on categorizing research uses of agricultural animals and defining standards for their care and use should be made by the IACUC based on both the researcher's goals and concern for animal well-being. … The *Guide* (P. 33) applies to agricultural animals used in biomedical research, including those maintained in typical farm settings."

Opin. A delicate balance is needed when considering the regulations that categorize the research and defining appropriate standards of care to support animal welfare in the context of the research. If the housing and environment are not relevant to the goals of the research, results may not be applicable, defeating the purpose of conducting the work. Similarly, if the animals are not in a good state of welfare, research results could have less value or become invalid. The IACUC is responsible for ensuring the appropriate standards are defined and implemented to maintain animal welfare and support the research goals. Knowledgeable agricultural animal scientists and veterinarians as members of the IACUC can help the committee find the right balance.

Surv. Does your IACUC allow simultaneous assignment of agricultural animals to more than one animal care and use protocol?

- No 12/27 (44%)
- Yes 15/27 (56%)

Surv. Does your IACUC allow sequential transfer of agricultural animals from one animal care and use protocol to another protocol?

- No 4/27 (15%)
- Yes 15/27 (56%)
- Yes, but only after review by a veterinarian 8/27 (30%)

21:8 If agricultural teaching and educational activities are taking place at a farm not owned by the institution, should

the institution obtain permission from the owner of the farm to conduct such activities?

Reg. There are no regulatory requirements for written permissions to conduct agricultural educational activities on a farm; however, written agreements between entities collaborating in farm animal research and teaching activities are expected when the activities are covered by an institutional Assurance with OLAW or as part of an institution's AAALAC, International accreditation. The *Guide* (p.15) indicates that "In cases of such collaboration involving animal use (beyond animal transport), the participating institutions should have a formal written understanding (e.g., a contract, memorandum of understanding, or agreement) that addresses the responsibility for offsite animal care and use, animal ownership, and IACUC review and oversight." If the activity is PHS funded, then it falls under the *Guide* or AWA regulations.

Opin. The implementation of a written agreement can protect both entities if there are concerns for animal welfare or adverse events associated with the educational activities. Institutions should consider consulting their legal counsel to ensure that risks are mitigated.

Surv. Does your IACUC review research or teaching activities involving agricultural animals if housed on private farms?

- No. 6/27 (22%)
- Yes. 12/27 (44%)
- Not applicable. None of our agricultural animals are housed on private farms. 9/27 (33%)

21:9 Should the IACUC conduct semiannual inspections of privately owned farm facilities where agricultural research or teaching activities are conducted?

Reg. The *Ag Guide* indicates that institutions must develop policies for animal care and use related to agricultural research conducted off-site as well as agricultural research using privately owned animals on- and off-site. The fact that research is conducted off-site does not lessen the responsibility of the institution to assure appropriate and humane animal care and use.

Opin. This question should be considered within the context of animal ownership. While the investigator and the institution can determine whether to accept a privately owned farm as a site for the animal use activity based on the standards of care and goals of the research, if the institution does not legally own the animals, the IACUC has no authority over the operations of a privately owned farm. The requirement in the *Ag Guide* for oversight does not mandate the IACUC to conduct physical inspections of those privately owned facilities, nor does it impart authority to the institution's veterinarian. However, in these situations, it is recommended the institution perform some form of due diligence and/or monitoring of the facility and animal care to ensure the welfare of the animals and the reputational protection of the institution. In situations where the institution is the legal owner of the animals, the IACUC should impart its authority to oversee the use and care of the animals at the privately owned facility in a manner they deem appropriate. Regular and direct communications between the Attending Veterinarian and the facility's veterinarian related to the health of the animals used in the research are highly advised. A written agreement between the owner, investigator, and institution is also advised and may be required if the study is federally funded.

21:10 Is it acceptable to rent/lease animals from a farm/institution for research or teaching use and house them in our facilities?

Reg. The *Ag Guide* indicates that institutions must develop policies for animal care and use related to research conducted off-site as well as research using privately owned animals on- and off-site. The fact that the animals housed at the institution are not owned by the institution does not lessen the responsibility of the institution to assure appropriate and humane animal care and use.

Opin. A relationship with a vendor or private commercial farm can be beneficial to both the institution and the commercial farm. Many teaching activities or agricultural research protocols do not involve invasive procedures, exposure to hazardous substances, or administration of investigational veterinary products. In these instances, using animals owned by a commercial farmer and then returning them to the farm of origin at the conclusion of the research or teaching need is a responsible use of animals. In these situations, collaboration between the Investigator, the Attending Veterinarian, the IACUC, and the institution's legal representation is critical to ensure appropriate protection of the institution. In addition to the scientific and welfare considerations of using the leased animals, the IACUC must also consider any risks to the animals and people of the institution's program.

21:11 What is the process for students to participate in educational activities if agricultural animals are used for educational purposes?

Reg. There are no required procedures for engaging students in activities using agricultural animals; however, when teaching goals are in the biomedical or basic sciences, the activity is regulated and therefore requires oversight by an IACUC. The Principal Investigator, who may also be the instructor, is responsible for compliance of activities with approvals by the IACUC. The institution is required to provide guidance in humane methods of animal maintenance, the 3Rs, the use of anesthetics and analgesics as appropriate, and methods by which deficiencies in animal care and treatment are reported.

Opin. While the teaching of standard agricultural practices not related to biomedical or basic sciences is not regulated by the AWA/R, all students who use animal models should be well informed of the ethical and regulatory considerations. The development of basic training requirements for all students who use animals can simplify institutional procedures for documenting required training. Even for institutions that do not engage in regulated teaching activities, providing basic instruction in humane and ethical animal use promotes responsible care and use of animals and supports a positive and responsible public reputation.

21:12 Are there animal facility construction guidelines for facilities that house agricultural animals?

Reg. There are no specific regulations detailing facility construction for agricultural animals. However, there are numerous guidelines published by the USDA Agricultural Research Service,[4] the Midwest Plan Service,[5] breed or species-specific organizations, and state agricultural extension programs that provide insights into facility construction. General information is also provided in the species-specific chapters of the *Ag Guide*.

Opin. Designers of agricultural facilities should incorporate many of the same considerations that biomedical facilities require during facility design: cleaning and sanitation; pest control; animal handling and transport; as well as separation of animals by species and age group. Additional factors of large impact on design of agricultural facilities include water and manure management, transport and removal of large animals which may require access to necropsy and carcass disposal areas, and, finally, security of outdoor facilities. These factors should be included in the design.

21:13 How should the IACUC review food and fiber production facilities as compared to biomedical research facilities?

Reg. Both the *Guide* and the *Ag Guide* instruct an IACUC to inspect animal facilities and study areas at least twice yearly. The basic premise for the inspection is the same for the different types of facilities.

Opin. Many agricultural facilities are modeled after production facilities found in industry or on small farms. There are relatively few published engineering standards for agricultural facilities, unlike biomedical facilities. When reviewing agricultural facilities, IACUC members should focus on the "performance standards" as described in the *Ag Guide* and pay particular attention to animal condition and comfort. While the term is not easily defined, a statement often referred to during inspections is: "the facility should represent a well-managed farm for the geographic area." IACUCs should have a scientist who is familiar with agricultural research as well as a veterinarian who is familiar with the species present. These individuals should be encouraged to participate in the facility inspections.

21:14 How is veterinary oversight and veterinary medical care managed in agricultural research settings?

Reg. The *Ag Guide* (p. 10) indicates, "Institutions should provide researchers and personnel with access to a veterinarian who has experience in the care of agricultural animals. The veterinarian can be full-time, part-time, or a private practitioner." This "program veterinarian" should also be involved in the development and oversight of the veterinary care program.

Opin. Institutions conducting biomedical research are familiar with the term "Attending Veterinarian," as the role is defined by the AWA and PHS Policy. To alleviate some confusion with veterinary roles, many agricultural programs have developed the position of Program Veterinarian, who may report to the institutional Attending Veterinarian. The Program Veterinarian is the individual with the experience and responsibility to lead the veterinary care program of an agricultural setting. The Attending Veterinarian may or may not be the Program Veterinarian, but the Program Veterinarian must be qualified and experienced in the species present. Certain components of delivering veterinary care may be assigned to other veterinarians or qualified husbandry managers. Regardless of how veterinary medical care is delivered to the animals, all parties involved have an obligation to provide regular and frequent communication to the Program Veterinarian so that ongoing issues can be addressed.

Surv. Is the Attending Veterinarian (AV) for agricultural animals the same person as the AV for the traditional laboratory animal program?

- No 5/27 (19%)
- Yes 22/27 (81%)

Surv. Does your institution use contracted veterinary services for agricultural animals?

- No. 9/27 (33%)
- Yes, full-time contracted veterinary services are provided. 3/27 (11%)
- Yes, part-time contracted veterinary services are provided. 11/27 (41%)

- Yes, contracted veterinarians provide back-up services only. 4/27 (15%)

Surv. If you have veterinarians providing veterinary care who are not associated with the animal care program, is there required communication with the Attending Veterinarian?

- No 4/27 (15%)
- Only in an emergency 2/27 (7%)
- Yes, every time they are on-site, consult on animal care, or initiate new animal treatments 9/27 (33%)
- Yes, at some regular interval 10/27 (37%)
- Other* 2/27 (7%)

*Other: The AV has access to all veterinary care records, including those generated by veterinarians not associated with the animal care program.

21:15 Should IACUCs be concerned about biosecurity when visiting agricultural facilities?

Reg. In an agricultural setting, biosecurity is defined as the security measures taken to prevent the unintentional transfer of pathogenic organisms between groups of animals (*Ag Guide*, p. 24). There are no current regulatory documents requiring a biosecurity program to be in place.

Opin. Husbandry managers, research personnel, and the program veterinarian should establish written guidelines detailing the biosecurity program for each location where agricultural animals are housed or transported. The guidelines should be comprehensive and describe how animals, facilities, and people all come together in the program. The IACUC should ensure a program of biosecurity is in place to protect the health and welfare of the animals.

21:16 How does an IACUC evaluate an agricultural animal health care program?

Reg. The *Ag Guide* (p. 10) indicates that adequate agricultural animal health care involves a written and implemented program for disease prevention, surveillance, diagnosis, treatment, and endpoint resolution.

Opin. A mechanism for direct, frequent, and regular communication is crucial between personnel who are responsible for daily animal care and observations, animal users, and the program veterinarian (*Ag Guide*, p. 10).

21:17 What should IACUCs know about transporting agricultural animals?

Reg. The interstate movement of animals is governed by the USDA and the impacted state's department of agriculture. Institutions must ensure that they are following all federal and state requirements when moving animals.

Opin. The movement of animals, whether long distances or within a farm, includes many aspects impacting the animals. Animal handling, mode and duration of transport, loading, unloading, social group disruption, potential isolation, change in food and water sources, and many other criteria can have an unpredictable impact upon groups of animals. These factors should be considered during the planning phases of transport in order to safeguard the welfare of the animals. Species-specific guidelines for transporting agricultural animals can be found in the *Ag Guide* (pp. 63–68).

21:18 Should IACUCs be concerned about standard procedures performed on agricultural animals to improve the animal's long-term welfare or help protect human handlers?

Reg. The *Ag Guide* (p. 23) provides a justification for performing common husbandry procedures to agricultural animals that may cause momentary pain or distress. Procedures such as beak trimming, castration, dehorning, and tail docking are common industry practices and are known to help reduce injuries to other animals or humans.

Opin. It is recommended that programs develop standardized written documents for performing these procedures and utilize input from the attending or program veterinarian. These documents should establish clear guidelines for ages and sizes of each species and determine if anesthetics or analgesics are required for each procedure. The IACUC has responsibility to consider if procedures may cause more than momentary pain and if the use of pain-relieving drugs would be appropriate. These documents can then be utilized as the basis of a training program.

Surv. Does your IACUC review standard agricultural practices such as dehorning, castration, needle teeth clipping, tail docking, beak trimming, early weaning, and regrouping of animals?

- No, we do not review standard agricultural practices unless they are the focus of the research. 6/27 (22%)
- We review standard agricultural practices only when they may cause pain (e.g., castration, dehorning, tail docking). 1/27 (4%)
- We review standard agricultural practices if performed on animals used for research and teaching, but not for animals used exclusively for production. 8/27 (30%)
- We review all standard agricultural practices for all agricultural animals regardless of their use for researching, teaching, or for production. 12/27 (44%)

Surv. If your institution's IACUC reviews standard agricultural practices, what are the methods of review? Select all that apply.

- Reviewed by the IACUC through inclusion in animal care and use protocols 19/27 (70%)
- Reviewed by the IACUC as standard operating procedures, separate from the animal care and use protocols 13/27 (48%)
- Reviewed by a subcommittee with agricultural animal experience as standard operating procedures 5/27 (19%)
- Reviewed by veterinarians with agricultural animal experience 16/27 (59%)

Surv. Who can perform surgeries in the agricultural animal program? Select all that apply.

- Animal technicians 12/27 (44%)
- Research technicians 14/27 (52%)
- Investigators 18/27 (67%)
- Visiting scientists 12/27 (44%)
- Graduate students 13/27 (48%)
- Undergraduate students 5/27 (19%)
- Visiting students 3/27 (11%)
- Farm managers 5/27 (19%)
- Private farm employees 1/27 (4%)
- Veterinarians, in addition to others 25/27 (93%)
- Veterinarians only 6/27 (22%)
- Other 2/27 (7%)

Surv. Does your institution require a veterinarian to perform standard agricultural practices such as castrations, dehorning, beak trimming, etc.?

- Yes, always 2/27 (7%)
- Certain practices, but not all 7/27 (26%)
- Only if the practice may cause pain and/or distress 1/27 (4%)
- No 16/27 (59%)
- Other 1/27 (4%)

21:19 Can a veterinarian who is the Principal Investigator for a project also serve as the veterinarian providing veterinary care to their animals?

Reg. While there is no regulation that specifically prohibits a qualified individual who is a veterinarian from providing veterinary care to their own research animals, IACUCs are reminded of the ***U.S. Government Principles for the Utilization and Care of Vertebrate Animals Used in Testing, Research, and Training,*** Principle IX, which states: "Where exceptions are required in relation to the provisions of these Principles, the decisions should not rest with the investigators directly concerned but should be made, with due regard to Principle II, by an appropriate review group such as an institutional animal care and use committee. Such exceptions should not be made solely for the purposes of teaching or demonstration."

Opin. The *Ag Guide* (p. 10) indicates that all programs utilizing agricultural animals must have a qualified veterinarian responsible for the agricultural animal health care program. If an institution were to allow a qualified investigator to provide routine veterinary care to their own animals, a clear mechanism for reporting to the Attending Veterinarian and regular communication between the two veterinarians must be established. The Program Veterinarian should regularly communicate animal health information to the Attending Veterinarian. If the Program Veterinarian is the PI of a project and is providing veterinary care to their own animals, the IACUC should be aware of the situation and ensure that a method for frequent communication to the AV is in place.

Surv. If there are researchers in the program who are veterinarians, can they provide clinical veterinary care for animals in the program?

- Yes, but only for animals participating on studies for which they are the Investigator 6/27 (24%)
- Yes, but not for animals participating on studies for which they are the Investigator 2/27 (8%)
- Yes, on any animal owned by the institution 9/27 (36%)
- No 4/27 (16%)
- Other 4/27 (16%)

*Other: noted that approval of the AV was required for veterinarians who are researchers to provide care.

21:20 How should IACUCs evaluate the euthanasia procedures required for agricultural animals?

Reg. Humane euthanasia of agricultural species is covered by the current version of the *AVMA Guidelines for the Euthanasia of Animals.*[6] For agricultural animals intended to enter the food chain, the *AVMA Guidelines for the Humane Slaughter of Animals* (AVMA, 2016)[7] should be consulted.

Opin. The AVMA Panel on Euthanasia describes many acceptable methods of euthanasia for larger animals. Human safety, availability of drugs in a remote situation, and the ultimate disposal method of the carcass are all factors that need to be considered when selecting a method of euthanasia for each possible situation that may arise in a farm setting. Physical methods such as gunshot or penetrating captive bolt are more commonly utilized in agricultural programs than in traditional biomedical research settings.

21:21 Does the IACUC have a responsibility to oversee the use of feed additives used in the agricultural program?

Reg. The *Ag Guide* states that agricultural animals should be fed in accordance with the requirements established for each species by the National Research Council (NRC) (1994, 2001, 2007a, b, 2012, 2016)[8] and as recommended for the geographic area.

Opin. The *Ag Guide* acknowledges that agricultural animals can vary considerably in body weight and condition during grazing and reproductive cycles. Feeding programs should allow animals to regain the body weight that is lost during the normal periods of negative energy balance. IACUCs should ensure that diets are properly formulated by qualified individuals and prepared in accordance with federal and state laws (e.g., Veterinary Feed Directive[9]).

Surv. Does your institution have a mechanism for assuring the nutritional quality of feedstuffs offered to agricultural animals?

- No 11/27 (41%)
- Yes 16/27 (59%)

21:22 Can veterinary products be used "off-label" in agricultural animals used for research and can the animals be allowed to subsequently enter the food chain?

Reg. The *Ag Guide* (p. 14) allows for the use of approved drugs in an extra label fashion and gives criteria for appropriate use and documentation. Such use is permitted only when it adheres to the regulations promulgated by the FDA under the Animal Medicinal Drug Use Clarification Act (AMDUCA) of 1994, Public Law 103-396 (U.S. Food and Drug Administration, 1994).[10] . The Food Animal Residue Avoidance Databank (FARAD)[11] or FDA should be contacted whenever guidance is needed.

Opin. The use of approved products off-label is, at times, needed to provide appropriate veterinary care to agricultural animals. IACUCs should not discourage this when done in accordance with AMDUCA and appropriate documentation is maintained. IACUCs should ensure a process is in place that prevents agricultural animals from entering the food chain prior to expiration of the applicable withdrawal time.

21:23 What should IACUC's consider when reviewing proposals to use non-pharmaceutical grade veterinary products and experimental compounds?

Reg. There is no specific regulation or guidance regarding the use of non-pharmaceutical grade drugs used in agricultural research. However, the *Ag Guide* (p. 2) states that the IACUCs responsibility when performing protocol review includes aspects of the proposed experiment or demonstration having to do directly with animal care and use. When agricultural animals are included under an institutional Assurance, OLAW provides an FAQ[12] that offers additional guidance on the use of non-pharmaceutical grade drugs.

Opin. IACUCs should follow the guidance provided for biomedical research programs and develop policies for the use of non-pharmaceutical grade drugs. Experimental compounds are inherently non-pharmaceutical grade, but the justification for their use is directly related to the intent of the research. IACUCs should develop policies that address purity, concentration, and other factors that could impact animal welfare when experimental compounds are used.

Surv. Does your institution release animals or animal byproducts for entry into the food chain?

Yes, animals only	11/27 (41%)
Yes, animal byproducts only	0/27 (0%)
Yes, animals and animal byproducts	12/27 (44%)
No	4/27 (15%)

Surv. Does your institution have a process for managing meat/milk withdrawal periods?

No	1/27 (5%)
Yes	19/27 (95%)

Surv. Can animals or their products that received experimental compounds be released into the food chain (meat, milk, eggs, etc.)?

Yes, after completing a withdrawal period	8/27 (30%)
Yes, with regulatory approval only (e.g., FDA INAD permits required)	5/27 (19%)
No	14/27 (52%)

21:24 How does an IACUC evaluate animal husbandry procedures for agricultural animals?

A. Animal Observations

Reg. The *Ag Guide* (p. 22) indicates that animals housed in intensive accommodations should be observed and cared for daily by trained and experienced caretakers.

Opin. In extensive environments such as rangeland pastures covering thousands of acres, it may be impossible to observe animals daily, similar to fish in a production pond that normally cannot be visualized because of the murky water. In both situations, managers can make indirect observations by paying attention to such things as feeding patterns, current weather patterns, and season of the year.

Surv. For animals that live in a natural environment such as a pasture or a pond, how often are they observed?

Once a day	12/27 (44%)
Twice a day	4/27 (15%)
Less than once a day	4/27 (15%)
Other*	7/27 (26%)

*Other: a respondent indicated a weekly observation interval.

Surv. How does your IACUC ensure the animals have an appropriate standard of husbandry? Select all that apply.

Husbandry SOP's	23/27 (85%)
Routine on-farm visits by veterinarians	14/27 (52%)
Semiannual facility inspections	24/27 (89%)
Facility inspections at an interval	6/27 (22%)
Other	2/27 (7%)

Surv. Does your institution allow compatible livestock of different species to be pastured together?

No	17/27 (63%)
Yes	10/27 (37%)

B. Animal Handling

Reg. Animal handling is defined as how agricultural animals are touched, moved, and interacted with during husbandry procedures and transport between pastures or housing locations. There are no specific regulations covering this activity; however, Chapter 5 of the *Ag Guide* provides detailed instructions for safe, effective handling of all agricultural species and should be consulted for further advice.

Opin. Agricultural programs should invest in a training program to ensure that all personnel who will be handling animals receive species-specific training on proper and safe animal handling practices. There are numerous resources available for each species, usually through species-specific associations or quality assurance programs.

C. Feed and Bedding Storage

Reg. The *Ag Guide* (p. 21) indicates all feedstuffs should be maintained in a manner so as to prevent contamination by chemicals or pests.

Opin. Feed containers should have tightly fitting lids and be maintained appropriately to prevent vermin entry. Feed storage areas for bulk feed should be regularly inspected for damage or evidence of pests. While the *Guide* (p. 66) specifies temperature and humidity conditions for feed storage, there are no such guidelines for agricultural feed. Bagged and bin stored feed should be labeled appropriately to prevent confusion as to its contents.

D. Provision of Food and Water

Reg. The *Ag Guide* (p. 21) indicates animals must be provided with feed and water in a consistent manner, on a regular schedule in accordance with established NRC requirements for each species.

Opin. Agricultural animals readily adapt to a routine feeding schedule. Once this happens, feeding time can be the best opportunity to observe animals for health issues or experimental effects. IACUCs should be aware of the feeding schedule and use this as an opportunity to evaluate the appropriateness for the feeding and observation schedules for each facility. There is no specified cleaning interval for feeding devices and equipment, but they should be cleaned as needed to prevent contamination of feed and buildup of waste material. It is recommended that feed, hay, and non-municipal water sources all be periodically tested to ensure appropriate quality.

Surv. What is the source of water for your agricultural animals held on pasture? Select all that apply.

Standard automatic livestock waterer	24/27 (89%)
Hose-filled tanks from deep wells	16/27 (59%)
Ponds/streams	10/27 (37%)
Other	3/27 (11%)

E. Vermin Control Considerations

Reg. The *Ag Guide* (p. 23) indicates that programs should be instituted to control infestation of animal facilities by vermin (e.g., flies, cockroaches, mosquitoes, lice, mites, ticks, grubs, rodents, skunks, and pest birds such as starlings, pigeons, and sparrows).

Opin. Prevention is the key to controlling vermin. Vermin are free-roaming; therefore, comprehensive plans should be developed that include the entire farm or campus in the control strategies. Facility design and maintenance are integral components of a vermin control program. Development of management plans for stormwater drainage and manure management should include a discussion of their impact on pest control. Pesticides and rodenticides may be a part of the control program but their usage should follow label recommendations and include a discussion of their downstream impacts and the possible impacts on feed supplies, impact on animals in the area, and impact on existing research programs.

Surv. Does the IACUC oversee animals used to protect livestock on pasture or animals used as part of the pest control program (e.g., cats)?

No.	10/27 (37%)
Yes.	7/27 (26%)
Not applicable. We do not use animals to protect livestock on pasture or as part of the pest control program.	10/27 (37%)

F. Agricultural facilities are complex operations that require expertise from multiple disciplines. IACUCs are encouraged to establish collaborative relationships with veterinarians, farm management, and other agricultural professionals. Subjects such as pest and predator control, waste management, appropriate housing and fencing, and building maintenance should be discussed among these stakeholders.

21:25 What role does the IACUC play in hazard identification in agricultural programs?

Reg. The *Ag Guide* (pp. 4–5) discusses the importance of hazard identification for personnel and details the hazards that are unique to agricultural facilities.

Opin. Agricultural programs have unique hazards that can represent serious and life-threatening health concerns. The institution should ensure sufficient expertise exists in the Occupational Health and Environmental Safety program to protect individuals from injuries related to the use of heavy equipment, large animals, chemical exposure, inclement weather, and zoonotic disease. Additional guidance for identification and mitigation of on-farm hazards is available at the National Ag Safety Database[13] and from numerous state agricultural extension agencies.

Surv. What unique elements are included in your institutional OHSP for individuals working with or exposed to agricultural animals? Select all that apply.

Chemical hazards – routine farm chemicals such as disinfectants, herbicides and insecticides, veterinary labeled drugs such as prostaglandins and tilmicosin, etc.	6/27 (22%)
Physical hazards – heat stress, cold exposure, power-take-off rotating shaft and auger safety, equipment rollover protection, etc.	6/27 (22%)
Biological hazards – zoonotic agents such as swine influenza, Q-fever, Salmonella, anthrax, etc.	10/27 (37%)
Animal allergies – hay fever, chick dander, etc.	8/27 (30%)
All of the above	16/27 (59%)
None of the above	2/27 (7%)

Surv. Does your institutional OHSP include a hearing conservation program that addresses the unique auditory hazards of a farm setting?

No	13/27 (48%)
Yes	14/27 (52%)

Surv. What auditory hazards are evaluated by your institutional OHSP? Select all that apply.

Tractor or other heavy equipment use.	11/27 (41%)
Working with swine.	11/27 (41%)
Using a captive bolt or rifle.	10/27 (37%)
Milking parlor activities.	4/27 (15%)
Use of a hydraulic chute.	6/27 (22%)
Our OHSP does not include evaluation of farm related auditory hazards.	10/27 (37%)
Other.	4/27 (15%)

Surv. Which category of individuals working with or exposed to agricultural animals should be included in your institutional OHSP? Select all that apply.

Paid employees that are not students	27/27 (100%)
Students, but only if they are a paid employee	19/27 (70%)
Students enrolled in coursework that exposes them to agricultural animals	16/27 (59%)
Personnel that may perform maintenance or repairs in the agricultural animal facilities	16/27 (59%)
Guests, visiting scientists, or volunteers	13/27 (48%)
None of the above	0/27 (0%)

Surv. How are individuals not enrolled in the institutional OHSP informed of the risks associated with working or visiting a farm environment? Select all that apply.

Required in-person training	12/27 (44%)
Required computer-based training	10/27 (37%)
Given a pamphlet or other materials to read	5/27 (19%)
No awareness is given	1/27 (4%)
Other	9/27 (33%)

21:26 What aspects of animal waste management should the institution review?

Reg. The *Ag Guide* (p. 20) describes the goals of the waste management program along with additional resources. State and local laws may apply and should be regularly consulted.

Opin. IACUCs should be informed of the waste management practices for agricultural programs and ensure appropriate expertise and resources are applied to this function. Waste management can directly impact worker health, animal health, pest control, groundwater contamination, and public perception of the operation, and it can create legal vulnerabilities.

Surv. Does your institution's IACUC review farm manure and waste management programs?

No	24/27 (89%)
Yes	3/27 (11%)

21:27 Is dedicated work clothing required for agricultural animal workers?

Reg. While no specific requirement for dedicated clothing exists, the *Ag Guide* (pp. 24–25) discusses the importance of biosecurity and the role of dedicated clothing and footwear in the biosecurity program.

Opin. Clothing is in contact with animals, chemicals, excreta, etc., as well as the skin of personnel. Clothing can act as a fomite and be a source of allergens and contamination of vehicles, offices, cafeterias, and other animals. The IACUC should be aware of the management of clothing and make recommendations that will protect the animals and individuals of the program. Dedicated clothing is highly recommended.

Surv. Do you allow farm workers to wear their work clothing home at the end of the day?

No, we do not allow our farm workers to wear work clothing home. They are required to change clothing before leaving the farm.	13/27 (48%)
Yes, we allow our farm workers to wear their work clothing home.	14/27 (52%)

21:28 Should institutions require zoonotic disease training as part of the occupational health program?

Reg. The *Ag Guide* (p. 3) states: "The program veterinarian, working in collaboration with scientists or instructors of record, should establish appropriate preventive medicine programs and husbandry practices to decrease the likelihood of transmission of potential zoonotic agents. Each institution must have an appropriate occupational health and safety program for evaluating human health risks associated with animal contact and must take steps to ensure that health risks for each individual are assessed and managed at an acceptable level."

Opin. Agricultural animals can serve as a source of zoonotic disease transmission. In research programs, agricultural animals may be sourced and transported to research facilities that are geographically distant from the animal source, and they can thus introduce diseases that are not endemic to the research facility's region. The IACUC should ensure the biosecurity, preventive medicine program, occupational health program, and training for employees are sufficiently robust to protect the animals and people from zoonotic disease transmission.

21:29 What vaccines should an agricultural program's occupational health program consider for employees?

Reg. The *Ag Guide* (p. 4) states: "It is important that all researchers, students, agricultural animal caretakers, and others be immunized against tetanus every 10 years based on the institution's risk assessment. Prophylactic vaccinations should also be considered when research is being conducted on infectious diseases for which effective vaccines are available."

Opin. Agricultural animals used in research, especially food and fiber research, have a high probability of being sourced from commercial production operations vs. vendors that supply purpose-bred animals for research. Given this, the risk of zoonotic disease exposure can be greater. IACUCs should ensure there is sufficient partnership between the animal program and the occupational health group to identify and assess the risks and determine when individuals should be offered vaccinations.

Surv. Does your institutional OHSP require vaccinations/titers or serum-banking for individuals working with or exposed to agricultural animals? Select all that apply.

Vaccinations/titers are required	8/27 (30%)
Serum-banking is required	0/27 (0%)
Vaccinations/titer screening are available upon request	11/27 (41%)
Serum-banking is available upon request	2/27 (7%)
Neither vaccinations/titer screening nor serum-banking are required or available upon request	11/27 (41%)

Surv. If your institutional OHSP requires vaccinations or serum-banking, which of the following are required? Select all that apply.

Rabies vaccination/titer screening	7/27 (26%)
Influenza (specifically H1N1) vaccination	4/27 (15%)
Tetanus vaccination	16/27 (59%)
Q-fever serology/titer screening	4/27 (15%)
Other	13/27 (48%)

REFERENCES

1. Guide for the Care and Use of Agricultural Animals in Research and Teaching, 4th Edition, 12020. American Dairy Science Association, the American Society of Animal Science, and Poultry Science Association.
2. Office of Laboratory Animal Welfare. Frequently Asked Questions G.27. May Institutions Utilize the Guide for the Care and Use of Agricultural Animals in Research and Teaching (Ag Guide) if their Programs Include Traditional Farm Animals? https://olaw.nih.gov/faqs#/guidance/faqs?anchor=50384.
3. U.S. Government Principles for the Utilization and Care of Vertebrate Animals Used in Testing, Research and Training. U.S. Government Principles for the Utilization and Care of Vertebrate Animals Used in Testing, Research, and Training | OLAW (nih.gov).
4. United States, Department of Agriculture, Agricultural Research Service (ARS). https://www.ars.usda.gov/
5. Midwest Plan Service. https://www.mwps.iastate.edu/
6. American Veterinary Medical Association. 62020. AVMA Guidelines for the Euthanasia of Animals. https://www.avma.org/resources-tools/avma-policies/avma-guidelines-euthanasia-animals.
7. American Veterinary Medical Association. 72016. AVMA Guidelines for the Humane Slaughter of Animals. https://

www.avma.org/sites/default/files/resources/Humane-Slaughter-Guidelines.pdf.
8. National Research Council (NRC) (1994, 2001, 2007a, b, 2012, 2016). https://animalnutrition.org/nrc_reportshttps://animalnutrition.org/nrc_reports
9. United States Food and Drug Administration, Veterinary Feed Directive (VFD). https://www.federalregister.gov/d/2015-13393
10. United States Food and Drug Administration, Animal Medicinal Drug Use Classification Act of 1994, (AMDUCA). https://www.fda.gov/animal-veterinary/guidance-regulations/animal-medicinal-drug-use-clarification-act-1994-amduca
11. Food Animal Residue Avoidance Databank (FARAD). https://www.farad.org/
12. Office of Laboratory Animal Welfare. Frequently Asked Questions F.124. May Investigators Use Non-Pharmaceutical-Grade Substances in Animals? https://olaw.nih.gov/faqs#/guidance/faqs?anchor=50361
13. National Ag Safety Database. https://nasdonline.org/

22

IACUC Considerations for Research Using Invertebrates

Stuart Leland and Alan Gelperin

Introduction

Living things are taxonomically divided into six kingdoms: animal, plant, fungi (yeasts, molds, mushrooms), protista (algae and protozoa), eubacteria, and archaebacteria. The animal kingdom can be divided into two main groups, vertebrates and invertebrates. Vertebrate animals have backbones while invertebrates do not. Over 95% of the animal kingdom consists of invertebrates and their unique genera and species number in the millions.

The use of invertebrates in research is common.[89] While a number of countries have regulations regarding the use of invertebrates in research, they are not regulated in the United States. Unfortunately, however, no data are available reporting on the use of invertebrates in research, and, therefore, it is impossible to accurately determine the number or proportion of their use in relation to other commonly used research animals. In general, but certainly not always, their small size, short life cycle, and simpler anatomy make them excellent animal models to answer many basic research questions. They are often easy to acquire and maintain (e.g., fruit flies and crayfish), making the costs associated with their use nominal in comparison to other research species.[1] Some researchers have turned to invertebrate species in an effort to implement the replacement, refinement, and reduction (3Rs) of animals in research. Specifically, based on current scientific thinking, invertebrates may not be capable of experiencing suffering, which makes their use less controversial than a vertebrate animal.[2] Some would consider the fact that most invertebrates are not regulated to be an advantage, negating many husbandry, facility, recordkeeping, and protocol regulatory stipulations.

This chapter explores institutional responsibilities, best practices, and ethical considerations when using invertebrates in research, testing, and teaching. The chapter is not meant to cover all aspects of invertebrate use in research, but rather to provide a framework for consideration when IACUC's are presented with the unique situation in which invertebrate animals are the proposed animal model of choice. Several highly relevant book-length treatments dealing with the welfare and medicine of invertebrate animals have recently appeared.[3–6]

22:1 What is an invertebrate and how are they classified?

Invertebrates are animals without backbones. The taxonomy of invertebrates is a field unto itself and, with recent advances in DNA, RNA, mitochondrial, amino acid, and protein sequencing to name a few common techniques, the classification of invertebrates has undergone a major revamping in the past decade. The major classes of invertebrates include annelids (earthworms and leeches), mollusks (clams, snails, squid, octopuses), crustacea (lobsters, crayfish), cnidaria (jellyfish, anemones, corals), insects (cockroaches, fruit flies, preying mantids), and protozoa.

22:2 What are the major research areas in which invertebrates are used?

While not an exhaustive list, invertebrates are used in neuroscience (crayfish for synaptic physiology[7, 8]), genetics (fruit flies for gene identification studies[9, 10]), drug testing (planaria for anesthetics testing[11, 12]), pharmacology (clam hearts for neuromodulator [e.g. peptides] testing[13]), regeneration (starfish, cephalopods[63, 64]), ecology (moths for predator-prey interactions[14, 15]), ecotoxicology,[71] and linking gene activity to behavior (*Caenorhabditis elegans* learning.[16, 17] Many species of crustacean decapods have been utilized over the past decades as biomedical models to understand neural network and physiology due to their large neurons and relatively simple nervous circuits.[67, 68, 78]

The use of *Caenorhabditis elegans* as a model system for linking gene activity to behavior was greatly facilitated by the complete definition of the CNS connectome in this species.[18, 19] Interest in this species was further heightened by the demonstration of both associative and nonassociative learning.[20] The study of learning with *Drosophila melanogaster* was initiated by Seymour Benzer and William Quinn[21] and was followed by identification of a single gene mutation, called dunce, that degraded the fly's learning ability.[22] Subsequent work has greatly expanded on these initial studies.[23–26] Crayfish neuromuscular synapses have provided a particularly favorable environment for the study of how presynaptic action potentials trigger the release of neurotransmitters,[27, 28] in addition to seminal studies of motor network function and postural control.[29–31]

Invertebrates are also used to study a number of potential contaminants in aquatic environments. Their simple anatomic structures and often transparent outer membranes make them compelling sentinels to examine the effects of many contaminants on neural development. Specifically, marine pollutants, including heavy metals (copper, iron, manganese, zinc, lead, mercury, etc.), microplastics, nanoparticles, pharmaceuticals, personal care products (drugs, chemicals, etc.), pesticides, insecticides, herbicides, fungicides, and endocrine disrupting

DOI: 10.1201/9781003272861-22

compounds can have neurotoxic effects on marine invertebrates. These marine pollutants can have sub-lethal, acute, or chronic effects on a single invertebrate, or an entire population or geographic area. As sentinels, invertebrates can alert humans of potential environmental risk factors, especially given trophic transfer and the fact that humans often represent the final consumer in a lengthy food chain.

Surv. Does your institution use invertebrates for teaching, research, or testing?

- No 141/189 (75%)
- Yes 48/189 (25%)

Surv. If invertebrate animals are used in research, specify the species being used. Species name(s):

- Please specify 49/49 (100%)

  - Mosquitos
  - Sea hare/sea slug
  - Cephalopods (including octopus, squid, cuttlefish)
  - Crayfish
  - Fruitfly
  - Snails, mussels
  - Urchins
  - Grasshoppers
  - Shrimp
  - Hermit crabs
  - Cockroaches
  - *Caenorhabditis elegans*
  - Ticks
  - Sea anemone
  - Starfish

22:3 What species-level considerations are given to invertebrates regarding the ability to experience pain and distress?

Nociception refers to a body's processing of noxious stimuli, which may include chemical, thermal, or physical stimuli. Nociception can occur strictly through the peripheral nervous system, in which case the stimuli might not be experienced consciously as pain, or it may occur both peripherally and centrally, in which case one experiences the noxious stimuli as a painful event. In order to perceive pain, an animal must process the painful event centrally. There must be anatomic structures and physiologic processes that allow the conception of pain followed by behavioral manifestations that allow the expression of pain, distress, or fear, including escape or withdrawal activity.[32] Note that withdrawal activity resulting from stimulation may occur only as a peripheral or reflex event and does not necessarily mean that central processing of the stimulus occurred.

Nociception can occur in the absence of pain. However, there remains active debate about the definition of perception and the neural processing necessary for that perception to occur centrally. This is especially true with animal species as numerous and diverse as invertebrates. It was not too long ago that invertebrates were believed to lack the anatomic and physiologic capacity to perceive pain similarly to vertebrate animals. For example, some invertebrates do not have a nervous system (e.g., sponges) or centralized ganglia associated with the processing of painful experiences (e.g., starfish, but see reference[33]). However, other invertebrates contain well-developed brains and are capable of advanced learning behaviors (e.g., octopuses, cuttlefish). The *AVMA Guidelines for the Euthanasia of Animals* also state that spiders, honeybees, butterflies, and ants are species that have similar brain characteristics as octopuses and cuttlefish.[32] While it is a fact that most invertebrates respond to noxious stimuli, it remains debatable about their ability to process such stimuli centrally and thus experience the stimuli as painful. Robert Elwood[34] does a nice job of discussing many of the anatomic structures and physiologic processes necessary to experience nociception, perception, reflexes, and pain and provides a helpful table to elicit further consideration of 15 key criteria for pain in animals as described by Sneddon in 2014.[35] Using these criteria and citing research, Elwood makes a compelling argument that many classes of invertebrates experience pain in manners similar to vertebrates and are therefore deserving of protections.

The definition of pain continues to evolve and change. In 2020, the International Association for the Study of Pain (IASP) revised its definition of pain as, "An unpleasant sensory and emotional experience associated with, or resembling that associated with, actual or potential tissue damage." This definition is accompanied by a number of important notes, including that pain can be learned and influenced by one's experiences in life, and that it is a "personal experience" affected by biological, psychological, and social factors.[83]

Opin. The default assumption should be that invertebrate species can experience pain and distress, as evidenced by the triggering of defensive and escape behaviors. This opinion is further supported by U.S. Government Principles for the Utilization and Care of Vertebrate Animals Used in Testing, Research, and Training IV, which states that investigators should assume that procedures causing pain or distress in humans may cause pain or distress in other animals.[84] This would include invertebrates.

Observing an invertebrate that looks back at you and follows your movements within its visual field presents a uniquely meaningful communication event. This visual interaction can provide a window into the Umwelt.[36] The human Umwelt is dominated by vision and thus visual interactions with invertebrates with visual systems that can focus on and track an observer provide a unique communication channel between observer and subject. Those who have watched an octopus observing the observer cannot help but be affected by this interaction. Cephalopods contain what is known as a lens eye, though it is different anatomically and functionally from a vertebrate eye. They focus using an aperture-type opening like the lens of a camera, as opposed to vertebrates that focus by changing the shape of the lens.[37] The same is true of watching a praying mantis watching you and following your movements with its head. These visual interactions make it hard to escape the inference of sentience on the part of the invertebrate during this interaction. Insects with highly developed visual systems raise the same set of issues as do decapod crustacea such as crayfish and lobsters. The morally defensible position is to assume that an invertebrate is sentient until proven otherwise, particularly for invertebrates with highly developed visual systems.

22:4 What are cephalopods, and why are they addressed as part of the accreditation process by AAALAC, International?

Cepahlopods are a set of closely related species of mollusks and include cuttlefish, squid, and octopuses. Octopuses have large brains and considerable behavioral flexibility, including advanced learning abilities. Squid and cuttlefish share these properties, including highly developed visual systems, which provide a uniquely effective communication channel with human observers. The list of additional species will expand as our understanding of learning ability and neural complexity for other invertebrates increases, particularly insects. As this occurs, we anticipate that more and more invertebrate species will receive attention from regulatory bodies.

AAALAC considers all animals, including farm animals, wildlife, aquatics, and invertebrates, to be included and evaluated as part of an accredited program where they are relevant to an institution's mission. AAALAC offers different approaches an institution may take to include the use of invertebrates as part of an overall program.[85]

22:5 What are decapod crustaceans and should they be considered higher level?

Decapod crustaceans belong to the class Malacostraca, and include crabs, lobsters, crayfish, shrimp, and prawns. Many decapod crustaceans have well-developed visual systems and eyes mounted on mobile eyestalks that follow salient visual stimuli and may guide body orientation to potential mates, predators, or prey. Crayfish also exhibit sleep behavior with brain activity during sleep divided into multiple phases.[38] Lobsters can live for 50 years and can attain weights of more than 40 lbs.

Opin. Crayfish and lobsters have well-developed central nervous systems, including a brain containing thousands of neurons. Many more central neurons are contained in the chain of ganglia running from the thorax to the tail. This central computing machinery is supplied by sensory receptors involving vision, olfaction, taste, pressure waves, and tactile input, among others. Decapod crustacea exhibit a number of complex behaviors, including homing behavior, navigation, and formation of dominance hierarchies. The complexity of behavior, brain organization, sensory systems, and motor outputs support the possession of the ability to feel pain.

22:6 What is the current regulatory landscape in the United States concerning the use of invertebrates in research?

In the United States, the Animal Welfare Act (AWA), the Animal Welfare Regulations (AWAR), and the Public Health Service Policy on Humane Care and Use of Laboratory Animals (PHS Policy) do not regulate invertebrate species nor make any mention of them. However, the *Guide for the Care and Use of Laboratory Animals*, Eighth Edition (*Guide*) mentions invertebrates, suggesting that many of the general principles governing humane care and ethical considerations provided for vertebrates also could be applied to invertebrates.[39]

In 2023, the National Institutes of Health (NIH) Office of Laboratory Animal Welfare (OLAW) put out a request for information on proposed guidance on cephalopod care and use. The guidance includes, among other things, that IACUCs review and approve the use of cephalopod research activity; that IACUC's perform semiannual evaluations of cephalopod programs and facilities; that only trained and qualified personnel be allowed to use and care for cephalopods; that users follow the *AVMA Guidelines for Euthanasia of Animals* unless there is scientific justification approved by the IACUC. The guidance also encourages the development of institutional policies and standard operating procedures for cephalopod care and use.[86]

In 2023, the NIH also put out a notice encouraging the use of the ARRIVE Essential 10 Checklist in all publications reporting on the results of vertebrate animal and cephalopod research.[87]

22:7 What is the regulatory landscape internationally concerning the use of invertebrates in research?

Reg. Sentience can be defined as "the capacity to have feelings, such as feelings of pain, pleasure, hunger, thirst, warmth, joy, comfort and excitement."[72] To be sentient is to perceive and respond to one's senses, including sight, sound, taste, touch, or smell.[69, 70] Interestingly, this word is infrequently used in the United States regulatory landscape when discussing animal welfare. However, many in the international community embrace the word "sentience," suggesting that it may be used to define which invertebrate species shall be covered through animal welfare or ethical regulation. In New Zealand the Animal Welfare Act recognizes sentience at the outset of the Act and includes in the definition of animal: "any octopus, squid, crab, lobster, or crayfish (including freshwater crayfish)."[40] The United Kingdom Animals (Scientific Procedures) Act 1986 does not currently regulate invertebrates nor mention sentience though it is clearly stated that invertebrates may be included in the definition of animal at the order of the Secretary of State. Sentience may also be addressed by the United Kingdom in the near future.[41] The Australian AWA states that "animals are sentient beings that are able to subjectively feel and perceive the world around them.'[42] It provides the following definition of animal: *animal means—(a) a live member of a vertebrate species, including—(i) an amphibian; and (ii) a bird; and (iii) a fish; and (iv) a mammal (other than a human being); and (v) a reptile; or (b) a live cephalopod; or (c) a live crustacean intended for human consumption.* The Australian AWA does not define sentience and so it is difficult to ascertain why some invertebrates are considered sentient and therefore covered species, while others are not.

In Canada, the federal government is prohibited from regulating experiments involving animals (Constitution Act of 1867). Instead, this responsibility is delegated to provincial governments. The Criminal Code of Canada[43] contains some general anti-cruelty language that protects animals in general but never defines animal and does not mention invertebrates.

The Canadian Council on Animal Care (CCAC) is an independent, nonprofit organization that serves in a capacity similar to that of AAALAC, International. In its fundamental principle document describing the categories of invasiveness in animal experiments, the CCAC specifically refers to invertebrates and instructs users that they "must adhere to humane principles, and take cognizance of the CCAC's policy statement on ethics of animal investigation." Invertebrate use must be assigned to a category of invasiveness (range A least invasive to E most invasive) and reviewed by an appropriate review committee if the category is B–E. The principle further goes on to state: "Cephalopods and some other higher invertebrates have systems as well developed as in some vertebrates, and may therefore warrant inclusion in Category B, C, D, or E." Invertebrates are not mentioned elsewhere in the Standards utilized by CCAC in its accreditation process.

The EU Directive 2010-63 refers to "the protection of vertebrate animals used for experimental and other scientific purposes." [44] On page 2 (item 8) it says: "In addition to vertebrate animals including cyclostomes, cephalopods should also be included in the scope of this Directive, as there is scientific evidence of their ability to experience pain, suffering, distress and lasting harm." During its deliberations, the European Union (EU) gave serious consideration to adding decapod crustaceans to this definition but ultimately decided not to do so. Switzerland and Norway are not part of the EU but also regulate the use of some invertebrates.[73]

The Japanese Act on Welfare and Management of Animals (Act No. 105 of 1973) does not mention invertebrate species nor make any reference to sentience.[45] The Korean Animal Protection Act does not mention invertebrate species nor make any reference to sentience.

Note that in the U.S., neither the AWA, AWAR, nor PHS Policy mention the word "sentient" or "sentience." The *Guide* mentions sentience once but only via a reference on p. 192 under "Amphibians, Reptiles, and Fish.'[46]

Opin. A number of international countries, including New Zealand, Australia, the European Union, and other European countries, actively define some species of invertebrates as potentially sentient in their regulations. The United Kingdom, Japan, and Korea do not currently regulate invertebrates. As more and more research becomes available demonstrating sentience of some invertebrate species, and if countries adopt sentience as a condition in need of protection, it appears likely that some of them will be added to law, regulation, or policy going forward.

Balancing the field of animal research compliance with the addition of a requirement to oversee invertebrates would require a monumental paradigm and resource shift in oversight. Unless a standard is created, either through regulation or policy in the United States or elsewhere, IACUCs will be left to develop their own mechanism for providing oversight on the care and use of invertebrates. The ethical use of invertebrates and moral obligation to consider sentience and treat accordingly with appropriate anesthetics and analgesics still requires investigation, which may ultimately result in legal and regulatory protections for this very large group of animals.

22:8 What should an IACUC consider when determining if the use of invertebrates should be overseen by the IACUC?

Reg. The following statement is from Section 1 (Definitions) in the Rules of Accreditation accessed on the AAALAC website[47]:

> All animals used or to be used in research, teaching or testing at accreditable units are to be included and evaluated in accordance with the standards set forth in Section 2 of these Rules. This includes traditional laboratory animals, farm animals, wildlife, and aquatic animals. Nontraditional animals, inclusive of invertebrate species, are also included where they are relevant to the unit's mission.

Within the AAALAC FAQ section under *Animals included in the AAALAC International Accredited "Unit"* is a section on invertebrate animals in which it states:

"Of particular concern to Council, vis-a-vis invertebrate colonies, is the inclusion of their review when such inclusion would benefit the program as a whole, or their exclusion would potentially jeopardize the overall program mission. Also, Council acknowledges the need to evaluate management of invertebrate colonies when those colonies are within a core animal facility and therefore subject to creating, or being subject to, potential adverse influences from activities in adjacent spaces. Possible areas of evaluation could include:

1. Training of personnel caring for and handling the animals.
2. Husbandry practices, including sanitation, diet, and adequate space (especially if contiguous with vertebrate holding areas).
3. Health monitoring and recordkeeping.
4. Species interactions for those held in the same room or area.
5. Physical plant provisions (e.g., emergency power availability for species whose environment would be negatively impacted by a power outage).
6. Occupational health and safety risks (e.g., GFI circuitry, zoonotic diseases, containment of mosquitoes used in malaria research, etc.)."

While AAALAC does require that invertebrates used as a part of the research mission of the institution be included as part of the animal care and use program, there are many ways in which that can be accomplished. Many institutions do not have a formal protocol for invertebrate work but notify the Institutional Animal Care and Use or Ethics Committee or comparable oversight body (IACUC/OB) via letter about the work they will be doing with these species. Or the oversight may be a Standard Operating Procedure (SOP), policy statement, guideline, etc. that is reviewed by the IACUC/OB and delineates the committee's responsibility over the use of invertebrates and provides guidance to investigators using invertebrates. Some of the issues the document may want to include are listed above. The intensity of oversight by the IACUC/OB may depend on such things as the species of invertebrate being used and the type of procedures being performed. Site visit teams will generally not review research activities involving lower level invertebrates such as zooplankton, sea slugs, nematodes, or mosquitoes, but they will review research with higher level invertebrates such as cuttlefish, squid, octopi, or decapod crustaceans. For example, if procedures are being conducted on a higher-level invertebrate, an IACUC/OB approved protocol may be an appropriate requirement."

Surv. Does your IACUC review activities involving invertebrates? Select all that apply.

- The IACUC conduct a formal review. 14/49 (29%)
- The IACUC is informed of the activities but does not conduct a formal review. 10/49 (20%)
- The Attending Veterinarian/veterinary staff are consulted and/or oversee animal husbandry and care. 7/49 (14%)
- None of the above. 22/49 (45%)

Surv. What activities involving invertebrates does your institution's IACUC review? Select all that apply.

- Scientific merit/ethical review 11/49 (22%)
- Housing locations and environmental parameters (e.g., temperature, humidity, water quality) 21/49 (43%)
- Husbandry procedures 17/49 (35%)
- Veterinary care 14/49 (29%)
- Any procedure with the potential for pain and/or distress 13/49 (27%)
- Any surgical procedure details 12/49 (24%)
- Any non-surgical procedure details, including restraint 12/49 (24%)
- Any anesthetic, analgesic, or experimental agent administered 13/49 (27%)
- Euthanasia 16/49 (33%)
- None of the above 27/49 (55%)

Surv. How frequently does your IACUC inspect locations where invertebrates are housed, held or used in research?

- More than semiannually 0/49 (0%)
- Semiannually 24/49 (49%)
- Annually 1/49 (2%)
- More than annually 0/49 (0%)
- Less than annually 0/49 (0%)
- Never 24/49 (49%)

22:9 When might the inclusion of invertebrates benefit the overall program or when might their exclusion potentially jeopardize a program of animal care?

Opin. The advantages and disadvantages of including invertebrate species in an institutional program of animal care should be based on the professional opinion of the AV, IACUC, and research community considering risks to existing vertebrate colony health and human safety. Such decisions will be influenced by the size of the invertebrate research program and the relevant invertebrate expertise available. If an institution remains unsure whether they should include invertebrate colonies as part of their overall program description, outside expertise may be sought to provide an objective opinion.

22:10 What are the expectations of an IACUC if it chooses to oversee an IACUC protocol involving invertebrates?

Opin. If an IACUC decides to review and approve a research protocol involving invertebrates, the expectations are similar to the review process that currently exists for vertebrate species with a number of additional considerations that may be unique for the species involved.[48] Many of these concepts are explored in depth in other chapters in this *Handbook*, including

- A justification for the use of animals and the species chosen (see 22:11)
- A justification for the number of animals to be used (see 22:11)
- A review of how the animals will be acquired and transported to their final destination (see 22:12)
- A complete description of the experimental procedures used, study design, and the animals' final disposition
- A description of any surgical procedures, including appropriate pre- and postoperative care, anesthesia, analgesia, and veterinary input into the design and implementation of surgical procedures (see 22:13)
- A description of any pain and distress that may be experienced by the invertebrates and how or whether it will be alleviated. This should include a description of humane endpoints
- A description of the method(s) of euthanasia and the method(s) used to confirm death (see 22:14)
- A description of animal housing, husbandry practices, space allocation, HVAC, and access security that may be unique to the invertebrate species (see 22:15–16, 22:18–20)
- Any research-specific training or species-specific training of research personnel for animal handling, restraint, and experimental use (see 22:17)
- Any potential occupational or health and safety concerns (see 22:21–22)

22:11 How might an IACUC review a proposal for appropriate species and animal numbers?

As with all experimental application, the species must be justified and the number of animals used experimentally should be designed to achieve statistically relevant results while abiding by the reduction concept as established by Russell and Burch's 3Rs.[54] There are a number of characteristics of unique invertebrate species that may require specialized training and knowledge, which will be covered in the next sections. If needed, IACUCs are encouraged to use *ad hoc* consultants as unique issues may be identified during protocol review.

Given that many invertebrates live in a group setting, such critical mass necessary for survival of the colony must be considered when building and supporting colony animals. If animals will be captured in the wild, IACUCs must consider bycatch, for example, during the drag of a pond, seascape, or other waterscape or netting of a geographic region, and what will be done with those bycatch species.

22:12 What are the considerations when acquiring and transporting animals?

Reg. The U.S. Department of Commerce, Bureau of Industry and Security (BIS) promulgates regulations affecting the exportation and importation of identified commodities, software, and technology that falls under the jurisdiction of the Export Administration Regulations (EAR).[57] For example, some arthropods (mosquitoes) may be affected by export control permits, especially if they are relevant to certain zoonotic diseases like encephalitis.

Invertebrate species that are not native to a locale may require special permits or licensure to import. State and local agencies may require collection or other permits for certain native wild species. Invertebrates that may be endangered or on the verge of extinction require special handling. The introduction of new species into new locales may threaten existing biodiversity or be a threat to humans, water sources, aquafers, agriculture, or livestock. The Convention on International Trade in Endangered Species of Wild Fauna and Flora (CITES) is an international agreement between governments.[55] Its aim is to ensure that international trade in specimens of wild animals and plants does not threaten the survival of the species.[56]

According to the EU Directive 2010-63, the National Competent Authority (NCA) must authorize the catch of animals from the wild if they are to be used for scientific purposes. Standardized studies exploring specific protocols used for transportation are lacking in most species. Training events have developed around invertebrate research, including biology and care, through the European Cooperation in Science & Technology (COST).[61]

Opin. Because many species used in research may be purchased from outside vendors or shipped from collaborators, IACUCs should consider the animal's macro- and microenvironment during transportation between facilities. Given that most invertebrates are ectotherms (cold-blooded), attention to temperature throughout the journey is especially important. International travel requires additional attention by adding potentially long wait periods, transfer procedures, possible permitting (see above), pressurization, humidity, and feed/watering issues to an already complicated process. This is especially important for those species that live in aquatic environments that may need monitoring and treatment during transportation.

22:13 What veterinary considerations are needed for procedures that have the potential for pain and/or distress? How is veterinary oversight managed?

Opin. For USDA-covered species, veterinarians are required to have input on surgical procedures, including pre- and postoperative care and the use of appropriate anesthetics and analgesics. Although there is no regulation requiring veterinary input for invertebrates, many of the skillsets used to administer and monitor anesthesia, perform surgery, and provide postoperative pain relief are transferrable to invertebrate species.

Most invertebrates are ectotherms and therefore rely on the external environment with minimal adaptability (burrowing or heat seeking) to regulate body temperature. The temperature of the environment, be it water or air, must therefore be a factor when considering anesthesia and analgesia for invertebrates.

Properly administered anesthesia has two primary purposes: the alleviation of pain and memory loss of the event associated with pain. Given the debate existing about invertebrates' ability to experience pain (see above), the moral imperative to administer anesthesia and analgesia remains controversial. However, given the increasing evidence that at least some invertebrates are likely capable of experiencing pain, it remains an important topic for consideration when using invertebrates in a research setting. Cooper[3] discusses the use of the halogenated inhalational anesthetic agents (isoflurane, halothane, or sevoflurane) up to 10% concentration or CO_2 as the most commonly used anesthetic agents for terrestrial invertebrates. Molluscs and other aquatic invertebrates are more commonly anesthetized with water-soluble agents such as tricaine methanesulfanate (MS-222) or benzocaine. Carbon dioxide has also been bubbled through the water, although its use remains controversial due to its lack of anesthetic properties and the potential acidity it causes when bubbled through solutions given the sensitivity of invertebrate gills and skin to changes in pH. More recently, magnesium chloride and ethanol, either separate or combined, have been described as effective anesthetics for cephalopods.[49, 74] Other agents have been described for invertebrate anesthesia, including chloretone, ethanol, or chloroform.[3] Isobutanol and lidocaine have been used in crustaceans.[50] Hypothermia may be used to facilitate handling and for temporary noninvasive procedures but its use as an anesthetic agent for invasive surgery is no longer recommended.[3] Recommendations for specific anesthetic and analgesic regimes for a number of commonly studied invertebrates can be found elsewhere.[5, 60]

The assessment of depth of anesthesia is poorly described in invertebrates. Immobility and the loss of the righting reflex appear to be the most reliable indicators of a surgical plane of anesthesia. Other parameters that may be useful to assess depth of anesthesia depend upon the species and may include the ability to remain attached to a substrate, muscle tone, speed and forcefulness of movement, decoloration or decrease in chromtophores in the case of cephalopods, and response to stimulation (touching, pin-prick). Neural recordings support the use of magnesium chloride and ethanol in cephalopods as these agents have further demonstrated depression of both efferent and afferent nerve activity believed to be associated with CNS depression and unconsciousness.[61, 75, 76]

Postoperative care of cephalopods includes using suitable recovery tanks that are supplied with clean, good quality water and free of obstacles. Postoperative recovery monitoring should involve monitoring the animal's respiratory rate, sucker condition, coloration, and reaction to stimuli. The use of antibiotics is not recommended for animals that will be released into the wild.[77]

As with other species under review by an IACUC, descriptions of the assessment of pain and distress, their alleviation, pre- and post-recovery plans (including the administration of analgesics), humane endpoints, and documentation should be reviewed and approved by the IACUC prior to approval of the research protocol. The IACUC is also responsible for ensuring appropriate training and competence of the research team exists to perform the experimental procedures, including the administration of anesthetics and analgesics and the assessment of pain and distress.

While the use of anesthetics for invertebrates is not well described in the literature, the use and administration of analgesics is even less well described. This may be due, in part, to the long-running debate about invertebrates' ability to experience pain and suffering. Given the advancement of invertebrate sentience research, it is hoped that the field of pain and distress relief in invertebrates will follow. At the current time, however, very little is described on the preemptive use and use of analgesics either to relieve anticipated pain and distress, such as that experienced during a surgical procedure, or to treat existing pain or distress. A helpful drug formulary providing a number of referenced doses for anesthetics, analgesics, antimicrobials, antifungals, and anti-parasitic agents can be found elsewhere.[51]

22:14 What method(s) are used to euthanize invertebrates?

Reg. The 2020 *AVMA Guidelines for the Euthanasia of Animals* (Euthanasia Guidelines) defines euthanasia as, "A method of killing that minimizes pain, distress, and anxiety experienced by the animal prior to loss of consciousness, and causes rapid loss of consciousness followed by cardiac or respiratory arrest and death."[32] The Euthanasia Guidelines address invertebrate euthanasia in two sections; one involving aquatic invertebrates and the other involving captive or terrestrial invertebrates.

The Euthanasia Guidelines assume that all animals are worthy of consideration for minimizing pain or distress and negative effects to the animal. In addition, animal handling and the euthanasia technique should minimize distress experienced by the animal prior to loss of consciousness. The physical environment the animal experiences leading up to the euthanasia procedure can influence the outcome. Given the extraordinary number of invertebrate species and the wide variety of environments with which they may be used in a research setting, it may not always be possible to achieve the complete absence of pain and distress.

The Euthanasia Guidelines acknowledges that because knowledge about invertebrate physiology and anatomy may be limited, identifying and applying principles associated with humane death may be difficult. The Euthanasia Guidelines work to balance the goals of euthanasia to minimize pain and distress with the reality of the many environments in which euthanasia of invertebrates may occur.

As for all species undergoing euthanasia procedures, training on species-specific behaviors, physiology, and anatomy is important to properly assess the effectiveness during the administration, observation, and assessment of final disposition of death in invertebrate species. Consultation with experienced veterinarians and researchers familiar with invertebrates should further corroborate the best available options for euthanasia and ensure proper training and assessment of final disposition.

The Euthanasia Guidelines classify specific techniques as acceptable, acceptable with conditions, and unacceptable:

Acceptable methods are those that consistently produce a humane death when used as the sole means of euthanasia. Methods acceptable with conditions are those techniques that may require certain conditions to be met to consistently produce humane death, may have greater potential for operator error or safety hazard, are not well documented in the scientific literature, or may require a secondary method to ensure death. Methods acceptable with conditions are equivalent to acceptable methods when all criteria for application of a method can be met. Unacceptable techniques are those methods deemed inhumane under any conditions or that the Panel on Euthanasia found posed a substantial risk to the human applying the technique. The Guidelines also include information about adjunctive methods, which are those that should not be used as a sole method of euthanasia, but that can be used in conjunction with other methods to bring about euthanasia.

Adjunctive methods are often recommended as the second part of a two-step euthanasia procedure. For invertebrates, the second step typically involves a technique to destroy the brain or major ganglia. Such techniques often involve physical methods, including pithing, freezing, boiling, or chemical methods, including alcohol or formalin.

Note that the varied and complicated life stages of invertebrates should also factor into the determination of the best method of euthanasia for invertebrates since not all stages of an invertebrate's life cycle are equally susceptible to euthanasia methods. Consultation with experienced veterinarians or researchers on preferred methods of euthanasia will help determine the most effective means. Euthanasia methods for invertebrates in developmental stages have not been determined.

Potential routes of administration of euthanasia methods include intrahemolymphatic for those terrestrial invertebrate species with a closed circulatory system (earthworms), intra-abdominal or intracoelomic for those terrestrial species with an open circulatory system (insects), or via immersion for aquatic invertebrates.

For both aquatic and terrestrial invertebrates, confirmation of death may be difficult. Therefore, the Euthanasia Guidelines recommend a two-step euthanasia procedure for most invertebrate species. For aquatic invertebrates, submersion in magnesium salts or ethanol is an acceptable first step of a two-step method.[75, 76] Note, however, that magnesium salts are ineffective for crustaceans.[32] Acceptable second steps include immersion in 70% alcohol and neutral-buffered 10% formalin. Other acceptable second steps include physical methods such as pithing freezing and boiling. Unacceptable methods of euthanasia for aquatic invertebrates include removing them from water, leaving them in improperly aerated water, or exposing them to caustic chemicals or trauma without first inducing unconsciousness.[32]

The Euthanasia Guidelines state that formaldehyde is an acceptable method of euthanasia for *Porifera* species (sponges). Formaldehyde use carries significant risks for personnel using it, and it is a known human carcinogen.[32]

As with aquatic invertebrates, the Euthanasia Guidelines for terrestrial or captive invertebrates recommend a terminal anesthesia procedure followed by a secondary method to ensure inactivation of the nervous system. Acceptable methods of euthanasia include the use of non-inhalational agents like pentobarbital injected into the coelomic cavity. Where injectable methods are not acceptable, other methods considered acceptable with conditions include the use of inhaled anesthetic agents followed by a secondary chemical or physical method to ensure death. The use of CO_2 may be useful for some terrestrial invertebrates, but there is no definitive research demonstrating its efficacy and therefore secondary methods are recommended if CO_2 is used. Many of the two-step methods and precautions described above for aquatic invertebrates also apply to captive or terrestrial invertebrates.

Assessing death in invertebrates is not as straightforward as it is for most vertebrate species. Consider assessing for the loss of movement, loss of reactivity to any stimulus (especially painful), and flaccidity as indicators of death in aquatic invertebrates. Due to the difficulty in properly assessing death in some invertebrates, it may be necessary to consult with others more experienced with the species in question or to consider performing a secondary means to ensure death as a matter of routine when euthanizing aquatic invertebrates.

22:15 How are animal husbandry needs determined for invertebrates?

Opin. Many IACUCs require that non-standard methods of housing and animal care be described in an IACUC protocol and invertebrate animal use often falls into this category. Husbandry of individual invertebrate species will vary based upon the species under study. For an excellent reference about husbandry and welfare of invertebrates commonly used in research, the reader is referred to a number of recent general texts on the matter. For example, guidelines and texts describing the welfare and care of a number of invertebrates used in research, testing, and training, including insects, honeybees, spiders, coral, cnidarian (jellyfish, hydra, sea anemone), crustaceans, cephalopods, and other invertebrates, have been published.[4, 5, 78]

22:16 How do you assess normal vs. abnormal health of invertebrate animals?

Opin. Much of the ability to observe and monitor the health of invertebrate animals rests on a firm understanding of normative behavior both of an individual and in a group setting. Before bringing a new species into a laboratory setting, husbandry staff should spend time learning how the species acts in the wild and captivity during different times of the day and other behaviors, including those that occur during feeding, hunting, predation, resting, play, sleep, with young, with mates, as well as including normal responses to threats and injury. Only when normal behavior is fully understood can one identify abnormal behavior that warrants further observation or treatment.

22:17 How are invertebrates handled and restrained?

In addition to learning how to properly handle animals during routine husbandry procedures, animal care staff and researchers must learn how to restrain animals. It is critical that animals be handled with adequate restraint to allow the experimental procedure to be performed without injury to the animal and while maintaining safety for the handler. Such handling can be especially challenging for those invertebrate species that have a mucinous slime coat used in nature as a protective feature.

Many cephalopods exhibit autotomy (voluntary separation or self-amputation) in natural environments or as a result of stress during research procedures, including surgery.[64] Autotomy has also been described in spiders in response to noxious stimuli and is a known response to predation.[88]

The skin of cephalopods plays an important role in concealment, communication, and as a barrier to protect the animal. Such a slime coat makes handling and restraint particularly challenging. When handling cephalopods during surgery, attention must be paid to avoid damaging the mucous layer/skin. Techniques used to help maintain the integrity of the skin include using moistened cotton soaked in sterile marine solution, using correctly diluted antiseptics, keeping the animal moist at all times, and to avoid using pure alcohol solutions, as this can act as a desiccant and damage the skin.[79] It is important that researchers, especially new investigators or those with little experience using these animals, are made aware of these species-specific requirements during training.

Pilot studies for newer procedures or to refine husbandry, handling, and restraint techniques may be a viable method for understanding species care and for practicing procedures that are new for a research laboratory and to allow for additional oversight early in the experimental model development. Usually, these are of shorter duration, with limited numbers of animals, and they utilize a check-in or reporting point with the IACUC to assess progress and identify and mitigate potential issues.

22:18 What considerations are needed for the animals' micro- and macro-environment?

Opin. Some specific considerations relative to invertebrate husbandry include physical plant (sealing, ante-rooms, special lighting, curtains), HVAC considerations to maintain optimal temperature and humidity, primary and secondary containment options. For example, one might consider whether the use of a hood or smaller sealed chamber is appropriate. Many invertebrates have unique housing and husbandry needs and may require that animal care personnel obtain specialized training and expertise in order to deliver and monitor critical husbandry parameters properly. These might include, for example, salinity, pH, intra-habitat structural modifications to meet species needs for hiding places, substrates needed for burrowing, normal feeding substrates and patterns, normal methods to obtain and retain water quality, acclimation, and appropriate light-dark cycles. In addition, husbandry staff should understand the different life cycles for the species involved and consider that all of the above parameters may be different depending upon the life stage (larva, nymph, polyp, etc.) of the invertebrate.

Within a laboratory setting, an example for discussion purposes is the fact that cephalopods are obligate predators. Their normal diet consists of crustaceans, fish, bivalves, and jellyfish, and they have been known to exhibit cannibalism. Byproducts of such feeding habits include shells, exoskeletons and other debris that represent waste that must be physically removed from holding tanks. Such debris may clog existing filters if not removed in a timely manner. When threatened, cephalopods may ink their environment, inhibiting the ability to observe them as required to assess health or research procedures and potentially leading to injury to the cephalopod who also cannot see and may traumatize themselves on the primary enclosure or physical substrate located within. Such inking episodes may require the ability to change out large volumes of tank water on short notice.[37]

The composition of materials used for primary and secondary enclosures and other substrates within an enclosure (e.g., environmental enrichment objects) may contain products that can leach into the water of aquatic vertebrates, thereby endangering the quality of the water and causing potential toxicity. Sanitation methods, and chemicals used to clean primary and secondary enclosures, should be evaluated to avoid potential toxicity or longer-term environmental or water contamination.

22:19 How does an animal care and use program determine animal enrichment needs for invertebrates?

Opin. Elements to enhance normative behavior within a primary enclosure should be a consideration for all animal species. Environmental enrichment for invertebrates may require an understanding of species-typical behavior in the wild that must be balanced with the necessity to observe animals on a daily basis to assess their health, welfare, or effects of the experimental procedure. Colony animals may require special enrichment needs to provide protection against self-predation, aggressive, and even cannibalistic behaviors. Animal husbandry, veterinary, and research personnel need to understand the normal behavior for the species, including isolated and social/group behavior so that enrichment needs for the species may be met. Physical structures such as huts, rocks, sand or pebbles, marbles, tubes, shelves, faux-walls, and plants (real or artificial) are all considerations for invertebrate species. Such items must be nontoxic, sanitizable, or disposable, and they must not leach harmful materials into the water. Some contouring to the environment by which higher regions (mounds) may be established with lower-level water features may be appropriate for amphibious species. Consideration for species-typical behaviors that encourage burrowing, or use of preferred vertical or ceiling surfaces may be important for some species. Five distinct enrichment categories for consideration with cephalopods have been described, including sensory, physical, olfactory, social, and food.[37]

22:20 What are the biosecurity considerations when working with invertebrates?

Opin. Biosecurity refers to those procedures and processes used to protect human and animal health against the introduction of pathogens that may cause harm to humans or animals or negatively impact research results. In a research setting, biosecurity is typically referred to at a physical "herd" level including facilities, rooms, colonies, or tanks. For example, one might refer to a "colony" of nematodes or a "tank" of lobsters and consider their viral, bacterial, or fungal status relative to the health of the colony/tank and their potential impact on human safety or impact on research results. Much of the research involving invertebrates relates to aquaculture and the potential impact of pathogens of aquatic invertebrates (e.g., Taura syndrome virus (TSV) in shrimp;[52] bees as carriers of *Varroa destructor*, a parasitic mite, and the endoparasitic fungus *Nosema ceranae*).[53] There is limited literature describing specific pathogens relative to research colonies outside of our knowledge that many invertebrates are known vectors of human disease (e.g., ticks), or contain toxins or poisons that are directly harmful to humans (stings, bites, poisons, or venoms).

22:21 What hazards and environmental health and safety concerns should an animal care and use program evaluate with respect to invertebrates used in research?

Opin. Collaborating closely with colleagues in environmental health and safety, who are familiar with the species being used, is an important part of working with invertebrate species. Environmental health and safety colleagues will have valuable and important safety information and input into housing, handling, and restraint, as well as insights on what to do in the event of an exposure. As with other species, hazard identification is an important consideration when working with invertebrates. For invertebrates, one should carefully consider the choice of cleaning, disinfectant, and topical medications to ensure there are no residual chemicals or ingredients that may cause injury to skin or breathing apparatus of invertebrate species, especially aquatic species. Enclosure or tank construction and furnishings should be safe and non-toxic for the species being housed.

Some invertebrates may have external needles, spicules, spikes, or hairs, some that can cause inflammatory dermatitis requiring special gloves to handle.[58] Handlers need to understand what personnel protective equipment should be worn and where the species should be handled (within its home cage, out in the open, within a hood, etc.). For those invertebrates that may contain toxins or poisons, special training should be implemented to inform staff and researchers of proper technique to minimize exposure as well as emergency actions to take in the event of an exposure, which may include reporting of the incident.

22:22 What occupational health items should be considered when working with invertebrates?

Opin. Collaborating closely with colleagues in occupational health, who are familiar with the species being used, is an important part of working with invertebrate species known to secrete poisons, venoms, or toxins, or whose physical external skeleton or may cause dermatitis upon exposure. Occupational health professional will have valuable knowledge on what to do in the event of an exposure. A number of invertebrates are poisonous, venomous, or secrete toxins, e.g., wasps, mosquitoes, box jellyfish or sea wasps (*Chironex sp.*). A sting from one of these species can cause severe inflammation and pain, hyper- or hypotension, impaired consciousness, cardiac arrythmias, shock, and even death, depending upon the species and dose of toxin received. Other problematic species include blue-ringed octopuses (*Hapalochlaena sp.*), Portuguese-man-of-war (*Physalia sp.*), and some sea urchins (e.g. flower urchins (also called trumpet urchins; *Toxopneustes sp.*).[59]

Proper primary and secondary enclosures should be carefully considered when working with venomous or toxin-excreting species. Dedicated personal protective equipment, including gloves, gowns, waterproof aprons, shoes or shoe covers, protective eyewear, and hair coverings, may all be considered when working with dangerous invertebrate species. Many of the poisons or venoms may have antidotes or antivenins. Ensuring timely access to such antidotes, antivenins, or other treatments may be of the essence to ensure a successful outcome.

Many zoonotic diseases use invertebrates as vectors. For example, ticks can carry Babesia (a protozoa), tick-borne encephalitis virus, heartland virus, Lyme disease, Tularemia, and a number of Ricketsial diseases (e.g., typhus, Rocky Mountain spotted fever). Cullicoid midges can carry the orbivirus associated with bluetongue in cattle or African horse sickness. Mosquitoes are common vectors of dengue virus, heartworm disease, malaria, filariasis, West Nile fever, yellow fever, and Zika. Snails serve as intermediate hosts or transmitting vectors for a multitude of agents that affect millions of people across the globe. For example, the rat lungworm *Angiostrongylus cantonensis* requires snails or slugs as intermediate hosts to develop through second and third state

TABLE 22.1

Advantages and disadvantages of including invertebrates within an overall program of animal care

Factor	Advantages	Disadvantages
Location	Controlling colonies located within a pre-existing vivarium allows one to protect other species and research that may occur within the vivarium	Colonies located outside a facility may have limited impact on other species or the research program
Species	Poisonous invertebrates or those that are vectors of human disease may warrant coverage as part of an occupational health and safety program or based on hazard identification and risk assessment	Non-poisonous, non-vector carrying invertebrates may not warrant coverage
Species	Octopus, squid, cuttlefish, crayfish, lobster, crab	Other invertebrate species
Endangered	Rare, protected or endangered species may warrant coverage	Common invertebrates may not warrant coverage
Use in research	Invertebrates used in major procedures or those deemed painful or distressful procedures in other species may warrant coverage	Invertebrates used in minor procedures or for observation purposes or non-painful procedures may not warrant coverage
Pathogen and microbial status	Invertebrates with known pathogens or those in which monitoring such status may be unknown, costly or difficult may warrant extra protection	Invertebrate colonies that do not pose a risk to human or other colony health may not warrant coverage
Source	Invertebrates from international sources may require coverage for shipping, licensing, permitting, or export control purposes	Invertebrates from local or national sources may not require or warrant coverage
Funding	Federal or private funding sources may demand more stringent oversight, including IACUC review	Internally funded research projects have no regulations that dictate oversight but potential effects on regulated species, especially if housed within a pre-existing vivarium, must be considered. See also whether nonregulated species may be voluntarily covered by an institution's PHS Assurance.

larvae in order to complete their lifecycle. Humans become infected by ingesting raw snails as a food source or indirectly through ingestion of contaminated produce. Although this disease causes major socioeconomic issues in many topical and subtropical countries, it has also been described within the continental United States and Hawaii.[80] Other human infectious agents that rely on snails as part of their life cycle cause angiostrongyliasis, clonorchiasis, fascioliiasis, fasciolopsiasis, opisthorchiasis, paragonimiasis, and schistosomiasis.[81]

REFERENCES

1. About Animal Testing. Murnaghan, I. Invertebrates Used for Testing. https://www.aboutanimaltesting.co.uk/ (accessed 8/4/22).
2. NC3Rs. https://www.nc3rs.org.uk/ (accessed 8/4/22).
3. Cooper JE. Anesthesia, analgesia, and euthanasia of invertebrates. *ILAR J* 52:196–204, 2011.
4. Carere C, and Mater J. *The Welfare of Invertebrate Animals.* Switzerland: Springer, 2019.
5. Lewbart GA. *Invertebrate Medicine.* Hoboken, NJ: John Wiley, 2022.
6. Linzey A, and Linzey C. *Animal Ethics and Animal Law.* Lanham, Maryland: Lexington Books, 2022.
7. Bilkey J, Nahirney PC, and Delaney KR. Time and exposure to serotonin affect releasability of recycled vesicles at crayfish claw opener muscle synapses. *Synapse* 74: e22136, 2020.
8. Lnenicka GA. Crayfish and Drosophila NMJs. *Neurosci Lett* 732:135110, 2020.
9. Yoon W, Min S, Ryu HS, Chung SJ, and Chung J. Discovery of levodopa-induced dyskinesia-associated genes using genomic studies in patients and Drosophila behavioral analyses. *Commun Biol* 5:872, 2022.
10. Munnik C, Xaba MP, Malindisa ST, Russell BL, and Sooklal SA. Drosophila melanogaster: A platform for anticancer drug discovery and personalized therapies. *Front Genet* 13:949241, 2022.
11. Ireland D, Bochenek V, Chaiken D, Rabeler C, Onoe S, Soni A, and Collins ES. Dugesia japonica is the best suited of three planarian species for high-throughput toxicology screening. *Chemosphere* 253:126718, 2020.
12. Pestana JLT, and Ofoegbu PU. Ecotoxicity assays using freshwater planarians. *Methods Mol Biol* 2240:125–137, 2021.
13. Krajniak KG. Invertebrate FMRFamide related peptides. *Protein Pept Lett* 20:647–670, 2013.
14. Ratcliffe JM, Fullard JH, Arthur BJ, and Hoy RR. Tiger moths and the threat of bats: decision-making based on the activity of a single sensory neuron. *Biol Lett* 5:368–371, 2009.
15. Neil TR, Shen Z, Robert D, Drinkwater BW, and Holderied MW. Thoracic scales of moths as a stealth coating against bat biosonar. *J R Soc Interface* 17:20190692, 2020.
16. Rankin CH. But can they learn? My accidental discovery of learning and memory in C. elegans. *J Neurogenet* 34:251–254, 2020.
17. Charlesworth AG, Seroussi U, and Claycomb JM. Next-gen learning: The C. elegans approach. *Cell* 177:1674–1676, 2019.
18. Wright EAP, and Goltsev AV. Statistical analysis of unidirectional and reciprocal chemical connections in the C. elegans connectome. *Eur J Neurosci* 52:4525–4535, 2020.
19. Taylor SR, Santpere G, Weinreb A, Barrett A, Reilly MB, Xu C, Varol E, Oikonomou P, Glenwinkel L, McWhirter R, Poff A, Basavaraju M, Rafi I, Yemini E, Cook SJ, Abrams A, Vidal B, Cros C, Tavazoie S, Sestan N, Hammarlund M, Hobert O, and Miller DM, 3rd. Molecular topography of an entire nervous system. *Cell* 184:4329–4347 e4323, 2021.
20. De Fruyt N, Yu AJ, Rankin CH, Beets I, and Chew YL. The role of neuropeptides in learning: Insights from C. elegans. *Int J Biochem Cell Biol* 125:105801, 2020.
21. Quinn WG, Harris WA, and Benzer S. Conditioned behavior in Drosophila melanogaster. *Proc Natl Acad Sci U S A* 71:708–712, 1974.
22. Dudai Y, Jan YN, Byers D, Quinn WG, and Benzer S. dunce, a mutant of Drosophila deficient in learning. *Proc Natl Acad Sci U S A* 73:1684–1688, 1976.
23. Bertin F, Moya-Alvarado G, Quiroz-Manriquez E, Ibacache A, Kohler-Solis A, Oliva C, and Sierralta J. Dlg is required for short-term memory and interacts with NMDAR in the drosophila brain. *Int J Mol Sci* 23:2022.
24. Bourouliti A, and Skoulakis EMC. Anesthesia resistant memories in drosophila, a working perspective. *Int J Mol Sci* 23:2022.
25. Flores-Valle A, and Seelig JD. A place learning assay for tethered walking Drosophila. *J Neurosci Methods* 378:109657, 2022.
26. Menzel R. In search for the retrievable memory trace in an insect brain. *Front Syst Neurosci* 16:876376, 2022.
27. Delaney K, Tank DW, and Zucker RS. Presynaptic calcium and serotonin-mediated enhancement of transmitter release at crayfish neuromuscular junction. *J Neurosci* 11:2631–2643, 1991.
28. Zucker RS, Delaney KR, Mulkey R, and Tank DW. Presynaptic calcium in transmitter release and posttetanic potentiation. *Ann N Y Acad Sci* 635:191–207, 1991.
29. Fields HL, and Kennedy D. Functional role of muscle receptor organs in crayfish. *Nature* 206:1235–1237, 1965.
30. Kennedy D, and Evoy WH. The distribution of pre- and postsynaptic inhibition at crustacean neuromuscular junctions. *J Gen Physiol* 49:457–468, 1966.
31. Calabrese RL. Motor coordination: a local hub for coordination. *Curr Biol* 24: R274–275, 2014.
32. American Veterinary Medical Association. *American Veterinary Medical Association Guidelines for the Euthanasia of Animals.* Schaumburg, IL: American Veterinary Medical Association, pp. 1–111, 2020.
33. Freas CA, and Cheng K. Neuroecology beyond the brain: learning in Echinodermata. *Learn Behav* 50:20–36, 2022.
34. Elwood R. Pain and suffering in invertebrates. *ILAR J* 52:175–184, 2011.
35. Sneddon L. Defining and assessing animal pain. *Animal Behaviour* 97:201–212, 2014.
37. Cooke G, Tonkins B, and Mather J. Care and enrichment for captive cephalopods. In: *The Welfare of Invertebrate Animals*, edited by Carere C, and Mather J. New York: Springer, 2019.
38. Osorio-Palacios M, Montiel-Trejo L, Oliver-Dominguez I, Hernandez-Falcon J, and Mendoza-Angeles K. Sleep phases in crayfish: relationship between brain electrical activity and autonomic variables. *Front Neurosci* 15:694924, 2021.

39. National Research Council of the National Academies. *Guide for the Care and use of Laboratory Animals.* Washington, DC: The National Academies Press, 2011.
40. Wells N, and Nicholson J. Five plus three: legislating for the Five Freedoms and the Three Rs--Animal Welfare Act 1999 (New Zealand). *Altern Lab Anim* 32 Suppl 1B:417–421, 2004.
41. Flecknell P. Replacement, reduction and refinement. *ALTEX* 19:73–78, 2002.
42. Englefield B, Blackman SA, Starling M, and McGreevy PD. A review of Australian animal welfare legislation, regulation, codes of practice, and policy, and their influence on stakeholders caring for wildlife and the animals for whom they care. *Animals (Basel)* 9:2019.
43. The Criminal Code of Canada. https://www.justice.gc.ca/eng/csj-sjc/ccc/index.html (accessed 9/7/22).
44. Directive 2010/63/EU on the Protection of Animals Used for Scientific Purposes. https://eur-lex.europa.eu/LexUriServ/LexUriServ.do?uri=OJ:L:2010:276:0033:0079:en:PDF (accessed 9/7/22).
45. Act on Welfare and Management of Animals. Japanese Law Translation. https://www.japaneselawtranslation.go.jp/en/laws/view/3798/en (accessed 9/7/22).
46. Chandroo KP, and Duncan IJH. Can fish suffer? Perspectives on sentience, pain, fear and stress. *Appl Animal Behaviour Sci* 86:225–250, 2004.
47. AAALAC, Intl. https://www.aaalac.org (accessed 8/11/22).
48. Harvey-Clark C. IACUC challenges in invertebrate research. *ILAR J* 52:213–220, 2011.
49. Scimeca J. Cephalopods. In: *Invertebrate Medicine*, edited by Lewbart G. New York: Blackwell Publishing, 2006.
50. Noga E, Hancock A, and Bullis R. Crustaceans. In: *Invertebrate Medicine*, edited by Lewbart G. Ames IA: Blackwell Publishing, pp. 179–194, 2006.
51. Cabot M, and Lewbart G. Drug Formulary. In: *Invertebrate Medicine*, edited by Lewbart G. Hoboken, NJ: John Wiley, 2022.
52. Moss S, Moss D, Arce S, Lightner D, and Lotz J. The role of selective breeding and biosecurity in the prevention of disease in penaeid shrimp aquaculture. *J Invert Pathology* 110:247–250, 2012.
53. Garrido C, and Nanetti A. Welfare of managed honey bees. In: *The Welfare of Invertebrate Animals*, edited by Carere C, and Mather J. New York: Springer, 2019, p. 248.
54. Russell W, and Burch R. *The Principles of Humane Experimental Technique*. London, UK: Methuen & Co., 1959.
55. Sollund R. Wildlife trade and law enforcement: a proposal for a remodeling of CITES incorporating species justice, ecojustice, and environmental justice. *Int J Offender Ther Comp Criminol* 66:1017–1035, 2022.
56. Convention on International Trade in Endangered Species of Wild Fauna and Flora. https://cites.org/eng (accessed 8/28/22).
57. Bureau of Industry and Security, Export Administration Regulations. bis.gov/regulations (accessed 9/7/22).
58. Haddad V, Jr. Environmental dermatology: skin manifestations of injuries caused by invertebrate aquatic animals. *An Bras Dermatol* 88:496–506, 2013.
59. The Fisheries Blog. Sackett DK. 2106. Who Needs a Backbone When You Have Toxins: Our Top 5 Most Venomous Ocean Invertebrates. https://thefisheriesblog.com/2016/04/17/who-needs-a-backbone-when-you-have-toxins-our-top-5-most-venomous-ocean-invertebrates (accessed 8/28/22).
60. Winlow W, Polese G, Moghadam HF, Ahmed IA, Di Cosmo A. Sense and Insensibility - An Appraisal of the Effects of Clinical Anesthetics on Gastropod and Cephalopod Molluscs as a Step to Improved Welfare of Cephalopods. *Front Physiol* 9:1147, 2018.
61. Lopes VM, Sampaio E, Roumbedakis K, Tanaka NK, Carulla L, Gambús G, Woo T, Martins CPP, Penicaud V, Gibbings C, Eberle J, Tedesco P, Fernández I, Rodríguez-González T, Imperadore P, Ponte G, Fiorito G. Cephalopod biology and care, a COST FA1301 (CephsInAction) training school: anaesthesia and scientific procedures. *Invert Neurosci* 17(3):8 2017.
62. Diggles BK. Review of some scientific issues related to crustacean welfare. *ICES J Marine Sci* 76:66–81, 2019.
63. Ben Khadra Y, Farrario C, Benedetto CD, Said K, Bonasoro F, Candia Carnevali MD, Sugni M. Re-growth, morphogenesis, and differentiation during starfish arm regeneration. *Wound Rep Regenerat* 23(4):623–634, 2015.
64. Imperadore P, Fiorito G. Cephalopod tissue regeneration: consolidating over a century of knowledge. *Front Physiol* 9:593, 2018.
65. Rodríguez-Romero A, Viguri JR, Calosi P. Acquiring an evolutionary perspective in marine ecotoxicology to tackle emerging concerns in a rapidly changing ocean. *Sci Total Environ* 764:142816, 2021.
66. Rodrigues ET, Pardal MÂ. The crab *Carcinus maenas* as a suitable experimental model in ecotoxicology. *Environ Int* 70:158–182, 2014.
67. Beltz BS, Kravitz EA. Physiological identification, morphological analysis, and development of identified serotonin-proctolin containing neurons in the lobster ventral nerve cord. *J Neurosci* 7:533–546, 1987.
68. Harzsch S, Dirksen H, Beltz BS. Development of pigment-dispersing hormone-immunoreactive neurons in the American lobster: homology to the insect circadian pacemaker system? *Cell Tissue Res* 335:417–429, 2009.
69. Bronfman ZZ, Ginsburg S, Jablonka E. The transition to minimal consciousness through the evolution of associative learning. *Front Psychol.* 7:1954, 2016.
70. Reimert I, Webb LE, van Marwijk MA, Bolhuis JE. Review: towards an integrated concept of animal welfare. *Animal* Suppl 4:100838, 2023.
71. Deidda I, Russo R, Bonaventura R, Costa C, Zito F, Lampiasi N. Neurotoxicity in marine invertebrates: an update. *Biology (Basel)* 10:161, 2021.
72. Birch J., Burn C, Schnell A, Browning H, Crump A. Review of the Evidence of Sentience in Cephalopod Molluscs and Decapod Crustaceans. LSE Consulting. LSE Enterprise Ltd. The London School of Economics and Political Science. 2021. https://www.lse.ac.uk/business/consulting/reports/review-of-the-evidence-of-sentiences-in-cephalopod-molluscs-and-decapod-crustaceans?_gl=1*1gmxpfy*_ga*MTMyNjcyNTQ5NC4xNzEyOTI3NTg0*_ga_LWTEVFESYX*MTcxMjkyNzU4NC4xLjEuMTcxMjkyNzcxMS42LjAuMA. (accessed 12 April 2024).

73. Smith JA, Andrews PL, Hawkins P, Louhimies S, Ponte G, Dickel L. Cephalopod research and EU Directive 2010/63/EU: requirements, impacts and ethical review. *J Exp Marine Biol Ecol* 447:31–45, 2013.
74. Escanez A, Lozano-Bilbao E, Paz S, Hardisson A, Gonzalez-Weller D, Rubio C, Lozano G, Gutierrez AJ. Assessments of metallic contents in rare cephalopods from the Canary Islands: relationships with depth habitat and body size. *Environ Sci Pollut Res Int* 28:54161–54169, 2021.
75. Butler-Struben HM, Brophy SM, Johnson NA, Crook RJ. In vivo recording of neural and behavioral correlates of anesthesia induction, reversal, and euthanasia in cephalopod molluscs. *Front Physiol* 9:109, 2018.
76. Abbo LA, Himebaugh NE, Demelo LM, Hanlon RT, Crook RJ. Anesthetic efficacy of magnesium chloride and ethyl alcohol in temporate octopus and cuttlefish species. *J Am Assoc Lab Anim Sci* 60:556–567, 2021.
77. Da Costa MR, Diana A. A systematic review on the link between animal welfare and antimicrobial use in captive animals. *Animals (Basel)* 12:1025, 2022.
78. Fiorito G, Affuso A, Basil J, Cole A, de Girolamo P, D'Angelo L, Dickel L, Gestal C, Grasso F, Kuba M, Mark F, Melillo D, Osorio D, Perkins K, Ponte G, Shashar N, Smith D, Smith J, Andrews PL. Guidelines for the care and welfare of cephalopods in research -a consensus based on an initiative by CephRes, FELASA and the Boyd group. *Lab Animal* 49:1–90, 2015.
79. Voss KM, Mehta RS. Asymmetry in the frequency and proportion of arm truncation in three sympatric California Octopus species. *Zoology (Jena)* 147:125940, 2021.
80. Cowie RH, Ansdell V, Panosian Dunavan C, Rollins RL. Neuroangiostrongyliasis: global spread of an emerging tropical disease. *Am J Trop Med Hyg* 107:1166–1172, 2022.
81. Lu L, Loker ES, Adema CM, Zhang SM, Bu L. Genomic and transcriptional analysis of genes containing fibrinogen and IgSF domains in the schistosome vector Biomphalaria glabrata, with emphasis on the differential responses of snails susceptible or resistant to Schistosoma mansoni. *PLoS Negl Trop Dis* 14:e0008780, 2020.
82. Panchin YV, Arshavsky YI, Selverston A, Cleland TA Lobster stomatogastric neurons in primary culture. I. Basic characteristics. *J Neurophysiol* 69(6):1976–1992, 1993.
83. https://www.iasp-pain.org/ International Association on the Study of Pain (accessed 04/04/24).
84. https://olaw.nih.gov/policies-laws/gov-principles.htm/ (accessed 04/04/24).
85. https://www.aaalac.org/accreditation-program/faqs/ B2. Invertebrate Animals (accessed 04/04/24).
86. https://grants.nih.gov/grants/guide/notice-files/NOT-OD-23-176.html/ (accessed 04/04/24).
87. https://grants.nih.gov/grants/guide/notice-files/NOT-OD-23-057.html/ (accessed 04/04/24).
88. Escalante I, Ellis VR, Elias DO. Leg loss decreases endurance and increases oxygen consumption during locomotion in harvestmen. *J Comp Physiol A Neuroethol Sens Neural Behav Physiol* 207(2):257–268, 2021.
89. Wilson-Sanders SE. Invertebrate models for biomedical research, testing, and education. *ILAR J* 52:126–152, 2011.

23

Occupational Health and Safety

Robin Lyn Trundy and Susan Stein Cook*

Introduction

Animal allergens, zoonotic diseases, and physical injuries are only some of the issues that need to be addressed in a general occupational health program associated with animal research. Although regulations and guidelines regarding human health and protection are a valuable resource for establishing a comprehensive occupational health program, their scope and level of detail will always be limited. Fortunately, most current health and safety regulations are performance oriented, focusing on the goal (i.e., a safe work environment) rather than describing how to get there, which allows institutions to tailor their safety programs to protect employees in the most effective way.

General Occupational Health and Safety

23:1 What is the IACUC's responsibility to ensure a safe work environment for persons working with laboratory animals?

Reg. The need to protect the health and safety of personnel involved in animal-based research is addressed in the *Guide*[1] (p. 17) and the PHS Policy[2] (IV,A,1,f). The PHS Policy (IV,A,1) requires institutions to use the *Guide* as the basis for their institutional program for activities involving animals. The *Guide*[1] (p. 17) states that institutions must establish an occupational health and safety program (OHSP) as part of the overall animal research program. The *Guide*[1] (p. 17) specifically references the National Research Council publication, *Occupational Health and Safety in the Care and Use of Research Animals*,[3] to be used as a tool for establishing a comprehensive occupational health and safety program (OHSP). While the main responsibility of the OHSP lies with the institution, the IACUC is specifically charged with the oversight and evaluation of the animal care and use program (*Guide*, p. 17). The institutional OHSP should identify potential hazards in the work environment and conduct a critical assessment of the associated risks (*Guide*, p. 18).

Although targeted to animal health and safety, the Animal Welfare Regulations require that research facilities ensure that all personnel involved in the animal care and use program are qualified to perform their duties (§2.32).[38]

Opin. The applicable referenced guidelines and regulations provide limited detail specifically addressing the IACUC's responsibility for occupational health and safety. Therefore, the institution must manage and assign responsibilities covering the various programs required by the *Guide*.[1] The quality of the OHSP will be impacted significantly by the quality of interaction of the IACUC with relevant institutional groups responsible for functions such as the research program, environmental health and safety, and occupational health services. Institutions are required to address OHSP issues and assign responsibilities to various groups or individuals inside the institution for management and oversight of the different components. For example, occupational health requires the input of medical professionals, while physical, chemical, biological, and radiological hazards are best addressed by specific committees or safety professionals with expertise in these areas. Ideally, the IACUC assumes the role of a facilitator, interacting with the various programs, committees, and individuals to ensure the existence of a comprehensive OHSP and compliance with the *Guide*.[1]

Surv. Does your IACUC or program have a means of monitoring your institution's occupational health and safety program to ensure that it is performing effectively?

• No, we do not monitor the program.	27/189 (14%)
• Yes, we routinely monitor the program to help ensure effectiveness.	117/189 (62%)
• We review the program only if a complaint is raised.	25/189 (13%)
• I don't know.	12/189 (6%)
• Other*	8/189 (4%)

* *The authors thank Kimyli Recca for her contribution to this edition of the chapter.

DOI: 10.1201/9781003272861-23

*Other: Respondents noted that monitoring can be accomplished as part of the post-approval monitoring process and the semiannual program review.

Surv. Are personnel who have regular direct animal contact required to participate in your institution's occupational health and safety surveillance program (OHSP)?

- Yes. Personnel with regular direct animal contact must participate in the OHSP and MAY NOT decline certain activities. 76/189 (40%)
- Yes. Personnel with regular direct animal contact must participate in the OHSP but MAY decline certain activities. 97/189 (51%)
- No. Personnel with regular direct animal contact are given the opportunity to participate in the OHSP but are not required. 16/189 (8%)

23:2 What is the IACUC's responsibility to ensure a safe work environment for persons who may have access to an animal facility but are not working with laboratory animals (e.g., facility support staff, students, visitors)?

Reg. As stated in 23:1, the institution at large is primarily responsible for the establishment of an OHSP. The PHS Policy[2] (IV,A,1,f) requires a health program for personnel who have frequent contact with animals, as well as all persons who work in laboratory animal facilities. According to the *Guide*[1] (pp. 18–19): "The extent and level of participation of personnel in the OHSP should be based in the hazards posed by the animals and materials used, the exposure intensity, duration, and frequency; to some extent, the susceptibility of the personnel; and the history of occupational illness and injury in the particular workplace."

In addition, the National Research Council publication, *Occupational Health and Safety in the Care and Use of Research Animals*[3] (pp. 12, 17) as referenced by the *Guide*,[1] suggests including not only animal caretakers, technicians, students, volunteers, investigators, and veterinarians, but also facility maintenance personnel, housekeepers, security, and other staff who must perform job duties in animal research environments.

Opin. Ideally, the IACUC facilitates the close interaction among institutional groups dealing with employee health and safety as it relates to animal research. For example, the institution's environmental health and safety (EHS) office oversees the health and safety of maintenance personnel working in areas with research hazards. The institutional occupational health group and the medical professionals should identify any necessary surveillance criteria for those who may enter animal holding areas under all anticipated scenarios. The assistance of these and other health and safety groups should be requested by the IACUC to perform risk assessments, project review, and make health and safety recommendations for the range of personnel who may interface with research animal environments regardless of the frequency or duration of exposure.

Surv. Which of the following groups are required to participate in some or all parts of your occupational health and safety (OHS) program? Select all that apply.

- Not applicable 7/189 (4%)
- Research personnel who handle animals 174/189 (92%)
- Animal facility personnel (e.g., husbandry, veterinary staff) who may handle animals 179/189 (95%)
- Animal facility office personnel who do not have to use animals in their work 56/189 (30%)
- Program personnel who enter the animal facility or laboratories where animals may be held/used for non-animal use purposes (e.g., maintenance, security) 119/189 (63%)
- IACUC members who do not use live animals in their work but may participate in semiannual inspections (e.g., nonaffiliated member, non-scientist member) in the animal facility and/or laboratories where animals are held/used. 117/189 (62%)
- Students, residents, post-docs, and others working only once with animals 113/189 (60%)
- Visitors with direct animal contact (e.g., outside company representatives or demonstrating techniques) 92/189 (49%)
- Visitors with no direct animal contact (e.g., observing others perform procedures) 30/189 (16%)
- Non-animal users who work in the same lab or in close proximity as those who do use animals 39/189 (21%)

Surv. What information is provided to individuals who are not enrolled in the institutional OSHP to inform them of the risks associated with entering the vivarium (e.g., inspectors)? Select all that apply.

- Animal allergens education and/or screening 149/189 (79%)
- Species-specific briefing (including zoonoses as applicable) 120/189 (63%)
- Research hazard safety practices if entering areas under biosafety, radiation safety, or chemical safety restrictions 141/189 (75%)
- Life safety/emergency response procedures 79/189 (42%)
- None of the above 16/189 (8%)

23:3 What is the IACUC's responsibility to ensure a safe work environment for research teams performing field studies or working with wild-caught species?

Reg. The PHS Policy[2] (IV,A,1,f) is applicable to all institutional programs involving animals, not only in animal facilities. Therefore, the OHSP must address all risks to investigators, technicians, and students associated with PHS-supported animal activity, regardless of where it occurs. The risks associated with unusual experimental conditions, such as those encountered in field studies or wildlife research, should also be addressed under the hazard identification and risk assessment responsibilities of the OHSP (*Guide*, p. 18).

Opin. General and specific regulatory health and safety requirements (i.e., Occupational Safety and Health Administration [OSHA] regulations) apply to worker protection both inside and outside an animal facility environment. Hazards associated with animal research outside a facility can be of great concern. One example is field research involving wild animals. The incorrect use of equipment and chemical agents to trap or immobilize wild animals can pose a significant physical hazard to animals and humans. Environmental hazards may be present. Zoonotic diseases, such as rabies or hantavirus pulmonary syndrome, may be found in certain wildlife species and can be life threatening. Field sites may be in areas with a possible seasonal risk of exposure to vector borne diseases (e.g., Lyme disease, West Nile virus). The IACUC should partner with knowledgeable institutional safety groups, state public health divisions, departments of natural resources, or other individuals with specific expertise to ensure that hazards unique to planned field activities are identified and appropriate health and safety measures are taken.

23:4 What elements should the OHSP include?

Reg. The *Guide*[1] (pp. 17–23) outlines an occupational health and safety program, as required in the PHS Policy (IV,A,1,f). This was summarized by NIH/OLAW[4] as follows: Minimally, the program must include:

- Pre-placement medical evaluation
- Identification of hazards to personnel and safeguards appropriate to the risks associated with the hazards
- Appropriate testing and vaccinations
- Training of personnel regarding their duties, any hazards, and necessary safeguards
- Policies and facilities that promote cleanliness
- Provisions for treating and documenting job-related injuries and illnesses
- Facilities, equipment, and procedures should be designed, selected, and developed to reduce the possibility of physical injury or health risk to personnel
- Good personal hygiene practices, prohibiting eating and drinking, use of tobacco products, and application of cosmetics and/or contact lenses in animal rooms and laboratories
- Personal protective equipment (PPE)

Opin. The minimal components outlined above are necessary for an effective OHSP. Safety professionals and occupational health professionals should work collaboratively with veterinary and animal care management staff to address all of these components. While different groups may be responsible for administration of certain components, it is important for the groups to communicate openly with the other groups when developing policies and procedures to build a cohesive program that truly supports the safety needs of the end users. Although published in 1997, Chapters 6 and 7 of *Occupational Health and Safety in the Care and Use of Research Animals*[3] still provide valuable foundational information for those seeking to establish and maintain a successful OHSP regardless of program size.

Surv. Which of the following does your institution use for occupational health and safety training? Select all that apply.

• An institutionally affiliated office/program	167/189 (88%)
• A private, for-profit organization	33/189 (17%)
• A state affiliated organization (e.g., workers compensation, loss control insurance)	16/189 (8%)
• An individual's personal physician	26/189 (14%)
• I don't know	5/189 (3%)
• Other	1/189 (1%)

Surv. Does your IACUC or program require an annual (or more frequent) health evaluation for people who routinely work directly with laboratory animals (other than tuberculosis testing)?

• Yes, and this includes a physical examination.	20/189 (11%)
• Yes, but this requires only filling out a questionnaire.	95/189 (50%)
• No, we do not require an annual health evaluation.	48/189 (25%)
• I don't know.	5/189 (3%)
• Other*	21/189 (11%)

*Other:

- The assessment interval depended on job role. For example, in some cases, annual screening is required for animal care personnel and triennial screening for others.
- Assessment interval of two years.

Surv. For animal facilities that usually house Old World nonhuman primates: How often are animal facility personnel tuberculosis tested?

• Not applicable	140/189 (74%)
• Never	1/189 (1%)
• Once a year	35/189 (19%)
• Twice a year	9/189 (5%)
• More than twice a year	1/189 (1%)
• I don't know	2/189 (1%)
• Other	1/189 (1%)

23:5 What institutional committees and departments should the IACUC interface with regarding personnel safety?

Reg. Per the *Guide*[1] (p. 17): "An effective OHSP requires coordination between the research program (as represented by the investigator), the animal care and use program (as represented by the AV, IO, and IACUC), the environmental health and safety program, occupational health services, and administration (e.g., human resources, finance, and facility maintenance personnel). Establishment of a safety committee may facilitate communication and promote ongoing evaluation of health and safety in the workplace. In some cases, there is a regulatory requirement for such a committee."

Opin. The IACUC has an oversight responsibility (see 23:1) related to personnel safety issues associated with animal protocols. To ensure that personnel safety issues are effectively addressed as part of the review process, the IACUC must seek input and communicate with the committees and departments specifically charged with regulatory compliance, worker safety, and environmental health. These groups and individuals interact directly with agencies such as OSHA, the Nuclear Regulatory Commission (NRC), and the Centers for Disease Control and Prevention (CDC). These committees and departments usually include the following:

- Environmental Health and Safety/ Occupational Safety provides assistance related to common physical hazards, personal protective equipment assessments, and chemical use, including anesthetics, analgesics, and disinfectants.
- Biological Safety provides assistance related to biological containment practices, human-derived material use, safe handling of sharps, biological safety cabinet use, and biohazardous waste management.
- Radiation Safety provides assistance with radiation safety training, signage, personnel exposure monitoring, and area surveys for radioactive contamination. The scope of controls for radiological hazards is determined by the Institutional Radiation Safety Committee. When radiological hazards are used in conjunction with animals, the IACUC must maintain open communication with the Radiation Safety Committee, or a radiation safety officer, to assure that radiation safety compliance items have been addressed before IACUC approval is granted.
- Occupational Health provides medical screening of personnel who are enrolled in the occupational health program, as well as information on allergies and hygiene.
- A Chemical Hygiene Committee may, or may not, be established at an institution. If such a committee is not established, a chemical hygiene officer with oversight for animal research facilities should be included in the IACUC review for protocols involving hazardous chemicals, including anesthetics and euthanasia agents.
- An Institutional Biosafety Committee (IBC) is required for an institution if it receives any NIH funds for recombinant DNA research. This committee is charged with review of research involving recombinant DNA molecules (see 23:22), although many IBCs also review the use of biological hazards, including potentially infectious agents and toxins. When recombinant DNA molecules are used in conjunction with animals, the IACUC and the IBC should work jointly to assure that all regulatory compliance items have been addressed before either committee approves a project (see 23:24).

Surv. What mechanisms are in place to ensure that IACUC protocols involving the use of research hazards are properly assessed and receive other institutional committee approval (i.e., Institutional Biosafety Committee, Radiation Safety Committee)? Select all that apply.

• Safety professionals are involved in the IACUC protocol review process.	132/189 (70%)
• Safety professionals participate in IACUC meetings.	99/189 (52%)
• IACUC representatives participate in IBC and/or RSC meetings.	110/189 (58%)

- Safety and/or IACUC administrative staff communicate (within or outside of electronic systems). 141/189 (75%)
- Other* 17/189 (9%)
- None of the above. 5/189 (3%)

*Other: The animal facility will not allow access until all safety committee approvals have been secured.

Surv. Does your IACUC include any of the following parties as voting or *ex officio* members? Select all that apply.

- Occupational health professional 51/189 (27%)
- Industrial hygienist 12/189 (6%)
- Biosafety professional 103/189 (54%)
- Radiation safety professional 29/189 (15%)
- Risk management professional 22/189 (12%)
- None of the above 70/189 (37%)

23:6 What circumstances may require medical surveillance beyond routine occupational health examinations for those who work with animals?

Reg. Some OSHA standards require a specific medical surveillance activity if an employee is exposed to a hazard at levels or durations above those that are considered to be safe. Examples include the Occupational Noise Exposure Standard,[5] the Respiratory Protection Standard,[6] and the Bloodborne Pathogens Standard.[7]

Under the Occupational Noise Exposure Standard[5] employees who are exposed to noise levels at or above 85 decibels for an eight-hour time-weighted average (TWA) must have baseline and annual audiometric testing through the occupational health provider. These levels may be exceeded in some facilities as a result of noise from animals or equipment.

The Respiratory Protection Standard[6] requires that any employee who is required to wear a respirator for protection against an airborne hazard on the job be medically evaluated before wearing a respirator; additional ongoing surveillance requirements are determined by the occupational health provider. This standard may apply to the use of certain chemical products in the research facility and for work with airborne infectious agents.

The Bloodborne Pathogens Standard[7] requires that employees who are exposed to certain human-derived materials (which include human cells under most circumstances) be included in the bloodborne pathogens exposure control program. These personnel must be offered the hepatitis B vaccination and receive specific post-exposure services if an employee sustains an exposure to such materials through a splash to the eyes, nose, or mouth; absorption through broken skin; or a cut with a contaminated object. In the animal research environment, personnel who handle human cells during the introduction of these materials into an animal model or harvest these materials from the animal model are likely to be affected by these requirements. The *Guide*[1] (p. 22) also states: "Vaccination is recommended if research is to be conducted on infectious diseases for which effective vaccines are available." Agent-specific recommendations are available in *Biosafety in Microbiological and Biomedical Laboratories*, sixth edition (BMBL).[8]

Opin. In addition to the OSHA requirements, an institution should ensure that the occupational health–medical surveillance activities include additional components to protect personnel if infectious disease research is under way in the animal research environment or laboratories. The BMBL[8] is considered a *de facto* standard by many agencies and organizations. This document includes agent summary statements under Section VIII for environments where an infectious disease risk is present. Additionally, the document provides information on modes of infection, laboratory safety and containment recommendations, vaccinations, medical surveillance information, and considerations for work with specific agents and biological materials. Updates in the BMBL (sixth edition) of relevance for IACUCs include an expanded Section VII on occupational health support for biomedical research and an expanded Appendix D that now includes information regarding biocontainment strategies for BSL-2 studies involving agricultural and loose-housed animal models.

The institution should also ensure that zoonotic diseases risk has been fully assessed when nonrodent species will be involved in studies and medical surveillance programs expanded to follow the latest public health guidance for addressing specific exposure risk accordingly. Common examples include Q fever risk associated with sheep[9] and herpes B virus exposure risk associated with macaques.[10]

23:7 What types of hazards are inherent in animal research settings?

Opin. Animal research settings involve unique hazards related to working with animals, but such common workplace hazards as cuts, burns, slips, trips, or falls also occur. In addition, the research might involve unique physical, chemical, biological, or radiological hazards. An overview of relevant physical, chemical, biological, and protocol-related hazards has been published in Chapter 3 of the document entitled *Occupational Health and Safety in the Care and Use of Research Animals.*[3]

In general, physical hazards in the animal research environment can include bites, scratches, and kicks; sprains and strains caused by the moving of equipment; operational hazards involving electricity, machinery, and noise; or protocol-induced hazards such as radioactivity or the use of lasers. Chemical hazards can be directly related to a specific agent or procedure. Examples include chemicals used for processing tissues and cleaning or disinfecting research equipment. The accidental inhalation of waste anesthetic gases provides another route of chemical exposure. Research protocols might require the application of toxins, carcinogens, and other hazardous chemicals. Biological hazards, such as infectious pathogens, can be introduced through naturally occurring or experimentally infected research animals or include the application of recombinant DNA technology. Another significant hazard is manifested in the increasing development of animal-related allergies or occupation-related asthma.[11] Repetitive motion injuries,[12] compassion fatigue,[13] and increased activities by individuals or organizations opposed to the use of animals in research are additional hazards relevant to the animal research setting.

Physical and Chemical Hazards

23:8 What is meant by a *physical hazard* that may be present in animal research settings?

Reg. According to *Occupational Health and Safety in the Care and Use of Research Animals*[3] (pp. 32–43) physical hazards are those generally associated with:

- Flammable materials
- Pressurized vessels
- Machinery
- Noise
- Temperature
- Vibration
- Electricity
- Nonionizing and ionizing radiation (also considered a radiation hazard)
- Ultraviolet radiation
- Lasers
- Illumination
- Sharp objects

Occupational exposure limits and protective measures for some of these hazards are covered under specific OSHA standards in the U.S.:

- Occupational noise exposure[5]
- Nonionizing radiation[14]
- Ionizing radiation[15]
- Electrical[16,17]

In addition, a hazardous chemical may have properties that will pose a danger to personnel and the environment such as being flammable, explosive, reactive with water and other incompatible materials, etc. These properties are fully described in the OSHA Hazard Communication Standard.[18]

Opin. Excessive noise can be produced by machinery and animals, especially pigs and dogs. Exposure to intense noise over time will result in hearing loss. If normal talking or phone conversation is not possible because of excessive noise, the noise level should be assessed by an industrial hygienist or safety professional. Electricity is of concern if electrical equipment is out of date, poorly maintained, or improperly grounded. Compliance with electrical code requirements is necessary, as is equipment maintenance. (See 23:31–23:36 for specific information on the management of radiological hazards.) The IACUC should seek the assistance of occupational health and safety professionals to assess the potential physical hazards associated with animal protocols and animal holding facility operations.

23:9 Should physical methods of euthanasia be considered a physical risk?

Reg. Physical methods of euthanasia as defined by the American Veterinary Medical Association (AVMA) *Guidelines for the Euthanasia of Animals*[19] include captive bolt, gunshot, cervical dislocation, decapitation, electrocution, focused beam microwave irradiation, exsanguination, maceration, stunning, and pithing. According to the 2020 AVMA *Guidelines*,[20] "Since most physical methods involve trauma, there is inherent risk for animals and humans. Extreme care and caution should be used." PHS Policy[2] (IV,C,1,g) requires euthanasia methods to be consistent with the AVMA recommendations unless a deviation for scientific reasons is justified, in writing, by the investigator.

Opin. As with all physical procedures and practices, the skill and expertise of the person performing the procedure determine the degree of risk. Prior to approval of such methods, the IACUC should evaluate the proficiency and level of training of individuals performing these procedures and be assured that the equipment is maintained in good working order.

23:10 What sharps safety practices should the IACUC consider as a way to reduce animal research-related injuries?

Reg. The OSHA Bloodborne Pathogens Standard[7] (which applies to human-derived contaminants such as human tissues, cells, and some viral vectors), the BMBL[8] (which applies to all levels of biohazard), and the *NIH Guidelines for Research Involving Recombinant or Synthetic Nucleic Acid Molecules (NIH Guidelines)*[20] (which generally refers to materials containing DNA from two different species) all prohibit or restrict actions with contaminated sharp devices that would likely lead to a needlestick or laceration injury. These prohibitions or restrictions include:

- The use of sharps devices is eliminated; needle-less or safety-engineered systems are adopted when possible; needle-locking syringes or syringes with integral needles are selected (required for BSL-2 activities under the *NIH Guidelines*[20]).
- Bending, breaking, recapping, or removal/separation of needles is prohibited unless deemed necessary and safety procedures are used to offset risk.
- Used devices are placed in a puncture-resistant sharps container immediately after use.
- Non-disposable sharps are placed in hard-walled containers after use.

Opin. The IACUC should partner with the safety professional who has oversight for the institutional bloodborne pathogens exposure control program to ensure that all nuances of the sharps restrictions and prohibitions are understood when developing practice expectations for those working with research animals and for those performing terminal procedures. For example, it is not valuable for the IACUC to cite a full syringe with a capped needle in a sharps container without asking for an explanation of what was observed. Note that the common practice restrictions mentioned in the biosafety standards focus on the device but not necessarily on the skill of the end user or the safe handling of an animal. When using a sharp for *in vivo* procedures, it is critical for the individual to be qualified in animal restraint techniques as well as be properly trained by senior lab and animal care personnel on sharps safety and disposal practices that will protect all parties.

23:11 How can the IACUC evaluate physical hazard risk?

Opin. Potential physical hazards for animals and research personnel working with them should be considered as part of the protocol review process. The IACUC may also use required animal facility inspections as opportunities to observe practices and procedures involving potential physical hazards, including those that fall outside of the scope of protocol activities, such as cage wash, feed handling, facility cleaning, etc. However, the IACUC should involve institutional occupational health and safety professionals in the evaluation process because control of such hazards is within the purview of these individuals' responsibilities.

23:12 What is meant by a *chemical hazard*?

Reg. A *hazardous chemical* is defined by OSHA[18] as any chemical which is classified as a physical hazard or a health hazard, a simple asphyxiant, combustible dust, pyrophoric gas, or hazard not otherwise classified. See Table 23.1. Per the OSHA Hazard Communication Standard,[18] a safety data sheet needs to be available and accessible for personnel for all products and agents that meet the criteria of a hazardous chemical regardless of their end use in the animal research or animal care environment.

Opin. The IACUC and AV should seek the assistance of the institution's industrial hygiene professionals or chemical safety officer when questions arise related to the hazard status and assessment of an experimental compound or previously unused chemical on an animal protocol.

23:13 What chemical hazards are commonly associated with animal research activities and environments?

Opin. Researchers commonly use a wide variety of chemicals. Toxic agents, carcinogens, euthanasia agents, and chemotherapeutic and other test agents can be used to induce disease, study metabolic processes, euthanize animals, or reduce and reverse the effects of disease. Other chemicals are used in analytic techniques. Most laboratories have and use

TABLE 23.1

Examples of Chemical Hazards

Physical Hazards	Health Hazards
Chemicals that pose one of the following hazardous effects:	Chemicals that pose one of the following hazardous effects:
• Explosive	• Acute toxicity (any route of exposure)
• Flammable (gases, aerosols, liquids, or solids)	• Skin corrosion or irritation
• Oxidizer (liquid, solid, or gas)	• Serious eye damage or eye irritation
• Self-reactive	• Respiratory or skin sensitization
• Pyrophoric (liquid or solid)	• Germ cell mutagenicity
• Self-heating	• Carcinogenicity
• Organic peroxide	• Reproductive toxicity
• Corrosive to metal	• Specific target organ toxicity (single or repeated exposure)
• Gas under pressure	• Aspiration hazard
• In contact with water emits flammable gas	

flammable solvents, corrosive liquids, and a variety of toxic chemicals. Common animal facility cleaning supplies include solvents, acids, disinfectants, and sanitizers such as bleach or quaternary ammonium compounds. Some of these are used for manual cleaning tasks and others in mechanical devices such as cage washers or foamers. These cleaning agents can be an exposure hazard in several scenarios, including improper mixing or dispensing technique, mechanical equipment failure, or improper storage of cleaning compounds resulting in container failure.

23:14 What is the difference between a chemical hazard and a particularly hazardous substance?

Reg. Per OSHA's Occupational Exposure to Hazardous Chemicals in Laboratories Standard (the Lab Standard),[21] particularly hazardous substances (PHS) are a subclass of hazardous chemicals that include select carcinogens, reproductive toxins, and substances which have a high degree of acute toxicity as defined by that standard. Under that standard, the employer must have a chemical hygiene plan (CHP) that outlines all procedures, equipment, personal protective equipment, and work practices intended to protect workers from exposures to hazardous chemicals used in laboratories. While the CHP encompasses all hazardous chemicals, standard operating procedures (SOPs) must be available for those designated as PHS that include established designated area use, use of containment devices, procedures for safe removal of contaminated waste, and decontamination procedures.

Opin. Many of the chemical hazards used in daily animal care and research operations (see 23:13) are likely to meet the criteria of a PHS. Therefore, it is strongly recommended that the IACUC partner with the institution's designated Chemical Hygiene Officer (or an industrial hygienist with lab chemical safety experience) to determine the applicability of this standard to the animal care and research activities. The Lab Standard is not written with the animal care and research setting in mind, and it is somewhat performance-based. Fortunately, articles addressing this setting are available on this topic, including an example of one institution's approach to addressing the performance-based expectations of this standard.[22,23]

23:15 Should all approved anesthetic and euthanasia agents be considered hazardous?

Reg. Approved anesthetic and euthanasia agents are considered hazardous under OSHA regulations if they meet one or more of the criteria identified for hazardous chemicals (see 23.12).

Opin. The degree of hazard varies from agent to agent. Factors that determine the degree of hazard include the specific agent and its properties, exposure concentrations, work practices, engineering controls, personal protective equipment used, and route of administration. The institution's designated Chemical Hygiene Officer (or an industrial hygienist with lab chemical safety experience) should be consulted regarding the hazard status of such agents. This individual should also assess the administrative practices for these agents to determine the need for specific practices and safety equipment to reduce personnel exposure. These assessments should include exposure monitoring when available to be used to establish procedures and training related to use, storage, and disposal of anesthetic and euthanasia agents. Training in proper techniques involving inhalation and non-inhalation agents should be provided.

23:16 How can an IACUC evaluate a chemical risk?

Opin. As previously stated, an IACUC should partner with the institution's designated Chemical Hygiene Officer, an industrial hygienist with lab chemical safety experience, or any other health and safety professional with specific knowledge and experience (e.g., toxicologist) to review and evaluate chemical risks. Factors to consider include the following:

- Chemical agents to be administered and hazardous properties
- Dosage and format of chemical to be administered
- Method of administration in animal model (feed, gavage, aerosol, injection, immersion, topical application)
- Available methods of restraint for administration and sample collection post-challenge
- Route of excretion
- Persistence of chemical hazard in animal and its tissues
- Housing configuration
- Level of training and experience of personnel

Based on these factors, injury/exposure prevention measures can be determined to include:

- Engineering controls (ventilation equipment, safety engineered sharps, etc.)
- Study-specific personal protective equipment
- Precautions for handling animals exposed to chemical agent
- Animal disposal procedures
- Bedding disposal procedures
- Cage decontamination procedures
- Administrative controls, including hazard signage, transport procedures, study-specific safety briefing for those with access to the animal room.

23:17 What institutional resources can the IACUC access to determine whether there is a need for personnel respiratory protection for a specific animal protocol?

Reg. The OSHA Respiratory Protection Standard[6] requires the institution to develop and implement a written respiratory protection program with worksite-specific procedures and elements for required respirator use. This program, and the use of respirators, are necessary if employees are exposed to air contaminated with harmful dusts, fogs, fumes, mists, gases, smokes, sprays, or vapors, and these contaminants cannot be controlled or eliminated by other means (e.g., engineering controls). Animal care–related examples of contaminants and their applications may include:

- Disinfection of animal rooms with aerosolized chemicals
- Project-specific feed preparations containing carcinogens
- Working with or near animals *and* being identified as having animal allergies

If an employer is required to have a respiratory protection program, the employer or institution must designate a program administrator who is qualified by appropriate training or experience to administer or oversee the respiratory protection program and conduct the required evaluations of program effectiveness. This individual is the key resource for the IACUC for matters associated with respiratory protection.

Opin. Effective respiratory protection programs require a knowledgeable program administrator, an occupational health professional who can provide medical evaluations, and an understanding, at the institutional level, of the restrictions and requirements for use of respirators by employees. The IACUC should confer with the respiratory protection program administrator (usually a member of environmental health and safety or the occupational health department) whenever the need for respiratory protection is identified or suggested, as part of an animal protocol.

23:18 What factors should the IACUC consider to ensure laboratory animal allergen exposure risk is effectively minimized for those working with or around animals?

Reg. Concerns regarding the effects of animal allergens received regulatory recognition in 1998 with the release of the NIOSH Alert[11] addressing occupational asthma in animal handlers. In that alert, several controls to reduce exposure to animal allergens were put forth and included:

1. Modifications to ventilation and filtrations systems
   - increasing ventilation rate and humidity in animal housing areas
   - separating ventilation of animal housing and handling areas from the rest of the facility
   - directing airflow away from workers and toward the back of animal cages
   - use of ventilated animal cage racks or filter-top cages
2. Animal selection and density
   - Use animal species or sex known to be less allergenic
   - Decrease animal density
3. Allergen containment practices
   - Perform animal manipulations in ventilated hoods or safety cabinets when possible
   - Keep cages and animal areas clean using care to control exposures during cleaning
   - Use bedding that will minimize dusts/dry debris
   - Avoid wearing street clothes while working with animals; leave work clothes at the workplace to avoid exposure of others to allergens
4. Personnel protection and surveillance
   - Provide training to educate workers about allergens and exposure control practices
   - Use personal protective equipment that will reduce skin contact with animal products/wastes
   - Provide health monitoring, counseling, and medical follow-up for workers who have become sensitized or have developed allergy symptoms.

According to the *Guide*[1] (p. 22), the medical surveillance program should promote the early diagnosis of allergies and evaluate an individual's medical history for preexisting allergies. In addition, personnel training should include information about laboratory animal allergies, preventive control measures, early recognition and reporting of allergy symptoms, and proper techniques for working with animals.

Opin. Regardless of the size of one's animal research program, exposure to laboratory animal allergens is a universal hazard that the IACUC should ensure is adequately addressed in their OHSP. In addition to measures to protect those who handle animals or their byproducts directly, consideration must be given to those who may be indirectly exposed to animal dander, urine, or other allergenic wastes. This can occur when support personnel who do not routinely work in the vivarium are brought into that environment to service equipment. It can also occur when animals are taken outside the vivarium for procedures, or when animals are maintained in a satellite facility in a lab research area. Once the animal is outside of the vivarium, exposure control largely relies on the researcher and their adherence to hygiene practices that will minimize the allergen exposure risk for those who are not included in medical surveillance activities for animal handlers. Therefore, a comprehensive assessment of all activities where direct or indirect exposure to animals or their products may occur should be completed to determine appropriate controls to protect those with incidental exposure. The occupational health professionals administering the medical surveillance program and the safety professionals with oversight for the impacted personnel (including skilled trades, custodial and research staff) should participate in this determination. Detailed resources on animal allergen control are referenced at the end of this chapter and will complement the assessment process.[3,24]

23:19 May outer garments that are worn in animal rooms (e.g., scrub suits) also be worn outside of the animal room?

Reg. According to the *Guide*[1] (p. 20), outer garments worn in animal rooms should not be worn outside the animal facility unless they are covered by other outerwear. However, if scrub suits are routinely covered by other outerwear, such as a disposable gown or jumpsuit while working in animal areas, the scrub suit can be worn outside of the animal facility once the disposable outerwear is removed.

Opin. Outerwear serves to limit contact of outer garments with allergens and other contaminants in the facility environment, thus reducing the probability of transferring those contaminants out of the facility. The same outerwear may also be prescribed as personal protective equipment to limit the wearer's exposure to a hazard (such as in an animal biosafety level 2 room). When outerwear is worn as PPE, it is important to assure that it is worn as prescribed, and removal and replacement of the outerwear is performed upon exit from the area where the hazardous agent is present to avoid spreading associated contaminants to other areas of the facility.

Biological and Radiological Hazards

23:20 What is meant by a *biological hazard*?

Reg. Currently there are no regulatory-based definitions for biological hazards. A regulatory definition for bloodborne pathogens has been published in the OSHA standards.[7] Recombinant DNA molecules are defined in the *NIH Guidelines for Research Involving Recombinant or Synthetic Nucleic Acid Molecules*.[20] Certain infectious materials are defined by the Department of Health and Human Services (HHS),[25,26] USDA,[27,28] and Department of Transportation regulations[29] related to their transfer, acquisition, and transportation.

Opin. The term *biological hazards* (or *biohazards*) commonly refers to agents or materials of biological origin that are potentially hazardous to humans, other animals, or plants. Although there is no standardized definition for biohazards, domestic institutions often include infectious agents, materials containing recombinant DNA molecules, toxins of biological origin, and human—or nonhuman primate—derived materials in the definition.

23:21 What key biosafety standards should the IACUC be aware of that apply to animal research facilities and activities?

Reg. Biosafety standards that apply to animal research in the United States include:

1. *CDC/NIH Biosafety in Microbiological and Biomedical Laboratories (BMBL)*[8] – Section V of this document outlines the animal biosafety level criteria for vivarium species; Appendix D outlines containment requirements for pathogens affecting agricultural animals and animals in open pen or loose-housing environments.
2. *NIH Guidelines for Research Involving Recombinant or Synthetic Nucleic Acid Molecules (NIH Guidelines)*[20] – Appendix G outlines containment practices for research involving recombinant DNA use in whole animals; Appendix M supersedes Appendix G if the species are of a size or have growth requirements that preclude the use of containment for laboratory animals (i.e., cattle, swine, sheep, goats, horses, poultry, aquatic species).

Opin. While the BMBL and *NIH Guidelines*[20] are the cornerstone biosafety standards that an IACUC should be aware of, expertise in understanding and interpretation of these standards and peripheral standards that may apply (such as the OSHA Bloodborne Pathogens Standard, import permit requirements, select agent regulations) is often needed. The IACUC should engage their institutional Biosafety Officer (or the Institutional Biosafety Committee Chair if no officer) to effectively address animal biosafety–related questions.

23:22 What is meant by *recombinant or synthetic nucleic acid molecules* and what are some examples?

Reg. *Recombinant or synthetic nucleic acid molecules* are defined by the *NIH Guidelines*[20] as

1. molecules that (a) are constructed by joining nucleic acid molecules and (b) that can replicate in a living cell, i.e., recombinant nucleic acids
2. nucleic acid molecules that are chemically or by other means synthesized or amplified, including those that are chemically or otherwise modified but can base pair with naturally occurring nucleic acid molecules, i.e., synthetic nucleic acids, or
3. molecules that result from the replication of those described previously

Under the *NIH Guidelines*,[20] common animal research–related examples of recombinant or synthetic nucleic acid molecule use that require registration with, and approval by, the IBC would include:

- Creation of transgenic animals (including rodents)
- Acquisition, breeding, or use of transgenic animals other than rodents
- Introduction of microorganisms or viruses (including viral vectors) containing recombinant or synthetic nucleic acid molecules in an animal model
- Introduction of cells or tissues containing recombinant or synthetic nucleic acid molecules in an animal model

Opin. By using recombinant DNA techniques, genes can be isolated, expressed, and transferred, resulting in a multitude of possible applications. The development of transgenic and knockout animals (e.g., rodents) has proved valuable for studying a variety of diseases. Other applications involve the use of genetically modified infectious agents for vaccine development. Transgenic technology has also been used to alter certain aspects of the biochemistry, hormonal balance, and protein products in livestock and to develop recombinant human insulin, growth hormone, and blood clotting factors.

Genetic engineering technologies continue to evolve and enhance the researcher's ability to isolate, manipulate, and express genetic material to support research goals. Unfortunately, not all genetic engineering applications are definitively addressed in the *NIH Guidelines*.[20] When in doubt, the NIH Office of Science Policy should be consulted for guidance regarding applicability.

23:23 Is research involving the use of transgenic animals subject to the *NIH Guidelines*?

Reg. The answer to this question depends on the species and other factors. As outlined in Appendix C-VII of the *NIH Guidelines*,[20] the purchase or transfer of transgenic rodents (i.e., mice, hamsters, guinea pigs, rats) for experiments that require Biosafety Level 1 (BL1) containment are exempt from the *NIH Guidelines*.[20] However, crossing conditions which prohibit exemption are based upon the use of certain viruses used in the creation of the transgene. The details of the exemption application can be found at Appendix C-VIII of the *NIH Guidelines*.[20] Additionally, creation of transgenic rodents is not exempt and must be registered with and approved by the institutional biosafety committee (IBC) in conjunction with initiating this activity.

Regarding nonrodent transgenic species, there are no exemptions under the *NIH Guidelines*.[20]

Opin. Any research use of nonrodent transgenic species requires registration with and approval by the IBC before initiation of the study per Section III-D-4 of the *NIH Guidelines*.[20] This is an important nuance for IACUCs to understand and ensure that their protocol review process intersects with the IBC to avoid inadvertent creation of an NIH reportable incident.

IACUCs can benefit greatly by referencing the NIH Office of Science Policy's *Animal Activities Table*[30] and frequently asked questions document entitled *Animal Experiments under the NIH Guidelines*[31] when educating animal researchers and committee members on this topic and building their review processes.

23:24 What animal research activities involving biohazards require IBC review and approval before they are initiated?

Reg. Any recombinant DNA research activity that falls under Section III-D of the *NIH Guidelines*[20] requires registration with, and review and approval by, the IBC before the activity is initiated. Activities involving whole animals are specifically included in Section III-D-4. The NIH Office of Science Policy's *Animal Activities Table*[30] summarizes these activities and containment levels.

Opin. Animal research activities that warrant BSL-2 or greater containment practices for research agent administration (such as biological toxins or human cells) or involve administration of an infectious agent should all be reviewed and approved by the IBC before the IACUC approves these activities on an animal protocol. These circumstances will require both research and animal care staff to receive study or agent-specific training prior to these activities being carried out to ensure personnel safety and proper management of the hazard in the vivarium or lab environment.

Surv. An IACUC protocol involves the use of research hazards that require review and approval by the Institutional Biosafety Committee (IBC). Does your institution require IBC approval before IACUC approval may be granted?

- Yes. IBC approval is always required prior to IACUC approval. 87/189 (46%)
- Yes. But under special circumstances, IACUC approval may be granted prior to IBC approval. 34/189 (18%)
- No. IBC approval is not required prior to IACUC approval. Other control points are in place to ensure IBC approval prior to use of the hazards. 61/189 (32%)
- No. IBC approval is not required prior to IACUC approval. There are no control points in place to ensure IBC approval prior to use of the hazards. 7/189 (4%)

23:25 How can the IACUC evaluate biological hazard risk?

Opin. The IACUC should partner with the institutional Biosafety Officer (or the IBC Chair if no biosafety officer is in place) to ensure that the spectrum of biohazards that may be used in studies or encountered via the species to be handled on protocols are fully recognized and safe study design is initiated before the work begins. Key references have been mentioned in this chapter, and a wealth of information can also be found via the American Biological Safety Association International website.[32] A listing of all registered and/or certified biosafety professionals is also available on this site if an IACUC wishes to contact a credentialed biosafety subject matter expert.

It is important to note that biosafety level and animal biosafety level definitions and criteria are not universal across biosafety standards or animal research settings. Therefore, it is essential to involve institutional biosafety representatives in developing institutional policies related to any research (or teaching) activities involving animals and biohazards. It is also equally important to recognize that risk assessments are specific in nature and not necessarily translational to other settings or institutions.

23:26 What is a Select Agent?

Reg. Biological agents and toxins that are considered by the Department of Health and Human Services (HHS), or the USDA, as having the potential to pose substantial harm or a severe threat to human, animal, or plant health or plant products are regulated as Select Agents. Under the HHS[25] and USDA[27] regulations, anyone who wishes to transfer or possess any quantity of a Select Agent pathogen must successfully complete an intensive registration and approval process through the appropriate regulatory agency. This process includes a security clearance component for all who will have access to the Select Agent. Possession of a Select Agent–listed toxin does not require this intensive registration process with the federal agencies if specified quantities of the Select Agent or toxin are not exceeded by the individual Principal Investigator (PI).

Opin. All researchers (and research-related committees such as the IACUC) should be aware of which agents and toxins are currently regulated as Select Agents. Current lists, including specific exemptions, are available through the applicable federal agency websites. Institutions should be vigilant in implementing policies to ensure that Select Agents are not unknowingly introduced to, and possessed by, the institution. The Biosafety Officer (BSO) and Chemical Hygiene Officer can contribute significantly in this regard.

23:27 How does the use of a Select Agent in an animal study impact the IACUC review process?

Opin. The review process should not change for protocols involving animals. However, this process should include direct communication with the PI and the Responsible Official for the Select Agents. Unique access restrictions and emergency response requirements for Select Agent work can be best clarified by these individuals. The Responsible Official is the individual who is charged by the institution with the authority and responsibility of ensuring compliance with HHS and/or USDA Select Agent regulations. In some instances, this individual will be the BSO. In all instances, the BSO will be affiliated with Select Agent regulatory compliance activities. Animal studies involving the use of a Select Agent should also be reviewed by the IBC.

23:28 Should a protocol involving macaques be reviewed and approved by the Institutional Biosafety Committee (IBC) before receiving final IACUC approval?

Opin. In lieu of specific regulatory requirements, guidelines and recommendations for implementing a safety and health program specific to herpes B virus and nonhuman primates have been established by the CDC[8,10,33] and others.[34,35] None of these guidelines or recommendations requires a separate IBC review or approval of research involving animals considered to be infected with herpes B virus prior to final IACUC approval. However, an institution that is working with or housing animals of the *Macaca* genus should have an established program to control this zoonotic hazard, including training, specific SOPs, and exposure response procedures as recommended by the references cited. Development and oversight of the program should be verified as part of an IACUC review of protocols involving macaques or other Old World primates. The BSO or the Institutional Biosafety Committee Chair (if there is no BSO) and the AV should be consulted as part of this verification.

Herpes B virus is one of the examples for which specific guidelines are recommended to control a zoonotic disease hazard associated with a given species. When a protocol under review involves animals that are known, or strongly suspected, to be infected with agents transmissible to humans, the IACUC must consult the BSO or the IBC regarding applicable institutional programs or reviews related to biological safety. Research animals should be housed and handled in accordance with the animal biosafety level practices outlined in the BMBL.[8]

23:29 What elements should be included in a herpes B control program to protect persons who must enter a macaque-holding area to provide facility support services?

Reg. The OSHA General Duty Clause[36] states: "each employer shall furnish to each of his employees, employment and a place of employment which are free from recognized hazards that are causing or are likely to cause death or serious physical harm to his employees." When an institution has macaques, a herpes B exposure control program is established because of the recognized hazard. This program needs to include provisions to protect all parties with exposure risk, not just those who work directly with the macaques. This is addressed in *Occupational Health and Safety in the Care of Nonhuman Primates*,[35] which recommends the development of qualification standards for support staff such as maintenance personnel, pest control services, housekeeping staff, and emergency response personnel who may enter the macaque-holding areas. This source recommends conducting a facility orientation and providing an overview of operationally specific hazards, safety practices, and emergency response procedures.

Opin. Support staff who enter macaque-holding facilities routinely and without being escorted should be thoroughly trained on herpes B exposure control measures and bite/scratch/fluid exposure protocols. When support staff enter the area only incidentally and infrequently, these individuals should be escorted by qualified vivarium staff who can ensure that they do not encounter macaques or items that are contaminated with macaque body fluids and dander while carrying out their task in the facility. Under this scenario, the escort should provide a safety briefing prior to entry explaining herpes B exposure risk, measures that have been taken to minimize the potential for exposure, specific safety practices that the support staff must follow including PPE requirements, and emergency response procedures.

23:30 Is it necessary or advisable for an IACUC to request documentation that cell lines, tumors, or non-sterile biologic fluids are free of adventitious agents before a protocol receives final IACUC approval?

Opin. The *Guide*[1] (p. 113) states that "transplantable tumors, hybridomas, cell lines, blood products, and other biologic materials can be sources of both murine and human viruses that can contaminate rodents or pose risks to laboratory personnel," and that appropriate testing should be considered to detect these agents in biological materials. Introduction of adventitious agents into an animal colony can alter research results, cause disease and death of animals, and waste animal life and time. Some agents have zoonotic potential under certain conditions. Therefore, it is strongly advised that the IACUC request documentation to ensure such agents are not introduced into the animal facility via cell lines, tumors, or other biological materials. Responsibility for oversight of this effort is frequently assigned to the AV as part of the veterinary care program. To support personnel post-exposure response activities, available documentation should be made accessible to the occupational health group that supports the animal research program.

Surv. Which of the following statements are true for your program about testing cell lines for adventitious murine viruses? Assume the cell lines are NOT derived from in-house animals. Select all that apply.

• Not applicable.	55/189 (29%)
• Our IACUC requires PCR or similar testing of all incoming cell lines.	37/189 (20%)
• Our IACUC requires PCR or similar testing of rodent cell lines only.	23/189 (12%)
• Our IACUC accepts a statement from an investigator that a cell line is "clean."	16/189 (8%)
• We treat all cell lines coming in from the American Type Culture Collection (ATCC) as "clean."	33/189 (17%)
• An investigator cannot use a cell line until the IACUC gives its approval to do so.	36/189 (19%)
• A veterinarian must approve the PCR or similar test results from a cell line before a study can use those cells.	33/189 (17%)
• We have no policy.	37/189 (20%)
• I don't know.	10/189 (5%)
• Other*	9/189 (5%)

*Other: Several respondents indicated that while there is no requirement via policy, the IACUC offers guidelines.

Radiological Hazards

23:31 What is meant by a radiological hazard?

Reg. According to Title 10, Code of Federal Regulations, Part 20.1003[37] the definition is as follows: "Radiation (ionizing radiation) means alpha particles, beta particles, gamma rays, x-rays, neutrons, high-speed electrons, high-speed protons, and other particles capable of producing ions."[31] Radiation may be machine produced (x-ray machines, accelerators), by-product materials (e.g., radioisotopes produced by a reactor), or naturally occurring atoms that emit radiation, such as uranium, radium, or other naturally radioactive atoms.

Opin. The level of hazard is determined by the amount of radiation, type of radiation, chemical form, method of use (procedures and protocols), and other factors. Radiological hazards must be assessed carefully by a qualified radiation protection professional. Many radiation safety issues are more public relations and regulatory compliance than real risk problems.

23:32 What are typical sources of radiological hazards in the research animal environment?

Opin. Typical sources of ionizing radiation in the animal research environment are radioactive materials administered as part of a research protocol or machine-produced radiation, all of which can be used for treatment, diagnostic, or research tools. Some commonly used radionuclides include H-3, C-14, F-18, Cu-64, Ga-68, Zr-89, Tc-99_m, In-111, I-131, Lu-177, and Ac-225. These radioisotopes may be administered to animals as radiolabeled antibiotics, chemical toxins, and blood flow tracers for trauma and injury studies. X- or neutron radiation may be administered to animals to study brain or other physiological characteristics, to treat cancers, or to pursue other areas of research or treatment. Sophisticated imaging technologies in common use, such as positron emission tomography and computed tomography, present unique radiation hazards. Theranostics studies combine diagnostic imaging and therapies with the same molecule which is radiolabeled differently; these treatments have proven to be very effective, and it is expected that this will be prominent in animal studies for some time. The use of such radiologic technology tools in animal studies can present exposure hazards for personnel handling these animals during and after exposure. Careful review and oversight by the radiation safety staff are needed.

23:33 Is external beam radiation, such as X-irradiation, considered a radiation hazard of concern to an IACUC?

Reg. External beam radiation is regulated as a radiation hazard and therefore should be recognized as such by the IACUC. Sources of external radiation beams could include sealed radioactive material (e.g., irradiators), X-ray devices, or accelerators.

Opin. External beam radiation effects should be considered a radiation hazard. Therapeutic or research uses involving high doses of X-radiation or proton irradiation that potentially can cause somatic effects should be considered similar to other procedures that cause pain and suffering for animals. The IACUC should communicate with the radiation safety department to assure that protocols involving X-irradiation, as well other forms of radiologic technologies, are properly planned, documented, and monitored.

The use of external beam radiation in the animal research environment is normally limited to therapeutic purposes (diagnostic and treatment). Nevertheless, it should be included in the overall animal care program oversight of the IACUC, as it can be a hazard. If a research protocol requires the use of X-radiation, that use requires IACUC review and approval. Therapeutic or research uses involving high doses of X-radiation that potentially can cause somatic effects should be considered similar to other procedures causing pain and suffering for animals.

All human health aspects related to external beam radiation must be addressed by institutional safety professionals in the radiation safety and occupational health areas. This area is usually beyond the IACUC's expertise and scope.

23:34 How can an IACUC evaluate radiological hazard risk?

Reg. Radiological risk potential is reflected in the rigor of the applicable federal, state, and local regulations for radiation. In general, the regulations require that each use of radiation must be authorized by the Radiation Safety Officer and, for larger programs, the Radiation Safety Committee, providing a built-in mechanism for knowledgeable review of the radiological hazards involved in any proposals that come before the IACUC.

Opin. The IACUC should evaluate only the effects to the animal of experiments involving radiation. The Radiation Safety Officer and the institutional radiation safety committee must assess radiation risk, determine precautions and management practices, and document the radiation protection evaluation. They also must inspect and ensure safety and compliance with the regulations. In certain cases, precautions must be taken by animal care staff. The instructions for precautions are often posted on the door of the room or on the cages of the animals for which the precautions are necessary. Animal care staff who will handle or assist with the management of animals that have been administered radioactive materials, or who will handle or manage the radioactive waste or bedding, must be trained in radiation safety at their institution.

23:35 What kind of documentation relative to an investigator's approved radioisotope use may be provided to the IACUC from a radiation safety office?

Reg. Regulations require the radiation safety office to maintain sufficient records to document compliance with the license conditions and applicable regulations. This will include records of the program content and implementation, audits and radiation surveys, individual radiation monitoring results, radioactive material inventory, dose to members of the public, waste records, etc. The authorization documentation (i.e., which animal users are authorized to use radioactive material or radiation emitting devices) will generally be of most interest to the IACUC. Coordination with the Radiation Safety Officer is needed to ensure that all radiation use associated with animals is authorized and that the appropriate documentation is maintained.

Opin. The safe use of radiation and radioactive materials entails a very comprehensive and strict program of safety and compliance performed by qualified radiation safety professionals. IACUCs should initiate and maintain a close and friendly working relationship with the radiation safety managers at their institutions and defer to their findings. The idea of combining radiation and animal use authorization documents may appear to offer an appealing opportunity to reduce duplication of effort, but generally the requirements and business rules of the IACUC and radiation safety programs are too different to make this approach practicable.

23:36 How does IACUC maintain compliance with the use of ionizing radiation?

Reg. Radioactive material must be obtained, stored, used, and processed for disposal according to the constraints of both the applicable regulations and the site's radioactive material license, but the regulatory agency, and hence the applicable regulations, varies by location. Most states are agreement states, meaning those states have an agreement with the Nuclear Regulatory Commission (NRC) to set forth and enforce their own state regulations for radiation protection and issue state radioactive material licenses. Radioactive material users in non-agreement states, nuclear reactor facilities, and all federal facilities in any state (e.g., military bases, Veterans Administration medical centers, etc.) are always regulated directly by the NRC, except for Department of Energy (DOE) sites, which have their own regulations. X-ray devices and accelerators are regulated by the states. The regulatory controls imposed by this regulatory framework are generally commensurate with the associated hazard, and the IACUC must coordinate with the site's Radiation Safety Officer to ensure that the applicable regulatory requirements have been addressed prior to approving such radiation use.

REFERENCES

1. National Research Council. 2011. *Guide for the Care and Use of Laboratory Animals.* Washington, DC: National Academy Press.
2. National Institutes of Health, Office of Laboratory Animal Welfare. 2015. *Public Health Service Policy on the Humane Care and Use of Laboratory Animals.* Bethesda, MD: U.S. Department of Health and Human Services.
3. National Research Council. 1997. *Occupational Health and Safety in the Care and Use of Research Animals.* Washington, DC: National Academy Press.
4. National Institutes of Health, Office of Laboratory Animal Welfare. Frequently asked questions G.2, What is required for an occupational health and safety program? http://grants2.nih.gov/grants/olaw/faqs.htm#instresp_2.
5. Office of the Federal Register, Code of Federal Regulations, Title 29, Part 1910.95. Occupational Noise Exposure, Washington, D.C.
6. Office of the Federal Register, Code of Federal Regulations, Title 29, Part 1910.134, Respiratory Protection. Washington, DC.
7. Office of the Federal Register, Code of Federal Regulations, Title 29, Part 1910.1030, Bloodborne Pathogens. Washington, DC.
8. Centers for Disease Control and Prevention and National Institutes of Health. 2019. *Biosafety in Microbiological and Biomedical Laboratories*, 6th ed. Washington, DC: Government Printing Office.
9. Centers for Disease Control and Prevention. Diagnosis and Management of Q Fever – United States. *MMWR* 2013;62(3).
10. Cohen, J.I., Davenport, D.S., J.A. Stewart et al. 2002. Recommendations for Prevention of and Therapy for Exposure to B virus (Cercopethicine Herpesvirus 1) *Clin. Infect. Dis.* 35:1191–1203.
11. National Institute of Occupational Health Alert. 1998. Preventing Asthma in Animal Handlers, U.S. Department of Health and Human Services, Publication No. 97–116.
12. Institute for Laboratory Animal Research. 2000. *Strategies That Influence Cost Containment in Animal Research Facilities*, p. 28. Washington, DC: National Academy Press.
13. American Association for Laboratory Animal Science. *Cost of Caring—Human Emotions in the Care of Laboratory Animals.* https://www.aalaslearninglibrary.org/pdf/course2593/cost_of_caring.pdf (accessed 8/6/2025).
14. Office of the Federal Register, Code of Federal Regulations, Title 29, Part 1910.97. Nonionizing Radiation, Washington, DC.
15. Office of the Federal Register, Code of Federal Regulations, Title 29, Part 1910.1096, Ionizing Radiation. Washington, DC.
16. Office of the Federal Register, Code of Federal Regulations, Title 29, Part 1910 Subpart S, Electrical. Washington, DC.
17. Office of the Federal Register, Code of Federal Regulations, Title 29, Part 1910.301–1910.399, Electrical. Washington, DC.
18. Office of the Federal Register, Code of Federal Regulations, Title 29, Part 1910.1200, Hazard Communication. Washington, DC.
19. American Veterinary Medical Association. 2020. AVMA Guidelines for the euthanasia of animals: 2020 edition. https://www.avma.org/sites/default/files/2020-02/Guidelines-on-Euthanasia-2020.pdf (accessed 8/6/2025).

20. National Institutes of Health. 2024. *NIH Guidelines for Research Involving Recombinant or Synthetic Nucleic Acid Molecules*. Bethesda, MD: U.S. Department of Health and Human Services.
21. Office of the Federal Register, Code of Federal Regulations, Title 29, Part 1910.1450, Occupational Exposure to Hazardous Chemicals in Laboratories. Washington, DC.
22. Thomann, W.R. 2003. Chemical safety in animal care, use and research. *ILAR J.* 44(1): 13–19.
23. Lee, V.K, Hubble, L.M. and S.W. Thomaston. 2018. Chemical use in animal models. *Appl. Biosafe* 23(3):153–161.
24. Harrison, D.J. 2001. Controlling exposure to laboratory animal allergens. *ILAR J.* 42(1):17–36.
25. Office of the Federal Register, Code of Federal Regulations, Title 42, Parts 72 & 73, Possession, Use, and Transfer of Select Agents and Toxins. Washington, DC.
26. Office of the Federal Register, Code of Federal Regulations, Title 42, Part 71.54, Import Regulations for Infectious Biological agents, Infectious Substances, and Vectors. Washington, DC.
27. Office of the Federal Register, Code of Federal Regulations, Title 7, Part 331, and Title 9, Part 121, Agricultural Bioterrorism Protection Act of 2002, Possession, Use and Transfer of Biological Agents and Toxins, Final Rule. Washington, DC.
28. Office of the Federal Register, Code of Federal Regulations, Title 9, Animal and Animal Products, Parts 92–95, 122. Washington, DC.
29. Office of the Federal Register, Code of Federal Regulations, Title 49, Parts 171–180, Hazardous Materials Regulations. Washington, DC.
30. https://osp.od.nih.gov/wp-content/uploads/2022/11/Animal_Activities_Table.pdf (accessed 8/06/2025).
31. https://osp.od.nih.gov/policies/biosafety-and-biosecurity-policy/faqs-for-research-on-genetically-modified-transgenic-animals-april-2024/ (accessed 8/06/2025).
32. https://absa.org/ (accessed 8/06/2025).
33. Centers for Disease Control and Prevention. 1987. Guidelines for the prevention of herpes virus simiae (B-virus) infection in monkey handlers. *MMWR* 36:680–682, 687–689.
34. National Institute for Occupational Safety and Health. April 1998. Health Hazard Evaluation Report: 98-0061-2687, Yerkes Primate Research Center. Lawrenceville, GA.
35. Institute for Laboratory Animal Research. 2003. *Occupational Health and Safety in the Care and Use of Nonhuman Primates*. Washington, DC: National Academy Press.
36. Office of the Federal Register, 29 U.S.C 654, Section 5, Occupational Health and Safety Act of 1970. Washington, DC.
37. Office of the Federal Register, Code of Federal Regulations, Title 10, Part 20.1003, Standards for Protection against Radiation. Washington, DC.
38. Animal Welfare Act and Animal Welfare Regulations. APHIS 41-35-076. 2020.

24

Personnel Training

Melissa C. Dyson, Tara Martin, and Kiirsa Pokryfke

Introduction

The AWAR, the PHS Policy, and the *Guide for the Care and Use for Laboratory Animals* (Guide) direct institutions to provide training for personnel engaged in animal research. Numerous methods ranging from one-on-one sessions to online tutorials have arisen to provide training. In response to the demand for increased training, many institutions have created specific positions for trainers as part of the IACUC and/or animal care staff. Virtual or online training resources and learning management systems have become widely available and frequently utilized for compliance and technical educational programs. IACUCs frequently utilize training and retraining as a means to achieve compliance and improve animal welfare, especially in response to episodes of repeated non-compliance encountered with research personnel.

A survey was conducted by the editors and authors of this book, and institutions responded to questions on personnel training. In this chapter, the survey results are expressed as the number of responses per number of respondents and percent of respondents. Thus, if more than one answer was provided, the sum of all responses will be greater than the number of respondents. If no answer was provided by some respondents, the number of respondents will be less.

24:1 Is there any requirement in either the AWAR or the PHS Policy for general training in laboratory animal care and use?

Reg. The AWAR require that research facilities include specific topics in their institutional training programs in addition to a general requirement that staff involved in the animal care and use program are qualified to perform their duties.[1] The *Guide* provides recommendations on the content of personnel training programs.[2] The PHS Policy requires that Assured institutions use the *Guide* as a basis for their animal care and use program and that individuals receive training to ensure appropriate practices are humane and minimize distress, and training or instruction in research methods that minimize the number of animals used in research.[3]

The relevant sections of the AWAR are §2.32,c,1–5 as shown below.[4]

c. Training and instruction of personnel must include guidance in at least the following areas:

1. Humane methods of animal maintenance and experimentation, including:
   i. The basic needs of each species of animal;
   ii. Proper handling and care for the various species of animals used by the facility;
   iii. Proper pre-procedural and post-procedural care of animals; and
   iv. Aseptic surgical methods and procedures;
2. The concept, availability, and use of research or testing methods that limit the use of animals or minimize animal distress;
3. Proper use of anesthetics, analgesics, and tranquilizers for any species of animals used by the facility;
4. Methods whereby deficiencies in animal care and treatment are reported, including deficiencies in animal care and treatment reported by any employee of the facility. No facility employee, Committee member, or laboratory personnel shall be discriminated against or be subject to any reprisal for reporting violations of any regulation or standards under the Act;
5. Utilization of services (e.g., National Agricultural Library, National Library of Medicine) available to provide information:
   i. On appropriate methods of animal care and use;
   ii. On alternatives to the use of live animals in research;
   iii. That could prevent unintended and unnecessary duplication of research involving animals; and
   iv. Regarding the intent and requirements of the Act.

In addition to the training requirements in §2.32,c,1–5, the AWAR (§2.38,I,1–3) set

DOI: 10.1201/9781003272861-24

requirements for development and implementation of institutional contingency (emergency) plans. A component of this requires employee training in such plans (See 25:2 and 25:3). Specifically, "the facility must provide and document participation in and successful completion of training for its personnel regarding their roles and responsibilities as outlined in the plan."

The PHS Policy does not specify training program content to the same degree as the AWAR; nevertheless, PHS Policy (IV,C,1,f) requires the IACUC, in its review of protocols, to determine that personnel conducting procedures are appropriately qualified and trained. In addition, the PHS Policy (IV,A,1,g) requires the institution to include in its NIH/OLAW Assurance a "synopsis of [the] training or instruction in the humane practice of animal care and use, as well as training or instruction in research or testing methods that minimize the number of animals required to obtain valid results and minimize animal distress, offered to scientists, animal technicians, and other personnel involved in animal care, treatment, or use."

The *Guide* (pp. 15–17) specifies that "all research groups should receive training in animal care and use legislation, IACUC function, ethics of animal use and the concepts of the Three Rs, methods for reporting concerns about animal use, occupational health and safety issues pertaining to animal use, animal handling, aseptic surgical technique, anesthesia and analgesia, euthanasia, and other subjects, as required by statute." In contrast, the *Guide* does not specify the content of training programs for animal care personnel. Instead, it states only that "[animal care] staff should receive training and/or have the experience to complete the tasks for which they are responsible."

Opin. The U.S. Government Principles, PHS Policy, and AWA emphasize that animals should not be subjected to unnecessary pain or distress. Animal care and research personnel who are well trained can perform animal research techniques with greater skill and fewer adverse outcomes for the animals, leading to less pain and distress. Consequently, it is important for institutions to provide training to animal care and research personnel to ensure humane care for animals used in research.[5]

24:2 Is there any requirement in either the AWAR or the PHS Policy for additional training in the care and use of a particular species?

Reg. The AWAR (§2.32,c,1,i; §2.32,c,1,ii; §2.32,c,3) require training on the basic needs of the relevant species; proper handling and care for various species used by the facility; and the proper use of anesthetics, analgesics, and tranquilizers for any animals used by the facility. Research facilities utilizing marine mammals are required to provide and document participation in and successful completion of a facility training program (§3.108).

The PHS Policy (IV,A,1) does not specify any type of species-specific training beyond being qualified for conducting procedures on the species being maintained or studied, but it does require that Assured institutions use the *Guide* as a basis for their animal care and use.

The *Guide* (p. 16) states that research personnel should have the necessary knowledge and expertise for the specific animal procedures proposed and the species used.

Opin. There is a vast diversity of species used in research, teaching, and testing. Although rats and mice account for the vast majority of animals used in research, numerous other species have been utilized and will continue to be utilized, depending on the suitability of particular models for the research being conducted.[6] It is impossible for any one person to be familiar with all species that are or might be used in research. Therefore, institutions must provide animal care and research personnel with access to training when experiments are planned to use new species.

24:3 Do the AWAR or the PHS Policy state whether or not training is required for specific research procedures (e.g., performing an arterial cut-down)?

Reg. Procedure-specific training specified in the AWAR (§2.32,c,1,iii; §2.32,c,1,iv) includes training on pre- and post-procedural care of animals and aseptic surgical methods and procedures. Additionally, research facilities are responsible for ensuring that all staff are qualified to perform their duties, which may require training for specific research procedures depending on the experience of individual staff members (§2.32,a). The PHS Policy does not require procedure-specific training, and as noted previously, the *Guide* (p. 16) indicates that procedure-specific training may be necessary to ensure that research personnel who perform experimental manipulations on animals can conduct their research humanely.

Opin. Innumerable animal research techniques are described in the scientific literature for use in the many disciplines in biomedical research. It is impossible for anyone to be skilled in all techniques commonly used in animal research. Training personnel should be prepared to teach common animal research techniques to individuals who conduct animal research. Development of skills and new techniques can be done in collaboration with training or veterinary personnel with experience in animal support and diverse technical skills. In addition, it can be valuable to identify individuals at the institution who work in research laboratories and have unique skills in performing specific research procedures with animals. These individuals should be cultivated as ancillary training staff who can be called upon to help train personnel when training requests for their particular skills are received. Engaging research personnel to participate in the training effort can be a fruitful and rewarding means to expand the training provided at an institution while conserving fiscal resources.

24:4 Who is responsible for ensuring that research and animal care personnel working with animals are adequately trained?

Reg. According to the AWAR (§2.32,a), the research facility is responsible for ensuring that research personnel are qualified to perform their duties. To accomplish this, the research facility must make training available to personnel engaged in animal care and use. The HREA (§495,c,1,b) and the PHS Policy (IV,A,1,g) require the awardee institution to assure that all personnel involved in animal care, treatment, or use have training available to them. Both the AWAR (§2.31,d,1,viii) and the PHS Policy (IV,C,1,f) specify that the IACUC, in its review of activities involving animals, must determine that personnel are appropriately qualified and trained. The *Guide* (p. 15) states that the institution is responsible for providing the resources to support training of animal care and use personnel and the IACUC is responsible for oversight of training and evaluation of the training program.

Opin. The responsibility for ensuring that personnel are adequately trained must be shared by the institutional officers, the IACUC, the animal care and veterinary staff, and the Principal Investigator (PI). Overall, it is an institutional responsibility to ensure that animal care and research personnel are adequately trained to care for research animals and conduct animal research procedures. The institution must be willing to commit the personnel and financial resources to accomplish this objective. Nevertheless, in practical terms, the responsibility for assuring that training is available lies with the IACUC, as oversight of the animal care and use program is an IACUC responsibility as an agent of the research facility. The IACUC is able to identify the personnel who should be trained and coordinate the resources and activities necessary to provide the training. The responsibility for actually providing the training usually rests upon the veterinary staff or the staff that supports the IACUC. These are the individuals who possess the scientific, clinical, and technical skills that the animal care and research staff need to perform their duties. Certainly, some specialized training may have to be provided by research personnel. Finally, the PI has a responsibility to convey to her or his staff the importance of receiving proper training in order to conduct humanely the animal studies in which they will be participating. The PI's attitude toward training sets the tone for laboratory personnel with regard to their participation in the institution's training program.

24:5 Should personnel be adequately trained before a research project begins, or can they be trained during the course of a project?

Opin. The AWAR and PHS Policy do not provide guidance on when or how training is provided; however, the AWAR do require the IACUC determine that personnel are adequately trained prior to being approved to work with animals. The *Guide* (p. 17) does state that all animal users should have adequate training before beginning animal work. Some types of training might be so specialized that they can be provided only by other members of the research staff in a mentor–trainee setting. It would be wise to provide training in a stepwise fashion, starting with the most basic techniques and progressing to the most complex and specialized and prior to the use of animals in experimental procedures.

Surv. At your institution, must personnel be adequately trained before a research project begins, or can they be trained during the course of a project via observations, assisting a fully trained person, etc.?

- In most instances, the person must be fully trained before the project begins. 59/188 (31%)
- In most instances, the person must have the basic skills needed before the project begins, but more detailed skills can be gained during the course of the project. 100/188 (53%)
- In most instances, all hands-on training can be gained during the course of the project. 23/188 (12%)
- I don't know. 0/188 (0%)
- Other. 6/188 (3%)

Most institutions represented by survey respondents require that personnel be given either full or basic training before working with animals.

24:6 Is training in animal care and use mandatory for personnel who will be working with animals?

Opin. Training for animal care and research personnel is certainly desirable in order to ensure the highest level of humane care. The AWAR (§2.32,a) state that it is the responsibility of the research facility to ensure that personnel are qualified to perform their duties and that "this responsibility shall be fulfilled in part through the provision of training and instruction to those personnel." The *Guide* (p. 15) emphasizes that all personnel must be adequately educated, trained, and/or qualified in basic principles of laboratory animal science. This implies that training, *per se*, is not necessarily mandatory, but that personnel have a particular knowledge base if they are involved in animal care and use. For example, an institution could choose not to require specific skills training for a new employee who had been trained at another institution or had acquired the necessary skills to perform his duties through experience. These skills (e.g., biomethodologies) can be documented via competency assessment as opposed to training, allowing the program to ensure that personnel are qualified to perform procedures without additional training. However, in practice, there is a great deal of institution-specific education that is important to transmit to individuals (e.g., policies) beyond technical skills that are essential for successful research implementation. Thus, some amount of training is typically necessary for new personnel, regardless of technical skill or experience at prior institutions.

24:7 What actions can an IACUC take if members of the research team do not complete their training?

Opin. Enforcement of institutional policies that require training is essential to maintaining a compliant animal care program. Methods to ensure compliance with training policies must be achievable and should limit the access of users to animals. Common methods of ensuring completion of required animal use training include limiting protocol approval until training completion, withholding animal facility access, or restriction from animal research until training is completed.

Surv. What are the consequences of failure to complete required training by the Principal Investigator or members of the research team? Select all that apply.

- Not applicable as we have no formal training program. 6/188 (3%)
- There are no consequences. 2/188 (1%)
- Protocol or amendment approval is contingent upon completion of training. 102/188 (54%)
- Employment is contingent upon the completion of training. 8/188 (4%)
- Access to the animal facility is contingent on completion of training requirements. 120/188 (64%)
- Personnel are restricted from performing animal research procedures until training is completed. 133/188 (71%)
- We report non-compliance with training requirements to the IACUC. 81/188 (43%)
- I don't know. 3/188 (2%)
- Other* 5/188 (3%)

*Other: Personnel who fail to complete training would be removed from the protocol.

24:8 Should an IACUC approve a protocol with novel procedures based on the surgical or other procedural experience of the investigator or other personnel involved?

Reg. Although the PHS Policy and the AWAR are silent on this matter, the *Guide* (p. 28) provides some insight into this issue. If novel procedures are encountered in protocols under review, the *Guide* suggests that the IACUC consider the use of pilot studies to identify and define humane endpoints.

Opin. In addition to pilot studies, the IACUC can require that novel procedures be observed by a veterinarian or other qualified staff member to ensure that they are performed in a humane and painless manner.

24:9 Person A wants to learn a new technique for Protocol A but will use animals on Protocol B under the supervision of a person from Protocol B. Should Person A be listed only on Protocol A or on both protocols?

Opin. The IACUC should consider the degree of involvement and the invasiveness of the procedure in determining whether to require Person A to be listed on Protocol B. For example, if someone needs to observe only a minimally invasive technique used in another laboratory, the IACUC would be justified in not requiring Person A to be listed on Protocol B. On the other hand, if someone must learn a complicated surgical procedure that would take extensive training by a skilled mentor, inclusion on the protocol should be required. Additionally, if training is regularly provided under Protocol B, the training should also be incorporated into the protocol and approved by the IACUC.

24:10 Would principal investigators who do not themselves perform procedures on animals have to receive animal care and use training?

Opin. Investigators who will not be conducting procedures on animals themselves should still be required to take basic introductory training in order to become familiar with federal, state, and institutional policies and procedures. As supervisors of research personnel who will be performing animal research procedures, they have a responsibility to convey the importance of utilizing the most humane methods in animal research and of regulatory compliance.

Surv. At your institution, do Principal Investigators, who will not be performing any hands-on animal work, have to partake in the institutional animal care and use training program?

- We have no training program. 2/188 (1%)
- Yes, but only basic training in animal care and use. More detailed training is not required. 85/188 (45%)
- Yes, they are required to complete the same training as staff who perform animal research procedures. 72/188 (38%)
- No, no additional training of any kind is required. 21/188 (11%)
- I don't know. 1/188 (1%)
- Other* 7/188 (4%)

* Other included the following responses:

- Must complete online training, but not hands-on training
- Must also complete occupational health and safety training

24:11 If a person has successfully completed an animal care and use training course at a different institution, should the IACUC at a new institution require that she retake a similar course at that institution or otherwise demonstrate his or her capabilities?

Reg. See 24:1–24:3.

Opin. As the IACUC examines the credentials of animal use personnel in the course of conducting its review of proposed activities, any training that has occurred at another institution can be considered. In order to do this, the IACUC may request that the individual provide detailed information on the nature of training at the previous institution. The amount of "credit" awarded by the IACUC will likely vary with the institution's policies and the content of the training at the previous institution. Regardless of the nature of the training at the previous institution, the individual should take any orientation and program specific training offered to acquire institution-specific information that would not be acquired in any other way. Additionally, assessment of technical skills of an experienced individual may be sufficient to replace requirements for completion of a specific course (see 24:12).

Surv. Does your IACUC waive training requirements for persons having documented training at another institution?

- Not applicable. We have no training program. 2/188 (1%)
- It varies with circumstances, but we can waive some training requirements. 100/188 (53%)
- We offer alternative means of satisfying training requirements, such as competency assessment. 23/188 (12%)
- We never waive training requirements. 52/188 (28%)
- I don't know. 2/188 (1%)
- Other* 9/188 (5%)

*Other responses included some of the following comments:

- If training was conducted at another institution, documentation to establish equivalency is required.
- Waivers would be granted only to short-term visitors, including those on campus for training or teaching purposes.

24:12 Can someone with experience in surgery or anesthesia in human patients perform these procedures on animals without demonstrating their competence to do so?

Reg. The *Guide* (p. 16) and the PHS Policy (IV,C,1,f) state that research personnel should have the necessary knowledge and expertise for the specific animal procedures proposed and the species used. In addition, the *Guide* (p. 115) indicates that researchers trained in human surgery may need additional training before performing surgery on animals. The AWAR require the IACUC to ensure the PI is qualified for his/her duties (see 24:1,24:2). Experience with humans does not necessarily qualify a person with regard to the same procedures in animals. Also, the AWAR (§2.32,c,3) require training specifically for the proper use of anesthetics for the species of animals being used.

Opin. Humans and animals differ greatly in anatomy, physiology, and patient compliance. Although similarities exist and can make training easier for personnel experienced in human techniques, it is not appropriate to assume that these techniques will be the same for animals.

Surv. Does your IACUC allow persons experienced with human patients (such as physicians who are anesthesiologists or surgeons) to perform anesthesia or surgery on laboratory animals without providing evidence of having acquired similar skills with laboratory animals?

- Yes, we usually do. 3/146 (2%)
- Yes, we might do so, but we evaluate each request on a case-by-case basis. 42/146 (29%)
- No, we do not. 96/146 (66%)
- I don't know. 3/146 (2%)
- Other* 2/146 (1%)

*Other responses included some of the following comments:

- Rodents – No exceptions; Large Animals – no for anesthesia, yes for surgery (under direct veterinary supervision).
- We would typically require training on a cadaver for the animal species involved in the experiment prior to allowing any live animal work.

24:13 Should all personnel performing animal care and use procedures receive the same training?

Opin. All personnel should be competent in performing animal care and use procedures. This can be achieved through a variety of training and mentorship opportunities that do not necessarily need to be the same for each type of work or individual. Animal care staff often require a great breadth of knowledge in order to provide consistent care to all of the animals they interact with. A combination of training methods can be very effective. For example, staff may be provided foundational information as part of a basic online training course, SOP review, and/or formal training by a trainer.

This can be followed by mentorship, where staff can be observed for true competency and provided with additional support as needed. Training expectations should reflect expected tasks and responsibilities in regard to animal care and use. Visitors or other personnel who may be observing or have limited work with animals may require less training prior to approval for activities.

Surv. How do the training methods differ between laboratory personnel and animal care staff? Select all that apply.

- Animal care staff is provided on the job training. 157/187 (84%)
- Animal care staff is provided online training. 130/187 (70%)
- Animal care staff is provided with SOP review. 139/187 (74%)
- Animal care staff complete hands-on in room training. 144/187 (77%)
- Animal care staff completes hands-on animal technique training. 133/187 (71%)
- Lab staff is provided online training. 136/187 (73%)
- Lab staff is provided just-in-time in person training. 85/187 (45%)
- Lab staff is required to complete hands-on animal technique training. 113/187 (60%)
- I don't know. 11/187 (6%)
- We don't have a training program. 4/187 (2%)

Surv. How do we ensure appropriate training for animal care staff? Select all that apply.

- Online training. 86/187 (46%)
- On the job training. 90/187 (48%)
- Mentoring. 73/187 (39%)
- Formal training by a trainer. 67/187 (36%)
- SOP review. 77/187 (41%)
- A combination of these. 160/187 (86%)
- I don't know. 4/187 (2%)
- We don't have a training program. 3/187 (2%)

Surv. Does your IACUC require that short-term personnel (e.g., summer students, visiting professors, etc.) fulfill the same training requirements as full-time investigators/educators and their staff?

- Not applicable. 9/187 (5%)
- Yes, short-term employees have to fulfill the same training requirements. 162/187 (87%)
- No, short-term employees have somewhat modified training requirements 14/187 (7%)
- No, short-term employees have no training requirements. 1/187 (1%)
- I don't know. 0/187 (0%)

24:14 If a person from another institution or a commercial organization is to demonstrate how to perform a procedure on a live animal, how can the IACUC assure that person is qualified to perform the procedure?

Opin. Visiting experts or trainers may be asked to perform or assist with procedures on animals in many types and sizes of institutions. When institutions encounter such requests, it is important to develop mechanisms to manage them effectively rather than forbid them outright so as not to inhibit education and training. These procedures are often an important part of the institution's educational programs—whether as formalized continuing medical education or outside of it. In addition, visiting professionals may be providing expertise otherwise unavailable to users within a given institution. Appropriate management of such requests depends on the nature of the procedure and on institutional resources, but often involves review of qualifications, monitoring of the procedure by or demonstration of skills to qualified personnel, or a combination.

Surv. At your institution, if a representative of a commercial company is to demonstrate the proper use of a new surgical product (e.g., for intestinal anastomosis) on a live animal, how does your IACUC assure that the representative is properly trained in the use of the product on the species to be used? Select all that apply.

- Not applicable. 70/187 (37%)
- We accept a written statement from the Principal Investigator that the person is qualified. 42/187 (22%)
- We review the written qualifications of the person. 75/187 (40%)
- We have a veterinarian watch the person perform the procedure. 34/187 (18%)
- We have a veterinarian watch the person perform the procedure, and the veterinarian has the authority to stop the procedure. 65/187 (35%)
- The person must first demonstrate the procedure to our IACUC representative before training others. 4/187 (2%)
- I don't know. 8/187 (4%)
- Other* 10/187 (5%)

*Other responses included some of the following comments:

- Such personnel would be required to undergo all the training that would be expected for institutional personnel.
- Requirements vary with species.
- The veterinarian meets with the individual prior to the procedure as a way to assess qualifications.v

24:15 Should students working with live animals during a class be required to participate in an animal use training program?

Opin. Students working in research laboratories should receive the same training as all other research personnel. Students who utilize animals in classroom settings only could receive abbreviated training specific to the procedures being conducted. This may be provided by the IACUC staff, institutional trainers, veterinary staff, or by the course instructor (who should have received all training required of research personnel). If training is provided by the course instructor, it is wise to periodically review to ensure it is up to date with institutional policies.

Surv. If your institution has courses in which students use live animals for any reason or length of time, are they required to participate in an animal use training program? Select all that apply.

• Not applicable.	70/188 (37%)
• Students in courses do not receive training.	11/188 (6%)
• Course instructors provide training to students in the course.	70/188 (37%)
• Students in courses receive the same or similar training as that provided to research personnel.	55/188 (29%)
• Students in courses receive their training from the same individuals who train research personnel.	32/188 (17%)
• Students in a single training lab get less (or no) training compared to students who will be using animals throughout the course.	4/188 (2%)
• Other*	7/188 (4%)

*Other responses included: Modified training may be provided by a veterinarian, IACUC Chair, or by means of a brochure.

24:16 Can a high school student perform animal research if he or she works under the guidance of an investigator with an IACUC-approved protocol?

Opin. It has become increasingly common for high school students to work as volunteers, students, or employees in animal research settings at colleges and universities. There is no inherent reason that a high school student cannot perform research procedures using animals if he or she is under the guidance and supervision of an investigator with approval to conduct the research. High school students should receive whatever training is necessary for them to perform their duties. Their training should be the same as that provided to college-level students and regular employees of the institution.

24:17 Is there any lower age limit below which the IACUC should disallow student participation in animal care and research?

Opin. There is no specific reason to place an arbitrary age limit on student participation in animal care and use. However, it is certainly uncommon for students in elementary and middle school to volunteer or seek employment in research settings. IACUCs are advised to treat any such request on a case-by-case basis. Any student, regardless of age, must be adequately supervised by an investigator with approval to conduct the animal studies. Other factors such as labor laws and institutional policies on employment and volunteers may limit the participation of younger students in animal research.

24:18 In terms of animal care and use, who is ultimately responsible for the activities of a student?

Reg. Although the PI is responsible for a proposal to conduct research and for the design and implementation of research involving animals (§1.1), the research facility is ultimately responsible for compliance with the AWAR (§2.30,b).

Opin. While every member and body within the research community shares responsibility for appropriate animal care and use, the PI bears the responsibility for any and all personnel, including students working in his or her laboratory. In the course of review and approval of a protocol, the PI should provide assurance to the IACUC that all personnel working with animals will be adequately trained to perform their duties. The investigator must understand that this responsibility extends to students working in the laboratory. Ultimately the IACUC (*Guide*, p. 15) and the research facility (AWAR §2.32,a) are responsible for ensuring that staff are appropriately qualified to work with animals in research.

24:19 Does the unaffiliated member of an IACUC need additional or supplemental training beyond that which is provided to other committee members?

Reg. According to the *Guide* (p. 17), IACUC members should be provided formal orientation training covering the animal care program, relevant legislation, regulations, guidelines and policies, animal facilities and laboratories, processes or protocols, and program review. Similarly, the AWAR (§2.32,a) state that the institution is responsible for ensuring that all scientists, research technicians, animal technicians, and other personnel involved in animal care, treatment, and use are qualified to perform their duties. This includes unaffiliated Committee members.

Opin. It is advisable to provide the same training to all IACUC members, not just the unaffiliated member(s). Even those experienced with animal research may have gaps in knowledge which training can address. However, depending on their own experience and expertise, unaffiliated members may desire additional training or orientation to the institution and to animal research in general. It is advisable to develop a mechanism through which this additional training can occur, such as lists of reference material, referrals to outside meetings or training opportunities, or designated personnel to contact for additional training within the institution.

Surv. Does your IACUC provide any additional training to the unaffiliated member of your IACUC that is not provided to other committee members?

- Yes, routinely 33//187 (33%)
- Yes, but only upon request of the unaffiliated member 53/187 (28%)
- No, we do not 93/187 (50%)
- I don't know 3/187 (2%)
- Other 5/187 (3%)

24:20 At what interval is periodic retraining of animal users required?

Reg. The AWAR require that training and instruction be made available, and the qualifications of personnel reviewed, with sufficient frequency to fulfill the research facility's responsibility to ensure that all personnel are qualified to perform their duties (§2.32,b).

Opin. Periodic retraining can help ensure personnel remain qualified—especially for tasks or duties not performed routinely. Retraining also provides a good opportunity for ensuring staff are up to date with recent policy changes. However, providing full retraining to all staff is labor and time intensive and may not be feasible. In many cases, abbreviated retraining that focuses on critical tasks selected by the institution may be sufficient. These will often include tasks that require attention to detail or that have a high risk of human or animal harm if performed incorrectly, such as performing surgery or working with hazardous agents.

Surv. Does your IACUC require routine periodic retraining of persons working with animals? Select all that apply.

- No 61/186 (33%)
- Required for Principal Investigator and research team 111/186 (60%)
- Required for animal care and veterinary staff 107/186 (58%)
- Required for IACUC members 82/186 (44%)

Surv. If your institution requires repeated training of animal users, how often does this retraining occur?

- Not applicable 38/187 (20%)
- Every other year 10/187 (5%)
- Every third year 69/187 (37%)
- Every fourth year 2/187 (1%)
- Every fifth year 9/187 (5%)
- Less frequent than every fifth year 4/187 (2%)
- Varies with the species or procedures to be performed 44/187 (24%)
- Other 11/187 (6%)

24:21 Should retraining be utilized as remediation for instances of protocol noncompliance?

Opin. Retraining can be an effective tool in addressing gaps in knowledge, understanding, or skills. However, protocol non-compliance can be caused by a combination of many factors, including poor or insufficient training, inadequate oversight and staff management or other factors. While training can be an effective adjunct in many of these cases, it may not be sufficient alone when addressing non-compliance. It is important to assess each non-compliance situation to determine the potential causes and match remediation efforts to the needs of each unique situation. However, if common mistakes are noted among multiple personnel, these could be incorporated into onboarding training, or routinely scheduled retraining.

Surv. Does your IACUC require research personnel to undergo retraining as remediation for instances of protocol non-compliance?

- Not applicable 6/187 (3%)
- Yes, occasionally 80/187 (43%)
- Yes, frequently 88/187 (47%)
- No 4/187 (2%)
- I don't know 4/187 (2%)
- Other 5/187 (3%)

24:22 Which categories of personnel should provide animal care and use training?

Opin. Institutions utilize a variety of personnel to provide training, including veterinarians, veterinary technicians, training personnel, laboratory animal technicians and technologists, and research personnel. No doubt the choice of individuals to fill these roles is affected by the size of the institution, its mission (academic, industrial, etc.), the organization of the IACUC and animal care services, the physical resources, the animal population and species maintained, and the training and skill of personnel (*Guide*, p. 15).

Surv. At your institution, who provides training courses in the care and use of laboratory animals? Select all that apply.

- Not applicable. We have no training program. 1/187 (1%)
- An outside private company. 18/187 (10%)
- The veterinary staff, including veterinarians and/or veterinary technicians. 158/187 (84%)

- The animal care staff, including animal care technicians, supervisors, and/or managers. 141/187 (75%)
- The IACUC staff. 54/187 (29%)
- Research personnel. 111/187 (59%)
- Designated training program personnel. 75/187 (40%)
- Other institutional personnel. 17/187 (9%)
- Other* 14/187 (7%)

*Other responses included:

- Use of web-based training

In the survey, veterinary personnel (84%), animal care personnel (75%), and research personnel (59%) were most often enlisted to provide training. A significant number of institutions (40%) have designated training program personnel. This is the case at the authors' institution where a Training Core has been established that is responsible for training of all animal care and animal research personnel.[5] The category "Other" included a wide variety of responses such as online resources (e.g., AALAS Learning Library and the Collaborative Institutional Training Initiative [CITI]), PIs, and outside consultants.

24:23 What departments or individuals should be responsible for providing training of personnel who are potentially exposed to hazards as part of their job duties?

Opin. Many individuals from existing groups at an institution, such as the IACUC staff, animal care program staff, and members of the research team, may participate in providing training to personnel exposed to hazards. However, personnel in these areas are not experts or authorities in occupational health issues, which may lead to unsafe practices due to gaps in training. As a result, development and guidance of the training program should be the responsibility of, or in partnership with, the institution's Occupational Safety and Risk Management Office (or its equivalent). However, monitoring and assuring the effectiveness of the program is the IACUC's responsibility.

Surv. Who is responsible for providing training of personnel potentially exposed to hazards? Select all that apply.

- Not applicable 2/187 (1%)
- The IACUC 14/187 (7%)
- The Principal Investigator 118/187 (63%)
- Animal facility personnel 64/187 (34%)
- The Occupational Safety and Risk Management Office (or its equivalent) 139/187 (74%)
- Other* 13/187 (7%)

*Other responses included: some respondents indicated this to be a responsibility of the IBC.

24:24 What subject matter should be included in animal care and use training programs?

Reg. See 24:1.

Surv. Which of the following topics are included in your training and education programs for animal users? Check as many boxes as appropriate.

- Not applicable, no training and education program in place 2/187 (1%)
- Purpose and general functions of the IACUC 171/187 (91%)
- Organizational structure of the institutional animal care and use program 142/187 (76%)
- The politics and ethics of animal experimentation 131/187 (70%)
- The 3 Rs and means of implementing them 171/187 (91%)
- Overview of the occupational health and safety program 160/187 (86%)
- Experimental endpoints and humane endpoints 161/187 (86%)
- Overview of institutional policies and procedures 143/187 (76%)
- Additional information on anesthesia and analgesia (if appropriate to your program) 146/187 (78%)
- Acceptable euthanasia methods 162/187 (87%)
- Record keeping requirements 143/187 (76%)
- How to report concerns about animal care and use 172/187 (92%)
- How and when to request veterinary help 146/187 (78%)
- Basic care and use of common laboratory animals 160/187 (86%)
- Basic surgery information (if applicable to your program) 132/187 (71%)
- Hands-on training in basic research techniques (e.g., animal handling, injections) 149/187 (80%)
- Hands-on training in surgical techniques 114/187 (61%)
- I don't know 0/187 (0%)
- Other* 11/187 (6%)

*Other responses included: Management of breeding colonies, the protocol submission process, and conflict of interest as additional training topics.

The survey results indicate several common themes that should be included in institutional training programs, namely institution-specific information on the IACUC, the animal care and use program, and policies; the 3Rs and humane endpoints; the occupational health and safety program; experimental procedures (anesthesia and analgesia, euthanasia; methods); basic care and use of common species; the provision of veterinary care; and reporting concerns. Hands-on training in basic research techniques is provided at the majority of institutions and a growing number of institutions (61%) also provide hands-on surgical training.

24:25 What type of information might the IACUC consider as essential to basic required training?

Opin. Basic training will vary somewhat from one institution to another, but certain common elements will most assuredly be incorporated. These include historical perspectives on animal use, the current public perspective of animal research, animal care and use regulations and policies, the importance of humane use of animals, institutional policies and procedures, the institutional animal care and use application, the institutional occupational health and safety program, and the procedures for registering concerns about animal use. Many institutions also include species-specific information on the basics of normal and abnormal behavior, signs of illness, and research techniques. As noted in 24:1, the AWAR (§2.32,c,1–5) define specific training topics that must be included in the training program. Similarly, the *Guide* (pp. 15–17) includes topics that should be included in training of animal care and use personnel.

24:26 What information is useful for training of IACUC members?

Opin. The AWAR do not explicitly identify IACUC members as needing training, but the requirement for training of IACUC members can be inferred from §2.32,a, which states that all individuals involved in animal care, treatment, and use be qualified to perform their duties. The *Guide* (p. 17) states that IACUC members should be provided formal orientation training covering the animal care program, relevant legislation, regulations, guidelines and policies, animal facilities and laboratories, and processes or protocol and program review, as well as be provided opportunities for ongoing education as needed. Training on routine and minor procedures (handling, restraint, injections, etc.) might be valuable for some members who have limited experience with animals, but training on complex procedures such as surgery is probably not warranted. In addition to material provided by the institution, a variety of online resources are available such as the American Association for Laboratory Animal Science (AALAS) Learning Library, AALAS' IACUC Central, the Collaborative Institutional Training Initiative, and NIH/OLAW. NIH/OLAW offers online seminars on various topics. Attendance at meetings sponsored by national or regional organizations, such as AALAS's national and local branches, may be helpful. Public Responsibility in Medicine and Research (PRIM&R) is a national organization that promotes ethical standards in research. They sponsor training for professionals in human and animal research protections, research ethics, and relevant regulations via both in-person meetings and webinars. IACUC 101, 201, and 301 are educational workshops sponsored by NIH/OLAW that provide training to IACUC members, administrators, veterinarians, animal care staff, researchers, regulatory personnel, and compliance officers. Scientists Center for Animal Welfare offers IACUC Training Workshops similar to IACUC 101. The IACUC Administrators Association sponsors Best Practice Meetings where topics are determined by the participants and varied institutional practices are discussed. The National Association for Biomedical Research periodically offers webinars to members on topics related to IACUC oversight and related regulation and legislation. The Federation of American Societies for Experimental Biology offers webinars and fact sheets about animal research. Americans for Medical Progress has a list of professional resources and position statements and provides public outreach training. In addition, numerous local and regional organizations focused on biomedical research and animal welfare may provide training opportunities, such as state AALAS branches, state veterinary medical associations, or state societies for medical research.

Surv. When training new IACUC members, what topics are covered? Check all appropriate responses. Select all that apply.

• We have no training program for new IACUC members.	4/187 (2%)
• Training is via on-the-job observations only. There is no formal program.	19/187 (10%)
• Animal use justification.	132/187 (71%)
• What constitutes an appropriate literature search.	99/187 (53%)
• Animal number justification.	128/187 (68%)
• Means of alleviating pain and distress.	120/187 (64%)
• Euthanasia techniques.	110/187(59%)
• Means of conducting semiannual inspections.	149/187 (80%)
• General policies of the IACUC.	164/187 (88%)
• Training and skill qualification standards for animal users.	88/187 (47%)
• Means of reporting and responding to concerns about animal welfare at our institution.	145/187 (78%)
• Expectations for housing and general care of animals.	103/187 (55%)
• Other*	22/187 (12%)

*Other responses included:

- Relevant regulatory expectations

- Principles of confidentiality
- Aspects of occupational health and safety
- Approaches to protocol review, IACUC meeting procedures, approaches to non-compliance

24:27 What strategies can be used to deliver training to animal care and use personnel?

Opin. Numerous strategies are available for training animal care and use personnel. Instructor-led training such as lectures and wet labs allow trainees to ask questions and instructors to provide corrections or tailor lessons in real time. The wet lab setting, where individuals or small groups are provided with hands-on training, provides optimal student-teacher interaction and is excellent for fostering skill development. This type of training is ideal for teaching animal care or research techniques such as handling, restraint, anesthesia, dose administration, blood collection, and surgery, as well as health assessment and operations such as cage changing. Real-time web-based training can be an effective means of teaching SOPs or practicing techniques that do not require the trainee to be present in the same place as the instructor, such as filling out surgical records or cage cards appropriately. On the other hand, these types of training strategies require scheduling ahead of time and require staff to be present. Self-paced learning completed by the student may be a more efficient use of time or personnel when presenting introductory material such as federal laws and national standards, institutional policies and procedures, the institutional animal care and use program, and institutional occupational health programs. For this type of material, pre-recorded videos, written material, and computer-based training may be more appropriate and efficient. This assertion is borne out by the results of the following survey, which demonstrate a range of training strategies at the surveyed institutions, most of which use a combination of training methodologies.

Surv. At your institution, what approaches are used to provide education and training to animal care and use personnel? Select all that apply.

- Not applicable. We have no training program. 1/187 (1%)
- We use live lectures (with or without slides). 97/187 (52%)
- We use live individual or small group hands-on training. 151/187 (81%)
- We subscribe to a commercial training program (e.g., CITI). 130/187 (70%)
- We use video presentations. 68/187 (36%)
- We use computer-based training (e.g., autotutorials). 83/187 (44%)
- We use written materials (e.g., handouts, institutional manuals, newsletters). 131/187 (70%)
- I don't know. 1/187 (1%)
- Other. 12/187 (6%)

24:28 What strategies can be used to organize training opportunities and communicate those to trainees?

Opin. A training program should be methodically structured for animal care staff and laboratory personnel to access and locate their training classes quickly. Training classes can be organized by species, audience, task, subject or training titles, or a combination of these organization methods. Personnel should have an easy way to (a) know what training they have assigned, and (b) be able to search for and find their training with minimal steps. Communication is often necessary to instruct animal program personnel on how to locate assigned training and on how to complete their classes and in what order. Communication can be through many channels, including targeted e-mails at various intervals, newsletters, webpages, posters within a facility, face-to-face instruction, or other forms of animal program communication.

Surv. How are the training courses structured and organized so the user can easily access them? Select all that apply.

- Training courses are categorized by species. 88/187 (47%)
- Training courses are categorized by learner. 57/187 (30%)
- Search tool using training class subjects and or titles. 60/187 (32%)
- Learning is automatically assigned with a learning management system. 67/187 (36%)
- Direct contact with training personnel is required to access training classes. 74/187 (40%)
- Training courses are not organized or structured. 14/187 (7%)
- I don't know. 6/187 (3%)
- We don't have a training program. 2/187 (1%)

Surv. How does the program determine and communicate training classes to different groups based on assignment and completion? Select all that apply.

- Training is assigned manually based on protocol review. 69/187 (37%)
- Training courses are assigned automatically as determined by planned activities. 49/187 (26%)
- Training courses are communicated via individual e-mail. 111/187 (59%)
- Training courses are communicated via mass e-mail. 42/187 (22%)

- Training courses are communicated using newsletters or posters in the vivarium. 27/187 (14%)
- Training courses are communicated by the training department. 33/187 (18%)
- A combination of these methods are used. 98/187 (52%)
- I don't know. 6/187 (3%)
- We don't have a training program. 5/187 (3%)

Surv. How does the training audience register for training courses? Select all that apply.

- The audience registers via a learning management system. 101/187 (54%)
- The audience registers for training classes via e-mail. 69/187 (37%)
- The training department registers the audience for training classes manually. 43/187 (23%)
- The audience contacts the IACUC office to register for training. 50/187 (27%)
- The audience contacts the training dept to register for training. 51/187 (27%)
- I don't know. 4/187 (2%)
- We don't have a training program. 5/187 (3%)

24:29 What technologies are available to support training assignment, provision, and documentation?

Opin. Technologies to support training programs are diverse and range from simple spreadsheets to complex learning management systems. In all cases, a plan must be in place for periodic review of the technology and any data captured by it to ensure the technology is meeting the needs of the institution. Microsoft Excel and Google Sheets can be useful for creating spreadsheets that help track coursework, but functions such as managing training assignments, training enrollment, and transcripts must still be performed manually by staff. Still, these offer the benefit of being readily accessible, inexpensive (except in personnel time), and relatively easy to use. While more expensive, learning management systems, such as AALAS Learning Library or CITI, offer many benefits, such as premade and customizable courses, assignment of training, automatic tracking of transcripts, and scheduling capability. Institutions may also have their own in-house learning management systems that the animal care and use program can use to assign training, allow for training enrollment, and provide completed training transcripts. In the case of large or complex training programs, a combination of technologies is likely to be used.

Surv. What kind of technology do you use at your institution to provide, assign, or document training? Select all that apply.

- Learning Management System (LMS) 92/187 (49%)
- Microsoft Excel sheets 88/187 (47%)
- Microsoft Access database 12/187 (6%)
- Microsoft Word 32/187 (17%)
- Google sheets 11/187 (6%)
- Other 45/187 (24%)
- I don't know 8/187 (4%)
- No training documentation system in place 5/187 (3%)

24:30 What strategies can be used to design effective training programs?

Opin. Many strategies can be utilized to design effective training. Organized training programs typically have various class types, including online classes, virtual classes, in-person workshops or wet labs, just-in-time training classes, etc. Well-designed training classes should be learner-focused. Learner-focused strategies for developing and designing training classes typically begin with development of training topics, goals, and associated objectives to fulfill training needs and assess learning throughout the class and beyond. Various instructional design models can be utilized to provide structure to the development of effective learner-center educational programs.

Surv. What data collection methods are used to identify knowledge, skill, or ability gaps that determine training needs? Select all that apply.

- Training evaluation surveys. 27/187 (14%)
- Discussion with personnel and supervisors. 108/187 (58%)
- Technical assessments of trainee. 85/187 (45%)
- Quizzes or tests. 50/187 (27%)
- On the job evaluations. 83/187 (44%)
- IACUC non-compliance data. 85/187 (45%)
- I don't know. 7/187 (4%)
- We don't use any data collection methods to identify knowledge, skill, or ability gaps. 30/187 (16%)

Surv. What tools and methods do you use to design and develop training? Select all that apply.

- Student evaluation surveys. 16/187 (9%)
- Instructional design models. 20/187 (11%)
- Google sheets. 3/187 (2%)
- PowerPoint. 83/187 (44%)
- Learning Management systems. 60/187 (32%)
- Articulate/captivate. 17/187 (9%)
- Other learning design tools. 11/187 (6%)
- I don't know. 12/187 (6%)
- We don't use any tools or methods for designing training. 48/187 (26%)

Surv. Do you use instructional design tools to help develop training? Select all that apply.

- ADDIE Model. 4/187 (2%)
- Backwards design model. 4/187 (2%)
- Bloom's Taxonomy of learning objectives. 5/187 (3%)
- Other* 10/187 (5%)
- I don't know. 24/187 (13%)
- We don't use an instructional design model. 146/187 (78%)

"Other" responses included:

- Canvas
- ICARE Train the Trainer (active learning techniques)
- PEAK

Surv. How are training goals and measurable objectives created when developing a training course? Select all that apply.

- Discussions with lab managers/animal care managers/supervisors 115/187 (61%)
- Utilizing other learning resources 72/187 (39%)
- Utilizing the outcome/result of training to develop goals and objectives 88/187 (47%)
- Not applicable; no training program in place 8/187 (4%)
- Other* 10/187 (5%)
- I don't know 26/187 (14%)

*Other responses included:

- Based on Guide requirements.
- DOPS (directly observed procedural skills)
- No specific goals or measurable objectives are used.

24:31 What are effective strategies for the development of training personnel?

Most subject matter experts in the field of laboratory animal care and use do not have expertise in pedagogy or education. Training programs are most effective when the personnel developing and providing educational programs understand both subject matter and educational programs design and delivery. The onboarding, development, and continuing education of training personnel should include provision of tools for effective training, training delivery, and educational class design and development. These opportunities can include supervision, mentorship, self-study, and assessment, as well as formal programs, certifications, and conferences designed to provide training in educational methodologies.

Surv. What onboarding or continuing education resources are available for personnel that provide training to others? Select all that apply.

- Structured onboarding training program for new trainers. 45/187 (24%)
- Resources for instructional design and educational practices. 45/187 (24%)
- Mentorship from experienced trainers. 94/187 (50%)
- Assessment of training techniques. 71/187 (38%)
- Other. 14/187 (7%)
- None. 13/187 (7%)
- I don't know. 29/187 (16%)
- We don't have training personnel in our program. 18/187 (10%)

Surv. How are training personnel evaluated to ensure consistency and effectiveness in training delivery and methods? Select all that apply.

- Evaluation or assessment by manager/supervisor. 106/187 (57%)
- Evaluation or assessment by peers. 47/187 (25%)
- Students evaluate the trainers via anonymous post- training surveys. 19/187 (10%)
- Pre- and post-tests provided to students 17/187 (9%)
- Other. 20/187 (11%)
- I don't know. 36/187 (19%)
- We don't have a training program. 12/187 (6%)

24:32 How should the IACUC inform the research community of changes in regulatory guidance, institutional policies, and standards of practice?

Opin. While changes in regulations happen slowly, changes in IACUC or veterinary and animal husbandry and care policies and guidelines may occur frequently. Communication of these changes in policies, practices, or expectations of animal users is essential for animal care program compliance. It can be difficult to communicate changes to the research community. Practices employed often include e-mail, newsletters, websites or town hall meetings, required continuing education, protocol renewals, or post-approval monitoring visits.

Surv. How does your IACUC provide updates to animal care and use personnel on changes in regulatory guidance, institutional policies, and standards of practice? Select all that apply.

- We do not provide any such updates. 5/187 (3%)
- We have required retraining or continuing education classes. 60/187 (32%)
- We use e-mail, newsletters, website, and town hall meetings. 150/187 (80%)
- We provide this information at post-approval monitoring visits. 82/187 (44%)
- We provide this information when protocols are being renewed. 104/187 (56%)
- I don't know. 5/187 (3%)
- Other. 16/187 (9%)

24:33 How should training be documented and what data should be maintained?

Reg. While the PHS Policy does not require that training records be maintained, the *Guide* (p. 15) states that all program personnel training should be documented.

Opin. It is important to maintain training records to demonstrate compliance with the responsibility of the IACUC to ensure that personnel are trained and qualified. Logically, it is difficult to meet these obligations without a mechanism to track the training and experience of individuals engaged in animal research. The most common methods of documentation are filing the sign-in sheets from training classes, maintaining an electronic database of personnel training. or utilizing a learning management system to document training completion. The most common types of records maintained include the personnel identification and organizational affiliation, class dates, class name or type, topics reviewed during training, and the trainer's identity. Documentation and scores of class pre- and post-tests, class quizzes and tests, and scored skill competency and proficiency may enhance the training documentation and can be used to discover knowledge, skill, and ability gaps.

Training data is also important to provide to the IACUC. Training activity metrics and evaluations capture trends to inform IACUC, and the IACUC makes decisions regarding training resources and requirements based on these metrics. Data collected may include items such as training completion data, learner evaluation of training, number of people who completed and passed competency or proficiency training, the pre- and post-test comparison scores, test score averages, and the number of classes which are offered to the research and animal care community. Having an effective, efficient form of data collection can also be important when evaluating annual trends.

Surv. How, and to what extent, does your institution document training and education efforts? Select all that apply.

- We do not document our training and education efforts. 7/187 (4%)
- We maintain the sign-in sheets from training classes. 105/187 (56%)
- We maintain electronic records from online training. 149/187 (80%)
- We maintain an electronic database of personnel training records. 131/187 (70%)
- We provide trainees with certificates of training. 73/187 (39%)
- Other* 4/187 (2%)

*Other responses included: Documentation via digital or hard-copy methods.

Surv. What records does your institution keep relative to training and education of animal use personnel? Select all that apply.

- Not applicable as we do not keep such records. 3/187 (2%)
- We maintain the person's name and organizational affiliation. 128/187 (68%)
- We maintain the names of classes or training sessions. 154/187 (82%)
- We maintain the dates of training sessions. 153/187 (82%)
- We maintain the names of the trainers. 88/187 (47%)
- We maintain a description of each person's training and experience in the IACUC approved protocol. 99/187 (53%)
- Other. 7/187 (4%)

Surv. What kind of data is captured regarding the training program? Select all that apply.

- Training effectiveness evaluation data 28/187 (15%)
- Pre- and post-training evaluation or test data 22/187 (12%)
- Training attendance or completion records 155/187 (83%)
- Proficiency or competency assessments 78/187 (42%)
- Total number of classes taught 40/187 (21%)
- Total number of classes taught by each method (online, in person, proficiency) 33/187 (18%)
- Total time spent on training activities 23/187 (12%)
- Total number of training classes per trainer 17/187 (9%)
- Total number of classes or time required versus requested training activities 8/187 (4%)

24:34 What resources are available for provision and documentation of training for animal users?

Opin. The most common methods for documenting and/or providing training include learning management system software, general database software, protocol management software, and web-based training management programs from an external vendor or organizations.

Surv. Does your institution utilize any of the following to document or provide training to animal users? Select all that apply.

- Learning management system software that is supported by your institution and provides class registration, online web modules, transcripts, and training documentation 64/187 (34%)
- Protocol management software that can track and document a person's training 71/187 (38%)

- Web-based training management programs from an external vendor or organization (e.g., AALAS learning Library, CITI) 141/187 (75%)
- General database software such as Microsoft Excel, Access, etc. 84/187 (45%)
- No use of any of the above methods 15/187 (8%)

24:35 Should the IACUC be provided with regular reports on training activities?

Opin. The PHS Policy and the AWAR do not specifically require reports on training activities. However, the *Guide* (p. 15) places the responsibility for providing oversight and for evaluating the effectiveness of the training program with the IACUC. Therefore, it would be important to provide training reports to the IACUC at regular intervals, perhaps as part of the semiannual program evaluation. Even if training information is incorporated into the protocol, summary reports to the IACUC could be very useful in evaluating training effort and personnel staffing levels. Training documentation is not only important for tracking training activities, but also for reporting on training activity metrics for the IACUC. Metrics can include instances of training, number of users identified as being assigned training, number of classes completed within a time period, and others. Learning management systems and other electronic database systems collecting this data can be used to provide these metrics. The reports can also be utilized to provide the IACUC with a summary of the training classes, topics covered within each, and how each training class is utilized to enhance the knowledge and skills of the learner. According to the survey, most institutions provide some type of reports to their IACUCs.

Surv. What documentation of training activities must be provided to the IACUC? Select all that apply. Select all that apply.

- We do not provide reports of training activities to our IACUC. 41/187 (22%)
- We provide the IACUC with a summary report on training activities as part of the semiannual report to the Institutional Official. 40/187 (21%)
- We provide reports on training activities only upon request from the IACUC. 60/187 (32%)
- We update the IACUC at its regular meetings. 41/187 (22%)
- Other. 32/187 (17%)

24:36 What are some effective ways to provide and document continuing education?

Reg. The *Guide* (pp. 16–17) recommends that institutions provide opportunities and support for continuing education for animal care and use personnel.

Opin. Continuing education is a wise investment for institutions because it improves personnel engagement in the institution's success, improves animal welfare, and improves employee morale. Continuing education opportunities can be provided by existing staff (trainers, supervisors, veterinarians, etc.), by institutional human resource and development personnel, by outside trainers, or by off-site training sessions at local, regional, or national meetings. Continuing education can be offered through online classes, refresher classes, SOP reviews, staff meetings, and in-person training classes. Specific animal technical skills practice beyond basic training requirements can also be utilized as continuing education events.

24:37 What forms of assessment can be used to evaluate personnel that have received training?

Opin. Some form of assessment is necessary to ensure that personnel have sufficient mastery of the procedures that they will be performing such that the welfare of animals being used will not be compromised. Evaluation methods and strategies can be utilized to provide a comprehensive picture of an individual's competency in performing various animal procedures. Formative assessments are usually provided during training and can capture knowledge and skill retention. These formative assessments may include online tests/quizzes, knowledge checks, demonstration of procedures during training, and pre- and post-training tests/quizzes. Other assessments, sometimes called summative assessments, may include on-the-job evaluation of procedures, written tests, post-approval monitoring, observation by other appropriately trained personnel, and utilizing standardized scoring rubrics while observing and monitoring procedures. Other more passive methods of evaluation may include reviewing noncompliance data, laboratory notebooks, training records, and semiannual facility inspection notes.

Surv. As part of the training for your animal care and use program, is any form of evaluation, assessment, or testing utilized? Select all that apply.

- Not applicable. We have no training program. 8/186 (4%)
- No, testing is not required as part of our training program. 28/186 (15%)
- Yes, written examinations. 48/186 (26%)
- Yes, demonstration of competency during training sessions. 104/186 (56%)
- Yes, worksite evaluation of competency by training or IACUC personnel. 64/186 (34%)
- Yes, worksite evaluation of competency by supervisory personnel. 89/186 (48%)
- Other. 13/186 (7%)

Surv. How do you assess an audience's retention of learning? Select all that apply.

- Formal assessments 40/187 (21%)
- Pre- and post-tests 39/187 (21%)
- Formative and summative assessments 15/187 (8%)
- Informal assessments such as in-person discussions 81/187(43%)
- On the job assessments 101/187 (54%)
- Discussions with supervisor 54/187 (29%)
- Other* 15/187 (8%)
- I don't know 13/187 (7%)
- No training program in place 11/187 (6%)

*Other responses included:

- Non-compliance data
- Post-approval monitoring results

24:38 How can training versus qualifications be effectively evaluated by the IACUC?

Reg. The *Guide* (p. 16) and the PHS Policy (IV,C,1,f) state that research personnel should have the necessary knowledge and expertise for the specific animal procedures proposed and the species used. The AWAR (§2.32,a) require that personnel be qualified, and that training can be used, in part, to ensure the necessary qualifications.

Opin. In order for the IACUC to evaluate personnel training and qualifications, investigators should provide the IACUC with specific information on their prior experience and training with the species and procedures proposed in their protocol as well as the experience and training of their staff members. The IACUC then should determine, in the course of reviewing the investigator's protocol, whether the research personnel have sufficient training or experience to conduct the proposed procedures. For example, an investigator may be trained as a human surgeon but, without specific training or experience in animal surgery, may not be qualified to perform surgical procedures in animals. Such an individual may need additional training on species-specific anatomy, physiology, behavior, anesthesia, and analgesia. Prior training or experience of the research staff can also be taken into consideration.

When semiannual inspections of animal care and use facilities are performed, the IACUC also can gain some insight into the adequacy of training by interviewing research personnel during these inspections. This approach not only identifies inadequacies in training after protocol approval has taken place, but also can be useful as an audit of the adequacy of the review process in identifying personnel in need of training. Additionally, institutions may require newly added personnel to a protocol to complete training regardless of prior experience at other institutions. Some performance standards, guidelines, and policies may differ, and it is an opportunity to observe and provide feedback to ensure their qualifications are adequate (see 24:39).

24:39 How can the IACUC determine when an individual requires additional training?

Opin. Ideally, additional training needs would be identified during basic or primary training, formative or workplace assessments. However, in practice it can be difficult to predict the competency of personnel in the workplace when they leave the training environment. Often inappropriate work practices are noted during laboratory inspections, post-approval monitoring visits, and these observations can result in the assignment of additional training requirements. Likewise, additional training requirements may be assigned as remediation for protocol noncompliance or animal use concern reports to the IACUC.

Surv. What criteria are used by your institution to indicate to the IACUC that an individual requires more training? Select all that apply.

- Not applicable 7/186 (4%)
- Poor performance on peri- or post-training evaluation process (e.g., tests, competency evaluation, worksite assessment) 93/186 (50%)
- Supervisor or PI recommendations 123/186 (66%)
- Non-compliance reports to the IACUC 156/186 (84%)
- I don't know 3/186 (2%)
- Other* 9/186 (5%)

- *Other responses included:
- Self-identified needs and requests.
- Issues, including adverse events, that are not strictly non-compliances.
- Observations during post-approval monitoring.

24:40 What metric(s) can be used to evaluate the effectiveness of a training program?

Reg. The AWAR place responsibility on the research facility to ensure that all personnel are qualified to perform their duties and to provide training and instruction and review the qualifications of personnel with sufficient frequency to fulfill this responsibility (9 CFR §2.32,a; §2.32,b). The *Guide* (p. 15) places the responsibility for providing oversight and for evaluating the effectiveness of the training program with the IACUC. Further, because the training program is part of the facility's animal care and use program, it must be evaluated semiannually, and personnel qualifications must be reviewed as part of the IACUC's process for reviewing and approving animal use activities (AWAR §2.31,c,1; §2.31,c,3; §2.31,d,1,viii; PHS Policy IV,B,1).

Opin. Evaluation of training should occur at several levels extending from evaluations of the training course by attendees, evaluating trainee knowledge and skills during or immediately after classes, and evaluating the effects of long-term application and change in personnel behaviors and skills. Effective long-term behavior change is difficult to instill and difficult to evaluate. This requires routine post-approval monitoring and/or quality assurance programs and personnel dedicated to performing this oversight. Additionally, appropriate workplace practices are affected by education and training, personnel management and oversight, and institutional culture. It is difficult to separate the effects of all of these issues to gain a final conclusion on the effectiveness of training. Data captured throughout the training process can include class evaluation scores, test scores, and post-training procedural assessment scores. Collection of this information can be utilized to assess quality and effectiveness or course and areas for improvement.

Surv. What metric(s) are used to evaluate the effectiveness of your training program? Select all that apply.

- Not applicable 13/186 (7%)
- We do not use any metrics 73/186 (39%)
- Data or reports from our post-approval monitoring assessments 70/186 (38%)
- Data or reports from worksite assessments by trainers, IACUC, animal care and veterinary staff 61/186 (33%)
- Data or reports from supervisors and Principal Investigators' assessments of trainees 37/186 (20%)
- Data from compliance reports to the IACUC 74/186 (40%)
- Other* 4/186 (2%)

*Other responses included: results of AAALAC, international site visits, and evaluations by the training core as means to evaluate the effectiveness of training.

Surv. How should the animal program evaluate the return on investment (ROI) for the training program and the trainers? Select all that apply.

- Assessing rates of incidence of compliance or animal welfare concerns 152/186 (82%)
- Student assessment of training value 54/186 (29%)
- Pre- and post-training assessment activities 79/186 (42%)
- Cost and effort comparison of training versus non-compliance management 52/186 (28%)

Acknowledgment

In preparation of this chapter for this fourth edition, the authors wish to acknowledge Howard Rush for his efforts in the preparation of Chapter 24 for the third edition of this text

REFERENCES

1. US Department of Agriculture. Animal Welfare Act (Public Law 89-544). 1966 (Last amended 2023).
2. National Research Council. 2011. *Guide for the Care and Use of Laboratory Animals.* Washington, DC: National Academy Press.
3. National Institutes of Health, Office of Laboratory Animal Welfare. 2015. *Public Health Service Policy on the Humane Care and Use of Laboratory Animals.* Bethesda, MD: U.S. Department of Health and Human Services.
4. US Department of Agriculture. Animal Welfare Regulations 9 CFR (US Department of Agriculture, 2023).
5. Dyson, M.C. and H.G. Rush. 2012. Institutional training programs for research personnel conducted by laboratory-animal veterinarians. *J. Vet. Med. Educ.* 39:160–168.
6. Carbone, L. 2021. Estimating mouse and rat use in American laboratories by extrapolation from Animal Welfare Act-regulated species. *Sci. Rep.* 11:493. https://doi.org/10.1038/s41598-020-79961-0

25

Disaster and Emergency Response Planning

Kevin Prestia, JoAnn Henry, and Jennifer Pullium

Introduction

Disaster preparedness can be a formidable task that requires a holistic approach to ensure continued operations in the face of a disaster or emergency. Understanding the interactions and partnerships within groups is essential to maintaining core functions. There is no standard, "one size fits all plan," but providing a framework that incorporates core concepts can help institutions become better equipped to assess and respond to these adverse incidents. While most institutions focus on the role of the veterinarian and animal care staff when developing a disaster plan, the IACUC members and support staff can be a valuable resource for ensuring compliance in the event of a disaster. This assistance can come in the form of background information for writing the plan as well as direct assistance with investigators during the disaster recovery process.

25:1 What is a crisis, disaster, or emergency?

Reg. According to the Canadian Council on Animal Care (CCAC), the term "crisis" has been defined as any unplanned event which triggers a real, perceived, or possible threat to the life, health, or safety of animals and/or personnel, or to the institution's credibility.[1] According to the *Guide for the Care and Use of Laboratory Animals* (*Guide*), disasters are "unexpected conditions that result in the catastrophic failure of critical systems or significant personnel absenteeism, or other unexpected events that severely compromise ongoing animal care and well-being."[2] While other agencies and organizations, such as the U.S. Department of Agriculture (USDA), Office of Laboratory Animal Welfare (OLAW), and AAALAC International provide expectations regarding response planning, they do not specifically define what constitutes a crisis or emergency. The USDA gives examples of circumstances that might invoke implementation of a contingency plan (https://www.aphis.usda.gov/sites/default/files/ac-tech-note-guide-contingency-plans-emergencies.508.pdf).[3] OLAW defines adverse events as those unexpected incidents that lead to harm or endanger the well-being of animals and humans at a research facility. OLAW considers large-scale damage to the area, the institution, or the entire animal facility to be an extensive adverse event, and events that cause damage to a specific area with an animal facility to be a contained adverse event.[4] An emergency, as defined by the Stafford Act, is any occasion or instance for which, in the determination of the president, federal assistance is needed to supplement state and local efforts and capabilities. As used in the Stafford Act, a "major disaster" is "any natural catastrophe ... or, regardless of cause, any fire, flood, or explosion, in any part of the United States, which in the determination of the President causes damage of sufficient severity and magnitude to warrant major disaster assistance under the Act to supplement the efforts and available resources or States, local governments, and disaster relief organizations in alleviating the damage, loss, hardship, or suffering caused thereby."[5]

Opin. Crisis, disaster, and emergency definitions may be established on the basis of the number of animals that may be affected and the magnitude of loss or damage to the facility. "Disaster," "incident," and "contingency plan" can be defined according to dictionary definitions; "an event causing great harm, damage or suffering," "an event that is either unpleasant or unusual," and "a plan that is made for dealing with an emergency, or with something that might possibly happen and cause problems in the future," respectively.

25:2 What is the regulatory requirement and guidance for a research institution to develop a disaster/contingency plan?

Animal Welfare Act – USDA, Animal and Plant Health Inspection Service (APHIS)[6]

Reg. The Animal Welfare Regulations require research facilities to develop, document, and follow an appropriate plan to provide for the humane care, treatment, transportation, and housing, and care of animals in the event of an emergency or disaster. Although "emergency or disaster" is not defined, the plan must address emergencies or disasters that can reasonably be anticipated and expected to be detrimental to the health and well-being of the animals in the research facility's possession. Contingency plans must list situations that would

DOI: 10.1201/9781003272861-25

trigger the need for the measures identified in a contingency plan to be put into action, outline specific tasks required to be carried out in response to specific emergencies or disasters, and identify a chain of command, including who will be responsible for fulfilling the listed tasks. The plan must address response and recovery efforts, highlighting the materials, resources, and training required, and list by name or position title the individual(s) responsible for fulfilling the specific tasks outlined in the contingency plan.

Training for employees regarding their roles, as detailed in the plan, must be provided. Research facilities must review the contingency plan on at least an annual basis and maintain documentation of their annual reviews, including documenting any modifications made to the plan during the previous year. Any substantive changes to the plan as a result of the annual review must be communicated to employees through training, conducted within 30 days of making the changes. Contingency plans, and all annual or more frequent review documentation, must be made available to APHIS and any funding federal agency representative.

PHS Policy on Humane Care and Use of Laboratory Animals[7]

Reg. No activity involving animals may be conducted or supported by the PHS until the institution conducting the activity has negotiated a written Assurance acceptable to the PHS. The Assurance fully describes the institution's program for the care and use of animals in PHS-conducted or PHS-supported activities and requires institutions to use the *Guide for the Care and Use of Laboratory Animals* (*Guide*) as a basis for developing and implementing an institutional program for activities involving animals. OLAW FAQ G3 "Do awardee institutions need animal facility disaster plans?"[8] details requirements and considerations for the program in developing a comprehensive and effective plan.[7]

Opin. The *Guide (p. 35)* states that facilities must have a disaster plan, and the well-being of animals and personnel during unexpected events should be considered as disaster plans are developed by institutions.[2] OLAW provides a Disaster Planning and Response Resources webpage (https://olaw.nih.gov/resources/disaster-planning.htm)[9] to assist institutions in planning and responding to natural and other disasters affecting animal facilities. In addition, OLAW FAQ F6 "How can institutions assure animal welfare when HVAC systems malfunction or fail?"[10] confirms that the institution has a responsibility to ensure animal welfare and is obligated to protect federal investments in PHS-supported research, testing, or training. OLAW addresses adverse events and defines these as unexpected incidents that lead to harm or endanger the well-being of animals and humans at a research facility. Conditions that jeopardize the health or well-being of animals, including natural disasters, accidents, and mechanical failures resulting in harm to animals, are reportable to OLAW (https://olaw.nih.gov/faqs/#/guidance/faqs).

***Guide for the Care and Use of Laboratory Animals*, eighth edition**[2]

Reg. A disaster plan that considers both personnel and animals should be prepared as part of the overall safety plan for the animal facility. The *Guide* notes that all facilities are subject to unexpected conditions that compromise ongoing animal care and well-being, and, as such, animal facilities must have a disaster plan. The plan should address the preservation of animals that are necessary for critical research or irreplaceable, and the plan should define the actions necessary to prevent animal pain, distress, and death. Responsible investigator(s) should be engaged during disaster plan development, balancing the priorities for triaging animal populations and the institutional needs and resources, but realizing that animals that cannot be relocated or protected must be humanely euthanized. The disaster plan should identify trained, essential personnel and ensure personnel safety, realizing that special facilities or operations may require additional detail and training in emergency procedures. The institution should approve the disaster plan, incorporate it into the overall institutional disaster response plan, and provide a copy to law enforcement and emergency personnel. A communication strategy that ensures coordination between institutional security personnel and fire or police officials should be outlined in the plan, and posting emergency procedures, names, and/or telephone numbers at critical locations has been recommended.

4th edition of the *Guide for the Care and Use of Agricultural Animals in Research and Teaching* (*Ag Guide*)[11]

Reg. The *Ag Guide* requires the development of a site-specific emergency plan for observing animals and providing care during emergency weather or health situations to care for agricultural animals. Unless the conditions are so unusual and extreme, the plan should ensure proper management and care for animals regardless of the conditions and define proper animal management and care and parameters to ensure employee safety. Consistent with the regulatory guidance from APHIS, emergency plans must name employees or positions that are considered essential for providing animal management and care.[3] Emergencies that occur in the geographic area of the research facility should be addressed in the plan and should include animal evacuation plans specific to the research.

Canadian Council on Animal Care[1]

Reg. The CCAC policy statement, Terms of Reference for Animal Care Committees, states that the institution and its Animal Care Committee must have in place "a crisis management program for the animal facilities and for the animal care and use program, in conjunction with any general institutional crisis management plan(s)." Plans should address power outages, work stoppages, incidents that interfere with the ability to come to work, fires, natural disasters, and large chemical spills, and they should include a communications plan for addressing public and media inquiries about animal use. Elements of the program should include a crisis management team, and a communications plan that includes emergency contact information and an effective strategy to communicate information effectively and efficiently to members of the institution and the public.[1]

Directive 2010/63/EU[12]

Opin. There is no statement from the European Parliament and of the Council of 22 September regarding emergencies, disasters, and contingency planning. The only mention in the Directive is that clear instructions on emergency procedures shall be prominently displayed.

AAALAC, International

Opin. As per the AAALAC, International website, in 2011 AAALAC International's Council on Accreditation began using three primary standards to evaluate animal care and use programs. These reference materials include the Eighth Edition of the *Guide*, the *Ag Guide*, and the *European Convention for the Protection of Vertebrate Animals Used for Experimental and Other Scientific Purposes.* As described previously in this chapter, the *Guide* and the *Ag Guide* require and provide extensive guidance on disaster and emergency planning.

25:3 How can institutions develop a disaster or emergency preparedness plan to maximize capabilities and minimize risk?

Opin. The National Preparedness Goal developed by the Federal Emergency Management Agency (FEMA) is the cornerstone for the implementation of Presidential Policy Directive 8 and identifies five mission areas: prevention, protection, mitigation, response, and recovery, in creating a framework for organizing preparedness activities.[13]

- Prevention refers to the capabilities necessary to avoid, prevent, or stop a threat or hazard.
- Protection refers to the capabilities necessary to secure against acts of terrorism and manmade and natural disasters.
- Mitigation refers to the capabilities necessary to reduce loss of life and property by lessening the impact of disasters.
- Response refers to the capabilities necessary to save lives, protect property and the environment, and meet the basic human needs after an incident.
- Recovery refers to the capabilities necessary to restore and rebuild the community following a disaster.

A comprehensive plan identifies and assesses risks, identifies capabilities to address these risks, is integrated with broader institutional and governmental plans, incorporates preparedness training to optimize capabilities, and is subject to ongoing review and modification. Continuity of operations, although not explicitly defined in the national planning framework, should be addressed in the contingency plan to ensure that mission essential activities continue to be performed during a wide range of emergencies. Plans may include communication strategies, alternate worksites and animal facilities, stockpiled supplies, alternate supply chains, and alternate sites for critical animal lines and documents.

Surv. Which of the following are included in your emergency response plan? Select all that apply.

• Protocols for the management of animals exposed to research hazards	141/189 (75%)
• Protocols for entering areas where research hazards may be present or stored	141/189 (75%)
• Protocols for dealing with hazardous equipment	107/189 (57%)
• Protocols for emergency responder interface	140/189 (74%)
• None of the above	14/189 (7%)

Surv. How does the institutional leadership demonstrate commitment to the emergency management program at your institution? Select all that apply.

• Existence of an Emergency Management, Enterprise Resilience Group, or similar entity.	104/185 (56%)
• Institution-wide training activities.	76/185 (41%)
• The animal program's disaster/ emergency plan is included in the overall institutional emergency plan.	149/185 (81%)
• Personnel from the Animal Care Program are participants in the Incident Command Center.	93/185 (50%)
• None of the above.	11/185 (6%)
• Other.	3/185 (2%)

Surv. Please select the external entities that your institutional contingency plan is shared with, so that if the need arises, they may assist with recovery efforts. Select all that apply.

- Local police 138/185 (75%)
- Local fire 126/185 (68%)
- Other local municipal departments 53/185 (29%)
- None of the above 37/185 (20%)

Surv. Who at your program receives training on how to deal with the press and public? Select all that apply.

- Specific individuals within a Communications or Media Relations type office 152/185 (82%)
- Institutional Official / CEO 94/185 (51%)
- Attending Veterinarian / Veterinary Staff 36/185 (19%)
- IACUC Chair, IACUC members, or IACUC administrative staff 32/185 (17%)
- Other senior leadership (e.g., Provost, President, Dean) 59/185 (32%)

25:4 What is the role of the recovery phase of emergency management, and when does it occur?

Opin. The recovery phase of emergency management begins immediately after the response to the adverse event, once critical function is restored and stability is achieved or imminent. The recovery phase may occur concurrent with the response phase. The recovery phase ends when normality or steady state has been achieved even when different from the pre-emergency condition. Ideally, an end to the emergency is declared based on evaluations, using agreed standards and communicated to the stakeholders. The post-event analysis should then begin.

The individual divisions within the program, as well as the institution, have different perspectives and priorities in both acute and long-term stages of response and recovery. Therefore, it is as important to ensure the contingency plans of divisions are aligned with the institutional plan as it is to ensure the plenary plan has considered all of the divisions. As an addition, or when lacking a plan, diligence toward communication practices will aid recovery efforts. The contingency plan should include a process for a post-incident analysis, based on the nature and severity of the incident, as the final part of the recovery process. If not performed immediately after, individuals should record notes for later analysis as details may be lost with time.

Surv. Who do you look to for guidance after a major disaster/incident? Select all that apply.

- The IACUC 66/185 (36%)
- Attending Veterinarian, veterinary staff 131/185 (71%)
- Institutional Official 133/185 (72%)
- Emergency Management + Emergency Resilience Group 135/185 (73%)
- Local law enforcement and/or local government agency 91/185 (49%)
- Other 8/185 (4%)

Surv. Does your institution have a designated chief resilience officer or department?

- No 120/185 (65%)
- Yes. Please provide their functional title at your institution (do not provide their name)* 65/185 (35%)

*Yes:

- Security Office Director
- Incident Commander
- Chief Executive Officer
- Campus Chief of Police
- Chief Operating Officer
- Chief Resilience Officer
- Chief Preparedness and Continuity Officer
- Director of Emergency Management
- Director, Disaster and Community Crisis
- Director of Facilities and Engineering Services
- Emergency Management Officer/Specialist
- Emergency Management/Preparedness Coordinator
- Emergency Policy Executive
- Emergency and Business Continuity Coordinator
- Executive Director, Environmental Health and Safety
- Human Resources Safety Official
- Manager of Resilience Center
- Medical Director
- Public Safety Officer
- Regional Dean

Surv. If you have a resiliency officer or department, approximately how frequently do they meet with members of the animal care program?

- Annually 9/64 (14%)
- Quarterly 10/64 (16%)
- Monthly 0/64 (0%)
- As needed 42/64 (66%)
- Not at all 3/64 (5%)

25:5 What should the IACUC and IACUC support team focus on prior to, or at the start of, the recovery phase?

Opin. Individuals who are not responsible for enacting the initial response can prepare for expected or potential consequences within their own division and help maintain normal business function when possible. These individuals can also serve as a communication hub, especially if located or equipped differently from first responders. Even if it is not possible to answer questions, IACUC members and staff can still be a source of support, continuity, or comfort to the research and animal care communities.

Qualified IACUC personnel may be asked to assist other groups engaged in the initial response as appropriate, such as IACUC specialists trained as veterinary technicians or experienced scientist-member researchers. IACUC staff and members have familiarity with the species housed, facility locations, and animal procedures performed based on the IACUC's protocol review and inspection responsibilities. Those participating in IACUC inspections will also already have received training and occupational health enrollment based on their specific exposures during the process.

A risk-based, carefully considered audit of the current state of the program and facilities should be performed as soon as safety and animal care are secured. The IACUC support team can assist in the first efforts of recovery by helping provide essential information, including animal census, equipment inventory, location of and points of contact for decentralized housing locations, and on-site conditions. Additionally, having a list of protocol expiration dates can help prevent unapproved animal activities and expenditure of funds on protocols that expire during an emergency.

25:6 How can the IACUC support team assist with the animal inventory?

Opin. Depending on the organization of the animal care and use program, the IACUC support team may or may not have direct access to current animal census information. However, the IACUC will have records of animal housing and other location information from the IACUC protocols and semiannual inspection process. IACUC records may be the only source of information available, even if incomplete, until animal care is secured and the veterinary and animal care division is able to allocate resources to reporting.

IACUC personnel may be asked to participate in census taking when appropriate. Confirming animal identification, location, and quantifying animal numbers is time consuming and can be laborious. Many factors contribute to the amount of effort needed, including the nature of the incident, the conditions under which the census is being recorded, and the type and location of the housing. Animal numbers may need to be estimated in outdoor enclosures or for large groups. Animal numbers may also need to be estimated for incident-specific reasons, including personnel safety. Any identified animal fatalities should be documented, and sick and dead animals managed according to the contingency plan. This process of determining animal identification and quantity should not be at the risk of personnel or impede the provision of routine and veterinary care.

25:7 What can the IACUC do before an emergency to improve the effectiveness of the animal inventory?

Opin. Partnering with the veterinary and animal care teams, the IACUC should consistently emphasize and demonstrate proper identification practices and recordkeeping standards throughout the program. As with any important data, animal inventory and colony records should be protected and accessible. Researchers should be reminded of the potential risk of reliance on physical, written notes (such as on cage cards.) This conversation can occur as part of orientation, during training, during post-approval monitoring activities, and during protocol submission. In addition to advising researchers, the IACUC may consider providing templates, guidelines, or detailing requirements for some or all animal records.

25:8 How can the IACUC support team assist with the equipment inventory?

Opin. In addition to the other program participants (e.g., facilities management; environment, health and safety; and veterinary and animal care), the IACUC support team may have knowledge of the location, quantity, and uses of equipment necessary for personnel safety and animal care and use, especially in laboratory spaces located outside of centralized animal facilities. Equipment necessary for safe animal care and emergency response can include fume hoods, central gas lines for oxygen and carbon dioxide, central vacuum and scavenging, and empty waste and sharps disposal containers. The emergency management response team should communicate the current functionality of equipment and building services. The veterinary/animal care division will communicate the status of ventilated caging and other housing, containment, and life support systems. Individuals responsible for decentralized animal housing locations need to be included in the plan to ensure comprehensive oversight and communication (see 25:14).

The IACUC support team can assist by utilizing information gathered about these locations from the protocol and during IACUC inspections and other post-approval monitoring visits. Equipment identified in this way might be utilized in an emergency situation to replace nonfunctional equipment. Similarly, the IACUC support team might identify locations with equipment that has not yet been, but should be, evaluated.

In addition to equipment, the IACUC support team may have knowledge of the storage location of controlled substances, anesthetics, or other drugs with the potential for abuse or theft. These locations may warrant additional security or relocation depending on the nature of the emergency.

25:9 How can the IACUC support team assist with ensuring access?

Opin. IACUC staff and members participating in the inspection process are familiar with how to access the buildings and areas in which housing facilities, satellites, and laboratories are located. Depending on the institution and program, individuals on the IACUC and staff may be provided with unrestricted access to these locations as a business practice, or as part of the response to the emergency. This may require individuals to obtain and produce physical or electronic credentials or institutional documentation to security for access.

Inter-building and intra-building transportation routes for live animals should also be known to the IACUC from the location approval and inspection process. In addition, the IACUC should be aware of locations where animals are frequently, or in quantity, maintained outside of housing. These routes and areas, as well as any vehicles used for transportation, should be considered for reliability and continuity of access as part of the evaluation.

25:10 What can the IACUC and Program do to support the mental and physical well-being of personnel during an emergency?

Opin. It is important to recognize that personnel and personal needs will change during the course of the emergency. Methods to support physical and emotional well-being should be tailored to the incident and in consideration for the personnel involved. Honest bidirectional communication is important to ensure that those providing support can adjust in response to changing needs or an increase in complexity, severity, or duration of the emergency. Personnel may be simultaneously impacted in the workplace and in their personal lives, either as a result of, or unrelated to, the emergency. Stressors can be divided into categories to help aid in systemic consideration, including:

- *Basic Stressors* – unavailable or unreliable transportation; school closings; lack of child care, elder care, or pet care; inability to easily address biological needs like hunger, civil unrest, or targeting
- *Physical Stressors* – working conditions that are uncomfortable, such as hot, cold, slippery, or dark; risk of exposure to pathogens or other hazardous materials; increased physical labor; longer work hours or commute
- *Emotional Stressors* – uncertainty; impediments to communication; isolation; risk to or loss of animal life; risk to or loss of human life; compassion fatigue
- *Professional Stressors* – real or perceived disparity in the treatment of different groups of employees; belief that colleagues are not as concerned or involved; supervisor's concerns regarding staff safety

One method of reducing stress during difficult circumstances is to provide individuals with meaningful ways to contribute to recovery. Programs should consider ways they can provide any volunteers with an opportunity to contribute in a purposeful way whenever possible.

Surv. Does your contingency plan include the following? Select all that apply.

- Detailed text regarding mass euthanasia 110/185 (59%)
- Care for employees' dependents 24/185 (13%)
- Care for employees' pets 11/185 (6%)
- Employee transportation, food, and shelter 73/85 (39%)
- Employee psychological support 48/185 (26%)
- No contingency plan in place 8/1185 (4%)
- None of the above 44/185 (24%)

25:11 What actions should the IACUC consider in the second phase of recovery?

Opin. The IACUC support team and members should perform an initial evaluation of the state of the program. The OLAW checklist, the institutional Animal Welfare Assurance, and AAALAC program description can be used as a framework. The IACUC can also consider developing a checklist specific for recovery as part of their contingency plan. Using a framework ensures all aspects of the program are considered, even if the level of detail varies based on circumstances, available resources, and timing. If the IACUC staff is coordinating the evaluation, it is essential to communicate results regularly with the members of the Committee and the Institutional Official (IO). Participation by the members of the IACUC should reflect their responsibility; it should not be expected that the staff can, or should, assume responsibilities for all aspects of the program.

25:12 What should the IACUC do after the initial evaluation of the facilities and program?

Opin. Requirements for reporting should be considered prior to an emergency and reviewed regularly as part of the program review. Requirements for reporting are based on multiple factors, including: animal species and source; type of research conducted; facility location; sources of funding; accreditation status; and regulation. Requirements for reporting, reporting method, and contact information should be compiled and available to the appropriate responsible individuals via multiple methods.

Once personnel safety and animal needs are identified and addressed, prompt reporting to OLAW and AAALAC International may be required.[14–16] The program should document details of reports made to capture the date, method of report (e.g., phone call, email, online portal), individual receiving the report, and details of the information reported. USDA Animal Care can provide technical expertise and, as part of a national community, facilitate assistance to regulated research facilities, so communication with USDA can support the facility's recovery. Moving forward, the IACUC should continue to document the recovery process. It is important to verify continuity of normal operations in addition to any changes or disruptions; and records and communications should reflect this. Periodic update reports may be needed and should be tracked and documented to ensure the program is meeting obligations for reporting to various entities.

25:13 What can the IACUC do to ensure every possible emergency situation is considered as part of the contingency plan?

Opin. No one can ensure that every potential adverse event is considered, especially in a rapidly changing, complex, and unpredictable world. However, the IACUC can focus on potential challenges personal to the program and emphasize the need for ongoing and planning with staff. Triaging animal populations in conjunction with investigators, having disaster plans coordinated with the IO or other senior level administrators, and having the plan approved at the institutional level are all essential. The IACUC can also develop, apply, and refine strategies for decision making that can be readily reapplied, even in novel situations. There are many shared experiences and plans available (including OLAW's Disaster Planning and Response Resources webpage)[9] that can help facilitate and expand brainstorming regarding aspects specific to different types of emergencies.

25:14 What challenges should the IACUC consider with respect to decentralized housing facilities?

Opin. Decentralized animal facilities can include facilities managed by the institution's veterinary and animal care division, facilities managed by researchers or departments, and off-site facilities, including field study sites, regardless of management. The increase in the number of and distance between locations, as well as the number of personnel responsible, increase the effort required for planning and recovery. One important way in which the IACUC can help mitigate these challenges includes keeping location and contact information up to date and accessible, especially for housing locations and essential personnel. In addition, consolidation of animals to fewer housing locations in the likelihood or event of an emergency is a practice that should be considered as part of planning and training.

25:15 What challenges should the IACUC and animal care team consider related to transportation?

Opin. Decentralized facilities can result in substantial intra-institutional transport of animals and supplies. Transportation routes should be checked for abandoned animals or supplies. In addition, the program's contingency plan should address when deliveries from vendors and importations from other sources should be postponed or canceled. Regular training should include the practice of ensuring the ability to complete the transport prior to departure and any alternative routes or methods, even for normal business occurrences such as elevator maintenance or fire drills. Animals may also need to be relocated to another institution or contract facility/off-site facility during an emergency. The IACUC can help with sharing personnel training, experience, and protocol information with the receiving institution as needed.

25:16 What challenges should the IACUC consider regarding ongoing experimental manipulations?

Opin. Considerations and actions for an emergency during an animal experiment should be part of initial, ongoing, and emergency planning and training. Including this topic in multiple training opportunities is important as the potential responses will vary greatly based on species, procedure, location, and the nature of the emergency. For example, rodents may be left secured in isolator cages on the benchtop or cart, allowing for immediate evacuation of personnel. Procedures involving hazards may require specific actions for personnel to take before they evacuate. The IACUC and program may consider training provided to researchers for procedures under anesthesia, with restraint or when an animal is not otherwise securely housed. A holistic approach that includes these considerations and actions as part of all routine and ongoing training provides researchers with information and practice before an actual emergency.

25:17 What other challenges should the IACUC consider as part of the contingency plan?

Opin. Other potential emergencies or secondary consequences from the primary emergency include: human intruders; protests; civil unrest; targeting;

students in emergency situations; damage to high-profile equipment; theft of drugs or supplies (including PPE); data breach or loss; falsification of records; research sabotage; chemical, biological, and/or radiation spills; theft or loss of valuable animals or lines; animal evacuation or escape; and/or climate change. Since there are so many possibilities, the IACUC may consider a potential incident or type for each program review as part of the ongoing evaluation of the contingency plan.

25:18 What are the challenges of intermittent work vs. work stoppage?

Opin. Intermittent work can be more challenging than a work stoppage. With intermittent work, each iteration may include changes to rules and requirements due to dynamic circumstances and needs. Personnel will need to remain vigilant to ensure understanding of current requirements. Constantly changing circumstances can lead to change fatigue even in the best of circumstances. Intermittent work may also increase the need for work coverage. If personnel access is restricted, either in quantity or in frequency, researchers may ask for, or need, assistance in monitoring their animals or performing procedures. The IACUC can help mitigate any complications by facilitating the process of training, personnel approval, and facility access whenever possible to ensure care is continued and personnel effort conserved.

25:19 What can an IACUC and IACUC support team do during an emergency in absence of a contingency plan?

Opin. Even in the absence of a predetermined contingency plan, a robust and tested communication strategy will aid response and recovery efforts. The IACUC support team should ensure the ability to maintain continued communication with IACUC personnel; IACUC members; IACUC leadership, including the Chair, the IO, and the Attending Veterinarian; and other key points of contact in the program and institution during an emergency. Ideally, multiple methods of contact are available for essential personnel, and this contact information is shared and periodically confirmed. Reliance on one method (including institutional personnel management systems) or one person for contact information is imprudent. Both normal business operations and the institutional contingency plan can, and should, be used as a basis for the contact list. The IACUC contact list should also include information for external reporting to government and other regulatory and oversight bodies as needed. It is important to consider all communication methods that might be useful in a particular situation, including atypical or imperfect methods, such as social media. One should not underestimate the value of hard copies of printed information as one method of sharing and securing contact information.

In addition to maintenance of contact information, communication methods and strategies can be developed and tested prior to an emergency. The emergency contact list should indicate the role and responsibilities for essential personnel for communication triage. The IACUC should anticipate communication and guidance from regulatory agencies such as OLAW and USDA in response to significant adverse events, major disasters and other large-scale events. The IACUC should have multiple methods documented and shared for contact and information sources for these agencies. Key requirements for external reporting can be compiled for reference. Instructions for mass communication methods such as blast e-mails, website changes, and other institutional messaging should also be available. The institution should consider a secondary, siloed list of contact information for researchers secured in case the protocol management software or database itself is compromised.

25:20 What can an IACUC and IACUC support team do before an emergency to ensure continuity of the program?

Opin. The animal care and use program and the institution are best served when all the operating units are functioning effectively, during both normal business and emergencies. Therefore, the IACUC and IACUC support team best serve the program by supporting the IACUC personnel that safeguard the business function of the IACUC. Fortunately, best practices for emergency management and contingency planning are similar to best practices for IACUC management. OLAW's definition of an "unexpected" incident is meant as "unplanned," not unprecedented. IACUC review processes require researchers to anticipate pain or distress and then develop a specific predetermined plan for monitoring and intervention, including endpoints. Contingency planning follows a similar cadence. Both also emphasize the use of objective, standardized, and timely evaluation; risk-benefit analysis; training and documentation; and ongoing review.

Ideally, the IACUC contingency plan is congruent with the institutional plan and harmonized with the facility response plan. Further, the IACUC component should consider both programmatic responsibilities and business requirements and needs.

Surv. How are personnel trained on the contingency plan at your institution? Select all that apply.

- Personnel are not trained on the contingency plan. 28/185 (15%)
- Personnel review a copy with their supervisor, manager, or other staff member. 125/185 (68%)
- Personnel conduct table-top exercises. 43/185 (23%)

- Personnel are trained using web-based programs. 29/185 (16%)
- Personnel are trained at institution-wide sessions. 41/185 (22%)
- Personnel are trained in isolation of other departments/divisions. 37/185 (20%)

25:21 How can an IACUC and IACUC support team create a contingency plan?

Opin. There are a multitude of effective and tested methods and strategies for emergency management and contingency planning. One straightforward method to develop a contingency plan for an IACUC and IACUC support team, especially in the absence of a model plan or template, is to use familiar IACUC processes and cadences.

Considering the regulatory assigned IACUC and institutional responsibilities as one axis of an array, and various types of emergency situations as the other, the converging points provide areas to detail specific responses. Those details can be described using a risk-based paradigm such as "MoSCoW" (M: Must have, S: Should have, C: Could have, W: Won't have).

The responsibilities to be assigned can be collated at varying levels of granularity from source documents, including Greene and Wright's Institutional and IACUC Requirements,[17] the OLAW Checklist[18]; any institutional framework for business operations or program review; the institution's PHS Animal Welfare Assurance (Assurance); and the AAALAC Program Description. Potential emergency circumstances can also be defined and characterized, especially initially, at various levels of specificity and culled from local and professional experiences and emergency management resources.

As an example, the review of activities involving animals is a responsibility assigned to the IACUC as part of normal operations. The institution has both programmatic processes (as described in the Assurance) and business processes to conduct this review. Flexibilities to consider for contingency planning include incorporating OLAW's Guidance on the "Use of Electronic Communications for IACUC Functions," a document that addresses general issues with respect to the Assurance, distribution of information, polling, security, and signatures; methods for full committee and designated member review of protocols; suspension of an activity; reporting; and semiannual review and inspection.[19] NOT-OD-24-067 details flexibilities for conducting the semiannual program and facility inspections, capturing details required for institutions covered by the Animal Welfare Regulations.[20] The contingency plan should also include plans in response to circumstances such as a local/social emergency (e.g., public transit workers strike) and an institutional emergency (e.g., targeted attack). For these intersections using a method such as MoSCoW, an institution might make the following plans[21]:

- During a public transit workers strike, the IACUC review MUST continue as before including electronic and off-site methods that have been and are regularly validated. Protocol review consultations SHOULD continue via virtual methods when possible. Reviews WOULDN'T be postponed due to transit.
- For an institutional emergency, such as a targeted (potentially electronic) attack: Protocol review MUST NOT occur until the technology has been confirmed to be secure.

25:22 How can an IACUC and IACUC support team add value to the contingency plan?

Opin. The APHIS requirements for contingency planning provide a reasonable framework that can be adapted for additional uses and include identifying triggers for when the plan is enacted; specific tasks as a specific response to a specific emergency; and a reminder for regular evaluation of the plan and prompt socialization for significant changes. The simple FEMA scheme of "prevention, protection, mitigation, response and recovery" is another cadence that can be applied to contingency planning. FEMA includes the ability to respond to the risk as a separate factor for consideration, which should be incorporated during planning and transparently communicated for maximum benefit. The allowance to consider and modify the response as needed acknowledges the individual institution's ability to utilize expertise and resources identified and available. It is expected that the response will be impacted by changing circumstances, additional information, and time. The ability to adapt should, therefore, be supported and considered an integral aspect of planning, allowing for agile responses. Communicating this commitment to all stakeholders will forewarn them of change and encourage trust in the plan and foster resiliency during a crisis. With increased changeability, communication assumes greater importance. Resources for communicating during a crisis are available and continue to be expanded and refined. Avoiding arbitrary escalation and embracing transparency can minimize speculation and counter misinformation among stakeholders. Downplaying the severity of the situation or withholding information can erode trust and have lasting consequences. Multiple, progressive communication strategies should be considered for each step of the contingency plan. Both APHIS and FEMA highlight preparedness and the focus on continuity, including developing strategies and considering alternatives.

25:23 How can an IACUC and IACUC support team support compliance during times of change?

Opin. It may help for the IACUC and IACUC support team to remind themselves of the intent ("spirit") of the law or standard when considering emergency changes to ongoing, especially long-standing, practices. Whenever a practice or standard or requirement is not or cannot be met, it should be acknowledged and documented for record and considered in context.

For example, the basing of IACUC review activities in Parliamentary Procedure and Robert's Rules of Order ensures representation and facilitates participation from the required individuals while also curtailing intimidation. These methods are generally accepted and applied readily in face-to-face meetings. When considering new or adjunctive methods of meeting, such as virtual meetings, it is important to consider if the new process meets the intent of the requirement or standard. Will all members have the same ability to freely participate? Does the IACUC have an adapted business process in place to document attendance, quorum, voting, etc.?

There may also be benefits to forced change. The support for and use of live-video or photography for certain inspection events may allow for more or different participants.

It is also important to consider the acute burden to specific groups during the emergency when considering the overall benefit of participation in lieu of prioritizing activities to ensure animal welfare. Consider the use of "hidden experts" and other *ad hoc* support especially in lieu of first responders, such as veterinary and safety personnel, and/or when the ability to be on-site is severely limited.

REFERENCES

1. Crisis Management Program, Canadian Council on Animal Care. https://ccac.ca/Documents/Assessment/Crisis-Management-Program.pdf
2. National Research Council. 2011. Guide for the Care and Use of Laboratory Animals. Washington, DC: National Academy Press. National Institutes of Health, Office of Laboratory Animal Welfare. 2015.
3. Development Guide: Contingency Plans for Emergencies, Animal and Plant Health Inspection Service, U.S. Department of Agriculture. https://www.aphis.usda.gov/sites/default/files/ac-tech-note-guide-contingency-plans-emergencies.508.pdf
4. OLAW NOT-OD-05-034. Guidance on Prompt Reporting to OLAW under the PHS Policy. http://grants.nih.gov/grants/guide/notice-files/NOT-OD-05-034.html
5. Robert T. Stafford Disaster Relief and Emergency Assistance Act, Public Law 93-288, as amended, 42 U.S.C. 5121 United States Code, Title 42.
6. US Department of Agriculture. Animal Welfare Act (Public Law 89-544). 1966 (Last amended 2023). U.S. Department of Agriculture. Animal Welfare Regulations 9 CFR (U.S. Department of Agriculture, 2023).
7. Public Health Service Policy on the Humane Care and Use of Laboratory Animals. Bethesda, MD: U.S. Department of Health and Human Services.
8. OLAW FAQ G3, "Do awardee institutions need animal facility disaster plans?" https://olaw.nih.gov/faqs#/guidance/faqs?anchor=10597
9. Disaster Planning and Response Resources. https://olaw.nih.gov/resources/disaster-planning.htm
10. OLAW FAQ F6 "How can institutions assure animal welfare when HVAC systems malfunction or fail?" https://olaw.nih.gov/faqs#/guidance/faqs?anchor=50386
11. The American Dairy Science Association, he American Society of Animal Science, and the Poultry Science Association. 2020. Guide for the Care and Use of Agricultural Animals in Research and Teaching, 4th Edition.
12. Directive 2010/63/EU of the European Parliament and of the Council of 22 September 2010 on the Protection of Animals Used for Scientific Purpose.
13. National Preparedness System. U.S. Department of Homeland Security November 2011.
14. https://grants.nih.gov/grants/guide/notice-files/NOT-OD-05-034.html
15. https://www.aaalac.org/accreditation-program/faqs/#H1
16. https://www.aaalac.org/accreditation-program/faqs/#H2
17. Office of Laboratory Animal Welfare; Institutional Animal Care and Use Committee Guidebook (nih.gov)
18. https://olaw.nih.gov/sites/default/files/Program%20Review%20cheklist.pdf
19. Use of Electronic Communications for IACUC Functions. *ILAR Journal* 37(4):190–192, 1995. Issues for IACUCs. https://olaw.nih.gov/guidance/articles/ilar95.htm
20. https://grants.nih.gov/grants/guide/notice-files/NOT-OD-24-076.html
21. Clegg, Dai, Barker, Richard 1994. Case Method Fast-Track: A RAD Approach. Addison-Wesley.
22. AVMA . Guidelines-on-Euthanasia-2020.pdf

26

Confidential and Proprietary Information

Tracy M. Heenan and Thomas A. Leach

Introduction

The purpose of this chapter is to provide a basic framework for understanding how the concepts of confidential information and proprietary information relate to IACUC activities. In providing institutional review and approval for research activities using animals, IACUCs might review information considered confidential or proprietary by the researchers. Even information that may not seem confidential or proprietary, such as a researcher's name, may necessitate protective treatment under certain circumstances.

IACUCs at government-supported institutions, such as public universities and federal research institutions, are subject to state and federal laws for open meetings and Freedom of Information acts for "public records"; however, even among such IACUCs, the application of these access-to-information laws varies. Some IACUCs have open meetings, and, therefore, whatever information is discussed by the committee becomes public information. Others have closed meetings and may not have the automatic release of information into the public domain. Similarly, state laws on access to public records may lead to radically different results, as some jurisdictions require release. Some jurisdictions may require release but may also provide exemptions that preserve confidentiality under certain circumstances. It is important to seek legal advice in determining jurisdictional requirements.

In reading the court decisions cited in the following sections, it is important to keep in mind that a court's decision is based on the set of facts before it. Thus, a different factual situation might lead to a different legal decision. This is one reason why courts within the same jurisdiction may appear to reach contradictory decisions. It also is important to remember that a published decision provides binding precedent only within the court's jurisdiction, such as within one state, and only as to the issue addressed. Of course, a broad and well-written decision might provide persuasive authority in other jurisdictions and to other related issues. Thus, for example, a decision by the Vermont Supreme Court, the state's highest court, on the applicability of that state's public records act to an IACUC is binding only within Vermont. The Vermont decision, however, might provide persuasive reasoning that would guide judges in other states when determining the applicability of their state public records laws to an IACUC in their state. Readers are encouraged to check with their institution's legal counsel or state attorney general for specific laws and interpretations in their particular state.

26:1 What are the sources of law that can apply to an IACUC's activities?

Reg. IACUCs at institutions using any live or dead warm-blooded animals (other than rats of the genus *Rattus* and mice of the genus *Mus*, bred for use in research) for research, testing, teaching, or experimentation purposes must comply with the federal AWA, found at 7 U.S.C. §§2131 through 2159, particularly §2143,B (setting forth the establishment and duties of the IACUC) and §2157 (prohibiting release and wrongful use of confidential information by members of the IACUC). IACUCs at such institutions must also comply with the federal regulations promulgated under the AWA (the AWAR, found at 9 C.F.R. §1.1–4.11), particularly AWAR §2.31 (detailing IACUC functions). The requirements specified in AWAR §2.32 (personnel qualifications), AWAR §2.33 (AV and adequate veterinary care), AWAR §2.35 (record-keeping requirements), and AWAR §2.36 (annual report) also are relevant to IACUC activities. The AWA has been in effect since 1966, with multiple amendments over the years and subsequent promulgation of regulations and standards in response to the AWA and its amendments in the AWAR.

IACUCs at federal or state institutions or institutions receiving federal or state funding are additionally subject to the laws, regulations, and policies applicable to the funding source or the institution. For example, 42 U.S.C. §289,d, created by the Health Research Extension Act of 1985 (HREA), sets forth requirements of an IACUC at institutions receiving funds from the PHS. In addition to the AWA and AWAR, institutions receiving PHS funding for research must conform to the PHS Policy, which is applicable to any PHS-conducted or PHS-supported use of a live vertebrate animal (whether warm blooded or cold blooded) for research training, experimentation, or biological testing, wherever that activity is performed. PHS Policy II requires every individual researcher receiving PHS support for an activity involving animals to be affiliated with or sponsored by an institution which can and does assume responsibility for compliance with this Policy, unless

DOI: 10.1201/9781003272861-26

other arrangements are made with the PHS. U.S. Department of Veterans Affairs facilities must follow VHA Handbook 1200.07, *Use of Animals in Research*,[1] which requires compliance with the PHS Policy even if PHS funds are not received. Institutions receiving funding from certain sources and all institutions accredited by AAALAC International must comply with the *Guide for the Care and Use of Laboratory Animals*[2] and the Federation of Animal Science Societies' *Guide for the Care and Use of Agricultural Animals in Research and Teaching*.[3] Although AAALAC International accreditation is voluntary and not mandated by law, the governing body of an institution may require accreditation or conformance with the standards described in these guides. The *U.S. Government Principles*, for example, which forms the basis for the PHS Policy, refers to the *Guide* for guidance in following its principles. It should be noted that PHS Policy does not affect applicable state or local laws or regulations that impose more stringent standards for the care and use of laboratory animals.

IACUCs at institutions with facilities in other countries or that review protocols to be performed in other countries must ensure that those protocols comply with all applicable laws of those countries. For example, one such set of standards that may apply is that of the *European Convention for the Protection of Vertebrate Animals Used for Experimental and Other Scientific Purposes* (CETS No. 123).[4]

State laws and regulations also may apply to IACUCs at institutions required or allowed by their state to be licensed or inspected by state authorities and may be more stringent than federal laws. For example, state or local laws or ordinances may impose restrictions on obtaining animals from certain sources, such as municipal animal control facilities or shelters. Likewise, the City of Cambridge, Massachusetts, employs a Commissioner of Laboratory Animals within the Public Health Department. The Commissioner is a licensed veterinarian who oversees the care and use of laboratory animals in Cambridge, as mandated by Ordinance 1086 of the city's municipal code. Cambridge is one of the few cities in the United States with such a program.

Additionally, IACUCs should be cognizant of state anti-cruelty laws which could apply to activities at their institutions despite statutes in many states that attempt to exempt research activities from animal cruelty laws. All states have laws prohibiting animal cruelty, and many states have significantly modified those statutes in recent years to increase penalties for violations. The AWA (§2145,b) expressly authorizes the USDA to cooperate with state and local authorities in animal welfare matters.

Statutes, regulations, and policies governing other activities of an institution may apply to activities involving animals used for research, teaching, or testing, and thus apply to an IACUC's activities as well. These could include laws of the Environmental Protection Agency (EPA), the Fish and Wildlife Service, or the Food and Drug Administration (FDA), such as the Good Laboratory Practices regulations,[5] among others. Certain protocols, such as studies involving biohazardous agents or radioactive materials, may implicate other federal, state, and local laws, regulations, and policies. An IACUC's activities also may be impacted by regulations of the Centers for Disease Control and Prevention (CDC) and Occupational Safety and Health Administration as they affect the safety and health of persons working with or near animals. The Endangered Species Act[6]; the Chimpanzee Health Improvement, Maintenance, and Protection Act[7]; the Migratory Bird Treaty Act[8]; or similar laws protecting certain animals may also occasionally factor into decisions by an IACUC.

IACUCs also should know that under the Animal Enterprise Terrorism Act,[9] it is a federal crime to damage or interfere with the operations of an enterprise using animals for research, education, or testing. This law prohibits conduct, including threats, vandalism, property damage, criminal trespass, harassment, and intimidation. Additionally, many states have laws protecting animal research facilities from criminal acts such as arson, economic sabotage, and similar acts of intimidation.

Finally, federal, state, local, and institutional laws not specific to animal-related activities can apply to an IACUC's activities, such as laws pertaining to patents or trade secrets, laws requiring public disclosure of records or meetings open to the public, and laws protecting personal privacy. The federal Freedom of Information Act[10] (FOIA) requires the federal government to provide access to records in its possession, which may include institutions' annual reports and inspection reports filed with APHIS/AC pursuant to the AWA. All states have laws providing for the public disclosure of state agency activities and records, which may apply to state universities. IACUCs at universities, other nonprofit organizations, and small businesses receiving federal government funding for research should be aware of the Bayh-Dole Act (Patent and Trademark Act Amendments of 1980),[11] which allows these institutions to retain title to and file for patents on inventions developed under that funding as well as promote those inventions commercially.

Opin. Many sources of law can bear on an IACUC's activities. They include, in order of hierarchy, federal and state constitutions, federal and state statutes, federal and state regulations, federal and state case law, and institutional policies.

Constitutions set out the fundamental, broad legal principles that govern the nation or the state that enacted them. The principle of academic freedom, for example, is often considered to be grounded in the First Amendment to the U.S. Constitution. State constitutions are sometimes found to be more protective of certain rights (such as the right of privacy) than is the federal Constitution, even where the wording in the two documents is similar.

Statutes are laws enacted by the federal Congress, a state legislature, or, in some states, by a vote of the people. In a particular state, both federal laws and the laws of that state apply. Laws are codified, or collected and organized systematically, by dividing them into titles or chapters by subject matter. The AWA, for example, is codified in the U.S. Code at Title 7, which includes laws pertaining to agriculture. The HREA is codified in various places within the U.S. Code, most particularly with respect to animal research at Title 42, which collects laws protecting the public health and welfare, and was implemented by the PHS Policy. Most federal and state codes and statutes are available for viewing online.

Regulations are rules enacted by a federal or state agency pursuant to requirements or authority granted in a statute. As with statutes, these rules are collected and typically organized by topic. Pursuant to the AWA, for example, the USDA has enacted specific rules (the AWAR) that are found in the Code of Federal Regulations (C.F.R.) at Title 9, which sets forth the regulations governing activities of the APHIS and other rules pertaining to animals and animal products. Strict administrative procedures govern the adoption of regulations. Typically, the agency is required to publish a notice of proposed rulemaking and to provide an opportunity for comment. Federal notices of this sort are published in the Federal Register; states have similar publications for the rulemaking activities of their agencies. Regulations tend to be much more detailed than statutes.

Case law is the set of written opinions by judges interpreting constitutions, statutes, regulations, and other sources of law when deciding a particular case. Not all lawsuits result in written case law. Only published decisions of appellate-level judges (those who review trial-level decisions) can be cited as binding precedent in other lawsuits. Comparatively few lawsuits result in this kind of decision. The vast majority of filed lawsuits settle prior to trial; only a few that go to trial are appealed, and only some of those that are appealed result in a decision written for publication. The citation to a case indicates where to find the published decision, which court made the ruling, and in what year.

26:2 What sources of authority other than those noted in 26:1 might apply to an IACUC's activities?

Reg. The *U.S. Government Principles for the Utilization and Care of Vertebrate Animals Used in Testing, Research, and Training* address compliance with the AWA and other applicable federal laws, guidelines, and policies and generally provide a set of overarching principles for ensuring that the use of research animals is justified and humane. These principles are incorporated in the PHS Policy. The PHS Policy applies to the use of live vertebrate animals in any activity supported or conducted by any PHS agency, including the NIH, FDA, and CDC. Also of note, OLAW has established Memoranda of Understanding with the National Aeronautics and Space Administration (NASA), National Science Foundation (NSF), Biomedical Advanced Research and Development Authority (BARDA), and the U.S. Department of Veterans Affairs (VA). As such, these agencies are subject to the same requirements and obligations as the government agencies outlined in the PHS Policy. The PHS Policy mandates compliance with the AWA and AWAR, as applicable, the *Guide*, and the *AVMA Guidelines for the Euthanasia of Animals*. VHA Handbook 1200.07, *Use of Animals in Research*,[12] applies to research, testing, and teaching activities involving laboratory animals in the Department of Veterans Affairs and mandates compliance with the PHS Policy, the *Guide*, and the *AVMA Guidelines for the Euthanasia of Animals*. Institutions that are accredited by AAALAC International or receive funding from certain sources must comply with the *Guide* and the Federation of Animal Science Societies' *Guide for the Care and Use of Agricultural Animals in Research and Teaching*.[13] The *AVMA Guidelines for the Euthanasia of Animals*,[14] prepared by the American Veterinary Medical Association, is an essential reference for IACUCs; deviations from these recommendations should be extremely rare and only when scientifically justified, with any such justifications documented in writing and approved by the IACUC.

Other sources of authority will apply to activities conducted in countries other than the United States.

Opin. In addition to the types of legal authority described, other sources of authority may apply, with varying degrees of force, to IACUC activities, even in a legal setting such as a lawsuit. The PHS Policy, the *Guide*, and the *AVMA Guidelines for the Euthanasia of Animals* carry a great deal of authority as the standards for research animal care and use. In addition, a court is likely to consider as presumptively appropriate other standards or references set out or condoned by AAALAC International or similarly influential professional organizations. Several government agencies have guidelines or requirements applicable to animal activities that should be followed in IACUC-approved protocols when appropriate and possible. Also, some professional scientific associations publish authoritative guidelines relevant to animal activities relating to their areas of interest. Courts often look for evidence, such as these references, of a "community standard" among similar facilities if there is a dispute as to whether an institution acted properly.

Other important sources of authority are any internally adopted policies that apply to the IACUC. These can be policies adopted by the animal facility, university, occupational health and safety department, or any entity with authority to set policy for the IACUC, including the IACUC itself. If an IACUC is to be subject to an internal policy that transcends legal requirements, however, it is important that the IACUC be able to comply with that policy. Having a policy with standards that are not met may place the IACUC in a more vulnerable legal position than having no policy at all.

Contractual obligations may apply to an IACUC's activities. Research involving third parties (such as collaborations or sponsored research) and research that involves transfers of third-party information or materials will likely be governed by contracts that restrict the use and disclosure of information. Such contracts may be identified as confidentiality agreements, memoranda of understanding, collaborative research agreements, sponsored research agreements, or material transfer agreements. Agreement names vary widely and although agreements may share the same name, they may contain differing obligations, restrictions, and rights. The terms and conditions of each agreement must be carefully read and understood. Research-related agreements usually contain language allowing the use and disclosure of information within an institution and to regulatory bodies in compliance with law; this permits an IACUC to perform its necessary functions. However, any disclosures beyond those strictly necessary to comply with law may violate such third-party agreements. Violations of contractual confidentiality restrictions can give rise to claims of breach and civil damages.

Finally, IACUC activities also may need to comply with an institution's policies on conflicts of interest or other policies relating generally to committee activities.

26:3 What is proprietary or confidential information?

Reg. Federal law defines certain types of proprietary information, including copyrights, patents, trademarks, and trade secrets. Under federal law, *copyright* protection can attach to "original works of authorship fixed in any tangible medium of expression."[15] A *patent* can be obtained for a "new and useful process, machine, manufacture, or composition of matter, or any new and useful improvement thereof."[16] A *trademark* is something that distinguishes the goods or services of the owner from those of others.[17] A *trade secret* is information that has economic value by virtue of its not being generally known to or readily ascertainable by the public.[18] *Confidential information* may be defined by federal or state law or by private agreements or policies. Confidential information can include trade secrets, personal information, individually identifiable health information, personnel records, information protected by a statutorily defined privilege such as the attorney–client or doctor–patient privilege, or other types of information.

Several federal laws protect individuals' privacy. For example, the federal Privacy Act limits the type of information that federal agencies, the military, and other government institutions may collect, maintain, use, and disclose on individuals.[19] Most states have additional laws protecting individuals from disclosure of their confidential information by others, through constitutional provisions, statutes, or regulations.

Opin. *Proprietary information* is information that has the legal status of personal property, that is, information that someone exclusively owns, exercises control over, or uses. The term is usually used for trade secrets, confidential business or financial information, or information about an invention that has not yet been disclosed in a filed patent application. IACUCs must be very careful not to disclose such information inappropriately or inadvertently. As such, legal counsel should be consulted when protected proprietary information could be discussed. Proprietary information also includes any intellectual property, including patents, trademarks, copyrights, and proprietary software. Patents, trademarks, and copyrights are publicly known, but they are considered personal property because unauthorized use or copying by someone other than their owner constitutes infringement. Other types of proprietary information are not known by the public and their public disclosure can cause irreparable harm to their owner.

A *trademark* or *service mark* is a word, name, symbol, device, or any combination, used or intended to be used to identify and distinguish the goods or services of one seller or provider from those of others, and to indicate the source of the goods or services. Although federal registration of a mark is not mandatory, it has several advantages, including notice to the public of the registrant's claim of ownership of the mark,[20] legal presumption of ownership nationwide, and exclusive right to use the mark on or in connection with the goods or services listed in the registration.[21] Marks may also be registered with states, but a trademark need not be registered at all to be valid and subject to infringement. An unregistered trademark may be identified by a ™ symbol, while a registered mark can be identified by a ® symbol. The Nike swoosh design and the Kleenex® name for facial tissues are examples of trademarks. Rights in a mark cease if the mark is not actively used for a period of time and registrations must be periodically renewed.[22]

Copyright is an author's exclusive right to copy or modify his or her original expression.[23] Copyright protection can attach to original works of authorship fixed in any tangible medium of expression, including literary works, pictorial and graphic works, and sound recordings.[24] Journal articles (including drafts), laboratory notebooks, drawings, and photographs are examples of works with copyright protection. Ideas, procedures, processes, principles, and the like cannot be protected by copyright. The term of a copyright typically runs until 70 years after the author's death.[25]

A *trade secret* may be any secret, commercially valuable information or compilation of information used in a business that gives the owner an advantage over competitors who do not know or use it and is the end product of either innovation or substantial effort. The term "use" includes "all the ways one can take advantage of trade secret information to obtain an economic benefit." This includes assisting, accelerating research, or even development.[26] It may include any form or type of financial, business, scientific, technical, economic, or engineering information, including a compilation, plan, formula, pattern, design, prototype, device, method, technique, process, procedure, program, or code, whether tangible or intangible, and whether or however stored, compiled, or memorialized.[27]

A person who asserts that information qualifies as a trade secret must show that reasonable efforts were made to keep the information confidential, the information has economic value, and the information is not generally known to or readily ascertainable by the public. When properly safeguarded, a trade secret may remain valuable for an indefinite period of time. One of the best-known examples of a well-kept trade secret is the formula for Coca-Cola®.

A U.S. *patent* gives its owner the right to exclude others from making, using, selling, or offering to sell in the U.S., or importing into the U.S.,[28] the "process, machine, manufacture, or composition of matter"[29] claimed in the patent. The term of a U.S. patent generally extends from the date of issuance until 20 years from the date of filing of the patent application.[30] A patent document includes a description of the invention, drawings (if necessary), and one or more claims. The claims particularly point out and distinctly claim the subject matter the applicant regards as the invention and define the scope of the protection afforded by the patent. A patentable invention must be useful, novel, and nonobvious.[31] Under the novelty requirement, no U.S. patent will be issued if the invention was in public use or described in any printed publication more than one year before the application for the patent was filed[32]; thus, disclosure to the public of information describing an invention can forever preclude the inventor from receiving a patent. In most other countries, this grace period is shorter or does not exist at all. In the early stages of a research project, investigators may not be able to identify reliably all the information that may be important to a future patent application and know which information must be protected to prevent compromising the value of the research, so all potentially patentable information must be kept confidential. Laboratory notebooks, drawings, and dated laboratory photographs are common evidence of inventions and thus should generally be treated as confidential. Even live animals may be evidence of a patentable invention.[33]

Under the Patent and Trademark Act Amendments of 1980, popularly known as the Bayh–Dole Act,[34] ownership of patentable inventions arising from federally funded research can be vested in the entities performing the research, including universities, other nonprofit organizations, and small businesses.[35] The holder of the patent rights can license those rights to private companies for further development and commercialization and collect royalties if the invention is sold or used. Thus, inappropriate disclosure by an IACUC could result in the foreclosure of an opportunity for a significant income stream to the IACUC's institution. Funding agreements require that if the grantee elects not to file an application for patent, the grant recipient must disclose inventions to the government in time for the government to file a patent application, and that the government be licensed to practice the invention if the grantee obtains a patent.[36] Federal agencies are authorized to maintain the confidentiality of information disclosing an invention the federal government funded for a reasonable period to permit a patent application to be filed.[37] If the grantee patents the invention but does not appropriately utilize it, the Bayh–Dole Act permits the government to step in and grant licenses to other companies to make the products or use the methods it helped finance for the benefit of the public.[38]

Confidential information is any information that is intended to be held in confidence or kept secret pursuant to either state or federal law or by private agreement or policy. Confidential information can include trade secrets and patentable information or educational, health, and personnel records. It can also include research information subject to a confidentiality agreement or nondisclosure agreement. Some information is considered confidential as a result of a statutorily defined privilege such as the attorney–client or doctor–patient privilege; an attorney should be consulted regarding the limits and requirements of these types of privileges because disclosure to another person or to the IACUC can destroy the confidentiality of such privileged information.

Marking something "confidential" is sometimes a condition of protecting information from disclosure under state and federal laws, as well as private agreements. But that action alone has no legal significance and does not protect information from disclosure. Legal protection is provided either through the law or under a contractual agreement between parties.

26:4 Under what circumstances might proprietary or confidential information disclosed to an IACUC become available to the public?

Reg. IACUCs of public institutions may be required to open their meetings to the public under state open meetings laws. Agencies of the federal government must disclose records in compliance with the federal FOIA.[39] Agencies of a state or local government must disclose records in accordance with the public records laws of their particular state or municipality. Documents and records in the possession or control of a government agency must be made available to the public for inspection and copying, unless some compelling justification exists for keeping them confidential.

Under AWA §2157, it is unlawful for any member of an IACUC to use or release to the public or reveal to another person any confidential information of the research facility, including "any information that concerns or relates to (1) the trade secrets, processes, operation, style of work, or apparatus; or (2) the identity, confidential statistical data, amount or source of any income, profits, losses, or expenditures, of the research facility." Such wrongful use or disclosure may result in removal from the IACUC, fines, and/or imprisonment. In addition, any institution or individual who is injured by such disclosure or use may bring a lawsuit against the IACUC member and recover damages and the costs of the suit from the IACUC member.

§2.35,f of the AWAR requires APHIS/AC inspectors to maintain the confidentiality of information reviewed for inspection purposes and prohibits them from removing materials from research facilities' premises "unless there has been an alleged violation, they are needed to investigate a possible violation, or for other enforcement purposes." The public release of any information or materials copied or removed from a research facility by an APHIS/AC inspector is governed by FOIA.

The federal FOIA provides nine exemptions to disclosure of records. One of the nine exemptions from disclosure under the federal FOIA is for "trade secrets and commercial or financial information that is obtained from a person and is privileged or confidential."[40] This exemption is intended to protect the interests of both the government and submitters of information. Privileged information is information that would ordinarily be protected from disclosure in civil litigation by a recognized evidentiary privilege, such as the attorney–client privilege. Confidential information is defined by FOIA as information that, if disclosed by the government,

1. May impair the government's ability to obtain necessary information in the future
2. Would substantially harm the competitive position of the person who submitted the information
3. Would impair other government interests, such as program effectiveness and compliance, or
4. Would impair other private interests.[41]

Another FOIA exemption prohibits release of "personnel and medical files and similar files the disclosure of which would constitute a clearly unwarranted invasion of personal privacy."[42] Internal government communications are also exempted from disclosure under FOIA.[43]

IACUCs should give special attention in circumstances involving research integrity,[44] particularly if the circumstances may lead to law enforcement actions against those involved. If research misconduct is committed in connection with research subject to IACUC oversight, IACUC records can ultimately become evidence in a law enforcement action. Under certain circumstances, FOIA provides an exemption for records or information compiled for law enforcement purposes.[45] State or local public records laws may have similar exemptions. Legal advice should be obtained prior to responding to FOIA requests relating to research involving potential research misconduct.

Opin. Information disclosed to an IACUC might become available to the public if it is discussed in an open meeting, if the information is included in a document or report submitted by the IACUC to a government agency, or if the information is disclosed in response to a public request because the IACUC is considered an agency of the local, state, or federal government. Whether a document or report was required by law or was submitted or maintained voluntarily is irrelevant to whether it will be disclosed. In litigation, confidential or proprietary information may be required to be disclosed via the discovery process to an adverse party when the information is relevant and necessary to the action, although it may be withheld from public disclosure under a protective order.

The confidentiality of proprietary or confidential information must be scrupulously maintained by an IACUC and each individual member of an IACUC. Inappropriate disclosure of confidential or proprietary information can destroy the value of years of research efforts and have devastating effects on the careers of researchers or the profits of a company or institution. Additionally, if investigators lack confidence that their information will be handled appropriately, they may be reluctant to provide information that the IACUC needs to evaluate proposals properly and meet its legal obligations. Moreover, inappropriate disclosures may lead to researchers becoming the object of harassment or terrorism by persons opposed to the use of animals in research, potentially endangering their lives or laboratories. Unauthorized disclosure or use of confidential or proprietary information by an IACUC member may also lead to removal from the IACUC, payment of a fine, imprisonment, and/or a civil lawsuit brought by the individual or institution injured by the use or disclosure.

If confidential or proprietary information is released to a government agency, even when that submission is required by law or involuntary, the information can become available to the public under FOIA or state or local public records laws. Only information created or obtained by a government agency or government contractor and actually in the agency's or contractor's possession or control can be disclosed; information the government merely has a right to obtain or may have relied on but did not obtain is not available to the public. Receiving a grant from a government agency does not make a private organization a government agency subject to FOIA, unless the government extensively supervises the private organization on a day-to-day basis; similarly, receiving a federal grant does not make a state university subject to the federal FOIA.[46] A private organization is not necessarily an agency of the government when designated by the government to perform certain functions, even if it has decision-making authority.[47] Any such arrangement must be examined individually, considering several factors, including the type and degree of control exercised by the government over the private entity as well as the portion of the entity's business which is performed by contract with the government.[48]

Each state has its own version of a FOIA or public records act that applies to its own government entities.[49] Additionally, each federal and state government agency operates under a specific set of rules or regulations governing its disclosures of records to the public.[50] Many states and local governments also have laws requiring that meetings of government boards, commissions, councils, committees, and the like, be open to the public.[51] The FOIA, public records laws, and open meetings laws are based on the principle that public officials and institutions are accountable to the people. Under this principle, the public is entitled to know how its government makes decisions and to review the documents and data behind those decisions.

According to FOIA and these other laws, "any person" regardless of motive, purpose, or need may seek government documents. Pursuant to these laws, every government agency must make available to the public, upon request, any information in its possession or control unless the information falls within the scope of one of several specified exemptions. Courts interpret these exemptions very narrowly. To prevent information from being disclosed under these laws, sufficiently specific arguments must be made regarding the exemption under which the information falls and the substantial harm that disclosure would cause. The institution wanting to withhold the information has the burden of proving that an exemption protects the information from disclosure.

Open meetings or public meetings laws, sometimes called "open door" or "sunshine" laws, require that meetings of "public bodies" be open to the public. These laws usually include a provision that the committee or other entity may adjourn into an executive session closed to the public when discussing privileged or confidential information.

The few courts that have considered any of these laws in relation to IACUCs have reached differing conclusions, although it must be understood that they were interpreting different laws. Under the federal FOIA, grant applications submitted to a government agency have been required to be disclosed by that agency, under the reasoning that the research designs of scientists not engaged in profit-oriented research are not commercial information or trade secrets.[52] Courts have also required disclosure of IACUC records, reports, and minutes not actually in the government's possession when a government-owned animal research facility is operated by a contractor.[53]

Some courts have ruled that an IACUC at a state university is a public body subject to the state's open meeting law and a public agency subject to the state's public records act.[54] These courts have required the release, upon request under the state's public records law, of applications for IACUC approval,[55] including unfunded grant proposals.[56] However, other courts have reached the opposite conclusion, reasoning that even an IACUC at a state university was created by order of the federal government and thus is accountable only to the USDA and federal funding agencies, not the state, even though the IACUC may be monitoring a state institution. Those courts conclude that disclosure of IACUC records is governed by the federal FOIA, and the state open meetings laws have no application to IACUC meetings.[57]

Institutions have generally, but not always, been successful in withholding from disclosure the names and contact information of researchers and staff under exemptions in the public records laws relating to privacy or safety. One court held that the names and work addresses of the animal researchers "serve to document the organization, functions, and operations" of a state university's activities and must be disclosed as public records,[58] but another court held that its state law allowed any information "concerning research" to be withheld from disclosure.[59]

Confidential and proprietary information such as trade secrets is always exempt from public disclosure, provided it is adequately proven to actually be confidential and proprietary, because exemptions are construed narrowly. For example, an Oregon court held that the names of experimental drugs and the names of companies for which research was performed constituted proprietary information exempt from disclosure under the state's public records law, but required the release of "daily logs" pertaining to the care of individual animals housed at a regional primate research center.[60]

Another possible circumstance under which proprietary or confidential information disclosed to an IACUC might become available to the public is illustrated by a case in which documents received in response to a FOIA request, including IACUC minutes and correspondence between the IACUC and an investigator, were used to support a *qui tam* ("whistleblower") action under the federal False Claims Act.[61] The False Claims Act imposes liability on those who defraud the government.[621] It encourages the uncovering of such fraud by permitting private persons (i.e., "whistleblowers") to bring *qui tam* actions on behalf of the government and allowing them to share in any monetary recovery.[63] The action alleged that a research scientist submitted a fraudulent grant application to the NIH.

In light of the varying court opinions as to what information is confidential or proprietary and what information the public has a right to see, IACUCs should seek guidance from their institutional legal counsel regarding preparation of reports, meeting minutes, and other documents, as well as how to appropriately conduct meetings open to the public. The general guideline should be that reports include everything requested in sufficient detail to be responsive to the purpose, but nothing more. If provided, guidance from the agency requesting a report should be followed.

26:5 How specific should IACUC protocols be when proprietary information related to techniques, compounds, or devices is at stake?

Reg. The AWA (§2143,A,6,B) states that neither the AWA nor the AWAR should be "construed to require a research facility to disclose publicly or to the Institutional Animal Care and Use Committee during its inspection, trade secrets or commercial or financial information which is privileged or confidential." AWA §2157 prohibits any member of an IACUC from using, releasing to the public, or revealing to another person any confidential information of the research facility. The AWAR (§2.35,f) require APHIS/AC inspectors to maintain the confidentiality of any information they obtain in connection with an inspection of a research facility and to disclose it only when necessary, in accordance with FOIA, which exempts the release of privileged and confidential information. Similarly, the HREA (§495,e) states that none of its guidelines or regulations "may require a research entity to disclose publicly trade secrets or commercial or financial information which is privileged or confidential."[64]

However, under the AWAR (§2.31,d; §2.31,e) and PHS Policy (IV,C,1), the IACUC is obligated to review proposed activities related to the care and use of animals and to determine that the proposed activities are in accordance with the AWA and, when applicable, the PHS Policy. Therefore, IACUC protocols must be sufficiently detailed to permit this required review.

Opin. Protocols submitted to an IACUC must be sufficiently detailed to permit the IACUC to evaluate adequately the proposed procedures involving animals, including understanding any potential physical, physiological, and behavioral discomfort, pain, and/or distress the animals may experience as a result of the procedures and related husbandry methods. Protocols must include enough detail to assure the IACUC that no alternatives to the use of animals are available and that the procedures will not unnecessarily duplicate previous experiments. IACUC reviewers must be provided sufficient information to understand whether proposed procedures will be performed with appropriate sedatives, analgesics, or anesthetics, or that the withholding of such agents is scientifically justified. Reviewers must also be able to evaluate whether surgery will be performed on animals and, if so, whether appropriate preoperative, operative, and postoperative care will be provided and whether any animal will be used in more than one surgical procedure.

If a proposed activity related to the care or use of animals utilizes a proprietary method, compound, or device, disclosure in the protocol of that proprietary information may be necessary for the

IACUC to adequately conduct its review of the proposed activity.

Protocols also must include sufficient specificity to permit the IACUC to understand whether proposed procedures may cause conditions for which animals may need medical care, including the provision of anesthetics, analgesics, or tranquilizers, and whether that medical care will be provided. The method of euthanasia of each animal associated with or involved in the proposed study must also be adequately explained in a protocol submitted to an IACUC.

The IACUC must be provided with sufficient information to be able to evaluate the adequacy of the training and other qualifications of all personnel who will conduct the proposed procedures involving animals.

Including descriptions and explanations of proposed procedures in a protocol submitted to an IACUC that are sufficiently detailed to permit the IACUC to evaluate the protocol adequately in accordance with the IACUC's legal responsibilities as just outlined may require the disclosure of confidential or proprietary information. Techniques, compounds, devices, and animal models may be patentable or may have significant commercial value if not publicly known, but the IACUC may need information about them to be able to understand properly any potential discomfort, distress, or pain the animals involved may experience as a result of the procedures or related husbandry methods.

Protection of confidential or proprietary information is no excuse for an inadequate review by an IACUC of proposed procedures involving animals. The IACUC must have sufficient facts to be able to perform its legal and ethical responsibilities and make informed decisions about the welfare of the research animals. Accordingly, the IACUC, and each individual IACUC member, must follow appropriate procedures to provide assurance to investigators that their secrets will not be revealed in ways that could cause harm to the researchers or loss of the value of their intellectual property. Additionally, IACUCs must diligently protect information revealed and entrusted to them and not permit disclosure to anyone not needing to know that information. See 26:8 below for procedural mechanisms an IACUC may adopt to protect confidential and proprietary information of researchers. Also see 8:19 for survey responses to who has access to IACUC protocol forms.

Surv. Does your IACUC make PIs aware of the need to protect confidential or proprietary information (e.g., techniques, compounds, or devices)?

- No 71/184 (39%)
- Yes 100/184 (54%)
- Not applicable 13/184 (7%)

Surv. If yes, what method does your IACUC use to identify or separate out confidential information? Select all that apply.

- Separate section of the protocol 8/100 (8%)
- Separate document or policy 21/100 (21%)
- PI identifies this content during review and redaction processes 37/100 (37%)
- Not provided with the protocol 28/100 (28%)
- I don't know 10/100 (10%)
- Other 12/100 (12%)

26:6 Must a sponsor of research be identified on the IACUC protocol forms?

Reg. Institutions may require their employees to disclose any outside financial support they receive or anticipate receiving in connection with research or other activities so they can identify any potential conflicts of interest or other ethical issues. Disclosure of sponsorships to the IACUC on protocol forms may be included in these institutional policies.

Government employees are required to comply with applicable laws on ethical conduct and financial disclosure and the implementing guidance and policies of their employer. Employees of the Department of Health and Human Services (HHS) and uniformed service officers in the Public Health Service Commissioned Corps on active duty, for example, are bound by the Standards of Ethical Conduct for Employees of the Executive Branch contained in 5 C.F.R. Part 2635. These employees are also subject to the executive branch-wide financial disclosure regulations at 5 C.F.R. Part 2634, the Employee Responsibilities and Conduct regulations at 5 C.F.R. Part 735, and the HHS regulations regarding conduct at 45 C.F.R. Part 73.[65]

Institutions and individual investigators that apply for or receive research funding from PHS granting agencies, including the NIH, must comply with Financial Conflict of Interest (FCOI) regulations in 42 C.F.R. Part 50 Subpart F. Similar regulatory requirements apply to institutions and investigators performing research under PHS contracts and are found at 45 C.F.R. Part 94. Note that intellectual property rights and interests, unless assigned to the institution, are included within the definition of financial interest in these regulations. Under these regulations, significant financial interests must be disclosed to the institutional designated official(s) when an application is submitted to the NIH for funding, upon discovering or acquiring a new significant financial interest, and on an annual basis. If a significant financial interest is determined to constitute a financial conflict of interest, the institution must make the investigator's name, the sponsor's name, and information regarding the significant financial interest publicly accessible through a website or written response to a request for the information.[66]

Opin. There is no *per se* legal requirement to identify a sponsor on protocols submitted to an IACUC. However, protocols for research involving funding from a government agency, a government contract, or a cooperative agreement with a government agency require disclosure of financial sponsors to ensure that the design, conduct, and reporting of funded research is not biased by any conflicting financial interest of the investigators. Therefore, institutions having investigators with such government involvement may adopt IACUC protocol forms for the convenience of the investigators that comply with applicable government regulations, so investigators need not prepare multiple forms. The identification of a federal funding agency is also necessary for the IACUC to know in order to fulfill its responsibilities under the PHS Policy for prompt reporting of non-compliance. However, work funded by private entities which results in non-compliance may not be required to be reported.

Conflict of interest policies require researchers to disclose the identity of sponsors and the nature and extent of any financial relationship between the researcher and sponsor, in order to preserve the integrity of research and prevent potential researcher bias. Institutional policies may require disclosure of a financial sponsor. Scientific journals, where researchers may wish to publish the results of their IACUC-approved animal studies, also frequently require authors to disclose financial support from entities other than their employers, as well as financial relationships with entities that could be perceived to influence the submitted work.

Although not required, IACUCs may choose to request identification of financial sponsors on IACUC protocol forms, as well as any management plan for a financial conflict of interest, to aid their review of activities involving animals. AWAR §2.31,d requires IACUCs to review proposed activities to ensure that all procedures involving animals will avoid or minimize discomfort, distress, and pain to the animals, as well as ensure that the activities do not unnecessarily duplicate previous experiments and that the PI considered alternatives to procedures that may cause more than momentary or slight pain or distress to the animals. A financial sponsor may, intentionally or inadvertently, negatively affect animal welfare through its influence on how a proposed activity is conducted or its hope for certain experimental outcomes. Investigators may believe, perhaps naively, that their actions are not biased by their desire to please their sponsors. Fully informed examinations of protocols, animal activities, and animal living conditions by IACUC members who are concerned primarily with the animals' welfare and not beholden to sponsors would seem to better conform to the purpose of IACUC review.

Even if a research sponsor's name is not listed on IACUC documents, sponsor information may be obtainable by a public records or FOIA request to a public research institution or about a taxpayer-funded activity.

Some agreements concerning the performance of research may require a sponsor's identity to be maintained as confidential to prevent the sponsor's competitors from learning about the research. This situation weighs against disclosure of sponsor information on protocol forms to the extent that it is not necessary for regulatory compliance or compliance with institutional policy. When the name of a sponsor, in conjunction with other information in the protocol, could reveal confidential or proprietary information of value to the sponsor's competitors, the sponsor's name should be deleted before the protocol form is released to anyone outside the IACUC.[67]

Surv. Does your IACUC protocol form ask for the identification of the financial sponsor of the study (if any)?

• Not applicable	8/184 (4%)
• Yes	147/184 (80%)
• No	27/184 (15%)
• I don't know	2/184 (1%)

26:7 Can researchers protect their own identities on IACUC protocol forms?

Reg. AWAR §2.31,c,6 and PHS Policy IV,B,6 give the IACUC the authority to accept, require modifications of, or withhold approval of animal activities. To be able to approve a proposed animal activity, an IACUC must know the identities of the personnel who will be performing each part of a protocol to determine whether they are adequately and appropriately qualified and trained.

Scientists, technicians, and other personnel involved with animal care, treatment, or research must be appropriately qualified and trained in humane practices and research methods; this is required by the AWA (§2143,d); AWAR (§2.31,d,1,viii), and the PHS Policy (IV,C,1,f). The HREA mandates that applicants for grants, contracts, or cooperative agreements involving animal research which are administered by the NIH or any national research institute provide assurances that "scientists, animal technicians, and other personnel involved with animal care, treatment, and use by the applicant have available to them instruction or training in the humane practice of animal maintenance and experimentation, and the concept, availability, and use of research or testing methods that limit the use of animals or limit animal distress."[68]

In addition to the identities of those persons performing procedures, the identities of other personnel who may be involved with a protocol may need to be disclosed for purposes of identifying conflicts of interest. Institutions and individual investigators that apply for or receive research funding from PHS granting agencies, including the NIH, must comply with FCOI regulations in 42 C.F.R. Part 50 Subpart F. Similar regulatory requirements apply to institutions and investigators performing research under PHS contracts and are found at 45 C.F.R. Part 94. Note that intellectual property rights and interests, unless assigned to the institution, are included within the definition of financial interest in these regulations. Under these regulations, significant financial interests must be disclosed to the institutional designated official(s) when an application is submitted to the NIH for funding, upon discovering or acquiring a new significant financial interest, and on an annual basis. If a significant financial interest is determined to constitute a financial conflict of interest, the institution must make the investigator's name, the sponsor's name, and information regarding the significant financial interest publicly accessible through a website or written response to a request for the information.[69]

It should be noted that PHS states: "Institutions may, at their discretion, represent the names of members other than the chairperson and veterinarian with program authority by using numbers or other symbols in submissions to OLAW. Sufficient information for OLAW to determine that all appointees are appropriately qualified must be provided. The identify of each member must be readily ascertainable by the institution and available to authorized OLAW or other PHS representatives upon request."[70]

Opin. The IACUC is required to determine that the personnel conducting animal research are properly qualified by their education and training. The identity of all personnel conducting procedures on animals must be disclosed in order for the IACUC to make this determination. Additionally, the identity of the PI overseeing the work is important to the evaluation of the protocol and should be disclosed to the IACUC. Identities may also be relevant for purposes of evaluating conflicts of interest in accordance with HHS regulations and institutional policies. If a researcher has a significant financial interest that is determined to be a financial conflict of interest, the institution must make the researcher's name, title, and role in the research publicly accessible through a website or written response to a request for such information.[71]

If IACUC records are requested by the public under FOIA or a state's public records law, or if IACUC meetings are open to the public, the identities of the researchers may or may not be able to be protected from disclosure, depending on the applicable law. Names and contact information of researchers are usually held to be exempt from disclosure, but biographies, credentials, and other information that could identify specific individuals may be required to be disclosed if such information is publicly available from other sources[72] (see 7:12).

Surv. Does your institution protect the identities of Principal Investigators on protocols?

- No 123/184 (67%)
- Yes 54/184 (29%)
- I don't know 7/184 (4%)

26:8 What procedural mechanisms may an IACUC adopt to identify and protect information considered proprietary by the researchers?

Reg. Under §2157 of the AWA, it is unlawful for any member of an IACUC to release to the public any confidential information of the research facility. Similarly, subsection (e) of the HREA states that none of its guidelines or regulations "may require a research entity to disclose publicly trade secrets or commercial or financial information which is privileged or confidential."[73] Confidential information includes trade secrets, processes, operations, style of work, apparatus, the identity, confidential statistical data, amount or source of any income, profits, losses, or expenditures of a research institution. The AWA (§2157) also prohibits any member of an IACUC from using or attempting to use to his or her advantage, or revealing to another person, any information entitled to protection as confidential information of the research facility.

Opin. While serving on an IACUC, members receive information from researchers. Some of the information will be confidential or proprietary. AWA §2157 imposes a duty on IACUC members not to disclose to anyone outside the IACUC any confidential information they learn through service on the IACUC and not to use any such information in their own research without the express permission of the owner of the information. Thus, it is important for an IACUC to implement mechanisms to first identify potentially confidential and proprietary information and then protect it from inappropriate disclosure.

Flagging. Although all information submitted to and created by an IACUC should be handled securely, researchers' confidential and proprietary information should be especially protected. Investigators could be requested to flag or mark confidential or proprietary information specially, including a designation on the title page that such information is included. Such information could be included as an attachment to a standard protocol form. In an electronic document, confidential or proprietary information could be included in a separate document that is encrypted or protected with an additional password. Under FOIA and state public records laws, documents containing both confidential and non-confidential information are subject to disclosure. When such documents are requested, the confidential information should be redacted prior to disclosure. Flagging or marking information considered confidential or proprietary, or including such information in an attachment, would facilitate identification and segregation of the information by the IACUC or institution in the event of a request by the public for inspection of IACUC records or the preparation of an IACUC report or submission to a government agency. However, unless information falls within the scope of a defined exemption in the applicable public records law, the information will nevertheless have to be disclosed upon request.

Investigators should be advised that although all information submitted to an IACUC will be treated confidentially to the extent permitted by law, some information may become available to the public through state public records requests or discussion in open meetings. Only information within the scope of a legal exemption to disclosure can be assured of protection; flagging or labeling portions of a submission as confidential or proprietary cannot guarantee nondisclosure but can assist the IACUC or the institution in reviewing a record in the event of a public records request or open meeting. Investigators should be instructed to identify and flag selectively only information that is truly confidential or proprietary.

Confidentiality or Nondisclosure Agreements. As a general rule, any information provided by researchers to an IACUC should be distributed only to those persons who have a need for the information. No information received by an IACUC member from a researcher should be forwarded, discussed, or otherwise shared with anyone not essential to the review of the procedures or related husbandry, whether or not the information seems confidential or proprietary. To emphasize IACUC members' duty not to disclose information, requiring that they sign confidentiality or nondisclosure agreements is a good practice.

Security. Basic security measures should always be taken to protect information about the use of animals in research or persons conducting such research because of the controversial nature of such activities and the ease with which they can be misinterpreted. These security measures include, at a minimum, building access controls, escorting visitors, and restrictions on photography.

Additional security measures should be taken to protect documents, data, and other written or recorded information provided by researchers to an IACUC, as well as IACUC meeting minutes in which such information is discussed. These measures should include storage in a locked cabinet in an area with restricted access and shredding or similar destruction of duplicative or otherwise unnecessary records. IACUC members should be encouraged to destroy their personal copies of IACUC-related documents or return them to the IACUC secretary when they are no longer needed. A clear and concise policy for the retention and destruction of records should be followed, and materials regularly destroyed when the retention period expires.

Electronic Security. Electronic materials are readily copied and distributed, so electronic protocol submissions, distribution, and reviews must be tightly controlled. The computers used for electronic storage, transmission, or discussion of researchers' information should be access-controlled and password-protected, including a requirement to enter a password after a period of nonuse, and encryption of IACUC-related documents should be considered. In establishing security policies and procedures for the protection of researchers' electronic confidential or proprietary information, institutions may find useful the specifications for the protection of electronic health information in the security regulations promulgated under the Health Insurance Portability and Accountability Act of 1996 (HIPAA).[74] These regulations describe administrative, physical, and technical safeguards to prevent access to sensitive information by unauthorized persons. Institutions that include or are affiliated with a medical center may have personnel and technical resources experienced in working under the HIPAA regulations available to assist in developing appropriate procedures and policies for the protection of proprietary information.

26:9 What responsibilities do IACUCs and IACUC members have to protect records? What guidelines should be followed to protect confidential and proprietary information?

Reg. IACUCs must, under AWAR §2.35 and PHS Policy IV,E, maintain certain records and reports for at least three years. AWAR §2.35,f and PHS Policy IV,E,2 require records that relate directly to proposed activities involving animals and proposed significant changes in ongoing activities reviewed and approved by the IACUC to be maintained for the duration of the activity and for an additional three years after the completion of the activity. These records must be available for inspection and copying by authorized USDA or funding federal agency representatives. In addition, an institution has a legal duty to identify, preserve, and produce any records relating to a potential or actual legal claim.

Under AWA §2157, it is unlawful for any member of an IACUC to use or release to the public or reveal to another person any confidential information of the research facility, including "any information that concerns or relates to (1) the trade secrets, processes, operations, style of work, or apparatus; or (2) the identity, confidential statistical data, amount or source of any income, profits, losses, or expenditures of the research facility." Such wrongful use or disclosure may result in removal from the IACUC, fines, and/or imprisonment. In addition, any institution or individual who is harmed by such disclosure or use may bring a lawsuit against the IACUC member and recover damages and the costs of the suit from the IACUC member.

Opin. The IACUC is required by law, under AWAR §2.35 and PHS Policy IV,E, to retain a copy of every approved protocol and to record and maintain minutes of all committee meetings and deliberations for at least three years. These laws further require that IACUC records that relate directly to proposed activities and proposed significant changes in ongoing activities involving animals must be retained for three years after the completion of the activity. These records must be available for inspection and copying by authorized representatives of the USDA or the funding federal agency (if applicable). The IACUC should serve as the custodian for such records and should maintain those records in a secure location, whether physically, electronically, or both. There is no legal requirement for individual members to maintain separate copies of protocols or other similar records. Maintenance of separate records increases the risk of disclosure of confidential and proprietary information they may contain. In addition, in the event the institution is required to produce such records in response to a legal claim, it will be required to produce all existing copies of such records. If an institution has implemented policies and procedures that (1) require original records to be maintained by a designated custodian and (2) require destruction of all other copies of such records once no longer needed for their intended use, it will be required to produce only the original record. However, the institution must be able to show that the policies and procedures are actually followed. Thus, if an IACUC does not require individual members to destroy such records at the end of meetings or after final action, or if an IACUC has not implemented policies and procedures to retain official records and destroy unnecessary copies, it may be required by law to identify, preserve, and produce both original records and all copies.

In today's paper and electronic environment, the task of identifying and producing all existing copies can require extensive resources and financial cost. Thus, IACUCs should have policies and procedures that include collecting copies of protocols and other documents reviewed by committee members, either at the end of the relevant meeting or when the committee's review is completed. Such practices not only protect confidential and proprietary information contained in the records, but also reduce legal and financial risk to the IACUC and institution. These practices should be governed by institutional record management policy. The IACUC practices should not conflict with the institution's other record retention requirements, including preserving records related to a potential or pending legal claim or retaining such records in compliance with other federal laws and regulations pertaining to animal-based research with which the IACUC must comply.

Surv. Does your IACUC require that protocols or other sensitive documents be turned in, destroyed, or otherwise removed at the end of an IACUC meeting or once a final action has been taken by the IACUC?

• Not applicable	40/184 (22%)
• Yes	50/184 (27%)
• No	90/184 (49%)
• I don't know	4/184 (2%)

Surv. What mechanism(s) does your institution use to protect the identities of Principal Investigators, IACUC members, and staff on documentation? Select all that apply.

• Coding	52/184 (28%)
• Initials	41/187 (22%)
• Redact names per state records law	78/187 (42%)
• Instructions provided indicating the need to limit personal information in document	37/187 (20%)
• Other	44/187 (24%)

26:10 For what records and related issues should an institution have a records retention policy?

Reg. The PHS Policy (IV,E,1) indicates that the "awardee institution shall maintain:

- a copy of the Assurance which has been approved by the PHS
- minutes of IACUC meetings, including records of attendance, activities of the committee, and committee deliberations
- records of applications, proposals, and proposed significant changes in the care and use of animals and whether IACUC approval was given or withheld
- records of semiannual IACUC reports and recommendations (including

minority views) as forwarded to the Institutional Official
- records of accrediting body determinations

PHS Policy (IV,E,2) states: "[a]ll records shall be maintained for at least three years; records that relate directly to applications, proposals, and proposed significant changes in ongoing activities reviewed and approved by the IACUC shall be maintained for the duration of the activity and for an additional three years after the completion of the activity. All records shall be accessible for inspection and copying by authorized OLAW or other PHS representatives at reasonable times and in a reasonable manner."[75]

AWAR §2.35(a) states: "The research facility shall maintain the following IACUC records:

- Minutes of IACUC meetings, including records of attendance, activities of the Committee, and Committee deliberations;
- Records of proposed activities involving animals and proposed significant changes in activities involving animals, and whether IACUC approval was given or withheld; and
- Records of semiannual IACUC reports and recommendations (including minority views), prepared in accordance with the requirements of § 2.31(c)(3) of this subpart, and forwarded to the Institutional Official."

All records and reports shall be maintained for at least three years (AWAR §2.35,f) . Records that relate directly to proposed activities and proposed significant changes in ongoing activities reviewed and approved by the IACUC shall be maintained for the duration of the activity and for an additional three years after completion of the activity. All records shall be available for inspection and copying by authorized APHIS or funding federal agency representatives at reasonable times.

Opin. IACUCs generate numerous documents associated with the many different IACUC responsibilities. Animal use protocols, facility and laboratory inspection reports, semiannual reviews, and investigation of non-compliance or animal welfare are just a few of the IACUC-associated records. The submission of a protocol alone results in the creation of subsequent documents including but not limited to the following: IACUC reviewer comments and veterinary consultations; verification of personnel training; occupational health risk evaluations; approval notices; protocol modifications; and the e-mail correspondence of individuals involved with the protocol review and approval process. Meeting minutes (written or recorded) are created to document member attendance and IACUC deliberations at convened meetings. The IACUC creates reports for review by the Institutional Official and external agencies such as the Semiannual Program Review and Facility Inspection Report (including minority views), as well as accounts of non-compliance or animal welfare concerns. Additional records include reports from accrediting bodies, Drug Enforcement Agency records and drug logs, financial disclosures, and purchasing records. Legal counsel should be consulted about the proper form and content of retained documents. Public institutions and documents submitted to agencies subject to open public records laws, such as OLAW and USDA, should be drafted with the assumption the document will be released to members of the public.

While federal requirements dictate the maintenance and disposition of certain documents, it is best practice for an institution to have a record retention policy to ensure proper maintenance, archival, and disposition of IACUC records. Expertise should be sought regarding time for retention and proper disposal. Some institutions have a formal Record Retention Office or record retention experts. In the absence of these resources, obtaining institutional legal input is strongly advised. For public institutions subject to state public records laws, it may be best practice to institute retention times no longer than federal and state requirements. IACUC offices should consider developing a plan and devote attention to regular assessment of records to ensure they are not maintained for longer than required. E-mail is commonly used to rapidly share information and receive feedback. IACUC members and staff should refrain from including sensitive information in e-mails as e-mails may be subject to Public Records requests in certain states. In addition, an institution has a legal duty to identify, preserve, and produce any records relating to a potential or actual legal claim.

Surv. Does your IACUC or institution have a records retention policy?

- No 17/184 (9%)
- Yes 167/184 (91%)

Surv. If yes to the above question, what dictates the time length for the record retention policy? Select all that apply.

- State regulation 71/167 (43%)
- Federal regulation 132/167 (79%)
- Institutional requirements 96/167 (57%)
- I don't know 8/167 (5%)
- Other* 2/167 (1%)

*Other: Records are retained for as long as the platform exists.

Surv. How long does your IACUC maintain protocols before they are shredded or purged from the system?

- <3 years 1/184 (1%)
- 3 years 35/184 (19%)
- 6 years (3 year protocol lifecycle plus an additional 3 years) 63/184 (34%)
- >6 years 57/184 (31%)
- I don't know 5/184 (3%)
- Other 23/184 (13%)

26:11 May an IACUC be sued for breach of proprietary confidence?

Reg. AWA §2157 addresses disclosures or wrongful uses of proprietary information by a member of an IACUC. This statute prohibits the release by an IACUC member of any confidential information that "concerns or relates to (1) the trade secrets, processes, operations, style of work, or apparatus; or (2) the identity, confidential statistical data, amount or source of any income, profits, losses, or expenditures, of the research facility." This statute also makes it illegal for any member of an IACUC "(1) to use or attempt to use to his advantage; or (2) to reveal to any other person any information which is entitled to protection as confidential information" under the statute. Penalties for violation of this law include a fine of up to $10,000 and imprisonment of up to three years, as well as removal from the IACUC; the violator also will be required to pay the damages, costs, and attorneys' fees of the injured investigator and/or the research institution. Thus, IACUC members have a duty not to disclose to anyone outside the IACUC any confidential information they learn through service on the IACUC and not to use or attempt to use any such information to their own advantage without the express permission of the owner of the information. In addition, any institution or individual who is injured by such disclosure or use may bring a lawsuit against the IACUC member and recover damages and the costs of the suit from the IACUC member.

Opin. An IACUC can be sued for breach of proprietary confidence. With respect to animals regulated by USDA, Animal Care, IACUCs are created under and obtain their authority from the AWA, which explicitly requires that IACUC members maintain the confidentiality of proprietary information they receive in the course of performing their responsibilities, specifies penalties for violations of confidentiality, and provides remedies for injured persons. Violation of this law could be the basis for a lawsuit against the IACUC or an individual member of the IACUC. The HREA also provides authority for IACUCs at institutions receiving funding from certain sources; it states that it should not be interpreted to require the IACUC or research entity to disclose trade secrets or commercial or financial information which is privileged or confidential. The HREA thus could potentially provide another basis for a lawsuit against an IACUC that releases such information, although the HREA does not specifically provide for rights and remedies of persons harmed by disclosures.

Additionally, a legal claim could be based on the improper disclosure of confidential proprietary information by an IACUC during an open meeting or in response to a request under a state public records act, or in a submission by the IACUC to a government agency. Further, a claim for breach of proprietary confidence also could be supported by inadvertent disclosure of confidential proprietary information. Inadvertent disclosure might occur if IACUC records or an IACUC member's notes are not stored securely or are disposed of in a way that permits viewing of confidential information by unauthorized persons.

A state university successfully avoided breaching proprietary confidence in *Mississippi State University v. People for the Ethical Treatment of Animals, Inc.*[76] The university had performed research for a company under contracts that included nondisclosure agreements requiring secrecy of information and no disclosure of confidential information. PETA sought IACUC records under the state public records law for projects, tests, and experiments funded by the company. The court held that the university was a public body as defined by state law and therefore the IACUC protocol forms were public records, but the data and information recorded on IACUC protocol forms was not public property because it was owned, furnished, or required by the company. The completed protocol forms in this case contained confidential commercial, proprietary, and trade secrets information about the company's research, so the university was able to convincingly argue that they fit within a statutory exemption permitting nondisclosure.

26:12 Must an IACUC at a state institution open its meetings to the public?

Reg. To promote transparency in government, every state has some version of law requiring all government business to be conducted in open meetings to which the public has access. The court's opinion in *Oklahoma Association of Municipal Attorneys* states why open meetings laws exist: "If an informed citizenry is to meaningfully participate in government or at least understand why government acts affecting their daily lives are taken, the process of decision making as well as the end results must be conducted in full view of the governed."[77]

However, AWA §2157 prohibits IACUCs from releasing to the public any confidential information of their research facilities.

Opin. Whether an IACUC at a state institution is required to open its meetings to the public depends on the wording and interpretation of the state's open meetings law. Open meetings laws, also referred to as "open door" or "sunshine" laws, require that meetings of public bodies be open to the public. If either the IACUC or the state institution is considered a public agency or public body as defined under the open meetings law of its state, then the IACUC may be required to hold its meetings open to the public. Nevertheless, these laws usually include a provision that the committee or other entity may adjourn into an executive session closed to the public when discussing confidential or proprietary information or information defined by law as privileged, such as communications between an attorney and client or doctor and patient.

The few state courts that have considered open meetings laws in relation to IACUCs have reached different conclusions, although it must be understood that they were interpreting different state laws (see also 26:4).

In *Animal Legal Defense Fund, Inc. v. Institutional Animal Care & Use Committee of University of Vermont*,[78] the Supreme Court of Vermont ruled that the IACUC at a state university was a committee of the university and thus was subject to the state's open meetings law.

In *Dorson v. State of Louisiana*,[79] a court in a different state held that the IACUC at a state university was created by the order of the federal government and thus was accountable only to the USDA and NIH, not the state, even though the entity monitored by the IACUC was a state institution. That court concluded that the state open meetings law had no application to IACUC meetings. The highest court in New York also held, in *In re American Society for the Prevention of Cruelty to Animals v. Board of Trustees of the State University of New York*,[80] that meetings of the IACUC of a state university did not need to be open to the public because the IACUC's powers and functions derived solely from federal law.

Surv. Are your IACUC meetings open either in full or in part to the general public?

- Not applicable. 6/184 (3%)
- Yes, and we are a public institution. 29/184 (16%)
- Yes, and we are a private institution. 5/184 (3%)
- No, and we are a public institution. 66/184 (36%)
- No, and we are a private institution. 78/184 (42%)

26:13 Are there special factors that should be considered when IACUC meetings are conducted on an electronic meeting platform?

Reg: While the AWA and PHS Policy are silent with respect to meetings held on electronic platforms, Notice Number: NOT-OD-06-052 issued by OLAW on March 24, 2006, related to the use of telecommunications when a convened meeting of a quorum of the IACUC is required under PHS Policy indicates that

> traditional convened meeting, physically attended by IACUC members, provides the optimal forum in which to conduct full committee review of proposals and consider potential suspensions. Introduction and integration of new members to the Committee is also most effectively accomplished during physically-convened meetings. However, OLAW recognizes that some forms of telecommunications facilitate the conduct of business, reduce regulatory burden, are standard practice in many forums, and enhance flexibility without compromising the quality of deliberation and interaction. The IACUC chairperson, as the appointed leader of the committee, bears some responsibility for holding committee meetings in a manner that encourages participation and facilitates interaction among members.[81]

Moreover, Notice Number NOT-OD-06-052 indicates that methods of telecommunications (e.g., telephone or video conferencing) are acceptable for the conduct of official IACUC business requiring a quorum, provided the following criteria are met:

- Members are given notice of the meeting.
- Documents normally provided to members during a physically convened meeting are provided to all members in advance of the meeting.
- All members have access to the documents and the technology necessary to fully participate.
- A quorum of voting members is convened when required by PHS Policy.
- The forum allows for real-time verbal interaction equivalent to that occurring in a physically convened meeting (i.e., members can actively and equally participate and there is simultaneous communication).
- If a vote is called for, the vote occurs during the meeting and is taken in a manner that ensures an accurate count of the vote. A mail ballot or individual telephone polling cannot substitute for a convened meeting.

- Opinions of absent members that are transmitted by mail, telephone, fax or e-mail may be considered by the convened IACUC members but may not be counted as votes or considered as part of the quorum.
- Written minutes of the meeting are maintained in accord with the PHS Policy, IV.E.1.b.

It should also be noted that USDA APHIS, Animal Care concurred with this Notice.[82]

Opin. Institutions are increasingly utilizing electronic meeting platforms to conduct meetings, including IACUC meetings. Electronic platforms are widely available, convenient, and save travel time and expense. The trend is likely to continue in the future. Use of these platforms to hold convened meetings include most, if not all, of the requirements applying to open (public) in-person meetings. IACUC meetings conducted via teleconference or via online meeting platforms, such as Zoom, Microsoft Teams, or Google Meet, are considered functionally equivalent to physically attended meetings. While the views or opinions expressed by absent members submitted in advance of an IACUC meeting can be considered by IACUC members at a convened meeting, a mail-in ballot or individual telephone or email poll is not a substitute for participation and those views or opinions cannot be counted as votes.

Institutions subject to "open meeting" requirements may also be subject to specific requirements for electronic meetings such as recording, identifying attendees, publicizing meeting dates, and admitting members of the public who wish to attend. Recorded meetings may be subject to similar requirements as written documentation such as record retention and availability through public records requests. Institutions must also make informed decisions about whether to record audio only or audio and video, who is responsible for recording the meeting, where recordings are stored, which institutional offices have copies, and how long the recordings are retained.

Surv. What methods does your IACUC use to conduct virtual meetings? Select all that apply.

- Teleconference 30/184 (16%)
- Videoconference using a virtual meeting platform (e.g., Zoom, Teams, etc.) 181/184 (16%)
- Not applicable 2/184 (1%)

Surv. If your institution uses tele- or videoconferencing for IACUC meetings, will this trend likely continue for future IACUC meetings?

- Not applicable. 0/182 (0%)
- Yes. 126/182 (69%)
- Not likely. 6/182 (3%)
- We have done this only because of the pandemic. 27/182 (15%)
- We have always met or had the option to meet virtually or by teleconference. 20/182 (11%)
- Other. 3/182 (2%)

Surv. For public institutions: does your institution consider an IACUC meeting as being open to the public?

- Yes 26/104 (25%)
- No 68/104 (65%)
- I don't know 6/104 (6%)
- Other 4/104 (4%)

Surv. If yes, how is the public notified and included with in-person, tele- or videoconference formats? Select all that apply.

- Newsletter 0/26 (0%)
- Posting on website 17/26 (65%)
- Inform Public Relations Office 2/26 (8%)
- Other* 8/26 (31%)

*Other:

- No specific notification is provided.
- Information is provided only in response to submitted requests.

Surv. Are your IACUC meetings recorded?

- No 134/184 (73%)
- Yes 50/184 (27%)

Surv. If yes, in what format are your meeting recorded?

- Audio/Video 31/49 (63%)
- Audio only 18/49 (37%)

Surv. If yes, who records and maintains the meeting recording?

- IACUC administration 46/50 (92%)
- Office of Legal Affairs (or similar) 1/50 (2%)
- Public Information Office 0/50 (0%)
- Not sure 0/50 (0%)
- Other* 3/50 (6%)

*Other: Respondent indicated that the information technology group records and maintains the recording.

Surv. If yes, how long are the recordings maintained?

- Erased soon after minutes are approved 32/50 (64%)
- < 3 years 2/50 (4%)

- > 3 years but less than 6 years 8/50 (16%)
- > 6 years 3/50 (6%)
- Other* 5/50 (10%)

*Other:

- On average, 60 days
- One year after the meeting date
- Recordings not destroyed, maintained in perpetuity

Surv. Does your institution prohibit audio and/or video recording of IACUC meetings by attendees?

- No 68/184 (37%)
- Yes 78/184 (42%)
- I don't know 27/184 (15%)
- Other* 11/184 (6%)

*Other: The institution does not have a specific policy or stated position on this matter.

Surv. If yes, what methods does your institution utilize to prohibit recordings? Select all that apply.

- Confidentiality policy 36/74 (49%)
- Signed confidentiality form 33/74 (45%)
- Announcements/reminders to not record 27/74 (36%)
- Prohibition on cellphones and recording devices in the meeting room 5/74 (7%)
- I don't know 5/74 (7%)
- Other* 5/74 (7%)

*Other: Respondent noted that recording is restricted by state party consent law

26:14 Is an IACUC at a state institution required to make its records available to the public?

Reg. The federal FOIA, state public records acts, and state open meetings laws require public bodies, such as government agencies and state universities, to make their information and records available to the public upon request.[83] Other IACUC records disclosed to the government include the Assurance, Annual Reports to OLAW, and noncompliance reports

Opin. Whether an IACUC at a state institution is required to make its records, including records not submitted to another entity, available to the public varies by state. This issue depends on the wording and interpretation of the state's public records law, also referred to as the state's FOIA. Public records laws generally require that records in the possession or control of a public body be accessible for inspection and duplication by the public except for specific exceptions enumerated in the law. Courts generally construe state public records acts strongly in favor of disclosure.

The determination of whether an IACUC must disclose records to the public is a multi-step analysis. The first piece of the analysis is whether the IACUC is a public body. If either the IACUC or the state institution is considered a public agency or public body, as defined in the public records law, then the IACUC may be required to make its records available to the public. Second, it must be determined whether any of the specific exemptions found in the public records law apply. Finally, disclosure depends on whether the records have already been disclosed to a federal government agency, which may make the records available to the public under the federal FOIA.

Few courts have considered whether IACUCs are public bodies subject to records disclosure. Those courts that have considered public records laws in relation to IACUCs have reached different conclusions, although it must be understood that they were interpreting different state laws (see also 26:4). Some courts have ruled that an IACUC at a state university is a public agency subject to the state's public records act.[84] These courts have required the release, upon request under the state's public records law, of applications for IACUC approval,[85] including unfunded grant proposals.[86] However, other courts have reached the opposite conclusion, reasoning that even an IACUC at a state university was created by order of the federal government and thus is accountable only to the USDA and federal funding agencies, not the state, even though the IACUC may be monitoring a state institution. Those courts conclude that disclosure of IACUC records is governed only by the federal FOIA[87] and the IACUC is not an entity "performing a governmental or proprietary function for the state."[88]

State public records laws, like the federal FOIA, typically include a list of exemptions. If one or more of the enumerated exemptions applies, the otherwise presumptively public document, or portions of it, may be withheld from disclosure. The institution seeking to withhold the information has the burden of proving that the information fits within an exemption. Exemptions are interpreted very narrowly. See 26:16 for more information on exemptions.

Institutions have generally, but not always, been successful in withholding from disclosure the names and contact information of researchers and staff under exemptions in the public records laws relating to privacy or safety. One court held that its state law allowed any information "concerning research" to be withheld from disclosure,[89] but another court held that the names and work addresses of the animal researchers "serve to document the organization, functions, and operations" of a state university's activities and must be disclosed as public records.[90] The court in *In re*

Physicians Committee for Responsible Medicine v. Hogan stated that public employees have a diminished expectation of privacy with regard to details of their public employment.[91] That court found that the public has a legitimate interest in being granted access to records detailing the research activities of state employees in order to assess the scientific value and social utility of the research being performed. The court permitted contact information of individual researchers to be withheld but required disclosure of information identifying researchers and describing details of their work in response to a request under the state's freedom of information law. It should be noted that PHS Policy states: "Institutions may, at their discretion, represent the names of members other than the chairperson and veterinarian with program authority by using numbers or other symbols in submissions to OLAW. Sufficient information for OLAW to determine that all appointees are appropriately qualified must be provided. The identify of each member must be readily ascertainable by the institution and available to authorized OLAW or other PHS representatives upon request."[92]

Confidential and proprietary information such as trade secrets is always exempt from public disclosure, provided it is adequately proven to actually be confidential and proprietary. For example, the court in *In Defense of Animals v. Oregon Health Sciences University*[93] found the names of drug company sponsors and experimental drugs to be exempt from disclosure because they constituted sensitive business records that would not ordinarily be disclosed to the companies' competitors. That court also held that staff member names were exempt from disclosure because the public interest in disclosure was not outweighed by the university's interest in nondisclosure.

Although not directly involving IACUC records, the opinion of the Supreme Court of Ohio in *State ex rel. Physicians Committee for Responsible Medicine v. Board of Trustees of Ohio State University*[94] is instructive. An advocacy organization had filed a writ of mandamus to compel a state university to release certain photographic and videotaped records created by the college of medicine in connection with its research on spinal cord injuries using laboratory mice and rats. The applicable public records statute exempted from public disclosure "intellectual property records" created during research, but only if those records had not been "publicly released, published, or patented."[95] In this case, the state university had lent some of the requested records to scientists and research trainees and shown a few of them to scientists at medical conferences. Acknowledging its responsibility to construe the public records act strictly in favor of disclosure, the court nevertheless held that the intellectual property records exception to disclosure was available in this case because of the efforts the state university took to protect the confidentiality of the records. The state university kept the records in secure cabinets within a locked office accessible only to members of the research laboratory, lent them only to persons who signed nondisclosure agreements, and showed them only to small groups of scientists and researchers under controlled circumstances. The court found that this limited sharing, for purposes related to the research, did not constitute prior public release, so the advocacy organization was not entitled to any of the records under the state public records act.

In the event an IACUC or state institution is not considered a public body under its state public records laws, or it is considered a public agency but exemptions apply, the public could still access IACUC records if the records were ever disclosed to the federal government. Once IACUC records are disclosed to the federal government or an agency of the federal government, such as semiannual inspection, reports of non-compliance, or annual reports, such records, which were safe under state public records laws, may be considered public records as defined under FOIA and may then be available to the public, unless they fall within a FOIA exemption. Information the federal government merely has a right to obtain or may have relied on but never actually obtained is usually not available to the public.

Responses to survey question 6:22 show the impact of FOIA and similar state laws on IACUC minutes.

26:15 Is an IACUC at a private institution subject to FOIA or a state's public records or open meeting laws and therefore required to make its records and meetings open to the public?

Reg. The federal Freedom of Information Act[96] states public records acts and state open meetings laws require public bodies, such as government agencies and state universities, to make their information and records available to the public upon request. These laws do not apply to private nongovernmental institutions.

Opin. As a general rule, private institutions are not required to make their records or meetings available to the public. However, if a private institution is supported by, affiliated with, or considered a state-aided, state-owned, or public institution, the state open meeting and public records laws may apply. Simply receiving a grant from a government agency does not make a private organization a public agency subject to these laws; nor does a private organization necessarily become a public body by performing specific functions for the government. But a government cannot delegate away its statutory responsibility to govern transparently by "outsourcing" its governmental functions.

IACUC records from a government-owned animal research facility operated by a private contractor must be disclosed under FOIA because the government is considered to own those records.[97] This may be true also under state public records acts.[98] Arrangements between private entities and governments must be examined individually, considering several factors, including the type and degree of control exercised by the government over the private entity as well as the portion of the entity's business which is done by contract with the government.[99]

When a private institution or IACUC at a private institution discloses any of its records, such as semiannual inspection or annual reports, to the federal government, such records become available to the public under FOIA, because they now constitute public records. Similarly, reports prepared by an APHIS/AC representative following inspection of a private research facility are public records. Information the government merely has a right to obtain or may have relied on but never actually obtained is usually not available to the public.

Even involuntary disclosure of IACUC records to the government may make those records available to the public. In *In Defense of Animals v. U.S. Department of Agriculture*, records relating to the USDA's investigation of a contract research organization, including IACUC records, were required to be disclosed under FOIA.[100] The research organization argued that disclosure would be perceived as a breach of its confidentiality agreements with its customers and cause it to lose study sponsors, but the court held that the only relevant harm of disclosure was whether a competitor could reap a commercial windfall from the use of disclosed proprietary information and the organization had failed to carry its burden of proving that harm.

26:16 How should IACUCs respond to FOIA or state public records act requests?

Reg. The federal FOIA[101] requires a federal agency to provide access to records in its possession or control to requesters. This may include providing access to an institution's annual reports and inspection reports filed with the APHIS/AC pursuant to the AWA, as well as an institution's Assurance statement and other reports made to NIH/OLAW. All states have their own public records laws, also referred to as state freedom of information acts. These state laws provide for the public disclosure of state agency activities and records, which may apply to state universities.

AWA §2157 states that it is unlawful for any member of an IACUC to release to the public any confidential information of the research facility. Confidential information includes the trade secrets, processes, operations, style of work, and apparatus of the facility, as well as the identity, confidential statistical data, amount or source of any income, profits, losses, or expenditures of the facility. It is also unlawful for any member of an IACUC to use or attempt to use to his or her advantage, or to reveal to another person, any information entitled to protection as confidential information of the research facility.

Opin. Many freedom of information requests for IACUC documents originate from animal rights activist groups and the number of requests from such groups is increasing.[102] Requests for information should be made directly to the federal or state agency, but an institution and its IACUC should take certain steps to protect confidential, protected, or proprietary information in anticipation of such requests.

Responding to FOIA Requests: Facts and Resources[103] is a helpful reference guide for researchers about FOIA and state open records laws developed jointly by the National Association for Biomedical Research (NABR), Society for Neuroscience (SfN), and Federation of American Societies for Experimental Biology (FASEB).

Most institutions' legal counsels maintain procedures for responding to governmental requests such as subpoenas. Institutions and their IACUCs should process freedom of information requests under a similar procedure. Individuals knowledgeable in FOIA and related state laws with an understanding of what information is subject to disclosure should process such requests. IACUC members and administrative staff should be trained on and become knowledgeable of FOIA and similar state laws. Guidelines or best practices should be established for developing records giving due consideration to the records most likely to be requested. Records used for governmental submissions should provide only necessary information and should not include extraneous information. If responding in connection with an alleged non-compliance situation, references to particular individuals by name should be avoided. Rather, titles or descriptions of individuals should be used. Finally, disclosure of confidential or proprietary information should be strictly avoided.

In a 2019 decision, the United States Supreme Court ruled that the term "confidential" under Exemption 4 of the Freedom of Information Act regarding the protection of confidential trade secrets and commercial or financial information must be given its "ordinary" meaning instead of using a substantial competitive harm test.[104] This ruling has the potential to change the way IACUCs at federal institutions protect such information, which could include the institution's own IACUC records or confidential information provided to the government by a private entity. The Department of Justice developed this step-by-step guide to determine whether commercial or financial information provided by a private entity is "confidential" under Exemption 4 :

1. Does the submitter customarily keep the information private or closely held? (This inquiry may in appropriate contexts be determined from industry practices concerning the information.)
   - If no, the information is *not* confidential under Exemption 4.
   - If yes, answer question 2.
2. Did the government provide an express or implied assurance of confidentiality when the information was shared with the government?
   - If no, answer question 3.
   - If yes, the information is confidential under Exemption 4 (this is the situation that was present in *Argus Leader*).
3. Were there express or implied indications at the time the information was submitted that the government would publicly disclose the information?
   - If no, the information is "confidential" under Exemption 4 (the government has effectively been silent—it hasn't indicated the information would be protected or disclosed—so a submitter's practice of keeping the information private will be sufficient to warrant confidential status).
   - If yes, and no other sufficient countervailing factors exist, the submitter could not reasonably expect confidentiality upon submission and so the information is *not* confidential under Exemption 4.[105]

Some freedom of information requests made under state public records laws can be denied due to an exemption. State public records laws typically include a list of exemptions. If one or more of the enumerated exemptions applies, the otherwise presumptively public document, or portions of it, may be withheld from disclosure. Courts have generally construed state public records acts strongly in favor of disclosure. The entity wishing to withhold information has the burden of proving the applicability of an exemption. In addition to the enumerated exceptions, federal law or constitutional principles might preclude disclosure in a particular situation.

This section discusses some of the more common enumerated exceptions, as well as the other sources of legal authority that might be relevant. Note that more than one exception or other law might apply and that state laws differ significantly, making it difficult to generalize.

Some public records acts include an exception for research-related documents. They differ widely in their wording and interpretation. In *Robinson v. Indiana University*, an Indiana court held that university records on animal use in research were "information concerning research" and thus exempt under that state's Public Records Act.[106] By contrast, in *Progressive Animal Welfare Society v. University of Washington*, the Washington Supreme Court held that unfunded grant proposals, as a whole, were not "valuable formulae" or "research data" as defined in Washington's public records act, although much of the material in the grant proposals did fall within that exemption.[107]

Public records acts typically include an exception aimed at protecting the privacy of employees. Whether an IACUC can delete (redact) names and addresses of researchers or others when responding to a public records request has been the subject of litigation, with differing results. In *State ex rel. Thomas v. Ohio State University*, the court required a state university to release the names and work addresses of animal research scientists. The university had, for security reasons, redacted that information from research-related documents provided in response to a public records request. The court acknowledged the likelihood of increased harassment but noted that there did not appear to be the "high potential … for victimization" required to establish a constitutionally based exemption and if criminal conduct resulted from the disclosure it "should be punished by criminal sanctions."[108] The court in *In re Physicians Committee for Responsible Medicine v. Hogan* similarly refused to protect the identity of researchers and the details of their research "merely because the passions of unknown terrorists or criminals might be inflamed" and stated that "public employees have a diminished expectation of privacy with regard to details of their public employment."[109] That court permitted redaction of phone numbers, e-mail addresses, building addresses, and similar information under a personal privacy exemption.

In contrast, in *S.E.T.A. UNC-CH, Inc. v. Huffines*, the court ruled that under that state's public records act, the identity of researchers and staff members could be withheld. The court reasoned that "public policy does require that any information contained in the [research] applications relating to the names of the researcher and staff members, their telephone numbers, addresses, their experience, and the department name be redacted from the IACUC applications."[110] That court also ruled that applications that are not approved need not be made public.

Again, PHS Policy states: "Institutions may, at their discretion, represent the names of members other than the chairperson and veterinarian with program authority by using numbers or other symbols in submissions to OLAW. Sufficient information for OLAW to determine that all appointees are appropriately qualified must be provided. The identify of each member must be readily ascertainable by the institution and available to authorized OLAW or other PHS representatives upon request."[111]

Many states have exceptions for "preliminary" or "draft" information, and these exceptions might preclude release of unapproved or unfunded proposals. In addition, intellectual property and public policy arguments often have great force when the research is at this early stage.

It is important to keep in mind that other state statutes outside the state public records laws may provide exemptions from disclosure for research records. Statutes protecting trade secrets as well as any anti-harassment statutes specifically relevant to animal researchers may not protect research records in their entirety but may protect various portions of the records from disclosure.

Freedom of information requests made under FOIA can also be denied due to an exemption. Some state courts have determined that the federal FOIA rather than the state public records act applies to the records of an IACUC.[112] The federal FOIA provides nine exemptions to disclosure of records. One of the nine exemptions from disclosure under the federal FOIA is for "trade secrets and commercial or financial information that is obtained from a person and is privileged or confidential." Privileged information is information that would ordinarily be protected from disclosure in civil litigation by a recognized evidentiary privilege, such as the attorney–client privilege. Confidential information is defined as information that, if disclosed by the government, (1) may impair the government's ability to obtain necessary information in the future; (2) would substantially harm the competitive position of the person who submitted the information; (3) would impair other government interests, such as program effectiveness and compliance; or (4) would impair other private interests.[113]

Another exemption under FOIA is for internal government communications[114] and yet another important exemption is for "records about individuals if disclosure would constitute a clearly unwarranted invasion of their personal privacy."[115] In deciding whether to release records that contain personal or private information about a person, the foreseeable harm of invading that person's privacy must be weighed against the public benefit that would result from the release.

The federal FOIA includes no exemption from disclosure for information qualified under the so-called privilege of academic freedom, which proponents trace to the First Amendment of the United States Constitution. The few cases that have considered the protection of material under an academic freedom claim in circumstances such as those an IACUC might encounter have rejected claims that this "privilege" prohibits disclosure.[116]

Surv. Does your institution have a Public Records Office or other trained personnel who handle public record requests or state-specific freedom of information act requests?

Yes, they manage the entire process (receive requests, review responsive documents, apply applicable redactions, etc.).

• Yes, they manage the entire process (receive requests, review responsive documents, apply applicable redactions, etc.).	44/183 (24%)
• Yes, they manage portions of the process and coordinate with other units.	68/183 (37%)
• No.	36/183 (20%)
• Not applicable.	21/183 (11%)
• I don't know.	14/183 (8%)

Surv. If yes to above question, what types of training do the individuals reviewing record requests have? Select all that apply.

• Law degree	48/64 (75%)
• Public records training	32/64 (50%)
• On the job training	28/64 (44%)
• No training	1/64 (2%)
• Other*	3/64 (5%)

*Other: Individual reviewing record requests has a DVM and PhD.

26:17 How does the principle of academic freedom apply to IACUC activities?

Reg. The Supreme Court of the United States has recognized the principle of academic freedom.[117] It has said: "The essentiality of freedom in the community of American universities is almost self-evident. No one should underestimate the vital role in a democracy that is played by those who guide and train our youth. To impose any straitjacket upon the intellectual leaders in our colleges and universities would imperil the future of our Nation. No field of education is so thoroughly comprehended by man that new discoveries cannot yet be made. … Teachers and students must always remain free to inquire, to study and to evaluate, to gain new maturity and understanding; otherwise, our civilization will stagnate and die."[118]

Opin. Researchers are often aware of the principle of academic freedom and are interested in whether it protects from public disclosure information related to their research. Academic freedom is a multifaceted principle, invoked in support of allowing scholars to study, discuss, and publish ideas free of inappropriate restraints. This principle is invoked in a variety of contexts, from debates over tenure to challenges over publication rights. This principle is explicitly reflected in the policies of numerous universities and academic associations. In addition, it is generally acknowledged to be firmly grounded in the free speech protections of the federal Constitution and state constitutions.[119]

One facet of academic freedom is often referred to as research scholars' privilege or academic privilege. Some have argued that this privilege is rooted in the First Amendment[120] and that it protects documents that would otherwise be available under public records statutes, because releasing the information would chill research, the unfettered discussion of ideas, and other activities that underlie academic freedom.[121] This is the context in which IACUCs are most likely to face the potential applicability of the principle of academic freedom.

Unfortunately, in this context, the principle of academic privilege has generally not fared well in the courts. In *University of Pennsylvania v. Equal Employment Opportunity Commission* (EEOC), the Supreme Court of the United States firmly rejected the university's attempts to expand academic freedom to protect confidential peer review materials from disclosure.[122] The university argued that peer review materials from tenure review files deserved protection under First Amendment academic freedom principles, claiming that the disclosure of such documents to the EEOC and ultimately the public, pursuant to a subpoena, would have a "chilling effect" on the tenure evaluation process and affect the "quality of instruction and scholarship" at the university. The Court disagreed, finding the university's reliance on past academic freedom cases misplaced since under the facts of the case, there were no attempts by the government to control university speech.

A year later, in *S.E.T.A. UNC-CH, Inc. v. Huffines*, the North Carolina Court of Appeals reviewed the same issue, but this time in the context of IACUC disclosure of records pursuant to state public record laws. That court, following *University of Pennsylvania v. EEOC*, rejected the academic privilege argument and ordered the IACUC records to be disclosed.[123] Similar decisions were made by the Supreme Courts of Ohio and Washington, which both refused to expand academic freedom to protect IACUC records from disclosure to the public.[124]

So, it appears that the role academic freedom plays in IACUC activities is rather small, at least as far as protecting research records is concerned. Until the Supreme Court rules differently, IACUCs would be better off understanding their state public record laws well in addition to any statutory exemptions that their legislature has seen fit to provide.

The University of California, Davis prevailed in a civil lawsuit brought by activists under the California Public Records Act. People for the Ethical Treatment of Animals (PETA), filed the suit in January 2019, seeking access to unpublished research data, in the form of video recordings, from work by two researchers at the California National Primate Research Center. In an unpublished decision released on January 11, 2022, the Superior Court of California, County of Yolo, ruled that releasing the material did not serve the public interest and would undermine academic freedom and the scientific process while increasing the risk that researchers could face physical harm and harassment from activists.[125]

26:18 What mechanisms are available to protect against harassment and vandalism?

Reg. To intentionally damage or cause the loss of any property, including animals or records, used by an animal research facility, or to conspire to do so, is a federal crime under 18 U.S.C. §43.[126] Many states have enacted similar laws making it a crime to enter an animal facility with the intent to commit prohibited acts. The prohibited acts vary from state to state and may include: (1) interfering with facility operations; (2) defacing or damaging the facility; (3) releasing, stealing, or killing animals; (4) taking photographs; and/or (5) committing acts to defame the facility. Punishment for violation of these federal and state laws can include a fine, imprisonment for a term of years or life, and a requirement for restitution, depending on the extent of the damage or loss caused.

Opin. The use of animals in research is a privilege granted to those investigators and programs that commit to meeting the highest ethical and regulatory standards. However, some opponents of animal research have engaged in criminal activity such as vandalism and harassment. Although comparatively infrequent, notable incidents have involved bombings, arson, threats of bodily harm, and publication of private information on the Internet.

An IACUC cannot prevent criminal activity by third parties, but it can adopt practices that minimize risk to researchers. For example, researchers' home addresses, telephone numbers, and social security numbers should never be requested or used in IACUC submissions. Files, offices, computers, and animal housing areas and laboratories should be locked and/or password protected. Documents containing confidential information should be shredded before disposal or recycling. Computer records should be backed up in case of theft or destruction of computers, but all computer hard drives should be erased, and any physical recording media should be appropriately destroyed before disposal or recycling. Academic institutions, in particular, abhor secrecy in favor of the free exchange of information; however, they might be able to learn from their counterparts in industry about appropriate security measures. On a personal level, researchers and IACUC members would be well advised to adopt practices designed to thwart identity theft, such as shredding personal financial information, maintaining unlisted home telephone numbers, and regularly changing passwords.

Another way to protect against crimes such as harassment and vandalism is to educate the public about the nature and purpose of animal research, the care and effort devoted to protocol review, animal care, the prevention or minimization of pain, the fact that animal studies are usually necessary to satisfy FDA requirements for safety and efficacy data as a prerequisite to human trials, and the benefit that animal research has given to humankind through the introduction of new drugs, devices, and therapies. (Although FDA regulations and guidance suggest that animal studies should be performed in most cases to demonstrate safety and efficacy prior to human clinical trials of new drugs and medical devices, validated *in vitro* tests can substitute for certain whole animal tests.[127] In limited situations, when definitive human efficacy studies cannot ethically be conducted because doing so would expose healthy human volunteers to lethal or permanently disabling toxic substances, studies in animals alone can serve as the basis for FDA approval of a new drug for human use.[128])

The institution's legal counsel should be educated on the risks to the institution and its employees so that counsel can be prepared if an incident occurs. Counsel should also be encouraged to join the "animal law" committees of the American Bar Association and state and local bar associations, which continue to deliberate proposed legislation that would impair or eliminate animal research. These associations tend to be overly represented by opponents of animal research, with little balance to any debate. Participation by more institutions' legal counsel could help ensure that legislation considers all sides of the issues.

Research facilities should evaluate the applicable state criminal and civil laws regarding trespassing, harassment, and property destruction and establish institutional policies to prevent and respond to such activities. If a public facility is involved, the constitutional right to free speech also must be considered, for example, in establishing boundaries for a protest. In some states, an individual may obtain an injunction based on a reasonable belief that he or she may be injured by the commission of animal research–related vandalism or harassed because of animal research activities. If an injunction is issued to prevent anticipated vandalism or harassment, the violation of the injunction itself gives rise to legal action. Although this type of statutory provision might be valuable in some circumstances, its practical benefits are limited by the fact that it requires researchers to file court documents identifying themselves, the harassers, and the anticipated unlawful action sufficiently in advance of that action. A number of states have enacted laws that classify some of the tactics employed by animal rights groups as terrorism. The benefit of these statutes is that they impose greater criminal penalties and might therefore deter opponents from engaging in illegal behavior. Another benefit is that terroristic threats, a tactic employed by some opposition groups, are actionable. Institutions can protect researchers by lobbying state lawmakers in support of such legislation.

Research associations are often excellent resources for practical steps that can be taken to reduce the level of confrontation over animal research. The National Association for Biomedical Research and its state affiliates, for example, have helpful resources regarding community outreach, relations with animal rights groups, and response to protests.

26:19 What mechanisms should be used by an IACUC and institution to ensure information collected from other units or other institutions is handled in a compliant manner and confidential information is protected?

Reg. While working with information collected from other units within an institution or with information collected from outside institutions, IACUC members and staff must remain aware of proprietary information, including copyrights, patents, trademarks, and trade secrets. As such, the provisions of AWA §2157 and §2.35,f of the AWAR are applicable.

Under AWA §2157 it is unlawful for any member of an IACUC to use or release to the public or reveal to another person any confidential information of the research facility, including "any information that concerns or relates to (1) the trade secrets, processes, operation, style of work, or apparatus; or (2) the identity, confidential statistical data, amount or source of any income, profits, losses, or expenditures, of the research facility." Such wrongful use or disclosure may result in removal from the IACUC, fines, and/or imprisonment. In addition, any institution or individual who is injured by such disclosure or use may bring a lawsuit against the IACUC member and recover damages and the costs of the suit from the IACUC member.

§2.35,f of the AWAR requires APHIS/AC inspectors to maintain the confidentiality of information reviewed for inspection purposes and prohibits them from removing materials from research facilities' premises "unless there has been an alleged violation, they are needed to investigate a possible violation, or for other enforcement purposes." The public release of any information or materials copied or removed from a research facility by an APHIS/AC inspector is governed by FOIA.

Opin: The submission of a protocol involves many different institutional units and subsequent actions including, but not limited to the following: veterinarians conducting protocol pre-review, consultation, and animal care; individuals responsible for providing and assessing personnel training; occupational health and safety professionals providing personnel risk assessment related to animal and hazardous agent use; units preparing contingency (disaster) planning; and departments performing grant congruency and assessing collaborations with other institutions. Many of the above generate records that may contain sensitive information and may be considered confidential, such as personal medical information.

Institutions should not only account for safe handling of their own institutional records but also safeguard records that may be obtained from other institutions. Private institutions or nonprofit agencies (such as AAALAC International) may not be subject to open public records laws. Institutions should keep in mind, however, that records received by a public institution from a private or nonprofit entity may become subject to the public institution's requirements; thus, potentially jeopardizing confidentiality of those documents.

Refer to prior sections of this chapter for a more complete review of these topics. In particular, section 26:3 *What is proprietary or confidential information?* And section 26:4 *Under what circumstances might proprietary or confidential information disclosed to an IACUC become available to the public?*

Surv. Does your institution have a mechanism to protect confidentiality when receiving documentation from collaborators, CROs, other institutions, subcontracted entities?

- No 22/183 (12%)
- Yes 93/183 (51%)
- Not applicable 9/183 (5%)
- I don't know 57/183 (31%)
- Other 2/183 (1%)

Surv. Does your IACUC review animal care, training, OSHA, contingency planning, or other non-study protocols or SOPs?

- No 35/183 (35%)
- Yes 142/183 (78%)
- I don't know 6/183 (3%)

Surv. If yes, how is the confidentiality of those documents and review secured? Select all that apply.

- Separate section of protocol 18/141 (13%)
- Secure electronic files 79/141 (56%)
- Label as 'Confidential' 28/141 (20%)
- I do not know 18/141 (13%)
- Not applicable 10/141 (7%)
- Other 14/141 (10%)

26:20 What mechanisms should be used by an IACUC and institution to ensure research collaborations between institutions are conducted in a compliant manner?

Reg. Although the *Guide* is not a regulation, the PHS Policy requires that Assured institutions base their programs of animal care and use on the *Guide*. The *Guide* (p. 15) states that institutions collaborating in research should enter into a formal written understanding (e.g., a contract, memorandum of understanding, or agreement) that addresses each party's responsibilities regarding off-site animal care and use, animal ownership, and IACUC review and oversight. In addition, NIH Grants Policy Statement on Written Agreements "requires that awardees have a formal written agreement with consortium participants that addresses the negotiated scientific, administrative, financial, and reporting requirements of the grant. This written agreement must include incorporation of applicable public policy requirements, including agreement for meeting the PHS Policy requirement for review and approval of proposed animal activities, significant changes to animal activities, and semiannual facilities review by an IACUC (IV.B.2.)"[129] Research collaborations are not addressed in the AWAR; however, the Animal Care Inspection Guide states that "it is the responsibility of the registrants to determine which party is responsible for the functions of the IACUC, animal care and handling, and reporting of the animals on the Annual Report" when registered research facilities contract research out to be conducted at another facility. If the research facilities do not delegate these functions, both facilities are responsible for non-compliance.[130]

Opin. Generally, any research collaborations between institutions should be memorialized in a written agreement to not only address responsibilities with regard to research, but also to allocate legal rights and obligations among the parties.[131] Any research collaboration agreement (such as a contract or memorandum of understanding) should:

1. Designate which institution's IACUC will be responsible for reviewing and monitoring the protocol(s)

2. Designate which institution's policies will apply to the researcher and the research
3. Designate which institution will report the animals on its Annual Report
4. Establish guidelines for animal care and use activities that require cooperation and coordination among the institutions, such as animal transport
5. Establish guidelines for the sharing of information regarding the research, program(s), as well as issues arising from the research (such as lapses, terminations, suspensions, or noncompliance)
6. Establish guidelines for handling non-compliance and, specifically, whether the institutions will investigate separately or jointly[132]

Depending on the nature of the research, collaborating institutions may choose to address other legal rights and obligations such as treatment of confidential and proprietary information, intellectual property, indemnification, insurance, publications, and use of names and trademarks. Institutions may also want to establish guidelines on how to manage certain media or public requests and/or inquiries.

By taking the time to establish a written agreement that clearly outlines each party's rights and responsibilities, collaborating institutions will create a productive relationship and environment that facilitates their research. In addition, the institutions are better situated to identify and address any non-compliance as well as to effectively communicate with governmental agencies and the public.

Surv. Which office at your institution is responsible for reviewing materials associated with collaborations, subcontracts, or sub-awards with other institutions? Select all that apply.

• Sponsored Programs Office/Office of Sponsored Research	139/176 (79%)
• Research integrity	27/176 (215%)
• IACUC	68/176 (39%)
• Not applicable	3/176 (2%)
• Other*	29/176 (16%)

*Other:

- Business development
- Business operations
- Chief Executive Officer or Chief Financial Officer
- Institutional Official
- Principal Investigator
- Intellectual property department
- Provost's office
- Quality Assurance unit

Surv. Does your institution have a mechanism to protect confidentiality when receiving documentation from collaborators, CROs, other institutions, subcontracted entities?

• No	22/183 (12%)
• Yes	93/183 (51%)
• Not applicable	9/183 (5%)
• I don't know	57/183 (31%)
• Other	2/183 (1%)

26:21 How should an IACUC protect confidential information and the rights of a person against whom an allegation of protocol non-compliance or animal mistreatment has been made?

Reg. Under the PHS Policy (IV,C,6) and the AWAR (§2.31,d,6), an IACUC may suspend a previously approved activity involving animals for reasons of protocol non-compliance. If the IACUC suspends an activity, the IO must, in consultation with the IACUC (AWAR §2.31,d,7; PHS Policy IV,C,7), review the reasons for suspension, take appropriate corrective action, and, if APHIS/AC regulated animals are involved, report that action with a full explanation to APHIS/AC (and any federal agency funding that activity). If only PHS Policy regulated animals are involved, it is reported to NIH/OLAW with a full explanation. The PHS Policy and AWAR are silent on whether an investigator can appeal the decision by the IACUC (see 33:47, 33:48).

Defamation is a false statement about a person which causes that person to suffer harm such as mental anguish or loss of reputation. Slander is a spoken defamatory statement. Libel is a written or otherwise recorded defamatory statement. Most states recognize *per se* defamation in which the alleged defamatory statement itself is presumed to have caused damage, eliminating any need to prove that the statement caused harm to the person. Statements against a person's professional character or standing are considered *per se* defamation in most jurisdictions.

Opin. Allegations of protocol non-compliance or animal mistreatment can have a significant impact on the accused person. Such allegations can negatively affect the person's current employment as well as future professional opportunities. If allegations are not kept confidential, the person may suffer reputational harm. Reputational harm may impact the manner in which the accused person is treated by colleagues, the professional community, and even the general public if the allegations are particularly egregious. Even worse, the person may face retaliation from animal extremist groups.

IACUCs should have a policy that clearly outlines the process of investigating allegations and issuing determinations resulting from such investigations. The policy should clearly define the rights and responsibilities of the IACUC, the IO, and the accused person. The accused person should be afforded an opportunity to defend against the allegations. For example, the accused person should be allowed to provide information in the form of records, testimony, or statements. In addition, the process should allow the accused person an opportunity to appeal any corrective actions. An appeals process will give an investigator the opportunity to introduce or reintroduce information that may not have been adequately considered in determining the corrective actions.

IACUC policy should require that allegations and investigations of protocol non-compliance and animal mistreatment be treated in confidence. Allegations that are not kept in confidence can lead to reputational harm of the accused person within the institutional community or professional community. If the allegations are particularly egregious, the accused person can suffer reputational harm with the general public. If the accused person suffers significant reputational harm that impacts his or her professional future, the IACUC runs the risk of a *per se* defamation claim. In today's electronic environment, information, without any consideration for its veracity, can be distributed worldwide within seconds and retraction of such information is nearly impossible. Therefore, IACUCs should maintain and distribute documentation and communications regarding allegations and investigations in a secure, confidential manner (see 29:23). For IACUC documentation related to the allegation, attention to avoid the use of names and laboratory identification and the use of gender-neutral language when referring to the individual involved in the allegations may be advisable.

Surv. What mechanisms does your institution have in place to allow for anonymity in reporting animal welfare concerns or non-compliance?

- Anonymous electronic route 127/183 (69%)
- Ethics hotline phone number 120/183 (66%)
- Physical drop box 47/183(26%)
- Other 36/183 (20%)

Surv. What mechanisms does your institution employ to ensure the reports of protocol non-compliance or animal mistreatment are maintained as confidential and anonymity is maintained? Select all that apply.

- Make reports gender neutral 68/183 (37%)
- Avoid use of personnel names or laboratory locations 126/183 (69%)
- Avoid specifics that would directly point to the Principal Investigator or laboratory 108/183 (59%)
- Use of third-party vendor to receive, log, and retain any reporter information (if provided) 25/183 (14%)
- Other 38/183 (21%)

Surv. Does your IACUC allow for appeals of any sanctions imposed on an animal user? Assume that any such appeals are made to the IACUC. See survey 33:47.

Surv. During an investigation of an allegation, does your IACUC have a formal policy about protecting the confidentiality of research or other records? See survey 33:20.

Surv. Does your IACUC have a formal policy about protecting the rights and confidentiality of the person against whom an allegation has been made? See survey 33:20.

26:22 What role should legal counsel play in IACUC activities?

Reg. The AWAR do not specifically require an institution's legal counsel or any attorney to serve as a member of the IACUC. Under PHS Policy (IV,A,3,b), the IACUC must comprise at least five members including, at a minimum, a veterinarian, a practicing scientist experienced in animal research, an individual "whose primary concerns are in a nonscientific area (for example, ethicist, lawyer, member of the clergy)," and a member of the community who is not affiliated with the institution or a family member of a person affiliated with the institution. Thus, it is not necessary to have an institution's legal counsel serve on the IACUC.

Opin. An IACUC should maintain a good working relationship with the institution's legal counsel. Whether or not legal counsel is a voting member of the IACUC, the IACUC should seek advice from counsel regarding its legal obligations. Legal counsel can provide training to new IACUC members regarding the legal responsibilities discussed in this chapter. Also, it is particularly important for an IACUC to seek legal advice when research poses potential compliance issues. Attorneys are trained to read and interpret the law and can serve as a valuable resource in understanding regulations, particularly when application of the regulations and guidance do not provide a clear resolution. An IACUC should seek legal advice in conducting investigations. Legal counsel can advise on how to properly conduct an investigation in accordance with the IACUC's policies and procedures, frame questions for interviews to obtain the necessary information to assess the non-compliance and assist in determining whether acts under investigation constitute non-compliance based on the findings. In addition, legal counsel should be consulted whenever responding to third-party or governmental inquiries such as FOIA requests and subpoenas.

REFERENCES

1. Department of Veterans Affairs. (2013). *Use of Animals in Research.* https://www.research.va.gov/programs/animal_research/documents.cfm.
2. Institute for Laboratory Animal Research, National Research Council of the National Academy of Sciences. (2011). *Guide for the Care and Use of Laboratory Animals*, 8th edition. Washington, DC: The National Academies Press. http://www.nap.edu/catalog.php?record_id=12910#toc.
3. Committees to revise the Guide for the Care and Use of Agricultural Animals in Research and Teaching. (2020). *Guide for the Care and Use of Agricultural Animals in Research and Teaching* (4th edition). Champaign, IL: https://www.asas.org/docs/default-source/default-document-library/agguide_4th.pdf?sfvrsn=56b44ed1_2.
4. Council of Europe. (1991). http://conventions.coe.int/treaty/Commun/QueVoulezVous.asp?CL=ENG&NT=123.
5. 21 C.F.R. §§ 58.1–58.219.
6. 16 U.S.C. §§ 1531–1544.
7. 42 U.S.C. § 283m.
8. 16 U.S.C. §§ 703–712.
9. 18 U.S.C. § 43.
10. 5 U.S.C. § 552.
11. 35 U.S.C. §§ 200–212, with implementing regulations at 37 C.F.R. Part 401.
12. Department of Veterans Affairs (2013). *Use of Animals in Research.* https://www.research.va.gov/programs/animal_research/documents.cfm
13. Guide for the Care and Use of Agricultural Animals in Research and Teaching, Fourth Edition
14. Committees to Revise the *Guide for the Care and Use of Agricultural Animals in Research and Teaching* (2020). *Guide for the Care and Use of Agricultural Animals in Research and Teaching* (4th edition). Champaign, IL: Federation of Animal Science Societies. https://www.asas.org/docs/default-source/default-document-library/agguide_4thpdf?sfvrsn=56b443e1_2
15. 17 U.S. Code § 102(a)
16. American Veterinary Medical Association. (2020). *AVMA Guidelines for the Euthanasia of Animals: 2020 Edition.* https://www.avma.org/sites/default/files/2020-02/Guidelines-on-Euthanasia-2020.pdf.
17. 17 U.S.C. § 102.
18. 35 U.S.C. § 101 et seq.
19. 15 U.S.C. § 1051 et seq.
20. 18 U.S.C. § 1839.
21. 5 U.S.C. § 552a.
22. 15 U.S.C. § 1072.
23. 15 U.S.C. § 1065.
24. 15 U.S.C. § 1059.
25. 17 U.S.C. § 106.
26. 17 U.S.C. § 102.
27. 17 U.S.C. § 302.
28. Oakwood Labs. LLC v. Thanoo, 999 F.3d 892 (3d Cir. 2021).
29. 18 U.S.C. § 1839(3).
30. 35 U.S.C. § 154.
31. 35 U.S.C. § 101.
32. 35 U.S.C. § 154.
33. 35 U.S.C. § 103.
34. 35 U.S.C. § 102.
35. See, e.g., Transgenic Animal Expressing Alzheimer's Tau Protein, U.S. Patent No. 8,288,608 (filed July 9, 2003).
36. 35 U.S.C. § 200 et seq.
37. 35 U.S.C. § 202.
38. 35 U.S.C. § 202(c).
39. 35 U.S.C. § 205.
40. 35 U.S.C. § 203.
41. 5 U.S.C. § 552.
42. 45 C.F.R. § 5.65; 5 U.S.C. § 552,b,4.
43. 45 C.F.R. § 5.65.
44. 5 U.S.C. § 552(b)(6); 45 C.F.R. § 5.67.
45. 5 U.S.C. § 552(b)(5); 45 C.F.R. § 5.66.
46. See, e.g., 42 C.F.R. Part 93 (applicable to misconduct in research supported by PHS).
47. 5 U.S.C. § 552(b)(7).
48. See *Forsham v. Harris*, 445 U.S. 169 (1980).
49. *Public Citizen Health Research Group v. Dep't of Health, Educ. & Welfare*, 668 F.2d 537, 543 (D.C. Cir. 1981).
50. *Public Citizen Health Research Group v. Dep't of Health, Educ. & Welfare*, 668 F.2d 537, 545 (D.C. Cir. 1981) (Tamm, J., concurring).
51. See, e.g., FOIAdvocates, State Public Record Laws, http://www.foiadvocates.com/records.html (accessed July 9, 2013).
52. See, e.g., 7 C.F.R. Part 1 (USDA); 21 C.F.R. Part 20 (FDA); 45 C.F.R. Part 5 (HHS).
53. See, e.g., http://sunshinereview.org/index.php/Portal:WikiFOIA.
54. *Washington Research Project, Inc. v. Dep't of Health, Educ. & Welfare*, 504 F.2d 238 (D.C. Cir. 1974) (summary statements and site visit reports of initial review groups consisting of nongovernmental consultants with no legal decision-making authority held to be exempt from public disclosure as intra-agency memoranda).
55. *In Defense of Animals v. Nat'l Inst. of Health*, 543 F. Supp. 2d 83 (D.D.C. 2008); *Sangre de Cristo Animal Protection, Inc. v. U.S. Dep't of Energy*, 1998 U.S. Dist. LEXIS 23505 (D.N.M. Mar. 10, 1998) (names and identities of contractor employees permitted to be redacted).
56. E.g., *Animal Legal Defense Fund, Inc. v. Institutional Animal Care & Use Comm. of Univ. of Vermont*, 616 A.2d 224 (Vt. 1992).
57. *S.E.T.A. UNC-CH, Inc. v. Huffines*, 399 S.E.2d 340 (N.C. Ct. App. 1991) (permitting redaction of names and contact information, as well as training and experience, of researcher and staff; applications not approved could be withheld); *In re Am. Soc'y for the Prevention of Cruelty to Animals v. Bd. of Trs. of the State Univ. of N.Y.*, 556

N.Y.S.2d 447 (N.Y. Sup. Ct. 1990) (permitting redaction of names, contact information, and grant number or application number of funding source).

58. *Progressive Animal Welfare Soc'y v. Univ. of Washington*, 884 P.2d 592 (Wash. 1994) (certain information held exempt from disclosure, including personal information of researchers; raw research data and hypotheses; budget breakdowns; and peer reviewers' comments).
59. *Dorson v. State of Louisiana*, 657 So.2d 755 (La. Ct. App. 1995); *In re Am. Soc'y for the Prevention of Cruelty to Animals v. Bd. of Trs. of the State Univ. of N.Y.*, 591 N.E.2d 1169 (N.Y. 1992).
60. *State ex rel. Thomas v. Ohio State Univ.*, 643 N.E.2d 126 (Ohio 1994).
61. *Robinson v. Indiana Univ.*, 659 N.E.2d 153 (Ind. Ct. App. 1995) (withholding animal care and use applications and other information requested about individual research projects).
62. *In Defense of Animals v. Oregon Health Sci. Univ.*, 112 P.3d 336 (Or. Ct. App. 2005).
63. *United States v. Catholic Healthcare West*, 445 F.3d 1147 (9th Cir. 2006).
64. 31 U.S.C. § 3729.
65. 31 U.S.C. § 3730.
66. Codified at 42 U.S.C. § 289d(e).
67. 5 C.F.R. § 5501.101.
68. 42 C.F.R. § 50.605.
69. See *Mississippi State Univ. v. People for the Ethical Treatment of Animals, Inc.*, 992 So.2d 595 (Miss. 2008); *In Defense of Animals v. Oregon Health Sci. Univ.*, 112 P.3d 336 (Or. Ct. App. 2005).
70. 42 U.S.C. § 289d(c).
71. 42 C.F.R. § 50.605.
72. See PHS Policy, IV.A.3.b, Footnote 6. and PHS Policy, IV.F.1.c, Footnote 12.
73. 42 C.F.R. § 50.605.
74. See *Sangre de Cristo Animal Protection, Inc. v. U.S. Dep't of Energy*, 1998 U.S. Dist. LEXIS 23505 (D.N.M. 1998) (names and identities of contractor employees were allowed to be redacted); *In re Physicians Comm. for Responsible Med. v. Hogan*, 918 N.Y.S.2d 400 (N.Y. Sup. Ct. 2010) (allowing redaction of contact information); *In re Am. Soc'y for the Prevention of Cruelty to Animals v. Bd. of Trs. of the State Univ. of N.Y.*, 556 N.Y.S.2d 447 (N.Y. Sup. Ct. 1990) (allowing redaction of names and contact information); *S.E.T.A. UNC-CH, Inc. v. Huffines*, 399 S.E.2d 340 (N.C. Ct. App. 1991) (allowing redaction of names and contact information, as well as training and experience, of researcher and staff); *In Defense of Animals v. Oregon Health Sciences University*, 112 P.3d 336 (Or. Ct. App. 2005) (holding staff member names exempt from disclosure); *Progressive Animal Welfare Soc'y v. Univ. of Washington*, 884 P.2d 592 (Wash. 1994) (holding exempt from disclosure the personal information of researchers). *Contra State ex rel. Thomas v. Ohio State Univ.*, 643 N.E.2d 126 (Ohio 1994).
75. Codified at 42 U.S.C. § 289d(e).
76. See 45 C.F.R. § 164.302 et seq.
77. PHS Policy IV.E.1&2.
78. *Mississippi State Univ. v. People for the Ethical Treatment of Animals, Inc.*, 992 So.2d 595 (Miss. 2008).
79. *Oklahoma Ass'n of Municipal Attorneys*, 577 P.2d 1310 (Okla. 1978).
80. *Animal Legal Defense Fund, Inc. v. Institutional Animal Care & Use Comm. of Univ. of Vermont*, 616 A.2d 224 (Vt. 1992).
81. *Dorson v. State of Louisiana*, 657 So.2d 755 (La. Ct. App. 1995).
82. *In re Am. Soc'y for the Prevention of Cruelty to Animals v. Bd. of Trs. of the State Univ. of N.Y.*, 591 N.E.2d 1169 (N.Y. 1992).
83. *Notice Number: NOT-OD-06-052*, Released March 24, 2006, https://grants.nih.gov/grants/guide/notice-files/NOT-OD-06-052.html.
84. See *USDA Animal Welfare Inspection Guide, Chapter 7.1.8. Telecommunications for IACUC Meetings*, Available at https://www.aphis.usda.gov/animal_welfare/downloads/Animal-Care-Inspection-Guide.pdf.
85. 5 U.S.C. § 552.
86. *Animal Legal Defense Fund, Inc. v. Institutional Animal Care & Use Comm. of Univ. of Vermont*, 616 A.2d 224 (Vt. 1992).
87. *In re Am. Soc'y for the Prevention of Cruelty to Animals v. Bd. of Trs. of the State Univ. of N.Y.*, 556 N.Y.S.2d 447 (N.Y. Sup. Ct. 1990) (permitting redaction of names, contact information, and grant number or application number of funding source); *S.E.T.A. UNC-CH, Inc. v. Huffines*, 399 S.E.2d 340 (N.C. Ct. App. 1991) (permitting redaction of names and contact information, as well as training and experience, of researcher and staff; applications not approved could be withheld).
88. *Progressive Animal Welfare Soc'y v. Univ. of Washington*, 884 P.2d 592 (Wash. 1994) (certain information held exempt from disclosure, including personal information of researchers; raw research data and hypotheses; budget breakdowns; and peer reviewers' comments).
89. *Dorson v. State of Louisiana*, 657 So.2d 755 (La. Ct. App. 1995).
90. *In re Am. Soc'y for the Prevention of Cruelty to Animals v. Bd. of Trs. of the State Univ. of N.Y.*, 591 N.E.2d 1169 (N.Y. 1992).
91. *Robinson v. Indiana Univ.*, 659 N.E.2d 153 (Ind. Ct. App. 1995) (withholding animal care and use applications and other information requested about individual research projects).
92. *State ex rel. Thomas v. Ohio State Univ.*, 643 N.E.2d 126 (Ohio 1994).
93. See PHS Policy, IV.A.3.b, Footnote 6. and PHS Policy, IV.F.1.c, Footnote 12.
94. *In re Physicians Comm. for Responsible Med. v. Hogan*, 918 N.Y.S.2d 400 (N.Y. Sup. Ct. 2010).
95. *In Defense of Animals v. Oregon Health Sci. Univ.*, 112 P.3d 336 (Or. Ct. App. 2005).
96. *State ex rel. Physicians Comm. for Responsible Med. v. Bd. of Trs. of Ohio State Univ.*, No. 2005-0612, 2006 WL 508325 (Ohio Mar. 15, 2006).
97. Ohio Rev. Code § 149.43.
98. 5 U.S.C. § 552.
99. *In Defense of Animals v. Nat'l Inst. of Health*, 543 F. Supp. 2d 83 (D.D.C. 2008); *Sangre de Cristo Animal Protection, Inc. v. U.S. Dep't of Energy*, 1998 U.S. Dist. LEXIS 23505 (D.N.M. Mar. 10, 1998).

100. See *Clarke v. Tri-Cities Animal Care & Control Shelter*, 181 P.3d 881 (Wash. Ct. App. 2008) (privately-run corporation providing animal control services under contract with municipal agency subject to public disclosure act).
101. *Public Citizen Health Research Group v. Dept of Health, Educ. & Welfare*, 668 F.2d 537, 545 (D.C. Cir. 1981) (Tamm, J., concurring).
102. *In Defense of Animals v. U.S. Dep't of Agric.*, 656 F. Supp. 2d 68 (D.D.C. 2009).
103. 5 U.S.C. § 552.
104. See, e.g., Margaret Snyder, *Freedom of Information and Openness*, http://www.iom.edu/~/media/Files/Activity%20Files/Research/NeuroForum/2011-JUL-26/Snyder.pdf.
105. *Food Marketing Institute v. Argus Leader Media*, 139 S. Ct. 2356 (2019).
106. See *Step-by-Step Guide for Determining if Commercial or Financial Information Obtained from a Person is Confidential Under Exemption 4 of the FOIA*. Available at https://www.justice.gov/oip/step-step-guide-determining-if-commercial-or-financial-information-obtained-person-confidential.
107. NABR, SfN, & FASEB, *Responding to FOIA Requests: Facts and Resources*, available at http://www.nabr.org/uploadedFiles/nabrorg/Content/Animal_Law/Responding_to_FOIA_Requests.pdf.
108. *Robinson v. Indiana Univ.*, 659 N.E.2d 153 (Ind. Ct. App. 1995).
109. *Progressive Animal Welfare Soc'y v. Univ. of Washington*, 884 P.2d 592 (Wash. 1994).
110. *State ex rel. Thomas v. Ohio State Univ.*, 643 N.E.2d 126 (Ohio 1994).
111. *In re Physicians Comm. for Responsible Med. v. Hogan*, 918 N.Y.S.2d 400 (N.Y. Sup. Ct. 2010).
112. See PHS Policy, IV.A.3.b, Footnote 6. and PHS Policy, IV.F.1.c, Footnote 12).
113. *S.E.T.A. UNC-CH, Inc. v. Huffines*, 399 S.E.2d 340 (N.C. Ct. App. 1991).
114. E.g., *Dorson v. State of Louisiana*, 657 So.2d 755 (La. Ct. App. 1995); *In re Am. Soc'y for the Prevention of Cruelty to Animals v. Bd. of Trs. of the State Univ. of N.Y.*, 584 N.Y.S.2d 198 (N.Y. App. Div. 1992).
115. 45 C.F.R. § 5.65.
116. 45 C.F.R. 5.66; see, e.g., *Washington Research Project, Inc. v. Dep't of Health, Educ. & Welfare*, 504 F.2d 238 (D.C. Cir. 1974).
117. 45 C.F.R. § 5.67.
118. See *Univ. of Penn. v. Equal Employment Opportunity Comm'n*, 493 U.S. 182 (1990); *S.E.T.A. UNC-CH, Inc. v. Huffines*, 399 S.E.2d 340 (N.C. Ct. App. 1991).
119. See *Sweezy v. New Hampshire*, 354 U.S. 234 (1957); *Keyishian v. Bd. of Regents*, 385 U.S. 589 (1967).
120. *Sweezy v. New Hampshire*, 354 U.S. 234, 250 (1957).
121. See *generally* Chang, A.W., *Resuscitating the Constitutional "Theory" of Academic Freedom: A Search for a Standard beyond Pickering and Connick*. 2001. Stan. Law Rev. 53:915; Byrne, J.P. 1989. Academic Freedom: A "Special Concern of the First Amendment." *Yale Law J.* 99:251.
122. *Keyishian v. Bd. of Regents*, 385 U.S. 589, 603-04 (1967).
123. See, e.g., Levinson-Waldman, R. (2011). *Academic Freedom and the Public's Right to Know: How to Counter the Chilling Effect of FOIA Requests on Scholarship.* http://www.acslaw.org/sites/default/files/Levinson_-_ACS_FOIA_First_Amdmt_Issue_Brief.pdf (suggesting possible ways to change laws to protect academic documents from disclosure requests).
124. *Univ. of Penn. v. Equal Employment Opportunity Comm'n*, 493 U.S. 182 (1990).
125. *S.E.T.A. UNC-CH, Inc. v. Huffines*, 399 S.E.2d 340 (N.C. Ct. App. 1991).
126. *State ex rel. Thomas v. Ohio State Univ.*, 643 N.E.2d 126 (Ohio 1994); *Progressive Animal Welfare Soc'y v. Univ. of Washington*, 884 P.2d 592 (Wash. 1994).
127. People for the Ethical Treatment of Animals, Inc. v. The Regents of the University of California, CV-2019-172, Ruling (Jan. 11, 2022).
128. See, e.g., *United States v. Fullmer*, 584 F.3d 132 (3rd Cir. 2009).
129. See, e.g., 21 C.F.R. § 312.23(a), § 812.27(a).
130. 21 C.F.R. § 314.610.
131. NIH GPS, chapter 15.2.1.
132. See *USDA Animal Welfare Inspection Guide, Chapter 7.4. Contracted Research or Projects that Involve Multiple Registrants.* Available at https://www.aphis.usda.gov/animal_welfare/downloads/Animal-Care-Inspection-Guide.pdf.

27

General Concepts of the Program Review and Facility Inspection

Joseph D. Thulin and Kenneth P. Allen

Introduction

The AWAR, PHS Policy, and *Guide* all require the IACUC to periodically review the animal care and use program and inspect the animal facilities. While the *Guide* (p. 25), which was not written solely for a U.S. audience, recommends that these activities occur at least annually, the AWAR and PHS Policy both require them to be conducted at least every six months. Although the program review and facilities inspection are listed as separate IACUC responsibilities in the regulatory documents, they are integrally related. Indeed, at many institutions, and perhaps at the vast majority, the two are orchestrated as a set of coordinated, semiannual events. Hence, the single term "semiannual" is used frequently to express the collective activities of both of these IACUC functions.

The semiannual review of the program and inspection of facilities may be the most important responsibilities of the IACUC. Admittedly, the fact that these are listed first among the several requisite functions of the IACUC in both the AWAR and the PHS Policy does not necessarily indicate that the framers of these documents considered them to be the most important. Nevertheless, it is through a properly conducted "semiannual" that the IACUC most comprehensively fulfills its purpose of evaluating the care, treatment, housing, and use of animals, and for assessing the institution's conformance to regulatory requirements and adherence to adopted standards.

In this chapter, we address implementation of the program review and facilities inspection as required by the AWAR and the PHS Policy and recommended by the *Guide*; however, readers are cautioned that state and local laws may add other requirements.

27:1 What is meant by reviewing the program of animal care and use?

Reg. Both the PHS Policy (IV,B,1) and the AWAR (§2.31,c,1) require the IACUC to review the institution's program of animal care and use every six months. Neither of these documents provides a succinct definition of "program," although the PHS Policy (IV,A,1) does list the elements that must be fully described in an institution's Animal Welfare Assurance with NIH/OLAW. The *Guide* (p. 6) defines the animal care and use program as "the policies, procedures, standards, organizational structure, staffing, facilities, and practices put into place by an institution to achieve the humane care and use of animals in the laboratory and throughout the institution." The PHS Policy (IV,A,1) specifies that the institution is to use the *Guide* as the basis for evaluation, whereas the AWAR (§2.31,c,1) require the evaluation to be based on the standards of the AWAR. None of these documents prescribe a specific methodology for conducting the program review. Instead, the IACUC may determine the best means of conducting evaluations of the research facility's programs (§2.31,c,3).

Opin. The definition of the animal care and use program in the *Guide* expresses well the scope of elements to include in the program review—regardless of whether the institution is required to use the *Guide* as the basis of its evaluation for regulatory or accreditation reasons. The review should consist of a comprehensive evaluation of each of those elements listed in the above definition. In so doing and while referring to the standards for the program set forth in the PHS Policy, AWAR, and the *Guide*, as applicable, the IACUC should determine the extent to which their program meets those standards. Institutions and their IACUCs have a great deal of flexibility in determining the specific methods used to conduct the program review; however, to ensure all the facets of the program are adequately and consistently considered, it is advisable to use as guidance for the review, an outline or checklist derived from the content of PHS Policy, AWAR, and the *Guide*. The outline should list the various program facets and standards derived from the applicable standards. Institutions may develop their own outlines or use one of the many that are readily accessible through the web. NIH/OLAW provides an excellent version of a program review checklist on its website.[1] This document provides a window into the scope of review that NIH/OLAW considers appropriate for the reviews.

27:2 What are the similarities and differences between the program review and the facility inspection?

Reg. In addition to the program review (see 27:1), the PHS Policy (IV,B,2) and the AWAR (§2.31,c,2) also require the IACUC to inspect all of the institution's animal facilities, including satellite

DOI: 10.1201/9781003272861-27

facilities/animal study areas, at least once every six months. The *Guide* (p. 25) recommends inspection of facilities and animal use areas at least annually. In addition to the frequency with which they must occur, the PHS Policy and AWAR have other requirements that are the same for both the program review and facility inspection. For example, the AWAR require both the reviews and inspections be conducted by a subcommittee of at least two committee members (see 27:8), and the PHS Policy and the AWAR have the same requirements for both the program review and facilities inspection in terms of categorizing identified deficiencies and reporting results to the IO. The PHS Policy (IV, B,3, footnote 8) states that the IACUC may determine the best means of conducting an evaluation of the institution's programs and facilities.

Opin. The program review and facility inspection are two discrete yet interrelated duties of the IACUC. On the one hand, the program review encompasses all the policies, plans, standard procedures, and systems under which the institution fulfills its obligation to care for and use animals in a lawful and ethical manner. On the other hand, the facility inspection is a physical and visual assessment of the animals, the environments in which the animals are housed and used, buildings, and related equipment. During the program review, the IACUC evaluates the expectations for animal care and use, while in the facility inspection, the committee validates the degree to which those expectations are effectively implemented. The outcomes or conditions observed during the facility inspection can indicate deficiencies that should be addressed in the program review. For example, if all the rodent cages are found to be poorly cleaned during a facility inspection, that condition may be due to a suboptimal program element such as an inappropriate Standard Operating Procedure (SOP) or a poor training program. Together, facility inspection and program review provide a comprehensive internal audit of the institution's animal care and use program. See Table 27.1.

27:3 What is the purpose of the program review and the facility inspection?

Reg. The semiannual evaluation of the animal care and use program and facilities is a primary means by which the IACUC fulfills its oversight responsibility, particularly in terms of assessing the extent to which the institution adheres to the provisions of the PHS Policy, AWAR, and *Guide*, as applicable, and developing plans to address identified deficiencies.

Opin. Most animal research facilities in the U.S. (depending on the source of funding and the species of animals used) fall under the jurisdiction of the PHS Policy and/or the AWAR. Both documents mandate semiannual program and facility evaluations (see 27:1 and 27:2). Therefore, an institution could take the view that these activities are necessary merely to meet regulatory expectations; however, such a narrow view would not meet the spirit and intent of the requirements.

The program review and facility inspection process should be considered a central element of post-approval monitoring (PAM, see Chapter 34). The review helps ensure the program is appropriately designed and organized to conform with the regulatory requirements and applicable professional standards. The inspections give the IACUC the means to confirm that the intended outcomes of the written program are indeed achieved. This is accomplished through the direct observations of the animals, procedures, holding facilities, records, and laboratories. In addition, a well-managed and documented animal care and use program helps improve public confidence in the research conducted at the facility and engenders increased support for government-sponsored research in general. Commercial entities may find that clients will seek out laboratories that adhere to the highest standards of animal care, and clients may be willing to pay a premium for high-quality work. In other words, a well-managed animal care and use program is a vital component of good scientific research.

Surv. How useful do you think the semiannual program review is to your IACUC?

• Not applicable, we do not do a semiannual program review.	3/182 (2%)
• Of little value; we do it for compliance purposes	13/182 (7%)
• Of moderate value; it sometimes helps us set policies, correct problems, and assure the quality of our program.	84/182 (46%)
• Very valuable; it often helps us develop new policies, correct problems, and is a key part of our quality assurance program.	82/182 (45%)
• I don't know.	0/182 (0%)

27:4 What laws, guidelines, and policies delineate what to identify when an IACUC inspects an animal facility and reviews the program of animal care and use?

Reg. The first law passed by Congress to regulate animal research was the Laboratory Animal Welfare Act in 1966. The second law was the Health Research Extension Act passed in 1985. Subsequent regulatory documents regarding research animal care and use, including IACUC inspections, were derived from these two laws. Two basic documents mandate and govern the administrative aspects of the facility inspection and the program review: PHS Policy (IV,B) and the AWAR (§2.31,c) (see

TABLE 27.1

Similarities and Differences in the Program Review and Facility Inspection

Feature	Program Review	Facility Inspection
Directed by	As assigned by IACUC Chair	As assigned by IACUC Chair
Primary focus of review	Policies, plans, procedures, organization, staffing, etc.	"Bricks and mortar," equipment, animals, and outcomes
General references in the *Guide*	Chapter 2–Animal Care and Use Program Chapter 3–Environment, Housing and Management Chapter 4–Veterinary Care	Chapter 3–Environment, Housing and Management Chapter 4–Veterinary Care Chapter 5–Physical Plant
General reference in the AWAR	Part 2, Subpart C–Research Facilities	Part 3–Standards
How review is conducted	• Talking around a tableAsking questions of management; persons who know and supervise the programReviewing written policies, plans, and proceduresReviewing documentationReceiving oral/ written reportsReporting on what is read or spoken	• Walking around the animal and research facilitiesAsking questions of the animal care and research staffExamining records typically kept in the animal facility or lab, e.g., animal health/use recordsVisual observation and physical assessmentGenerally reporting on what you see
Minimal required participation	• At least two members of the IACUCGenerally, need more people to conduct the review, including representative persons familiar with the policies, plans, and procedures and how they are used:VeterinariansIACUC administratorsMaintenanceAnimal care managementOccupational health and safetyResearch administrators	• At least two members of the IACUCFewer people needed; will talk with persons encountered during the review:Animal husbandry and veterinary careMaintenance personnelPersons doing the work in research laboratoriesEnvironment, health and safety
Inspection of animals	Generally not	Always
Physical review of laboratories	No	Yes
Ratification of findings and reports	Majority of IACUC members' signatures	Majority of IACUC members' signatures

27:1 and 27:2). It should be noted that PHS Policy (IV,B) requires the IACUC to use "the *Guide* as a basis for evaluation" when conducting the program review and facility inspection. In addition, if agricultural research is conducted, the 4th edition of the *Guide for the Care and Use of Agricultural Animals in Agricultural Research and Teaching* (*Ag Guide*)[2] should be consulted (see Chapter 21).

Opin. Other references also should be used to evaluate an animal care and use program. The primary standard for conducting an animal care and use program is the *Guide*. It is customary for research facilities to follow the *Guide* even if they have no legal obligation to do so. Both the *Guide* and *Ag Guide* contain extensive references; a few of the more frequently used references are the following:

- Semiannual program review and facility inspection checklist[1]
- U.S. Government Principles for the Utilization and Care of Vertebrate Animals Used in Testing, Research, and Training (part of the PHS Policy)
- Occupational Health and Safety in the Care and Use of Research Animals[3]
- Biosafety in Microbiological and Biomedical Laboratories[4]
- Guide to the Care and Use of Experimental Animals[5]
- Institutional Animal Care and Use Committee Guidebook[6]
- Tutorial: PHS Policy on Humane Care and Use of Laboratory Animals[7]

27:5 How frequently should program reviews and facility inspections be conducted?

Reg. PHS Policy (IV,B,1; IV,B,2) and the AWAR (§2.31,c,1; §2.31,c,2) require inspection of the animal facilities and review of the program of animal care and use at least once every six months. The *Guide* (p. 25) states that these activities "should occur at least annually or more often as required (e.g., by the Animal Welfare Act and PHS Policy)." Subsequent to the provisions of the 21st Century Cures Act, Section 2034(d), the USDA and OLAW agreed "that the timing of facility inspections can include flexibility of within 30 calendar days of the six-month interval from the last inspection, provided there is no forward drift of the date from year to year. To avoid forward drift, the IACUC should consider scheduling facility inspections during the same calendar month from year to year. This 30-day flexibility allows extra time for IACUCs without changing the current or future dates of subsequent inspections."[8]

Opin. In general, programs should strive for an interval between successive program and facility evaluations that is as close to six months as feasible. However, there should be no hesitation in employing the 30-day flexibility afforded by USDA and OLAW guidance. Major departures from the six-month interval normally should be prevented to avoid the necessity to reset the IACUC calendar and potential regulatory non-compliance (see 28:7).

27:6 Can the twice yearly inspections occur in November and again during the following January?

Reg. See 27:5.

Opin. No (see 27:5). The regulatory language dictates an inspection every six months, not twice a year. The IACUC may choose to conduct inspections more frequently, but not less frequently. Consequently, if the committee followed a November inspection with another the following January, the next inspection would be due six months later, i.e., in July. It is advisable, therefore, to plan on conducting inspections at six-month intervals, in this case either May and November or January and July.

27:7 Must the program review and facility inspection be conducted at the same time?

Opin. No. At small institutions, the IACUC often will conduct both the program review and the facility inspection on the same day. At larger facilities, however, time constraints typically make it necessary to conduct the reviews over several days or even weeks during the reporting period. At very large facilities, inspections may be accomplished by inspecting one sixth of the facility each month during a 6-month period. It can be advantageous to conduct the facility inspection prior to the program review in that some program deficiencies might come to light only during the inspection (see 27:14). Alternatively, the IACUC might elect to discuss and compare at a convened meeting the results of both the program review and the inspection prior to preparation of the final report to the Institutional Official. A sample semiannual schedule is shown in Table 27.2.

TABLE 27.2

Sample IACUC Inspection and Program Review Schedule

Event/reporting period[a]	April 1–September 30	October 1–March 31
Program review	1st week of October	1st week of April
Facility inspection	2nd week of October	2nd week of April
Semiannual report	November	May

[a] Do not confuse this reporting period with that for the annual APHIS/AC or AAALAC report. This reporting period is merely a sample six-month period for the semiannual report to the IO. Any six-month period could be used.

27:8 Does the entire IACUC need to participate in the program review and facility inspection?

Reg. To comply with the AWAR (§2.31,c,3), a subcommittee participating in the facility inspection and program review must be composed of at least two committee members and "no Committee member wishing to participate in any evaluations … may be excluded." Under PHS Policy (IV,B,3, footnote 8, 2015 reprint), the IACUC "may, at its discretion, determine the best means of conducting an evaluation of the institution's programs and facilities." The AWAR (§2.31,c,3) require that the majority of the members of the IACUC review and sign the report.

Opin. While it is the practice at some institutions, particularly those with small animal care and use programs, for the entire IACUC to conduct the program review and facility inspection, this is not necessary. At many institutions, one or more IACUC subcommittees are charged to conduct the reviews and facility inspections. This is particularly apt for large institutions in that spreading the workload among multiple subcommittees can reduce the overall burden borne by individual committee members. It should be noted that regardless of the method chosen, the IACUC retains responsibility for the animal care and use program. As such, the full IACUC must review the final results of the review and inspection and approve the final report before submission to the IO.

27:9 Minimally, who should be included on each program review and inspection team?

Reg. See 27:8.

Opin. There are large variations in how institutions form teams/subcommittees to conduct the program review and inspection. Ideally, a typical team would consist of several IACUC members with at least one scientist, one veterinarian, and, where possible, one unaffiliated member. Other useful team members (not necessarily from the IACUC) are the animal care supervisor, a person with occupational health experience, and a member of the building operations/maintenance department (see Table 27.1). These internal consultants can round out the committee's subject matter expertise in their evaluation of the program and facility and provide additional perspective in the facility's operation. No IACUC member who wishes to participate may be excluded from participation (see 27:8). A useful practice at large facilities is to have one member of the team who has prior knowledge of the facility through participation in the last evaluation and one new member who can provide a fresh or unbiased review (see 27:21). Another practice is to train IACUC members through mock inspections in which deficiencies are staged and members are challenged to identify the deficiencies. In any case, it is important to keep the inspection and review process vibrant so that the process does not become routine and just another walk around the facility or a "rubber stamping" of the program.

Surv. Who participates in the semiannual program review and inspection at your institution? Select all that apply.

• Teams consisting of at least two IACUC members for all areas reviewed	108/182 (59%)
• Teams consisting of at least two IACUC members for areas where USDA-regulated animal species are housed/used; Teams inspecting areas where non-USDA regulated animal species are house/used may include one IACUC member	53/182 (29%)
• The entire IACUC at all times	34/182 (19%)
• External or internal consultants as necessary	33/182 (18%)
• Other	12/182 (7%)

27:10 Can outside consultants be used by the IACUC during inspection and review?

Reg. The PHS Policy (IV,B,3, footnote 8, 2015 reprint) states that the "IACUC may invite ad hoc consultants to assist in conducting the evaluation. However, the IACUC remains responsible for the evaluation and report." The AWAR (§2.31,c,3) have similar language allowing *ad hoc* consultants in addition to the two committee members to conduct the inspection and review.

Opin. Yes; however, two members of the IACUC must participate in the inspection and review to satisfy the AWAR (see 27:8). Most institutions possess ample expertise within their own organization and *ad hoc* consultants typically are not necessary for facility inspection and review purposes. In practice, outside consultants (other than veterinary consultants retained by some institutions) are not commonly used. Yet, in situations where internal expertise is limited or additional validation is sought, the use of an outside consultant can be valuable. For example, an *ad hoc* consultant might be employed when evaluating ABSL-3/4 laboratory space if the IACUC does not possess the necessary expertise to evaluate the facility. The use of internal consultants may also be desirable (see 27:9).

27:11 Can a recent AAALAC site visit be used in lieu of a semiannual program review and facility inspection?

Reg. According to NIH/OLAW Notice NOT-OD-24-076,[8] a site visit by AAALAC can be used as the semiannual program review and inspection of the animal facilities, provided the requisite elements of PHS Policy (IV,B,3) are met (see 27:8). USDA concurred with this guidance, as long as the requirements under AWAR §2.31,c and §2.35,a,3 are fulfilled such as:

- The IACUC adheres to the requisite frequency of program reviews and facility inspections (see 27:5; 27:6).
- At least two members of the IACUC must participate in both the program review and facility evaluation components of the site visit.
- All members of the IACUC must have the opportunity to participate with the AAALAC site visitors in the program review and facility evaluation.
- A properly convened quorum of the IACUC should review a summary of the findings from the site visit, determine whether any identified deficiencies are minor or significant, decide on suitable corrective action, and set a schedule for corrections to be completed.
- Committee actions relative to the review of the site visit findings need to be properly recorded in the official minutes of the IACUC.
- Minority opinions must be included if submitted.
- Relevant sections of the AAALAC site visit report are available to the USDA inspector.

Opin. No. It is not accurate to say that an AAALAC International site visit may be used **in lieu or instead of** the IACUC's review of the program and inspection of facilities. Rather, the IACUC may perform its evaluation of the program and facilities **in conjunction with** an AAALAC site visit. In such cases, the AAALAC site visitors are essentially acting as external consultants providing information about the program and facilities for consideration by the IACUC. Along these lines, it is important to note that while the AAALAC site visit entails a comprehensive review of the animal care and use program and facilities, in some cases, particularly at larger institutions, the AAALAC site visitors might not evaluate all the facilities requiring a semiannual inspection by the IACUC, e.g., all laboratories where survival surgery is conducted. In such cases, the committee will need to ensure those areas are inspected during or after the site visit. It also should be noted that the IACUC may identify deficiencies that were not identified by the AAALAC site visitors, and, conversely, might decide to exclude from their evaluation items identified by AAALAC. The IACUC should consider the findings presented by the AAALAC site visitors shortly after the site visit rather than waiting for the formal communication from AAALAC conveying the program's accreditation status. It is advisable to convene a regular IACUC meeting soon after the AAALAC site visit for the purpose of discussing the AAALAC findings and any other issues/deficiencies identified by the IACUC and endorsing the report as a formal IACUC review. If too much time is allowed to elapse, appropriate action may be delayed, including the timely correction of deficiencies. Timely committee action will also aid in writing the post-AAALAC visit response, which usually is due to AAALAC within 10 business days after the site visit.

Surv. Did your IACUC conduct a semiannual review of the program and/or facility inspection in conjunction with your last triennial AAALAC International site visit?

- No 60/182 (33%)
- Not applicable, my institution is not AAALAC-accredited 46/182 (25%)
- Yes, only the facility inspection 19/182 (10%)
- Yes, only the program review 6/182 (3%)
- Yes, both the program review and the facility inspection 46/182 (25%)
- I don't know 5/182 (3%)

27:12 What is an effective method for conducting the program review?

Reg. Footnote 8 of the PHS Policy (August 2015 reprint) states that the IACUC may, at its discretion, determine the best means of conducting an evaluation of the institution's programs and facilities.

Opin. The AWAR and PHS Policy do not specify the method to be used for conducting the semiannual program review. As such, there are large variations in the methods employed among institutions. However, there are two accepted formats in common use. One involves a discrete review at six-month intervals of the entire program, typically at a single meeting or group of meetings scheduled within a relatively short window of time. The other involves an ongoing or rolling review process in which a portion of the program, e.g., program management (*Guide*, pp. 13–24) is reviewed as part of normal (generally monthly) IACUC meetings, such that the entire program is evaluated in the meetings held over the course of six months. The authors prefer the former in that a discrete review conducted over a short period of time can simplify planning and documentation processes. However, we acknowledge that the rolling review method can help with time management for some committees. In all instances, either the full committee or a subcommittee may conduct the review.

Regardless of the format used, the committee's review process should address all programmatic aspects described in the *Guide* (p. 11). As mentioned previously, checklists based on the outline of the *Guide* can be very helpful in this regard (see 27:1). Often, all the information required to determine whether a particular standard is being met is not at hand when merely reviewing a checklist. As such, many IACUCs have implemented procedures to ensure relevant information is presented to the committee for evaluation. Commonly, these methods include interviews of or presentations/reports by key program personnel, and review of program documents, such as IACUC policies/procedures and animal facility standard-operating procedures, as well as research records.

Surv. Which of the following methods/tools does your IACUC use to conduct its semiannual program review of humane care and use of animals? Select all that apply.

- The NIH/OLAW checklist that is on the OLAW website 152/182 (84%)
- A checklist prepared by our own institution 66/182 (36%)
- Interviews with key personnel (e.g., veterinarian, IACUC Chair, occupational health service representative, etc.) 51/182 (28%)
- A list of items to discuss prepared by the IACUC Chairperson or IACUC office staff 61/182 (34%)
- Special presentations regarding various aspects of the animal care and use program 23/182 (13%)
- Review of a portion of our program at every full committee meeting 27/182 (15%)
- Subcommittees to review different elements of the program 39/182 (21%)
- Other 9/182 (5%)

27:13 What sorts of documentation should be reviewed during the program review and inspection of the facilities? (see 27:2; 27:7)

Reg. The AWAR, the PHS Policy, and the *Guide* are silent on what documents must be reviewed as part of the semiannual program review and facilities inspection.

Opin. An important aspect of the facility inspection is to look at records/documents associated with the care and use of animals such as animal medical records, laboratory records of animal use, animal room service logs, cage washer maintenance records, etc. Real-time observations of the animals and facilities during the inspection provide a snapshot view of the conditions at that specific time. The review of key records can provide a longitudinal perspective of conditions and practices. For example, examination of animal room service logs provides evidence of conditions maintained over time rather than just during the inspection. Similarly, examination of animal health/surgical records provide valuable information about delivery of veterinary medical care over time, which augments the animal observations made during the inspection. As such, the examination of program documentation helps the committee form a complete view of the state of the animal care and use program.

Surv. What documentation does your IACUC consistently review during its program review? Select all that apply.

- Institutional/IACUC policies 133/182 (73%)
- IACUC meeting minutes 61/182 (34%)
- Animal resource SOPs 72/182 (40%)
- Animal resource disaster/emergency response plan 94/182 (52%)
- Animal behavioral management/ environmental enrichment program records 46/182 (25%)
- AAALAC International program description 34/182 (19%)
- Personnel training records 46/182 (25%)
- Medical/surgical records for USDA-regulated species 33/182 (18%)
- Medical/surgical records for mice, rats, and other non-USDA-regulated species 35/182 (19%)
- Animal morbidity and mortality records 41/182 (23%)
- Animal disease surveillance results 34/182 (19%)
- Animal room environmental monitoring logs/documents 47/182 (26%)
- Animal room care logs/documentation 50/182 (27%)
- Documentation of cage washer/ autoclave function 47/182 (26%)
- Other 40/182 (22%)

Surv. What documentation does your IACUC consistently review during facility inspection? Select all that apply.

- Institutional/IACUC policies 41/182 (23%)
- IACUC meeting minutes 18/182 (10%)
- Animal resource SOPs 45/182 (25%)
- Animal resource disaster/emergency response plan 39/182 (21%)
- Animal behavioral management/ environmental enrichment program records 66/182 (36%)
- AAALAC International program description 10/182 (5%)
- Personnel training records 58/182 (32%)
- Medical/surgical records for USDA-regulated species 94/182 (52%)
- Medical/surgical records for mice, rats, and other non-USDA-regulated species 110/182 (60%)
- Animal morbidity and mortality records 73/182 (40%)
- Animal disease surveillance results 57/182 (31%)
- Animal room environmental monitoring logs/documents 159/182 (87%)
- Animal room care logs/ documentation 169/182 (93%)
- Documentation of cage washer/ autoclave function 145/182 (80%)
- Other 19/182 (10%)

27:14 Can the review of animal care and use programs be based on observations made during the inspection of the facilities? (see 27:2; 27:7)

Opin. While observations made during the facility inspection certainly do provide information relevant to the program, facility inspection findings should not be used as the sole basis for the program review. Conditions in the animal facility and laboratories where animals are used can reflect the overall quality of the program, and the IACUC should determine whether deficiencies identified during the inspection have programmatic implications. However, assessment of the facility inspection results would not likely result in the evaluation of all aspects of the program. More than a visual inspection of the facilities is required to determine whether the animal care and use program is operating in accordance with federal requirements. For example, it is not likely that problems with the composition of the IACUC or documentation of the occupational health program will be identified during a facility inspection. Therefore, the IACUC should have discrete processes for both the program review and the facility inspection.

27:15 Can the review of the animal care and use program be conducted while inspecting the facilities?

Opin. Although conceivable, especially at institutions with very small programs, it is not advisable for a committee/subcommittee to carry out the program review and inspection simultaneously. As suggested in 27:14, it would be highly unlikely that a physical inspection of the facility would provide information sufficiently representative of the state of all aspects of the program. In addition, it generally would be impractical to attempt to go through a list of program elements and determine the extent to which the standards for those elements are met while simultaneously viewing and assessing the facility. For these reasons, separate (either temporally or by different subcommittees) sessions for the program review and facility inspections are recommended.

27:16 Should the IACUC evaluate the animal facility disaster plan as part of the program review? If so, what are the requirements for a disaster plan? (see 25:2)

Reg. The *Guide* (pp. 35, 74–75) requires facilities to have a disaster plan that considers both animals and humans. The animal facility plan should be part of the overall institutional disaster plan. Some key elements of the plan include the actions necessary to prevent animal pain, distress, and deaths due to loss of critical systems/infrastructure; how the facility will preserve animals that are necessary for critical research activities or are irreplaceable; and preparation and training of key personnel, e.g., the AV and colony manager, having responsibilities in the event of a disaster; and how significant personnel absences will be handled. The AWAR require a documented contingency plan (§2.38,l). The contingency plan requirements are generally similar to the *Guide* recommendations and mandate the identification of situations that would trigger the activation of the plan; outline specific tasks to be carried out; identify the chain of command and the concomitant responsibilities of personnel; address how response and recovery would be handled; and provision of training of personnel regarding their roles and responsibilities. The research facility is required to review the plan on at least an annual basis to ensure that it adequately addresses these criteria. NIH/OLAW has published guidance on this topic.[9,10] Since the disaster/contingency plan is a part of the animal care and use program, it is implicitly within the scope of the IACUC's review of the program.

Opin. Since it is an important component of the overall animal care and use program, the IACUC should evaluate the animal facility disaster plan as part of the semiannual program review. It is a relatively straightforward task to compare the disaster/contingency plan content against the guidance listed above. However, it is advisable for IACUCs to assess the currency and readiness of the plan. To this end, the committee should ask questions of the animal facility director/manager and perhaps institutional safety personnel about training staff concerning the plan. Questions might include:

- What training is given to the animal care staff regarding the disaster plan?
- Does management ever conduct drills or tabletop exercises to simulate disaster situations and practice responses?
- What processes are in place to ensure that all aspects of the plan are kept current, e.g., lists of responsible personnel, emergency notification contact lists/calling trees, etc.?

Questions such as these will help the IACUC determine whether the disaster plan is merely a document or a set of processes that are ready to be implemented should the need arise. An NIH/OLAW webinar on disaster planning can be viewed at https://olaw.nih.gov/education/educational-resources/webinar-2013-03-07.htm.

27:17 How much time should the IACUC expect to devote to conducting the program review?

Opin. Many factors influence the amount of time the IACUC devotes to the program review. Most important are the specific methods and level of detail chosen to perform the review. If an IACUC already has an in-depth familiarity with their program and decides to have a single meeting in which they use a checklist of program components to guide their review, that committee might complete its work in as little as one hour. However, such a quick and perhaps superficial review may not yield very valuable results for many programs and could result in an inadequate assessment. An appropriate and complete program review requires the input of individuals and groups having the proper expertise to conduct an evaluation of the program as it exists and determine the extent to which it meets (or deviates from) the regulatory requirements and other standards. In well-functioning programs, the IACUC can rely on program area experts, e.g., the AV, IACUC support staff, behavior management staff, animal care managers, and occupation health and safety personnel, to provide the committee information and preliminary assessments of the various aspects of the program. To permit greater depth of evaluation and spread effort, multiple subcommittees may be formed to separately review subcomponents of the program, e.g., institutional policies and responsibilities, occupational health and safety, veterinary medical care, etc. This preliminary work by subcommittees can save time for the committee as a whole. Assuming there is adequate preparation and depending on the size and complexity of the program, it is reasonable to expect that a formal subcommittee meeting to review the entire program can usually be completed within a period of about two hours. In the event multiple subcommittees are used to evaluate individual program components, considerably less time would be required for each subcommittee meeting. However, the state of the program and identification of deficiencies can lengthen that time. Full committee review of the subcommittee reports takes additional time.

It is worth noting that the amount of time required to complete a thorough program review is not necessarily proportional to the size of the institution. Small institutions might lack certain program elements, such as biocontainment facilities, that add complexities to the review; however, all institutions regardless of size are subject to the same general requirements for program organization, implementation, and oversight. Table 27.3 illustrates a typical review process.

Surv. Approximately how much time is devoted by the IACUC membership to the semiannual program review? Do not include time to conduct the facilities inspection. Estimate the total time as an entity, not the cumulative time of each individual member.

- Not applicable, semiannual reviews not conducted 2/182 (1%)
- Approximately 1–2 hours 73/182 (40%)
- Approximately 2–4 hours 65/182 (36%)
- Approximately 4–6 hours 18/182 (10%)
- Over 6 hours 22/182 (12%)
- Don't know 2/182 (1%)

27:18 Can alternate members of the IACUC perform semiannual inspections that involve PHS Policy and AWA-regulated species?

Reg. The AWAR and PHS Policy do not address this issue specifically. As stated previously, the AWAR (§2.31,c,3) require the inspection team to consist of a subcommittee of at least two IACUC members. NIH/OLAW guidance on this topic states: "An alternate may only contribute to a quorum and function as an IACUC member if the regular member for whom they serve as alternate is unavailable to participate in IACUC business, whether it is because that member is unable to attend the meeting, has to leave the meeting early or arrive late, or is recused from participating due to a real or potential conflict of interest."[11] As such, if the alternate functions as an IACUC member during semiannual program review and facility inspections, they may do so only if the regular member for whom they serve is unavailable. Using alternates to conduct inspections solely to reduce the workload of regular members of the IACUC (for example, using an alternate to serve on a different inspection group from the regular member) is not consistent with the cited guidance. However, the guidance does not preclude alternate members from participating as *ad hoc* consultants (see 27:10) or in addition to the two regular members required by the AWAR. USDA concurs with this guidance.[12]

Opin. Alternates should participate in the semiannual program reviews and facility inspections. Clearly, an alternate member may participate instead of the regular IACUC member when the regular member is unable to take part in the inspection. Beyond that, however, the participation of alternate members should be viewed as an important element in their development. Alternate members should be appropriately trained and experienced to fulfill their role, i.e., to substitute for an unavailable regular member. Given the centrality of the semiannual program review and inspection to the IACUC's function, it follows that alternate members should be thoroughly familiarized with this process. In instances where alternate members are inexperienced, this is best achieved by their participation as *ad hoc* consultants or "extras" with teams of at least two regular voting members. But even for experienced alternates, e.g., former regular members, this form of participation is a good way for them to maintain familiarity with the program and process. Alternates developed in this way will be better prepared to step in for a regular member when required (see 5:24).

Surv. Does your IACUC allow an alternate member of the committee to perform inspections if the IACUC position for which they are the alternate is performing inspections at the same time but in

TABLE 27.3

Typical Steps for the Program Review Process

Time Frame	Activity	Who	What	Time Required
30 days or less prior to full IACUC review	Subcommittee review(s)	Each subcommittee should include at least two members of the IACUC as well as area experts as indicated	Collectively reviews the program in terms of the combined and accumulated knowledge of the area experts. Makes recommendations to the full committee	Up to 2 hours
Within 30 days of subcommittee review	Full IACUC review	Members of the IACUC and invited area experts as desired	Review findings of the subcommittee; establish corrective action and schedules of completion. Significant deficiencies may need more expeditious handling	Up to 2 hours; with good preparation at previous levels meeting may be much shorter
Within 30–45 days of full committee review; some significant deficiencies may need more timely action	Post any administrative paperwork	IACUC Administrator Note: At small facilities a committee member may serve as the Administrator	Complete minutes, send out notice of required corrective action, and complete other required documentation. Significant deficiencies may need more expeditious handling	As long as needed; should generally complete process in 30–45 days
See above	Final report	IACUC Chair and IACUC Administrator	Signed by committee and forwarded to the IO	See above

a different location? Assume this question is for USDA-covered species.

- Not applicable 72/182 (40%)
- Yes, we do 51/182 (28%)
- No, we do not 52/182 (29%)
- I don't know 4/182 (2%)
- Other 3/182 (2%)

27:19 What areas should the IACUC include in the semiannual facility inspection?

Reg. Both the PHS Policy (IV,B,1; IV,B,2) and the AWAR (§2.31,c,1; §2.31,c,2) require the IACUC to inspect the institution's animal facilities once every six months (see 16:1). The PHS Policy (III,B) further defines animal facility to include "Any and all buildings, rooms, areas, enclosures, or vehicles, including satellite facilities, used for animal confinement, transport, maintenance, breeding, or experiments inclusive of surgical manipulation," and requires the inspection to include "satellite facilities," which are defined in the Policy (III,B) as those areas that are "outside of a core facility or centrally designated or managed area in which animals are housed for more than 24 hours." NIH/OLAW has further clarified the definition to include "areas where any form of surgical manipulations (minor, major, survival, nonsurvival) are performed."[13] The AWAR (§1.1) require the inclusion of "animal study areas," which are defined as areas "outside of a core facility or centrally designated or managed area in which animals are housed for more than 12 hours." The *Guide* (p. 25) recommends that in addition to the animal facility the inspection include "animal use areas"; however, the term is not further qualified.

Opin. The semiannual facility inspection should include at a minimum all areas, whether they be for the purpose of animal housing, conducting animal procedures, or for support functions, e.g., cage wash, that are within the "envelope" of the central animal housing facility(ies), those areas that meet the definitions of satellite facility and/or animal study area, and performance areas such as imaging, behavioral testing units, specialized housing, and transport vehicles. In our opinion, however, best practice would be for the committee to include not only those areas specified above, but also all areas where animals are used—even areas where animals are taken for noninvasive procedures or euthanasia and tissue collection. Yet, it could be exceedingly difficult for some IACUCs, particularly at large institutions, to visit every six months every single laboratory/area where animals are used. In such cases, it is essential that the IACUC implement alternative means of assessing those activities and spaces. NIH/OLAW[14] has stated: "When considering IACUC responsibilities for semiannual review, it is important to keep in mind that the institution, usually acting through the IACUC and/or the facility veterinarian, is responsible for all animal-related activities of the institution, regardless of where animals are maintained or the duration of their stay." NIH/OLAW FAQ E.1 states: "Institutions have discretion with regard to how they oversee areas used for routine weighing, dosing, immunization, or imaging, but should monitor such areas on a random or fixed schedule to effectively oversee activities at the institution." Therefore, IACUCs have broad discretion in the way they oversee animal work that is conducted in laboratories. The degree, frequency, and method of IACUC oversight of these other animal use areas may depend on the nature of the activity or the species involved. The methods may involve, for example, random or scheduled visits (which may be less frequent than semiannual) by IACUC members, the veterinary staff, or PAM staff. Whatever method is chosen, the authors recommend that documentation/reports of the oversight activities be presented to the full committee for its consideration by the committee and inclusion in the semiannual report to the Institutional Official (IO) (see 24:12).

Surv. What areas does your IACUC include in your semiannual facility inspection? Select all that apply.

- Central animal holding areas 179/182 (98%)
- Satellite animal holding areas 132/182 (73%)
- Surgery and procedure rooms within the animal facility 171/182 (94%)
- Service areas such as cage wash, food storage, cage storage 176/182 (97%)
- Personnel areas within the animal facilities, including locker and break rooms 78/182 (43%)
- Animal facility loading dock(s) 102/182 (56%)
- All laboratories holding animals for over 12 or 24 hours, depending on the species 134/182 (74%)
- All laboratories to which animals are brought, regardless of length of stay or procedures performed 117/182 (64%)
- Laboratories where survival surgery is performed 144/182 (79%)
- Laboratories where non-survival surgery is performed 140/182 (77%)
- Laboratories where noninvasive procedures such as behavioral testing are performed 133/182 (73%)
- Human clinical areas where animals are brought for procedures such as imaging 61/182 (34%)

27:20 But what about remote, outdoor agricultural animal sites and field study sites? Does the IACUC have to visit those in person too?

Reg. The PHS Policy and *Guide* do not specifically address the inspection of outdoor, non-confinement agricultural animal sites such as pastures or field study sites, although OLAW has stated that IACUC inspections of field study sites are not required. While the AWAR (§2.31,c,2) do not specifically consider non-confinement agricultural animal sites, field study sites "containing free-living wild animals in their natural habitat" are exempt from the inspection requirement. Of course, if animals are captured and/or confined at the field site for more than 12 hours, it meets the AWAR definition of "study area" (§1.1) and will need to be included in the semiannual inspection. While not defining the terms, the *Ag Guide* (p. 2) states that the IACUC should include "active agricultural animal facilities and study areas" in its semiannual inspections.

Opin. The IACUC should evaluate all animal housing areas regardless of location on a semiannual basis. Outdoor housing of livestock may range from small paddocks to pastures of hundreds or even thousands of acres. In large pasture settings, the observation of animals and conditions may be challenging not only because of size of the land, but also because of its topography, e.g., wooded areas, ravines, etc. In addition, lands on which animals are pastured as well as field study areas often are long distances from the campus/administrative home of the IACUC, which creates further logistical challenges. While it may not be practical to observe every animal at these locations, the committee should assess the adequacy of the environment for the species kept. Specific considerations by the IACUC should include the appropriateness of shelter from sun, wind, and precipitation; state of repair fences, erected shelters, and other structures; adequacy of forage/feed; appropriateness of animal access to water during all seasons; and exclusion/protection from predators.[15] Ideally, the IACUC inspections should be in person; however, the use of contemporary technical tools such as video-equipped drones and devices with live video streaming capability can be useful in augmenting or in some cases replacing the committee's on-site evaluation (see 27:21).

Surv. Does your IACUC include remote, non-confinement agricultural animal sites (e.g., pastures) and field study sites in its semiannual inspections?

• Not applicable, we don't have those types of sites in our program.	124/182 (68%)
• Yes; We visit all remote, non-confinement agricultural animal sites but don't inspect field study sites.	22/182 (12%)
• Yes; We evaluate video recordings of remote sites and interview personnel who work or conduct research activities at those sites. There is no on-site inspection.	7/182 (4%)
• Yes; We interview only personnel who work or conduct research activities at those sites.	7/182 (4%)
• No; We don't evaluate remote sites.	22/182 (12%)

27:21 Can aspects of the program review and semiannual inspection of animal facilities be performed virtually?

Reg. See 27:1, 27:2. NIH/OLAW and USDA/APHIS offer guidance on using alternative methods for IACUC meetings. "Methods of telecommunications (e.g., telephone or video conferencing) are acceptable for the conduct of official IACUC business requiring a quorum, provided the following criteria are met:

- All members are given notice of the meeting.
- Documents normally provided to members during a physically-convened meeting are provided to all members in advance of the meeting.
- All members have access to the documents and the technology necessary to fully participate.
- A quorum of voting members is convened when required by PHS Policy.
- The forum allows for real time verbal interaction equivalent to that occurring in a physically convened meeting (i.e., members can actively and equally participate and there is simultaneous communication).
- If a vote is called for, the vote occurs during the meeting and is taken in a manner that ensures an accurate count of the vote. A mail ballot or individual telephone polling cannot substitute for a convened meeting.
- Opinions of absent members that are transmitted by mail, telephone, fax or e-mail may be considered by the convened IACUC members but may not be counted as votes or considered as part of the quorum.
- Written minutes of the meeting are maintained in accord with the PHS Policy, IV.E.1.b."[16,17]

Opin. During the COVID-19 pandemic, many research institutions adopted the use of virtual platforms for communication purposes. As a result, IACUC, and other meetings related to the care and use of laboratory animals, were often performed virtually at institutions. NIH/OLAW offered flexibility regarding program review and facility inspection(s) during this time.[18] The notice was rescinded in 2022. That said, many institutions have decided to continue using virtual platforms for IACUC business where allowed. There is some flexibility regarding alternatives to in-person inspections of animal facilities using virtual platforms. According to NIH/OLAW,

> IACUCs may use videos, photographs, written descriptions, or other appropriate remote methods to inspect facilities housing non-AWA regulated species. These methods may be useful, for example, where no qualified individual has access to an inspection site (e.g., high containment and remote satellite facilities) or when one qualified individual provides a prerecorded or real-time virtual tour to IACUC members wishing to participate but who are unable to do so safely. There is no requirement for IACUCs to retain these materials once the semiannual report has been submitted to the IO. However, for areas housing AWA-regulated species, inspections must be physically performed; videos, photographs, written descriptions, or other remote methods are not appropriate. [19]

For areas housing AWA-regulated species, live feed is allowed for inspections using video.

Surv. Does your IACUC regularly use virtual platform(s) for the conduct of any portions of the semiannual program review and inspection?

- No. 81/182 (45%)
- We use virtual platform(s) as part of the program review but not the inspection. 66/182 (36%)
- We use virtual platform(s) as part of the inspection but not the program review. 6/182 (3%)
- We use virtual platform(s) as part of the program review and inspection. 26/182 (14%)
- I don't know. 3/182 (2%)

27:22 Can an IACUC inspection team inspect areas more often than twice a year?

Reg. See 27:5.

Opin. Yes, facilities may be inspected as often as the IACUC deems necessary. The PHS Policy and the AWAR do not preclude IACUC inspections more frequently than once every six months. An IACUC may "reset the clock" if it wishes to perform an inspection early and then perform the next inspection six months later. More frequent inspections may be required to verify that previously identified concerns have been corrected. Some IACUCs conduct inspections on an ongoing basis. For example, an IACUC may set aside a time each month to visit laboratories where animal research is conducted. Under this system, an individual laboratory may be inspected more frequently. It is also a common practice to evaluate new areas before housing animals.

Members of the animal care staff, such as directors and managers, are often members of the IACUC. In addition, at least one veterinarian must be a member. These individuals are usually in the animal facility on a regular basis between formal IACUC inspections and, therefore, have the opportunity to evaluate and report to the IACUC more frequently than every six months. Individual(s) performing PAM visits on behalf of the committee also should report findings to the IACUC on a frequent basis and many animal holding/use areas will be evaluated more than twice per year as a result of such inspections combined with the semiannual inspections.

27:23 Should animal facilities be inspected only if animals are present at the time of inspection?

Reg. See animal facility definitions in 16:1.

Opin. Not necessarily. Within the central animal facility there may be holding rooms that are empty or procedure/surgery rooms that are not in use at the time of the inspection. These areas still should be evaluated for acceptability for the intended use. Similarly, animals might not be present in satellite facilities or other animal use areas outside of the central facility at the time the IACUC inspection. In these cases, the committee should still evaluate the spaces and review records and interview staff about the activities in order to assess their adequacy. If it is likely that animals will never again be housed or used in a particular area, the area can be removed from the list of those locations to be inspected (see 27:26).

27:24 What information should be provided to the facility inspection team, either before or during the inspection, about the animal activities that are approved for each area they inspect?

Opin. To conduct a proper evaluation, the team needs to have some knowledge about the animal activities that are approved for the areas they inspect. At the outset of the inspection, the team should have knowledge of the approved protocols, species used, and general categories of procedures performed, e.g., survival surgery, in each location that is to be inspected. This information usually can be included in a list provided to the inspection team. During the inspection, it is advisable for the committee to ask questions of the investigators and research staff concerning the details of how animal procedures are conducted. In addition, the team should have the capability of referencing in real time the currently approved protocols. This permits the comparison of what is observed with what has been approved by the IACUC. Methods to accomplish this may include bringing hard copies of the approved protocols along on the inspection or ensuring access to electronic documents at computer stations, laptops, or mobile computing devices.

Surv. For the semiannual facility inspection, how are the inspectors informed as to what procedures the IACUC has approved for any particular investigator/educator?

Select all that apply.

- We expect the IACUC inspectors to review protocols prior to the inspection. 49/182 (27%)
- The inspector(s) may bring a copy of the approved protocol(s) with them on the inspection. 45/182 (25%)
- The inspector(s) obtains that information from the investigator/educator. 57/182 (31%)
- The inspector(s) already are thoroughly familiar with all the procedures approved and don't need additional information for the inspection. 48/182 (26%)
- IACUC office personnel communicate that information to inspectors. 85/182 (47%)
- We provide hard copies of the investigator's protocols to the inspectors. 15/182 (8%)
- We provide electronic access to the investigator's protocols via a mobile device (e.g., laptop, tablet, etc.). 56/182 (31%)
- Don't know. 2/182 (1%)
- Not applicable. 10/182 (5%)
- Other. 12/182 (7%)

27:25 When doing semiannual inspections, is it a good idea to have some IACUC members "specialize" in certain areas year after year or is it a better idea to have members rotate among areas?

Opin. There is a benefit to having inspection teams consisting of a combination of people who have inspected an area, or areas, previously in addition to new members. Retaining members on a team that have inspected area(s) previously is particularly valuable as it facilitates a historical perspective. Such individuals are available to offer their previous experience regarding the areas and to answer questions the new members may pose regarding the previous inspection. Conversely, changing or adding new members to an inspection team will add a "new set of eyes." The new members may offer fresh perspectives or different expertise to the inspection team. This scenario is beneficial to all members involved and should result in a more thorough evaluation than scenarios where members repeatedly inspect the same areas or rotate every inspection.

Surv. For semiannual inspections, do you always send the same people to the same places or do you rotate the responsibilities?

- We are a small institution and everybody goes to the same places during every inspection. 64/182 (35%)
- We are large enough to rotate assignment areas but we usually do not do that. 10/182 (5%)
- We usually try to rotate assignment areas. 99/182 (54%)
- I don't know. 1/182 (1%)
- Other. 8/182 (4%)

27:26 Should the IACUC allow a previously unused area to be used for live animal–related work without a formal IACUC inspection?

Reg. Both the PHS Policy (IV,B,6) and the AWAR (§2.31,c,6) indicate that the IACUC should "review and approve, require modifications in (to secure approval) or withhold approval of those activities related to the care and use of animals." According to NIH/OLAW, "Institutions are responsible for oversight of all animal-related activities regardless of how long or where the activity occurs."[13] The *Guide* (p. 25) indicates that IACUC oversight functions include "regular inspection of facilities and animal use areas."

Opin. No research project involving live animals may be conducted prior to a review and approval by the IACUC. As a component of that review, the IACUC should assess the suitability of the proposed space for planned animal-related work. The methods for inspecting the area should be similar to those used for conducting the semiannual inspection, for example, forming an inspection team of at least two members of the IACUC to evaluate the appropriateness of new animal housing or study area. A common approach is to have the AV (or other program veterinarian) and one additional IACUC member conduct the inspection and report back to the full committee. In any case, work on the protocol should not commence until the area has been evaluated and approved by the IACUC (see 29:4).

Surv. Does your IACUC allow a previously unused area to be used for live animal–related work (e.g., survival rodent surgery) without a formal IACUC inspection?

- Yes, routinely without any additional conditions. 3/182 (2%)
- Yes, but the PI must notify the IACUC prior to use. 13/182 (7%)
- Veterinary review and approval are required by our IACUC. 50/182 (27%)
- One or more IACUC members must review and approve the area. 96/182 (53%)
- I don't know. 9/182 (5%)
- Other. 11/182 (6%)

27:27 Is there a conflict of interest when the director of the animal facility (an IACUC member) participates on a team that inspects the core animal facility?

Reg. The PHS Policy (IV,C,2), AWAR (§2.31,d,2), and *Guide* (p. 26) all contain language indicating the need for the recusal of IACUC members from deliberations on animal use proposals for which they have a conflict of interest, i.e., are personally involved. However, none of these documents have a similar statement with regards to the facility inspection. To the contrary, with respect to the semiannual program review and facility inspection the AWAR (§2.31,c,3) state that "no committee member wishing to participate in any evaluation conducted … may be excluded."

Opin. There might be a perception of a conflict of interest because the director, who often is the AV, would be in a position of having to criticize his or her own area of responsibility. However, in practice, this is seldom a problem. Animal program directors desire a well-managed and regulation-compliant animal care facility/program since they are faced with annual unannounced inspections by APHIS/AC and periodic review by AAALAC. Using this same logic, animal program directors often seek input from others regarding the animal facility since programmatic improvements are in their best interest. Directors seldom hide problems; rather, they are the ones to point out problems and potential problems to newer members of the team. Due to his or her background, the director is often the most knowledgeable individual on the IACUC regarding animal care standards. Thus, the director, as well as others involved in management of the animal facility, are valuable resources for other committee members and should be available to answer questions during the facility inspection of the animal facility.

Nevertheless, if there remains concern about the perception of a conflict of interest, the IACUC may consider assigning members other than the facility director to formally conduct the review and permit the director to participate as an observer and to provide information. If the IACUC believes that the director, particularly if the director also is the AV, must be excluded for fear of interfering or otherwise corrupting the inspection process, this is a sign of more serious programmatic problems that are beyond the scope of this text.

Surv. To avoid potential conflicts of interest, does your IACUC attempt to assure that during semiannual inspections the attending veterinarian or other member(s) of the IACUC who are responsible for the daily operations of your central animal facility(s) do not inspect the central animal facility(s)?

- Not applicable. 14/182 (8%)
- Yes, we try to prevent this from happening. 26/182 (14%)
- Yes, but they may participate in the inspection in an informal capacity to provide information to the people performing the inspection. 76/182 (42%)
- No, we make no particular effort to prevent this from happening. 65/182 (36%)
- I don't know. 0/182 (0%)
- Other. 1/182 (1%)

27:28 Is there a conflict of interest when an investigator who also is an IACUC member participates on a team that inspects their own laboratory?

Reg. See 27:27.

Opin. As with the scenario in 27:27 regarding the animal facility director, this situation can be, and typically is, considered a conflict of interest by the IACUC. In this case, the individual should not perform the primary evaluation of their own research laboratory; rather, other committee members should serve in that capacity. But just as the facility director should not be precluded from participating in the evaluation of the animal facility, neither should an investigator be excluded from the inspection of their laboratory because they are an IACUC member. Just as the facility director is a valuable resource during the inspection of the animal facility, the researcher/investigator, whether or not they are an IACUC member, should be present during the laboratory inspection so that they may answer questions and provide information to the inspection team as necessary.

Surv. To avoid potential conflict of interest, does your IACUC attempt to ensure that during semiannual inspections investigators do not inspect their own laboratories?

- Not applicable. 15/182 (8%)
- Yes, we try to avoid this from happening. 63/182 (35%)
- Yes, but they may participate in the inspection in an informal capacity to provide information to the people performing the inspection. 83/182 (46%)
- No, we make no particular effort to avoid this from happening. 21/182 (12%)
- I don't know. 0/182 (0%)
- Other. 0/182 (0%)

27:29 Should facility inspections be announced or unannounced?

Opin. Opinions differ, but most institutions favor announced inspections. There are no requirements in the AWAR, *Guide*, or PHS Policy stipulating announced or unannounced inspections. The advantages of both have been briefly discussed.[6] An unannounced inspection gives the inspection team a clear picture of what is routinely occurring in the animal facility and laboratories where animals are used. On the other hand, announced inspections allow essential persons to be available to answer questions. Announced visits also give investigators the incentive to review their areas and procedures with their lab personnel; they may motivate some PIs to elevate the standards within their labs, especially if they are provided with a listing by the IACUC of what some common types of infractions are. These higher standards will ideally prevail through the entire six-month period between inspections. As there are benefits from both announced and unannounced inspections, an IACUC may want to try both approaches to determine what best fits their particular institution. In any case, the choice is that of the individual IACUC (see 29:6).

In addition to semiannual IACUC inspections, IACUC oversight of animal activities may be enhanced by a PAM program and these visits may be announced and/or unannounced. PAM is discussed in the *Guide* (pp. 33–34). PAM visits may supplement semiannual inspections, so such inspections do not need to be unannounced. In addition, PAM visits are generally targeted for one specific area or laboratory and a more thorough evaluation of the animal care/procedures can be performed than during a semiannual inspection. PAM visits offer additional oversight that supplements semiannual facility inspections (see Chapter 34).

Surv. Are your semiannual facility inspections announced or unannounced?

- Not applicable; we do not do semiannual facility inspections. 2/182 (1%)
- Usually announced. 148/182 (81%)
- Usually unannounced. 26/182 (14%)
- We leave it to the discretion of the inspectors. 6/182 (3%)
- Other. 0/182 (0%)

27:30 Should the IACUC inspectors speak with individuals participating in research using animals?

Opin. Yes, this is a generally accepted practice and is highly desirable. There are two main groups of people with whom the committee should interact during the inspection: the research staff and the animal resource staff. Conversations with individuals from these groups often yield valuable information about the program and facilities. For example, through informal interviews at the time of the inspection, it is possible to verify that SOPs are being followed, that persons are appropriately trained, and that anesthesia and euthanasia methods are appropriate and in accordance with approved protocols. This allows the inspection team to be more efficient and obtain the information they seek in a timely manner.

Some institutions require that each laboratory have a representative present to answer questions from the inspection team. Other IACUCs schedule inspections at times when persons are normally working and question those persons available on a less formal basis. The inspection process can and should be a learning experience for both the IACUC and the staff. Free and open communication facilitates this learning process.

The IACUC should take time at the end of a visit to give feedback to the persons involved. When warranted, positive as well as negative comments should be made. IACUCs may consider having an "exit interview" with interested persons as a means of exchanging information.

Surv. How do your IACUC teams interact with animal care and laboratory personnel during facility inspections?

- We always speak with both animal care and laboratory personnel about their activities as an integral part of our inspections. 126/182 (69%)
- We always speak with laboratory personnel during inspections of their areas; however, we generally don't speak with animal care personnel unless we have a specific question. 39/182 (21%)
- We generally don't interact with either animal care or laboratory personnel during inspections. 15/182 (8%)
- I don't know. 2/182 (1%)

27:31 What is a deficiency, what criteria are used to categorize a deficiency as minor or major (significant), and who makes that determination?

Reg. Neither the PHS Policy nor the AWAR explicitly define the term "deficiency." However, the implicit definition is any unapproved departure from the provisions of the *Guide*, PHS Policy, or AWAR (see PHS Policy IV,B,3; AWAR §2.31,c,3). Regarding the criteria that distinguish a significant deficiency from a minor, the AWAR and PHS Policy are in concordance: "A significant deficiency is one which … in the judgment of the IACUC and the Institutional Official, is or may be a threat to the health or safety of the animals" (AWAR §2.31,c,3; PHS Policy IV,B,3).

Note: While many IACUCs use "major" and "significant" interchangeably, the term "major" is not present in the regulatory lexicon as a qualifier or category of deficiencies. NIH/OLAW offers guidance regarding minor deficiencies and notes: "Generally, a minor deficiency refers to a problem for which an immediate solution is not necessary to protect life or prevent distress (e.g., peeling or chipped paint)."[20]

Opin. Given that a deficiency is any unapproved departure from the provisions of the regulatory standards,[21] it follows that a thorough knowledge of those standards is necessary to identify a deficiency. One could argue that for any deficiency the IACUC should be able to specifically identify by chapter and verse the regulatory provisions from which there has been an unapproved departure. Once the committee has determined that a finding represents an unapproved departure from the regulatory standards, i.e., that it is a deficiency, the IACUC can then consider whether the finding represents an actual or potential threat to the health and safety of the animals. If the IACUC judges that it does, then the finding is categorized as significant. If not, the finding by default is considered minor.

While it often is straightforward to determine if a finding meets the threshold of being a deficiency and if the deficiency meets the criteria to be classified as significant, there are many cases when such discrimination is not easy. The IACUC has broad discretion in this regard. It is critical that the IACUC ensures that its membership includes individuals having in-depth knowledge of the regulatory standards, as well as an understanding of the nuances of their application. The authors recommend that all deficiencies be presented at a regularly convened meeting of the IACUC, so that the full committee can make the final determination as to the nature of deficiencies identified in the program review or facility inspection.

It should be noted that according to the regulatory guidance provided (AWAR §2.31,c,3; PHS Policy IV,B,3), both the IACUC and the IO have joint responsibility for deciding if a deficiency is significant. That being said, having the inspection team/subcommittee preliminarily classify findings as minor or significant at the time of the inspection can be beneficial. For example, if a finding is preliminarily called significant, a discussion with appropriate individuals in the animal facility or laboratory, as appropriate, can be initiated so that the deficiency is corrected as soon as possible. In addition, this initial determination by the inspection team will facilitate a subsequent review by the full committee. As noted above, the authors' advice is that the full committee should make the final determination as to whether a deficiency is registered as minor or significant in the report to the IO (see 27:35).

Surv. Once the IACUC determines a deficiency is significant, is the IO consulted to determine if they agree with the IACUC's determination? Select all that apply.

- The IACUC consults with the IO during the initial review process to determine if a deficiency is considered significant. 5/182 (3%)
- As part of a convened meeting, the IACUC determines if a deficiency is considered significant and informs the IO of their decision after the meeting but prior to submission of its semiannual report. 40/182 (22%)
- The IACUC, through its report, informs the IO about any deficiency that is considered significant. 130/182 (71%)
- I don't know. 2/182 (1%)
- Other. 5/182 (3%)

27:32 What are some examples of minor and significant deficiencies?

Reg. See 27:31. NIH/OLAW offers guidance.[20]

Opin. Any deficiency that could cause injury, death, or severe distress to animals or people is considered significant. Examples could include failures in heating, ventilating, and air-conditioning systems and the related electrical systems and power failures of sufficient duration to affect important areas such as surgical suites. Not only can program or facility deficiencies be included as significant (if they meet the definition of the term), but also accidents and natural disasters can result in a reporting of a significant deficiency. For example, if a hurricane tore off a roof of an area where animals were kept, and injury, death, or severe distress to the animals resulted, it can be reported as a significant deficiency due to the resulting conditions of the facility and animals. (Note that such events would likely also require reporting to OLAW; see 27:48.) Additional examples of significant deficiencies would include deviations from an approved protocol such as withholding analgesics without IACUC approval or performing unapproved surgery, particularly if such procedures result in clinical signs of illness in research animals. If significant deficiencies are identified at an NIH/OLAW Assured institution, and are either programmatic or PHS supported, they must be reported to NIH/OLAW.

Although the PHS Policy and AWAR distinguish between significant and minor deficiencies only in terms of the actual or potential threat to the health or safety of the animals and people, some institutions elect to classify as significant other serious deficiencies that may not directly affect animals. One example of this might be a deficiency in the occupational health and safety program that is judged to threaten the health or safety of personnel. The authors concur that there are some specific situations, such as the example noted, when it would be reasonable for the IACUC to classify as significant a serious deficiency that does not impact animal health or welfare; however, we caution against a broad expansion of the classification; so, for example, there is a complete blurring between a significant deficiency and an event that requires reporting to a regulatory agency (see 27:40). NIH/OLAW offers guidance on reportable incidents.[17]

Examples of minor deficiencies include expired food, unsealed wood in an animal surgery area, missing ceiling tiles, a dripping faucet, and small cracks in the wall. These deficiencies do not directly impact the health of research animals or pose a direct threat to their safety. They are, however, problems that should be addressed.

27:33 Should deficiencies and possible corrective actions be discussed at the time of the inspection?

Opin. In general, yes, although opinions vary. If the inspection team identifies a deficiency it considers to be significant, the responsible individual (investigator, facility manager, or veterinarian) should be informed immediately of the problem so they can correct it as soon as possible. Delaying notification about a potentially significant deficiency until the entire committee is informed only prolongs the problem. The IACUC can always request additional action be taken to correct a deficiency at a later date. By discussing the deficiency with the appropriate persons at the time of the inspection, the inspecting subcommittee can make sure that their recommended course of action and date for correction are workable before presenting them to the entire IACUC for review.

There are times when it might be prudent to withhold discussion of findings at the time of the inspection, pending a full IACUC discussion. If the reviewing subcommittee is unsure of the nature of the deficiency, it is advisable to withhold judgment. On the other hand, if a problem is serious but corrective action is not immediately required, then consideration by full committee review is warranted. For example, while reviewing animal records the inspecting subcommittee discovers that an investigator had performed major survival surgical procedures on rodents but did not administer postoperative analgesics as specified in the IACUC-approved protocol; however, all the animals had been euthanized prior to the inspection. Although this is a serious non-compliance issue, and the health and the well-being of the animal was affected at the time of the surgery and during recovery, no immediate intervention may be required by the reviewers. Nevertheless, the reviewers should immediately report their findings so that the IACUC can assess the situation and, if warranted, take immediate action. In accordance with NIH/OLAW's guidelines, the IACUC would also determine whether the violation warranted reporting to NIH/OLAW, depending on the funding source for the activity, its impact on PHS-supported work, or if it was found to be a program-wide problem.

27:34 An IACUC has a policy of requiring investigators to euthanize animals by two methods (e.g., carbon dioxide followed by a physical method) to assure the animals are dead. During a semiannual inspection, it is determined that a PI is using only one of the two approved methods. Is this considered a minor or a significant deficiency?

Reg. See 27:30. NIH/OLAW offers guidance on this issue.[20,22]

Opin. Minor deficiencies are problems for which immediate resolutions are generally not necessary to protect life or prevent distress. If the IACUC considered it necessary to require two methods of euthanasia to ensure death, it would follow that if only one method were used, there would be a possibility that the animal would be in distress if a quick and painless death did not occur. Thinking in this mode could lead one to list it as a significant deficiency. There would be no question about this if during the inspection animals were found in a disposal refrigerator/freezer and conscious. If animals were euthanized using only one method and no distress was found by the inspection team, either by direct observation or from speaking to personnel working in the lab, the committee could decide to classify it as a minor deficiency as an immediate solution was not necessary to protect lives or prevent distress. The infraction would be the investigator's lack of adherence to the methodology described in the protocol only. That being said, NIH/OLAW guidance on reporting includes, "Failure to adhere to IACUC-approved protocols" and this would be considered a reportable event based on NIH/OLAW guidance.[22]

27:35 How are results of inspections and reviews recorded and processed and do semiannual inspection reports need to be signed by a majority of IACUC members?

Reg. The PHS Policy (IV,B,3; IV,E,1,d), AWAR (§2.31,c,3), and *Guide* (p. 25) all require that upon completion of the review and inspection a report is to be forwarded to the IO. However, only the AWAR (§2.31,c,3) require the report to be signed by a majority of the IACUC members. The HREA of 1985 (§495,b,3,C) references a requirement for certification that the review has been conducted but does not specify how this is to be done (see 7:1; 27:38–27:41). NIH/OLAW guidance specifies: "Final reports of the semiannual evaluations and inspections are considered full committee actions and should be reviewed and endorsed by a majority of the IACUC."[23]

Opin. Three methods are often used to document the facility inspection and program review:

- Completed review/inspection forms or checklists
- IACUC minutes
- Summary reports

The method or format used depends on the complexity of the institution and the preference of the IACUC. The results/findings of reviews and inspections should be recorded at the time the evaluations are conducted. Checklists such as those provided by NIH/OLAW (see 27:1) or customized forms are valuable tools for recording outcomes. Raw data from the inspections and reviews should be compiled as needed to facilitate IACUC review of the results, tracking of follow-up actions, and inclusion in the semiannual report to the IO. The use of spreadsheet or database software is highly useful in this regard and is strongly recommended. In cases where hard copy raw data subsequently are transcribed into an electronic format, e.g., spreadsheet or database, the original hard copy documents should be retained. The final report to the IO may include an executive summary of the IACUC's findings. However, the report should in all cases include the details of any programmatic or facility deficiency identified by the committee. Often such details, particularly for minor facility deficiencies, are provided in the form of an appendix to the report. The written report must be certified, and the most effective way to document certification is via signatures of the IACUC members at a convened meeting that involves a discussion regarding the facility inspection and program review.

27:36 How should the IACUC decide whether a deficiency item identified during the facility or program review is actually included on the final report?

Reg. The PHS Policy (IV,B,3) and the AWAR (§2.31,c,3) both state that the report to the IO "must identify specifically any departures from the provisions" of the respective regulatory standards, i.e., the PHS Policy, the *Guide*, or the AWAR. NIH/OLAW offers some guidance.[23]

Opin. All items that the IACUC determines to be deficiencies, whether minor or significant, must be included in the report. It is important for the IACUC to have a robust process for determining whether program review and inspection findings are judged to be deficiencies and whether deficiencies are classified as minor or significant (see 27:33). IACUCs should avoid the temptation to exclude items because they think they will reflect poorly on the institution. The semiannual report to the IO must include identification of any deficiencies in the program or facility, including classification as either significant or minor and a reasonable and specific plan and schedule for correction. Reports with minor items listed demonstrate that the inspection process is working.

27:37 Should findings deemed to be insignificant or suggestions for improvement be reported to the IO?

Reg. See 27:30, 27:31, and 27:35. AWAR (§2.31,c,5) and PHS Policy (IV, B, 6) both indicate that a function of the IACUC is to "Make recommendations to the Institutional Official regarding any aspect of the institution's animal program, facilities, or personnel training."

Opin. The threshold for reporting to the IO is whether the IACUC considers a finding to be a deficiency or not. If so, it must be included in the report and its correction tracked. If the IACUC determines that the finding does not represent a departure from the provisions of the regulatory standards, it should not be considered a deficiency. That said, a function of the IACUC, as determined by PHS Policy and AWAR, is to make recommendations to the IO. This could be interpreted to mean that recommendations or suggestions for improvement that are not deemed to be deficiencies by the IACUC should be included on the report to the IO.

IACUCs frequently encounter items that are of some import but do not meet the criteria for a programmatic deficiency. Such could be the case in the discovery of one overcrowded mouse cage in a room of 1,000 cages. Depending on the specifics, an IACUC may or may not deem this to be a deficiency. But even if it is not, it would be prudent to point out the overcrowded cage to the animal care and/or investigative staff so they can attend to it. Similarly, a single burnt-out light fixture might not have sufficient impact on the room environment to rise to the level of a deficiency, but it still is worthwhile to take note of it so that it can be addressed. While isolated items like these might not end up as deficiencies and accounted in the report, there may be other non-deficiency items that do warrant a mention in the report. Examples could include concern about aging equipment that has not yet failed or outmoded procedures or drugs for which better options are available. The IACUC may wish to have these kinds of findings included in the report under headings such as, "Other Comments," "Notes," or "Recommendations," or "Suggestions for Improvement (SFI)."

Surv. In addition to deficiencies, does your IACUC log suggestions or recommendations during the semiannual evaluation and include them in the report to the IO?

- No, we record deficiencies only in the report to the IO and don't log suggestions/recommendations during the review and inspection. 10/182 (5%)
- No, we record deficiencies only in the report to the IO, but we do also log suggestions/recommendations report those directly to the responsible party. 41/182 (23%)
- Yes, we include both deficiencies and suggestions/recommendations in the report to the IO. 129/182 (71%)
- I don't know. 2/182 (1%)

27:38 How does your IACUC follow up on significant deficiencies to ensure they have been corrected within the time frame specified by the IACUC?

Reg. In accordance with the PHS Policy (IV,B,3) and AWAR (§2.31,c,3), "a reasonable and specific plan and schedule" must be given for correction of each deficiency. It does not matter whether the deficiency is designated as minor or significant. Furthermore, "any failure to adhere to the plan and schedule that results in a significant deficiency remaining uncorrected shall be reported in writing within 15 business days by the IACUC, through the IO, to APHIS/AC and any Federal agency funding that activity" (AWAR §2.31,c,3). In accordance with the PHS Policy (IV,F,3), "The IACUC, through the Institutional Official, shall promptly provide OLAW with a full explanation of the circumstances and actions taken with respect to … any serious or continuing noncompliance with this Policy [or] any serious deviation from the provisions of the *Guide*." Furthermore, a recent policy notice from NIH/OLAW states that prompt reporting is required if there is a "failure to correct deficiencies identified during the semiannual evaluation in a timely manner."[17]

Opin. IACUCs should devise plans that are attainable within the timetable established. Significant deficiencies must be handled immediately and plans formulated to ensure the health and well-being of the animals. The criteria for determining appropriate deadlines for correction are set in concert with a reasonable timetable for the type of correction required. Thus, for significant deficiencies, there may be two timetables, one to ensure the animal's health and well-being immediately and a second to correct the deficiency. The type of deficiency discovered will determine the appropriate timetable for correction. For example, a facility issue (bricks and mortar) may take longer to correct than a programmatic issue in which a correction can be put into motion just by changing a policy. An issue that can be corrected by the IACUC, animal care staff, or investigative staff may be addressed more rapidly than a problem that involves outside contractors and facility issues, which can involve a long procurement process or allocation of funding. Temporary "quick fix" solutions may be used while planning and implementing the final solution.

Surv. How does your IACUC follow up on significant deficiencies to assure that they have been corrected within the time frame specified by the IACUC? Select all that apply.

- Not applicable. 2/182 (1%)
- We have no follow-up procedure. 5/182 (3%)
- The original inspectors perform the follow-up and report back to the IACUC. 35/182 (19%)
- The IACUC administrative office performs the follow-up. 126/182 (69%)
- The IACUC Chairperson designates the person(s) who will perform the follow-up. 51/182 (28%)
- I don't know. 2/182 (1%)
- Other* 17/182 (9%)

*Other:

- Anmal facility manager.
- IACUC Chair.
- Principal Investigators provide written attestation that deficiencies in their areas have been corrected.

27:39 Who decides the appropriate resolution of specific deficiencies? How is this done?

Reg. Both the PHS Policy (IV,B,3) and the AWAR (§2.31,c,3) require the IACUC to report on the deficiencies found during the facility and program review and to categorize them as either minor or significant (major). The IACUC is required to develop a "reasonable and specific plan and schedule" for correcting each deficiency. A deficiency is resolved when the IACUC determines that it has been corrected.

Opin. The IACUC has the ultimate responsibility for determining what actions are needed to correct a deficiency. However, it is best to involve the stakeholders who will be affected by the committee's decision. As appropriate, the IACUC should consult the AV, the investigator, and the maintenance department. If major funding is needed to correct an item, it will also be necessary to consult the IO or other administrative persons to set a reasonable plan for correction. Such involvement makes for a more collegial atmosphere and ultimately a better solution.

27:40 What action is taken if a significant deficiency is not corrected in accordance with the established schedule?

Reg. In some cases, institutions are required to report uncorrected significant deficiencies. The AWAR

(§2.31,c,3) are explicit in this regard, requiring that "Any failure to adhere to the plan and schedule that results in a significant deficiency remaining uncorrected shall by reported in writing within 15 business days by the IACUC, through the Institutional Official, to APHIS and any Federal agency funding that activity." While not specifically citing significant deficiencies, the PHS Policy (IV,F,3) requires prompt reporting of any serious or continuing non-compliance with the PHS Policy, any serious deviation from the provisions of the *Guide*, or any suspension of an activity by the IACUC. If a significant deficiency meets any of these criteria, NIH/OLAW must be notified.[17]

In addition to the reporting requirement specified above, the IACUC has the option of suspending the activity associated with the significant deficiency. The AWAR (§2.31,c,8; §2.31,d,6) authorize the IACUC to suspend an activity involving animals if it is not being conducted in accordance with the description of that activity provided by the PI and approved by the committee. This may occur only at a convened meeting of a quorum of the IACUC with the suspension vote of a majority of the quorum present. The PHS Policy (IV,C,6) allows the IACUC to "suspend an activity that it previously approved if it determines that the activity is not being conducted in accordance with applicable provisions of the Animal Welfare Act, the *Guide*, the institution's Assurance … the IACUC may suspend an activity only after review of the matter at a convened meeting of a quorum of the IACUC and with the suspension vote of a majority of the quorum present." As applicable, suspensions must be reported to APHIS/AC and any federal funding agency as per AWAR §2.31,d,7.

Opin. Because a significant deficiency has, or may have, an adverse effect on animal health and safety it must be corrected quickly. In some instances, the initial action need not result in complete resolution, but the action must protect the animals. For example, if a dog run is found with a broken fence with potential to cause an injury, the immediate action would be to remove the dog to a suitable enclosure. The action will correct the significant deficiency. The dog run can then be fixed as time permits.

Another option that may be taken by the IACUC in response to a lack of attention regarding correction of a significant deficiency is the suspension of an animal activity. Typically, this is reserved only for situations involving an immediate threat to animal health or well-being that is not resolved or a continued lack of compliance in regard to a significant deficiency or deficiencies, and most IACUCs use this only in situations with continued non-compliance or egregious violations. IACUCs should use due diligence to ensure that corrective action is taken in a timely manner.

27:41 For how long should facility inspection and program review records be maintained?

Reg. PHS Policy (IV,E,2) states: "All records shall be maintained for at least three years; records that relate directly to applications, proposals, and proposed significant changes in ongoing activities reviewed and approved by the IACUC shall be maintained for the duration of the activity and for an additional three years after completion of the activity." The AWAR (§2.35,f) use similar wording.

Opin. In practice, there are two opinions regarding recordkeeping. Many institutions maintain records indefinitely because of the difficulty in segregating records into activities that are either complete or ongoing. At these institutions, records are kept as long as storage is available. However, some institutions have concerns about storing unnecessary records because of freedom of information laws. These institutions generally store minutes and other general records for only three years and specific protocols for the duration of the project plus three years. It is unusual, in the authors' opinion, for regulatory bodies to request records older than three years. It should be noted that many institutions currently use, or are currently considering, electronic protocol management systems. With such systems, records that date beyond three years may be identified with minimal effort and deleted from the system. Conversely, such records can be archived and maintained indefinitely depending on institutional preference (see 26:14 and 26:15).

27:42 How should the IACUC address minority opinions relative to the semiannual inspection and program review?

Reg. Both PHS Policy (IV,F,4) and AWAR (§2.31,c,3) mandate that minority reports, if any, be included with the semiannual report and PHS Policy requires minority views to be included in the annual report to NIH/OLAW.

Opin. Minority reports are infrequent because most IACUCs can devise reasonable compromises to difficult problems. Dissenting members generally do so by voting "no" on an issue and requesting that a notation be made in the regular committee minutes. However, a member(s) may write a full minority opinion if so desired. This opinion must be included as part of the official report.

27:43 Who receives the findings from the program review and facility inspection?

Reg. In accordance with PHS Policy (IV,B,3) and the AWAR (§2.31,c,3), the IACUC must submit a report of the inspection and program review to the IO.

Opin. It also is useful to submit a copy of the report to those who are responsible for correcting deficiencies, such as the facility management, the AV, and the facilities maintenance department. Members of the IACUC should receive a copy of the report as well.

27:44 Other than the IO and the IACUC, who has legal access to the IACUC's semiannual reports to the IO?

Reg. According to NIH/OLAW, semiannual reports must be submitted to NIH/OLAW if requested.[17] PHS Policy (IV,E,2) requires that "All records shall be accessible for inspection and copying by authorized OLAW or other PHS representatives at reasonable times and in a reasonable manner." In addition, PHS Policy (IV,A,2) requires institutions lacking AAALAC accreditation must submit its most recent semiannual report from the IACUC to NIH/OLAW when the institution's PHS Animal Welfare Assurance is renewed. According to the AWAR (§2.31,c,3,a), reports of the semiannual inspection and program review "shall be maintained by the research facility and made available to APHIS and to officials of funding Federal agencies for inspection and copying upon request."

Opin. The regulatory verbiage used may be interpreted to mean that NIH/OLAW and APHIS/AC may request semiannual reports as part of the inspection process or the Assurance negotiation. Legal access for the public depends on whether the facility is private or public and depends on applicable federal regulations and state laws. Legal counsel should be sought to address this issue (see 26:4).

27:45 Is the IACUC responsible for compliance oversight for the use and proper storage of federally controlled drugs used in animal-based research?

Reg. PHS Policy (IV,B,2) indicates that the IACUC must inspect all of the institution's animal facilities once every six months using the *Guide* as a basis for evaluation. The *Guide* (p. 34) suggests that the "handling and use" of controlled substances be addressed during semiannual inspections. The sample semiannual program review checklist from NIH/OLAW includes a section on drug storage and control.[1] Even though some guidance is given, there is no regulatory requirement that the IACUC serve as the regulatory body for DEA-regulated drugs that are used in research animals.

Opin. No. The IACUC shares with the AV the responsibility for ensuring the use of all drugs (particularly anesthetics, analgesics, and anti-infective agents) in animal subjects is consistent with current accepted laboratory animal veterinary practice and congruent with the activities approved by the IACUC. However, this responsibility does not imply that the IACUC's regulatory compliance role should include enforcement of federal, state, or local controlled substances laws and regulations. The institution should provide resources/support for a separate program to monitor storage and use of controlled substances. For example, a controlled drug officer could provide oversight for the ordering, use, and destruction of controlled substances. During the semiannual inspection, an IACUC indeed should look at controlled and noncontrolled drugs and how they are stored. However, the committee's primary focus should be on making sure that the drugs are used in accordance with the approved research protocols and within established expiration dates. That said, if the committee observes deviations from controlled substances storage or recordkeeping requirement, it may still be considered a minor deficiency, and the appropriate institutional personnel should be notified.

Surv. During semiannual inspections, do the inspectors check for expired drugs (including controlled substances, if applicable)?

• Not applicable.	5/182 (3%)
• Yes, they do for all drugs.	155/182 (85%)
• Yes, they do for all drugs except for controlled substances.	13/182 (7%)
• No, they do not.	6/182 (3%)
• I don't know.	2/182 (1%)
• Other.	1/182 (1%)

27:46 Can the IO overrule a significant deficiency finding of the IACUC on a semiannual report?

Reg. The PHS Policy (IV,B,3) and AWAR (§2.31,c,3) both state that "a significant deficiency is one which … in the judgment of the IACUC *and* [emphasis added] Institutional Official, is or may be a threat to the health or safety of the animals."

Opin. The regulatory language could be interpreted to mean that a deficiency is significant only if both the IACUC and IO judge it to be so. In practice, however, the IO typically does not participate in the IACUC's process for determining whether a deficiency is significant, and the IO is merely informed of the committee's designation, e.g., via the semiannual report, after the fact. Rarely, a situation may arise in which the IO does indeed disagree with the IACUC with respect to the designation of a significant deficiency. In such a case, it would not mean that the IO is "overruling" the committee. It may just mean that the deficiency might not technically be considered significant because both parties did not designate it so. If this situation were to occur, it would be in the best interest of the institution if the IACUC and IO would jointly consider the matter and come to an agreement in terms of the severity of the deficiency. If the IO and the IACUC could not come to an agreement, consultation with the regulatory agencies (i.e., NIH/OLAW and/or APHIS/AC, as applicable) may be warranted.

27:47 Do all conditions that jeopardize the health or well-being of animals have to be reported to NIH/OLAW?

Reg. PHS Policy (IV,F,3) indicates that "the IACUC, through the IO, shall promptly provide OLAW with a full explanation of the circumstances and actions taken with respect to (a) any serious or continuing noncompliance with this Policy; (b) any serious deviation from the provisions of the *Guide*; or (c) any suspension of any activity by the IACUC." NIH/OLAW offers guidance regarding reporting.[22,24] A number of situations that would require reporting are included such as any "conditions that jeopardize the health or well-being of animals, including natural disasters, accidents, and mechanical failures, resulting in actual harm or death to animals." There are situations where animal morbidity and mortality are noted but reporting is not warranted such as, "death of animals that have reached the end of their natural life spans … death or failures of neonates to thrive when husbandry and veterinary medical oversight of dams and litters was appropriate … infrequent incidents of drowning or near-drowning of rodents in cages when it is determined that the cause was water valves jammed with bedding (frequent problems of this nature, however, *must* be reported promptly along with corrective plans and schedules)."

Opin. No. The IACUC should utilize professional judgment with respect to reporting while adhering to the guidance provided by NIH/OLAW. If a committee cannot come to a consensus regarding a decision on reporting, NIH/OLAW may be consulted for guidance in such situations.

27:48 In the event of a naturally occurring disaster such as a hurricane, burst water pipes, etc. in which animals are not hurt or distressed, is it necessary to report the same to NIH/OLAW or AAALAC (for accredited institutions)?

Reg. NIH/OLAW offers specific guidance on this issue and states: "inoperable HVAC, electrical or watering systems, failure of such systems sufficient to affect critical housing and operational areas, and situations such as natural disasters that cause injury, death, or severe distress to animals"[15] and "conditions that jeopardize the health or well-being of animals, including natural disasters, accidents, and mechanical failures, resulting in actual harm or death to animals"[17] are reportable. The *Guide* (pp. 74–75) offers some verbiage regarding requirements for an emergency plan and required response in the event of an emergency (see 27:16).

Opin. Natural disasters that cause severe distress, injury, or death are considered significant deficiencies. As a significant deviation from the *Guide*, and based on the guidance from NIH/OLAW, such deficiencies must be reported. This issue has been addressed in the literature and as a significant deviation from the *Guide*, it must be reported to NIH/OLAW.[17] AAALAC requires accredited institutions to report adverse events as a result of "natural disasters" and the results of "OLAW/USDA investigations."[25]

REFERENCES

1. National Institutes of Health, Office of Laboratory Animal Welfare. 2012. *Semiannual Program Review and Facility Inspection Checklist*. http://grants.nih.gov/grants/olaw/sampledoc/cheklist.pdf.
2. American Dairy Science Association, the American Society of Animal Science, and the Poultry Science Association. 2020. *4th edition of the Guide for the Care and Use of Agricultural Animals in Research and Teaching*. https://www.asas.org/docs/default-source/default-document-library/agguide_4th.pdf?sfvrsn=56b44ed1_2.
3. National Research Council. 1997. *Occupational Health and Safety in the Care and Use of Research Animals*. Washington, DC: National Academy Press.
4. U.S. Public Health Service, Centers for Disease Control and Prevention. 2020. *Biosafety in Microbiological and Biomedical Laboratories*, 6th edition, Washington, DC: U.S. Government Printing Office.
5. Canadian Council on Animal Care. 1993/1984. *Guide to the Care and Use of Experimental Animals*. Vols. I and II. Ottawa: Canadian Council on Animal Care.
6. Applied Research Ethics National Association/Office of Laboratory Animal Welfare. 2002. *Institutional Animal Care and Use Committee Guidebook*, 2nd edition. Bethesda: National Institutes of Health.
7. National Institutes of Health, Office of Laboratory Animal Welfare. 2012. Tutorial: *PHS Policy on Humane Care and Use of Laboratory Animals*. http://grants.nih.gov/grants/olaw/tutorial/index.htm.
8. National Institutes of Health, Office of Laboratory Animal Welfare. 2024. Notice NOT-OD-24-076, *Guidance on Flexibilities for Conducting Semiannual Inspections of Animal Facilities*. https://grants.nih.gov/grants/guide/notice-files/NOT-OD-24-076.html.
9. National Institutes of Health, Office of Laboratory Animal Welfare. 2023. Frequently Asked Questions, G.3, *Do Awardee Institutions Need Animal Facility Disaster Plans?* http://grants.nih.gov/grants/olaw/faqs.htm#instresp_3.
10. National Institutes of Health, Office of Laboratory Animal Welfare. 2023. *Disaster Planning and Response Resources*. http://grants.nih.gov/grants/olaw/disaster_planning.htm.
11. National Institutes of Health, Office of Laboratory Animal Welfare. 2011. Notice NOT-OD-11-053, *Guidance to Reduce Regulatory Burden for IACUC Administration Regarding Alternate Members and Approval Dates*. http://grants.nih.gov/grants/guide/notice-files/NOT-OD-11-053.html.
12. Brown, P., Mullaney, R. A Word from OLAW and USDA. *Lab Anim* 52, 138 (2023). https://doi.org/10.1038/s41684-023-01200-6.

13. National Institutes of Health, Office of Laboratory Animal Welfare. 2012. Frequently Asked Questions, E.1. *Should the IACUC Inspect Laboratories or Other Sites Where Investigators Use Animals?* https://olaw.nih.gov/faqs#/guidance/faqs?anchor=question50347.
14. National Institutes of Health, Office of Laboratory Animal Welfare. 2008. *The Public Health Service Responds to Commonly Asked Questions.* http://grants.nih.gov/grants/olaw/references/ilar91.htm.
15. Thulin JD and Underwood WJ. IACUC considerations for the use of livestock in translational research. *ILAR J* 56(1), 139–146 (2015). doi: 10.1093/ilar/ilv003.
16. National Institutes of Health, Office of Laboratory Animal Welfare. 2006. Notice NOT-OD-06-052, *Guidance on Use of Telecommunications for IACUC Meetings under the PHS Policy on Humane Care and Use of Laboratory Animals.* https://grants.nih.gov/grants/guide/notice-files/NOT-OD-06-052.html.
17. APHIS/USDA. 2021. USDA Animal Welfare Inspection Guide. 7.1.8. Telecommunications for IACUC Meetings. https://www.aphis.usda.gov/animal_welfare/downloads/Animal-Care-Inspection-Guide.pdf.
18. National Institutes of Health, Office of Laboratory Animal Welfare. 2020. Notice NOT-OD-20-088, Flexibilities for Assured Institutions for Activities of Institutional Animal Care and Use Committees (IACUCs) Due to COVID-19. https://grants.nih.gov/grants/guide/notice-files/NOT-OD-20-088.html?utm_campaign=+42055075&utm_content=&utm_medium=email&utm_source=govdelivery&utm_term= [Rescinded, no replacement].
19. National Institutes of Health, Office of Laboratory Animal Welfare. 2021. NOT-OD-21-164. *Guidance on Flexibilities for Conducting Semiannual Inspections of Animal Facilities.* https://grants.nih.gov/grants/guide/notice-files/not-od-21-164.html [Rescinded. New NOT-OD-24-075]
20. National Institutes of Health, Office of Laboratory Animal Welfare. 2023. Frequently Asked Question, E.2, *How Does the IACUC Distinguish between Significant and Minor Deficiencies?* https://olaw.nih.gov/faqs#/guidance/faqs?anchor=question50348.
21. National Institutes of Health, Office of Laboratory Animal Welfare. 2012. Departures from the *Guide.* https://olaw.nih.gov/education/educational-resources/webinar-2012-10-04.htm.
22. National Institutes of Health, Office of Laboratory Animal Welfare. 2005. Notice NOT-OD-05-034, *Guidance on Prompt Reporting to OLAW under the PHS Policy on Humane Care and Use of Laboratory Animals.* http://grants.nih.gov/grants/*Guide*/notice-files/not-od-05-034.html.
23. National Institutes of Health, Office of Laboratory Animal Welfare. 2013. *Semiannual Report of the Program Review and Facility Inspection.* https://olaw.nih.gov/resources/documents/ioreport.htm.
24. National Institutes of Health, Office of Laboratory Animal Welfare. 2023. *Reporting Noncompliance.* https://olaw.nih.gov/guidance/reporting-noncompliance.htm.
25. AAALAC International. 2022. Frequently Asked Questions, I.2. Maintaining Accreditation. *Managing and Reporting Adverse Events.* https://www.aaalac.org/accreditation-program/faqs/#H2.

28

Inspection of Animal Housing Areas

Christine A. Boehm and Carolyn M. Malinowski*

Introduction

Animal housing facilities must be inspected by the IACUC at least once every six months, according to the AWAR[1] (§2.31,c,2) and PHS Policy[2] (IV,B,2). Nevertheless, aside from this simple directive, a few subsequent policy clarifications, and the detailed standards for animal housing and care provided in the AWAR[1] and the *Guide*,[3] these regulatory authorities allow institutions the flexibility to develop inspection methods that function effectively and efficiently for their specific research program. An IACUC may approach implementation of the inspection requirement in a variety of ways. The information provided in this chapter is intended to assist IACUCs in developing an effective animal housing facility inspection program tailored to the needs of their individual institutions.

Information, ideas, and opinions expressed in this chapter include a compilation of results from a survey of IACUCs at institutions located throughout the United States. The first survey, conducted in 1997, gathered information from 93 IACUCs on how their inspections were managed and administered. The most recent survey included information from over 290 IACUCs from institutions of varying size, type, and culture.

While the information provided in this chapter may represent the way in which many, or even most, IACUCs conduct animal housing and care inspections, the common practices described should not be interpreted as the best practices for every individual institution. The methods that an institution's IACUC employs depend on the size, type, and culture of the institution. The "best" inspection methods are likely to be very different for large versus small facilities, centralized versus decentralized facilities, corporate organizations versus educational institutions, and so forth. IACUCs are encouraged to exercise their professional judgment and creativity in developing an effective animal housing facility inspection program tailored to the particular needs of their individual institutions.

28:1 What is the definition of an animal housing facility?

Reg. See 16:1; 28:13.

28:2 What is the purpose of conducting IACUC inspections of animal housing facilities?

Reg. The IACUC, as an agent of the research facility, must fulfill its regulatory responsibility to inspect animal housing areas at least once every six months to evaluate the facility's compliance with applicable regulations, standards, and guidelines (AWAR[1] §2.31,c,2; PHS Policy[2] IV,B,2; HREA[4] §495,b,3,A). The recommendations of the *Guide*[3] (p. 25) specify that the IACUC should inspect the animal facilities at least annually.

Opin. The IACUC must represent its institution both by advancing the institutional mission and by serving as the institutional conscience. By inspecting animal housing areas every six months, the IACUC can ensure that the animal housing and care program is furthering the institutional mission by providing high-quality animal care (through proper attention to animal welfare), maintaining a suitable environment for research activities, and safeguarding the health and safety of its personnel. In addition to ensuring the institution's compliance with applicable regulations, the IACUC must be confident that the animal housing and care program is accomplishing these goals in a scientific, humane, and ethical manner.

28:3 Should an institutional guideline, SOP, or policy be developed for IACUC inspection of animal housing facilities?

Reg. There is no regulatory requirement for IACUCs to establish guidelines, Standard Operating Procedures (SOPs), or policies regarding inspection of animal housing facilities. The AWAR[1] (§2.31,c,3) state that the IACUC may determine the best means of conducting evaluations.

Opin. Many IACUCs have found the creation of guidelines, SOPs, or policies regarding inspection of animal housing facilities to be helpful in the administration and management of the inspection process. Clarifying expectations for IACUC members conducting inspections, veterinary and administrative staff supporting inspections, and managers of facilities undergoing inspection can promote consistent and thorough inspections and enhance rapport among all parties. Issues addressed in the guidelines, SOPs, or policies can include the following:

* This chapter builds upon the contributions of Patricia A. Ward and Debra L. Hickman, who were authors of previous editions of this chapter in *The IACUC Handbook*.

DOI: 10.1201/9781003272861-28

- Who conducts the IACUC inspections
- The orientation or training inspectors receive
- How inspections are scheduled
- Whether inspections are to be announced or unannounced
- How inspectors should prepare for each inspection
- How inspections will be facilitated
- Which sites are to be inspected
- Which general issues are considered during inspections
- Findings of previous inspections as examples
- How findings will be communicated, documented, and distributed
- How the IACUC will follow up deficiencies
- How long the inspection documents will be retained
- Expectations of timing to correct deficiencies

28:4 Who conducts the IACUC inspections of animal housing areas?

Reg. While the PHS Policy[2] (IV,B,2), the *Guide*[3] (pp. 25, 34), and the AWAR[1] (§2.31,c,3) specify animal facility inspection as an IACUC function, the AWAR[1] offer the most explicit "minimal" answer to this question: "No Committee member wishing to participate … may be excluded. The IACUC may use subcommittees composed of at least two Committee members and may invite ad hoc consultants to assist."

Opin. At many institutions, particularly those with smaller or centralized animal facilities, the entire IACUC participates in the animal housing facility inspection. However, in larger or decentralized facilities, the IACUC may divide into subcommittees or appoint a single subcommittee to tackle the job. Such subcommittees must include at least two IACUC members if housing areas for AWA-regulated species will be inspected (AWAR[1] §2.31,c,3). Duly appointed and qualified alternate members of the IACUC may conduct the inspections in place of corresponding regular members if the regular member is unavailable.[5] A site visit from the AAALAC may be utilized for the semiannual animal facility inspection, as long as the participation and endorsement of the IACUC meets the requirements of the AWAR and the PHS Policy.[6] Members of the veterinary or IACUC staff also may participate in the inspection to provide expertise, continuity, and support. During the COVID-19 pandemic, many IACUCs split the inspection personnel into smaller groups to maintain physical distancing; for further information on this topic, please see Chapter 25.

Surv. At your institution, which of the following inspects animal housing areas? Select all that apply.

• Voting members only	35/181 (19%)
• Voting members	148/181 (82%)
• Alternate members	91/181 (50%)
• Consultants	31/181 (17%)
• IACUC staff who are not appointed IACUC members	80/181 (44%)
• Husbandry staff who are not appointed IACUC members	31/181 (17%)
• Laboratory staff who are not appointed IACUC members	19/181 (10%)
• Environmental Health and Safety staff	66/181 (36%)
• Facilities and Maintenance staff	27/181 (15%)

28:5 Should IACUC members receive orientation or training in preparation for conducting facility inspections?

Reg. The AWAR[1] require that all personnel involved in the animal care and use program be qualified to perform their duties (§2.32,a). The *Guide*[3] (p. 24) states that "it is the institution's responsibility to provide suitable orientation, background materials, access to appropriate resources, and, if necessary, specific training to assist IACUC members in understanding and evaluating issues brought before the committee."

Opin. Most IACUCs arrange for some form of orientation or training for incoming members that includes information about inspection of animal housing facilities. This training is especially beneficial to the nonscientist and nonaffiliated members of the IACUC. Many IACUCs will pair up new members with more senior members to facilitate the training and information transfer during IACUC inspections. Information presented in the inspection training session or packet can include the following:

- An overview of the size and scope of the institution's animal housing facilities, including maps or floor plans
- Institutional guidelines, SOPs, or policies regarding IACUC inspections
- IACUC policies
- SOPs from the animal housing and care program
- Copies of regulatory, professional, and institutional animal housing and care standards
- Guidelines for evaluating animal well-being
- The institutional inspection checklist, if used
- Sources of additional information (e.g., websites or information about contacting key laboratory animal organizations)

Most IACUCs find it worthwhile to spend a significant portion of the training effort reviewing the various established standards that apply to the areas to be inspected. These include the animal housing and care standards provided in the AWAR,[1] the *Guide*,[3] and the *Guide for the Care and Use of Agricultural Animals in Agricultural Research and Teaching* (*Ag Guide*).[7] As applicable, IACUCs may also consider reviewing the various animal management documents produced by, among others, the Institute of Laboratory Animal Resources and the National Research Council.[8] A working understanding of these standards not only enables IACUC members to evaluate regulatory compliance issues in the facilities inspected, but also helps to develop their professional judgment with regard to the more important issues of animal welfare, research integrity, and personnel health and safety.

Surv. What methods of training does your institution use for training IACUC inspectors? Select all that apply.

- Available online modules (i.e., CITI, AALAS Learning Library) 117/181 (65%)
- Institution-designed online modules 26/181 (14%)
- In-person training 137/181 (76%)
- Other* 17/181 (9%)
- None of the above 10/181 (6%)

*Other:

- Discussion at IACUC meetings
- Industry conferences
- Printed checklist
- Pairing with experienced IACUC member as mentor
- Printed references and resource material
- Webinars

28.6: Should animal facility inspectors participate in the institutional occupational health program?

Reg. There is no regulatory requirement for IACUC members to participate in the institutional occupational health program. Many IACUC members are already enrolled in their institution's occupational health program by virtue of being involved with animal research.

Opin. Every institution will approach this question differently. However, as many institutions are now asking for housekeeping, security, and maintenance personnel to enroll in the occupational health program, it would be a natural extension of the program to include animal facility inspectors. A recommended approach is to perform a risk-based assessment to determine if an IACUC member who physically enters a room requires inclusion in the occupational health program.

Surv. Do all of your IACUC inspectors participate in the institutional occupational health program?

- No 49/181 (27%)
- Yes 132/181 (73%)

28:7 What periods are appropriate for IACUC inspections of animal facilities?

Reg. Both the AWAR[1] (§2.31,c,2) and the PHS Policy[2] (IV,B,2) specify that animal facilities be inspected at least once every six months (see 27:5–27:7). The *Guide*[3] (p. 25) states that facilities should be inspected at least annually or more often as required by the AWA and PHS Policy. A notice published by the NIH (NOT-OD-21-164) provides additional guidance in conducting semiannual inspections by allowing a 30-day flexibility from the original date of inspection.[9]

Opin. For the most part, meeting this requirement is not difficult for institutions where the animal facilities can be inspected in a relatively short time. For larger or decentralized facilities, however, meeting this requirement can be more difficult. If completing the inspection will take several days, the IACUC is faced with management of a large time commitment. Under these circumstances, most IACUCs will establish regular "inspection windows" at six-month intervals and break the inspection up over several days within each window. Alternatively, an IACUC can opt to distribute inspection trips throughout the entire six-month period. In this case, or if the inspection window spans more than a few weeks, care must be taken to inspect the various sites in approximately the same order for each six-month period. Otherwise, the inspection interval for some sites might exceed the six-month maximum. This could occur if a particular site is inspected early in one six-month period and late in the next. If the 30-day flexibility allowance is implemented, care must be taken to prevent forward drift from year to year.[9]

Surv. Has your institution ever requested an exemption from OLAW to delay an IACUC inspection?

- No 149/181 (82%)
- Yes 32/181 (18%)

28:8 Should inspections be announced or unannounced?

Opin. See 27:29 and 29:6.

28:9 How should IACUC inspectors prepare for each inspection?

Opin. Before embarking on each inspection, most IACUC inspectors want to review the sites to be visited with particular respect to entry and exit restrictions, previous inspection findings or deficiency history, and experimental activities being conducted. Entry and exit restrictions may affect the way inspectors dress for the inspection, the order in which various areas are visited, and the extent to which exposure of inspectors to certain other animals is prohibited for a specified period before or after the inspection. Reviewing the previous inspection findings or deficiency history of the sites to be inspected may help IACUC inspectors focus attention on problem areas or remind them to express appreciation for improvements made. Some IACUC inspectors may wish to review experimental protocols approved for use in the areas to be inspected. Doing so can alert them to animal use procedures of particular concern, such as those requiring special consideration during protocol review (*Guide*,[3] p. 25) and allow IACUC inspectors to make pertinent observations or inquiries during the inspection visit. Alternatively, some IACUC inspectors choose to wait until after the inspection to review the experimental protocols of animals noted during the inspection.

28:10 What strategies can be used to facilitate inspections?

Opin. Every IACUC looks for ways to facilitate the inspection process. Including administrative or veterinary staff on the inspection team is one popular strategy used to provide expertise, continuity, and support. Another is the use of the previous inspection report or an inspection checklist to use as a reference or to record findings during the inspection. NIH/OLAW has made a sample checklist available on its website,[8] and institutions may download and modify it to suit their own programs. Some IACUC inspection teams carry copies of pertinent regulations and standards, and equipment for recording findings (photo, audio, or video) or for validating conditions (temperature or humidity reader, light meter, smoke bottle, etc.) (see 28:11).

28:11 What electronic technology might the IACUC use to aid in the conduct of the semiannual program review and facility inspection?

Reg. The regulations are flexible on what technology can be used to inspect animal care and housing facilities. Electronic technologies are allowed for any and all sites to be inspected. However, for AWA-regulated species, only live feeds are allowed, rather than pre-recorded videos, photos, or written descriptions.[10]

Opin. The advent of technologies such as iPads®, tablet computers, smartphones, and data management systems has streamlined the IACUC inspection process. As a result of the COVID-19 pandemic, many institutions incorporated these technologies into their semiannual inspection processes. These technologies may include FaceTime, Zoom, GoPro, and other live stream technologies that allow the user real-time interaction and visualization of the area, including the ability to zoom in on specific areas as the inspection is conducted.

Surv. Which of the following technologies does your IACUC employ to facilitate the conduct and documentation of the semiannual program review? Select all that apply.

• Mobile devices (e.g., laptop, tablet device, etc.) for recording findings	41/181 (23%)
• Commercial IACUC/animal resource data management product(s)	10/181 (6%)
• Internally developed data management system	20/181 (11%)
• Written inspection forms or checklists that are compiled electronically after the inspection	162/181 (90%)
• None of the above	10/181 (6%)
• Other*	7/181 (4%)

*Other: Respondents also indicated the use of handwritten notes taken by either the inspector or the IACUC coordinator.

Surv. Which of the following technologies does your IACUC employ to facilitate the conduct of the facility inspection? Select all that apply.

• Video recordings	14/181 (8%)
• Go Pro	0/181 (0%)
• Facetime	15/181 (8%)
• Zoom	27/181 (15%)
• Teams	14/181 (8%)
• WebEx	4/181 (2%)
• None of these; inspections conducted in person	143/181 (79%)
• Other	13/181 (7%)

*"Other
Box

- Facetime
- Photographs of animals at remote locations

28:12 Should the program review necessarily precede the facility inspection during the semiannual review?

Opin. While there is no right or wrong way with respect to the sequence of events for the program review and facility inspection, it is advantageous for members of the inspection team to review all of the SOPs and policies associated with the animal program before inspecting the animal facility. This allows the inspection team to be confident in their evaluation of the animal facility.

Surv. Which of the following best describes the sequencing of your semiannual program review and facility inspection?

• Program review prior to the facility inspection	32/181 (18%)
• The facility inspection prior to the program review	95/181 (52%)

- Program review while conducting the facility inspection 12/181 (7%)
- Program review at the same time as the facility inspection, but by different pairs/groups of IACUC members 10/181 (6%)
- It varies; no set format 24/181 (13%)
- Other 8/181 (4%)

28:13 What sites should be included during the inspection of animal housing facilities?

Reg. The AWAR[1] (§1.1, Animal) define an animal as a live or dead warm-blooded animal, thus does not include all vertebrates. The AWAR[1] also specifically exclude birds, rats of the genus *Rattus*, and mice of the genus *Mus* bred for use in research; horses not used for research purposes; and other farm animals such as livestock or poultry used or intended for use as food or fiber or the improvement of animal nutrition, breeding, management, or production efficiency. Animal areas containing free-living wild animals in their natural habitat need not be inspected (§2.31,c,3). The PHS Policy[2] (III,A) applies to any live vertebrate animal used in research, research training, experimentation, or biological testing according to the applicability of the institution's Animal Welfare Assurance (e.g., where PHS funding is involved). Most IACUCs resolve this discrepancy simply by including the housing areas of all vertebrate animals, regardless of species or type of use, in their inspections. Some IACUCs elect to inspect the housing areas of invertebrate species as well, especially when the inclusion of such areas benefits the program as a whole.[11]

What constitutes an animal housing area is also a matter of regulatory definition (see 16:1). The AWAR[1] (§2.31,c,2) specify that animal facilities, excluding wild habitats but including study areas, must be inspected. While a definition of an animal facility is not provided in the AWAR,[1] a housing facility (AWAR[1] §1.1, Housing Facility) is defined as any land or shelter intended to house animals, divided into indoor housing facility (structure or building with environmental controls), sheltered housing facility (facility that provides protection from the elements), and outdoor housing facility (housing that does not meet any other definition of housing). The AWAR[1] (§1.1, Study Area) further define a study area as any area where animals are housed for more than 12 hours.

The PHS Policy[2] (III,B) defines an animal facility as "any and all buildings, rooms, areas, enclosures, or vehicles, including satellite facilities, used for animal confinement, transport, maintenance, breeding, or experiments inclusive of surgical manipulation. A satellite facility is any containment outside of a core facility or centrally designated or managed area in which animals are housed for more than 24 hours." The PHS Policy[2] definition is quite comprehensive, including support and experimental surgery areas along with housing areas (see 27:19; 29:1).

Opin. Many IACUCs, particularly those wishing to obtain or maintain accreditation by AAALAC, want to comply with the animal housing inspection requirements of the AWAR,[1] PHS Policy,[2] and the *Guide*.[3] To accomplish this, IACUCs will visit all areas where animals are housed or kept (including study areas, satellite facilities, and laboratories) for more than 12 hours (AWAR-regulated species) or 24 hours (all other vertebrate species) during their semiannual inspections. Typically, areas where rats and mice are kept are not inspected if those animals are held there for less than 24 hours, unless survival or non-survival surgery is conducted at that location (see NIH/OLAW FAQ E.1[12]). Temporarily unoccupied animal housing areas also are visited by most IACUCs. Corridors and anterooms contiguous with these areas are included in most inspections. Field research locations are not usually inspected by IACUCs unless animals are held in captivity for more than 12 hours or the IACUC is particularly interested in observing the field research procedures or setting. Cage washing and other sanitation facilities are included in the semiannual inspections, as are diet preparation and food and bedding storage areas. Most IACUCs also inspect areas where caging, equipment, and supplies are stored. Loading docks and transport vehicles, particularly those dedicated to traffic of animals and animal-related supplies and equipment, are usually inspected, as are carcass storage and disposal areas. Animal procedure areas contiguous with the animal housing facilities, including surgical suites, procedure rooms, and euthanasia stations, are also usually included in the semiannual inspection of the animal housing and support areas (see 27:19).

Surv. Does your IACUC inspect temporarily empty animal holding rooms?

- No 26/181 (14%)
- Yes 155/181 (86%)

28:14 What general issues should be considered when inspecting animal housing facilities?

Reg. The AWAR[1] (§2.31,c,1) require the IACUC to use the AWAR as a guide for semiannual inspections. PHS Policy[2] (IV,B,1) requires the IACUC to use the *Guide* as a basis for evaluation.

Opin. Paramount among inspection issues are animal well-being, research integrity, and personnel health and safety. Proper management of animal care and use programs promotes all three, and IACUC inspectors seek to evaluate the success of the program in this context. To do this during an inspection, most IACUCs try to organize the effort and attention they give to the various aspects of the animal care and use program. Many IACUCs will utilize currently available tabulations, such as the table of contents of the *Guide* or the Facility Inspection Checklist[9] available from NIH/OLAW, to provide such structure. Some IACUCs prefer to focus their attention on issues of particular concern in the type of animal facilities at their institutions (see also 28:9).

No matter how the inspection is structured, the many issues that should be considered by IACUC inspectors can be overwhelming once they actually enter an animal housing area to conduct an inspection. One popular approach is for IACUC inspectors to begin with the condition of the animals and expand their focus to the following general categories in any convenient order. Note that contingency plans for emergency situations (AWAR[1] §2.38,l; *Guide*,[3] p. 35) invariably have a physical component (e.g., animal rack movements, emergency supply storage, evacuation routes), and these areas should be included during the inspection of animal facilities. This is particularly true if there is an intent to shelter animals in place during an emergency situation (AWAR[1] §2.38,l,1).

Issue to consider:

- Animal health and well-being
- Primary enclosures, environmental conditions including environmental enrichment, and physical plant
- Housekeeping, disinfection, and sanitation
- Water, food, bedding, and other supplies (including expiration date of drugs and appropriate marking of secondary containers)
- Animal euthanasia, carcass storage, and carcass disposal
- Labels, signage, and records
- Personnel, training, and occupational safety
- Transportation vehicles
- Documentation of the above as appropriate, including contingency plans and disaster plans (AWAR[1] §2.38,l,2)

28:15 How should inspection findings be communicated, documented, and distributed?

Reg. Both the AWAR[1] (§2.31,c,3) and the PHS Policy[2] (IV,B,3) require that reports of inspection findings be forwarded to the IO at the close of every six-month inspection period. The AWAR require the report to be signed by a majority of IACUC members. Although not an explicit requirement of PHS Policy, it is best practice and recommended by OLAW that the report is signed by a majority of IACUC members. This is usually done in conjunction with the IACUC's semiannual evaluation of its institutional program for humane use and care of animals. Inspection findings may be summarized in the semiannual report to the IO, or copies of detailed site-by-site inspection records may simply be attached to the report. Sample reports are available on NIH/OLAW's website.[13] In either case, the inspection findings must distinguish significant deficiencies from minor deficiencies (significant deficiencies represent a threat to the health or safety of animals; see 27:32) and include a plan and schedule for correction (AWAR[1] §2.31,c,3; PHS Policy[2] IV,B,3). The records and reports of the inspections must be maintained for a period of three years (AWAR[1] §2.35,a,3 and f).

Opin. In addition to reporting inspection findings to the IO, IACUCs generally communicate their findings to animal facility managers, veterinary staff, and laboratory personnel in order to assist them in correcting deficiencies and to obtain full compliance. This communication can occur through many means, depending on the culture of the institution. If inspections are arranged with animal facility managers and laboratory staff in advance, most IACUC inspectors ask questions and discuss concerns with those personnel throughout the inspection visit. Alternatively, and particularly if appropriate personnel are not present during the inspection, IACUC inspectors may arrange for a verbal interview with the personnel either at the conclusion of the inspection visit or at a later time. In addition to verbal communication, most IACUCs will present responsible personnel with a written report of the inspection findings. This may be a simple list or table of deficiencies or a completed checklist presented at the conclusion of the inspection or a more formal inspection report sent at a later date. In either case, these written inspection reports generally contain the following information:

- Date and site location
- Names of IACUC inspectors
- Deficiencies observed and whether each deficiency was significant or minor
- A plan and date for correction

Additionally, many IACUCs also will include one or more of the following in their inspection reports:

- The administrative unit responsible for the facility inspected
- Names of facility representatives present at the inspection
- A list, including description and contents, of all rooms inspected

- Whether deficiencies observed were first time or previously cited
- The specific standard or regulation violated by each deficiency
- A general summary, comment, or commendation

28:16 How should the IACUC follow up on deficiencies?

Reg. Both the AWAR[1] (§2.31,c,3) and the PHS Policy[2] (IV,B,3) require that program and facility deficiencies be included in the semiannual report to the IO, that a reasonable and specific plan and date for correction be indicated for each deficiency, and that minor deficiencies be distinguished from significant deficiencies (deficiencies that may threaten the health or safety of animals). Additionally, the AWAR[1] require that failure to adhere to the correction schedule for significant deficiencies be reported within 15 days by the IACUC through the IO to APHIS/AC and any federal agency funding the cited activity (AWAR[1] §2.31,c,3). PHS Policy[2] (IV,F,3,a) requires prompt reporting to NIH/OLAW of serious or continuing non-compliance with the Policy. (See 29:12.)

Opin. When an IACUC encounters a situation that represents a threat to the health or safety of animals, prompt corrective action must be taken and a thorough follow-up conducted in order to ensure that the situation does not recur. The vast majority of deficiencies cited by the IACUC on inspection are minor, and the follow-up action depends on the culture of the institution. Depending on the nature of the minor deficiency (e.g., painting, repairs, etc. that are reliant on the schedule of others or availability of goods), corrections may require regular check-ins with the facility and rescheduled due dates. Many IACUCs re-inspect the cited facility at the correction deadline to verify resolution of the deficiency. Most IACUCs specifically assess the correction of previously cited deficiencies at subsequent inspections. Occasionally, an IACUC encounters previously cited minor deficiencies that either have not been corrected or have recurred. These deficiencies usually do not become "elevated" to major deficiencies unless the uncorrected minor deficiency has become a threat to the health or safety of animals. When there are recurrent or uncorrected minor deficiencies, the IACUC, using authority granted to it by its institution and federal regulations, can employ a variety of strategies to obtain compliance. Examples include the following:

- Increased frequency of IACUC inspections
- Utilizing Post-Approval Monitoring programs
- IACUC interview of responsible persons
- Notification of superiors in the chain of command
- IACUC-issued verbal or written warnings
- Withdrawal of IACUC approval to house or use animals
- Discipline of personnel in accordance with institutional policy

These actions are generally taken after discussion and approval by the IACUC and with the engagement of the IO. Care should be taken with repeated or ongoing minor deficiencies as this could be interpreted as a lack of engagement or oversight on behalf of the IO by regulatory authorities.

28:17 For how long should inspection documents be retained?

Reg. Both the AWAR[1] (§2.35,a,3 and f) and PHS Policy[2] (IV,E,2) require that semiannual reports to the IO, which include the IACUC's animal facility inspection findings, be retained by the institution and be available for inspection for at least three years.

Opin. If additional inspection records are created, the length of time they are retained is at the discretion of the IACUC. Many IACUCs retain these records for the same three years as the semiannual reports to the IO. Semiannual reports to the IO and additional inspection records are a valuable resource of documentation of animal facility management history. As such, IACUCs are likely to retain these records for an extended period. IACUCs at institutions subject to federal or state freedom of information acts or open records laws should consult appropriate persons at their institution for advice on their records retention program (see 26:14–26:16).

28:18 How should the IACUC assess the well-being of animals during inspection?

Reg. IACUC inspectors should endeavor to assess the physical and psychological well-being of all animals. According to the *Guide*[3] (p. 41), "An appropriate Program ... provides environments, housing, and management that are well suited for the species or strains of animals maintained and takes into account their physical, physiologic, and behavioral needs, allowing them to grow, mature, and reproduce normally while providing for their health and well-being." To accomplish this, many factors that affect the micro- and macro-environments should be considered, including the structural environment; social environment; animal activity patterns and behaviors; habituation and training for routine husbandry and experimental procedures; food and water; environmental factors, such as temperature, light cycles, and air flow; bedding and nesting materials; environmental enrichment; and sanitation and disinfection procedures. The standards of the AWAR[1] (§3.8; §3.81; §3.154) include specific provisions for the exercise of dogs and environmental enhancement of nonhuman primates and birds not bred for use in research.

The *Guide*[3] (p. 63) notes: "High levels of repetitive, unvarying behaviors (stereotypies, compulsive behaviors) may reflect disruptions of normal behavioral control mechanisms due to housing conditions or management practices." Identification of these types of behaviors can indicate to IACUC inspectors that the psychological well-being of the animals demonstrating the behavior, and perhaps the colony in general, has not been adequately addressed.

Opin. During animal housing facility inspections, the first thing to capture the attention of IACUC inspectors is likely to be the physical condition of the animals. By virtue of training or experience, most IACUC inspectors will readily distinguish normal healthy animals from those exhibiting signs of illness, injury, pain, or distress. Discovery of animals in the latter group should prompt further inquiry into the cause of the condition and the way it is being managed. If the condition is known to be experimentally induced, IACUC inspectors may wish to verify that the condition and its management are consistent with information provided in the IACUC-approved protocol. If the condition is not related to the experimental procedure, inquiry into the colony health status and other recent illnesses or injuries may be in order, and management practices questioned if undesirable patterns are identified. In either instance, IACUC inspectors should be assured that all animals are checked daily, that veterinary care is available around the clock if needed, contact information is current, the staff is aware of how to contact the veterinarian when needed, and that the veterinary staff is promptly notified of animal health problems and closely monitors the effectiveness of the implemented treatment and management plan (*Guide*,[3] pp. 112–113; AWAR[1] §2.33,b).

If the animal housing and care program is adequately providing for the psychological well-being of animals, IACUC inspectors should expect to see primary enclosures with features that accommodate the animals' species-specific postures and motor activities, management practices that cater to species-specific needs for cognitive stimulation and social interaction, and compatible social groupings where appropriate (see 32:2).

28:19 How should the IACUC evaluate animal enclosures during inspection?

Reg. The *Guide*[3] (pp. 57–63), AWAR[1] (§3.6; §3.15 ;§3.28; §3.53; §3.80; §3.101; §3.104; §3.128; §3.153), and *Ag Guide*[7] provide specific standards for primary enclosures.

Opin. To evaluate compliance with these standards, IACUC inspectors generally ask themselves the following questions when examining an animal enclosure:

- Are social species socially housed according to institutional policies and/or IACUC protocols?
- Is the enclosure of appropriate design for the comfort, security, and observation of the animals housed?
- Are animals able to engage in normal postural positions?
- Are animals able to engage in species-appropriate behaviors?
- Is it constructed of durable and sanitizable materials?
- Is space adequate for the size and number of animals contained?
- Is the enclosure clean and in good repair?
- Are food and water readily accessible?
- Does the enclosure in any way jeopardize the well-being of the animals housed therein?
- Are appropriate enrichment devices present in the enclosure?

The *Guide*[3] (p. 41) has clear expectations of the performance outcomes that are expected for animal housing environments. It is understood that the laboratory environment will not fully approximate the natural environment, but the institution should take steps to ensure that animals are able to engage in species-appropriate behaviors, move through normal postural adjustments, and have social interactions that are appropriate for the species. A basic understanding of the natural history of the species in use (e.g., social or isolated, burrowing, foraging, etc.) will assist the inspectors in their evaluation of the appropriateness of the animal enclosures for meeting the well-being of animals. The inspectors should note the condition of the animals, balance that with the institutional policies and guidelines, and evaluate them in the context of current literature regarding natural history and husbandry recommendations.

The same general questions listed above apply to non-traditional housing systems utilized in agricultural research. Additionally, the methods and systems utilized for protecting animals from heat and cold stress as well as predators should be evaluated. Further parameters to consider when evaluating an extensive agricultural system includes the sufficiency of geographic features of the area to provide shelter and protection from the elements as well as the provision of protective artificial shelters, windbreaks, and sunshades. The *Ag Guide*[7] should be referred to for specific standards for both extensive and intensive agricultural animal management systems.

28:20 What environmental parameters should the IACUC evaluate during inspection of an animal facility?

Reg. Specific standards for proper animal housing environmental conditions, including temperature, humidity, lighting, ventilation, air quality, noise, and vibration, are provided in the *Guide*[3] (pp. 43–45) and AWAR[1] (Part 3). The PHS Policy[2] (IV,A,1) uses the *Guide* as a basis for the development and implementation of an animal care and use program. The *Ag Guide*[7] provides specific environmental parameters for agricultural animal species.

Opin. IACUC inspectors generally find some environmental parameters, such as extremes of temperature and humidity, inadequate ventilation, excessive odor, and chronically loud noise, to be easily assessable the moment they enter an animal housing room. Other parameters require a little more investigation. For example, adequate assessment of the lighting cycle and magnitude and frequency of temperature fluctuation require more than momentary sensory perception by the IACUC inspectors. This may be addressed by reviewing facility records of these parameters. Additionally, depending on the type of primary enclosure present, the environmental conditions experienced by the animals within the enclosure may be very different from those perceived by the IACUC inspectors in the larger room environment. Records of HVAC function at the room and facility level should be reviewed as part of the semiannual inspection process to ensure that the macro-environment consistently meets the intended standards. In situations where vibration may be a problem, most commonly from construction activities in surrounding areas, facility managers should be aware and have a plan for monitoring and tracking vibration levels. Most IACUC inspectors interview animal facility managers about the monitoring methods used for such situations and review the monitoring documentation to ensure that the appropriate environmental conditions are being provided. The results of tests or actual experience with emergency alarms and response systems should be included in the discussion and records review. For animals housed in extensive agricultural management systems, the climatic and regional weather events should be considered. IACUC inspectors should also assess the adequacy of hazard (e.g., chemical, biological, and radiological) containment (see Chapter 23 for details).

Surv. Does your IACUC report deviations from environment parameters to OLAW?

- No 7/181 (4%)
- Yes 17/181 (9%)
- Depends on the effect on animal welfare 124/181 (69%)
- Not applicable 24/181 (13%)
- Other 9/181 (5%)

28:21 How should the IACUC evaluate the animal facility's physical plant during inspection?

Reg. As indicated in 28:17, IACUC inspectors should consult the *Guide*[3] and AWAR[1] for specific standards regarding the animal housing facility physical plant.

Opin. In addition to ensuring compliance with these standards, IACUC inspectors generally evaluate the facilities in terms of design, construction, building systems, fixtures and equipment, security, maintenance, and housekeeping. Room and facility level records may be evaluated to ensure that these expectations are being met appropriately. To satisfy most IACUC inspectors, the facility's physical plant should meet the following criteria:

- The design of the structure should be conducive to achieving the goals of animal well-being, research integrity, and personnel health and safety.
- Construction should be structurally sound, surface finishes should be sanitizable, and physical barriers should be in place to prevent entry of vermin.
- Building systems (electrical, plumbing, heating, ventilation, air conditioning, etc.) should be reliable and sufficient to support facility demands.
- The facility should be outfitted with fixtures and equipment appropriate to the housing, care, and use of the species housed. The surfaces of these items should be sanitizable.
- Equipment requiring periodic service or certification should be properly maintained.
- Appropriate security should include provisions for preventing both escape of animals and entry of unauthorized personnel.
- The facility structure and building systems should be maintained in good repair.
- In general, the facilities should be clean and free of unnecessary clutter.

For the average IACUC inspector, the evaluation of the physical plant for agricultural animals may be difficult. It is recommended to include someone with professional expertise in extensive and intensive agricultural animal management systems.

28:22 How should the IACUC assess the adequacy of housekeeping, disinfection, and sanitation practices during animal facility inspections?

Reg. The *Guide*[3] (pp. 69–73) recommends frequent bedding changes and cleaning and disinfection of primary enclosures and animal housing rooms. The AWAR[1] (§3.11; §3.31; §3.56; §3.84; §3.106; §3.107; §3.131; §3.158) provide specific minimal requirements for cleaning, sanitization, housekeeping, and pest control in animal housing facilities.

Opin. During inspections, most IACUC members evaluate these issues through direct observation and examination of records. Terrestrial animals are examined to ensure that they are clean and dry. Bedding should not be excessively soiled. Primary enclosures should not exhibit an accumulation of soil. Likewise, animal housing rooms should neither exhibit an accumulation of soil or waste nor be excessively cluttered (most IACUC inspectors discourage storage of equipment and supplies that are not used in the routine care and use of the animals). All surfaces should be sealed, intact, and sanitizable, and agents used for sanitation and disinfection should be appropriate for the purpose, be properly stored, and be used according to the manufacturers' directions. IACUC inspectors should see evidence of an effective vermin control program. Waste material should be removed from the animal rooms in a timely fashion and handled in a manner that minimizes the potential to introduce contaminants or attract vermin to the animal facility. Many IACUC inspectors also evaluate the effectiveness of an animal housing facility's housekeeping, disinfection, and sanitation program in terms of colony health. They may inquire about recent spread of disease and identify questionable housekeeping, disinfection, or sanitation practices that may contribute to colony health problems (see 28:19; 28:24).

Inspection of facilities for washing and sanitizing caging and equipment also are included in the IACUC's evaluation of sanitation practices. Most IACUC inspectors want to see that these facilities do not exhibit an accumulation of soil and that caging and equipment are processed in such a way as to minimize the recontamination of clean items. Chemical agents and automatic washing equipment used for sanitation and disinfection of caging and equipment should be appropriate for the purpose used, be properly stored, and be used according to manufacturers' directions. Most IACUC inspectors want to verify the effectiveness of the sanitation process. This can be done by inspecting records of validation test results (e.g., cage washer temperature indicator strips and cultures of sanitized surfaces). If the process calls for sterilization of caging or equipment, the performance of sterilizers is similarly evaluated by most IACUC inspectors.

The nature of agricultural housing systems requires similar evaluation parameters. Additional parameters to evaluate, particularly for intensive systems, include excreta management to prevent pollution of water, air, and soil and the spread of parasites as well as to minimize dust. There may be local, state, and federal regulations, policies, and laws that need to be considered (see Chapter 21).

28:23 How should the IACUC evaluate the adequacy of the water, food, bedding, medications, and other supplies provided to or used for animals?

Reg. Prior to inspection, IACUC inspectors are advised to consult the *Guide*[3] (pp. 65–69), *the Ag Guide*,[7] and AWAR[1] (Part 3) for specific standards regarding the quality, storage, and provision of water, food, bedding, medications, and other supplies. Of note, the AWAR[1] specifically require potable water to be continuously available to dogs unless restricted by the attending veterinarian or the IACUC (§3.10,a).

Opin. IACUC inspectors should be assured that all supplies that have contact with animals are appropriate and safe. Food, bedding, medications, and so on can usually be evaluated by direct observation during inspection. Acceptable animal food must be appropriate for the species, within its recommended shelf life, and stored under conditions that minimize the potential for premature deterioration, spoilage, contamination, or infestation. Food products should be labeled and dated with the expiration date. It should be available to animals *ad libitum* or provided in an appropriate ration and available to all individuals in a group unless veterinary or IACUC-approved research protocols dictate otherwise. Bedding should be absorbent, non-nutritive, nontoxic, and stored in a manner that minimizes the potential for spoilage, contamination, or infestation. Food and bedding should be free of foreign objects and normal in appearance, texture, and odor. Medications, treatments, and other veterinary or experimental supplies should be appropriate for the type of animal and research intended, medical grade, within designated expiration dates, and stored according to package directions (IACUC inspectors should be aware that special storage and record keeping requirements are imposed by the regulations of the Drug Enforcement Agency for controlled substances).[14] Unless veterinary or IACUC-approved research protocols dictate otherwise, drinking water should be available to animals *ad libitum*, potable, and free of contaminants. To assess the latter, most IACUC inspectors interview facility managers about the source of the water provided (municipal, well, purified, etc.) and the integrity of the system that delivers the water to the animals (plumbing systems, manifold flushing practices, backflow prevention, etc.). Inspectors will also usually examine the results of any water analysis conducted and inquire about colony health problems in which animal drinking water may have played a role.

28:24 What aspects of animal euthanasia, carcass storage, and carcass disposal should concern the IACUC during animal facility inspections?

Opin. Most IACUC inspectors visit areas for animal euthanasia, carcass storage, and carcass disposal in conjunction with their inspection of animal housing facilities. Euthanasia areas should be clean, and the methods employed should be consistent with the recommendations of the current edition of the American Veterinary Medical Association's *AVMA Guidelines for the Euthanasia for Animals: 2013 Edition*[15] (see Chapter 18).

Carcasses are generally stored in dedicated refrigerators or freezers, with special provisions for the labeling and storage of carcasses contaminated with infectious, radioactive, or chemical hazards, including ether. Whether carcass disposal is handled by the institution internally or contracted to a professional waste management company, IACUC inspectors should be apprised of the disposal methods and be assured that these methods comply with all applicable regulations, including local requirements and restrictions. IACUC inspectors also may wish to review any facility records associated with animal euthanasia and carcass disposal, particularly those required by the AWAR[1] (§2.35,b) for documentation of the final disposition of dogs and cats. Standards on animal euthanasia, carcass storage, and carcass disposal also can be found in the *Guide*[3] (pp. 73–74, 123–124, 142).

The slaughter of animals may be part of an agricultural research program. Methods should be consistent with the current edition of the *AVMA Guidelines for the Humane Slaughter of Animals*[16] (see Chapter 21).

28:25 What labels, signage, and records should be evaluated by the IACUC during animal facility inspections?

Opin. In an animal housing and care facility managed in accordance with the AWAR,[1] PHS Policy,[2] and the recommendations of the *Guide*,[3] most IACUC inspectors can expect to see labels, signage, and records associated with nearly every aspect of the program. Animal enclosures should be labeled with pertinent information (see the *Guide*,[3] p. 75) about the animals and their intended use (e.g., PI and IACUC-approved protocol number). Information for reporting concerns regarding animal welfare should be clearly posted (*Guide*,[3] p. 24). Animal food should be labeled with the species formulation and the milling or expiration date. Various cleaning, husbandry, veterinary, and experimental substances found in the animal facility should be labeled with the name and opening or expiration date of each item. Usage records should accompany controlled substances as required by the regulations of the Drug Enforcement Administration.[13] Hazardous substances and contaminated animals or materials should be clearly identified when the hazard is present and labeled with precautions. Special instructions should be prominently posted to assist personnel to comply with non-routine or critical procedures. Animal housing areas should be posted with current emergency contact information. Husbandry activities should be recorded on log sheets. Environmental readings should be logged or recorded on automatic monitoring system printouts. Animal monitoring records (post-procedural, food and water consumption, enrichment, exercise, etc.) are usually present at the animal housing location and should be reviewed. Animal health, clinical and veterinary treatment records, as well as animal receiving and disposition records should be maintained in an appropriate location and inspected by the IACUC. Records of sanitation and disinfection, including cage washer and sterilizer performance, also should be available for review by IACUC inspectors.

While not all of these labels, signs, and records are specifically required by the AWAR,[1] PHS Policy,[2] and recommendations of the *Guide*[3]—and certainly not every activity in the animal housing facility need be recorded—many of these records and documents will prove to be useful animal facility management and inspection tools. These may include signage for PPE requirements for entry, emergency call lists, or emergency egress on cage washers. Some IACUCs choose to require that extensive records be kept of animal housing activities and parameters, particularly to document to outside authorities that there is compliance with standards. Others rely on the state of their facilities and animal care programs to demonstrate compliance and forgo recordkeeping not specifically required by applicable regulations and standards.

28:26 What personnel issues should concern the IACUC during animal facility inspections?

Opin. One of the highlights of the IACUC animal housing facility inspection process is the opportunity for dialogue between IACUC members and personnel working in the animal facility. Such personnel can include members of the research staff, as well as facility managers and animal care and support personnel. In their conversations, most IACUC inspectors try to evaluate the extent to which personnel are qualified, through either experience or training, to perform their duties (AWAR §2.32), and whether or not experimental procedures and work practices are consistent with IACUC-approved animal use protocols, institutional guidelines, policies, and SOPs. As APHIS/AC mandates training of all staff regarding the implementation of the contingency plan and their role (AWAR §2.38,l,3), the IACUC inspectors should discuss this issue with key personnel.

Some of the most important personnel issues IACUC members consider during their inspections of animal housing facilities concern the occupational health and safety of animal care, use, and support personnel. IACUC inspectors should look for evidence that personnel are informed about and protected from the various health and safety risks associated with exposure to animals and any associated hazardous experimental agents or procedures. Examples of such evidence include elements of facility design and management (such as space delineation and air pressure differentials) that separate personnel areas and activities from those of animals, especially where hazardous agents are present. Compliance of personnel with appropriate personal hygiene and protective equipment recommendations also can indicate an effective occupational health and safety program (see Chapter 23).

Additionally, it is important that personnel who are working with animals understand that they are able to anonymously report concerns about animal welfare to the IACUC for investigation. Because personnel who are reporting animal welfare concerns have an expectation of whistle-blowing protection, ensuring that the reports can be made anonymously and that all personnel know how to make those anonymous reports is of the utmost importance (see 33:8-33:10).

28:27 What methods are useful for assessing remote animal holding locations?

Opin. Off-site animal holding study locations can be very challenging to inspect as they may require a significant time commitment from IACUC members for travel to and from the facility. Some facilities may have multiple facilities spread out over multiple geographic locations, thus making the logistics of arranging the IACUC inspections difficult as well. Off-site locations also may take more time to inspect as the holding areas are very different from those IACUC inspectors are accustomed to, and thus there is a large deal of education that must also happen before and during the site visit. While video recordings may be acceptable, it is up to each individual IACUC to develop a policy regarding off-site animal housing areas that is acceptable for their institution and still ensures the health and safety of the animals. It may be beneficial for the IACUC to enlist the help of *ad hoc* subject matter experts to aid in the evaluation of specialized or unusual animal facilities.

Surv. How often does your IACUC visit remote animal holding/study locations in person?

• Every six months	59/181 (33%)
• Once a year	13/181 (7%)
• Every three years	2/181 (1%)
• Other	13/181 (7%)
• Not applicable, no remove animal holding/study locations	94/181 (52%)

28:28 Should the IACUC include high containment areas in the semiannual inspection?

Reg. The AWAR[1] require that the IACUC inspect, at least once every six months, all of the research facility's animal facilities, including animal study areas (§2.31,c,2).

Opin. Yes. All animals in the research facility's animal facilities areas must be included in the semiannual inspection. Animal Biosafety Level (ABSL) 3 or 4 animal laboratories present special challenges to IACUCs because of the advanced level of training, personal protective measures, and security clearance that may be required to enter these spaces. However, with careful consideration, IACUCs can select members of a subcommittee to enter these areas, such as employee health and safety personnel who are familiar with these practices or who may already have approved clearances. An additional method that can be utilized is to inspect these areas during planned shutdowns.

Surv. As part of the semiannual facility inspection, how does your institution's IACUC evaluate high containment animal holding/study locations (i.e. ABSL-3/4 animal laboratories)?

• We do not have any high containment facilities.	122/181 (67%)
• The committee or subcommittee enters the facility(s) to conduct the inspection.	47/181 (26%)
• Evaluation of records from the facility(s), but the committee does not enter the high containment facility.	3/181 (2%)
• We use video recordings from the facility.	2/181 (1%)
• We do not inspect or evaluate our high containment facility(s).	2/181 (1%)
• Other.	5/181 (3%)

28:29 How should the IACUC evaluate training of the staff members involved in animal care?

See 24:33–39.

Surv. How should the IACUC evaluate the training of staff members involved in animal care? Select all that apply.

• Review of training records	146/181 (81%)
• Observation of techniques and procedures	120/181 (66%)
• Interview staff	81/181 (45%)
• Other	25/181 (14%)

28:30 How should IACUC's inspect animal facilities with approved housing exemptions and exceptions to required standards?

Reg. The AWAR stipulate that the IACUC shall conduct reviews of proposed activities or significant changes to ongoing activities to determine if they are in accordance with the subchapter unless acceptable justification for a departure is presented in writing and meet the requirements outlined in the section (§2.31,d.). Additionally, the PHS Policy[2] stipulates that the IACUC will confirm that a research project was conducted in accordance with the AWAR and is consistent with the *Guide*[3] unless acceptable justification for a departure is presented (IV, C, 1).

Opin. The IACUC should have a method for keeping track of approved housing exemptions and provide relevant information to inspectors who are auditing those areas. These housing areas should be inspected within the scope of the approved exemption.

28:31 Should the IACUC inspect temporarily empty animal holding rooms?

Opin. As part of a thorough and complete inspection, the IACUC should audit temporarily empty animal holding rooms.

28:32 How should the IACUC inspect vehicles used to transport animals?

Reg. The AWAR[1] stipulate that the cargo spaces of primary conveyances must be designed to protect the health and well-being of animals and to ensure their safety and comfort during transport. Fresh and/or auxiliary air, protection from temperature extremes, mechanically sound vehicles, methods to prevent the ingress of exhaust fumes, and clean interior spaces are required for transport vehicles. For details on transporting covered species, refer to the appropriate sections of the AWAR.[1]

While the *Guide*[3] does not provide specific information regarding the inspection of transport vehicles, it does state that planning for transport should occur to ensure animal safety and well-being as well as to provide an appropriate level of animal biosecurity. Additionally, zoonotic risks should be minimized; animals should be protected from environmental extremes; overcrowding avoided; provisions made for animal physical, behavioral, and psychologic needs and comfort; as well as protecting the animals and personnel from physical trauma (p. 107). The use of private vehicles for the transport of research animals is discouraged (p. 108).

Opin. Provision of an appropriate thermal environment is one of the primary cornerstones of safe and humane animal transport. Ensuring the vehicle is able to provide a stable ambient temperature and identifying how that temperature is monitored is of primary concern for IACUC inspectors. Mechanically sound, clean vehicles with methods to secure animals during transport are essential (see 16:4–5). For more information surrounding the transport of research animals, refer to the *Guidelines for the Humane Transportation of Research Animals*.[17]

Surv. Does your IACUC inspect facility-owned vehicles used to transport animals?

- No 72/181 (40%)
- Yes 109/181 (60%)

Surv. Does your IACUC inspect personal vehicles used to transport animals?

- No 123/180 (68%)
- Yes 57/180 (32%)

REFERENCES

1. United States Department of Agriculture, Animal Plant and Health Inspection Service. Animal Welfare Act and Animal Welfare Regulations. 2017. APHS 41-35-076.
2. US Department of Health and Human Services, National Institutes of Health, Office of Laboratory Animal Welfare. Public Health Service Policy on Humane Care and Use of Laboratory Animals. 2015.
3. The National Academies, National Research Council. 2011. *Guide for the Care and Use of Laboratory Animals*, 8th edition. Washington, DC: The National Academies Press.
4. Health Research extension Act of 1985. Public Law 99-158.
5. National Institutes of Health, Office of Extramural Research. 2011. Notice NOT-OD-011-053, Guidance to Reduce Regulatory Burden for IACUC Administration Regarding Alternate Members and Approval Dates. http://grants.nih.gov/grants/guide/notice-files/not-od-00-007.html.
6. Committees to revise the Guide for the Care and Use of Agricultural Animals in Research and Teaching. 2010. *Guide for the CARE and Use of Agricultural Animals in Agricultural Research and Teaching*, 3rd edition. Savoy: Federation of Animal Science Societies. https://www.fass.org/images/science-policy/Ag_Guide_3rd_ed.pdf.
7. Institute for Laboratory Animal Research. 2013. http://dels.nas.edu/ilar/.
8. National Institutes of Health, Office of Laboratory Animal Welfare. 2012. Sample Semiannual Facility Inspection Checklist Animal Housing and Support Areas. http://grants.nih.gov/grants/olaw/sampledoc/cheklist.htm.
9. Guidance on Flexibilities for Conducting Semi Annual Facility Inspections of Animal Facilities.. https://grants.nih.gov/grants/guide/notice-files/NOT-OD-24-075.html.
10. Association for the Assessment and Accreditation of Laboratory Animal Care International. 2012. Frequently Asked Questions B.2. http://www.aaalac.org/accreditation/faq_landing.cfm#A2.
11. National Institutes of Health, Office of Laboratory Animal Welfare. Frequently Asked Questions. https://olaw.nih.gov/faqs#/guidance/faqs?anchor=50347.
12. National Institutes of Health, Office of Laboratory Animal Welfare. 2013. Semiannual Report to the Institutional Official. http://grants.nih.gov/grants/olaw/sampledoc/ioreport.htm.

13. Office of the Federal Register. 1988. Controlled Substances Act, amended May 1, 1987, Title 21, Food and Drugs, Chapter 13, Drug Abuse Prevention and Control, 21 CFR §1301.11–§1308.15. Schedule of Controlled Substances, Washington, DC.
14. American Veterinary Medical Association. 2020. *AVMA Guidelines for the Euthanasia for Animals*: 2020 edition. Schaumburg: American Veterinary Medical Association. https://www.avma.org/sites/default/files/2020-02/Guidelines-on-Euthanasia-2020.pdf.
15. American Veterinary Medical Association. 2016. *AVMA Guidelines for the Humane Slaughter of Animals*: 2016 edition. Schaumburg: American Veterinary Medical Association. https://www.avma.org/sites/default/files/resources/Humane-Slaughter-Guidelines.pdf.
16. National Research Council of the National Academies, Institute for Laboratory Animal Research. 2006. *Guidelines for the Humane Transportation of Research Animals*. The National Academies Press. Washington, D.C.
17. Prescott, M. Guidelines for the Humane Transportation of Research Animals Committee on Guidelines for Humane Transport of Laboratory Animals, xvi+376 pp., 2006, National Research Council Press, Washington, D.C.

29

Inspection of Individual Laboratories

Neil S. Lipman and Scott E. Perkins

Introduction

One of the IACUC's principal roles is to ensure that animals used in research, testing, and teaching are used humanely by trained staff. Although IACUC review of proposals describing planned animal activity is an important mechanism for meeting this role, observation of animal use provides the most direct assurance that these goals are attained. At some institutions, animals are never removed from the animal facility as all activities are performed within animal holding or procedure rooms. Nevertheless, it is common for animals to be transported from the animal facility to an investigator's laboratory for experimental use. While animal resource personnel can readily observe the appropriateness of activities conducted within the animal facility, this task becomes significantly more difficult, especially at large institutions, when animal research is conducted in an investigator's laboratory.

The specific regulatory directive requiring IACUCs to inspect animal use areas is based upon the type of activity occurring, regardless of whether the location is physically within a centralized facility or a decentralized location such as an investigator's laboratory. The IACUC is charged with ensuring that animal use is conducted humanely by appropriately trained personnel and that these activities have been approved, in advance, by the IACUC. It is this chapter authors' opinion that laboratories in which animals are used should be visited by the IACUC in order to meet applicable regulations, policies, and the *Guide*.

29:1 Is it necessary to inspect investigators' laboratories where research with animals is conducted?

Reg. See 27:19. The AWAR (§1.1; §2.31,c,2) require that all of the research facility's animal facilities, including *animal study areas* (defined as any building room, area, enclosure, or other containment outside a core facility or centrally designated or managed area in which animals are housed for more than 12 hours) be inspected at least once every six months by a subcommittee of at least two IACUC members. Although the two committee members need not inspect the areas simultaneously, both must physically inspect the entire animal facility at least once within the six- month period.

The PHS Policy (III,B) defines an *animal facility* as "any or all buildings, rooms, areas, enclosures, or vehicles, including satellite facilities, used for animal confinement, transport, maintenance, breeding, or experiments inclusive of surgical manipulation. A satellite facility is any containment outside of a core facility or centrally designated or managed area in which animals are housed for more than 24 hours." All facilities, including satellite facilities, must be inspected at least once every six months by the IACUC (PHS Policy IV,B,2). (See 16:11 for comment on field study sites.) The *Guide* (p. 25) does specify that the IACUC, as part of its oversight functions, should conduct regular inspections of animal activity areas at least annually, or more often if required by the AWAR or PHS Policy.

Opin. The AWAR and PHS Policy mandate inspection of investigators' laboratories where research with animals is conducted when the activities performed meet their definition of an animal facility. This includes locations where animals are maintained in the laboratory longer than a specified period, as noted (see below) or with respect to the PHS Policy, surgery is performed. Additionally, the *Guide* (p. 25) specifies that the IACUC, as part of its oversight functions, should conduct regular inspections of animal activity areas at least annually or more often if required by the AWAR or PHS Policy. The *Guide* also suggests in this same chapter (p. 34) that laboratory inspections be considered as one component of a post-approval monitoring program. It is also noteworthy that AAALAC requires institutions seeking or renewing accreditation to identify protocols in which procedures are conducted on animals outside animal resource program managed areas as a component of the program description. These areas may be visited during the site visit based upon the nature of the procedures conducted at the site and the site visitors' discretion. Therefore, many institutional IACUCs assume this responsibility in order to meet their regulatory and institutional charge of ensuring that animals are used humanely by trained personnel. Laboratory visitation frequently provides the added benefit of enhancing dialog between the committee and investigative staff, frequently improving the

DOI: 10.1201/9781003272861-29

IACUC's understanding of proposals (see 27:19; 27:23).

NIH/OLAW, in its FAQ E.1[1] has further clarified the IACUC inspections of laboratories where investigators use animals as follows:

> Institutions are responsible for oversight of all animal-related activities regardless of how long or where the activity occurs. Satellite facilities (defined by PHS Policy as a containment outside a core or centrally managed area in which animals are housed for more than 24 hours) and areas where any form of surgical manipulations (minor, major, survival, non-survival) are performed must be inspected at least once every six months by the IACUC as part of the semiannual evaluation. Institutions have discretion with regard to how they oversee areas used for routine weighing, dosing, immunization, or imaging, but should monitor such areas on a random or fixed schedule to effectively oversee activities at the institution. USDA requires semiannual inspection of "animal study areas" defined as areas where USDA covered animals are housed for more than 12 hours.

The AWAR require that all of the research facility's animal facilities, including animal study areas, be inspected. These inspections are a key component of self-regulation and the intent is to ensure that the IACUC reviews animal use, including procedures that may cause pain and distress. As a result, IACUC inspection of laboratories is encouraged to the extent that it helps the IACUC and the institution fulfill its responsibilities under the AWA to oversee all animal care and use activities at the institution and ensure compliance with the regulations and standards.

Surv. During your semiannual inspections, which laboratories that hold or house animals are inspected by your IACUC? Select all that apply.

- Not applicable. 10/181 (6%)
- We inspect all laboratories that hold or house animals for any length of time. 134/181 (74%)
- Only those holding or housing animals for more than 12 hours (USDA covered species). 28/181 (15%)
- Only those holding or housing animals for more than 12 hours (non-USDA covered species). 15/181 (8%)
- Only those holding or housing animals for more than 24 hours (non-USDA covered species). 22/181 (12%)
- We inspect a random selection of laboratories. 6/181 (3%)
- We inspect only those laboratories deemed by the IACUC to be of increased concern. 8/181 (4%)
- Other. 2/181 (1%)

29:2 Is it necessary to inspect and evaluate every laboratory where animals are used at the time of each IACUC inspection?

Reg. See 27:19, 29.1.

Opin. As addressed in 29 :1, many IACUCs conduct routine laboratory inspections in order to meet their oversight responsibility of the institution's animal care and use program. It should be noted that the AWAR (§2.31,c,2) require that the IACUC inspect all animal facilities , including animal study areas, at least once every six months. Other animal facilities include surgical rooms, food storage, and cage wash areas.[2] Some of these areas may not house animals for more than 12 hours or may not house animals at all. Nevertheless, APHIS/AC will inspect these areas and cite any non-compliance; therefore, it is appropriate to correct any non-compliance found during IACUC inspections before they are cited by APHIS/AC inspectors.

It is also the authors' experience that AAALAC site visit teams will often request to visit laboratories outside of the centrally managed animal resource program where survival surgical procedures on any species are performed due to the invasive nature of the procedures and the potential for post-procedural complications. As we noted in the introduction to this chapter and in 25:1, it is our opinion that such inspections are appropriate.

Surv. When performing a semiannual laboratory inspection, which areas do the inspectors visit (other than laboratories and remote field sites)? Select all that apply.

- Not applicable 13/181 (7%)
- Survival surgical areas 162/181 (90%)
- Non-survival surgical areas 158/181 (87%)
- Animal transport vehicles 99/181 (55%)
- Non-surgical procedure areas (e.g., behavior testing, euthanasia stations) 156/181 (86%)
- Only areas we can easily access 2/181 (1%)

29:3 How frequently should each laboratory where animals are used be inspected?

Reg. Research laboratories in which APHIS/AC-covered species are maintained for more than 12 hours must be inspected by a subcommittee of at least two IACUC members every six months (AWAR §1.1, Animal; §1.1, Study Area; §2.31,c,2–3). Additionally, laboratories that are considered part of the research institution's animal facilities must also be inspected. IACUCs serving institutions required to comply with the PHS Policy (III,B; IV,B,2) must inspect, at least once every six months, all laboratories where surgery is performed and laboratory spaces in which animals are maintained for more than 24 hours. As stated in 29:1, the *Guide* (p. 25) does specify that the IACUC as part of its oversight functions should conduct regular inspections of animal activity areas at least annually or more often if required by the AWAR or PHS Policy. (See 27:19, 29:1, 29:2)

Opin. As there are no other regulatory requirements mandating laboratory inspection other than what is recommended in the AWAR and PHS Policy, it is at the individual IACUC's discretion whether or not and how often they should inspect animal research laboratories at their institution that do not meet these time or use requirements. It is the authors' opinion that inspections of at least select animal research laboratory inspections be conducted by the IACUC at least annually. It may be prudent to inspect select areas more frequently, for example every six months, as part of the semiannual inspection, based on the nature of the activities conducted at the site (see 29:1). Additionally post-approval monitoring (PAM) sessions, which can ensure compliance and assist research staff, conducted at the site of animal use may also supplement scheduled or unscheduled inspections of animal use areas.

29:4 Should a laboratory be approved by the IACUC before permitting an investigator to use the laboratory to conduct procedures on animals?

Reg. The AWAR (§2.31,d,1) require the IACUC, as part of the protocol review process, to ensure that all components of animal care and use are in compliance with the AWAR prior to approval of the protocol. PHS Policy (IV,C,1) requires that the project be conducted in accordance with the AWA and the *Guide* (see 29:1). If a laboratory is designated as a satellite facility, it would be necessary for the IACUC to visit to ensure that it can meet the requirements of the *Guide*.

Opin. Other than as noted for the AWAR and the PHS Policy, there are no regulatory requirements mandating the approval of laboratories by the IACUC. Nevertheless, it is the IACUC's responsibility to ensure that the proposed activity is conducted at a suitable location. It is the authors' opinion that there are significant advantages of having the IACUC inspect laboratories before providing approval for work with animals. Inspection of the laboratory provides the IACUC the opportunity to determine whether the facilities and equipment are suitable for the proposed activity. In addition, it provides an opportunity for the investigative staff to gain a better understanding of IACUC functions and provides IACUC members with a more thorough understanding of the proposed project. At institutions where IACUC approval of sites is not feasible because of time constraints, the IACUC can delegate this responsibility to a member of the institution's animal resource program's staff.

29:5 How does the IACUC determine the location of laboratories used for research with animals?

Opin. The most direct method for determining which laboratories are used to conduct research on animals is to request the information in the animal use proposal reviewed by the IACUC. This information is frequently used to generate a database. Additionally, laboratories may be identified on the basis of information provided by the animal resources staff or the institution's administration.

Surv. How does your IACUC determine the location of laboratories used for research with animals? Select all that apply.

- Not applicable 14/181 (8%)
- From information on the IACUC protocol form 142/181 (78%)
- From a separate database which we periodically update 32/181 (18%)
- From requests to investigators to identify their laboratory's location 39/181 (22%)
- From the knowledge of animal care personnel 62/181 (34%)
- Other* 9/181 (5%)

*Other: Facilities were of limited size and scope, thus allowing close knowledge of locations where animal work is performed.

29:6 Should the IACUC inspection be announced or unannounced? If an IACUC inspection team misses a laboratory during an inspection or if a member of the investigative staff is not present in the laboratory at the time of the inspection, should they go back at an alternative time to complete the inspection?

Opin. There are advantages and disadvantages of both methods. An announced site visit generally ensures that a knowledgeable staff member is available to provide laboratory access, discuss animal research activities, and answer questions. Staff availability permits dialog, which is advantageous to both the IACUC members and research staff, who frequently gain a better understanding of the IACUC's functions and activities. However, when conducting announced site visits, IACUC members may be exposed to an artificial environment not reflective of the normal state of the laboratory, personnel conduct, or animal use. Further, it is not uncommon for laboratories to cancel animal research activities when expecting a site visit (see 27:29).

Unannounced inspections permit the IACUC members to view the laboratory and its activities under normal operating conditions. However, research staff may not be present, potentially limiting laboratory access, or if they are present, ongoing research responsibilities may limit them in interfacing with IACUC members.

Institutions conducting announced inspections may also conduct unannounced inspections if the IACUC has specific concerns that may not be uncovered if inspections are announced.

The determination as to whether a laboratory should be revisited if it is missed is dependent on federal and state regulatory requirements and the regulatory and institutional charge of the IACUC. The AWAR (§2.31,c,2) and the PHS Policy (III,B; IV,B,2) require that animal facilities and study areas be inspected at least once every six months by the IACUC to maintain compliance (see 29:1; 29:2). Therefore, if the area is mandated for inspection and access cannot be obtained, the IACUC must reschedule a visit within the required inspection timetable.

Surv. During a semiannual laboratory inspection, if a laboratory cannot be inspected (for any reason) does your IACUC make an attempt to revisit the lab during the same inspection cycle or is the inspection deferred until the next semiannual inspection?

- Not applicable. 21/181 (23%)
- We typically attempt to revisit the lab during the current semiannual inspection time frame. 142/181 (78%)
- We typically wait until the next semiannual inspection. 2/181 (1%)
- We leave the decision to the inspection team. 5/181 (3%)
- We leave the decision to the IACUC office or IACUC Chairperson. 9/181 (5%)
- Other. 2/181 (1%)

29:7 Is it necessary for a PI or their designee to be present during the inspection?

Opin. Although not mandated by the AWAR or PHS Policy, there are clear advantages in having a knowledgeable research staff member present during the inspection (see 29:6). If the inspections are announced, the IACUC may stipulate that the Principal Investigator (PI) or designee be present during the inspection. With unannounced visits, the IACUC cannot ensure that access can be gained to the area.

29:8 Prior to the semiannual inspection, should the IACUC members who are visiting the laboratory be briefed on procedures conducted in the laboratory?

Opin. It is advantageous for IACUC members to be briefed on procedures occurring in the laboratory prior to the inspection, by the AV, a knowledgeable IACUC member, or by consulting the protocol. Alternatively, the PI or his or her designee can provide this information at the time of the inspection.

29:9 What areas and issues should the IACUC focus on during inspection of the laboratory?

Reg. The intent of the laboratory inspections is to ensure compliance with applicable regulations. The AWAR stipulate that the regulations and standards found in title 9, Chapter I, subchapter A-Animal Welfare be used as a basis for inspecting animal facilities (§2.31,c,2).

Opin. IACUC members should

- Ensure appropriate facilities, personnel, and equipment are present to provide adequate veterinary care
- Observe the laboratory and equipment used for cleanliness and adequate sanitation
- Assess human safety issues such as the following:
- Food or drink consumption within the laboratory
- Anesthetic gas scavenging
- Appropriate use, labeling, storage, and handling of hazardous substances
- Use of appropriate animal handling procedures
- Sharps as well as syringes and needles are handled appropriately
- Hazard and carcass waste disposal
- Accuracy of posted signage
- Observe the condition of animals, if they are present
- Confirm compliance with approved IACUC protocols, including ascertaining that procedures and staff have been described and included in the approved protocol
- Ensure staff understand the requirement for gaining IACUC approval in advance of utilizing a different species and/or changing/adding an experimental procedure(s) in accordance with institutional policy
- Evaluate the adequacy of training
- If surgery is performed, determine adherence to aseptic technique, appropriate instruments are utilized and they are in appropriate condition, suitability and understanding of anesthesia monitoring, and postoperative care, including pain relief
- Review methods, equipment, and/or materials used for euthanasia and restraint
- Verify that all pharmaceuticals are handled and stored appropriately and are not expired
- If using controlled substances, ensure their storage, use, and recordkeeping is in

compliance with federal, state, and/or institutional policies

- Review records, especially when conducting surgical procedures (see 29:10)
- Confirm that equipment requiring regular maintenance or certification, e.g., anesthetic vaporizers, biological safety cabinets, and guillotines, are maintained/certified as appropriate
- If the laboratory is approved for animal housing, determine that the animals' husbandry needs are met, environmental conditions are both stable and suitable and appropriate records are maintained, and the animals are transported appropriately to/from the laboratory
- Consider using NIH/OLAW's Semiannual Program Review and Checklist[3]

Surv. If your IACUC inspects areas (other than a central laboratory animal facility) in which animals are permitted to be housed for more than one day, which of the following are typically evaluated by the inspectors? Select all that apply.

• Not applicable	58/181 (32%)
• Light levels	93/181 (51%)
• Temperature	116/181 (64%)
• Humidity	108/181 (60%)
• Air changes/hour	71/181 (39%)
• Daily veterinary rounds (if daily rounds are made)	55/181 (30%)
• Daily husbandry rounds	104/181 (57%)
• Other	10/181 (6%)

29:10 What records should the IACUC inspection team review when visiting the laboratory?

Opin. Record evaluation is influenced by the activities that are conducted in the laboratory. If the laboratory is approved for the housing and care of animals, the PI's staff should maintain, and the IACUC members should periodically examine, those records indicating the frequency of cage changing, environmental monitoring (e.g., temperature, humidity, and light cycles), laboratory sanitation, and daily animal observations (including weekends and holidays). Records should be similar to those maintained by animal resource personnel in the animal facility. The following records, if applicable to the use of the laboratory, should be maintained and made available to the IACUC members during an inspection:

- Numbers of animals used per unit time (period in which the IACUC can compare actual animal usage to the number of animals requested in the protocol)
- Procedures conducted
- Treatment or drug administration
- Anesthetic, surgical, and postoperative care
- Equipment maintenance, e.g., guillotines and anesthesia vaporizers
- Controlled substances log
- Dosimetry for radiation exposure if using radionuclides or exposed to ionizing radiation
- Any institutionally required documentation when using hazardous substances
- Copies of institutionally required policies, handbooks, or guidelines
- Training records, especially for euthanasia methods that require training documentation
- Weight records for food or fluid restricted animals

29:11 What criteria can the inspection team use to make sure that acceptable standards are being met in the laboratory?

Opin. There are many criteria that can and should be utilized to determine that acceptable standards are used in the research laboratory. IACUC members should utilize the *Guide* and the AWAR, as well as IACUC and institutional policies, as the basis for determining the suitability of the laboratory for conducting research with animals. Depending on the nature of the research conducted and the institution, the Good Laboratory Practices Act,[4] the Occupational Safety and Health Act,[5] the Bloodborne Pathogen Standard,[6] Standards for Protection against Radiation,[7] Lab Safety Standard,[8] Controlled Substances Act,[9] NIH Recombinant or Synthetic Nucleic Acid Molecules Guidelines,[10] and Centers for Disease Control and Prevention's *Biosafety in Microbiologic and Biomedical Laboratories*, 6th edition[11] also may be used as criteria. OLAW's Semiannual Program Review and Facility Inspection Checklist[3] serves as a valuable resource for evaluating special facilities, including aseptic surgery, non-survival surgery, rodent surgery, and hazardous agent containment.

29:12 How should the IACUC proceed after identifying an item of deficiency in a laboratory?

Reg. The AWAR (§2.31,c,3) require the IACUC to identify deficiency items as significant or minor and develop a plan and timetable for correction. The AWAR (§2.31,c,3) define a *significant deficiency* as "one which, with reference to Subchapter A, and, in the judgment of the IACUC and the Institutional Official, is or may be a threat to the health or safety of the animals." If the established timetable is not adhered to and a significant deficiency remains uncorrected, the deficiency must be reported to APHIS/AC and any federal agency funding the activity within 15 days (AWAR §2.31,c,3). Similarly, the PHS Policy (IV,B,3) requires the IACUC to draft reports that are submitted to the Institutional Official (IO) of its inspection findings. As with the AWAR, identified deficiencies must be distinguished as significant or minor on the basis of whether they are or may be a threat to the health or safety of the animals. If program or facility deficiencies are noted, the report to the IO must contain a reasonable and specific plan and schedule for correction of each deficiency. Non-compliance must be promptly reported to NIH/OLAW (PHS Policy IV,F,3,a). Serious deviations from the provisions of the *Guide* must be promptly reported to OLAW (PHS Policy IV,F,3,b). See also NIH/OLAW guidance on reporting departures from the *Guide*, eighth edition at https://grants.nih.gov/grants/olaw/departures.htm (see 28:16; 30:16).

Opin. Although not specifically addressed in the regulations, the authors recommend expanding the criteria for a significant deficiency to include those that may be, or are, a threat to the health or safety of personnel in addition to animals. The authors find it most effective for deficiencies to be discussed with the responsible individuals or their designee, if available, at the time of the site visit. The ensuing dialog may clarify misinterpretations. Deficiencies should be reviewed by the IACUC and categorized as described. A plan and timetable for correction should be established. A written notice should be sent from the IACUC to the responsible individual, describing the stated deficiency and a plan and timetable provided for correction. The authors advise that the IACUC develop a mechanism by which the responsible individual can contest the deficiency if there is disagreement. Further, the responsible individual should be requested to provide written documentation that the deficiency has been corrected. The IACUC may wish to re-inspect to ensure the correction is effective.

Alternatively, the authors are aware of institutions that effectively use a multicopy form that is completed by the site visitors at the time of the site visit. The form contains a checklist for the inspectors to use to identify compliance or lack thereof. If deficiencies are noted, a timetable for correction based on preapproved criteria established by the IACUC is provided at the time of the site visit. The form is signed by the responsible individual or designee at the time of the site visit, and a copy is retained by the laboratory. The form stipulates that a formal response is required from the responsible individual for all non-compliant items.

Surv. As part of your semiannual laboratory inspection, how does your IACUC inform an investigator of a significant deficiency? Select all that apply.

- Not applicable. 8/181 (4%)
- The inspection team notifies the laboratory during the inspection. 108/181 (60%)
- The IACUC sends a written notification after the inspection. 160/181 (88%)
- The IACUC calls the laboratory after the inspection. 34/181 (19%)
- I don't know. 1/181 (1%)
- Other. 6/181 (3%)

All communications should convey the date of expected resolution. The authors recommend that the IACUC either re-inspect the area to determine whether the item was corrected by the specified date or confirm correction during the next scheduled inspection for non-compliant items categorized as minor. Depending on the severity and repetition of the item of non-compliance, the IACUC may impose sanctions on the investigator (see Chapter 33).

29:13 What criteria should be used for evaluation of biological safety cabinets (BSCs), laboratory hoods (also referred to as chemical or fume hoods), and other devices providing engineering controls (e.g., snorkels and downdraft tables) ?

Reg. The CDC's *Biosafety in Microbiological and Biomedical Laboratories*, sixth edition,[11] *Appendix A, Part 7-Certification of BSCs*, indicates that a BSC be certified, preferably by an accredited field certifier, before it is placed in service, after it has been repaired or relocated or annually, using the National Sanitation Foundation (NSF) Standard 49 (NSF/ANSI Standard 49-2018, Annex F) specifications. NSF 49 states that it is customary for the person conducting the designated tests to affix to the cabinet a certificate of adequate performance when the cabinet meets all field test criteria. Performance of laboratory hood or other types of engineering controls is addressed in OSHA's Laboratories Standard[8] (OSHA CFR Title 29 Part 1910.1450) requiring each institution to establish methods for evaluating hood performance in their required Chemical Hygiene Plan.

Opin. Current certification of each BSC should be confirmed during a laboratory inspection by examining the certification label on the cabinet. The IACUC inspection team should be acquainted with their institution's laboratory hood and engineering controls certification process and verify the devices are appropriately certified. If the device is in use, the authors recommend observing its use to ensure safe operational procedures are followed.

29:14 What standards should be applied to animals being housed in laboratories?

Reg. The AWAR (§2.31,c,2) require the same standards to be used (Title 9, Chapter I, subchapter A-Animal Welfare) for evaluating the animal study area as are used for the core facility or centrally designated or managed area. The PHS Policy (IV,B,2), as with the AWAR, makes no distinction in evaluation criteria between the animal facility and laboratory if the 24-hour housing requirement is met; therefore, the *Guide* is used as the basis for laboratory evaluation.

Opin. In the authors' experience, laboratory housing for extended periods is frequently necessitated by several factors: the need for the animals to be maintained in close proximity to unique equipment that is not found in or would be difficult or impossible to relocate to the animal facility; the need to house the animals under unique environmental conditions which cannot be replicated in the animal facility; and/or the transportation of animals to and from the animal facility would confound experimental results. When there is a need for frequent or prolonged laboratory housing, the expectation, in the authors' opinion, is that the laboratory would be dedicated to this purpose and would be constructed similarly to an animal holding room or procedure laboratory in the centralized facility. There may be situations in which the essential need for laboratory housing is for a limited time and infrequent. In this circumstance, the IACUC may consider giving greater latitude with respect to laboratory conditions; however, the IACUC must ensure that the animal's welfare and safety are not at risk, there is no risk to personnel who will access the laboratory, and the environmental conditions and stability are conducive with obtaining repeatable and reliable results. Routine monitoring of animal welfare by appropriately trained staff is essential for all laboratory housing.

REFERENCES

1. National Institutes of Health, Office of Laboratory Animal Welfare. 2021. Frequently Asked Questions. http://grants.nih.gov/grants/olaw/faqs.htm.
2. U.S. Department of Agriculture, Animal and Plant Health Inspection Service. 2022. *Animal Welfare Inspection Guide*. https://www.aphis.usda.gov/animal_welfare/downloads/Animal-Care-Inspection-Guide.pdf .
3. National Institutes of Health, Office of Laboratory Animal Welfare. 2021. Semiannual Program Review and Facility Inspection Checklist. http://grants.nih.gov/grants/olaw/sampledoc/cheklist.htm.
4. Office of the Federal Register. 1978. Code of Federal Regulations, Title 43, Good Laboratory Practice Regulations. Washington, DC. https://www.ecfr.gov/current/title-21/chapter-I/subchapter-A/part-58.
5. Office of the Federal Register. 1984. Code of Federal Regulations, Title 29, Part 1910, Occupational Safety and Health Standards, Subpart G, Occupational Health and Environmental Control. Washington, DC. http://www.osha.gov/pls/oshaweb/owastand.display_standard_group?p_toc_level=1&p_part_number=1910.
6. Office of the Federal Register. 1991. Code of Federal Regulations, Title 29, Part 1910.1030, Bloodborne Pathogens, Occupational Health and Safety Act. Washington, DC. http://www.osha.gov/pls/oshaweb/owadisp.show_document?p_table=STANDARDS&p_id=10051.
7. Office of the Federal Register. 1992. Code of Federal Regulations, Title 10, Chapter 1 Part 20, Standards for Protecting Against Radiation Nuclear Regulatory Commission Regulations. Washington, DC. http://www.nrc.gov/reading-rm/doc-collections/cfr/.
8. Office of the Federal Register. 1990. Code of Federal Regulations, Title 29, Part 1910, Department of Labor, Occupational Safety and Health Administration, Occupational Exposures to Hazardous Chemicals in Laboratories: Final Rule. Washington, DC. https://www.ecfr.gov/current/title-29/subtitle-B/chapter-XVII/part-1910/subpart-Z/section-1910.1450.
9. U.S. Department of Justice. 2022. Pharmacist's Manual: An Informational Outline of the Controlled Substances Act. Diversion Control Division. https://www.deadiversion.usdoj.gov/GDP/(DEA-DC-046R1)(EO-DEA154R1)_Pharmacist%27s_Manual_DEA.pdf.
10. U.S. Department of Health and Human Services. 2019. Guidelines for Research Involving Recombinant or Synthetic Nucleic Acid Molecules (NIH Guidelines). https://osp.od.nih.gov/wp-content/uploads/NIH_Guidelines.pdf.
11. U.S. Department of Health and Human Services. 2020. *Biosafety in Microbiological and Biomedical Laboratories*, 6th edition. HHS Publication No. (CDC) 300859. Washington, DC: U.S. Government Printing Office. https://www.cdc.gov/labs/pdf/CDC-BiosafetyMicrobiologicalBiomedicalLaboratories-2020-P.pdf.

30

Inspection of Surgery Areas

Scott E. Perkins and Neil S. Lipman

Introduction

The AWAR (§2.31,c,2) require IACUCs to inspect "all of the research facility's animal facilities, including animal study areas" for species regulated by the AWAR. The PHS Policy (IV,B,2) requires any and all animal facilities and areas to be inspected at least every six months, including satellite facilities and surgical areas. The PHS Policy (IV,A,1) also requires adherence to the *Guide* (p. 25), which recommends inspection of the animal facilities and animal care and use areas at least annually or more often as required (e.g., by the AWA and the PHS Policy). The *Guide* and AWAR specify requirements for surgical facilities and procedures for conducting survival surgery on rodents and AWAR-regulated species. The IACUC must approve all surgical procedures involving animals to ensure that appropriately trained individuals conduct the procedures in a humane manner. This chapter is intended to help the reader to comply with these regulations and recommendations.

30:1 Should all surgical sites be visited and evaluated at the time of each IACUC inspection?

Reg. There are federal requirements for IACUCs to inspect areas used to conduct surgical procedures on AWAR-regulated species[1] regardless of whether or not the site is located within an animal facility (*Guide*, p. 25). The AWAR (§2.31,c,2) require that animal facilities, including study areas, be physically inspected at least once every six months by a subcommittee of at least two IACUC members. More than one subcommittee may be utilized, such that different subcommittees may inspect different areas of the facility. However, both members of any subcommittee must each physically inspect all of their designated areas.

The PHS Policy (IV,B,2) also requires any and all animal facilities to be inspected at least every six months. Specifically, the PHS Policy (III,B) includes "any and all buildings, rooms, areas, enclosures, or vehicles, including satellite facilities, used for animal containment, transport, maintenance, breeding, or experiments inclusive of surgical manipulation." This includes areas used for non-survival surgery.[1] PHS Policy (IV,A,1) also requires adherence to the provisions of the *Guide* (p. 25), which states: "Program review and facilities inspections should occur at least annually or more often as required (e.g., the Animal Welfare Act and PHS Policy)."

Opin. The AWAR (§2.31,c,2) and PHS Policy (IV,B,2) require semiannual inspection of surgical areas.[1] Therefore, all surgical areas, including surgical facilities outside a core animal facility, should be inspected semiannually by the IACUC to maintain compliance with and meet the spirit of the AWAR, PHS Policy, and *Guide*. Both the AWAR (§2.31,d,1,ix) and the *Guide* (pp. 115–120, 144–145) detail facility and procedural requirements for conducting survival surgical procedures on rodents or USDA-regulated species.

Surv. Which surgical sites does your IACUC visit as part of its semiannual inspection? Select all that apply.

- Not applicable 10/181 (6%)
- Survival surgical sites for USDA-covered species 63/181 (35%)
- Survival surgical sites for all species 168/181 (93%)
- Non-survival surgical sites for USDA-covered species 61/181 (34%)
- Non-survival surgical sites for all species 150/181 (83%)

Surv. During the semiannual inspection of surgical facilities, are your inspectors given a checklist of items to evaluate?

- Not applicable. 9/181 (5%)
- Yes, we use the NIH/OLAW aseptic surgery checklist. 87/181 (48%)
- Yes, we use an in-house developed checklist. 58/181 (32%)
- No. 27/181 (15%)

30:2 What is the minimal frequency for evaluating surgical sites?

Reg. See 30:1.
Opin. See 30:1.

DOI: 10.1201/9781003272861-30

30:3 What facilities are required for conducting survival surgery on non-rodent and rodent species?

Reg. The IACUC must differentiate between major and non-major surgical procedures when evaluating facilities for performing survival surgical procedures, as facility requirements depend on the procedure conducted (see 30:5 for definitions).

Both the AWAR (§2.31,d,1,ix; §2.31,d,1,x) and the *Guide* (pp. 144–145) specify facility requirements for conducting major survival surgery on nonrodent species. The AWAR (§2.31,d,1,ix) require that "major operative procedures on non-rodents will be conducted only in facilities intended for that purpose which shall be operated and maintained under aseptic conditions." The AWAR do not require dedicated facilities when conducting major operative procedures at field sites; however, aseptic technique must be followed (§2.31,d,1,ix).

The *Guide* (pp. 144–145) provides significantly more detail pertaining to facility requirements, recommending that nonrodent surgical facilities include a variety of functional components, usually but not always separated by physical barriers. These include areas for surgical support, animal preparation, surgeon's scrub, operating room, and postoperative recovery. The need to provide all of these areas depends on the size and scope of the surgical program. With respect to ventilation, the *Guide* (p. 139) states that "areas of surgery should be kept under relative positive pressure with clean air."

Opin. It is the chapter authors' opinion that the minimal requirements for most operative suites are:

- Area for animal preparation
- Area for surgical scrub
- Operating room
- Area for animal recovery

The surgical suite should be designed and operated in a manner to control microbial contamination. This can be accomplished by

- Minimizing traffic in the area
- Adhering to rigorous sanitation and disinfection programs
- Limiting equipment present within the operating room to only essential movable equipment (such as surgical and support tables and anesthesia and monitoring equipment)
- Outfitting the surgical suite with a ventilation system that provides directional airflow maintaining the operating room at positive pressure with respect to the animal preparation and surrounding areas (under most circumstances)

Facility requirements for conducting non-major surgery on nonrodents or major survival surgery on rodents are less stringent. The AWAR (§2.31,d,1,ix) do not require dedicated facilities for conducting these procedures but do specify the use of aseptic technique. An area used for such procedures can be a room or portion of a room, which is easily sanitized and is dedicated for that purpose at the time of surgery. The laboratory bench or surgical table should be disinfected prior to and after use, be free of ancillary equipment, be located in an area free of drafts, and have a suitable temperature. If conducting surgery (non-survival or survival) in an area used by or accessed by individuals not directly involved in the procedure, consideration should be given to occupational health risks associated with exposure to the animals, e.g., laboratory animal allergens, or waste anesthetic gases. The authors' recommend that surgical procedures performed on rodents be performed in an isolated location to minimize these risks.

30:4 What facilities are required for conducting non-survival surgery on animals?

Opin. There are no facility-specific requirements when conducting non-survival surgical procedures. The area used for non-survival surgery should be a room or portion of a room, which is easily sanitized and at the time of surgery, dedicated for that purpose. The laboratory bench or surgical table should be disinfected prior to and after use, be free of ancillary equipment, and be located in an area free of drafts with a suitable temperature. In general, adherence to aseptic technique is unnecessary. The surgical site should be clipped and the instruments clean, and the surgeon should wear gloves. The IACUC should evaluate the duration of non-survival surgery as procedures can occasionally last for sufficient periods for sepsis to develop. In these cases, adherence to aseptic technique is advisable. Additionally, the IACUC should evaluate the need for aseptic technique or "aseptic" facilities, if certain procedures are being performed (e.g., harvesting an organ for transplantation) (see 19:21).

30:5 How does one differentiate a major from a minor surgical procedure?

Reg. The AWAR (§1.1) define a *major operative procedure* as "any surgical intervention that penetrates and exposes a body cavity or any procedure which produces permanent impairment of physical or physiological functions" (see 19:4). The PHS Policy itself provides no definition of major or minor surgical procedures, but it does require adherence to the provisions of the AWA for regulated species (PHS Policy II) and of the *Guide* (PHS Policy IV,A,1).

The *Guide* (pp. 117–118) defines a major survival surgical procedure as one that "penetrates and exposes a body cavity, produces substantial impairment of physical or physiological functions, or involves extensive tissue dissection or transection." It defines *minor survival surgery* as one in which the procedure "does not expose a body cavity and causes little or no physical impairment."

NIH/OLAW provides the following guidance in FAQ F.13:

> "OLAW recognizes the authority of the IACUC to determine whether specific manipulations used in research are major operative procedures. The IACUC's determination must be based on a detailed description of the procedure and the anticipated or actual consequences, as characterized by the investigator. In some cases, the classification by the IACUC of a procedure as major or minor may be readjusted post-procedurally depending on clinical outcome. If the IACUC, after thorough review, determines that the surgical procedure only penetrates but does not expose a body cavity and that the procedure does not produce substantial impairment, the IACUC may conclude that it is not a major operative procedure. Any laparoscopic surgery that produces substantial impairment of physical or physiological function must be considered a major operative procedure."

Opin. Although both the AWAR and the *Guide* provide a definition of a major operative procedure, the classification of some surgical procedures requires IACUC interpretation, which is based on the procedure's extent, effects, and potential for development of complications. Major operative procedures include laparotomy, thoracotomy, and limb amputation. However, APHIS/AC acknowledges that the IACUC has latitude to determine whether a laparoscopic procedure is a major or minor operative procedure.[2] Minor operative procedures include wound suturing, peripheral vessel cannulations, and certain routine farm animal procedures. The *Guide* (p. 117) indicates that "laparoscopic surgeries and some procedures associated with neuroscience research, (e.g., craniotomy, neurectomy) may be classified as major or minor depending on their impact on the animal" (see 19:8).

30:6 Assume an animal is regulated by the AWAR and the PHS Policy. Under what circumstances can a major survival surgical procedure be performed in an area other than a dedicated surgical suite approved by the IACUC?

Reg. The AWAR (§2.31,d,1,ix) state that "major operative procedures on non-rodents will be conducted only in facilities intended for that purpose which shall be operated and maintained under aseptic conditions." The AWAR (§2.31,d,1,ix) do not require dedicated facilities when conducting major operative procedures at field sites, although aseptic procedures are required. Additionally, the AWAR (§2.38,k,1) allow the IACUC to approve exceptions to the regulations and standards "when such exceptions are specified and justified in the proposal to conduct the activity."

PHS Policy (IV,A,1) requires adherence to the *Guide* (p. 62), which states: "Unless an exception is specifically justified as an essential component of the research protocol and approved by the IACUC, aseptic surgery should be conducted in dedicated facilities or spaces." Nevertheless, the *Guide* (pp. 117–118) does provide exceptions for emergencies and field sites: "Emergency situations sometimes require immediate surgical attention under less-than-ideal conditions," and "Generally, agricultural animals maintained for biomedical research should undergo surgery with techniques and in facilities compatible with the guidelines set forth in this section. However, some minor and emergency procedures commonly performed in clinical veterinary practice and in commercial agricultural settings may take place under field conditions. Even when conducted in an agricultural setting, however, these procedures require the use of appropriate aseptic technique, sedatives, analgesics, anesthetics, and conditions commensurate with the risk to the animal's health and well-being."

Opin. It is the authors' opinion, in accordance with the *Guide*, that major operative procedures can be performed in a non-dedicated surgical suite when scientifically justified (e.g., unique instrumentation in the laboratory) and approved by the IACUC. In the event of an emergency when the animal cannot be transported to the dedicated operating room, this should be considered a clinical medical requirement, determined by veterinary medical judgment, and does not require IACUC approval.

30:7 Can non-survival surgery on nonrodent species, or survival surgery or non-survival surgeries on rodents, be performed in survival surgical facilities used for nonrodents? If yes, are there any special procedures that should be followed?

Opin. IACUCs may permit the conduct of survival rodent surgery or non-survival surgical procedures in survival surgical facilities if either of the following conditions is met:

- Non-survival procedures that are conducted utilize the same standards that apply to survival procedures in accordance with the standards of the *Guide* (p. 118).
- The operating room is thoroughly sanitized and disinfected prior to conduct of a survival surgical procedure in nonrodents (*Guide* p. 118).

30:8 What areas and issues should the IACUC focus on during inspection and evaluation of surgical areas?

Reg. PHS Policy (IV,B,2) requires any and all animal facilities (as defined in PHS Policy III,B) to be inspected at least every six months, including "any and all buildings, rooms, areas, enclosures, or vehicles, including satellite facilities, used for animal confinement, transport, maintenance, breeding, or experiments inclusive of surgical manipulation." PHS Policy (IV,A,1) requires compliance with the *Guide* (pp. 115–120), which includes IACUC evaluation of aseptic procedures, postoperative evaluation and care, anesthetic monitoring, and other elements. Other sections of the *Guide* provide further information on OHS (*Guide*, pp. 17–23), design of facilities for aseptic surgery (*Guide*, pp. 144–145), and other considerations. The AWAR (§2.32), PHS Policy (IV,C,1,f), U.S. Government Principle VIII, and the *Guide* (p. 115) all address the need for proper qualifications and experience of personnel working with animals (see 30:19).

Opin. The AWAR (§2.31,d,1,ix) and the *Guide* (pp. 116–120, 144–145) describe facility and procedural requirements for conducting survival surgical procedures in nonrodents. In the authors' opinion, minimal evaluation of standards for surgical areas should include the following:

- Assessment of personnel training
- Adherence to aseptic technique
- Scavenging of waste anesthetic gases
- Correct use of unexpired pharmaceuticals and materials
- Appropriate documentation of controlled substances use
- Scientifically justified and IACUC approved use of non-pharmaceutical-grade chemicals and other substances
- Proper maintenance and use of equipment and procedures for peri-operative monitoring and care
- Maintenance of peri-operative records
- Availability of appropriate equipment and pharmaceuticals for emergencies
- Proper disposal of sharps

Surv. During the semiannual inspection of surgical areas, which of the following items are routinely evaluated by the inspectors? Select all that apply.

• Not applicable	10/181 (6%)
• No policy on this issue in place	12/181 (7%)
• The general condition of the physical plant	128/181 (71%)
• Personnel training and surgeons' competence	62/181 (34%)
• Scavenging of anesthetic waste gases, if used	139/181 (77%)
• Maintenance of gas anesthesia machines, if used	146/181 (81%)
• Availability of appropriate surgical instruments	98/181 (54%)
• Availability of appropriate surgical equipment	107/181 (59%)
• Drug storage and drug inventory	150/181 (83%)
• General cleanliness	163/181 (90%)
• Records of disinfecting the area	83/181 (46%)
• Procedures used for instrument and supply sterilization	109/181 (60%)
• Methods of testing autoclaves and appropriate test records	105/181 (58%)
• Proper preparation of the animal (i.e., skin prep)	91/181 (50%)
• Adherence to aseptic technique during surgery	96/181 (53%)
• Use of appropriate surgical garb	83/181 (46%)
• Posting of emergency contact information	124/181 (69%)
• Peri-operative monitoring and records	122/181 (67%)

30:9 What reference sources are available to help evaluate the issues in question 30:8?

Opin. The IACUC should utilize the AWAR, the *Guide* and its associated references, applicable state regulations, and institutional policies to develop acceptable standards for surgical facilities and their operation. The IACUC can also consult references, as well as expert consultants, to assist in the development of standards for conducting surgery. Additional references include veterinary and human surgery and anesthesia texts, texts and references on experimental surgery, the Good Laboratory Practices Act,[3] the Occupational Safety and Health Act,[4] the Blood Borne Pathogen Standard,[5] Standards for Protection against Radiation,[6] Lab Safety Standard,[7] information on the Controlled Substances Act,[8] Recombinant DNA Guidelines,[9] the Centers for Disease Control and Prevention (CDC) publication *Biosafety in Microbiologic and Biomedical Laboratories*,[10] NIH/OLAW Semiannual Program Review and Facilities Inspection Checklist,[11] U.S. Department of Justice's Researcher's Manual: An Informational Outline of the Controlled Substances Act,[8] and references on disinfection and sanitation control.[12]

30:10 What records should the inspection team review when visiting nonrodent and rodent surgical areas? (see 19:38)

Opin. The IACUC should confirm that procedures and staff training, experience, and qualifications have been accurately described and included in the approved animal care and use proposal. The IACUC should review pertinent animal health and experimental records, including the following:

- Peri-operative records describing the surgical procedure performed
- Experimental agents administered
- Dose, route, and frequency of anesthetics and analgesics administered
- Required records for controlled substances
- Frequency and type of intra-operative monitoring procedures performed
- Postoperative monitoring and observations
- Quality assurance records for sterilization of materials (periodically reviewed)
- All records should be made available for IACUC review whether they are maintained with the animal, in the surgical facility, or by the investigator

30:11 How detailed should surgical and postoperative records be? What are the essential elements of an adequate record?

Reg. The *Guide* (pp. 75–76) states: "Medical records for individual animals can also be valuable, especially for dogs, cats, nonhuman primates, and agricultural animals (Suckow and Doerning 2007). They should include pertinent clinical and diagnostic information, date of inoculations, history of surgical procedures and postoperative care, information on experimental use, and necropsy findings where applicable." The AWAR (§2.33,b,5) require "adequate pre-procedural and post-procedural care in accordance with current established veterinary medical and nursing procedures."

The HREA (§495,a,2,B) requires "appropriate pre-surgical and post-surgical veterinary medical and nursing care for animals" used in biomedical and behavioral research (see 19:38; 19:39).

Opin. It is the opinion of the authors that the AWAR and HREA regulations could be interpreted as surgical records being a required component of adequate veterinary care. Records should be detailed and contain all relevant information, including, but not limited to

- Animal identification
- Project identification
- Detailed description of the surgical procedure
- Experimental agents administered
- Dose, route, and frequency of anesthetics and analgesics administered
- Frequency (e.g., time interval) and type (e.g., blood pressure, heart rate) of intraoperative monitoring procedures performed
- Peri-operative administrations and treatments (e.g., fluids, other medications)
- Postoperative monitoring and observations recorded
- Postoperative treatments administered

Surv. Does your IACUC or animal facility provide investigators with specific forms to document the surgical information typically required by your IACUC (such as drugs used, vital signs monitoring, and observations)?

• Not applicable.	11/181 (6%)
• Yes, and the form must be used.	45/181 (25%)
• Yes, either the provided form or an investigator designed facsimile is required to be used.	67/181 (37%)
• The use of a form is suggested but not required.	28/181 (15%)
• No.	30/181 (17%)

30:12 Should surgical records be kept for all species or only certain ones?

Reg. The AWAR require that activities that involve surgery include appropriate provision for preoperative and postoperative care of the animals in accordance with established veterinary medical and nursing practices (§2.31,d,1,ix). Research facilities are also required to establish and maintain programs of adequate veterinary care (§2.33,b). Additionally, the AWAR include specific requirements for medical records for dogs, including surgeries and treatments (§3.13,b).

Opin. The *Guide* (pp. 75–76) states that records are especially valuable for dogs, cats, nonhuman primates, and farm animals. Although there are no specific surgical record requirements in the AWAR or PHS Policy, anesthetic, surgical, and postoperative records are components of established veterinary medical practice and adequate veterinary care. PHS Policy also requires adherence to the requirements of the AWAR for regulated species (PHS Policy II) and the *Guide* (PHS Policy IV,A,1).

It is the opinion of the authors that surgical records should be maintained for all species to provide documentation that appropriately trained individuals conducted the research in a humane manner (see 19:38–19:39).

Surv. For which species does your IACUC require that adequate surgical records be maintained?

• Not applicable	6/181 (3%)
• No policy in place	11/181 (6%)
• For all vertebrate species	123/181 (68%)
• For all mammalian species	24/181 (13%)

• For USDA covered species only (and we perform surgery on more than just USDA covered species)	13/181 (7%)
• For USDA covered species (and we perform surgery ONLY on USDA covered species)	1/181 (1%)
• I don't know	1/181 (1%)
• Other	2/181 (2%)

30:13 Should surgical records be kept for each individual animal or can they be maintained by groups?

Opin. There are no regulatory requirements for having individual surgical records for each individual animal. The *Guide* (pp. 75–76) notes that clinical records for individual animals can be useful and states that they should include the history of surgical procedures and postoperative care information. It is the opinion of the authors that records can be maintained by groups for rodents, but nonrodents should have surgery records maintained for individual animals (see 30:10; 31:18; 31:20).

30:14 Do records for survival surgery differ from those maintained for animals undergoing non-survival surgical procedures?

Opin. Neither the AWAR nor the *Guide* stipulates that recordkeeping should be less stringent or different for animals undergoing non-survival procedures. Records should be maintained in accordance with established veterinary medical standards as noted in 30:10 and 30:12 (see 19:38–19:39).

Surv. Does your IACUC have different intraoperative recordkeeping requirements for survival vs. non-survival major surgical procedures (assume the same species would be used)?

• Not applicable.	9/181 (5%)
• We have no policy.	27/181 (15%)
• Our intraoperative record keeping requirements are the same.	96/181 (53%)
• Our intraoperative record keeping requirements are less rigid for non-survival procedures.	34/181 (19%)
• I don't know.	15/181 (8%)

30:15 What are the recordkeeping requirements when performing surgery on rodents?

Reg. The *Guide* (pp. 75–76) states that clinical records for animals are valuable and describes the kind of information to be maintained. Rodents are not excluded from these statements.

Opin. At the authors' institutions, records are required for rodents undergoing surgical procedures. The frequency of monitoring procedures and details for intraoperative monitoring procedures are described in the IACUC protocol. Records may be maintained for groups depending on the nature of the surgical procedure and the scope of peri-operative care. Rodent records should, at minimum, include the following:

- A description of the surgical procedure performed
- Experimental agents administered
- Dose and route of anesthetics and analgesics administered
- Postoperative monitoring and observations
- Treatments administered during the peri-operative period

30:16 How should the inspection team proceed after identifying an item of non-compliance in a surgical area?

Reg. The IACUC should identify the item as significant or minor and develop a timetable for correction as specified in the AWAR (§2.31,c,3) and the PHS Policy (IV,B,3) (see 29:12).

Opin. The policy for contacting the appropriate individual varies depending on the magnitude of the item. IACUC options for notification include the following:

- Briefing the responsible individual from the laboratory at the time of inspection, and/or later by telephone or e-mail
- Sending a copy of an inspection report to the responsible individual
- Sending a formal letter documenting the non-compliant item

All communications should convey the date of expected resolution. Depending on the severity and repetition of the item of non-compliance, the IACUC may impose sanctions on the investigator such as a letter of reprimand from the IO, cessation of research for a specified time or permanently, mandatory (re)training, restriction of specific staff from conducting procedures, or increased inspection frequencies. The IACUC may revisit the area prior to the next site visit to confirm correction (see 29:12 and Chapter 33).

30:17 Can unrelated major survival surgeries be conducted simultaneously in a single operating room on multiple non-rodent animals of the same or different species? Does the response differ for rodent species?

Opin. Neither the AWAR nor the *Guide* has restrictions on the performance of more than one surgery, on the same or different species, in the same operating room. There are unique situations (e.g., inter- and intraspecies transplantation) when conducting surgery on more than one animal in the same operating room is justifiable.

Because of the potential for infectious agent transmission and behavioral incompatibility, it is not advisable to conduct survival surgical procedures on more than one species simultaneously unless the species are of similar microbiological status; for example, specific pathogen-free rodents. Conducting simultaneous multiple survival surgeries on different animals of the same species is acceptable as long as the nature of each surgical procedure conducted does not place the other animals undergoing surgery at increased risk (e.g., different microbiological status of animals or an infectious disease study).

30:18 Can a major survival and a non-survival operative procedure be performed simultaneously on nonrodent species in the same operating room?

Opin. When justifiable, as described in 30:17, survival and non-survival surgical procedures can be conducted in the same operating room. However, both surgeries should be conducted with aseptic technique, as if they were survival procedures, because strict aseptic technique must be followed in accordance with the regulations and with the purpose of providing the highest level of sanitation for the animal undergoing the survival procedure.

30:19 What level of training should the IACUC require for individuals performing survival surgical procedures? Should the IACUC verify whether individuals have adequate training and skills to perform major operative procedures?

Reg. The IACUC has the responsibility for ensuring that individuals conducting surgery, irrespective of species used, have the necessary skills to execute the surgical procedure successfully. Both the AWAR (§2.32) and the PHS Policy (IV,c,1,f) require that all personnel involved in animal care, treatment, and use are qualified and trained to perform their duties. U.S. Government Principle VIII stipulates that "investigators and other personnel shall be appropriately qualified and experienced for conducting procedures on living animals." In addition to the conduct of the surgery, individuals associated with the activity must be skilled at administering and monitoring anesthesia and have the capability of providing suitable peri-operative care.

Opin. Although the IACUC's responsibility is clear, neither the PHS Policy nor the AWAR describe implementation methods. There are a variety of mechanisms that the committee can use to meet this responsibility (see Chapter 24). IACUCs frequently garner information pertaining to an investigator's training and skills via required descriptions of them in the initial proposal submitted to the IACUC. Although not all individuals on a specific project may have suitable skills to carry out the project, the IACUC must determine that there is sufficient skilled staff available to meet the project's goals. Clearly, the variety and scope of surgical procedures conducted in a research or training setting vary considerably, requiring the IACUC to make this determination on a case-by-case basis. The IACUC may request additional information or invite the investigative staff to meet with them to review their experience and training.

At many institutions, the animal resource program personnel are responsible for the oversight of a centralized surgical facility. The committee frequently relies on the animal resource program's veterinary or paraprofessional staff for information pertaining to an individual's skill or training. Alternatively, the committee may request that one or several of its members observe procedures on a select number of animals before giving the project full approval. Adequacy of training should be ascertained prior to the start of the project; however, it is important for continuing assessment to be conducted. An infrequently used option is that the committee may request the aid of a consultant to make an assessment for projects of a sensitive nature or requiring highly specialized staff.

30:20 What constitutes aseptic technique for major operative procedures? (see 19:18)

Reg. The AWAR (§2.31,d,1,ix) and the *Guide* (pp. 115–120) require that major survival surgical procedures on all animals be conducted utilizing aseptic procedures, including surgical gloves, masks, sterile instruments, and aseptic technique.

Opin. Aseptic technique is a body of practices employed to prevent microbial contamination of living tissues or sterile materials by excluding, removing, or killing microbial organisms. Aseptic technique requires strict adherence to practices in several areas. The laboratory bench or surgery table should be clean, disinfected, and free of ancillary equipment. Devices or equipment (e.g., animal restraining devices, monitoring equipment, stereotaxic devices) that will be required in the surgical field also should be disinfected. These practices reduce or eliminate potentially infectious organisms and the substrates on which they grow.[13,14] All surgical instruments, implantable devices, and materials that will contact the surgical site or are implanted in the animal must be sterilized. The sterilization method selected is dependent on time considerations, specialized equipment available, and the composition of the material to be sterilized. Sterilization monitoring devices should be routinely utilized to validate sterilization techniques. Surgical instruments and supplies should be prepared prior to sterilization so that they can be opened without being touched.[15]

The regulatory requirements for preparation of the surgeon for rodent surgery are less rigorous than those required for nonrodents. Either a clean or disposable laboratory coat or a surgical scrub shirt should be worn by the surgeon and assistants.

A surgical mask (which serves to prevent contamination of the wound by droplets of saliva from the surgical team) and a surgical cap, for individuals who have long hair, which can potentially fall into the surgical field, should be worn. The surgeon should wash his or her hands with a disinfectant and don sterile surgical gloves by a method that maintains sterility (*Guide*, p. 118).[16]

Preparation of the surgical site involves removal of fur by clipping or using a depilatory and disinfection of the skin by scrubbing with suitable agents. The surgical site should be covered with a sterile drape. The goal of aseptic technique is to prevent the surgeon, the surgical site, and all instruments and equipment utilized from causing infection and subsequent disease.[16]

30:21 If performing major operative procedures on multiple rodents, can a single set of surgical instruments be used on multiple animals? (see also 19:20)

Opin. Groups of non-USDA regulated rodents are frequently surgically manipulated during a single session. Care must be taken to prevent contaminating one animal from another. Sterilized instruments should always be used when initiating a surgical session. New sterile gloves should be donned any time they become contaminated; preferably, separate sterile surgical instruments should be provided for each animal subject. When a limited number of surgical instruments are available and separate sterile surgical packs are not feasible, the following options may be considered:

- Use two sets of instruments. One set is used for incising and manipulating the skin, which is considered a potentially contaminated site because of microbial flora. Once the initial incision and skin manipulation are complete, these instruments are set aside and not used again without prior sterilization. A second set of instruments is used to manipulate deeper, sterile tissues. The second set of instruments (which is shared) should be thoroughly rinsed with sterile saline solution between animals. Care must be exercised to maintain the sterility of the surgical gloves by avoiding contacting tissues with fingers when utilizing this option.
- Utilize a glass bead sterilizer to sterilize instrument tips between animals. Recognize that only the instrument tips are sterilized. Care must be taken to maintain sterility of the tips within the surgical field and prevent touching tissues with instrument handles or other non-sterile instrument parts. Because non-sterile instrument handles are held with gloved fingers, contact of tissues with the fingers is prevented.
- Have two or more sets of sterile instruments available so that the contaminated set can soak in cold sterilant solution for the minimal time required by the solution manufacturer, to kill vegetative bacteria prior to reuse. Instruments must be rinsed thoroughly with sterile saline or water prior to reuse.

REFERENCES

1. National Institutes of Health, Office of Laboratory Animal Welfare. 2021. Frequently Asked Questions E.1, Should the IACUC Inspect Laboratories or Other Sites Where Investigators Use Animals? http://grants.nih.gov/grants/olaw/faqs.htm#prorev_1.
2. Petervary, N. 2013. Best practices for IACUCs in the evaluation of multiple major operative procedures. *Lab Anim. (NY)* 42:330–333.
3. Office of the Federal Register. 1978. Code of Federal Regulations, Title 43, Good Laboratory Practice Regulations, Washington, DC. https://www.ecfr.gov/current/title-21/chapter-I/subchapter-A/part-58.
4. Office of the Federal Register. 1984. Code of Federal Regulations, Title 29, Part 1910, Occupational Safety and Health Standards, Subpart G, Occupational Health and Environmental Control, Washington, DC. http://www.osha.gov/pls/oshaweb/owastand.display_standard_group?p_toc_level=1&p_part_number=1910.
5. Office of the Federal Register. 1991. Code of Federal Regulations, Title 29, Part 1910.1030, Bloodborne Pathogens, Occupational Health and Safety Act, Washington, DC. http://www.osha.gov/pls/oshaweb/owadisp.show_document?p_table=STANDARDS&p_id=10051.
6. Office of the Federal Register. 1992. Code of Federal Regulations, Title 10, Chapter 1 Part 20, Standards for Protecting against Radiation. Nuclear Regulatory Commission Regulations, Washington, DC. http://www.nrc.gov/reading-rm/doc-collections/cfr/.
7. Office of the Federal Register. 1984. Code of Federal Regulations, Title 29, Part 1910, Occupational Safety and Health Standards, Subpart G, Occupational Health and Environmental Control, Washington, DC. http://www.osha.gov/pls/oshaweb/owastand.display_standard_group?p_toc_level=1&p_part_number=1910.
8. U.S. Department of Justice. 2022. Researcher's Manual: An Informational Outline of the Controlled Substances Act. Diversion Control Division. https://www.deadiversion.usdoj.gov/GDP/(DEA-DC-046R1)(EO-DEA154R1)_Pharmacist%27s_Manual_DEA.pdf.
9. U.S. Department of Health and Human Services. 2011. Guidelines for Research Involving Recombinant DNA Molecules (NIH Guidelines). http://oba.od.nih.gov/oba/rac/Guidelines/NIH_Guidelines.pdf.
10. U.S. Department of Health and Human Services. 2009. *Biosafety in Microbiological and Biomedical Laboratories*, 5th edition. HHS Publication No. (CDC) 21-1112. Washington, DC.: U.S. Government Printing Office. http://www.cdc.gov/biosafety/publications/bmbl5/BMBL.pdf.

11. National Institutes of Health, Office of Laboratory Animal Welfare. 2012. Semiannual Program Review and Checklist. http://grants.nih.gov/grants/olaw/sampledoc/cheklist.htm.
12. McDonnell, G. and Hansen J. (eds.). 2020. *Block's Disinfection, Sterilization, and Preservation*, 6th edition. Philadelphia: Wolters Kluwer.
13. Cockshutt, J. 2003. Principles of surgical asepsis. In *Textbook of Small Animal Surgery*, 3rd edition, ed. D. Slatter, Chapter 10. Philadelphia: W.B. Saunders.
14. Hobson, H.P. 2003. Surgical facilities and equipment. In *Textbook of Small Animal Surgery*, 3rd edition, ed. D. Slatter, Chapter 13. Philadelphia: W.B. Saunders.
15. Mitchell, S.L. and J. Berg. 2003. Sterilization. In *Textbook of Small Animal Surgery*, 3rd edition, ed. D. Slatter, Chapter 11. Philadelphia: W.B. Saunders.
16. Shmon, C. 2003. Assessment and preparation of the surgical patient and operating team. In *Textbook of Small Animal Surgery*, 3rd edition, ed. D. Slatter. Philadelphia: W.B. Saunders.

31

Assessment of Veterinary Care

Carolyn M. Doerning and Bernard J. Doerning

Introduction

While any animal may experience spontaneous illness, animals used in research or educational programs are at risk for possible husbandry related problems, potential adverse events due to specific procedures, and genetic manipulations that may result in susceptibility to diseases or specific deleterious phenotypes. For these reasons, the availability of adequate veterinary care is essential to an institutional animal care and use program. Specific care and treatment of all animals should always be guided by the professional judgment of an appropriately qualified veterinarian.

The overall goal of the program of veterinary care is to ensure that the animals remain healthy and are used humanely. In this regard, the Attending Veterinarian (AV) has a responsibility to the animals to ensure animal well-being, to the institution to promote regulatory compliance, and to the investigator to facilitate research. Of these, the well-being of the animals is paramount.

A challenge for the IACUC is to identify means by which to evaluate the scope and adequacy of the program of veterinary care. The IACUC should recognize that the implementation of a sound program of veterinary care relies on the professional guidance and judgment of the AV and veterinary staff who have training and experience in laboratory animal medicine, or the species being attended. Useful parameters for the IACUC to consider include the following:

1. The provisions for preventive, clinical and emergency veterinary care
2. The completeness and accuracy of medical records that allow for retrospective evaluation of the program
3. The training and number of staff available to assist the AV or investigators in delivery of acceptable care
4. The mechanism for daily observation of the animals and reporting of animal health concerns and the processes put in place by the AV and their staff to respond to such information

31:1 What is meant by adequate veterinary care?

Reg. Adequate veterinary care is an essential part of every animal care program. The Animal Welfare Act Regulations (AWAR) (§2.33, a) state: "Each [registered] research facility shall have an attending veterinarian who shall provide adequate veterinary care to its animals."[1] The PHS Policy (IV, C,1, e) requires the IACUC to assure that "medical care for animals will be available and provided as necessary by a qualified veterinarian."[2] Specific descriptions of proper veterinary care are not contained in the PHS Policy; however, the *Guide for the Care and Use of Laboratory Animals* (*Guide*), which is used as a basis for evaluation by the PHS Policy, lists the following as components of adequate veterinary care (p. 105)[3]:

- Animal procurement and transportation
- Preventive medicine
- Clinical disease, disability, or related health issues
- Protocol-associated disease, disability, and other sequelae
- Surgery
- Pain and distress
- Anesthesia and analgesia
- Euthanasia

In addition to indicating that the program of veterinary care is the responsibility of the AV, the *Guide* (p. 14) endorses the American College of Laboratory Animal Medicine's (ACLAM) position statement on *Adequate Veterinary Care*.[3,4] This position statement requires the veterinarian to be provided with appropriate authority and resources to ensure "the adequacy of veterinary care and animal well-being during all phases of an animal's life."[4]

Opin. *Adequate veterinary care* implies a minimal acceptable standard; however, an AV, by virtue of specialty training and/or experience, should strive to develop their institution's program of veterinary care to a level that exceeds this minimum. General aspects of adequate veterinary care encompass three overlapping areas: animal husbandry, animal health, and study coordination.

Animal husbandry is much more than cleaning cages and feeding the animals. Prevention of disease and injury is a key objective of the animal husbandry program. The following aspects

DOI: 10.1201/9781003272861-31

of animal husbandry are areas of concern as they relate to veterinary care:

Animal procurement: Potential vendors should be carefully evaluated because newly acquired animals can introduce disease into established colonies. A health surveillance program to screen or monitor incoming animals and/or potential vendors is key to selecting appropriate sources of animals. All animals should be observed upon arrival at the facility for signs of illness.

Acclimation and quarantine: In general, animals should be allowed a period of acclimation and physiologic stabilization after arrival at the facility to recover from shipping stress and adapt to their new surroundings. Animals known or suspected of being from contaminated sources should be quarantined in such a way that risk to other animals is minimized. Quarantine often involves special precautions, including restriction of personnel access, maintenance of air pressure differentials between quarantine and other facility areas, strict disinfection procedures, and diagnostic testing to determine pathogen status.

Separation of species and sources: Different species and animals from different sources are best maintained in separate housing units. Separating animals by species is useful to reduce anxiety due to interspecies conflict and to minimize the possibility of interspecies disease transmission (pp. 11–12), and in some cases is required (e.g., rabbits under §3.58). Similarly, animals from different sources should be separated to control possible transmission of infectious diseases between groups. Separation can be achieved by housing animals in separate rooms, in various forms of isolators, or in specifically designed individually ventilated cage (IVC) systems.

Daily animal care: All animals should be housed in cages of appropriate size and style for the species and cleaned and maintained in accordance with the standards of the AWAR and the *Guide*. In addition, it is a regulatory imperative that all animals are observed every day to ensure that their health and living conditions remain acceptable. A Standard Operating Procedure (SOP) describing methods and equipment to be used in animal care is often useful to maintain consistency across all areas of the facility. A written or electronic check-off documenting animal husbandry activity provides a permanent record that can be used to evaluate the consistency of animal care over time.

Animal health is intricately related to veterinary care. Ill or injured animals must receive prompt medical attention. Important issues to address include the following:

Observation: All animals must be observed daily (*Guide*, p. 74; AWAR §2.33, b, 3) and any unexpected deaths and deviations from normal behavior or appearance must be reported to the attending veterinarian. Sick or injured animals require a plan for alerting the veterinary staff to the need for prompt attention.[1,3]

Reporting of animal health problems: Many different mechanisms exist to report animals in need of veterinary attention. Whether it is electronic recordkeeping, cage cards, or "animal health forms," a process must be defined to capture and record progress of care (see 31:29). Documentation of findings by the veterinarian and/or trained animal care staff, treatment recommendations, and the progression of the animal's condition is often a legal requirement and an effective way to assist the IACUC in evaluation of the adequacy of veterinary care.

Anesthesia and analgesia: According to the AWAR (§2.31, d, 1, iv, A), "Procedures that may cause more than momentary or slight pain or distress to the animals will be performed with appropriate sedatives, analgesics or anesthetics, unless withholding such agents is justified for scientific reasons."[1] The PHS Policy (IV, C, 1, b) is essentially the same as the AWAR.[2] Similarly, the *Guide* (p. 120) states: "An integral component of veterinary medical care is prevention or alleviation of pain associated with procedural or surgical protocols."[3] It is the duty of the veterinarian to advise the investigator on proper use of these drugs and to ensure that animals are being adequately treated. The use of anesthetics and analgesics should be carefully documented in writing. The IACUC may find evaluation of such records to be useful in confirming adequate veterinary care.

Survival surgery: A major responsibility of the AV is to assure that survival surgical procedures are conducted in proper facilities by trained individuals using acceptable methods. Again, records of procedures performed and documentation of qualifications of personnel can be used by the IACUC to evaluate this aspect of veterinary care.

Peri-operative and post-procedural care: Additional care and attention are often critical to the successful outcome of surgical or other invasive procedures that involve anesthesia. For example, intraoperative hemodynamic and body temperature maintenance; care immediately after surgery, including administration of supportive fluids, analgesics, and other drugs; monitoring of vital signs such as temperature, pulse, respiration, and pain scores; provision of supplemental heat to prevent hypothermia; and continuous monitoring of the animal until it is awake and ambulatory. Longer-term care might include administration of antibiotics or other supportive care, monitoring of the surgical site for signs of dehiscence or infection, and overall continued attention to the general medical needs of the animal. For all peri-operative and post-procedural care, it is important that accurate records be maintained for retrospective evaluation if complications arise. This record also allows the IACUC to evaluate this aspect of veterinary care. The IACUC should also be aware of the requirements set forth by the AWAR for procedural care (§2.31, d, 1, ix; §2.33, b, 5).[1]

Pre-study veterinary consultation. The AWAR require the IACUC to ensure that the investigator has consulted with the AV during the planning of research proposals that may result in more than momentary pain or distress to the animal (§2.31, d, 1, iv, B). The AWAR (§2.33, b, 4) state that the program of veterinary care must include "guidance to principal investigators and other personnel involved in the care and use of animals regarding handling, immobilization, anesthesia, analgesia, tranquilization, and euthanasia."[1]

Training and assessment of skills: Adequate veterinary care is greatly augmented by training and evaluation of skills for anyone planning to conduct animal research, and appropriate qualifications and training are required as per the PHS Policy (IV, C, 1, f).[2] As part of the program of veterinary care, the AV has an integral role in developing and evaluating relevancy and accuracy of the facility's training program curriculum. The ongoing oversight and evaluation of the training program's efficacy is a core responsibility of the IACUC (*Guide*, p. 15, §2.31, d, 1, viii).[3]

31:2 Must the AV be a laboratory animal specialist?

Reg. The AWAR (§2.31, b, 3, i) and the PHS Policy (IV, 3, b, 1) indicate that the IACUC must include at least one member who is a Doctor of Veterinary Medicine with training or experience in laboratory animal science and medicine who has direct or delegated responsibility for activities involving animals at the research facility.[1,2] The AWAR further require that the AV be a voting member of the IACUC, but facilities with more than one veterinarian may appoint another veterinarian with delegated program responsibility for activities involving animals to the IACUC (§2.33, a, 3). The AWAR (§1.1) also mandate that the AV be a person who has graduated from a veterinary school accredited by the American Veterinary Medical Association's Council on Education, or has a certificate issued by the American Veterinary Medical Association's Education Commission for Foreign Veterinary Graduates, or has received equivalent formal education as determined by the USDA.[1] Furthermore, the AV must have received training and/or experience in the care and management of the species being attended. The PHS Policy (IV, C, 1, e) states that the IACUC must determine that "medical care for animals will be available and provided as necessary by a qualified veterinarian," although the Policy does not specifically define *qualified*.[2]

The *Guide* (p. 106) indicates that the AV must be an individual who has a veterinary degree and has training or experience in laboratory animal science and medicine. However, the *Guide* does not specifically require the AV to serve as the veterinary member of the IACUC.[3]

Opin. The answer to this question is yes, if a *specialist* is defined as someone who has specialized training or experience. Individuals may have several areas of specialization, including laboratory animal science and medicine. The American Veterinary Medical Association recognizes laboratory animal medicine as a specialty practice area of veterinary medicine.

Completion of an internship at a laboratory animal facility as part of the veterinary professional curriculum is one means by which individuals might gain minimal experience in laboratory animal medicine. Individuals may become specialists as a result of on-the-job experience under the supervision of an experienced AV. Some individuals choose to complete a formal residency in laboratory animal medicine at a training program recognized by ACLAM. Such residencies allow for concentrated study and experience in the specialty. An individual who has completed an approved residency or has approved practical experience, has an approved publication in the field of laboratory animal medicine, and has successfully passed the appropriate examination is eligible for certification as an ACLAM Diplomate. Diplomate status within ACLAM is widely accepted as evidence of significant training and experience in laboratory animal medicine.

Membership and involvement in relevant professional organizations can be interpreted as evidence of specialized interest in laboratory animal science and medicine. For example, activity in the American Association for Laboratory Animal Science (AALAS) or the American Society of Laboratory Animal Practitioners (ASLAP) can contribute to the professional knowledge and continuing education of individuals active in the profession.

A veterinary practitioner who lacks specific training in laboratory animal medicine might still ably fill the role of AV. For example, if the species maintained at the facility are those with which the practitioner routinely works, their experience would be appropriate. In instances when a veterinary practitioner serves as a consulting veterinarian for a research facility, it is important for that individual to supplement their knowledge base where needed. This might be done by enrollment in professional associations such as AALAS or ASLAP, attendance at professional meetings that focus on laboratory animal medicine, completion of appropriate continuing education, or access to appropriate texts or other reference material for the veterinarian to expand their knowledge base.

Surv. In general, who makes immediate decisions regarding the veterinary care provided to animals at your institution?

- The Attending Veterinarian or veterinarian designee for all species 152/181 (84%)
- The Attending Veterinarian or veterinarian designee for USDA-covered species, but someone else for rodents or aquatics (e.g., a veterinary technician) 8/181 (4%)
- Veterinary technicians for all species 7/181 (4%)
- Animal care technicians for all species 5/181 (3%)
- Principal investigator or study director has the final say 3/181 (6%)
- Other* 6/181 (3%)

*Other: Noted the use of a veterinarian-approved rubric to assist technical personnel who are authorized to make such decisions.

31:3 Is it necessary for all organizations housing or using laboratory animals to have an AV?

Reg. Any institution that is under the purview of AAALAC, the AWAR, or the PHS Policy is required to have an AV (see 31:1). Other organizations that fall outside the purview of those standards may still choose to employ an AV on a full-time or consulting basis.[1–3]

Opin. The AV typically has several functions within the institution (AWAR §2.33, a; and §2.33, b; *Guide*, pp. 14, 27, 30, 34). These functions may be delegated to others in the institution, but the AV maintains regulatory oversight responsibility for their implementation. For example, it is the duty of the AV to ensure that the husbandry of the animals is appropriate for the species. The AV is responsible for overseeing the development and management of programs for preventive medicine to minimize the likelihood of animals' developing clinical illness or injury. Furthermore, the AV directs the diagnosis and treatment of clinical disease affecting the animals. In addition, the AV typically advises investigators and the IACUC on methods to relieve or alleviate pain and distress, peri-operative care, and surgical procedures. The AV must also be familiar with applicable state veterinary acts, inclusive of licensure requirements, particularly in the discharging of certain official duties such as signing interstate health certificates or verifying rabies vaccination or tuberculosis status of animals. In short, the role of the AV is to ensure that animals remain healthy and are used in a judicious and humane manner.

Several types of facilities are not required under the existing regulations to have an AV. For example, institutions that use only rats of the genus *Rattus* and mice of the genus *Mus*, bred specifically for use in research or teaching, are not required to have an AV if they lack AAALAC accreditation and do not receive PHS or other federal funds for research or teaching using animals. Such facilities should still develop a written plan for animal health in the event of unexpected illness or injury. In addition, if animals are to be used in surgical or other invasive procedures, it is advisable to identify an individual who can serve as the AV.

31:4 Under what circumstances would a full-time versus a part-time versus a consultant veterinarian be adequate?

Reg. The AWAR (§2.33, a, 1) require that institutions using a consultant veterinarian have a written program of veterinary care, that the AV makes regularly scheduled visits to the facility, and appropriate care, including emergency care, is available.[1] The PHS Policy in IV.A.1.c notes that the Assurance should state the qualifications, authority, and responsibility of the veterinarian(s) who will participate in the program and the percent of time each will contribute to the program. OLAW does acknowledge the concept of a full-time versus part-time veterinarian. The Assurance requires a backup veterinarian to ensure adequate care.[2] PHS Policy (IV, A, 3, b) and the AWAR (§2.33, a, 1) also require IACUC membership to include a veterinarian with direct or delegated program authority and responsibility.

Opin. PHS Policy (IV, C, 1, e) implies that veterinary staffing must be sufficient to provide appropriate medical care whenever the need arises. In this regard, the IACUC is charged with confirming that the level of veterinary service is adequate for the institution's research project(s) and, when applicable, the PHS Assurance.

The need for the services of the AV or other ancillary staff varies extensively with the size and scope of animal research–related operations of each institution. In this sense, it is not possible to outline specific standards for the need of a full-time or other veterinarian.

ACLAM requires that a formal arrangement for veterinary care exists in all instances, while recognizing that the specific arrangement may vary with the number and species of animals involved and the nature of the experimentation at the institution.[4]

Institutions where research involving survival surgery or other invasive procedures are performed on a routine basis should seek greater veterinary involvement than institutions where this does not occur. Similarly, other conditions that merit increased veterinary involvement are shown in Table 31.1.

The IACUC should consider the factors listed in Table 31.1 and determine whether the available veterinary staff can provide the necessary service

TABLE 31.1

Factors That Increase Need for Veterinary Services

Animal-Related Factors	Research-Related Factors
• Large number of animals in the colony • Use of phylogenetically higher mammals such as dogs, cats, and nonhuman primates • Use of immunodeficient animals • Use of animal models requiring specialized husbandry or medical care	• Use of biohazards • Use of chemical hazards • Use of radiologic hazards • Procedures requiring the use of anesthetics or analgesics, particularly survival surgery • Survival surgical procedures involving extensive peri-operative care • Anticipated pain or distress

and oversight. Each facility has its own unique characteristics and needs. Some small facilities may need a full-time veterinarian, while some larger ones may require only a part-time or consulting veterinarian. The authors' experience has shown that institutions characterized by three or more factors listed can generally justify the need for a full-time veterinarian. Facilities that are characterized by few or none of the listed factors may find a part-time veterinarian (employed by the institution) or even a consultant veterinarian (one who provides veterinary care on a fee-for-service basis to the institution) to be sufficient.

Veterinary support in addition to the AV may be needed as the number of relevant factors increase, particularly the animal census or the number of complex research studies. Such support can be in the form of additional veterinarians or, if appropriate, ancillary veterinary staff such as veterinary technicians or trained animal care technicians (see 31:9). Veterinary activities performed by technicians should be conducted under the direction of a qualified veterinarian. Communication between the veterinarian and technicians should occur regularly to ensure that the veterinarian is aware of any animal health or care issues and can advise the technical staff accordingly (AWAR §2.33, b, 3.; *Guide*, pp. 106, 114).[1,3] Under any arrangement, veterinary services must be available at all times, including evenings, weekends, and holidays (AWAR §2.33, b, 2; and see 31:6). For Assured institutions with a single veterinarian, OLAW requires the institution to also provide provision for a backup veterinarian.[2] It is important that methods for contacting the veterinarian are up to date and communicated to all research and animal care personnel.

31:5 If a veterinary consultant is used when the AV is unavailable, what criteria might the IACUC use to assess the adequacy of their credentials?

Reg. As discussed in section 31:2, the AWAR, the PHS Policy, and the *Guide* describe qualifications for the AV. The requirements/qualifications discussed for the AV are applicable to the backup vet, as noted in the PHS Policy (IV.A.1.c). The AWAR (§2.31, d, 1 vii.) and the PHS Policy (IV, C, 1, e) state that medical care for animals shall be available and provided by a qualified veterinarian; "qualified."[1–3] OLAW utilizes the AVMA definition of veterinarian for all veterinarians working independently within an institution's animal care and use program.

Opin. The AWAR require that each facility have an attending veterinarian who provides adequate veterinary care, and that all personnel are qualified to perform their duties. Provided that the facility has an AV who satisfies the regulatory requirements of an AV, veterinarians who do not meet those requirements should provide veterinary care under the supervision or at the direction of the AV. When evaluating qualifications of veterinary consultants, it would be reasonable for the IACUC to use the criteria described in Section 31:2, which define a qualified AV as a reference point. The professional judgment of the AV, as the person with the regulatory responsibility for providing adequate veterinary care, should be deferred to in this evaluation process. A consultant should have at least some familiarity with laboratory animal species in terms of handling and common medical issues, and the IACUC should seek to assess the consulting veterinarian's background for evidence of this experience. More importantly, consultants should understand what resources are available for the prevention, diagnosis, and treatment of disease in laboratory animal species. In this regard, the IACUC should ensure that such resources are familiar to the consulting veterinarian and are readily available. This might mean that the institution provides funds or access to relevant texts, online resources, professional organizations, and/or continuing education opportunities. Importantly, the IACUC should evaluate the consulting veterinarian's availability and accessibility. In other words, does the consultant have sufficient time away from other commitments to be able to respond to concerns or emergencies involving animals at the facility? Are they available via in-person or virtual means to quickly respond or advise on the needs of ill animals? Though there are no specific criteria available for answering these questions, the IACUC should formulate a plan to assure knowledgeable and prompt veterinary care and consultation can be provided when needed.

31:6 If a consultant is used as the AV, how frequent should visits be and how can emergency care be provided if the consultant is not available?

Reg. The frequency of visits to a facility by a consultant can vary significantly with the size and scope of animal use activities (*Guide*, p. 14).[3] The AWAR state that "regularly scheduled" visits shall be made to the facility (§2.33, a, 1).[1] As previously stated, OLAW requires availability of a backup vet. The expectations for this individual's skills are that they are trained in the species being used.

Opin. Visits should be frequent enough to provide the same degree of oversight to each study as would be provided to studies in a larger facility employing full-time veterinary staff. The experience of the authors is that on-site visits that occur less frequently than once a month diminish the effectiveness of the AV if animals are housed continuously at the facility. As the scope and complexity of the facility increase, so should the frequency of on-site visits by the consultant. Visits by the consulting veterinarian should be regularly scheduled and supplemented with additional visits as needed.

The availability of consultants to respond to emergency situations may be limited by time or distance constraints. Nonetheless, emergency veterinary care must be provided (AWAR §2.33, b, 2.; *Guide*, p. 114).[1,3] Trained veterinary or animal care technicians can sometimes provide such care under the direction of the veterinarian through instructions communicated by such means as a virtual meeting, phone call, e-mail, or prior written instructions and guidelines. In contrast, some facilities make prior arrangements with a local practicing veterinarian to provide occasional emergency care on-site. In general, this arrangement meets the OLAW requirement for a backup veterinarian when an Assured institution has a consultant or a single veterinarian in the program. Any veterinarian employed in this capacity should be provided with initial training regarding the species of animals and the type of research at the facility. For example, calling in an inexperienced local veterinarian to perform a complex nonhuman primate surgery could be considered malpractice. Both the veterinarian and the facility should be comfortable with each other's expectations before entering this relationship.

31:7 What staff, other than veterinarians, are useful in providing adequate veterinary care? What level of support staff is considered adequate?

Reg. PHS Policy (IV, A, 3, b, 1) and the AWAR (§2.33, b) require that each research facility shall establish and maintain a program of adequate veterinary care.[1,2]

Opin. Ancillary personnel such as veterinary technicians, research technicians, Principal Investigators (PIs), and trained animal care technicians often provide valuable support to the program of veterinary care (see Section 31:4). Such individuals usually interact closely with the animals on a daily basis and can serve as a primary point for observation of animal health. The IACUC through oversight, program review, and semiannual inspections contribute to the integrity of the institution's standards of veterinary care.

The size of the staff needed to provide adequate veterinary care is completely dependent upon the size and complexity of the animal facility. Unreported or untreated illness or injury, insufficient feed or water, or signs of insufficient husbandry can all indicate inadequate observation of (or response to) the animals and may signal a need for additional staff.

Any animal housed in a research facility must be observed at a minimum of once per day, every day (AWAR §2.33, b, 3.; *Guide*, p. 74).[1,3] This need can increase with the invasiveness or potential for untoward outcomes of the study. It is within the scope of the IACUC's animal welfare oversight to suggest that more frequent observations are integrated into a research protocol.

31:8 What type of training should be provided to animal care and research staff with respect to veterinary care issues?

Reg. The AWAR require training of personnel in the basic needs of each species, proper handling and care for the various species of animals used, proper pre- and post-procedural care, aseptic surgery methods and procedures, and proper use of pain-relieving drugs (§2.32).[1]

Opin. There are numerous resources available to augment in-house training programs. Conferences, webinars and online courses can help personnel, as described in 31:9, that might potentially examine animals exhibiting signs of illness. It is critical for the IACUC to ensure that these individuals are familiar with both normal and abnormal appearances and behaviors of animals (Table 31.2). Training provided or evaluated by the veterinary staff should ensure that personnel are able to recognize an animal in need of veterinary attention and should understand the procedures for notifying the veterinary staff if an animal is even suspected of being ill. Though a clear mechanism to report sick or injured animals to the AV must exist, the rigor of reporting may vary with the nature of the specific study; thus, a dermal abrasion in an otherwise normal animal should be called to the attention of the veterinary staff but does not likely constitute an emergency. For certain animal models, it is important that personnel also learn to recognize medical nuances specific to that model.

TABLE 31.2

Some Indications of Illness in Animals

Body System	Abnormal Clinical Signs
Respiratory system	Nasal discharge, labored breathing, blue tint to mucosal membranes
Gastrointestinal system	Vomiting, diarrhea, thin appearance
Integumentary system	Bleeding or seeping wounds, alopecia, observed persistent scratching, closed wounds
Nervous system	Inability to move normally, overly aggressive behavior, lethargy
Circulatory system	Edema in extremities, ascites, labored breathing

For example, in an animal model of hemophilia it would be important for the staff to understand the urgent nature of even a seemingly minor wound. Similarly, in an animal model of Parkinson's disease, it would be important to instruct personnel that tremor of limbs is an expected outcome.

Some staff may also benefit from training related to the provision of veterinary care. In this regard, it may be useful to train personnel with respect to administration of drugs by various routes, bandaging of surgical wounds, or other common veterinary care procedures. The IACUC should verify the method of documenting proficiency for ancillary personnel expected to provide veterinary care.

31:9 What role do PIs play in the provision of adequate veterinary care?

Opin. Investigators play a significant role in veterinary care. It is critical to recognize that veterinary care is not exclusively the treatment of sick animals. It includes the proper care and use of animals in such a way that animals remain healthy, are treated humanely, and recover normally from experimental manipulations. Beginning with protocol development, an appropriate animal model must be chosen to ensure a project is successful. Potential adverse reactions or outcomes should be identified by the PI and plans made to address them. Indeed, the IACUC should establish a mechanism which allows PIs to easily report unanticipated clinical outcomes related to novel genotypes or procedures performed on animals (see 31:27). Investigators and their technicians will observe animals frequently during the procedure and are, therefore, in an excellent position to report potential problems to the veterinary staff. For example, weight loss in a cage of group-housed animals may not be apparent to an intermittent observer, whereas review of the investigator's records of daily individual body weights of animals could detect the problem.

Depending on the staffing of the facility, PIs or their research staff may be responsible for aspects of husbandry or administering treatments prescribed by the veterinarian. As part of this, they should be instructed on procedures to contact the veterinary staff when animals are found in poor clinical condition.

For these reasons, it is critical to view the role of the PI and research staff, along with their prior experience and training with the respective species and procedures, as providers or facilitators of veterinary care. Although the AV is ultimately responsible for provision of proper veterinary care, the IACUC should recognize that there are others who have an ethical, if not a regulatory, responsibility to impact the health and welfare of research animals positively.

31:10 What aspects of a veterinary care program can and should be addressed under Standard Operating Procedures (SOPs)?

Reg. The AWAR require a written program of veterinary care if the AV is a consultant/part-time employee (2.33,a,1).[1] The PHS Policy (IV, A, 1) requires that the Assurance statement fully describe the institution's program for animal care and use, using the *Guide* as a basis for the development of that program.[2] The *Guide* (p.12) encourages the establishment and periodic review of written procedures that ensure consistent application of *Guide* standards, one of which is the provision for adequate veterinary care.[3]

Opin. It is worthwhile for institutions to have a written description of the veterinary care program which can be reviewed by the IACUC during the semiannual program review. While the contents of this document should be customized for the facility, it should address regulatory standards such as daily assessment of animal health and well-being; prevention, control, diagnosis, and treatment of animal disease and injury; training of veterinary and research personnel on proper handling, restraint, anesthesia, analgesia, and euthanasia methods; training of research personnel in appropriate surgical techniques and procedures; emergency preparedness; and guidelines for monitoring surgical procedures and providing peri-operative care. Additionally, a facility may utilize or augment their program of veterinary care with documents typically referred to as SOPs. In general terms, the function of an SOP is to provide written guidance in sufficient detail so that a person with the appropriate education, training, and experience can perform the procedure described in the SOP. Research facilities that wish to conduct animal studies under Good Laboratory Practices (GLP) conditions are required by the Food and Drug Administration to maintain SOPs (21 CFR Part 58), specifically, Section 58.81 requiring the following.[5]

(a) A testing facility shall have standard operating procedures in writing setting forth nonclinical laboratory study methods that management is satisfied are adequate to ensure the quality and integrity of the data generated in the course of a study.

(b) Standard operating procedures shall be established for, but not limited to, the following:

1. Animal room preparation
2. Animal care
3. Receipt, identification, storage, handling, mixing, and method of sampling of the test and control articles
4. Test system observations
5. Laboratory tests
6. Handling of animals found moribund or dead during study
7. Necropsy of animals or postmortem examination of animals
8. Collection and identification of specimens
9. Histopathology
10. Data handling, storage, and retrieval
11. Maintenance and calibration of equipment
12. Transfer, proper placement, and identification of animals

Non-GLP facilities often operate under SOPs as well. Alternatively, and sometimes additionally, both GLP and non-GLP facilities may develop guidelines or policies. These are documents that help define the conduct of procedures or situations that may have been a source of controversy or earlier deliberation for which the IACUC has reached a consensus. For example, the IACUC might approve a policy to allow with certain restrictions the controversial practice of chronic housing of rodents in wire bottom cages.

Surv. For what types of protocols do veterinarians pre-review/review protocols? Select all that apply.

• No review of protocols by veterinarians at my institution	8/180 (4%)
• All protocols	147/180 (82%)
• All surgical protocols	15/180 (8%)
• All protocols with pain or distress	19/180 (11%)
• All USDA-species protocols	9/180 (5%)
• All protocols with federal funding	4/180 (2%)
• Other*	11/180 (6%)

*Other:

- Any protocols for which the PI requests such review
- All new protocols
- Any protocols in which animals are used in procedures that would involve USDA category D or E procedures
- Only when time permits
- All technical procedures
- As determined by the IACUC Chair

Surv. Does your institution authorize the Attending Veterinarian to use the Veterinary Verification and Consultation process to authorize certain protocol changes in lieu of the approval of a significant modification by the IACUC?

• Yes, routinely	91/181 (50%)
• Yes, but only in a clinical emergency	38/181 (21%)
• No	52/181 (29%)

31:11 How often should you evaluate equipment and pharmaceuticals considered essential items for an adequate program of veterinary care?

Reg. The AWAR (§2.33, b, 1-2) state that, among other considerations, appropriate facilities and equipment must be available to prevent, control, diagnose, and treat diseases and injuries.[1] Furthermore, the AWAR require the IACUC to review the research facility's program for humane care and use of animal and inspect all of the animal facilities at least once every six months (§2.31, c, 1 and §2.3, c, 2). PHS Policy also requires evaluation of the institutions' program and inspection of all institutional animal facilities (including satellite facilities) at least once every six months, in accordance with IV.B.1. and 2. and reports prepared in accordance with IV.B.3. The *Guide* (p. 105) describes components of an effective program of veterinary care to include preventive medicine; surveillance, diagnosis, treatment, and control of disease; management of protocol-associated disease or disability; anesthesia and analgesia; surgery and post-surgical care; assessment of well-being; and euthanasia.[3] Thus, these documents do not specify the essential items required for adequate veterinary care and allow the AV and the institution to decide which items are essential to achieve the programmatic objectives described.

Opin. The specific items needed to provide adequate veterinary care will vary widely with the species housed within the facility and the type of research being conducted. For example, a program in which specific pathogen-free rodents are used and euthanized for collection of tissues within several days of arrival will have a much different need for specialized equipment and supplies than a program in which nonhuman primates undergo survival surgery and are maintained for several years afterward. In general, the facility should rely upon the advice of the AV in deciding which equipment and supplies are necessary to the needs of the animals and the research in a way that meets the objectives stated in the AWAR and the *Guide*. Most veterinary care programs find that routine supplies include items such as topical or injectable antibiotics, bandage materials, materials for euthanasia and necropsy, material and equipment for surgical repair of wounds in an aseptic manner, supplies for microscopic examination of samples, sterile fluids, antiseptics, syringes and needles, and means to provide supplemental warmth, such as an incubator or circulating water heating pad. These items are only examples of the kinds of supplies that would be useful to address the typical clinical needs of animals in a research facility. The list is by no means complete and each veterinary care program should have supplies to meet its specific needs. For example, radiographic and other imaging equipment is essential to some programs, but not for others. Though availability of needed supplies is essential, the veterinary care program need not necessarily have all equipment and supplies on hand but rather only have ready access to such resources. Adequate veterinary care is facilitated to a far greater extent when the veterinary care program has direct ownership of these resources.

31:12 If a drug not approved in the protocol is used in an animal for experimental purposes, should the IACUC regard this as an item of non-compliance? Should it be reported to OLAW?

Reg. The AWAR (§2.31, e, 3) require a complete description of the use of the animal. The description is incomplete if it does not include all experimental drugs.[1] The AWAR (§2.31, d, 4.; §2.31, e, 1–5) require IACUC approval of significant changes in the use of animals. Adding a drug is a significant change and would be considered a non-compliance if done without IACUC approval.[1] Both the AWAR (§2.31, c, 6) and the PHS Policy (IV, B, 6) indicate that the IACUC is expected to review and approve activities related to animal care and use.[1,2] The *Guide* (pp. 25–26) indicates that an animal use protocol should provide a clear and concise sequential description of procedures involving the use of animals; information on the impact of the proposed procedures on the animals' well-being; and consideration of the use of hazardous materials and provision of a safe working environment.[3] The AWAR (§2.31, d) similarly describe the need for consideration for procedures that might result in pain or distress to animals.[1]

Opin. Failure to specify a drug in the protocol limits the ability of the IACUC to determine if these provisions of the *Guide*, the PHS Policy, and the AWAR have been met; thus, this situation would be considered an item of non-compliance.

NIH/OLAW states[6] that failure to adhere to an IACUC-approved protocol is one example of a reportable situation. If a drug was used to provide veterinary clinical care to an individual animal, such use would not be reportable. In contrast, it would be regarded as a significant, reportable change if a drug not indicated in the approved protocol is used for experimental purposes. If changes in drugs are made by a veterinarian for humane reasons during a study, such use can be initiated through an institutional veterinary verification and consultation (VVC) policy.

31:13 Is it necessary to use only pharmaceutical quality drugs when they are available for animal use?

Reg. The *Guide* (p. 31) states that pharmaceutical-grade chemicals or substances should be used, when available, for all animal-related procedures, and that use of non-pharmaceutical grade compounds should be approved by the IACUC as part of a protocol.[3] NIH/OLAW is in agreement on this point and emphasizes that the use of pharmaceutical-grade compounds is expected and will help avoid toxicity or side effects that might result from use of non-pharmaceutical grade materials.[7]

Opin. NIH/OLAW defines the use of a pharmaceutical-grade compound as a drug that is approved by the U.S. Food and Drug Administration for which a chemical purity standard has been established by the United States Pharmacopeia-National Formulary or British Pharmacopeia. For many test compounds, pharmaceutical-grade material is not available, thereby requiring, by default, the use of non-pharmaceutical-grade compound. The *Guide* (p. 31) and NIH/OLAW recognize that there may be circumstances in which pharmaceutical-grade materials may not be available for scientific purposes.[2,7] In such cases, use of non-pharmaceutical-grade compounds may be approved for use in a protocol if scientifically justified and approved by the IACUC. Use of such agents is not automatic and cannot be justified for reasons of convenience or cost. When a specific compound is not available as a pharmaceutical-grade material, the PI might also consider using other, similar compounds available as pharmaceutical-grade material if such use would not be expected to alter the scientific integrity of the study. NIH/OLAW states that when non-pharmaceutical grade materials are used the IACUC is expected to evaluate the possibility of adverse events by considering such factors as grade, purity, sterility, acid-base balance, pyrogenicity, osmolality, stability, site and route of administration, compatibility of components, side effects and adverse reactions, storage, and pharmacokinetics.[7] In this regard, it would seem reasonable that the IACUC might develop a plan for monitoring animals for adverse events or side effects when non-pharmaceutical grade materials are used.

Surv. Under what conditions would your IACUC approve non-pharmaceutical quality drugs to be used in animals? Select all that apply.

- Under no circumstances 10/181 (6%)
- When pharmaceutical quality drugs are not available 132/181 (73%)
- When scientific need requires the use of non-pharmaceutical quality drugs 151/181 (83%)
- No specific justification is required by our IACUC 6/181 (3%)
- Other* 10/181 (6%)

*Other:

- When a pharmaceutical quality drug is not available in the specific formulation that is needed.
- When compounding, preparation, storage, and disposal information is provided.
- Novel test compounds.

31:14 Can expired drugs, biologics, or supplies be used for terminal procedures?

Reg. NIH/OLAW states that use of expired materials is not consistent with adequate veterinary care, and that expired euthanasia agents, anesthetics, and analgesics should not be used for terminal procedures.[8] Allowance for use of expired materials is given for circumstances in which the manufacturer verifies efficacy of the material beyond the expiration date or if the investigator can convincingly demonstrate to the IACUC that use of expired materials would neither harm animal welfare nor compromise the scientific integrity of the study. The AWAR and *Guide* do not specifically address the use of expired materials in animals; however, it is stated that surgical activities be performed in accordance with established veterinary medical practices (2.31, d, 1, x) and that institutions must provide a veterinary program that offers a high quality of care (p. 105).[1,3]

Opin. It is widely accepted that the use of expired materials in survival procedures is inconsistent with proper veterinary care. Though some institutions have allowed use of expired materials only in nonsurvival procedures, it is clear that both APHIS/AC and NIH/OLAW consider this to be acceptable only when such materials do not include those that would be used for pain relief (anesthetics or analgesics), euthanasia, or physiologic stability (emergency drugs). Although these agencies allow the possibility to use such expired materials, it would be under only a very unusual circumstance in which an investigator could justify to the IACUC that the materials could not possibly pose any threat to animal well-being or to the scientific integrity of the study. Doing so would represent a substantial task, as the investigator might be expected to demonstrate that no physiological stress would occur even in the anesthetized animal. The *Guide* supports the positions of NIH/OLAW and APHIS, though in less specific terms.

The IACUC must consider whether using expired materials is consistent with high quality of care. As described by NIH/OLAW, appropriate documentation from the manufacturer and/or investigator should demonstrate efficacy and lack of negative impact to animal welfare or study outcomes to the satisfaction of the IACUC.[8] Without this validation, the use of expired materials is discouraged, as described in 31:16.

Surv. Does your IACUC allow the use of expired drugs, biologics, or supplies?

- No, not in any situation 88/181 (49%)
- Yes, but only for terminal procedures, and not expired anesthetics or analgesics 85/181 (47%)
- Yes, but only for terminal procedures, and including expired anesthetics or analgesics 3/181 (2%)
- Yes, for both survival and terminal procedures, but not expired anesthetics or analgesics 4/181 (2%)
- Yes, for both survival and terminal procedures, and including expired anesthetics or analgesics 1/181 (1%)

31:15 Under what circumstances can the AV permit changes to an approved protocol in lieu of approval by the IACUC?

Reg. The AWAR state that the AV has direct or delegated authority for activities involving animals (§1.1) and that medical care for animals will be available and provided as necessary by a qualified veterinarian (§2.31, d, 1, vii).[1] Further, the AWAR state that the AV shall have authority to ensure the provision of adequate veterinary care (§2.33, a, 2) and that research facilities shall have programs of adequate veterinary care that include the use of appropriate methods to prevent, control, diagnose, and treat diseases and injuries (§2.33, b, 2).[1] However, the AV does not have authority to change experimental procedures that have been approved by the IACUC.

Opin. It is apparent from the AWAR, the PHS Policy, and the *Guide* that the AV is granted latitude to exercise professional judgment in the delivery of adequate veterinary care when there is an immediate clinical need. Modifications to the protocol beyond those circumstances require adherence to the institutional VVC policy. If such a policy does not exist or the request exceeds the scope of VVC, approval of the IACUC would be needed for this change. It is an important distinction to note that although the AV does not have blanket authority to permit protocol changes, they do maintain authority to halt protocol-approved activities resulting in significant animal health, welfare, or compliance concerns until an IACUC can be convened to formally discuss the matter.[9]

Surv. Does your Attending Veterinarian have the authority to approve a change in an anesthetic or analgesic drug without the approval of the IACUC?

- Yes, routinely 64/181 (35%)
- Yes, but only in a clinical emergency 84/181 (46%)
- No 28/181 (15%)
- I don't know 5/181 (3%)

31:16 Are there any circumstances under which an external veterinary specialist (i.e., board-certified ophthalmologist or surgeon) can provide veterinary services for research studies or clinical care? If so, what should be the criteria for IACUC approval and oversight?

Reg. The *Guide* (p. 112) states that methods of disease prevention, diagnosis, and therapy should be those currently accepted in veterinary practice.[3] The AWAR (§2.33, b, 2–3) state that a program of veterinary care includes the availability of appropriate facilities, personnel, and equipment to prevent, control, diagnose, and treat diseases and injuries.[1] Though no specific mention is made of utilizing off-site resources, these passages imply that the importance to animal health offered by certain equipment, methods, and expertise is substantial. One might infer, then, that if the necessary resources are not available at the institution, the use of external resources or expertise is a practical alternative.

Opin. At some institutions, particularly small ones, the expertise or equipment needed to make accurate diagnoses or to provide proper treatment may not exist. For example, a small institution that maintains rabbits producing a valuable polyclonal antibody may not have radiographic equipment or orthopedic expertise to diagnose and treat a rabbit suspected of fracturing its leg. Though euthanasia is an option in such cases, that approach may not represent judicious use of animals if the condition is one that could be reasonably treated so that the animal regains health as well as research usefulness. In such cases, diagnostic and treatment options at other facilities, such as nearby academic institutions, zoos, or private veterinary clinics, are an acceptable means for providing veterinary care if the animal can be humanely and safely transported to such a site. In the authors' experience, it is often more practical from a biosecurity and logistical standpoint for veterinary professionals from external sites to visit the research facility to provide specialized care, whether it be advanced clinical procedures, such as dentistry or orthopedic surgery, or for research-driven needs, such as pre-study ophthalmic examinations.

The IACUC should focus on several oversight aspects when it is determined that the medical needs of an animal are best served by an external practitioner. The IACUC should evaluate the credentials and experience and ensure that any institutional requirements for visiting personnel are met. Regarding the use of off-site resources, it should first be clear that there is benefit to the animal in instances when transportation is required. For example, the need for specialized imaging equipment needed for the diagnostic effort of a nonhuman primate may be evident, but if the animal is moribund and has a poor prognosis, there may be limited advantage to such an effort. Second, the IACUC should be certain that transport regulations are followed, and that transport does not place the animal or personnel at risk. Transport of animals infected with a biohazardous agent, for example, would be unwise. Third, the specific methods of transport, the location and business entity of the off-site facilities, and the overall infrastructure, including expertise, being made available should be understood by the IACUC. Though specific instances of need for use of off-site facilities cannot always be anticipated, the IACUC should expect the veterinary staff to generate a written plan if the possibility exists for use of such facilities. If use of an off-site facility is a routine part of the provision of health care, the facility should be included in the semiannual inspection of facilities. If off-site facilities are to be used only as part of a contingency plan for veterinary care, the IACUC could, instead, evaluate such sites by written description from the AV after their visit to the site.

31:17 Can a facility veterinarian provide treatment of employee pets or other veterinary services (i.e., filling prescriptions)? Is a policy and IACUC oversight required for these activities?

Opin. There is nothing in the AWAR, the PHS Policy, or the *Guide* that prohibits a research facility from providing care to private pets. Because these animals are not research animals, we can only assume that IACUC involvement should center on the program of adequate veterinary care under which the facility would be operating. Even though pet animals may not be covered under regulatory requirements set forth by APHIS, there are a number of other issues that should be addressed, including the following:

- The appropriate state and federal licenses of those providing the care

- Liability to the individual and institution in the event of complications
- The biosecurity and behavioral impacts on the in-house research animals
- A fair and equitable program describing whose pet animals will be treated and which services can or cannot be provided
- The time required for the veterinary staff to provide such care and the possibility that other duties of the veterinary staff might not be completed as a result
- The funding source used to procure the supplies or equipment to provide private pet care, including the billable hours for facility staff
- The legal considerations for the individual and institution to provide for-profit or non-profit services to staff

The IACUC may be asked to monitor issues such as these to provide assurances and safeguards for the research facility.

31:18 Are cage cards an acceptable location for recording veterinary medical information?

Reg. There is no specific regulatory passage in either the AWAR or the PHS Policy that specifically forbids notation of veterinary medical information on cage cards.[1,2] The *Guide* (p. 75) states only that "identification cards should include the source of the animal, the strain or stock, names and locations of the responsible investigators, pertinent dates (e.g., arrival date, birth date, etc.), and protocol number, where applicable."[3] The AWAR specifically require medical records only for dogs (§3.13, b) and marine mammals (§3.110, d), although medical records may be considered an important component of adequate veterinary care.

Opin. Though use of cage cards for recording veterinary medical information is not specifically mentioned in the AWAR, the PHS Policy, or the *Guide*, the amount and scope of information that can be noted on a cage card are substantially limited by the size of the card. As described in 31:21, the expectations of APHIS in terms of information included in a medical record are broad enough to rule out use of a cage card generally for animals under AWAR purview and maintained for extended periods; still, notes made on a cage card in those cases might serve as a useful summary of relevant clinical events that are recorded in greater detail in a more formal clinical record. In contrast, for animals such as rodents or others that will not be maintained for long periods and are generally free of illness, notations made directly on the cage card may be an efficient way to record veterinary medical information; with large colonies of such animals, it can be efficient to have pertinent clinical information available with the animal.

31:19 Who should be allowed access to veterinary care records and where should the records be maintained?

Reg. The *Guide* (p. 76) indicates that medical records for animals should be readily accessible to investigators, veterinary staff, and animal care staff.[3] The AWAR refer more specifically to records pertaining to covered animal activities being available upon request for inspection by authorized APHIS representatives.[1]

Opin. Veterinary care records exist to guide the clinical care of an individual or group of animals. For this reason, such records should be readily accessible to members of the veterinary staff, including veterinary technicians. When animal health records are maintained by the PI, in a research notebook, for example, some mechanism should be established so that the veterinary staff can easily access clinical information. However, it is difficult to ensure seamless flow of information when health records are maintained in research notebooks or research labs. For obvious reasons, it is critical that both the IACUC and representatives from regulatory or accrediting agencies be allowed easy and complete access to veterinary care records; without such access, it is impossible to determine accurately whether adequate veterinary care is being provided. Furthermore, it is important for animal care and research personnel to be provided with access to veterinary care records relevant to animals under their charge. Such personnel often are the initial point at which ill or injured animals are identified, and these same personnel typically are both curious about and have some degree of responsibility to ensure that sick or injured animals are receiving appropriate care and veterinary attention. For all these reasons, it is logical for most veterinary records to be physically maintained within the veterinary services unit.

Surv. Which of the following would be expected to be found on an animal health record of any species at your institution? Select all that apply.

• Identification of the animal	172/181 (95%)
• Medical or surgical treatments that are not research related	133/181 (73%)
• Medical or surgical treatments that are research related	152/181 (84%)
• General clinical observations	145/181 (80%)
• Surgical records	134/181 (74%)
• Anesthesia/analgesia records	135/181 (75%)
• Dates of treatments or observations	166/181 (92%)
• Times of treatments or observations	123/181 (68%)
• Initials or signature of person making the observations	149/181 (82%)

- Enrichment activities 52/181 (29%)
- Exercise times 34/181 (19%)
- Other* 11/181 (6%)

*Other:

- Behavioral or physical testing
- Characteristics of the animal such as age, sex, and strain

Surv. What format does your veterinary care unit use for its medical records?

- Paper medical records 59/181 (33%)
- Electronic medical records 20/181 (11%)
- A combination of paper and electronic medical records 93/181 (52%)
- Other* 8/181 (4%)

*Other: Respondent noted that cage cards are used for recording medical records for rodents.

Surv. At your institution who has access to animal medical records at your institution? Select all that apply.

- The Attending Veterinarian 154/180 (86%)
- The veterinary staff, in general 127/180 (71%)
- Animal care staff (if separate from the veterinary staff) 115/180 (64%)
- The investigator/educator and their research staff 114/180 (63%)
- The IACUC Chairperson 63/180 (35%)
- The IACUC (all committee members) 71/180 (39%)
- IACUC administrators 67/180 (37%)
- The Institutional Official 50/180 (28%)
- Legal counsel 15/180 (8%)
- Any person who requests to see the records 25/180 (14%)
- Other* 21/180 (12%)

*Other:

- Any employee when the need is justified
- Associate Director of Research Compliance and Integrity
- Quality assurance unit

31:20 When should an animal have an individual health record, as opposed to a cage, cohort, or colony record? (see also 31:11; 31:12)

Reg. The AWAR (§2.35, §3.13) cover recordkeeping requirements related to veterinary care.[1] Although the AWAR specifically require medical records only for dogs and marine mammals, there is an expectation that health records sufficiently comprehensive to demonstrate the delivery of adequate health care consistent with established veterinary practice are required for species regulated under the AWAR. Though the regulations stop short of requiring entirely separate records for each animal (other than dogs and marine mammals), it is established veterinary practice for notations for individual treatment of an animal to be made as entries specific to that animal.

Opin. In general, all observed illnesses or abnormalities should be entered into a clinical record; however, an alternate view holds that many animals, such as rodents, can be successfully maintained without formal medical records.[10] Records are frequently more detailed for some species than for others. For example, individual health records for dogs, cats, nonhuman primates, and agricultural animals are useful (*Guide*, p. 75). Regarding recordkeeping for species not specifically addressed in the AWAR, there is an expectation that medical records are maintained to the standard of veterinary care. The ACLAM position statement on Medical Records for Animals Used in Research, Teaching, and Testing provides recommendations for best practices.[10]

A useful rule of thumb is that individual records should be maintained for those animals that receive routine periodic individual health evaluations. For example, individual records must be maintained for dogs, detailing vaccinations, parasite screenings, any treatments administered and annual physical examinations, while routine husbandry records performed on all animals in a group may be kept on a single record. Separate records are not needed for animals that may have periodic evaluation by examination of several representative individuals from the colony. Rather, the health information of such animals can be recorded as a colony health record. An alternative approach is that individual records should be maintained for long-lived large animals (*Guide*, p. 75).[3] In either case, individual records are useful for species such as dogs, cats, nonhuman primates, rabbits, and farm animals, and of less value for animals maintained in colonies such as rodents or fish. In large-scale colony management, a standard practice is when an animal in a colony receives treatment as an individual (e.g., for an injury), a record of that treatment is maintained. Individual surgical records are also useful for animals that have had invasive procedures, regardless of species.

Surv. At your institution, for which species are individual animal health records prepared, even if a health problem has not arisen?

- No maintenance of individual health records 22/180 (12%)
- All vertebrate species at our institution 49/180 (27%)
- All warm-blooded species at our institution 3/180 (2%)

• Only mammalian species at our institution	10/180 (6%)
• Only species covered by the Animal Welfare Act regulations	73/180 (41%)
• Other*	23/180 (13%)

*Other:

- All species for which individuals can be identified
- Any animal which presents with a health issue

Surv. For what length of time does your institution maintain animal health records?

• Until the final disposition of the animal	12/180 (7%)
• For three years after the final disposition of the animal	46/180 (26%)
• For three years after the study is completed	65/180 (36%)
• We never discard animal health records	26/180 (14%)
• Other*	31/180 (17%)

*Other:

- 5 years
- 6 years after the study is completed.
- 7 years after the study is completed.
- Length of time depends on the funding, species, or specific regulatory requirements.
- For the calendar year following completion of the study.

31:21 What records and information should the IACUC review if a PI proposes to create an animal having a novel genotype with an unknown phenotype?

Reg. The *Guide* (p. 28) recognizes that whether genetic manipulation is targeted or random, the resulting phenotype may be unpredictable and can affect the animal's well-being and that the IACUC should be informed when characterization of a novel phenotype reveals a condition that negatively affects an animal's well-being.[3] The PHS Policy includes no specific reference to this issue, though it does state (IV, A, 1) that institutions having an Assurance are required to use the *Guide* as a basis for developing and implementing an institutional program for activities involving animals.[2] The AWAR do not directly address this issue, but general guidance might be taken from a statement (§2.31, d, 1, v) which emphasizes that animals experiencing severe, chronic pain or distress that cannot be alleviated will be euthanized at the end of the procedure or earlier.[1]

Opin. For the IACUC to properly engage its charge as described in the *Guide* and the regulations, it must understand the impact on clinical health and well-being of phenotypes which result from genetic manipulation. In this regard, the IACUC should identify a mechanism by which unexpected outcomes and phenotypes counter to well-being are reported to the IACUC. Such a mechanism might create a process for the investigator to inform the IACUC, perhaps by a standard form, by e-mail notification, or other documented means. This idea might be extended to any unexpected outcome of experimental manipulation of an animal, even those not involving genetic manipulation. Indeed, the *Guide* (p. 28) indicates that genetically modified animals are an example of a model for which increased monitoring for unexpected outcomes could be implemented, suggesting an expectation that unexpected adverse outcomes of any experimental model should be reported to the IACUC.[3]

When an investigator proposes to create an animal having a novel genotype, the IACUC would likely need to review certain information in order to adequately understand the scope of the phenotype effect on animal well-being. For instance, if the genetic alteration is expected to result in a specific disease phenotype, the investigator should indicate so as part of the protocol approval process. Similarly, any steps that will be taken to mitigate distress to affected animals, such as an increased monitoring frequency and euthanasia at defined endpoints, should be described. If the genetic alteration is intended to mimic a naturally occurring aberration of humans or other animals, the IACUC might choose to review relevant literature to assess the likelihood and degree of clinical disease and distress. Further, the effect of alteration on similar genes might be considered by the IACUC in assessment of likely effect. After approval and once the work has begun, the IACUC might utilize a post-approval monitoring (PAM) program to review any reports of animal illness or deaths of the animals, or review colony production records to assess if there is any evidence of fetal or neonatal mortality. The veterinary staff can also conduct PAM by making a clinical assessment of animals with the novel genotype to assess the degree of clinical disability, if any, and effect on animal well-being. Similarly, the IACUC might ask the investigator to provide a summary, written or verbal, of the phenotypic characterization of the animals, with emphasis on traits that could impact animal health and well-being.

31:22 Are there any circumstances under which an investigator may rightfully refuse veterinary care for their research animals?

Reg. The AWAR state that the AV must have appropriate authority to ensure adequate veterinary care (§2.33, a, 2).[1] Similarly, the *Guide* (p. 105) states that a veterinary program that offers a high quality of care and ethical standards must be provided regardless of the number of animals or species maintained.[3] The *Guide* also states that for animals on research protocols, the veterinarian or their designee should make every effort to discuss any problems with the PI or project director to jointly determine the most appropriate course of treatment or action.[3] Because the PHS Policy (IV, A, 1) requires adherence to the AWAR and principles in the *Guide*,[2] the expectations with respect to veterinary care are the same.

Opin. It is unequivocal that the AV has been granted authority by APHIS and the PHS Policy to make decisions regarding the veterinary care provided to animals. Clearly, the AV has the final say in matters related to veterinary care, but it is preferable that agreement between the veterinarian and the investigator be reached. This principle must not be interpreted as endowing the AV with authority to make veterinary decisions that contradict veterinary parameters outlined in an approved protocol. For example, if the IACUC has approved a procedure involving unrelieved pain and distress (supported with scientific justification), the AV does not have the authority to administer treatments to mitigate that pain and distress. In instances in which the animal is in immediate peril, the AV may need to make clinical decisions without consulting the investigator.

It is conceivable that an investigator may choose to euthanize animals rather than pursue the treatment recommended by the veterinarian. Depending upon the clinical circumstances and the nature of the research, such a decision may be practical and reasonable. Similarly, it is within the authority of the veterinarian to euthanize an animal for clinical reasons (*Guide*, p. 114).[3]

31:23 If an animal is clearly suffering, can the IACUC allow a veterinarian to euthanize the animal if the research laboratory personnel cannot be contacted, or if the PI refuses veterinary care?

Reg. The AWAR state that the AV must have the authority to ensure adequate veterinary care and that animals experiencing severe or chronic pain that cannot be relieved will be euthanized.[1] The *Guide* (p. 114) states that adequate veterinary care must be provided to animals and that the veterinary care program is the responsibility of the AV and that the veterinarian must have the authority to treat the animal, remove it from the experiment, institute appropriate measures to relieve severe pain or distress, or euthanize the animal if necessary.[3] Because the PHS Policy (IV, A, 1) requires adherence to the AWAR for AWAR-regulated species and the principles in the *Guide*, the expectations with respect to veterinary care are the same.[2]

Opin. Under the AWAR, the program of veterinary care includes guidance to PIs regarding euthanasia (§2.33, b, 4).[1] Likewise, the *Guide* (p. 105) considers euthanasia to be a component of the veterinary care program for which the AV has responsibility.[3] The *Guide* (p. 14) also endorses the ACLAM Position Statement on *Adequate Veterinary Care*, which requires the AV to have "final authority on decisions regarding animal disposition," and authority and resources to assure that "appropriate methods of handling, restraint and interventional therapeutics are being used prior to euthanasia, and that appropriate delivery of euthanasia is performed when required."[4] It is apparent that the intent of these documents is to assign to the AV authority to euthanize animals that are suffering and for which the IACUC has not approved procedures that involve suffering or death as an endpoint.

Surv. If an animal is clearly suffering, does your IACUC allow a veterinarian to euthanize the animal if the research laboratory personnel cannot be contacted?

- Yes, the veterinarian has that authority. 167/180 (93%)
- No, the veterinarian does not have that authority, they must wait for PI/study director input. 10/180 (6%)
- Other* 3/180 (2%)

*Other: PI may proactively delegate that authority to the veterinarian but is not required to do so.

31:24 How often should notations be made in the animal health records for an animal reported to be ill? Should a medical record require stated frequency of treatment or observation?

Reg. The AWAR require that all regulated animals be observed daily to assess their health and well-being (§2.33, b, 3). A similar rule for daily observation is recommended in the *Guide* (p. 112) with more frequent observations occurring based on professional judgment and risk of illness, injury, or abnormal behavior.[1,3]

Opin. Clinically ill or abnormal animals should be followed until resolution, either through return to clinical normalcy or through euthanasia or determination that the relevant condition is clinically minor with little likelihood of increased severity. Records should document follow-up care and subsequent evaluations, up through resolution or euthanasia.

The exact frequency at which follow-up evaluations should be made varies considerably with the clinical condition and depends upon the sound professional judgment of the veterinarian. Animal suffering, pain, or distress indicates severe conditions that might necessitate evaluation several times daily, while other conditions might be appropriately evaluated once every several weeks. In general, follow-up evaluations should occur and be documented in the records at a frequency such that significant increases in the severity of the condition could reasonably be expected to be detected. For most conditions, this usually involves at least daily evaluation with more frequent evaluations if necessary. For less threatening or improving conditions, weekly or biweekly evaluations are often sufficient.

Repeated instances of ill animals' progressing to a more severe disease status in the absence of corresponding notations in the record may indicate that animals are not being examined frequently enough. Conversely, if notations exist for such animals without adjustments to treatment, that could suggest a decline in the animal's clinical condition is duly noted but appropriate action not being taken be either veterinary or research staff. In either circumstance, the IACUC must further investigate the situation as potentially indicative of inadequate veterinary care.

31:25 How can the IACUC determine that the veterinary response to clinical monitoring and findings is reasonable and adequate?

Opin. Having determined that the program for preventive medicine and health monitoring is properly designed and implemented, the IACUC must further ensure that results of these efforts are interpreted correctly, and that appropriate action is taken when potential problems are identified. To accomplish this, the IACUC must evaluate the response of the veterinary staff, in particular the AV. The semiannual program reviews are a good place to start. How has the program changed in the last six months? Are there more or fewer clinical problems? Are the same issues coming up repeatedly and, if so, is there a plan proposed to address the issue? Is the staff proactive or reactive?

In general, the IACUC should see evidence of a veterinary staff that responds expeditiously to identified problems related to animal health, monitoring, and surveillance. The precise response can vary and can include continued monitoring of animals, isolation of suspect animals, specific treatment of animals, or euthanasia of all affected or exposed animals. Again, the appropriateness of the response depends on the type and species of animals, the nature of the research, and the professional judgment of the AV and the veterinary staff.

Additionally, the IACUC should routinely examine health records, specifically during semiannual inspections, to determine whether the responses of the veterinary staff have been appropriate. If not, the IACUC should inform the AV, encouraging the development of other strategies to resolve the problem. The IACUC must be assured that problems are not being ignored, and that the veterinary program has adequate resources (i.e., equipment, training or staffing levels) to provide prompt and appropriate animal care.

31:26 How can the IACUC ensure that veterinary care records are adequate?

Opin. Records for veterinary care should be evaluated for completeness and for evidence of an adequate response to reported problems (see 31:26). For example, records should clearly identify the specific animals involved, the clinical history, the source of the animal, the investigator, the age and gender of the animal, the date of animal receipt, and the types and dates of experimental or therapeutic procedures that have been performed on the animals (*Guide*, pp. 75–76).[3]

Appropriate diagnostic efforts and prescribed treatments should be documented. In addition, the record should indicate each time the animal is medicated or otherwise treated. The clinical progress of the animal should be noted until the clinical problem has been resolved, or the animal has been euthanized. Evidence that an animal languished until it died, or the root cause of the clinical problem was not addressed, can be considered as indicative of inadequate veterinary care (see 31:33). In addition, proper records should reflect the substance of any discussions with investigators or other staff on ways to minimize problems in the future. Individuals examining animals or providing treatment or other actions should sign or initial their notations.

Records can be considered to be inadequate if the IACUC cannot readily understand the steps taken to identify, report, diagnose, discuss, treat, and continue to evaluate an animal until the humane resolution of the case.

31:27 For what type of studies does the IACUC require direct veterinary involvement (e.g., requiring a veterinarian to check Category E animals or new surgical procedures)? For these studies, what is an appropriate period for this oversight?

Reg. The IACUC is charged with ensuring that "personnel conducting procedures on the species being maintained or studied will be appropriately qualified and trained in those procedures" (AWAR §2.31, d, 1, viii.; PHS Policy IV, C, 1, f).[1,2] The *Guide* (p. 114) states that the veterinarian or their designee should discuss with the investigator any problems so that a course of action may be determined.[3]

Opin. A commonly used system employed by many IACUCs is to require direct veterinary involvement in evaluating pain and in training and documenting the surgical skills of the researchers. It is important to remember that this requirement for training as specified in the AWAR applies to all procedures being conducted on animals. Because of expected pain, potential pain, and the skill required for surgical procedures, an IACUC may impose additional means of oversight for Category E protocols and those requiring new surgical procedures and in this way guarantee that those conducting the procedures are adequately trained and competent. Although direct veterinary involvement is effective, other methods of performing this task should not be overlooked. Experienced in-house investigators or external resources may be more familiar with the specific procedures and better suited for training or qualifying the competency of the person performing the technique. They may also be more attuned to the subtle signs of pain that may occur. The duration of this oversight should be directly related to the skill and competence of the individuals performing the procedures. Some leeway must be given during the training period to those learning the procedures, but, as a rule, individuals learning techniques should have a documented experienced trainer nearby, preferably in the same room, who can provide guidance, suggestions, and assistance as necessary. The role of the veterinarian or their designee should conclude the process by assessing and documenting the appropriate skill levels and reporting this to the IACUC.

Surv. For which of the circumstances described below does your IACUC require direct veterinary involvement? Select all that apply.

- Usually for observing major survival surgical procedures that have not been performed previously at our institution, but on Animal Welfare Act– regulated species only — 43/180 (24%)
- Usually for observing major survival surgical procedures that have not been performed previously at our institution on any vertebrate species — 86/180 (48%)
- Usually for observing procedures performed on Animal Welfare Act–regulated species with unrelieved pain/distress — 32/180 (18%)
- Usually for observing procedures performed on any species with unrelieved pain/distress — 55/180 (31%)
- Only if specifically requested by the IACUC — 84/180 (47%)
- None of the above — 9/180 (5%)
- Other* — 17/180 (9%)

*Other:

- For any newly trained surgeon.
- Upon request by the Principal Investigator.
- The vivarium manager observes new procedures and may contact the veterinarian if needed.

31:28 What is an appropriate organizational relationship among the IACUC office, the veterinary care staff, and the IO?

Reg. The AWAR (§2.31, a) state that the Chief Executive Officer (CEO) of the research facility must appoint the IACUC and the IACUC reports its evaluations to the Institutional Official (IO) (§2.31, c, 3).[1] The *Guide* (p. 13, 25) likewise states that the IACUC reports to "the IO about the status of the program " and further indicates that a shared responsibility relationship should exist between the IO, the AV, and the IACUC for the overall direction of the animal care and use program.[3] NIH/OLAW states that there should be direct and clean lines of responsibility and authority so that the organization is able to respond quickly to situations that may arise; and that key elements include the IO, the IACUC, and the AV.[11] Further, the IO should have authority to allocate resources to maintain the program in accordance with recommendations from the IACUC and the AV; and the IO should clearly define and assign responsibilities for program aspects such as training and occupational health.[3]

Opin. Often it is difficult to distinguish between the roles of the IACUC, veterinary staff, facility operations, and even the IO. Some smaller facilities with limited staffing may require one individual to fill multiple roles. To guard against possible conflict of interest—where those inspecting and assessing the operations are also those in charge of them—the trend in larger facilities has been to create an obvious separation of responsibilities. This system provides for the checks and balances similar to those seen in government. Thinking about these relationships as being analogous to the U.S. federal government may clarify the separate of roles. The executive branch (CEO/IO) appoints and supports the IACUC and facility. The judicial branch (IACUC) interprets and enforces regulations, and the legislative branch (veterinary care and facilities operations) oversees the daily operational needs. The trick then becomes one of effectively maintaining good communications and close working relations. The overall goal of all groups is to provide optimal animal care while facilitating the progress of research. Keeping this shared mission in mind can help maintain a collaborative working relationship with all stakeholders.

31:29 What types of veterinary issues might the IACUC commonly find during the semiannual programmatic review and facility inspection?

Opin. The IACUC is required to review the program of veterinary care to ensure that it is adequate. This means that programs, personnel, and other resources are available for the diagnosis, treatment, control, and prevention of animal disease and injury. Veterinary records detail the delivery of this program and should therefore be reviewed to assess the adequacy of implementation by the veterinary and research staff. Records, particularly those maintained by research staff, are frequent sources of concern as insufficient detail or data entry indicates a lack of attention to the provision of animal health support and/or a lack of attention to proper recordkeeping. If the IACUC believes that veterinary care is inadequate, such a finding should be cited as a deficiency which may be either minor or major depending on the potential impact on the animals. Under such circumstances, it would be reasonable for the IACUC to discuss such concerns with the veterinarian. It is important to note that the membership of the IACUC, if the AV and designees are removed from the deliberation, may not possess the expertise to adequately and fairly evaluate the veterinary care program. In this case, the IACUC might consider enlisting the service of a consultant veterinarian with appropriate expertise managing a program of similar scope to offer an additional opinion with respect to the rigor of veterinary care.

Surv. During the semiannual program review, what types of programmatic veterinary-related findings have you ever identified? Select all that apply.

- No review of the program of veterinary care during semi-annual program review — 10/180 (6%)
- Inadequate veterinary care as provided by our laboratory animal veterinarians — 9/180 (5%)
- Inadequate veterinary care as provided by veterinary technicians — 10/180 (6%)
- Inadequate veterinary care provided by research teams — 30/180 (17%)
- Inadequate medical records kept by veterinary services — 12/180 (7%)
- Inadequate medical records kept by research groups — 67/180 (37%)
- None of the above — 86/180 (48%)
- Other* — 11/180 (6%)

*Other:

- Use of expired anesthetics, analgesics, or surgical prep materials in animals.
- We assess the IACUC's ability to evaluate veterinary care as part of our semiannual program review.
- SOPs relating to communication of animal health issues.

31:30 What veterinary care considerations are required when an experimental protocol requests a deviation from standard practice, such as restriction of food and water ?

Reg. The AWAR (§2.38, k, 1) require research facilities to comply with the standards of care as set forth in Part 3-Standards, but the IACUC may approve exceptions when they are specified and justified in the proposal to conduct the activity.[1] The *Guide* (p. 31) states that animals should be closely monitored to ensure that food and fluid resources meet their nutritional needs, bodyweight should be recorded at least weekly (more frequently if the restrictions are greater), and written records should be maintained for each animal to document daily food and fluid consumption as well as clinical or behavioral changes.[3]

Opin. Food or fluid restriction may be required to achieve the desired results of various research protocols. These protocols should be planned to require the least restriction necessary to achieve the desired scientific results. The IACUC should evaluate the scientific justification for requested deviations from standard husbandry practices in light of alternative methods. For example, positive reinforcement training using a highly desired food reward may provide similar results without restricting an animal's basic nutritional needs. In proposals where such restriction is justified, the IACUC should expect a monitoring plan tailored to the duration and severity of restriction. For restrictions greater than 24 hours or procedures that limit access to food or water to specific time periods, the *Guide* (p. 31) suggests special attention be paid to the level of restriction, potential adverse consequences of regulation, and methods for assessing animal health and well-being, which is generally the role of the veterinarian.[3] Depending on the severity of the restrictions, monitoring of food and fluid intake as well as body weights and condition should be documented.

Surv. In a study in which an animal's food or water is restricted (other than for pre-surgical fasting), which of the following parameters are typically monitored at your institution? Select all that apply.

- Not applicable, we do not have food/fluid restriction studies. — 35/180 (19%)
- The animal's body weight. — 138/180 (77%)
- The animal's hydration status. — 113/180 (63%)
- The animal's behavior. — 117/180 (65%)
- We do not monitor any parameters. — 3/180 (2%)
- Other* — 9/180 (5%)

*Other:

- Body condition score
- Food intake and animal outputs (fecal, urine)
- Grimace scale
- Clinical condition and clinical signs

31:31 What provisions for veterinary care should the IACUC expect in the case of field studies?

Reg. The AWAR (§1.1) define field study as a study conducted on free-living wild animals in their natural habitat and exempts such defined field studies from the regulations; however, this term excludes any study that involves an invasive procedure, harms, or materially alters the behavior of an animal under study.[1] This distinction must not be overlooked. Studies conducted in the field that involve an invasive procedure, harm, or materially alter the behavior of an animal under study fall under IACUC purview to ensure they comply with the AWAR and institutional animal care and use program. The *Guide* (p. 32) advocates that many issues associated with field studies are similar to those for species maintained and used in the laboratory, suggesting that provisions for veterinary care during field studies should be included in the overall program of veterinary care.[3] The AWAR (§2.31, d, 1, ix) do provide for one exception related to procedures conducted in the field: operative procedures conducted at field sites need not be performed in dedicated facilities, but must be performed using aseptic procedures.[1] NIH/OLAW has provided the following guidance on IACUC oversight of field studies: while semiannual IACUC inspections of field study sites are not required and in many circumstances are impractical, IACUCs should be apprised of the circumstances under which studies are conducted so that they can consider risks to personnel and impact on study subjects.[13] This may be partially accomplished by written descriptions, photographs, or videos that document specified aspects of the study site. The IACUC should also ensure that appropriate permits are in place. It is acknowledged that "The role of the attending veterinarian is altered because veterinarians rarely accompany researchers to field sites, especially not on a routine basis to provide veterinary care."[13] When capture, handling, confinement, transportation, anesthesia, euthanasia, or invasive procedures are involved, the IACUC must ensure that proposed studies are in accord with the *Guide* (p. 32). The IACUC must also ensure compliance with the regulations and permit requirements of pertinent local, state, national, and international wildlife regulations." [13]

Opin. The PHS Policy requires a "health program for personnel who work in laboratory animal facilities or have frequent contact with animals" (IV,A,1,f). The *Guide* states: "Each institution must establish and maintain an occupational health and safety program as an essential part of the overall program of animal care and use. The nature of the OHSP will depend on the facility, research activities, hazards, and animal species involved" (*Guide*, pp. 17–23). Investigators preparing for field studies should consult with veterinarians and other subject matter experts in the design of their protocols. While the goal of the veterinary care program should be similar to that of laboratory animal operations, differences in capture, restraint, identification, surgery drug dosages, and other aspects of the work may be significantly different in their implementation. The IACUC should also ensure that the occupational health and safety issues are addressed so that the health and safety conditions of the animals and the persons in the field are not compromised. OLAW provides further guidance for developing an occupational health and safety program in their FAQ section.[14]

REFERENCES

1. Animal Welfare Act Regulations (AWAR) (§2.33(a)). https://www.nal.usda.gov/animal-health-and-welfare/animal-welfare-act.
2. PHS Policy on Humane Care and Use of Laboratory Animals. https://olaw.nih.gov/policies-laws/phs-policy.htm.
3. Guide for the Care and Use of Laboratory Animals (Guide). https://olaw.nih.gov/resources/publications/guide-care-2011.htm.
4. American College of Laboratory Animal Medicine. Position Statements: Adequate Veterinary Care. https://www.aclam.org/media/70f4a175-c820-4ea4-91fe-a71afd07e1d8/87UFqA/ACLAM/About%20Us/Position%20Statements/ACLAM%20Position%20Statement%20-%20Adequate%20Veterinary%20Care-Update%202022%20_FINAL%20JAALAS%20format.pdf. Accessed February 5, 2023.
5. 21 CFR Part 58—Good Laboratory Practice for Nonclinical Laboratory Studies. Food and Drug Administration. Amended 1999.
6. National Institutes of Health, Office of Laboratory Animal Welfare. 2005. Notice NOT-OD-05-034. Guidance on prompt reporting to OLAW under the PHS Policy on Humane Care and Use of Laboratory Animals. http://grants.nih.gov/grants/guide/notice-files/NOT-OD-05-034.html.
7. National Institutes of Health, Office of Laboratory Animal Welfare. 2012. Frequently Asked Questions F.4, May Investigators Use Non-Pharmaceutical-Grade Compounds in Animals? https://olaw.nih.gov/faqs#/guidance/faqs?anchor=question50361.
8. National Institutes of Health, Office of Laboratory Animal Welfare. 2012. Frequently Asked Question F.5, May Investigators Use Expired Pharmaceuticals, Biologics, and Supplies in Animals? https://olaw.nih.gov/faqs#/guidance/faqs?anchor=question50362.
9. National Institutes of Health, Office of Laboratory Animal Welfare. *ARENA/OLAW Institutional Animal Care and Use Committee Guidebook*, 2nd Edition, p 161. 2002. https://olaw.nih. gov/sites/default/fles/GuideBook.pdf.
10. American College of Laboratory Animal Medicine. Position Statements: Medical Records for Animals Used in Research, Teaching, and Testing. https://www.aclam.org/media/858c1f18-6eda-49c0-b36b-47ab1a16d7e3/mNlnLw/ACLAM/About%20Us/Position%20Statements/position_medical%20records%202020.pdf. Accessed May 14, 2024.
11. National Institutes of Health, Office of Laboratory Animal Welfare. 2012. Frequently Asked Questions G.4, What Kind of Administrative Organization Works Best for Ensuring Compliance? https://olaw.nih.gov/faqs#/guidance/faqs?anchor=question50379.

12. National Institutes of Health, Office of Laboratory Animal Welfare. 2022. Frequently Asked Questions E.4., Is the IACUC Required to Inspect Field Study Sites? https://olaw.nih.gov/faqs#/guidance/faqs?anchor=question651.
13. Ellen Paul, Robert S. Sikes, Steven J. Beaupre, John C. Wingfield. Animal welfare policy: implementation in the context of wildlife research—policy review and discussion of fundamental issues, *ILAR Journal*, 56(3), 312–334, 24 February 2016. https://doi.org/10.1093/ilar/ilv073.
14. National Institutes of Health, Office of Laboratory Animal Welfare. 2012. Frequently Asked Questions G.2, What Is Required for an Occupational Health and Safety Program? https://olaw.nih.gov/faqs#/guidance/faqs?anchor=50377.

32

Research Animal Enrichment

Jennifer Camacho and Kathryn Bayne

Introduction

In the United States, the concept of addressing the *behavioral needs* of research animals first came to the forefront through the 1985 amendments to the Animal Welfare Act (PL 99–198, "The Improved Standards for Laboratory Animals Act"), which directed the Secretary of Agriculture to develop regulations to ensure an adequate physical environment to promote the psychological well-being of nonhuman primates and to provide for the exercise of dogs. These stipulations were considered progressive at the time. The contemporaneous *Guide for the Care and Use of Laboratory Animals*[1] did not yet recognize this important aspect of the welfare of research animals, stating that meeting the biological needs of the animals involved "maintenance of body temperature, urination, defecation, and, if appropriate, reproduction." However, the resulting regulations pertaining to exercise for dogs and psychological well-being for nonhuman primates prompted a significant amount of research into laboratory animal behavior, and, more specifically, environmental enrichment. The output of this research effort included significant changes in the way in which nonhuman primates were housed in laboratories and consideration of promoting physical activity in dogs.

This evolution in the consideration of the behavioral needs (i.e., not just the biological needs) of research animals was reflected in the seventh edition of the *Guide*,[2] which devoted an entire section to the "behavioral management" of research animals. While the principal target species of this section were nonhuman primates and dogs, the recommendations were sufficiently broad to be applicable to a variety of research animal species. A behavioral management program was described as including three elements: (1) the structural environment; (2) the social environment; and (3) activity (both physical and cognitive). Methods to enhance each of these areas were described to improve the welfare of the animal. The eighth edition of the *Guide*[3] affirmed the application of environmental enrichment to diverse research animals. Concomitantly, the accrediting organization, AAALAC International,* published Position Statements[4] on "Cage or Pen Space" and "Social Housing" as well as Frequently Asked Questions (FAQs)[5] on "Environmental Enrichment," "Social Housing and Social Experience," and subsequently on "Dairy Calf Housing and Social Enrichment."

As the momentum increased for providing enrichment to nonhuman primates, additional guidance was developed and offered by the federal government. For example, the USDA and NIH/OLAW collaborated on publishing a series of six booklets that provide an introduction to "Enrichment for Nonhuman Primates"[6] and produced a seminar recording on the subject, "AC and OLAW Perspective on Nonhuman Primate Enrichment and Social Housing."[7] In addition, the USDA published a bibliography of environmental enrichment articles specific to nonhuman primates, which was updated in 2015 and again in 2021, titled "Environmental Enrichment for Nonhuman Primates Resource Guide"[8] and includes general references pertaining to environmental enrichment and behavioral management, primate behavior and abnormal behavior, as well as species-specific references.

In the last decade or so, the scope of environmental enrichment has expanded to encompass numerous other species of animals used in research and a large number of research studies have evaluated the impact of enrichment on both the animals and the research. The newest edition of the *Guide*[3] echoes further maturation of the field in a section titled "Environmental Enrichment" that more broadly addresses enrichment techniques for both large and small research animals, as well as aquatic animal species. NIH/OLAW has developed FAQs to further clarify their position on social housing for nonhuman primates,[9] positive reinforcement for nonhuman primates, the use of performance standards in determining housing issues, the use of performance standards in determining environmental enrichment issues, and the use of performance standards in determining rabbit housing practices. The provision of enrichment is generally considered a key element of addressing the principle of refinement in the 3Rs and thus continues to gain momentum in implementation across species when assurance is provided that neither the health of the animal nor the scientific goal of the study is compromised by its provision.

It should be noted that the provision of environmental enrichment to agricultural animals used in research and teaching has been well established for many years. For both production and research purposes, it has long been recognized that both animal management and animal health are enhanced through the use of species-specific enrichment strategies and techniques. The *Guide for the Care and Use of Agricultural*

* The authors wish to make clear that AAALAC International is a private, nonprofit accreditation organization with institutions participating voluntarily. It is not a regulatory agency and its perspective shared throughout this chapter is intended to aid those institutions that participate in AAALAC's accreditation program.

DOI: 10.1201/9781003272861-32

Animals in Research and Teaching (*Ag Guide*, 2020)[10] is a useful resource for effective enrichments for agricultural animals commonly used in research.

The authors recognize that the term "environmental enrichment" has been critiqued as being too narrow in scope, with the preferred term being "behavioral management program" to reflect the numerous and diverse aspects of the research animal's life that can be improved to foster high quality welfare and science.[11] While federal regulations use the term "environmental enhancement," OLAW and the *Guide* refer to "environmental enrichment." The decision was made to retain use of the term "environmental enrichment."

Despite the numerous advances in improving research animal welfare through behavioral management programs, the authors agree with the statement from the Preface of the *Guide* (p. xiv) that "space and housing needs of laboratory species and the need and best methods for providing enrichment, exercise, and human contact merit further study."

32:1 Excluding IACUC-approved exemptions, does your IACUC require enrichment to be provided to all species used?

Reg. The AWAR (§3.8) require that research institutions "develop, document, and follow an appropriate plan to provide dogs with the opportunity for exercise" through group housing in cages, pens, or runs that provide at least 100% of the required space for each dog, by maintaining singly housed dogs in twice the minimum floor space, or by providing access to a run or open area at a specified frequency and duration. The AWAR (§3.81) also specify that a research institution must "develop, document, and follow an appropriate plan for environment enhancement adequate to promote the psychological well-being of nonhuman primates." The regulations stipulate that the plan must address social housing, environmental enrichment (such as cage complexities, food items, and human interaction), special considerations (e.g., infant and young juvenile animals, animals showing signs of psychological distress, animals whose activity is restricted, primates unable to see and hear other animals of their own or compatible species, and great apes over 110 pounds), restraint devices, and exemptions. The AWAR include similar provisions for birds not bred for use in research (§3.154).

The *Guide*, which applies to all vertebrate animals (p. 2), describes possible enrichment techniques for a variety of animals, ranging from rodents to nonhuman primates. The *Guide* (p. 4) establishes the 3Rs as the foundation of the recommendations contained in the document and notes that refinement includes modifications to husbandry procedures (p. 5). In addition, the *Guide* (p. 53) describes enrichment strategies for a variety of species as well as potential negative consequences of some enrichment approaches for different species. Taken together, the *Guide* affirms that environmental enrichment should be provided to all vertebrate species used in research.

AAALAC International states: "Environmental enrichment is an important method of improving the well-being of many laboratory animal species and may be accomplished by the provision of stimuli, structures and resources that facilitate the expression of species-appropriate behaviors. Environmental enrichment, as one component of an IACUC (or comparable oversight body)-approved behavioral management program, should be provided in a consistent manner across the animal program." AAALAC International's expectation that the environment of research animals be enriched extends beyond vertebrates. Specifically, the AAALAC International Reference Resource, "Guidelines for the Care and Welfare of Cephalopods in Research[12]—a consensus based on an initiative by CephRes, FELASA and the Boyd Group" —states: "Enriched environments must be provided, to allow the animals to express their normal behaviour."

Opin. The corpus of peer-reviewed literature regarding environmental enrichment methods for the diversity of species used in research has increased dramatically in recent years, and it frequently includes impact on the animal (e.g., physiology, neuroanatomy, etc.) and various types of studies. Thus, investigators and veterinarians can often determine the best methods for enriching the environments of their research subjects while minimizing confounding influences on their studies, and IACUCs can foster appropriate attention to the provision of environmental enrichment by prompting investigators to describe planned enrichment, regardless of the species used.

Surv. Excluding IACUC-approved exemptions, does your IACUC require enrichment to be provided to all species used?

Survey participants report that the IACUC mandates enrichment for all species in 81% of cases. However, 19% of respondents note that extra enrichment is not required when animals are housed socially. This is particularly notable for aquatic and reptile species, where minimal or no enrichment is typically provided (4%, 7/180).

- Yes, to all species 146/180 (81%)
- If animals are social/group housed, additional enrichment is not required 27/180 (15%)
- No 7/180 (4%)

32:2 How does your IACUC interpret the term "research animal enrichment"?

Reg. Although it does not define the term, the *Guide* (p. 52) states that "the primary aim of environmental enrichment is to enhance animal well-being by providing animals with sensory and motor stimulation, through structures and resources that facilitate the expression of species-typical behaviors and promote psychological well-being through physical exercise, manipulative activities, and cognitive challenges according to species-specific characteristics." The *Guide* (p. 53) also notes that "well-conceived enrichment provides animals with choices and a degree of control over their environment, which allows them to better cope with environmental stressors." The AWAR (§3.81 and §3.154) address enrichment only for nonhuman primates and birds and, like the *Guide*, do not define the term. However, the term "environment enhancement" is used in the regulations to refer to the social, structural, and activity-related aspects of behavioral management. Requirements include social grouping and providing a means of expressing non-injurious species-typical activities through cage complexities, manipulanda, foraging opportunities, etc.

Opin. Environmental enrichment is an approach to animal housing that includes both social and non-social stimuli that are provided to foster the expression of species-typical behaviors and reduce the occurrence of abnormal behaviors.[13] It is a component of animal husbandry that is designed to enhance the quality of captive animal care and improve animal welfare by identifying and providing the environmental stimuli necessary for optimal behavioral, physiological, and physical well-being. In practice, this covers a multitude of innovative and imaginative techniques, devices, and practices aimed at offering the animals the opportunity to express species-appropriate behaviors in a safe and controlled manner.[14]

Care should be taken to enrich the physical environment in the primary enclosure by providing means of expressing only non-injurious species-appropriate activities and ensuring that any enrichment provided does not pose the potential for harm to the animal.[15] Species differences should be considered when determining the types or methods of enrichment, and thus a basic understanding of the normal behavior of the species in question is requisite to a successful enrichment program. Examples of environmental enrichments include perches, swings, mirrors, nesting material, hiding places, and other cage complexities; providing objects to manipulate; social opportunities; varied food items; using foraging or task-oriented feeding methods; providing control over some aspect of the environment, such as visual access to other animals, music, or video; and providing interaction with the caregiver or other familiar and knowledgeable person consistent with personnel safety precautions. Of course, the provision of the presumed enrichment must be well thought out and with a sound knowledge of the species' normal behavior, constraints imposed by the research, safety considerations, and the caregiver's level of experience.[16]

Surv. How does your IACUC interpret the term "research animal enrichment"?

Many respondents provided multiple definitions of enrichment, demonstrating a good understanding of its various aspects. They equally emphasized two main definitions: one focusing on improving animal welfare through promoting physical and psychological well-being or facilitating species-typical behaviors, and another emphasizing the inclusion of items or structures in the cage. Additionally, many agreed that enrichment should help reduce abnormal behavior. A large number of respondents, between 44% and 72%, indicated that caging complexity housing practices to promote positive social interactions fit into the definition of enrichment. The following data presents the distribution of interpretations of "laboratory animal enrichment" from the 180 unique respondents to the question.

• Care practices that promote physical and psychological well-being	155/180 (86%)
• Care practices that promote species-typical behavior	154/180 (86%)
• Care practices that provide caging complexity	80/180 (44%)
• Care practices that provide objects to the animals (toys, treats, nesting material, perch, etc.)	154/180 (86%)
• Care practices that are aimed at decreasing aberrant behavior	112/180 (62%)
• Housing practices to promote positive social interaction	129/180 (72%)
• Other	4/180 (2%)

32:3 Does your IACUC have an institutional document that details the research animal enrichment plan?

Reg. The AWAR (§3.8) require research institutions to document the plan to provide dogs with the opportunity for exercise. Further, "The plan must include written standard procedures to be followed in providing the opportunity for exercise." The AWAR also stipulate what information must be described in the plan for individually housed dogs and group housed dogs, including the methods and period of providing the exercise opportunity and any exemptions. Similarly, the AWAR (§3.81) require research institutions to document the plan to provide environmental enrichment "adequate to promote the psychological well-being of nonhuman primate." Also, "The plan must be in accordance with the currently accepted professional standards as cited in appropriate professional journals or reference guides, and as directed by the attending veterinarian." The plan must address the social needs of nonhuman primates that form social groups in nature; describe the methods used to enrich the physical environment; address special considerations (i.e., infants and young juveniles, those animals showing signs of psychological distress, those primates with IACUC approved restricted activity, those primates unable to see and hear other primates of their own or compatible species, and great apes weighing more than 110 lbs.); the use of restraint devices; and identify exemptions. The AWAR contain a similar provision for birds not bred for use in research (§3.154).

Although the *Guide* states that enrichment programs should be reviewed by the IACUC, researchers and veterinarians "on a regular basis" and "updated as needed," the means by which that review and update occur is not specified. A written environmental enrichment plan is specified for other common research animals "Species-specific plans for housing and management ... should also include strategies for environmental enrichment" (p.58), "psychological enrichment" for avian species (p. 60) and appears to be implied for all species.

Opin. A written canine exercise plan and environmental enrichment plan that encompasses all species becomes a useful tool for the IACUC during the semiannual program review. The exercise program and enrichment strategies described in the document can be compared to what the IACUC visualizes during the semiannual facility inspection, thereby providing the IACUC with an objective assessment of the implementation of the plans. In addition, the IACUC can judge if the plan has been updated to conform with the evolving state of knowledge for the species and for the particular type of research.

Surv. Does your IACUC have an institutional document that details the research animal enrichment plan?

Only 68 out of 180 respondents (38%) reported having an IACUC policy at their institution. The majority, 123 out of 180 respondents (68%), primarily rely on Standard Operating Procedures (SOPs) to document the specifics of their research animal enrichment plan. Some respondents mentioned that the plan's details might also be included in the IACUC protocol or detailed in the AAALAC International Program Description.

• Not applicable, we do not provide enrichment.	4/180 (2%)
• We have an IACUC policy for our institution.	68/180 (38%)
• There are Standard Operating Procedures (SOPs) that detail the environmental enrichment plan.	123/180 (68%)
• Other.	23/180 (13%)

32:4 If animals are exempt from all or part of your enrichment program, what are the criteria used to justify the exemption?

Reg. The AWAR stipulate that nonhuman primates and birds may be exempted from the enrichment program and dogs may be exempted from the exercise program for scientific, health, or behavioral reasons (§3.81,e; §3.154,e; and §3.8,d), respectively). Specific examples offered as an acceptable justification for exempting a primate from social housing include overly aggressive behavior/incompatibility, a contagious disease, debilitation because of the age of the animal, or other conditions such as arthritis. Unless the basis for the exemption is a permanent condition, exemptions made by the AV must be documented and reviewed at least every 30 days (§3.8,d,1; §3.81,e,1; §3.154,e,1). Records of exemptions must be maintained by the research facility and must be made available to USDA officials or officials of any pertinent funding federal agency upon request (AWAR §3.81,e,3.; §3.154,e,3; §3.8,d,3.).

PHS Policy (II; IV,A,1,) requires institutions to comply with the AWAR and the *Guide*. Additional guidance for the social housing requirement of nonhuman primates is offered by the NIH/OLAW in the context of NIH/OLAW FAQ F.14, which states: "Exemptions to the social housing requirement must be based on strong scientific justification approved by the IACUC or for a specific veterinary or behavioral reason. Lack of appropriate caging does not constitute an acceptable justification for exemption."

Opin. The management of an enrichment program includes, of necessity, provisions for allowing exemptions for select animals from participating in aspects of the program. Exemptions may be based on concerns related to the health of the animal (either related to the experiment or due to natural causes) or the scientific goals of the study. Exemptions from an enrichment program should require strong scientific justification carefully scrutinized by the IACUC and not be considered as "all or nothing" participation by the animal. An animal may be exempted from only a portion of the program, while still enjoying the benefits of other elements of the program (e.g., the animal may be precluded from social housing but offered a variety of nonsocial enrichments, such as food treats or toys). For example, an animals' role in research involving infectious diseases, atypical rearing conditions, physical restraint, surgery, pain, substance abuse, or aggression should not be excluded *a priori* from the enrichment program.[17] Rather, protocols that may pose restrictive environments for animals should be reevaluated periodically to determine if new technologies or strategies could be implemented to reduce the restrictions placed on the animal.

Four criteria are proposed for assessing animal well-being in addition to general physical health[18]:

- The animal's ability to engage in beneficial species-typical behavior
- The animal's ability to cope effectively with environmental changes
- The absence of maladaptive or pathological behavior
- The presence of a balanced temperament

When evaluating an animal for an exemption based on health or behavior, it is important to understand fully what is "normal" for an animal in that condition. For example, an aged animal frequently has reduced perceptual and locomotor capabilities,[19] and arthritis has been shown to alter behavior.[20] Thus, the enrichment program may need to be modified to accommodate limitations imposed by the health or condition of the animal.

Surv. If animals are exempt from all or part of your enrichment program, what are the criteria used to justify the exemption?

While 19% of respondents (34/180) stated that no animals are exempted from receiving environmental enrichment, the majority of those indicating exemptions cited scientific justification determined by the IACUC based on research needs, accounting for 78% (141/180). Exemptions due to veterinary criteria, behavioral reasons, or social incompatibility were mentioned by 98 out of 180 respondents, with these determinations made by a veterinarian. A small number of respondents identified alternative reasons for exemption, such as animal attrition (e.g., if only one animal remains in a social group) or specific to zebrafish species.

- No animals exempted from receiving environmental enrichment — 34/180 (19%)
- Scientific justification based on research needs, as determined by the IACUC — 141/180 (78%)
- Veterinary medical exemption, as determined by a veterinarian — 98/180 (54%)
- Behavioral exemption or social incompatibility as determined by veterinarian or dedicated animal care staff — 98/180 (54%)
- Other — 7/180 (4%)

32:5 How are enrichment supplies and resources (including personnel and space, such as food prep and storage) funded?

Opin. Since the provision of an enrichment program is mandated by federal law (see 32:1) for some species, it is critical that adequate resources be identified to ensure compliance with this institutional responsibility. An ongoing budgetary commitment to the enrichment program will allow for the program to evolve as new information becomes available and to have a meaningful impact on the welfare of the research animals. Inadequate funding can cause a program to be ineffective and not meet regulatory requirements or recommendations of the *Guide*. The NIH states in the *Institutional Administrator's Manual for Laboratory Animal Care and Use*[21] (p. 13) that "sustained and visible support from institutional officials is absolutely essential to establishing and maintaining a high-quality animal care and use program. … They can assure sufficient monetary and personnel resources are allocated to the institution's program." The enrichment program should not be viewed as separate or an "extra" program but rather as an integral component of the entire animal care and use program. In addition to funding enrichment through *per diem* charges, institutional personnel often implement low-cost and innovative approaches to providing environmental enrichment. In addition to funding supplies used to enrich the animals' housing environments, the institution will typically need to commit space for the storage of supplies as well as a food preparation area for some species that benefit from the provision of fresh food items (e.g., providing nonhuman primates with fruit).

Surv. How are enrichment supplies and resources (including personnel and space, such as food prep and storage) funded?

The survey results indicate various sources of funding for enrichment supplies and resources. The majority, 47% (85/180), include enrichment costs as part of the *per diem* charge. While many others, 34% (62/180), obtain funding from the animal facility budget, separate from the *per diem* charge. None of the respondents reported funding from the IACUC; however, smaller numbers of respondents receive an independent institutional budget (12/180) or rely on donations (6/180) or enrichment supplies provided by the principal investigator (15/180).

- From the animal facility budget, but independent of a *per diem* charge — 62/180 (34%)
- As part of the *per diem* charge — 85/180 (47%)
- From the IACUC's budget — 0/180 (0%)
- From an independent institutional budget — 12/180 (7%)
- No specific budget; enrichments are donated or collected by the animal care staff — 6/180 (3%)
- Other — 15/180 (8%)

32:6 What personnel resources are allocated for the implementation and oversight of your enrichment program?

Reg. The *Guide* (p. 53) states: "Enrichment programs should be reviewed by the IACUC, researchers, and veterinarian on a regular basis to ensure that they are beneficial to animal well-being and consistent with the goals of animal use. They should be updated as needed to ensure that they reflect current knowledge." The FAQ on Environmental Enrichment issued by AAALAC International states: "AAALAC site visitors will focus their attention on the IACUC's review of the enrichment program, documentation of the review, and implementation of the program. AAALAC expects that the enrichment program will be reviewed regularly by the IACUC, and that the IACUC adequately represents the research community and veterinarian(s) at the institution in the review of the enrichment program." While the AWAR require enrichment/exercise only for NHPs, dogs, and birds, all personnel involved in the animal care and use program must be adequately qualified to perform their duties (§2.32) for all regulated species, so if a person is responsible for providing enrichment to other regulate species, that person must be trained or qualified to do so.

Opin. The way in which institutions allocate human resources for the implementation of the enrichment program often varies with the size of the institution, the species used at the institution, and whether it is an academic institution. For example, large programs may have the funds to employ a staff member dedicated to the behavior and enrichment program, whereas smaller programs may exclusively rely on the animal care staff to create, implement, monitor, and update the enrichment program. Institutions that use only rodents are more likely to rely on husbandry staff to manage the enrichment program. Academic institutions have the opportunity to involve students in the support of aspects of the program. Regardless of the type of individual who manages the program, the IACUC must ensure that personnel involved in animal care and use of regulated species are adequately qualified to perform their duties, including the provision and management of environmental enrichment, through training or experience.

Surv. What personnel resources are allocated for the implementation and oversight of your enrichment program? Select all that apply.

Only 14% of the respondents (26/180) have dedicated personnel specifically hired to implement and oversee the animal enrichment program. The majority of the survey respondents (83/180) indicate oversight of enrichment programs primarily divided between the research person caring for the animals and a responsibility rotated between animal or veterinary cares staff. Others in the survey indicated other methods for personnel allocation, including utilizing students whenever possible (10/180).

- We have one or more persons specifically hired to implement the animal enrichment program. 26/180 (14%)
- We rotate responsibilities among our animal care or veterinary staff. 83/180 (46%)
- The research person caring for the particular animals provides the enrichment for those animals. 83/180 (46%)
- We use students whenever we can. 10/180 (6%)
- Other. 25/180 (14%)

32:7 Who receives training regarding environmental enrichment at your institution?

Reg. The AWAR (§2.32,a; §2.32,b; §2.32,c) require that all personnel involved in animal care and use are qualified to perform their duties and that training and instruction of personnel include "humane methods of animal maintenance." The PHS Policy (IV,A,1,g) stipulates inclusion of a "synopsis of training or instruction in the humane practice of animal care and use" in the NIH/OLAW Animal Welfare Assurance document. Personnel qualifications and training also are emphasized in the *Guide* (p. 16), which states: "Personnel caring for animals should be appropriately trained … and the institution should provide for formal and/or on-the-job training to facilitate effective implementation of the Program and the humane care and use of animals." The *Guide* (p. 53) also notes: "Personnel responsible for animal care and husbandry should receive training in the behavioral biology of the species they work with to appropriately monitor the effects of enrichment as well as identify the development of adverse or abnormal behaviors." NIH/OLAW emphasizes in its FAQ F.14, pertaining to social housing, that "Staff performing nonhuman primate socialization should be trained and competent in the procedure and knowledgeable about the animals."

Opin. The value of providing training on enrichment, and the recognition of normal and abnormal behavior, may vary with the species used at the institution because of the inherent complexities of observing behavioral changes and of providing enrichment to nonhuman primates versus mice and rats. Regarding nonhuman primate enrichment, the National Research Council[22] notes (p. 44) that the success of an enrichment program is dependent on personnel having knowledge of, and experience with, nonhuman primate behavior, and that "periodic training of staff to acquaint them with advances in the field is essential." Safety training also may be considered an essential topic in a behavioral management training program (pp. 19–20). Hazards to staff associated with forming or separating social pairs or groups of nonhuman primates due to possible resulting aggression, or associated with animals reaching out of cages to grab (and possibly scratch) personnel as they attach or remove enrichment items affixed to the cage front should be addressed with relevant personnel. Training should be provided to the variety of personnel at the institution involved in implementing and monitoring the enrichments. This could include research staff, veterinary medical personnel, husbandry staff, and members of the IACUC. Because the IACUC is responsible for oversight of the enrichment program and is evaluating it during the semiannual program reviews and facility inspections, it is important that committee members receive training so that they can adequately assess the success of the program performing these activities.

Surv. Who receives training regarding environmental enrichment at your institution?

Among the respondents, 22% (40/180) do not offer specific training on environmental enrichment. However, the majority (72% or 130/180) indicated that training is mainly provided to animal care staff. Multiple responses were allowed, with some indicating training for Principal Investigators, research staff, and veterinary medical staff. Only a small fraction (6% or 11/180) mentioned hiring personnel with prior enrichment experience.

- We do not provide any specific training on environmental enrichment. 40/180 (22%)
- Training is provided to IACUC members. 37/180 (21%)
- Training is provided to Principal Investigators and their research staff. 72/180 (40%)
- Training is provided to animal care staff. 130/180 (72%)
- Training is provided to the veterinary medical staff. 73/180 (41%)
- Specific enrichment personnel are hired with prior enrichment experience. 11/180 (6%)
- Other. 4/180 (2%)

32:8 What is the most commonly used form of training regarding environmental enrichment offered to your staff?

Reg. The *Guide* states: "Personnel responsible for animal care and husbandry should receive training in the behavioral biology of the species they work with to appropriately monitor the effects of enrichment as well as identify the development of adverse or abnormal behaviors" (p. 53). On-the-job training is recognized as the primary means of training and should be "supplemented with institution-sponsored discussion and training programs and reference materials." Additionally, it is recommended: "Personnel caring for laboratory animals should also regularly engage in continuing education activities" and recommends attendance in local and national laboratory animal science meetings, and provides suggested resources, including the Animal Welfare Information Center (AWIC), the Laboratory Animal Welfare and Training Exchange (LAWTE), AALAS, and ILAR (NRC 1991), among others.

Opin. Seminars at local, regional, or national meetings; scientific publications; online tutorials and webinars; consultants who provide training at the institution; and on-the-job training are good means for providing staff training.

Surv. What is the most commonly used form of staff training regarding environmental enrichment?

The survey results reveal varied approaches to training on environmental enrichment. The majority, 51% (92/180), utilize online resources such as webinars and electronic forums, while 48% (87/180) integrate training into their Standard Operating Procedures (SOPs). Additionally, 31% (56/180) send staff to conferences or workshops, and 29% (52/180) provide formal in-house training sessions. A smaller percentage, 14% (26/180), offer books and journals on enrichment topics, and 4% (8/180) hire external experts for training. Meanwhile, 19% (34/180) do not provide any specific training, and 5% (9/180) indicated other methods for training provision, such as on the job training.

- Not applicable, we do not provide training. 34/180 (19%)
- We provide formal training sessions in-house. 52/180 (29%)
- We send people to various conferences, workshops, etc. that address enrichment topics. 56/180 (31%)
- We use online resources (e.g., webinars, electronic forums, AALAS Learning Library, etc.). 92/180 (51%)
- We use outside experts to provide training. 8/180 (4%)
- Training is intrinsic in our Standard Operating Procedures (SOPs). 87/180 (48%)
- We provide books and journals that include enrichment topics. 26/180 (14%)
- Other. 9/180 (5%)

32:9 Are your institution's enrichment program and associated records reviewed by the IACUC as part of the semiannual program review or at another frequency?

Reg. The AWAR (§2.31,c,1) state that the IACUC shall "review at least once every six months the program for humane care and use of animals using Title 9, Chapter I, subchapter A-Animal Welfare, as the basis for evaluation." This reference includes the sections pertaining to dog exercise and environmental enhancement to promote psychological well-being. The *Guide* (p. 25) ascribes to the IACUC the responsibility for "oversight and evaluation of the entire Program and its components as described in other sections of the *Guide*," which includes the environmental enrichment and social housing sections. Specifically, the *Guide* (p. 53) states: "Enrichment programs should be reviewed by the IACUC, researchers, and veterinarian on a regular basis to ensure that they are beneficial to animal well-being and consistent with the goals of animal use." The PHS Policy states (IV,B,1) that the IACUC must "review at least once every six months the institution's program for humane care and use of animals, using the *Guide* as a basis for evaluation." Furthermore, The AWAR (§2.33,a,2) specify that "each facility shall assure that the attending veterinarian has appropriate authority to ensure the provision of adequate veterinary care and to oversee the adequacy of other aspects of care and use" and it can be implied that this include evaluation of enrichment provisions to ensure health and safety.

Opin. In accordance with the AWAR, institutions that have an enrichment plan for nonhuman primates and/or birds, reviewed and approved by the veterinarian, and/or a canine exercise program must review those aspects of the animal care and use program semiannually. Similarly, if the institution holds a PHS Assurance or is accredited by AAALAC, the behavioral management program described in the *Guide* should be reviewed as part of the semiannual program review. Some elements of these programs may also be evaluated during the facility inspections, for example, assessing the degree of implementation of the plan, how consistently it is implemented across the institution, whether the implementation conforms to institutional guidance, and gauging its effectiveness in terms of the animals' behavior or other measures of well-being. The IACUC has a responsibility to review the animal care and use program semiannually and to ensure its completeness, appropriateness, and currency.

Surv. Are your institution's enrichment activities and associated records reviewed by the IACUC as part of the semiannual program review or at another frequency? (e.g., reviewing for updates, exemptions, and consistency of implementation and species included).

The majority of respondents, 43% (77/180), indicated that the IACUC reviews enrichment activities and associated records during its semiannual program review. Additionally, 25% (27/180) stated that the review occurs at a frequency other than semiannually. However, 35% (63/180) reported that the IACUC does not review enrichment activities or records. A smaller percentage, 7% (13/180), mentioned other review methods, such as annual reviews, informal discussions for USDA-regulated species, or reviews during facility inspections.

- Yes, the IACUC does review the enrichment activities and records during its semiannual program review. 77/180 (43%)
- Yes, at a frequency other than a semiannual program review. 27/180 (25%)
- No, the IACUC does not review the enrichment activities or records. 63/180 (35%)
- Other. 13/180 (7%)

32:10 What information is maintained in your enrichment records?

Reg. The AWAR (§3.81) require that dealers, exhibitors, and research facilities develop, document, and follow an environment enhancement plan that is adequate to promote the psychological well-being of nonhuman primates. The plan must include specific provisions to address the social needs of nonhuman primates that are social in nature, a description of the environmental enrichment provided, and special considerations (i.e., primates requiring special attention). They further require that records of exemptions of animals from the plan be maintained (AWAR §3.81,e,3) and be made available to USDA officials or representatives of federal funding agencies upon request. A similar plan must be developed to provide dogs with the opportunity for exercise and must include written standard procedures to be followed in providing the opportunity for exercise. It must address dogs housed individually, dogs housed in groups, methods and periods of providing the opportunity to exercise, and exemptions (AWAR §3.8).

PHS Policy (II; IV,A,1) requires institutions to comply with the AWAR as applicable and to follow the *Guide*. The *Guide* (p. 75) notes that records of rearing and housing histories are useful for the management of many species. Institutions must keep records of departures from the *Guide* and IACUC approval of any departures (PHS Policy IV,B,3).

Opin. Documentation not only serves as a verifiable record of enrichment and socialization activities, but also validates the effectiveness of ongoing enrichments or provides evidence of the need to modify the enrichment and socialization program. The National Research Council[22] (p. 26) suggests that nonhuman primate enrichment programs should include a mechanism whereby there are "protocols for diagnosing the cause of physical impairments and abnormal behavior, determining when remediation is necessary, developing remediation plans, assessing the effectiveness of remediation, and maintaining appropriate records."

Surv. What information is maintained in your enrichment records?

The majority of respondents, 40% (72/180), indicated that they do not maintain enrichment records. For those who do keep records, 35% (63/180) document the provision of food treats, and 32% (57/180) maintain a distribution or rotation schedule of manipulative objects used for enrichment. Additionally, 31% (55/180) record details on social housing, and 29% (52/180) include information on animal behavior. Observations of enrichment usage and effects were noted by 24% (43/180), while 40% (72/180) document exemptions from receiving enrichment. Other information recorded by 9% (16/180) of respondents includes specific research-related changes, presence or absence of enrichment materials, identification of single housed animals, and time spent in playpens.

- Enrichment records not kept — 72/180 (40%)
- Distribution or rotation schedule of manipulative objects used for enrichment — 57/180 (32%)
- Provision of food treats — 63/180 (35%)
- Observation of usage and/or effects of the enrichment — 43/180 (24%)
- Exemption from receiving the enrichment — 72/180 (40%)
- Information on animal behavior (normal or if aberrant behaviors) — 52/180 (29%)
- Details on social (pair or group) housing — 55/180 (31%)
- Other — 16/180 (9%)

32:11 How does the IACUC ensure the efficacy of the enrichment activities (e.g., animal usage, rate or occurrence of aberrant behavior, etc.)?

Reg. The AWAR (§2.31,c,1) and PHS Policy (IV,B,1) require that the IACUC review at least once every 6 months the program for humane care and use of animals. The *Guide* (p. 9) states that the IACUC has oversight and evaluation responsibility for the animal care and use program and the components of the program described in the *Guide*. The *Guide* (p. 25) also recommends program review and facilities inspections occur at least annually, or more often as required.

Opin. Morgan *et al.*[23] reviewed various methods of assessing enrichment programs. Methods included reliance on others to inform the IACUC, common sense, and empiricism. They noted that a significant disadvantage of reliance on others is that acceptance is often based on the authority of the other person (the tendency to accept that a point of information is true simply because it is that of an authority figure). They suggested that common sense is a good starting point for evaluating enrichment, although its main disadvantage is that common sense is grounded on individual interpretations and biases. They strongly argued that empirical testing of enrichment strategies is necessary to validate their use both from the standpoint of the value to the animal and for judicious expenditure of limited resources.

An assessment of the enrichment program has the greatest relevancy when it is considered in the context of the welfare of the research animal. Specifically, the question should be asked whether the enrichment used (within any constraints imposed by the research goals) is fostering research animal welfare or is it of null or minimal value to the animal. This question broadens the scope of the evaluation of the enrichment program to more properly focus on the animal rather than the enrichment itself, such as use of the toy, consumption of a food treat, etc. For many research animal species, a reduction in abnormal behaviors (e.g., obsessive-compulsive behaviors) or an increase in the breadth of expression of normal behaviors (e.g., nest building, foraging, etc.) are indicators of improved well-being and the literature is rich with reports of ways to achieve these for numerous species. Also, on a pragmatic level, expenditure of limited fiscal resources for enrichment methods that are not achieving the goal of improving research animal welfare is an unsound management practice.

Surv. How does the IACUC ensure the efficacy of the enrichment activities (e.g., animal usage, rate or occurrence of aberrant behavior, etc.)?

The majority of respondents, 68% (122/180), rely on veterinary medical, animal care, or dedicated enrichment staff to advise the IACUC about any issues with enrichment activities. Additionally, 41% (74/180) depend on research staff to inform the IACUC of problems. For program assessment, 31% (56/180) have the IACUC evaluate enrichment activities during semiannual reviews or inspections. Meanwhile, 21% (38/180) consider the program successful if AAALAC is satisfied during its site visit. A smaller portion, 26% (46/180), do not assess the efficacy of their enrichment activities, and 3% (6/180) cited other methods for ensuring efficacy such as presentations to the IACUC by a staff behaviorist.

- We do not assess the efficacy of our enrichment activities. — 46/180 (26%)
- The IACUC assesses the enrichment activities during the semiannual (or other frequency) program review and/or inspections. — 56/180 (31%)
- We consider the program successful if AAALAC is pleased with it during its site visit. — 38/180 (21%)
- We depend on the veterinary medical, animal care, or dedicated enrichment staff to advise the IACUC of any problems with the enrichment activities. — 122/180 (68%)

- We depend on the research staff to advise the IACUC on any problems with the enrichment activities. 74/180 (41%)
- Other. 6/180 (3%)

32:12 If your investigators have concerns about using animal enrichment, what are their objections?

Reg. The *Guide* (p. 53) also states: "Not every item added to the animals' environment benefits their well-being." Enrichment can serve as a vector for disease transmission, pose a risk of injury if entangled, entrapped, ingested, or in the case of food enrichment (treats) can lead to obesity in animal models.[24] In addition, some specific enrichment items have been reported to increase stress in certain species. For example, marbles are used as stressors in mouse anxiety studies.[25] The *Guide* (p. 53) elaborates on the potential problems of providing enrichment in some cases, "In some strains of mice, cage dividers and shelters have induced overt aggression in groups of males, resulting in social stress and injury (Bergmann *et al.* 1994; Haemisch *et al.* 1997)." Emphasis is placed on the IACUC, researchers, and veterinarians to ensure that the environmental enrichment provided is safe and beneficial to the animals' well-being and consistent with the animal-use goals.

The *Guide* (p. 54) acknowledges: "Some scientists have raised concerns that environmental enrichment may compromise experimental standardization by introducing variability, adding not only diversity to the animals' behavioral repertoire but also variation to their responses to experimental treatments," but that "A systematic study in mice did not find evidence to support this viewpoint (Wolfer *et al.* 2004), indicating that housing conditions can be enriched without compromising the precision or reproducibility of experimental results." The *Guide* (p. 54) notes that "Further research in other species may be needed to confirm this conclusion." But to strengthen the case for providing enrichment, the *Guide* (p. 54) concludes the discussion by stating, "However, it has been shown that conditions resulting in higher-stress reactivity increase variation in experimental data (e.g., Macrì *et al.* 2007). Because adequate environmental enrichment may reduce anxiety and stress reactivity (Chapillon *et al.* 1999), it may also contribute to higher test sensitivity and reduced animal use (Baumans 1997)."

Opin. Both proponents of enrichment and scientists who work with animals living in enriched environments recognize that the presence of enrichment can influence the data generated.[15] In some species, the effects of providing enrichment on the research may not be as profound as they are in other species (e.g., dogs or nonhuman primates vs. rodents). Regardless, a prudent course of action is to evaluate the type of enrichment being provided to the animal on a study-by-study basis and assess with the investigator whether it may confound research results. If so, then perhaps another type of enrichment could be considered, or the animal may need to be exempted from the enrichment program, with approval by the IACUC.[26] A recent review of mouse enrichment illustrates the point that enrichment may influence the scientific data by introducing an independent variable, or it may facilitate scientific inquiry by stimulating new research directions.[27] Toth *et al.*[28] argue in favor of standardizing the research rodent's environment to maximize the sensitivity of the test and to minimize the number of animals used that will achieve appropriate power of the statistical test. They also make a strong case that the type of enrichment available to the animals in the study, and when it was made available, should be reported by the researcher in the published study. Others have argued equally persuasively that standardization can lead to inaccurate results and poor reproducibility of the data.[29] Standardization as a necessary component of behavioral phenotyping of mutant mice has been adjured[30]; however, others suggest that systematic variation of genetic and environmental backgrounds is a sound approach to augmenting the validity of study results, including behavioral phenotyping,[31] and have gone so far as to suggest that environmental standardization is a cause of poor reproducibility of research results.[32] However, even basic procedures in the animal facility, such as cage change and transporting rodents in a cage from the animal holding room to a procedure room, can affect the physiology of the animals,[33] and there are numerous environmental variables such as noise, temperature, light level, and others that contribute to a lack of standardization across studies.[31] Thus, it appears that as researchers examine more closely various housing and husbandry aspects of their research subjects, there are multiple entry points for variability in a study and researchers should be aware of these and assess them in the context of the specifics of the study.

Baumans[34] has suggested that with the provision of enrichment, the mouse can exhibit more species-appropriate behavior and thus may be able to better cope with unexpected changes to its environment and respond more uniformly to different challenges. Abnormal behavior, stress, fear and anxiety, and impaired thermoregulation are potentially confounding variables that may adversely affect the outcome of animal-based studies and consequently increase variation in the data. However, neurological, physiological, and immunological effects have been described in rodents and other research animal species offered different types of enrichment (for example, see a brief review of the effects on mice by Bayne and Würbel[26]). Therefore, it behooves investigators and IACUCs to be aware of the potential influence, or lack thereof, enrichment may have on the study animals and thus the resulting data.

Concern about possible harm to the animal has merit as reports of injury to animals[35–39] or death[40] resulting from various social or non-social types of enrichment appear in the literature despite the clear intent of offering these enrichments as a means to enhance animal well-being. In addition, Bayne et al.[41] observed that Kong® toys presented to nonhuman primates were not completely sanitized after going through a mechanical cage washer; therefore, some attention should be given to how toys are distributed across a primate colony, and further evaluation of the efficacy of sanitation of enrichment devices should be undertaken.

Despite these cautions, there is a large body of evidence to support the many positive effects of enrichment on promoting the welfare of numerous species of research animals, when the enrichment technique is chosen carefully, and the entire enrichment program is monitored appropriately. The introduction of new enrichment methodologies should be done carefully, and possibly incrementally, to assess any possible negative consequences associated with its use. Additionally, if one form of enrichment is precluded due to the nature of the study, alternative enrichment devices or techniques should be considered before removing the animal(s) from the enrichment program completely. Publishing information regarding the types of enrichment used successfully in various experimental paradigms is encouraged to further refine enrichment programs across institutions.

Surv. If your investigators have concerns about using animal enrichment, what are their objections?

The majority of investigators (56%) reported that they do not usually have concerns about using animal enrichment or that these concerns are infrequent. Among the concerns raised, the most common were worries about unforeseen effects introducing experimental variables and affecting scientific data (41%), followed by concerns about social housing potentially leading to injury to the animals (22%). Other concerns included worries about harm to the animals (17%), problems with sanitization introducing pathogens (7%), and the ease of access, visibility, or handling of the animals (8%). Fewer respondents cited the cost of items and/or labor (5%) and other miscellaneous concerns (3%) as reasons for their objections to using animal enrichment.

- We do not usually have concerns, or they are infrequent. 100/180 (56%)
- Concern that there will be an unforeseen effect that will introduce and experimental variable and effect scientific data. 74/180 (41%)
- Cost of items and/or labor to clean and/or replace. 9/180 (5%)
- Problems with sanitization that may introduce pathogens. 13/180 (7%)
- Concern that harm may occur to the animal (e.g., injury, ingestion, entanglement). 30/180 (17%)
- Concern that social housing may result in injury to the animals. 40/180 (22%)
- Ease of access, visibility or handling of the animals 15/180 (8%)
- Other. 6/180 (3%)

32:13 For institutions that exclusively use rodents, please indicate the environmental enrichment provisions offered.

Reg. The NIH/OLAW website[7] provides resources regarding environmental enrichment provisions, including suggestions on types of enrichment for mice and rats[42] The *Guide* recommends the use of enrichment (p. 68) based on studies where rodents had exhibited a clear preference for certain materials, such as nesting material,[43] which encourages the expression of species-specific behaviors, fosters control of body temperature,[44] and provides an opportunity to retreat in case of disturbances.[45] Additionally, the *Guide* references "wooden chew sticks for some rodent species" (p. 53) and indicates that the addition of enrichment items should be considered "in the absence of other animals" (p. 64). The *Guide* refers to the use of enrichment as "complex," cautioning that the use of enrichment "affects animal phenotype and may affect the experimental outcome" (pp. 53–54). Variables that affect how the animals respond to enrichment include species, strain, phenotype, age, gender, housing density, cage size and quality of the space, including vertical access and structures. However, the *Guide* notes that "it is difficult to compare these studies due to the study design and experimental variables that have been measured. These issues remain complex and should be carefully considered when housing rodents" (p. 58). A systematic study conducted by Wolfer *et al.* 2004,[46] referenced by the *Guide* (p. 54), indicates that "housing conditions can be enriched without compromising the precision or reproducibility of experimental results."

Opin. Environmental enrichment in research rodents is generally defined as the inclusion of items or design features in the animals' housing environment with the aim of increasing the expression of natural behaviors and improving animal welfare. Enrichment is intended to have a positive effect on reducing stereotypic behavior, improving breeding performance, and reducing aggression. Further, it should not negatively impact animal health or scientific outcomes.[47] Enrichment studies have yielded contradictory findings about the potential benefits and unintended consequences. Enrichment has been reported to have different effects depending on the strain, age, gender, social construct, scientific study, and housing types, among other parameters.[48] These differences lead to concern for variation in the experimental outcomes and stagnate the use of enrichment. Arguments opposing the use of environmental enrichment are drawn from concern that it could be a confounding variable and adversely affect the validity of experimental outcomes when considering historic models, usually conducted when the lack of enrichment was commonplace. Counterarguments suggest that lack of enrichment results in an inadequate environment that might similarly produce variability and concern for animal welfare, stress, social isolation, and the occurrence of abnormal behaviors; all which may not translate directly to the normal human or animal condition.[49,50]

Methods of enrichment described vary greatly by type, composition, and disposability. Enrichment types include nests, structures, running wheels, chew manipulanda and nutritional enrichments, or combinations of these items (particularly nesting supplemented by additional item(s)) and transfer of nesting material. Other aspects of rodent care may be considered enriching and include cognitive enrichment, such as animal training, human–animal interactions, behavioral testing apparatus; enclosure space; and environmental parameters such as airflow, temperature, and lighting.

- ***Nesting*** is commercially available in a variety of materials including cotton, wood, and paper, and in diverse substrate forms such as compressed, tissue, paper, cellulose, wool, and shavings. Mice have a strong motivation to build nests and have been documented to work for access to nesting material.[51] Nesting activity is well documented to produce affiliative behavior for both males and females in numerous background and transgenic strains of mice.[52] It may have a positive effect on reproduction[53] and may provide mice with the ability to modulate environmental parameters such as temperature and lighting. Nesting behavior can even be used as an indicator of welfare to assess general condition, pain, and discomfort after research procedures.[54] As such, nesting material has mostly positive associations and has become emphasized as an industry standard to promote animal welfare. The use of nesting material supplemented with enrichment provides more opportunities for the rodents to express diverse behaviors, and as noted in the *Guide* (p. 43), and thermoregulate. Similar to the provision of any enrichment, such choices should be made with consideration of the scientific study aims.
- ***Structures*** are commercially available in a variety of materials, including metal, plastic, wood, or paper products that are molded into shelters, such as hiding structures, tunnels, and resting surfaces that may or may not have cut-outs for animal egress and airflow. Structures can have a mitigating effect on the animals' fear and aggression by providing them a place to flee from aggressive cage mates and they also can reduce the stress response associated with cage change if animals are transferred in a tunnel.[55] However, structures can also have varying effects on studies, depending on the strain of the mouse and whether they are combined with nesting material.[56]
- ***Chew*** devices consist of natural cuts of wood or synthetic plastic or fiber composites. They can encourage species-typical gnawing behavior and may foster dental health by reducing the development of malocclusion.[57]
- ***Exercise*** (e.g., a running wheel) is generally accepted as a research tool to gauge the effects of physical activity on various systems, including cardiovascular, immune, and endocrine and neuro-biological parameters such as neurodegenerative disease, learning and cognition[58] Running wheels do not appear to have significant effects on food intake and body weight[59]; however, running wheels may prompt stereotypic behavior and the exclusion of other behaviors.[60]
- ***Nutritional*** enrichment is most commonly provided as food "treats," to expand dietary options, to increase natural foraging behaviors or as rewards paired with procedures, husbandry, or handling.[61] The provision of treats for enrichment should not be confused with nutritional items that are offered as a part of a scientific study for diet supplementation or operant conditioning. Some studies have identified benefits associated with the use of treats, for example as supplements to improve reproductive outcomes,[62] reduce food grinding behavior,[63] and promote stress resilience.[64] Nutritional enrichment should be provided only under the guidance of the attending veterinarian to assure that low-value foodstuffs do not negatively impact the health of the animal.

- ***Nest transfer*** is considered the movement of some or all of the formed nest to a clean cage during routine husbandry. This practice may be beneficial in reducing signs of stress and aggression.[65]
- ***The provision of environmental enrichment to singly housed animals*** should be considered, as rodents have been shown to be highly motivated to interact with conspecifics[66]; studies examining the effects of social isolation have documented negative outcomes on breeding performance, ability to cope and cognition.[67]

The use of enrichment is supported by positive outcomes, with evidence that barren environments may seriously impair animal welfare, while enrichment is thought to normalize the animal model.[68] The type of enrichment used should be relevant for, and not interfere with, the scientific measures. Peer-reviewed literature (e.g., PubMed, utilizing the services of the institutional librarian) can be helpful in evaluating enrichment for different scenarios, and consideration should be given to validating the data produced by the enriched animals to ensure valid and reproducible outcomes.

Surv. For institutions that exclusively use rodents, please indicate the environmental enrichment provisions offered.

The majority of institutions exclusively using rodents for research reported offering a combination of environmental enrichment provisions, with 78% providing nesting material plus other provisions and 79% offering structural enhancements like tunnels, igloos, or shelters. Additionally, 61% provided chew manipulanda such as wood or nylabones, and 58% added environmental enrichment specifically for single-housed animals. Nutritional enrichment, like sunflower seeds or mini-treats, was offered by 39% of the institutions. Some facilities incorporated exercise equipment, such as running wheels or swings, for 29% of the rodents. Nest transfer during cage changes and weaning was practiced by 45% of the institutions. A smaller percentage, 18%, offered only nesting material, and 8% mentioned other forms of enrichment, with one respondent highlighting the use of group housing in an enrichment cage.

- Nesting material only 33/180 (18%)
- Nesting material plus other provisions 141/180 (78%)
- Structure (i.e. tunnel, igloo, shelter) 142/180 (79%)
- Chew manipulanda (e.g., wood, nylabone, other plastic or wood item) 110/180 (61%)
- Exercise equipment (e.g., running wheel, swing) 53/180 (29%)
- Nutritional (e.g., sunflower seeds, mini-treats) 71/180 (39%)
- Addition of environmental enrichment for single housed animal 104/180 (58%)
- Nest transfer during cage change and/or during new cage set up and/or weaning 81/180 (45%)
- Other 15/180 (8%)

32:14 Which of the following species do you typically house individually when there is no scientific, veterinary, medical, or behavioral reason not to socially house the animals?

Reg. The AWAR (§3.81,a) state: "The environment enhancement plan must include specific provisions to address the social needs of nonhuman primates of species known to exist as social groups in nature." The AWAR allow exemption from social housing if the animal has an infectious disease, shows aggressive behavior, or is incompatible with other animals. The AWAR (§3.81,a,2) state that individually housed nonhuman primates must be able to see and hear nonhuman primates of their own or compatible species, unless the veterinarian determines that this contact would compromise the well-being of the animal. The AWAR include similar provisions for birds (§3.154).

The Overview of the *Guide* (p. xviii) states: "With respect to nonhuman primates (NHPs), the Committee endorses social housing as the default and has provided some species-specific guidance." The *Guide* (p. 64) states: "Single housing of social species should be the exception and justified based on experimental requirements or veterinary-related concerns about animal well-being. In these cases, it should be limited to the minimum period necessary, and where possible, visual, auditory, olfactory, and tactile contact with compatible conspecifics should be provided." The authors of the *Guide* recognized that social housing is not always possible for experimental, health, and behavioral reasons. The *Ag Guide* (2020) states that all agricultural animals are social, with the exception of the adult boar, and social enrichment for common agricultural animals used in research and education are described.

Regarding the social housing of nonhuman primates, NIH/OLAW states in FAQ F.14: "There is universal agreement among oversight agencies that nonhuman primates should be socially house. … Single housing of social primates is considered an exception. ... Exemptions to the social housing requirement must be based on strong scientific justification approved by the IACUC or for a specific veterinary or behavioral reason. Lack of appropriate caging does not constitute an acceptable justification for exemption. When necessary, single housing of social animals should be limited to the minimum period necessary."

AAALAC has issued a Position Statement[69] that "Social housing will be considered by AAALAC International as the default method of housing unless otherwise justified based on social incompatibility resulting from inappropriate behavior, veterinary concerns regarding animal well-being, or scientific necessity approved by the IACUC. … The institution's policy and exceptions for single housing should be reviewed on a regular basis and approved by the IACUC … and/or veterinarian."

Opin. The provision of opportunities for social housing of compatible animals is perhaps the most effective means of environmental enrichment for nonhuman primates and other social research animal species. A compatible social partner is an ever-changing stimulus that can also provide comfort through mutual grooming and physical proximity, aid in thermoregulation, serve as a playmate, etc.

Because social relationships among some species of animals, such as primates, are frequently based on hierarchical relationships that require caution when forming social pairs or groups of animals and can change with time, ongoing monitoring of the animals' compatibility is essential to prevent harm resulting from aggressive encounters. However, social interactions are considered essential to both the welfare of the primates and their use as a research model. Indeed, "Social interactions are considered to be one of the most important factors influencing the psychological wellbeing of nonhuman primates. Knowing that most primates benefit from social interactions, it should be obvious that they can be harmed by a lack of social interaction."[70]

Some strains of rodents engage in barbering behavior or more overt aggression that results in the wounding of cage mates, possibly due to territorial behavior expressed over an enrichment item.[71] A crowdsourcing study[72] illustrated that in male mice, strain, the number of mice per cage, cage cleaning procedures, and transfer of nesting material impacted aggression. Some strains of mice provided only nesting material as an enrichment can be housed compatibly.[73] However, it is suggested that when rodents are housed with compatible conspecifics, they experience a social buffering effect that can reduce the impact of experimental stressors.[74]

As rabbits are often observed in social groups in nature, social housing of rabbits has recently received greater attention. Efforts to evaluate the effects of social housing on rabbits have supported the challenge associated with this type of housing. Nevalainen et al.[75] confirmed that female rabbits were compatible in a social housing setting only until sexual maturity, while Vervaecke et al.[76] identified linear hierarchies being established as early as 10 weeks of age. Love[77] provided a thorough review of the behavioral biology of the rabbit in the wild with a well-reasoned extrapolation of approaches and obstacles to social housing these animals in laboratories. DiVincenti and Rehrig (2016) discuss the implications of social housing rabbits used in research and offer guidance on assessing the potential for success.[78] Therefore, personnel should be familiar with the behavior of the animal (at the species as well as individual level), the research study, and the possible consequences of social housing on the research objectives before commencing social housing.

Surv. Which of the following species do you typically house individually when there is no specific scientific, veterinary, medical, or behavioral reason not to socially house the animals (e.g., for fighting or due to concern of injury)?

The majority of respondents indicated that rabbits (24%) and mice (21%) were the species they typically house individually when there are no specific scientific, veterinary, medical, or behavioral reasons not to socially house the animals. Rat species followed closely behind at 14%. Reptile and amphibian species were individually housed by 12% of the institutions, while agricultural animal species and avian species were least likely to be individually housed at 6% and 4%, respectively. A significant 48% of respondents selected "Other," indicating a variety of species, including guinea pigs, hamsters, opossums, and lemurs, among others, as animals typically housed individually.

• Nonhuman primates	7/180 (4%)
• Dog	9/180 (5%)
• Cat	3/180 (2%)
• Rabbit	43/180 (24%)
• Agricultural animal species	10/180 (6%)
• Rat	26/180 (14%)
• Mouse	38/180 (21%)
• Fish	8/180 (4%)
• Reptile and amphibian species	21/180 (12%)
• Avian species	7/180 (4%)
• Other	87/180 (48%)

32:15 Does your institution supplement the animals' standard diet by providing food treats as part of an enrichment program for the following species?

Reg. With regard to nonhuman primates, the AWAR (§3.81,b) state: "Examples of environmental enrichments include … varied food items; using foraging or task-oriented feeding methods." The *Guide* (p. 67) notes: "In some species (e.g., nonhuman primates) and on some occasions, varying nutritionally balanced diets and providing 'treats,' including fresh fruit and vegetables, can be appropriate and improve well-being. Scattering food in the bedding or presenting part of the diet in ways that require the animals to work for it (e.g., puzzle feeders for nonhuman primates) gives the animals the opportunity to forage, which, in nature, normally accounts for a large proportion of their daily activity." The *Guide* (p. 53) offers the caution that "foraging devices can lead to increased body weight (Brent 1995)." The *Ag Guide*[10] offers examples of nutritional enrichment, defined as involving "either presenting varied or novel food types or changing the method of food delivery," for a variety of agricultural animals used in research. The *Ag Guide* cautions that nutritional enrichment might be contraindicated when it is a part of the research.

Opin. Creating opportunities for nonhuman primates to express foraging behavior has been proposed as a beneficial enrichment technique for many years, though the degree of benefit is variable in terms of the length of time stereotypic behavior is mitigated[79–82] or based on the type of foraging device used.[83] Food enrichment with rabbits has also been evaluated, with the finding that rabbits spent more time engaged with the food enrichment than non-food enrichment items and that a specific type of food enrichment was preferred. Scattering food in a pen containing socially housed rabbits has been recommended to reduce boredom and decrease incidents of aggression.[84] Chemically defined food items are commercially available, which can be especially useful for rabbits used in toxicology studies.[85] Food enrichment has been recommended in cats[86] and dogs[87] to reduce boredom and stimulate activity that simulates natural behaviors of searching for and obtaining food. Cautions regarding monitoring these animals for weight gain and for destruction of toys and other devices in which the food items are placed are offered. Food enrichment for rodents has been suggested as a means of stimulating natural foraging behaviors and reducing boredom.[88] Novel food enrichments are also proposed for rodents,[89,90] though potential contamination of feed provided on the cage floor should be a consideration. Nutritional enrichment has been suggested as one means of reducing oral stereotypic behavior in cattle.[10]

Surv. Does your institution supplement the animals' standard diet by providing food treats as part of an enrichment? If so, for which of the following species?

The majority of institutions supplement their animals' standard diet with food treats as part of enrichment, with 39% providing treats for rabbits and similar percentages for mice (33%) and rats (32%). Nonhuman primates also received treats from 26% of the institutions. For other species, agricultural animals, dogs, and cats received treats at 17%, 17%, and 6%, respectively. Avian species and reptile and amphibian species received treats from 12% and 9% of the institutions, respectively. A diverse range of species was indicated in the "Other" category, including guinea pigs, ferrets, bats, and wildlife, among others, with treats being provided by 17% of the institutions for these animals.

- We do not provide treats to supplement the normal diet 46/180 (26%)
- Nonhuman primates 46/180 (26%)
- Dog 31/180 (17%)
- Cat 11/180 (6%)
- Rabbit 70/180 (39%)
- Agricultural animal species 31/180 (17%)
- Rat 58/180 (32%)
- Mouse 60/180 (33%)
- Fish 15/180 (8%)
- Reptile and amphibian species 16/180 (9%)
- Avian species 21/180 (12%)
- Other 30/180 (17%)

32:16 Does your institution provide manipulanda (e.g., toys/puzzles/chains/chewing material) as part of your enrichment program?

Reg. The AWAR (§3.81,b) state: "Examples of environmental enrichments include providing perches, swings, mirrors, and other increased cage complexities; providing objects to manipulate." The *Guide* (p. 53) recommends offering manipulative activities and cognitive challenges as part of the enrichment program, specifically "manipulable resources such as novel objects and foraging devices for nonhuman primates; manipulable toys for nonhuman primates, dogs, cats, and swine; wooden chew sticks for some rodent species; and nesting material for mice." The *Ag Guide*[10] describes "occupational enrichment" and "physical enrichment' as two of five types of enrichment. Both involve objects that the animal can control, which provides cognitive stimulation, or increases the complexity of the environment or available resources.

Opin. Species for which the provision of enrichment has a broader literature base (e.g., nonhuman primates, agricultural animals, rodents) or for which the behavior of the animals is well understood (e.g., dogs and cats) appear to receive enrichment more commonly than those species for which consideration of enriching the environment is a relatively new husbandry practice (e.g., fish, amphibians). But even this trend may be changing, as the increased use of fish in research has led to more reports in the literature regarding the influence of enrichment on fish behavior. For example, studies have demonstrated more flexible behavior in fish reared in structurally complex environments, the promotion of social learning, and the expression of dominance displays that match those expressed in wild counterparts.[91–93] Of note, as has been observed with rodents, the effects (or lack of effect) of enrichment varies with the species of fish.[94] As more scientific studies are published regarding the effects of enrichment on aquatic species, husbandry practices may more routinely incorporate suitable enrichment methods for these animals as well.

The National Agricultural Library's Animal Welfare Information Center has links to several databases pertaining to enrichment techniques for a variety of research animals.[95] However, it is not all-inclusive and thus literature searches for current information on enrichment for a specific species of animal (e.g., through PubMed) is an excellent adjunct approach to designing a meaningful enrichment program that is based on the successful experience of others, with a reduced risk of harm to the animals. Further, those with an institutional librarian should rely on their expertise to assist scientists and the IACUC in literature searches, whether to search for alternatives or for potential improvements to the animal care and use program, e.g., environmental enrichment.

Surv. Does your institution provide manipulanda (e.g., toys/puzzles/chains/chewing material) as enrichment? If so, for which of the following species?

The majority of institutions provide manipulanda as enrichment for their animals, with rats at 52% and mice at 46% being the most commonly enriched species. Rabbits followed closely at 38%, while agricultural animal species were enriched by 22% of the institutions. Nonhuman primates received manipulanda enrichment from 25% of the respondents, and dogs and cats were enriched at 19% and 11%, respectively. Avian species received this type of enrichment from 13% of the institutions. Fish, reptile and amphibian species, and the "Other" category, which included chinchillas, ferrets, and swine, and others, had lower rates of manipulanda provision at 2%, 3%, and 17%, respectively.

- Manipulanda as enrichment not provided 33/180 (18%)
- Nonhuman primates 45/180 (25%)
- Dog 35/180 (19%)
- Cat 19/180 (11%)
- Rabbit 68/180 (38%)
- Agricultural animal species 39/180 (22%)
- Rat 93/180 (52%)
- Mouse 82/180 (46%)
- Fish 4/180 (2%)
- Reptile and amphibian species 5/180 (3%)
- Avian species 23/180 (13%)
- Other 31/180 (17%)

32:17 Does your institution provide structural enrichment to the animals' cage/enclosure (e.g., huts, tubes, boxes, shelves, perches, beds, nesting material) as part of an enrichment program?

Reg. The AWAR (§3.81,b; §3.154,b) recommend that primate and bird enrichment include "providing perches, swings, mirrors, and other increased cage complexities." The AWAR also require elevated resting surfaces for cats (§3.6,b,4). The *Guide* (p. 53) states: "Examples of enrichment include structural additions such as perches and visual barriers for nonhuman primates (Novak *et al.* 2007); elevated shelves for cats (Overall and Dyer 2005; van den Bos and de Cock Buning 1994) and rabbits (Stauffacher 1992); and shelters for guinea pigs (Baumans 2005)." The *Guide* (p. 53) further notes: "Well-conceived enrichment provides animals with choices and a degree of control over their environment, which allows them to better cope with environmental stressors (Newberry 1995). For example, visual barriers allow nonhuman primates to avoid social conflict; elevated shelves for rabbits and shelters for rodents allow them to retreat in case of disturbances (Baumans 1997; Chmiel and Noonan 1996; Stauffacher 1992); and nesting material and deep bedding allow mice to control their temperature and avoid cold stress during resting and sleeping (Gaskill *et al.* 2009; Gordon 1993, 2004)." The *Ag Guide* describes physical enrichment, "which can involve ... objects, substrate, or permanent structures." Calves that are offered more complex environments exhibit more play behavior[96] and a similar finding has been observed in swine.[97,98] Swine also show reduced aggression when visual barriers are included in the enclosure.[99]

Opin. The commercial production of various enrichment devices for several types of research animals has increased significantly over the last several years. Many of these items increase the structural complexity of the animal's cage environment. Examples for rodents include round, square, arched, or tented tunnels; huts/shacks (which are either destructible or indestructible); balls; igloos/domes (with or without a running wheel); and a variety of types of nesting material such as wood, paper, and cotton. Dogs housed in runs or larger caging are often provided with raised beds. Felines housed in rooms or large caging are provided with climbing structures and shelves. Similarly, nonhuman primates are provided opportunities for perching, climbing, hiding, swinging, and numerous other activities through the use of added cage "furniture," the specific type depending on the species of primate and quantity of space available. For many species of agricultural animals, structural elements, such as dividers or climbing structures, can reduce agonistic behaviors.[97–99]Aquatic animals, such as zebrafish, show a preference for habitats with vegetation,[100] including artificial vegetation[101] Stevens *et al.* (2021) have provided a review of the benefits and challenges associated with enrichment for zebrafish.[102]

Surv. Does your institution provide structural enrichment to the animals' cage/enclosure (e.g., huts, tubes, boxes, shelves, perches, beds, nesting material)? If so, for which of the following species?

The majority of institutions provide structural enrichment to their animals' cages or enclosures, with mice and rats being the most commonly enriched species at 80% and 74%, respectively. Rabbits followed at 36%, while reptile and amphibian species and avian species were both enriched at 27%. Fish received structural enrichment from 19% of the institutions. Among domestic animals, cats and dogs were enriched at 12% and 11%, respectively, while agricultural animal species had an 11% enrichment rate. Nonhuman primates received structural enrichment from 20% of the respondents. The "Other" category, which included a variety of species such as chinchillas, ferrets, pigs, opossums, and octopus had a 15% rate of structural enrichment provision. Only 8% of the institutions did not provide any form of structural enrichment.

• Structural enrichment not provided	14/180 (8%)
• Nonhuman primates	36/180 (20%)
• Dog	19/180 (11%)
• Cat	21/180 (12%)
• Rabbit	65/180 (36%)
• Agricultural animal species	20/180 (11%)
• Rat	133/180 (74%)
• Mouse	144/180 (80%)
• Fish	34/180 (19%)
• Reptile and amphibian species	48/180 (27%)
• Avian species	48/180 (27%)
• Other	27/180 (15%)

32:18 Does your institution provide periodic access to increased space as enrichment (e.g., for exercise, behavior, or social opportunity)?

Reg. The AWAR (§3.8) require that dogs be provided the opportunity to exercise regularly if they are kept individually in cages, pens, or runs that provide less than two times the required floor space for that dog or if the dogs are housed in groups if the enclosure does not provide in total at least 100 percent of the required space for each dog if housed separately. The *Guide* echoes this recommendation, stating: "Dogs and cats, especially when housed individually or in smaller enclosures (Bayne 2002), should be allowed to exercise."

Opin. The *Guide* (p. 55) states: "An animal's space needs are complex and considerations of only the animal's body weight or surface area may be inadequate." The *Guide* lists several criteria that should be considered when addressing space needs of research animals (e.g., age, natural behavior, special needs, etc.). The fundamental principle for offering part-time access to a larger space is the provision of an opportunity to express more diverse species-typical behaviors. These might involve compatible social interactions, locomotor or postural behaviors that cannot be expressed in the primary enclosure, and/or engaging with novel structures, objects, or substrates not available in the primary enclosure (i.e., physical or sensory enrichment).

For example, the European Convention for the Protection of Vertebrate Animals Used for Experimentation and Other Scientific Purposes (ETS 123, Appendix A),[103] a primary standard used by AAALAC International, states: "Outside runs provide an environmental enrichment opportunity for cats … and should be provided where possible." Institutions that house cats in indoor/outdoor runs can provide part-time access to additional space that offers physical and sensory enrichment. Alternatively, cats may be removed from their home cages and provided access to a larger space for access to climbing, scratching, toy-directed play, and/or compatible social opportunities. Dogs may be introduced to a play area with other compatible dogs, which may be further enriched with climbing and hiding structures. Part-time access to larger indoor caging or outdoor enclosures is also a means of enriching nonhuman primates.

Surv. Does your institution provide periodic access to increased space as enrichment (e.g., for exercise, behavior, or social opportunity) to any of the species?

The majority of institutions, 46%, do not provide periodic access to increased or enriched space as a form of enrichment for their animals. Among the species that do receive increased space, dogs and rabbits lead the way with 23% and 22%, respectively. Agricultural animal species and nonhuman primates follow at 12% and 17%. Cats, rats, and mice have lower rates at 6%, 9%, and 8%, respectively. Avian species and the "Other" category, which includes ferrets, guinea pigs, swine, and marine mammals, are provided with increased space by 4% and 9% of the institutions, respectively. Fish and reptile and amphibian species have the lowest rates at 2%.

• Access to increase/enriched space not provided	82/180 (46%)
• Nonhuman primates	31/180 (17%)
• Dog	41/180 (23%)
• Cat	10/180 (6%)
• Rabbit	40/180 (22%)
• Agricultural animal species	22/180 (12%)
• Rat	17/180 (9%)
• Mouse	14/180 (8%)

- Fish 3/180 (2%)
- Reptile and amphibian species 4/180 (2%)
- Avian species 8/180 (4%)
- Other 17/180 (9%)

32:19 What is your institution's approach to mitigate aberrant behaviors in your research animals?

Reg. The AWAR direct the use of enrichment for animals showing signs of psychological distress evidenced in their behavior of physical appearance (§3.81,c,2). The AWAR (§3.81) require facilities "develop, document, and follow an appropriate plan for environmental enhancement adequate to promote the psychological well-being of nonhuman primates." The plan should also address the social needs of the primates. In this manner, the regulations take both a preventive and a mitigating approach. The *Guide* recognizes that appropriate enrichment that provides animals with choices and some control over their environment can lead to better coping and reduced stress for the animals.[104,105] The *Ag Guide* specifically acknowledges that environmental enrichment can "reduce the incidence or severity of undesirable or abnormal behaviors."

Opin. Reduction (though not necessarily elimination) of some abnormal behaviors has been achieved through the use of environmental enrichment techniques targeted to specific behaviors.[106] Some success has been achieved in mitigating overgrooming behavior in nonhuman primates with the provision of a grooming board, a simple deflection of the behavior rather than elimination of the cause of it.[106] Similarly, excessive pacing in nonhuman primates may be interrupted by distracting the animal with a high-value food item in a foraging device.[13] However, many abnormal behaviors in nonhuman primates are deeply entrenched and reduction of their expression, let alone their elimination, is often not successful.[107] Such behaviors may not be a reflection of compromised welfare due to current housing conditions, but rather an indication of a previous period of the animal's life when welfare was compromised, such as early weaning or inadequate socialization. Therefore, while consideration should be given to targeting some abnormal behaviors with specific enrichment strategies, taking into consideration the species and strain, this approach should be buttressed with a fully implemented enrichment program that encompasses a diversity of methods and encompasses both prevention and mitigation strategies.[108] The importance of social housing nonhuman primates to prevent and mitigate many abnormal behaviors cannot be overstated.

The positive effects of enrichment for reducing the expression of undesirable behaviors in a diverse range of research animals have been documented for several years. Early examples include the provision of artificial turf as a foraging substrate for nonhuman primates significantly reducing the expression of four types of abnormal behaviors[13] and the addition of a cardboard tube in the home cage resulting in a 40% reduction in wire-gnawing in male ICR mice.[109] The body of literature has continued to grow, illustrating the benefits of a variety of enrichment techniques for numerous types of research animals. For example, in female C57BL/6NCrl mice, barbering has been shown to be eliminated by the provision of "nest pucks," paper strips woven and compressed into a puck shape.[110] Customizing the enrichment approach for various strains of mice has been recommended, as a more generalized approach was not found to be as effective unless the mice were "super enriched."[111] Food enrichment was shown to significantly reduce the expression of abnormal behaviors (e.g., stereotypic head movement) in individually housed New Zealand rabbits (Ceballos et al. 2016). Adult zebrafish exposed to classical music were less anxious in a novel tank test and more active compared to nonenriched controls, evidenced not only by their behavior by also by whole-body cortisol assay,[112] while zebrafish larvae manifest reduced anxiety-like behavior when housed in a tank with structural complexities (Lego® bricks of different shapes and colors[113]).

Surv. What is your institution's approach to mitigate aberrant behaviors in your research animals?

The majority of institutions, at 61%, evaluate animals showing signs of aberrant behavior and consider the continuation of the study in consultation with the Principal Investigator, veterinary staff, and/or IACUC (Institutional Animal Care and Use Committee) members. Enrichment is provided for animals with concerning behaviors by 42% of the institutions, while 27% offer enrichment and document a change in behavior. About 41% of the respondents indicated that they have no specific approach for addressing aberrant behaviors, treating them as clinical cases if they arise. A small percentage, 2% (4/180), mentioned pharmacological intervention as an "Other" approach to mitigate aberrant behaviors.

- No specific approach, they are treated as clinical cases, if they arise. 74/180 (41%)
- Enrichment is given for animals with concerning behaviors. 75/180 (42%)
- Enrichment is given and a change in behavior is documented. 49/180 (27%)
- Animals showing signs of aberrant behavior are evaluated and continuation on study evaluated by the PI, veterinary and/or IACUC staff. 110/180 (61%)
- Other. 4/180 (2%)

32:20 Does your institution use operant conditioning or positive reinforcement training (PRT)?

Reg. Animal training is included in the AWAR, Part 1: Definition of Terms, for "Handling, which include: "petting, feeding, watering, cleaning, manipulating, loading, crating, shifting, transferring, immobilizing, restraining, treating, training, working and moving, or any similar activity with respect to any animal" (§1.1). While AWAR specifically refer to the use of training only for marine mammals, emphasizing the importance of personnel experience and supervision (§3.108), they include interaction with caregivers as one method of environmental enrichment for nonhuman primates and birds (§3.81,b; §3.154,b). OLAW's FAQ F15 states: "Should positive reinforcement training be used for nonhuman primates? In situations where it is safe and feasible, nonhuman primates should be given positive reinforcement training to perform desired cooperative activities involved with research and husbandry. This type of training may also aid in reducing stress from capture and restraint and the need for chemical darts." The *Guide* offers a more expanded recommendation, stating that animals "placed in restraint devices should be given training (with positive reinforcement) to adapt to the equipment and personnel" (p. 29). The *Guide* further recommends "highly preferred food or fluid as positive reinforcement, instead of restriction" (p.31). Additional training references are made for "habituating animals to routine husbandry or experimental procedures" (p. 64), indicating that the "principles of operant conditioning may be employed during training sessions, using progressive behavioral shaping, to induce voluntary cooperation with procedures" (p. 65).

Opin. The terms positive reinforcement training (PRT), operant conditioning and animal training are often used interchangeably to describe purposeful human-animal interactions that encourage an animal to cooperate with a task on a routine basis in return for a reward. Usually high-value food or fluid rewards are used to shape the desired behavior.[114] By pairing negative experiences (e.g., an injection) with a positive reward, the animal is desensitized to the negative experience and instead a positive association is formed. Techniques used to encourage an animal to participate cooperatively are described by subject matter experts through publications (e.g., Magden 2017[115]), through web-based or in-person training (e.g., Perlman 2021[116]), or through professional training organizations dedicated to the training of domestic animals (e.g., Pryor 2014[117]). The most prevalent and perhaps the most important application of PRT is procedural training, where the animals are rewarded for cooperating in diagnostic and treatment procedures such as blood collection and injection to reduce the need for chemical immobilization.[118] PRT may have applications in experimental procedures that currently rely on food or fluid restriction, instead of cooperative training using highly valued rewards, such as cognitive research. It is also important to note that PRT can readily be applied to daily activities, as any time a staff member interacts with an animal there is some level of communication occurring that can be leveraged as training.[119] Training can be used as enrichment, to reduce stereotypic behavior,[120] to gain cooperation with husbandry procedures, to reduce aggression,[121] and to adjust social hierarchy.[122] Despite the potential benefits, training is not widely incorporated in research animal care programs, likely due to the perceived time investment and availability of skilled personnel. Training professionals contend that the time invested in training is saved during animal cooperation with procedures, and it is recommended to incrementally encourage the application of training for a single animal, process, or procedure.[123]

Surv. Does your institution use operant conditioning or positive reinforcement training?

The majority of institutions, 66%, use operant conditioning or positive reinforcement training as a part of the experimental design for certain studies. Additionally, 39% employ this approach for husbandry or veterinary applications, such as blood collection with minimal restraint. For enrichment or behavioral shaping purposes, 26% of the institutions use operant conditioning or positive reinforcement training. A smaller portion, 17%, indicated "Other" uses for this training method but do not specify.

- As a part of the experimental design for certain studies — 118/180 (66%)
- For husbandry or veterinary applications (e.g., blood collection with minimal restraint) — 70/180 (39%)
- For enrichment or behavioral shaping purposes (e.g., to respond to aberrant behavior) — 47/180 (26%)
- Other — 31/180 (17%)

32:21 Are social, exercise, and/or behavioral needs of the species considered in the purchase of animal housing enclosures or construction and design of animal facilities?

Reg. The AWAR indicate that primary enclosures be constructed to allow sufficient space to make "normal postural adjustments with freedom of movement," for nonhuman primates, rabbits, guinea pigs, hamsters, birds, and other warm-blooded animals regulated under the AWA (§ 3.8,c; 3.53,c,1; 3.28,b,1; 3.153,b; 3.128). It is further specified for dogs and cats to "provide sufficient space to allow each dog and cat to turn about freely, to stand, sit and lie in a comfortable, normal position, and to walk in a normal manner" (§ 3.6,a,2,xi.). The *Guide* speaks to the design of the animal enclosures, stating that "cage height should take into account the animal's typical posture and provide adequate clearance" (p. 56), and for social species; enclosures should be built to "facilitate the avoidance of social conflicts" (p. 64) and provide "enough feeding points to minimize competition for food" (p. 66). Additionally, specific mention is made for the design of housing for brachiating species that "cage design should enhance brachiating movement" (p. 61). The lack of appropriate caging is not sufficient justification for single housing of social species.[124]

Opin. Facility design should include specific components for social, exercise, and behavioral needs of the species early in the design process. By not doing so, challenges with accommodating the species' natural behavior, effective sanitization of enrichment substrates, and safety of personnel moving the species may arise. Retrofitting existing facilities to incorporate these components at a later date can be difficult. Enrichment items, while they may not have a direct impact on facility design, at a minimum should be considered for storage, preparation space, and refrigeration. Consideration should be given during the design of facilities to addressing the species-specific behavioral needs; requirements for increased space; complex caging systems; floor drains/troughs designed to prevent enrichment items from obstructing waste systems; and flexible housing options, including adaptations to the primary enclosure to accommodate individuals, social pairs, and larger groups.[125] An optimal facility should incorporate features to minimize animal stress and maximize the efficiency of movement.[129] Facility design and housing components can leverage the natural behaviors of the species to hide, flee, or flock, for example, to eliminate the need for physical handling during transitions for husbandry and research procedures. Inclusion of windows as a form of enrichment for some species (e.g., primates), indoor and outdoor housing as part of behavioral management program for large animals, or exercise runs in order to meet exercise requirements for dogs should be addressed in the planning process.

Surv. Are social, exercise, and/or behavioral needs of the species specifically considered in the purchase or construction and design of animal housing enclosures?

For the majority of institutions, 60%, involve veterinary medical, IACUC and/or enrichment staff in determining the needs for housing and space, considering social, exercise, and/or behavioral needs of the species in the purchase or construction and design of animal housing enclosures. Additionally, 51% of the institutions make enrichment and/or social housing considerations when purchasing or constructing enclosures. For any new construction or significant renovations, 32% always consider providing an enriched environment, while 16% prioritize it. Only 14% indicated that housing is typically selected to house as many animals as possible within limited space. A total of 4% of respondents selected "Other" approaches to considering animal needs in housing design.

- No, housing is typically selected to house as many animals as possible is the limited space. 25/180 (14%)
- Enrichment and/or social housing considerations are made in the purchase or construction of enclosures. 91/180 (51%)
- Veterinary, IACUC, and/or enrichment staff are involved in determining needs the needs for housing and space. 108/180 (60%)
- For any new construction or significant renovations, providing an enriched environment is always usually considered. 57/180 (32%)
- For any new construction or significant renovations, providing an enriched environment is always prioritized. 29/180 (16%)
- Other. 7/180 (4%)

REFERENCES

1. National Research Council. 1985. *Guide for the Care and Use of Laboratory Animals.* U.S. Department of Health and Human Services, Public Health Service, National Institutes of Health.
2. National Research Council. 1996. *Guide for the Care and Use of Laboratory Animals.* Washington, DC: National Academies Press.
3. National Research Council. 2011. *Guide for the Care and Use of Laboratory Animals.* Committee for the Update of the Guide for the Care and Use of Laboratory Animals. Washington, DC: National Academies Press.
4. Association for the Assessment and Accreditation of Laboratory Animal Care, International. *Position Statements "Cage and Pen Space" and "Social Housing"* In Accreditation Program. Online: www.aaalac.org/accreditation-program/position-statements/. Accessed May 16, 2024.
5. Association for the Assessment and Accreditation of Laboratory Animal Care, International. 2022. In *Accreditation Program.* FAQs. Online: www.aaalac.org/accreditation-program/faqs/.
6. U.S. Department of Agriculture and Office of Laboratory Animal Welfare. 2005. *Enrichment for Nonhuman Primates.* U.S. Public Health Service. Online: http://grants.nih.gov/grants/olaw/request_publications.htm.

7. National Institutes of Health, Office of Laboratory Animal Welfare. 2012. AC and OLAW perspective on nonhuman primate enrichment and social housing. Online: http://grants.nih.gov/grants/olaw/educational_resources.htm#special-seminars.
8. U.S. Department of Agriculture, National Agricultural Library, Animal Welfare Information Center. 2012. *Environmental Enrichment for Nonhuman Primates.* Online: http://www.nal.usda.gov/awic/pubs/Primates2009/primates.shtml.
9. National Institutes of Health, Office of Laboratory Animal Welfare. 2012. Frequently Asked Questions: PHS Policy on Humane Care and Use of Laboratory Animals. Online: http://grants.nih.gov/grants/olaw/faqs.htm.
10. Tucker, C.B., M.D. MacNeil, G. Delta, and A.B. Webster. 2020. *Guide for the Care and Use of Agricultural Animals in Research and Teaching.*
11. Turner, P.V. and K. Bayne. 2023. Research animal behavioral management programs for the 21st century. *Animals,* 13(12):1919.
12. Fiorito, G., A. Affuso, J. Basil, A. Cole, P. de Girolamo, L. D'angelo, and P.L. Andrews. 2015. Guidelines for the care and welfare of cephalopods in research–a consensus based on an initiative by CephRes, FELASA and the Boyd Group. *Lab Anim,* 49(2 suppl):1–90.
13. Bayne, K., S. Dexter, H. Mainzer, *et al.* 1992. The use of artificial turf as a foraging substrate for individually housed rhesus monkeys (*Macaca mulatta*). *Anim Welf,* 1:39–53.
14. Fillman-Holliday, D. and M. Landi. 2002. Animal care best practices for regulatory testing. *ILAR J,* 43:49–58.
15. Bayne, K. 2005. Potential for unintended consequences of environmental enrichment for laboratory animals and research results. *ILAR J,* 46:129–139.
16. Wolfle, T.L. 1991. Psychological well-being: The billion-dollar solution. American Psychological Association. *Through the Looking Glass: Issues of Psychological Well-Being in Captive Nonhuman Primates.* (pp. 119–128).
17. Novak, M.A. and A.J. Petto. 1991. American Psychological Association. *Through the Looking Glass: Issues of Psychological Well-Being in Captive Nonhuman Primates* (pp. xiii–285).
18. Tobey, E. and K. Adams. 2021. Environmental Enrichment for Nonhuman Primates Resource Guide. Animal Welfare Information Center, National Agricultural Library. Series No. 47.
19. Bayne, K. 1985. Qualitative observations of idiosyncratic behavior in old monkeys. In *Behavior and Pathology of Aging in Rhesus Monkeys,* eds. R.T. Davis, and C.W. Leathers, 201–222. New York: Alan R. Liss.
20. Rothschild, B. 2012. Interpretation of primate behavior, ambulation and biomechanics: Caveat arthritis. *J Primatol,* 1(3):1000e112.
21. National Institutes of Health. 1996. Institutional administrator's manual for laboratory animal care and use. *NIH Publication,* 88–2959.
22. National Research Council. 2003. *Occupational Health and Safety in the Care and Use of Nonhuman Primates.* Washington, DC: National Academies Press.
23. Morgan, K.N., S.W. Line, and H. Markowitz. 1998. Zoos, enrichment, and the skeptical observer: The practical value of assessment. In *Second Nature: Environmental Enrichment for Captive Animals,* eds. D.J. Shepherdson, J.D. Mellen and M. Hutchins, 153–171. Washington, DC: Smithsonian Institution Press.
24. Brent, L. 1995. Feeding enrichment and body weight in captive chimpanzees. *J Med Primatol,* 24:12–16.
25. De Boer, S.F. and J.M. Koolhaas. 2003. Defensive burying in rodents: ethology, neurobiology and psychopharmacology. *Eur J Pharmacol,* 463:145–161.
26. Bayne, K. and H. Würbel. 2014. The impact of environmental enrichment on the outcome variability and scientific validity of laboratory animal studies. *Revue scientifique et technique-Office international des épizooties,* 33(1):273–280.
27. Bayne, K. and H. Wurbel. 2012. Mouse enrichment. In *The Laboratory Mouse,* 2nd edition, ed. H. Hedrich, 545–564. New York: Elsevier.
28. Toth L.A., K. Kregel, L. Leon, et al. 2011. Environmental enrichment of laboratory rodents: The answer depends on the question. *Comp Med,* 61:314–321.
29. Würbel, H. 2001. Ideal homes? Housing effects on rodent brain and behaviour. *Trends Neurosci,* 24:207–211.
30. Richter S.H., J.P. Garner, B. Zipser, et al. 2011. Effect of population heterogenization on the reproducibility of mouse behavior: A multi-laboratory study. *PLoS One,* 6(1):e16461.
31. Van der Stay, F.J. and T. Steckler. 2002. The fallacy of behavioral phenotyping without standardization. *Genes Brain Behav,* 1:9–13.
32. Würbel, H. 2002. Behavioral phenotyping enhanced - Beyond (environmental) standardization. *Genes Brain Behav,* 1:3–8.
33. Richter, S.H., J.P. Garner, and H. Würbel. 2009. Environmental standardization: Cure or cause of poor reproducibility in animal experiments? *Nat Methods,* 6:257–261.
34. Gerdin, A.-K., N. Igosheva, L.-A. Roberson, et al. 2012. Experimental and husbandry procedures as potential modifiers of the results of phenotyping tests. *Physiol Behav,* 106:602–611.
35. Baumans, V. 2010. The laboratory mouse. In *The UFAW Handbook on the Care and Management of Laboratory and Other Research Animals,* 8th edition, eds. R. Hubrecht, and J. Kirkwood, 276–310. Oxford: Wiley-Blackwell.
36. Shomer, N.H., S. Peikert, and G. Terwilliger. 2001. Enrichment-toy trauma in a New Zealand White rabbit. *Contemp Top Lab Anim Sci,* 40(1):31–32.
37. Bazille, P.G., S.D. Walden, B.L. Koniar, *et al.* 2001. Commercial cotton nesting material as a predisposing factor for conjunctivitis in athymic nude mice. *Lab Anim, (NY)* 30(5):40–42.
38. Rowan, K.E.K. and L. Michaels. 1980. Injury to young mice caused by cottonwool used as nesting material. *Lab Anim,* 14:187.
39. Hahn, N.E., D. Lau, K. Eckert, et al. 2000. Environmental enrichment-related injury in a macaque (*Macaca fascicularis*): Intestinal linear foreign body. *Comp Med,* 50:556–558.
40. Wolfle, T.L. 2005. Environmental enrichment. *ILAR J.* 46(2):79–82.
41. Bayne, K, S.L. Dexter, J.L. Hurst, *et al.* 1993. Kong® toys for laboratory primates: Are they really an enrichment or just fomites? *Lab Anim Sci,* 43:78–85.

42. Winnicker, C., B. Gaskill, J. Garner, and K.R. Prichett-Corning. 2012. *A Guide to the Behavior & Enrichment of Laboratory Rodents* (No. 636.02 GUI).
43. Hess SE, S. Rohr, B.D. Dufour, B.N. Gaskill, E.A. Pajor, and J.P. Garner. 2008. Home improvement: C57BL/6J mice given more naturalistic nesting materials build better nests. *JAALAS,* 47:25–31.
44. Gaskill BN, S.A. Rohr, E.A. Pajor, J.R. Lucas, and J.P. Garner. 2009. Some like it hot: Mouse temperature preferences in laboratory housing. *Appl Anim Behav Sci,* 116:279–285.
45. Baumans V. 1997. Environmental enrichment: Practical applications. In *Animal Alternatives, Welfare and Ethics,* eds. L.F.M. Van Zutphen and M. Balls, 187–197. Elsevier.
46. Wolfer, D.P., O. Litvin, S. Morf, R.M. Nitsch, H.P. Lipp, and H. Würbel. 2004. Cage enrichment and mouse behaviour. *Nature,* 432(7019):821–822.
47. Northrup, E., N. Held, H.J. Hedrich, and A. Bleich. 2012. Mice may become caught in nesting material. *Lab Anim,* 41(6):147–147.
48. Van Loo, P.L., J.A. Mol, J.M. Koolhaas, B.F. Van Zutphen, and V. Baumans. 2001. Modulation of aggression in male mice: influence of group size and cage size. *Physiol Behav,* 72(5):675–683.
49. Bayne, K. 2018. Environmental enrichment and mouse models: Current perspectives. *Anim Model Exp Med,* 1(2):82–90.
50. Würbel, H. and J. Garner. 2007. Refinement of rodent research through environmental enrichment and systematic randomization. National Centre for the Replacement, Refinement and Reduction of Animas in Research. Online: www.nc3rs.org.uk/sites/default/files/documents/Refinementenvironmentalenrichmentandsystematicrandomization.pdf.****
51. Olsson, I.A. and K. Dahlborn. 2002. Improving housing conditions for laboratory mice: a review of "environmental enrichment." *Lab Anim,* 36:243–270.
52. Hutchinson, E., A. Avery, and S. VandeWoude. 2005. Environmental enrichment for laboratory rodents. *ILAR J,* 46(2), 148–161.
53. Sherwin C.M. 1997. Observations on the prevalence of nest-building in non-breeding TO strain mice and their use of two nesting materials. *Lab Anim,* 31:125–132.
54. Gallo, M.S., A.Z. Karas, K. Pritchett-Corning, J.P. Garner, G. Mulder, and B.N. Gaskill. 2020. Tell-tale TINT: does the time to incorporate into nest test evaluate postsurgical pain or welfare in mice? *J Amer Assoc Lab Anim Sci,* 59(1):37–45.
55. Gouveia, K. and J.L. Hurst. 2013. Reducing mouse anxiety during handling: effect of experience with handling tunnels. *PloS One,* 8(6):e66401.
56. Van de Weerd H.A., P.L.P. Van Loo, L.F.M. Van Zutphen, J.M. Koolhaas, and V. Baumansn. 1997. Preferences for nesting material as environmental enrichment for laboratory mice. *Lab Anim,* 31:133–143.
57. Froberg-Fejko, K. 2014. Give a rat a bone: satisfying rodents' need to gnaw. *Lab Anim,* 43(10):378.
58. Pawlowicz A, A. Demner, and M.H. Lewis. 2010. Effects of access to voluntary wheel running on the development of stereotypy. *Behav Proc,* 83:242–246.
59. Harri, M., J. Lindblom, H. Malinen, M. Hyttinen, T. Lapveteläinen, S. Eskola, and H.J. Helminen. 1999. Effect of access to a running wheel on behavior of C57BL/6J mice. *Comp Med,* 49(4):401–405.
60. Richter, S.H., P. Gass, and J. Fuss. 2014. Resting is rusting: a critical view on rodent wheel-running behavior. *Neurosci,* 20(4):313–325.
61. Baumans, V. 2005. Environmental enrichment for laboratory rodents and rabbits: Requirements of rodents, rabbits and research. *ILAR J,* 46:162–170.
62. Lecker, J. and K. Froberg-Fejko. 2016. Using environmental enrichment and nutritional supplementation to improve breeding success in rodents. *Lab Animal,* 45(10):406–408.
63. Pritchett-Corning, K.R., R. Keefe, J.P. Garner, and B.N. Gaskill. 2013. Can seeds help mice with the daily grind? *Lab Anim,* 47(4):312–315.
64. Toyoda, A. 2020. Nutritional interventions for promoting stress resilience: Recent progress using psychosocial stress models of rodents. *Anim Sci J,* 91(1):e13478.
65. Van Loo, P.L.P., L.F.M. Van Zutphen, and V. Baumans. 2003. Male management: coping with aggression problems in male laboratory mice. *Lab Anim,* 37(4):300–313.
66. Olsson, I.A.S. and K. Westlund. 2007. More than numbers matter: The effect of social factors on behaviour and welfare of laboratory rodents and non-human primates. *Appl Anim Behav Sci,* 103(3–4):229–254.
67. Arakawa, H. 2018. Ethological approach to social isolation effects in behavioral studies of laboratory rodents. *Behav Brain Res,* 341:98–108.
68. Sherwin, C.M. 2004. The influences of standard laboratory cages on rodents and the validity of research data. *Anim Welf,* 13(1):9–15.
69. Association for Assessment and Accreditation of Laboratory Animal Care International. 2012. Social housing. Online: http://www.aaalac.org/accreditation/positionstatements.cfm#social.
70. National Research Council. 1998. *The Psychological Well-Being of Nonhuman Primates.* Washington, DC: National Academies Press.
71. Hannibal, D.L., E. Bliss-Moreau, J. Vandeleest, B. McCowan, and J. Capitanio. 2017. Laboratory rhesus macaque social housing and social changes: Implications for research. *Amer J Primatol,* 79(1):e22528.
72. Lidster, K., K. Owen, W.J. Browne, and M.J. Prescott. 2019. Cage aggression in group-housed laboratory male mice: an international data crowdsourcing project. *Sci Rep,* 9(1):1–12.
73. Howerton, C.L., J.P. Garner, and J.A. Mench. 2008. Effects of a running wheel-igloo enrichment on aggression, hierarchy linearity, and stereotypy in group-housed male CD-1 (ICR) mice. *Appl Anim Behav Sci,* 115(1):90–103.
74. Denommé, M.R. and G.J. Mason. 2022. Social buffering as a tool for improving rodent welfare. *J Am Assoc Lab Anim Sci,* 61(1):5–14.
75. Nevalainen, T.O., J.I. Nevalainen, F.A. Guhad, *et al.* 2007. Pair housing of rabbits reduces variances in growth rates and serum alkaline phosphatase levels. *Lab Anim,* 41:432–440.

76. Vervaecke, H., L. De Bonte, L. Maertens, F. Tuyttens, J.M.G. Stevens, and D. Lips. 2010. Development of hierarchy and rank effects in weaned growing rabbits (Oryctolagus cuniculus). *World Rabbit Sci*, 18(3):139–149.
77. Love, J.A. 1994. Group housing: Meeting the physical and social needs of the laboratory rabbit. *Lab Anim Sci*, 44:5–11.
78. DiVicenti, L. and A. Rehrig. 2016. The social nature of the European Rabbit (*Oryctolagus cuniculus*). *J Am Assoc Lab Anim Sci*, 55:729–736.
79. Byrne, G.D. and S.J. Suomi. 1991. Effects of woodchips and buried food on behavior patterns and psychological wellbeing of captive rhesus monkeys. *Am J Primatol*, 23:141–151.
80. Bayne, K., H. Mainzer, S. Dexter, *et al.* 1991. The reduction of abnormal behaviors in individually housed rhesus monkeys (*Macaca mulatta*) with a foraging/grooming board. *Am J Primatol*, 23:23–35.
81. Novak, M.A., J.H. Kinsey, M.J. Jorgensen, *et al.* 1998. Effects of puzzle feeders on pathological behavior in individually housed rhesus monkeys. *Am J Primatol*, 46:213–227.
82. Gottlieb, D.H., S. Ghirardo, D.E. Minier, *et al.* 2011. Efficacy of 3 types of foraging enrichment for rhesus macaques (*Macaca mulatta*). *J Am Assoc Lab Anim Sci*, 50:888–894.
83. Harris, L.D, L.B. Custer, E.T. Soranaka, *et al.* 2001. Evaluation of objects and food for environmental enrichment of NZW rabbits. *Contemp Top Lab Anim Sci*, 40(1):27–30.
84. Morton, D.B., M. Jennings, G.R. Batchelor, *et al.* 1993. Refinements in rabbit husbandry. Second report of the BVAAWF/FRAME/RSPCA/UFAW joint working group on refinement. *Lab Anim*, 27:301–329.
85. Bayne, K. 2003. Environmental enrichment of nonhuman primates, dogs and rabbits used in toxicology studies. *Toxicol Pathol*, 31(Suppl):132–137.
86. McCune, S. 2010. The domestic cat. In *The UFAW Handbook on the Care and Management of Laboratory and Other Research Animals*, 8th edition, eds. R. Hubrecht, and J. Kirkwood, 453–472. Oxford: Wiley-Blackwell.
87. MacArthur Clark, J. and C.J. Pomeroy. 2010. The laboratory dog. In *The UFAW Handbook on the Care and Management of Laboratory and Other Research Animals*, 8th edition, eds. R. Hubrecht and J. Kirkwood, 432–452. Oxford: Wiley-Blackwell.
88. Baumans, V. 2005. Environmental enrichment for laboratory rodents and rabbits: Requirements of rodents, rabbits and research. *ILAR J*, 46:162–170.
89. Brown, C. 2009. Novel food items as environmental enrichment for rodents and rabbits. *Lab Anim (NY)*, 38:119–120.
90. Brown, C. 2010. Organic wheatgrass as environmental enrichment. *Lab Anim (NY)*, 39:74–75.
91. Moberg, O., V.A. Braithwaite, K.H. Jensen, *et al.* 2011. Effects of habitat enrichment and food availability on the foraging behaviour of juvenile Atlantic Cod (*Gadus morhua L*). *Environ Biol Fish*, 91:449–457.
92. Strand, D.A., A.C. Utne-Palm, P.J. Jakobsen, *et al.* 2010. Enrichment promotes learning in fish. *Ma Eco-Prog Ser*, 412:273–282.
93. Berejikian, B.A., E.P. Tezak, S.C. Riley, *et al.* 2001. Competitive ability and social behaviour of juvenile steelhead reared in enriched and conventional hatchery tanks and a stream environment. *J Fish Biol*, 59:1600–1613.
94. Brydges, N.M. and V.A. Braithwaite. 2009. Does environmental enrichment affect the behaviour of fish commonly used in laboratory work? *Appl Anim Behav Sci*, 118(3):137–143.
95. National Agricultural Library's Animal Welfare Information Center. 2021. AWIC Resource Series. "Environmental Enrichments for Nonhuman Primates: A Resource Guide."
96. Jensen, M.B., K.S. Vestergaard, and C.C. Krohn. 1998. Play behavior in dairy calves kept in pens: the effect of social contact and space allowance. *Appl Anim Behav Sci* 56:97–108.
97. Dudink, S., H. Simonse, I. Marks, F.H. de Jonge, and B.M. Spruijt. 2006. Announcing the arrival of enrichment increases play behaviour and reduces weaning-stress-induced behaviours of piglets directly after weaning. *Appl Anim Behav Sci*, 101:86–101.
98. Casey, B., D. Abney, and E. Skoumbourdis. 2007. A playroom as novel swine enrichment. *Lab Anim (NY)*, 36:32–34.
99. McGlone, J.J. and S.E. Curtis. 1985. Behavior and performance of weanling pigs in pens equipped with hide areas. *J Anim Sci*, 60:20–24.
100. Barcellos, H.H., G. Koakoski, F. Chaulet, K.S. Kirsten, L.C. Kreutz, A.V. Kalueff, and L.J. Barcellos. 2018. The effects of auditory enrichment on zebrafish behavior and physiology. *Peer J*, 6:e5162.
101. Gatto, E., M. Dadda, M. Bruzzone, E. Chiarello, G. De Russi, M.D. Maschio, A. Bisazza, and T. Lucon-Xiccato. 2022. Environmental enrichment decreases anxiety-like behavior in zebrafish larvae. *Dev Psychobiol*, 64(3):e22255.
102. Stevens, C.H., Reed, B.T., and Hawkins, P. 2021. Enrichment for laboratory zebrafish—A review of the evidence and the challenges. *Animals*, 11(3):698.
103. Conseil de l'Europe, C. 1986. *European Convention for the Protection of Vertebrate Animals Used for Experimental and Other Scientific Purposes/Convention européenne sur la protection des animaux vertébrés utilisés à des fins expérimentales ou à d'autres fins scientifiques:[Strasbourg, 18. III. 1986]*. Conseil de l'Europe Section des publications.
104. Haskell, M.J., K. Masłowska, D.J. Bell, D.J. Roberts, and F.M. Langford. 2013. The effect of a view to the surroundings and microclimate variables on use of a loafing area in housed dairy cattle. *Appl Anim Behav Sci*, 147:28–33.
105. Ramirez, A., A. Quiles, M. Hevia, and F. Sotillo. 1995. Observations on the birth of goats. *Can J Anim Sci* 75:165–167.
106. Bayne, K., H. Mainzer, S. Dexter, *et al.* 1991. The reduction of abnormal behaviors in individually housed rhesus monkeys (*Macaca mulatta*) with a foraging/grooming board. *Am J Primatol*, 23:23–35.
107. O'Neill, P. 1989. A room with a view for captive primates: Issues, goals, related research and strategies. In *Housing, Care and Psychological Wellbeing of Captive and Laboratory Primates*, ed E.F. Segal, , 135–182. Park Ridge, NJ: Noyes Publications.
108. Würbel, H. and M. Stauffacher. 1998. Physical condition at weaning affects exploratory behaviour and stereotypy development in laboratory mice. *Behav Proc*, 43(1):61–69.

109. Bailoo, J.D., E. Murphy, M. Boada-Saña, J.A. Varholick, S, Hintze, C. Baussière, K.C. Hahn, C. Göpfert, R. Palme, B. Voelkl, and H. Würbel. 2018. Effects of cage enrichment on behavior, welfare and outcome variability in female mice. *Front Behav Neurosci*, 232.
110. Moody, C.M., E.A. Paterson, D. Leroux-Petersen, and P.V. Turner. 2021. Using paper nest pucks to prevent barbering in C57BL/6 Mice. *J Amer Assocr Lab Anim Sci*, 60(2):133–138.
111. Bayne, K. and L.C. Anderson. 2015. Laws, regulations, and policies affecting the use of laboratory animals. In *Laboratory Animal Medicine,* 23–42. Academic Press.
112. Engeszer, R.E., L.B Patterson, A.A. Rao, and D.M. Parichy. 2007. Zebrafish in the wild: a review of natural history and new notes from the field. *Zebrafish*, 4(1):21–40.
113. Spence, R. and C. Smith. 2008. Innate and learned colour preference in the zebrafish, Danio rerio. *Ethology*, 114(6):582–588.
114. Laule G.E., M.A. Bloomsmith, and S.J. Schapiro. 2003 The use of positive reinforcement training techniques to enhance the care, management, and welfare of primates in the laboratory. *J Appl Anim Welf Sci,* 6:163–173.
115. Magden, E.R. 2017. Positive reinforcement training and health care. In *Handbook of Primate Behavioral Management,* pp. 201–215. CRC Press.
116. Perlman, J.E., M.A. Bloomsmith, M.A. Whittaker, J.L. McMillan, D.E. Minier, and B. McCowan. 2012. Implementing positive reinforcement animal training programs at primate laboratories. *Appl Anim Behav Sci*, 137(3–4):114–126.
117. Pryor, K. and K. Ramirez. 2014. Modern animal training. In *The Wiley Blackwell Handbook of Operant and Classical Conditioning* (p. 455).
118. Graham, M.L. 2017. Positive reinforcement training and research. In *Handbook of primate behavioral management,* 187–200. CRC Press.
119. Laule, G. 2010. Positive reinforcement training for laboratory animals. In *The UFAW Handbook on the Care and Management of Laboratory and Other Research Animals,* 206–218.
120. Coleman, K. and A. Maier. 2010. The use of positive reinforcement training to reduce stereotypic behavior in rhesus macaques. *Appl Anim Behav Sci*, 124(3–4):142–148.
121. Minier, D.E., L. Tatum, D.H. Gottlieb, A. Cameron, J. Snarr, R. Elliot, *et al.* 2011. Human-directed contra-aggression training using positive reinforcement with single and multiple trainers for indoor-housed rhesus macaques. *Appl Anim Behav Sci*, 132(3–4):178–186.
122. Veeder, C.L., M.A. Bloomsmith, J.L. McMillan, J.E. Perlman, and A.L. Martin. 2009. Positive reinforcement training to enhance the voluntary movement of group-housed sooty mangabeys (*Cercocebus atys atys*). *J Amer Assoc Lab Anim Sci*, 48(2):192–195.
123. National Research Council. 1998. *The Psychological Well-Being of Nonhuman Primates.* Washington, DC: National Academies Press.
124. Hessler, J. and N. Lehner. 2011. *Planning and Designing Research Animal Facilities*. Academic Press.
125. Grandin, T. 1980. Livestock behavior as related to handling facilities design. *Int J Study Anim Probl,* 1(1):33–52.

33

Adverse Events and Protocol Non-compliance

Stacy L. Pritt and Madeline L. Budda

Introduction

IACUCs are charged by various regulatory mandates, as well as AAALAC International accreditation requirements, to review non-compliance and adverse events (also known as unanticipated outcomes, but with a less negative connotation). Non-compliance may be self-reported by researchers or identified during routine IACUC operations, the performance of animal husbandry or veterinary duties, or from allegations made by anyone within or outside of the animal care and use program. Adverse events are those events resulting in negative outcomes that are not described or anticipated within an approved IACUC protocol, including operational or equipment failures, natural disasters, or accidents associated with such failures or disasters that result in negative effects on animals. Oftentimes, adverse events are reported by researchers or veterinary staff and may or may not represent non-compliance depending on the circumstances. Collectively, non-compliance and adverse events are often called "animal welfare concerns," even though there may be non-compliance or adverse events that do not involve animal procedures or care. Limiting available IACUC and institutional reporting mechanisms, as discussed later in this chapter, to "animal welfare concerns" may inadvertently preclude the reporting of adverse events or non-compliance that do not directly involve research animals but may still fall within the IACUC's purview. Animal care and use programs are encouraged to adopt terminology that promotes appropriate reporting of adverse events and protocol non-compliance and that is readily understood by the institution's research community.

Many IACUCs refer to reviews of adverse events and incidents of non-compliance as "investigations," even upon the immediate receipt of an allegation and/or at the time of gathering facts about the situation, but that is not necessary. Such reviews could be labeled as "inquiries," allowing the IACUC to determine the actual extent or severity as well as any intent underlying the potential non-compliance with only those that meet certain criteria subsequently being called "investigations." A system like this, which is loosely modeled after the Health and Human Services Office of Research Integrity (HHS ORI) research misconduct allegation review process, can provide for the prioritization of reviews and represent an efficient utilization of institutional resources that is potentially less threatening or accusatory to the research community.[1]

Whatever mechanism is used for the evaluation of adverse events and non-compliance, evaluations must be done carefully and with the utmost concern for due process and confidentiality as careers can be severely damaged by unfounded accusations and unsubstantiated rumors.[2] Throughout all of this, animal welfare cannot be compromised. The result is a complicated and nuanced activity required by the animal research oversight regulations and entities.

The intent of this chapter is to provide the reader with guidance on many of the challenges that IACUCs encounter when confronted with adverse events and protocol non-compliance. Because of the seriousness and sensitivity of this IACUC role, it is important that processes are crafted and adopted ahead of time and then employed with consistency from review to review. The time to hunt for the most appropriate operating procedure is not when the IACUC finds itself in the midst of a complex review, possibly from an anonymous allegation. Although IACUC policies can never account for all contingencies, it is wise to have an agreed upon outline that has been previously approved by the IACUC to conduct reviews. Otherwise, the IACUC risks creating further non-compliance by not adequately handling non-compliance, creating confusion among its members and the research community as to how non-compliance is addressed, and losing its ability to effectively oversee animal research due to inconsistency, unreliability, and a loss of trust.

Also important is a complete understanding of the authority your IACUC has under the AWAR, PHS Policy, AAALAC accreditation standards, other funding agency requirements (e.g., Department of Defense), its own policies, and the institution. Even though the AWAR and PHS Policy mandate the IACUC's role in reviewing adverse events and protocol non-compliance, they are silent on the process for handling the reviews up until a suspension of previously approved activities is sought.

No IACUC works in a vacuum. Rules, regulations, policies, and the like require a conscientious IACUC and strong institutional support. Without the latter, even the most conscientious IACUC can be hampered in its effort to review adverse events and protocol non-compliance, potentially leading to additional regulatory and policy concerns.

33:1 What is meant by adverse events and protocol non-compliance?

Reg. Neither the PHS Policy nor the AWAR directly define adverse events; however, U.S. Government Principle IV states that the proper use of animals

DOI: 10.1201/9781003272861-33

includes the avoidance or minimization of discomfort, distress, and pain when consistent with sound scientific practices. The AWA (§2131.1) states that a purpose of the Act is "to insure that animals intended for use in research facilities or for exhibition purposes or for use as pets are provided humane care and treatment." The AWA (§2143,A,3,a) states that animal care, treatment, and practices in experimental procedures will minimize animal pain and distress. The *Guide* (p. 2) notes that its basic goal is to promote the humane care and use of laboratory animals.

Likewise, protocol non-compliance is not directly defined. However, both the PHS Policy (IV,B,6) and the AWAR (§2.31,c,6; §2.31,c,7) require IACUC approval for animal use activities and proposed significant changes. Therefore, for the limited scope of this chapter (i.e., adverse events and protocol non-compliance), any component of any procedure or activity that occurs but is not approved by the IACUC constitutes protocol non-compliance. In recent years, OLAW has provided numerous examples of their strict interpretation of protocol non-compliance, so the reader is referred to their references for further guidance.[3]

Opin. Given the name, adverse events can be construed as those events that have resulted in animals having negative experiences or outcomes that are not described or anticipated within the approved IACUC protocol. Adverse events are those that influence animal health, welfare, and research animal comfort, along with the level of pain and distress animals experience. Animal mistreatment, either deliberate or accidental, defined as physical or psychological wrongful or abusive treatment of an animal, could be included in the scope of adverse events.[4] Other adverse events include equipment malfunction, compromised animal housing facilities, and inadvertent or unplanned lack of animal husbandry care that led to the harm or death of animals, which encompasses human error and natural disasters.

Protocol non-compliance indicates that procedures approved by the IACUC are not being followed.[4] If approved policies are incorporated into a protocol, then protocol non-compliance may include policy non-compliance. Non-compliance with IACUC or other relevant institutional policies, for all intents and purposes, can be handled the same as protocol non-compliance. Examples include the performance of unapproved surgeries, participation of an unapproved person in animal projects, or the administration of unapproved drugs to research animals. It is not unusual to have some overlap between adverse events and protocol non-compliance. For example, the unauthorized restraint of a nonhuman primate can potentially entail both an adverse event (outcome) for the animal and protocol non-compliance.

When faced with protocol non-compliance, an IACUC should ascertain the extent and scope of the non-compliance and then determine remediation. Remediation, described as the imposition of corrective and/or punitive actions within this chapter, can include the retraining of individuals, protocol changes to clarify activities or add options for procedures, policy revisions, policy creation, equipment and facility improvements, personnel changes, and other actions that the IACUC and veterinary staff feel are appropriate to ensure that the non-compliance does not recur. Some situations, such as animal mistreatment, severe and unrelenting post-procedural pain that has not been addressed, and surgical complications, may require immediate intervention by the IACUC and/or veterinary staff to provide for the needs of the animals along with subsequent punitive actions.

IACUCs are also charged with oversight of activities not directly related to animals, such as occupational health and safety. Discussions in this chapter concerning adverse events and protocol non-compliance will be limited to issues directly affecting animals in order to provide a common element given the scope of these topics.

NIH/OLAW routinely comments on departures from the *Guide* in terms of the standards that must, should, and may be met.[5] In many of the examples provided by NIH/OLAW, these departures may not represent non-compliant activities in an individual IACUC-approved protocol. Thus, the relation between departures to the *Guide* and adverse events as well as protocol non-compliance will not be discussed in this chapter.

Select scenarios regarding protocol non-compliance are presented at the end of this chapter.

33:2 What role does an institution have in defining the concepts of adverse events and protocol non-compliance?

Reg. The Institutional Official (IO), by signing an Animal Welfare Assurance with NIH/OLAW, commits the institution to compliance with the PHS Policy (IV,4). Non-compliance and the response to adverse events is ultimately the responsibility of the IO who represents the institution's administration (PHS Policy III,G; IV,A). The IO has a similar responsibility for ensuring compliance with the AWAR (AWAR §1.1, Institutional Official). The IO has the right to stop previously approved activities, independent of the IACUC, but cannot authorize or approve an animal activity to occur without IACUC approval (AWAR §2.31,d,8). Although the IO bears ultimate responsibility for compliance, the IO and the IACUC share the responsibility for the review and investigation of animal welfare concerns (*Guide*, p. 23).

Opin. To avoid confusion, the words "suspend" and "suspension" will be used in this chapter to indicate a temporary or permanent cessation of a previously approved animal activity that is imposed by the IACUC in accordance with federal requirements (AWAR §2.31,d,6; PHS Policy IV,C,6). Words such as "stop" and "halt" will be used to indicate a temporary or permanent cessation of a previously approved animal activity by an individual, such as the AV or the IO, who has the appropriate authority granted by the institution along with the applicable federal regulations and guidelines.

The IACUC is, ultimately, a regulatory committee that functions as an agent of the institution (PHS Policy IV,B; AWAR §2.31,c). The institution must make it known that the IACUC not only helps to assure regulatory compliance that results in good animal welfare but also is an indispensable link in protecting the researcher and institution against unwarranted accusations. All of this goes hand in hand with ensuring the integrity of the institution.

Because regulatory committees are often disparaged by investigators, it must be understood that an IACUC cannot properly function without general institutional ethical and administrative support, including financial and infrastructure resources. The institution, by actions and words, must support the IACUC. The IO should be a powerful ally and should have a clear understanding of their responsibilities and authority under the AWAR and PHS Policy and routinely communicate this support to the IACUC and investigators.

33:3 How do IACUCs typically identify adverse events and protocol non-compliance?

Reg. The AWAR (§2.32,c,4) state that research facility personnel must receive training on how to report alleged deficiencies in animal care and treatment, but there are no specific statements in the AWAR as to how to comply with this requirement. The *Guide* (pp. 23–24) states that institutions must develop methods for reporting animal welfare concerns and mechanisms for reporting concerns should be posted in prominent locations in the animal facility and on applicable institutional websites. The *Guide* (p. 24) also notes that multiple points of contact (e.g., IO, IACUC Chair, and AV) are recommended.

Opin. Absent from the above regulatory documents is the notion that institutions can proactively monitor for non-compliance and adverse events through post-approval monitoring (PAM), although the *Guide* does provide a brief description of PAM (pp. 33–34). Institutions routinely employ procedures developed to proactively identify adverse events and non-compliance, typically embodied in a PAM program. Voluntary reporting programs for adverse events, separate from PAM, are also frequently encountered within animal programs.

If allegations of non-compliance are made, they must eventually reach the Chair of the IACUC, unless the allegation is against the Chair, or if the Chair has a conflict of interest with the allegation. In those cases where the Chair is unable to participate in a review, a previously adopted policy should state who will help adjudicate the issue. As an example, this might be the Vice Chair of the IACUC, the AV, the leader of the IACUC administrative office, or the IO.

Typical communication methods for adverse events and protocol non-compliance include direct verbal conversations combined with written notifications, usually through e-mail or through an electronic system, to the AV, IACUC Chair, IO, IACUC members, college deans, IACUC office director or similar individual, compliance officer, and others. This information must be efficiently sent to the IACUC Chair. It is helpful to have a written IACUC policy on how adverse events and protocol non-compliance notifications should be communicated.

Surv. At your institution, what are the methods by which the IACUC learns of possible adverse events or protocol non-ompliance? Select all that apply.

- Self-reports from investigators 151/180 (84%)
- Concerns or allegations reported by faculty, staff, students, visitors, or the public 140/180 (78%)
- Findings identified by the institution's veterinary or animal care team 162/180 (90%)
- Findings identified during continuing review mechanisms (e.g., post-approval monitoring, semiannual inspections, annual progress reports) 157/180 (87%)
- Other* 7/180 (4%)

*Other methods noted by respondents included AAALAC site visits and routine monitoring of surgical activities by the AV.

33.4 What responsibilities do IACUCs have in response to adverse events or protocol non-compliance?

Reg. The AWAR (§2.31,c,4) state the IACUC shall "review and, if warranted, investigate concerns involving the care and use of animals at the research facility resulting from public complaints received and from reports of noncompliance received from laboratory or research facility personnel or employees." Similarly, the PHS Policy (IV,B,4) requires IACUCs to "review concerns involving the care and use of animals at the institution." The *Guide* (p. 23) expands upon this topic by stating: "the institution must develop methods for reporting and investigating animal welfare concerns, and employees should be aware of the importance of and mechanisms for reporting animal welfare concerns. In the United States, responsibility for review and investigation of these concerns rests with the IO and the IACUC." The Frequently Asked Question (FAQ) G.8. on the

NIH/OLAW website states that "the IACUC must consider allegations of noncompliance or animal welfare issues as concerns that must be addressed in accordance with relevant PHS Policy provisions and Animal Welfare Regulations (AWAR)."[6]

Opin. Adverse events and/or instances of protocol non-compliance left unaddressed can potentially threaten animal welfare and scientifically sound research outcomes. IACUCs have a fundamental responsibility to examine reports of adverse events and protocol non-compliance. A key phrase in the AWAR is "if warranted," as not all reports or allegations of adverse events or protocol non-compliance necessarily require an investigation by the IACUC. Institutions should therefore develop and maintain a plan for how to review adverse events and non-compliance, or policies and procedures outlining the expectations and performance standards for the IACUC to determine when investigation of concerns, adverse events, or protocol non-compliance is necessary. For example, a complaint that "the dogs prefer Brand A dog food to Brand B" should require far less action from the IACUC than a complaint that "Dr. X is performing unapproved activities on dogs." The IACUC Chair or appropriate designee(s) may perform an initial inquiry to ascertain the depth and scope of the adverse event or protocol non-compliance, but the IACUC should make the final determination according to its policies and procedures. In reference to language in the AWAR and PHS Policy, it is appropriate for the IACUC to interpret "institution" and "research facility" as any site where the IACUC has jurisdiction. The IACUC, in consultation with the IO, is responsible for defining any corrective action and response timeline it deems appropriate and communicating through the IO any reportable events to applicable external oversight entities (e.g., APHIS/AC, NIH/OLAW, AAALAC International, or the sponsoring agency).

Surv. At your institution, who decides whether to inform the IACUC of an allegation or report of an adverse event or protocol non-compliance?

• The Committee is informed of all allegations or reports.	130/180 (72%)
• The IACUC administrative staff makes the decision.	20/180 (11%)
• The IACUC Chair makes the decision.	41/180 (23%)
• Another IACUC-affiliated person makes the decision (e.g., veterinarian, Institutional Official).	31/180 (17%)
• A non-compliance "subcommittee" of representatives. from multiple groups.	17/180 (9%)
• Other*	6/180 (3%)

*Additional entities noted by respondents included other animal program leadership roles.

Surv. At your institution, how quickly is an allegation or report of an adverse event or protocol non-compliance brought to the attention of the entire IACUC? Select all that apply.

• We have never had any allegations or reports.	13/180 (7%)
• It varies with the severity of the report.	67/180 (37%)
• The full IACUC is notified as soon as possible.	60/180 (33%)
• The full IACUC is notified at its regularly scheduled meetings.	93/180 (52%)
• An IACUC subcommittee is promptly notified (e.g., Chair, plus others).	55/180 (31%)
• A subcommittee is notified and resolves the issue and the full IACUC is rarely or is not notified.	3/180 (2%)
• Other*	6/180 (3%)

*Additional frequencies noted by respondents included quarterly summary reports and urgent meetings called when needed to address critical incidents.

33:5 Should the IACUC develop and support a mechanism for researchers to self-report adverse events and protocol non-compliance?

Reg. The AWAR (§2.32,c,4) require the institution to provide training and instruction to personnel on "methods whereby deficiencies in animal care and treatment are reported, including deficiencies in animal care and treatment reported by any employee of the facility." The *Guide* states (pp.16–17) that "the institution should provide appropriate education and training to members of research teams," and that the "training should be tailored to the particular needs of research groups; however, all research groups should receive training in … methods for reporting concerns about animal use."

Opin. The institution must have methods for personnel to report concerns about animal activities and deliver training on those reporting methods. Yet, neither the AWAR, PHS Policy, nor the *Guide* specifically comment on the types of methods an institution should offer. The expectation for the IACUC to self-report to external oversight and funding entities under proscribed circumstances outlined in the AWAR (§2.31,c,3; §2.31,d,7) and PHS Policy (IV,C,7) suggest self-reporting methods developed within the institution are also important. Guidance on institutional self-reporting is also described in OLAW's notice NOT-OD-05-034 and AAALAC International's FAQ I.2.[7,8] Specifically in notice NOT-OD-05-034, OLAW states that "the underlying foundation of the PHS Policy is one of institutional self-evaluation, self-monitoring, and self-reporting."[8] This supports the notion that methods for personnel to self-report adverse events and protocol non-compliance to the IACUC align with the spirit of those federal regulations and policies governing institutional animal care and use programs.

Offering a variety of effective reporting methods written to maximize understanding among the members of the institution's animal research community, including methods to self-report, may increase the likelihood personnel will inform the IACUC of adverse events and protocol non-compliance. Training personnel to recognize adverse events and protocol non-compliance, and to take responsibility for their own actions by notifying the IACUC, may result in more collaborative efforts between the PI, IACUC, AV, and IO to promptly alleviate any negative effects on animals and prevent similar events from happening in the future. Effective training and self-reporting methods implies investigators trust the committee to handle self-reports in a fair and collegial manner (see 33:37).

Surv. At your institution, what mechanisms have been established for researchers to self-report adverse events or protocol non-compliance to the IACUC?

- We have no mechanisms for researchers to self-report to the IACUC. 11/180 (6%)
- Researchers are expected to promptly self-report to the AV, the IACUC Chair, the IACUC office, and/or the Institutional Official as soon as the event occurs. 157/180 (87%)
- Researchers may wait to self-report at the next scheduled continuing review event (e.g., progress report, post-approval monitoring event, semiannual inspection). 6/180 (3%)
- We do not expect researchers to self-report, the expectation is that adverse events or non-compliance will be identified via post-approval monitoring or incident reports from animal resource staff. 3/180 (2%)
- Other 3/180 (2%)

Surv. At your institution, does the IACUC handle self-reported adverse events or protocol non-compliance by researchers differently from allegations reported by someone else?

- We have never received a self-report of an adverse event or protocol non-compliance. 22/180 (12%)
- We handle all reports the same, regardless of the source. 120/180 (67%)
- If a researcher self-reports and provides a reasonable resolution, consequences may be reduced to encourage future self-reporting. 34/180 (19%)
- Other* 4/180 (2%)

*At least one respondent noted that any report from outside the PI's group would be handled similarly to an anonymous report.

33:6 The PHS Policy speaks of providing NIH/OLAW with a full explanation of the circumstances and actions taken with respect to any serious or continuing non-compliance with PHS policy or any serious deviation from the provisions of the *Guide*. What are some examples of serious or continuing protocol non-compliance?

Reg. PHS Policy (IV,F,3,a–b) requires the IACUC, through the IO, to "promptly provide OLAW with a full explanation of the circumstances and actions taken with respect to any serious or continuing non-compliance with this Policy [or] any serious deviation from the provisions of the *Guide*." Protocols are often used by the IACUC in part to document the manner in which an investigator commits to adhere with the PHS Policy, the AWAR, and the *Guide*; as such, protocol non-compliance may be viewed as a subcategory of noncompliance with the PHS Policy or the *Guide*. Neither the PHS Policy nor the *Guide* provide specific examples of what incidents may constitute serious or continuing non-compliance or serious deviations under PHS Policy (IV,F,3,a-b). However, the *Guide* (p. 13) does note that the general evaluation of the animal care and use program is left to the judgment of the IACUC, AV, and IO.

The phrase "serious or continuing non-compliance" is not mentioned in the AWAR. However, the proceeding discussion references statements from the AWAR (§2.31,c,3) as a regulatory basis for interpreting and defining serious or continuing non-compliance. With specific reference to semiannual deficiencies, the AWAR (§2.31,c,3) state that "a significant deficiency is one which, with reference to Subchapter A, and, in the judgment of the IACUC and the Institutional Official, is or may be a threat to the health or safety of the animals. If program or facility deficiencies are noted, the reports must contain a reasonable and specific plan and schedule with dates for correcting each deficiency. Any failure to adhere to the plan and schedule that results in a significant deficiency remaining uncorrected shall be reported in writing within 15 business days by the IACUC, through the Institutional Official, to APHIS and any Federal agency funding that activity."

Opin. General examples of protocol non-compliance include initiating animal activities without IACUC approval; deviating from the manner in which an animal activity was described in an approved animal protocol; and failing to monitor animals or observe humane endpoints as described in an approved animal protocol. As some instances of protocol non-compliance may also constitute non-compliance with the PHS Policy and/or a deviation from the *Guide*, it is necessary in those circumstances for the IACUC to determine whether the non-compliance meets its threshold for classification as serious or continuing and thus reportable to NIH/OLAW in accordance with PHS Policy (IV,F,3,a–b). However, there is no regulatory or policy mandate that IACUCs must classify every instance of protocol non-compliance as serious, continuing, both, or neither. Indeed, discussing whether a non-reportable instance of protocol non-compliance should be classified as serious or non-serious, continuing or non-continuing, may divert limited time and resources from more pertinent decisions the IACUC must make. Additionally, since the IACUC, AV, or others may disagree with what OLAW considers to be serious or continuing, for institutions with a Animal Welfare Assurance, the prevailing wisdom is still to contact OLAW for guidance on what is reportable unless it is otherwise clear from OLAW's published guidance.

No definitions for "serious" and "continuing" with respect to non-compliance are offered in the PHS Policy or the AWAR, but descriptions associated with deficiencies identified during semiannual reviews may be applied. While protocol non-compliance is not always synonymous with semiannual deficiencies, the concept of a "significant deficiency," that is, any action that "is or may be a threat to the health or safety of animals" in reference to semiannual reviews (PHS Policy IV,B,3; AWAR §2.31,c,3) could also describe a serious protocol non-compliance. Similarly, continuing non-compliance may be defined as protocol non-compliance for which the PI fails to resolve in accordance with the plan and correction schedule defined by the IACUC.

Whereas some types of protocol non-compliance are clearly significant enough for the IACUC to classify them as serious (e.g., performing surgery without IACUC approval), others are not quite as clear (e.g., forgetting to replace a watering device immediately after an animal recovers from anesthesia). Even relatively minor protocol non-compliance that does not threaten animal well-being can still pose a challenge for the IACUC, especially when the same PI or research team commits multiple instances of minor protocol non-compliance over time. AAALAC International's FAQ I.2. states that "institutions are encouraged to evaluate whether the single incidents over time collectively indicate a more significant concern."[7] Accredited institutions should maintain a written plan to help the IACUC determine whether non-compliance should be considered serious and/or continuous.

NIH/OLAW provides several examples of serious (i.e., reportable) non-compliance in its notice NOT-OD-05-34.[8] Some of the examples from this notice that may be classified as protocol non-compliance include:

- conduct of animal-related activities without appropriate IACUC review and approval
- failure to adhere to IACUC-approved protocols
- implementation of any significant change to IACUC-approved protocols without prior IACUC approval as required by PHS Policy (IV,B,7.)
- conduct of animal-related activities beyond the expiration date established by the IACUC
- participation in animal-related activities by individuals who have not been determined by the IACUC to be appropriately qualified and trained as required by PHS Policy (IV,C,1,f)
- failure to maintain appropriate animal-related records (e.g., identification, medical, husbandry)
- failure to ensure death of animals after euthanasia procedures (e.g., failed euthanasia with CO_2)

33:7 Should the IACUC investigate all allegations of adverse events or protocol non-compliance?

Reg. The AWAR (2.31,c,4) state that IACUCs shall "review, and, if warranted, investigate concerns involving the care and use of laboratory animals at the research facility from public complaints received and from reports of noncompliance received from laboratory or research facility personnel or employees." The PHS Policy (IV,B,4) uses the same language as the AWAR in that IACUCs are required to "review concerns involving the care and use of animals at the institution."

Stronger language is used in the *Guide* (p. 23), where the "investigating and reporting" of animal welfare concerns is described, but the requirement is directed at the institution. In other words, it is the institution that "must develop methods for reporting and investigating animal welfare concerns."

Opin. Building upon other federal guidance, such as the multi-tiered approach incorporated into the HHS ORI research misconduct regulations, institutions may consider that not every concern received about an adverse event or protocol non-compliance should be "investigated."[1] Given the negative connotation associated with the word "investigate," institutions may elect to not use that term when reviewing adverse events and non-compliance. Synonyms for "investigate" can include "review," "assess," "evaluate," and "consider."

Additionally, institutions could develop their own multi-level review process whereby all concerns are still received and reviewed, but only those meeting certain requirements (e.g., reportable to a federal agency, concern is received as an anonymous allegation, or additional information and direct observation is needed) deserve to be scrutinized at a level akin to an investigation. With such a process, the term "investigation" could be reserved for only those concerns warranting a high level of evaluation. Often, when these types of multi-level systems are adopted, the investigation level becomes a "formal investigation," indicating that prior levels of review have already occurred before the IACUC moves into an investigatory mode.

Such an approach could serve to provide thoughtful deliberations about the most serious of adverse events and non-compliance and distribute the review of cases accordingly. PIs may be more receptive to voluntarily reporting issues to the IACUC because they will not immediately be investigated, and the institution may be more supportive of an IACUC that does not appear to add to administrative burden by investigating every concern that comes its way.

Surv. At your institution, how often do allegations of adverse events or protocol non-compliance lead to formal investigations by the IACUC or designees (e.g., PAM staff)?

- Every allegation leads to a formal investigation. 53/180 (29%)
- All allegations are reviewed, and more than 75% of allegations lead to a formal investigation. 26/180 (14%)
- All allegations are reviewed, and between 50%–74% of allegations lead to a formal investigation. 16/180 (9%)
- All allegations are reviewed, and between 25%–49% of allegations lead to a formal investigation. 20/180 (11%)
- All allegations are reviewed, and less than 25% of allegations lead to a formal investigation. 43/180 (24%)
- We've never received an allegation. 22/180 (12%)

33:8 Who can bring allegations of adverse events or protocol non-compliance to the IACUC?

Reg. The AWAR (§2.31,c,4) specify that the IACUC must review concerns received from members of the general public and institutional employees. The PHS Policy does not expressly mention the type of individuals who can bring allegations to the IACUC; however, it does state that "all institutions are required to comply, as applicable, with the Animal Welfare Act and other Federal statutes and regulations relating to animals" (PHS Policy II). There is nothing in the AWAR or PHS Policy that precludes the IACUC from acting upon concerns raised from any source.

Opin. The IACUC should be prepared to receive and encourage the reporting of adverse events or protocol non-compliance from anyone. Free and open communication is the first step in building the trust necessary for people to feel comfortable voicing their concerns. Any limitations on who can present allegations to the IACUC may engender the perception that the institution has something to hide or that it does not have a genuine commitment to its animal care and use program.

The most common sources of information regarding adverse events or protocol non-compliance are veterinary and animal husbandry teams, PAM, and institutional employees. Nevertheless, methods to receive reports from individuals outside of the institution are also essential, as members of the public may possess information or viewpoints of which the institution would otherwise be unaware. In addition, it is typical for APHIS/AC veterinary medical officers as well as site visitors from AAALAC International to communicate any evidence or findings of adverse events or protocol non-compliance to the IACUC Chair, the IO, AV, IACUC office director or other similar individual, and compliance officer. See 33:16; 33:32; 33:33; 33:34 for additional information.

33:9 What communication pathways should be followed in order to bring allegations of adverse events or protocol non-compliance to the attention of the IACUC?

Reg. Institutions are required to develop methods for reporting animal welfare concerns and train personnel on those methods (AWAR §2.32,c,4; *Guide*, pp. 17, 23), but guidance on specific communication pathways is limited. The *Guide* (p. 24) does recommend having multiple points of contact (e.g., IO, IACUC Chair, and AV), and mechanisms for reporting concerns should be posted in prominent locations, such as in the animal facility and on institutional websites.

Opin. Institutions have developed various procedures to ensure allegations of adverse events and protocol non-compliance reach the IACUC. Such procedures are often developed with a focus on providing timely and efficient communication. Examples include direct verbal conversations and written letters to the AV, IACUC Chair, IO, IACUC members, college deans, IACUC office director or other similar individual, and other members of institutional leadership. Many institutions also offer a hotline or similar resource by which anyone may submit an anonymous allegation through a variety of approaches (e.g., phone call, written note, online form, or quick response (QR) code message). Hotline services may be hosted by the institution or through a third party, but ultimately the appropriate points of contact (e.g., IO, IACUC Chair, AV, IACUC office director or other similar individual) must be notified of the allegation in a timely manner. It may be helpful to have a written policy on how to do this, but the institution should minimally ensure the IACUC office's contact information is easily available to anyone who may receive an initial allegation. If the allegation is against the IACUC Chair, the policy should state who will help adjudicate the complaint. This could be the vice chair of the IACUC, the AV, the IO, or an IACUC administrator.

Surv. At your institution, how are concerns about adverse events and protocol non-compliance typically received? Select all that apply, but limit to the most common sources.

• Via the Attending Veterinarian	116/180 (64%)
• Via animal care or veterinary technical staff	129/180 (72%)
• Via the IACUC Chair	73/180 (41%)
• Via any IACUC member	47/180 (26%)
• Via the Institutional Official	23/180 (13%)
• Via an IACUC administrator, trainer, or PAM specialist	98/180 (54%)
• Via an institutional leader (e.g., dean, department chair)	11/180 (6%)
• Via an institutional hotline or other compliance reporting process	72/180 (40%)
• Other*	16/180 (9%)

*Other methods noted by respondents included AAALAC site visits, veterinary technicians, and animal husbandry personnel.

33:10 How are individuals trained and informed that they have the right to bring complaints to the IACUC about animal care and use?

Reg. Institutions are required to develop methods for reporting animal welfare concerns and train personnel on those methods (AWAR §2.32,c,4; *Guide*, pp. 17, 23). The *Guide* (p. 24) recommends having multiple points of contact (e.g., IO, IACUC Chair, and AV), and mechanisms for reporting concerns should be posted in prominent locations, such as in the animal facility and on institutional websites.

Opin. It is beneficial for all institutional personnel to receive training on how to bring complaints to the IACUC about animal care and use. No single approach will be universally appropriate. Oftentimes, this information can easily be incorporated into a general training resource for personnel that covers multiple aspects of the institution's animal care and use program. Additional approaches may include focused orientation or training sessions, newsletter articles, PAM, and/or signage with reporting instructions posted in the animal facility and on an institutional website. Personnel most likely to encounter or participate directly in animal activities may be targeted to receive more frequent training on this topic.

It may also be beneficial for the institution to provide the IO and IACUC members with training to help these individuals understand their role in receiving complaints and investigating adverse events and protocol non-compliance (see Chapter 24 Personnel Training for additional discussions on training for IACUC members). Nowadays, IACUC administrators play a pivotal role in the success of IACUCs, part of which is their provision of training to IACUC members, IOs, and the research community. Unclear institutional policies, general lack of expertise among the IACUC members, and intrapersonal dynamics within the committee can potentially derail the whole process and threaten the institution's commitment to resolving and minimizing the recurrence of adverse events or protocol non-compliance.[9] The institution may avoid these risks through training exercises for the IO and IACUC that cover relevant policies and expectations from the institutional, state, and federal levels; special considerations for other entities, such as AAALAC International or study sponsors may also be included (see 33:5; 33:9; 33:32; 33:33).

Surv. At your institution, how are people notified of the means available to them to bring their concerns to the attention of the IACUC? Select all that apply.

• No formal mechanism in place	6/180 (3%)
• Required in-person or online training	102/180 (57%)
• From handouts to investigators and their research teams	48/180 (27%)
• Through posted signage	154/180 (86%)
• Through periodic newsletters or updates through the IACUC and/or animal facility	53/180 (29%)
• Through an IACUC or animal facility website	127/180 (71%)
• Other*	11/180 (6%)

*Other methods of notification noted by respondents included listserv announcements, discussions during PAM visits, hands-on training, review of IACUC policies, orientation for new staff members, and protocol documentation procedures.

33:11 Must a complainant be identified to the IACUC or the IACUC Chair?

Reg. There are no federal laws, regulations, policies, or guidelines requiring the identification of a complainant. The *Guide* (p. 24) states that "the process [of reporting a concern] should include a mechanism for anonymity." The AWAR (§2.32,c,4) state that "no facility employee, Committee member, or laboratory personnel shall be discriminated against or be subject to any reprisal for reporting violations of any regulation or standards under the [Animal Welfare] Act."

Opin. Knowing the identity of a complainant can be valuable during an IACUC investigation. Occasionally, a complainant may desire to remain anonymous to avoid reprisal or similar reactions, and some institutions maintain reporting processes where individuals can report concerns to an independent third party with purported complete anonymity. IACUCs should proactively ask complainants when their identity is known whether they wish to keep their anonymity as the IACUC moves forward with its review, as well as whether or not they would like to be notified of the outcome of the review. When a complainant requests anonymity, they should be advised of the possibility that complete anonymity cannot be guaranteed due to certain legal, regulatory, and institutional requirements.

Requiring complainants to identify themselves to the IACUC or the IACUC Chair as a standard practice is not recommended as it may deter individuals from making valid complaints. It is important for the IACUC to be familiar with any relevant institutional policies and laws regarding confidentiality, whistleblower protections, fair notice, and due process. When consistent with institutional policy, the identity of a complainant may be withheld unless otherwise required under state public records laws.[2] On the other hand, when the institution maintains the position that the complainant's identity must be documented before the IACUC can investigate an allegation, this could create conflict between the IACUC and the institution. Under such circumstances, the committee may be able to bypass such a requirement by documenting the IACUC itself as the official complainant in order to mask the identity of the true complainant.

On occasion, allowing the IACUC access to the identity of a complainant (at least on a restricted, as-needed basis) may be critical to the outcome of a resulting investigation. Situations in which the complainant's testimony is the only evidence supporting an allegation of an adverse event or protocol non-compliance are examples in which observing strict anonymity could prove detrimental. For practical purposes, the IACUC may be unable to obtain a full accounting of the incident if the complainant is unavailable for questioning during the course of an investigation. Members may also be less inclined to trust an anonymous complainant, simply because they have no basis to evaluate the complainant's motives or credibility.

In cases where a complainant contacts a program leader directly, if consistent with IACUC policy, that program leader could consider leading the inquiry and, if necessary, any formal investigation to maintain the complainant's anonymity. Often, the animal research community, animal care technicians, veterinary technicians, and similar groups develop a sense of trust in certain program leaders (e.g., veterinarians, IACUC administrative specialists) and thus may report a concern directly to them and specifically request anonymity. As such, it is important that multiple individuals in the program have sufficient training to lead an investigation of concerns reported to the IACUC according to not only the regulatory requirements, but also IACUC and institutional policies.

33:12 Is there protection against repercussions if a person expresses a concern to the IACUC; that is, are "whistle-blowers" protected?

Reg. The AWAR (§2.32,c,4) prohibit discrimination against or reprisal to employees, IACUC members, or laboratory personnel for reporting violations of the AWAR or the AWA itself. As a result, legal action can potentially be taken against the facility in the event a reprisal can be proven. The employee of the facility, however, has no private right of action under the AWA or AWAR.

The PHS Policy states that all institutions are required to comply, as applicable, with the AWA and other federal statutes and regulations (PHS Policy II; U.S. Government Principle I). The *Guide* (p. 24) states that there should be compliance with applicable whistleblower policies, nondiscrimination against the concerned or reporting party, and protection from reprisals.

Opin. Although the PHS Policy does not offer the specific protection against whistleblowers that is written into the AWAR, the statement in the *Guide* (p. 24) is a standard required for PHS Assured institutions (see PHS Policy IV,A,1). This provides a greater degree of protection to whistleblowers than was previously offered. Nevertheless, NIH/OLAW has written, "as the PHS Policy does not contain explicit whistle blower protections, whistle blower anonymity will be honored only to the extent allowed under the FOIA [Freedom of Information Act], or other applicable laws, regulations or IACUC/institutional policies."[10]

33:13 What responsibilities do IACUCs have to ensure reporting and whistleblower policies and procedures align with institutional, state, and federal expectations?

Reg. The AWAR and *PHS Policy* are silent on this topic. The *Guide* notes (p. 24) that the process outlined by the institution for reporting concerns should take into consideration "applicable whistleblower policies."

Opin. IACUCs should be mindful of additional requirements set upon them by institutional policies, as well as educational system policies and/or state or local policies or statutes. While whistleblower protection is generally a vanguard of higher education in the United States, it is the responsibility of the IACUC to ensure that their policies and actions align with all relevant requirements. Not doing so can jeopardize the IACUC's ability to fulfill its mandate to review adverse events and non-compliance, and its integrity within the institution.

33:14 How frequently do IACUCs investigate allegations of adverse events or protocol non-compliance?

Reg. No federal regulations, policies, or guidelines address the frequency of IACUC investigations.

Opin. The frequency in which an IACUC investigates allegations of adverse events or protocol non-compliance undoubtedly varies with the size and culture of the institution. In particular, the IACUC should anticipate adverse events to occur at a frequency responsive to the volume of animals, species of animals included in the animal care and use program, and the invasiveness of animal activities supported by the institution. Institutional culture may also influence how many adverse events and instances of protocol non-compliance are communicated to the IACUC. Survey responses from the previous edition of this book suggest most IACUCs rarely investigate allegations (i.e., less than two or three times a year).[11] Data acquired from this edition of the survey indicate that the frequency at which IACUCs investigate allegations may be more variable between institutions, although the majority of respondents indicated allegations are evaluated less than six times a year.

Allegations of adverse events or protocol non-compliance should first undergo an initial inquiry by a representative of the IACUC; this could be the IACUC Chair, AV, subcommittee, or a designated IACUC administrator. The results of an initial inquiry should help the IACUC to assess the significance and validity of the allegation. In the authors' experience, it is not uncommon for initial inquiries to reveal allegations that are based upon honest misunderstandings and ultimately do not merit formal investigation for that reason. Similarly, the initial inquiry may reveal substantial evidence to support an allegation, yet the nature of the non-compliance can be fully resolved without the need for more in-depth investigative efforts. For these reasons, the IACUC is likely to receive more allegations of adverse events or protocol non-compliance than it formally investigates. In order to comply with AWAR (§2.31,c,4) and PHS Policy (IV,B,4), the IACUC should be informed of all allegations and determine which may warrant investigation, regardless of who performs the initial inquiry.

Surv. At your institution, how frequently does the IACUC or designee (e.g., PAM staff) investigate/fact-find allegations/reports of adverse events or protocol non-compliance?

- Never 18/180 (10%)
- Less than twice a year 52/180 (29%)
- Two or three times a year 41/180 (23%)
- About every other month 29/180 (16%)
- Monthly 16/180 (9%)
- Weekly 10/180 (6%)
- Other* 14/180 (8%)

*At least one respondent indicated this only occurred once at their institution. Others selecting this option indicated frequencies of less than once per year, and at least weekly.

33:15 To what extent should allegations of adverse events and protocol non-compliance be documented?

Reg. The PHS Policy (IV,E,1,b) requires the institution to maintain minutes of IACUC meetings, activities of the IACUC, and IACUC deliberations. The *Guide* (p. 24) states that reported concerns and corrective actions taken should be documented. Reported allegations that are discussed at IACUC meetings, or acted upon by the IACUC, must be documented. The AWAR (§2.35,a) also have a requirement for maintaining records of IACUC meetings. The research facility must maintain documentation of the committee's reviews and investigations conducted in response to complaints received in order to demonstrate its compliance with these regulations. However, the research facility may determine the form of and method of such documentation.

Opin. Allegations of adverse events and protocol non-compliance can be documented in many ways to satisfy regulatory and institutional requirements. Any actual documents, electronic or otherwise, received that contain allegations should be retained according to an institution's record retention policy, assuming that the records retention policy is consistent with regulatory requirements found in the AWAR (§2.35,f) and the PHS Policy (IV,E,2). Oftentimes, such policies call for the discarding of physical documents once they have been converted into an electronic format. Original versions of documents containing allegations do not have to be reviewed or produced during IACUC meetings. When individuals receive allegations verbally, they should document the conversation in writing as completely as possible. Since IACUC policies and procedures may outline multiple methods for adverse event and protocol non-compliance review, allegation document review and documentation of such reviews may vary and no established standard exists.

If the review takes place during a convened IACUC meeting, the essence of the review should be included in the meeting minutes. This could include a brief and succinct description of "who, what, when" along with the committee's decision. Detailed, sentence by sentence, transcription of the review is not necessary. When reviews take place outside of convened IACUC meetings, such as subcommittee meetings, documentation would be needed only as called for by the IACUC. Otherwise, conclusions of those deliberations may be sent on to the IACUC for review.

Documentation of any review should be organized chronologically to ensure maximal understanding of the situation. Complainants who did not indicate their desire for anonymity at the time they made an allegation should be asked if they wish to remain anonymous and if they wish to receive any updates from the IACUC as to the outcome of their review.

IACUCs should also consider any open records laws, also known as sunshine laws, their institutions may be subject to, as well as institutional policies for non-compliance, when determining how to document allegations and their reviews.

Surv. At your institution, does the IACUC have a policy regarding the retention and/or disposition of documents collected or created in the course of investigating allegations of protocol non-compliance, adverse events, or animal mistreatment? Select all that apply.

- The institution has a policy on record retention, which includes IACUC records. 114/180 (63%)
- The institution does not have a policy, but the IACUC does have a record retention policy. 34/180 (19%)
- Neither the institution nor the IACUC has a policy, but the IACUC must follow local or state public records laws (e.g., Freedom of Information Act, sunshine laws). 20/180 (11%)
- Neither the institution nor the IACUC has a policy, and the IACUC is not required to follow local or state public records laws (e.g., Freedom of Information Act, sunshine laws); 6/180 (3%)
- Other. 6/180 (3%)

33:16 Must the IACUC investigate allegations of non-compliance identified by external agencies (e.g., during an APHIS/AC inspection or AAALAC site visit)?

Reg. Yes. One of the IACUC functions under the AWAR (§2.31,c,4) is to "investigate concerns involving the care and use of animals at the research facility resulting from public complaints received" as well as from personnel within the institution. Although the AWAR do not expressly define the term "public," it would be reasonable to interpret its usage within this context to include members of federal agencies and other external oversight entities. If the IACUC investigation leads to the suspension of an animal activity, the IACUC, through the IO, must report that action to APHIS and any federal funding agency (AWAR §2.31,d,7).

Further, confirmed non-compliance should be reported to NIH/OLAW if it meets the reporting requirements of the PHS Policy (IV,F,3; IV,B,4). In notice NOT-OD-05-034, OLAW and the Office of Extramural Research state that institutions with PHS Assurances are not exempt from reporting under PHS Policy IV.F.3 when qualifying situations "are identified by external entities such as the United States Department of Agriculture or [AAALAC International], or by individuals outside the IACUC or outside the institution."[8]

Opin. One of the functions of the IACUC is to investigate concerns involving the care and use of animals at the institution, and this is not limited to concerns raised only within the institution. With regard to allegations raised during a facility inspection conducted by APHIS/AC, the IACUC should also be cognizant of the Memorandum of Understanding (MOU), which allows NIH/OLAW and APHIS/AC to share information pertinent to the enforcement of the AWAR and PHS Policy.[12] Furthermore, site visits of the institution conducted by representatives of AAALAC International may produce concerns described as either suggestions for improvement (SFIs) or mandatory items in an exit briefing or in a report from AAALAC's Council on Accreditation. AAALAC International may not require the institution to investigate or correct SFIs, but mandatory items (i.e., serious problems that do not conform to requirements/expectations, reference resources, or otherwise have the potential to harm people or animals) must be corrected before full accreditation is granted.[13] It is notable that while institutions have some latitude on whether to address an SFI for the purposes of AAALAC accreditation, the IACUC must be informed of any allegation of adverse events or non-compliance and consider the allegation in accordance with the AWAR (§2.31,c,4) and PHS Policy (IV,B,4). The IACUC should therefore be prepared not only to review a finding of non-compliance identified during an APHIS/AC inspection or site visit from AAALAC International, but also (depending on the nature of the non-compliance) to report the outcome of any investigation and corrective actions to NIH/OLAW, APHIS/AC, and/or AAALAC, as applicable (see 33:58).

33:17 How should institutions handle allegations against the IACUC for non-compliance with federal regulations, policies, or guidelines?

Reg. Federal regulations, policies, and guidelines do not delineate expectations on how institutions should handle allegations against the IACUC for non-compliance. However, these resources do define institutional positions other than the IACUC who share responsibility for the animal program. The

AWAR (§1.1) define the IO as "the individual at a research facility who is authorized to legally commit on behalf of the research facility that the requirements of 9 CFR parts 1, 2, and 3 will be met." Similarly, The *Guide* (p. 13) states the IO "bears ultimate responsibility for the Program, although overall Program direction should be a shared responsibility among the IO, AV, and IACUC."

Opin. Based upon the authority vested in the IO under the AWAR, and the PHS Policy through the *Guide*, the IO is responsible for developing and implementing processes to handle allegations of non-compliance against the IACUC. Depending upon the nature of the allegation, it may be reasonable for an initial inquiry to be performed by a senior member of the IACUC administrative office, a subcommittee of the IACUC, or even the IACUC itself under the IO's oversight. The IACUC should have experience in self-assessment and reflection through its performance of semiannual programmatic reviews as required by the AWAR (§2.31,c,1) and PHS Policy (IV,B,1). Therefore, the IACUC can perform a self-review if acceptable to the IO and identify any problems and implement solutions.

The *Guide* (p. 24) states that offering "multiple points of contact, including senior management, the IO, IACUC Chair, and AV, are recommended" for personnel to report concerns about animal welfare. The institution should therefore establish a process for the IACUC Chair to promptly report to the IO any allegations against the IACUC and, if necessary, for the IACUC Chair to recuse themselves from further participation that may bias any resulting inquiry or investigation. If an IO decides to delegate the investigation to another entity or body of institutional representatives (e.g., the IACUC administrative office, other research compliance specialists, college deans, or faculty), the IO must ensure delegates are knowledgeable on federal laws, regulations, policies, and guidelines germane to IACUC functions and responsibilities. The IACUC administrative office or the AV may provide an important perspective during such an investigation. Of note, the AV is also a member of the IACUC in compliance with membership requirements in the AWAR (§2.31,b,3,i) and PHS Policy (IV,A,3,b,1). The IO may therefore need to consider the details and context of the specific allegation to determine whether the AV can participate in the investigation without bias or conflict.

It is recognized, however, that there may be times where the allegation against the IACUC is so severe and/or there is apparent bias within the IACUC that the IACUC cannot reasonably review a complaint made against it. In this case, the IO should appoint another individual or group to review the complaint with the IACUC's cooperation. Ideas for other groups to serve such a purpose include compliance, administration, legal counsel, or quality assurance departments that do not have functional oversight of the IACUC. Alternatively, qualified outside consultants may also be engaged to conduct the review.

33:18 How should IACUCs handle allegations against the IO, an IACUC member, or the Attending Veterinarian?

Reg. The AWAR (§1.1) and the PHS Policy via the *Guide* define institutional positions responsible for the animal program. In particular, the *Guide* (p. 13) states the IO "bears ultimate responsibility for the Program, although overall Program direction should be a shared responsibility among the IO, AV, and IACUC." The *Guide* also states (p. 24) that offering "multiple points of contact, including senior management, the IO, IACUC Chair, and AV, are recommended" for personnel to report concerns about animal welfare.

Opin. The institution's policy outlining the handling of concerns regarding the animal program should include provisions that help ensure all allegations are evaluated fairly and consistently. The IO, the AV, and the IACUC typically share leadership responsibility for the institution's animal care and use program. Many institutions also have an administrative office that supports the IACUC and provides additional leadership to the program. The leader or director of the IACUC administrative office may also contribute to leadership responsibilities. One or more of these entities is likely to handle most allegations of adverse events or protocol non-compliance. When the accused party is the IO, the AV, or an IACUC member, it may be appropriate for that party to recuse themselves from any resulting investigation. Doing so helps eliminate any actual or perceived bias in the process and allows the institution to uphold its commitment to reviewing concerns involving the care and use of animals on behalf of the IACUC (AWAR §2.31,c,4; PHS Policy IV,B,4).

When a complainant makes an allegation against the AV or an IACUC member other than the Chair, it may be possible for the IACUC administrative office, Chair, and/or other non-accused members of the IACUC to help conduct the resulting investigation, with oversight from the IO. If an IO determines it is necessary to delegate an investigation to another entity or body of institutional representatives (e.g., research compliance specialists, college deans, legal counsel, or faculty), they must ensure delegates are knowledgeable on federal laws, regulations, policies, and guidelines germane to IACUC functions and responsibilities. The AV, IACUC Chair, or an IACUC administrator who is capable of assuming a neutral position with respect to the allegation may serve as an important resource during such an investigation.

If the AV, IACUC administrator, and/or IACUC Chair report directly to the IO within the institution's hierarchical chain of command, it may be difficult, but not impossible, for them to handle an allegation against the IO. For this reason, and because organizational structure varies considerably between institutions, the authors strongly recommend that institutional policy address procedures for handling allegations against the IO. Institutions may obtain an outside subject matter expert as a consultant to review its animal care and use program in such cases. If possible, allegations should be managed or overseen by a member of the institution (or a consultant) that can act as a neutral party. This could be the IACUC's vice chair, or a research compliance specialist (e.g., biosafety officer, research integrity officer) within the institution who does not report directly to the IO. If the individual serving as the IO does not also serve as the Chief Executive Officer (CEO), or the highest operating official of the institution, then an individual serving in a position hierarchically above the IO should be notified (PHS Policy IV,3,a, footnote 5)[14] (see 33:17).

33:19 What types of documentation might an IACUC need to perform an initial inquiry or investigate a complaint fully?

Reg. The AWAR task the IACUC, as an agent of the research facility, to "review, and, if warranted, investigate concerns involving the care and use of animals" (§2.31,c,4) and authorizes the IACUC "to suspend an activity involving animals" (§2.31,c,8) if the investigator fails to perform the activity in accordance with the description of the activity approved by the IACUC (§2.31,d,6).

The AWAR, the PHS Policy, and the *Guide* outline the types of records that must be maintained by the institution. Under the AWAR (§2.35,a), the IACUC must document meeting minutes (including attendance, activities, and deliberations); approval (or the withholding of approval) of proposed animal activities and significant changes in animal activities; and semiannual reports and recommendations as described in the AWAR (§2.31,c,3). The PHS Policy (IV,E,1) also requires the institution to maintain a copy of the PHS-approved Assurance and records of accrediting body determinations. The *Guide* (p. 24) also states that "reported concerns and any corrective actions taken should be documented."

Opin. The overall purpose of an investigation is to determine whether an adverse event or protocol non-compliance occurred, and to identify any corrective actions that may be necessary to resolve the incident and/or prevent its recurrence. Documentation collected during an initial inquiry or investigation should support this purpose. The types of documentation needed for the IACUC to adjudicate allegations of adverse events or protocol non-compliance should therefore facilitate the committee's understanding of the incident. Once the IACUC understands the full scope and circumstances regarding the incident, it can reasonably determine whether an animal activity was conducted in accordance with IACUC approval or institutional standards, and what (if any) corrective actions are indicated. Ideally, a written policy with procedures that align with the institution's expectations and requirements for documentation should be used to guide the investigation. Because the IACUC does not have any authority to summon people or records, institutional support is crucial to the investigative process. However, minimum documentation could include a description of the initial allegation; the name(s) and contact information of individuals involved; the plan and steps of the investigation; and any relevant evidence discovered during the course of the investigation.

A summary should ultimately be presented and discussed during an IACUC meeting and captured in the resulting minutes. Allegations or reports that do not require formal investigation may not entail any additional documentation. A previous survey indicated that a majority of IACUCs only considered initiating an investigation if there was enough documentation to suggest a significant problem exists.[15]

When multiple methods of inquiry are needed (e.g., personnel interviews and records reviews), it is advisable to document what was reviewed and learned. Any documents retained for the investigation should be directly related to the allegation made. All investigative documents should be kept confidential and made available only to the IACUC, the IO, the respondent(s) for whom an allegation is made against, and persons directly involved in the process or as otherwise required under state public record laws and institutional policies.[2]

(see 33:2; 33:15).

Surv. At your institution, who is primarily responsible for identifying, collecting, and distributing documentation to the IACUC during an investigation?

• The Principal Investigator or accused party	10/180 (6%)
• The party who made the allegation	2/180 (1%)
• Animal care or veterinary staff	4/180 (2%)
• A subcommittee of the IACUC	26/180 (14%)
• The IACUC office, PAM staff, etc.	123/180 (68%)
• Other*	15/180 (8%)

*Other responses included the AV and the IACUC Chair.

33:20 Should the IACUC let all concerned persons know the basis of a complaint and the procedures to be followed? Who might be "concerned persons"?

Reg. According to the *Guide* (pp. 23–24), the IACUC's response to reports of an animal welfare concern "should include communication of findings to the concerned employee(s), unless such concerns are reported anonymously; corrective actions if deemed necessary; and a report to the IO of the issue, findings, and actions taken."

Opin. In the event the IACUC receives an allegation or report of an adverse event or protocol non-compliance, the committee should follow the IACUC's and institution's policy regarding notification procedures of concerned persons. Individuals who may be considered "concerned persons" under such a policy include the IACUC, the IO, the complainant or person(s) making the allegation or report, the respondent or person(s) against whom an allegation is made, and possibly the immediate supervisor of the respondent. Clearly, individual circumstances may dictate who qualifies as a "concerned person" and thus needs to be notified. For example, the AV may need to also be classified as a concerned person if an adverse event resulted from a structural or procedural problem within the animal facility. Identifying the appropriate individuals will help ensure transparency and due process and will help protect the legal and ethical rights of all concerned persons.[2]

The IACUC may also need to consider the appropriate time in which concerned persons should first be formally notified. For example, if the IACUC submits an official notice to a respondent and their supervisor of a complaint as soon as it is received, this may impose unnecessary distress on the respondent if the committee subsequently concludes through an initial inquiry that the complaint was wholly unfounded. Identifying the appropriate time in which to notify concerned persons may vary depending on the nature and circumstances of the complaint. All concerned persons should be notified of the complaint and the IACUC's investigative procedures by the time the IACUC determines a formal investigation into the complaint is warranted.

Surv. At your institution, how/when is a person notified of an allegation of adverse events or protocol non-compliance against them?

- We have never had any report of adverse events or non-compliance. 18/180 (10%)
- We provide notification as soon as the fact-finding process begins. 78/180 (43%)
- We provide notification as soon as the IACUC is informed of the allegation. 42/180 (23%)
- We provide notification only if the IACUC decides to proceed with a formal investigation. 33/180 (18%)
- All such decisions are made by the legal office. 1/180 (1%)
- Other* 8/180 (4%)

*Other responses noted that the determination is made on a case-by-case basis; following a discussion about the allegation at a convened IACUC meeting; and following the notification of the IO.

Surv. Does your institution or IACUC have a formal policy for protecting the rights and confidentiality of the person against whom an allegation has been made?

- We have no such formal policy. 39/180 (22%)
- We have a formal policy. 137/180 (76%)
- Other* 4/180 2%)

*At least one respondent noted they were uncertain of the existence of a relevant policy.

33:21 Once a complaint has reached the IACUC, what is the responsibility of the IACUC Chair?

Reg. Regulatory and other guidance documents do not outline any specific role for the IACUC Chair when it comes to dealing with adverse events and protocol non-compliance.

Opin. The Chair typically oversees the efforts described in this chapter. The Chair should ensure that the review moves forward and follows the required processes. If the review involves the Chair or if the Chair has a conflict of interest regarding the event or non-compliance, then a Vice Chair, another IACUC member designated by the IO, or leader of an IACUC administrative unit, could oversee any activities surrounding the review. In other cases, the Chair may be informed of findings and how the investigation is progressing, while the leader of an IACUC administrative unit or post-approval monitoring staff may be collecting details constituting the investigation.

33:22 How should an IACUC conduct an investigation when the Chair has a conflict of interest?

Reg. There is no regulatory guidance for how to handle a review of adverse events or non-compliance that involve the IACUC Chair.

Opin. IACUCs should have policies and processes established to identify the typical conflicts of interest they may encounter, such as when IACUC members are participating in the research being discussed by the IACUC (AWAR §2.31,d,2; PHS Policy IV,C,2). Additionally, for situations involving non-compliance, the Chair and other IACUC members involved with the review should not have any conflicts of interest in the matter, including financial conflicts of interest. Examples of such conflicts include research involving a drug or medical device when the Chair has an interest in the manufacturer of the drug or device, or research involving intellectual property for which the Chair receives income from licensing fees or royalties. If IACUCs do not have access to that type of information, they can work with other administrative units within the institution to find it. Much like following its own policies, if the IACUC Chair or IACUC member were to participate in an adverse event or non-compliance review and a conflict of interest were identified after the fact, the IACUC's integrity could be called into question.

As mentioned in 33:29, a vice chair, if one exists, another member of the IACUC, or a senior leader of the IACUC administrative unit could lead a review if necessary due to a Chair's involvement or conflicts. The IO is the most logical person to make that designation, although the IACUC could already have the process outlined in a policy. The designated individual should assume the responsibilities of effectively leading the review as per all relevant requirements but would not assume any other Chair duties. In these instances, the Chair recuses themselves from the review but should be available in case the IACUC has questions of them, similar to what happens when an IACUC member recuses themselves from protocol review when a conflict exists.

33:23 Should the IACUC approach each concern on an individual basis, or should the IACUC have a policy to handle concerns in a consistent manner?

Reg. Although each complaint is likely to have its own unique needs, the AWAR and PHS Policy require the IACUC to review all complaints (see 33:4; 33:7) but are silent on the process of how to do so.

Opin. It is suggested that the IACUC have a policy or other guidance document that states the general procedures to be used for most allegations reaching the committee. This helps prevent against the potential for non-compliance to relevant requirements and promotes consistency in the reviews and corrective actions.

33:24 Can the IACUC Chair appoint an investigative subcommittee?

Reg. Responsibility for the review and investigation of animal welfare concerns rests with the IO and the IACUC (*Guide*, p. 23; AWAR §2.31,c,4, respectively), but the applicable regulations and guidelines are silent on the use of any subcommittees.

Opin. Yes. There is nothing in the AWAR or PHS Policy that details how an IACUC should proceed with an investigation if the IACUC determines one is needed. The choice of procedures is entirely at the discretion of the IACUC, with input, as needed, from the IO.

Regulatory committees in research often leverage subcommittees to review adverse events and non-compliance. These subcommittees can be standing or appointed on a case-by-case basis.

In addition to an investigative subcommittee, the Chair can do their own investigation, the Chair can ask the AV, leader of an IACUC administrative unit, or post-approval monitoring staff to perform an initial investigation, and so forth. It is strongly recommended that an IACUC policy on this topic be established well before a potential problem arises (see 33:23; 33:25).

Surv. At your institution, who typically conducts investigations/fact-finding? Select all that apply, but limit to the most common options

- The IACUC Chair or Vice Chair typically conducts investigations/fact-finding. 79/180 (44%)
- The IACUC Office leadership typically conducts investigations/fact-finding. 66/180 (37%)
- An IACUC subcommittee typically conducts investigations/fact-finding. 69/180 (38%)
- A research liaison/PAM staff typically conducts the investigation/fact-finding. 34/180 (19%)
- Our program has never performed any kind of investigation/fact-finding. 15/180 (8%)
- High-profile cases (e.g., significant animal welfare issues, anonymous reports) are typically managed by IACUC leadership (e.g., Chair, Vice Chair, Office Director, AV, etc.). 67/180 (37%)
- Other* 12/180 (7%)

*Respondents also noted the institution's AV, legal office, research integrity office, the full IACUC, and veterinary technicians.

Surv. At your institution, how are investigative subcommittees typically formed?

- PAM staff start the investigation and report findings to IACUC leadership. 25/180 (14%)
- The IACUC Chair (or designee) appoints IACUC members +/- veterinarians to investigate/fact-find the issue. 114/180 (63%)
- Our IACUC has never needed to conduct an investigation of non-compliance/adverse events. 22/180 (12%)
- Other* 19/180 (11%)

*Other responses indicated that investigative subcommittees are typically formed by the institution's AV, research integrity office, or that a standing subcommittee is already in place to handle such needs.

33:25 What might be the composition of an appropriate investigative subcommittee?

Opin. Since there is no regulatory guidance on this matter, there are no federal regulations or policies defining the composition of an investigative subcommittee of an IACUC. As is typical of other research compliance committees, investigative subcommittees can be composed of one or more persons, including non-IACUC members. Some IACUCs assign one person (often a veterinarian or IACUC administrative office staff member) to do all (or at least preliminary) information gathering. Some IACUCs and IOs attempt to assure that a scientist serve on an investigative subcommittee if the allegation is against a scientist. It is suggested that in a large IACUC, there be broad enough representation on a subcommittee to help ensure due process and acceptance of its recommendations.

Subcommittees can be appointed on an as needed or *ad hoc* basis or standing subcommittees can be used to address adverse events and non-compliance on an ongoing basis. Standing subcommittee members may gain significant knowledge about how to deal with adverse events and non-compliance that over time could prove to be valuable.

33:26 Can the IACUC use non-IACUC members as part of an investigative subcommittee?

Opin. Yes. There is no federal regulation or policy prohibiting this. However, since the IACUC is charged with reviewing all claims of adverse events and protocol non-compliance (see 33:7), it is suggested that only the IACUC or IO appoint such member(s) and that they report to the IACUC directly or through the IACUC subcommittee of which they are a part. For example, post-approval monitoring staff may conduct fact-finding regarding a complaint and report their findings to the Chair, AV, and administrative leader, then describe the incident to the full IACUC at the next convened meeting with or without proposed corrective actions. At the meeting, the committee may accept the proposed corrective actions, develop corrective actions, require the investigation to continue, or require additional corrective actions beyond those proposed.[16] The IACUC and IO retain the responsibility for investigative subcommittees under their auspices and any subsequent IACUC actions or determinations. It is strongly recommended that an IACUC policy on this topic be established well before the need arises (see 33:23; 33:25).

33:27 If an IACUC investigation reveals evidence of non-compliance outside the committee's purview (e.g., COI, professional misconduct, biosafety non-compliance), how should the IACUC proceed?

Reg. The federal regulations are silent on this topic.

Opin. IACUCs should have established processes for determining if a complaint is relevant to the IACUC and, if not, how to appropriately forward it on to another individual or group within the institution that would best handle the complaint. Sometimes, this involves work to determine the purview of the IACUC's scope, such as its oversight of biohazardous material use, conflict of interest of committee members, and the publication of inaccurate material, and can extend beyond the scope given to the IACUC by the regulations.

Complaints should always be reviewed by the appropriate individual or group, otherwise it calls into question the authority and legitimacy of the review. IACUCs should be careful to ensure that they follow all institutional policies and reporting lines when forwarding complaints that are not clearly within their purview, so good communication with other offices is key (see Chapter 37).

33:28 Should the IACUC keep a complainant informed of the *progress* of an initial inquiry or formal investigation?

Reg. The PHS Policy via the *Guide* (pp. 23–24) states the IACUC's response to reports of an animal welfare concern "should include communication of findings to the concerned employee(s), unless such concerns are reported anonymously; corrective actions is deemed necessary; and a report to the IO of the issue, findings, and actions taken." The AWAR contain no comparable statement.

Opin. Assuming the complainant is known, the institution should determine whether the IACUC needs to keep complainants informed of its progress in assessing the allegation (see 33:11). In making this determination, the institution should be aware of and consider the potential consequences of the IACUC sharing information with a complainant during an active inquiry or investigation. Under most circumstances, it may be sufficient to provide the complainant with limited feedback such as general status updates and an anticipated date by which the IACUC plans to conclude the investigation. Detailed communications with a complainant during an active investigation is not recommended, and, in some cases, it may violate institutional or state requirements concerning fair notice and due process.[2] Other consequences may include unwarranted pressure on the IACUC or investigative committee, and leaks of confidential information.

Surv. At your institution, is a known complainant informed of the progress of any ongoing inquiry or investigation?

- Our IACUC has never needed to conduct an investigation of non-compliance/adverse events. 23/180 (13%)
- We do not keep the complainant informed. 17/180 (9%)

- We provide the complainant with all information. 7/180 (4%)
- We usually provide the complainant with all information, unless there is sensitive/confidential information that is uncovered. 67/180 (37%)
- We provide the complainant with information only if they request to be informed. 43/180 (24%)
- Other* 23/180 (13%)

*Other responses indicated that the complainant is provided with a summary outcome upon the conclusion of an inquiry or investigation; this decision depends on the situation; and the complainant is provided periodic status updates.

33:29 Should the IACUC keep a complainant informed of the *results* of an initial inquiry or formal IACUC investigation?

Reg. According to the *Guide* (pp. 23–24), the IACUC's response to reports of an animal welfare concern "should include communication of findings to the concerned employee(s), unless such concerns are reported anonymously; corrective actions is deemed necessary; and a report to the IO of the issue, findings, and actions taken." The AWAR contain no comparable statement.

Opin. Yes, providing limited results to a complainant once an investigation is complete reinforces the credibility of the IACUC to both the institution and the public. An earlier survey indicated that most IACUCs communicate at least some of the investigative findings to the complainant.[11] Data acquired from this edition of the survey suggests this continues to be the case. However, a potential downside to full disclosure is the complainant may not be satisfied with the IACUC's determination. The complainant may then feel compelled to respond by initiating actions directly against the institution or an individual. It is also possible that other parties involved in the investigation may claim defamation of character if they learn or perceive that the IACUC provided the complainant with confidential or proprietary details gathered during the investigation. The institution should be aware of these risks and mitigate them by outlining expectations and procedures for the IACUC to inform complainants of the results from a resolved inquiry or investigation.

The results of an investigation should be provided to the IACUC in a timely manner and documented in the meeting minutes in alignment with PHS Policy (IV,E,1,.b) and AWAR (§2.35,a,1). The level of detail to be documented in the minutes is a matter of institutional policy. In the authors' opinion, the level of detail regarding the results of an investigation provided to the complainant need not exceed that documented in the meeting minutes (see 33:11).

33:30 Is legal or other representation on behalf of the accused person or the institution allowed during the course of an IACUC investigation?

Reg. PHS Policy and the AWAR are silent regarding this issue.

Opin. Past survey responses to this question suggest most IACUCs have not addressed this issue.[11] Depending on the nature of the allegation, it may be reasonable to involve representation from the institution's legal team, human resources, or research compliance specialists for the accused and/or the IACUC during an investigation, especially in an advisory capacity. Allowing legal, human resources, or other types of institutional representation to actively participate in an IACUC investigation could potentially undermine the IACUC's authority under AWAR (§2.31,c,4; PHS Policy IV,B,4), although institutional policy will ultimately dictate what level of involvement other units may have in this process. An exception to this may be in the case where the institution formally designated its legal team to conduct investigations on behalf of the IACUC. Ultimately, the institution should ensure its position on external representation during IACUC investigations aligns with any applicable federal, state or institutional policies.

33:31 What can the IACUC do to ensure confidentiality of research or other records that are reviewed during the course of an investigation?

Reg. The AWA prohibits IACUC members from releasing any confidential information of the research facility and establishes penalties for violations (AWA §2157), but no directions are given as to how to ensure confidentiality. Various state and local regulations and statutes, along with some federal funding agency requirements for sensitive research, may require that research and other records reviewed be kept confidential.

Opin. To ensure confidentiality, IACUCs may elect to keep no records of investigative proceedings (not recommended as this is not compliant with federal regulatory requirements) or keep limited records, and many instruct people involved with the proceedings to maintain confidentiality following IACUC or institutionally established policies. IACUCs may require that members and others assisting with adverse event or protocol non-compliance review sign confidentiality agreements.

It must be remembered that allegations remain allegations until proven otherwise, and unintentional leaks of information can adversely affect professional careers and individual well-being. It is suggested that records be kept as they may be crucial for further reference, particularly if APHIS/AC or NIH/OLAW needs to conduct their own investigation. The availability of records should be restricted to those people who have a legitimate need to have access to them. Use of locked cabinets, secured computers, or restricted access software may be necessary (see 33:15).

Surv. During an investigation of an allegation of noncompliance, does your institution or IACUC have a formal policy about protecting the confidentiality of research or other records?

- We have no such formal policy. 74/180 (41%)
- We do have a formal policy. 98/180 (54%)
- Other* 8/180 (4%)

*At least one respondent noted they were uncertain of the existence of a relevant policy.

33:32 How might the AV report to the IACUC or PI animal activities that occurred but were not included in an approved IACUC protocol?

Reg. The AWAR (§2.31,c,4) state the IACUC must "review and, if warranted, investigate concerns involving the care and use of animals at the research facility resulting from public complaints received and from reports of noncompliance received from laboratory or research facility personnel or employees." Similarly, the PHS Policy (IV,B,4) requires IACUCs to "review concerns involving the care and use of animals at the institution." The *Guide* (p. 24) states that there should be multiple methods for reporting concerns to the IACUC.

Opin. At many institutions, the AV and animal care team are uniquely aware of the activities taking place within the animal facility. This perspective, in addition to the training and qualifications these personnel typically possess, makes them a valuable resource for identifying protocol non-compliance. It is therefore important that the institution establish reasonable methods for the AV and animal care team members to communicate findings of protocol non-compliance to the PI and the IACUC.

For practical reasons, the nature of the noncompliance may dictate the approach to reporting. If a member of the animal care team finds that the PI's technician placed an unapproved toy in an animal's enclosure, for example, that team member may communicate the finding to the AV, who, in turn, discusses acceptable forms of environmental enrichment directly with the investigator. These types of findings may not necessarily require a direct report to the IACUC if the institution allows the AV to exercise their discretion on such matters.

For significant activities performed without IACUC approval (e.g., a surgical procedure), the most direct method of reporting is for the AV to personally contact the IACUC Chair or IACUC administrators to explain the situation, followed by a written description of the alleged violation. The AV may also contact the IO or request another IACUC member contact the Chair. It is not necessary for the AV to be identified as the source of the complaint in the IACUC's documentation of the investigation (see 33:8).

33:33 How might the AV report to the IACUC or PI adverse events that impacted animal well-being?

Reg. On the topic of post-approval monitoring, the *Guide* (p. 34) states that "institutions may also consider the use of veterinary staff and/or animal health technicians to observe increased risk procedures for adverse events (e.g., novel survival surgeries, pain studies, tumor growth studies) and report their findings for review by the IACUC."

Opin. As previously mentioned, the AV and animal care team may possess valuable insight into activities which take place within the animal facility. Reasonable pathways for the AV to report adverse events that impact animal well-being to the IACUC and the PI in a timely and effective manner are therefore essential. The AV should have means to communicate findings of adverse events to the PI immediately, so that any ongoing negative impacts to animal well-being are corrected as soon as possible. To facilitate this, PIs may need to provide after-hours and emergency contact information to the AV and the animal care team at the onset of an animal activity. The institution must grant the AV sufficient authority to access all animals in order to provide veterinary care (*Guide*, p. 14). This authority must be sufficient to allow the AV to initiate immediate action in the event that animal well-being is in jeopardy and efforts to communicate with the PI are unsuccessful. The most direct method of reporting an adverse event may be for the AV to contact the IACUC Chair or an IACUC administrator, as well as the PI as soon as possible, followed by a written description of the event and any actions taken to prevent or mitigate the potential or ongoing threats to animal well-being (see 33:8; 33:32).

33:34 How might findings of adverse events or protocol non-compliance identified through the post-approval monitoring (PAM) program be reported to the IACUC?

Reg. The AWAR (§2.31,c,4) state the IACUC must "review and, if warranted, investigate concerns involving the care and use of animals at the research facility resulting from … reports of noncompliance received from laboratory or research facility personnel or employees." Similarly, the PHS Policy (IV,B,4) requires the IACUC to "review concerns involving the care and use of animals at the institution." The *Guide* (p. 24) states that there should be multiple methods for reporting concerns to the IACUC. The *Guide* (p. 33) also states that "continuing IACUC oversight of animal activities is required by federal laws, regulations, and policies. … Postapproval monitoring (PAM) is considered here in the broadest sense, consisting of all types of protocol monitoring after the IACUC's initial protocol approval."

Opin. A robust PAM program is likely to identify a relatively large proportion of unintentional or otherwise unreported instances of protocol non-compliance and adverse events within the institution. The *Guide* (p. 34) suggests a PAM program needs to focus on education and cultivating partnerships with investigators in order to be successful. This type of instructive approach could be especially useful in situations where relatively minor instances of protocol non-compliance are discovered. In such instances, personnel performing the PAM might bring identified non-compliance to the attention of the investigator and instruct them on how to draft an appropriate amendment to resolve the issue. However, the PAM program must also extend IACUC oversight and should therefore have a means of promptly notifying the IACUC when non-compliance is discovered. It is important for the institution to establish clear and consistent lines of reporting between the PAM program and the IACUC, the AV, and investigators.

If the PAM program is led by the AV or a member of the IACUC, then reporting mechanisms discussed elsewhere in this chapter may be sufficient. If, however, the PAM program is managed through an office or entity that is independent of the AV and the IACUC, additional mechanisms may need to be developed to ensure timely and appropriate communication. It is strongly recommended that all PAM findings and outcomes be made accessible to the IACUC (both positive and negative), so that the committee can investigate or respond to specific negative findings and understand general programmatic trends.

33:35 What is meant by corrective or punitive actions?

Reg. There are no regulatory definitions of corrective or punitive actions within the AWAR, but corrective actions are discussed in terms of how IACUCs must address suspensions of previously approved activities (AWAR §2.31,d,7). Likewise, corrective actions are mentioned in the PHS Policy as those actions taken to address suspensions of activities (IV,C,7) while they appear to be required of an entity when reporting of non-compliance as per the Health Research Extension Act of 1985 (§495,d,2).[14]

Corrective actions are directly called out in the *Guide* (p. 24) as being potentially appropriate to deal with animal welfare concerns. Furthermore, the *Guide* states that animal welfare "concerns and any corrective actions taken should be documented."

Opin. Sanctions are often mentioned on the NIH/OLAW FAQ website but without a specific definition of the word.[6] It may be useful for IACUCs to refer to "corrective actions" and avoid words implying punishment in its deliberations, policies, and communications because of the more positive connotations "corrective actions" may engender. This is also the phrase referenced within the AWAR and PHS Policy. Corrective actions imply that a researcher is not being punished but is being shown a path forward and an opportunity to improve through the correction of select issues. Focusing on the IACUC's responsibilities to protect animal welfare and educating personnel on appropriate animal care and use—rather than punishing researchers—may foster better relations with the research community and institutional leadership. IACUCs should use terms such as "punitive action" sparingly, and only when true punishment is necessary to manage egregious events. Punitive actions, then, refer to any punishing action taken by the IACUC to help ensure compliance with the AWAR or PHS Policy.

It is important that corrective and punitive actions are just two of the many tools IACUCs can use to help address adverse events and protocol non-compliance.

The IACUC always has the authority to make recommendations to the IO regarding any aspect of the institution's animal care and use program (PHS Policy IV,B,5; AWAR §2.31,c,5). If the IACUC believes that the institution should impose punitive actions at the institutional level (e.g., a formal reprimand for repeated violations), the IACUC can make those recommendations to the IO, although the IO can initiate such actions on their own based on IACUC findings.

33:36 What corrective or punitive actions may the IACUC impose if allegations of adverse events or protocol non-compliance are verified?

Reg. The AWAR (§2.31,c,8) and PHS Policy (IV,B,8) allow only for suspension of an activity previously approved by the IACUC. Thus, any other corrective or punitive actions are imposed by actions of the IACUC, IO, or the institution (see 33:35; 33:49; 33:60-62).

Opin. Suspending previously approved animal activities can have significant downstream consequences for researchers, the IACUC, veterinary staff, animal husbandry staff, the animals, and the institution. The IACUC therefore should use suspensions sparingly to address egregious adverse events or protocol non-compliance. Suspensions of activities may be most appropriate when animal welfare is in question, or when time is needed to concentrate on retraining, revising protocols, and other corrective actions deemed necessary by the committee. In the absence of egregiousness, the vast majority of IACUCs prefer to work with non-compliant researchers in a manner that sets them up for future success through the implementation of corrective actions. IACUCs, therefore, have a wide range of options when it comes to potential corrective actions, examples of which are provided throughout this chapter and include retraining, formal counseling with the IACUC Chair or other authoritative figure, review of animal activities to determine if changes (e.g., protocol amendments) are needed to facilitate compliance, veterinary observations of animal procedures, heightened PAM, restrictions on animal facility access, specific personnel restrictions, monetary fines, etc.[2,17]

33:37 Should the IACUC develop and support a mechanism for researchers to self-correct —subject to IACUC oversight—adverse events and protocol non-compliance?

Reg. The AWAR (§2.31,c,8; §2.31,d,6) state that the IACUC is authorized to suspend animal activities "if it determines that the activity is not being conducted in accordance with the description of that activity provided by the principal investigator and approved by the Committee." Similarly, the PHS Policy (IV,C,6) states: "the IACUC may suspend an activity which it previously approved if it determines that the activity is not being conducted in accordance with applicable provisions of the Animal Welfare Act, the guide, the institution's Assurance, or IV.C.1.a–g of this policy." The *Guide* (pp. 23–24) only addresses the IACUC's response to animal welfare concerns within the context of communicating to the concerned employee(s) and to the IO, and that the response should be documented.

Opin. The AWAR and PHS Policy limit discussion on corrective actions to the IACUC's authority to suspend animal activities. However, if the IACUC supports the notion that many non-compliance issues stem from lack of awareness or knowledge, the AWAR statement (§2.31,d,1,viii) that IACUC must ensure "Personnel conducting procedures on the species being maintained or studied will be appropriately qualified and trained in those procedures" should also be considered in its development and support of such a mechanism.

The institution may grant its IACUC the authority to implement additional types of corrective actions in response to adverse events and protocol non-compliance. In some cases, the IACUC may request the researcher—or personnel responsible for the adverse event or protocol non-compliance—propose a corrective action plan for IACUC consideration. This strategy would not be appropriate if the researcher denies culpability or is otherwise uncooperative during the investigation. But when the researcher admits to wrongdoing (or self-reported the event or non-compliance to the IACUC), and has expressed a serious commitment to improve, they may be well positioned to assess their situation and identify the most appropriate actions to effectively avoid the recurrence of the event or non-compliance. This approach may also be helpful when there are internal problems present in the research group, including personnel-related issues, over which the IACUC has no authority. When the IACUC decides to allow researchers an opportunity to self-correct adverse events or protocol non-compliance, the committee must maintain oversight during the process. Any self-corrections should first be reviewed and approved by the IACUC and documented as with any other corrective action plan. (See 33:5.)

33:38 Can the IACUC suspend a previously approved animal activity or a person?

Reg. An activity, as defined in the AWAR (§1.1, Activity), is any element of research, testing, or teaching procedures that involve the care and use of animals. The PHS Policy (I) defines activities as research, research training, and biological testing. The AWAR (§2.31,d,6) and PHS Policy (IV,B,8; IV,C,6) specifically allow the IACUC to suspend a previously approved activity.

Opin. The IACUC can suspend a previously approved animal activity in whole or in part after review of the matter at a convened quorum of the IACUC, and with a suspension vote of a majority of the quorum present. See 33:40 for a discussion of partial versus complete suspension of an animal activity.

Suspension of an individual, either temporary or permanent, specific to procedure or from animal research all together, is also an option for the IACUC. NIH/OLAW recognizes this and has clarified that the IACUC can suspend a person, not just an animal activity, and such a suspension is reportable to NIH/OLAW as per the relevant reporting guidelines[18] (see 33:60). Based on one author's conversations with NIH OLAW, suspensions of personnel are not the same as suspensions of previously approved activities and do not require a convened meeting of the IACUC with a quorum of the members present.

33:39 Is there any regulatory requirement to have a "face-to-face" meeting of a quorum of the IACUC in order to vote on the suspension of an animal activity?

Reg. PHS Policy (IV,C,6) and the AWAR (§2.31,d,6) state that a vote to suspend an animal activity can occur "only after review of the matter at a convened meeting of a quorum of the IACUC and with the suspension vote of a majority of the quorum present."

Opin. The regulatory requirement noted previously suggests, but does not explicitly state, that a face-to-face committee meeting is required. NIH/OLAW has provided guidance for virtual meetings for the conduct of official IACUC business requiring a quorum, provided that certain criteria are met.[19] APHIS/AC has endorsed the NIH/OLAW guidance.[20]

There may be unique circumstances in which there is an immediate need to halt an animal activity without a meeting of the full IACUC because of a significant danger to animal welfare. This is discussed in 33:44-46.

33:40 Must the IACUC suspend an entire previously approved activity, or can it suspend parts of a previously approved activity?

Reg. An activity, as defined in the AWAR (§1.1, Activity), is any element of research, testing, or teaching procedures that involve the care and use of animals. As defined by PHS Policy (I), an activity includes research, research training, and biological testing.

Opin. The AWAR and PHS Policy are silent as to whether the IACUC has the authority to suspend only portions of a previously approved activity. The AWAR (§2.31,d,6) and the PHS Policy (IV,B,8) both speak of an "activity involving animals." Neither state that an entire IACUC protocol must be suspended. It therefore may be interpreted to allow the IACUC the authority to suspend either a full activity or a portion of such an activity. This opinion has been supported by NIH/OLAW and APHIS/AC,[21] and NIH/OLAW also has written that "a portion of a protocol may be suspended and the rest of the activity may continue to [remain] to be approved."[18] As an example, an IACUC can determine that only a surgical component of an ongoing research protocol need be suspended or that no new activities, or certain activities, can be initiated (see 33:61). IACUCs may utilize only partial activity suspensions, or the non-allowance for the initiation of new activities, in order to prioritize and address those research activities that are resulting in harm to the animals and/or non-compliance without stopping all research for a PI.

33:41 Does the IACUC, the IO, or both decide on corrective or punitive actions to be taken?

Reg. The AWAR (§2.31,d,7) and the PHS Policy (IV,C,7) state that the IO, in consultation with the IACUC, must take appropriate corrective action after a suspension of a previously approved activity (see 33:35).

Opin. The wording of the AWAR and PHS Policy as stated above suggests that the IO has somewhat more authority than the IACUC in the development of corrective actions after suspensions. In practice, most IACUCs decide what corrective and punitive actions are to be taken based on delegated authority from the IO or institution through policy.[11] IOs can implement corrective or punitive actions separately from the IACUC, either at the same time as the IACUC's actions or afterwards, although coordination with the IACUC would be preferred in these cases.

33:42 If corrective or punitive actions are imposed by the IACUC, do they occur via a formal full committee vote or can the IACUC Chair take action on behalf of the committee?

Reg. A formal committee vote at a convened meeting of a quorum of the IACUC is required for the suspension of a previously approved activity (AWAR §2.31,d,6; PHS Policy IV,C,6).

Opin. Neither the AWAR nor the PHS Policy address how corrective actions are implemented other than suspension of a previously approved activity. The method of approving corrective or punitive actions (other than a suspension) is left up to the IACUC, but a policy should be established to ensure consistency in IACUC processes and decisions. Any such policy should align with other related institutional policies.

33:43 If an allegation of an adverse event or protocol non-compliance is verified, must the IACUC apply corrective or punitive actions?

Reg. PHS Policy (IV,C,6) states that the IACUC may suspend an activity if it determines that the activity is not being conducted in accordance with applicable provisions of the AWA, the *Guide*, the institution's Animal Welfare Assurance, or PHS Policy (IV,C,1,a–g). The AWAR (§2.31,d,7) and the PHS Policy (IV,C,7) state that the IO, in consultation with the IACUC, must take appropriate corrective action in the event of a suspension of a previously approved activity.

Opin. If, in the opinion of the IACUC, corrective or punitive actions are not needed or are not appropriate, they need not be applied. A clearly minor and unintentional misinterpretation of an IACUC policy that has created no problem for an animal is an example of a verified protocol non-compliance that might lead to the IACUC revising its own policy for clarification to improve compliance.

For corrective or punitive actions other than a suspension of a previously approved activity, institutional and IACUC policies should be established and followed. The authors strongly suggest that any authority that the IACUC, IO, or any other institutional representative has to impose corrective or punitive actions, in addition to suspensions, be unambiguously communicated to the institution's research community.

33:44 Can the IACUC Chair apply corrective or punitive actions (other than suspension of a previously approved protocol) without full IACUC approval if the IO is in agreement with the actions?

Reg. Sections 33:35–43 of this chapter outline the requirements for corrective or punitive actions.

Opin. The only corrective or punitive actions expressly described in the AWAR and PHS Policy is the suspension of a previously approved activity, and a vote for suspension must occur at a meeting of a quorum of the full committee, even though corrective actions are referenced in both sections of the documents, along with the *Guide* (see 33:36). Thus, the application of any corrective or punitive actions by the Chair, with or without full IACUC approval, is a matter of IACUC or institutional policy (see 33:47). There may be unique circumstances under which a Chair might unilaterally impose a corrective or punitive action, pending review by the IACUC as per the IACUC's procedures (see 33:45, 33:66).

33:45 Can the IACUC Chair, the AV, or a veterinarian other than the AV suspend (halt) an ongoing IACUC-approved research project or related activity?

Reg. The AWAR (§1.1 Attending Veterinarian; §2.33,a,2) state that the AV must have direct or delegated authority for activities of animal care and use and must be able to provide adequate veterinary care. While they may not approve an activity that has not been approved by the IACUC, officials of the research facility may attach other conditions to animal use activities (§2.31,d,8). The PHS Policy has a similar statement (IV,A,3,b,1) regarding veterinary care. The *Guide* (p. 114) requires that the veterinarian have the authority to remove an animal from an experiment in the case of a pressing health problem.

Opin. The IACUC is identified in both the AWAR (§2.31,c,8; §2.31,d,5) and the PHS Policy (IV,B,8) as possessing the authority to suspend an animal activity. However, nothing in these documents precludes the institution from formally extending or delegating this authority to other officials of the research facility, including the IACUC Chair, AV, or another veterinarian. Ideally, any terms and conditions associated with the delegation of IACUC authority should be described in writing and endorsed by the IO and/or the IACUC, and the resulting document should be disseminated to investigators. It should also be described in the institution's Animal Welfare Assurance approved by NIH/OLAW, the veterinary care statement for APHIS/AC (for consulting AVs), and the program description submitted to AAALAC International (if applicable). In the authors' opinion, the IACUC Chair, the AV, and other veterinarians with delegated responsibility to the animal program should have the authority to halt an animal activity, subject to subsequent rapid notification and review by the IACUC. Halting research in response to non-compliance with PHS Policy, the AWAR, the *Guide*, or the institution's Animal Welfare Assurance must be reported to applicable external oversight entities; however, if the IACUC finds no such noncompliance, there is no requirement to report this action[8,22] (see 33:66).

Surv. At your institution, does anyone have the authority to halt an animal activity without an official vote of the IACUC to suspend a protocol/activity? Select all that apply.

- Yes, our IACUC Chair or Vice Chair has the authority to suspend an animal use protocol. 28/180 (16%)
- Yes, our Attending Veterinarian has the authority to suspend an animal use protocol. 69/180 (38%)
- Yes, our IACUC Office Director has the authority to suspend an animal use protocol. 4/180 (2%)
- No, only the IACUC has the authority to suspend an animal use protocol. 89/180 (49%)
- Other* 43/180 (24%)

*Respondents disclosed that the AV, the IO, and/or a college dean within their institution may also have the authority to halt an animal activity without an official vote of the IACUC for suspension.

Surv. Does your institution give your Attending Veterinarian the authority to suspend an animal activity, subject to subsequent timely review by the IACUC? Select all that apply, but limit to the most common options.

- Yes, our attending veterinarian has the authority to suspend any animal activity/protocol. 82/180 (46%)
- Yes, in some cases; we have a policy that outlines when the AV can suspend an activity/protocol and when the IACUC can suspend an activity/protocol. 17/180 (9%)
- No, but the AV has the authority to provide a treatment or euthanize the animal that is experiencing pain or distress. 35/180 (19%)
- No, only the IACUC has the authority to suspend an animal activity/protocol. 21/180 (12%)
- Other* 25/180 (14%)

*At least one respondent clarified that their AV may halt an animal activity but does not have authority to suspend an animal protocol.

33:46 Can the Principal Investigator (PI) voluntarily halt a study in order to facilitate the IACUC's investigation of an adverse event or protocol non-compliance?

Reg. The AWAR (§1.1) define *principal investigator* as the person responsible for a proposal to conduct research and for the design and implementation of research involving animals. The AWAR (§2.31,c,4) charge the IACUC to "review and, if warranted, investigate concerns involving the care and use of animals at the research facility resulting from public complaints received and from reports of noncompliance received from laboratory or research facility personnel or employees." Similarly, the

PHS Policy (IV,B,4) requires that the IACUC "review concerns involving the care and use of animals at the institution." The *Guide* (p. 23) says the "institution must develop methods for reporting and investigating animal welfare concerns, and employees should be aware of the importance of and mechanisms for reporting animal welfare concerns. In the United States, responsibility for review and investigation of these concerns rests with the IO and the IACUC."

Opin. There may be times when a PI decides to voluntarily and temporarily halt their animal activities while the IACUC investigates an adverse event or protocol non-compliance. Such a decision may be driven by a recognition of unanticipated outcomes and/or a desire to comply with the IACUC's investigation or to avoid the risk of committing continued non-compliance if the inciting cause is not yet known. Following the statements mentioned above from the *Guide* (p. 23), collaborating with the PI on a voluntary suspension could be an effective method for the institution to prevent additional harm to animals while animal welfare concerns are investigated.

An example in which the IACUC may find this method appropriate is when an animal unexpectedly dies a few hours after receiving a dose of test drug (an adverse event). The definitive cause of death is initially unclear, and the IACUC may feel it inappropriate to suspend the activity; perhaps the animal was advanced in age or was experiencing other morbidities which could have caused the death regardless of whether the animal received the test drug on that day. The PI may want to halt the study while the research team reviews records to rule out problems with the source, batch, diluent, dosage, or other characteristics of the test drug, in conjunction with veterinary staff and/or the IACUC. This type of review could take several days, especially if collaborators from other institutions need to be consulted and a sample of the drug needs to be submitted to a diagnostic lab for analysis and may necessarily require additional dosing activities be put on hold in the interim to avoid risking the loss of additional animals and valuable research resources. If the IACUC views the PI's decision as reasonable and appropriate, and the PI's activities do not obstruct the committee's review efforts, this could be an effective method for the two parties to work together to diagnose the adverse event more efficiently than if the IACUC acted as the sole review unit. If, after a review of the incident, the IACUC recommends reporting the adverse event to NIH/OLAW's Division of Compliance Oversight, the PI's decision to voluntarily halt the activity may be viewed positively by NIH/OLAW, as the intervention signals a willingness for the PI to work with the IACUC.

Ultimately, if an adverse event or protocol non-compliance is significant enough to warrant a suspension, the IACUC should formally vote to suspend the activity in accordance with the AWAR (§2.31,d,6) and PHS Policy (IV,C,6), regardless of whether the PI decided to initially halt the activities of their own accord.

33:47 Can an appeal be made to the IACUC or other institutional authority if corrective or punitive actions are imposed?

Reg. Neither the AWAR, PHS Policy, nor the *Guide* mention this topic. However, any appeal process implemented by the institution cannot allow participants to approve an animal activity that has not been approved by the IACUC (AWAR §2.31,d,8; PHS Policy IV,C,8).

Opin. The IACUC is not infallible. Prerequisites for IACUC membership do not typically include expertise in investigational conduct or concepts like fair notice and due process. Furthermore, the negative impact an individual may sustain as the target of IACUC-imposed corrective or punitive actions may extend beyond the scope of the committee's original decision. The IACUC should therefore be mindful of the repercussions (intentional or unintentional) of its actions and take precautions to ensure the committee does not target individuals unfairly by codifying expectations in a written policy which outlines an appeal process. Such a policy or set of procedures can help the IACUC apply corrective or punitive actions in a consistent manner. This is especially important when long periods of time may elapse between cases in which the IACUC is compelled to impose corrective or punitive actions, or when cases tend to differ widely in scope or impact.

However, only the IACUC can reverse its suspension of an animal activity (AWAR §2.31,d,8; PHS Policy IV,C,8; and see 33:56). It may therefore be appropriate for an institutional appeal process to handle the suspension of animal activities distinctly from all other corrective or punitive actions. If appeals to IACUC actions other than the suspension of an animal activity are permissible, then the IACUC should carefully consider whether or not this type of an appeal can also be made to the IO or another institutional authority. If the institution decides to allow such appeals, then the IACUC should be prepared to have its action overridden.

For the IACUC to run smoothly and accomplish its task, it is strongly recommended that all policies concerning corrective or punitive actions be clearly understood by the IACUC, IO, and all animal users.

Surv. Does your IACUC allow appeals of any corrective or punitive actions imposed on an animal user?

Assume that any such appeals are made to the IACUC.

- We do allow for appeals. 140/180 (78%)
- We do not allow for appeals. 21/180 (12%)
- Other* 19/180 (11%)

*More than one respondent clarified that provisions established at their institution allow for such appeals under specific circumstances.

33:48 Can corrective or punitive actions imposed by the IACUC be subsequently appealed by an APHIS/AC administrator or to the NIH/OLAW?

Reg. Neither the AWAR nor the PHS Policy address this question. The PHS Policy (V,A,5) allows NIH/OLAW to waive provisions of its policy, but that section refers to general programmatic decisions, not decisions on a given corrective or punitive action.

Opin. No. There are no external oversight entities that disclose provisions for such appeals.

33:49 Can an institution or IO override the IACUC and impose different corrective or punitive actions on a Principal Investigator or animal user?

Reg. The only responsive action an IACUC may perform under the authority of the AWAR (§2.31,d,6) and the PHS Policy (IV,B,8; IV,C,6) is to suspend an activity involving animals. The AWAR state (§2.31,d,7) that when the IACUC suspends an animal activity, "the Institutional Official, in consultation with the IACUC, shall review the reasons for suspension, take appropriate corrective action, and report that action." In the event of an IACUC suspension, the IO is required to report the suspension with a full explanation to NIH/OLAW (PHS Policy IV,C,7) or APHIS/AC (AWAR §2.31,d,7), and any federal agency funding the activity.

Opin. Stipulations for an IACUC to suspend an animal activity as described in the AWAR (§2.31,d,6) include reviewing "the matter at a convened meeting of a quorum of the IACUC and with the suspension vote of a majority of the quorum present." The AWAR (§2.31,d,7) go on to require that the IACUC consult with the IO on the reasoning for suspension and take appropriate corrective action. It would therefore seem that although the authority to suspend an animal activity should primarily rest with the IACUC, the IO can influence the stipulations of a suspension or other corrective action. If aligned with the institution's policy on this topic, the IO may have authority to override IACUC decisions regarding corrective or punitive actions (with the exception of reinstating an activity the IACUC has decided to suspend). This authority by the IO could serve as the basis for an institutional appeals process concerning IACUC decisions.

This question highlights the important relationship between the IACUC and the IO, and the need for both parties to communicate regularly. The level of engagement practiced by the IO, starting from the onset of an adverse event or crisis, can directly impact the success (or failure) of interventions, outcomes, and repeated occurrences.[9] The IACUC Chair should therefore notify the IO at the time a serious allegation or concern is received and keep the IO apprised throughout the investigation. This allows the IO the opportunity to provide counsel, if appropriate, and engenders trust in the IACUC to make reasonable and appropriate judgements on matters concerning the animal care and use program. When these elements of communication and mutual understanding are in place, the possibility of the IO overriding the IACUC's decisions, and thus potentially undermining the committee's authority, is minimized (see 33:47).

33:50 Can an IACUC impose corrective or punitive actions if the vertebrate animals used are not currently covered by the AWAR (e.g., common laboratory rats, *Rattus norvegicus*) and funding for the study is from nonfederal sources?

Reg. The AWAR (§1.1, Animal) and PHS Policy (III,A) clearly state which nonhuman animals are regulated by their respective agencies. No federal laws, regulations, or policies expressly apply to activities funded exclusively through nonfederal sources and involve only species exempt from the AWAR.

Opin. The institution is left to decide appropriate conduct where gaps in laws, regulations, and federal policies exist. Many institutions choose to extend the IACUC's oversight responsibilities to all vertebrate animals used in research, teaching, and testing activities conducted within the organization, regardless of whether those animals fall under the auspices of the AWAR or PHS Policy.[23] Extension of IACUC oversight to nonfederally funded activities is beneficial if data based on research involving animals is to be published, as most journals expect, if not require, IACUC approval of such studies. The institution may clarify its position to include all vertebrate animal use in its Animal Welfare Assurance to NIH/OLAW, and not just those of studies funded by funding agencies requiring an Assurance. If, as part of an institutional policy, an IACUC is empowered to apply corrective or punitive actions in addition to the suspension of a previously approved activity, then in many instances those actions could be applied to all vertebrate animals.

The AAALAC International Rules of Accreditation (sections 1–2) indicate that all animals used in research, teaching, or testing at accredited units should be included and evaluated by the institution based on standards outlined in the *Guide* and any applicable statutes and governmental regulations.[24] An AAALAC-accredited institution may therefore authorize the IACUC to impose corrective or punitive actions against any animal-related activity, regardless of species or funding source, to ensure its obligations as an accredited unit are met.

33:51 An investigator performs an animal activity that does not have IACUC approval. Must the IACUC take a formal vote in order to suspend an activity that it never approved?

Reg. The purpose of the IACUC is "to assess the research facility's animal program, facilities and procedures" (AWAR §2.31,a). Similarly, the PHS Policy (IV,A,3,a) states the CEO must appoint an IACUC "to oversee the institution's animal program, facilities, and procedures." The AWAR (§2.31,c,6) and the PHS Policy (IV,B,6) authorize the IACUC to approve activities involving the care and use of animals. Suspension of an ongoing animal activity requires a formal vote of a quorum of the full committee (AWAR §2.31,d,6; PHS Policy IV,C,6) (see 33:36).

Opin. The IACUC is responsible for overseeing all animal activities at the institution. If an investigator performs an animal activity without IACUC approval, then the activity must stop as soon as the non-compliance is discovered (or realized by the PI) and reported to the IACUC. In FAQ B.9, NIH/OLAW addresses this situation by stating, "The PHS Policy, *Guide*, and the USDA Animal Welfare Regulations presume that all ongoing animal activities have received the required prospective review and approval. An activity that has been undertaken without prior approval should be halted and subsequently reported to OLAW because it constitutes serious noncompliance."[6] Given this, the IACUC has the regulatory authority to halt (or suspend) an activity it never approved. A quorum of the IACUC would need to vote in favor of suspension in accordance with the AWAR (§2.31,d,6) and the PHS Policy (IV,C,6). Alternatively, some IACUCs maintain a policy that extends the committee's authority to halt any animal activity that is performed without IACUC approval. Such a policy should have unanimous support from the IO and all committee members and would allow the IACUC's decision to be implemented as soon as the non-compliance is discovered, without the need to convene a full committee meeting.

Another option would be for the institution to assign an individual, such as the IACUC Chair, the AV, or IO, the authority to halt animal activities conducted without IACUC approval. Such authority should have strict limitations, as it can override an important function of the IACUC. The IO, in consultation with the IACUC, would still need to notify the appropriate external oversight entities of the suspension, such as USDA, OLAW, AAALAC International, and/or the funding agencies[8] (see 33:45; 33:69).

33:52 An IACUC voted to suspend a researcher's animal use activity if any instance of protocol non-compliance were to be confirmed during a certain period. Another instance of non-compliance was confirmed during that period. Is it necessary for the IACUC to take another vote to suspend the protocol?

Reg. A formal vote of a majority of the quorum present at a convened meeting of the IACUC is required for the suspension of a previously approved activity (AWAR §2.31,d,6; PHS Policy IV,C,6).

There is nothing in the AWAR or PHS Policy that prohibits the IACUC from reversing a decision. Indeed, the AWAR (§2.31,d,4) and PHS Policy (IV,C,4) allow for the IACUC to reconsider its decision in reference to withholding approval of an animal activity. Reconsidering a previous action is therefore a reasonable and proper action of the IACUC. If there is to be a reversal or other change to a previous action, the IACUC must do so by holding a formal vote on the matter.

Opin. When the IACUC votes to suspend an animal activity or initiate any other type of corrective or punitive action in response to non-compliance, it is essential that any stipulations, concessions, or other caveats associated with the suspension be recorded in detail with the IACUC's decision and communicated to both the IO and the researcher. In this case, the provisional suspension of a researcher's activity will not necessarily take effect until (or unless) the researcher commits another instance of protocol non-compliance within a defined period of time in the future. This decision conceivably raises a number of concerns that the IACUC should be prepared to address at the time of its initial vote. One set of concerns relate to the institution's responsibility to ensure the researcher receives fair notice of any allegations of non-compliance, and the IACUC's obligation to uphold a fair and unbiased investigation.[2] Regarding the latter, it may be difficult for the IACUC to conduct a nonbiased investigation when the committee has already voted for suspension before the instance of protocol non-compliance even occurred.

Another set of concerns are related to the type or nature of non-compliance that will trigger the IACUC's decision to suspend the activity. Non-compliance can range from a very minor incident that has no impact on animal well-being or research outcomes to a significant incident that endangers the safety of animals and/or personnel. Further, circumstances outside of the researcher's control may impact the significance of a non-compliant event that the IACUC would have otherwise taken into consideration before imparting corrective actions. It would therefore be appropriate for the IACUC to revote on the decision to suspend the activity after the next instance of protocol non-compliance is confirmed (see 33:47; 33:56).

33:53 How long should corrective actions or a suspension of a previously approved activity last?

Reg. The applicable federal regulatory requirements are silent on this topic.

Opin. The duration or any corrective or punitive action, as well as suspension of a previously approved activity, is at the sole discretion of the IACUC, but it is unusual for an activity to be permanently suspended. Many activities are suspended for a length of time that the IACUC believes will allow for corrective actions, such as retraining or the submission of protocol amendments to occur. Any suspension must include the enactment of corrective actions to help prevent future problems (AWAR §2.31,d,7; PHS Policy IV,C,7). It is highly suggested that any time an IACUC suspends a previously approved activity the suspension deliberations include discussion and establishment of the criteria to be met for reinstatement of the activity be considered at a later date. That way, as the researchers work through the corrective actions, the IACUC can monitor that progress and prepare for reinstatement based upon predetermined criteria.

33.54 If an animal activity is suspended, can an investigator continue to breed their animals but not use them for research?

Reg. See 33:35 for a definition of a suspension. An activity is any element of research, testing, or teaching procedures that involve the care and use of animals (AWAR §1.1, Activity; PHS Policy I).

Opin. A suspension (AWAR §2.31,d,6; PHS Policy IV,C,6) includes either the temporary or the permanent stop of a previously approved animal activity, including the initiation of any new activities. Since animal breeding is an animal activity, suspensions may or may not include animal breeding. If all activities described in an IACUC protocol are suspended, then animal breeding would also be suspended unless otherwise stated in the suspension. IACUCs do have the option of suspending only select activities, so animal breeding could continue if not included in the suspension. Alternatively, the suspension may require the transfer of animals to a "holding protocol," which may allow for animal breeding if not prohibited by the suspension.[25] It is highly suggested that if the adverse event or protocol non-compliance leading to the suspension involved animal breeding, that it is not allowed until such time that the IACUC is assured that the issues have been resolved and will not recur.

33:55 If an animal activity is suspended, who is financially responsible for the upkeep of animals that are affected by the suspension?

Reg. The IACUC is authorized to suspend an animal activity under the AWAR (§2.31,c,8; §2.31,d,6) and the PHS Policy (IV,C,6–7). Neither the AWAR, PHS Policy, nor the *Guide* address financial responsibility of animals that may be affected by an IACUC-imposed suspension.

Opin. USDA/APHIS is silent on this issue, but NIH holds the position that the grantee institution is responsible for ensuring the care and upkeep of animals affected by a suspension. This is described in several documents. First, the Animal Welfare Requirements section of the NIH Grants Policy Statement (NIHGPS) specifies that "no costs for activities with live vertebrate animals may be charged to the NIH if there is not a valid Animal Welfare Assurance and IACUC approval of the activity."[26] In NOT-OD-07-044, NIH clarifies that the "absence of IACUC approval includes failure to obtain IACUC approval, expiration, or suspension of IACUC approval."[27] This document goes on to hold the grant recipient (i.e., the institution receiving the grant) responsible for continuing to maintain and care for animals during periods when animal activities do not comply with NIHGPS. This includes costs associated with equipment and supplies related to animal activities, paying *per diem* costs, salary support for persons doing technical work associated with the animals.[18] However, on a case-by-case basis, the NIH granting institute or center may allow expenditures of grant funds for the care and maintenance of animals.[26] Finally, OLAW and the Office of Extramural Research offer additional clarification in NOT-OD-10-081 that grantees must provide "certification that no unallowable costs were charged to NIH grant funds during a period of noncompliance."[28] A detailed accounting of the unallowable charges made to each grant should be provided to the funding component (see 33:58).

33:56 If a protocol is suspended, how is it reactivated?

Reg. The AWAR and the PHS Policy are silent on this issue. NIH/OLAW and APHIS/AC have provided guidance on reactivation reiterating that the regulations do not include any specific reinstatement procedure and stating: "APHIS and OLAW expect that the suspended protocol, along with any amendments added as a result of the suspension, will be reinstated only by review and approval according to the IACUC protocol approval process."[29] Also included in this guidance is the acknowledgment that the IACUC can establish criteria and procedures stricter than those employed during the typical IACUC protocol review process for reinstatement of an activity suspension.

Opin. The IACUC should lift a suspension only after it has been determined that the concerns and non-compliance that led to the suspension have been resolved and that the activity moving forward will be in full compliance with the PHS Policy, the NIH/OLAW Animal Welfare Assurance, AWAR, the *Guide*, IACUC policies, and any other relevant guidelines. Conceptually, reactivating a suspended protocol (in whole or in part) is analogous to approving a new protocol or approving a major amendment to a protocol. Either requires approval via full committee or designated member review. The IACUC should establish a policy to provide for reactivation since a suspension can be for a finite period or have an undefined endpoint unless otherwise noted by the IACUC at the time of the suspension. Such a policy may also describe additional criteria and procedures that an IACUC may utilize when reinstating a suspended activity.

33:57 If the IACUC imposes restrictions or limitations on a previously approved activity, how might they be lifted?

Reg. The regulations are silent on this topic.

Opin. IACUCs have the authority to impose restrictions and/or limitations on previously approved activities to not only correct a previous non-compliance, but also to provide mitigation for activities that pose a risk to animal welfare. Examples of such conditions include veterinary oversight of procedures (direct veterinary staff observation of procedures), IACUC oversight of procedures (direct IACUC observation of procedures, many time accomplished through a formal PAM program, also called "heightened" or "focused" or "enhanced" monitoring), limitations on the number of animals that can be used, provision of data to the IACUC or veterinarians before additional work can proceed, etc. It is up to the IACUC to determine the nature of imposed restrictions and limitations, but they should be focused on supporting the research while balancing the need for careful animal use given situations with a higher risk for an adverse outcome, including non-compliance. One should expect that at the time of establishing restrictions and limitations to be imposed, that the IACUC also determine the criteria to be used to determine when those restrictions and limitations will be lifted, while reserving the right to continue restrictions and limitations if the criteria are not met. IACUCs could set up an automatic lifting of restrictions or limitations when criteria are met, or there could be further IACUC deliberation on the restrictions or limitations and a vote taken to lift them.

33:58 Apart from corrective or punitive actions imposed by an IACUC, what actions can NIH/OLAW and APHIS/AC impose on an institution for non-compliance?

Ref. The Secretary of Agriculture is authorized under the AWA (§2146,a) to investigate whether a research facility "has violated or is violating any provision of this chapter or any regulation or standard issued thereunder" and to "permit inspectors to confiscate or destroy in a humane manner any animal found to be suffering as a result of a failure to comply with any provision of this chapter or any regulation or standard issued thereunder if ... such animal is held by a research facility and is no longer required by such research facility to carry out the research, test, or experiment for which such animal has been utilized." Similarly, the Director of NIH is authorized under the Health Research Extension Act (Public Law 99-158, §495,d) to "suspend or revoke such grant or contract under such conditions as the Director determines appropriate" if the grant or contract recipient fails to provide conditions of animal care and use that meet the standards in the PHS Policy and subsequently fails to correct the issue after notification and opportunity to do so.[14]

Opin. APHIS/AC and NIH/OLAW may impose corrective actions on an institution that is found to be non-compliant with the AWAR and PHS Policy, respectively. APHIS/AC may request APHIS Investigative and Enforcement Services (IES) to investigate an allegation of violation of the AWAR. AWAR violations are acted on in a variety of ways, depending on multiple factors. Infractions may be settled with an official notice of warning or a stipulation offer. Official letters of warning notify the institution that further infractions can result in more stringent enforcement actions. Stipulations allow alleged violators to pay a penalty in lieu of formal administrative proceedings. In cases of serious or chronic violations, consequences become more substantial.[30] A research facility found in violation of the AWAR may be assessed a civil penalty of not more than $10,000 for each such violation. Each violation and each day during which a violation continues is considered a separate offense.[31]

NIH/OLAW evaluates all allegations or indications of non-compliance with the PHS Policy or an institution's Animal Welfare Assurance. When an assured institution informs NIH/OLAW of reportable events promptly and an accordance with notice NOT-OD-05-034, OLAW gives the institution the opportunity to take corrective action; only rarely is a grant or award terminated due to an institution's failure to implement appropriate corrective action.[8,32] If after an appropriate investigation NIH/OLAW determines that non-compliance has occurred, it may restrict its approval of an institution's Assurance. This may include limiting the scope of an institution's Assurance applicability to restrict some or all projects until specified corrections have been implemented; requiring prior NIH/OLAW review of some or all research conducted under the Assurance; requiring additional training for some or all investigators; or requiring special reporting to OLAW. If deemed appropriate, NIH/OLAW may ultimately withdraw its approval of an institution's Assurance. If that should happen, no research can move forward using PHS financial support until an appropriate Assurance is approved by NIH/OLAW.[10] Office of Management and Budget Circular A-21 informs educational institutions receiving federal grants and contracts that any costs attributable to any federal sponsored agreement (such as a grant or contract) cannot be shifted to any other federally sponsored agreement to avoid restrictions on their use caused by circumstances such as an expired IACUC protocol or suspension of an animal activity.[33]

On a practical level, corrective action from NIH/OLAW will likely result in the institution needing to direct nonfederal funds to cover animal maintenance costs while activities are suspended or restrictions on its Animal Welfare Assurance are imposed. It should also be recognized that NIH/OLAW and APHIS/AC work together under a Memorandum of Understanding which allows each agency to provide the other with information that is pertinent to the enforcement of the AWAR and PHS Policy[12] (see 33:55).

33:59 If corrective or punitive actions are imposed, which persons or agencies must be informed and who does the informing?

Reg. The AWAR (§2.31,d,7) require the IO, in consultation with the IACUC, to "report that action [to suspend an activity involving animals] with a full explanation to APHIS and any Federal agency funding that activity." The IACUC should record any corrective or punitive actions it imposes in its meeting minutes (AWAR §2.35,a,1; §2.35,a,2). An APHIS/AC representative may review these minutes during the course of a facility inspection (AWAR §2.35,f). PHS Policy (IV,C,7) requires the IO, in consultation with the IACUC, to "review the reasons for suspension, take appropriate corrective action, and report that action with a full explanation to OLAW." The PHS Policy (IV,F,3) also requires the IACUC, through the IO, to report any serious or continuing non-compliance with the PHS Policy or serious deviations from the provisions of the *Guide* to NIH/OLAW.

Opin. When the IACUC imposes corrective or punitive actions, it generally does so in response to a confirmed incident such as a serious adverse event or instance of protocol non-compliance. Oftentimes the committee may need to inform certain individuals, agencies, or other entities of suspected or confirmed incidents of adverse events or protocol non-compliance prior to the time in which it imposes corrective or punitive action. Once the IACUC formally decides to impose corrective or punitive action, personnel that should be informed include the IO and the personnel against whom the committee's actions are directed. It is essential that the IACUC's decision, including any stipulations or other caveats associated with it, be clearly communicated to these individuals. Additional personnel that may also need to be notified, depending upon institutional policy or precedence, include the individual(s) who initially reported the concern to the IACUC (if applicable) and the supervisor of the personnel facing corrective or punitive action. The IACUC Chair or an IACUC administrator typically communicates the committee's decision to these parties.

External oversight entities and/or funding agencies may also need to be informed, depending upon the nature of the inciting event or incident, species of animal involved, and the type of responsive action imposed by the IACUC. Expectations for receiving reports vary among different external oversight entities and funding agencies, so the institution should be familiar with its reporting responsibilities to ensure reports are made in a timely and appropriate fashion. In cases where animal activities are conducted through a collaborating institution or as part of a consortia effort, an inter-institutional memorandum or agreement should discern what types of incidents need to be communicated between the institutions, and which institution is responsible for reporting the actions to applicable oversight entities and funding agencies.

When the IACUC suspends an animal activity that is regulated under the AWA (§2132,e; §2132,g), the IACUC, through the IO, must report the action with a full explanation to APHIS/AC. If the suspended activity is funded by a federal agency, that agency must also be notified of the IACUC's action as outlined in agency requirements. Notably, if the IACUC imposes corrective or punitive action other than the suspension of an animal activity, then notification of APHIS/AC and federal funding agencies is not required under the AWAR.

Suspension of an animal activity that falls under the auspices of the PHS Policy, or is otherwise included under the institution's Animal Welfare Assurance, should be reported to NIH/OLAW (PHS Policy IV,F,3,c). Under the NIHGPS, the institution must also notify any applicable NIH institute or center supporting the grant award.[26] It is the responsibility of the PI or the authorized organization representative (also known as the signing official in the NIH ERA Commons), and not the IACUC, to report the suspension to the appropriate NIH grants management officer.[18] Reports to NIH/OLAW about non-compliance findings that did not incur an action of suspension but nonetheless qualify as reportable situations described in NOT-OD-05-034 should typically include details of any corrective or punitive actions that were applied by the IACUC via institutionally granted authority.[8] These reports should be submitted by the IO, in consultation with the IACUC.

If the institution's animal care and use program is accredited with AAALAC International, the institution may also need to inform AAALAC of the results of the investigation and any corrective or punitive actions, as described in AAALAC's FAQ C.7[7] The IO, in consultation with the IACUC, may submit reports through the institution's unit contact to the AAALAC International Executive Office or the Regional Director responsible for the institution's geographic area. Alternatively, the institution may inform AAALAC through the unit's annual report (see 33:60).

33:60 Other than suspending an animal activity, what types of IACUC-related problems are to be reported to the NIH/OLAW, APHIS/AC, AAALAC, and/or federal funding agencies?

Reg. Aside from the suspension of an animal activity, the AWAR (§2.31c,3) require the IO, in consultation with IACUC, to report to APHIS/AC (and any federal agency funding the activities) any significant programmatic or facility deficiencies not corrected in accordance with the plan and schedule defined by the IACUC in its semiannual report. PHS Policy (IV,F,3) requires the IO, in consultation with the IACUC, to report to NIH/OLAW any serious or continuing non-compliance with the PHS Policy or serious deviations from the provisions of the *Guide*. Details can be found in NIH Notice NOT-OD-05-034.[8] The NIHGPS requires grantee institutions to report instances of serious non-compliance with PHS Policy (IV.F.3) to the NIH institute or center supporting the grant award.[26]

Opin. NIH/OLAW offers examples of actions that require reporting to the Division of Compliance Oversight in NOT-OD-05-034.[8] They include (but are not limited to) the following.

- IACUC suspension or other institutional intervention that results in the temporary or permanent interruption of an activity due to non-compliance with the PHS Policy, AWA, the *Guide*, or the institution's Animal Welfare Assurance
- Failure to adhere to IACUC-approved protocols
- Serious or continuing non-compliance with the PHS Policy
- Serious deviations from the provisions of the *Guide*, such as shortcomings in programs of veterinary care, occupational health, or training
- Failure to maintain appropriate animal-related records
- Failure to ensure the death of an animal after euthanasia
- Failure of animal care and use personnel to carry out veterinary orders (e.g., treatments)

Note that the suspension of a person (i.e., suspending an individual's privileges to conduct animal activities) would be reportable to NIH/OLAW only if the reason for the suspension involved serious non-compliance.[34] Regarding situations demonstrative of programmatic failure (e.g., unalleviated stress within mouse breeding colonies caused by excessive noise and vibrations during nearby building construction), institutions maintaining an Animal Welfare Assurance with OLAW must report the incident whether the research is supported with NIH funds or not.[35] For activities supported by the PHS, NIH Notices NOT-OD-07-044 and NOT-OD-10-081 state that performing an animal activity without IACUC approval (including performing activities on a suspended protocol), performing an animal activity without a valid Assurance on file with NIH/OLAW, or performing an animal activity *after* the expiration of an IACUC approval must be reported to the institute or center supporting the award.[6,27,28] Any such report is an institutional requirement and is not a mandated responsibility of the IACUC.

If the institution is accredited with AAALAC International, the unit contact must also inform AAALAC of any protocol violations which had the potential to compromise animal welfare in the annual report. Additionally, an adverse event report must be promptly submitted to AAALAC under the following circumstances (as described in AAALAC's FAQ I.2.).[7]

- Inadequate veterinary care
- Conditions that resulted in unexpected animal harm or deaths (due to accidents or errors, equipment failure, or natural disaster)
- Significant animal right activities (e.g., protests, break-ins, property damage, FOIA and other public records requests that include AAALAC International documents)
- Inappropriate euthanasia techniques and/or failure to confirm euthanasia
- Substantiated complaints or reports regarding animal welfare concerns
- Internal or external reviews/inspections or other similar reports that document significant adverse events or non-compliance that resulted in animal harm or death; investigations by national oversight bodies; and other serious incidents or concerns that negatively impact animal well-being (e.g., failure to follow the approved protocol which resulted in

- compromised animal welfare; death during transport)
- Significant human health issue directly related to the animal care and use program

Surv. How frequently does your IACUC inform relevant federal or international oversight entities (e.g., NIH/OLAW, USDA/AC, NSF, NASA, DOD, VA, or AAALAC) about activities that require reporting?

- We have never informed oversight entities of any such actions. 31/180 (17%)
- We typically inform oversight entities of such activities less than 5 times a year. 123/180 (68%)
- We typically inform oversight entities of such activities between 6–10 times a year. 16/180 (9%)
- We typically inform oversight entities of such activities between 11–20 times a year. 7/180 (4%)
- We typically inform oversight entities of such activities more than 20 times a year. 3/180 (2%)

33:61 If only parts of a previously approved activity are suspended by the IACUC (e.g., the surgical portion of a large project), must the IACUC notify the NIH/OLAW, APHIS/AC, or AAALAC?

Reg. The AWAR (§2.31d,6) authorize the IACUC to "suspend an activity that it previously approved if [the IACUC] determines that the activity is not being conducted in accordance with the description of that activity provided by the principal investigator and approved by the Committee." When reporting the suspension to APHIS/AC and any federal agency funding that activity, a full explanation of the action is required (§2.31,d,7).

PHS Policy (IV,F,3) requires the IACUC, through the IO, to promptly report any serious or continuing non-compliance with PHS Policy, any serious deviation from the provisions of the *Guide*, or any suspension of an activity to OLAW, along with a full explanation of the circumstances and actions taken. An "activity" is defined as any element of research, testing, or teaching procedures that involve the care and use of animals (AWAR §1.1, Activity; PHS Policy I).

Opin. The language on suspension used by the AWAR, PHS Policy, and NIH/OLAW notice NOT-OD-05-034 refers to an animal activity and therefore need not apply to an entire IACUC protocol.[8] If the IACUC suspends one or multiple elements within an animal protocol, the action is reportable (as applicable) to APHIS/AC, NIH/OLAW Division of Compliance Oversight, the federal funding agency, and AAALAC International even if all other elements within the protocol were allowed to proceed uninterrupted. In some instances (e.g., animal morbidity or mortality that exceeds the anticipated levels described in the approved protocol), the IACUC might suspend part of an ongoing activity until more information can be gathered in an investigation. If all or part of an animal activity is suspended by the IACUC, applicable external oversight entities should be notified (see 33:40; 33:60; 33:66).

33:62 Must a federal funding agency be informed if allegations of adverse events or protocol non-compliance are verified, but corrective or punitive actions are not imposed by the IACUC?

Reg. Federal funding agencies must be informed if the IACUC suspends an activity (AWAR §2.31,d,7). The NIHGPS requires instances of serious non-compliance to be reported to NIH/OLAW and the NIH institute or center supporting the grant award.[26]

Opin. Federal funding agencies must be informed when the IACUC suspends an activity involving species covered under the auspices of the AWAR. These funding agencies would also need to be informed in the event that a major deficiency is identified during a semiannual program review and not corrected within the appropriate timeframe (§2.31,c,3).

Items 33:59 and 33:60 describe conditions under which an institution receiving PHS funds must inform federal funding agencies of problems concerning animal use activities. Notifications of this nature are a responsibility of the institution, and not necessarily the IACUC. It is the responsibility of the PI or the institution's authorized organization representative (also known as the Signing Official in the NIH ERA Commons) to report the suspension to the appropriate NIH grants management officer.[18]

33:63 Must nonfederal funding agencies be informed of full or partial suspensions of a previously approved activity?

Reg. The regulations are silent on this topic.

Opin. Agencies and entities not associated with the U.S. federal government often sponsor animal research through grants, contacts, or other agreements. When that type of funding has been identified as funding animal research that has been suspended, the institution (e.g., IACUC office) or researcher should review any applicable funding agreements to determine what exactly should be reported to the funder. Institutions, including IACUC administrative offices and researchers, should be given the opportunity to review these types of nonfederal funding agreements to evaluate the ability to fulfill any reporting requirements contained in such agreements. Of note, some agreements can contain sweeping reporting requirements such as being notified anytime any deficiency is identified within the entire animal care and use program. Ideally, when nonfederal sponsors want to receive reports of adverse events or non-compliance, such reports should be limited to the animal research that is being directly supported by the sponsor and reporting triggers should be unambiguous.

Surv. Has your IACUC ever informed a nonfederal funding agency of an activity in which the IACUC suspended or imposed corrective actions?

- No, we have never informed a nonfederal funding agency. 139/180 (77%)
- Yes, we have informed a nonfederal funding agency, but very rarely. 39/180 (22%)
- Yes, we inform nonfederal funding agencies with relative frequency. 2/180 (1%)

33:64 If an IACUC suspension of a previously approved activity occurs, what information should be provided to NIH/OLAW, APHIS/AC, AAALAC, and funding agencies?

Reg. The AWAR (§2.31,d,7) require the IO, in consultation with the IACUC, to report the suspension "with a full explanation to APHIS and any Federal agency funding that activity." Any corrective or punitive actions imposed by the IACUC should be recorded in the minutes of the IACUC meeting (AWAR §2.35,a,1; §2.35,a,2) and are thus subject to review by APHIS/AC during an inspection of the research facility (AWAR §2.35,f).

Similarly, PHS Policy (IV,C,7) requires the IO, in consultation with the IACUC, to "review the reasons for suspension, take appropriate corrective action, and report that action with a full explanation to OLAW." The PHS Policy (IV,F,3) also requires the IACUC, through the IO, to report to NIH/OLAW any serious or continuing non-compliance with the PHS Policy or serious deviations from the provisions of the *Guide*. The conduct of animal activities when a protocol has been suspended, expired, or does not have IACUC approval must be reported to the relevant NIH institute or center supporting the grant award[26] (see 33:6; 33:59–61).

Opin. When the IACUC suspends an animal activity that is regulated under the AWA (§2132,g), the IACUC, through the IO, should report the action with a full explanation to APHIS/AC. The level of detail needed to provide a "full explanation" may be open to interpretation but should generally summarize the inciting adverse event or protocol non-compliance, its impact on the research facility, and the IACUC's corrective action plan. NIH/OLAW has provided guidance to institutions on this topic, and information that should be included in a report to OLAW's Division of Compliance Oversight includes the following.[8]

- The institution's Animal Welfare Assurance number
- Relevant PHS grant or contract numbers
- Effect(s) on PHS-supported activities
- Explanation of the cause for suspension
  - What happened, including the species of animal(s) involved
  - When and where the situation took place
  - Personnel involved
- The institution's response to address the situation
- Corrective plans (short term and long term) and implementation schedule

These guidelines for reporting to NIH/OLAW could also serve as the basis for reporting to APHIS/AC and AAALAC International, as applicable. If the institution's animal care and use program is accredited with AAALAC International, the institution should consult with its policy and AAALAC's FAQ C.7, which offers guidance on the nature and timing of reporting.[7]

33:65 An institution has an IACUC but it does not house or use animals that fall under the auspices of the AWAR or PHS Policy. Must the institution inform APHIS/AC and NIH/OLAW of any suspensions of a previously approved activity?

Reg. No. If the suspended animal activity does not include species covered under the AWA (§1.1, Animal), there is no reporting obligation to APHIS/AC. Similarly, if the institution does not conduct activities that apply to the PHS Policy, there are no reporting obligations to NIH/OLAW.

Opin. If the institution does not house or use animals that fall under the auspices of the AWAR or the PHS Policy, the institution is not required to inform APHIS/AC or NIH/OLAW of any corrective or punitive action imposed by its IACUC. Any reporting responsibilities described within the AWAR and PHS Policy would not apply to such an institution. Use of the phrase "Institutional Animal Care and Use Committee" does not guarantee the institution is accountable to all the requirements codified within the AWAR and PHS Policy. Nevertheless, if the institution maintains an active Animal Welfare Assurance with NIH/OLAW, reporting a suspension to the OLAW Division of Compliance Oversight may be required if the inciting event or incident was indicative of a programmatic failure, even if the activities were not directly supported by PHS.[35]

If the institution is accredited with AAALAC International, the IACUC should refer to its policy on adverse event assessment and reporting to determine if the suspension merits reporting to this accrediting body. If the inciting event or incident and/or IACUC's responsive action meets the standards for reporting under its policy, the unit contact should promptly inform AAALAC International (see 33:50).

33:66 Should NIH/OLAW or the APHIS/AC be informed that a project has been suspended pending the completion of an IACUC investigation?

Reg. One of the IACUC's functions under the AWAR (§2.31,c,4) is to "investigate concerns involving the care and use of animals at the research facility." The AWAR (§2.31,d,6) allow the IACUC to "suspend an activity that it previously approved if it determines that the activity is not being conducted in accordance with the description" approved by the IACUC. When reporting the suspension to APHIS/AC and any federal agency funding that activity, a full explanation should be included (§2.31,d,7).

PHS Policy (IV,F,3) requires the IACUC, through the IO, to promptly report any serious or continuing non-compliance with PHS Policy, any serious deviation from the provisions of the *Guide*, or any suspension of an activity to OLAW, along with a full explanation of the circumstances and actions taken (see 33:44-45).

Opin. The intent of the AWAR in authorizing the IACUC to suspend animal activities is to provide the committee with a mechanism to return the institution to a state of compliance with the regulations and standards when it determines that an activity is not being conducted in a manner consistent with its description at the time of approval. Thus, a majority vote for suspension suggests that the evidence gathered during a review is strong enough to conclude that some degree of non-compliance occurred. Once the IACUC votes to suspend an activity or protocol, it must then report the suspension, through the IO, to APHIS/AC and/or NIH/OLAW (as applicable), regardless of the status of the IACUC's review.

Conversely, if the review is in progress and no suspension by the IACUC has yet occurred, the threshold for reporting under the AWAR (§2.31,d,6) has not yet been achieved. That is, the IACUC may not yet have determined an activity was conducted in a manner non-compliant with its description at the time of IACUC approval. If all the details of an incident are not yet known, the IACUC cannot properly comply with the AWAR (§2.31,d,7) or PHS Policy (IV,C,7) on reporting the suspension along with a *full explanation* of the problem and corrective actions. There is nothing in the AWAR or PHS Policy that expressly requires the institution to notify APHIS/AC or NIH/OLAW about an activity that is under investigation. However, both NIH/OLAW and APHIS/AC have invited institutions to voluntarily advise each agency of any incidents resulting in adverse outcomes to allow proactive identification, correction, and prevention of issues that may negatively impact animal welfare.[36] NIH has explicitly requested early notification in the form of a preliminary report in its notice NOT-OD-13-044, stating: "institutions should notify OLAW promptly of reportable situations. As such, it is appropriate to submit a preliminary report prior to the completion of a full investigation and implementation of a corrective plan."[37]

When responding to a scenario in which an investigator voluntarily halts a study while the IACUC investigates an adverse event, the Deputy Administrator of APHIS/AC has indicated there was no regulatory obligation for the institution to report the event to APHIS/AC because the IACUC did not formally suspend the activity.[36] In notice NOT-OD-05-034, OLAW and the Office of Extramural Research state: "Institutions should notify OLAW of matters falling under [PHS Policy IV,F,3] promptly, i.e., without delay. Since PHS Policy (IV,F,3) requires a full explanation of circumstances and actions taken, and the time required to fully investigate and devise corrective actions may be lengthy, OLAW recommends that an authorized institutional representative provide a preliminary report to OLAW as soon as possible and follow up with a thorough report once action has been taken."[8] A potential benefit of the institution submitting a preliminary report of a reportable event before the investigation is complete is that NIH/OLAW may offer the IACUC guidance as it decides what actions are necessary to correct the issue and prevent recurrence[32] (see 33:67).

33:67 How soon after a previously approved activity is suspended must that suspension be reported to APHIS/AC or NIH/OLAW?

Reg. The AWAR (§2.31,d,7) do not provide a time frame for notifying APHIS/AC of a suspension of an activity involving animals. The PHS Policy (IV,F,3,c) requires the IACUC, through the IO, to promptly notify NIH/OLAW in the event of "any suspension of an activity by the IACUC" and with a full explanation of the circumstances and actions taken. The same reporting expectation goes for any serious or continuing non-compliance with the PHS Policy and serious deviation from the provisions of the *Guide* (PHS Policy IV,F,3,a-b).

Opin. NIH/OLAW describes in its notice NOT-OD-05-034 that an authorized institutional representative should provide a preliminary report to the OLAW "promptly, i.e., without delay" once the IACUC suspends an activity (either temporarily or permanently).[8] Therefore, a preliminary report to NIH/OLAW may be necessary prior to the completion of a full investigation and implementation of a corrective plan. The final report should be signed by the IO and include a detailed explanation of the incident and actions taken by the institution to resolve the issue or prevent recurrence.[37]

Although the AWAR do not specify a time frame by which a suspension should be reported, it would seem reasonable for the IO to notify APHIS/AC once all reporting requirements have been achieved (§2.31,d,7). Delaying this notification unnecessarily for an extended period could compromise the accuracy of the report or engender distrust by APHIS/AC in the institution's ability to adhere to this and other IACUC functions outlined in the AWAR (§2.31). On the other hand, the IO could also notify APHIS/AC prior to the preparation of a final report. This would be somewhat analogous to submitting a preliminary report to NIH/OLAW. Although preliminary reporting is not a regulatory obligation under the AWAR, the IO may choose to inform both APHIS/AC and NIH/OLAW at the same time when the suspended activity will ultimately require reporting to both entities. In this way, the institution can ensure both oversight entities are promptly informed and prevent (if desired) APHIS/AC from primarily learning of an incident secondhand through communications with NIH/OLAW.[12]

If the institution is accredited with AAALAC International, the IACUC should consult its Adverse Event Assessment and Reporting Plan to determine whether the suspension meets the institution's standards for prompt reporting. Notably, the IACUC's response (e.g., suspension) may not necessarily constitute definitive criteria for reporting to AAALAC as described in its FAQ C.7; rather, it is the nature of the event or circumstance (e.g., conditions that resulted in unexpected harm to animals or humans) that should guide the institution's decision to inform AAALAC.[7] As with NIH/OLAW, AAALAC requests adverse event reports be submitted "promptly," and the reporting plan should be available for review by the site visit team (see 33:64; 33:66).

33:68 Does the IACUC have to report to APHIS/AC or NIH/OLAW that an investigator has voluntarily halted part or all of an animal activity in response to an adverse event, protocol non-compliance, or IACUC investigation?

Reg. The AWAR (§2.31,d,6) authorize the IACUC to suspend an activity that it previously approved if it determines the activity is not being conducted in accordance with IACUC approval. Suspension is determined following IACUC review during a convened meeting with a quorum of the committee present and a majority vote in favor of the action. The IACUC must then review its decision with the IO and report the action with a full explanation to APHIS/AC and to any federal agency funding that activity (AWAR §2.31,d,7). The AWAR define an activity as any element of research, testing, or teaching procedures that involve the care and use of animals (§1.1). The PHS Policy (IV,F,3) requires that the IACUC, through the IO, inform NIH/OLAW of any suspension of an activity by the IACUC, any serious deviation from the *Guide*, or any serious or continuing non-compliance with the PHS Policy.

Opin. There may be occasions when an investigator responds to an adverse event or protocol non-compliance by voluntarily halting their animal activities for a period of time, typically to avoid recurrence of an adverse event or non-compliance until the source or inciting factor can be identified and/or addressed. The investigator initiates this action voluntarily, or without the IACUC imposing a formal action to halt animal use. For example, a researcher may determine an amendment to the existing animal activity approved by the IACUC is necessary. Voluntarily halting animal activities while the researcher waits to secure IACUC approval of the amendment is not synonymous with suspension. The IACUC may therefore need to take into consideration the investigator's intentions as well as the significance of the inciting incident to determine whether the action is reportable to APHIS/AC or NIH/OLAW.

While responding to a hypothetical scenario in which an investigator voluntarily halted a study until the IACUC and veterinarians could investigate the death of an animal, the Deputy Administrator of APHIS/AC indicated that the research facility was under no regulatory obligation to report the event to APHIS/AC.[36] Reporting a suspension to APHIS/AC is necessary only if a quorum of the IACUC were to suspend the activity at a convened meeting and by a majority vote (AWAR §2.31,d,7) or if the adverse event or protocol non-compliance were part of a significant deficiency not corrected in accordance with a semiannual inspection (§2.31,c,3). An activity that is halted or suspended by the IO or an individual authorized by the institution to act on behalf of the IACUC (e.g., IACUC Chair or AV) is also reportable.[22]

Regardless of who decides to halt an animal activity, an adverse event or protocol non-compliance that results in animal morbidity or mortality is considered reportable to NIH/OLAW.[36] Information about the incident that is recorded in the IACUC meeting minutes or semiannual reports must be made available to APHIS/AC, OLAW, or other PHS representatives upon request (AWAR §2.35,a,1–3; PHS Policy IV,D,1–2). Guidance from AAALAC International on prompt reporting also includes "conditions that resulted in unexpected animal harm or deaths" due to accidents or errors, equipment failure, or natural disaster[7] (see 33:2; 33:45).

Surv. Does your institution report an animal user's voluntary halt or stoppage of an animal activity to NIH/OLAW?

- Yes, we report any voluntary stoppage of any animal activity. 14/180 (8%)
- Maybe; we usually do not report a voluntary stoppage of an activity unless there are extenuating circumstances. 58/180 (32%)
- No, we report only if the IACUC officially suspends an activity, as required. 86/180 (48%)
- Other* 22/180 (12%)

*At least one respondent noted that reporting to NIH/OLAW was not applicable to their program.

33:69 An IO stops an animal activity in response to protocol non-compliance or adverse events, even though the IACUC voted not to suspend the same animal activity. Is this action reportable to the NIH/OLAW or APHIS/AC since it did not arise from an IACUC vote?

Reg. Language in the AWAR and PHS Policy does not preclude the IO from imposing corrective or punitive actions with regard to animal activities. In order for the IACUC to suspend an animal activity, a majority vote by a quorum of the committee is required (AWAR §2.31,d,6; PHS Policy IV,C,6).

Opin. Although the IO cannot approve an animal activity that the IACUC has not approved (AWAR §2.31,e,8; PHS Policy IV,C,8), there is nothing in the AWAR or PHS Policy that prohibits the IO from halting an animal activity. According to the response to NIH/OLAW's FAQ B.10., the IO "(or another authorized official) may unilaterally halt, terminate, or impose sanctions on any animal activity, as this is not prohibited by the PHS Policy."[6] Through its notice NOT-OD-05-034, NIH/OLAW would consider such an action reportable[8] (see 33:49).

33:70 Assume a study is initiated without IACUC approval, the problem is discovered, and when the PI is told to halt all animal work, they respond by asserting that halting the work will mean that all the data collected to date will be meaningless and the lives of the animals will have been wasted. What should the IACUC do?

Reg. The AWAR (§2.31,c,6) authorize the IACUC to "review, approve, require modifications in (to secure approval), or withhold approval of those components of proposed activities related to the care and use of animals." Similarly, the PHS Policy (IV,B,6) authorizes the IACUC to "review and approve, require modifications in (to secure approval), or withhold approval of those components of PHS-conducted or supported activities related to the care and use of animals." Beginning an animal activity without IACUC approval constitutes serious non-compliance with both the AWAR and the PHS Policy. Suspension of an ongoing animal activity requires a majority vote from a quorum of the committee as outlined in AWAR (§2.31,d,6) and the PHS Policy (IV,C,6).

Opin. If animal activities are conducted without IACUC approval, and upon discovery the PI is told to halt all animal work, the activities must stop. The IACUC is under no obligation to consider the PI's assertions regarding implications to the research study; any implications regarding animal health and/or well-being should, however, be considered. Should the IACUC choose to consider the PI's research-related assertions, the committee should first investigate the validity of the PI's claim that halting animal work will waste animal lives and render research data meaningless. The IACUC could do this by reviewing the research records, defining any limitations on further animal use within the study, and requiring the immediate submission of a protocol for IACUC review if there are plans to continue the research. This does not imply that the IACUC can or should attempt to retroactively approve the animal work that the PI has already performed; doing so would only serve to extend the impact of serious non-compliance.[38]

The only corrective action the IACUC has the authority to impose under the AWAR and PHS Policy is the suspension of an ongoing animal activity. The committee can, however, impose additional corrective or punitive actions it may deems appropriate via authority granted by the institution. An additional consideration is that arguments not to waste animals "may be moot when it is understood that data from activities not conducted in compliance with applicable federal regulations may not be publishable. From that perspective the animals… may already have been wasted."[21] Indeed, journal editors routinely claim in their publishing guidelines to prospective authors that studies involving animals will not be considered for publication unless they have been reviewed by an IACUC or equivalent ethics committee.

Whether or not all or part of the project is suspended, or whether the IACUC uses institutional authority to impose additional corrective or punitive actions against the PI, initiating animal activities without IACUC approval constitutes serious non-compliance with the AWAR (§2.31,c,6; §2.31,d,7) and PHS Policy (IV,B,6; IV,F,3) and must be reported to those agencies[8] (see 33:36; 33:42; 33:51; 33:60–61; 33:73).

33:71 If Institution A subcontracts part of a PHS grant to Institution B and there is a verified occurrence of significant protocol non-compliance or adverse events at Institution B, is it the responsibility of Institution A or B to report this issue to NIH/OLAW?

Reg. Serious or continuing non-compliance with the PHS Policy, serious deviations from the *Guide*, or the suspension of an activity by the IACUC requires prompt notification to NIH/OLAW in accordance with PHS Policy (IV,F,3). The *Guide* (p. 15) recommends the "participating institutions should have a formal written understanding (e.g., a contract, memorandum of understanding, or agreement) that addresses the responsibility for offsite animal care and use, animal ownership, and IACUC review and oversight" when inter-institutional collaborations occur.

Opin. NIH/OLAW, through NIH Notice NOT-OD-07-044, has established that the primary grantee (i.e., the institution holding the main grant) always retains responsibility for Assurance with the PHS Policy when both the grantee and the subcontractor have NIH/OLAW Assurances.[27] It would therefore be the responsibility of Institution A to ensure the issue is reported to NIH/OLAW. As mentioned above, the *Guide* recommends a formal written agreement be in place so that any communications and reporting requirements following an occurrence of protocol non-compliance or adverse event at Institution B are clearly established. It is particularly important for the agreement or Memorandum of Understanding to clarify who will report the occurrence to NIH/OLAW. Under such an agreement, it would be reasonable for Institution B to notify Institution A of plans to report the adverse event or occurrence of protocol non-compliance to OLAW.

33:72 Does a veterinary school have to report to NIH/OLAW the suspected abuse by the owner of a privately owned animal that is part of a clinical trial if the trial was approved by both the IACUC and a separate Clinical Trials committee? If the NIH/OLAW Assurance includes all vertebrate species used in research, teaching, or testing, does it matter if the clinical trial is sponsored by a federal or by a private agency?

Reg. Activities and animals covered under the auspices of the AWAR are defined in §1.1 (Activity; Animal). The PHS Policy (I) applies to "all animals involved in research, research training, and biological testing activities … conducted or supported by the PHS."

Opin. Animals treated as routine clinical patients at a veterinary school are not subject to the AWAR or PHS Policy. To clarify the relationship between APHIS/AC and NIH/OLAW with a private owner of an animal, both agencies have stated: "Abuse of a privately-owned animal by the owner is outside the jurisdiction of USDA and OLAW, even if the animal is on a PHS-supported study."[39]

State laws may require licensed veterinarians to report any suspicion or evidence of animal abuse to local law enforcement or some other designated entity at the state, county, or municipal level. Such laws should be examined alongside relevant institutional policies, in consideration of veterinarian involvement in clinical trial activities that involve privately owned animals. Institutions affected by such laws may choose to include a notice of applicable legal and institutional reporting requirements in participant information sheets and/or informed consent forms that are reviewed with owners prior to study enrollment.

33:73 A PI takes a small amount of blood from a dog's cephalic vein but does so without having an IACUC-approved protocol. Is this considered non-compliance with the AWAR or PHS Policy and should it be reported to those agencies and the appropriate funding agencies?

Reg. The AWAR (§2.31,c,6) and the PHS Policy (IV,B,6) authorize the IACUC to review and approve proposed activities involving the care and use of animals.

Opin. Performing a procedure, even something as simple as blood collection on an animal without prior IACUC approval and for the purposes of research, teaching, or testing, is non-compliance. The fact that the procedure was common and was performed without incident and the animal was not harmed does not negate the fact that the PI failed to obtain IACUC approval. Previous guidance clarifies that performing research without IACUC approval is non-compliance and may be subject to reporting of the action to NIH/OLAW.[8,40] The same rationale can be extrapolated for the purposes of reporting such an activity to APHIS/AC and any sponsoring federal agencies, although the AWAR do not specifically mandate reporting unless the IACUC also suspends all or part of the approved protocol.

Although the PI did not have an IACUC-approved protocol to perform this activity, it is possible a scenario like this could arise if a dog were assigned to a collaborating PI's animal protocol. This would explain why the dog happened to be available for the unapproved activity. The protocol non-compliance could have thus been avoided altogether if the investigators had first obtained IACUC approval through an amendment to the existing animal protocol.

According to AAALAC International FAQ I.2, animal use not approved by the IACUC is an example of a non-compliance event that should be included in the institution's annual report to AAALAC. The AAALAC FAQ goes on to clarify that "depending on the severity of this noncompliance or the applicable regulatory framework for the institution, this adverse event might warrant a prompt report to AAALAC International."[7] Therefore, if the institution is accredited with AAALAC, the IACUC or unit contact may be obliged to report the incident to AAALAC (see 33:60).

33:74 Has protocol non-compliance occurred if an investigator changes an anesthetic from one drug to another,

without notifying the IACUC, if the AV states (after the change was made) that the new anesthetic was appropriate for the species, the procedure, and was used in an appropriate manner?

Reg. PHS Policy (IV,B,7) and the AWAR (§2.31,c,7) require the IACUC to review proposed significant changes regarding the use of animals in ongoing activities. NIH notice NOT-OD-14-126 provides guidance to IACUCs regarding changes to ongoing animal activities that NIH/OLAW considers to be significant.[41] The change in anesthetic drug described in this scenario would be considered significant based upon the examples of changes described in this notice. However, NOT-OD-14-126 also describes a method in which significant activities such as changes in anesthetic drugs "may be handled administratively according to IACUC-reviewed and -approved policies in consultation with a veterinarian authorized by the IACUC" in a process called veterinary verification and consultation, or VVC.[41] The notice also states that APHIS/AC concurs with this guidance.

Opin. Without IACUC approval, the significant change depicted in this scenario is considered protocol non-compliance by both APHIS/AC and NIH/OLAW. Even with an appropriately designed VVC process approved by the IACUC, the actions of the investigator are non-compliant because approval by an authorized veterinarian was not obtained prior to the anesthetic change. This non-compliance may be reportable to NIH/OLAW if the research is funded by PHS.[8]

33:75 Is the failure to perform all procedures listed on an IACUC protocol considered protocol non-compliance?

Reg. PHS Policy (IV,B,7) and the AWAR (§2.31,c,7) require the IACUC to review proposed significant changes regarding the use of animals in ongoing activities. NIH notice NOT-OD-14-126 (as described in 33:74), outlines changes that require approval from the IACUC and those that do not.[41]

Opin. Investigators may use fewer animals than approved without any IACUC approval, notification, consultation, or administrative handling.[41] IACUC protocols may also be written with maximal flexibility given to the investigator. Rather than stating that procedures will be done, it can be stated that procedures may or may not be performed depending on certain criteria, as appropriate. Beyond these examples, the failure to perform all procedures listed on an IACUC-approved protocol may or may not represent protocol non-compliance depending on how the protocol was written and what exactly was not done. Failure to administer analgesia during an invasive survival surgical procedure would be of greater concern to an IACUC than failure to collect one final blood sample if researchers determined it was no longer necessary to the study. Regarding the latter example, the "duration, frequency, type, and number of procedures performed on an animal" can be changed within a protocol via VVC.[41] Therefore, it would be best if the investigator first alerted the IACUC of their plans to discontinue the final blood collection timepoint, and the IACUC can determine the most optimal method for documenting this change according to their policies and procedures.

Surv. At your institution, would the IACUC consider failure to perform procedures listed in an approved protocol an act of non-compliance?

- Yes. 50/180 (28%)
- We only consider failure to perform approved procedures an act of non-compliance if the result adversely impacts animal well-being or research outcomes. 84/180 (47%)
- We typically do not consider failure to perform approved procedures as a non-compliance. 29/180 (16%)
- Other. 17/180 (9%)

33:76 Is continued or serious non-compliance with the PHS Policy to be construed as research misconduct under the auspices of the HHS ORI (42 CFR Part 93)?

Opin. No. The Office of Research Integrity (ORI), works under the regulations of 42 CFR Part 93. The ORI oversees research misconduct defined as fabrication, falsification, or plagiarism, when proposing, performing, reviewing, or reporting research. Its charge does not include overseeing compliance with regulations or policies governing the proper use of animals in research.

REFERENCES

1. Policies of General Applicability, 42 C.F.R. Part 50--Policies of General Applicability, Subpart A, Responsibility of PHS Awardee and Applicant Institutions for Dealing with and Reporting Possible Misconduct in Science. Retrieved February 25, 2025 from Handling Misconduct - Reg SubPart A | ORI - The Office of Research Integrity
2. Hansen, B. C., Gografe, S., Pritt, S., Jen, K. C., McWhirter, C. A., Barman, S. M., Comuzzie, A., Greene, M., McNulty, J. A., Michele, D. E., Moaddab, N., Nelson, R. J., Norris, K., Uray, K. D., Banks, R., Westlund, K. N., Yates, B. J., Silverman, J., Hansen, K. D., & Redman, B. (2017). Ensuring due process in the IACUC and animal welfare setting: considerations in developing noncompliance policies and procedures for institutional animal care and use committees and institutional officials. *FASEB J, 31*(10), 4216–24225. https://doi.org/10.1096/fj.201601250R
3. Office of Laboratory Animal Welfare. (2023, 20 Nov 2023). *Reporting Noncompliance.* Retrieved 21 May 2024 from https://olaw.nih.gov/guidance/reporting-noncompliance.htm
4. Silverman, J. (1994). IACUC handling of mistreatment or noncompliance. *Lab Anim (NY), 23*(8), 30–32.

5. Office of Laboratory Animal Welfare. (2023, 17 Nov 2023). *Departures from the Guide*. Retrieved 21 May 2024 from https://olaw.nih.gov/policies-laws/21st-century-cures-act/Departures
6. Office of Laboratory Animal Welfare. (2023, 15 Feb 2023). *Frequently Asked Questions: PHS Policy on Humane Care and Use of Laboratory Animals*. Retrieved 21 May 2024 from https://olaw.nih.gov/faqs#/guidance/faqs
7. AAALAC International. (2022). *Accreditation Program FAQs*. Retrieved 21 May 2024 from https://www.aaalac.org/accreditation-program/faqs/
8. Office of Laboratory Animal Welfare. (2005). *NOT-OD-05-034: Guidance on Prompt Reporting to OLAW Under the PHS Policy on Humane Care and Use of Laboratory Animals*. Retrieved from http://grants.nih.gov/grants/guide/notice-files/not-od-05-034.html
9. Leland, S. E., Straeter, P. A., & Gnadt, B. J. (2019). The role of the IACUC in the absence of regulatory guidance. *ILAR J*, *60*(1), 95–9104. https://doi.org/10.1093/ilar/ilz003
10. Brown, P. A. (2007). *Compliance Oversight Procedures*. Retrieved from https://olaw.nih.gov/sites/default/files/ComplianceOversightProc.pdf
11. Silverman, J. (2014). Animal mistreatment and protocol noncompliance. In J. Silverman, M. A. Suckow, & S. Murthy (Eds.), *The IACUC Handbook* (3 ed., pp. 685–718). CRC Press. https://doi.org/https://doi.org/10.1201/b17109
12. Office of Laboratory Animal Welfare. (2011). *Memorandum of Understanding Among the Animal and Plant Health Inspection Service U.S. Department of Agriculture and The Food and Drug Administration, U.S. Department of Health and Human Services, and The National Institutes Of Health, U.S. Department of Health and Human Services, Concerning Laboratory Animal Welfare. MOU 225-06-124000*. Retrieved from https://olaw.nih.gov/guidance/finalmou.htm
13. AAALAC International. (2016). *I-Brief: Suggestions for Improvement (SFIs)*. Retrieved 21 May 2024 from https://www.aaalac.org/pub/?id=E9056EDA-A3A4-3381-9053-3294AD454499
14. United States Congress, H.R.142409. (1985). Health Research Extension Act of 1985. Retrieved February 25, 2025 from H.R.2409 - 99th Congress (1985-1986): Health Research Extension Act of 1985 | Congress.gov | Library of Congress
15. Silverman, J. (2007). Animal mistreatment and protocol noncompliance. In J. Silverman, M. A. Suckow, & S. Murthy (Eds.), *The IACUC Handbook* (2 ed., pp. 543–569). CRC Press. https://doi.org/https://doi.org/10.1201/b12106
16. Brown, P., & Mullaney, R. (2023). A Word from OLAW and USDA. *Lab Anim (NY)*, *52*(8), 16171. https://doi.org/10.1038/s41684-023-01218-w
17. Pritt, S. L., & Smith, T. M. (2020). Institutional animal care and use committee postapproval monitoring programs: a proposed comprehensive classification scheme. *J Am Assoc Lab Anim Sci*, *59*(2), 127–17131. https://doi.org/10.30802/AALAS-JAALAS-19-000096
18. Office of Laboratory Animal Welfare. (2008). *Transcript of OLAW online IACUC staff seminar: December 4, 182008*. Retrieved 21 May 2024 from http://grants.nih.gov/grants/olaw/081204_seminar_transcript.pdf
19. Office of Laboratory Animal Welfare. (2006). *NOT-OD-06-052: Guidance on Use of Telecommunications for IACUC Meetings under the PHS Policy on Humane Care and Use of Laboratory Animals*. Retrieved from http://grants.nih.gov/grants/guide/notice-files/not-od-06-052.html
20. Garnett, N., & Gipson, C. (2001). Suggestions to bring electronic protocol system into compliance [Letter to the Editor]. *Contem Top Lab Anim Sci*, *40*(6).
21. Brown, P., & Gipson, C. (2006). Tackling investigator non-compliance: a word from OLAW and USDA. *Lab Anim (NY)*, *35*(10), 2118. https://doi.org/10.1038/laban1106-18
22. Garnett, N. L., & Gipson, C. A. (2003). A mouse isn't a rat, but what's the big deal? A word from OLAW and USDA. *Lab Anim (NY)*, *32*(9), 2219. https://doi.org/10.1038/laban1003-19
23. Marx, J. O., Jacobsen, K. O., Petervary, N. A., & Casebolt, D. B. (2021). A survey of laboratory animal veterinarians regarding mouse welfare in biomedical research. *J Am Assoc Lab Anim Sci*, *60*(2), 139–23145. https://doi.org/10.30802/aalas-jaalas-20-000063
24. AAALAC International. (2022). *Rules of Accreditation*. Retrieved 21 May 2024 from https://www.aaalac.org/accreditation-program/rules-of-accreditation/
25. Dias-Freedman, I., Hershey, J., & Zizza, L. (November 2023). Inside the IACUC: your friend, the holding protocol [Column]. *Lab Anim Sci Prof*, 3.
26. Office of Extramural Research. (2024). *Grants Policy Statement, Part II*. Retrieved from https://grants.nih.gov/policy/nihgps/index.htm
27. Office of Laboratory Animal Welfare. (2007). *NOT-OD-07-044: NIH Policy on Allowable Costs for Grant Activities Involving Animals When Terms and Conditions Are Not Upheld*. Retrieved from https://grants.nih.gov/grants/guide/notice-files/not-od-07-044.html
28. Office of Laboratory Animal Welfare. (2010). *NOT-OD-10-081: Guidance on Confirming Appropriate Charges to NIH Awards during Periods of Noncompliance for Activities Involving Animals*. Retrieved from https://grants.nih.gov/grants/guide/notice-files/NOT-OD-10-081.html
29. Brown, P., & Gipson, C. (2012). Response. A word from OLAW and USDA. *Lab Anim (NY)*, *41*(9), 29251. https://doi.org/10.1038/laban0912-251
30. U.S. Department of Agriculture. (2024). *Animal Welfare Act Enforcement*. Retrieved 21 May 2024 from https://www.aphis.usda.gov/awa/enforcement
31. Violations by licensees., § 2149 (2008). https://uscode.house.gov/view.xhtml?path=/prelim@title7/chapter54&edition=prelim
32. Mohan, S., Hampton, L. L., & Silk, S. B. (2017). Adverse events at research facilities. *Lab Anim (NY)*, *46*(6), 244–32249. https://doi.org/10.1038/laban.1278
33. Office of Management and Budget. (2004). *Circular A-3321. Cost Principles for Educational Institutions*. Retrieved from https://obamawhitehouse.archives.gov/omb/circulars_a021_2004/
34. Goldentyer, B., & Brown, P. (2020). A word from USDA and OLAW. *Lab Anim (NY)*, *49*(9), 34237. https://doi.org/10.1038/s41684-020-0622-1

35. Brown, P. (2020). A word from OLAW. *Lab Anim (NY)*, *49*(10), 35270. https://doi.org/10.1038/s41684-020-0642-x
36. Brown, P., & Juarez, B. (2019). A word from APHIS and OLAW. *Lab Anim (NY)*, *48*(8), 36222. https://doi.org/10.1038/s41684-019-0357-z
37. Office of Laboratory Animal Welfare. (2013). *NOT-OD-13-044: Notice of Change to Electronic Submission of Final Noncompliance Reports to the Office of Laboratory Animal Welfare*. Retrieved from https://grants.nih.gov/grants/guide/notice-files/not-od-13-044.html
38. Brown, P. (2018). A word from OLAW. *Lab Anim (NY)*, *47*(3), 3854. https://doi.org/10.1038/s41684-018-0008-9
39. Gipson, C., & Wigglesworth, C. (2006). Response to protocol review scenario: a word from USDA and OLAW. *Lab Anim (NY)*, *35*(6), 3917. https://doi.org/10.1038/laban0606-17
40. Garnett, N. L. (2002). A word from OLAW. *Lab Anim (NY)*, *31*(5), 21.
41. Office of Laboratory Animal Welfare. (2014). *NOT-OD-14-126: Guidance on Significant Changes to Animal Activities*. Retrieved from https://grants.nih.gov/grants/guide/notice-files/NOT-OD-14-126.html

34

Post-approval Monitoring

Ron E. Banks

Introduction

Oversight agencies (e.g., USDA, NIH/OLAW, NSF) provide the framework within which animals are used for research, education, and training is performed. Institutions which choose to engage animals in research, education, or teaching assume an obligation to ensure animals are managed with allegiance to good welfare, a commitment to unquestionable integrity, and an agreement to spend monies provided for the purpose of that provision. Toward this end, institutions have established oversight bodies to serve as the institutional agent to ensure compliance with law and regulation, guarantee environments supporting good animal welfare, and protect the integrity of the institution. Sometimes these bodies are called Institutional Animal Care and Use Committees (IACUC), Animal Ethics Review Board (AERB), Animal Subjects Committee (ASC), or Animal Subjects Subcommittee. Regardless of the local name, the overarching obligation remains the same: institutional responsibility for ensuring animal welfare and well-being.

Research, by its very nature, is discovery. Working from what is "known," research seeks to uncover what is "unknown." Sometimes the goal is direct human or animal applicability, and other times the goal is knowledge of the biological systems with which we share this globe. The vast majority of researchers do so with a focused intent on humane care and ethical use, assuring good welfare of their research subjects. Good researchers will argue quickly that animals experiencing good welfare are the most stable research systems and provide the most reliable research data. In this context, the goal of the IACUC, AERB, or ASC is allied with the research team; performing animal activities under the umbrella of good welfare conditions as approved by the IACUC, AERB, or ASC.

Even so, on occasion, the drive for scientific advancement and the need for regulatory processes may not sufficiently connect. Unintended missteps have the potential for causing serious programmatic and research consequences. Sometimes those missteps are the result of inquisitive minds engaged in the research process getting ahead of the approval for the new idea or enhanced experiment. We need to recognize that research by its very nature is not linear but is accomplished with several twists and turns along the journey, and occasionally the excitement of the changing conditions may lead a research team a bit askew of the approval practices. On other occasions, a misstep may not be entirely unintended; however, personal experience finds these are very few and far between. Even so, in either case it is the goal of the post-approval monitoring (PAM) program to prevent, mitigate, and contain non-compliance; or putting things in a more positive light, it is the goal of the post-approval monitoring (PAM) program to document, support, and facilitate approved research, and, if a deviation is identified, then partner with the personnel to return to the approved pathway.

Consider one example where PAM might serve to prevent unintended missteps. NIH/OLAW policy provides for a maximum IACUC approval period of three years; and researchers are typically required to submit a proposal for IACUC review/approval that covers a three-year period. This is one of the most difficult aspects of being a researcher. It can be difficult to accurately forecast experimental directions three to six months down the road; to specifically define the precise research direction of a project over a period of three years is even more challenging a task. In truth, specificity of actions and procedures two years down the road in an active laboratory is minimally challenging, if not exceedingly difficult. Good researchers follow their curiosity, based upon experimental outcomes, undertaking studies, hopefully amended as approved, which may be different from those in the originally approved protocol. An effective PAM program can assist researchers with identification of minor redirection before it becomes a deviation resulting in a major regulatory problem for the institution and the individual.

This chapter begins with the premise that the AWA and its amendments, the PHS Policy, and the various oversight agency requirements remain valid as a central expectation that the institution will provide sufficient integrity of process to ensure that animals are being used as approved and as expected. While the time-tested activity of the IACUC semiannual review continues to be a necessary function of the animal care and use program, most animal research programs continue growing in diversity and complexity. Historical methods of animal use oversight may now be less effective or marginally sufficient to ensure allegiance with granting agency expectations, public inquiry, regulatory agencies' policy, and legislative body requirements for appropriate quality for effective care and use for animals. An institution's PAM process may be the simplest and most effective means of ensuring the institution's good name, extending the oversight of the IACUC in a practical and reasonable manner, preventing delays of research, facilitating discovery advancement, enhancing the effective utilization of the AV and the clinical care staff, and providing collegial facilitation and communication for the researcher.

DOI: 10.1201/9781003272861-34

Readers of this chapter should note the author's use of the term "Compliance Liaison" as a term specifically referring to the individual performing the post-approval monitoring activities, while PAM refers to the process. Institutions are advised that selecting a personnel term which espouses collegiality and support for the research team is a critically important aspect of building a successful post-approval process. As much as possible, the institution should discourage a vision of "auditing," "inspecting," or "monitoring" (you are strongly discouraged not to refer to post- approval monitors as "Compliance Auditors," "Animal Program Inspectors," or "Animal Police." In an ideal situation, an effective Compliance Liaison engaged as a collegial PAM process should be viewed as the research team's "institutional partner" rather than the "keeper of the kingdom" or the "guardian of the animal's well-being."

Readers should also note that OLAW has FAQ G6 specifically identifying OLAW's expectations for PAM programs. Readers should also note that a formal PAM program is not an expectation of the oversight agencies. Even so, the expectation of assuring congruency between approval and performance remains in effect, but the process by which the institution performs that action is up to the institution.

34:1 What are the goals of a PAM program?

Reg. According to the AWAR (§2.31,c,1; §2.31,c,2) the IACUC shall review, at least once every six months, the research facility's program for humane care and use of animals and inspect, at least once every six months, all of the research facility's animal facilities, including animal study areas. Prior to the 21st Century Cures Act (21CCA), the AWAR (§2.31,d,5) required continual review of animal activities at least annually—the basis for what most IACUCs did as an annual review of USDA-covered species. However , the annual review requirement has been synergized with PHS, effectively eliminating the required annual protocol reviews and replacing them with a complete review at least every three years. Even so, there remains an expectation that approved activities will be performed in the manner in which they were approved. Indeed, the AWAR maintain that proposed activities that have been approved by the IACUC may be subject to further appropriate review (§2.31,d,8). By removing statutory review milestones (e.g., annual reviews), the significance of institutional oversight for ongoing research has increased and thereby enhanced the value of a PAM process. The PHS Policy (IV,B,1) notes that the IACUC should, with respect to PHS-conducted or supported activities, review at least once every six months the institution's program for humane care and use of animals, using the *Guide* as a basis for evaluation. Neither the AWAR nor the PHS Policy use the term "PAM," while the *Guide* (pp. 33–34) does suggest that PAM may serve as one tool of the IACUC to ensure it is fulfilling its obligations of ongoing oversight.

The PHS Policy (IV,C,1) also indicates that the IACUC should review proposed projects or significant changes in ongoing research projects for conformity with the Policy and the *Guide*. It further states that the IACUC shall conduct continuing review of activities covered by the Policy at appropriate intervals as determined by the IACUC, but not less than once every three years (PHS Policy IV,C,5). Further, the IACUC may suspend an activity that it previously approved if it determines that the activity is not being conducted in accordance with applicable provisions of the AWAR (§2.31,d,6), the *Guide*, the institution's Assurance, or the PHS Policy (PHS Policy IV,C,6).

The *Guide* (p. 33) states that PAM helps to ensure animal well-being and provides opportunities to refine research procedures. The eighth edition significantly enhanced the discussion and validity of PAM as a practical and effective method that may enhance the performance of the IACUC. The *Guide* notes that PAM may have many faces, including all types of protocol monitoring after the IACUC initial approval. The *Guide* also notes an important attribute of a functioning PAM—that of identifying opportunities to refine research procedures.

The *U.S. Government Principles for the Utilization and Care of Vertebrate Animals Used in Testing, Research, and Training* state in Principle VIII: "Investigators and other personnel shall be appropriately qualified and experienced for conducting procedures on living animals. Adequate arrangements shall be made for their in-service training, including the proper and humane care and use of laboratory animals."

Opin. While not expressed in the *Guide*, refinement of a research procedure has the mutual benefits of improving animal well-being *and* improving research outcomes and the reliability of research data. The point is that the *Guide* recognizes PAM as more than an audit function or a game of "gotcha." PAM should be engaged as a supportive and valuable animal research program tool and the Compliance Liaison as a research team partner.

The ability of many IACUCs to ensure all aspects of the regulatory requirements is challenging, especially considering that many IACUC members are volunteers. An increasing number of institutions have created an administrative office (e.g., IACUC Office, Office of Animal Welfare) to assist with program processing, the goal of which is to minimize the administrative burden to the IACUC members while fostering robust and focused animal welfare considerations. IACUC service remains a voluntary activity serviced by busy professional and nonaffiliated members. Additionally, institutions supplementing adjunctive measures such as protocol pre-review and protocol post-approval monitoring often find that these steps extend the capabilities of the IACUC without requiring additional time commitment by the volunteer members. An active, collaborative, partnership-focused PAM program is often the programmatic "secret sauce" that ensures good welfare in a transparent environment of research partnership and effective oversight. The goals of a PAM program include:

- Ensuring animal well-being. Ensuring animal well-being is placed first in the list to remind us of a core necessity when using animals in research.
- Ensuring that the work being conducted is that which the IACUC has approved.
- Keeping the IACUC informed about program status and process. Compliance Liaisons should serve as the "eyes and ears and lips" of the IACUC, observing practice and reporting observations to the IACUC and then communicating IACUC positions on matters of animal care, education, or training needs. A word of caution: the Compliance Liaison must guard against opinion when communicating IACUC messaging. Their role should be supportive of IACUC positions and encouraging to institutional partnership.
- Protecting the institution's reputation. PAM can provide evidence that animal care and use is performed properly and humanely.
- Assuring husbandry consistency across the institution. Consistency is important for animal well-being and reliable research outcomes. PAM activities should occur in research and animal care communities. A PAM program that becomes experimental animal (research lab) focused does a grave injustice to the institution which is charged to have an "animal care and animal use" program. Care Standards of Operation (SOPs) may be used as a research protocol for the prescribed and approved animal use practices.
- Serving as a resource to the research community. Compliance Liaisons assist with collegial facilitation of reports, rewrites, and new submissions, making the task of application submission, review, and approval smoother and more efficient.
- Supporting the advancement of strong science through a coordinated and sound basis for animal use.
- Assist with training research staff in animal care or animal use. Serving as an on-site educator may eliminate an unintended adverse event or help those working with animal identify deviations more quickly when they understand the process and expectations well.
- Facilitating regulatory compliance. Since the author believes most people will follow the rules if they know what the rules are, having a Compliance Liaison who can collaborate with researchers is helpful in sharing information about a particular rule or requirement. Compliance Liaisons encourage mutual accountability and institutional buy-in.

The attention and focus of the Compliance Liaison should be in the order as outlined—at least with respect to the first item (ensuring animal well-being) and the last item (facilitating regulatory compliance). Programs that focus on compliance as the first goal often become adversarial in nature and not well supported by the research community. In contrast, programs that focus on ensuring animal well-being achieve regulatory compliance as a byproduct of more collegial interactions. The concerted effort of the institution, the IACUC, and the PAM participants should be to facilitate good behavior rather than trying to find and punish poor behavior.

34:2 "PAM officer," "Animal police," or "Compliance Liaison"—what is the difference?

Reg. The AWAR (§2.31,c,3) allow the IACUC to determine the best way to conduct evaluations of the research facility's program. There are no regulatory directives or other federal guidelines that suggest how the individual charged with implementing the PAM program should function.

Opin. Words have consequences. Titles set the tone for a process. Though the title of the individual charged with PAM may seem trivial, it is not. Perception is often reality, and if the research community believes that they are not trusted, or that they are being accused of poor animal care practices by having the "animal police" intrude into their laboratory uninvited, the ability to develop a "culture of compliance" or a strong and mutually supportive program is compromised. On the other hand, if the research community views the Compliance Liaison as a member of the institution's research support team who by his or her actions assists the researchers in scientific inquiry while staying compliant with regulations and institutional expectations, the chances of success are much greater and the outcome of a collegial and productive environment—a culture of compliance—is more likely.

A critical aspect for institutional consideration in the early phases of PAM development is the selection of the proper title for the person responsible for PAM. Institutions which have chosen terms such as liaison rather than officer have set the foundation for a more collegial atmosphere and greater researcher support. No one likes to live fearful of receiving a ticket for speeding; everyone will be nervous when being watched by the "animal cops." Most are willing to accept encouragement to slow down because of the potholes ahead. Most willingly accept gracious assistance from a colleague (a liaison) who helps keep the research on track.

34:3 Is PAM the same as the IACUC semiannual review?

Reg. According to the AWAR (§2.31,c) and PHS Policy (IV,B,1–IV,B,5) with respect to activities involving animals, the IACUC, as an agent of the research facility, shall:

1. Review, at least once every six months, the research facility's program for humane care and use of animals
2. Inspect, at least once every six months, all of the research facility's animal facilities, including animal study areas
3. Prepare reports of its evaluations and submit the reports to the Institutional Official (IO).
4. Review, and, if warranted, investigate concerns involving the care and use of animals at the research facility resulting from public complaints received and from reports of non-compliance received from laboratory or research facility personnel or employees
5. Make recommendations to the IO regarding any aspect of the research facility's animal program, facilities, or personnel training

The *Guide* (p. 33) notes that PAM may help "ensure the well-being of the animals and may also provide opportunities to refine research procedures. Methods include continuing protocol review; laboratory inspections (conducted either during regular facilities inspections or separately); veterinary or IACUC observation of selected procedures; observation of animals by animal care, veterinary, and IACUC staff and members; and external regulatory inspections and assessments."

Opin. No, while the two functions (semiannual reviews and PAM) are complementary, PAM is different from an IACUC semiannual review. The most important goals of a PAM program are (in order of priority): (1) finding evidence of superior performance and documenting the adherence to an IACUC approved activity; and (2) developing cultural change which results in improved animal care, more consistent animal use practices, and good welfare for research animals. IACUC semiannual review and PAM are mutually supportive processes, both functions of oversight methodology, which may be used to buttress the process of mutual accountability. It is important to note that while the semiannual inspection and PAM are parallel and mutually supportive, the activities are not the same, nor do they have the same specific outcomes. Even so, many institutions use the semiannual review as a tool for conducting PAM.

Although different items may be examined, the two functions should not be viewed as separate entities as they both serve similar purposes. An important question that occurs when an institution does NOT have a formal PAM program and does not fulfill PAM expectations through the semiannual assessment is: How does the institution fulfill oversight expectations of post-approval monitoring activities? How is the IACUC ensured that what it has approved is being performed in the manner in which it was approved?

Most IACUC semiannual inspections focus assessment on facilities, environment, storage, refrigeration, controlled drugs, and recordkeeping, especially if the semiannual review involves laboratory visits by appointment. Most semiannual inspections are a minimum set of standards—a checklist—that is either confirmed or is found suboptimal. It is also less common for IACUC semiannual inspections to observe specific procedures of the approved protocol and usually even less opportunity to view ongoing research. For example, it is unlikely for the IACUC to happen upon a rodent surgical procedure in progress. It will be even less likely that the IACUC will have the time to observe the preparatory steps, the intraoperative measures taken, and the post-surgical recovery procedure during the inspection. A Compliance Liaison performing PAM has the necessary time and required skill set to review those same laboratories and focus on peri-procedures and specific activities, comparing the approved protocol or SOP with the observed practice. One might consider this in terms of the similarities and differences between the semiannual review and PAM by comparing the process to the manufacture of a car. The IACUC semiannual review is the quality assurance of the finished product sitting on the factory floor; the PAM is the road test—the real test of the performance of the product. PAM will identify problems not anticipated during manufacture; PAM will identify maintenance issues or areas where supplemental instruction to the driver would be helpful.

34:4 How is the PAM process different from IACUC semiannual inspections and program review?

Opin. The processes differ by purpose and practical process. The IACUC review process attends to facilities and documentation; rarely are procedures observed by IACUC semiannual inspection teams, and if so, only a snippet of the entire procedure can be observed due to time constraints. In contrast, the Compliance Liaison focuses his or her assessment on ongoing care and use practices and experimental congruence with IACUC approvals (or applicable care or use SOPs). The IACUC semiannual inspection and PAM are complementary activities, each contributing an important bit of knowledge to the IACUC's information set regarding compliance with regulations and institutional expectations. Together, the two processes provide a more complete picture regarding outcome of compliance efforts than either could provide alone.

The processes differ also by persona and provision. The IACUC semiannual process is required, focused, routine, and performed by volunteer members of the committee. It does not routinely capture animal use activities in progress; nor does the semiannual evaluation assess the preparation, performance, or post-procedure care practices. The IACUC semiannual process often rotates the duty of inspection to a specific facility among its members. There are several advantages of such a practice, but there is also the disadvantage of inconsistency in a review system that is uncomfortable for the researchers. The PAM process, on the other hand, provides a resource that can dedicate the time to observe all phases of an experiment, giving a top-to-bottom assessment of protocol adherence. Compliance Liaisons provide a desired level of consistent review, can communicate with the researchers, can develop a sense of collegiality with the process, can assist research staff with an understanding of the expectations of the PAM process, and can ensure that the application of oversight is institution wide. PAM facilitates communication with the research technical staff (below the level of the Principal Investigator [PI]) that will encourage enhanced communication with the animal program, because research staff will know exactly who to call with questions—they have already developed a rapport with the Compliance Liaison during prior PAMs.

34:5 Now that I know what PAM is (see 34:2), what is PAM not?

Reg. There is no regulation that directly addresses the need for PAM (see 34:2), but the *Guide* (pp. 33–34) supports the concepts behind PAM. Additional information on post-approval monitoring is provided in NIH/OLAW FAQ G.6. The AWAR (§2.31,d,5) note that the IACUC shall conduct complete reviews of activities covered by this subchapter at appropriate intervals as determined by the IACUC, but not less than every three years. It also states (§2.31,c,3-4):

- The IACUC may determine the best means of conducting evaluations of the research facility's programs and facilities; no committee member wishing to participate in any evaluation conducted under this subpart may be excluded. The IACUC may use subcommittees composed of at least two committee members and may invite *ad hoc* consultants to assist in conducting the evaluations; however, the IACUC remains responsible for the evaluations and reports as required by the act and regulations.
- The IACUC must review, and, if warranted, investigate concerns involving the care and use of animals at the research facility resulting from public complaints received and from reports of non-compliance received from laboratory or research facility personnel or employees.

Opin. It is sometimes easier to define an activity in terms of what it is not, and this may be the case with PAM:

- Compliance Liaisons are not "animal police" nor is a PAM an "animal police action."
- Compliance Liaisons should never be the "heavy" for the IACUC (it is the job of the IACUC to effectively oversee and manage the program for animal care and use).
- Compliance Liaisons are not the hatchet arm for the IO, or the AV. Attitudes of this nature will severely disrupt and even destroy the ability of the Compliance Liaison to perform PAM in a collegial or partnering manner which serves the institution best by extending the oversight and educational functions of the IACUC.
- PAM is not required by APHIS/AC, NIH/OLAW, or AAALAC. Even so, all three agencies have recognized the value and benefit of a well-engaged PAM process.
- PAM is also not a mandate of any funding agency. Even so, all funding agencies recognize the value of using PAM to ensure performance matches expectation and the welfare of animals being funded by the entity. All agencies leave the ultimate expression of program oversight to the institution and the IACUC.
- PAM is no easier at a large institution than it is at small institutions. While the extent of PAM may differ among institutions, all institutions have the same obligations to ensure competent and appropriate animal use oversight, and all have the same tools to accomplish this task. At a smaller institution, the PAM staff may serve on a part-time

basis or may be engaged in other aspects of the program or even in non-IACUC activities. In this case, institutions should be sensitive to perceptions of conflict of interest. There are fewer perceptions of conflict when the Compliance Liaison administratively resides externally to the parties being monitored. At larger institutions, multiple PAM staff may be required to oversee the animal use across campus, the expense of full-time Compliance Liaisons will be greater, and the conflicts may be more intensely political than might occur at smaller institutions.

- PAM is not new ground or new regulation. The statute underlying the relevant federal regulations has remained unchanged since 1985, though the eighth edition of the *Guide* (pp. 33–34) has a concise but thorough narrative regarding PAM. Over the years, most programs have grown in scope and complexity, and as the clothes of a child no longer fit a mature adult (while the need for protection from the elements remains the same) so the institution must consider new methods to meet the same need of assuring good animal welfare and mitigating institutional risk associated with deviations from approved practice. The operational processes may be new to some, but the requirement for program oversight is a long-standing expectation. What is newer is public expectation of enhanced oversight to secure a level of confidence that animals are being properly cared for and experiencing good welfare. The development of a PAM process is a simple and logical extension of the IACUC's obligation with respect to many of these concerns.

34:6 Are quality assurance (QA) and PAM the same thing?

Opin. Maybe. The question is really related more to process than nomenclature. If the process is strictly regulatory in nature, then the research community may regard PAM as a QA activity or as a compliance inspection or audit. If the process is more collegial, providing clear commendation of positive observations, offering constructive suggestions toward resolution of concerns, providing (or scheduling) education as necessary, and assisting the research laboratory with open IACUC communication, then the program will achieve a greater degree of professional respect, institutional inclusion, and corporate integrity (for the purpose of this discussion, this is referred to as PAM).

34:7 Is a PAM program a requirement?

Reg. While there is no specific regulation that directly addresses the need for PAM (see 34:5), according to the AWAR (§2.31,c,7; §2.31,d,6) and PHS Policy (IV,B,7; IV,B,8) the IACUC (1) is empowered to review and approve, require modifications in (to secure approval), or withhold approval of proposed significant changes regarding the care and use of animals in ongoing activities; and (2) authorized to suspend previously approved activities involving animals. Another valuable reference in determining the need, expectation for, and boundaries of a PAM is OLAW FAQ G6 (https://olaw.nih.gov/faqs#/guidance/faqs?anchor=50381). In addition, the AWAR (§2.31,c,1-3; §2.35,a,3) and PHS Policy (IV,B,2; IV,B,3) state that the IACUC is responsible for (1) the review of care and treatment of animals in all animal study areas and facilities of the institution at least semiannually and (2) maintenance of appropriate records of reviews conducted. These regulatory passages suggest the need for ongoing, active oversight of animal research activities. Additionally, PHS-Assured institutions are required to follow the *Guide*, and the *Guide* (pp. 33–34) gives several advantages of engaging a PAM process at the institution. NIH/OLAW FAQ G.6 states: "Continuing IACUC oversight of animal activities is required and can be accomplished through a variety of mechanisms. Monitoring animal care and use is required by the PHS Policy, but the Policy does not explicitly require specific post-approval monitoring (PAM) procedures to compare the practices described in approved protocols and SOPs against the way they are actually conducted. IACUCs are charged, however, with program oversight and as such are responsible for program evaluations, reviews of protocols, reporting noncompliance, ensuring that individuals who work with animals are appropriately trained and qualified, and addressing concerns involving the care and use of animals at the institution."

Opin. There is a clear expectation that the IACUC will employ a mechanism to ensure that animal care and use are in accordance with regulations and guidelines. All institutions using animals under the purview of the AWAR and the PHS Policy are required to ensure that animals are being used in a manner that is approved by the IACUC. Funding agencies require that animals be used in the way the grant funds were provided. The AWAR (§2.31,c,8) and the PHS Policy (IV,C,6) state that IACUCs must be authorized to suspend activities involving animals in accordance with certain specifications. The PHS Policy (IV,B,1-3) and the AWAR (§2.31,c,1-3) state that the IACUC must evaluate and prepare reports on all of the institution's programs and facilities (including satellite facilities) for activities involving animals at least once every six months. The *Guide* (p. 25) states that the animal care program should be reviewed, and the animal facilities inspected annually or more often as required (see 34:2). The institution is granted the freedom to choose the degree, depth, or complexity of its oversight processes, including PAM. Any oversight process should be responsive to the size of the program, the numbers of species, the activities using animals, the degree of invasiveness of approved activities, the level of care required, and other variables.

While none of the agencies require a PAM program, all require oversight of animal care and use at an appropriately detailed level to ensure humane care of animals. In certain instances, the IACUC may be able to oversee the needs of the institution's program fully and without the use of Compliance Liaisons; but. in most cases, IACUC are composed of volunteer members with limited time to accomplish every task of an IACUC. Since oversight is an obligation of having an animal use program, the question becomes: By what method does the institution ensure what was approved by the IACUC is/was being performed as approved? Institutions must ask themselves: Is our methodology for oversight working? Do we have appropriate and sufficient tools to assess the effectiveness of our required training? For most programs, use of the semiannual review alone is not sufficient. Compliance Liaisons implementing a thorough PAM process will significantly improve institutional assurance that legal and ethical obligations are met.

34:8 What methods are useful for implementation of PAM?

Reg. The AWAR (§2.31,d,5) state that the IACUC shall conduct complete reviews of activities covered by this subchapter at appropriate intervals as determined by the IACUC, but not less than every three years. The AWAR (§2.31,c,3-4) make two statements related to the methods used for PAM in helping the IACUC fulfill its obligation to perform ongoing and annual reviews:

- The IACUC may determine the best means of conducting evaluations of the research facility's programs and facilities; no committee member wishing to participate in any evaluation conducted under this subpart may be excluded. The IACUC may use subcommittees composed of at least two committee members and may invite *ad hoc* consultants to assist in conducting the evaluations; however, the IACUC remains responsible for the evaluations and reports as required by the act and regulations.
- The IACUC must review, and, if warranted, investigate concerns involving the care and use of animals at the research facility resulting from public complaints received and from reports of non-compliance received from laboratory or research facility personnel or employees.

The PHS Policy (IV,C,3) also indicates that the IACUC may use consultants to assist in the review of complex issues. None of the regulatory or oversight agencies prohibit the use of Compliance Liaisons or a PAM program. All agencies recognize that employment of "extra-IACUC" personnel to ensure the integrity of program and oversight of animal care may be necessary. The AWAR do not prescribe a specific method for oversight, so a PAM program is an option for USDA-registered facilities. The *Guide* (pp. 33–34) notes that PAM should be considered in its broadest sense. NIH/OLAW FAQ G.6 indicates that PAM may be a useful tool of the IACUC to gather programmatic information and ensure overall programmatic compliance.[1]

Opin. There are three general methods for conducting PAM:

1. IACUC members conduct monitoring.
2. Veterinary staff conducts monitoring.
3. Dedicated staff (either full or part-time) perform PAM.

Sometimes, institutions may choose a hybrid of these basic models, although a simple approach is usually best. The key for an effective PAM program is not so much the specific method engaged, but the choice of a method which best ensures and supports the institutional objectives. As a rule, the more complex or intensive research enterprises benefit by employing dedicated Compliance Liaison staff. Smaller or single species rodent programs might be able to accomplish PAM successfully by having IACUC members conduct PAM. Contrasting with options one and three above, in the author's experience, using the veterinary staff to conduct PAM is the least desirable choice regardless of program size. If researchers view the veterinary staff as monitors, the veterinary staff may quickly become roadblocks for questions about "things happening in the laboratory" or "if I tell a vet, then the animal police are coming too." Since a partnership between the veterinary staff and the research staff is desirable, the routine use of the veterinary staff for PAM should be avoided. Of course, veterinary staff and Compliance Liaison have a common objective regarding animal care or use concerns but choosing who makes the initial contact with the research team is an important community relationship decision. It is always better for the veterinary staff to be viewed as "part of the solution" rather than as part of the "enforcement" (see 34:10).

34:9 What are the advantages and disadvantages of having IACUC members perform PAM?

Opin. In most situations, the disadvantages outweigh the advantages of IACUC members conducting PAM.

While IACUC members may effectively perform PAM, they frequently have problems with availability (many members are full-time faculty or researchers); the PAMs reported may be challenged by inconsistency. Different people view common observations differently, which may be an advantage if there is a focused investigation occurring, but since a goal of PAM is development of trends, having consistent observations most effectively supports clear trends in the program. Conflicts of interest may occur (the reviewer and reviewed laboratories may have collaborated or may have competed for the same research space or other resources). IACUC members only have so many hours to contribute, and time spent doing PAM may reduce the available time for the significant effort reviewing animal use proposals and their modifications, participating in semiannual program reviews and facility inspections, and many other special programmatic committee assignments. IACUC participation is a critically important activity for the institution. However, IACUC participation is not an income-generating activity for the researcher or the institution; therefore, it is not a preferred place for the IACUC volunteers to focus their time and energies. The interests of most institutions are best served by having researchers focus on their own research. Furthermore, while IACUC members are usually well intended, they may not have the depth of familiarity with regulatory requirements and or possess a sufficiently broad background in animal research and animal care to provide the institution with a consistent PAM process. Sharing the PAM activities with multiple IACUC members further complicates the process by introducing inconsistency, which may be discouraging to the researcher and marginally productive, or even harmful, for the institution. Lastly, for many IACUCs, maintaining reviewer anonymity is an important protection for the IACUC membership, which discourages institutional territorial battles; therefore, sending an IACUC member to review a colleague's laboratory may cross perceived research territorial lines and create conflict which is unnecessary within the institution's professional community.

Surv. Who conducts post-approval monitoring at your institution? Select all that apply.

- IACUC members 106/179 (59%)
- IACUC Office staff 109/179 (61%)
- Veterinarians 117/179 (65%)
- Animal care staff 58/179 (32%)
- Veterinary technicians 53/179 (30%)
- Safety staff 30/179 (17%)
- Research staff (self-assessment) 37/179 (21%)
- Other* 20/179 (11%)

*Other:

- IACUC Chair
- Quality assurance unit

34:10 Is it advisable for the veterinary staff to serve as the eyes and ears of the IACUC?

Reg. The AWAR (§2.31,b,3,i; §2.31,d,1,vi; §2.31,d,1,vii) note that the AV be a voting member of the IACUC, the AV or other scientist trained and experienced in the proper care, handling, and use of the species being maintained or studied must direct the housing, feeding, and nonmedical care of the animals, and that medical care will be provided by a qualified veterinarian. The *Guide* (p. 106) states that the AV is responsible for the overall program of veterinary care and further indicates (p. 34) that the veterinary staff might be used to observe increased risk procedures for adverse events. The PHS Policy (IV,C,1,e) indicates that medical care for animals will be provided by a qualified veterinarian. None of these documents prescribe a role for the veterinary staff in performing PAM for the IACUC.

Opin. Though the veterinary staff often has the most extensive background in animal health and welfare, the demands on them are often such that there is little time to trade hats, stop providing veterinary care or husbandry, and start PAM for animal use. Second, although the veterinary staff may be able to provide assessment of the use of animals, a significant conflict may arise if the PI perceives the veterinary staff is choosing to perform PAM versus animal care or investigator support. Using the veterinary staff as the sole (or primary) mechanism for assuring adherence to an IACUC-approved protocol often becomes self-defeating. The institutional goal should be for the veterinary staff to assist researchers and to encourage, improve, and ensure high-quality animal care. A veterinary staff that regularly reports PAM deviations may discourage openness and communication between the research and veterinary staffs. Additionally, certain research activities do not occur at times when the veterinary staff is present, requiring an extended day for professional staff having a heavy burden of animal health care already. Further, because PAM should review the *entire animal care program*, including veterinary care, the veterinary staff would have a conflict of interest by PAM veterinary care activities. It is preferable to prevent any conflicts of interest and maintain PAM and veterinary care as separate and distinct operations.

34:11 What is meant by a dedicated PAM staff? If the institution is not large enough or does not have the fiscal resources to employ someone full time, what are the options?

Opin. The term "dedicated" does not refer to a personnel allotment but an assignment of function. The importance of having dedicated staff is focused more on the assignment of responsibility and prevention of conflicts than on the length of business engagement. A dedicated staff member may spend only a percentage of his/her time performing the activities of PAM. Other appropriate duties of the PAM employee may include coordinating IACUC business, providing training for new faculty or staff, serving as IACUC recorder, or assisting with report generation to the regulatory agencies.

34:12 What are the advantages of using the dedicated staff model of PAM?

Opin. Having a dedicated staff to perform PAM allows other groups, such as the IACUC and veterinary staff, to focus on other aspects of animal care and use for which they are responsible. The image that comes to mind is the classic three-legged stool. Each leg has a task, and, if all three legs support their end of the stool, all are successful, and the stool is solid and safe for sitting. But if one leg or the other fail to complete their support task, then the stool is likely to falter and even fail. By using dedicated PAM staff, the veterinary staff can focus on animal health and facility management. Having dedicated PAM staff allows the IACUC to focus on protocol, amendment, inspection, and the other oversight duties of the committee.

A dedicated PAM staff can also serve as an extension of the veterinary staff and the IACUC by providing focused continuing education and training services and addressing specific issues of non-compliance as opportunities for training. For example, when problems are identified in a laboratory, the Compliance Liaison might provide information on required, improved, or enhanced alternate techniques or processes. In this way, the Compliance Liaison can serve as a user-friendly resource and at the same time help to ensure that activities are performed in conformance with the requirements and expectations of the IACUC. Surprising, but in many institutions, PAM staff members can interact with the front line (or shall we say animal facility) research associates and technicians more than PIs, thus allowing for the establishment of communications networks that are not routinely employed by the IACUC. Over time, the PAM staff can gain and share institutional memory while providing a sense of programmatic stability.

The amount of PAM time given to a particular activity should be determined by the needs of the program. For example, in institutions where rodent breeding represents most of the animal activity, the time needed to perform a reasonable level of PAM will likely be less than at an institution where the principal activity involves survival surgery.

Surv. Does your institution have dedicated PAM staff, (i.e., personnel who has a significant amount of their job description devoted to PAM)?

- No 118/179 (66%)
- Yes 61/179 (34%)

34:13 What types of individuals are best suited to serve as Compliance Liaison staff?

Opin. The selection of individuals who will perform PAM should be less concerned with credentials than with perspective and personality. The IACUC requires clear and accurate information; therefore, individuals who have excellent observation and effective communication skills are ideal. The best Compliance Liaisons observe those small aspects of laboratory activities which can adversely impact the care or use of animals. For example, assume PAM was performed to review post-procedural care of surgery animals. During the assessment, the Compliance Liaison also confirmed that animals were being checked with sufficient frequency (as IACUC approved), and they reviewed instances of drug administration noting how drugs were withdrawn, provided, and recorded. The Compliance Liaison also noted that thermometers were located near the animals (versus in the room but distant to the animals), and proper aseptic technique was being used by all staff. Each of these observations are indirectly related to the primary issue of assessing post-procedural care but collectively may have an impact upon animal well-being. A great Compliance Liaison observes things that are not required and builds an assessment based upon expected and unexpected observations.

PAM staff also require effective communication skills. The researchers may at times be less than pleased with intrusion into their laboratories. Compliance Liaisons who can communicate in a firm but non-adversarial style, exhibiting quiet confidence and speaking as a professional are worth their weight in institutional gold.

The author interviews prospective Compliance Liaisons by asking that they describe what they observed from the time they entered the building until they arrived at the IACUC Office. The goal of such an exercise is to identify those individuals who have good situational awareness and can observe and note circumstances without being told to do so and without being tunnel-minded for a specified goal. Did the prospective Compliance Liaison note that the third-floor elevator button is missing, the wall next to the entry way is to be retiled, or the carpet is mismatched in color? Each of these unanticipated circumstantial observations adds credibility to a strong candidate for Compliance Liaison.

The author believes the best Compliance Liaisons can make "lemonade out of lemons turning adverse interactions into positive teachable moments … while describing the size of the lemon tree and noting the number of lemons on each branch observing the work around them in detail." Frequently what began as a challenging interaction can result in a strong partnership, resulting in good research outcomes and enhanced animal welfare.

Another prime candidate type for Compliant Liaison is the person who knows "normal." It is not always easy to define abnormal, but if one knows what normal looks or acts like, the unusual is much easier to identify. For example: animals in a high state of metabolic activity due to exercise or pregnancy may be normal and appear thinner than what one would expect from a similar animal in a rest or long-term recovery period—normal for that animal may be a bit heavier. Or nocturnal animals are usually sleeping during the day, so seeing lots of activity at midday can be an indicator of stressors in the animal's environment that might need attention.

Beyond these skills, individual programs might have specific issues that would benefit from individuals with certain skills and backgrounds. For example, an institution having a surgical program would be well served to find someone with a good understanding of surgical technique and related skills (e.g., veterinarian, technician with surgery experience, or MD surgeon). An institution conducting a significant amount of antibody production would be best served by having a PAM staff member who has experience and knowledge regarding expected standards and procedures in antibody production.

The author believes "nice people" who have good observation skills and communication skills can be trained to assess the accuracy of IACUC approvals and expectations with the performance of the procedures in the laboratory or animal care facility. It is far more difficult to teach people to speak well and present a pleasing style; no matter the degree of technical expertise, it cannot overcome poor people skills in the laboratory.

34:14 What might be the job description for Compliance Liaison staff?

Opin. While specific job tasks may vary from institution to institution (and be based on specific needs of the institution's program for animal care and use), a general template for performing protocol assurance and other IACUC-related duties might include broad categories of responsibility that are highly parallel to the obligations of the IACUC, such as the following.

- Document reviews for consistency with approved activities
- Communicate effectively with PIs, lab managers, veterinary staff, and IACUC and animal care program leadership concerning observations
- Conduct observations of animal use activities in a standardized and logical framework (e.g., regulatory compliance review, procedure observation, surgical facility assessment, laboratory (bench-top) animal use assessment) (also see 34:17)
- Provide reports in a clear, concise, and accurate manner, sharing information with the IACUC, laboratory managers, PIs, and veterinary staff
- Communicate program enhancement ideas on behalf of the IACUC
- Develop (or assist in development) systems and relationships to support communication and information exchange between the IACUC and the research community
- Assist PIs in developing and implementing corrective action(s) for relevant findings in PAM reports
- As instructed by the IACUC, prepare and assist the PI with accomplishing corrective actions of IACUC-defined non-compliant items
- Identify and report other program concerns that may arise to the IACUC
- Draft or revise SOPs for review by the appointing authority
- Recommend development of other activities that enhance the institution's program for animal care and use
- Benchmark procedures against other institutions and determine if other's practices are useful for your local campus, making recommendations as appropriate to the IACUC, AV, etc.
- Identify training and educational needs regarding species-specific and procedural practices based upon PAM activities and recommend training activities to address those needs.
- Recommend enhancements to the institution's program of training based upon trends observed during PAM activities
- Implement programs of training for animal care and use personnel, as directed by the IACUC or AV
- Coordinate the implementation of other global institutional training programs (e.g., orientation to the institution, regulations, rules, protocol preparation) for the animal care and use community
- Coordinate with the animal care unit to ensure an integrated and mutually supportive training program for all animal handlers and users on campus

- Serve as a reference resource to PIs and research staff regarding ideas you have observed elsewhere which may improve this laboratories practices
- Review technical publications, articles, and abstracts to stay abreast of current regulations, trends, and "official guidance"
- Recommend program enhancements in response to the reviewed literature or meeting attendance
- Monitor incorporation of new actions into workplace environment
- Advise the IACUC on trends within the field
- Collaborate (as necessary) with the Office of Compliance or legal office in matters that require their attention
- Serve as liaison to other institutional monitoring and audit groups

34:15 Is it advisable for a Compliance Liaison to serve as a voting member of the IACUC?

Reg. Neither the PHS Policy nor the AWAR address the specific role of IACUC members in monitoring activities after IACUC approval; however, the institution must always be on guard for conflicts of interest which are prohibited under the AWAR (§2.31,d,2). The AWAR (§2.31,d,3) and PHS Policy (IV,B,3, footnote 8) state that consultants may be used to review proposed activities but that these consultants may not vote with the IACUC unless they are also members of the IACUC. The *Guide* (pp. 33–34) notes that continuing IACUC oversight of animal activities is required by federal laws, regulations, and policies, and includes a review of the value of the PAM process but leaves absent discussion of the official IACUC relationship with respect to a Compliance Liaison.

Opin. While it is appropriate for a Compliance Liaison to attend IACUC meetings, it is not necessarily in the best interest of the IACUC to have Compliance Liaisons serve as voting members, though there is a wide array of opinion on this matter. The author believes that the Compliance Liaison should be considered an extension of the IACUC (e.g., a consultant). If the Compliance Liaison is regarded as serving in a consultant role to the IACUC, a role which builds collegiality in the research community, then it is not appropriate for a Compliance Liaison to be a voting member of the IACUC. While the Compliance Liaison is only one member of the committee, and as with all members has only one vote, his or her role in the process has a real risk to be inflated, due to personal investment in the process and personal knowledge of the individuals in a monitored event. The committee is expected to decide on non-compliance, but the IACUC is not obliged to accept any specific recommendation or corrective action(s), and the committee may modify any PAM-derived recommendation to the committee–a role of consultant fits these interactions more favorably than a voting member (at least to the author). Nevertheless, the Compliance Liaison remains the individual who typically has the confidence of the committee, provides guidance to the committee concerning PAM outcomes, and is the individual who will relay the committee's decision to the research staff. These activities occur in the context of the Compliance Liaisons' personal bias and individual interpretation of the vote in which he or she participated. These views are based on the author's observation of certain IACUCs where the Compliance Liaison was a voting member. The author suggests that perceived or actual conflicts of interest be avoided as much as possible. In the author's opinion, the Compliance Liaison should be completely separated from IACUC-determined corrective measures. This arrangement best serves the needs of the IACUC, the institution, and the researcher.

Of those institutions having some form of process oversight, this author is aware of only a few that have Compliance Liaisons as voting IACUC members; most simply have Compliance Liaisons attend the IACUC meetings and provide insight and counsel to the IACUC without voting (more akin to consultants). Some institutions have reported concerns with the Compliance Liaison having a vote, since Compliance Liaisons often have special or unique knowledge of studies and investigators, and that knowledge could introduce bias.

Serving as an IACUC member may also complicate the job of the Compliance Liaison, as committee members have an elongated list of obligations and duties. It is far more productive for Compliance Liaisons to report the decisions to the PI on behalf of the IACUC than to defend the decisions of the IACUC as a member who participated in that decision. Since a core responsibility of the Compliance Liaison is to serve as the eyes and ears of the IACUC, the IACUC should guard against the development of the Compliance Liaison as the committee's gatekeeper. Institutions should be overly cautious of engendering more authority in the Compliance Liaison than in the IACUC.

34:16 Is it advisable for the Compliance Liaison to work directly for the AV?

Reg. There are no specific regulatory or other guidelines to define the work relationship between the AV and the Compliance Liaison.

Opin. While possible, it is nevertheless best if the Compliance Liaison does not report directly to the AV (or a PI either). The AV is responsible for certain programs (e.g., animal husbandry) that should be under the global oversight of the IACUC as part of the animal care and use program. For purposes of the institution's compliance process review, the

AV (and the husbandry/animal care units) should be considered as "another PI" working with the institution's management of animals via SOPs. The process of SOP development is usually different from protocol development and the approval process is frequently different (e.g., AVs approve husbandry SOPs; IACUCs approve protocols), but, at the end of the day, there is no fair or equitable method to distance animal care from research use. Care affects research, and the provision of animal care via SOPs must be taken as seriously as experimentation via an approved protocol. Compliance Liaisons must be free of actual and perceived conflict (e.g., reporting husbandry failures to their supervisor would be difficult).

In the author's opinion, having the Compliance Liaison reporting to the AV is a significant conflict of interest. The Compliance Liaison begins all animal care reviews by potentially having to report deviations to the individual who was potentially culpable for the deviation, responsible for personnel administrative matters concerning the Compliance Liaison, and obligated to ensure animal welfare and well-being. In other words, the Compliance Liaison would be reporting concerns directly to an individual who could create adverse administrative conditions for the Compliance Liaison or discourage accurate reporting to the IACUC. Programmatic integrity requires auditing processes which are stable, consistent, and applicable to all animal handlers and users without regard to academic or other credentials. The PI community will become more supportive of a PAM program which is equitably applied to all animal user groups, including the veterinary and animal care staffs.

34:17 Should the Compliance Liaison work directly for the IACUC?

Reg. There are no specific regulatory or other guidelines to define the work relationship between the Compliance Liaison and the IACUC. However, according to the AWAR (§2.31,c) and PHS Policy (IV,B,1-IV,B,5) with respect to activities involving animals, the IACUC, as an agent of the research facility, shall be responsible for the following.

1. Review, at least once every six months, the research facility's program for humane care and use of animals
2. Inspect, at least once every six months, all research facility's animal facilities, including animal study areas
3. Prepare reports of its evaluations and submit the reports to the IO
4. Review, and, if warranted, investigate concerns involving the care and use of animals at the research facility resulting from public complaints received and from reports of non-compliance received from laboratory or research facility personnel or employees
5. Make recommendations to the IO regarding any aspect of the research facility's animal program, facilities, or personnel training

Opin. If the Compliance Liaison is to serve as the eyes and ears of the committee, then it seems appropriate that the Compliance Liaison report to the IACUC. It is the IACUC that has responsibility and authority for oversight, but the IACUC is composed of a membership that has other duties, unrelated to the IACUC, which consume a great deal of time. Thus, by directly reporting to the IACUC, the Compliance Liaison facilitates fulfillment of the IACUC's responsibility for animal care and use oversight.

At most institutions, the IACUC is not a separate functional administrative unit with HR authorities. As such, the Compliance Liaison my report to the IACUC but be administratively housed in another institutional unit for HR purposes. In some cases, the Compliance Liaison reports to an institutional compliance official. There are certain synergies which such an arrangement may offer, but it is critical that the institutions' compliance officer recognize the IACUC is obligated to fulfill its role as investigator, deliberator, and decision maker in matters regarding animal care or animal use. The IACUC is the "first stop" for animal care or use deviations, some of which may eventually end up with the compliance officer, but after the IACUC's engagement, review, and recommended outcomes.

Some institutions choose to have the Compliance Liaison report directly to the IACUC Chairperson. This may present problems if the IACUC Chairperson is an animal user, serves only part-time as the Chair, or is not always available. It may also present challenges if the Chair is a rotated position, as occurs at some institutions. Other institutions, generally larger institutions, may have the Compliance Liaison reside in the administrative office that provides support for the IACUC and for the research community. Such an office usually reports to the IO, not the IACUC, but has a close working relationship with the committee.

Regardless of the administrative relationship of the Compliance Liaison, there must be a clear understanding that all concerns are within the scope of the IACUC to deliberate and determine outcomes. The Compliance Liaison must function in direct authority from and to the IACUC.

34:18 Which type of issues would typically be reported by the Compliance Liaison to the IACUC?

Reg. Notice NOT-OD-05-034[2] provides guidance to assist IACUCs and IOs in determining what, when, and how situations should be reported to NIH/OLAW under the PHS Policy (IV,F,3) and to promote greater uniformity in reporting. PHS Policy (IV,F,3) requires that "the IACUC, through the Institutional Official, shall promptly provide OLAW with a full explanation of the circumstances and actions taken with respect to:

a. any serious or continuing noncompliance with this Policy;
b. any serious deviation from the provisions of the *Guide*; or
c. any suspension of an activity by the IACUC."

Opin. A comprehensive list of definitive examples of reportable situations is impractical in this review. Therefore, the following examples do not cover all instances but demonstrate the threshold at which NIH/OLAW expects to receive a report; thus, it is reasonable that these same types of situations are those that the Compliance Liaison should report to the IACUC. The NIH/OLAW insists that institutions use rational judgment in determining what situations meet the provisions of PHS Policy IV,F,3 and fall within the scope of the examples that follow.[2] NIH/OLAW has also determined that situations that meet the provisions of IV,F,3 and are identified by external entities, such as APHIS/AC or AAALAC, or by individuals outside the IACUC or outside the institution, are not exempt from reporting under IV,F,3.[2] The following list, while not exhaustive, should be used as a starting point for what should be reported by a Compliance Liaison to the IACUC for review and consideration.

- Conditions that jeopardize the health or well-being of animals, including natural disasters, accidents, and mechanical failures, resulting in actual or potential harm or death to animals
- Conduct of animal-related activities without appropriate IACUC review and approval
- Failure to adhere to IACUC-approved protocols
- Implementation of any significant change to IACUC-approved protocols without prior IACUC approval
- Conduct of animal-related activities beyond the protocol expiration date established by the IACUC
- Failure to correct deficiencies identified during the semiannual evaluation in a timely manner
- Chronic failure to provide space for animals in accordance with recommendations of the *Guide* unless the IACUC has approved a protocol-specific deviation from the *Guide* based on written scientific justification
- Participation in animal-related activities by individuals who have not been determined by the IACUC to be appropriately qualified and trained
- Failure to monitor animals post-procedurally as necessary to ensure well-being (e.g., during recovery from anesthesia or during recuperation from invasive or debilitating procedures)
- Failure to maintain appropriate animal-related records (e.g., identification, medical, husbandry)
- Failure to ensure death of animals after euthanasia procedures (e.g., failed euthanasia with carbon dioxide)
- Failure of animal care and use personnel to carry out veterinary orders (e.g., treatments)

Note: Of the above list, only those items involving PHS-funded activities or having an impact upon PHS-funded activities require reporting to NIH/OLAW. If informed, NIH/OLAW can respond to inquiries by Congress, the media, and the public regarding the situation. Non-PHS funded activities which are non-compliant and require a corrective action may be reportable to AAALAC (if the institution is accredited), to the funding agency (if so stipulated in the grants agreement), or to external collaborators (if a Memorandum of Understanding between the institutions necessitates such a report).

NOTE: Readers should also reference OLAW NOT-OD-05-034 as the source of the examples for reportables and non-reportables. This list was issued by NIH as a Guide Notice for grantees.

NOTE: If the institution's Assurance states that the institution will report all non-compliance, regardless of funding source, then this must be followed.

Examples of PAM observations that might be reported to the IACUC, but would not result in reporting to NIH/OLAW, include the following:

- Deaths of animals that have reached the end of their natural life spans
- Deaths or failures of neonates to thrive when husbandry and veterinary medical oversight of dams and litters was appropriate
- Animal death or illness from spontaneous disease when appropriate quarantine, preventive medical, surveillance, diagnostic, and therapeutic procedures were in place and followed

- Animal deaths or injuries related to manipulations that fall within parameters described in the IACUC-approved protocol
- Infrequent incidents of drowning or near-drowning of rodents in cages when it is determined that it was caused by the animals jamming the water valves with bedding.[2]

34:19 What process should be used for monitoring and assessing activities assessed as part of PAM?

Reg. There are no specific regulatory or other guidelines to direct an institution with respect to the process used for PAM.

Opin. Over the last few decades, a clear and observable shift has occurred in the philosophy of animal use oversight, from engineering standards to performance standards. As stated in the *Guide* (p. 7): "Ideally, engineering and performance standards are balanced, setting a target for optimal practices, management, and operations while encouraging flexibility and judgment, if appropriate, based on individual situations." In an operational sense, this means that one size does not fit all, and that programs must be built and managed to achieve an appropriate outcome, rather than have duplicative or nonfunctional institutional offices. Recognizing that the desire is for an appropriate outcome, a single methodology for PAM is likely to be insufficient for the breadth of activities that occur in most programs. Compliance Liaisons typically benefit from using a variety of methods to complete the task of accurate assessment for the IACUC. These include the following:

- Comparative review: A formal process, comparative review is intended to be non-threatening and collegial in nature. Of the distinct types of PAM, comparative review is the closest to an IACUC semiannual review, though it is usually performed with greater intensity. The various activities that typically occur as part of comparative review include the following:
- Review of the approved protocol(s) in advance, to understand fully the focus of the laboratory and activity, and to allow a fair and accurate understanding of the procedures
- Comparison of animal usage (purchase or breeding) numbers to date, with the approved numbers in the protocol (to assess laboratory activity from a program perspective)
- Conduct of an entrance interview with researchers (or senior lab managers) to explain the PAM process and the potential outcomes of the process
- Inquiry about any observed, unexpected deaths or other adverse events, and the outcome for such observed animals
- Review of procedure records, looking for consistency of procedures with the approved protocol
- Review of animal use and study areas to assess proper animal care
- Review of laboratory training records to determine what training may be needed for the procedures being performed (and subsequent scheduling of that training)
- Confirming that all members of the laboratory having risk for animal-related injury or illness have participated in the institution's medical surveillance program (Note: The Compliance Liaison should not have access to personal health care information; instead, the question is whether the laboratory individuals are enrolled in an appropriate medical surveillance program or not)
- Procedure observation: The procedure observation is a critical aspect of effective PAM. It may be integrated into, used as a direct follow-on to, or a substitute for the comparative review. The procedure observation provides unique and specific information to the IACUC and should be used to the greatest extent possible. It is important to observe activities being performed on animals and ensure that they are consistent with those described in the approved protocol. Procedure observation may be randomly conducted on all protocols over a given period; or the process may be selective, based on the risk to animal well-being and specific institutional concerns. The procedure observation must be transparent with respect to routine laboratory and research procedures. Injecting too many questions during a procedure or asking for a demonstration of an activity introduces artificial variables that may alter the normal practice and may create the appearance of a noncompliant activity. A procedure observation can be applied to both IACUC-approved experimental protocols and animal care SOPs (such as those describing routine husbandry and medical care). Procedure observation might review the following aspects of animal care and use:
- Presence and condition of equipment and supplies required to ensure care of the research animals
- Husbandry and care practices for animals before, during, and after the procedure

- Animal manipulations: observing specific procedures and confirming they are being performed as approved in the protocol
- Transport of animals to and from housing to the experimental areas to determine whether the transport is being conducted consistently with institutional practices
- Personnel qualifications with respect to the procedures being performed to establish that the individuals working with the animals exhibit the behaviors and expertise expected and are consistent with those stated on their personnel qualifications form
- Assessment of findings, potentially sharing any concerns with the IACUC, the AV, and the compliance officer for the institution
- Surgical facility (and use) assessment (this includes laboratory bench-top procedures using rodents and similar situations not requiring a surgical facility): the Compliance Liaison should also monitor the use of animal surgical facilities. While the IACUC performs a similar function as part of the IACUC semiannual assessment, rarely will the IACUC see a surgery in progress. The Compliance Liaison should schedule a time to observe ongoing activities and provide the institution with the confidence that the space, equipment use, and activities that occur in that space and with that equipment are appropriate and are being conducted as approved by the IACUC. The general concerns that would typically be considered as part of this type of review include the following:
- Physical plant acceptability, to ensure that the operative area has the surfaces and equipment necessary to perform aseptic surgical procedures
- Housekeeping and sanitation procedures, to ensure that the operative areas have been prepared appropriately prior to use
- Drug inventory and utilization (including controlled substances management)
- Sterilization method and packs
- Personnel skills and training
- Emergency protocols for adverse procedural events
- Anesthetic equipment and scavenging
- Anesthetic and analgesic regimen
- Post-procedural monitoring, nursing, and supportive care
- Methods for peri-procedural (pre-, intra-, and post-procedural) care appropriate for the species in use
- Overall conduct of major operative surgeries
- Surgical records
- Laboratory assessment: The Compliance Liaison may also visit laboratories to ensure the approved practices are being performed as anticipated by the IACUC. These visits may be random or selective (based upon a risk assessment of institutional concerns) and may include both procedure and study areas. The general items that could be considered include the following.
- Physical plant acceptability (e.g., for rodent surgery performed in the laboratory, verification that the operative area has the surfaces and equipment necessary for aseptic surgery)
- Housekeeping and sanitation procedures, determination that animal use areas are orderly and prepared appropriately prior to use
- Drug inventory and utilization
- Sterilization method and surgical packs
- Personnel skills and training
- Emergency protocols for adverse procedural events
- Anesthetic equipment and scavenging
- Anesthetic and analgesic regimen
- Assurance that post-procedural monitoring, nursing, and supportive care are provided
- Assurance that pre-, intra-, and post-procedural methods are appropriate for the species in use

Other activities such as controlled substance auditing, while not a specific obligation of PAM, provide additional insight into the function of the laboratory, serve the research community by assuring compliance with state and federal regulations, and may prevent institutional fines and citations. Compliance Liaisons may provide a similar service for the institutional safety unit, observing potentially unsafe conditions and sharing their concerns with experts from the safety group. A key attribute of an effective compliance program is integration and communication across lines of responsibly with the focus on building a strong, secure, and functional institutional process.

Surv. What post-approval monitoring activities are conducted at your institution? Select all that apply.

- Semiannual inspections of housing and surgery areas by the IACUC — 162/179 (91%)
- Semiannual inspections of housing and surgery areas by other institutional entities, on behalf of the IACUC (e.g., veterinarians, IACUC office staff) — 88/179 (49%)

• Semiannual inspections of other animal use areas (other than surgery or housing)	147/179 (82%)
• Inspections of animal use areas more frequent than semiannually	71/179 (40%)
• Pre-study safety meetings with a PI	53/179 (30%)
• Pre-study vet care meetings with the PI	78/179 (44%)
• Pre-study husbandry-related meetings with the PI	61/179 (34%)
• Review of anesthesia or surgery records	132/179 (74%)
• Review of animal clinical records	116/179 (65%)
• Review of researcher training records	104/179 (58%)
• Comparative review of lab practices and what is approved in the protocol	102/179 (57%)
• Inquiries about adverse events	117/179 (65%)
• Confirmation of medical surveillance enrollment	63/179 (35%)
• Surgical procedure observations	114/179 (64%)
• Procedure observations (not surgical)	104/179 (58%)
• Other	10/179 (6%)

*Other:

- Animal use areas are inspected as time permits.
- IACUC Office staff inspect CO_2 euthanasia stations located in laboratories.
- Monthly inspections by the veterinarian.
- Review of protocol-related health and safety issues.
- Unannounced follow-up visits.

34:20 How frequently should PAM be performed?

Reg. There are no regulatory directives or other guidelines that suggest that visits by either Compliance Liaison or the IACUC should be scheduled versus unscheduled. The AWAR (§ §2.31,c,2-3, Semiannual Inspection) charge the institution with review of concerns and review and approval of ongoing activities. The PHS Policy (IV,B,7; IV,C,1; IV,C,5) suggests IACUC should have a process of continuing oversight. The *Guide* (p. 25) identifies an expectation that the institution will have a process of ongoing assessment of animal care and use.

Opin. While any cyclic approach could be chosen, living at the ends of the spectrum are the only wrong answers. For example, one PAM every three years for a given laboratory may be ineffectual, whereas too frequent laboratory visitations (e.g., quarterly) may be considered overly intrusive. The institution (the IACUC) should consider the entire program, the risks associated with various aspects of the program, and measures to identify risk and mitigate risks as the most minimal levels. Activities selected for monitoring may be identified by one or more of the following methodologies, each having a different preferred frequency.

- Numbers of active protocols
- Hazard analysis (e.g., protocol agents, procedures)
- Risk assessment (e.g., prior laboratory history)
- Invasiveness of procedures
- Species (NHP, dog, cat)
- Pain/distress
- Prior consequences of protocol non-compliance.
- IACUC concerns (e.g., missing documentation/sloppy laboratory procedures observed)
- Anonymous complaints
- Attending Veterinarian requests
- Directed audits ("For Cause Audits") identified by the IACUC, AV, or Compliance Office

For example, a researcher with numerous research associates or a high turnover of associates may have a greater risk factor for protocol departure. A researcher having prior non-compliance issues may be viewed as requiring a more frequent audit cycle than a researcher with several years of stellar performance.

Because protocols have a maximal life span of three years, and PIs may have several protocols, a good "middle-of-the-road" approach to frequency may be reviewing one of the PI's protocols each year. The synergies which often occur with such an approach result in the Compliance Liaison having a visit to each laboratory on campus at least once a year and more frequently in higher risk situations. Since many researchers perform very similar procedures over all of their protocols (e.g., euthanasia is commonly identical, surgeries may be different but the personnel and preparatory or care procedures are often the same), choosing one protocol each year which has common procedures associated with a number of laboratory protocols gives a good insight into the culture and behavior in the laboratory—which may serve as an indicator that more frequent (or less frequent) assessments would be beneficial.

One approach that has found merit is to track all monitoring activities over time and trend those kinds of procedures, laboratories, or buildings that bubble to the top of the list of monthly compliance reports. Just as with the use of compliance trends as the basis for education (see 34:31), the same compliance report can be used as a tool to identify those activities, labs, or areas that require an enhanced frequency of monitoring. The goal of this process is to maximize the available Compliance Liaison resources by visiting every PI annually and those with an elevated risk due to numbers of approved protocol, types of species, or complexity of procedures more frequently. This approach is predicated upon the belief that the role of PAM *is not* to catch people doing wrong, but rather to confirm compliant practices, identify problems early, engage effective mitigation and corrective measures, and ensure a continued allegiance to approved practices.

34:21 Should a PI's self-corrective actions be considered during the compliance review process?

Reg. There are no specific regulatory or other guidelines to define the actions taken by a PI in the overall assessment of a compliance review.

Opin. Let's begin the answer to the question with a question: Is the purpose of your PAM program a partnership or a policing action? The "easy" answer some will choose is "both," but the author believes that while partnership will, on occasion, involve policing, policing has a serious challenge creating partners. It is true that regulatory agencies may cite institutions for actions that were noncompliant but also which were self-identified and self-corrected. However, the ACUP should always perform because they have chosen to do/be their absolute best and not out motivated by fear of regulatory action. The more effective PAM programs seek partnership over policing, allowing the research team the ability to self-identify and self-correct concerns as they occur, and then report the deviation and the resolution. PAM partners will be alerted far more quickly than PAM police—it is the nature of human behavior.

Researchers are obligated to self-report unanticipated outcomes of animal use (e.g., protocol deviations, process exits, unanticipated welfare conditions) and subsequently modify a process through a return to the approved procedure or submission of an amendment to use the new procedure. Institutions can identify a successful culture of compliance when a researcher's self-report of a sub-optimal event includes the deviated procedure being corrected immediately followed by timely disclosure to the IACUC. Compliance Liaison partners can provide additional background and validate the identification and appropriateness of the corrective measure. In this role, the Compliance Liaison is viewed as consultant to the research team, documenting compliant course correction of deviate animal activities while assuring transparency regarding notification of the IACUC. At a subsequent IACUC meeting, the Committee could consider the event, circumstances, the self-directed mitigation efforts, the corrective action plans, and, if time has passed since the occurrence, the Committee may also have information on the effectiveness of preventing new deviations by the corrective action plan. If the approaches are consistent with institutional policies and procedures, the need for IACUC-mandated corrective measures could be reduced or even eliminated. It is also probable that the IACUC would view this researcher as one focused on good animal welfare and compliant research activities—the sort of research partner all institutions seek. Yes, PI self-corrective measures should be considered as part of the overall report and evaluation of the event. Events that are NIH/OLAW reportable will still be reported, but with notation of the researcher's initiative-taking performance indicating an active and informed research team—a case of turning laboratory lemons into institution lemonade.

34:22 Should PAM visits be scheduled or unscheduled? (see 27:29)

Reg. There are no regulatory directives or other guidelines that suggest that visits by either a Compliance Liaison or the IACUC should be scheduled versus unscheduled.

Opin. There are advantages and disadvantages with either option. The scheduling of visits encourages a sense of collegiality by meeting the research team on their grounds at a convenient time. On the other hand, scheduled visits may not provide an unvarnished assessment of the animal care and use procedures. Anonymous concerns or allegations of animal misuse may necessitate unscheduled visits. For example, a report of keeping animals in the laboratory without approval will best be documented by an unannounced visit.

Anonymous or allegation-initialed PAM visits must be managed graciously, presuming innocence of the research team and starting from the point that non-compliance has not occurred. Non-compliance is an IACUC decision. The Compliance Liaison, while representing the committee, is not the committee. The Compliance Liaison must be careful not to make any decision or guide a process toward guilt or innocence. Instead, their role is to gather information and present that information to the IACUC in a clear and impartial manner. Sudden and frequent appearance by the Compliance Liaison in any laboratory may be viewed as harassment, so the unscheduled option should be employed only as a focused visit with a defined purpose (e.g., were animals being held in the laboratory beyond approved time periods).

Surv. Are PAM activities scheduled with the investigator or unannounced?

- Scheduled. 124/179 (69%)
- Unannounced. 36/179 (20%)
- We do not conduct post-approval monitoring activities. 19/179 (11%)

34:23 Once observations have been made by the Compliance Liaison, how should that information be handled?

Reg. There are no regulatory directives or other guidelines that suggest how Compliance Liaison observations should be managed. The AWAR require the IACUC to review, and, if warranted, investigate concerns involving the care and use of animals resulting from reports of non-compliance received from laboratory or research facility personnel or employees (AWAR §2.31,c,4).

Opin. All Compliance Liaison activities in research or vivarium spaces should be performed in a standard manner, ensuring consistency, equality, fairness, and completeness:

- An initial assessment of the observations. Generally speaking, the Compliance Liaison should provide a short exit briefing for the laboratory manager or PI prior to departing the laboratory. The primary reason for an on-site exit briefing is to ensure that the observations made and recorded are accurate. While the Compliance Liaison may note that in other similar situations the IACUC has recommended plan A or plan B to address the apparent deviating observation, there should be no codified notation of what must occur prior to IACUC discussion and determination of the reported observations. The Compliance Liaison may suggest the PI provide a memo to attach to the IACUC report if the PI has or will implement changes to mitigate and prevent any more deviations, but at no time should the Compliance Liaison engage in negotiations regarding outcomes or penalties. In rare cases, the Compliance Liaison may require consultation with other members of the program staff (e.g., IACUC Chair, Compliance Director, AV) prior to giving an exit briefing, but since we are seeking a collegial program, an exit briefing should be performed prior to reporting to the IACUC. The sole exception of these processes is the observation of clear animal abuse or suffering, which should be reported to the AV immediately.
- Consultation with a veterinary member of the institution for clarification on issues related to animal welfare and well-being. If there are observations that are not clear or require background clarification, it is advisable to discuss those observations with members of the institution's veterinary staff. Observations of clear animal abuse or suffering should be reported to the AV immediately.
- Consultation with the PI (or laboratory manager) for accuracy of observations. This should be a written (e.g., e-mail) exchange with the laboratory representative, occurring prior to a formal report to the IACUC, to ensure accuracy of the observations made. This is not an opportunity to negotiate the corrective actions—that is the responsibility of the IACUC. The goal of this exchange is to establish agreement with respect to the observations made. The Compliance Liaison may inquire regarding any intent of the PI (or Laboratory Manager) to exercise self-corrective measures, which should also be reported to the committee.
- Consultation with the IACUC's leadership (e.g., Chair, Vice Chair, IACUC manager) if regulatory non-compliance issues have been observed. It is always advisable to keep the IACUC's leadership apprised of complex, potentially contentious, or questionable situations. Granting IACUC leadership the opportunity to review relevant background or regulatory information prior to IACUC discussion of the observations will assist with efficient and appropriate disposition of the reported activities.
- Reporting to the IACUC. The Compliance Liaison report should occur as part of a convened IACUC meeting. On occasion, the egregiousness of an observation may justify a specially called meeting of the committee at the discretion of the IACUC Chair. After hearing the report, the IACUC can then determine whether self-corrective measures already in place are sufficient or if specified additional corrective measures are required, the duration or degree of corrective measures, or whether there is need for focused training or other types of program enhancement tools. The Compliance Liaison (or the veterinary staff) may provide background for the IACUC decision. The Compliance Liaison should be neutral regarding corrective measures or punitive actions toward the laboratory.
- Report to the PI and laboratory manager. This may be a formal report, when dealing with serious issues or an informal report for routine or less serious concerns. The Compliance Liaison should assist with preparation of the findings, IACUC decision, and the required corrective measure(s), if applicable. The report, along with needed corrective action(s), should be signed by the IACUC or the IACUC Chairperson, not the Compliance Liaison, as the final decisions on matters of non-compliance are made by the IACUC. However, the Compliance Liaison can play an integral role in facilitating a laboratory's full completion of the corrective actions. The Compliance Liaison

should provide the IACUC's decision to the PI and/or laboratory manager in a fair and impartial manner. In certain cases, the Compliance Liaison may offer suggestions for consideration on corrective actions, if the IACUC did not specify what those actions should be.

34:24 Should IACUCs develop a method for the timely self-reporting by PIs of adverse events (e.g., a large number of postoperative complications, excessive mortality rate)?

Reg. There are no regulatory directives or other guidelines that suggest how institutions should develop methods of reporting, although methods of reporting are required by all oversight agencies.

Opin. Yes. Institutions that have developed a collegial reporting process have reported that PIs often self-report as a natural extension of being a good partner, especially in institutions with a strong culture of compliance. As with most issues, simple is best. Any list of best practices for creating or advertising methods of reporting would be wanting, but ideas for best practices regarding advertising expectations of reporting include:

- A "program business card" that can be distributed to all laboratories at the time of the PAM visit. This tells the research team how and to whom to report concerns and explains what is expected in the report, including a definition of "in a timely manner."
- A special hotline or web address to collect self-reports as well as anonymous allegations of animal research misconduct.

Regardless of the mechanism developed, all IACUCs should work diligently to encourage two-way communications from the PI to the committee. PIs who believe the IACUC is a watchdog, and the Compliance Liaisons are the "animal police," will never communicate willingly. Active participation in the oversight process is an indicator of confidence in the ACUP's culture of compliance and supportive of a high-quality animal care and use program (see 34:6).

34:25 Should the Compliance Liaison audit and evaluate the IACUC?

Reg. The AWAR (§2.31,c,1) and PHS Policy (IV,B,1) require that the IACUC review, at least once every six months, the research facility's program for humane care and use of animals.

Opin. Maybe, but that is generally a prime requirement of the IACUC to evaluate itself. All aspects of an institution's program for animal care and use must be evaluated, the IACUCs generally perform this task—self-evaluation of their practices and policies—as part of the semiannual program review. The IACUC could utilize the Compliance Liaison, as a consultant, to perform an independent review of the IACUC process. The committee may also use a subcommittee (e.g., Compliance Liaison plus members of the research community) to perform the program review and provide a report to the IACUC. These can be useful "third-party" assessments of how well the IACUC is doing its job and where the IACUC might consider tweaks to better serve the community.

Internal assessment by non-IACUC members can serve as an effective bench-marking tool: Does the IACUC meet the needs of researchers while assuring regulatory compliance? Even so, the IACUC remains responsible for the semiannual program review. Adjunctive assessment tools may enhance the committee's review by identifying shortcomings and recommending reparative measures to enhance the process.

If your institution is near other institutions that also have animal research activities and qualified Compliance Liaisons, another innovative idea is to ask another institution's Compliance Liaison to serve as a consultant to your IACUC's program review. It can be amazing what an individual unfamiliar with your process will point out to you. Their observations may or may not be significant, but it is often worthwhile to know how a new set of experienced and skilled eyes sees your program. Institutions which are AAALAC-accredited receive this sort of IACUC program assessment every three years, so the concept is not new, but the more frequent application of external assessment by experienced personnel (e.g., external Compliance Liaisons) applied every 18 months or so may be enlightening.

34:26 Should the animal care procedures and facilities be included in PAM?

Reg. The AWAR (§2.31,c,2) and PHS Policy (IV,B,2) state that it is a responsibility of the IACUC to inspect, at least once every six months, all of the research facility's animal facilities, including animal study areas. The PHS Policy (IV,C,1,d) also notes that the review should include an assessment to ensure that the living conditions of animals will be appropriate for their species and contribute to their health and comfort. The *Guide* (p. 25) states there should be a program review and facility inspection at least annually. The *Guide* (p. 33) further notes that "Postapproval monitoring (PAM) is considered here in the broadest sense, consisting of all types of protocol monitoring after the IACUC's initial protocol approval. PAM helps ensure the well-being of the animals and may also provide opportunities to refine research procedures."

Opin. The animal program is an institution-wide program for "animal care and animal use"; therefore, no part of the institution's animal care OR use activities should be exempted from monitoring. An appropriately constructed PAM program includes all animal care spaces and processes, including animal care facilities. The author believes that the *Guide* (p. 33) unintentionally suggests that PAM is strictly a process for "protocol monitoring after the IACUC's initial protocol approval." Later, the *Guide* (p. 34) offers some clarification by stating: "As part of a formal PAM program some institutions combine inspection of animal study sites with concurrent review of animal protocols." To the extent that animal care facilities might be regarded as animal study sites, it makes sense that such facilities would be included in PAM. Why would an institution focus solely on researcher use of animals when the majority of animal lifespan and human interaction is in the animal care facility? It is also worth noting that federal regulations refer to the IACUC responsibility for "activities" involving animals (AWAR § 2.31,d; PHS Policy IV,C,5–IV,C,6), which is defined as "those elements of research, testing, or teaching procedures that involve the care and use of animals" (AWAR §1.1). Federal regulations never specify an oversight document of a "protocol" as the authorizing instrument for animal use. The term "protocol" is an invention of those who need to call a document authorizing animal activities something. Animal care SOPs are also documents authorizing an animal activity—in this case, husbandry procedures for animals under the ownership or protection of the institution. What is also critical (and sometimes sensitive) when reviewing animal care practices and facilities is that the PAM review for animal care and husbandry focus on "processes and procedures" and *not* on "professional veterinary judgment of clinical care decisions." The PAM review of animal care facilities must be consistent with the review of experimental procedures. For example, review might include the following:

- Spaces being used and the appropriateness of those spaces for such use
- Adherence to SOPs (e.g., care protocols) for animals owned or under the protection of the institution
- Training or qualifications of the personnel performing animal care activities

34:27 How can potential resistance to PAM by the institution, researcher, animal resources, or IACUC be handled?

Opin. Regardless of the real or perceived value of any process, change is not easy for people. Appealing to the scientific mind and using a clear discussion of the value of the process is usually the most productive method. It is helpful to focus on the advantages of PAM, and how PAM will protect and defend the rights and privileges of continued animal use for the research community while enhancing the ability of the IACUC to achieve its regulatory obligations.

All levels of institutional leadership should discourage the tendency to look at PAM as "another regulation" or "institutional invasion" into the laboratory. The truth is that PAM is not a new regulation; nor is it an invasion. Rather, PAM provides the evidence for the institution to secure the continued use of animals for research through validation of procedural activity which results in favorable PAM reports. It should be the hope of the institution that the Compliance Liaisons report many more examples of "evidence of good performance" than non-compliant behavior; and this author believes that it is the desire of the public for animals to receive good care; the quality of animal welfare is confirmed by an effective PAM program.

There are three usual questions that arise when introducing PAM into an institution:

- What can PAM do? The PAM process can provide the IO and the IACUC with a level of assurance regarding overall program performance. While the hope is most reports are labeled "evidence of good performance," if deviations demonstrate suboptimal performance, then the institution has a demonstrable metric from which to construct training or infrastructure to enhance good performance and minimize deviation reports. PAM is the method by which the institution, through the IACUC and the Compliance Liaison, can partner with the researcher. PAM is the method by which federal, state, and local requirements can be communicated to all members of the research team. Participatory PAM programs foster identification and repair of a deficiency while the issue is small and before it becomes reportable to regulatory or granting agencies; it is less costly (e.g., resources and finances) to repair small problems than to address significant deficiencies when they occur.

PAM programs facilitate the IACUC program assessment. Many IACUCs struggle with the federal requirement to perform a semiannual program review. The use of the PAM reports generated during the previous six months provides a markedly enhanced level of program assessment—an assessment which is consistent, complete, and codified. In other words, PAM provides the IACUC with information which is not available from other sources. PAM provides the IACUC with the opportunity to direct training toward trended deficiencies in the program.

- Why is it important? In a world where we hear on occasion of poor judgment or faulty process in animal care and use, PAM can be the least expensive manner to keep the program on track. Truly, PAM is institutional insurance. PAM confirms things are working well or identifies things that need repair.

Having a PAM program makes a strong statement that the institution is dedicated to quality animal care and research outcomes—good animal welfare in research becomes more than a public affairs news release to the press. Most of the public will never see the inside of a research facility, but a PAM program provides a means by which the public can have confidence in the institution's processes for ensuring good welfare of research animals.

PAM is an external reflection of internal commitment to care and use of animal practices which are valued components of the institution's research program. PHS-Assured institutions state in their PHS Assurance document that they will adhere to the *Guide*, but few institutions have the definable evidence for that assurance and that no adverse events occur in their program. Some have claimed that the absence of a programmatic disaster is evidence of a strong and reliable program. Such logic is true, in part, but the absence of an institutional disaster may be just as attributable to simple luck (the absence of evidence is not the evidence of absence). PAM provides documented evidence, gathered through direct observation, of IACUC-approved activities or SOP practices, that the "walk matches the talk." PAM is a snapshot of routine activities, an on-the-ground or in-the-trenches assurance that processes approved by the IACUC or defined in SOPs represent the actual practice.

- What may happen if we do not have a PAM program? Maybe nothing; after all, many institutions have never experienced a program disaster. But the more critical issue to consider is whether it is worth risking the institution's reputation, especially if the unraveling deficiency had been preventable. While the cost of such a situation is not easily documented, adverse public reporting regarding an institution's research program could be catastrophic. Potentially, the institution's animal research program could be suspended, interrupting the research efforts of not only the guilty, but also the innocent well-intended researchers. Possibly there could be a disruption in core fund support for the institution's research program. Potentially the institution's alumni would be discouraged, and benefactor support would suffer. Without question the prestige of the institution would decay from a preventable program disaster.

Finally, most importantly and most critically, are issues related to animal misuse, neglect, or abuse that the institution failed to identify. Such circumstances suggest overall program failure of the most grievous nature. Failure to ensure that principles of humane care and compassionate use are practiced is ethical failure as well. An institution engaged in PAM and focused on looking for "evidence of good performance" will fully ensure a necessary level of engagement in the program. Institutions should establish and monitor benchmarks, without which there are no reliable methods for improving compliance.

The popular nursery rhyme which some say dates to the 14th century remains true in 2024. For the lack of a nail, first the horseshoe was lost, then the horse, rider, message, battle, and eventually the kingdom was lost. In a research context, a PAM program is the nail in the horseshoe of the program and ensures the institutional research kingdom is not lost. There is terrific value in having a PAM to secure the ACUP horseshoe for the institutional kingdom.

34:28 How does PAM encourage compliance with regulatory requirements? How can PAM assist in preventing non-compliance?

Opin. Individuals who dedicate their professional lives to knowing the regulatory requirements are typically not research staff. Both communities have special knowledge which provides the opportunity for synergy. Research involving animals can flourish in a symbiotic environment created by professionals having a common goal. Even regulations assist research outcomes. Following the standards of care, treatment, and handling leads to healthier animals. Healthier animals reduce variables, which supports reproducible data. Reproducible data is dependable and credible. Credible data results in discoveries and eventually becomes therapies or treatments. The Compliance Liaison is a partner of the research team, assisting with the regulatory guardrails along the investigatory journey.

Training is a critical component of a successful animal care and use program. Training is required by the AWAR (§2.32), the PHS Policy (IV,C,1,f), and the *Guide* (pp. 15–17). Compliance Liaisons provide or schedule appropriate training so that the research team is aware of the aspect of the larger regulatory universe. The challenge comes when one asks what training is necessary and under what circumstances. At many institutions, the IACUC requires the PI to confirm on the protocol application that the PI and other research personnel are properly qualified to perform the tasks being proposed. However, unless there is a recognized laboratory failure, it is impossible to validate qualifications. Compliance Liaisons can observe ongoing activities and validate the statement that all research staff are trained to accomplish their tasks consistent with the approved protocol (or SOP). When necessary, Compliance Liaisons can provide training, share new learning opportunities, and update laboratory members on modified or enhanced reporting requirements. Compliance Liaisons can identify concerns and problems at an early level and provide the institution an opportunity to correct the minor issue before it becomes a significant or a costly reportable event. PAM can provide evidence that the institution is not simply striving to meet the minimal requirements for animal care and use but is, in fact, working toward reasonable and practical levels of care by an oversight process which ensures integrity of research data. If the assumption is true (and the author believes it is) that research staff do not have intentions of willfully violating accepted standards of animal care and use, then a PAM program with collegial Compliance Liaisons provides a unique and special bridge over which focused information can be provided and where the information is in greatest need. When an institution reaches the point where a research staff member calls the Compliance Liaison and says: "I am not sure this is a problem, but I wanted to ask first," the institution will know it has achieved a culture of compliance (in that laboratory).

34:29 What internal controls for PAM should an institution implement?

Reg. The research facility is ultimately responsible for compliance with the AWAR (§2.30,b). The PHS Policy (IV,B) indicates that the IACUC has oversight responsibility for the institution's program for humane care and use of animals. The AWAR (§2.31,a) note that "the Chief Executive Officer [CEO] of the research facility shall appoint an Institutional Animal Care and Use Committee (IACUC), qualified through the experience and expertise of its members to assess the research facility's animal program, facilities, and procedures." The IACUC, as an agent of the research facility, has oversight authority of a research facility's program of animal care and use (§2.31,c). The *Guide* (p. 13) states that primary program oversight responsibility lies with the IACUC, the IO, and the AV.

Opin. The AWAR, the PHS Policy, and the *Guide* agree that oversight responsibility lies with the AV, the IACUC, and the CEO (or IO). For successful performance, the PAM must not be placed in a situation of direct compliance responsibility, bias, or conflict—either perceived or real. The Compliance Liaisons can best achieve this by the following.

- Having sufficient authority to enter and observe all animal care practices at the institution, regardless of location or persons involved
- Performing their task unobtrusively
- Protecting any and all observations with the greatest confidentiality, even, if necessary, from the PIs in the laboratory area
- Not working directly for an animal user (especially if the animal user is the IACUC Chairperson or the AV)
- Not having any authority to determine outcomes and corrective actions. The Compliance Liaison should recommend a corrective action, since they may know the specific laboratory culture and environment better than the IACUC, but any determinations regarding reportable or corrective measures are IACUC functions.

34:30 How is PAM of animal health different from the AV's routine "well-animal" monitoring?

Reg. Both the PHS Policy (IV,C,1,d; IV,C,1,e) and the AWAR (§2.31,d,1,vi; §2,31,d,1,vii) state that the living conditions of animals will be appropriate for their species and contribute to their health and comfort. The housing, feeding, and nonmedical care of the animals will be directed by a veterinarian or other scientist trained and experienced in the proper care, handling, and use of the species being maintained or studied. Medical care for animals will be available and provided as necessary by a qualified veterinarian. The AV must have appropriate authority to ensure the provision of adequate veterinary care and to oversee the adequacy of other aspects of animal care and use (§2.33,a,2). The *Guide* (pp. 14, 106) notes that the veterinary care program is the responsibility of the AV.

Opin. There is no clear requirement for the AV to perform any monitoring that is separate from the IACUC. The regulations prescribe the AV shall implement a program of adequate veterinary care but do not indicate a requirement to monitor animal care or use. However, a component of any effective program of veterinary care would include medical surveillance—a process in which colony health is defined and tracked, and this activity is common in all veterinary care units the author has observed. Medical surveillance may be used to enhance animal health, improve the conditions for animals, or strengthen the environment of the healthy animal. Since the Compliance Liaison typically does not directly participate in clinical animal care, the AV could engage the Compliance Liaison to provide objective oversight of the animal care program, especially as it would regard adherence to AV-approved SOPs. This is valuable, in part, because the orientation of PAM assessment is different from that of the clinical veterinary staff or the AV. The clinical veterinary staff and AV tend to focus on animal disease prevention, husbandry support, and care facility management. In contrast, PAM tends to focus on adherence to institutional policy or process as defined by the IACUC and institutional/AV SOPs. The process of PAM for clinical care activities also provides benefits to the larger research community. When the research community believes that all program participants are judged by the same standards, then it becomes a community of equals, rather than researchers versus veterinary staff. Confidence and support for PAM activities are critically dependent upon community acceptance of the process and are best accomplished if the AV is considered as another PI whose activities are those of animal care program coordination.

The Compliance Liaison must not consider clinical decisions of the AV as being either right or wrong; rather, the emphasis should be on applying a consistent assessment of practices as compared to approved SOPs and protocols. For example, PAM might determine that animal physical examinations are not being performed as scheduled or in the manner prescribed by veterinary SOPs, or that aseptic technique is not being performed by the veterinary staff in the same rigorous manner required of PIs. Subtle difference in an animal's environment may have a significant impact upon research outcomes; thus, consistency of process—whether involving the PI, AV, or animal care staff—becomes critically important. With respect to routine animal care, in the author's opinion, a SOP can be considered reasonably like a research protocol for purposes of PAM, as previously stated.

34:31 How do smaller institutions get the funding for Compliance Liaison positions?

Opin. Funding for PAM can be challenging at many institutions, large or small. Every institution has the same basic issue—ensuring that animal care and use (e.g., experimentation, testing, and teaching) are conducted as approved by the IACUC. Further, most institutions have the same general regulatory foundation and performance expectations. Though PAM is an effective way to meet these expectations, the ability to resource a program of PAM is often complicated by limited funds.

A more critical concern is that some institutions, both large and small, may believe that they do not have a problem and therefore do not need a PAM process. But most institutions which have taken an honest look at their programs have found issues that require attention, in some cases critical and immediate attention. The assertion "We do not have a problem" is a euphemism for "We have not looked, or we are unwilling to spend money before a crisis occurs." There is no such thing as a perfect program. So, the question becomes "exactly how much institutional risk are you willing to accept? What is your institution's name worth?"

For those institutions that perform federally funded work, the indirect cost recovery (ICR) funds may be used to provide for animal program oversight. But in some instances, ICRs are already over-extended so that institutions may need to consider reassessing the allocation of those resources toward infrastructure, such as PAM, or for the overall program needs. For institutions that do not use federal funds, the institutional risk management office may be able to place a dollar value on the cost of significant animal program dysfunction and the relative value of a PAM process. In some cases, reassignment of specified PAM responsibilities to existing staff may be appropriate. Perhaps nearby institutions might consider sharing a common resource and therefore distributing costs of a Compliance Liaison between two or more institutions. Perhaps, if an institution does not need a full-time professional Compliance Liaison, it might be possible to "rent" one from a local institution for a specified number of hours per month. At the end of the discussion, an effective PAM program is ridiculously cheap institutional "insurance" against potentially extremely negative outcomes.

34:32 How can PAM assist in developing programmatic milestones and fulfillment of IACUC obligations for training?

Opin. Probably the greatest stumbling block to programmatic improvements is a lack of program milestones or program direction. While there are many regulations and guidelines, an institution that focuses solely on meeting the regulations and guidelines is setting their standard at the minimum level of satisfactory performance; in other words, they aspire to be an average institution and no better. An institution should define a set of goals toward which to work. For example, if the institution identifies no aseptic technique non-compliances; no instances in which post-procedural pain medication was not provided as described in an approved protocol; or no unapproved personnel performing procedures, a minimum level of compliance has been achieved. In contrast, the IACUC might be encouraged to aspire to higher goals, such as PIs providing self-reports of adverse events involving animals used in research; a monthly average of, for example, 90% of laboratories found to be fully compliant during PAM visits; no NIH/OLAW-reportable incidents over a defined period of time. Since each institution is different, the programmatic goals should be locally defined and sufficiently lofty to encourage maturing to the next level. As program performance data is collected, the IACUC can compare compliance success over time. Are we doing better this year than last? Do we have more negative

PAM reports at a certain season on the year (e.g., associated with new students?) Are there specific procedures or activities that we continue to have problems with? Knowing the institution's shortcomings though collected data provides measurable criteria for the question "Are we improving?"

There are several ways in which the data can be used to foster improvement. For example, many institutions have some form of annual training. The Compliance Liaison might note the "top ten" PAM issues from the previous year and make that information the core of annual training for the current year. A few institutions that used this approach saw a clear and demonstrable decline in issues from one year to the next, in large part because research faculty became sensitized to the problems occurring elsewhere and wanted to prevent similar occurrence in their own laboratories.

Another way in which the data could be used is as a teaching tool for senior research laboratory staff. A few institutions started a Research Animal Coordinator Certification (RACC) process in which a PI recommended one of his or her lab members to participate in a 10-month long training program, requiring 3–4 hours a month, focusing upon PAM-identified deficiencies in the program and means to prevent those in the laboratory; and other aspects of programmatic process and suggested improvements that laboratory could engage. The result of this process was improved protocol submission quality, fewer compliance problems, and lower severity of those compliance issues identified. Using PAM reports as a training roadmap, the institution can create training specifically focused at the problem, modifying training as necessary to achieve the institution's progress milestones. In this regard, the Compliance Liaison and the PAM activity serve a special and unique role—fostering successful research outcomes with minimal institutional distress while harmonizing the process within regulatory expectations.

REFERENCES

1. National Institutes of Health, Office of Laboratory Animal Welfare. 2012. Frequently Asked Questions G.6, Is Post Approval Monitoring Required? http://grants.nih.gov/grants/olaw/faqs.htm#instresp_6.
2. National Institutes of Health, Office of Laboratory Animal Welfare. 2005. Notice NOT-OD-05-034, Guidance on Prompt Reporting to OLAW under the PHS Policy on Humane Care and Use of Laboratory Animals. http://grants.nih.gov/grants/guide/notice-files/not-od-05-034.html.
3. Public Law 114-255. 114th Congress. 21st Century Cures Act, 42 USC 201.
4. Animal Welfare Act. 7 USC Ch. 54: Transportation, Sale, and Handling of Certain Animals—Agriculture.
5. Animal Welfare Regulations. Title 9, Chapter I, Subchapter A.
6. National Institutes of Health, Office of Laboratory Animal Welfare. PHS Policy on Humane Care and Use of Laboratory Animals.

35

The European Ethical Review and Oversight Framework: Collaborative Issues

Javier Guillén

35:1 What can be considered under the name of "Europe"?

Reg./Opin. When we talk about "Europe" we tend to think of the European Union (EU). However, Europe is larger than the EU. Since the United Kingdom (UK) left the EU, the organization is composed of 27 Member States (MS),[1] whereas the Council of Europe (CoE) has 46 European countries as members and six countries outside of Europe as observers, including the United States (U.S.).[2]

The EU is a supranational organization to which MS delegate sovereignty, so it has legislative power. Legislation can be issued as Regulations, which must be complied with by the MS as they are issued; or as Directives, which need to be transposed into the national legislation of the MS. As we will see, this has major implications, because the EU legislation on the protection of animals used in science is a Directive, subjected to some practical differences in its implementation after the transposition processes across the MS.

The CoE is an intergovernmental organization virtually covering the entire European continent, including all of the EU MS but also many others. It has no legislative power and seeks voluntary cooperation within its members through Recommendations, Agreements, and Conventions, which can be (or not) signed and ratified by the member countries.

We will be focusing on the European countries having more developed legislation, which are those belonging to the EU and a few others, such as the UK, Norway, and Switzerland, that have comparable legislative frameworks.

35:2 How is the use of animals in science regulated in Europe?

Reg. The 27 countries of the European Union (EU) are covered by the European Directive 2010/63/EU on the protection of animals used for scientific purposes which took effect on January 1, 2013 (this was the deadline for the transposition process into national legislation) and further amendments issued in 2019 and 2024.[3,4,5]

The publication of this Directive had the following three main objectives:

- Implementation of the 3Rs
- Levelling the playing field for science across the EU
- Promoting transparency

All Member States (MS) transposed the Directive into national legislation, and they cannot implement stricter measures unless they were in place before the enforcement of the Directive. Important to note is that implementation of practical matters of the Directive (and its transposition) in many MS is delegated to regional competent authorities (as the regulatory agencies are called in the EU), which, as we will see below, has some important implications.

Some other countries have similar legislation. The United Kingdom (UK) had to adapt it initially to the Directive (in fact, the Directive adopted significant parts of the existing UK legislation).[6] Norway voluntarily follows the Directive standards, and Switzerland has a different but highly comparable legislation though independent from the Directive.[7–9] Other European countries have legislation of variable quality.

The Directive has 66 Articles, including important items such as the Scope; the 3Rs; Oversight, Evaluation, Authorization and Retrospective Assessment of Projects (the main topic of this chapter); Key Named Personnel and Institutional Responsibilities; Animal Care and Accommodation, Installation and Equipment; Competence of Personnel; Killing of Animals; Nonhuman Primates; Inspections; and the Statistical Reporting and Non-technical Project Summaries. It also has eight Annexes with more specific requirements relating to several articles. One of them refers to the requirements for care and accommodation of animals, which is based in a Convention issued years before by the Council

DOI: 10.1201/9781003272861-35

of Europe (CoE), known as ETS 123, and, more specifically, its Appendix A.[10,11] This part of the EU legislation concerning housing requirements is particularly important, since much more housing space is required for most of the species than in the United States (US) legislation. For example, this difference is very significant for dogs and nonhuman primates. The European Commission (which is like the "government" of the EU) has published several documents intended as guidance to assist MS and others affected by this Directive to arrive at a common understanding of the provisions contained in the Directive.[12] A complete description of the European legal framework is available elsewhere.[13]

Opin. The EU and a few other European countries have one of the strictest regulated environments for animal research in the world. However, although one of the intentions of the Directive is to establish consistency across all MS, there are differences in the way it is implemented at the national and even the regional level. This is due to the different practical processes developed after it was transposed into national legislation and also by the diverse interpretations by applicable competent authorities in each country or region. Also, one has to bear in mind that Europe is not only the EU and that other European countries may have different requirements. This means that in terms of research collaborations between both sides of the Atlantic, it is essential to pay attention not only to the Directive's requirements, but also to the national (or even regional) transposition and implementation of the Directive (for EU members) or to the specific legislation of non-EU countries.

As stated above, one important difference between EU requirements and those in other countries such as the U.S. refers to the housing space conditions, especially significant for species such as dogs or nonhuman primates. This may have implications for certain type of collaborations between EU and U.S. institutions. For example, housing in the EU must comply with EU standards regardless of the funding source; or some EU organizations may require implementation of EU standards outside the EU. Other more specific differences concerning ethical review and oversight will be highlighted in the subsequent sections of this chapter. Many U.S. institutions rely on the AAALAC International accreditation to decide on collaborations with foreign institutions. They have to acknowledge that one of the AAALAC Primary Standards applied in Europe is the cited ETS 123 (including its Appendix A), which was adopted as such because it is not regulatory in nature, is based more on performance standards than the legislation, and can be applied all across Europe. However, the ETS 123 does not offer specific recommendations on IACUC-like bodies, so the descriptions of and discussions on ethical review and oversight processes in this chapter will mainly focus on the EU Directive and other comparable European legislation.

35:3 Which are the IACUC-like bodies in Europe and what are their composition and function?

Reg. According to Art. 26 of the Directive,[3] in the European Union (EU) each breeder, supplier, and user must set up an Animal-Welfare Body (AWB). In the United Kingdom (UK) the equivalent is called Animal Welfare and Ethical Review Body (AWERB).

The AWB shall include at least the person or persons responsible for the welfare and care of the animals and, in the case of users, also a scientific member. This same article does not require the official participation of the Designated Veterinarian (the equivalent of the Attending Veterinarian in the United States) as a *member*, though the AWB shall receive *input* from the Designated Veterinarian. It is important to point out that the position of the Designated Veterinarian in an establishment can be held by a "suitably qualified expert where more appropriate" (Art. 25), an addition that was based on certain kinds of establishments (e.g., fish research) where other professionals with specific expertise could play that role. This minimum composition does not preclude institutions to have a more complex composition, and, in several countries, the presence of the Designated Veterinarian has been enforced during the transposition process of the Directive into some Member States legislation. Normally, the AWB may include representation from researchers, persons taking part in the experiments, veterinarians, animal care personnel, lay persons, and others.

The tasks of the AWB according to the Directive (Art. 27) are as follows:

a. Advise the staff dealing with animals on matters related to the welfare of animals in relation to their acquisition, accommodation, care, and use
b. Advise the staff on the application of the requirement of replacement, reduction, and refinement, and keep it informed of technical and scientific developments concerning the application of that requirement
c. Establish and review internal operational processes as regards monitoring, reporting, and follow-up in relation to the welfare of animals housed or used in the establishment
d. Follow the development and outcome of projects, taking into account the effect

on the animals used, and identify and advise as regards elements that further contribute to replacement, reduction, and refinement

e. Advise on rehoming schemes, including the appropriate socialization of the animals to be rehomed

One can easily realize that the ethical evaluation of the protocol (in the EU they are called projects) is not included as one of the basic functions of the AWB. As evidenced by the tasks assigned to the AWB, this body has mainly advisory and 3Rs promotion roles. The Directive assigns the project evaluation task to the competent authority (Art. 36,2). However, the Member States (MS) may designate bodies other than public (government) competent authorities for the implementation of specific tasks laid down in the Directive only if there is proof that they have the necessary expertise and infrastructure and are free from conflicts of interest (Art. 59.1). This means that MS can potentially designate AWBs or other types of IACUC-like bodies as competent authorities if they fulfill these conditions, which, in practice, means that the project evaluation can be delegated to these bodies. Therefore, what has happened in most of the MS is that the public competent authorities have delegated the project evaluation to these institutional (or in some countries also external) bodies, as far as their composition ensures compliance with the expertise required for the project evaluation (Art. 38). Therefore, many AWBs in the EU are involved in the ethical review process performing the project evaluation. In some countries the AWB acts as an Ethics Committee (EC), while in others (e.g., France) the AWB and EC must be different entities that can coexist in the same institution.

The expertise required for an IACUC-like body to perform the project evaluation is as follows.

a. The areas of scientific use for which animals will be used, including replacement, reduction, and refinement in the respective areas
b. Experimental design, including statistics, where appropriate
c. Veterinary practice in laboratory animal science or wildlife veterinary practice, where appropriate
d. Animal husbandry and care, in relation to the species that are intended to be used

When comparing with specific functions performed by the typical IACUC and supporting office in the United States, such as the Program review, facilities inspection, training oversight, policy management, and post-approval monitoring, the tasks assigned to the EU AWB are more vague, although it could be argued that those same functions fall under the advisory and follow-up tasks listed in Art. 27 (e.g., Follow the development and outcome of projects). A European Commission Expert Working Group published guidance for AWB.[14]

In addition to the AWB, institutions must appoint a person or persons to be responsible for overseeing the welfare and care of the animals in the establishment (sometimes called Animal Welfare Officer, or in the UK called the Named Animal Care and Welfare Officer [NACWO]); ensure that the staff dealing with animals have access to information specific to the species housed in the establishment; and be responsible for ensuring that the staff are adequately educated, competent, and continuously trained, and that they are supervised until they have demonstrated the requisite competence (Art. 24.1). Also, institutions must appoint a Designated Veterinarian (in the UK the Named Veterinary Surgeon [NVS]) with expertise in laboratory animal medicine, or a suitably qualified expert where more appropriate, charged with advisory duties in relation to the well-being and treatment of the animals (Art. 25).

In countries outside of the European Union we find diverse approaches, from countries with no IACUC-like bodies to others with different versions of an EC. In some countries with poor or non-existing legislation it is possible to find oversight bodies at committed research institutions mirroring the U.S. IACUC composition and function, which in some cases are also called an IACUC.

Opin.

The minimum composition required by the Directive for the AWB is so short that, in theory, breeder establishments could legally have the AWB comprising one member only and only two members in user establishments. This minimum composition differs from the requirements in the U.S. AWAR (§2.31,a; §2.31,b) and the PHS Policy (IV,A,3,a.; IV,A,3,b) that require three or five members, respectively. This difference may be explained by the fact that the Directive does not assign the protocol (project) evaluation to the AWB, whereas this is an essential function of the IACUC in the U.S. The Directive initially assigns that function to the public (government) competent authority, although making possible the delegation of this function to other bodies, as described above. Several Member States require a more complex composition so they can fulfill correctly the tasks assigned to the AWB. A major controversy arose over why the Directive does not mandate the Designated Veterinarian to be a regular member of the AWB but only requires the AWB to "receive input from the designated veterinarian or the expert referred to in Article 25." There is no indication on what "input" means in practical terms. Other more complex compositions of the AWB that have been implemented by some of the MS require official veterinary participation. As pointed out above, there is room for establishments not having a veterinarian, but rather a "suitable qualified expert." Although the intention in the Directive is that this should happen only in establishments using non-conventional species only where other specialists may have a better knowledge of them (e.g., some fish or amphibians), this is not clearly stated and could be potentially challenged from the legal point of view by establishments using conventional species. Therefore, there could be potentially some establishments with no veterinarian and, as a consequence, no veterinary input to the Animal Welfare Body. However, this will very rarely happen in practice, and, in general, MS have ensured that all establishments have a veterinarian. A more complex composition than the minimum required by the Directive should be necessary also to comply with all the tasks assigned to the AWB, and even more when the competent authorities delegate other functions to it. When collaborating with European partners, one must be ready to find different IACUC-like bodies and the ethical review process in general (see 35:4).

35:4 How are European IACUC-like bodies and competent authorities (government regulatory agencies) organized for the evaluation/authorization/inspection/oversight of animal use?

Reg. In the European Union (EU), according to the Directive[3] the Member States (MS) have to designate a so-called competent authority (or authorities) to carry out the obligations arising from the Directive (Art. 3,7; Art. 59,1) and ensure that the competent authorities carry out regular inspections of all breeders, suppliers, and users, including their establishments, to verify compliance with the Directive requirements (Art. 34). There are MS who have designated a single central national competent authority (e.g., at the Ministry level) and others that have designated multiple competent authorities, usually at regional level. In all cases, a national authority must serve as the contact point for the purposes of the Directive to the European Commission (Art. 59.2). As we have seen above, the same Article 59 allows MS to designate bodies other than public competent authorities for the implementation of specific tasks laid down in the Directive. This can happen only if the designated body has the expertise and infrastructure required to carry out the tasks and is free of any conflict of interests regarding the performance of the tasks. Therefore, there are MS that have designated IACUC-like bodies at institutional level (or external) as competent authorities which have the task of project evaluation (e.g., France, Spain, and the Netherlands). In theory even the project authorization could be performed by other designated competent authorities, although this last task is, except in Belgium, kept at the public/government competent authority level. The possibility of this delegation of authority and the potential conflict of interest of an IACUC-like body are discussed further in the opinion part below. A European Commission Expert Working Group published guidance on project evaluation.[15]

Outside of the European Union, cases may be diverse. For example, in Switzerland the public authority lies at regional (cantonal) level. In the UK, the central government Home Office Inspectorate is the competent authority.[16]

In general:

- The initial evaluation of projects is performed in most countries by IACUC-like bodies, though in some cases it can be performed by external bodies (authorized by the public competent authorities) or directly by the public competent authorities.
- Authorization of animal use must be granted by public/government competent authorities either at central or regional level (except in Belgium, where institutional IACUC-bodies have been delegated this task). Member States shall ensure that the decision regarding authorization is taken and communicated to the applicant no more than 40 working days from the receipt of the complete and correct application. This period shall include the project evaluation.
- Inspections are performed by the same public competent authorities.
- Other oversight activities are mainly performed by the IACUC-like bodies, with potential participation of the public competent authorities depending on their level of engagement.
- Member States have national committees to advise the competent authorities and AWB on matters dealing with the acquisition, breeding, accommodation, care, and use of animals in procedures and ensure sharing of best practice.

Needless to say, any activity related to collaborative research between European and other institutions (e.g., from the U.S.) that involve the use of animals in a European establishment will be subjected to the applicable competent authority evaluation, authorization, inspection, and oversight.

Opin. Due to different national administrative systems, there is wide diversity in the organization of European public competent authorities. Only in a few countries do the competent authorities have laboratory animal science specialized full-time professionals (as, for example, the Home Office Inspectorate in the United Kingdom). In most cases, competent authorities for animal research matters also have to attend to other non-related areas (e.g., farms, zoos, etc.) and may not have the resources to address in detail all the requirements of the applicable legislation for animal research. A typical scenario may be represented by a competent authority at the regional level (local government) with little time to dedicate to animal research institutions and, moreover, with little chances to get specialized training in laboratory animal science. This may result in a very basic oversight being performed, usually based on engineering rather than performance standards. The heterogeneous character of the competent authorities enhances the importance, in terms of collaborative research, of identifying, and getting a good knowledge of the competent authority in the area to see how the legislation is applied. The European Commission holds regular meetings with national representatives trying to harmonize the implementation of the legislation.

The Directive's intention was to create the institutional AWB with advisory and follow-up duties for the development of a culture of care based on the principles of the 3Rs at the establishment, whereas the public competent authorities would be responsible for the evaluation and authorization of projects to ensure objectivity and independency in these processes and the inspections. However, many Member States have delegated the project evaluation process to bodies other than the public competent authorities. In most cases this is performed by IACUC-like bodies, that in some countries (e.g., Spain or the Netherlands) can be even external. This has created a heterogeneous landscape in the way the project evaluation is performed across countries, with potential participation, alone or in combination, of IACUC-like bodies, external bodies, and public competent authorities.[17] It could be argued that potential conflicts of interest or lack of independency during the project evaluation process could arise when this is performed by IACUC-like bodies when comparing with public competent authorities. But at least these last ones retain the responsibility for the final authorization. The reason for the delegation of the project evaluation has been in most cases pragmatic, as many public competent authorities lack the expertise and human resources to perform an appropriate evaluation of all project proposals. Also, local IACUC-like bodies may have a much better knowledge of the actual situation in the establishment with regard to the applicant, the personnel (training and competence) involved, and other resources and facilities, which surely impact the efficiency of the evaluation.

A significant difference from the U.S. system is that the positive evaluation by any of these bodies is not sufficient to start the experiments involving animals, because there is an additional step of authorization by the public competent authorities. Although there is an established maximum time for the evaluation/authorization of projects, there may be significant differences in the timeliness of the response, which can create differences contrary to the intention of establishing consistency across the EU. In fact, this happened frequently in the first years after the implementation of the Directive, and it still happens to a certain extent. Also, differences can exist within a single country, between different regional competent authorities, not only referring to the timelines, but also to the decision on the approval of the same type of projects (e.g., those involving species like nonhuman primates).

The role and efficiency of national committees vary significantly across countries, from some very active to others that only exist on paper. When active, they may be consulted for implementation of the legislation or to make a decision on ethically complicated projects.

35:5 What activities are covered by the legislation and require evaluation, authorization, and oversight by the IACUC-like bodies and/or the competent authority?

Reg. Under the European Union (EU) Directive,[3] the scope includes all activities "where animals are used or intended to be used in procedures or bred specifically so that their organs or tissues may be used for scientific purposes" (Art. 1.2).

There is a difference with the U.S. framework with regard to the animals covered that has to be taken into consideration: the Directive (Art. 1.3) applies to live nonhuman vertebrate animals, including independently feeding larval forms and fetal forms of mammals from the last third of their normal development; and also live cephalopods.

The concept of a procedure is crucial, being defined in Article 3.1 as "any use, invasive or non-invasive, of an animal for experimental or other scientific purposes, with known or unknown outcome, or educational purposes, which may cause the animal a level of pain, suffering, distress or lasting harm equivalent to, or higher than, that caused by the introduction of a needle in accordance with good veterinary practice." This includes the creation and maintenance of a genetically modified animal line in any such condition, but the Directive (and the UK legislation), excludes the killing of animals solely for the use of their organs or tissues. A project can involve one or more procedures.

Under Article 5 of the Directive, procedures may be carried out only for the purposes of

a. Basic research
b. Translational or applied research with any of the following aims:
i. The avoidance, prevention, diagnosis or treatment of disease, ill-health or other abnormality, or their effects in human beings, animals, or plants
ii. The assessment, detection, regulation, or modification of physiological conditions in human beings, animals, or plants
iii. The welfare of animals and the improvement of the production conditions for animals reared for agricultural purposes
c. For any of the aims in point (b) in the development, manufacture, or testing of the quality, effectiveness, and safety of drugs, foodstuffs, and feed-stuffs, and other substances or products
d. Protection of the natural environment in the interests of the health or welfare of human beings or animals
e. Research aimed at preservation of the species
f. Higher education, or training for the acquisition, maintenance, or improvement of vocational skills
g. Forensic inquiries

Interestingly, the MS can implement a simplified administrative procedure for projects containing procedures classified as "non-recovery," "mild," or "moderate" and not using nonhuman primates that are necessary to satisfy regulatory requirements, or which use animals for production or diagnostic purposes with established methods.

Activities that fall out of the scope of the Directive (Art. 1.5) and therefore are not subjected to oversight, evaluation, and authorization are as follows.

a. Non-experimental agricultural practices
b. Non-experimental clinical veterinary practices
c. Veterinary clinical trials required for the marketing authorization of a veterinary medicinal product
d. Practices undertaken for the purposes of recognized animal husbandry
e. Practices undertaken for the primary purpose of identification of an animal
f. Practices not likely to cause pain, suffering, distress, or lasting harm equivalent to, or higher than, that caused by the introduction of a needle in accordance with good veterinary practice

Opin. First, it is important to note that in the Directive, not only vertebrate animals, but also live cephalopods are included. The inclusion of cephalopods may not have a big impact because the use is very limited (if any) in most European countries, but it raises the issue of the general lack of knowledge about these animals that may affect the quality of the activities by the IACUC-like bodies. The consideration of independently feeding larval forms may be more important, especially due to the increasing use of aquatic species such as the zebrafish. For example, when is a zebrafish embryo considered to be an independent larval form? There is a general common agreement that they be considered an independent larval form from five days post-fertilization under normal laboratory conditions. The implication is huge if we consider that Member States have to collect and make publicly available, on an annual basis, statistical information on the use of animals in procedures (Art. 54.2). Counting such small animals may be impossible and the statistics will have to be approximations. The fetal forms of mammals from the last third of their normal development were already covered in some countries (e.g., UK) but since the implementation of the Directive this is a requirement in the entire EU and something to consider in collaborative projects.

Second, the concept of procedure is even more important as it will determine if and how many activities by AWB and competent authorities are to be covered. If every single action at the severity level of the introduction of a needle had to be considered a procedure, it would be practically impossible to work from the administrative burden point of view. For this reason, it was agreed by the European Commission that, in practical terms, a procedure be considered as a combination of one or more technical acts carried out on an individual animal for an experimental or other scientific purpose and which may cause that animal pain, suffering, distress, or lasting harm. For example, if an experiment involves gavage, injections, laparotomy, and sampling, all actions together are to be considered a single "procedure."

One of the more important discussions held in the EU refers to the concept of procedure and the activities that fall out of the scope of the Directive, such as practices undertaken for the purposes of recognized animal husbandry. More specifically, the discussion focused on deciding if tissue sampling for genotyping of genetically altered lines is to be considered husbandry (and therefore exempted of all Directive requirements) or not, with all the administrative consequences. Very importantly, the European Commission's interpretation is that tissue sampling for genotyping cannot be considered as a recognized husbandry practice because its primary purpose is to either fulfil scientific needs or confirm scientific suitability. Only the sampling procedures that are below the severity threshold (introduction of a needle) should be exempted, which has promoted the implementation of noninvasive methods. This means that not only the creation of genetically altered lines and the breeding/maintenance of lines with harmful phenotypes are to be considered as procedures, but also that evaluation and authorization of the maintenance of lines with not-harmful phenotypes will be needed, unless the methods used for tissue sampling are noninvasive (e.g., saliva, feces) or are used in conjunction with identification practices, which are explicitly excluded as a procedure. An example is using the ear tissue for both identification and genotyping purposes. The implication of this interpretation is also large, not only for evaluation and authorization issues, but also for the statistics.

Also under discussion is that the killing of animals solely for the use of their organs or tissues is not considered a procedure and therefore no evaluation and authorization by IACUC-like bodies or public competent authorities is required for projects involving only the euthanasia of animals, if this is performed according to any of the established euthanasia methods in Annex IV of the Directive. But aren't these animals being used to fulfill scientific needs? Is not killing an animal to be considered as above the threshold of introducing a needle? The same reasons given to include sampling for genotyping as a procedure seem not to apply to killing animals to obtain tissues and vice versa. Regardless of these practices being considered as procedures or not, the IACUC-like bodies should establish institutional policies to be followed in each case.

35:6 Are there special considerations for the use of nonhuman primates and endangered species?

Reg. Yes. According to the Directive[3] nonhuman primates (NHP) cannot be used for all the aims listed in the Article 5 of the European Union Directive already described in 35:5. Their use is restricted (Art. 8,1) to procedures meeting the following conditions:

a. The procedure has one of the purposes referred to in (i) points (b)(i) or (c) of Article 5 and is undertaken with a view to the avoidance, prevention, diagnosis, or treatment of debilitating or potentially life-threatening clinical conditions in human beings; or (ii) points (a) or (e) of Article 5.
b. There is scientific justification to the effect that the purpose of the procedure cannot be achieved by the use of species other than nonhuman primates. A debilitating clinical condition for the purposes of this Directive means a reduction in a person's normal physical or psychological ability to function.

The definitions in Article 5 include basic research, translational or applied research (related to human disease), safety (regulated) studies, and preservation of the species. Severe restrictions apply to using NHP endangered species (Art. 8.2) and, importantly, the use of great apes is forbidden (Art. 8.3). The latter is subject to a safeguard clause (Art. 55.2) that states that great apes can be used only when a Member State has justifiable grounds for believing that action is essential for the preservation of the species or in relation to an unexpected outbreak of a life-threatening or debilitating clinical condition in human beings. A Member State may adopt a provisional measure allowing the use of great apes in procedures having one of the purposes referred to in Article 5, provided that the purpose of the procedure cannot be achieved by the use of species other than great apes or by the use of alternative methods. Also, it is important to remember that projects involving the use of NHP are must undergo retrospective assessment (see Question #10).

In addition, NHP are not only included in the list of animals that can be used only if they have been bred to be used in procedures (Art. 10.1 and Annex I), but also NHP will be allowed to be used in procedures only where they are the offspring of NHP which have been bred in captivity (F2) or where they are sourced from self-sustaining colonies. These conditions apply since November 2022, after two feasibility studies performed in 2017 and 2022. A self-sustaining colony is defined in the Directive as "a colony in which animals are bred only within the colony or sourced from other colonies but not taken from the wild, and where the animals are kept in a way that ensures that they are accustomed to humans" (Art. 10,1). More recently, the European Commission has indicated a more specific definition of self-sustaining colonies with regard to the 2022 feasibility study and statistical reporting:

- – established colonies in which no wild-caught animals introduced after November 2017; and
- – colonies established after that date by using only F1 or higher generation animals obtained

from other self-sustaining colonies and having never introduced wild-caught animals.

Similar restrictions apply to other endangered species (Art. 7), with the main difference that they cannot be used for basic research and no safeguard clause exists.

Opin. There are three key points to consider in the Directive about the use of NHP. First, the use of great apes is basically forbidden in the European Union (EU). They could still be used in theory in other countries where they are not regulated. The second point refers to the potential different interpretation of the definition of a "debilitating or potentially life-threatening clinical conditions in human beings" included in the Directive as "a reduction in a person's normal physical or psychological ability to function" (Art. 8,1.b). Apparently, this definition covers the large majority of the current NHP use, although doubts may arise with regard to some types of research, such as reproductive studies. The third point, related to the exclusive use of NHP only where they are F2 or sourced from self-sustaining colonies, is worrying the research community, as it may impact severely the already reduced availability of animals for research purposes. The COVID-19 pandemic led China to stop the exportation of NHP, which, in addition to the increasing demand of animals for COVID-related research, created the "perfect storm" where, due to the limited availability of animals, their cost has skyrocketed and many institutions, especially those with less financial resources (i.e., academic institutions), are having great difficulties to obtain animals for research. This situation is aggravated in the EU due to the, as described, strict requirements. It is a matter at issue for the EU Member States to use the possibility of an exemption to these requirements based on scientific grounds. However, the European Commission will be reviewing this, and, in theory, an exemption cannot be based just on lack of availability or supply of F2 animals.

With regard to other endangered species, their use is even more restricted, as no basic research is allowed regardless of its aim and no safeguard clause is offered. Authorization from competent authorities to use endangered species, even in the cases considered in the Directive, will likely be difficult to obtain.

35:7 In Europe, what is the engagement of IACUC-like bodies in the implementation of Standard Operating Procedures?

Reg. The tasks assigned to the Animal Welfare Body (AWB) in the Directive,[3] as described previously, include to "establish and review internal operational processes as regards monitoring, reporting and follow-up in relation to the welfare of animals housed or used in the establishment" (Art. 27.1.c).

Opin. It could be argued that this task apparently restricts the requirement of Standard Operating Procedures (SOPs) to those directly involved in animal welfare aspects, which leaves much room for interpretation about the extent of procedures that should be established by the AWB. While the assignment of this task seems to be appropriate, the feasibility of correctly performing this task (and the other tasks assigned to the AWB) may depend on the composition of the AWB. Complex institutions that do not go beyond the minimum requirements of the Directive and do not ensure the AWB has the necessary human resources and expertise will never be able to appropriately perform all the tasks assigned. When collaborating with European institutions, the composition and performance of the AWB may serve to evaluate the level of commitment of the institution to animal care and use.

35:8 Are European IACUC-like bodies involved in post-approval monitoring activities?

Reg. Yes, at least partially in the European Union (EU). One of the tasks of the Animal Welfare Body (AWB) is to "follow the development and outcome of projects, taking into account the effect on the animals used, and identify and advise as regards elements that further contribute to replacement, reduction and refinement" (Art. 27.1.,d). In addition to this, for all authorized projects there must be "persons responsible for the overall implementation of the project and its compliance with the project authorization" (Art. 40.2.b), and these persons must "ensure that the projects are carried out in accordance with the project authorization" (Art. 24.2.b).

Opin. The IACUC-like bodies, such as the AWB in the EU, should be clearly involved in post-approval monitoring. The persons referred to in Articles 24 and 40 are meant to be the persons legally responsible for the compliance with the project authorization (i.e., the Principal Investigator) but not the persons performing an independent post-approval monitoring. However, the authority of the AWB is not clear as its role is mainly to give advice. Therefore, it will remain at the institution's discretion to establish an AWB with enough resources and authority to perform a complete follow-up of the projects and give appropriate advice on the implementation of the 3Rs, as well as to respond positively to that advice. At the very least, records of any advice given by the Animal Welfare Body and decisions taken regarding that advice are to be kept for at least three years (Art. 27.2). These records may serve as one of the indicators of the performance of the Animal Welfare Body and of the institutional support of its activities.

The post-approval monitoring activities are very important because of the Directive requirement to provide, in the statistical information, the *actual* severity of the procedures performed (Art. 54.2). This means that for all uses of animals, the severity caused to each animal is to be included and not only the prospective categorized severity. How can the collection of all this information be coordinated? Who is supposed to collect the information? How can the assessment criteria be harmonized? The participation of the researchers seems to be crucial to collecting information, but the AWB should play a decisive role too. This is further discussed in 35:10.

35:9 Is there a severity classification of experimental procedures?

Reg. Yes. The European Union (EU) Directive indicates that the procedures included in projects have to be classified as "non-recovery," "mild," "moderate," or "severe" on a case-by-case basis (Art. 15).[3] To harmonize the classification across Europe, Annex VIII offers examples for each of the categories. Unless a safeguard is applied (Art. 15.2 and 55.3), procedures that involve severe and long-lasting pain, suffering, or distress are prohibited. Furthermore, one of the Expert Working Groups promoted by the European Commission published more detailed guidance on interpretation of the severity, including examples, and numerous workshops on severity assessment are being carried out in many countries with the purpose of harmonization.[18,19]

Other different severity categories may be found in other European countries (e.g., Switzerland, ranging 0–3).

Opin. The harmonization of the prospective classification of severity constitutes one important challenge, as the classification assigned has several important implications, including practical requirements for researchers (see 35:10) and societal ones concerning the public statistical reports. The European Commission has promoted numerous workshops on this matter trying to harmonize the classification of severity with the participation of recognized experts. The workshops reveal the initial disparity of interpretation and serve as a very good tool for harmonization based on the discussions held. Professionals from many EU Member States have been trained to lead subsequent workshops in their respective countries, after agreeing on the criteria used for the severity classification. Although there will always be different opinions, the practical discussions are essential, as well as the way the professionals attending the training convey the information to their respective IACUC-like bodies.

35:10 What is the retrospective assessment of projects?

Reg. All projects using nonhuman primates and projects involving procedures classified as "severe" (Art. 39.2) shall undergo, by the competent authority, a retrospective assessment to evaluate:

(a) whether the objectives of the project were achieved;
(b) the harm inflicted on animals, including the numbers and species of animals used, and the severity of the procedures; and
(c) any elements that may contribute to the further implementation of the requirement of replacement, reduction, and refinement

Only projects classified as "mild" or "non-recovery" may be exempted; the competent authority may decide to perform the above evaluation for the projects classified as "moderate."

Although this retrospective assessment is, in theory, the responsibility of the public competent authority (Art. 39.1), it appears as one of the activities that could be delegated to the local IACUC-like body, as one of its recognized tasks is to follow the development and outcome of projects, taking into account the effect on the animals used (Art. 27.1.d). The European Commission Expert Working Group on the project evaluation has also offered guidance on the retrospective assessment.[15]

Opin. Although this task is assigned to the competent authority, the role of the IACUC-like bodies, especially the Animal Welfare Body, and the researchers is essential to comply with this requirement. First, because, as has been noted, many IACUC-like bodies have been delegated the task of the project evaluation, and also one of their tasks is to follow up the development of projects taking in consideration the effects on the animals used. Second, because, in practice, the researchers are the most capable of providing the required information on the actual severity experienced by each animal, which is the main factor to be considered for the retrospective assessment. When collaborating with European Union institutions, the possibility of being subjected to a retrospective assessment should be considered.

35:11 Are IACUC-like body members required to follow specific training?

Reg. There is no reference in the European Union Directive for specific training of Animal Welfare Body (AWB) members, though it could be implied in some articles (see opinion below).[3]

Opin. One can argue that there is a legal obligation to implement this training, depending on the interpretation of one of the requirements for personnel of authorized establishments (Art. 24.1.c), which states that there must be person(s) "responsible for ensuring that the staff are adequately educated, competent and continuously trained and that they are supervised until they have demonstrated the requisite competence." Although this requirement is clearly related to Article 24, which states that "the staff shall be adequately educated and trained before they perform any of the following functions: (a) carrying out procedures on animals; (b) designing procedures and projects; (c) taking care of animals; or (d) killing animals," the same principle could be applied to the members of the AWB or other IACUC-like body. The expertise to be considered for the project evaluation, listed in 35:3, should be taken in account for the training of IACUC-like bodies. In practice, many institutions do not have a formal training program for IACUC-like body members but for some basic initial training on the committee operations.

35:12 To whom does the IACUC-like body report at the typical research institution?

Reg. There is no clear indication about this in the European Union Directive.[3]

Opin. Authorized establishments have to communicate to the competent authority the person who is responsible for ensuring compliance with the provisions of the Directive (Art. 20.2). That person can be interpreted as the equivalent to the Institutional Official in the United States. Having the IACUC-like body report directly to this person would seem to be a reasonable approach, although it is not specified in the legislation. If IACUC-like bodies, such as the Animal Welfare Body (AWB), do not report to this person directly, they should report to someone delegated by that individual who has enough authority within the institution to implement actions as a response to the AWB advice.

In practice, IACUC-like bodies report to different individuals/positions, such as Deans or Vice Principals of research in academic institutions and Site Heads or persons at research management level in private companies and public research institutions.

Also, the IACUC-like body can communicate with the competent authorities concerning the project evaluation (in the countries where they have been assigned this task) to continue with the authorization process.

REFERENCES

1. European Union. Countries. https://european-union.europa.eu/principles-countries-history/country-profiles_en. Accessed July 11, 2022.
2. Council of Europe. Members. https://www.coe.int/en/web/about-us/our-member-states. Accessed July 11, 2022.
3. The European Parliament and the Council of the European Union. 2010. Directive 2010/63/EU of the European Parliament and of the Council of 22 September 2010 on the Protection of Animals Used for Scientific Purposes. *Official Journal of the European Union:* L 276:33–79. http://eur-lex.europa.eu/LexUriServ/LexUriServ.do?uri=OJ:L:2010:276:0033:0079:en:PDF.
4. The European Parliament and the Council of the European Union. 2019. Regulation (EU) 2019/1010 of the European Parliament and of the Council of 5 June 2019 on the alignment of reporting obligations in the field of legislation related to the environment, and amending Regulations (EC) No 166/2006 and (EU) No 995/2010 of the European Parliament and of the Council, Directives 2002/49/EC, 2004/35/EC, 2007/2/EC, 2009/147/EC and 2010/63/EU of the European Parliament and of the Council, Council Regulations (EC) No 338/97 and (EC) No 2173/2005, and Council Directive 86/278/EEC (Text with EEA relevance)
5. Commission Delegated Directive (EU) 2024/1262 of 13 March 2024 amending Directive 2010/63/EU of the European Parliament and of the Council as regards the requirements for establishments and for the care and accommodation of animals, and as regards the methods of killing animals.
6. Animals (Scientific Procedures) Act (1986). London: HMSO. https://www.legislation.gov.uk/ukpga/1986/14/contents. Accessed 11 July 2022.
7. Swiss Federal Department of Home Affaseverityirs, Federal Food Safety and Veterinary Office. French version: 455 Loi fédérale sur la protection des animaux (LPA) du 16 décembre 2005 (Etat le 1er janvier 2022). German version: Tierschutzgesetz (TschG) vom 16. Dezember 2005 (Stand am 1. Januar 2012). English version: 455 Animal Welfare Act (AniWA) of 16 December 2005 (Status as of 1 January 2022).
8. Swiss Federal Department of Home Affairs, Federal Food Safety and Veterinary Office. French version: 455.1 Ordonnance sur la protection des animaux (OPAn) du 23 avril 2008 (Etat le 1er juin 2022). German version: 455.1 Tierschutzverordnung (TschV) vom 23. April 2008 (Stand am 1. Juni 2022). No English version.
9. Swiss Federal Department of Home Affairs, Federal Food Safety and Veterinary Office. French version: 455.163 Ordonnance de l'OSAV concernant la détention des animaux d'expérience, la production d'animaux génétiquement modifiés et les méthodes utilisées dans l'expérimentation animale (Ordonnance sur l'expérimentation animale) du 12 avril 2010 (Etat le 1er mai 2010). German version: 455.163 Verordnung des BLV über die Haltung von Versuchstieren und die Erzeugung gentechnisch veränderter Tiere sowie über die Verfahren bei Tierversuchen(Tierversuchsver ordnung) vom 12. April 2010 (Stand am 1. Mai 2010). English version: 455.163 FSVO Ordinance on Laboratory Animal Husbandry, the Production of Genetically Modified Animals and Methods of Animal Experimentation (Animal Experimentation Ordinance) of 12 April 2010 (Status as of 1 May 2010).
10. European Convention for the Protection of Vertebrate Animals used for Experimental and Other Scientific Purposes. 1986. http://conventions.coe.int/treaty/en/treaties/html/123.htm.
11. Council of Europe. 2006. Appendix A of the European Convention for the Protection of Vertebrate Animals Used for Experimental and Other Scientific Purposes (ETS No. 123). Guidelines for Accommodation and Care of Animals (Article 5 of the Convention). Approved by the Multilateral Consultation. Cons 123 (2006) 3. http://conventions.coe.int/Treaty/EN/Treaties/PDF/123-Arev.pdf
12. European Commission Consensus Documents. https://ec.europa.eu/environment/chemicals/lab_animals/interpretation_en.htm

13. Guillén, J., Prins, J-B., Howard, B., Degryse, A-D., Gyger, M. (2017) The European Framework on Research Animal Welfare Regulations and Guidelines. In J. Guillén (Ed.) Laboratory Animals: Regulations and Recommendations for the Care and Use of Animals in Research (pp. 117–202). Cambridge: Academic Press.
14. National Competent Authorities for the implementation of Directive 2010/63/EU on the Protection of Animals Used for Scientific Purposes (2014). A Working Document on Animal Welfare Bodies and National Committees to Fulfil the Requirements under the Directive. https://ec.europa.eu/environment/chemicals/lab_animals/pdf/endorsed_awb-nc.pdf (Accessed 15 July 2022).
15. National Competent Authorities for the Implementation of Directive 2010/63/EU on the Protection of Animals Used for Scientific Purposes (2013). Working Document on Project Evaluation and Retrospective Assessment. https://ec.europa.eu/environment/chemicals/lab_animals/pdf/Endorsed_PE-RA.pdf. Accessed 15 July 2022.
16. Home Office, UK. Research and Testing Using Animals. http://www.homeoffice.gov.uk/science-research/animal-research/. Accessed October 17, 2012.
17. Guillen, J., Robinson, S., Decelle, T., Exner, C. & Fentener van Vlissingen, M. (2015). Approaches to animal research project evaluation in Europe after implementation of Directive 2010/63/EU. *Laboratory Animals*, 44:1, 23–31.
18. National Competent Authorities for the implementation of Directive 2010/63/EU on the Protection of Animals Used for Scientific Purposes (2012). Working Document on a Severity Assessment Framework. https://ec.europa.eu/environment/chemicals/lab_animals/pdf/Endorsed_Severity_Assessment.pdf. Accessed 14 July 2022
19. National Competent Authorities for the Implementation of Directive 2010/63/EU on the Protection of Animals Used for Scientific Purposes (2013). Examples to Illustrate the Process of Severity Classification, Day-to-Day Assessment and Actual Severity Assessment. https://ec.europa.eu/environment/chemicals/lab_animals/pdf/examples.pdf. Accessed 14 July 2022.

36

Ethical Review Framework in Asia-Pacific Region: Collaborative Considerations

Sonja T. Chou and Dewi K. Rowlands

Introduction

The Asia-Pacific region is generally regarded as the part of the world near the western Pacific Ocean. Although the areas included in Asia-Pacific vary depending on context, the region generally includes East Asia, Oceania, South Asia, and Southeast Asia, which includes over 50 countries, regions, or territories. For the purposes of this chapter, we will focus on the countries or territories that are most active in animal research, those being Australia, India, Japan, People's Republic of China, Taiwan, Republic of Korea, Singapore, and Thailand. As the regulatory landscape in different countries evolves over time, readers are encouraged to conduct online searches for the most current legislative information.

36:1 How is animal research regulated in the Asia-Pacific region?

Reg. As a diverse region, there is no single entity that covers regulation of research animal use, with each country, territory, or region having their own laws, regulations, or guidelines. Each country recognizes ethical concerns over the use of animals in research, teaching, and testing, and each has developed and implemented their own regulatory framework for regulatory oversight and promotion of animal welfare following similar principles in North America and Europe. Although many of the jurisdictions are member organizations of trading and cooperation bodies, e.g. Asia-Pacific Economic Cooperation (APEC) and the Association of Southeast Asian Nations (ASEAN), no efforts have yet been made to harmonize research animal regulation.

Australia's research is regulated under separate legislation in each of Australia's six states and two major territories. It is united through a national code that is integrated into the legislation in each of these jurisdictions. This is known as "The Australian Code for the Care and Use of Animals for Scientific Purposes" (The Code).[1] Although the specific legislation differs between the states and territories,[2] all require prior licensing or registration of the scientific establishment that uses animals, and/or premises where animals are bred, held, or used for scientific purposes; prior ethical review and approval, and ongoing oversight, of the animal care and use by an appropriately constituted Animal Ethics Committee (AEC); and adherence to the principles and guidelines set out in the "The Code."[1,2]

In India, anyone engaged in conducting scientific experiments on animals must act in conformity with the provisions of the prevention of Cruelty to Animals Act, 1960, and the Breeding of and Experiments on Animals (Control and Supervision) Rules, 1998, as amended in 2001 and 2006.[3–5]

Animal research facilities in Singapore must comply with the Guidelines on Care and Use of Animals for Scientific Purposes (2022) set by the National Advisory Committee for Laboratory Animal Research (NACLAR), under the Animals and Birds Act, Chapter 7, the Animals and Birds (Care and Use of Animals for Scientific Purposes) Rules 2004.[6,7] The NACLAR Guidelines are presented in four parts; the Guiding Principles for the care and use of animals for scientific purposes; the IACUC, a guideline detailing how the IACUC should operate in carrying out its responsibilities; Training, a guideline providing the training scope and requirements for all personnel involved in the care and use of animals for scientific purposes; and Occupational Health and Safety in Animal Care and Use. Any facility that intends to keep or use animals for scientific purposes must obtain a license from the Animal and Veterinary Service (AVS) and must establish an IACUC, with the AVS maintaining regulatory oversight over the IACUC.[6–8]

Thailand has only recently enacted legislation for the use of animals for scientific purposes with the Animals for Scientific Purposes Act (2015).[9] There has been a voluntary Ethical Principles and Guidelines for the Use of Animals for Scientific Purposes (Ethical Principles) since 1999, later replaced by the Ethics in Procedures on Animals for Scientific Purposes (2015) as well as a Standard for Institutional Animal Care and Use Committee

DOI: 10.1201/9781003272861-36

(2012), issued by the National Research Council of Thailand (NRCT).[9–13]

In the People's Republic of China, laboratory animal affairs are regulated and administered within a multi-tiered government system. The State Council approved and issued the Statute on the Administration of Laboratory Animal Use in 1988 and designated the Ministry of Science and Technology (MOST) as the government agency responsible for setting regulatory requirements related to animal use in research and for overseeing the development of laboratory animal science in the country.[14,15] For the implementation of the regulations and further strengthening oversight by provincial level bureaus of science and technology, most provinces, municipalities, and autonomous regions will issue similar regulations and guidelines.[16] There are also over 100 national standard documents (e.g. GuoBiao or China National Standard; GB) for the purpose of laboratory animal quality control, issued by the China State Bureau of Technical Supervision, covering animal health profiles, genetic control, feed formulation, animal facility and housing environment, and testing methodologies.[16,17] Finally, for institutions seeking voluntary certification of their animal research facilities through China National Accreditation Service for Conformity Assessment (CNAS), institutions are expected to comply with GB/T 27416-2014 "Laboratory Animal Facilities – General Requirements for Quality and Competence."[18]

The Act on Welfare and Management of Animals is the main law that regulates the use of animals in research, testing, and education in Japan.[19] Originally issued as the Act on Protection and Management of Animals in 1973, it had gone through multiple amendments to incorporate considerations for animal welfare and mandated a self-regulation system for animal experimentation. Furthermore, three different ministries have issued fundamental administrative guidelines, including the Ministry of Education, Culture, Sports, Science and Technology; the Ministry of Health, Labour and Welfare; and the Ministry of Agriculture, Forestry and Fisheries.[20–22] Based on the nature of animal use, an institution will follow the guidelines of one of the three ministries and with the Standards Relating to the Care and Management of Laboratory Animals and the Alleviation of Pain issued by the Ministry of Environment.[23–25]

Animal experiments in the Republic of Korea are regulated under two main laws, the Animal Protection Act (APA), established by the Ministry of Agriculture Food and Rural Affairs (MAFRA) in 1991, and the Laboratory Animal Act (LAA), established by the Ministry of Food and Drug Safety (MFDS) in 2008, each with its own expectations for institutional animal use oversight.[26,27] The APA is applicable toward all institutions that conduct animal testing, while the LAA typically applies to institutions that breeds animals for research purposes or use animals in the development of, or in safety/quality control of food, drugs, medical devices, cosmetics, etc. An institution may follow APA or LAA or both, depending on the institution's registration under which Ministry.[24]

In Taiwan, the Animal Protection Act (1998), last amended in 2021, is the law that regulates the treatment of animals in research, teaching, testing, companion, exhibition, and transport, and is enforced by Council of Agriculture (COA) Executive Yuan.[28]

Opin. There is no single unified approach to how animal research is regulated within the region and the level of detail as well as what is covered varies between each jurisdiction. The general approach, however, for all major regions/countries that perform animal research do share similarities with *The Guide for the Care and Use of Laboratory Animals (Guide)* in that an IACUC-like body plays a significant role in overall oversight of animal research.[29] The need for an IACUC-like body is stipulated in law, but the degree to which the guidelines or regulations are enforced differ by region.

One feature that is very apparent is that independent accreditation by AAALAC International is relatively common, with the region having the most accredited units outside of North America. With a burgeoning biotech industry, and diverse regulatory framework, such accreditation has no doubt contributed to some degree of harmonization as well as making collaboration easier with international organizations.

36:2 How is animal research oversight accomplished in Asia-Pacific?

Reg. Animal research oversight in Australia is accomplished through a system of enforced self-regulation at the center of which is the institution's Animal Ethics Committee (AEC). Institutions and their AECs must undergo an independent external review at least once every four years to ensure ongoing compliance with the Code.[1,2]

In India these provisions are enforced by an independent Committee for the Purpose of Control and Supervision of Experiments on Animals (CPCSEA), a statutory body under the Prevention of Cruelty to Animals Act, 1960, under the Department of Animal Husbandry and Dairying (DAHD), Ministry of Fisheries, Animal Husbandry and Dairying (MoFAH&D).[3–5] CPCSEA appoints an Institutional Animal Ethics Committee (IAEC) to each institution for a period of five years that has the overall responsibility to oversee the conduct of animal experiments and functioning of the animal facilities. The IAEC has to ensure compliance with all regulatory requirements, applicable rules, guidelines, and laws and must hold a meeting every six months.[3–5]

Singapore's animal research facilities must be licensed as stated in the Animal and Birds (Care and Use of Animal for Scientific Purposes) Rules. Inspection of the animal research facility is conducted by Animal and Veterinary Services (AVS), National Parks Board (NParks) and an annual report is required by each licensed institution. The establishment of an IACUC is required by the licensee institution ensuring the use of animals for scientific purposes is in accordance with the Guiding Principles and relevant legislation. The appointed IACUC of each facility acts as an internal approval and auditing body to evaluate all proposed use of animals for scientific purposes in compliance with the NACLAR Guidelines, and it is overseen by the AVS.[6–8]

For Thailand, the NRCT appoints the National Committee for Research Animal Development (NCRAD) to be responsible for monitoring and promoting the ethical use of animals in research. The care and use of laboratory animals in all institutions are overseen by the Secretariat Office of NCRAD. The Committee has the authority to investigate the internal affairs of an institution, which is subject to public complaints or accusation by the people, public media, published documents, academic publications, and research funding organizations for violation of the "Ethical Principles." There are two national committees for the "Supervision and Promotion for Procedures on Animals for Scientific Purposes," and an "Ethics Committee," which, between them, formulate policies, ethics, standards, and procedures for animal care and use, as well as make appropriate decisions on non-compliances. According to the Animals for Scientific Purposes Act 2015, every institution that uses animals in research, testing, production of biological materials, and teaching is also obliged to establish at least one committee to manage and be accountable for the use of animals, so that the "Ethical Principles" are followed.[9–13]

Within mainland China, following the Regulation on the Management of Laboratory Animal Licenses, there is a mandatory licensure system for working with laboratory animals.[15] Institutions holding animal breeding and/or animal use licenses are expected to self-regulate by establishing an IACUC and comply with program oversight expectations as stated in the Guidelines on Humane Treatment of Laboratory Animals, the first national guideline with specific consideration for laboratory animal welfare.[30] While not mandated, institutions also typically comply with GB/T 35892-2018 "Laboratory Animal – Guideline for Ethical Review of Animal Welfare," the first national standard for quality management of laboratory animal welfare and ethics, as demonstration of their commitment to animal research program oversight.[31,32]

While there is no specific mandated oversight by regulatory agencies in Japan, institutions develop their own self-inspection system to determine if they are compliant with applicable regulations and institutional policies. Institutions will reference the Guidelines for Proper Conduct of Animal Experiments (2006), issued by the Science Council of Japan,[33] along with Ministry guidelines and standards such as the Standards Relating to the Care and Management of Laboratory Animals and the Alleviation of Pain,[23] and numerous voluntary guidelines provided by professional organizations, to self-regulate and promote animal welfare.

For animal research institutions in the Republic of Korea, in addition to the periodic internal inspection by their own IACUC, they are also subject to government oversight depending on the institutional registration and the regulation(s) followed. Specifically, institutions registered and following the Animal Protection Act are annually inspected by MAFRA. If the institution follows the Laboratory Animal Act (e.g. registered animal testing facilities conducting animal testing or raising laboratory animals), they will be annually inspected by representatives of the Korean Association for Laboratory Animals (KAFLA) on behalf of MFDS.[25]

Under the Taiwanese Animal Protection Act, the COA Executive Yuan sets expectations for periodic self-evaluation by an institutional panel for the care and use of lab animals (e.g. IACUC), as well as by an external inspection team with representation from municipal or county (city) government animal protection inspectors, laboratory animal-related scholars, experts, and representatives of relevant agencies and animal protection groups. Annually, for all registered institutions, the COA inspection team will either perform an on-site inspection or review institutional annual reports, as a form of government regulatory oversight of animal research institutions. COA published the Guidelines for the Care and Use of Laboratory Animals as a reference to research institutions to further clarify oversight expectations.[34] While very similar to the ILAR *Guide* in terms of content, it further defined the role and training requirement of an IACUC executive secretary; added special considerations for animal re-use, retirement, rehabilitation, and rehoming; required semi-annual program self-evaluation; and touched on the use of non-pharmaceutical-grade chemicals for euthanasia of animals.

Opin. Oversight of animal research within Asia-Pacific is a mix of government enforcement, self-regulation, and independent external review, as well as through voluntary accreditation. Similar to North America, in all jurisdictions an IACUC is in place that has the overall responsibility of ensuring that the institution provides appropriate animal care and abides with local regulations and requirements, along with periodic reporting requirements of animal use information, either directly to the governing oversight authorities or through public disclosure. For institutions that wish to build collaborative relationships across different jurisdictions, it is recommended that animal ownership and activity oversight responsibilities be clarified. Additional and/or periodic assessment through auditing or third-party program review may be necessary to confirm conformity to industry current best practice.

36:3 What IACUC-like bodies are required for compliance with regulatory expectations in Asia-Pacific and what is its expected composition?

Reg. For all the jurisdictions described, an IACUC-like body (e.g. Institutional Animal Care and Use Committee, Laboratory Animal Management Committee, etc.) is required under local legislation. In terms of composition, they are generally similar to what is required in the ILAR *Guide*,[29] in that there is a need for veterinary, scientific, and public/nonaffiliated member input. However, differences observed between jurisdictions include the need for specialty members (e.g. non-scientific member equivalent) and/or additional mandated members.

Australia does not stipulate the need to have a non-scientific member *per se*, but rather a person with demonstrable commitment to, and experience in, furthering the welfare of animals who is not otherwise associated with the institution and who is not currently involved in animal research. The public member however, must be someone who has never been involved in animal research; with this person being viewed, by the wider community as bringing a completely independent view. There is also a requirement that membership should include a person responsible for routine care of animals within the institution, and additional members with relevant skills (such as a statistician) may also be appointed.[1–2]

The IAEC in India is composed of five members proposed by the institution, including various scientists from within the institution and at least one veterinarian involved in the care of the animals; four members are appointed by CPCSEA, to include an external scientist and a socially aware member. There is no requirement, however, for the socially aware member to be a non-scientist or someone who has not been involved in animal research.[3]

For Singapore and Thailand, the composition is very similar to that of the *Guide*, with some small differences in that the Attending Veterinarian must be appointed to the IACUC rather than any veterinarian with training or experience in laboratory animal medicine as stipulated in the ILAR *Guide*.[6,10,29]

In mainland China, the Guidelines for the Humane Treatment of Laboratory Animals promotes 3Rs principles and requires each registered institution to establish an IACUC or ethics committee to oversee their animal care and use program.[30] Further defined under the China National Standards GB/T 27416-2014 and GB/T 35892-2018, the ethics committee shall be comprised of at least three people with background in veterinary medicine or as a laboratory animal specialist, have experience in research animal management or animal experimentation, be a lay member of the public, etc., but with no more than three members from the same department or branch of an institution or body on the committee at the same time.[18,31]

The IACUC composition expectation in Japan is different from others in that a veterinarian and a public member are not required. As such, the IACUC composition will typically comprise a member with knowledge of animal experimentation, a person with knowledge of laboratory animals (may be a veterinarian), and a person with academic experience.[24]

Under Article 27 of the Korean APA, registered institutions must establish an IACUC with membership of at least three but no more than 15 members, appointed by the head of each animal testing institution, and a Chairperson is elected by committee members.[26] Under Article 7 of the Laboratory Animal Act, the establishment of a Laboratory Animal Management Committee (LAMC) is expected in animal testing facilities, with committee membership of least four but no more than 15 members, appointed by the manager of the laboratory animal facility.[27] Due to the differences in composition and function expectations for the IACUC-like body under either APA or LAA, the MFDS and the Animal and Plant Quarantine Agency (QIA, affiliated with MAFRA) published the "IACUC Standard Operating Guidelines for an 'integrated IACUC'" to meet the requirements under both sets of regulation.[35,36] Essential committee members include a veterinarian or a PhD scientist with experience in animal experiments and a person referred by an animal protection organization.[25]

IACUC member composition in Taiwan must consist of at least three people, including one laboratory animal specialist, and one nonaffiliated member who is not a laboratory animal user or a veterinarian. The IACUC must also have an IACUC Executive Secretary position filled by an individual who has completed specific IACUC training recognized by the Central Authority of Taiwan.[34]

Opin. In general, in order to offer a well-balanced perspective in the pursuit of ethical and responsible animal use in scientific activities, the IACUC-like bodies will include representation from the scientific, veterinary, and non-scientific/public communities. While institution-specific self-governing approaches are developed to ensure compliance with regulatory requirements, institutions mostly follow guidelines issued by regulatory agencies and professional organizations. Continued program improvement is made possible through additional third-party auditing/inspection activities and animal program assessment entities to keep institutions abreast of current best practices.

36:4 What type of activities may require oversight, evaluation, and authorization?

Reg. For all jurisdiction, activities such as research or teaching that involve living vertebrate animals are included in oversight, evaluation, and authorization of research.

There may be differences in their oversight for the use of invertebrate animals as well as cells and fetuses. In Australia, the Code does extend to include invertebrates such as cephalopods and, increasingly, this is being extended to include decapod and crustaceans.[1] The Australian Code also requires that consideration be given to embryonic, fetal, and larval forms.[1] Thailand's definition is broader and includes invertebrates such as insect, shrimp, crab, and octopus, whereas Singapore's NACLAR Guidelines now include cephalopods in the definition of animals.[6,10]

Opin. While not all jurisdictions have the same definition for "laboratory animals" or "animal facilities," institutions are expected to apply the same standards of care to all animals used for scientific and teaching purposes.

36:5 Do IACUC-like bodies in Asia-Pacific have the same functions as US IACUCs, especially with regard to the written protocol (project) evaluation and proposed activity approval?

Reg. In Australia, depending on the nature of the work, some projects may require additional approvals or licensing to ensure compliance with other regulations, such as those relating to gene technology, quarantine, wildlife protection, or work health and safety. Projects involving collaboration between institutions or across state boundaries also require approval from the other AECs or state regulatory authorities.

In India, the IACUC (IAEC) has been empowered to approve project proposals on rodents and rabbits only, whereas any experimentation using larger animals such as cats, dogs, primates, sheep, and other larger animals must be submitted to the CPCSEA National Committee for approval.[3]

Singapore's NACLAR Guidelines for Institutional Animal Care and Use Committee (Chapter III) indicate that an IACUC has authority to approve protocols independent of the Institutional Official (IO); however, the institution is not obligated to conduct the research activities despite IACUC approval.[6]

According to the Thai Standard for Institutional Animal Care and Use Committee (2012), the IACUC must: review animal use protocols; perform oversight of the animal use protocol procedures; establish Standard Operating Procedures (SOPs) and practices in the care and use of animals at the institution and make a routine annual institutional report and submit it to the IO.[12]

In mainland China, the IACUC is tasked with semiannual inspection of the institution's animal program management, operational policies and processes; overseeing animal use projects before, during, and after completion; investigating any instances of non-compliance; and providing reports to the institutional management and senior administrative body supervising the institute.[18,32]

Under Japanese Ministry guidelines, the IACUC's functions are to (1) review written animal use proposals to ensure compliance with regulations and internal policies/guidance documents and report the results of deliberation as a recommendation to the institutional official or CEO for final approval; and (2) upon receiving approval status of animal use proposals, provide guidance on an as-needed basis.[24]

Within the Republic of Korea, under the APA, the main function of the IACUC is for the ethical management and protection of research animals. Each institution will establish one IACUC, and the heads of institutions that have animal facilities will submit research proposals to the IACUC. The IACUC will deliberate on the proposed animal experiment; direct animal experimentation in accordance with the principles of animal experiments; and make recommendations for improving animal welfare to the head of the institutions.[25] Under the LAA, the establishment of the LAMC is for the purpose of reviewing animal use proposals and ensuring safe and ethical treatment of animals as well as reliability of animal testing, similar to the responsibilities described under APA. Each animal facility is expected to establish a LAMC. However, if there are no animal experiments being conducted within the animal facility (e.g., breeding or teaching only), then the establishment of the LAMC is not mandated.[27,37]

As described in the Taiwanese Guide, the expected functions of the IACUC are similar to those described in the ILAR *Guide*, including reviewing written animal use protocols and amendments. For institutions that use nonhuman primates, dogs, and cats, a copy of the approved animal use protocol must also be submitted to the COA.[34]

Opin. All jurisdictions require registered institutions to develop an internal system of self-regulation and animal program oversight through an IACUC-like body. Although the extent of their oversight in different components within an animal care and use program may vary, the basic duties include review of proposed animal use, evaluation of animal facility management, and ensuring compliance with applicable laws, regulations, policies, and guidelines.

36:6 Are IACUC-like bodies in Asia Pacific involved in post-approval monitoring activities?

Reg. Australia's Code (Section 2) outlines the responsibilities of the Animal Ethics Committee (AEC) in conducting follow-up review of approved projects and activities. Such follow-up review may include proposing amendments to an approved project and review of annual progress and any unexpected adverse events.[1] The AEC must also "monitor the care and use of animals, including housing conditions, practices and procedures involved in the care of animals in facilities" by performing inspections of animal facilities and records of where animals are housed and laboratories where animal use occurs.[1,2]

For India, IAECs are expected to be involved in post-approval monitoring (PAM) activities. Reports of research protocols have to be submitted for review and consideration of IAEC of the establishment at prescribed intervals and at the end of the study. Any Serious Adverse Events (SAEs) need to be reported to IAEC and in cases where there are any protocol deviations involving large animals, they should be also reported to CPCSEA immediately. IAEC members are also expected to visit animal facilities before each meeting, which must be held at least twice a year, to oversee areas where animals are housed and experiments are performed to ensure compliance to ethical approvals and regulatory requirements.[3–5]

In Singapore, the Animal and Birds (Care and Use of Animal for Scientific Purposes) Rules (rule 8.1) include the need for facility inspection, review of the modification of the project; withdrawal approval for or suspend any project that is not conducted in accordance with the guidelines or approval.[7] NACLAR Guidelines for Institutional Animal Care and Use Committee (2022) (Section III 5) describes PAM as part of the IACUC oversight responsibilities, which should consist of at least an annual review of activities conducted under each protocol; monitoring and reporting of adverse events and unexpected outcomes; animal acquisition and tracking; and an annual inspection.[6]

GB/T 35892-2018 of mainland China expects regular inspection by the IACUC to monitor project activities and approve any significant study modifications that may have negative effect on the animals. Upon project completion, a final report from the project leader will be submitted to the IACUC for retrospective animal welfare and ethical conduct review.[30]

While there is no defined PAM under Japanese regulations and Ministry guidelines, animal users will submit reports from experimental results to the IO or department head.[33] Based on the reports and with input from the IO, IACUC will perform self-inspections. Current ministry guidelines encourage the IO to have the results of self-inspections/evaluations verified by a third party independent from the research institution.[20–22]

In Korea, PAM is carried out in accordance with IACUC Standard Operation Guidelines jointly published by the Animal and Plant Quarantine Agency (QIA) and the MFDS,[35] and generally performed in conjunction with animal facility checks.[38]

Following the Taiwanese Guide, Section 1.2.3, PAM includes facility/laboratory inspection, continuing protocol review, and observation of selected procedures. Institutions must disclose their PAM results in their annual report to COA.[34]

Opin. While not all jurisdictions specify how or when PAM should be performed, all have expectations that approved animal activities will be monitored by the IACUC to ensure compliance.

36:7 Is there a severity classification of experimental procedures?

Reg. Unlike the United States or Europe, there is no defined severity classification system in common across Asia-Pacific jurisdictions.

Australia's severity rating is not part of the Australian Code but rather within state reporting guidelines. For example, for New South Wales, potential impact on animals are reported according to observation involving minor interference; animal unconscious without recovery; minor conscious intervention; minor surgery with recovery; major surgery with recovery; minor physiological challenge; major physiological challenge and death as an endpoint.[39] In the State of Victoria any lethality study may only proceed following special approval from the Minister of Agriculture.[40]

In Singapore, the NACLAR Guidelines for Institutional Animal Care and Use Committee (Section II 9.3 – Annual report by the institution) specify the annual necessity for reports on animal number in each pain/ distress category. They must differentiate and report the numbers of animals used in scientific activities at each facility that involve (1) no more than momentary or slight pain or distress and no use of pain-relieving drugs; (2) accompanying pain or distress to the animals and for which appropriate anesthetics, analgesic, or tranquilizing drugs were used; (3) accompanying pain or distress to the animals and for which the use of appropriate anesthetics, analgesic, or tranquilizing drugs would have adversely affected the procedures, results, or interpretation of the teaching, research, experiments, surgery, or tests, with an explanation of the procedures producing pain or distress in these animals and the reasons such drugs were not used must be attached to the annual report; (4) animals being bred, conditioned, or held for scientific activities at each facility but not yet used for such purposes.[6]

Thailand and India do not have any severity classification, but those that are AAALAC International accredited follow the USDA pain categories. The Taiwanese Guidelines for Laboratory Animal Use and Care recommended a slight modification to the USDA pain category, where Pain Category D1 is for studies where animals are subjected to potential minor painful and stressful procedure for which they receive appropriate anesthetic and analgesics, such as intubation under anesthesia, minor survival surgery under general anesthesia, or exposure to nonlethal chemical or drugs without physical change. Pain Category D2 is for studies where animals are subjected to potential moderate to severe painful and stressful procedures in which they receive appropriate pain relief, such as major survival surgery, long-term physical restraint, induction of behavioral distress, deprivation of maternal care, introduction of aggression or predator/bait interaction, etc.[34]

The Japanese and Korean guidelines generally followed the Scientists Center for Animal Welfare (SCAW) five categories of pain scale, from experiments using dead organisms, plants, bacteria, protozoa, or invertebrates (Category A) to training or research activities in vertebrate animals with no pain or distress (Category B), mild pain or distress for a short period of time (Category C), moderate to severe pain or distress (Category D), or infliction of severe pain or distress above tolerance threshold of conscious animals (Category E).[24,33]

Opin. There is no unified system for categorizing animal pain and distress; however, it is commonly recognized that animal pain and distress from experimental manipulation should be minimized, and pain management should be considered during the review of animal use proposals, as well as animal monitoring expectations during experimental procedures, and setting of appropriate humane study endpoints. The use of appropriate anesthesia, analgesia, sedative, and euthanasia are subject to guidance by the designated institutional veterinary authority (e.g. Attending Veterinarian) and regularly reviewed as part of animal care and use program oversight by the IACUC-like body.

36:8 In what topics are IACUC-like body members required to be trained?

Reg. While most IACUC-like bodies in Asia-Pacific regions are tasked with the responsibility to oversee animal user training programs or to ensure personnel competency in qualification and skills, there is generally no specific requirement for training of IACUC members, except in the regions mentioned below.

In Australia, several states mandate that members of AECs are provided with an understanding of the ethical and humane care and use of animals with organizations such as the Australian and New Zealand Council for the Care of Animals in Research and Teaching (ANZCCART) to meet such requirements.

In India, IAEC members are expected to be knowledgeable of new guidelines and any amendments to relevant laws. The CPCSEA conducts training programs for updating existing CPCSEA nominees as well as for those individuals who are keen to apply to CPCSEA for becoming IAEC members, and the CPCSEA stipulates that IAEC members should be encouraged to attend national and international training programs/conferences.[3]

In Singapore, only qualified personnel shall be appointed as an IACUC member, with the NACLAR Training Guides (Section IV) specifying the training of IACUC members. All members must participate in the didactic components of "Responsible Care and Use of Laboratory Animals" (RCULA) and receive relevant in-house training that should include responsibilities and functions of IACUC members; performing protocol reviews; conducting program reviews and facility inspections; performing ongoing assessments of approved protocols; and addressing concerns involving animal care and use. Formal training must also be provided to at least 50% of IACUC members to emphasize the roles and responsibility of the IACUC and provide other relevant information necessary for IACUC members to discharge their responsibilities.[6]

Based on the Korean APA (Article 27), the IACUC members who are recommended from nonprofit organizations should complete training offered by MAFRA on animal protection, welfare, or animal research topics.[26]

Stated in the Taiwanese Guide, Section 1.2.1, animal research institutions are expected to provide IACUC members with training to understand their work and role. In addition, the IACUC Executive Secretary must complete the specific IACUC training program recognized by the Central Authority of Taiwan every three years.[34]

Opin. Regardless of regulatory requirements or expectations for IACUC member training, most institutions will identify their own approach to ensuring members of their IACUC-like body are adequately knowledgeable about their role and responsibilities in order to ensure institutional compliance with applicable laws and regulations overseeing animal care and use.

36:9 To whom does the IACUC-like body report to at a typical research institution?

Reg. There is no unified system for reporting structure for an IACUC-like body within the Asia-Pacific region.

Opin. Most guidelines specifically indicate the CEO, head of institution, or Institutional Official as the person the IACUC-like body reports to; however, the actual reporting line may depend on the institutional structure and, in some jurisdictions, additional regulatory reporting requirement (e.g. CPCSEA of India, AVS of Singapore, QIA or MFDS of Korea, COA of Taiwan, etc.).

36:10 Are there special considerations for the use of nonhuman primates and endangered species?

Reg. Nonhuman primates (NHP) are widely used and bred within the region, and, as with Europe and North America, most jurisdictions have additional regulations, expectations, and justifications needed for their use. The import and/or export of NHP is subject to country-specific implementation of the Convention on International Trade in Endangered Species of Wild Fauna and Flora (CITES).

Australia's major funding body for medical research, the National Health and Medical Research Council (NHMRC) publishes the "Principles and Guidelines for the Care and Use of Non-human Primates for Scientific Purposes," and does not fund research involving NHP unless the work is of high quality and significance as determined by NHMRC peer-review process.[40] NHMRC have stated that "Non-human primates must not be used for scientific purposes except when no alternative to the use of non-human primates is suitable to achieve the stated aims of the project, and the potential effects on the non-human primates are justified by the potential benefits."[2,40] There are also enhanced requirements for training and competency for NHP work in Australia.[41]

For India, all animal facilities where NHP are housed should have approval from CPCSEA and, as applicable for all large animals, projects involving NHP will first be reviewed by IAEC and then, if recommended by IAEC, will be submitted to CPCSEA for final approval. Rehabilitation of large animals has been recognized in India as the 4thR (rehabilitation) and has evolved as an official policy of the CPCSEA since 2004. Establishments conducting experiments on large animals (including NHP) are responsible for maintaining the animals post-experimentation until their natural death either at its own rehabilitation facility or at any Animal Welfare Organization (AWO) registered with the Animal Welfare Board of India (AWBI) and/or Government.[3,42]

In Singapore, NACLAR Guiding Principles for the Care and Use of Animals for Scientific Purposes Appendix IV: Additional Information on the Care and Use of Non-human Primates for Scientific Purposes, provides information regarding the complexity in the care and management of NHP and include consideration on the behavioral well-being, enclosure, zoonotic diseases, and safety aspects of the program.[6]

Thailand doesn't explicitly have special considerations for NHP use but does emphasize that the use of wildlife is restricted to scientific research that cannot be replaced by any other kind of animals, and researchers are to abide by the laws and policies for wildlife conservation.

Mainland China is a significant global supplier of laboratory NHP. The NHP are considered as rare and endangered wild animals and are protected under the Wildlife Protection Law of China and CITES.[16,43] Common species of laboratory NHP, including macaques and marmosets, are national key protected wild animals subject to additional oversight under the State Forestry and Grass Administration.[44,45] If their use for scientific research is necessary, aside from the need to have a valid animal use license for housing and working with the specific nonhuman primate species, an institution must follow additional administrative licensing procedures enforced under the provincial-level Department of Forestry to procure and transport NHP. Annually, each breeding facility is allowed to sell a limited number of NHP for domestic use and for exportation.

In Japan, the importation of NHP is restricted by the Invasive Alien Species Act and the Act on Prevention of Infectious Diseases and Medical Care for Patients of Infectious Diseases.[25] The Guidelines for Proper Conduct of Animal Experiments emphasizes compliance with applicable regulations when importing NHP, potential zoonotic disease risks, and occupational health and safety expectations. Additional care and use information may be found in guidance documents from the Primate Society of Japan.[33]

Regulated under the Ministry of Environment, breeding facilities of NHP (and other globally endangered species) in the Republic of Korea must comply with the registration requirements under the Wildlife Protection and Management Act, Article 16-2.[46] While there are no national guidelines for the care and use of laboratory NHP, Seoul National University Hospital has prepared and shared its own guideline for working with laboratory NHP.[25]

Oversight for the use of NHP for scientific purposes in Taiwan falls under the APA, but, depending on the species use, special registration, permitting, and reporting requirements may apply. For example, the use of cynomolgus macaques will require a special permit on use of wildlife and endangered species, issued by the Forestry Bureau under COA (H.C. Chen, personal communication, August 2, 2022). Similar to projects involving the use of dogs and cats, the institution working with NHP must submit a copy of the animal use protocol to the COA.[34]

Opin. Nonhuman primate species, especially specific pathogen-free laboratory macaques (e.g., rhesus and cynomolgus macaques) play an irreplaceable role in preclinical research and evaluation of biomedicine, neuroscience, and reproductive biology. In recent years, the rapid growth in the biopharmaceutical market has led to increased demand for this valuable animal resource, especially of cynomolgus macaques. However, unlike other common laboratory animal species, several important factors have contributed toward the global supply shortage. According to the China Laboratory Animal Information Network, within mainland China there are strict breeding practice and quality control measures stipulated under Chinese National Standards which must be followed by domestic breeding facilities; long production cycle of macaques (e.g. 6–7 years to produce suitable animals for experiments) and reported gradual decrease in birth rate due to aging breeding population could not keep up with short-term global market interest. To ensure their animal supplies, several major research labs have either procured small to mid-sized breeding farms or have established long-term partnerships with large domestic breeding facilities, thus further limiting the availability of NHP to smaller institutions. While there are calls for Southeast Asia countries to ease their import/export regulations in order to address production needs of NHP, there is concurrent concern over public health risks from inadvertently incorporating wild captured animals (or animals from substandard breeding farms) that are likely carriers of pathogenic microorganisms, which cause latent diseases such as *Mycobacterium tuberculosis* orsimian viruses.[47,48]

The current shortage of laboratory NHP further emphasizes the need to select and use laboratory animals scientifically and rationally, to abide by the ethics of animal experimentation, to follow the 3R principles, and to hold steadfast in our commitment to continuously improve the welfare of laboratory animals.

36:11 What are the typical expectations with respect to the 3Rs?

Reg. In Australia, the 3Rs are a mandatory requirement under the Governing Principles (Section 1) of the Code which states that replacement, reduction, and refinement (the 3Rs) are to be applied at all stages of research. Because the Code is integrated into state and territory legislation, the expectation to apply the 3Rs is considered legally binding.[1]

In the case of India, to IAECs the 3Rs principles are normally the basis of most of the discussions with a need to review the literature and declare whether the experiment has already been done *in vitro/in vivo* elsewhere and to justify why the experiment cannot be performed *in vitro*.[3]

For Singapore, Animal and Birds (Care and Use of Animal for Scientific Purposes) Rules (Rule 10 (2)) specify that IACUC approval of use of animals for scientific purposes complies with the requirements that include the 3R principles.[7] The NACLAR Guiding Principles for the Care and Use of Animals for Scientific Purposes (Section II.4) emphasize the Guiding Principles that includes the 3Rs, whereas Section III.4 stipulates the need for ensuring alternatives are considered during protocol review.[6]

The Thai Ethical Principles and Guidelines for the Use of Animals for Scientific Purposes (Ethical Principles) and the Ethics in Procedures on Animals for Scientific Purpose 2015 are based on the 3Rs, so they form the backbone of national guidelines and regulations on the ethical use of research animals.[10,11]

In mainland China, the promotion of animal welfare through emphasis of the 3Rs and/or assurance of the Five Rights to enjoy comfort, good health, and psychological well-being are specified in the Guidelines on Humane Treatment of Laboratory Animals, GB/T 27416-2014 (Attachment B), and GB/T 35892-2018 (Section 3.8 and 3.10).[18,31,32] These principles are universally accepted and followed by institutions breeding and using animals for scientific purposes.

As stated in Japan's Guidelines for Proper Conduct of Animal Experiments (Section 4), the 3Rs are the ideology behind both animal experimentation and the handling of laboratory animals.[33] Drafting of animal experiment protocols and experimental procedures should comply with the internationally accepted 3R principles of animal experimentation.

Korean Guidelines for the Standard Operations of the IACUC (Section 3) set the IACUC expectation for consideration of the 3Rs when deliberating on animal experimentation proposals, and examples for animal replacement, reduction, and refinement are provided.[35]

Under Taiwan's Guide for the Care and Use of Laboratory Animals (Section 1,3), ethics of animal use and principles of the 3Rs are required training topics for all animal users. Research teams are encouraged to design animal experiments following the spirit of the 3Rs.[34]

Opin. The 3Rs (replacement, reduction, and refinement) principle is widely accepted within the circle of laboratory animal professionals in the Asia-Pacific region. Along with the 5Fs (freedom from thirst and hunger; freedom from discomfort; freedom from pain, harm, and disease; freedom to carry out most natural behaviors; and freedom from fear and distress),[49] these basic animal welfare–related concepts are often covered during training for IACUC-like bodies and animal users. Depending on the country, reference to the 3Rs and/or 5Fs can be found in regulatory guidance documents and/or in publications issued by professional organizations to promote consideration and enforcement of these principles by the IACUC-like bodies.[13,25,32]

Country- or region-specific laws, regulations, and guidelines relevant to animal welfare oversight and recommended websites for further information

Australia	Animal Ethics Infolink https://www.animalethics.org.au/legislation
	Animal Research Act 1985 No. 123 https://legislation.nsw.gov.au/view/html/inforce/current/act-1985-123
	Animal Research Regulation 2021 (under the New South Wales Animal Research Act 1985) https://legislation.nsw.gov.au/view/html/inforce/current/sl-2021-0477
	Australian Code for the Care and Use of Animals for Scientific Purposes (2013) https://www.nhmrc.gov.au/about-us/publications/australian-code-care-and-use-animals-scientific-purposes
	Other Australian legislation: https://www.animalethics.org.au/legislation/other-australian-legislation
Hong Kong, China	Cap. 340 Animals (Control of Experiments) Ordinance. 1963. https://www.elegislation.gov.hk/hk/cap340
	Cap. 169 Prevention of Cruelty to Animals Ordinance 1935 https://www.elegislation.gov.hk/hk/cap169
	Code of Practice Care and Use of Animals for Experimental Purposes (2004) https://www.afcd.gov.hk/english/aboutus/abt_adv/files/Code_of_Practice_Care_and_Use_of_Animals_for_Experimental_Purposes_English.pdf
India	Committee for the Purpose of Control and Supervision of Experiments on Animals (CPCSEA) – Acts and Rules https://cpcsea.nic.in/Content/54_1_ACTSANDRULES.aspx
	The Prevention of Cruelty to Animals Act, No. 59 of 1960. Ministry of Fisheries, Animal Husbandry & Dairying https://cpcsea.nic.in/WriteReadData/userfiles/file/No_59.pdf
	Breeding of and Experiments on Animals (Control and Supervision) Rules, 1998 (amended in 2001 and 2006) https://cpcsea.nic.in/WriteReadData/userfiles/file/2006.pdf
	Committee for the Purpose of Control and Supervision of Experiments on Animals (CPCSEA) – All Guidelines https://cpcsea.nic.in/Content/55_1_GUIDELINES.aspx
	Compendium of CPCSEA 2018, Committee for the Purpose of Control and Supervision of Experiments on Animals (CPCSEA), Ministry of Environment, Forest and Climate Change, Government of India https://cpcsea.nic.in/WriteReadData/userfiles/file/Compendium%20of%20CPCSEA.pdf
	Standard Operating Procedure for Institutional Animal Ethics Committee. 2010. CPCSEA. https://cpcsea.nic.in/WriteReadData/userfiles/file/SOP_CPCSEA_inner_page.pdf
Indonesia	Regulation of the Government of the Republic of Indonesia Number 95 of 2012 Concerning Veterinary Public Health and Animal Welfare http://www.flevin.com/id/lgso/translations/JICA%20Mirror/english/4936_PP_95_2012_e.html
	Regulations of the Food and Drug Supervisory Agency Number 7 of 2014 concerning Guidelines for In Vivo Nonclinical Toxicity Testing (in Indonesian) https://jdih.pom.go.id/download/product/704/7/2014
	Regulation of the Food and Drug Supervisory Agency Number 18 of 2021 concerning Guidelines for Preclinical Pharmacodynamic Testing of Traditional Medicines (in Indonesian) https://jdih.pom.go.id/download/product/1282/18/2021
	The Head of National Nuclear Energy Agency of Indonesia Regulation Number 195/KA/XI/2011 about Ethical Guidelines for the use and Care of Experimental Animals (in Indonesian) https://jdih.batan.go.id/unduh/jdih/93175678607Perka_Etik_Final.pdf
	Regulation of the Minister of Health (PMK) Number 7 of 2016 concerning the National Research Ethics Commission and Development of National Health

Japan	Act on Welfare and Management of Animals (Act No. 105 of October 1, 1973) https://www.cas.go.jp/jp/seisaku/hourei/data/AWMA.pdf
	Basic Policies for the Conduct of Animal Experimentation in the Ministry of Health, Labour and Welfare 2006 and amended in 2015 (Japanese) https://www.mhlw.go.jp/file/06-Seisakujouhou-10600000-Daijinkanboukouseikagakuka/honbun.pdf
	Fundamental Guidelines for Proper Conduct of Animal Experiment and Related Activities in Academic Research Institutions under the jurisdiction of the Ministry of Education, Culture, Sports, Science and Technology Notice No. 71. June 1, 2006 https://www.lifescience.mext.go.jp/policies/pdf/an_material011.pdf
	Basic Guidelines for Conducting Animal Experiments at Research Institutes under the jurisdiction of the Ministry of Agriculture, Forestry and Fisheries, Notice No. 307. June 1, 2006 (Japanese) https://www.affrc.maff.go.jp/doubutsujikken/attach/pdf/doubutsujikken_kihonshishin-5.pdf
	Guidelines for Proper Conduct of Animal Experiments (2006) Science Council of Japan https://www.scj.go.jp/ja/info/kohyo/pdf/kohyo-20-k16-2e.pdf
	Standards Relating to the Care and Management of Laboratory Animals and the Alleviation of Pain. (2006) Ministry of Environment https://www.env.go.jp/nature/dobutsu/aigo/2_data/laws/nt_h25_84_en.pdf
Malaysia	Animal Welfare Act 2015 (Laws of Malaysia Act 772) https://extwprlegs1.fao.org/docs/pdf/mal176896.pdf
	Malaysian Code of Practice for the Care and Use of Animals for Scientific Purposes (MYCODE) 2019. Animal Welfare Board, Department of Veterinary Services https://www.dvs.gov.my/dvs/resources/user_1/2020/Kebajikan%20Haiwan/GP01.pdf
Philippines	Republic Act 8485 An Act to Promote Animal Welfare in the Philippines, otherwise known as "The Animal Welfare Act of 1998" https://www.officialgazette.gov.ph/1998/02/11/republic-act-no-8485/
	Republic Act No. 10631 An Act Amending Certain Sections of Republic Act No. 8485, otherwise known as "The Animal Welfare Act of 1998" https://www.officialgazette.gov.ph/2013/10/03/republic-act-no-10631/
	Rules and Regulations on the Conduct of Scientific Procedures Using Animals (1999), Department of Agriculture Bureau of Animal Industry https://www.bai.gov.ph/index.php/laws-issuances/administrative-order/category/79-ao-1999?download=934:ao40-s-1999-rules-and-regulations-on-the-conduct-of-scientific-procedures-using-animals
	PALAS Code of practice for the care and use of laboratory animals in the Philippines (COP)
People's Republic of China	Regulations for the Administration of Affairs Concerning Experimental Animals. 1988. Ministry of Science and Technology (English translation) https://english.calas.org.cn/index.php?m=content&c=index&a=show&catid=50&id=136
	Guidelines on Humane Treatment of Laboratory Animals, 2006. Ministry of Science and Technology (in Chinese) https://www.most.gov.cn/xxgk/xinxifenlei/fdzdgknr/fgzc/gfxwj/gfxwj2010before/201712/t20171222_137025.html
	China National Standard GB/T 27416-2014 – Laboratory animal facilities – General requirements for quality and competence, Appendix A, 2014. (in Chinese) https://openstd.samr.gov.cn/bzgk/gb/newGbInfo?hcno=75FCA517D0290A9A1FE5B928801537A5
	China National Standard GB/T 35892-2018 - Guidelines for the ethical review of laboratory animal welfare. 2018. https://www.aaalac.org/pub/?id=2193A985-E2E4-BC2F-4824-22F5BAFFA47C
Republic of Korea	Animal Protection Act (English translation) https://elaw.klri.re.kr/eng_service/lawView.do?hseq=50163&lang=ENG
	Laboratory Animal Act (English translation) https://elaw.klri.re.kr/kor_service/lawView.do?hseq=46430&lang=ENG
	Guidelines for the standard operations of the IACUC (in Korean). Gimcheon: Animal and Plant Quarantine Agency; https://ebook.qia.go.kr/20180621_173152/
Singapore	Animals and Birds Act (Ch. 7, Sec. 80) Animals and Birds (Care and Use of Animals for Scientific Purposes) Rules R 10 G.N. No. S558/2004. 2004. Agri-Food and Veterinary Authority of Singapore. https://sso.agc.gov.sg/SL/ABA1965-R10
	NACLAR Guidelines on the Care and Use of Animals for Scientific Purposes (2nd ed.) 2022. National Advisory Committee for Laboratory Animal Research https://www.nparks.gov.sg/-/media/naclar-guidelines-(second-edition)_v2.ashx
Taiwan	Animal Protection Act (1998) (Amended 2021) https://law.moj.gov.tw/ENG/LawClass/LawAll.aspx?pcode=M0060027
	Guide for the Care and Use of Laboratory Animals (2018) Council of Agriculture, Executive Yuan. https://animal.coa.gov.tw/download/file/181205-1.pdf

Thailand	Animals for Scientific Purposes Act, B.E. 2558 (A.D. 2015) https://labanimals.nrct.go.th/LinkClick.aspx ?fileticket=Ezbx7i1V9A8%3D&portalid=0
	The Ethical Principles and Guidelines for the Use of Animals for Scientific Purposes (1999) National Research Council of Thailand (English translation) https://www.aaalac.org/pub/?id=E905AF01-DA92-FCA3-9A57 -49F479FA27A3
	The Standard for Institutional Animal Care and Use Committee (2012) National Research Council of Thailand
	The Ethics in Procedures on Animals for Scientific Purposes (2015) Committee for Supervision and Promotion of Procedures on Animals for Scientific Purposes. National Research Council of Thailand.
	Guidelines on animal use protocol review. (2014) National Research Council of Thailand.
New Zealand	Animal Welfare Act 1999 https://www.legislation.govt.nz/act/public/1999/0142/latest/DLM49664.html
	Good Practice Guide for the Use of Animals in Research, Testing and Teaching (2021) Ministry for Primary Industries and the National Animal Ethics Advisory Committee https://www.mpi.govt.nz/dmsdocument/33585 -good-practice-guide-for-the-use-of-animals-in-research-testing-and-teaching

Abbreviations and Acronyms

AEC	Animal Ethics Committee
ANZCCART	Australian and New Zealand Council for the Care of Animals in Research and Teaching
APA	Animal Protection Act, Republic of Korea
APEC	Asia-Pacific Economic Cooperation
ASEAN	Association of Southeast Asian Nation
AWBI	Animal Welfare Board of India
AWO	Animal Welfare Organization
AV	Attending Veterinarian
AVS	Animal & Veterinary Services
CEO	Chief Executive Officer
CITES	Convention on International Trade in Endangered Species of Wild Fauna and Flora
COA	Council of Agriculture, Executive Yuan, Republic of China, Taiwan
The Code	The Australian Code for the Care and Use of Animals for Scientific Purposes
CPCSEA	Committee for the Purpose of Control and Supervision of Experiments on Animals
DAHD	Department of Animal Husbandry and Dairying, India
GB	GuoBiao, or China National Standard
IACUC	Institutional Animal Care and Use Committee
IAEC	Institutional Animal Ethics Committee
ILAR	Institute for Laboratory Animal Research
IO	Institutional Official
KAFLA	Korean Association for Laboratory Animals
KCLAM	Korean College of Laboratory Animal Medicine
LAA	Laboratory Animal Act, Republic of Korea
LAMC	Laboratory Animal Management Committee
MAFRA	Ministry of Agriculture Food and Rural Affairs, Republic of Korea
MFDS	Ministry of Food and Drug Safety, Republic of Korea
MoFAH&D	Ministry of Fisheries, Animal Husbandry and Dairying
MOST	Ministry of Science and Technology, People's Republic of China
NACLAR	National Advisory Committee for Laboratory Animal Research
NCRAD	National Committee for Research Animal Development
NHMRC	National Health and Medical Research Council
NHP	Nonhuman primates
NRCT	National Research Council of Thailand
PAM	Post-approval monitoring
QIA	Animal and Plant Quarantine Agency, Republic of Korea

RCULA	Responsible Care and Use of Laboratory Animals
SCAW	Scientists Center for Animal Welfare
SOP	Standard Operating Procedures
The 5Fs	The 5 Freedoms (freedom from hunger or thirst; freedom from discomfort; freedom from pain, injury, or disease; freedom to express normal behavior; freedom from fear and distress)
The 3Rs	Replacement, reduction, and refinement
US	United States
USDA	United States Department of Agriculture

Acknowledgment

We thank the following individuals for their invaluable input regarding regional practices: Dr. Shikha Yadav of National Institute of Biologicals, India; Dr. Mayu Uchihashi of Medtronic Japan Co., Ltd., Japan; Dr. Young-Shin Joo of Institute of Biomedical Industry at the Catholic University of Korea, Republic of Korea; Dr. Cheng-Fen, Lai of Keelung Chang Gung Memorial Hospital, Republic of China, Taiwan; Dr. Thattawan Kaewsakhorn of Chiang Mai University, Thailand; Dr. Hou-Chun Chen of Thammasat University, Thailand; Dr. Yasmina Paramastri of National University of Singapore, Singapore; Dr. Pooi-Fong Wong of Universiti Malaya, Malaysia; Dr. Imelda Winoto of Mochtar Riady Institute for Nanotechnology, Indonesia; and Dr. Malcolm France of the University of New South Wales, Australia.

REFERENCES

1. National Health and Medical Research Council. 2013. Australian code for the care and use of animals for scientific purposes. https://www.nhmrc.gov.au/about-us/publications/australian-code-care-and-use-animals-scientific-purposes.
2. Noonan D. and Williams V. 2018. Chapter 12. Laboratory Animals Regulations and Recommendations: Australia and New Zealand. In: *Laboratory Animals*, 2nd edition, ed. J. Guillén, 375–419. Amsterdam: Elsevier Academic Press.
3. Qadri S.S.Y.H., and Ramachandra S.G. 2018. Chapter 8. Laws, Regulations, and Guidelines Governing Research Animal Care and Use in India. In: Laboratory Animals, 2nd edition, ed. J. Guillén, 293–317. Amsterdam: Elsevier Academic Press.
4. The Prevention of Cruelty to Animals Act, No. 59 of 1960. Ministry of Fisheries, Animal Husbandry & Dairying. https://cpcsea.nic.in/WriteReadData/userfiles/file/No_59.pdf.
5. Breeding of and Experiments on Animals (Conrol and Supervision) Rules, 1998 (amended in 2001 and 2006). https://cpcsea.nic.in/WriteReadData/userfiles/file/2006.pdf.
6. National Advisory Committee for Laboratory Animal Research (NACLAR). 2022. *Guidelines on the Care and Use of Animals for Scientific Purposes* (2nd ed.). https://www.nparks.gov.sg/-/media/naclar-guidelines-(second-edition)_v2.ashx.
7. Agri-Food and Veterinary Authority of Singapore. 2004 Animals and Birds Act (Ch. 7, Sec. 80) Animals and Birds (Care and Use of Animals for Scientific Purposes) Rules R 10 G.N. No. S558/2004. https://sso.agc.gov.sg/SL/ABA1965-R10.
8. Retnam L., Chatikavanij P., Kunjara P., Paramastri Y.A. et al. 2016. Laws, Regulations, Guidelines and Standards for Animal Care and Use for Scientific Purposes in the Countries of Singapore, Thailand, Indonesia, Malaysia, and India. *ILAR J.* 57(3):312-323.
9. Animals for Scientific Purposes Act, B.E. 2558 (A.D. 2015). https://labanimals.nrct.go.th/LinkClick.aspx?fileticket=Ezbx7i1V9A8%3D&portalid=0.
10. The Ethical Principles and Guidelines for the Use of Animals for Scientific Purposes (1999) National Research Council of Thailand (English translation). https://www.aaalac.org/pub/?id=E905AF01-DA92-FCA3-9A57-49F479FA27A3.
11. The Ethics in Procedures on Animals for Scientific Purposes (2015) Committee for Supervision and Promotion of Procedures on Animals for Scientific Purposes. National Research Council of Thailand.
12. The Standard for Institutional Animal Care and Use Committee (2012) National Research Council of Thailand.
13. Gettayacamin M., Grant R., Davun H. Heidrich JE. et al.. Chapter 11. Laws, regulations, guidelines, and principles pertaining to laboratory animals in Southeast Asia. 2018. In: *Laboratory Animals*, 2nd edition, ed. J. Guillén, 319–373. Amsterdam: Elsevier Academic Press.
14. Ministry of Science and Technology. 1988. Regulations for the Administration of Affairs Concerning Experimental Animals. (English translation). http://english.calas.org.cn/index.php?m=content&c=index&a=show&catid=50&id=136.
15. Ministry of Science and Technology. 2001. Regulation on the Management of Laboratory Animal licenses. (in Chinese). http://www.jsdw.org.cn/uploadfile/20120715131345491.pdf.
16. Bayne K., Wang J., and Pang W. Chapter 9. Oversight of Animal Research in China. 2018. In: *Laboratory Animals*, 2nd edition, ed. J. Guillén, 263–291. Amsterdam: Elsevier Academic Press.
17. National Public Service Platform for Standards Information Website. National Standard Catalog Search. https://std.same.gov.can/go/gbQuery. Accessed June 20, 2022.
18. General Administration of Quality Supervision, Inspection and Quarantine of the People's Republic of China, Standardization Administration. 2014. China National Standard GB/T 27416-2014 Laboratory Animal Facilities

– General Requirements for Quality and Competence. (Chinese). https://openstd.samr.gov.cn/bzgk/gb/newGbInfo?hcno=75FCA517D0290A9A1FE5B928801537A5.

19. Act on Welfare and Management of Animals. (Act No. 105 of October 1, 1973, last revision 2014). https://www.env.go.jp/nature/dobutsu/aigo/1_law/files/aigo_kanri_1973_105_en.pdf.
20. Ministry of Education, Culture, Sports, Science and Technology, Notice No. 71. 2006. Fundamental Guidelines for Proper Conduct of Animal Experiment and Related Activities in Academic Research Institutions under the Jurisdiction of the Ministry of Education, Culture, Sports, Science and Technology. http://www.lifescience.mext.go.jp/policies/pdf/an_material011.pdf.
21. Ministry of Health, Labour and Welfare. 2006. Basic Policies for the Conduct of Animal Experimentation in the Ministry of Health, Labor and Welfare (Japanese). https://www.mhlw.go.jp/file/06-Seisakujouhou-10600000-Daijinkanboukouseikagakuka/honbun.pdf.
22. Ministry of Agriculture, Forestry and Fisheries, Notice No. 307. 2006. Basic Guidelines for Conducting Animal Experiments at Research Institutes under the Jurisdiction of the Ministry of Agriculture, Forestry and Fisheries (Japanese). https://www.affrc.maff.go.jp/doubutsujikken/attach/pdf/doubutsujikken_kihonshishin-5.pdf.
23. Ministry of Environment. 2006. Standards Relating to the Care and Management of Laboratory Animals and the Alleviation of Pain. https://www.env.go.jp/nature/dobutsu/aigo/2_data/laws/nt_h25_84_en.pdf.
24. Ogden BE, Pang W., Agui T., Lee BH. 2016. Laboratory Animal Laws, Regulations, Guidelines and Standards in China Mainland, Japan, and Korea. *ILAR J.* 57(3):301-311.
25. Kurosawa TM, Park JH, Hong CC. Chapter 10. Laws, Regulations, Guidelines, and Principles Pertaining to Laboratory Animals in Far East Asia. 2018. In: *Laboratory Animals*, 2nd edition, ed. J. Guillén, 293-317. Amsterdam: Elsevier Academic Press.
26. Ministry of Agriculture Food and Rural Affairs. 1991. Animal Protection Act (Amended 2011, 2013, 2014, 2015). https://law.go.kr/LSW/eng/engLsSc.do?menuId=2&query=animal%20protection%20act.
27. Ministry of Food and Drug Safety. 2009. Laboratory Animal Act Act. No. 9025 Mar 28, 2008 (Amended 2010, 2013, 2016, 2017). https://law.go.kr/LSW/eng/engLsSc.do?menuId=2&query=laboratory%20animal%20act.
28. Council of Agriculture, Executive Yuan. 1998. Animal Protection Act (Amended 2021). https://law.moj.gov.tw/ENG/LawClass/LawAll.aspx?pcode=M0060027.
29. National Research Council 2011. *Guide for the Care and Use of Laboratory Animals* (8th ed.). Washington, DC: The National Academies Press.
30. Ministry of Science and Technology. 2006. Guideline on the Humane Treatment of Laboratory Animals (Chinese). https://www.most.gov.cn/xxgk/xinxifenlei/fdzdgknr/fgzc/gfxwj/gfxwj2010before/201712/t20171222_137025.html.
31. General Administration of Quality Supervision, Inspection and Quarantine of the People's Republic of China, Standardization Administration. 2018. China National Standard GB/T 35892-2018 Guidelines for the Ethical Review of Laboratory Animal Welfare (in Chinese). https://openstd.samr.gov.cn/bzgk/gb/newGbInfo?hcno=9BA619057D5C13103622A10FF4BA5D14.
32. MacArthur Clark JA, Sun D. 2020. Guidelines for the Ethical Review of Laboratory Animal Welfare People's Republic of China National Standard GB/T 35892-2018 [Issued 6 February 2018 Effective from 1 September 2018]. *Animal Model Exp Med.* 3:10304–113. https://www.aaalac.org/pub/?id=2193A985-E2E4-BC2F-4824-22F5BAFFA47C.
33. Science Council of Japan. 2016. Guidelines for Proper Conduct of Animal Experiments. https://www.cas.go.jp/jp/seisaku/hourei/data/AWMA.pdf.
34. Council of Agriculture, Executive Yuan. 2018. Guideline for the Care and Use of Laboratory Animals (in Chinese). https://animal.coa.gov.tw/download/file/181205-1.pdf.
35. Animal and Plant Quarantine Agency. 2017. Guidelines for the Standard Operations of the IACUC (E-Book in Korean). http://ebook.qia.go.kr/20180621_173152.
36. Ahn N., Park J. and Roh S. 2022. "Operational Issues of the Institutional Animal Care and Use Committee in Korea. *J. Vet Sci.* 23(4): e59. https://doi.org/10.4142/jvs.22087.
37. Lee BH, Joo YS, and Gettayacamin M. Overview of Local Requirements for Site Visits in Republic of Korea (PowerPoint Presentation, Site Visit Guidance, AAALAC International).
38. Ahn N., Park J. and Roh S. 2021. "The Status and Issues of the Institutional Animal Care and Use Committee of Seoul National University: From Its Establishment to the Present Day. *Exp Anim.* 70(4): 532–540. https://doi.org/10.1538/expanim.21-0066.
39. Guidance for Researchers on Reporting Animals Used in Research - Secretary's Animal Care and Ethics Committee. New South Wales Government. 2021. https://www.dpi.nsw.gov.au/__data/assets/pdf_file/0003/1330905/Guide-for-reporting-animals-used-in-research-v3.pdf.
40. State of Victoria Prevention of Cruelty to Animals Regulations. 2019. https://www.legislation.vic.gov.au/in-force/statutory-rules/prevention-cruelty-animals-regulations-2019/001.
41. National Health and Medical Research Council. 2016. Principles and Guidelines for the Care and Use of Non-Human Primates for Scientific Purposes. https://www.nhmrc.gov.au/about-us/publications/principles-and-guidelines-care-and-use-non-human-primates-scientific-purposes.
42. Guidelines of CPCSEA for Reuse/ Rehabilitation of Large Animals Post Experimentation 2020. https://cpcsea.nic.in/WriteReadData/LnPdf/GuidelinesofCPCSEAforReuseRehabilitationofLargeAnimals.pdf
43. Law of the People's Republic of China on the Protection of Wildlife. 1988. Order No. 9 of the President of the People's Republic of China on November 8, 1988. http://www.china.org.cn/english/environment/34349.htm
44. State Forestry Administration on Strengthening Experimental Monkeys Management Notification 2014 (in Chinese). http://www.forestry.gov.cn/uploadfile/gfxwj/201336/aa52f73e-f9a4-4ded-8f70-fcbbfbe02f2e.pdf

45. Response to the recommendation of Deputies No. 0092 of the Fifth Session of the 15th Municipal People's Congress, Shanghai Forestry Bureau. 2021. https://lhsr.sh.gov.cn/dbjy/20210809/c1fea9ba-ab96-4685-a088-f70e7d3a5e8a.html.
46. Ministry of Environment. 2017. Wildlife Protection and Management Act. https://elaw.klri.re.kr/eng_service/lawView.do?hseq=46346&lang=ENG
47. China Laboratory Animal Information Network "Why Are Laboratory Nonhuman Primates So Scarce?" from Those Things about Experimental Animals (2021-02-09). http://www.lascn.net/Item/89791.aspx Accessed August 30, 2022.
48. China Laboratory Animal Information Network "Supply-side Factors in the Shortage of Laboratory Nonhuman Primates?" from Those Things About Experimental Animals (2022-07-05) http://www.lascn.net/Item/96975.aspx.
49. Farm Animal Welfare Council Press Statement issued December 5, 1979. "Five Freedoms." https://webarchive.nationalarchives.gov.uk/ukgwa/20121007104210/http://www.fawc.org.uk/pdf/fivefreedoms1979.pdf.

37

Operational Communications and Workflow

Natalie Mays, Tanise Jackson, and Jennifer A. Klahn

Introduction

The Institutional Animal Care and Use Committee (IACUC) is an independent committee mandated by the Animal Welfare Act and the Public Health Service Policy on Humane Care and Use of Laboratory Animals (PHS Policy). Despite this independent status, the IACUC best serves the animal care and use program when it partners with other institutional units. This chapter will discuss those units outside of the IACUC with which the IACUC and other animal program staff may engage in the process of handling animal care and use protocols, performing facility inspections, and conducting other business associated with the program. There are no regulatory mandates to direct these interactions; as such, the following guidance is based largely on author experience.

37:1 Which institutional groups should the IACUC and the IACUC Office regularly communicate with?

Principal Investigator (PI): The PHS Policy specifies that one of the functions of the IACUC with respect to PHS-conducted or supported activities is to review new or significant changes to ongoing research projects related to the care and use of animals and determine that the proposed projects are in accordance with the PHS Policy.[1] After making the determination, the IACUC is required to notify investigators and the institution in writing of its decision to approve or withhold approval or if modifications are required to secure IACUC approval.

Institutional Official (IO): The PHS Policy indicates that one of the responsibilities of the IO is resource planning for the animal care and use program. The IO also bears ultimate responsibility for the animal care and use program. The IACUC is required to communicate regularly with the IO. The IACUC administrator may assist the IACUC by drafting communications from IACUC leadership to the IO, arranging meetings with the IO, or inviting the IO to attend IACUC meetings where investigations of animal welfare concerns will be discussed. The IO may be asked to moderate when a researcher is in a dispute with the IACUC in order to help solve the disagreement.

Other institutional administrators and leadership: IACUC administrators should become acquainted with other research administrators in their institution to share and compare duties and responsibilities. Administrators are encouraged to be familiar with their institutional leadership to better understand how decisions are made and who to reach out to about matters that are related to the program but not expressly under the purview of the IO.

Institutional Biosafety Committee (IBC) and environmental health and safety: Researchers at institutions that receive federal funds for research involving the use of recombinant or synthetic nucleic acid molecules in research within the United States are required to adhere to the NIH Guidelines for Research Involving Recombinant or Synthetic Nucleic Acid Molecules. The institution fulfills this requirement by establishing an IBC to review and approve recombinant or synthetic nucleic acid molecule activities. Researchers proposing to conduct experiments that require IBC notification or approval are required to comply with the NIH Guidelines by submitting to the IBC. The IACUC and IBC must maintain regular communication when the research involves approval by both regulatory committees. Creation of some novel transgenic rodents is an example of animal research that requires IBC approval. Some institutions require IBC approval prior to IACUC approval. Environmental Health and Safety (EH&S) personnel are an integral part of the program's researcher safety / occupational health program and they typically serve on an IACUC or IBC. These representatives provide consultation on the safe use and handling of hazardous agents used in animal research, laboratories, and animal facilities. They also may perform personnel risk assessments in locations where anesthetic gases such as isoflurane are used. The IACUC office regularly communicates with EH&S during protocol review and IACUC inspections.

Institutional Review Board (IRB) and human protections group: When animal-based research also involves human subjects' research subject to IRB review, the IACUC and IRB review and approval processes must be synchronized so that each committee is aware of each other's actions. If the IACUC requires modifications, the IRB must be informed and vice versa. IACUC administrative staff must ensure that the research involving IRB approval is not initiated until both the IACUC and the IRB have granted approval.

Occupational health: Regulations require personnel with significant animal exposure to be enrolled in an occupational health program. IACUC administrators should verify that individuals are enrolled in the occupational health program prior to IACUC approval being granted for the individual to work with animals.

Veterinary clinical services group: IACUC membership includes at least one veterinarian with training or experience in

DOI: 10.1201/9781003272861-37

laboratory animal science and medicine. The IACUC administrator may convey veterinarian review comment responses from researchers to veterinarians. They may also regularly communicate with veterinarians if joint statements between the veterinary staff and the IACUC need to be released to the research community.

Husbandry management group: Research animals supported by federal funding sources must be associated with an active IACUC-approved animal protocol. When protocols are either expired or suspended and animals remain on census, the IACUC administrator may query the husbandry management group to verify the number of animals remaining and determine the disposition of those animals (e.g., transfer to another appropriate active protocol, euthanasia, holding protocol).

Library staff: Regulations require that researchers proposing the use of USDA-covered species in pain categories D or E perform a search for alternatives to painful procedures prior to submitting to the IACUC. IACUC administrators can assist researchers in collaborating with librarians in the initial stages of their research project to ensure searches are appropriate for the type of research being proposed.

Grants office: The NIH Grants Policy requires IACUC approval for animal procedures proposed in federal grants prior to funding such proposals. If a federal grant is funding the proposal, the procedures described in the IACUC protocol must be congruent with the procedures described in the grant. Verification of IACUC approval and grant/protocol congruency is the responsibility of institutions receiving federal funding for animal studies. Many institutions delegate these responsibilities to the IACUC or IACUC administrators. When applicable, IACUC administrators should be aware of grant funding cycles and coordinate with grants' personnel when funding sources for animal research have been identified.

Research sponsors: Sponsors of research studies involving the use of animals usually require that the proposed studies be approved by the IACUC when the proposal is submitted to the sponsor or prior to distribution of funds. IACUC administrators may be responsible for providing documentation of IACUC approval to sponsors. Industry sponsors may require negotiation of contracts involving the use of animals. Depending on institutional policies, IACUC administrators may be involved in these contract negotiations as well.

Regulatory or accrediting agencies: IACUC administrators may be the contact person for regulatory bodies or accrediting agencies. When applicable, they can submit documents such as annual reports on behalf of the institution. They may also be the initial contact person when agencies schedule inspections or site visits.

Public: One of the IACUC's responsibilities is to review any concerns involving the care and use of research animals at the institution. When concerns come from the public, the IACUC administrator must follow institutional policies regarding steps to forward those concerns to the IACUC. In addition, the IACUC administrator may be responsible for acknowledging receipt of the concern and reporting on the outcome of the IACUC's investigation.

Surv. Which groups at your institution do your IACUC and IACUC Office regularly communicate with? Select all that apply.

• Principal Investigators and their research staff	175/179 (98%)
• Institutional Official	156/179 (87%)
• IBC	113/179 (63%)
• Environmental health and safety	138/179 (78%)
• IRB and human research protections office	31/179 (17%)
• Occupational health group	102/179 (57%)
• Veterinary clinical services group	111/179 (62%)
• Husbandry management group	108/179 (61%)
• Library staff	13/179 (7%)
• Grants office	91/179 (51%)
• Research sponsors	17/179 (10%)
• Regulatory or accrediting agencies	92/179 (52%)
• Public Affairs/Communication	32/179 (18%)
• Other institutional administrators and leadership	66/179 (37%)
• Legal counsel	50/179 (28%)
• Foreign and domestic institutions where collaborative research occurs	31/179 (17%)
• Information Technology	58/179 (33%)
• Conflict of Interest office	37/179 (21%)
• Other*	6/179 (3%)

*Other:

- Building and grounds/facilities personnel
- Data practices/public records/FOIA office
- Privacy officer
- Stem cell research oversight office

37:2 What are the shared administrative processes between the IACUC, IRB, and the associated offices?

Opin. The IACUC oversees the animal care and use program by reviewing and monitoring research involving animals, and the IRB is responsible for overseeing, reviewing, and monitoring research involving human subjects. Administrators supporting the IACUC and/or IRB may provide the following assistance in support of their relevant committee(s) and the research community:

- Writing policies and procedures specific to regulations and institutional requirements for submitting to the IACUC and IRB
- Assisting researchers in creating IACUC and IRB submissions
- Routing protocols through the workflow from submission to approval
- Sharing committee members, such as including medical doctors on the IACUC
- Participation in research administration training specific to IACUC and IRB processes

- Representing the committee as an initial contact person and subject matter and regulatory expert

Federal regulations do not explicitly outline administrative procedures commonly shared among institutional regulatory committees. However, based on the author's experience, the policies and standard operating procedures of these committees tend to follow similar processes for review and oversight. These policies, guidelines, and standard procedures, which pertain to committees like the IACUC and IRB, are often crafted using official institutional templates. This approach allows other departments, such as research compliance or legal, to incorporate language applicable to all regulatory committees.

Research handbooks, which function as reference materials for researchers engaged in animal or human subjects research, can serve as valuable resources. These handbooks can also assist administrative personnel as they aid researchers in preparing IACUC or IRB submissions.

Institutional websites for IACUC and IRB typically exhibit a consistent appearance and functionality.

One effective strategy for ensuring transparency in the review processes of regulatory committees is to make the workflow from protocol submission to approval publicly accessible through policies, websites, and electronic protocol review software dashboards.

Federal regulations and guidance documents define the minimum composition requirements for both the IACUC and the IRB. Additionally, both committees may engage consultants with expertise in specific areas to assist in reviewing protocols that require specialized knowledge. Members qualified to serve on both committees may be shared, unless they are nonaffiliate members whose sole affiliation with the institution is membership on a single institutional regulatory committee.

The initial and ongoing training of regulatory committee members can be similar, except for topics directly related to animal or human subjects research. Training on certain policies, such as institutional conflict of interest policies, is relevant to both IACUC and IRB members. Members of these committees should abstain from participating in the review of research in which they have a conflicting interest, except when providing information requested by the committee.

Administrative staff for the IACUC and IRB can serve as the primary point of contact for researchers looking to submit studies to each committee. They can also function as subject matter experts in regulations and submission workflows.

Surv. What, if any, similarities are there in your institution's procedures to submit research protocols to the IACUC and IRB? Select all that apply.

- Shared institutional online submission system — 62/178 (35%)
- IACUC and IRB staff reside in the same office — 55/178 (31%)
- Shared membership between committees — 31/178 (17%)
- Both committees report to the same institutional leader — 120/178 (67%)
- Administrative process for offices that support the IACUC and IRB — 51/178 (29%)
- Administrative responsibilities for the offices of the IACUC and IRB — 44/178 (25%)

37:3 What are the shared administrative processes between the IACUC, the IBC, and the associated safety offices (Environmental Health and Safety, Chemical Safety, Radiation Safety, etc.)?

Opin. Administrative processes shared among these groups may include the following:

- Writing policies and procedures specific to regulations and institutional requirements for submitting to the IACUC and IBC
- Assisting researchers in creating IACUC and IBC submissions
- Shared committee members including safety personnel on the IACUC and veterinarians on the IBC
- Participation in research administration training specific to IACUC and IBC processes
- Verifying assignments for review of hazard use on IACUC protocols to safety IACUC members or their associated offices (biohazards-IBC, chemical hazards-chem safety group, radiation hazards-rad safety group)
- Confirming approval by IBC and other regulatory and non-regulatory safety committees such as chemical safety and radiation safety

Federal regulations do not explicitly outline the administrative processes typically shared among institutional regulatory committees. Based on the author's experience, the policies and standard operating procedures of these committees tend to follow similar processes associated with review and oversight. These written policies, guidelines, and standard procedures related to the IACUC and IBC, and extending to other safety committees, are typically drafted using official institutional templates. This approach allows other departments, such as research compliance or legal, to incorporate language that applies to all regulatory committees.

Research handbooks, serving as reference documents for researchers engaged in animal research involving hazardous agents, can be valuable resources. These handbooks provide guidance for researchers, and administrative personnel can use them when assisting in the creation of IACUC or safety committee submissions.

Institutional websites for IACUC, IBC, and other safety committees typically maintain a consistent appearance and functionality.

To enhance transparency in the review processes of regulatory committees, it is highly effective to publish workflows from submission to approval in policies, websites, and on dashboards of electronic submission and review software.

Federal regulations and guidance documents describe the minimum composition requirements for the IACUC and IBC. While OSHA mandates a safe workplace, it does not specify the composition of safety committees.[2] Institutions usually have policies outlining the required membership. The IACUC and IBC may engage consultants with expertise in specific areas to assist or support in the review of protocols requiring additional expertise. Members qualified to serve on both committees may be shared, except for nonaffiliate members, whose sole affiliation with the institution is membership on a single institutional regulatory committee.

IACUCs are required to have at least one veterinarian experienced in laboratory animal medicine,[1,3,4] while IBCs require at least one member with expertise in animal containment principles for research involving use of recombinant or synthetic nucleic acid molecules in animals.[5] The IBC "animal expert" is typically a veterinarian who may also serve on the IACUC. IACUCs may also include members of the IBC who are experts in research involving hazardous agents.

The initial and ongoing training of regulatory committee members can be similar, including training on policies like institutional conflict of interest policies, which are relevant to both IACUC and IBC members. Members of both committees should refrain from participating in the review of research in which they have a conflicting interest, except to provide information requested by the committee.

While regulations require the IACUC Chair to assign designated member reviewers, IACUC staff typically assign protocols involving hazardous or biological agents to the appropriate IACUC safety member, who may come from environmental health and safety, radiation safety, or chemical safety. If animal experiments involve recombinant or synthetic nucleic acid molecules, IBC review should be performed. Institutions should develop policies outlining the workflow for submissions requiring review by both the IACUC and IBC and/or other institutional safety committees, with IBC approval obtained before IACUC approval. Regardless of the situation, research subject to institutional policies and federal regulations for both IACUC and IBC review should obtain approval before research initiation, with administrative staff from both committees serving as gatekeepers by verifying approval from their respective committees.

37:4 What are the shared processes between the IACUC and Embryonic Stem Cell Research Oversight (ESCRO) Committee?

Opin. It is the authors' experience that there are not usually processes shared between an IACUC and the institution's ESCRO Committee. That said, the IACUC may use their own review process as an opportunity to remind researchers of their institutional responsibility to seek ESCRO Committee review of any embryonic or human pluripotent stem cell use in animal activities. This could be achieved by simply adding contact information for the institutional ESCRO on the IACUC website, adding a question about ESCRO approval on the IACUC protocol form, or actively inquiring of the PI whether their research meets ESCRO oversight requirements and providing a direct referral to the ESCRO Committee. If the IACUC sees an increase in this type of research or has increasing concerns about the responsible use of such materials, they make consider inviting the ESCRO administrator to an IACUC meeting and/or appointing that individual as an *ad hoc* member of the IACUC.

37:5 What are the shared processes between the IACUC and Occupational Health Professionals?

Reg. PHS Policy (IV,A,1,f) requires a "health program for personnel who work in laboratory animal facilities or have frequent contact with animals."[1] The *Guide* (pp. 17–23) further reinforces the institutional obligation to provide an occupational health and safety program for individuals who are part of the animal program.[3]

Opin. The IACUC is not responsible for establishing or managing the institutional occupational health program (OHP); however, the administrative staff can play an important role in identifying individuals who should participate in the program because of working with or near animals. This can be managed in the context of protocol review, or some other periodic record review, during which individuals named as working with animals are flagged for enrollment in the OHP. If tied to protocol review, the IACUC should determine whether failed or expired OHP participation should hold up approval. An online system that integrates with training and enrollment records can facilitate the IACUC administrator's assessment of OHP participation.

The institution should determine OHP vaccine, surveillance, and screening requirements specific to different species and risk groups, including whether an individual can decline vaccinations and still be allowed to work with or be near a given species. In most cases, minimal requirements (e.g., Tdap, Hepatitis B) are intended for protection of the human worker, while, in other cases, such requirements serve to protect the animals from workers (e.g., regular TB screening). OHP clearance should, therefore, be inclusive of all species with which an individual works, and the IACUC administrative staff can verify this during the protocol review process.

37:6 What is the role of the IACUC with respect to the review of hazards?

Reg. The *Guide* (p. 18) states: "A comprehensive OHSP [occupational health and safety program] should include a hierarchy of control and prevention strategies that begins with the identification of hazards and the assessment of risk associated with those hazards."[3]

Opin. The PHS Policy[1] and AWAR[4] are silent on IACUC review of hazards; however, the PHS Office of Laboratory Animal Welfare (OLAW) requires compliance with the *Guide*, which includes a lengthy consideration of the institutional OHSP in identifying, assessing, and mitigating potential hazards in the work environment, including "experimental hazards such as biologic agents (e.g., infectious agents or toxins), chemical agents (e.g., carcinogens and mutagens), radiation (e.g., radionuclides, X-rays, lasers), and physical hazards (e.g., needles and syringes)."[4] The IACUC does not have to directly evaluate such risks, but iy should ensure that such evaluation is occurring, for example, by involving institutional safety officers in IACUC discussions and cross-checking submissions to other safety committees. The authors recommend that protocol submissions be screened (e.g., by the IACUC administrative staff) for inclusion of hazardous agents and the appropriate safety representative pulled into the review process as an ancillary reviewer with the opportunity to provide comments on location of procedures, appropriate PPE, waste disposal, and more. The IACUC should be reasonably assured that the responsible investigator is seeking appropriate approval for use of hazardous agents in the proposed animal activities and must determine whether IACUC approval will be contingent upon approval from other committees.

37:7 What is the IACUC's role in assessing and evaluating (e.g., semiannually) the safety program?

Reg. According to the PHS Policy, it is the IACUC's responsibility to semiannually review the institution's program for humane care and use of animals, and this process includes review of the safety program associated with animal research, i.e., the occupational health and safety program. Similarly, the AWAR require review of all aspects of the animal program which could be interpreted to include the occupational health and safety program.

Opin. How the review of the animal care and use program will be conducted is unique to each institution. No formal guidance is provided for the review. OLAW provides a semiannual review checklist as guidance for completing the review. The IACUC is responsible for assuring that a program is in place which covers all personnel who work in laboratory animal facilities. The following are sections that should be evaluated:

1) Changing, washing, and showering facilities are available as appropriate
2) Hazardous facilities are separated from other areas and identified as limited access
3) Personnel training is provided based on risk (e.g., zoonoses, hazards, personal hygiene, special precautions, animal allergies)
4) Personal hygiene procedures are in place (e.g., work clothing, eating/drinking/smoking policies)
5) Procedures for use, storage, and disposal of hazardous biologic, chemical, and physical agents are in place
6) Personal protective equipment (PPE) for the work area is appropriate and available
7) Program for medical evaluation and preventive medicine for personnel includes:
   a) Pre-employment evaluation including health history
   b) Immunizations as appropriate (e.g., rabies, tetanus) and tests as appropriate
   c) Zoonosis surveillance as appropriate (e.g., Q-fever, tularemia, Hantavirus, plague)
   d) Procedures for reporting and treating injuries, including accidents, bites, allergies, etc.
   e) Promotes early diagnosis of allergies, including pre-existing conditions
   f) Considers confidentiality and other legal factors as required by federal, state, and local regulations
8) Waste anesthetic gases are scavenged
9) Hearing protection is provided in high noise areas

10) Respiratory protection is available when performing airborne particulate work
11) Special precautions for personnel who work with nonhuman primates, their tissues, or body fluids include:
    a) Tuberculosis screening provided for all exposed personnel
    b) Training and implementation of procedures for bites, scratches, or injuries associated with macaques
    c) Personal protective equipment (PPE) is provided, including gloves, arm protection, face masks, face shields, or goggles
    d) Injuries associated with macaques are carefully evaluated and treatment implemented
12) Occupational safety and health for those involved with field studies is reviewed by EH&S Committee or Office

The IACUC should conduct a semiannual evaluation of the safety program, which includes reviewing associated policies and procedures through review at an IACUC meeting or IACUC inspections. Subcommittees within the IACUC can be tasked with reviewing specific safety-related aspects of the animal care and use program. The findings from these subcommittee evaluations can then be presented to the full IACUC for discussion during a convened meeting of the full committee or its semiannual program review meeting.

IACUC inspection teams may include representatives from institutional safety departments. including experts from environmental health and safety, chemical hygiene, radiation safety, and/or occupational health and safety, depending on the facilities inspected and anticipated research risks present. These experts are well equipped to assess potential risks related to personnel working in animal facilities or animal study locations. Occupational health personnel should play a role in conducting risk assessments related to medical evaluations, including appropriate immunizations, zoonosis surveillance, treatment of injuries or accidents, and managing exposure to animal allergens.

Decisions made by the IACUC regarding the evaluation of the safety program must be effectively communicated to the Institutional Official in the IACUC's Semiannual Report to the IO.

37:8 Is there a requirement for groups to coordinate approvals, e.g., IACUC and rDNA/IBC?

Opin. The IACUC is responsible for the oversight of all animal-related work. If biohazards and/or radioactive materials are to be used in the proposed animal research, then these hazards must be approved by the IACUC as well as by the relevant safety committee. The authors recognize that there are different approaches to addressing these other reviews (e.g., IBC, safety) as they impact the IACUC review and approval process. With semiannual review of the occupational health and safety program being a responsibility of the IACUC, the IACUC should ensure that precautions are taken with handling, use, and storage of hazardous agents relative to the animals as well as the personnel working with animals. Shared membership on both the IACUC and the IBC aids in the assurance that all the necessary precautions are reviewed and in place for protection of the animals and personnel. Communication between the IACUC and relevant safety committees is critical for compliance. Policies and SOPs may be created to define the workflow for approvals between committees, outlining the review of each other committee before the IACUC gives final approval or procedures put in place to receive information from the safety programs on specific guidelines for use of hazards.

The organization structure at each institution may differ with one single goal in mind: that safety from hazards are identified and mitigated for animals and personnel alike.

Surv. How does your institution ensure that biohazards or radioactive compounds are not used in animals without appropriate approvals?

• IACUC does not approve a protocol or modification describing a hazard to be used in animals without documentation of associated hazard approval.	129/178 (72%)
• The IACUC may approve the animal work but does not notify the PI of the approval until the associated hazard approval is confirmed.	24/178 (13%)
• IACUC protocol approval focuses on animal use only and the IBC and environment health and safety group monitors hazard use.	43/178 (24%)
• Through the continuous monitoring program of IACUC protocols.	45/178 (25%)
• Researcher communicates the plan to use biohazards to animal facility personnel.	48/178 (27%)
• Other.	10/178 (6%)

37:9 What are some successful communication strategies between these groups?

Opin. Effective communication has always been and continues to be a cornerstone of a successful animal care and use program. Communication between committees can take various forms, particularly when both committees operate under the same administrative umbrella with shared administrative staff or shared committee members. This setup allows for the exchange of information related to research, reviews, and approval processes, and committee members can serve as advocates to address any questions or concerns that may arise during committee meetings.

In the context of electronic systems, information can be routed to multiple committees for approval before receiving final approval from the IACUC, and integration between systems will further enhance communication. The use of each review system can be instrumental in establishing the appropriate workflow to ensure that the necessary information is provided and approvals are obtained. The specific review workflow that is implemented may vary depending on the size and structure of your institution, and it may incorporate one or more of the aforementioned processes to facilitate efficient communication and approval processes.

37:10 What types of communication should or must the IACUC have with Principal Investigators and the animal research community?

Reg. PHS Policy (IV,C,4) states: "The IACUC shall notify investigators and the institution in writing of its decision to approve or withhold approval of those activities related to the care and use of animals, or of modifications required to secure IACUC approval. If the IACUC decides to withhold approval of an activity, it shall include in its written notification a statement of the reasons for its decision and give the investigator an opportunity to respond in person or in writing."[1]

Opin. In addition to those notifications required by PHS Policy, the authors recommend regular communications to the animal research community regarding important programmatic issues (e.g., unavailability of centralized equipment or disease outbreak), new and updated policies impacting the animal program, and general recommendations about the IACUC review and approval process. Such communications may be distributed via an e-mail distribution list or closed listserv, posted to the IACUC or research compliance website, or via some other means appropriate to the program (e.g., newsletter, alerts). Messages should be clear and concise, with important information highlighted up front, to ensure it is seen. Researchers are often overwhelmed with electronic communications, so the IACUC should be mindful of the frequency and mechanism(s) by which communications are distributed.

With electronic IACUC systems, the notifications of meeting dates and other deadlines may be communicated via e-mail or alert systems. Using the best form of communication for your institution (alerts, e-mail notifications, newsletters, etc.) is key to keeping the IACUC and research community informed of current changes needed to stay in compliance.

37:11 What services can the IACUC provide to investigators to support their work?

Opin. The IACUC, either directly or through the administrative staff, is well -positioned to guide investigators on institutional requirements for animal care and use activities. Scientist and veterinary members may wish to meet with new investigators to discuss their research program prior to protocol submission in order to cover institutional practices, procedures, and policies. The IACUC staff can support such meetings by providing "how to" and "quick tip" type of guidance, for example, "The committee generally asks for more detail about XYZ." As an alternative to individual meetings, the IACUC may choose to develop a new investigator orientation, either as an in-person or as online training; prepare a protocol review cheat sheet; expand resources on their website; and more.

The IACUC is advised to periodically consider their local requirements (i.e., those that are implemented on top of regulatory and accrediting body requirements) and the value added of each. In some cases, value may not be apparent or may appear trivial when compared to the burden it causes. An IACUC that can self-reflect on the service it provides the research community and the institution as a whole is an IACUC that may be more inclined to modify overly bureaucratic processes to better support research while retaining a strong commitment to animal welfare.

Opportunities the IACUC office should be pursue to interface with investigators may include:

- Regularly scheduled office hours
- Office hotline (phone/e-mail/website)
- Scheduled seminars and presentations
- Surveys to solicit comments on IACUC performance
- Short video training sessions, for example on the following topics:
- Protocol writing, including sections that are typically challenging for researchers to complete
- Exceptions and exemptions to IACUC policies, procedures, and review processes

REFERENCES

1. National Institutes of Health, Office of Laboratory Animal Welfare. 2015. PHS Policy on Humane Care and Use of Laboratory Animals. http://grants.nih.gov/grants/olaw/.htm.
2. U.S. Department of Labor, Occupational Safety and Health Administration, Law and Regulations, Standards

3. National Research Council. 2011. *Guide for the Care and Use of Laboratory Animals. Eight Edition.* Washington, DC: National Academies Press.
4. U.S. Department of Agriculture, Animal and Plant Health Inspection Service APHIS 41-35-076 Animal Care: *Animal Welfare Act and Animal Welfare Regulations.* https: www.aphis.gov/animal-welfare
5. National Institutes of Health, Office of Science Policy. 2024. NIH Guidelines for Research Involving Recombinant or Synthetic Nucleic Acid Molecules (NIH Guidelines). NIH_Guidelines.pdf

38

Reducing Regulatory Burden

Sally Thompson-Iritani, Elaine Kim, and Aubrey Schoenleben

Introduction

38:1 What is "regulatory burden" and why should it be reduced?

There are two types of burden that need to be considered when we are discussing oversight of our animal care and use programs: compliance burden and self-imposed regulatory burden (**Figure 38.1**).

Regulatory burden can be considered the combination of compliance burden and self-imposed regulatory burden and is the sum of all administrative costs (e.g., time, employees, and money) related to maintaining regulatory compliance for a single organization. In general, regulatory burden increases in magnitude and complexity as an organization grows and expands its scope of activities (**Figure 38.2**).

It is important to state that when we discuss regulatory burden, the impact of regulatory burden, and how we might address it, we must be thoughtful about how we are supporting animal welfare and scientific inquiries. It should not be misconstrued that when we talk about decreasing regulatory burden that we are suggesting that there should be weaker regulations on how we care for our animals and support our science. It is quite the opposite. Our suggestion for decreased regulatory burden is to redistribute resources and focus those resources on infrastructure and increased personnel time to improve animal welfare and spur scientific discoveries (**Figure 38.3**).

Compliance Burden

To enforce laws and regulations, the regulatory agencies impose rules and regulations on how organizations operate their animal care and use programs. This compliance burden's scope has become increasingly complex over the years as societal interest in animal welfare has increased. Animal care and use programs, as well as the larger research infrastructure they reside within, must navigate federal regulations as they relate to not only animal research, but also areas such as environmental health and safety, sponsored research, and, more recently, foreign influence (Figure 38.4). To comply with the laws and expectations, there is an associated compliance burden. Initially this takes the form of understanding new rules and implementing policies and procedures that ensure compliance, e.g., all the must, should, and may tables that followed the 2011 edition of the *Guide* and numerous iterations of flow charts on how to determine what is a departure from the *Guide*. The associated compliance burden can be quantified as the administrative costs associated with maintaining regulatory compliance. Several governmental agencies are required to post the estimated burden (in hours) associated with a specific task or regulation, which can assist an organization in understanding this cost (**Figure 38.4**).

Self-Imposed Regulatory Burden

When organizations implement policies and procedures that go beyond the requirements set forth by regulating and accrediting agencies, these actions are considered "self-imposed regulatory burden." Although the term may be connotated with negativity, there may be particularly good reasons for an organization to have policies and procedures in place in areas that are not explicitly required, such as safeguarding institutional reputation, avoiding negative scrutiny from the public and other external stakeholders, or protecting staff and students from harassment or physical harm. One commonly referred to example of self-imposed burden is AAALAC accreditation. AAALAC accreditation is a voluntary activity for an animal care program, and many programs choose to be accredited to ensure that they are being recognized as a pillar program that is being regularly reviewed and evaluated by an external agency of experts in the field. Obtaining and maintaining AAALAC accreditation, however, necessitates a significant investment of resources. Another common example is seen in policymaking. Policies and procedures are often put in place as a reaction to a near miss or an actual non-compliance. It may be important for an organization to respond to this with additional oversight. A key concept is to evaluate each of the Animal Care and Use Program's policies regularly and determine how they are supporting compliance, the animals, the science, and the organization. Developing a checklist and ensuring that they are meeting certain criteria is one way to support an organization's endeavor to remain compliant and responsive to increasing pressures on our research community.

38:2 What are the negative consequences of regulatory burden?

There are significant challenges associated with regulatory burden that neither contribute to compliance nor benefit animal welfare and/or science. The animal care and use program

DOI: 10.1201/9781003272861-38

Regulatory Burden	
Compliance Burden- the cost of maintaining compliance.	Self-Imposed: impose rules beyond the requirements.

FIGURE 38.1 Two types of burden associated with oversight of the animal care and use program.

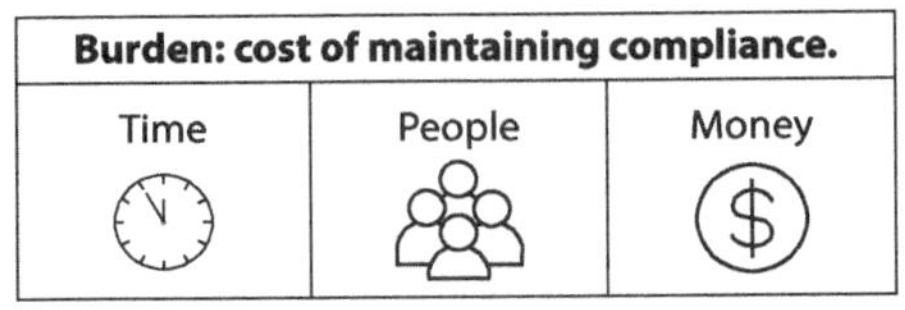

FIGURE 38.2 Regulatory burden is the sum of all administrative costs.

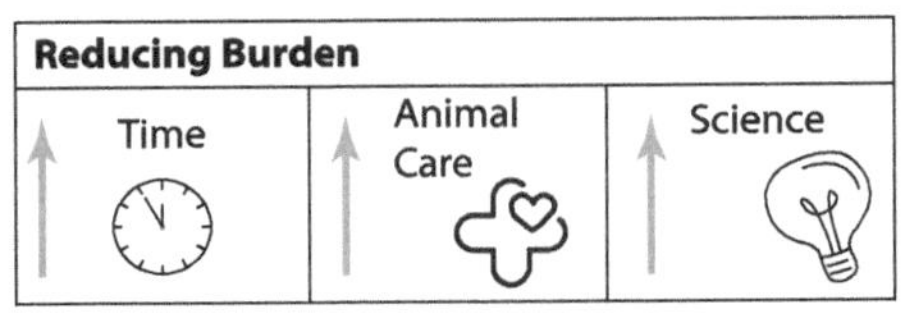

FIGURE 38.3 Reducing burden will support improved animal welfare.

oversight community needs to critically assess all activities to ensure a positive outcome associated with the established policy or procedure. As noted in Table 38.1, there has been a consistent burden on the scientific community since the early 2000s, and the most recent survey in 2018 showed that this burden is increasing despite efforts to decrease regulatory burden.

The 2018 Federal Demonstration Partnership (FDP) survey states: "Reducing the administrative workload associated with federally funded projects is one of the primary ways of increasing the **efficiency and effectiveness of research**."

Working with Regulatory Agencies

There is an opportunity to work with regulatory agencies to change compliance burden if it is shown that a regulation or policy is not supporting the goals as intended. The 21st Century Cures Act (2016) (govinfo 2016) directed regulatory agencies to work together and identify areas to decrease burden. Specifically, it mandates that "the NIH (National Institutes of Health) Director collaborate with the Agriculture Secretary and Food and Drugs Commissioner to complete a review of the policies and regulations for the care and use of laboratory animals. As appropriate, they shall revise and reduce burden placed on investigators, while protecting research animals and maintaining the integrity and credibility of research findings." The strategy regulatory agencies employed to review policy and the resulting actions are summarized in Section 38.4.

Balance

It is important to consider how best to balance compliance, animal welfare, and science. It is also essential to constantly rebalance to retain the "sweet spot" where risk acceptance/aversion is characterized in perspective with the need to move the science forward (**Figure 38.5**).

The examples in this chapter are intended to help provide you with tools to apply to the unique situations at an individual organization. No two organizations are identical, and it is important to establish who owns each process and how it is being assessed to ensure that it is meeting its originally intended purpose.

Survey. Does your institution require any of the following examples of self-imposed regulatory burden? Select all that apply.

- Assigning USDA (United States Department of Agriculture) pain/distress categories to protocols — 139/178 (78%)
- Reviewing most/all protocols by FCR instead of DMR — 54/178 (30%)
- Inspecting areas other than those used for housing or surgery — 125/178 (70%)
- Reviewing types of protocols not required by the regulations (e.g., invertebrates, observational field studies), etc. — 36/178 (20%)
- Literature searches for non-painful/non-distressful procedures or for non-regulated species — 72/178 (40%)
- IACUC review and approval of minor modifications (e.g., personnel, animal numbers, etc.) — 84/178 (47%)
- Comparing protocols to non-PHS funded grants (congruency review) — 34/178 (19%)
- Rewrite a protocol every three years — 128/178 (72%)

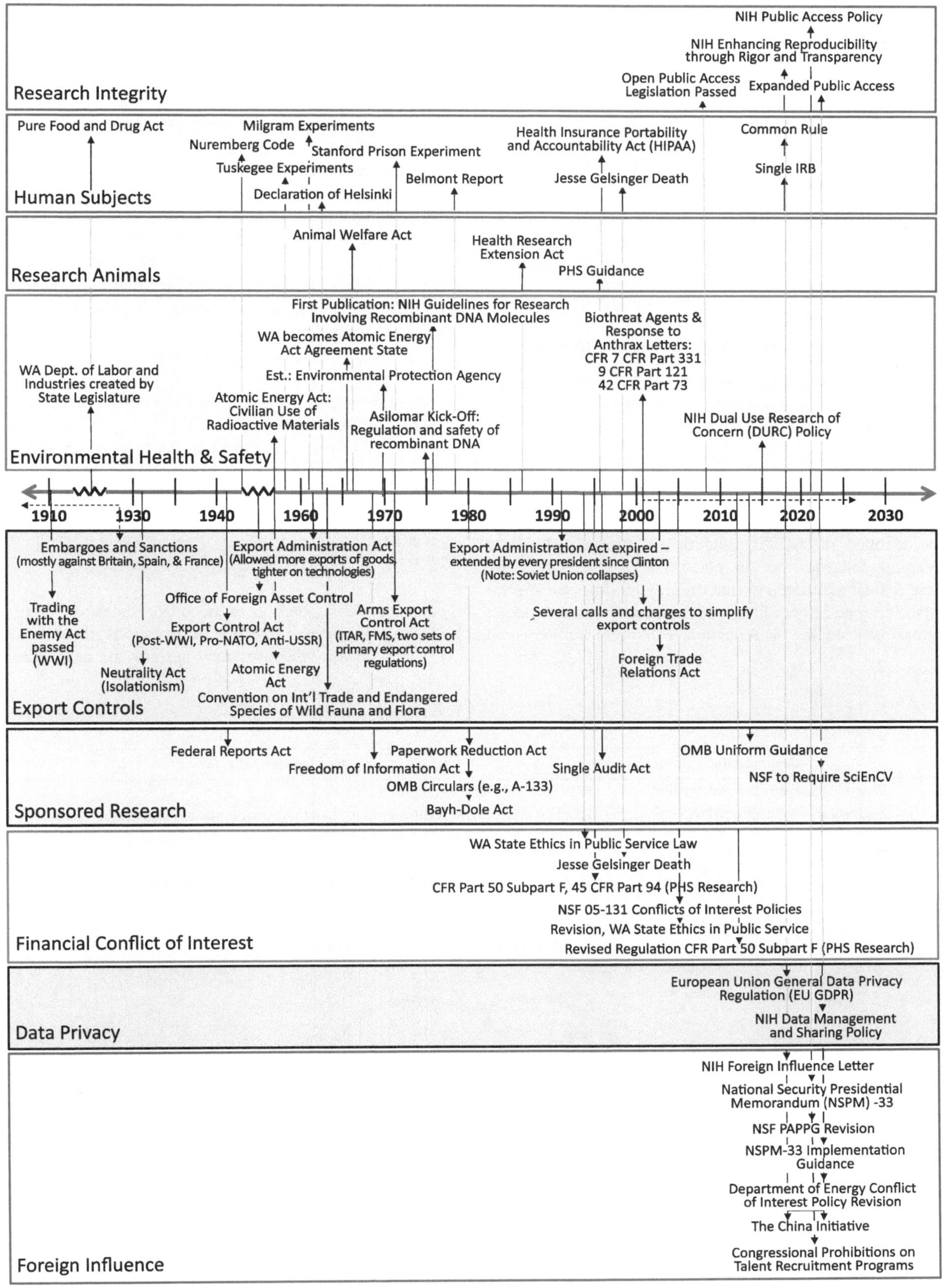

FIGURE 38.4 Federal Regulatory Landscape. The regulatory landscape that research organizations must operate within is complex and comes with a significant cost in terms of compliance burden. ©University of Washington.

- Annual reviews of protocols or submission of progress reports — 91/178 (51%)
- Inclusion of information in protocols not cited in the PHS Policy, Animal Welfare Act and Regulations, or the *Guide* — 53/178 (30%)
- Widely expansive PAM program and/or zero-tolerance approach to compliance — 18/178 (10%)

38:3 What are common examples of self-imposed burden at the programmatic level, and how can that regulatory burden be reduced?

Organizational Structure (Outlining the Burden)

The animal care and use program sits within a larger research infrastructure at an organization. When that research infrastructure is designed to be less flexible or has minimal risk tolerance, this may translate into more complex and time-consuming procedures to avoid compliance issues. For example, there can be multiple "hard stops" or approvals required prior to setting up funding accounts, obtaining animal facility access, or, once an IACUC protocol is approved, additional gatekeeping activities put in place where researchers are required to take additional online training modules for several reasons. These additional layers of complexity can counteract procedural efficiencies (or reduction of burden) implemented by one unit within an organization's larger system of research administration.

There is also the idea at the animal care and use program level that a committee such as the IACUC may be asked to weigh in on the type of research that should occur at the organization, which can be a slippery slope regarding committee purview and responsibilities. The federal regulations should be used as the starting point and foundation for any discussion regarding the IACUC's oversight and the approach to animal activities that the CEO and IO desire to implement at their organization. If there are additional duties that the IACUC is being asked to oversee, there should be clear and consistent documentation of this need, outlining what is being asked of the IACUC and why, acknowledging that the oversight is in addition to federal regulations.

Regarding animal activities in general—there is an ethical duty and there is also gauging the organization's risk aversion, opinion, and voice on certain types of activities where live animals are involved, and how various people in leadership positions choose to address animal activities and animal welfare.

Opin. Thus, there is a need for balance. Table 38.2 includes examples of questions that should be asked whenever new systems or procedures are

TABLE 38.1

Regulatory Burden for Scientists

Year	Organization	Burden (overall)
2007	Federal Demonstration Partnership	42% (Federal Demonstration Partnership n.d.)
2012	National Research Council (NRC)	Problem of excessive regulatory burdens on university research programs could cost "billions of dollars over the next decade." (National Academies Press. 2012)
2012	Federal Demonstration Partnership	42% (Federal Demonstration Partnership n.d.)
2013	Council on Governmental Relations' (COGR)	Ongoing increase in regulations affecting PIs and research institutions. (COGR 2013)
2014	National Science Board Report	42%
2018	Federal Demonstration Partnership	44.3% (Federal Demonstration Partnership n.d.)

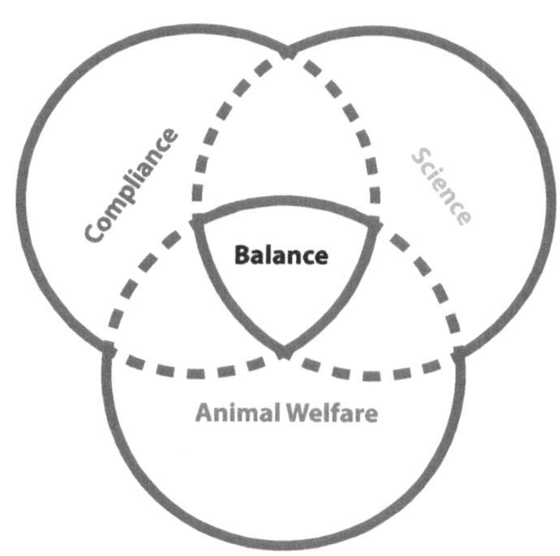

FIGURE 38.5 Balancing animal welfare, science, and compliance to determine the "sweet spot" where we can balance these adequately.

TABLE 38.2

Evaluating New Systems and Procedures

Questions to consider	Key Stakeholders (Academic)	Key Stakeholders (Non-Academic)[a]
Are the systems put in place benefiting animal welfare?	IACUC Administrators, IACUC Members,[b] Animal Care Staff, Principal Investigators (PIs), and Researchers.	Research Compliance Staff, IACUC Members, Animal Care Staff, Scientists, Research Associates
Are the systems/processes improving the quality of the science or activity (e.g., teaching)?	PIs and researchers, IACUC Members, Department Chairs, College Deans	Scientists, IACUC Members, Unit Leaders
Is the new system or proposed change required by regulation? Is it adding value (e.g., to animal welfare, the science, the organization)? Is there data to support the change? Is more documentation mitigating real risk?	Leadership and staff from key stakeholders (e.g., Sponsored Programs, EHS (Environmental Health and Safety), Training), relevant committee administrators (e.g., IACUC, IBC (Institutional Biosafety Committee), stem cells), IACUC Members, PIs, and researchers,	Research compliance staff, Grants/Accounting leadership and staff, EHS leadership and staff, any relevant trainers within these units
Does this system have a high barrier to entry, to pursue and continue research activities at the organization, thus contributing to a discouraging (or challenging) climate of research?	PIs and research staff responsible for doing this work; Executive Leadership at the organization, including the IO and CEO. Sponsored Programs leadership and EHS leadership responsible for various trainings and documentation (additional administrative burden on PIs and researchers)	Scientists, Unit Leaders (EHS, Grants/ Accounting), IO and CEO, Executive leadership at the organization
Does this system directly result in researchers and other personnel leaving the institution due to onerous research administrative processes?	Same as above, as well as faculty council representatives, College Deans, HR, IO, and CEO.	HR, Unit Leaders, IO and CEO, Executive Leadership

[a] There may be different names or organizational structures within private sector/industry compared to academic institutions; thus, this is included to provide alternative labels for similar staff/positions.

[b] IACUC Members here include all required members on the committee as stated in the regulations: scientist, non-scientist, nonaffiliated member, Chair, Attending Veterinarian.

considered for implementation in animal research administration. Leadership at an organization—not just the IO, but also the CEO and other executive leadership—should discuss these items. It is crucial that administrators, veterinarians, IACUC members, and other animal care and use staff be included in these conversations to provide their unique and valuable input on what is required to help animal research occur safely and to promote an effective and efficient system. We provide suggestions below for key stakeholders to include in each conversation.

Applying these questions to an example: Revising the IACUC Protocol Form. Who should be included in these discussions? To answer this question, examine which people and units are impacted by the protocol creation and review process (e.g., researchers, IACUC administrators, IACUC members, veterinarians). Consider the questions and sections included on the current form. Are they included based on regulations or because of the needs of various units (e.g., EHS)? Each of these stakeholders should be included in the discussion to determine which questions on the form or parts of the review process add value and which need revision or removal.

Grant Congruence and Inter-Institutional Agreements (Regulation)

The expectation for grant congruency is rooted in sponsor requirements, with NIH being the largest public funder of biomedical research. PHS Policy and the NIH Grants Policy Statement require awardee institutions to verify that animal activities described in a grant have been reviewed and approved by the IACUC prior to release of funding (see below for the specific requirements outlined in these policies) (NIH 2024).The purpose of this verification is to "ensure that public funds are used to promote the highest level of scientific integrity, public accountability, and social responsibility as reflected in humane animal care." Memoranda of Understanding between OLAW and other organizations, such as the National Science Foundation (NSF), Department of Veterans Affairs (VA), and National Aeronautics and Space Administration (NASA), extend these expectations to research funded by these organizations.*

- "Applications or proposals (competing and noncompeting) covered by this Policy from institutions that have an approved Assurance on file with OLAW shall include verification of approval (including the date of the most recent approval) by the IACUC of those

* The following entities are included within the scope of OLAW's purview: PHS (NIH, CDC, FDA), HHS BARDA, NSF, VA, NASA.

components related to the care and use of animals. For competing applications or proposals only, such verification may be filed at any time prior to award unless specifically required earlier by the funding component. If verification of IACUC approval is submitted after the application or proposal, it shall state the modifications, if any, required by the IACUC. The verification shall be signed by an individual authorized by the institution but need not be signed by the Institutional Official." (PHS Policy IV,D,2)

- "PHS awarding units may not make an award for an activity involving animals unless the prospective awardee institution … has provided verification of approval by the IACUC of those components of the application or proposal related to the care and use of animals...No award shall be made until.. .the institution or institutions have provided verification of approval by the IACUC of those components of the application or proposal related to the care and use of animals." (PHS Policy V,B)
- Recipient institutions must verify, before award, that the Institutional Animal Care and Use Committee (IACUC) has reviewed and approved the components of grant applications related to the care and use of animals. (NIH Grants Policy Statement 4.1.1.2)

(Outlining the Burden)

One source of administrative burden related to animal research and teaching within an academic institution is the federal requirement for institutions to perform grant-to-protocol congruency for certain sources of federal funds. An institution is responsible for this review process and despite this understandable language, some institutions either mandate that IACUC offices perform this review or the institution requires all funding, regardless of the source, to go through a grant-to-protocol congruency process. These processes may have been implemented in response to specific events where non-compliance was discovered or when a gap in communication resulted in a high-risk activity occurring without oversight or due to the lack of mechanism to accurately track the funding source(s) associated with a protocol. Thus, a new requirement was instituted to minimize risk and prevent future non-compliance. In these instances, it would be beneficial to gather relevant data, metrics, or other pertinent information within the organization's research administrative units to further investigate if the added procedures are truly having the intended impact or if there may be a more efficient and less onerous way to mitigate these organizational risks while improving programmatic efficiencies.

Similar processes and requirements may also be instituted within private organizations or companies, even if no federal funding is secured by the organization and all funding is secured by private investors. That may have a different layer of risk that the organization's leadership is trying to mitigate, related to concerns about public perception of animal research carried out by the company, or impact on investors funding future work. There is also the consideration of whether a company is private or publicly traded, which may impact the level of risk a company is willing to take in relation to animal research activities.

Opin. There are several strategies that institutions can consider when reducing self-imposed burden related to grant congruence. The first is to conduct congruency review only when required by the sponsor. Funding opportunities from sponsors, such as PHS, require institutions to verify that animal work has been approved before award. However, funding from many smaller or private organizations does not include this requirement. Furthermore, when sponsors do require a congruence review, institutions should conduct such a review only when stipulated. As an example, congruence review is required only at initial award for NIH grants. Ongoing review does not need to be performed with each annual progress report or as part of the triennial protocol review. Eliminating these unnecessary reviews significantly reduces burden without compromising animal welfare. It behooves the institution to develop a system to clearly identify which sponsored projects require congruence review and which do not. A tight partnership between the IACUC support office and sponsored projects office can facilitate such a system.

Second, introducing greater balance into the review process itself can help to reduce the time and effort required when conducting a congruence review. In response to the 21st Century Cures Act, OLAW has made a concerted effort to clarify expectations around an institution's responsibility for congruence review, highlighting several adjustments institutions can implement to make the process less burdensome (NOT-OD-22-005, NOT-OD-22-006). (NIH 2024) Most notably, a side-by-side review of the complete grant application and the protocol is not explicitly required. Institutions are encouraged to develop a procedure for evaluating key components of the grant, such as the research strategy or vertebrate animal section, to confirm consistency between the grant and the protocol. Also, only a brief description of activities in years four and five of a grant must be included in the protocol if the experimental details and procedures will be refined or amended later or during the three-year renewal. It is understood that research projects are dynamic and expected to evolve over their lifecycles as data is collected and innovative technologies and advances come online. There is no value in including a detailed description of studies anticipated toward the end of the grant. These experiments can be fully fleshed out and added to the protocol with an amendment once fully conceptualized. If there is a change in scope as the project evolves, it is expected that the PI (Principal Investigator) will talk with their Program Office or other Authorized Organization Representative (AOR). This requirement of the Grants Policy Statement should not be confused with the need for a congruence review. It is worth noting that there may be instances where congruence review will be required after initial award.

The institution may ask the IACUC office to support their researchers by reminding them to communicate with their AOR to comply with NIH requirements.

Last, there is value in exploring which unit within an institution is best poised to conduct a congruence review. Although the IACUC and IACUC support offices are most often tasked with this, it is not required. Confirmation of congruence is an institutional responsibility. When determining which unit within your organization should take on this responsibility, consider your institutional culture and structure, the expertise needed, the infrastructure of the institution, and the actual requirements of a congruence review. The questions provided in Table 38.2 can provide a starting point for these conversations.

Closely related to grant congruence review, inter-institutional agreements (or lack thereof) represent another area that is often ripe with self-imposed burden. Per the *Guide* (p. 15) (ILAR Committee 2011), collaborating institutions should have a "formal, written understanding" outlining each institution's responsibilities, including IACUC oversight, for the shared project. These written agreements most often take the form of a Memorandum of Understanding (MOU) or inter-institutional agreement (IIA). Having a well-constructed MOU or IIA template can be extremely valuable. Use of a template can help to ensure that standards are consistently applied across projects and eliminates the need for redundant reviews by both institutions. Key elements of a well-constructed MOU or IIA include specifying when each institution is responsible for (1) oversight of animal activities, including review and approval of related IACUC protocols and inspection of animal facilities, (2) reporting of adverse events or non-compliances, and (3) animal ownership. We also recommend that the MOU template addresses which institution will be responsible for congruence review of related grants. Specifically, we suggest that each institution performs congruence review for the work that is taking place at their facility, rather than requiring an exchange of protocols or amendments. Additionally, creating a template that is amendable can save time and effort overall. An amendable MOU allows institutions to negotiate the terms of their agreement once and is structured to cover all collaborative projects that occur between the two institutions. A list of covered projects can be maintained in an appendix that is updated and shared between the two institutions as new collaborations come online. The Federal Demonstration Partnership (FDP) recently developed a MOU template in support of these efforts (FDP 2023).

38.2 What are common examples of how IACUC oversight can contribute to self-imposed regulatory burden? What strategies can an animal program use to mitigate these?
IACUC oversight can also contribute to the overall regulatory burden within an animal care and use program. One common area where this may occur is the scope of the IACUC's purview as it relates to the type of animal activities that require review and approval. It is common for institutions to expand the committee's scope beyond what is required by federal regulations or oversight bodies.

Here are some examples:

- Review and approval of studies involving higher order cephalopods (oversight of IACUC is encouraged by AAALAC International, but not a regulatory requirement except for NASA), wildlife management activities, or observation only fieldwork (i.e., does not materially alter behavior of animal under study). Other agencies or groups providing oversight of wildlife activities include city, county, or state parks and wildlife agencies and federal agencies such as the National Park Service.
- Review and approval of studies involving privately owned animals enrolled in veterinary clinical studies, where scientific merit review and ethical oversight is provided by a veterinary clinical studies committee and state veterinary licensing board (and other accreditation groups). If the study is not supported by PHS funds, IACUC review is not required in addition to the clinical review (OLAW FAQ A7) (OLAW 2024).
- Review and approval of studies that use banked tissues (collected for purposes other than the proposed work), tissues collected from dead animals other than those covered by the AWAR, or embryonated eggs prior to hatching.

Programs with strict oversight of any research or teaching activities with animals or animal tissues or products have the potential to stifle science, innovation, and collaboration. When the administrative burden appears insurmountable, the animal care and use program has low risk tolerance, and there is a punitive approach to compliance, then the researcher is less likely to want to begin such an arduous administrative journey. Another common area that can contribute to programmatic burden is how the IACUC (and the organization) chooses to carry out their semiannual inspections, including determining which facilities/areas to inspect, e.g., housing facilities, animal study areas, satellite facilities, privately owned facilities that periodically house animals owned by the organization, etc. In this section, we will explore the burden associated with these two areas.

IACUC Purview

Reg. Both the PHS Policy* and the AWA (Animal Welfare Act)† define when an IACUC-approved protocol is required for animal activities. AAALAC International's Rules for Accreditation‡ also define which animal activities must be included within the IACUC's purview. IACUC approved protocols are required for:

- All PHS-conducted or supported activities that involve the use of any live vertebrate animals in research, research training, experimentation, biological testing, or related activities (PHS Policy II).
- Activities at a research facility involving any live dog, cat, nonhuman primate, guinea pig, hamster, rabbit, or any other warm-blooded animal, which is being used or is intended for use for research, teaching, testing, or experimentation (AWAR §1.1). Notably, the AWAR do not require IACUC review of activities involving birds, rats of the genus *Rattus*, and mice of the genus *Mus* bred for use in research; horses not used for research purposes; and other farm animals (e.g., livestock or poultry) used or intended for use as food or fiber, or livestock or poultry used or intended for use for improving animal nutrition, breeding, management, or production efficiency, or for improving the quality of food or fiber. Additionally, field studies are exempt from IACUC review under the AWAR. The AWAR define field study as a study conducted on free-living wild animals in their natural habitat, excluding any study that involves an invasive procedure, harm, or material alteration of behavior of animals under study.
- Though not a regulatory body, institutions accredited by AAALAC International are required to have an IACUC or similar oversight body. All animals used in research, teaching, or testing at accreditable units are included. This includes traditional laboratory animals, farm animals, wildlife, and aquatic animals, and nontraditional animals, such as invertebrate species when applicable.

Opin. In its simplest form, IACUC review and approval of animal activities would be limited to when it is required by regulatory agencies and oversight bodies. Under these circumstances, the IACUC would review research, teaching and testing activities that utilize only live vertebrate animals funded by PHS, NSF, VA, or NASA[4], or when those activities utilize USDA-regulated species. If an institution is AAALAC accredited, the scope of the IACUC purview is widened to include all vertebrate animals owned by the institution that are used for research, teaching, or testing, regardless of funding source. Refer to Chapter 2, "Circumstances Requiring an IACUC," for a more detailed discussion and examples of when a protocol is required. With the privilege of using animals comes an ethical duty to ensure good animal welfare that must be balanced against compliance and the performance of sound science. An institution must be thoughtful and intentional when extending the IACUC's oversight of animal activities beyond what is required, taking into consideration the impact of that change on the animals as well as the overall program for animal care and use.

As noted above, there are several animal activities that are explicitly exempt from IACUC review but are commonly included within the IACUC's purview for simplicity or ease of tracking, including observational field studies and the use of banked tissues. An alternate strategy for programs to consider is to create an institutional policy around these topics that is regularly reviewed and approved by the IACUC, outlining the institution's expectations, and providing guidance to researchers when working in these areas. For example, the IACUC could develop a policy outlining when a field study does and does not require an IACUC-approved protocol. Use of this strategy is supported by OLAW, as suggested in FAQ A.4 (OLAW 2024) on the use of embryonated eggs. When taking this approach, stakeholders impacted by the policy should be engaged to ensure that they understand the scope and motivation for the change. These discussions are also helpful for ensuring alternate points of view have been considered, clarifying points of confusion, and mitigating potential risks. Applying this to our field study example from above, an institution should connect with field researchers as part of implementing a new policy to ensure researchers understand when an IACUC protocol is and is not required, as well as defining a clear process for documenting this decision point (i.e., the IACUC determined that oversight is

* PHS Policy on Humane Care and Use of Laboratory Animals | OLAW (nih.gov) Accessed May 26, 2024.

† Animal Welfare Act | National Agricultural Library (usda.gov). Accessed May 26, 2024.

‡ Rules of Accreditation - AAALAC. May 26, 2024.

not required) to facilitate publication of research in academic journals.

Another approach programs might consider is borrowed from the world of human subjects' research. Human subjects' research is reviewed and approved by an Institutional Review Board (IRB), the human subjects' equivalent of an IACUC. However, research projects that pose no or minimal risk to subjects are considered exempt from IRB review (45 CFR 46.104). Exempt projects undergo an administrative review by IRB support staff. Institutions with a low-risk tolerance or that have a defined business need to document which research projects are exempt from IACUC review (e.g., to support publication of data generated from exempt projects) could apply a similar concept to animal research projects. Rather than requiring an investigator to submit a full protocol for exempt work, a brief description of the project (collected via e-mail or a short form) could be reviewed and approved by IACUC support staff or simply kept on file as notification of the work to be performed.

Semiannual Program Review and Facility Inspections

Reg. The HREA, PHS Policy, and AWA define which facilities within an animal care and use program need to be inspected and how frequently they need to be inspected.

Specifically:

- "Each animal care committee of a research entity shall review the care and treatment of animals in animal study areas and facilities of the research entity at least semiannually to evaluate compliance with applicable guidelines established under subsection (a) for appropriate animal care and treatment." (HREA Sec. 495(b)(3)(A))
- The IACUC shall review at least once every six months the institution's program for humane care and use of animals using the *Guide* as a basis for evaluation. (PHS Policy IV,B,1)
- The *Guide* (p. 6) defines the animal care and use program as "the policies, procedures, standards, organizational structure, staffing, facilities, and practices put into place by an institution to achieve the humane care and use of animals in the laboratory and throughout the institution."
- The IACUC shall inspect at least once every six months all the institution's animal facilities (including satellite facilities) where PHS-conducted or supported activities occur using the *Guide* as a basis for evaluation. (PHS Policy IV,B,2)
- The PHS Policy defines "animal facility" as "any and all buildings, rooms, areas, enclosures, or vehicles, including satellite facilities, used for animal confinement, transport, maintenance, breeding, or experiments inclusive of surgical manipulation (minor, major, survival, non-survival). A satellite facility is any containment outside of a core facility or centrally designated or managed area in which animals are housed for more than 24 hours."
- The AWAR require that the IACUC inspect, at least once every six months, all the research facility's animal facilities, including animal study areas, using the AWAR, 9 CFR Chapter 1, subchapter A – Animal Welfare, Parts 1–4, as a basis for evaluation. (AWAR §2.31,c,2)*
- The AWAR define "study area" as "any building room, area, enclosure, or other containment outside of a core facility or centrally designated or managed area in which animals are housed for more than 12 hours." (9 C.F.R. § 1.1)

Opin. The administrative burden associated with the conduct of semiannual facility inspections is significant (e.g., scheduling, visit coordination, effective follow-up, including resolution of inspection findings), and how an organization chooses to carry out these inspections can impact the degree of burden associated with this process. Notably, PHS Policy Footnote 8 and AWAR (§2.31c,3) allow IACUCs discretion in determining how best to conduct facility inspections. OLAW and USDA have codified flexibilities available to institutions when conducting inspections (NOT-OD-24-076). With regards to timing, inspections can take place within 30 calendar days of the six-month interval if there is no forward drift (i.e., inspections take place the same months every year, such as January and July). Staggered scheduling is also permitted, such that all facilities do not need to be visited on the same day or within the same month. A rotating schedule can be developed allowing for a manageable number of facilities to be visited within a given month so long as all facilities that must be inspected within the six-month period are visited during that time. Additional guidance was also provided with regards to who can perform the semiannual inspection. For housing or procedure areas used for species not regulated by USDA, an IACUC member or a qualified individual invited by the IACUC can perform the inspection. The qualified individual does not need to be an IACUC

* ac_bluebook_awa_508_comp_version.pdf (usda.gov). Accessed May 26, 2024

member or associated with the institution but must have a working knowledge of relevant regulations, including the *Guide*, to appropriately evaluate the facility.

Additionally, institutions have discretion with how some areas used for routine procedures such as weighing, dosing, and euthanasia are inspected. If used for non-USDA regulated species, these areas can be inspected on a random or fixed schedule. Institutions may also consider using a risk-based approach for inspecting these areas. For example, these low-risk spaces could be inspected annually if no deficiencies are identified. If a deficiency is identified, then the space's risk level increases, and the space is placed on a semiannual inspection schedule until that space can be maintained without deficiencies for a defined period. Alternatively, institutions can opt not to inspect these spaces at all if they do not meet the criteria defined in the regulations. Note, however, that OLAW expects some IACUC oversight of all areas where animal activities are performed at a frequency determined to be appropriate by the IACUC. These approaches may help reduce the administrative burden for IACUCs and investigators, without negatively impacting animal welfare.

Closely related to the facility inspections, the required semiannual program review is an evaluation of all policies, procedures, and systems that the institution has in place to support the humane care and use of animals. The program review is an ideal time for the IACUC to look at their program holistically to critically assess each of the components to ensure that they are working together to support compliance, the animals, the science, and the organization, and fine-tune the balance between these as needed.

Questions the committee may consider asking as part of this evaluation include:

- Do the policies and procedures in place add value to the program? How is value defined?
- Do they demonstrate a tangible benefit (to the animals, the organization, sound science)?
- Do they demonstrate an intangible benefit, adding to a healthy programmatic culture?
- Are there opportunities to combine or eliminate redundant policies or procedures, or to strengthen communication between units?

The regular collection of data or metrics related to research administration at an organization can aid an IACUC's assessment of its program and help it determine if the systems in place are providing the support that the program needs. This sort of regular, constructive assessment will help institutions retain the "sweet spot" between risk acceptance and risk aversion that supports strong animal welfare and allows science to move forward.

As suggested, developing a checklist is one mechanism an institution can use to ensure that they are performing an effective semiannual program review. NIH/OLAW provides a comprehensive checklist on their website to support program review and facility inspection. This checklist is editable and can be modified to support this more robust self-assessment. Like the semiannual facility inspections, it is prudent to take advantage of the flexibility offered by regulatory agencies with regards to timing. Semiannual program reviews can be completed within the same calendar month each year and can be completed on a staggered schedule. The review also does not need to be completed as part of a convened meeting or with a quorum. The IACUC can opt to complete the program review using subcommittees if the final report is available to all members of the committee for comment, signed by a majority of the committee members, includes any minority views, and is submitted to the IO.

38.5 What are common examples of how the protocol review and approval process can be burdensome? How can this burden be reduced?

(Outlining the Burden)

Many examples of protocol focused burden have been documented in published reports. Here we focus on protocol burden with the potential for broader, programmatic, and organizational impact. These examples can be used to start conversations to help committee members, IACUC administrators, and program leadership come to a better understanding of what is needed to cohesively support animal welfare and regulatory compliance, provide rigorous protocol review, and improve or support effective and efficient communication related to IACUC protocol review.

In this section, we discuss:

1) Mechanisms of protocol review
2) Frequency of protocol review
3) Protocol content

1. **Mechanisms for Protocol Review**

Reg.

The PHS Policy and AWAR stipulate the IACUC can use the following mechanisms to review proposed animal work:

- "Written descriptions of research projects that involve the care and use of animals shall be available to all IACUC members, and any member of the IACUC may obtain, upon request, full committee review of those research projects. If full committee review is not requested, at least one member of the IACUC, designated by the chairperson and qualified to conduct the review, shall review those research projects, and have the authority to approve, require modifications in (to secure approval), or request full committee review of those research projects. If full committee review is requested, approval of those research projects may be granted only after review at a convened meeting of a quorum of the IACUC and with the approval vote of a majority of the quorum present." (PHS Policy IV,C,2; AWAR 2.31,d,2)

Opin. For an IACUC to function effectively and efficiently, it must utilize the full range of approval mechanisms available to it. The IACUC can approve proposed animal activities or changes to animal activities via full committee review (FCR) or designated member review (DMR). OLAW stipulates that significant changes to the protocol, such as a change in study objectives or the degree of pain or distress, must be approved using one of these mechanisms. However, organizations have the flexibility to handle other types of changes administratively. Guidance from OLAW places these administrative changes into two categories:

(1) administrative changes that must be supported by IACUC-reviewed and IACUC-approved policy, and
(2) administrative changes that do not need to be supported by IACUC-reviewed and administrative changes that do not need to be supported by IACUC-approved policy.

Within the first category, the institution can develop a set of policies that will allow specific protocol changes to be made in consultation with a veterinarian authorized by the IACUC. This approval mechanism is called Veterinary Verification and Consultation (VVC). In this approval mechanism, the veterinarian "is serving as a subject matter expert to verify that compliance with the IACUC-reviewed and IACUC -approved policy is appropriate for the animals in this circumstance." Changes in animal activities that can be approved via VVC include changes in anesthesia, analgesia, and euthanasia methods. Importantly, veterinarians utilizing this method always have the option to refer the protocol change to the IACUC for review. See Chapter 11, "Amending IACUC Protocols," for a detailed discussion of VVC. Institutions can also develop an IACUC-reviewed and IACUC-approved policy allowing administrative handling of changes in previously approved animal numbers.

Within the second category, updates to personnel (other than the PI), contact information, or correction of typographical errors can be handled administratively without an IACUC-approved policy. There are often questions about the definition of the phrase "correction of typographical errors" and what types of errors may be corrected administratively. Most often, these changes are literal typos or grammatical errors in the protocol, but they can also be a change in protocol title or clarifying a response in the protocol that does not change the intent or procedure being performed. This is especially pertinent when the change was intended to be incorporated into the protocol prior to its approval after consultation with an IACUC member or the Attending Veterinarian. In general, a good approach to this type of administrative change is to ask if the typographical error changes the protocol significantly (which would need to go to the IACUC for DMR or FCR) or if the protocol change qualifies as a VVC (which would need to go to a veterinarian authorized by the IACUC).

Utilizing each of these approval mechanisms when appropriate can help institutions reduce the burden for their IACUC and investigators. It allows changes to research projects to be reviewed in the most efficient manner possible, supporting scientific progress without compromising the quality of review or animal welfare.

2. Frequency of Protocol Review

Reg. Regular review of ongoing animal activities is required by the PHS Policy and the AWAR. Specifically:

- The PHS Policy states that "the IACUC shall conduct continuing review of each previously approved ongoing activity … at appropriate intervals as determined by the IACUC, including a complete review… at least once every three years." (PHS Policy, IV,C,5)
- The AWAR state that "the IACUC shall conduct complete review of activities … at appropriate intervals as determined by the IACUC, but not less than every 3 years." (AWAR 2.31,d,5)

Opin. The continuing review of protocols—the annual renewal and triennial renewal—offers an opportunity for improvement from a burden reduction standpoint. Effective December 27, 2021, USDA replaced the requirement for IACUCs to review protocols annually with a requirement to conduct

TABLE 38.3

Summary of Burden Reduction Strategies

Topic Area	Burden Reduction Strategies
A. Animal Care Program	Programmatic risk assessment and aversion to risk discussions (less or more flexible, maintain rigor of review)
	Curate rather than collect institutional policies
	Balanced grant to protocol congruence review
	Use of standard MOU template
B. IACUC Purview	Intentional IACUC oversight
	Effective utilization of DMR, VVC, and administrative approval
	Risk based assessment of activities
C. Protocol Review and Approval	Eliminate annual review of protocols
	Complete review ≠ re-write
	Use of standard protocol and procedure templates

a complete review at least once every three years. This regulatory change was made in response to the 21st Century Cures Act and brought the AWAR into alignment with requirements of the PHS Policy. This change is intended to reduce the time and effort put into conducting annual reviews. However, elimination of this requirement at the institutional level may be more complex. For example, some institutions utilize the annual review process to track animal use activity, which is needed to complete required annual reports for USDA and AAALAC, collect information related to IACUC-approved enrichment exemptions, and receive updates on the status of active projects. It is recommended that the institution carefully evaluate the impact that this change could have on their program. Can other procedures be put into place to collect data from investigators each year related to animal use (e.g., electronic animal operations systems, annual surveys, post-approval monitoring visits)? Could enrichment exemptions authorized by the IACUC for USDA-regulated birds and primates be reviewed annually as part of a full committee meeting or as part of the semiannual program review? Key to this evaluation is weighing the time, effort, and resources required for any new procedures put in place relative to that currently dedicated to annual reviews.

Both the PHS Policy and the AWAR state that approved animal activities must undergo a complete review at least once every three years. Counter to widespread belief, a complete review does not equate to a complete re-write of the protocol. A protocol is best viewed as a living document that evolves and is modified alongside the research that it supports. At the time of triennial renewal, rather than re-writing the protocol, the investigator should review the current content and update it to reflect the lab's plans for the coming three-year period. This includes updating personnel, animal use locations, study plans, animal numbers, and literature search for alternatives, as applicable. Additionally, a summary of work completed during the triennial period and how the outcome of that work impacts plans for the next three years can assist the IACUC in doing a complete review of the protocol.

3. **Protocol Content**

As noted above, several published reports identify the preparation, revision, and review of IACUC protocols as a primary contributor to administrative burden. Protocol content offers institutions several opportunities to address this burden. We will discuss the literature search for alternatives, the protocol form, and use of standard templates and procedures.

The literature search for alternatives is a frequently cited source of burden in this area. Identifying and incorporating refinements and non-animal alternatives into a research program is a critical component of animal research. The literature search request included on many protocol forms, however, is often treated as a box checking exercise and does not provide substantive value to the investigator, the animals, or the science. Only the AWAR require the PI to provide "a written narrative description of the methods and sources, e.g., the Animal Welfare Information Center, used to determine that alternatives [are] not available" (9 CFR 2.31 (d)(1)(ii)). Thus, this specific requirement to include a narrative should be applied only to protocols that include USDA-regulated species, rather than all protocols. Generative AI (artificial intelligence) tools may provide an opportunity for balance in this area in the future by allowing investigators to efficiently summarize published literature relevant to their area of study, as well as identify possible alternative models and procedural refinements. The institution should look programmatically at how it can better identify and promote the use of available alternatives and refinements, encouraging their consideration earlier in the research lifecycle where they will be of greater impact.

The IACUC is also often viewed as a gatekeeper. In this role, the protocol form is used to collect information that is necessary (or desired) for related processes or partner departments, such as facility access or occupational health and safety review but is not actually required by regulation (refer to Chapter 8, "General Format of Protocol Form," for details on protocol form requirements). Although these other processes are important, when evaluating your program for burden it is essential to determine what pieces of information are being collected because they are required by regulations or have been identified as real risks by organizational leadership and what pieces of information are being collected to support best practices. For the latter, additional conversation and analysis with partner departments may be warranted to determine the rationale behind including that information on the protocol form and if there are other reasonable mechanisms to collect that data. OLAW has a sample study proposal template available at

https://olaw.nih.gov/resources/documents/animal-study-prop.htm

Standard Procedures

Use of standard templates or procedures is another mechanism that animal care programs can use to reduce the time and effort associated with the preparation and review of protocols, easing burden for researchers, IACUCs, and IACUC support staff. Creation of a standard procedure library allows the animal program to define the content of common procedures used by their research community (e.g., administration of anesthetics and analgesics, euthanasia methods), ensuring the procedures are aligned with federal and state regulations and institutional policy. Standard procedures can be drafted by veterinary or administrative staff and then reviewed and approved by the IACUC.

Once an institution has built a library of standard procedures, researchers can then insert those procedures directly into their protocol. This strategy saves researchers time during the review process by providing them with key elements of their protocol that are clearly written and at an appropriate level of detail. If the researcher's experimental plans do not allow for standard procedures to be used, the ability to use a similar procedure as a starting point will save them time when writing the protocol because they will have a better understanding of what procedural details are required in the protocol. Both contribute to a smoother review process, with less back and forth between the researcher, the veterinarians, IACUC, and IACUC support staff.

Using standard procedures saves the investigator time in preparing their protocol and the veterinarians, IACUC, and IACUC support staff time during the review process. Once approved, standard procedures can be treated like institutional policies or Standard Operating Procedures (SOPs) and be reviewed on a regular schedule (e.g., once every three years). In doing so, the specific details of a standard procedure do not need to be reviewed in depth each time it is inserted into a protocol; its use just needs to be evaluated within the experiment's context. This reduces the time required to complete administrative and veterinary pre-review and the effort required during committee review.

38:6 How does the 21st Century Cures Act contribute to reducing regulatory burden?

TABLE 38.4

Summary of 21CCA Planned and Implemented Changes

Burden Area	Burden Reduction Strategy
Agency Coordination	Use of AAALAC program description for portions of OLAW Domestic Assurance
	Implemented 60-day comment period for significant policy guidance
	Update OLAW disclaimer on policy guidance
	Coordination with BARDA, DOD (Department of Defense), FDA, NASA, NSF, USDA, and VA ongoing
Protocol Review	USDA removed the requirement for protocols to be reviewed annually, harmonizing expectations with NIH/OLAW
	Update guidance on use of DMR, DMR after FCR, and VVC
	Update guidance on animal activities that are exempt from IACUC review
	Update guidance on non-pharmaceutical grade substances
Registration and Reporting	Requirement to renew registration every three years was removed
	Timing for submission of annual reports to OLAW and USDA was harmonized (October 1–September 30); working to develop a shared submission portal
	Updated guidance on departures from the *Guide*
	Updated guidance on reporting non-compliances
Semiannual Inspections and Program Review	Updated guidance on flexibilities allowed for inspections
	Updated guidance on flexibilities allowed for program review
Training and Resources	Support industry led training resources, including IACUC best practices through the IACUC Administrators' Association and IACUC training through iCARE, SCAW (Scientists Center for Animal Welfare), etc.
	CUSP (Compliance Unit Standard Procedure) repository for standard procedures
	Universal Protocol Sample
	Updated guidance on grant-to-protocol congruence
	Updated guidance on use of zebrafish larvae

The 21st Century Cures Act (govinfo 2016) is a comprehensive set of legislation passed in December 2016 with the intent of facilitating biomedical research and drug development through the reduction in administrative burden for researchers. This legislation tasked the Director of the NIH to work cooperatively with the USDA and FDA (Food and Drug Administration) to review "all applicable regulations and policies for the care and use of laboratory animals and make revisions, as appropriate, to reduce administrative burden on investigators while maintaining the integrity and credibility of research findings and protection of research animals." In response, a working group was formed in 2017 with the goal of identifying "inconsistent, overlapping, and unnecessarily duplicative regulations and policies" and to take necessary steps to reduce burden associated with these regulations and policies, including improved coordination within and across agencies. The working group conducted an in-depth review of applicable laws, published reports and surveys, and gathered feedback at listening sessions from a wide range of stakeholders, including the regulated community and the public. The working group issued a preliminary report of proposed actions in March 2018, and requested additional feedback from all interested parties. A final report outlining the actions that the agencies would take to reduce administrative burden and improve coordination was published in August 2019. Table 38.4 summarizes the main action items to come out of the working group.

REFERENCES

COGR. (2013). Council of Government Relations. https://www.cogr.edu/

FDP. (2023). *Memorandum of Understanding (MOUs) for Collaborating Institutions.* https://olaw.nih.gov/resources/documents/IACUC_MOU

Federal Demonstration Partnership. (n.d.). *2007, 2012, 2018 Faculty Workload Survey. Federal Demonstration Partnership. 2020. FDP Foundation. Faculty Workload Survey - The Federal Demonstration Partnership* (thefdp.org) Accessed May 22, 2024. Retrieved May 26, 2024, from https://thefdp.org/demonstrations-resources/faculty-workload-survey/

govinfo. (2016). *21st Century Cures Act.* https://www.govinfo.gov/content/pkg/BILLS-114hr34enr/pdf/BILLS-114hr34enr.pdf

ILAR Committee (Ed.). (2011). *Guide for the Care and Use of Laboratory Animals.* National Research Council.

National Academies Press. (Ed.). (2012). *National Research Council, Global Affairs, Board on Higher Education, & Committee on Research Universities. Research Universities and the Future of America: Ten Breakthrough Actions Vital to Our Nation's Prosperity and Security.*

NIH. (2024). *NIH Grants Policy Statement.* https://grants.nih.gov/policy/nihgps/index.htm

OLAW. (2024). *FAQs: PHS Policy on Humane Care and Use of Laboratory Animals.* https://olaw.nih.gov/faqs#/guidance/faqs

39

Research Reproducibility and the Role of the IACUC

Esther J. Pearl, Mark J. Prescott, Nathalie Percie du Sert, and Penny S. Reynolds

Introduction

Institutional Animal Care and Use Committees (IACUCs) and equivalent oversight bodies, such as the United Kingdom (UK) Animal Welfare and Ethical Review Body (AWERB), the European Union's Animal Welfare Body (AWB), and Australia's Animal Ethics Committee (AEC), play a central role in ensuring appropriate standards of animal use at the institutional level. In addition to deciding from an ethical perspective whether proposed animal research protocols can be conducted and on what terms, the IACUC also has an important role as gatekeeper for building quality into experiments before animals are used. This helps ensure the research is rigorous and reproducible and data from animal research is useable, meaning animal lives are not wasted.[1, 2] Factors which affect rigor and reproducibility in bioscience research include the quality of experimental design (e.g., choice of model, use of appropriate controls, measures to reduce bias, and appropriate sample sizes), research practices, data analysis, and reporting—all of which IACUC members can scrutinize as part of protocol review.[3–5] The extent to which this information is reviewed, however, varies across committees. In assisting investigators to plan, design, conduct, and record research studies appropriately, the IACUC can also contribute to improve the reporting of studies in the published literature. Additionally, the IACUC's role in improving animal welfare and standardizing animal care also impacts the reliability and repeatability of *in vivo* experiments. Animals with compromised welfare show altered physiology, immunology, and behavior, which can lead to distorted results, unreliable conclusions, and unwanted variation in scientific output.[1, 6, 7] Most systems of animal research oversight globally require the IACUC, or equivalent, to conduct a harm-benefit analysis, weighing the potential benefits of the study against the animal welfare concerns, and to assess compliance with the 3Rs principles.[8–10] Research-funding bodies and accrediting organizations recommend oversight committees assess the rigor and reproducibility of proposed research protocols, although there are few explicit requirements in regulations for IACUCs to do so.[4] Rigor and reproducibility assessment is a key part of conducting harm-benefit analysis, helping to ensure minimization of both animal use and suffering, and increasing the likelihood that the research findings will truly add to the knowledge base and realize anticipated benefits.[11] In so doing, the IACUC can protect the reputation of the institution and those involved in the research. Finally, the IACUC is also uniquely positioned to influence the local research culture and practices, set standards, and ensure researchers are supported to do the best science. This includes disseminating resources developed to improve experimental design, reporting, and reproducibility, such as the Experimental Design Assistant and ARRIVE guidelines (Animal Research: Reporting of *in vivo* Experiments) from the NC3Rs (the United Kingdom's National Centre for the Replacement, Refinement and Reduction of Animals in Research).[12–14] This chapter outlines key considerations for IACUCs in their role in helping to improve rigor and reproducibility in animal research.

39:1 What guidelines are there regarding the role of the IACUC in promoting reproducibility?

The U.S. ILAR* *Guide for the Care and Use of Laboratory Animals* (*Guide*; p. 26)[8] notes that while responsibility for reviewing the scientific merit of proposed research normally lies outside of the IACUC, the committee members should evaluate scientific elements of the protocol as they relate to the welfare and use of the animals. Hypothesis testing, sample size, group numbers, and adequacy of controls are all given as examples that can relate directly to the prevention of unnecessary animal use or duplication of experiments. The *Guide* (p. 25) also stipulates that, whenever possible, the number of animals and experimental group sizes should be statistically justified in the protocol, for example by the provision of a power analysis.[8]

The most recent edition of the *Guide* was published in 2011.[8] In the intervening decade, awareness of poor reproducibility in bioscience research, its implications, and factors that can address the problem have increased dramatically. Accordingly, in 2021 AAALAC International (the premier body for accrediting organizations using animals in research, teaching, or testing worldwide) recommended that animal care and use programs should consider scientific rigor and experimental reproducibility when evaluating animal research activities.[4] Evaluation may be based on an initial review by a grants review panel, it may be assigned to the internal IACUC protocol review process, or, in some cases, it may be reviewed by both. In any of these scenarios, it was recommended that the IACUC review should confirm that the protocol contains pertinent study design elements and ensure adherence to the

* Now the Board on Animal Health Sciences, Conservation, and Research [BAHSCR] within the Division on Earth and Life Studies of the National Academies.

DOI: 10.1201/9781003272861-39

principles of the 3Rs. The review should include aspects such as the adequacy of the rationale for the use of animals, the purpose and design of the research, including randomization, masking (also known as blinding), and controls, and the justification of species and animal numbers proposed (e.g., via statistical power analyses, if appropriate). To increase awareness and competency in researchers, their staff, and IACUC members, the IACUC should confirm that the program provides education, training, and recommendations related to scientific rigor and experimental reproducibility.

In the United Kingdom the equivalent of the IACUC is the institutional Animal Welfare and Ethical Review Body (AWERB). Among other tasks, the AWERB reviews project license applications before these are considered for formal authorization by the national regulator, the Home Office. Broadly speaking, the AWERB's role in project review is to ensure that the project license application has been prepared to a satisfactory standard, consistent with local requirements relating to good practice in science, experimental design, animal welfare, ethics, and the 3Rs. Questions on fundamental aspects of good experimental design are included within the Home Office project license application form. Therefore, these aspects are commonly reviewed by AWERBs, together with justification for choice of experimental and control groups, measures to minimize bias, and determination of sample size. Guidance on principles of good practice for AWERBs from the Royal Society for the Prevention of Cruelty to Animals (RSPCA) and the Laboratory Animal Science Association (LASA)[15] notes that while not every AWERB may review each of these aspects in detail, the members need to feel comfortable that these issues have been addressed satisfactorily. For example, if there is a local statistician who has already reviewed the statistical design, then this may be considered sufficient.

Within the European Union, Directive 2010/63/EU on the protection of animals used for scientific purposes[10] does not mandate that the local Animal Welfare Body (AWB) assess scientific rigor and reproducibility. Emphasis instead is put on the competent authority carrying out the project evaluation having suitable expertise in experimental design and statistics to ensure animal numbers are minimized and the investigators themselves having been trained in the design of projects and procedures.[10, 16] Nonetheless, there is recognition from all stakeholders that assessing whether the proposed research plan meets contemporary quality criteria for experimental design and analysis is part of assessing the likelihood of it delivering robust findings and achieving the proposed benefits, which is fundamental to the harm-benefit analysis.[16–20]

39:2 What are the reasons for an IACUC to assess likely reproducibility?

Poor research reproducibility is an ethical, as well as a scientific, issue. Animal research is conducted with the understanding that present animal suffering should be mitigated as far as possible and may be offset by the potential for translation and future clinical benefit. However, recognition is increasing that animals are being wasted in poorly designed, inappropriately analyzed, and (ultimately) non-informative experiments. This must be considered as part of the harm-benefit analysis. The importance of experimental design and analysis has been emphasized in recent discussions of research quality criteria[21, 22] and assessment of research review processes.[22] If an experiment is not designed rigorously enough to yield reliable results, or if the methods and results are not published or shared fully and transparently, the research cannot be used by the scientific community and any benefit cannot materialize. Irreproducible research cannot contribute to improving the health of humans or other animals and therefore does not benefit science and society. Because the benefits do not offset the harms done to the animals, the research is unethical. Poor-quality research has been called out as fundamentally irresponsible, [23] negligent, [24] and unethical.[25] Therefore, as part of duty of care and ethical oversight, both investigators and the IACUC have the responsibility to ensure harms are minimized and benefits maximized by performing studies that meet the best-possible quality standards.

Poor quality research also involves costs to researchers, institutions, and funding agencies. The financial impact of poor data and irreproducible research have been previously estimated as exceeding US$28 billion in the United States (U.S.) alone.[26] Major funding agencies have been increasingly alarmed by the lack of translational success and reduced value for money resulting from poor quality research. To combat this, they now actively promote rigorous study design in funding applications. For example, over a decade ago the National Institutes of Health (NIH) recognized that key information indicating scientific rigor was missing from much of published government-funded pre-clinical research, and in 2012 presented core reporting standards for reproducibility.[27] The 2021 ACD-NIH working group report *Enhancing Rigor, Transparency, and Translatability in Animal Research*[5] highlighted the necessity for good study design, analysis, and results reporting. Recently, NIH has provided updated information on incorporating best practices for rigor and reproducibility into research plans and grant applications.[28] Likewise, the U.S. Department of Defense has endorsed both NIH core reporting standards and best practice reporting guidelines, such as the ARRIVE guidelines, to increase the rigor of experimental study design and reporting and to maximize reproducibility and translational potential.[5, 13, 14, 28]

Poor quality research erodes public confidence in the necessity for animals in scientific research, [29, 30] and poor research practice resulting in animal welfare violations has caused serious reputational damage to research institutions (e.g., [31–33]). A 2018 Pew Research Center poll indicated that just over half of the U.S. public opposes animals in research.[34] Recent Ipsos surveys showed that public acceptance of animal research in the United Kingdom is contingent on perceived purpose and welfare considerations.[35, 36] However, public perception that research animals are wasted results in increased calls to ban animal research outright or severely restrict animal use.[36]

A properly planned, designed, and reported study is a "force-multiplier." Good quality experiments enable rapid convergence on a solution, provide clear interpretable results, and dramatically reduce both the amount of testing and the resources required. Data that can be ensured to be reliable, valid, and reproducible has a longer shelf life. High-quality, well-reported data contribute reliable information to databases and systematic reviews, thereby informing further research. Findings from an informal survey of IACUCs conducted as part of this book are included in this chapter. They highlight the opportunity to improve practices and the scope for the IACUC to play a greater role.

39:3 What is the role of the IACUC in improving the reproducibility of research?

The IACUC can promote reproducibility throughout the research cycle. Good practice can be encouraged in the early stages during protocol review and introduced and reinforced by educational outreach.[37] Over the past decade, the perceived role of the IACUC has evolved beyond the traditional emphasis on animal welfare and compliance to include promotion of good science. The IACUC does not usually conduct peer review of the scientific and technical merit of research proposals. However, per the *Guide* and federal mandate,[8, 38, 39] the IACUC is responsible for evaluating scientific merit as it pertains to animal welfare and must incorporate 3Rs considerations into animal use protocol reviews. Institutions that place value on scientific excellence and "optimized animal model science" will promote reproducible studies and greater translational success.[1, 37, 40]

The 2021 ACD-NIH (Advisory Committee to the Director, National Institutes of Health) working group report[5] made no specific recommendations for the role of the IACUC in improving research quality. However, the report strongly recommended better pre-study planning to ensure that reproducibility is built into studies before they begin. IACUCs already have the obligation to review and confirm that the proposed research is scientifically and ethically justifiable, and that basic reproducibility items are addressed in the protocol. These include the choice and justification of the animal model, study and humane endpoints, experimental agents, animal care, housing and husbandry, analgesia, and euthanasia. In addition, personnel involved in the research must meet minimum training standards, and skill sets are expected to meet basic competency standards.[1, 8] The NIH has shifted the onus to the IACUC for animal numbers overview and justification.[39] Animal numbers justification is the top priority for research validity and reproducibility[41] and is a key feature of most IACUC protocols.

Harm-benefit analysis forms a key part of the ethical review of animal research protocols in countries around the world. AAALAC International expects this to be a primary consideration in IACUC protocol review, and the *Guide* outlines areas where the IACUC is "obliged to weigh the objectives of the study against potential animal welfare concerns" (p. 27).[8] In general, assessment of "harms" is chiefly the responsibility of veterinarians and animal care staff, and "benefits" rely almost entirely on self-reporting by investigators.[42] IACUCs rarely consider scientific importance of the research or the use of alternatives.[43] However, it has recently been suggested that review of harm-benefit criteria be extended to include scientific validity criteria such as study design and appropriate statistical analyses,[20–22] as initially proposed by Bateson.[44] The ACD-NIH working group report[5] prioritized the role of appropriate statistical study designs and development of investigator statistical competencies. A potential barrier is that relatively few IACUC committees include applied statisticians,[43] though PHS Policy (IV,C,3)[45] notes that the IACUC may invite consultants to assist in the review of complex issues, and committee members may consider themselves unqualified to assess statistical technicalities. However, the basics necessary for protocol review need not be highly mathematical and require only commonsense criteria for adequate evaluation.[42]

Educational outreach initiatives can also improve research reproducibility. Updating investigators with current best practice information can be done through existing (and free) online resources and tools, without adding to an already heavy administrative burden on the part of IACUC personnel. For example, the Federation of American Societies for Experimental Biology recommended that IACUCs and institutional animal facilities work together with professional societies to promote awareness and use of the guidelines from the NC3Rs (ARRIVE) and BAHSCR to improve the reporting and the reproducibility of animal research.[46] The ACD-NIH working group report also suggests award of non-monetary prizes or other forms of recognition to highlight exemplary work in study design and analysis.[5]

Surv. Does your IACUC regard assessing the strength of experimental design as part of its responsibility?

- No, it is outside of the IACUC's responsibility. 32/178 (18%)
- No, our researchers might think we are overstepping the IACUC's responsibility. 10/178 (6%)
- No, we do not have the expertise. 4/178 (2%)
- No, if the project is funded, we can assume the study design is appropriate. 18/178 (10%)
- Yes, good experimental design is important ethically as well as scientifically (so animals are not wasted on a study unlikely to yield reliable results). 114/178 (64%)

Surv. Does your IACUC educate researchers on best practice and guidelines for reporting animal experiments in the scientific literature such as Animal Research: Reporting of *in vivo* Experiments (ARRIVE)?

- No, it is outside our responsibility. 53/178 (30%)
- No, we do not have the knowledge. 13/178 (7%)
- No, researchers do not ask us for this. 62/178 (35%)
- Yes, it is part of training resources available to all our researchers. 35/178 (20%)
- Yes, it is part of training resources available to new researchers. 6/178 (3%)
- Yes, we highlight these resources to researchers, but only when their protocol indicates they need assistance. 9/178 (5%)

39:4 What key areas of expertise help an IACUC review the rigor of experimental design in protocols?

In the United Kingdom and European Union, knowledge, understanding, and expertise in statistics and experimental design are identified as key competencies to include within an AWERB or AWB and, accordingly, many involve a local statistician. This expertise is needed to advise adequately on the application of the 3Rs, and particularly to assess whether the number of animals is appropriate for the proposed study.[15, 44, 47] Including such expertise on the IACUC will ensure that researchers are not unduly pressured to reduce the number of animals to a level where the study does not have sufficient statistical power to produce reliable and reproducible results. This is also helpful to assess whether the design is efficiently tailored to the research question, if the statistical analysis plans are sound, and if there is appropriate accounting for sources of variability.[40]

If none of the committee members have statistical expertise, the IACUC can request that researchers seek external advice on the experimental design and statistical analysis plans, which can then feed into the IACUC's assessment. Of note, PHS Policy (IV,C,3) states that the IACUC may invite consultants to assist in the review of complex issues, and it seems that this may include a statistician.[45] Online tools to support researchers to design robust experiments, like the Experimental Design Assistant (EDA),[12] can also facilitate review of the experimental design. The EDA functionality includes a summary report of key design features, including experimental groups, sample size with justification, bias reducing measures, statistical analysis plans, and animal characteristics. This report can be submitted to the IACUC as supporting documentation to facilitate discussions with researchers.[48]

Surv. Does your IACUC include a statistician?

- No, we do not think we need a statistician. 31/178 (17%)
- No, we have not considered including a statistician. 63/178 (35%)
- No, we have had trouble recruiting a statistician but would like one on the IACUC. 51/178 (29%)
- Yes, we usually have a statistician (depending on their availability). 22/178 (12%)
- Yes, we make sure we always have a statistician. 11/178 (6%)

Surv. How does your institution train researchers and the IACUC in scientific rigor and reproducibility?

- No training is provided. 68/178 (38%)
- Our institution refers us to online resources. 46/178 (26%)
- Our institution requires completion of an online training module. 34/178 (19%)
- Our institution requires attendance at an in-person training lecture or webinar. 13/178 (7%)
- Other* 17/178 (10%)

*Other:

- Feedback is given at the time of protocol review, as necessary.
- Discussed during research orientation for new faculty.
- Departments are charged with training their faculty with respect to research rigor and reproducibility.

39:5 What are key ways to reduce bias and improve the reproducibility of research?

Bias threatens the validity of an animal experiment and whether its results will be reliable and reproducible. Bias is caused by inadequacies in the design, conduct, or analysis of the experiment, resulting in the introduction of error in the estimation of the effect of the intervention(s) the experiment was designed to test. Key measures to reduce the risk of bias include randomization, masking, and defining *a priori* inclusion and exclusion criteria, the primary outcome measure and the statistical analysis plan.[49] The role of these measures is briefly outlined here, and discussion of them in the context of protocols submitted for ethical review is addressed in subsequent questions in this chapter.

Using a randomization sequence to allocate group or treatment type to animals (or experimental units) reduces selection bias and the risk of systematic differences between experimental groups before the experiment begins. Randomization is also necessary to meet one of the key assumptions of the statistical analysis and allow inferential statistics (e.g., t-tests and ANOVAs) to be used.[50, 51] It is good practice to use randomization throughout the experiment to minimize other potential sources of bias such as the order of treatments, the order of measurements, or the position of cages on a rack. This ensures that any difference measured between the experimental groups can be attributed to the intervention being studied and not systematic differences in the way animals are treated, processed, or housed.

Masking (also known as blinding) is a technique to reduce performance and detection bias, which occurs when researcher expectations unconsciously influence their behavior toward the animals, the animals' behavior, and the results of the experiment. For example, knowing which animals have received a painful surgery might prompt researchers to handle these animals with additional care compared to the animals that received the sham surgery. This, in turn, might trigger differences in behavior between the groups or differential stress responses, which could be erroneously attributed to the intervention being tested. Another example would be when animals are injected with either a drug or saline; if the researcher does not manage to inject the entire contents of the syringe in one go, unconscious bias might prompt them to stop if they know the syringe contains only saline, whereas they would inject the remainder if they know the syringe contains an active drug.

Ideally, researchers should be unaware of the intervention(s) animals have received or which animals are in the same experimental groups from the start of the experiment until the data have been analyzed. This includes when animals are allocated to groups and the intervention administered, during the conduct of the experiment when the animals are handled and cared for, when measurements are collected, and during the data analysis. Sometimes it is not possible to mask every stage of an experiment, for example, when there are visible differences between the groups or when no additional staff are available to assist, but it should always be possible to conduct at least some of the stages masked. The importance of masking has been demonstrated in experiments using a range of animal species, including rodents, nematodes, and farm/agricultural animals.[52–55] Systematic reviews of animal studies published in the literature also show that experiments that make no mention of randomization and masking are more likely to report false positive results and large differences between experimental groups compared to more robustly designed studies that used randomization and masking.[56–58]

Other types of bias commonly seen in animal experiments include attrition bias and selective reporting. Attrition bias occurs when missing animals, samples, or data points create systematic differences between the experimental groups.[59] Selective reporting of outcomes or analysis, usually in favor of positive or statistically significant results, create a distortion in the published literature which overestimates the efficacy of treatments and hampers the translation of animal research into clinical benefits.[60, 61] Measures to mitigate against these risks include setting up precise inclusion and exclusion criteria in advance. These define what animals and/or samples are eligible for the study and the conditions that will need to be met for animals, samples, or data points to be excluded. Exclusion criteria may reflect the humane endpoints for the study but also encompass standards linked to quality control or data outliers.[14] Pre-specifying data analysis plans in the protocol ensures researchers have considered the variables and outcome measures needed to answer the research question. It prompts them to choose statistical analysis methods based on what is appropriate for testing the hypothesis and to calculate the appropriate sample size based on the changes expected in the primary outcome measure. Pre-specified data analysis plans demonstrate rigor, reduce the risk that different statistical analysis methods will be tried until a statistically significant result is obtained (this is known as p-hacking), and increase the likelihood that the most relevant findings will be reported as planned, regardless of the direction of the results.

The following questions in this chapter give more information about the above, except for considerations of sample size, which is covered in Chapter 14.

Surv. Does your IACUC regard reviewing the use of bias reducing strategies (e.g., randomization and blinding) as part of its responsibility? Select all that apply.

- No, it is outside the IACUC's responsibility. — 70/178 (39%)
- No, if the project is funded, we can assume the study design is appropriate. — 41/178 (23%)
- No, the protocol is an observational study so randomization cannot be performed. — 4/178 (2%)
- Yes, we expect protocols to mention what bias reducing strategies will be used. — 32/178 (18%)
- Yes, we expect detailed information on which steps will involve randomization and blinding (e.g., during allocation, when the intervention is given, during outcome assessment, during analysis). — 11/178 (6%)
- Yes, we expect detailed information on which steps will involve randomization and blinding, as well as how they will be implemented (e.g., block randomization, coding of individual animals, researcher is blinded when measurements are taken). — 6/178 (3%)
- Other* — 14/178 (8%)

*Other: used multiple strategies

39:6 What are key considerations when reviewing the use of randomization and masking in protocols?

While reviewing the design of animal experiments is not strictly within the remit of the IACUC, several relatively simple checks can be made to assess the quality of experimental design and prompt improvements to ensure animals will not be wasted in experiments that cannot yield reliable results. Randomization and masking are consistently identified as some of the top priority measures to use to improve the quality of animal research.[13, 27]

In the description of the experiment, a key piece of information to look for in the protocol or request from researchers is how different experimental groups (e.g., treatment and control) are allocated to the animals. A mention of animals being "randomized" or "randomly allocated" to groups is often insufficient. A true randomization is a formal procedure based on a computer-generated algorithm that cannot be influenced by the researcher, even subconsciously. This is not assigning animals to groups haphazardly or arbitrarily. Bespalov and colleagues have compiled a useful resource presenting different types of randomization, with practical examples and a list of tools for researchers to use.[51] An appropriate description of randomization should include how the random sequence will be generated (e.g., the randomization sequence will be created using the Experimental Design Assistant).

Descriptions of masking (blinding) in the protocol should include who will be aware of group allocations at different stages of the experiment.[62] Simply mentioning that masking will be used provides limited information, as masking is not a binary process. It may not be feasible to mask some parts of the experiment, and different strategies may need to be used throughout the experiment depending on the nature of the treatments or interventions the animals receive. For example, masking during data collection (i.e., during outcome assessment) may require an independent staff member to flip the labels on the cages so that the treatment and group information remains hidden while the researcher observes the animals or measures a particular behavior. Masking during the intervention may involve someone loading and coding syringes so that the researcher can inject the animals without knowing what dose is being injected. Plans for masking should consider the steps of the experiments that are at risk of subjective bias, the masking strategy to be used, personnel who will be involved, and those responsible for keeping the masking code secure so that the researcher can be unmasked at the end of the experiment. While masking increases the reliability of the results, it also increases the risk of error, which could lead to animals and resources being wasted; thus, it is important to check that the procedure has been carefully planned and de-risked.[51]

Surv. For some protocols blinding is not appropriate at all steps. What justification do you accept for not including blinding? Select all that apply.

- We accept any explanation a researcher provides for not blinding at allocation and when conducting the experiment, as long as the analysis will be conducted blind. 17/178 (10%)
- We accept any explanation a researcher provides, as long as they provide one. 29/178 (16%)
- The animals in the protocol are visually distinct (e.g., lean and obese mice) and analysis will be conducted blind. 7/178 (4%)
- The protocol is a dose escalation study, so all animals receive the same treatments in the same order and the analysis will be conducted blind. 8/178 (4%)
- The protocol is a longitudinal study, so measurements are taken at different times and the analysis will be conducted blind. 7/178 (4%)
- We do not expect researchers to justify whether or not blinding is to be performed. 106/178 (60%)
- Other* 24/178 (13%)

*Other: Noting that most studies conducted at that institution were observational and not conducive to blinding

As multiple answers were allowed percentages add up to more than 100%.

39:6.1 Can the IACUC insist that randomization and masking are used?

Although it is not within the remit of the IACUC to mandate use of bias-reducing measures, the IACUC can strongly encourage the use of bias-reducing measures. To lower the barriers to uptake the IACUC can fulfil an important educational role through promotion of training courses on this topic, making resources available for researchers to consult (see resources listed below in 39:20), and communicating the risks and consequences of not using such measures. Most funders and journals expect researchers to comply with the ARRIVE guidelines and report whether randomization and masking were used when the research is published.[5, 13, 63] Bringing up these issues with researchers at an early stage will help them to set up and conduct the experiment using rigorous methodology, which they are more likely to describe transparently in publications.

A common reason for not using masking in animal experiments is concern around the appropriate care for animals that have received a severe intervention. Animals in such experimental groups often receive differential care, with husbandry adjustments (e.g., wet food if the animals are expecting to have difficulties eating or chewing) and additional welfare checks compared to animals in control groups. This is problematic because a different diet or additional handling may create systematic differences between the experimental groups that are not related to the intervention being tested. As much as possible, it is recommended that all animals in the experiment receive the same husbandry and care to prevent the introduction of bias. It is therefore important for the IACUC to question and discuss the use of differential care between groups. In many situations, it will be possible to use masking and treat all animals in the same way. However, in some experiments, animals in one of the experimental groups may have transient welfare needs (e.g., following disease-inducing surgery) that would not be tolerated in control or sham-operated animals and may trigger humane endpoints. In these specific cases, the same care management plan cannot be applied to all groups and animal care staff must know the treatment each animal has received (or have ready access to this information), because understanding the effect of the intervention will be critical for care decisions. Risk of bias due to different handling must be balanced against the potential for additional harm to the animals if all staff are masked to group identity. In such circumstances, the role of the IACUC is to consider the options carefully, on a case-by-case basis, as part of the harm-benefit analysis.

39:6.2 What role can facility personnel play in implementing randomization and masking?

Implementing bias-reduction strategies such as randomization and masking often requires an extra pair of hands. Animal facility personnel can provide the extra support needed. If animal facility staff are involved in the study, for example, by recording clinical signs in animals, the researcher may need to conceal the intervention animals have received (i.e., the animal facility staff member is masked and the researcher is aware). In other cases, where the animal facility staff are not directly involved in the experiment, they could help implement the randomization and masking. They could randomize animals to specific interventions (based on a randomization sequence provided by the researcher) and code the animals so the researcher is unaware of which animals received which intervention. The "key" for uncovering what the code means (i.e., which intervention each animal received) should be kept by the person helping with the masking (in this case the animal facility personnel) until the study is completed. This allows the researcher to remain masked for the entire study and reduce the risk that subjective bias affects the findings of the study.

39:7 What are the key considerations when assessing if outcome measures are properly defined in a protocol?

"Outcome measures" usually refer to the dependent or response variables the investigator will measure to test the research hypotheses. The primary outcome is the most critical or important variable used to assess the effects of an intervention, and it is used to determine the sample size. Thus, the primary outcome measure needs to be identified clearly in the sample size justification.

Identification of what will be measured and how it will be measured is integral to descriptions of procedures/protocols. Therefore, all outcome measures must be clearly defined, specific, measurable, and relevant to the study hypotheses, aims, and objectives. Outcomes that are not clearly defined will be inconsistently measured and collected and will not provide reliable information. A working definition of all experimental outcomes to be measured must be provided, especially if the outcome has multiple definitions. For example, outcomes identified as "immune response" or "motor function" are too vague and contain no actionable information. Other outcomes such as "survival" have several competing definitions (e.g., an estimate of probability derived from counts and proportions or time to a pre-specified humane endpoint). The protocol must also contain a description of methods, how often outcome variables will be measured (once or many times per animal), frequency and timing of repeated sampling, and if sampling is invasive, terminal, or otherwise humane and acceptable.[42]

3:7.1 Can an IACUC insist on better defined outcome measure(s)?

Yes. The operative word here is "defined." The IACUC is mandated to review details of all procedures performed on animals, and procedures are dictated by what is going to be measured ("experimental outcomes"). How the primary experimental outcome is defined will also determine how effect size (and therefore sample size) will be estimated. "Unexpected" adverse outcomes must also be anticipated, and prospective plans put in place for early identification and mitigation, such as through the use of humane endpoints.

Surv. Does your IACUC assess whether outcome measures are defined properly?

• No, it is outside the IACUC's responsibility.	28/178 (16%)
• No, we do not have the expertise.	6/178 (3%)
• No, if the project is funded, we can assume outcome measures are defined properly.	32/178 (18%)
• Yes, but we do not do this systematically.	80/178 (45%)
• Yes, we have a system (e.g., checking that each outcome measure is feasible, specific, measurable, and quantifiable and that the timings and frequency of measurements are mentioned).	32/178 (18%)

39:8 How else can reproducibility be improved?

Both investigators and reviewers should be made aware of research strategies that improve overall performance standards and promote reproducibility. These strategies encompass *experimental design*, *study management*, and *animal health and welfare*.

Experimental design is more than the technical, methodological, and material descriptions routinely included in protocols. It also includes statistically based components that are not normally reported, such as correct specification of the experimental unit, relevant outcome measures, inclusion and exclusion criteria, and bias minimization methods (see question above 39:5). The design also encompasses descriptions of the groups being compared. Including controls in experiments allows researchers to tell if a treatment is having an effect, but only if the control is appropriate for the assay or experimental design. If the controls are not appropriate for the experiment the results could be inconclusive or even misleading.[42] Another key feature of statistically based experimental designs is the formal "structuring" of independent (or predictor) variables. Specific experimental designs account for diverse sources of variation efficiently and effectively and discriminate experimental signal from noise. Unfortunately, such designs are rarely described in the pre-clinical literature, although designs appropriate for animal-based studies have been available for almost a century. In practice, the majority of rodent model protocols describe multiple experiments involving two-group comparisons, which are more typical of clinical trials and usually inappropriate for assessing multiple interacting factors with small sample sizes.[12, 64–68] Design advice and templates are available for animal-based research that describe how to control experimental variation without increasing sample size, and incorporate design elements to increase validity.[12, 64–68] Using an appropriate sample size is also essential. Studies with inappropriately small sample sizes (i.e., underpowered studies) are not as likely to detect the effect of an intervention, even if the effect does exist. Small sample sizes can also lead to artificial inflation of the effect size, meaning that in studies that do detect an effect the results suggest that the effect is larger than it really is. Incorrect identification of the experimental unit (the entity exposed to the intervention independently of all other experimental units) can lead to unintentionally underpowered studies.[69] Conversely, overpowered studies waste animals and may give statistically significant results for differences with no biological relevance. For more detailed information regarding appropriate sample sizes, see Chapter 14.

Study management incorporates both planning and operational logistics. Protocols should present a focused blueprint of work that could realistically be accomplished in the given time with resources, budget, and personnel actually available and have best-practice care and welfare plans in place. Investigators should be discouraged from presenting large "wish-lists" of everything they hope to accomplish. Poorly planned and undirected study management can introduce additional uncontrolled and often unrecognized variation into experimental processes and seriously jeopardize research quality and reproducibility. In addition, there are serious legal issues involved with protocol non-compliance. In the U.S., deviations from protocol are reportable to the NIH Office of Laboratory Animal Welfare (OLAW). Common causes of protocol non-compliance includes include failure of investigators and technical staff to review and adhere to all procedures described in their protocol, conduct of studies where animal-based activities have not been described and approved, failure to provide appropriate post-procedural care and monitoring, and inadequate documentation. Variation in operator performance can be reduced by training all personnel to predetermined best-practice standards, frequent assessment and evaluation of compliance with simple relevant performance metrics, and incorporating ongoing quality control measures (such as checklists and visual aids) into the experimental workflow.[70] Animal models used for translational research must be adequately and routinely characterized and documented, as inheritance of genes associated with specific diseases may be affected by genetic drift, quiet mutations, environmental factors, and functional variants (e.g., [71]), and numerous sub-strain differences have been reported (e.g. [72]). Misidentification, mislabeling, and/or cross-contamination of mouse strains and cell lines[73, 74] are additional sources of management failure. Investigators should be prepared to include genotyping and cell line authentication into the protocol work description and thoroughly document all strain- and cell-related authentication procedures.[74]

Animal health and welfare. Confounders resulting from overlooked factors in animal signalment, care, and husbandry can induce phenotype changes in study animals, contribute to unpredictable effects on experimental results, and obscure true treatment effects. Rodents are especially sensitive to even subtle changes in environmental factors, such as social grouping, food, bedding, temperature, photoperiod, and circadian cues; such factors account for many irreproducible rodent experiments.[75] Pain management and the minimization of pain and distress are ethical imperatives but can pose methodological challenges. Unrelieved, unidentified, or inadequately treated pain, suffering, and stress are additional confounders of study outcomes. Protocol reviewers should insist on prospective comprehensive welfare plans that include recognition of pain and pain-associated behaviors, best-practice pain palliation measures, choice of analgesic agents, and clearly defined humane endpoints, with criteria for deployment.[76] There should be a very high scientific and ethical bar for withholding analgesia. It is not sufficient or scientifically justifiable for the investigator to opine that "Analgesia might affect experimental endpoints." This statement is ethically and scientifically problematic as often there are analgesic agents that can be given without affecting the specific experimental outcome(s) and the effect of unrelieved pain can itself be a confound.[77]

39:8.1 What role can animal facility personnel play to improve reproducibility?

Animal facility personnel can play a key role in introducing changes that improve reproducibility. The ACD-NIH working group report recommends increased liaison between investigators and institutional animal resource specialists and greater awareness of institutional practices involved in the delivery of animal care.[5] Facility personnel can assist with randomization and blinding (see Section 39:6.2), but more crucially assist in educating and training researchers in refinement techniques that improve animal welfare. Animals experiencing poor welfare may have altered behavior and physiology, introducing unwanted variation into experimental outcomes that may confound the ability to quantify changes in the biological parameters under investigation. Conversely, there is abundant evidence to show refinements improve, rather than compromise, internal validity. For example, traditional methods of picking up laboratory mice via the tail cause high aversion, high anxiety, and stress, which is avoided when non-aversive methods such as a handling tunnel or cupped hands are used instead.[78, 79] Rodent lifespan is shortened, and health and welfare adversely affected, by "standard" housing conditions that lack enrichment or opportunities for normal activity and behaviors,[80] whereas enriched environments improve animal well-being without compromising data.[81] Incorporating non-aversive handling and "planned heterogeneity" with enrichment significantly reduces both stress and extraneous and uncontrolled variation, thus promoting more reliable results.[82–84]

39:9 What are the key considerations when assessing the appropriateness of control groups/animals?

"Controls" are used to establish cause and effect relationships between interventions and responses. The effect of the test intervention is assessed by comparison of the mean differences between treatment and control groups. Controls provide a standard of comparison for evaluating if results can be attributed directly to the independent (or predictor) variables or are instead an artifact of some unknown confounder or time-dependency in the data. Controls therefore provide a method to distinguish experimental signal from noise. To minimize bias, experiments with controls should also include formal randomization and masking (see above).

Johnson and Besselsen[85] identify five types of controls commonly used in animal research: positive, negative, standard, sham, and vehicle. "Positive" controls assess validity of the test, that is, if the response can actually be measured and distinguished from background noise. "Negative" controls assess the magnitude of response that would be expected if the test intervention had no effect. "Standard" controls are procedures or agents for which the effect is already known and verified ("standard of care"), to be compared to the new intervention. "Sham" and "vehicle" controls are both a form of negative control. "Shams" are used to discriminate systemic effects induced by the test intervention from those resulting from procedures not of direct interest, but which are necessary to create the condition of interest. An example is when surgical procedures are required for instrumenting the animal so measurements can be obtained before the test intervention is administered. The sham control enables discrimination of inflammatory and physiological responses due to the surgery itself from the physiological effects of the test intervention. "Vehicle" controls assess potential physiological effects of the excipient, carrier, or solubilizing solution in which an active agent is suspended. Finally, studies can be designed so that each animal serves as its own control. Repeated-measures designs (paired, before-after, crossover, longitudinal) have more power than parallel arm trials because they account for variation due to individual differences and therefore minimize bias and variance.

Concurrent controls are the most relevant comparators for determining treatment-related effects and most appropriate for ensuring validity. Historical controls (i.e., non-concurrent controls for experiments completed in the past) have been suggested as a means of reducing the numbers of animals. Historical data can be useful for informing the planning and design of new experiments; for example, for variance and baseline mean estimation or calculating Bayesian priors for a plausible range of effect sizes.[86, 87] They are also useful in oncology studies for interpreting rare tumor incidence, severity, and/or proliferation patterns, or identifying evolving trends in tumor biology and behavior.[88] However, for most animal-based studies, comparison of new experimental data to historical controls can introduce serious biases. First, randomization is essential for valid statistical inference, but it is extremely difficult or impossible to incorporate historical controls into randomization plans. Second, historical controls will be appropriate only if stringent "Pocock criteria" can be met to ensure that data are comparable.[89, 90] For historical controls to be valid in animal studies, all experimental conditions (animals, animal housing and husbandry, experimental procedures, study design, endpoints, and methods of measurement) except for the treatment to be tested must be identical to the historical condition. However, these restrictions are unlikely to be met in practice.[86] In reality historical controls will be subject to sources of variation other than those found in the new experiment, including changes in animal husbandry, laboratory techniques, personnel and personnel skill levels, strain, vendor or source, animal eligibility criteria, routes of agent administration, chemical reagents, and suppliers. Therefore, historical controls are not recommended for most laboratory experimental studies.

Choice of control strains must consider both environmental and genetic backgrounds. There is increasing awareness that animals maintained under "standard" non-enriched housing and husbandry conditions will not be appropriate controls as compromised physical and behavioral health make them metabolically morbid and more susceptible to disease than controls raised in housing with appropriate environmental enrichment.[80] Genotype and genetic background can profoundly affect results and interpretation of results in comparative studies. For comparator strains, genetic heterogeneity in both outbred and inbred strains may occur from genetic drift and spontaneous mutation,[71, 72] poor breeding colony management, and/or variation in the environments in which animals are bred and then subsequently transferred.[91, 92] Differences between vendors, breeding locations, and even barrier facilities can profoundly affect both genotype and phenotype and confound both experimental results and attempts at replication.[92]

Surv. Does your IACUC check that experimental controls are included in the protocol? Select all that apply.

• No, it is outside of the IACUC's responsibility.	20/178 (11%)
• No, we assume that the researcher knows this best.	21/178 (12%)
• No, we do not have the expertise.	2/178 (1%)
• Yes, we check that at least one control group is mentioned.	47/178 (26%)
• Yes, we check that it is clear what kind of controls are in the protocol (e.g., positive, negative, animals as their own control).	94/178 (53%)
• Other*	10/178 (6%)

*Other: Respondents noted that some protocol reviewers will, on occasion, request information regarding experimental controls.

39:9.1 Can an IACUC insist on inclusion of control groups/more appropriate control groups?

It is part of the IACUC ethical oversight role to assess "adequacy of controls" (*Guide*, p. 26),[8] although there are no specific directions for doing so. The IACUC cannot mandate choice and type of controls, but it is certainly well within its remit to query any unusual numbers or omissions. Assessment of "adequacy" will be based on ethical considerations for animal numbers and groups. On the one hand, failure to include appropriate controls may result in an inconclusive experiment, which will essentially waste all the animals used. Conversely, requests for large numbers of animals in unnecessary or redundant control groups are equally wasteful and should be discouraged (see Section 39:9.2).

39:9.2 What are some situations when control groups are not necessary?

Some sort of comparator is almost always required in experimental studies. Exceptions to inclusion of a control condition are untreated iatrogenic injury (i.e., injury caused by a procedure rather than the disease, such as a surgical site infection) or disease when spontaneous recovery is not possible, and certain early-stage pilot or proof of concept studies. Sham surgery controls may be required if the chances are high for biased experiment management or assessment (if masking is not possible); otherwise, all effort should be made to standardize procedures to a uniformly high level of consistency and competence. Sham controls will be unnecessary for acute, well-established procedures with highly predictable experimental endpoints from which spontaneous recovery is not possible.[86] Examples include untreated spinal cord crush or transection,[86] severe hemorrhagic shock,[93] or sepsis.[94] Redundant controls will occur if studies are designed as numerous two-group trials, each with its own "control." This approach is inefficient, has low precision, and cannot determine any potential interactions, such as synergisms or inhibitory effects which are usually of most interest. The result is the waste of large numbers of animals for limited information. More appropriate designs are available and well described that enable screening of multiple predictors with small sample sizes.[64, 65]

Operational pilots used to determine, standardize, and finalize details of logistics, procedures, methods, resources, and processes will usually not need a comparator group (and may not need any animals at all). However, pilots designed to generate preliminary data for subsequent studies must include controls (and if possible be randomized and masked) to strengthen the evidence for or against proceeding with the investigation. It is strongly suggested that descriptive or observational cohort or cross-sectional studies include comparators or comparator groups for meaningful comparisons. These will consist of subjects matched on various attributes, for example, age, sex, and sample collection times. Study designs where animals act as their own control may not require a separate control group (see [64–66]).

39:10 What are the key considerations when assessing if protocols should include males and females?

Most animal research has been conducted on male animals. Results found in males are not necessarily applicable to females, meaning we have incomplete knowledge of many diseases and biological functions. Researchers using only one sex often erroneously assume their results also apply to the other sex, when, in fact, they can only draw conclusions for the sex they included in their study. Single-sex studies also miss any sex-based differences that might exist in biological functions, disease processes, and treatment responses.

To reduce sex bias in research the NIH now suggests (NOT-OD-15-102) that sex as a biological variable will be factored into research designs, analyses, and reporting in vertebrate animal and human studies, unless there is strong justification to perform a single-sex study (https://grants.nih.gov/grants/guide/notice-files/not-od-15-102.html). This does not necessarily mean researchers need to specifically look for sex differences. Instead, it is recommended that both sexes are included to ensure results are generalizable across both sexes. Sex differences are common—this was demonstrated recently in a large phenotyping study of over 2,000 mutant mouse lines which found that sex differences existed in phenotypes across a broad range of organs and tests.[95] There are also numerous studies where the sex of the animal affects key outcomes.[96–99] Whether or not researchers are specifically looking for sex differences, they must include how they will account for sex in their analysis plan and justify the sample sizes accordingly.

39:10.1 When is it appropriate to use only one sex in a protocol?

There are cases where it is appropriate to only use one sex in studies. This includes studying diseases that only occur in one sex, such as ovarian cancer, or biological processes that are sex-specific, such as pregnancy, embryo implantation, or sperm motility.[100, 101]

Single-sex studies can sometimes also be justified for practical and logistical reasons, such as studying rare conditions where samples are hard to get. These justifications should be assessed on a case-by-case basis. When only one sex is being used for practical or logistical reasons, researchers should be aware that their results may not generalize to the whole population, and they can draw conclusions only about the phenomenon in the sex they are studying.

Stating "there is no sex difference" (with or without evidence) is not an appropriate justification for a single-sex study. Without a strong justification both sexes should be used.

Surv. Does your IACUC request justification if only one sex is being used?

- No, it is outside of the IACUC's responsibility. 29/178 (16%)
- No, we do not have the expertise if doing so is essential to the experimental paradigm. 2/178 (1%)
- No, if the project is funded, we can assume using only one sex is appropriate. 27/178 (15%)
- Yes, we request justification but do not specify details. 41/178 (23%)
- Yes, we request justification that is in line with NIH policy. 62/178 (35%)
- Other* 17/178 (10%)

*Other: Noted that for some studies reviewers may request that information

39:10.2 Can an IACUC insist on inclusion of females and males?

The IACUC can ask for more information if a single-sex study is proposed without a strong justification, or if it is unclear which sex(es) is/are going to be used in a study. The IACUC can also remind researchers that using both sexes is expected by funders and journals, and that it is a requirement for NIH-funded researchers unless there is robust justification.

Resources to help formally review the inclusion of both sexes during protocol review include the Sex Inclusive Research Framework (SIRF) developed by Karp and colleagues[101] and a decision tree developed by Duffy and colleagues.[102]

39:11 How can an IACUC support researchers in improving reproducibility after protocol review?

The IACUC is uniquely positioned to support researchers locally with good experimental design/protocols. IACUCs can raise awareness and encourage the use of free resources, such as the NIH training modules on rigor and reproducibility[103] and the Experimental Design Assistant.[12] To help increase use of the ARRIVE reporting guidelines, see the action plans available from the NC3Rs with suggested steps tailored to universities and research institutes.[104]

Continuing IACUC oversight of animal activities is required by federal laws, regulations, and policies (*Guide*, p. 33).[8] Post-approval monitoring can provide researchers with further support after protocol review. By being aware of where a researcher is in the research process, the IACUC can direct them to the most useful resources for that stage of the process. Post-approval monitoring can also help the IACUC identify patterns among local researchers, which could identify specific local training needs.[105, 106]

Surv. Does your IACUC track the progress of protocols after review? Select all that apply.

- No, it is outside the IACUC's responsibility. 18/178 (10%)
- No, we do not have the resources. 24/178 (13%)
- Only if the protocol is submitted for amendment or renewal. 88/178 (49%)
- Yes, we follow up on protocol compliance, protocol deviation, and if deviations were justifiable. 73/178 (41%)
- Yes, we follow up which experiments have been completed. 22/178 (12%)
- Yes, we follow up which experiments are published. 6/178 (3%)

Surv. Does the IACUC check if experiment reporting is in accordance with guidelines such as ARRIVE?

- No, this is outside the IACUC's responsibility. 92/178 (52%)
- No, we do not have the resources. 71/178 (40%)
- Yes, we ask researchers if they report in accordance with guidelines. 11/178 (6%)

- Yes, we check the publications to see if they report in accordance with guidelines. 4/178 (2%)

39:12 What resources are available to support researchers in planning experiments and using bias-reducing strategies such as randomization and masking?
Most institutions do not train researchers in how to increase the scientific rigor and reproducibility of their experiments, either at undergraduate level or beyond, and many do not have the expertise to provide this training. However, free resources are listed below and can be communicated to researchers.

Resource	Description	Reference
General study planning resources		
EQIPD (Enhancing Quality In Preclinical Data) quality system https://quality-preclinical-data.eu	Quality system describing principles, strategies, and operational criteria to incorporate rigor into design, conduct, and analysis of pre-clinical early drug development research	[67]
FRAME (Fund for the Replacement of Animals in Medical Experiments) Experimental Design https://frame.org.uk/resources/experimental-design/	Web page includes a flowchart to plan research programs strategically	[68]
PREPARE (Planning Research and Experimental Procedures on Animals: Recommendations for Excellence) guidelines https://norecopa.no/PREPARE	Guidelines identifying major logistic considerations for early-stage planning of animal studies	[107]
Experimental design resources		
British Pharmacological Society Experimental Design eLearning Collection https://www.bps.ac.uk/education-engagement/research-animals/blinding-elearning-resource	E-learning resources on masking (blinding) and how to identify the experimental unit	
Experimental Design Assistant https://eda.nc3rs.org.uk/	Interactive online tool that helps researchers design robust experiments and provides bespoke feedback on experimental plans Creates randomization sequences that take account of any factors that need to be randomized separately (blocking factors or factors of interest) and facilitate the implementation of masking Produces a report summarizing key design elements of experiments that can be submitted to an IACUC committee to demonstrate rigorous design	[12]
ETPLAS (Education and Training Platform for Laboratory Animal Science) EU-10 & EU-11: Design of procedures and projects https://etplas.eu/learn/eu-10-experimental-design/ https://etplas.eu/learn/eu-11-1-design-of-procedures-and-projects/	E-learning modules on the design of procedures and projects (levels 1 and 2), aligned with the European Commission education and training framework	
NIH Rigor and Reproducibility – Training https://grants.nih.gov/policy-and-compliance/policy-topics/reproducibility/training	Online training modules in rigor and reproducibility	[103]
SIRF (Sex Inclusive Research Framework) https://openinnovation.astrazeneca.com/preclinical-research/sex-inclusive-research-framework.html	An evaluation framework to assess whether an *in vivo* research proposal is sex inclusive, has an appropriate justification for only including one sex, or includes (or is at risk of including) sex bias.	[101]

Randomization tools		
RandoMice https://github.com/RvE54/RandoMice	Desktop randomization tool	[108]
Random.org https://www.random.org/	Online randomization tool	
Sample size calculators		
Experimental Design Assistant – Power calculator for t-tests https://eda.nc3rs.org.uk/experimental-design-group #PowerCalc	Online power calculator with guidance to determine each of the parameters in the calculation	
G*power https://www.psychologie.hhu.de/arbeitsgruppen/ allgemeine-psychologie-und-arbeitspsychologie/ gpower	Desktop power calculators for many statistical tests	
Reporting guidelines		
ARRIVE guidelines https://arriveguidelines.org/	Guidelines for reporting research involving animals transparently and thoroughly. The Explanation and Elaboration of the guidelines is also helpful when designing studies.	[13, 14]

As well as the above resources there are many books and articles on the design of animal experiments including the following:

Resource	Reference
Books	
The design and statistical analysis of animal experiments Bate and Clark, 2014	[64]
Experimental design for laboratory biologists: Maximizing information and improving reproducibility Lazic, 2016	[65]
Experimental design for the life sciences Ruxton and Colegrave, 2010	[109]
The design of animal experiments: reducing the use of animals in research through better experimental design Festing, Overend, Cortina Borja, and Berdoy, 2016	[110]
Journal articles	
What is the optimum design for my animal experiment? Karp and Fry, 2021	[66]
Special issue of the *ILAR* journal: The design and statistical analysis of animal experiments Various authors, 2014	[59,111–120]
The design and statistical analysis of animal experiments Festing, 2002	[114]
Guidelines for the design and statistical analysis of experiments using laboratory animals Festing and Altman, 2002	[121]
Practical aspects of experimental design in animal research Johnson and Besselsen, 2022	[85]
Use of factorial designs to optimize animal experiments and reduce animal use Shaw, Festing, Peers, and Furlong, 2002	[122]

Collaborative research groups such as CAMARADES (Collaborative Approach to Meta Analysis and Review of Animal Data from Experimental Studies)[123] and SYRCLE (Systematic Review Centre for Laboratory animal Experimentation)[124] are excellent resources for specific research fields such as stroke, neuropathic pain, and toxicology. They also provide a number of e-training resources and tools for assessing research quality and conducting systematic reviews.[124–126]

REFERENCES

1. Everitt, J.I. and Berridge, B.R. (2017) The role of the IACUC in the design and conduct of animal experiments that contribute to translational success. *ILAR J* 58, 129–134. 10.1093/ilar/ilx003
2. Mohan, S. and Huneke, R. (2019) The role of IACUCs in responsible animal research. *ILAR J* 60, 43–49

3. The Academy of Medical Sciences (2015) Reproducibility and reliability of biomedical research: improving research practice. https://acmedsci.ac.uk/policy/policy-projects/reproducibility-and-reliability-of-biomedical-research
4. AAALAC (2021) AAALAC accreditation program FAQs – scientific reproducibility. https://www.aaalac.org/accreditation-program/faqs/#Reproducibility
5. ACD-NIH (2021) Enhancing rigor, transparency, and translatability in animal research. https://acd.od.nih.gov/working-groups/eprar.html
6. Prescott, M.J. and Lidster, K. (2017) Improving quality of science through better animal welfare: the NC3Rs strategy. *Lab Anim* 46, 152–156. 10.1038/laban.1217
7. Pritt, S.L. and Hammer, R.E. (2017) The interplay of ethics, animal welfare, and IACUC oversight on the reproducibility of animal studies. *Comp Med* 67, 101–105
8. National Research Council (2011) *Guide for the Care and Use of Laboratory Animals* 8th edn), The National Academies Press
9. United Kingdom Parliament 2014. Consolidated version of ASPA 1986. https://www.gov.uk/government/publications/consolidated-version-of-aspa-1986
10. European Commission (2010) Directive 2010/63/EU of the European Parliament and of the Council of 22 September 2010 on the protection of animals used for scientific purposes. *Official Journal of the European Union*
11. Graham, M.L. and Prescott, M.J. (2015) The multifactorial role of the 3Rs in shifting the harm-benefit analysis in animal models of disease. *Eur J Pharmacol* 759, 19–29. 10.1016/j.ejphar.2015.03.040
12. Percie du Sert, N. *et al.* (2017) The Experimental Design Assistant. *PLoS Biol* 15, e2003779. 10.1371/journal.pbio.2003779
13. Percie du Sert, N. *et al.* (2020) The ARRIVE guidelines 2.0: Updated guidelines for reporting animal research. *PLoS Biol* 18, e3000410. 10.1371/journal.pbio.3000410
14. Percie du Sert, N. *et al.* (2020) Reporting animal research: Explanation and elaboration for the ARRIVE guidelines 2.0. *PLoS Biol* 18, e3000411. 10.1371/journal.pbio.3000411
15. RSPCA and LASA 2015. Guiding principles on good practice for Animal Welfare and Ethical Review Bodies. https://www.lasa.co.uk/PDF/AWERB_Guiding_Principles_2015_final.pdf
16. European Commission (2013). National Competent Authorities for the implementation of Directive 2010/63/EU on the protection of animals used for scientific purposes: Working document on Project Evaluation and Retrospective Assessment.
17. Voipio, H.-M. *et al.* (2004) Nordic–European workshop on ethical evaluation of animal experiments. *Scand J Lab Anim Sci* 31, 251–267
18. Smith, J. *et al.* (2007) Principles and practice in ethical review of animal experiments across Europe: summary of the report of a FELASA working group on ethical evaluation of animal experiments. *Lab Anim* 41, 143–160
19. Animals in Science Committee 2021. Harm-benefit analysis review, recommendations for AWERBS: ASC report. https://www.gov.uk/government/publications/harm-benefit-analysis-review-animals-in-science-committee-report
20. Sena, E.S. and Currie, G.L. (2019) How our approaches to assessing benefits and harms can be improved. *Anim Welfare* 28, 107–115. 10.7120/09627286.28.1.107
21. Eggel, M. and Wurbel, H. (2021) Internal consistency and compatibility of the 3Rs and 3Vs principles for project evaluation of animal research. *Lab Anim* 55, 233–243. 10.1177/0023677220968583
22. Vogt, L. *et al.* (2016) Authorization of animal experiments is based on confidence rather than evidence of scientific rigor. *PLoS Biol* 14, e2000598. 10.1371/journal.pbio.2000598
23. Bronstad, A. *et al.* (2016) Current concepts of harm-benefit analysis of animal experiments - report from the AALAS-FELASA working group on harm-benefit analysis - part 1. *Lab Anim* 50, 1-20. 10.1177/0023677216642398
24. MacCallum, C.J. (2010) Reporting animal studies: good science and a duty of care. *PLoS Biol* 8, e1000413. 10.1371/journal.pbio.1000413
25. Altman, D.G. (1980) Statistics and ethics in medical research. Collecting and screening data. *Br Med J* 281, 1399-1401. 10.1136/bmj.281.6252.1399
26. Freedman, L.P. *et al.* (2015) The economics of reproducibility in preclinical research. *PLoS Biol* 13, e1002165. 10.1371/journal.pbio.1002165
27. Landis, S.C. *et al.* (2012) A call for transparent reporting to optimize the predictive value of preclinical research. *Nature* 490, 187–191. 10.1038/nature11556
28. Lauer, M., S 2022. Update following ACD recommendations on enhancing rigor, transparency, and translatability in animal research. https://acd.od.nih.gov/documents/presentations/06092022_Animal_Research.pdf.
29. Akhtar, A. (2015) The flaws and human harms of animal experimentation. *Camb Q Healthc Ethics* 24, 407–419. 10.1017/S0963180115000079
30. Ormandy, E.H. *et al.* (2019) Animal research, accountability, openness and public engagement: report from an international expert forum. *Animals (Basel)* 9. 10.3390/ani9090622
31. Kitroeff, N. (2015) USDA cites University of Oklahoma for abusing animals. https://www.bloomberg.com/news/articles/2015-02-20/usda-cites-university-of-oklahoma-for-abusing-animals#xj4y7vzkg
32. O'Dell, R. (2021) University of Arizona criticized over animal welfare violations, research misconduct
33. Spencer, C. (2021) USC's research labs accused of animal cruelty again. https://dailytrojan.com/2021/02/19/uscs-research-labs-accused-of-animal-cruelty-again/
34. Pew Research Center (2018) Most Americans accept genetic engineering of animals that benefits human health, but many oppose other uses. https://www.pewresearch.org/science/2018/08/16/most-americans-accept-genetic-engineering-of-animals-that-benefits-human-health-but-many-oppose-other-uses/.
35. Clemence, M. and Leaman, J. (2016) Public attitudes to animal research in 2016. https://www.ipsos.com/en-uk/attitudes-animal-research-2016.
36. Clemence, M. (2018) Public attitudes to animal research in 2018. https://www.ipsos.com/en-uk/public-attitudes-animal-research-2018.

37. Silverman, J. *et al.* (2017) The role of the IACUC in ensuring research reproducibility. *Lab Anim* 46, 129–135. 10.1038/laban.1213
38. US Government (2018) U.S. government principles for the utilization and care of vertebrate animals used in testing, research, and training. https://olaw.nih.gov/policies-laws/gov-principles.htm
39. National Institutes of Health 2015. Simplification of the vertebrate animals section of NIH grant applications and contract proposals. https://grants.nih.gov/grants/guide/notice-files/not-od-16-006.html.
40. Mohan, S. *et al.* (2018) Contribution of the breadth and depth of IACUC membership to experimental design as a factor in research reproducibility. *J Am Assoc Lab Anim Sci* 57, 104–109
41. Vollert, J. *et al.* (2020) Systematic review of guidelines for internal validity in the design, conduct and analysis of preclinical biomedical experiments involving laboratory animals. *BMJ Open Sci* 4, e100046. 10.1136/bmjos-2019-100046
42. Reynolds, P. (2021) Statistics, statistical thinking, and the IACUC. *Lab Anim* 50, 266–268. 10.1038/s41684-021-00832-w
43. Silverman, J. *et al.* (2015) Decision making and the IACUC: Part 1 - Protocol information discussed at full-committee reviews. *J Am Assoc Lab Anim Sci* 54, 389–398
44. Bateson, P. (1986) When to experiment on animals. *New Sci* 109, 30–32
45. United States Department of Health and Human Services, (2015) Policy on Humane Care and Use of Laboratory Animals. US Department of Health and Human Services. https://olaw.nih.gov/policies-laws/phs-policy.htm
46. FASEB (2016) Enhancing research reproducibility: recommendations from the Federation of American Societies for Experimental Biology. https://www.faseb.org/FASEB/media/PDF/News/Washington%20Update/FASEB_Enhancing-Research-Reproducibility_1.pdf.
47. European Commission (2014) Animals used for scientific purposes. https://ec.europa.eu/environment/chemicals/lab_animals/pubs_guidance_en.htm.
48. EDA (2015) Supporting funding applications and ethical review with the EDA. https://eda.nc3rs.org.uk/funding-applications-and-ethical-review.
49. Reichlin, T.S. *et al.* (2016) The researchers' view of scientific rigor-survey on the conduct and reporting of in vivo research. *PLoS One* 11, e0165999. 10.1371/journal.pone.0165999
50. Vorland, C.J. *et al.* (2021) Errors in the implementation, analysis, and reporting of randomization within obesity and nutrition research: a guide to their avoidance. *Int J Obes* 45, 2335-2346. 10.1038/s41366-021-00909-z
51. Bespalov, A. *et al.* (2020) Blinding and randomization. In *Good Research Practice in Non-Clinical Pharmacology and Biomedicine* (Bespalov, A. *et al.*, eds), pp. 81–100, Springer International Publishing
52. Cordaro, L. and Ison, J.R. (1963) Psychology of the scientist: X. Observer bias in classical conditioning of the planarian. *Psychol Rep* 13, 787–789. 10.2466/pr0.1963.13.3.787
53. Rosenthal, R. and Fode, K.L. (1963) The effect of experimenter bias on the performance of the albino rat. *Behav Sci* 8, 183–189. 10.1002/bs.3830080302
54. Tuyttens, F.A.M. *et al.* (2014) Observer bias in animal behaviour research: can we believe what we score, if we score what we believe? *Animal Behaviour* 90, 273–280. 10.1016/j.anbehav.2014.02.007
55. van Wilgenburg, E. and Elgar, M.A. (2013) Confirmation bias in studies of nestmate recognition: a cautionary note for research into the behaviour of animals. *PLoS One* 8, e53548. 10.1371/journal.pone.0053548
56. Bebarta, V. *et al.* (2003) Emergency medicine animal research: does use of randomization and blinding affect the results? *Acad Emerg Med* 10, 684–687. 10.1111/j.1553-2712.2003.tb00056.x
57. Vesterinen, H.V. *et al.* (2010) Systematic survey of the design, statistical analysis, and reporting of studies. *J Cereb Blood Flow Metab.* 10.1038/jcbfm.2010.217
58. Holman, L. *et al.* (2015) Evidence of experimental bias in the life sciences: why we need blind data recording. *PLoS Biol* 13, e1002190. 10.1371/journal.pbio.1002190
59. O'Connor, A.M. and Sargeant, J.M. (2014) Critical appraisal of studies using laboratory animal models. *ILAR J* 55, 405–417. 10.1093/ilar/ilu038
60. Neves, K. and Amaral, O.B. (2020) Addressing selective reporting of experiments through predefined exclusion criteria. *Elife* 9. 10.7554/eLife.56626
61. Tsilidis, K.K. *et al.* (2013) Evaluation of excess significance bias in animal studies of neurological diseases. *PLoS Biol* 11, e1001609. 10.1371/journal.pbio.1001609
62. Karp, N.A. *et al.* (2022) A qualitative study of the barriers to using blinding in in vivo experiments and suggestions for improvement. *PLoS Biol* 20, e3001873. 10.1371/journal.pbio.3001873
63. NIH Principles and Guidelines for Reporting Preclinical Research. https://www.nih.gov/research-training/rigor-reproducibility/principles-guidelines-reporting-preclinical-research.
64. Bate, S.T. and Clark, R.A. (2014) *The Design and Statistical Analysis of Animal Experiments.* Cambridge University Press
65. Lazic, S.E. (2016) *Experimental Design for Laboratory Biologists: Maximising Information and Improving Reproducibility* Cambridge University Press
66. Karp, N.A. and Fry, D. (2021) What is the optimum design for my animal experiment? *BMJ Open Sci* 5, e100126. 10.1136/bmjos-2020-100126
67. Bespalov, A. *et al.* (2021) Introduction to the EQIPD quality system. *Elife* 10. 10.7554/eLife.63294
68. FRAME Experimental Design. https://frame.org.uk/resources/experimental-design/
69. Lazic, S.E. *et al.* (2018) What exactly is 'N' in cell culture and animal experiments? *PLoS Biol* 16, e2005282. 10.1371/journal.pbio.2005282
70. Reynolds, P.S. *et al.* (2019) Informing efficient pilot development of animal trauma models through quality improvement strategies. *Lab Anim* 53, 394–404. 10.1177/0023677218802999
71. Mahajan, V.S. *et al.* (2016) Striking immune phenotypes in gene-targeted mice are driven by a copy-number variant originating from a commercially available C57BL/6 strain. *Cell Rep* 15, 1901–1909. 10.1016/j.celrep.2016.04.080

72. Simon, M.M. *et al.* (2013) A comparative phenotypic and genomic analysis of C57BL/6J and C57BL/6N mouse strains. *Genome Biol* 14, R82. 10.1186/gb-2013-14-7-r82
73. Uchio-Yamada, K. *et al.* (2017) Incorrect strain information for mouse cell lines: sequential influence of misidentification on sublines. *In Vitro Cell Dev Biol Anim* 53, 225-230. 10.1007/s11626-016-0104-3
74. Souren, N.Y. *et al.* (2022) Cell line authentication: a necessity for reproducible biomedical research. *EMBO J* 41, e111307. 10.15252/embj.2022111307
75. Reardon, S. (2016) A mouse's house may ruin experiments. *Nature* 530, 264. 10.1038/nature.2016.19335
76. Carbone, L. (2011) Pain in laboratory animals: the ethical and regulatory imperatives. *PLoS One* 6, e21578. 10.1371/journal.pone.0021578
77. Peterson, N.C. *et al.* (2017) To treat or not to treat: the effects of pain on experimental parameters. *Compar Med* 67, 469-482
78. Gouveia, K. and Hurst, J.L. (2019) Improving the practicality of using non-aversive handling methods to reduce background stress and anxiety in laboratory mice. *Sci Rep* 9, 20305. 10.1038/s41598-019-56860-7
79. Hurst, J.L. and West, R.S. (2010) Taming anxiety in laboratory mice. *Nat Methods* 7, 825-826. 10.1038/nmeth.1500
80. Martin, B. *et al.* (2010) "Control" laboratory rodents are metabolically morbid: why it matters. *Proc Natl Acad Sci USA* 107, 6127-6133. 10.1073/pnas.0912955107
81. Andre, V. *et al.* (2018) Laboratory mouse housing conditions can be improved using common environmental enrichment without compromising data. *PLoS Biol* 16, e2005019. 10.1371/journal.pbio.2005019
82. Richter, S.H. *et al.* (2011) Effect of population heterogenization on the reproducibility of mouse behavior: a multi-laboratory study. *PLoS One* 6, e16461
83. Voelkl, B. *et al.* (2018) Reproducibility of preclinical animal research improves with heterogeneity of study samples. *PLoS Biol* 16, e2003693. 10.1371/journal.pbio.2003693
84. Nakamura, Y. and Suzuki, K. (2018) Tunnel use facilitates handling of ICR mice and decreases experimental variation. *Journal of veterinary medical science*, 18-0044
85. Johnson, P.D. and Besselsen, D.G. (2002) Practical aspects of experimental design in animal research. *ILAR J* 43, 202–206. 10.1093/ilar.43.4.202
86. Kramer, M. and Font, E. (2017) Reducing sample size in experiments with animals: historical controls and related strategies. *Biol Rev* 92, 431–445
87. Bonapersona, V. *et al.* (2021) Increasing the statistical power of animal experiments with historical control data. *Nat Neurosci* 24, 470–477
88. Keenan, C. *et al.* (2009) Best practices for use of historical control data of proliferative rodent lesions. *Toxicol Pathol* 37, 679–693
89. Hatswell, A. *et al.* (2020) Summarising salient information on historical controls: a structured assessment of validity and comparability across studies. *Clin Trials* 17, 607–616
90. Pocock, S.J. (1976) The combination of randomized and historical controls in clinical trials. *J Chron Dis* 29, 175–188. 10.1016/0021-9681(76)90044-8
91. Löscher, W. *et al.* (2017) The relevance of inter-and intra-strain differences in mice and rats and their implications for models of seizures and epilepsy. *Epil Behav* 73, 214–235
92. Gileta, A.F. *et al.* (2022) Genetic characterization of outbred Sprague Dawley rats and utility for genome-wide association studies. *PLoS Genet* 18, e1010234
93. Tremoleda, J.L. *et al.* (2017) Modeling acute traumatic hemorrhagic shock injury: challenges and guidelines for preclinical studies. *Shock* 48, 610–623
94. Zingarelli, B. *et al.* (2019) Part I: minimum quality threshold in pre-clinical sepsis studies (MQTiPSS) for study design and humane modeling endpoints. *Shock* 51, 10
95. Karp, N.A. *et al.* (2017) Prevalence of sexual dimorphism in mammalian phenotypic traits. *Nat Commun* 8, 15475. 10.1038/ncomms15475
96. McCullough, L.D. *et al.* (2005) Ischemic nitric oxide and poly (ADP-ribose) polymerase-1 in cerebral ischemia: male toxicity, female protection. *J Cereb Blood Flow Metab* 25, 502-512. 10.1038/sj.jcbfm.9600059
97. Riley, A.L. *et al.* (2018) Sex as a biological variable: drug use and abuse. *Physiol Behav* 187, 79–96. 10.1016/j.physbeh.2017.10.005
98. Hodes, G.E. *et al.* (2015) Sex differences in nucleus accumbens transcriptome profiles associated with susceptibility versus resilience to subchronic variable stress. *J Neurosci* 35, 16362–16376. 10.1523/JNEUROSCI.1392-15.2015
99. Parodi, G. *et al.* (2022) The response of the rodent gut microbiome to broad-spectrum antibiotics is different in males and females. *Front Microbiol* 13, 897283. 10.3389/fmicb.2022.897283
100. National Institutes of Health. Consideration of sex as a biological variable in NIH-funded research. https://orwh.od.nih.gov/sites/orwh/files/docs/NOT-OD-15-102_Guidance.pdf.
101. Karp, N.A. *et al.* (2025) The Sex Inclusive Research Framework to address sex bias in preclinical research proposals. *Nature Communications* 16, 3763. 10.1038/s41467-025-58560-5
102. Duffy, K.A. *et al.* (2020) Filling the regulatory gap: potential role of institutional review boards in promoting consideration of sex as a biological variable. *J Womens Health (Larchmt)* 29, 868–875. 10.1089/jwh.2019.8084
103. National Institutes of Health (2022) Rigor and reproducibility training. https://grants.nih.gov/policy-and-compliance/policy-topics/reproducibility/training.
104. The National Centre for the Replacement Refinement and Reduction of Animals in Research (2022) ARRIVE action plans. https://arriveguidelines.org/supporters.
105. Collins, J. (2008) Postapproval monitoring and the institutional animal care and use committee (IACUC). *ILAR J* 49, 388–392
106. Banks, R.E. and Norton, J.N. (2008) A sample postapproval monitoring program in academia. *ILAR J* 49, 402–418
107. Smith, A.J. *et al.* (2018) PREPARE: guidelines for planning animal research and testing. *Lab Anim* 52, 135–141. 10.1177/0023677217724823
108. van Eenige, R. *et al.* (2020) RandoMice, a novel, user-friendly randomization tool in animal research. *PLoS One* 15, e0237096. 10.1371/journal.pone.0237096
109. Ruxton, G. and Colegrave, N. (2010) *Experimental Design for the Life Sciences*. Oxford University Press

110. Festing, M.F.W. *et al.* (2016) *The Design of Animal Experiments: Reducing the Use of Animals in Research through Better Experimental Design* (Revised and updated edition) SAGE Publications Ltd
111. Bailoo, J.D. *et al.* (2014) Refinement of experimental design and conduct in laboratory animal research. *ILAR J* 55, 383–391. 10.1093/ilar/ilu037
112. Festing, M.F. (2014) Randomized block experimental designs can increase the power and reproducibility of laboratory animal experiments. *ILAR J* 55, 472–476. 10.1093/ilar/ilu045
113. Festing, M.F. (2014) Evidence should trump intuition by preferring inbred strains to outbred stocks in preclinical research. *ILAR J* 55, 399–404. 10.1093/ilar/ilu036
114. Festing, M.F. (2002) The design and statistical analysis of animal experiments. *ILAR J* 43, 191–193
115. Fry, D.J. (2014) Teaching experimental design. *ILAR J* 55, 457–471. 10.1093/ilar/ilu031
116. Garner, J.P. (2014) The significance of meaning: why do over 90% of behavioral neuroscience results fail to translate to humans, and what can we do to fix it? *ILAR J* 55, 438–456. 10.1093/ilar/ilu047
117. Hooijmans, C.R. *et al.* (2014) Meta-analyses of animal studies: an introduction of a valuable instrument to further improve healthcare. *ILAR J* 55, 418–426. 10.1093/ilar/ilu042
118. Nevalainen, T. (2014) Animal husbandry and experimental design. *ILAR J* 55, 392–398. 10.1093/ilar/ilu035
119. Parker, R.M. and Browne, W.J. (2014) The place of experimental design and statistics in the 3Rs. *ILAR J* 55, 477–485. 10.1093/ilar/ilu044
120. Pearl, D.L. (2014) Making the most of clustered data in laboratory animal research using multi-level models. *ILAR J* 55, 486–492. 10.1093/ilar/ilu034
121. Festing, M.F. and Altman, D.G. (2002) Guidelines for the design and statistical analysis of experiments using laboratory animals. *ILAR J* 43, 244–258. 10.1093/ilar.43.4.244
122. Shaw, R. *et al.* (2002) Use of factorial designs to optimize animal experiments and reduce animal use. *ILAR J* 43, 223–232
123. The University of Edinburgh (2020) CAMARADES.
124. Radboudumc SYRCLE. https://www.syrcle.network/.
125. Systematic Review Facility 2022. SyRF. https://syrf.org.uk/.
126. The University of Edinburgh (2022) CAMARADES tools and resources.

Index

B

C

D

E

J

Q

R

S

T

U

V

W

Z